ENGINEERING
DRAWING AND DESIGN

Second Edition

ENGINEERING
DRAWING AND DESIGN

Second Edition

David A. Madsen

Department Chair Drafting Technology
AutoCAD Premier Authorized Training Center
Authorized Softdesk Civil Engineering Training Center
Former Member
Board of Directors
American Design Drafting Association
Clackamas Community College, Oregon City, OR

Terence M. Shumaker (Chapters 1 CADD, 2, 17, and 20)

Drafting/CADD Instructor
Manager
AutoCAD Premier Authorized Training Center
Softdesk Civil Engineering Training Center
Clackamas Community College, Oregon City, OR

J. Lee Turpin (Chapters 7 and 8)

Former Drafting Instructor/Department Chair
Vocational Counselor
Clackamas Community College, Oregon City, OR

Catherine Stark (Chapters 19 and 22)

Engineering Technician/CADD
City of Portland, Oregon

Delmar Publishers

I(T)P An International Thomson Publishing Company

Albany • Bonn • Boston • Cincinnati • Detroit • London • Madrid
Melbourne • Mexico City • New York • Pacific Grove • Paris • San Francisco
Singapore • Tokyo • Toronto • Washington

NOTICE TO THE READER

Cover Design: Bob Clancy/Spiral Design

Delmar Staff
Publisher: Robert Lynch
Administrative Editor: John Anderson
Developmental Editor: Kathleen Tatterson
Senior Project Editor: Christopher Chien
Production Manager: Larry Main
Art and Design Coordinator: Lisa Bower
Editorial Assistant: John Fisher

COPYRIGHT © 1996
By Delmar Publishers
a division of International Thomson Publishing Inc.

The ITP logo is a trademark under license.

Printed in the United States of America

For more information, contact:

Delmar Publishers
3 Columbia Circle, Box 15015
Albany, New York 12212-5015

International Thomson Publishing Europe
Berkshire House 168-173
High Holborn
London, WC1V 7AA
England

Thomas Nelson Australia
102 Dodds Street
South Melbourne, 3205
Victoria, Australia

Nelson Canada
1120 Birchmont Road
Scarborough, Ontario
Canada, M1K 5G4

Delmar Publishers' Online Services
To access Delmar on the World Wide Web, point your browser to:
http://www.delmar.com/delmar.html
To access through Gopher: **gopher://gopher.delmar.com**
(Delmar Online is part of "thomson.com", an Internet site with information on more than 30 publishers of the International Thomson Publishing organization.)
For information on our products and services:
email: **info@delmar.com**
or call **800-347-7707**

International Thomson Editores
Campos Eliseos 385, Piso 7
Col Polanco
11560 Mexico D F Mexico

International Thomson Publishing GmbH
Konigswinterer Strasse 418
53227 Bonn
Germany

International Thomson Publishing Asia
221 Henderson Road
#05-10 Henderson Building
Singapore 0315

International Thomson Publishing—Japan
Hirakawacho Kyowa Building, 3F
2-2-1 Hirakawacho
Chiyoda-ku, Tokyo 102
Japan

2 3 4 5 6 7 8 9 10 XXX 01 00 99 98 97

Library of Congress Cataloging-in-Publication Data
Madsen, David A.
   Engineering drawing and design / David A. Madsen [with] Terence
M. Schumaker, J. Lee Turpin, Catherine Stark.—2nd ed.
      p.   cm.
   Includes index.
   ISBN 0-8273-6720-1
   1. Mechanical drawing.   I. Title.
T353.M1965   1996
604.2—dc20   95-36318
                                              CIP

CONTENTS

PREFACE

Engineering Drawing and Design is a practical, in-depth textbook that is easy to use and understand. The content may be used as presented, or the chapters may be rearranged to accommodate alternate formats for traditional or individualized instruction. *Engineering Drawing and Design* may be used for the following courses, so your students need only one text for their entire curriculum:

○ COMPUTER-AIDED DESIGN DRAFTING (CADD)
  Hardware, software, and applications.
○ MECHANICAL DRAFTING
  Sketching, lettering, lines, geometric constructions, multiviews and auxiliary views, dimensioning, fasteners, sectioning, working drawings, details, assemblies, parts lists, and engineering changes.
○ DESCRIPTIVE GEOMETRY
  Beginning and advanced.
○ MANUFACTURING PROCESSES
○ WELDING PROCESSES
○ GEOMETRIC TOLERANCING
  Based on ASME Y14.5M—1994.
○ MECHANISMS
  Linkages, gears, cams, bearings, and seals.
○ BELT AND CHAIN DRIVES
○ FLUID POWER
○ PICTORIAL DRAWINGS
○ STRUCTURAL DRAFTING
○ CIVIL DRAFTING
○ INDUSTRIAL PIPE DRAFTING
○ HEATING, VENTILATING, AND AIR CONDITIONING (HVAC)
○ SHEET METAL DRAFTING AND PATTERN LAYOUT
○ ELECTRICAL DRAFTING
  Electrical power substation design.
○ ELECTRONIC SCHEMATIC DRAFTING
○ ENGINEERING CHARTS AND GRAPHS

▼ MAJOR FEATURES

Engineering Drawing and Design has these important features:

○ CADD throughout.
○ ANSI, ASME, and related standards emphasized.
○ Professional Perspectives.
○ Engineering design process.
○ Math Appendix and Applications.
○ Step-by-step layout methods.
○ Engineering layout techniques.
○ CADD techniques.
○ Practical appendices.
○ Real industry problems.

Engineering Drawing and Design provides a practical approach to drafting as related to the American National Standards Institute (ANSI) and the American Society of Mechanical Engineers (ASME) standards and common alternates that may be found in traditional industrial standards, such as the American Welding Society or the American Institute for Steel Construction. Also presented, when appropriate, are standards and codes related to specific engineering graphics fields. One excellent and necessary foundation to engineering drawing and design and the implementation of a common approach to graphics nationwide is the emphasis of standardization in all levels of drawing and design instruction. When students become professionals, this text will go along as a valuable desk reference.

Each chapter provides realistic examples, illustrations, problems, and related tests. The examples illustrate recommended design presentation based on ANSI/ASME standards and other related national standards and codes with actual industrial drawings used for reinforcement. The correlated text explains drawing techniques and provides professional tips for skill development. Step-by-step layout methods provide a logical approach to setting up and completing the drawing problems.

A typical engineering design process leads the content of every chapter. This gives the student an early understanding of the type of engineering project that is found in the specific design and drafting area discussed in the chapter.

There is a comprehensive math appendix that provides the easy-to-understand essentials of engineering drafting mathematics. Numerous examples illustrate how the math concepts relate to the practical applications. Math continues to be a part of every chapter, giving examples and instruction on how math is used in the specific discipline. Nearly every chapter contains math problems along with the practical drafting problems.

The appendices contain the types of charts and information that is used daily in the engineering design and drafting environment. These appendices include common fastener types and data, fits and tolerances, metric conversion charts, tap drill charts, and other manufacturing information.

▼ INDUSTRIAL APPROACH TO PROBLEM SOLVING

The drafter's responsibility is to convert the engineering sketch or instructions to formal drawings. The text explains how to prepare drawings from engineering sketches by providing the learner with the basic guides for layout and arrangement in a knowledge-building format. One concept is learned before the next is introduced. Problem assignments are presented in order of difficulty within each chapter and throughout the text. The concepts and skills learned in one chapter are used in subsequent chapters so that by the end of the text the student will have the ability to solve problems using a multitude of previously learned activities. The problems are presented as 3-D or actual industrial layouts in a manner that is consistent with the engineering environment. Early problems provide suggested layout sketches. It is not enough for students to duplicate drawings from given assignments; they must be able to think through the process of drawing development. The goals and objectives of each problem assignment are consistent with recommended evaluation criteria based on the progression of learning activities.

▼ COMPUTER-AIDED DESIGN AND DRAFTING

Computer-aided design and drafting (CADD) is presented throughout the text. CADD topics include:

○ CADD hardware.
○ CADD software.
○ CADD material requirements.
○ Specific CADD applications.
○ CADD template menus and symbol libraries for specific engineering drafting applications.
○ Increased productivity with CADD.
○ Parametrics in CADD applications.

▼ COURSE PLAN

The introduction to *Engineering Drawing and Design* provides a detailed look at drafting as a profession, and includes occupations, professional organizations, occupational levels, opportunities, computers, and the engineering design process.

Drafting Equipment, Media, and Reproduction Methods

The study of engineering drawing and design in this text begins with equipment, materials, and reproduction for manual drafting with specific instruction on how to use tools and equipment.

Introduction to CADD

Next you receive an in-depth introduction to CADD, including developing CADD skills, computer languages, drafting with the computer, using menus, CADD in industry, the CADD environment, the job market, computer drafting equipment, CADD drawing materials, and future CADD hardware and software.

Sketching, Lettering, Lines, and Geometric Construction

You then learn how sketching, lettering, lines, and geometric constructions may be properly used in both manual and computer-aided design in accordance with ANSI/ASME standards.

Multiviews and Auxiliary Views

A complete study of multiviews and auxiliary views, in accordance with ANSI/ASME standards, provides accurate and detailed instruction on topics such as view selection and placement, first- and third-angle projection, and viewing techniques. You are provided with a step-by-step example on how to lay out multiview and auxiliary view drawings.

Descriptive Geometry

This text provides all the descriptive geometry coverage that students need, including discussion of lines, planes, angles, slope, bearing, intersecting lines and planes, and vectors. Problems are presented as actual engineering situations. This is an easy-to-understand set of two chapters with related problems.

Manufacturing Processes

This is a complete introduction to manufacturing processes, including product development, manufacturing materials, material numbering systems, hardness and testing, casting and forging methods, design and drafting, complete machining processes, computer numerical control, computer-integrated manufacturing, machine features and drawing representations, surface texture, design of machine features, tool design, and statistical process control.

Dimensioning

This chapter is in accordance with ASME Y14.5M—1994 and provides complete coverage on dimensioning systems, rules, specific and general notes, tolerances, symbols, and dimensioning for CAD/CAM. You are provided with a step-by-step example showing how to lay out a fully dimensioned multiview drawing.

Fasteners and Springs, and Welding Processes

These two chapters cover the complete range of fastening devices and welding processes available. The fasteners chapter covers screw threads, thread cutting, thread forms, thread representations and notes, washers, dowels, pins, rings, keys and keyseats, rivets, and springs.

The welding chapter provides an in-depth introduction to processes, welding drawings and symbols, weld types, symbol usage, weld characteristics, weld testing, welding specifications, and prequalified welded joints.

Sections, Revolutions, and Conventional Breaks

This chapter explains in detail every type of sectioning practice available to the mechanical engineering drafter. Additionally, the chapter includes treatment of unsectioned features, conventional revolutions, and conventional breaks.

Geometric Tolerancing

This chapter provides a complete instruction to the reading and use of geometric tolerancing symbols and terms as presented in ASME Y14.5M—1994. It features excellent coverage on datums, feature control, basic dimensions, geometric tolerances, material condition, position tolerance, virtual condition, geometric tolerancing, and CADD.

Mechanisms, Belt and Chain Drives

These two chapters provide you with extensive coverage of linkage mechanisms, gears, cams, and belt and chain drives. You are given detailed information on the selection of bearings and lubricants. The use of vendors' catalog information is stressed in the design of belt and chain drive systems. Actual mechanical engineering design problems are provided for gear train and cam plate design. Working drawings include details, assemblies, and parts lists—the most extensive discussion on engineering changes available. A complete analysis on how to prepare a set of working drawings from concept to final product.

Pictorial Drawing

This chapter is a complete review of 3-D drafting techniques used in manual and computer-aided drafting. The content includes extensive discussions in isometric, dimetric, trimetric, perspective, exploded drawing, and shading methods.

Engineering Fields of Study

The balance of this text provides you with chapters that offer the most complete information available in specific engineering drawing and design fields. These chapters may be used individually for complete courses. They contain comprehensive instruction, problems, and tests. These fields include:

○ *Fluid power*—with complete coverage on forces, pressure, work, power, hydraulics, pneumatics, diagrams, symbols, equipment, systems, and operations.

○ *Industrial process piping*—with explanations and detailed examples of pipe and fittings, valves and instrumentation, pumps, tanks, and equipment, flow diagrams, piping plans and elevations, piping isometrics, and piping spools.

○ *Structural drafting*—of reinforced and precast concrete, truss and panalized framing, timbers, laminated beams, steel joists and studs, prefabricated systems, structural steel, structural welding, set of structural drawings.

○ *Civil drafting*—covers the complete discipline of mapping, including legal descriptions, survey terminology, and plot plans. Civil drafting includes road layout, cuts and fills, and plan and profile drawings.

○ *HVAC and sheet metal drafting*—provides complete coverage of HVAC systems, heat exchangers, HVAC symbols, single- and double-line ducted systems, working from an engineer's sketch, sections and details, common sheet metal developments with step-by-step layout procedures, sheet metal intersections, and chassis fabrication drawings.

○ *Electrical drafting and power substation design*—provides the only coverage of its type in electrical power transmission. The content is specific to electrical drafting of wiring diagrams, cable assemblies, one-line and elementary diagrams, electrical power system symbols, plot plans, bus layouts, ground layouts, conduit layouts, electrical floor plan symbols, power supply plans, and schematics.

○ *Electronic schematic drafting*—exclusively related to the electronics industry. This includes topics such as block diagrams, electronic components and symbols, engineer's sketch, component numbering, IC systems, logic diagrams, LSC schematics, SMT, electronics artwork, layers, marking, drilling, assembly drawings, photo drafting, and pictorial diagrams.

○ *Engineering charts and graphs*—with the most complete coverage on rectilinear, log and semilog scales, surface, column, pie, concurrency, alignment, organizational, production control process (SPC), distribution, and pictorial charts. The chapter includes an in-depth analysis of chart design.

The text contains CADD discussion and examples, engineering layout techniques, working from engineer's sketches, professional practices, and actual industrial examples. The problem assignments are based on actual real world products and designs.

▼ SECTION LENGTH

Chapters are presented in individual learning segments that begin with elementary concepts and build until each chapter provides complete coverage of each topic.

▼ APPLICATIONS

Special emphasis has been placed on providing realistic problems. Problems are presented as 3-D drawings, engineering sketches, and layouts in a manner that is consistent with industry practices. Most of the problems have been supplied by the industry. Each problem solution is based on the step-by-step layout procedures provided in the chapter discussions. Problems are given in order of complexity so that students may be exposed to a variety of engineering experiences. Early problems recommend the layout to help students have time. Advanced problems require the students to go through the same thinking process that a professional faces daily, including drawing scale, paper size selection, view layout, dimension placement, sectioning placement, and many other activities. Problems may be solved using manual or computer-aided drafting as determined by the individual course guidelines. All problems should be solved in accordance with recommended ANSI, ASME, or other related industry practices. Problems may not be presented in the best arrangement for a quality solution. Students should always approach a problem with critical analysis based on view selection and layout, and dimension placement when dimensions are used. Students should not assume that problem information is presented exactly as the intended solution. Advanced problems often require the student to evaluate the accuracy of provided information. Chapter tests provide a complete coverage of each chapter and may be used for student evaluation or as study questions.

ACKNOWLEDGMENTS

I would like to express my deepest appreciation to my wife, Judy, and my children, Michael, Britt, and David, for their support during the long and seemingly never-ending development of this text. Also, a special thanks to my contributing authors:

❍ Terence M. Shumaker, for his outstanding implementation of the introduction to CADD and related topics, industrial pipe drafting, and pictorial drawing chapters.
❍ J. Lee Turpin, for his experience in presenting descriptive geometry to students in a manner that is easy to follow and understand.
❍ Catherine Stark, for a professional coverage of fluid power and civil drafting.
❍ Alan Jefferis, my coauthor of *Architectural Drafting and Design,* for his professional support.
❍ Steve Brown, for his presentation of technical math in the appendix, chapter applications, and problems. The math instruction is given in an innovative and practical manner that immediately addresses the real-world issues.
❍ Rod Rawls, for the development of many engineering design processes based on years of engineering consultation.

I would like to give special thanks and acknowledgment to the many professionals who helped formulate this work. Also, thanks to those who reviewed the manuscript in an effort to help me publish the best possible text:

Richard Anderson
ITT Technical Institute

Tom Bledsaw
ITT Technical Institute

Gary Burggraff
Moorhead Technical College

Steve Clayden
CAD Institute

Richard Erickson
Eden Area R.O.P.

Gene Fosheim
Lake Washington Technical College

John Frostad
Green River Community College

Larry Gallo
Clearfield Vo-Tech

Ray Heikkila
State Technical Institute

Keith Jackson
ITT Technical Institute

Thomas Marsh
St. Augustine Technical Center

Milton Moore
N.W. Hillyard AVTS

Frank Morgan
CAD Institute

Dean Noble
ITT Technical Institute

Tim Riordan
ITT Technical Institute

Richard Staples
Eastern Maine Technical College

Dan Walker
Southeast Career Center

The quality of this text is also enhanced by the support and contributions from industry and vendors. The list of contributors is extensive, and acknowledgment is given at each figure illustration. (Many figures appearing in Chapters 20, 21, and 22 were reprinted from Jefferis and Madsen, *Architectural Drafting and Design,* 3rd edition, © 1996 by Delmar Publishers.) The following individuals and companies gave an extraordinary amount of support:

Accugraph Corporation
Ontario, Canada

Aerojet TechSystems Company
Sacramento, California

American Institute for Design and Drafting
Rockville, Maryland

American National Standards Institute
New York, New York

Berol USA RapiDesign
Burbank, California

Bishop Graphics, Inc.
Westlake Village, California

CALCOMP
Anaheim, California

Carven Community College

Chartpak
Northampton, Massachusetts

Computervision Corporation
Bedford, Massachusetts

Consul and Mutoh, Ltd.
Anaheim, California

Bill Curtis
Curtis Associates
Portland, Oregon

Hyster Company
Portland, Oregon

Doug Major
Mark Hartman
Professional Consultants

Parker-Hannifin
Cleveland, Ohio

Tom Pearce
Stanley Hydraulic Tools,
    A Division of The Stanley Works
Portland, Oregon

Jean K. Shirkoff, consultant
Portland, Oregon

Leonard Sosnovske
Bonneville Power Administration
Portland, Oregon

T & W Systems
Huntington Beach, California

Ricardo Wilkins
Michael Spiegel
Fluid Air Components
Portland, Oregon

Engineering Drawing and Design

The Short History of Engineering Drawing

J. H. Oakey

We are all aware of the marvelous drawings and the inventive genius of Leonardo da Vinci. It is naturally assumed that he was the originator of drafting and that after his work, all inventors and engineers carefully duplicated his designs in some form in their drawings. This is not so. History is not well documented in the realm of inventors and engineers. The only source for the detailed history of Leonardo's work is his own careful representation. His drawings were those of an artist. They were 3-D and they generally were without dimensional notations as shown in Figure I–1. Craftsmen worked from the 3-D representation, and each machine or device was one off; parts were not interchangeable. In fact, interchangeability was not realized except in special demonstrations until the development of the micrometer in the late 1800s, and even then it was not easily achieved.

Without the concept of interchangeability, accurate drawings were not necessary. Inventors, engineers, and builders worked on each product on a one-off basis, and parts were manufactured from hand sketches or hand drawings on blackboards. Coleman Sellers, in the manufacture of fire engines, had blackboards with full-size drawings of parts. Blacksmiths formed parts and compared them to the shapes on the blackboards. Cole-

man Seller's son, George, recalls lying on his belly using his arms as a radius for curves as his father stood over him directing changes in the sketches until the drawings were satisfactory. Most designs used through the 1800s were accomplished by first completing a hand sketch of the object to be built. These were then converted into wooden models (3-D modeling) from which patterns were constructed. This practice was followed well into this century by some. Most of us are familiar with the stories of Henry Ford and his famous blackboards. What is news is that these were also the Henry Ford "drafting tables." Henry would sketch cars and parts three dimensionally and have patternmakers construct full-size models in wood.

An effort to create a program to standardize drawing came when the Franklin Institute was founded in Philadelphia in 1824. It was founded "to advance the general interests" of mechanics and entrepreneurs "by extending a knowledge of mechanical science." One of the goals of the Institute was to establish a mechanical drawing school. Unfortunately, the academic program was never realized because two factions argued over whether to base it on classical academics (Latin and Greek) or on science and practical courses. Eventually the founders abandoned the academic purposes of the Institute.

Also in 1824, in upstate New York, Amos Eaton founded a school "for the purpose of instructing persons . . . in the application of science to the common purposes of life." This school has grown to be known today as Rensselaer Polytechnic Institute.

In 1862, with the coming of federal government assistance in the form of the Morrill Act, more technical schools emerged, and we can assume that mechanical drawing *slowly* became a part of the intellectual tool kits of trained mechanics, engineers, and inventors.

This can, in part, put to rest the common thought (as we look at pictures of pyramids, steam engines, and other engineering examples) that "it all started with a drawing." Certainly there were freehand sketches, cartoons, and other forms of graphic models. It is probably accurate to say that before most things were built, they were tested with a 3-D model. Just as mechanical drafting slowly took the place of hand sketches and wooden models, today computer design and drawing and 3-D

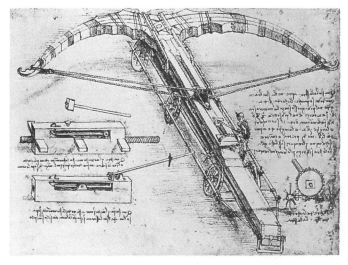

Figure I–1   The drawing of a design by Leonardo da Vinci.

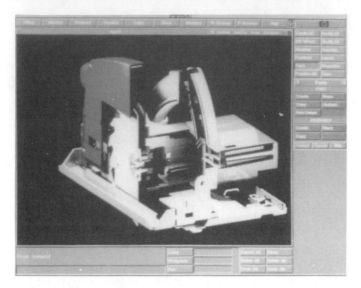

Figure I–2   3-D modeling is likely to replace board drafting.

computer modeling are likely to rapidly replace board drafting. An example is shown in Figure I–2. The eventual utility and order of computer modeling and design remain to be established as part of the developing graphic process.

▼ DRAFTING TECHNOLOGY

According to the *Dictionary of Occupational Titles,* published by the U.S. Department of Labor, drafting is grouped with professional, technical, and managerial occupations. This category includes occupations concerned with the theoretical and practical aspects of such fields of human endeavor as architecture; engineering; mathematics; physical sciences; social sciences; medicine and health; education; museum, library, and archival sciences; law; theology; the arts; recreation; administrative specialties; and management. Also included are occupations in support of scientists and engineers and other specialized activities such as piloting aircraft, operating radios, and directing the course of ships. Most of these occupations require substantial

educational preparation, usually at the college, junior college, or technical institute level.

Men and women employed in the drafting profession are often referred to as drafters. A general definition of *drafter,* as prepared by the Career Information System at the University of Oregon, follows:

Drafters translate data and sketches of engineers, architects, and scientists into detailed drawings that are used in manufacturing and construction. Their duties may include interpreting directions given to them, making sketches, preparing drawings to scale, and specifying details. Drafters may also calculate the strength, quality, quantity, and cost of materials. They utilize various drafting tools and computer equipment, engineering practices, and math to complete drawings.

▼ OCCUPATIONS IN ARCHITECTURE AND ENGINEERING DRAFTING FIELDS

There are several types of drafting technology occupations. While drafting in general has one basic description, specific drafting areas have unique conceptual and skill characteristics. The types of drafting occupations fall into three general professional areas; architecture, engineering, and surveying. The following are specific drafting areas as defined by the *Dictionary of Occupational Titles.*

Architectural Drafter

Draws artistic architectural and structural features of any class of buildings and like structures. Draws designs and details, using drawing instruments. Confirms compliance with building codes. May specialize in planning architectural details according to structural materials used. (See Figure I–3.)

Landscape Drafter

Prepares detailed scale drawings from rough sketches or other data provided by Landscape Architect. May pre-

Figure I–3(a)   Computer-generated elevations. *Courtesy Piercy & Barclay Designers, Inc.*

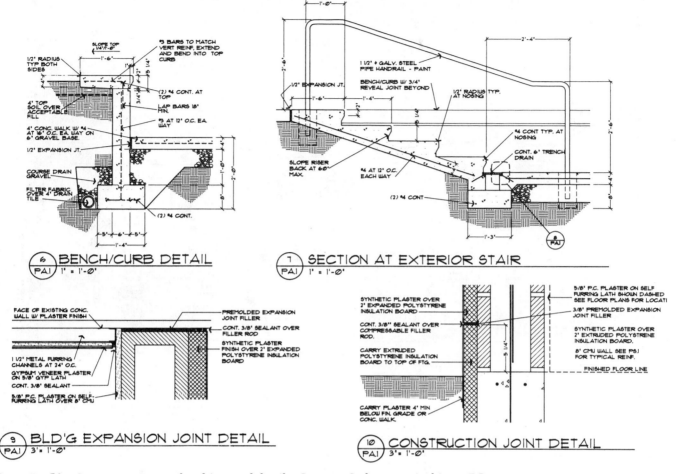

Figure I–3(b) Computer-generated architectural details. *Courtesy Soderstrom Architects PC.*

pare separate detailed site plan, grading and drainage plan, lighting plan, paving plan, irrigation plan, planting plan, and drawings and detail of garden structures. May build models of proposed landscape construction and prepare colored drawings for presentation to client. (See Figure I–4.)

Electrical Drafter

Prepares electrical-equipment working drawings and wiring diagrams used by construction and repair crews who erect, install, and repair electrical equipment and wiring in communications centers, power plants, industrial establishments, commercial or domestic buildings, or electrical distribution systems, performing duties described under Drafter. (See Figure I–5.)

Aeronautical Drafter

Specializes in preparing engineering drawings of developmental or production airplanes and missiles and ancillary equipment, including launch mechanisms and scale models of prototype aircraft, as planned by Aeronautical Engineer.

Electronic Drafter

Drafts wiring diagrams, schematics, and layout drawings used in manufacture, assembly, installation, and repair of electronic equipment such as television cameras, radio transmitters and receivers, audio-amplifiers, computers, and radiation detectors, performing duties as described under Drafter. Drafts layout and detail drawings of racks, panels, and enclosures. May conduct service and interference studies and prepare maps and charts related to radio and television surveys. May be designated according to equipment drafted. (See Figure I–6.)

Civil Drafter

This category is also known by the following titles: Drafter, Civil Engineering; Drafter, Construction; and Drafter, Engineering. Prepares detailed construction drawings, topographic profiles, and related maps and specification sheets used in planning and construction of highways, river and harbor improvements, flood control, drainage, and other civil engineering projects, performing duties as described under Drafter. Plots maps and charts showing profiles and cross sections indicating relation of topographical contours and elevations to buildings, retaining walls, tunnels, overhead power lines,

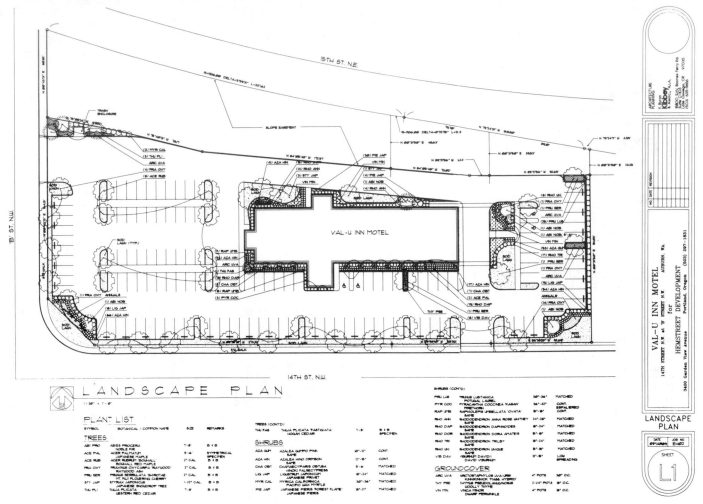

Figure I–4 Landscape plan. *Courtesy OTAK, Inc., for landscaping, and Kibbey & Associates, for site plan.*

and other structures. Drafts detailed drawings of structures and installations such as roads, culverts, fresh water supply and sewage disposal systems, dikes, wharfs, and breakwaters. Computes volume of excavations and fills and prepares graphs and hauling diagrams used in earth-moving operations. May accompany survey crew in field to locate grading markers or to collect data required for revision of construction drawings. May be designated according to type of construction. (See Figure I–7.)

Structural Drafter

Performs duties of Drafter by drawing plans and details for structures employing structural reinforcing steel, concrete masonry, wood, and other structural materials. Produces plans and details of foundations, building frame, floor and roof framing, and other structural elements. (See Figure I–8.)

Castings Drafter

Drafts detailed drawings for castings, which require special knowledge and attention to shrinkage allowances and such factors as minimum radii of fillets and rounds.

Patent Drafter

Drafts clear and accurate drawings of varied sorts of mechanical devices for use by Patent Lawyer in obtaining patent rights.

Tool Design Drafter

Drafts detailed drawing plans for manufacture of tools, usually following designs and specifications indicated by Tool Designer.

Mechanical Drafter

Drafts detailed working drawings of machinery and mechanical devices indicating dimensions and tolerances, fasteners and joining requirements, and other engineering data. Drafts multiple-view assembly and subassembly drawings as required for manufacture and repair of mechanisms. Performs other duties as described under Drafter. Mechanical drafting, in general, is the core of the engineering drafting industry. (See Figure I–9.)

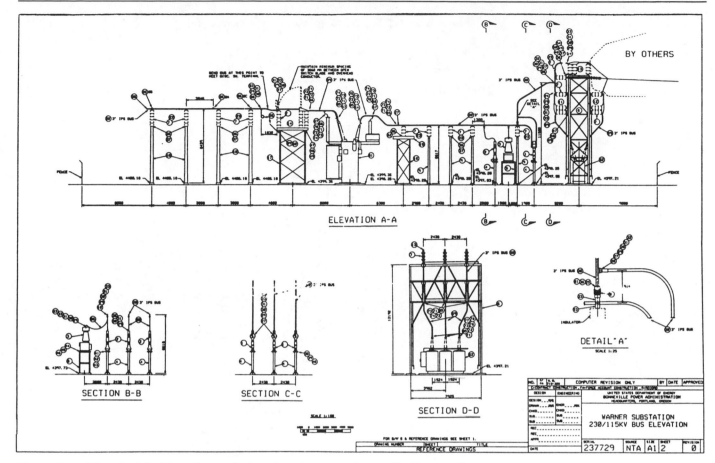

Figure I–5 Electrical drafting plan and elevation. *Courtesy Bonneville Power Administration.*

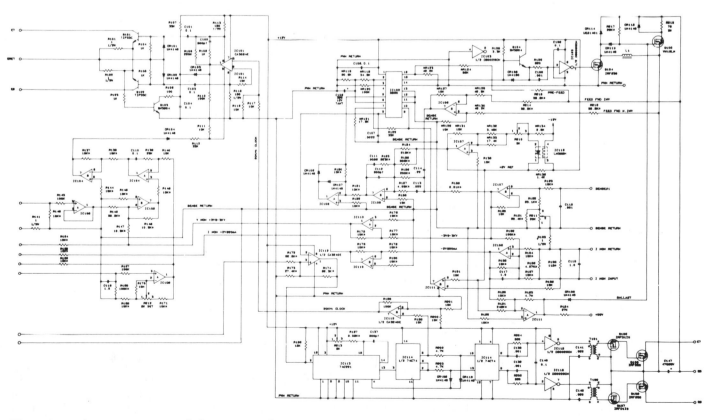

Figure I–6 Computer-generated electronics schematic. *Courtesy Versacad Corporation.*

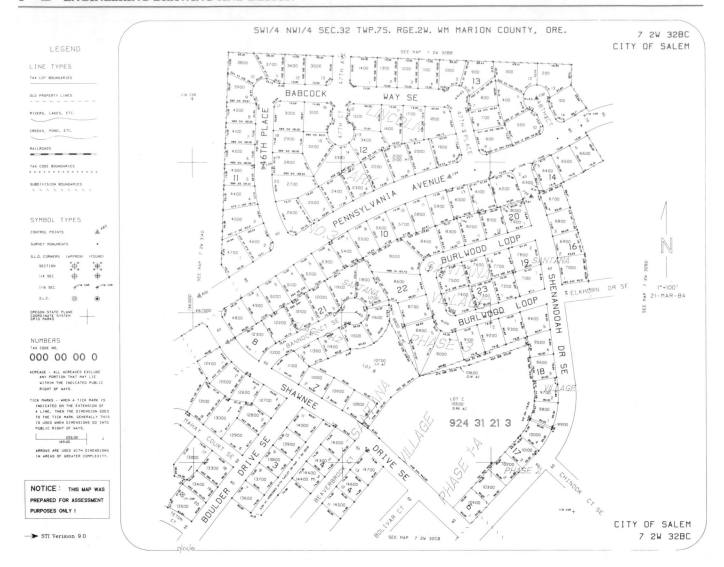

Figure I–7 Civil drafting, computer-generated drafting subdivision plat. *Courtesy Glads Project.*

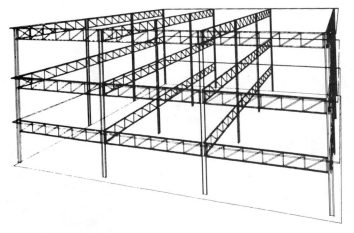

Figure I–8 Computer-generated structural perspective. *Courtesy Computervision Corporation.*

Directional Survey Drafter

Plots oil- or gas-well boreholes from photographic sub-surface survey recordings and other data. Computes and represents diameter, depth degree, direction of inclination, location of equipment, and other dimensions and characteristics of borehole.

Geological Drafter

Draws maps, diagrams, profiles, cross sections, directional surveys, and subsurface formations to represent geological or geophysical stratigraphy and locations of gas and oil deposits. Performs duties described under Drafter. Correlates and interprets data obtained from topographical surveys, well logs, or geophysical prospecting reports, utilizing special symbols to denote geological and geophysical formations or oil field installations. May finish drawings in mediums and according to specifications required for reproduction by blueprinting, photographing, or other duplication methods.

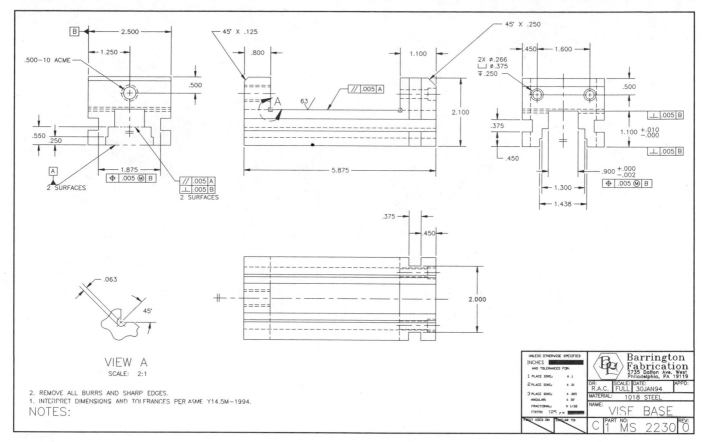

Figure I–9 Computer-generated mechanical drafting. *Courtesy Richard Clouser.*

Geophysical Drafter

Draws subsurface contours in rock formations from data obtained by geophysical prospecting party. Plots maps and diagrams from computations based on recordings of seismograph gravity meter, magnetometer, and other petroleum prospecting instruments and from prospecting and surveying field notes.

Heating and Ventilating Drafter

Also known as Heating Ventilating and Air Conditioning (HVAC) Drafter. Specializes in drawing plans for installation of heating, air-conditioning, and ventilating equipment. May calculate heat loss and heat gain for buildings for use in determining equipment specifications, following standardized procedures. May specialize in drawing plans for installation of refrigeration equipment. (See Figure I–10.)

Plumbing Drafter

Also known as Piping Drafter. Specializes in drafting plans for installation of plumbing and piping equipment for residential, commercial, and industrial installations. (See Figure I–11.)

Automotive Design Drafter

Designs and drafts working layouts and master drawings of automotive vehicle components, assemblies, and

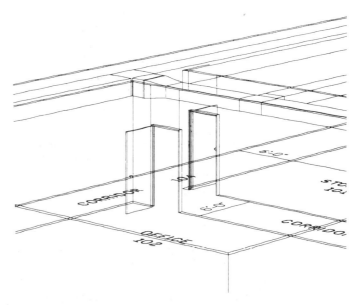

Figure I–10 Computer-generated HVAC pictorial. *Courtesy Computervision Corporation.*

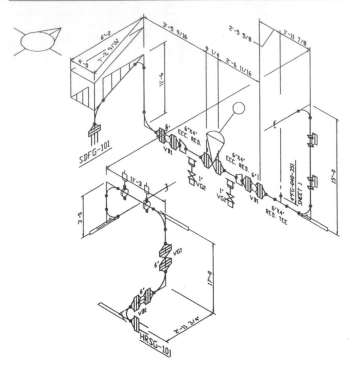

Figure I–11 CADD isometric piping layout. *Courtesy Autodesk, Inc.*

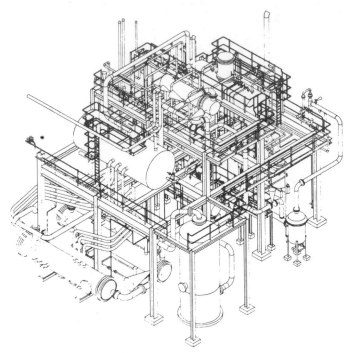

Figure I–12 CADD piping pictorial. *Courtesy Computervision Corporation.*

systems from specifications, sketches, models, prototype and verbal instructions, applying knowledge of automotive vehicle design, engineering principles, manufacturing processes and limitations, and drafting techniques and procedures, using drafting instruments and work aids. Analyzes specifications, sketches, engineering drawings, ideas, and related design data to determine critical factors affecting design of components based on knowledge of previous designs and manufacturing processes and limitations. Draws rough sketches and performs mathematical computations to develop design and work out detailed specifications of components. Applies knowledge of mathematical formulas and physical laws. Performs preliminary and advanced work in development of working layouts and final master drawings adequate for detailing parts and units of design. Makes revisions to size, shape, and arrangement of parts to create practical design. Confers with Automotive Engineer and others on staff to resolve design problems. Specializes in design of specific type of body or chassis components, assemblies or systems such as door panels, chassis frame and supports, or braking system.

Oil and Gas Drafter

Drafts plans and drawings for layout, construction, and operation of oil fields, refineries, and pipeline systems from field notes, rough or detailed sketches, and specifications. Develops detail drawings for construction of equipment and structures, such as drilling derricks, compressor stations, gasoline plants, frame, steel, and

masonry buildings, piping manifolds and pipeline systems, and for manufacture, fabrication, and assembly of machines and machine parts. (See Figure I–12.)

Technical Illustrator

Lays out and draws illustrations for reproduction in reference works, brochures, and technical manuals dealing with assembly, installation, operation, maintenance, and repair of machines, tools, and equipment. Prepares drawings from blueprints, designs, mockups, and photographs by methods and techniques suited to specified reproduction process or final use, such as diazo, photo-offset, and projection transparencies, using drafting and optical equipment. Lays out and draws schematic, perspective, axonometric, orthographic, or oblique-angle views to depict function, relationship, and assembly sequence of parts and assemblies, such as gears, engines, and instruments. Shades or colors drawing to emphasize details or to eliminate undesired background, using ink, crayon, airbrush, and overlays. Pastes instructions and comments in position on drawing. May draw cartoons and caricatures to illustrate operation, maintenance, and safety manuals and posters. (See Figure I–13.)

Cartographic Drafter

Draws maps of geographical areas to show natural and construction features, political boundaries, and other features. Analyzes survey data, survey maps and photographs, computer- or automated-mapping products, and other records to determine location and names of

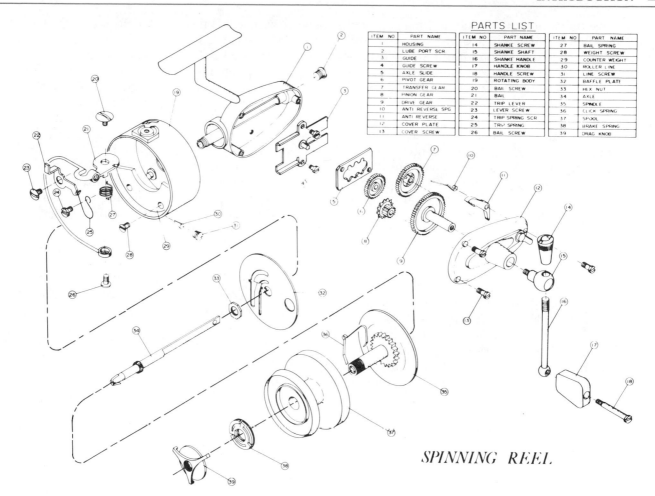

ITEM NO	PART NAME	ITEM NO	PART NAME	ITEM NO	PART NAME
1	HOUSING	14	SHANKE SCREW	27	BAIL SPRING
2	LUBE PORT SCR	15	SHANKE SHAFT	28	WEIGHT SCREW
3	GUIDE	16	SHANKE HANDLE	29	COUNTER WEIGHT
4	GUIDE SCREW	17	HANDLE KNOB	30	ROLLER LINE
5	AXLE SLIDE	18	HANDLE SCREW	31	LINE SCREW
6	PIVOT GEAR	19	ROTATING BODY	32	BAFFLE PLATE
7	TRANSFER GEAR	20	BAIL SCREW	33	HEX NUT
8	PINION GEAR	21	BAIL	34	AXLE
9	DRIVE GEAR	22	TRIP LEVER	35	SPINDLE
10	ANTI REVERSE SPG	23	LEVER SCREW	36	CLICK SPRING
11	ANTI REVERSE	24	TRIP SPRING SCR	37	SPOOL
12	COVER PLATE	25	TRIP SPRING	38	BRAKE SPRING
13	COVER SCREW	26	BAIL SCREW	39	DRAG KNOB

SPINNING REEL

Figure I–13 Technical illustration, exploded isometric assembly.

features. Studies records to establish boundaries of properties and local, national, and international areas of political, economic, social, or other significance. Geological and topographical maps are drawn by a Cartographic Drafter.

Photogrammetrist

Analyzes source data and prepares mosaic prints, contour-map profile sheets, and related cartographic materials requiring technical mastery of photogrammetric techniques and principles. Prepares original maps, charts, and drawings from aerial photographs and survey data and applies standard mathematical formulas and photogrammetric techniques to identify, scale, and orient geodetic points, estimations, and other planimetric or topographic features and cartographic detail. Graphically represents aerial photographic detail, such as contour points, hydrography, topography, and cultural features, using precision stereoplotting apparatus or drafting instruments. Revises existing maps and charts and corrects maps in various stages of compilation. Prepares rubber, plastic, or plaster three-dimensional relief models.

▼ PROFESSIONAL ORGANIZATION

The *American Design Drafting Association (ADDA)* is a nonprofit, professional organization dedicated to the advancement of design and drafting.

The basic knowledge and skill required to do board drafting is also helpful in preparing input data for or operating computerized drafting systems such as digitizers, plotters, and photo composition devices.

Whatever means are used to generate drawings, persons skilled in drafting techniques are in demand. Therefore, drafting can be a rewarding and challenging career to those who can visualize and portray ideas graphically.

▼ HOW TO BECOME A DRAFTER

Persons considering drafting as a career should be able to visualize what is to be drawn and be mechanically minded. Potential drafters should be willing to work hard and learn new things, and be able to adjust to varying working conditions. Above all, they should be neat and accurate in their work.

Today, most industries, to fill their drafting job openings, are employing only graduates who have received specialized training in design and drafting from technical institutes, junior or community colleges, and vocational schools. The specialized training includes basic and advanced drawing, mathematics (algebra, geometry, and trigonometry), physics or chemistry, English, humanities, technology courses, and courses in specific fields such as aeronautical, architectural, automotive, electrical, electronics, illustrative, mapping, mechanical, piping, structural, and sheet metal drafting. Computer-aided drafting is essential training for today's employment market.

▼ DRAFTING OCCUPATIONAL LEVELS

There are several levels of advancement for drafters. They generally are based on educational background and practical experience. For advancement, most companies require career employees to hold a two-year college or trade-school degree; however, persons with an equivalent or larger amount of practical experience in a specific job-related area can obtain advanced occupational levels also.

Occupational levels differ slightly from one industry to another although the general classifications are generally accurate nationwide. Suggested educational levels have been added for reference.

▼ DRAFTING JOB OPPORTUNITIES

Drafting job opportunities, which include all possible drafting employers, fluctuate with national and local economies. This is no different from most other employment in industry or construction. Drafting is tied closely to manufacturing and construction so that a slowdown or speedup in these industries nationally affects the number of drafting jobs available. This same effect upon drafting opportunities may be experienced at the local level or with specific industries. For example, construction may be strong in one part of the country, and slow in another, so the demand for drafters in those localities is strong or slow accordingly. Other examples are demonstrated when automobile manufacturers experience poor sales, and drafting opportunities decline, or when hi-tech industries expand, and more drafters are needed. Public and private indicators suggest that the demand for drafters will continue to be strong for graduates of two-year post-secondary drafting curriculums. The emphasis should be placed on drafting skills, computer-aided drafting, math, English, and oral- and written-communication skills.

The types of drafting jobs available are also controlled by local demands. A predominance of mechanical drafting jobs are found in metropolitan areas where manufacturing is strong, while in outlying areas there may be more civil or structural drafting jobs. Each local area has a need for more of one type of drafting skill than another. Also, drafting curriculums in different geographical areas generally specialize in the fields of drafting that help fill local employment needs. Some drafting programs offer a broad-based education so that graduates may have versatile employment opportunities. When selecting a school look into curriculum, placement potential and local demand. Talk to representatives of local industries for an evaluation of the drafting curriculum.

Opportunities for advancement for drafters are excellent, although dependent on the advancement possibilities of the specific employer. Advancement also depends on an individual's initiative, ability, product knowledge, and willingness to continue to be educated. Additional education for advancement usually includes increased levels of mathematics, pre-engineering, engineering, computers, and advanced drafting. Drafting has traditionally been an excellent stepping-stone to designing, engineering, and management.

▼ DRAFTING SALARIES AND WORKING CONDITIONS

Salaries in drafting professions are comparable to salaries of other professions with equal educational requirements. Employment benefits vary according to each individual employer. However, most employers offer vacation and health insurance coverage, while others include dental, life, and disability insurance.

▼ COMPUTERS IN DRAFTING

The industrialized world is wrestling with the dramatic and constant changes of the electronic revolution and its child, the computer. The speed-of-light capabilities of the computer have shaped the gathering, processing, storage, and retrieval of information into a major industry. Every day our lives are touched by the effects of computers. Contrary to science-fiction works of years past, computers have not grown to occupy the core of the earth but have shrunk to microscopic proportions. They now occupy tiny spaces inside our machines and tools.

The tools of the drafting trade have also felt the electronic touch of computers. The silicon chip, that maze of delicate circuitry, has found its way into drafting tools. The heart of modern drafting tools is now the computer. The core of a manual drafting workstation, parallel bars, arm- and track-drafting machines, are rapidly becoming obsolete. The tables they are attached to are being used as reference space and handy places for books, papers, and vendor catalogs.

Now the drafter, or operator, uses electronic hardware to construct drawings, not on paper, but on what looks like a television screen. The screen is called a monitor, video display screen, or video display terminal. The drafting process is now called *computer-aided design drafting,* or *CADD.* Most people say"computer-aided drafting" and use the acronym of CAD. Actually, CAD means computer-aided design.

▼ THE ENGINEERING DESIGN PROCESS

Creativity is very important in the world of our modern manufacturing and construction processes. Engineers, architects, and designers must solve many problems for the design to be functional, attractive, and competitively priced among the endless number of products available today. Creativity may be fostered by the following principles:

○ **New Methods:** Problems in industry and construction often must be solved in areas that require original ideas or revisions of existing designs.

○ **Determination:** A designer must have a lot of determination in order to keep working on the project until the problems are solved.

○ **Attitude:** The designer needs a positive attitude and must realize the possibilities of modern technology.

○ **Confidence:** A good designer must have confidence in his or her ability to solve a problem within the marketing and manufacturing or construction requirements.

○ **Experimentation:** Careful testing and recording of data is a key to successful design. Experimentation, prototyping, testing, and analysis are often used in the design process.

○ **Logic:** A designer may lose ideas and waste valuable time without a logical approach to problem solving. A step-by-step design process pays off when combined with the other qualities previously discussed.

Design Analysis

The design analysis process is a methodical approach that is used by engineers, architects, and designers to solve problems. Most designers use some process for problem solving. Most of these processes are based on a logical system of experimentation, data gathering, and record keeping. A typical design analysis process has these elements:

○ problem identification ○ analysis
○ preliminary ideas   ○ decision making
○ refinement   ○ implementation

This entire process is often referred to as research and development, or R & D.

The Six Steps of Design Analysis

Problem Identification. The first step of the design process is to identify everything that is known about the problem such as historical and background information, economic requirements, manufacturing or con-

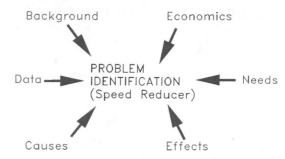

Figure I–14 The elements that go into problem identification.

struction processes available, and market information. The design process is often based on consumer demand for the product. A common way to begin this process is to put all of the known elements together in the following steps:

Step 1.   Identify the problem.

Step 2.   List the known limitations.

Step 3.   Specify the desirable features.

Figure I-14 illustrates the main elements of problem identification. This problem is the design of a gear reducer that reduces the revolutions per minute (rpm) of a motor for use in driving a winch.

Preliminary Ideas.   The designer records all ideas in the form of notes, written descriptions, and sketches. These preliminary ideas may be original or modifications of existing products. Figure I–15 shows the elements that make up typical preliminary ideas.

Problem Refinement.   Now the designer takes several preliminary ideas and reduces them to two or three that are the best. These best ideas are then refined with more detail. This additional detail may include formal sketches, manual or computer-aided preliminary drawings. The preliminary drawings often have the major dimensions, physical properties, shape, and weight characteristics as displayed in Figure I–16.

Analysis.   The analysis of the problem is a very important part of the design process. After the designer nar-

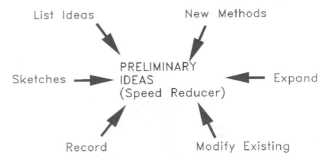

Figure I–15 The elements that make up typical preliminary ideas.

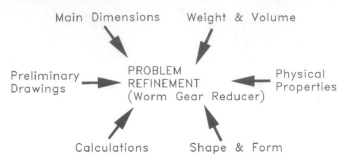

Figure I–16 Characteristics of preliminary drawings.

rows the design down to a few refined preliminary ideas, he or she is ready to consider cost, strength, function, and anything else that affects the design. Here the designer may set up test situations and record data in order to come up with the best result. Prototypes or models may be built. Three-dimensional computer-generated models may be created. These computer models may be used to provide stress analysis, weight calculations, or other needed calculations. They might even be used to generate an animation of the product in use. This also gives the designer a chance to consider physical properties such as material, shape, size, color, or texture. Some elements of analysis are shown in Figure I–17.

Decision.   It is time to pick the best design. This might be a joint decision between a team of designers, engineers, and manufacturing or construction experienced leaders. The positive and negative features of each preliminary design might be listed to help narrow down the decision. (See Figure I-18.)

Implementation.   This is the step in the design process in which formal drawings are made, specifications are

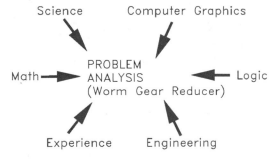

Figure I–17 Elements of analysis.

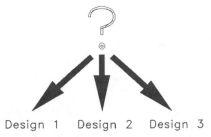

Figure I–18 Deciding on the preliminary design to use.

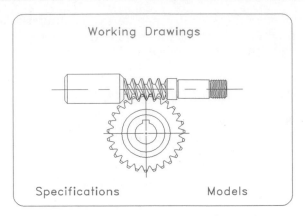

Figure I–19 The implementation stage of the design process.

written, and the product is manufactured. In some cases, actual manufacturing or construction may begin. In other situations, it may be necessary to manufacture a prototype for final testing before the product is released to mass production. Figure I–19 illustrates the implementation stage of the design process.

CADD Applications

ONCE IS ALWAYS ENOUGH WITH CADD

by Karen Miller
Graphic Specialist with Tektronix, Inc.

Reusability is one of the important advantages of CADD. With CADD it is never necessary to draw anything more than once. This advantage is even enhanced by developing a CADD symbols library. Building a parts library for reusability has increased productivity, decreased development costs, and set the highest standards for quality at the Test and Measurement Documentation Group at Tektronix, Inc.

The parts library began by reusing 2-D isometric parts created by the AutoCAD illustrator and saved as blocks (symbols) in a Parts Library Directory. By pathing the named blocks back to this library directory, each block becomes accessible to any directory and drawing file. This allows the CADD illustrator to insert library symbols into any drawing simply by typing the block name.

The illustrator adds new parts to the library as a product is disassembled and illustrated. Each part is given the next available number as its block name in the library, as shown in Figure I–20.

Originally, the AutoCAD drawings were combined with text (written using Microsoft Word) in Ventura Publisher to create technical publications.

Now, the AutoCAD drawings are added to document files in Interleaf technical publishing software. The entire parts library is available to both AutoCAD and Interleaf users (Figure I–21).

From the point of view of cost management, the parts library has saved hundreds of hours of work. From the illustrator's view, the parts library helps improve productivity and frees time for new or complex projects.

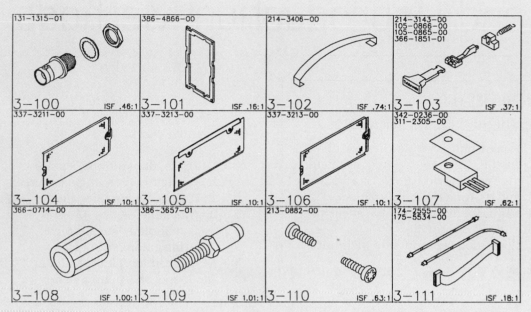

Figure I–20 CADD parts library. *Courtesy Karen Miller, Technical Illustrator, Tektronix, Inc.*

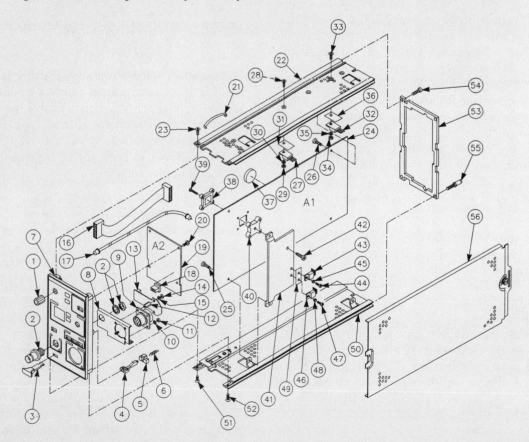

Figure I–21 Integrating the parts from the library for a complete exploded assembly drawing. Tektronix current probe amplifier. *Courtesy Karen Miller, Technical Illustrator, Tektronix, Inc.*

CHAPTER 1

Drafting Equipment, Media, and Reproduction Methods

LEARNING OBJECTIVES

After completing this chapter you will:

❍ Describe and demonstrate the use of various manual drafting tools and equipment.
❍ Read engineer's, architect's, and metric scales, and drafting machine verniers.
❍ Discuss and use drafting media, sheet sizes, and title block information.
❍ Explain common reproduction methods.

▼ DRAFTING EQUIPMENT

Drafting tools and equipment are available from a number of vendors that sell professional drafting supplies. For accuracy and long life, always purchase high-quality equipment. Local vendors can be found by looking in the yellow pages of your telephone book under headings such as: Drafting Room Equipment & Supplies, Blueprinting, Architects' Supplies, Engineering Equipment & Supplies, or Artists' Materials & Supplies.

Drafting supplies and equipment can be purchased in a kit or items can be bought individually. Whether equipment is purchased in a kit or by the individual tool, the items that are normally needed include the following:

❍ One 0.3 mm automatic drafting pencil with 4H, 2H, and H leads.
❍ One 0.5 mm automatic drafting pencil with 4H, 2H, H, and F leads.
❍ One 0.7 mm automatic drafting pencil with 2H, H, and F leads.
❍ One 0.9 mm automatic drafting pencil with H, F, and HB leads. Drafters may elect to purchase two or more pencils and use a different grade of lead in each. Doing so reduces the need to constantly change leads. Some drafters use a light-blue lead for layout work.
❍ 6-in. bow compass.
❍ Dividers.
❍ Eraser. Select an eraser that is recommended for drafting with pencil on paper.

❍ Erasing shield.
❍ 8-in. 30°–60° triangle.
❍ 8-in. 45° triangle.
❍ Irregular curve.
❍ Scales:
  1. Triangular architect's scale.
  2. Triangular civil engineer's scale.
  3. Triangular metric scale.
❍ Drafting tape.
❍ Circle template (small holes).
❍ Lettering guide (optional).
❍ Arrowhead template (optional).
❍ Sandpaper sharpening pad.
❍ Dusting brush.

▼ DRAFTING FURNITURE

Tables

There is a large variety of drafting tables available, ranging from economical models to complete professional workstations. Drafting tables are generally sized by the dimensions of their tops. Standard tabletop sizes range from 24 × 36 in. to 42 × 84 in.

The features to look for in a good-quality, professional table include:

❍ One-hand tilt control.
❍ One-hand or foot height control.
❍ The ability to position the board vertically.
❍ An electrical outlet.
❍ A drawer for tools and/or drawings.

Some manufacturers ship tables with tops that are ready to draw on. Others ship tables with tops of steel or basswood that need to be covered. Basswood can be a drawing surface; however, most offices commonly cover drafting-table tops with smooth, specially designed surfaces. The material, usually vinyl, provides the proper density for effective use under normal drafting conditions. After compass or divider points have pierced the surface, the small holes close and so provide a smooth surface for continued use. Drafting tape is commonly used to adhere drawings to the tabletop, although some drafting tables have magnetized tops and use magnetic strips to attach drawings.

Chairs

Better drafting chairs have the following characteristics:

○ Padded or contoured seat design.
○ Height adjustment.
○ Foot rest.
○ Fabric that allows air to circulate.
○ Sturdy construction.

▼ DRAFTING PENCILS, LEADS, AND SHARPENERS

Mechanical Pencils or Lead Holders

The term *mechanical pencil,* also referred to as lead holder, is applied to a pencil that requires a piece of lead to be manually inserted; by some physical action such as twist, push, or pull, the lead is mechanically, or semiautomatically, advanced to the tip. These have been replaced by automatic pencils for most use.

Leads for mechanical pencils are bought separately. Leads are graded by hardness and designated by a number and letter, or one or two letters. The designation is usually found along or on one end of the lead. Always sharpen the end opposite the designation. Figure 1–1a shows a mechanical pencil.

Automatic Pencils

The term *automatic pencil* refers to a pencil with a lead chamber that, at the push of a button or tab, advances the lead from the chamber to the writing tip and, when a new piece of lead is needed, advances the new piece to the tip. Automatic pencils are designed to holds leads of one width so you do not need to sharpen the lead. These pencils are available in several different lead sizes. Drafters have several automatic pencils. Each pencil has a different grade of lead hardness and is used for a specific technique. (See Figure 1–1b.)

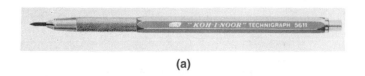

(a)

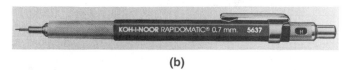

(b)

Figure 1–1   (a) Mechanical pencil, also known as lead holder. *Courtesy Koh-I-Noor, Inc.* (b) An automatic pencil. Lead widths are 0.3, 0.5, 0.7, and 0.9 mm. *Courtesy Koh-I-Noor, Inc.*

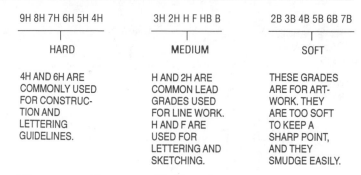

9H 8H 7H 6H 5H 4H	3H 2H H F HB B	2B 3B 4B 5B 6B 7B
HARD	MEDIUM	SOFT
4H AND 6H ARE COMMONLY USED FOR CONSTRUCTION AND LETTERING GUIDELINES.	H AND 2H ARE COMMON LEAD GRADES USED FOR LINE WORK. H AND F ARE USED FOR LETTERING AND SKETCHING.	THESE GRADES ARE FOR ART-WORK. THEY ARE TOO SOFT TO KEEP A SHARP POINT, AND THEY SMUDGE EASILY.

Figure 1–2   The range of lead grades.

Lead Grades

The leads that you select for line work depend on the amount of pressure you apply and other technique factors. Experiment until you identify the leads that give the best line quality. Leads commonly used for thick lines range from 2H to F, while leads for thin lines range from 4H to H, depending on individual preference. Construction lines for layout and guidelines are very lightly drawn with a 6H or 4H lead. Figure 1–2 shows the different lead grades.

Polyester and Special Leads

Polyester leads, also known as *plastic leads,* are for drawing on *polyester drafting film,* often called by its trade name, *Mylar®.* Plastic leads come in grades equivalent to F, 2H, 4H, and 5H and are usually labeled with a prefix and number. Pentel™, for example, uses P1 and P2. Some companies make a combination lead for use on both vellum and polyester film.

Colored leads have special uses. Red lead is commonly used for corrections. Red prints as a black line. Blue leads can be used on the original drawing by the supervisor to tell the drafter the corrections that need to be made. Blue lead will not show on the print when the original drawing with blue lines on it is run through a diazo machine. Some drafters use light-blue lead for all layout work and guidelines because it will not reproduce on a diazo printer; however, it may show on a photo copier.

Basic Pencil Technique

Automatic pencils do not require rotation, although some drafters feel rotating the pencil makes darker lines. The automatic pencil should be held near vertical. The full surface of the lead is used when the pencil is in a vertical position. Provide enough pressure and go over each line enough times to make a line dark and crisp. Take care not to make it too thick. Figure 1–3 shows some basic pencil motions.

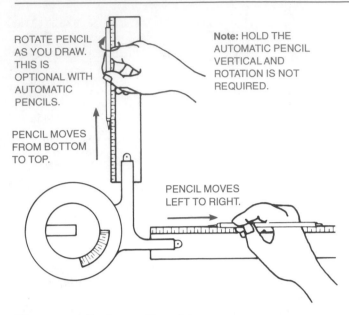

Figure 1–3 Basic pencil motions.

Sanding Block

Especially useful for sharpening compass lead, this is one of the simplest devices used for sharpening. Sandpaper stapled to a wooden paddle is called a sandpaper, or sanding block. Plastic lead fills sandpaper rapidly so several sheets are needed as compared to graphite lead. To avoid smudging your drawing, use the sanding block away from your drawing table and dispose of the graphite carefully.

Pocket Pointer

A portable sharpener for mechanical pencils is the pocket pointer. This pointer contains blades that sharpen lead to a conical point. The pocket pointer works with either graphite or plastic lead.

Cutting Wheel Pointer

The best type of conical-point sharpener for mechanical pencil holders is a mechanical lead pointer with a tool-steel cutting wheel. Use the slots provided in the top to expose the right length of lead to get a sharp or slightly dull point. The slightly dull point is used for lettering. This sharpener works on graphite or polyester leads equally well.

▼ TECHNICAL PENS AND ACCESSORIES

Technical Pens

Also known as technical fountain pens, technical pens have improved in quality and ability to produce excellent inked lines. These pens function on a capillary action where a needle acts as a valve to allow ink to flow from a storage cylinder through a small tube,

which is designed to meter the ink so a specific line width is created. Specially designed technical pens are used for liquid ink plotters in computer-aided drafting. A technical pen is shown in Figure 1–4. Technical pens may be purchased individually or in sets. The different tip sizes used to make various line widths range from a narrow number 6 × 0 (.005 in./0.13 mm) to a wide number 7 (.079 in./2 mm). Figure 1–5 shows a comparison of some of the different line widths available with technical pens.

Technical pens are available in different price ranges as determined by the kind of material used to make the point. Each kind of point has a recommended use. Stainless steel points are made for use on vellum. These points wear out very rapidly when used on polyester film. Tungsten carbide points are recommended for polyester film and can be used on vellum. Jewel points provide the longest life and can be used on vellum or polyester film with the best results. They also provide the smoothest ink flow on polyester film.

In addition to having the advantage of a constant line width, technical pens have a reservoir that allows the drafter to make inked lines for a long period of time before ink must be added. Technical pens may be used with templates to make circles, arcs, and symbols. Compass adapters to hold technical pens are also available. Technical pen tips are designed to fit into scribers for use with lettering guides. This concept is discussed further in Chapter 3.

Most pen manufacturers have designed pen holders or caps that help keep pen points moist and ready for use. When pens are used daily, cleaning may be necessary only periodically. Some symptoms to look for when pens need cleaning include the following:

○ Ink constantly creates a drop at the pen tip.
○ Ink tends to flow out around the tip holder.
○ The point plunger does not activate properly. If you do not hear and feel the plunger when you shake the pen, then it is probably clogged with thick or dried ink.

Figure 1–4 Technical pen. *Courtesy Koh-I-Noor, Inc.*

6x0	4x0	3x0	00	0	1	2	2½	3	3½	4	6	7
.13	.18	.25	.30	.35	.50	.60	.70	.80	1.00	1.20	1.40	2.00
005 in	007 in	010 in	012 in	014 in	020 in	024 in	028 in	031 in	039 in	047 in	055 in	079 in
13 mm	18 mm	25 mm	30 mm	35 mm	50 mm	60 mm	70 mm	80 mm	1 00 mm	1 20 mm	1 40 mm	2 00 mm

Figure 1–5 Technical pen line widths. *Courtesy Koh-I-Noor, Inc.*

○ Ink does not flow freely when drawing a line.
○ Ink flow starts with difficulty.

The ink level in the reservoir should be kept between one-quarter to three-quarters full. If the cylinder is too full, the ink may not flow well; if too little ink is present, then the pen may skip when in use.

Pen Cleaning

Read the cleaning instructions that come with the brand of pen that you purchase. Some pens require disassembly for cleaning while others should not be taken apart. Some manufacturers strongly suggest that pen tips remain assembled. The main parts of a technical pen include the cap, nib, pen body, ink cartridge (or reservoir), and pen holder as seen in Figure 1–6. The nib contains a cleansing wire with drop weight and a safety plug. Most manufacturers recommend that the nib remain assembled during cleaning as the cleansing wire is easily damaged. To fill the technical pen, unscrew the cap, holder, and clamp ring from the pen body. Remove the cartridge and fill it with ink to within ¼ of the top. Slowly replace the cartridge and assemble the parts. Pens should be cleaned before each filling or before being stored for a long period of time. Clean the technical pen nib, cartridge, and body separately in warm water or special cleaning solution.

Ultrasonic pen cleaners are available to clean points. Pens are placed in a tank where millions of energized microscopic bubbles, generated by ultrasonic action, carry cleaning solution into the smallest openings of the drawing point to scrub the tube inside and out.

Syringe pressure pen cleaners and point starters are also available for cleaning pens. These cleaners use pressure and suction for the cleaning action. These units are provided with a connector for use as a pen starter. Specially formulated pen cleaner should be used for best results in either the ultrasonic or syringe units. Pen cleaner can also be used to soak pen points for cleaning by hand.

Ink

Drafting inks should be opaque, or have a matte or semi-flat black finish that will not reflect light. The ink should reproduce without hot spots or line variation. Drafting ink should have excellent adhesion properties for use on paper or film. Certain inks are recommended for use on film in order to avoid peeling, chipping, or cracking. Inks recommended for use in technical pens also have nonclogging characteristics. This property is especially important for use in high-speed computer-graphics plotters.

When selecting an ink, be sure to purchase one for the job you want done. First, determine how the ink will be applied, that is, from a technical pen, computer plotter, air brush, fountain pen, calligraphy pen, or with a brush. Second, determine the surface the ink will be used on, such as paper (vellum or bond), polyester film, or acetate. Third, determine if your use requires the ink to be opaque, fast drying, waterproof, or erasable.

▼ ERASERS AND ACCESSORIES

The common shapes of erasers are rectangular and stick. The stick eraser works best in small areas. There are three basic types of erasers: pencil, ink, and plastic. The plastic eraser is used for plastic lead or ink on polyester film. These erasers are identified by their white or translucent color. Select an eraser that is recommended for the particular material used. When used with ink, apply very light pressure, and take the utmost care because the friction developed by the speed of erasure can easily damage the drafting film surface and prevent the adhesion of ink when redrawing over the erased area. Moistening the eraser during use on polyester film helps to reduce any damage to the drawing surface.

Erasing Tips

When erasing, the idea is to remove an unwanted line or letter. You do not want to remove the surface of the paper or polyester film. Erase only hard enough to remove the unwanted line. However, you must bear down hard enough to eliminate the line completely. If all the line does not disappear, ghosting results. A ghost is a line that seems to have been eliminated but still shows on a print. Lines that have been drawn so hard as to make a groove in the drawing sheet can cause a ghost too. To remove ink from vellum, use a pink or green eraser, or an electric eraser. Work the area slowly. Do not apply too much pressure or erase in one spot too long or you will go through the paper. On polyester film, use a vinyl eraser and/or a moist cotton swab. The inked line usually comes off easily, but use caution. If you destroy the matte surface of either a vellum or polyester sheet you will not be able to redraw over the erased area.

Lead may be picked up from the drawing board surface and transmitted to the back of the drawing surface. When this happens, the drawing must be

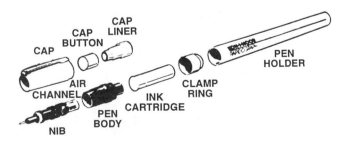

Figure 1–6 The parts of a technical pen. *Courtesy Koh-I-Noor, Inc.*

turned over and the graphite removed from the back surface of the drawing.

Electric Erasers

Professional drafters use electric erasers. Those with cords that plug in are best, but cordless, rechargeable units are also available. When working with an electric eraser, you do not need to use very much pressure because the eraser operates at high speed. The purpose of the electric eraser is to remove unwanted lines quickly. Use caution, these erasers can also remove paper quickly!

Erasing Shield

Erasing shields are thin metal or plastic sheets with a number of differently shaped holes. (See Figure 1–7.) They are used to erase small, unwanted lines or areas. For example, if you have a corner overrun, place one of the slots of the erasing shield over the area to be removed while covering the good area.

Eradicating Fluid

Eradicating fluid is primarily used with ink on film or for removal of lines from sepia (brown) prints. Sepia prints are often used to make corrections rather than correcting the original drawing. The eradicating fluid is most often applied with a brush or cotton swab. If in doubt about its application, follow the manufacturer's instructions; however, application is usually done by lightly moistening the area to be corrected. Then the solution is wiped with a tissue, being careful to remove all residue. Eradicating fluid is especially effective for removing aged ink lines and for erasure of large areas.

Cleaning Agents

Special eraser particles are available to help reduce smudging and to keep the drawing and your equipment clean. The particles also help float triangles, straightedges, and other drafting equipment to reduce line smudging. Use this material sparingly since too much of it can cause your lines to become fuzzy. Cleaning powders are not recommended for use on ink drawings, nor on polyester film.

Dusting Brush

Use a dusting brush to remove eraser particles from your drawing. Doing so helps reduce the possibility of smudges. Avoid using your hand to brush away eraser particles because the hand tends to cause smudges which reduces drawing neatness.

The brush should be cleaned regularly with soap and water because it picks up graphite particles and casts a slight film over the drawing.

▼ DRAFTING INSTRUMENTS
Kinds of Compasses

Compasses are used to draw circles and arcs. However, using a compass can be time consuming. Use a template, whenever possible, to make circles or arcs more quickly. A compass is especially useful for large circles.

There are several basic types of compasses:

1. Drop-bow compasses are used mostly for drawing small circles. The center rod contains the needle point and remains stationary while the pencil or pen leg revolves around it. (See Figure 1–8.)

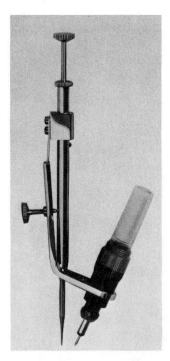

Figure 1–8 Drop bow compass, technical pen model. *Courtesy J. S. Staedtler, Inc.*

Figure 1–7 Erasing shield. *Courtesy Koh-I-Noor, Inc.*

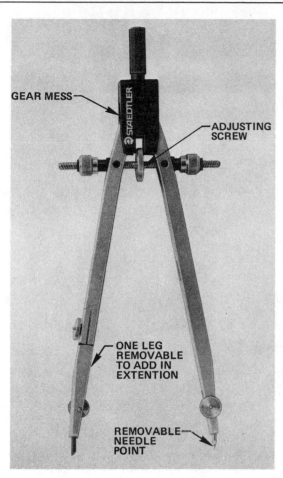

Figure 1–9 Bow compass. *Courtesy J. S. Staedtler, Inc.*

2. Bow compasses, shown in Figure 1–9, are used for most drawings. This type of compass is commonly used by professionals.
3. Beam compasses consist of a bar with an adjustable needle, and a pencil or pen attachment for swinging large arcs or circles. Also available is a beam that is adaptable to the bow compass. Such an adapter works only on bow compasses that have a removable break point.

Use of Compass

Keep both the compass needle point and lead point sharp. The points are removable for easy replacement. The better compass needle points have a shoulder on them. The shoulder helps keep the point from penetrating the paper more than necessary. Compare the needle points in Figure 1–10.

The compass lead should, in most cases, be one grade softer than the lead you use for straight lines because less pressure is used on a compass than a pencil. Keep the compass lead sharp. An elliptical point is commonly used with the bevel side away from the needle leg. Keep the lead and the point equal in length. Figure 1–11 shows properly aligned and sharpened points on a compass.

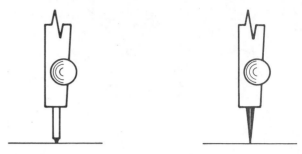

Figure 1–10 Compass points.

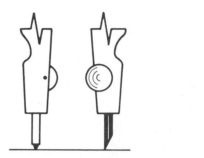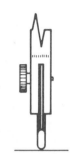

Figure 1–11 Properly sharpened and aligned elliptical compass point.

Use a sandpaper block to sharpen the elliptical point. Be careful to keep the graphite residue away from your drawing and off your hands. Remove excess graphite from the point with a tissue or cloth after sharpening. Sharpen the lead often.

Some drafters prefer to use a conical point in their compass.

Protecting the Sheet During Compass Use. If you are drawing a number of circles from the same center, you will find that the compass point causes an ugly hole in your drawing sheet. Reduce the chance of making such a hole by placing a couple of pieces of drafting tape at the center point for protection. There are small plastic circles available for just this purpose. Place one at the center point, then pierce the plastic with your compass. (See Figure 1–12.)

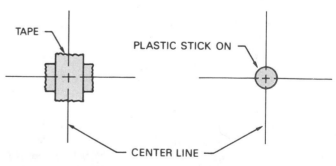

Figure 1–12 Drawing sheet protection from compass point.

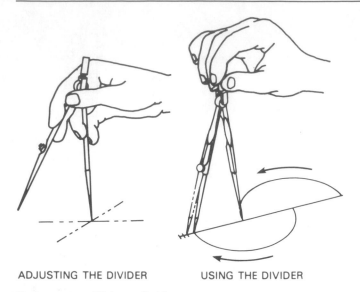

ADJUSTING THE DIVIDER   USING THE DIVIDER

Figure 1–13 Using a divider.

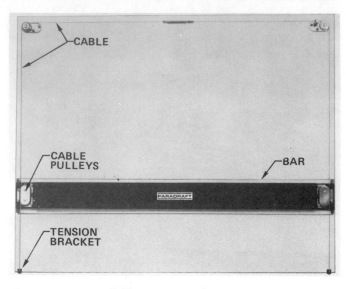

Figure 1–14 Parallel bar. *Courtesy Charvoz-Carsen Corporation.*

Dividers

Dividers are used to transfer dimensions or to divide a distance into a number of equal parts.

Note: Do not try to use dividers as a compass.

Some drafters prefer to use bow dividers because the center wheel provides the ability to make fine adjustments easily. Also, the setting remains more stable than with standard friction dividers.

A good divider should not be too loose or tight. It should be easily adjustable with one hand. In fact, you should control a divider with one hand as you lay out equal increments or transfer dimensions from one feature to another. Figure 1–13 shows how the divider should be handled when used.

Proportional

Proportional dividers are used to reduce or enlarge an object without the need of mathematical calculations or scale manipulations. The center point of the divider is set at the correct point for the proportion you want. Then you measure the original size line with one side of the proportional divider and the other side automatically determines the new reduced or enlarged size.

Parallel Bar

The parallel bar slides up and down the board to allow you to draw horizontal lines. (See Figure 1–14.) Vertical lines and angles are made with triangles in conjunction with the parallel bar. The parallel bar is commonly found in architectural drafting offices because architectural drawings are frequently very large. Architects often need to draw straight lines the full length of their boards and the parallel bar is ideal for such lines.

Triangles

There are two standard triangles. One has angles of 30°–60°–90° and is known as the 30°–60° triangle. The other has angles of 45°–45°–90° and is known as the 45° triangle. Figure 1–15 shows these popular triangles.

Some drafters prefer to use triangles in place of a vertical drafting machine scale as shown in Figure 1–16. The machine protractor or the triangle can be used to make angled lines. Drafters who use parallel bars rather than drafting machines also use triangles to make vertical and angled lines.

Triangles may also be used as straightedges to connect points for drawing lines without the aid of a parallel bar or machine scale. Triangles are used individually or in combination to draw angled lines in 15° increments. (See Figure 1–17.) Also available are adjustable triangles with built-in protractors that are used to make angles of any degree up to a 45° angle.

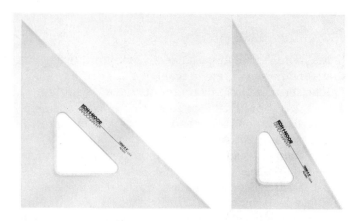

Figure 1–15 45° and 30°–60° triangles. *Courtesy Koh-I-Noor, Inc.*

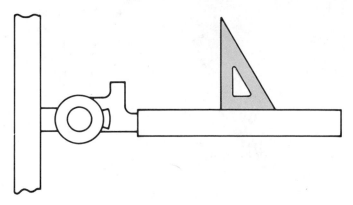

Figure 1–16 Using a triangle with a drafting machine.

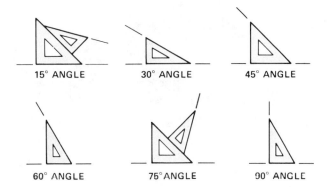

Figure 1–17 Angles that may be made with the 30°–60° and 45° triangles individually or in combination.

Templates

Circle Templates. Circle templates are available with circles in a range of sizes beginning with ¹⁄₁₆ in. The circles on the template are marked with their diameters and are available in fractions, decimals, or millimeters. Sample circle templates are shown in Figure 1–18. A popular template is one that has circles, hexagons, squares, and triangles.

Always use a circle template rather than a compass. Circle templates save time and are very accurate. For best results when making circles, try to keep your pencil or pen perpendicular to the paper. To obtain proper wide lines with a pencil, use a 0.9 mm automatic pencil.

To use a circle template properly, first draw the centerlines of your circle. Then exactly align the dashes on the template with the centerlines as shown in Figure 1–19. Proceed to trace the outline of the circle.

To draw arcs with a circle template, use one of two methods. One method is to draw the centerlines of the arc, align the template with the proper diameter, and draw the arc. Keep in mind that the template circles are marked in diameter while the size of an arc is given in radius. Remember to divide the template size in half to find the proper arc radius. The other method

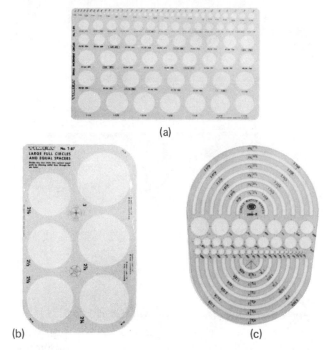

(a)

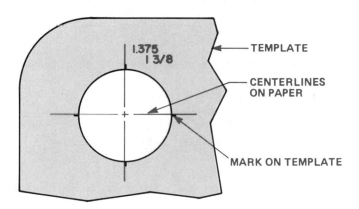

(b)   (c)

Figure 1–18 **(a)** Small circles. **(b)** Large, full circles. **(c)** Large half circles. *Courtesy Teledyne Post.*

Figure 1–19 Using a circle template to draw a circle.

of drawing arcs is to lightly draw outside construction lines, then fill in the arc to the points of tangency. This method is demonstrated in Chapter 4, Geometic Construction. Be sure the connection between the arc and the straight line is smooth.

When using a circle template and a technical pen, keep the pen perpendicular to the paper. Some templates have risers built in to keep the template above the drawing sheet. Without this feature there is a risk of ink running under a template that is flat against the drawing. If your template does not have risers, purchase and add template lifters, use a few layers of tape placed on the underside of the template (although tape does not always work well), or place a second template with a larger circle under the template you are using. (See Figure 1–20.)

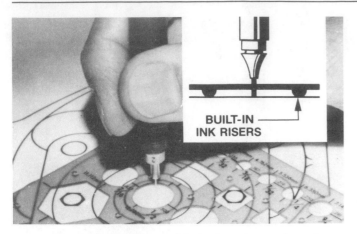

Figure 1–20 A template with built-in risers for inking. *Courtesy Chartpak.*

Ellipse Templates. Ellipses are circles seen at an angle. The parts of a circle are shown in Figure 1–21, while the parts of an ellipse are shown in Figure 1–22.

The kind of pictorial drawing known as isometric projects the sides of objects at a 30° angle in each direction away from the horizontal. Isometric circles are ellipses aligned with the horizontal, right, or left planes of an isometric box as in Figure 1–23. Isometric ellipse templates automatically position the ellipse at the proper angle of 35° 16'. (See Figure 1–24.)

Isometric drawing is discussed in detail in Chapter 18.

Use A Template

Never use a compass if you can use a template. Templates increase drafting speed and are very accurate.

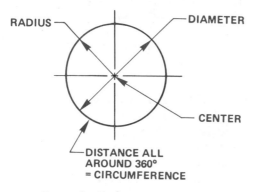

Figure 1–21 Parts of a circle.

Figure 1–22 Parts of an ellipse.

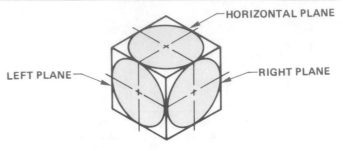

Figure 1–23 Ellipses in isometric planes.

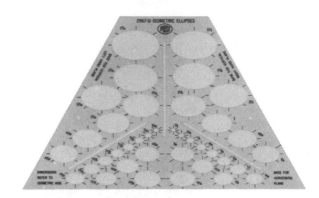

Figure 1–24 Isometric ellipse template. *Courtesy Teledyne Post.*

Irregular Curves

Irregular curves are commonly called *French curves.* These curves have no constant radii. An irregular curve is shown in Figure 1–25a while a flexible curve is shown in Figure 1–25b. Figure 1–26 shows a radius curvé, composed of a radius and tangent. The radius on these curves is constant and their sizes range from 3 ft to 200 ft, which are commonly used in highway drafting. In addition to these two kinds of curves there are available ship's curves. The curves in a set of ship's curves become progressively larger and, like French curves, have no constant radii. They are used for layout and development of ships' hulls.

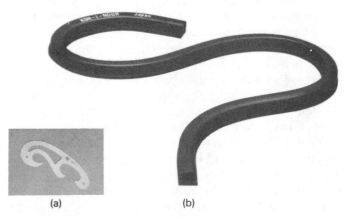

(a) (b)

Figure 1–25 **(a)** Irregular or French curve. *Courtesy Teledyne Post.* **(b)** Flexible curve. *Courtesy Koh-I-Noor, Inc.*

Figure 1–26   Radius curve.

In order to draw an irregular curve, points on the curve are developed or plotted. The points are then connected with a light line. Then a flexible curve is bent to fit, or a portion of an irregular curve is matched to the light line to include at least three plotted points. Care must be taken to make the curve flow smoothly. This smooth flow of the line is accomplished by never drawing the full length of the curve, but by overlapping each successive setting of the irregular curve. (See Figure 1–27.)

▼ DRAFTING MACHINES

Drafting machines, for the most part, take the place of triangles and parallel bars. The drafting machine maintains a horizontal and vertical relationship between scales, which also serve as straightedges. A protractor allows the scales to be set quickly at any angle. There are two types of drafting machines, arm and track. Although both types are excellent tools, the track machine has generally replaced the arm machine in industry. A major advantage of the track machine allows the drafter to work with a board in the vertical position. A vertical drafting surface position is generally more comfortable to use than a horizontal table. For school or home use, the arm machine may be a good economical choice.

Arm Drafting Machine

The arm drafting machine is compact and less expensive than a track machine. The arm machine clamps to a table and through an elbow-like arrangement of supports allows the drafter to position the protractor head and scales anywhere on the board. An arm drafting machine is shown in Figure 1–28.

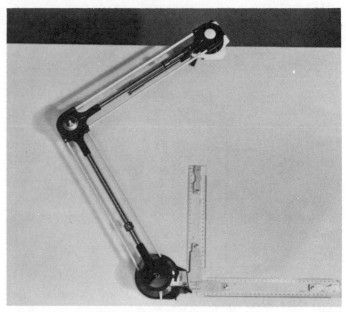

Figure 1–28   Arm drafting machine.

Track Drafting Machine

A track drafting machine has a traversing arm that moves left and right across the table and a head unit that moves up and down the traversing arm. There is a locking device for both the head and the traversing arm. The shape and placement of the controls of a track machine vary with the manufacturer although most brands have the same operating features and procedures. Figure 1–29 shows the component parts of a track drafting machine. The track drafting machine has the following advantages over the arm drafting machine:

1. The track machine is more stable and in some cases more accurate than an arm machine.
2. You can draw with your table inclined at a steep angle and the head will not slide down the table as it will with an arm machine.

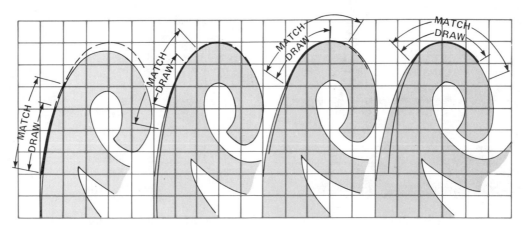

Figure 1–27   Irregular curve development.

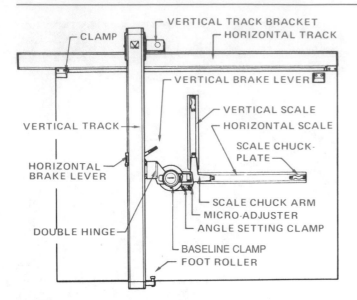

Figure 1–29 Track drafting machine and its parts. *Courtesy Consul & Mutoh, Ltd.*

3. Both the head and traversing arm may be locked in any position. This feature is important when using a lettering guide or other equipment that requires a stationary position.

As with the arm machines, track drafting machines have a vernier head allowing the user to measure angles accurately to 5'. A track drafting machine head protractor and vernier scale are shown in Figure 1–30. An optional dial head, available on some brands, allows for quick and easy angular measurements without the use of a vernier.

Sizes of Drafting Machine

When ordering a drafting machine, the specifications should relate to the size of the drafting board on which it is mounted. For example, a 37½ × 60 in. machine would properly fit a table of the same size.

Coordinate Reading and Processing Drafting Machine

There is a drafting machine available that offers a digital reading of angles, X and Y, and polar coordinates and has a resolution of 5'. A computer may be added to this machine that greatly speeds the time-consuming calculations that are done during layout and after a drawing is complete.

Controls and Machine Head Operation

Drafting machine heads contain the controls for horizontal, vertical, and angular movement. Although each brand of machine contains similar features, controls may be found in different places on different brands. (See Figure 1–31.) Most machines have the following controls:

1. Baseline adjustment—releases the scales so they can move but the protractor is not affected.
2. Index control—permits automatic stops every 15°. It can also be pushed in and locked to let you adjust the machine to any angle.
3. Indexing clamp—locks the protractor at angles other than 15° increments so you can draw an accurate line without the protractor moving.

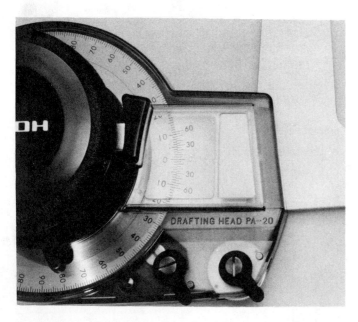

Figure 1–30 Drafting machine head protractor and vernier. *Courtesy Consul & Mutoh, Ltd.*

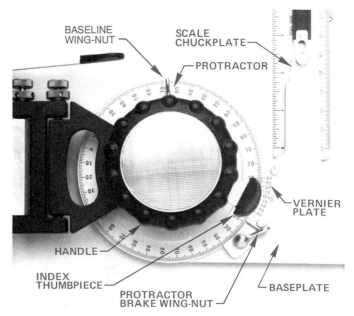

Figure 1–31 Drafting machine head controls and parts. *Courtesy Vemco Corporation.*

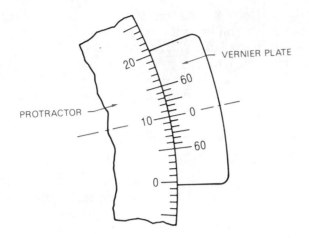

Figure 1–32 Vernier plate and protractor showing a reading of 10°.

To operate the drafting machine protractor head, place your hand on the handle and using your thumb, depress the index thumbpiece. Doing so allows the head to rotate. Each increment marked on the protractor is one degree with a label every 10°. As the vernier plate (the small scale numbered from 0 to 60) moves past the protractor, the zero on the vernier aligns with the angle that you wish to read. For example, Figure 1–32 shows a reading of 10°. As you rotate the handle, notice the head automatically locks every 15°. To move the protractor past the 15° increment, you must again depress the index thumbpiece.

Having rotated the protractor head 40° clockwise, the machine is in the position shown in Figure 1–33. The vernier plate at the protractor reads 40°, which

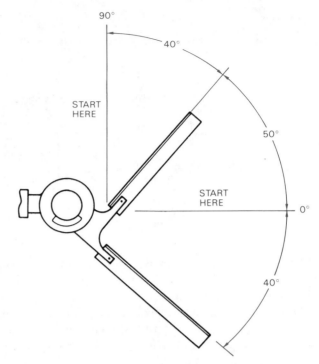

Figure 1–33 Angle measurement with the drafting machine.

means that both the horizontal and vertical scale have moved 40° from their original position at 0° and 90° respectively. The horizontal scale reads directly from the protractor starting from 0°. The vertical scale reading begins from the 90° position. The key to measuring angles is to determine if the angle is to be measured from the horizontal or vertical starting point. See the examples in Figure 1–34.

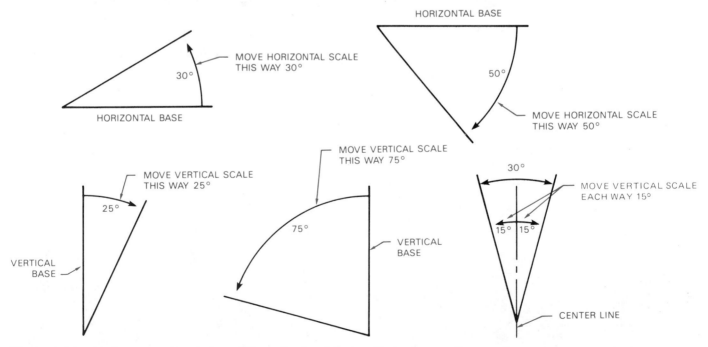

Figure 1–34 Angle measurements from either a horizontal or vertical reference line.

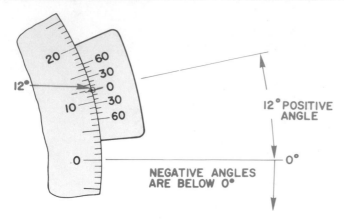

Figure 1–35 Measuring full degrees.

Measuring full degree increments is easy since you simply match the zero mark on the vernier plate with a full degree mark on the protractor. See the reading of 12° in Figure 1–35. The vernier scale allows you to measure angles as accurately as 5' (minutes). Remember, 1 degree equals 60 minutes (1° = 60') and 1 minute equals 60 seconds (1' = 60").

Reading and Setting Angles with the Vernier. To read an angle other than a full degree, assume the vernier scale is set at a *positive angle* as shown in Figure 1–36. Each mark on the vernier scale represents 5'. First see that the angle to be read is between 7° and 8°. Then find the 5' mark on the *upper* half of the vernier (the direction in which the scale has been turned) that is most closely aligned with a full degree on the protractor. In this example, it is the 40' mark. Add the minutes to the degree just passed. The correct reading then, is 7°40'. The procedure for reading *negative angles* is the same, except to read the minute marks on the *lower* half of the vernier.

Suppose you wished to set the angle 7°40' as shown in Figure 1–36. First release the protractor brake and disengage the indexing mechanism with the thumb control. Rotate the protractor arm counterclockwise

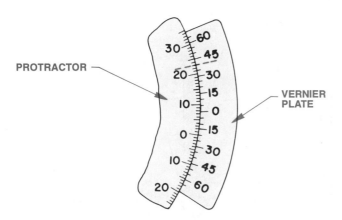

Figure 1–36 Reading positive angles with the vernier.

until the zero of the vernier is at 7°. Then slowly continue the rotation until the 40' mark on the upper half of the vernier aligns with the nearest degree mark on the protractor. Lock the protractor brake and draw the line. The procedure for setting negative angles is essentially the same except for turning the protractor head in a clockwise motion.

Machine Setup

To insert a scale in the baseplate chuck, place the scale flat on the board and align the scale chuckplate with the baseplate chuck on the protractor head. Firmly press, but do not drive the scale chuckplate into the baseplate chuck. To remove a scale, use the scale wrench as shown in Figure 1–37. Slip the wrench over screw C and turn clockwise, thus pressing curved section B strongly against section A of the baseplate chuck. Removing a scale by hand without the aid of a key could result in damage to the scale and/or machine.

Scale Alignment. Before drawing with any drafting machine, the scales should be checked for alignment and, if needed, be adjusted at right angles to each other. For best results with a track drafting machine, the scales should also be aligned with respect to the horizontal track. Both operations can be accomplished through the following procedure:

Step 1. Tighten the flat-head screw nearest the end of the scale on each scale chuckplate. Insert the scales in the baseplate chuck and press them firmly into place. Release the inner scale chuckplate lock-screw on the horizontal scale. Set the scale near the center of its angular range of adjustment and tighten the lock-screw.

Step 2. Draw a reference line parallel to the horizontal track by:
a. Locking the vertical brake and releasing the horizontal brake.
b. Placing a pencil point at zero on the horizontal scale and moving both the pencil and protractor head together laterally

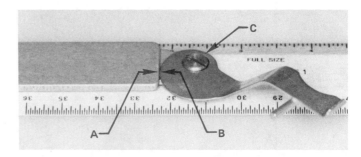

Figure 1–37 Scale removal. *Courtesy Vemco Corporation.*

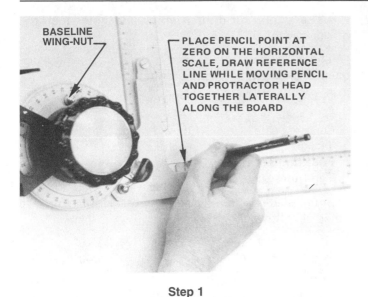

Step 1

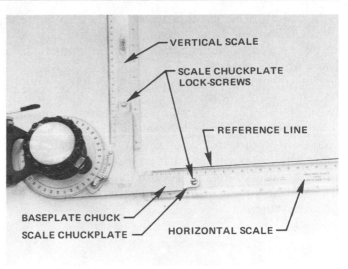

Step 2

Figure 1–38   Steps 1 and 2, scale alignment.

along the board. (See Figure 1–38.) *Caution:* Merely drawing the pencil along the scale will not assure the line being parallel to the horizontal track.

Step 3. Release the lock nut on the micrometer baseline screw and turn this screw until the scale is brought parallel to the reference line. (See Figure 1–39.) Tighten the lock nut firmly. Some machines have a baseline wing-nut. If yours does, release the baseline wing-nut and bring the scale parallel to the reference line. Tighten the baseline wing-nut. For those machines that have a baseline zero on the protractor (usually found to the left of the handle), first loosen the baseline wing-nut and align the arrow to 0°. Then lock the baseline wing-nut. Finally, loosen the horizontal scale chuckplate lock-screw and adjust the horizontal scale to the reference line, then retighten the screw.

Step 4. Remove the horizontal scale, turn it 180°, and replace it. Loosen the scale chuckplate lock-screw, adjust the scale parallel to the reference line, and tighten the lock-screw. Now the horizontal scale is properly aligned when inserted from either end.

Step 5. Move the head 90° clockwise and adjust both ends of the vertical scale in the same manner as the horizontal scale and along the same reference line drawn in step 2b. (See Figure 1–40.)

By following this procedure, you have established a reference line setting that is parallel to the horizontal track and adjusted the scales so they are parallel to the track and perpendicular to each other. For satisfactory results when drawing, the screws on the scales must be

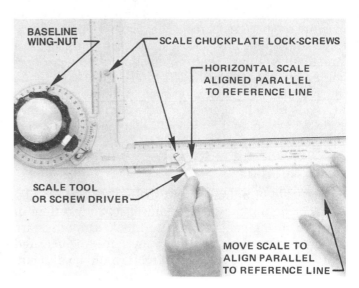

Figure 1–39   Steps 3 and 4, scale alignment.

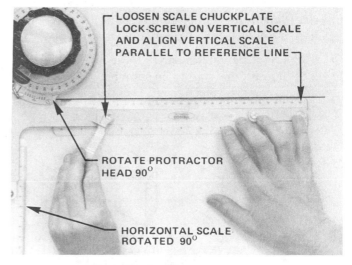

Figure 1–40   Step 5, scale alignment.

tight and the scale chuckplates pressed firmly into the chucks. Use good judgement when tightening any mechanism as too much force can cause components to break or wear out rapidly.

The alignment procedure of steps 1 and 2 should be checked periodically, even daily. Doing so may seem like a lot of work and trouble, but once you have gone through the procedure several times it becomes routine. By checking and adjusting the scale alignment often, you are sure that your drawings are accurate. One of a drafter's great frustrations is to prepare a layout and then discover that the machine is not properly aligned.

▼ SCALES

Scale Shapes

There are four basic scale shapes as shown in Figure 1–41. The two-bevel scales are also available with chuckplates for use with standard arm or track drafting machines. These machine scales have typical calibrations and some have no scale reading for use as a straightedge alone. Drafting machine scales are purchased by designating the length needed, 12, 18, or 24 in., and the scale calibration such as metric, engineer's full scale in 10ths and half scale in 20ths, or architect's scale ¼" = 1'–0" and ½" = 1'–0". Many other scales are available.

Scale Notation

The scale of a drawing is usually noted in the title block or below the view of an object that differs in scale to that given in the title block. Drawings are scaled so the object represented can be illustrated clearly on standard sizes of paper. It would be difficult, for example, to make a full size drawing of a Boeing 747, thus a scale that reduces the size of such a large object must be used. Machine parts are often drawn full size or even twice, four, or ten times larger than full size, depending on the actual size of the part.

The scale selected, then, depends on:

○ The actual size of the part.
○ The amount of detail to be shown.
○ The paper size selected.

○ The amount of dimensioning and notes required on the part.

The following scales and their notation are frequently used on mechanical drawings.

Full scale	=	FULL or 1:1
Half scale	=	HALF or 1:2
Quarter scale	=	QUARTER or 1:4
Twice scale	=	DOUBLE or 2:1
Four times scale	=	4:1
Ten times scale	=	10:1

Some scales used on architectural drawings are noted as follows:

⅛"	=	1'–0"	1"	=	1'–0"
¼"	=	1'–0"	1½"	=	1'–0"
½"	=	1'–0"	3"	=	1'–0"

Some scales used in civil drafting are noted as follows:

1"	=	10'	1"	=	50'
1"	=	20'	1"	=	60'
1"	=	30'	1"	=	100'

Metric Scale

ANSI/ASME   According to the American National Standards Institute (ANSI) and the American Society of Mechanical Engineers (ASME):

The commonly used SI (International System of Units) linear unit used on engineering drawings is the *millimeter*. On drawings where all dimensions are either in inches or millimeters, individual identification of units is not required. However, the drawing shall contain a note stating: UNLESS OTHERWISE SPECIFIED, ALL DIMENSIONS ARE IN INCHES (or MILLIMETERS as applicable). Where some millimeters are shown on an inch-dimensioned drawing, the millimeter value should be followed by the abbreviation *mm*. Where some inches are shown on a millimeter-dimensioned drawing, the inch value should be followed by the abbreviation *IN*.

Metric symbols are as follows:

millimeter	=	mm
centimeter	=	cm
decimeter	=	dm
meter	=	m
dekameter	=	dam
hectometer	=	hm
kilometer	=	km

Some metric-to-metric equivalents are the following:

10 millimeters	=	1 centimeter
10 centimeters	=	1 decimeter
10 decimeters	=	1 meter

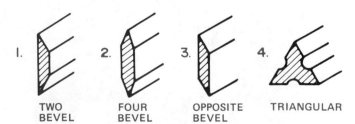

1. TWO BEVEL   2. FOUR BEVEL   3. OPPOSITE BEVEL   4. TRIANGULAR

Figure 1–41 Scale shapes.

10 meters   = 1 dekameter
10 dekameters = 1 kilometer

Some metric-to-U.S. customary equivalents are the following:

1 millimeter = .03937 inch
1 centimeter = .3937 inch
1 meter   = 39.37 inches
1 kilometer = .6214 mile

Some U.S. customary-to-metric equivalents are the following:

1 mile = 1.6093 kilometers = 1609.3 meters
1 yard = 914.4 millimeters = .9144 meter
1 foot = 304.8 millimeters = .3048 meter
1 inch = 25.4 millimeters = .0254 meter

To convert inches to millimeters, multiply inches by 25.4.

Figure 1–42 shows the common scale calibrations found on the triangular metric scale. One advantage of the metric scales is that any scale is a multiple of 10; therefore, any reductions or enlargements are easily performed. No mathematical calculations should be required when using a metric scale. Always select a direct reading scale. To avoid the possibility of error, avoid multiplying or dividing metric scales by anything but multiples of ten.

Civil Engineer's Scale

The triangular civil engineer's scale contains six scales, two on each of its sides. The civil engineer's scales are calibrated in multiples of ten. The scale margin displays the scale represented on a particular edge. The following table shows some of the many scale options available when using the civil engi-

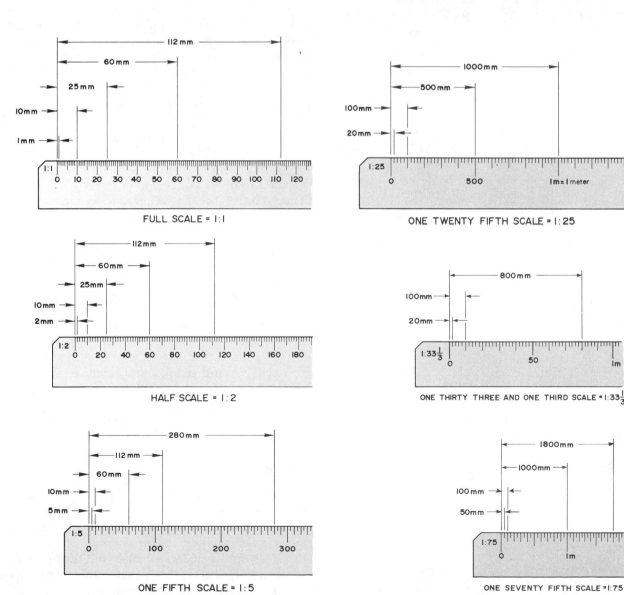

Figure 1–42   Metric scale calibrations.

neer's scale. Keep in mind that any multiple of ten is available with this scale.

CIVIL ENGINEER'S SCALE					
Divisions	Ratio	Scales Used with This Division			
10	1:1	1" = 1"	1" = 1"	1" = 10'	1" = 100'
20	1:2	1" = 2"		1" = 20'	1" = 200'
30	1:3	1" = 3"		1" = 30'	1" = 300'
40	1:4	1" = 4"		1" = 40'	1" = 400'
50	1:5	1" = 5"		1" = 50'	1" = 500'
60	1:6	1" = 6"		1" = 60'	1" = 600'

The 10 scale is often used in mechanical drafting as a full, decimal-inch scale, shown in Figure 1–43. Increments of ¹⁄₁₀ (.1) in. can easily be read on the 10 scale. Readings of less than .1 in. require the drafter to approximate the desired amount, as shown in Figure 1–43. Some scales are available that refine the increments to ¹⁄₅₀th of an inch. The 10 scale is also used in civil drafting for scales of 1" = 10' or 1" = 100' and so on. (See Figure 1–44.)

The 20 scale is commonly used in mechanical drawing to represent dimensions on a drawing at half scale (1:2). Figure 1–45 shows examples of half-scale decimal dimensions. The 20 scale is also used for scales of 1" = 2', 1" = 20', 1" = 200', as shown in Figure 1–46.

The remaining scales on the engineer's scale may be used in a similar fashion. For example, 1" = 5', 1" = 50', and so on. Figure 1–46 shows dimensions on each of the civil engineer's scale. The 50 scale is popular in civil drafting for drawing plats of subdivisions.

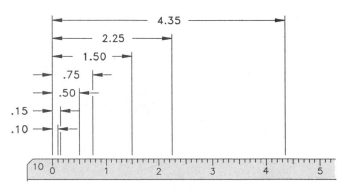

Figure 1–43   Full engineer's decimal scale (1:1).

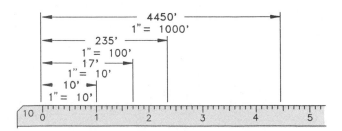

Figure 1–44   Civil engineer's scale, units of 10.

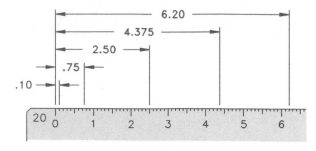

Figure 1–45   Half scale on the engineer's scale (1:2).

Architect's Scale

The triangular architect's scale contains 11 different scales. On ten of them each inch represents a foot and is subdivided into multiples of 12 parts to represent inches and fractions of an inch. The eleventh scale is the full scale with a 16 in the margin. The 16 means that each inch is divided into 16 parts and each part is equal to ¹⁄₁₆th of an inch. Look at Figure 1–47 for a comparison between the 10 engineer's scale and the 16 architect's scale. Figure 1–48 shows an example of the full architect's scale, while Figure 1–49 shows the fraction calibrations.

Look at the architect's scale examples in Figure 1–50. Note the form in which scales are expressed on a drawing. The scale notation may take the form of a word; full, half, double, or a ratio, 1:1, 1:2, 2:1; or an equation of the drawing size in inches or fractions of an inch to one foot, 1" = 1'–0", 3" = 1'–0", ¼" = 1'–0". The architect's scale commonly has scales running in both directions along an edge. Be careful when reading a scale from left to right. Do not confuse its calibrations with the scale that reads from right to left.

▼ CLEANING DRAFTING EQUIPMENT

Cleaning the drafting equipment each day helps keep your drawing clean and free of smudges. This is easily done with mild soap and water or with a soft rag or tissue. Avoid using harsh cleansers or products that are not recommended for use on plastic. It is also a good practice to clean your hands periodically to remove graphite and oil. Also, keep your hands dry.

▼ DRAFTING MEDIA

Several factors other than cost should influence the purchase and use of drafting media. The considerations include durability, smoothness, erasability, dimensional stability, and transparency.

Durability should be considered if the original drawing will have a great deal of use. Originals may tear or wrinkle and the images become difficult to see if the drawings are used often.

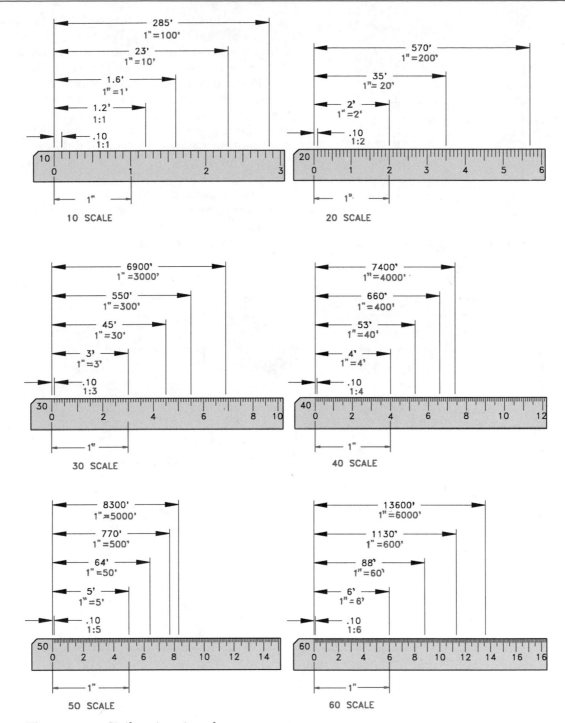

Figure 1–46   Civil engineer's scales.

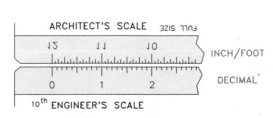

Figure 1–47   Comparison of full engineer's scale (10) and architect's scale (16).

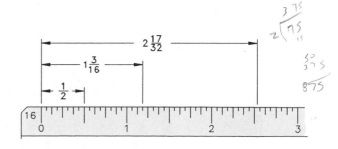

Figure 1–48   Full (1:1), or 12" = 1'–0", architect's scale.

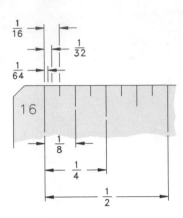

Figure 1–49 Enlarged view of architect's (16) scale.

Smoothness relates to how the medium accepts line work and lettering. The material should be easy to draw on so that the image is dark and sharp without a great deal of effort on the part of the drafter.

Erasability is important because errors need to be corrected and changes frequently made. When images are erased, ghosting should be kept to a minimum.

Ghosting is the residue that remains when lines are difficult to remove. These unsightly ghost images are reproduced in a print. Materials that have good erasability are easy to clean up.

Dimensional stability is the quality of the media to not alter size due to the effects of atmospheric conditions such as heat and cold. Some materials are more dimensionally stable than others.

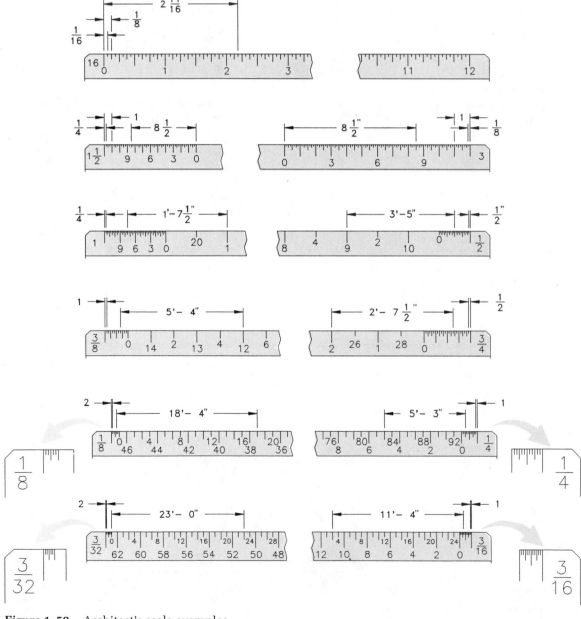

Figure 1–50 Architect's scale examples.

Transparency is one of the most important characteristics of drawing media. The diazo reproduction method requires light to pass through the material. The final goal of a drawing is good reproduction, so the more transparent the material the better the reproduction, assuming that the image drawn is professional quality. Transparency is not important when using the photocopy process.

▼ PAPERS AND FILMS

Vellum

Vellum is drafting paper that is specially designed to accept pencil or ink. Lead on vellum is the most common combination used in the drafting industry today. Many vendors manufacture quality vellum for drafting purposes. Each claims to have specific qualities that you should consider in the selection of your paper. Vellum is the least expensive material having good smoothness and transparency. Use vellum originals with care. Drawings made on vellum that require a great deal of use could deteriorate, as vellum is not as durable as other material. Some brands are better erased than others. Affected by humidity and other atmospheric conditions, vellum generally is not as dimensionally stable as other materials.

Polyester Film

Polyester film, also known by its brand name, Mylar, is a plastic material that offers excellent dimensional stability, erasability, transparency, and durability. Drawing on Mylar is best accomplished using ink or special polyester leads. Do not use regular graphite leads as they smear easily. Drawing techniques that drafters use with polyester leads are similar to graphite leads except that in general the polyester leads are softer and feel like a crayon when used.

Mylar is available with a single or double matte surface. *Matte* is a non-glossy, slightly textured surface. The double matte film has texture on both sides so drawing can be done on either side if necessary. Single matte film is the most common in use with the non-drawing side having a slick surface.

Protect the Mylar Surface

When using Mylar you must be very careful not to damage the matte by erasing. Erase at right angles to the direction of your lines and do not use too much pressure. Doing so helps minimize damage to the matte surface. Once the matte is destroyed and removed, the surface will not accept ink or pencil. Also, be cautious about getting moisture on the Mylar surface. Oil from your hands can cause your pen to skip across the material.

Normal handling of drawing film is bound to soil it. Inked lines applied over soiled areas do not adhere

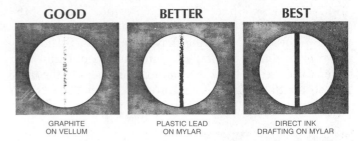

Figure 1–51 A magnified comparison of graphite on vellum, plastic lead on Mylar, and ink on Mylar. *Courtesy Koh-I-Noor, Inc.*

well and in time will chip off. It is always good practice to keep the film clean. Soiled areas can be cleaned effectively with special film cleaner.

Mylar is much more expensive than vellum; however, it should be considered when excellent reproductions, durability, dimensional stability, and erasability are required of original drawings.

Reproduction

The one thing most designers, engineers, architects, and drafters have in common is that their finished drawings are intended for reproduction. The goal of every professional is to produce drawings of the highest quality that give the best possible prints when they are reproduced.

Many of the factors that influence the selection of media for drafting have been discussed; however, the most important factor is reproduction. The primary combination that achieves the best reproduction is the blackest, most opaque lines or images on the most transparent base or material. Each of the materials mentioned makes good prints if the drawing is well done. If the only concern is the quality of the reproduction, ink on Maylar is the best choice. Some products have better characteristics than others. Some individuals prefer certain products. It is up to the individual or company to determine the combination that works best for their needs and budget. The question of reproduction is especially important when sepias must be made. (Sepias are second- or third-generation originals. See Sepias in this chapter.)

Look at Figure 1–51 for a magnified view of graphite on vellum, plastic lead on Mylar, and ink on Mylar. Judge for yourself which material and application provides the best reproduction. As you can see from Figure 1–51, the best reproduction is achieved with a crisp, opaque image on transparent material. If your original drawing is not good quality, it will not get better on the print.

▼ SHEET SIZES, TITLE BLOCKS, AND BORDERS

ASME/ANSI standard sheet sizes and format are specified in the documents ANSI Y14.1—1980 (Reaffirmed

1987) *Drawing Sheet Size and Format,* and ASME Y14.1M—1992 *Metric Drawing Sheet Size and Format.*

All professional drawings have title blocks. Standards have been developed for the information put into the title block and on the surrounding sheet next to the border so the drawing is easier to read and file than drawings that do not follow a standard format.

Sheet Sizes

ANSI Y14.1—1980 specifies sheet size specifications in inches as follows:

SIZE DESIGNATION	SIZE IN INCHES	
	Vertical	Horizontal
A	8½	11 (horizontal format)
	11	8½ (vertical format)
B	11	17
C	17	22
D	22	34
E	34	44
F	28	40

There are four additional size designations (G, H, J, and K); these apply to roll sizes.

The *M* in the title of the document Y14.1M—1992 means all specifications are given in metric. Standard drawing sheet sizes are designed as follows:

SIZE DESIGNATION	SIZE IN MILLIMETERS	
	Vertical	Horizontal
A0	841	1189
A1	594	841
A2	420	594
A3	297	420
A4	210	297

Longer lengths are referred to as elongated and extra-elongated drawing sizes. These are available in multiples of the short side of the sheet size.

Standard inch sheet sizes are shown in Figure 1–52a and metric sheet sizes are shown in Figure 1–52b.

Zoning

Some companies use a system of numbers along the top and bottom margins and letters along the left and right margins called *zoning.* Notice in Figure 1–52a that numbered and lettered zoning begins on C-size drawing sheets and in Figure 1–52b zoning is used on A3 sheet sizes and larger. Zoning allows the drawing to read like a road map. For example, the reader can refer to the location of a specific item as D4, which means that the item can be found at or near the intersection of D across and 4 up or down.

Title Blocks

Companies generally have title blocks and borders preprinted on drawing sheets to reduce drafting time and cost. Some companies use an adhesive title block so that one standard title block can be attached to any size drawing sheet.

Drawing sheet sizes and sheet format items such as borders, title blocks, zoning, revision columns, and general note locations have been standardized so that the same general relationship exists between engineering drawings internationally. The ANSI Y14.1—1980 and ASME Y14.1M—1992 documents specify the exact size and location for each item found on the drawing sheet. It is recommended that standard sheet sizes and format be followed to improve readability, handling, filing, and reproduction. Each company may use a slightly different design, although the following basic information is located in approximately the same place on most engineering drawings:

1. Title block. Lower right corner.
   a. Company name.
   b. Confidential statement.
   c. Unspecified dimensions and tolerances.
   d. Sheet size.
   e. Drawing number.
   f. Part name.
   g. Material.
   h. Scale.
   i. Drafter signature.
   j. Checker signature.
   k. Engineer signature.
2. Revision Column. Upper right corner, over or next to the title block.
   a. Revision symbol, number or letter.
   b. Description.
   c. Drafter.
   d. Date.
3. Border Line.
   a. With or without zoning.

Figure 1–53 shows a sample title block.

Title Block Definitions

○ *Tolerances* are discussed in detail in Chapter 10, although it is important to know that a tolerance is a given amount of acceptable variation in a size or location dimension. All dimensions have a tolerance.
○ *Millimeters* and *inches.* All dimensions are in millimeters (mm) or inches (in.) unless otherwise specified.
○ *Unless otherwise specified* means that, in general, all of the features or dimensions on a drawing have the relationship or specifications given in the title block unless a specific note or dimen-

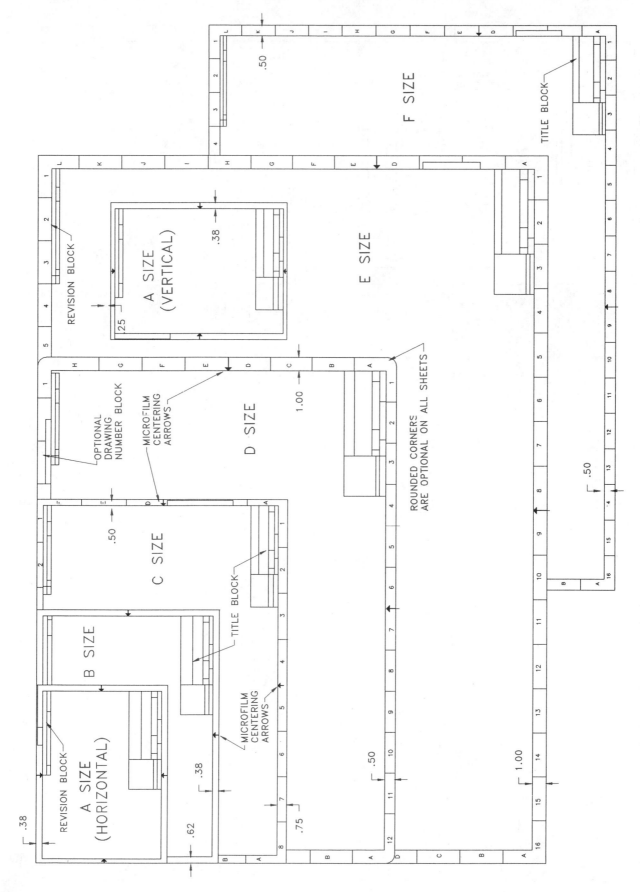

Figure 1-52(a) Standard inch drawing sheet sizes.

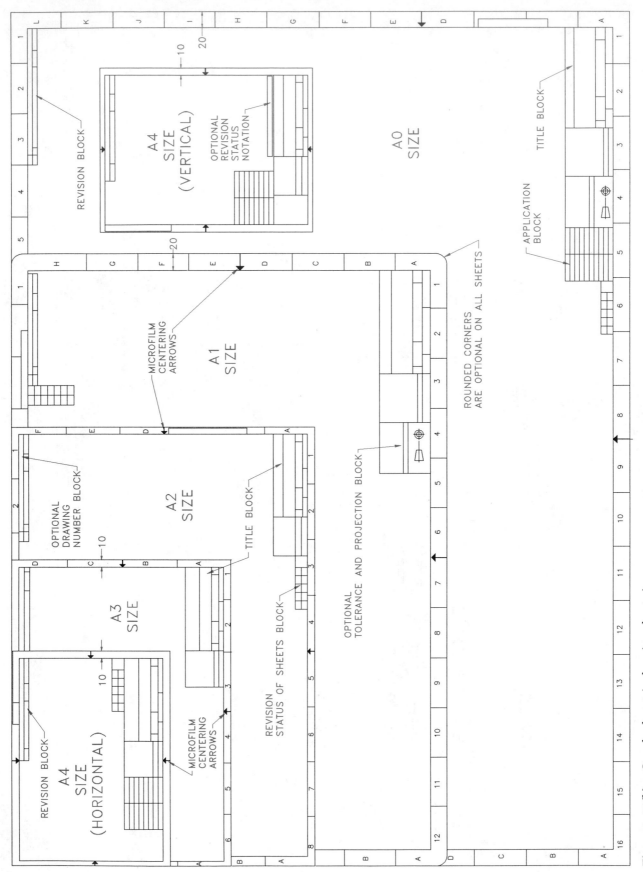

Figure 1–52(b) Standard metric drawing sheet sizes.

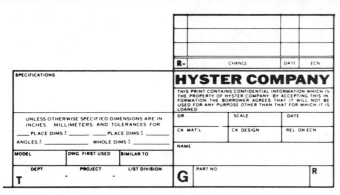

Figure 1–53 Sample title block. *Courtesy Hyster Company.*

sional tolerance is provided at a particular location in the drawing.

○ *Unspecified tolerances* refers to any dimension on the drawing that does not have a tolerance specified. This is when the dimensional tolerance required is the same as the general tolerance shown in the title block.

○ *Revisions.* When parts are redesigned or altered for any reason the drawing will be changed. All drawing changes are commonly documented and filed for future reference. When this happens the documentation should be referenced on the drawing so that users can identify that a change has been made. Before any revision can be made, the drawing must be released for manufacturing.

Title and Revision Block Instructions

The title block displayed in Figure 1–54a shows most of the common elements found in industrial title blocks. The revision block, Figure 1–54b, found in this

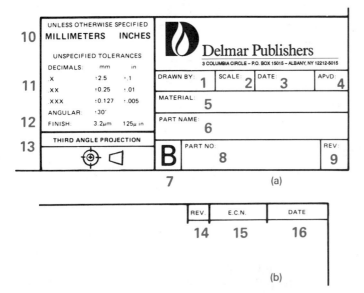

Figure 1–54 **(a)** Title block elements. **(b)** Revision block elements.

format is located in the upper right corner of the drawing sheet.

The large numbers shown on Figure 1–54 refer to the following instructions for completing the title and revision blocks. All lettering should be .125 or .188 in. high as indicated in the instructions. The lettering style should be vertical upper-case Gothic freehand, or mechanical lettering as specified by the instructor or company standards. Computer graphic systems are to use vertical upper-case letters unless otherwise specified.

1. DRAWN BY: .125 in. high lettering. Identify yourself by using all your initials such as DRC, DAM, JLT, unless otherwise specified.

2. SCALE: .125 in. high lettering. Examples of scales to fill in this block are FULL or 1:1, HALF or 1:2, DBL or 2:1, QTR or 1:4, NONE.

3. DATE: .125 in. high lettering. Fill in this block using the following order: day, month, and year such as 18 NOV 96.

4. APVD: .125 in. high lettering. This block is to be used by the instructor or checker who initials his or her approval of the drawing.

5. MATERIAL: .125 in. high lettering. Describe here the material used to make the part, for example, BRONZE, CAST IRON, SAE 4320. (The material is given in each problem.)

6. PART NAME: .188 in. high lettering. Insert here the name of the part, such as COVER, HOUSING. (The part name is given in each problem.)

7. B: .188 in. high if unprinted. Gives sheet sizes shown in Figure 1–52. When preprinted borders are used, this sheet size designation is usually printed. When adhesive title blocks are used, this area is often left blank to be filled in by the drafter.

8. PART NO.: .188 in. high lettering. Fill in the drawing number of the part. Most companies have their own part numbering system. While numbering systems differ, they are often keyed to categories such as the disposition of the drawing (casting, machining, assembly), materials used, related department within the company, or a numerical classification of the part. (Part numbers for drawing problems will be given.)

9. REV: .188 in. high lettering. Fill in the revision letter of the part or drawing. A new or original drawing is — (dash). The first time a drawing is revised the — changes to an A, for the second drawing change a B is placed here, and so on. The letters *I, O, S, X,* and *Z* are not used. When all of the available letters *A* through *Y* have been used, double letters such as *AA* and *AB,* or *BA* and *BB* are used.

10. UNLESS OTHERWISE SPECIFIED: Unless otherwise specified, drawing dimensions will be given in millimeters or inches. When the pre-

dominant dimensions are in MILLIMETERS neatly blacken out INCHES; when in INCHES, blacken out MILLIMETERS.

11. UNSPECIFIED TOLERANCES: One, two, and three place decimals are established as MILLIMETERS or INCHES in a manner the same as item 10. For example, if the drawing dimensions are predominately MILLIMETERS then INCH related tolerances are blackened out. Angular tolerances for unspecified angular dimensions are ±30'.

12. FINISH: In this block fill in the unspecified surface finish, for surfaces that are identified for finishing without a specific callout.

13. THIRD-ANGLE PROJECTION: The drawing assignments in this text are done using third-angle projection unless otherwise specified. A complete discussion of this topic is found in Chapter 5.

14. REV: Fill in the revision letter or number in this block, such as A, B, C, etc. Succeeding letters are to be used for each Engineering Change Notice (ECN) or group of related ECNs regardless of quantity of changes on an ECN. *Note:* the Rev: block in the title block must be changed to agree with the last REV. letter in the revision block.

15. E.C.N.: In this block, fill in the Engineering Change Notice (ECN) letter covering the Engineering Change Request (ECR), or group of ECRs that require the drawing to be revised.

16. DATE: Fill in the day, month, and year on which the ECN package is ready for release to production, such as 6 APR 96, 19 JUN 96, etc.

Drafting revisions are completed in chronological order by adding horizontal columns and extending the vertical column lines.

CADD Applications

PRODUCING A CADD DRAWING

The following discussion gives you an overview of how drawings are produced on a CADD system.

The term *producing* means to display the drawing in some fashion on one of several different pieces of hardware. Figure 1–55 shows a drawing displayed on the screen. This is done with a command such as OPEN. A quick print of the screen

Figure 1–55 Drawings can be displayed quickly on the screen. *Courtesy Hewlett-Packard Company.*

display is available to those with a system that has a hardcopy unit. A *hard copy* is a paper copy, while *soft copy* is video or audio format.

A second type of print hardcopy is produced by the printer. A PRINT command may handle this job. A printed copy is useful for checking purposes much the same as a hardcopy unit print. The size is limited with both types of prints and therefore most drawings produced in this manner are not to scale.

Quality reproduction of a CADD drawing is produced by a pen plotter using wet-ink pens as shown in Figure 1–56. Quick, true-scale plots of good quality are produced by electrostatic plotters. Plotted prints are started with the PLOT command at the computer. The operator can request a plot of a drawing presently in the computer memory, or of any drawing file that is on a disk. Such a request can be made from the workstation, or from other terminals in the office. Some companies that use large, wet-ink pen plotters, will leave the plotting of drawings to the night shift. Then only one shift has to load and clean ink pens and make sure the plot is running properly. Other companies use plotters that accept roll paper. A list of drawings to be plotted can be given to the computer just before the day shift is over, and the plotter will produce a continuous stream of drawings until the entire list is plotted. A take-up reel behind the plotter rolls up all

Figure 1–56 A pen plotter. *Courtesy Hewlett-Packard Company.*

the drawings, which are then cut and trimmed the next day.

Drafters with small firms usually request a plotted copy at any time they need one. Many plotters use felt-tip pens of varying widths and colors that can be left in their carousels and receptacles without fear of drying out. Using felt-tip pens makes it easy to get a plot whenever needed because the hassle of working with wet-ink pens is eliminated. Pen plotters are rapidly being replaced by new technology. See Chapter 2 for examples.

▼ DIAZO REPRODUCTION

Diazo prints, also known as ozalid dry prints, or blueline prints, are made with a printing process that uses an ultraviolet light that passes through a translucent original drawing to expose a chemically coated paper or print material underneath. The light does not go through the dense, black lines on the original drawing. Thus the chemical coating on the paper beneath the lines is not exposed. The print material is then exposed to ammonia vapor which activates the remaining chemical coating to produce blue, black, or brown lines on a white or clear background. The print that results is a diazo, or blue-line print, not a blueprint. The term *blueprint* is now a generic term that is used to refer to diazo prints even though they are not true blueprints.

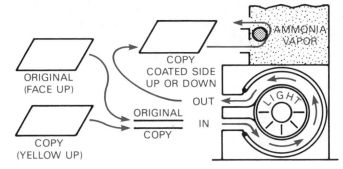

Figure 1–57 The diazo print process.

Making a Diazo Print

Before you run a print using a complete sheet, run a test strip to determine the proper speed setting.

To make a diazo print, place the diazo material on the feedboard, coated (yellow) side up. Position the original drawing, image side up, on top of the diazo material, making sure to align the leading edges. Use diazo material that corresponds in size with your original.

Using fingertip pressure from both hands, push the original and diazo material into the machine. The light passes through the original and exposes the sensitized diazo material except where images (lines and lettering) exist on the original. When the exposed diazo material and the original drawing emerge from the printer section, remove the original and carefully feed the diazo material into the developer section, as shown in Figure 1–57.

In the developer section the sensitized material that remains on the diazo paper activates with ammonia vapor and blue lines form, although some diazo materials make black or brown lines. Special materials are also available to make other colored lines, or to make transfer sheets and transparencies. Figure 1–57 shows the diazo print process.

Storage of Diazo Materials

Diazo materials are light sensitive so they should always be kept in a dark place until ready for use. Long periods of exposure to room light cause the diazo chemicals to deteriorate and reduce the quality of your print. If you notice brown or blue edges around the unexposed diazo material, it is getting old. Always keep it tightly stored in the shipping package and in a dark place such as a drawer. Some people prefer to keep all diazo material in a dark refrigerator. Doing so preserves the chemical for a long time.

Sepias

Sepias are diazo materials that are used to make secondary originals. A secondary or second-generation original is actually a print of an original drawing that can be used as an original. Changes can be made on

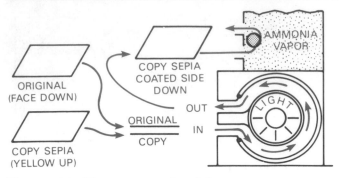

Figure 1–58 The reverse sepia print process.

the secondary original while the original drawing remains intact. The diazo process is performed with material called sepia. Sepia prints normally form dark brown lines on a clear (translucent) background; therefore, they are sometimes called brown lines. In addition to being used as an original to make alterations to a drawing without changing the original, sepias are also used when originals are required at more than one company location.

Sepias are normally made with the original face down on the sensitized sepia material. The balance of the print process is the same as just described except that the sepia should be fed into the developer coated-side down. Doing so allows the coating on the sepia material to come into direct exposure with the ammonia in the development chamber. Experiment both ways and observe the quality of your reproduction. Standard diazo copies may reproduce better with the coated-side down; it depends on your machine and materials.

Sepias are made in reverse (reverse sepias) so that corrections or changes can be made on the matte side (face side) and erasures can be made on the sensitized side. Sepia materials are available in paper or polyester. The resulting paper sepia is similar to vellum and the polyester sepia is Mylar. Drawing changes can be made on sepia paper with pencil or ink, and on sepia Mylar with polyester lead or ink. Figure 1–58 shows an illustration of the sepia print process. The use of sepia reproduction is being rapidly replaced by computer-aided drafting.

▼ SAFETY PRECAUTIONS FOR DIAZO PRINTERS
Ammonia

Ammonia has a strong, unmistakable odor. It may be detected, at times, while operating your diazo printer. Diazo printers are designed and built to provide safe operation from ammonia exposure. Some machines have ammonia filters while others require outside exhaust fans. When print quality begins to deteriorate and it has been determined that the print paper is in good condition, the ammonia may be old.

Ammonia bottles should be changed periodically, as determined by the quality of prints. When the ammonia bottle is changed it is generally time to change the filter also.

When handling bottles of ammonia or when replacing a bottle supplying a diazo printer, the following precautions are required.

Eye Protection. Extreme care should be taken to avoid direct contact of ammonia with your eyes. Always wear safety goggles, or other equivalent eye protection, when handling ammonia containers directly or handling ammonia supply systems.

Avoid Contact with Ammonia and the Ammonia Filter. Avoid contact of ammonia and the machine's ammonia filter with your skin or clothing. Ammonia and ammonia residue on the filter can cause uncomfortable irritation and burns when it touches your skin.

Avoid Inhaling Strong Concentrations of Ammonia Fumes. The disagreeable odor of ammonia is usually sufficient to prevent breathing harmful concentrations of ammonia vapors. Avoid prolonged periods of inhalation close to open containers of ammonia or where strong, pungent odors are present. The care, handling, and storage of ammonia containers should be in accordance with the suppliers' instructions and all applicable regulations.

First Aid. If ammonia is spilled on your skin, promptly wash with plenty of water, removing your clothing if necessary, to flush affected areas adequately. If your eyes are affected, as quickly as possible irrigate with water for at least 15 minutes.

Anyone overcome by ammonia fumes should be removed to an area of fresh air at once. Apply artificial respiration, preferably with the aid of oxygen if breathing is labored or has stopped. Obtain medical attention at once in event of eye contact, burns to the nose or throat, or if the person is unconscious.

Ultraviolet Light Exposure. Under the prescribed operating instructions, there is no exposure to the ultraviolet rays emitted from the illuminated printing lamps. However, to avoid possible eye damage, under no circumstances should anyone attempt to look directly at the illuminated lamps.

▼ HOW TO PROPERLY FOLD PRINTS

As you have already learned, prints may come in a variety of sizes ranging from small 8½ × 11 in. to 34 × 44 in. or larger. It is easy to file the 8½ × 11 in. size prints because standard file cabinets are designed to hold this size. There are file cabinets available called flat files that may be used to store full size unfolded prints. However, many companies use standard file cabinets. Larger prints must be properly folded before they can be filed in a standard file cabinet. It is also important to properly fold a print if it is to be mailed. Folding large prints is much like folding a road map.

Folding is done in a pattern of bends that results in the title block and sheet identification ending up on the front. This is desirable for easy identification in the file cabinet. The proper method used to fold prints also aids in unfolding or refolding prints. Look at Figure 1–59 to see how large prints are properly folded.

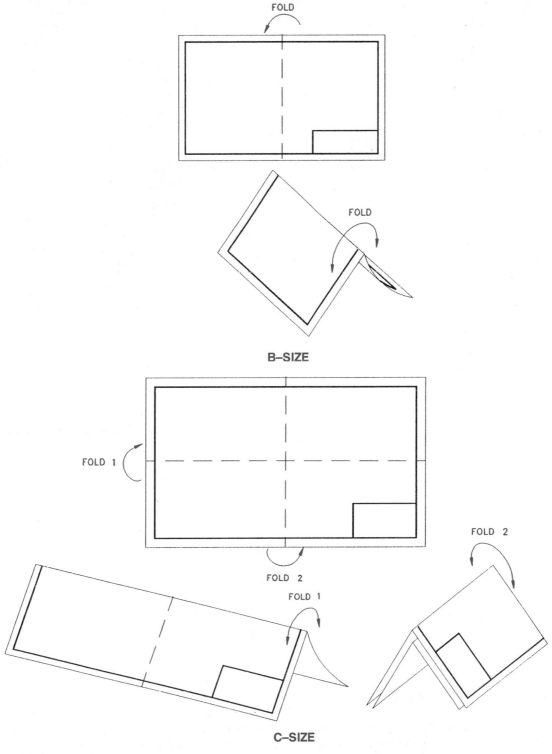

Figure 1–59 How to properly fold B-size, C-size, and D-size prints.

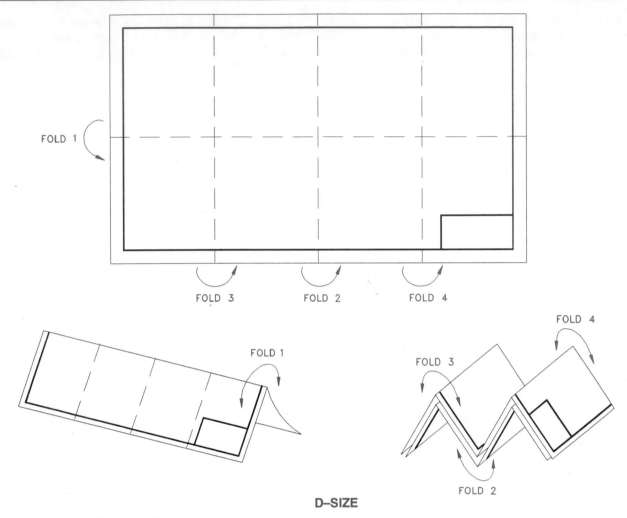

Figure 1–59 *(Continued)*

▼ PHOTOCOPY REPRODUCTION

Photocopy printers are also known as engineering copiers. Prints can be made on bond paper, vellum, polyester film, colored paper, or other translucent materials. The reproduction capabilities also include instant print sizes ranging from 45 to 141 percent of the original size. Larger or smaller sizes are possible by enlarging or reducing in two or more steps, that is by making a second print from a print.

Almost any large original can be converted into a smaller sized reproducible print, and then the secondary original can be used to generate diazo or photocopy prints for distribution, inclusion in manuals, or for more convenient handling. Also, a random collection of mixed-scale drawings can be enlarged or reduced and converted to one standard scale and format. Reproduction clarity is so good that halftone illustrations (photographs) and solid or fine line work have excellent resolution and density.

Engineering copiers are rapidly replacing diazo print machines in the engineering and drafting world because of their versatility and competitive cost. Also, engineer-

ing copiers do not use ammonia, which is an important consideration when purchasing a print machine.

▼ MICROFILM

Microfilm is photographic reproduction on film of a drawing or other document which is highly reduced for ease in storage and sending from one place to another. When needed, equipment is available for enlargement of the microfilm to a printed copy. Special care must be taken to make the original drawing of the best possible quality. The reason for this is that during each "generation," the process of "blowback" makes the lines narrower in width and less opaque than the original. The term *generation* refers to the number of times a copy of an original drawing is reproduced and used to make other copies. For example, if an original drawing is reproduced on sepia or microfilm, and the sepia or microfilm is used to make other copies, this is referred to as a *second generation* or *secondary original.* When this process has been done four times, the drawing is called a *fourth generation.* The term *blowback* means bringing the drawing from the microfilm back to a printed copy. A true test of the original drawing's quality is the ability to maintain good reproductions through the fourth generation of reproduction.

In many companies original drawings are filed in drawers by drawing number. When a drawing is needed, the drafter finds the original, removes it, and makes a copy. This process works well although, depending on the company's size or the number of drawings generated, drawing storage often becomes a problem. Sometimes an entire room is needed for drawing storage cabinets. Another problem occurs when originals are used over and over. They often become worn and damaged, and old vellum becomes yellowed and brittle. Also, in case of a fire or other kind of destruction, originals may be lost and endless hours of drafting vanish. For these and other reasons, many companies are using microfilm for storage and reproduction of original drawings. Microfilm is used by industry for photographic reproduction of original drawings into a film-negative format. Cameras used for microfilming are usually one of three sizes: 16, 35, or 70 mm. Microfilming processor cameras are now available that prepare film ready for use in a few seconds. ASME Y14.5M—1994 provides recommended microfilm reduction factors for different sheet sizes. These help ensure standardization.

Aperture Card and Roll Film

The microfilm is generally prepared as one frame or drawing attached to an aperture card, or as many frames in succession on a roll of film. The aperture card, shown in Figure 1–60, becomes the engineering document and replaces the original drawing. A 35 mm film negative is mounted in an aperture card, which is

Figure 1–60   Aperture card with microfilm. *Courtesy 3M Company.*

a standard computer card. Some companies file and retrieve these cards manually, while other companies punch the cards for computerized filing and retrieval. The aperture card, or data card as it is often called, is convenient to use. All drawings are converted to one size, regardless of the original dimensions. All engineering documents are filed together, instead of having originals ranging in size. Then for easy retrieval each card is numbered with a sequential engineering drawing and/or project number. Although the aperture card is more popular, some microfilm users prefer roll film, as seen in Figure 1–61. One disadvantage of roll film is that the reader has to look through a roll of film images to find the one drawing needed. This system is used in some government agencies for storage and retrieval of local maps and plats.

CADD Applications

CAD/CAM

The optimum efficiency of design and manufacturing methods exists without reproducing a single copy of a drawing of a part. Computer graphic systems may now directly link the engineering department by way of computer-aided design (CAD) to the manufacturing department machines operated by computer-aided manufacturing (CAM). Through the use of numerical control dimensioning and layout practices, the drafter or designer can prepare an engineering drawing on a computer graphics system and then have the com-

puter convert the drawing's coordinate data to a numerical control computer tape. The computer software for this translation is known as a post processor and is the same program that reproduces data for manufacturing with a numerically-controlled machine tool. When the electronics are connected directly from engineering to manufacturing, there is a direct link between CAD and CAM. A copy of the actual drawing is never generated, unless someone needs a hard copy (print) for review away from the computer terminal.

Figure 1–61 Roll microfilm. *Courtesy 3M Company.*

Microfilm Enlarger Reader-printer

Microfilm from aperture cards or rolls can be displayed on a screen for review, or enlarged and reproduced as a copy on plain bond paper, vellum or offset paper-plates. Most microfilm reader-printers can enlarge and reproduce prints ranging from A0 to A4 size. The prints are then available for disbursement to manufacturing, sales, or engineering department personnel.

Computerized Microfilm Filing and Retrieval

The next step in the automation of engineering-drawing filing systems is the computerized filing and retrieval of microfilm aperture cards. Large companies and independent microfilming agencies are able to take advantage of this technology. Small companies can send their engineering documents to local microfilm companies, known as job-shops, where microfilm can be prepared and stored. Companies that use computerized microfilm systems can have a completed drawing microfilmed, stored in a computer, and available on-line in the manufacturing department in a few minutes.

▼ DIGITIZING EXISTING DRAWINGS

Digitizing is a method of transfering drawing information from a digitizer into the computer. The digitizing function of automatically sending commands to the computer from a tablet menu is discussed in Chapter 2. A digitizing tablet may be used to take existing drawings and digitize the information into the computer. Digitizer tablets range in size from 6 × 6 in. to 44 × 60 in. Most school and industry CAD departments use 11 × 11 in. digitizer tablets to input commands from standard and custom tablet menus. In most cases companies do not have the time to convert existing drawings to CAD, because the CAD systems are heavily used for new product drawings. Therefore, there are an increasing number of businesses that digitize existing drawings for other companies.

▼ OPTICAL SCANNING OF EXISTING DRAWINGS

Scanning is a method of reproducing existing drawings in the form of computer drawings. Scanners work

CADD Applications

CADD VS. MICROFILM

Microfilm has become an industry standard for storage and access of drawings. Large international companies especially rely on the microfilm network to insure that all worldwide subsidiaries have the ability to reproduce needed drawings and related documents. One big advantage of microfilm is the ability to archive drawings. Archive means to store something permanently for safekeeping. The use of CADD in the engineering and construction industries has made it possible to create and store drawings on flexible disk, tape, or optical disk. This has made it possible to quickly retrieve drawings that have been stored. A big advantage of this advancement involves local use of these drawings. When the CADD-generated drawings are retrieved, they are of the same quality as when they were originally drawn. The CADD drawings can be used to make multiple copies, or to redesign the product. Yet, microfilm remains an important part of the engineering environment, because microfilmed drawings last forever and there is currently a broad base of microfilm use worldwide. The optical disk comes closest to achieving the archival quality of microfilm, and perhaps one day a satellite network will exist that allows drawings on disk to be transmitted anywhere instantly. Until storage on disk becomes archival quality, microfilm will remain as a long-term method of saving drawings.

much the same way as taking a photograph of the drawing. One advantage of scanning over digitizing existing drawings is the entire drawing, including dimensions, symbols, and text is transferred to the computer. A disadvantage is that some drawings, when scanned, require a lot of editing to make them presentable. The scanning process picks up visual graphic information from the drawing. What appears to be a dimension, for example, is only a graphic representation of the dimension picked up from the original drawing. What you get is an exact duplicate of the existing drawing. However, if you want the dimensional information to be technically accurate, the dimensions must be redrawn and edited on the computer. There are software packages available that make this editing process quick and easy.

After the drawing is scanned, the image is sent to a raster converter that translates information to digital or vector format. A *raster* is an electron beam that generates a matrix of pixels. *Pixels* make up the drawing image on the computer as explained in Chapter 2. A raster editor is then used to display the image for changes.

In the final analysis, companies involved in scanning can, in many cases, reproduce existing drawings more efficiently than companies that digitize existing drawings.

One economic product, CAD/Camera™, by Autodesk, Inc., is scanning software that runs on a microcomputer and works in conjunction with an electronic scanning camera. When an existing drawing has been transferred to the computer, it becomes an AutoCAD drawing that may be edited as necessary. Houston Instrument has an optical scanning accessory that attaches to their DMP-60 series of drafting plotters. This system provides an easy-to-use, cost-effective way to enter up to 36 × 48 in. drawings into the computer system. The scanning accessory can automatically scan drawings from paper, vellum, film, or blueline and convert the hardcopy image into a raster data file.

MATH APPLICATION

ANGLE MEASUREMENT IN RADIANS

Figure 1–62 It is important to convert from one type of angle measurement to another, because drafting machines work with degrees and minutes while many computers and calculator applications measure in degrees and decimal fractions of degrees. *Courtesy Megatek.*

Although drafting machines, surveying instruments, and many modern machine tools work with degrees and minutes, many computer and calculator applications require angles to be measured in degrees and decimal fractions of a degree.

It is important to be able to apply math to convert from one type of angle measurement to another because not all machines and computers do this automatically.

To convert from minutes of a degree to decimal degrees, divide the number of minutes by 60. For example, 7° 40' is the same as 7.666666°, or rounding, 7.67° because 40 ÷ 60 = .666666. To convert from decimal degrees to minutes, multiply the decimal portion of the angle by 60. For example, 18.75° is the same as 18° 45' because .75 × 60 = 45.

DRAFTING EQUIPMENT TEST

DIRECTIONS
Answer the questions with short complete statements or drawings as needed.

PART 1. MANUAL DRAFTING EQUIPMENT

1. Describe a mechanical drafing pencil.
2. Describe an automatic drafting pencil.
3. List three lead sizes available for automatic pencils.
4. Identify the lead hardness that is normally recommended for the following functions:
   Range of common drafting leads
   Sketching leads
5. Name the type of technical pen tip that will normally provide the longest pen life.
6. What precautions should be taken when disassembling a technical pen tip?
7. List three characteristics recommended for technical pen drafting inks.
8. Describe an advantage and a disadvantage of using electric erasers.
9. Name the type of compass most commonly used by professional drafters.
10. Why should templates, instead of compasses, be used whenever possible?
11. Why should a compass point with a shoulder be used whenever possible?
12. Identify two common lead points that may be used in compasses.
13. What is the advantage of placing drafting tape or plastic adhesive disks at the center of a circle before using a compass?
14. A good divider should not be too loose or too tight. Why?
15. List two divider uses.
16. List two basic types of manual drafting machines.
17. How many degrees are in the interval between the locking points for most drafting machine heads?
18. How many minutes are there is one degree, seconds in one minute?
19. What is the degree of accuracy achieved by a drafting machine vernier?
20. Why should drafting machine scales be checked for alignment daily?
21. Why should centerlines be established before drawing a template circle?
22. Describe the recommended procedure for drawing smooth irregular curves.

23. Show how the following scales are noted on a mechanical drawing:
    Full scale, half scale, quarter scale, twice scale, ten scale.
24. According to ANSI standards, identify the drawing note that indicates that all dimensions are in metric.
25. Why should a mechanical drafter have a working knowledge of precision measuring instruments?

PART 2. MEDIA AND REPRODUCTION

1. List five factors that influence the purchase and use of drafting materials.
2. Why is media transparency so important when reproducing copies using the diazo process?
3. Describe vellum.
4. Describe polyester film.
5. What is another name for polyester film?
6. Define matte, and describe the difference between single and double matte.
7. Which of the following combinations would yield the best reproduction: graphite on vellum, plastic lead on polyester film, or ink on polyester film?
8. What are the primary elements that will give the best reproduction?
9. What are the standard sheet sizes as defined by ANSI for the following sheet designations: A size; B size; C size; D size; E size?
10. Describe zoning.
11. Define the following title block terms: tolerances, millimeters and inches, unless otherwise specified, unspecified tolerances, and revisions.
12. Given the following title block, describe the numbered elements 1 through 13.

	UNLESS OTHERWISE SPECIFIED	◊ Delmar Publishers
10	MILLIMETERS INCHES	3 COLUMBIA CIRCLE – P.O. BOX 15015 – ALBANY, NY 12212-5015
	UNSPECIFIED TOLERANCES	
11	DECIMALS: mm in .X ±2.5 ±.1 .XX ±0.25 ±.01 .XXX ±0.127 ±.005 ANGULAR ±30' FINISH: 3.2µm 125µ in	DRAWN BY: 1 SCALE: 2 DATE: 3 APPD: 4
12		MATERIAL: 5
	THIRD ANGLE PROJECTION	PART NAME: 6
13	⊕ ◁	B PART NO: 8 REV: 9

7

13. What is another name for the diazo print?
14. Is a diazo the same as a blueprint? Explain.
15. Describe how the diazo process functions.
16. How does the speed of the diazo printer affect the resulting print?
17. Describe how to and the importance of running a test strip.
18. What should a good quality diazo print look like?
19. Describe the characteristics of slightly exposed or old diazo materials.
20. Define sepia.
21. Why are sepias generally made in reverse (reverse sepia)?
22. Discuss the safety precautions involved in handling ammonia.
23. Describe the recommended first aid for the following ammonia accidents: ammonia spilled on the skin; ammonia in the eyes; inhaling excess ammonia vapor.
24. What is the potential for hazard involving the ultraviolet light in the diazo process?
25. List two advantages of the photocopy reproduction method over the diazo process.
26. Define the microfilm process.
27. What is the big advantage of microfilm?
28. What are the two common microfilm formats?
29. Which microfilm format is the most popular, and why?
30. Discuss the function of a microfilm enlarger reader-printer.
31. Discuss computerized microfilm filing and retrieval.
32. Discuss how computerized reproduction can take place without generating a single copy.

CHAPTER 1

DRAFTING EQUIPMENT PROBLEMS

PART 1. READING SCALES AND DRAFTING MACHINE VERNIERS

1. Given the following engineer's scale, determine the readings at A, B, C, D, and E.

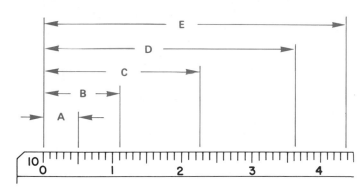

FULL SCALE = 1:1

2. Given the following engineer's scale, determine the readings at A, B, C, and D.

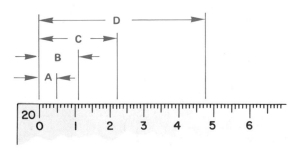

HALF SCALE = 1:2

3. Given the following architect's scale, determine the readings at A, B, C, D, E, and F.

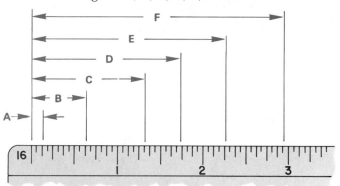

FULL SCALE = 1:1

4. Given the following metric scales, determine the readings at A, B, C, D, and E.

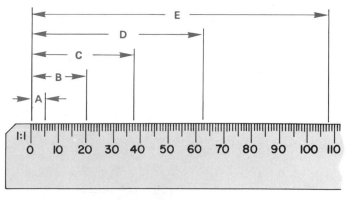

FULL SCALE = 1:1

4. *Continued.*

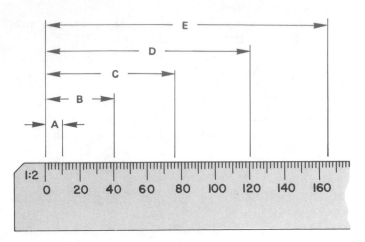

HALF SCALE = 1:2

5. Given the following drafting machine protractors, determine the angular readings.

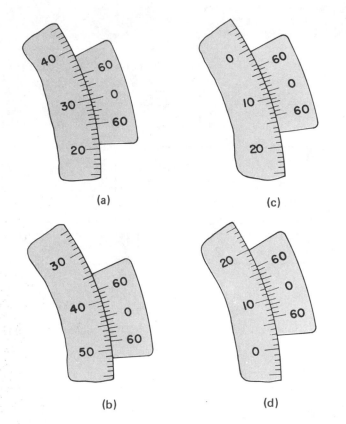

PART 2. READING A TITLE BLOCK

Given the following title block, describe the numbered elements 1 through 13.

UNLESS OTHERWISE SPECIFIED			
10 MILLIMETERS INCHES			

10

UNLESS OTHERWISE SPECIFIED	◆ Delmar Publishers
MILLIMETERS INCHES	3 COLUMBIA CIRCLE – P.O. BOX 15015 – ALBANY, NY 12212-5015

UNSPECIFIED TOLERANCES

11

DECIMALS:	mm	in
.X	±2.5	±.1
.XX	±0.25	±.01
.XXX	±0.127	±.005
ANGULAR	±30'	
FINISH:	3.2μm	125μ in

DRAWN BY: 1	SCALE: 2	DATE: 3	APPD: 4
MATERIAL: 5			

12

THIRD ANGLE PROJECTION	PART NAME: 6

13   ⊕ ◁ | B | PART NO: 8 | REV: 9 |

7

MATH PROBLEMS

Convert the following angle measurements to decimal degrees:

1. 15'
2. 7° 30'
3. 18° 5'
4. 200° 18'
5. −13° 42'

Convert the following angle measurements to degrees and minutes.

6. 60.4°
7. 9.5°
8. .27°
9. 177.8°
10. −45.1°

Introduction to Computer-Aided Design and Drafting (CADD)

LEARNING OBJECTIVES

After completing this chapter you will:

○ Describe the new motor and mental skills needed to be a CADD drafter.
○ Explain the CADD environment.
○ Discuss computers in drafting and how the industry uses CADD.
○ Evaluate CADD training and the job market.
○ Discuss and evaluate CADD equipment, including monitors, computers, input devices, output devices, storage, and software.
○ Explain the use of different CADD materials, supplies, and media.
○ Get started with CADD by exploring the use of the menu, using pointing devices, manipulating the drawing, drawing symbols, creating a symbol library, and storing a drawing.
○ Solve an engineering problem dealing with CADD.

THE ENGINEERING DESIGN PROCESS

Suppose you were asked to: a) Evaluate CADD software packages to determine the best choices for your company; and b) Evaluate microcomputers and peripherals to run the CADD software. What would you do? There are no "right" choices, only a lot of different ones, all of which could do the job.

One solution to this problem is to set out with an organized plan, and most important, stick to it. The following lists contain items that you should keep in mind when evaluating and recommending CADD software and hardware.

Software

○ What kinds of drawings does the company create?
○ Are the drawings the company's product?
○ Are the drawings used in the manufacture of a product?
○ How complex are the drawings?
○ Is 3-D required for a majority of the drawings?
○ Are client drawings used and edited?
○ Do the drawings have to be sent to a client or contractor?
○ If the previous two items were answered "yes," what kind of system is used by the clients?
○ How much money is available to spend?
○ How many packages must be purchased?
○ If more than ten packages must be purchased, does the software manufacturer provide site licenses?
○ How good is the software documentation?
○ Is quality training available? If so, is it local, or must you travel?
○ How often are software upgrades issued?
○ What is the average cost of software upgrades?
○ How good is the software support? Is there a hot-line number?
○ Are third-party software applications available?

Hardware

○ How much money is available to spend?
○ How many workstations does the company want?

(Continued)

ENGINEERING DESIGN PROCESS *(Continued)*

○ How many workstations does the company need?
○ How large are the typical drawings?
○ How much space is available for hardware?
○ How important is speed in the computer?
○ Where will archieve drawings be stored?
○ How many colors are required in the display?
○ Are digitizer tablet menus needed?
○ Will old drawings be converted to CADD?
○ What kind of paper copy is needed?
○ Will paper copies be reproduced? If so, how?
○ What size of flexible disk is needed?
○ If client drawings are used, how are they received—by disk or by phone modem?
○ How many people will be using the system?
○ Do people in the office work with overlay drafting now?
○ Do several people use the same drawing, or copies of the same base?
○ Is interoffice communication important?
○ Is it important that several employees be made aware of any updates to project drawings?
○ What are the hardware requirements of the CADD software?

These lists include a sampling of the many things to be considered and decisions that must be made before a CADD system is purchased. Unfortunately, many companies rush into the purchase of a system without answering even a few of these questions. This eventually leads to headaches that could have been avoided. Remember, the more planning that is involved at the beginning of any project, the less time will be required to execute the project.

▼ WHAT IS CADD?

The term *computer-aided design and drafting (CADD)* refers to the entire spectrum of drawing with the aid of a computer, from straight lines to color animation. An immense range of artistic capabilites resides under the heading of CADD, and drafting is just one of them. Three-dimensional industrial modeling and analysis is one of the specialized areas that has developed as a result of CADD. (See Figure 2–1.)

▼ DEVELOPING NEW SKILLS

CADD requires both manual and mental dexterity. Using tools involves motor skills. The CADD professional must also interpret, visualize, and achieve aesthetic layouts which are accomplished by using mental skills. Let's examine these two types of skills a little closer as they apply to computer graphics.

Figure 2–1 The screen display of this piping system was generated with a 3-D modeling software package. *Courtesy Applications Development, Inc.*

Motor Skills

Typing and digitizing are the principal motor skills you will be using while working with a CADD system.

CADD systems require data entry with a keyboard and a pointing device, such as a digitizer or a mouse. The digitizer, or menu tablet, may contain all of the commands used with the system plus a number pad and letters of the alphabet (alphanumeric key pads). Many commands and parameters are entered using the keyboard. Therefore, we find that the two skills of typing and digitizing are often used in roughly equal amounts depending on the configuration of the equipment.

A CADD professional using a digitizer is required to be able to pinpoint certain spots on a drawing, often looking through a set of crosshairs in a puck, and then pressing a button to *digitize* the point. Many systems use the pen-shaped stylus that is held in the hand like a pencil. The operator points to a spot and it is digitized.

Systems using flexible disks may require the operator to load disks into the disk drives as shown in Figure 2–2. This operation requires a steady hand and light touch as the disks may be damaged if inserted into the drive too hard or too fast.

Although some systems demand accuracy in locating points with the stylus or puck, many allow for sloppy pointing. If you are close when pointing to a specific item, the computer looks for it, or snaps to the nearest line, feature, or coordinate point. Such a feature is an advantage because it allows an operator to develop speed without concentrating on the exact location of points.

Manual drafting is actually much more demanding in the area of motor skills than is CADD. This fact may open drafting careers to people who do not possess certain motor functions. CADD systems using speech activated

Figure 2–2 Operator is loading a floppy disk into a disk drive.

Figure 2–3 The reduction in size of CADD equipment has enabled small engineering and architectural firms to purchase systems. *Courtesy TRI-STAR Computer Corporation.*

commands will open the field of drafting and design even further for persons with limited motor skills. Many physically challenged people possess unseen talents in design, drafting, and art that may be hidden behind a disability. The removal of certain barriers by the computer will enable many of these people to nurture and develop those talents despite certain physical limitations.

Mental Skills

As a computer drafter you need to approach problems and drawings in a systematic manner. Planning a drawing is especially important because your time at the computer may be limited. Much of the data you enter is based on the Cartesian coordinate system; therefore, a working knowledge of these coordinates is imperative.

Mental skills also involve the basic function of knowing which button to push and when. This is related to the ability to give verbal instructions. Remember, you are issuing commands to the computer and they must be given in the correct sequence. If they are not, you may encounter difficulty.

In addition, you must make a variety of decisions as you create drawings and designs at the computer: the location of views; which way to rotate an object; the best 3-D view of an object; the location of callouts and local notes; and the placement of notes, charts, and schedules for the best balance. These and many other decisions are based on your mental skills of perception, analysis, and synthesis.

▼ INDUSTRY AND CADD
Who Uses CADD?

Computer drafting systems are small enough to occupy just a few square feet of desk space. Individual contractors, designers, and architects can have one in

their home or office. (See Figure 2–3.) CADD has found its way into every corner of drafting, design, and engineering, and is used extensively in architecture, HVAC, structural, civil, process piping, landscape design, pattern making, machine design, and solid modeling.

Computers have found their way into the manufacturing process, too. Computerized welding machines, machining centers, punch-press machines, and laser-cutting machines are commonplace. Many firms are engaged in *computer-aided design/computer-aided manufacturing (CAD/CAM)*. In a CAD/CAM system, a part is designed on the computer and transmitted directly to the computer-driven machine tools that manufacture the part as seen in Figure 2–4. Within that process there are other computerized steps along the way.

Figure 2–4 Numerical control (NC) information required by the machining center in the background can be generated on the CAD/CAM system in the foreground. *Courtesy Computervision Corporation.*

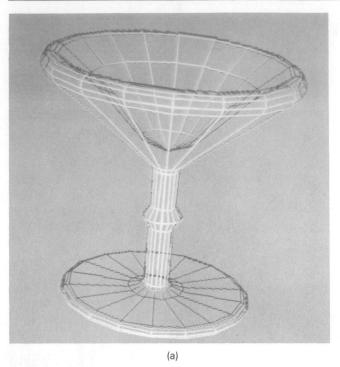

(a)

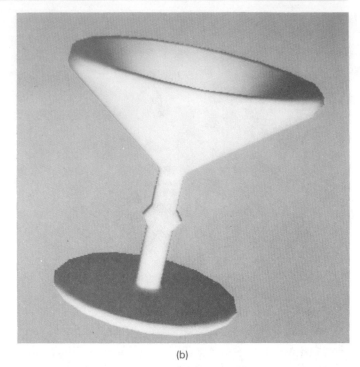

(b)

Figure 2–5 **(a)** All the points and vectors of an object are given to the computer to create a 3-D wire form. **(b)** A solid form can be created by filling in or shading and rendering a wire form. *Courtesy Megatek.*

The combination of the entire design, material handling, manufacturing, and packaging process is referred to as *computer-integrated manufacturing (CIM).* Within this process, the computer and its software controls most, if not all, portions of the manufacturing. A basic CIM system may include transporting the stock material from a holding area to the machining center at which several machining functions are performed. From there, the part may be moved automatically to another station at which additional pieces are attached, then on to an inspection station, and from there to shipping or packaging.

Designers and engineers are also using CADD. For example, a part can be designed as a model on the screen of a computer in a process called *solids modeling.* The model can first be drawn as a *wire form,* or outline, in which you can see all edges. (A wire form object is shown in Figure 2–5a.) Then the model outline can be filled in, or *rendered,* with various colors and shades on the monitor. In addition, the model may be rotated to any viewing angle desired, appear to be cut into and broken apart, then put through a series of simulated tests, and analyzed for strengths and defects in design. (See Figure 2–5a and b.) This process can be referred to as *CAE,* or computer-aided engineering. (See Figure 2–6.)

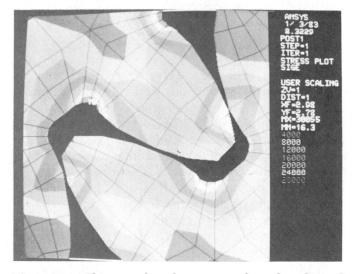

Figure 2–6 Objects such as these gear teeth can be subjected to simulated stress analysis on a computer screen. *Courtesy Swanson Analysis Systems, Inc.*

▼ COMPUTER DRAFTING EQUIPMENT

The electronic tools used for CADD are referred to as *hardware.* An individual CADD workstation relies on a computer for data processing, calculations, and com-

munications with other pieces of peripheral equipment. Peripheral equipment is any additional hardware item that uses the computer's files and performs functions that the computer cannot handle. A typical CADD workstation is shown in Figure 2–7.

The additional functions and services provided by the peripheral equipment fall into three categories: input, output, and storage. Input means to put information inside the computer to be acted on in some way. Input can come from the keyboard, a function board, or a

Figure 2–7 A typical CADD workstation with dual display screens and peripheral equipment. *Courtesy Calcomp.*

digitizer. Output refers to information that is sent from the computer to a receiving point. The receiving point may be a display on a video monitor, an inked drawing done by a plotter, or printed images and data provided by a printer. Storage refers to disk, tape and CD drives, which allow the operator to store programs, drawing files, symbols, and data.

We will examine all of the hardware mentioned here, and look closely at additional equipment in each of the categories described. All of the equipment is rendered functional by computer programs, or software, which is also discussed.

Monitors

The most obvious component of the CADD workstation is the video display screen, or monitor. We are all familiar with this piece of equipment because it looks much like a TV screen. The most common forms of monitors are vector and raster displays. Figure 2–8 is an example of a monitor. The monitor contains either an electron gun that sits behind the screen and shoots electrons, which strike the screen causing it to glow, or a flat-panel matrix display. Screens may show a single color (monochrome) or a full color display. The monitor comes in a variety of forms which we examine here.

Vector Monitor. This type of monitor is also known as stroke writing, vector writing, line drawing, cursive, or calligraphic writing. The beam of the electron gun is aimed at X and Y coordinate points (called pixels) and connects these points with a line, much like the way you draw a line with a pencil. Each line and feature on the screen is rewritten frequently to allow the operator to view the image as constant. This is called a refresh system because the image is refreshed many times per second.

If large amounts of data are displayed on a refresh screen, it is possible to get a flicker because it takes longer for the electron gun to return to an individual pixel to refresh it. During that lapse of time the pixel may begin to fade. When the electron beam strikes the fading pixel again, it brightens, which is seen as flicker. All of the data is stored as coordinates (vectors); therefore curved lines may appear to have a pointed shape because curves are drawn with short straight line segments connecting many vectors. Since any element on the screen can be deleted and redrawn without repainting the entire screen, the vector terminal can be used to animate images.

Raster Display. A raster display is similar to a television display. The screen is composed of a gridwork of tiny squares which appear as dots and are known as pixels. The word *pixel* is an acronym for picture (pix) element (el). The resolution of the screen, or the crispness of the image, is determined by the number of pixels. Each pixel is actually a *bit,* as shown in Figure 2–9. A bit is a binary digit and can be either on or off. A bit that is receiving a signal is on, and is lit. The absence of a sig-

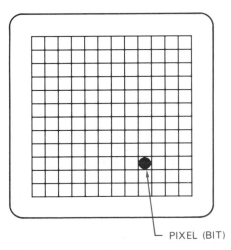

PIXEL (BIT)

Figure 2–9 Video display screens ar composed of a grid, or framework of pixels (bits), or dots, that can be turned on to display graphics or alphanumeric characters.

Figure 2–8 The monitor can display CADD system designs in full color. *Courtesy Nanao.*

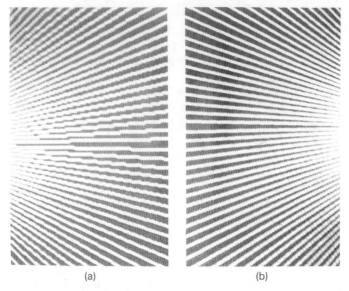

(a)   (b)

Figure 2–10 **(a)** A raster display produces jagged lines called "jaggies"; **(b)** but the jaggies can be reduced by high-resolution screens. *Courtesy Megatek.*

Figure 2–11 Portable microcomputer with a flat panel display. *Courtesy Toshiba.*

nal means the bit is off. A raster display is often called a bit map because it is a map composed of pixels. Low-resolution raster screens may produce lines that have jaggies, or a jagged look. This is especially true for angled lines and curved features. The lines in Figure 2–10a and 2–10b show how an image would appear on raster screens of different resolution if magnified. Individual lines and features can be deleted without repainting the screen because the image is constantly refreshed. The brightness of each individual pixel can be adjusted to show highlights and shading.

Color and Monochrome Monitor. The color monitor is similar to a color TV set. A pixel on a color screen is actually composed of three dots, red, green, and blue (RGB). Three electron guns supply these three primary colors. Monochrome screens can be purchased with a variety of single display colors such as green, amber, white, black on white, blue, and red.

Maxtrix Display. There are ways of producing a display other than shooting electrons at a screen. The general term, *matrix display,* refers to rows and columns of elements or components that can be turned on or off, much the same as pixels. Some forms of matrix displays are liquid crystal display (LCD), light-emitting diode (LED), electro-luminescent panels, and plasma display.

These matrix screen displays are also called *flat-panel displays.* (See Figure 2–11.) They do not require the deep cabinet needed by the monitor to house the electron gun. Present research and development is pointed toward flat-panel displays because of the reduction in physical size and the lower power requirements.

The Heart of a Computer

The processing function of the CADD workstation is handled by the *central processing unit (CPU).* The CPU is normally an integrated circuit (IC) chip, dedicated to a specific function. Additional chips contain specific instructions that allow a certain component of the CADD system to operate. These instructions can only be read by the computer, and then acted on. They are called *read only memory (ROM).* This is the memory that the computer never forgets. Other chips within the computer are empty. Information can be put into these chips and taken out again at any time. They are called *random access memory (RAM).*

The size of a computer's RAM is specified by the number of characters (letters or numbers) that it can hold. A character is formed by a group of eight bits (binary digits). When these eight bits are acted on as a whole unit it is called a *byte.* Thus the capacity of a computer's RAM is referred to by its number of bytes of memory, usually expressed in thousands or millions. One thousand bytes is called a K (kilo) and a million bytes is an "mb" (megabyte). A computer that contains 8 megs of RAM can hold over 8 million (mega) bytes of information in its internal memory.

The Mainframe Computer

The large computer that occupies its own climactically controlled room, such as the one shown in Figure 2–12, is commonly known as a mainframe computer. The term *mainframe* at one time referred to the size of the cabinet holding the computer. Most early CADD systems relied on a mainframe computer, and some

Figure 2–12 The mainframe computer can support many workstations and occupies its own climactically controlled room. *Courtesy Intergraph Corporation.*

Figure 2–13 A minicomputer, shown in the background, can serve several workstations. *Courtesy Computervision Corporation.*

still do. The mainframe computer may possess from 32- to 100-million bytes of memory. Mainframe computers often handle many other functions in addition to the graphics requirements of the CADD workstations. They are often referred to as host computers because they serve so many different terminals. They can support up to several thousand terminals on-line at one time.

The Minicomputer

The minicomputer represents a break from the restrictions of the centrally located mainframe computer. The minicomputer is smaller physically and in memory than a mainframe, but it can accomplish the same tasks. A plus for the minicomputer is its ability to support up to several hundred user terminals simultaneously (depending on the requirements of the users). The abilities of the minicomputer and its cost relative to the mainframe make it highly compatible with the requirements of many companies. CADD systems operating on a minicomputer are shown in Figure 2–13.

The Microcomputer

The microcomputer is also known as a desktop computer or a personal computer. (See Figure 2–14.) It possesses at least one microprocessor chip and normally between 1MB and 16MB of RAM. Many micros now provide for memory expansion up to 32 megabytes. Most micros serve one terminal, but the trend is turning toward networking, which is connecting several terminals to share resources such as a large hard disk. Networking gives a single microcomputer some of the same capabilities as the minicomputer.

The microcomputer is available in a variety of forms, from the basic desktop unit shown in Figure 2–15a, to the tower model with CD ROM player and multimedia

Figure 2–14 A microcomputer and peripherals are used by several CADD systems.

speakers shown in Figure 2–15b. The miniaturization of components has enabled laptop computers, Figure 2–16, to become nearly as powerful as microcomputers.

Input Devices

Keyboard. There are several methods of putting information into a computer. The most obvious is with the familiar alphanumeric keyboard (alpha-letters, numeric-numbers). (See Figure 2–17.) Written instructions can be given to the computer from the keyboard. Graphic instructions can also be entered from the keyboard in the form of X, Y, Z coordinates, circle diameters, number of polygon sides and curve radii. Using a keyboard

(a)

(b)

Figure 2–15 **(a)** A standard desktop microcomputer with dual disk drives. *Courtesy NEC Information Systems, Inc.* **(b)** A tower microcomputer with CD ROM and multimedia speakers. *Courtesy TriStar.*

Figure 2–16 This laptop computer has a flat-panel display and minitrackball pointer. *Courtesy Toshiba.*

Figure 2–17 This computer keyboard resembles that of a typewriter, with the addition of several keys and a calculator keypad. *Courtesy Capital Spectrum, Inc. for Dell Computers.*

could be a tedious method if all the drawing data were given to the computer in this fashion, but it isn't. Only specific instructions are issued to the computer from the keyboard.

The layout of the computer keyboard varies between different brands of CADD systems. One keyboard is shown in Figure 2–18. The standard QWERTY typewriter keys are arranged the same, and the additional keys can be found along the top, left, and right sides of the standard keyboard. Consider the following description of some of the most common keys found on a CADD system keyboard.

Return Key. The name of this key varies, but it is most commonly labeled RETURN, or ENTER, and is usually located at the right side of the keyboard. It is similar to the carriage return key on a typewriter. The computer responds to a line of data entry after this key is pressed.

(a)

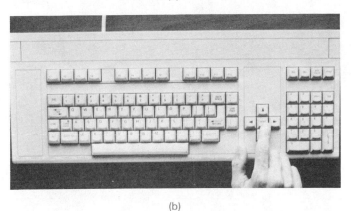

(b)

Figure 2–18 **(a)** Standard typewriter keyboard. *Courtesy Sharp.* **(b)** Typical alphanumeric computer keyboard. *Courtesy Texas Instruments, Inc.*

Escape Key. This key allows you to escape from the current command or activity that you're engaged in.

Cursor Keys. The function of these keys is to move the screen cursor. They are arranged in a group and appear as four arrows pointing in the four compass directions. They are often called arrow keys.

Home Key. This key is used in conjunction with the cursor keys. It moves the cursor to either the top left corner of the screen or to the beginning of the line. It can be found in the center of the cluster of cursor keys or at the upper left of the keyboard.

Backspace Key. If you make a mistake in typing, press this key and you can delete what you just typed. This key can also serve to move the screen cursor back along the line just typed.

Control Key. This key is similar to a typewriter shift key because it is usually pressed along with another key (or keys) to initiate a function or command within the program. The control key is used extensively in

word processing but its operation is often replaced with function keys in CADD systems.

Function Keys. These are additional keys located just about anywhere on the keyboard. (See Figure 2–18b.) A popular location is along the top of the keyboard. The number and grouping pattern varies by manufacturer. They are used by the program to activate specific functions or commands. Their use may change in the program, or from one program to the next, hence the need for function key masks or overlays. The overlay is usually a plastic template with written labels that fits around the keys. Some computers and programs display the labels of the function keys at the bottom of the screen, which eliminates the need for key overlays.

Calculator Keypad. This cluster of keys is usually referred to as the keypad and contains the numbers 0 through 9, a decimal point, and arithmetic function symbols.

Become intimately familiar with the keyboard you will be working with, and you will be rewarded with less wasted time searching for keys and smoother-running sessions at the computer overall.

Digitizer. The second method of entering drawing data into the computer is by pointing to it. This is done using an input device called a digitizer. The digitizer is often called a graphics tablet, or tablet for short. It is a low-profile piece of hardware that has a flat, smooth surface covering a gridwork of thin wires. (See Figure 2–19a and b.) A pointing cursor is attached to the digitizer by a wire.

The pointing cursor used with the digitizer can take two standard forms, *stylus* or *puck.* The stylus looks like a ballpoint pen with a wire coming out of the top. (See Figure 2–20a.) Another application of the stylus is in the 3-D digitizer shown in Figure 2–20b. This instrument enables you to digitize an existing object or model, and input the data into a CADD system as a 3-D model. This procedure is also known as *reverse engineering.*

The puck is usually rectangular in shape and many have from one to sixteen buttons on it. It too is attached to the digitizer by a wire. To locate points with the puck you must look through a small lens containing crosshairs. Two different types of pucks are shown in Figure 2–21a and b.

When the stylus or puck is brought near the active surface of a *digitizer tablet,* its position is determined by the electronics inside the tablet. As the stylus is moved across the tablet, crosshairs on the video display screen track its movement. By pressing the stylus to the tablet at the beginning of a line, the coordinates for that point are transmitted to the computer. Move the stylus to the end of the line and touch the tablet at that point. The final coordinates are then transmitted. A digitizer tablet operator can reduce an entire drawing to digits without ever stopping to calculate a single number.

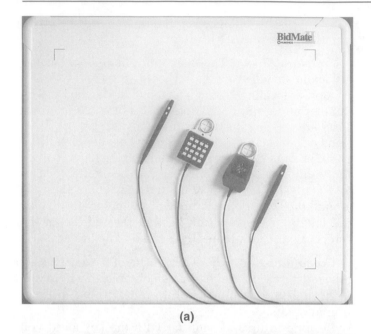

(a)

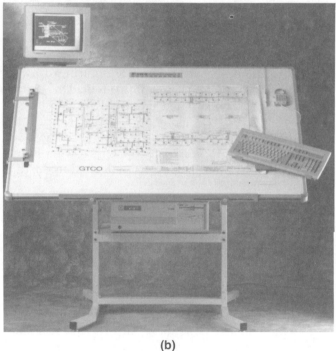

(b)

Figure 2–19 (a) Drawing data can be entered into the computer by using a digitizer. *Courtesy Numonics Corporation.* (b) A table-model digitizer. Note the stylus and puck. *Courtesy GTCO Corporation.*

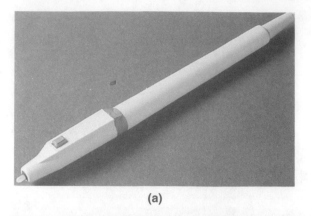

(a)

(b)

Figure 2–20 (a) A hand-held stylus resembles a ballpoint with a thin cable attached to it. *Courtesy Numonics Corporation.* (b) A 3-D digitizer enables objects to be input into a CAD system as a 3-D model. *Courtesy FARO Technologies, Inc.*

Digitizer tablets are also used to give instructions to a graphics system. This input method uses a menu having several graphics commands printed on it. A command is a specific instruction that is issued to a computer. The user simply presses the stylus on the desired command and the computer system accepts it. (See Figure 2–22.)

Several other input devices exist that perform the same function as the stylus and puck, but do not have to be used with a digitizer. The most well-known is famil-iar to anyone who has ever played a video game. It is a joystick, as seen in Figure 2–23. This little spring-loaded stick allows you to move the crosshairs on the screen by tilting the *joystick* in any direction. The stick always returns to the center of its receptacle, but the crosshairs remain in the last position they were placed.

The trackball is a ball mounted inside a box as seen in Figure 2–24. The operator simply pushes on the ball to make it rotate and the crosshairs on the screen move in the direction of rotation.

Turn the *trackball* over so that you are holding onto the box and rolling the ball on the table, and you have the principle of the mouse. The *mouse* may have a small roller ball on its underside, or two wheels mounted

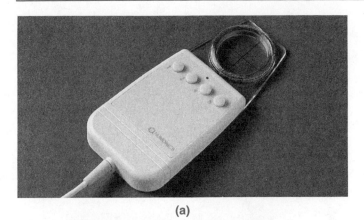

(a)

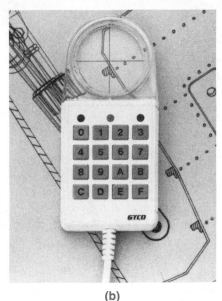

(b)

Figure 2–21 **(a)** A four-button puck. *Courtesy Numonics Corporation.* **(b)** A sixteen-button puck. *Courtesy GTCO Corporation.*

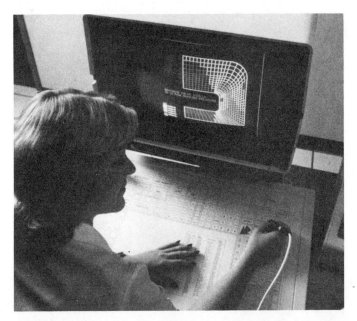

Figure 2–22 Using a stylus to select a command from the digitizer menu. *Courtesy Computervision Corporation.*

Figure 2–23 A joystick allows the operator to move the screen cursor. *Courtesy Chessell-Robocom Corporation.*

Figure 2–24 The ball of a trackball is rolled to move the cursor on the screen. *Courtesy ITAC Systems, Inc., Garland, Texas.*

at 90° to each other. The mouse, shown in Figure 2–25, is moved about on a flat surface in order to move the crosshairs on the screen. The optical mouse requires a special board on which to roll because it operates without moving parts. (See Figure 2–26.) The mouse and trackball are also attched to the terminal with a cable.

Output Devices

The primary output device is the video display screen which was discussed earlier. But there are several other

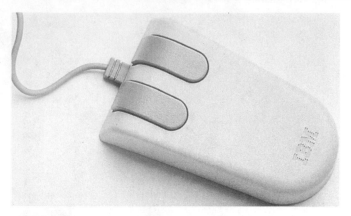

Figure 2–25   This two-button mouse is a pointing device for use with microcomputers. *Courtesy IBM.*

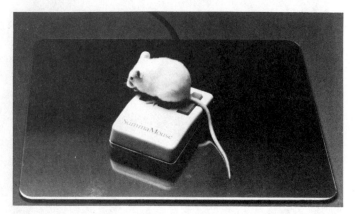

Figure 2–26   The optical mouse uses a reflective surface to control the movement of the screen cursor. *Courtesy Summagraphics, Inc.*

methods of producing a drawing. Sending it to the plotter is a popular option, be it pen or electrostatic plotter. A quick print can be transmitted to the hardcopy unit, which performs a function similar to a photostatic copier. The printer can be used to get a paper copy of drawing data, program listings, or even a print of the screen display. Not all CADD systems use each of these pieces of equipment.

Pen Plotters.   On a vector graphics display screen, lines and features are drawn using X and Y coordinates. This information is stored in the computer as a display list. The display list contains all the information that the plotter needs to function. The information stored on a raster screen (bit mapped) must first be converted to a display list before it is sent to the plotter because the plotter operates in two directions only. Movement of the pen is along the X or Y axes. Some plotters allow the pen to move in both directions, while others move the paper in one direction and the pen in the other. The paper size that can be used on this plotter is determined by the width and length of the bed.

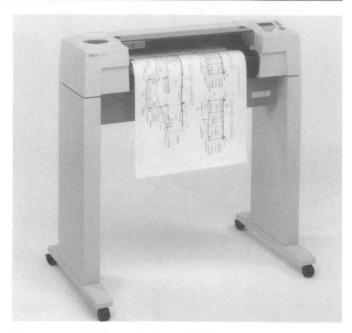

Figure 2–27   The pens and paper both move to produce a plotted drawing in this pen plotter. *Courtesy Hewlett-Packard Company.*

The type of pen plotter shown in Figure 2–27 is called a microgrip plotter. It uses grit-wheel paper-grippers. Some people refer to it as a pinch-wheel plotter. This plotter moves the pen in one direction and the paper in the other.

Electrophotographic Plotter.   The high-speed electrophotographic plotter is often referred to as a printer because of the nature of its process. As with pen plotters, the paper moves in one direction on an electrophotographic plotter. (See Figure 2–28.)

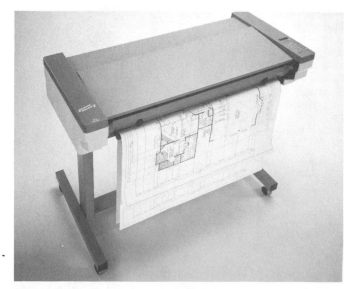

Figure 2–28   An electrophotographic plotter. *Courtesy JDL Company, Ltd.*

It uses a direct-imaging process similar to that of office copiers. A photosensitive drum is given a neutral charge by a light source. Negatively charged toner adheres to the neutral areas, and the toner is transferred to positively charged media. Heat and pressure then bonds the image to the paper. An electrophotographic plotter is shown in Figure 2–28.

Impact Printer.   A variety of printers fall into the category of impact printer. The action of this kind of printer is to strike a ribbon and paper to produce a character or image. The dot-matrix printer, seen in Figure 2–29, hammers tiny wires in specified locations to create a character composed of dots. A matrix is just an arrangement of rows and columns (Figure 2–30). Scoreboards in stadiums around the country flash scores using a matrix of lights.

Nonimpact Printer.   The nonimpact printer may be found in several different forms. The thermal printer uses tiny heat elements to burn dot-matrix characters into treated paper. Blue or black images can be produced. The electrophotographic printer has been described separately. The ink-jet printer, seen in Figure 2–31, squirts miniscule droplets of ink at the paper to produce a dot-matrix character. Some of these printers spray a continuous stream of ink at the paper.

The best quality image produced by a nonimpact printer comes from the laser printer. (See Figure 2–32.) The laser printer is more closely related to the electrostatic copier process than to the striking, poking, burning, and spraying printing methods. A laser draws lines on a revolving plate that is charged with a high voltage. The light of the laser causes the plate to discharge. An ink/toner will then adhere to the laser drawn images. The ink is then bonded to the paper by pressure or heat.

Figure 2–29   A dot-matrix printer. *Courtesy JDL Company, Ltd.*

Figure 2–30   A matrix is an arrangement of rows and columns.

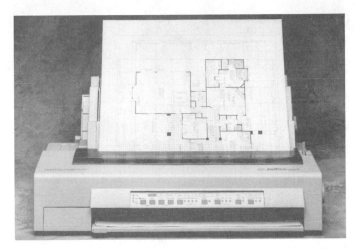

Figure 2–31   An ink jet printer. *Courtesy Summagraphics Corporation.*

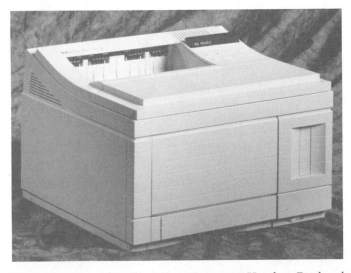

Figure 2–32   A laser printer. *Courtesy Hewlett-Packard Company.*

Storage

Storage can be either temporary or permanent. Temporary storage is often referred to as memory. This integrated circuit storage works fine when the power is on. But when the power goes off, the contents of temporary storage dissipate into the atmosphere. The temporary storage that is used to handle drawing data is random access memory (RAM). Drawings need greater protection than is available with RAM, so they are transferred to magnetic tape or disk. This transfer of data is handled by a tape drive, seen in Figure 2–33, or disk drive, shown in Figure 2–34. Once the drawing files have been secured on tapes or disks, they are in permanent storage. They may be filed in racks in the computer room, or sent to a vault as backup copies.

When both tape and disk drive units are used within a company, the current drawings are stored on magnetic disks. At the end of the working day, or whenever policy dictates, the contents of the disks are dumped (transferred) to the magnetic tape drives and saved on tape. This cleans off the disks for reuse. Copies can then be made of the tapes for backups.

Disk and tape drives are often referred to as mass storage devices because they are devices that store large amounts of data. Disks and tapes come in a variety of sizes and are discussed in greater detail later in this chapter.

Software

A computer program that provides specific instructions to enable a computer to do a certain task is called software. When CADD software is purchased from a manufacturer it is known as a systems program. This specialized software provides a service for a specific endeavor such as integrated circuit (IC) layout, process piping design, solids modeling, HVAC, structural, and architectural. The company that purchases a software package may eventually begin to modify the contents. It may even create some programs of its own. This can be referred to as applications software that the customers can apply to their own needs.

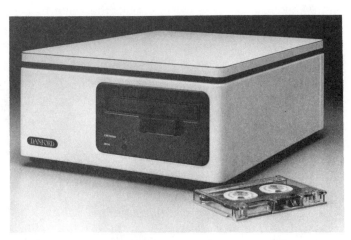

Figure 2–33 This tape-drive unit accepts a tape cartridge. *Courtesy Danford Corporation.*

▼ FUTURE CADD HARDWARE AND SOFTWARE

As the density of integrated circuits increases and memory capacities rise, larger programs will be able to operate faster with less power. Microcomputers are taking on the characteristics of mini and mainframe systems. Smaller CADD systems are able to store and handle more data because of their greater computing power.

The trend is toward total automation of the design, testing, and manufacturing process, and the integration of the disciplines of drafting and design. Eventually these tasks will be accomplished by a single computer system. These systems will have diagnostic abilities to pinpoint problems within the equipment, and redundant or auxiliary components that become active when another component fails.

Flat-panel and high resolution display screens will continue to increase in popularity and may eventually overshadow video display technology. But video display is still viable and continued research will produce screens of superior resolution. Raster screens are available with resolutions of approximately 150 × 200 pixels, up to 1000 × 1000 pixels. In the near future we will see displays of up to 4000 lines.

Color monitors are becoming standard, as are color hardcopy and color plotters. The increased use of three-dimensional (3-D) modeling will make color indispensable.

Figure 2–34 A disk pack is used in the disk drive. *Courtesy Computervision Corporation.*

Biocontrols

A variety of input methods may be developed but the voice method may soon be preferred. Systems already exist that accept voice commands. Voice synthesizer chips are being used commercially in a variety of applications to speak certain phrases or instructions. But CADD computers will recognize a certain number of spoken words (commands) and translate them into action. Another possible method of input may be the human body itself. The operator will wear an elastic band containing tiny electrodes. This band, placed anywhere on the body, will sense brain waves, muscle movements, and electro-optical signals (eye movement). These measured signals are then converted into digital format and used to control any number of processes. Current research in this area holds great promise for enabling physically challenged persons to realize their potential by tapping into their knowledge and insight while ignoring any perceived disabilities.

Laser Holography

Solids modeling and realistic shaded and rendered color models of parts and systems are becoming more commonplace as computer power increases. The use of 3-D may even evolve into the realm of holography. Lasers can be used to project a three-dimensional image in space. Designers and engineers could walk around the object and examine it from all sides. Animated screen displays enable us to view an object from all angles now, but the use of holography may expand those abilities. Projectors could be mounted around an open lot and the image of a new building could be projected onto the site at full scale. It could then be examined, tested, and changed on the very spot it is to be built.

Virtual Reality

The manner in which we work with computers and models will be dramatically changed by the technology called *virtual reality (VR).* The user wears a helmet or goggles containing two small flat-panel screens, one for each eye. On one hand is worn an *instrumented glove* that senses finger movement. The virtual reality user is actually inside the model and can see his or her hand in the screen. Not only can the user move around inside the model, but can interact with it.

Virtual reality technology is the next logical step in the design process. We have always lacked the ability to place ourselves inside the models we create and see them from a variety of viewpoints before they are built. A VR system gives us the capability to interact with a model of any size, from molecular to astronomical. Future surgeons will learn on virtual cadavers and will practice a real operation on a virtual body constructed from scanned images of the human to be operated on.

Home designers will walk around inside the house, stretching, moving and copying shapes in order to create the finished product.

Buildings will be designed and placed on virtual building sites. Clients will be able to take "walkthrough" tours of the building before it is built and make changes as they "walk." Scientists will be able to conduct experiments on a molecular level by placing themselves inside a model of chemical compounds. Using *telerobotics,* a person can "see" through a robot's eyes while in a safe virtual environment, in order to guide a robot into a hazardous situation. These possibilities are being researched and developed today using the variety of equipment shown in Figure 2–35.

The principal component of "immersive" VR is the *head-mounted display (HMD),* which may also include earphones. (See Figure 2–35a.) Using an HMD gives you the feeling of being totally inside the virtual world, because your eyes are not distracted by outside activity. Hence, the term "immersive." A second type of HMD, the *suspended display system,* is shown in Figure 2–35b. It is not as heavy to wear because it is suspended by the articulated boom, which contains all of the necessary cabling and sensors.

The HMD technology is moving toward smaller, less cumbersome units. The use of goggles, a special monitor, and sensing devices can enable the user to see in 3-D. (See Figure 2–35c.) An instrumented glove containing sensors can be worn that enables you to manipulate 3-D objects and navigate in a "virtual world." (See Figure 2–35d.)

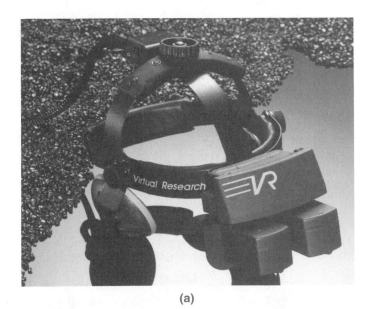

(a)

Figure 2–35   (a) This virtual reality head mounted display incorporates both stereo viewing with two small monitors, and stereo sound using attached headphones. *Courtesy Virtual Research.*

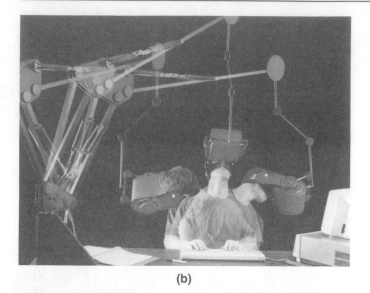

(b)

(c)

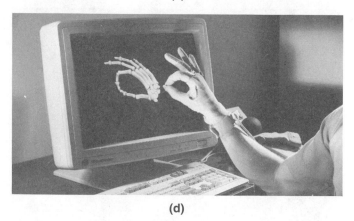

(d)

Figure 2–35 *(Continued)* **(b)** This HMD is a suspended display system that eliminates much of the weight, and provides six degrees of freedom of movement. *Courtesy Leep Systems, Inc.* **(c)** The stereo eyewear shown here enable the user to see 3-D images when used with a special infrared emitter and stereo-ready monitor. *Courtesy Stereo Graphics Corporation.* **(d)** The twenty-two sensor instrumented glove enables a user to manipulate objects and navigate in a virtual world. *Courtesy Virtual Technologies.*

A field of tremendous opportunities will develop around the creation of virtual worlds. These worlds will be detailed 3-D models of a wide variety of subjects. Worlds will need to be constructed for many different applications. Persons who can construct realistic 3-D models will be in great demand. The fields of VR and GIS (Geographic Information Systems) will be combined to create "intelligent" worlds from which an infinite wealth of data can be obtained while occupying the virtual world. No town or city will be without a virtual model of their municipality.

The implications for this technology are tremendous and the only limitation for its uses is our imaginations.

▼ CADD MATERIALS
Drawing Media

Drafting tools may have changed radically but the finished product is still rendered on paper (vellum) or drafting film (Mylar). The same paper sizes that are used in manual drafting are used in computer drafting. Those sizes are discussed in Chapter 1. The size of the plotter will, of course, determine the size of the paper used. Drafters occasionally must create drawings that are larger than standard paper sizes. This is accomplished by using roll paper or Mylar that is available in standard widths. Figure 2–36 shows a plotter using roll paper.

Other papers used in computer drafting are printer papers and hardcopy unit papers. Printer paper used with impact printers is basically a bond paper available in sev-

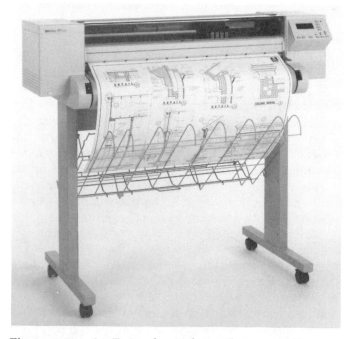

Figure 2–36 An E-size drum plotter that uses roll paper. *Courtesy Hewlett-Packard Company.*

eral weights (thicknesses) and sizes, and may have trac-tor-feed holes along the sides. Treated papers that are sen-sitive to heat are used in thermal printers. Hardcopy units can use a variety of mediums, depending on the process, such as electrostatic, ink jet, or photo. A dry silver paper is used with the photo plotter type of hard-copy unit.

Drawing Pens

The types of pens used in plotters are liquid ink, wet ink, liquid roller, felt tip, ceramic tip, and ballpoint. Some examples are shown in Figure 2–37. The best quality lines are obtained with the wet-ink pen. This type of pen is seen in Figure 2–38, and is basically an adap-tation of the technical pen. Wet-ink pens have a variety of tip widths, and a reservoir for ink. Most plotters can hold several pens and automatically switch pens to change line width or color. The main problem with using multiple wet-ink pens is preventing the tips from drying between use. Since the use of wet-ink pens is time consuming and costly, most preliminary plots are done with felt-tip pencil or ballpoint pens. A plotter using felt-tip pens is seen in Figure 2–39.

Figure 2–39　Felt-tip pens are being used in this A-size plot-ter. *Courtesy Summagraphics Corporation.*

Storage Media

The storage file for the computer drafting system is the magnetic tape or disk. The tape or disk drive holds the drawing file as magnetic impulses called *bits.* The drive can also read the data from the tape or disk and send it to the computer. The tape or disk drive unit is often referred to as the mass storage device. The term *mass storage* technically means a large-capacity storage device but it is commonly understood to mean any external storage beyond that inside the computer.

Magnetic Disks

The most popular form of storage media is the mag-netic disk. The disk itself is made of either a rigid or flex-ible material, usually plastic, which is coated with a magnetic oxide. The disk resembles a brown phonograph record without grooves. The read/write heads of the disk drive skim just above the surface of the disk and deposit or remove magnetic impulses. Disks come in a variety of sizes and forms.

Mainframe and minicomputer systems often use the removable hard-disk module or disk pack. The *disk pack* contains several rigid disk platters 14 in. in diam-eter connected with a spindle. (See Figure 2–40.) When the disk pack is not being used (off line) it is stored in a plastic case with a handle on top. The bottom of the plas-tic case is removed before the pack is inserted into the disk drive. Once in the drive the plastic cover is removed to provide room for the access arms to get between the disks. The access arms are the read/write heads. When not inside the disk drive, the disk packs must be securely attached to the bottom and cover of the plastic case. This protects them from dust and other contamination. Care should be taken not to drop or otherwise mishandle the disk pack.

A second type of hard disk is the *fixed hard disk,* which cannot be removed from the disk drive. This is a rigid disk made of plastic or metal that is permanently sealed inside the disk drive. A fixed hard disk is faster and holds more data than its flexible cousin, and is usually offered with small computers such as micros and minis.

Figure 2–37　Examples of various types of plotter pens. *Courtesy Mars Plot, J. S. Staedtler, Inc.*

Figure 2–38　Wet-ink plotter pens. *Courtesy Mars Plot, J. S. Staedtler, Inc.*

Figure 2–40 Disk packs. *Courtesy Uarco, Inc.*

The most well-known of the magnetic disks is the flexible disk. It is also called a *floppy disk* or diskette. The floppy disk is much thinner than a hard disk and thus is floppy. It is encased in a paper cover and comes with a dust jacket that must be removed before the disk is inserted into the drive. Floppies come in two standard sizes. The standard diskette has a diameter of 5¼ in., and the micro floppy is 3½ in. The micro floppy is encased in rigid plastic and the surface of the disk is not exposed. When the disk is inserted into the disk drive, a shutter opens to expose the disk to the read/write head. Examples of the various forms of disks are shown in Figure 2–41.

Disks and disk jackets should never be written on with pencil or ballpoint pen. Labels should be written first, then attached to the disk. Disks are magnetic and can be ruined if placed too near magnetic materials. Any magnetic field can rearrange the information stored on a disk. Motors, television sets, loud-speakers, and ringing phones can damage the information stored on nearby floppy disks. Temperature extremes can damage disks, so try to store them in a room with a fairly comfortable temperature range. Dust, hair, and assorted atmospheric debris can contaminate a disk and render it worthless. See that your computer room is clean and that disks do not lie around outside their jackets and storage boxes.

Floppy disks must be handled carefully. Never touch the exposed areas of the disk. When inserting a floppy into the disk drive, do so gently, in the manner shown in Figure 2–42. If a disk is inserted into the disk drive too hard, it may bend or crease.

New disks must be initialized or formatted. The INITIALIZE or FORMAT command is issued to the computer after inserting a new disk into the drive. The disk is then divided into sectors and tracks to accommodate the storage of files or records. An area for the disk directory is also established by the drive. The sectors of a disk resemble slices of a pie, and the tracks are like the grooves on a phonograph record. (See Figure 2–43.)

Figure 2–42 The proper way to insert a flexible disk into a disk drive.

Figure 2–41 Examples of various types of flexible disks. *Courtesy Devoke Data Products.*

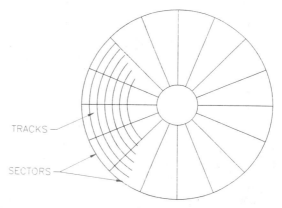

TRACKS

SECTORS

Figure 2–43 A disk is divided into pie-shaped sectors and tracks that circle the disk-like concentric rings.

Magnetic Tape

A less expensive type of magnetic storage media is magnetic tape. It is used primarily for backup and off-line storage of drawings and data. Companies using computers with tape drives and disk packs usually use disks for daily work, and then transfer the drawings to tape as backup storage on a regular basis. Disks are used during working hours because they are faster and hold more data.

Floppy disks are faster than tape because they are direct-access media. The disk is divided into sectors, tracks, and bytes. A specific location on a disk is much like the numbers of the houses on a street or apartments in a building. In fact a specific location on a disk is also called an address. When the disk drive discovers the address, it can go directly to that location. Hence the term direct access.

Magnetic tape, on the other hand, is said to have sequential access. The information on tape is recorded on parallel tracks. There are eight tracks, each containing one bit. Therefore the width of a tape can hold one byte and the bytes are recorded in blocks. A ninth track on the tape is used for automatic checking purposes. (See Figure 2–44.) Because tape is a long, thin strip, it must be wound and rewound to find the correct address, a process that takes time. The location of your drawing may be at one end of the tape. Once the tape drive learns the address, it must then wind the tape until it finds the location. All the addresses are located in sequence and must be accessed sequentially.

Magnetic tape also comes in a variety of forms. The reel-to-reel tape that is used in mainframes and some minis looks quite similar to the reel-to-reel tape that is used on tape decks. The standard size is ½ in. wide. (See Figure 2–45.) Tapes used for storage must be recopied periodically. If they are not recopied, the coils of tape may contaminate each other. Tape is tightly coiled and if left that way for several years, the magnetism of the individual bits may begin to merge with others or adjacent parts of the tape, thus creating an unreadable tape. This problem is not encountered with disks.

Magnetic tape is also available in cartridge and cassette form. These come in various lengths, much the same as audio tape. All tape is coated on one side, whereas disks can be coated on one or both sides and be

Figure 2–45   The reel-to-reel tape drive unit on the right can send stored drawings to the drum plotter on the left. *Courtesy Gerber Scientific Instrument Company.*

designated as single sided or double sided. The same precautions hold for magnetic tape as were mentioned for disks; always keep them inside their storage cases when not in use, and never touch the exposed tape.

Optical Disks

Optical disks offer an economical, high-speed method of storing computer-generated drawings. Some advantages of optical storage disks over magnetic media include increased storage volume, transfer of information to a standard output device or directly to microfilm. One single optical disk can store a trillion bytes of information, which is accessible in only a few seconds. Optical disks are available in 5¼-, 12-, and 14-in. sizes.

CD ROMs

One of the fastest growing storage technologies is the *compact digital read only memory,* or *CD ROM.* This is the same technology that is used for audio CD ROMS. The thin plastic platter is treated on one side with tiny pits and a reflective surface. Digital impulses are read by a laser and transformed into either audio or digital information. Many companies are providing parts catalogs and drawing libraries of components such as doors, windows, fasteners, etc., on CD ROM. CD ROM recorders will enable the disks to be used much as floppy disks are used now, for temporary and permanent storage of data.

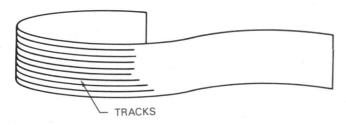

Figure 2–44   Magnetic tape stores data on parallel tracks.

▼ GETTING STARTED WITH CADD

This section contains a general discussion of the basic aspects of CADD systems such as AutoCAD. Software documentation and specialized textbooks provide the detailed information you need to learn the software and work productively with it. The following discussions provide an overview of topics such as screen and tablet menus, use of pointing devices, the rectangular coordinate system, and basic drawing, editing, and viewing procedures.

Starting a CADD Drawing

When your CADD software is loaded you are presented with a drawing screen, or *graphics area,* and menus. The graphics area is where you construct your drawing, and the menus contain items that you can select to perform specific functions. Regardless of the software you use, it is important to provide some time for planning your work before using the computer. Some of the items to consider before beginning work on any CADD project are:

○ Size of the object being drawn.
○ Size of the final plotted drawing sheet.
○ Scale of the plotted drawing.
○ Text heights and styles to be used.
○ Type of dimensioning needed for the drawing.
○ Format of the drawing (2-D, 3-D, or isometric).
○ Types of views required.
○ Special settings or values needed.

The more planning time you provide, the smoother your drawing session will go.

Using the Menu

The *menu* is a list of related items (commands, numbers, symbols, etc.) from which you can select, similar to a menu in a restaurant. An example of a typical menu is seen in Figure 2–46. You select the options on the menu by using some sort of pointing device. The menu is usually a sheet of paper or a plastic overlay that is printed with commands used in the particular CADD system. The menu is attached to an electronically sensitive surface called a tablet, or digitizer. A menu can also appear on the monitor. Commands may be selected using function keys, cursor keys, a stylus, mouse, or by touch. The screen-menu commands may appear anywhere around the periphery of the screen, depending on the CADD system.

Menu Layout. An important aspect of the menu is the manner in which it is arranged. There is no ideal arrangement for a menu, and that is why the menus that come with commerical CADD systems are often customized by the user company to suit its needs.

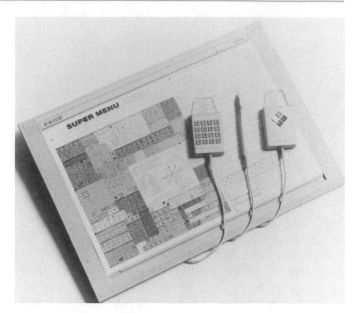

Figure 2–46 A typical tablet menu showing commands and functions. *Courtesy GTCO Corporation.*

The most important aspect of the menu is the grouping of items in related clusters. The item groups are defined by borders, space, or colors, and are usually identified by a heading that defines their relationship. A good example of a typical menu layout is seen in Figure 2–47.

A good method to use when learning the menu is to begin with general items and work toward specific items. First learn what the command groups are, then where the command groups are located. Then as you

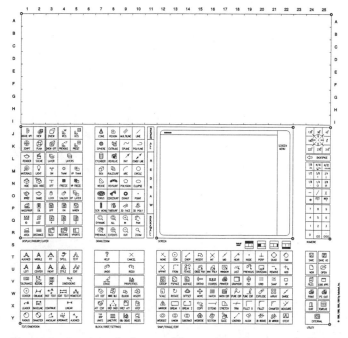

Figure 2–47 A typical menu layout. *Courtesy Autodesk, Inc.*

begin to use the system, try to remember which commands are in each group. You will do better by first learning the commands in the groups you use most often, then adding those used the least.

Pointing Devices. The most popular method of selecting items on a menu tablet is with a pointing device, which is a general term for a variety of instruments that are attached to the digitizer or computer terminal by a cable. The most common pointing devices are the puck, stylus and mouse. These pointing devices were discussed earlier in this chapter.

▼ DRAWING WITH THE COMPUTER

Creating good drawings with a CADD system hinges on just a few things: your ability to learn and remember, hand-to-eye coordination between the digitizing surface and the screen, your ability to undertand and visualize coordinate systems, and your typing ability, although this varies among CADD systems.

In this section we will look at how you use the above-mentioned skills to create drawings. The intention is not to discuss the techniques used with a particular brand of CADD system, but to describe the general concepts, techniques, and commands used when working with any computer drafting system. A familiarity with the basics of CADD will allow you to more readily grasp the particulars of any one system.

Cartesian Coordinate System

Absolute Coordinates. The *Cartesian coordinate system* is a rectangular coordinate system that locates a point by its distances from intersecting, perpendicular planes. Drawings done with a computer are based on the Cartesian (rectangular) coordinate system, so it is imperative that you develop a thorough understanding of it.

The two-dimensional rectangular coordinate system is illustrated in Figure 2–48. The point of intersection of the

dark vertical and horizontal lines (planes) is called the *origin*. The origin has a value of zero. Values along the horizontal increase as you move to the right of the thick vertical line. This is called the *X axis*. The *Y axis* is the horizontal line and values increase as you move upward from the origin. The values just mentioned are positive. Note in Figure 2–48 that each axis can also have negative values. Negative X values are to the left of the vertical plane, and negative Y values are below the horizontal plane. Study this figure carefully before continuing.

The best way to understand the rectangular coordinate system is to work with a piece of 10 lines to the inch graph paper. Locate the intersection of any two heavy perpendicular grid lines on your graph paper. Mark this point as the origin. Both X and Y have values of zero at the origin. At each inch mark on the horizontal line to the right of the origin, write the values 1, 2, 3, 4, etc. Do the same for the inch lines above the origin along the vertical axis line. (See Figure 2–49.) Now locate the point of X = 2, Y = 2. First count two inches to the right on the X axis, then count two inches straight up on the Y axis. This point has the value of X = 2, Y = 2. Only one point in this coordinate system can have that value.

Not all points will have even inch values. Locate the point X = 4.5, Y = 2.9 on your grid paper. This point is located directly from the origin and not from any other point. Coordinate values are always written with the X value given first. Therefore, this point is also written 4.5, 2.9. Count the grid squares along the X and Y axes to find this point, or use a scale and measure. Remember that every point must have at least two coordinate values. A point of just 4.5 could be any location 4.5 in. from the plane.

Relative or Incremental Coordinates. Objects can also be drawn using a moving or relative origin. In this method, each point becomes the origin and the next point is located in relation (relative) to the previous point. This method is called *relative* or *incremental coordinates*. If each point becomes the origin for the next

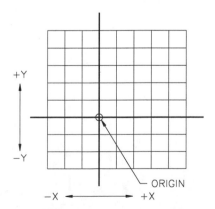

Figure 2–48   Two-dimensional Cartesian (rectangular) coordinate system.

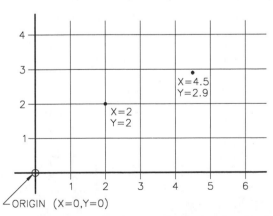

Figure 2–49   Points located on a rectangular coordinate grid.

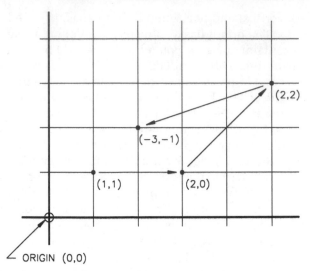

Figure 2–50 Points located using relative (incremental) coordinates.

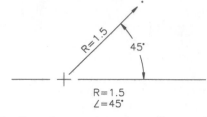

Figure 2–52 Locating a point using polar coordinates.

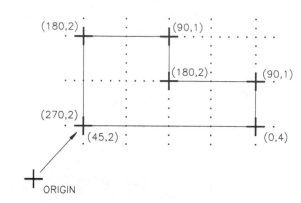

Figure 2–53 Object drawn using polar coordinates.

point, then numbers to the left and below the origin are negative. See Figure 2–50 for an example of this method.

The relative coordinate method may be useful when using the dimensions of each line to lay out a part. It is important that you keep in mind the positive and negative aspects of this method, because each point becomes the origin of a coordinate system. For example, moving to the left of a point gives the next point a negative X value. The object in Figure 2–51 shows the relationship of points on an object constructed with relative coordinates.

Polar Coordinates. Not all points can be located properly with a rectangular coordinate system. For example, a point may be at a 45° angle from horizontal at a radius distance of 1.5 in. from a reference point. Such a location defines a point by *polar coordinates*. Note in Figure 2–52 that the horizontal line has a value of 0°. Angles are measured in a counterclockwise direction from horizontal. The radius distance is measured from a specified reference, or center, point.

Polar coordinates are relative and can be used to create a drawing if all the angles of the part are known. Figure 2–53 is the same object drawn in Figure 2–51, but

the values given are polar coordinates. The first number at each corner of the object is the angle from the previous point (origin), and the second number is the radius from the origin. Study this object until you are familiar with the concept of polar coordinates.

Three-Dimensional Coordinates. To construct a three-dimensional coordinate system, all we need do is add a third dimension, the *Z axis*. The third axis, Z, projects out toward the viewer, Figure 2–54. The intersection of these three axes forms the origin, which has the numerical value of X = 0, Y = 0, and Z = 0. Any point that is behind the origin on the Z axis has a negative value.

The object shown in Figure 2–55 is shown in wire form, and the X, Y, Z coordinate values of each point are indicated. The first number of each set is X, the second

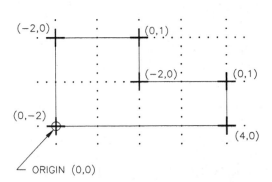

Figure 2–51 Object drawn using relative coordinates.

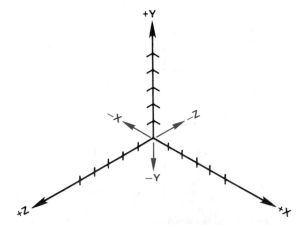

Figure 2–54 Three-dimensional coordinate system.

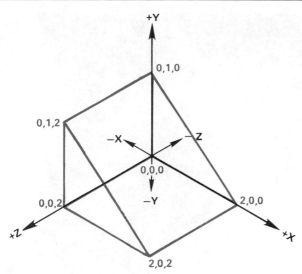

Figure 2–55   Object plotted using 3-D coordinate system.

Y, and the third Z. Study the object in Figure 2–55 until you understand the technique to plot points in three dimensions (3-D).

Using Coordinates

Now try your hand with the object given in the table on this page. You are provided only the coordinate values for each point on the object. Locate each point given on the chart and draw a line as you go. Lay out a 3-D axis like the one shown in Figure 2–54. Use a scale to measure the coordinate values given and plot them as shown in the example in Figure 2–56. The solution to this exercise is found at the end of this chapter.

Three-Dimensional Coordinate System.   The addition of the *Z axis* adds depth to the object and allows the CADD professional to create a model form. A popular misconception about computer graphics is that the

entire object can be shown in three dimensions (3-D) just by entering coordinates for a couple of views. This is not always true. The important thing to realize is that every point on the drawing must be defined by numerical locations, or coordinate points, for the computer to generate a 3-D form.

THREE-DIMENSIONAL COORDINATE EXERCISE			
Point	X	Y	Z
1	0	0	0
2	3	0	0
3	3	1	0
4	1	3	0
5	0	3	0
6	0	3	4
7	1	3	4
8	3	1	4
9	3	0	4
10	0	0	4
11	0	3	4
12	1	3	4
13	1	3	0
14	3	1	0
15	3	1	4
16	3	0	4
17	3	0	0
18	0	0	0
19	0	0	4

The process of entering three-dimensional data into a system is relatively simple. The principal component is your ability to visualize three dimensions and distinguish and separate any number of planes in one view. When entering data in 3-D, you should establish an origin point. (See Figure 2–55.) You can then draw using XYZ coordinates based on that origin. Programs such as AutoCAD allow you to move and rotate the origin to create new coordinate systems. You can place the origin and coordinate system on any plane of an object by using the UCS command in AutoCAD to create a new *user coordinate system*. This allows you to draw any shape, using its actual dimensions, from any angle or viewing point you choose.

Note in Figure 2–55 that the coordinate system can accommodate negative coordinates. An object can be drawn in any or all of the four quadrants. An object can just as easily be drawn using all negative coordinates as positive ones. Most drafters work with positive coordinates unless it is more convenient to locate the origin in the center of the part or a feature.

A CADD 3-D model becomes amazingly versatile when all of its vertices have been entered. Once the computer knows the location of all of the points of an object it can turn and rotate the object to any angle, as with the robot arms shown in Figure 2–57. And that is one of the features of CADD that makes it so appealing in engineering, design, testing, and manufacturing.

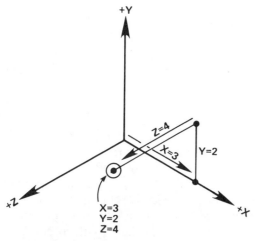

Figure 2–56   Locating a point using 3-D coordinates.

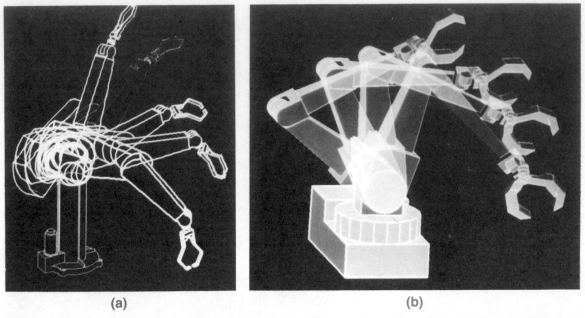

(a)   **(b)**

Figure 2–57   **(a)** These robot arms can be rotated in any direction once all of the points have been entered into the computer. *Courtesy Computervision Corporation.* **(b)** Another example of robot arm movement. *Courtesy Spectragraphics Corporation.*

Manipulating the Drawing

The subject of manipulating the drawing covers a wide range of capabilities within the CADD system, capabilities that may vary from one system to another. Manipulating the drawing begins with editing functions and moves into other aspects of creating a drawing, such as layout, design, additions, initial set up, examining microscopic detail, standing far away from the object, and looking at the object from any angle. Let's look at a few of these manipulative aspects.

ZOOM.   A popular command that enables the operator to get close to the work is a command called *ZOOM.* Selecting the ZOOM command allows the drafter to draw a rectangle around an area that needs to be enlarged. After the rectangle is drawn, the entire screen is filled with the area that was inside the rectangle. It is as if a window were placed around an area, and then the viewer walked closer to the window so that the area inside the frame was all that could be seen. This is a great command to use when working in a detailed area of a drawing. Once inside a window, the command can be used again and again to get even closer to the object or feature. Figure 2–58 illustrates how the ZOOM command works.

MIRROR.   The *MIRROR* command can give reverse copies of an entity or symbol by simply specifying an axis on which to reflect the object. The image can be duplicated at any distance from the original, or appear to be attached to it, such as the object in Figure 2–59.

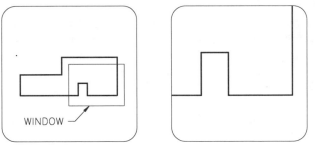

Figure 2–58   The ZOOM command brings the object, or a feature of the object, closer.

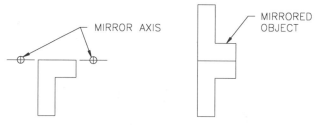

Figure 2–59   The MIRROR command can be used to reflect a feature.

COPY.   Once a feature or symbol has been drawn or located on a drawing, it can be duplicated with the *COPY* command. With the COPY command you need only identify the item to be copied, then indicate where it is to be copied. A similar command, called ARRAY, enables you to copy an object in a row and column arrangement, or in a circular pattern copied about a center point (polar array).

Figure 2–60   Example of grid snap.

HATCH.   By selecting a command called *HATCH,* you can choose a sectioning pattern, point to an enclosed polygon or circular shape, and automatically the area inside the polygon is filled with the pattern.

GRID.   The *GRID* command can be used to do a couple of things. The first is to paint a grid of dots on the screen in any increment needed, be it millimeters, decimal inches, or feet. A grid snap or grid roundoff can also be set which snaps any digitized point, line, or feature to a grid intersection.

Imagine that you have grid lines on the screen that are 2 in. apart. Any point that you pick on the tablet will be moved to the nearest 2-in. grid intersection. Figure 2–60 illustrates this process. Try to digitize a point along a horizontal line from point A. Purposely move the pointer up somewhat from the intended line to point B1. If your new point is less than 1 in. vertically from the previous point it will be snapped to the previous Y coordinate, point B. If your second point, B2, was greater than an inch above the previous Y coordinate, the new point will be snapped to the next grid intersection, point C. This is a great aid in drawing straight lines and aligning features.

Viewpoint.   After an object has been drawn, we may want to stand back and view it from a different angle. The view may be either orthographic or three-dimensional. Of course, you remember that a 3-D view can only be displayed if all of the X, Y, & Z coordinate points of the object have been entered into the computer. Systems that have the ability to rotate objects in 3-D open an exciting world of possibilities for the CADD professional. It's not only exciting, but fun to watch an object rotate on the screen in front of you.

These are but a few of the specialized commands and functions found in CADD systems. Each system has commands similar to the ones discussed.

▼ CREATING AND USING SYMBOLS

One of the great time-saving features of CADD technology is the ability to draw a feature once, save it, and then use it over and over without ever having to draw it again. That is why one of the first things a company does after purchasing a CADD system is to begin developing a symbol library, cell library, or symbol directory. This is a file of symbols that can be called up, located on the menu tablet, and used on a drawing. The area for a symbol library is shown on the menu in Figure 2–61.

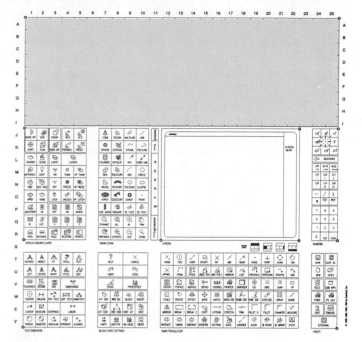

Figure 2–61   The upper portion of this menu overlay houses the symbol library. *Courtesy Autodesk, Inc.*

Drawing a Symbol

The process of creating a symbol is a simple one. Begin a drawing with the name of the symbol as the file name. When you start the actual drawing of the symbol, begin by placing the origin of your coordinate system at a spot on the symbol that would be convenient to use as a point of reference or location on a drawing. Note the symbols in Figure 2–62. They all have small circles at their location points. This is the spot that you point to on your drawing to locate the symbol.

When the symbol has been drawn it must then be stored, but in a slightly different manner from a regular drawing. It is a symbol and not a drawing, and must be stored as such. A command such as BLOCK enables you to save the drawing as a symbol. This allows the symbol to be used in ways that a regular drawing cannot be used.

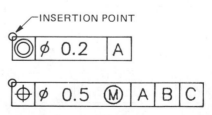

Figure 2–62   Location or insertion points on symbols.

Creating a Symbol Library

After all of the required symbols have been drawn they can be used in a variety of ways. It is important that all the symbols or blocks be located in a common area, such as a floppy disk, or a directory on the hard disk drive. The simplest is to just insert or merge them with an existing drawing, calling the symbol up by its file name. If you have the ability to customize the software's menus, the symbols can be added to a new screen or pull-down menus for easy access. In addition, they can be added to an area of the tablet menu as discussed previously. These methods of locating the symbols in menus allow them to be picked and then placed in the drawing, thus eliminating the need to type file names.

If symbols are placed on a digitizer or tablet menu, the final task would be to have all the symbols on one template plotted in a pattern that matches their layout on the template. This can then be used as an overlay, or mask on the menu, which would look similar to those shown in Figure 2–63.

Using Symbols

Most of the editing features that are used to revise drawings can be used to alter symbols. In addition, there are several commands that enable you to work with the symbols. The entire symbol template can be selected for use by choosing a command such as MENU. Another method of choosing a specific symbol template is to have all of the template names displayed on the menu tablet. Then it is just a matter of pointing to the name of the template that is needed.

A tablet menu symbol can be selected by pressing the pointer inside the particular symbol box. The symbol can then be located on the drawing by simply pressing the pointer to the spot where it is desired. Some systems may automatically display the symbol at the crosshairs location. As the puck is moved, the symbol moves with it. When the symbol is in the desired location, simply press a button on the puck and the symbol is placed in that spot.

Symbols are fluid just like the rest of the drawing, and can be changed, moved, renamed, and copied. Commands such as ROTATE, COPY, MIRROR, RENAME, and MOVE can all be used to work with symbols.

▼ SCALING AND SCALE FACTORS

Drawings created using manual techniques are laid out using a scale, whereas drawings created with CADD are drawn full scale and plotted to a specific scale. Mechanical drawings utilize scales such as ½" = 1", ¼" = 1", and ⅛" = 1". Architectural drawings are representations of buildings and use scales such as ½" = 1', ¼" = 1', and ⅛" = 1'. Some of the smallest scale drawings are in the civil engineering field. A *small scale drawing* shows a large area, while a *large scale drawing* shows a small area. These drawings represent areas of land measured in feet or miles, and some of the scales used are 1" = 10', 1" = 20' and 1" = 50'.

The most important point to remember when using AutoCAD is that you always draw full size, and plot at the proper scale. But this also means you must plan ahead and know what scale the final plotted drawing is to be. This is discussed in detail later. Drawing in AutoCAD at full scale means that if a part measures 4.35", you draw it 4.35" long. Even if the final drawing will be plotted on a B-size sheet of paper at ¼" = 1", it is still drawn at full size in AutoCAD. A house that measures 48'–6" along one wall must be drawn that exact size in AutoCAD. You will soon realize that it is critically important to plan your work carefully so that you set up the correct drawing limits to work in. Establishing the correct drawing limits will enable you to draw anything full size within those limits—from a microscopic computer circuit to a map of the solar system.

The planning that should occur prior to beginning the drawing involves, but is not limited to, the following items related to scaling:

○   Size of paper for final plot or print.
○   Size of the object to be drawn.
○   Height of plotted text.
○   Type of dimensioning to be used.

Planning the scale at which a drawing is to be plotted is important because the plotted scale dramatically affects

SYMBOL LIBRARY: PIPING

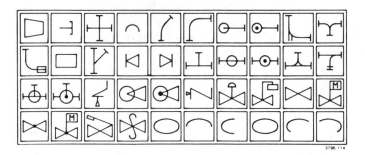

SYMBOL LIBRARY: SCHEMATIC

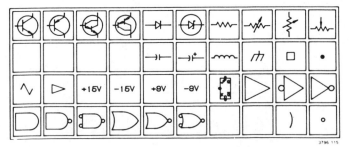

Figure 2–63   Typical symbol template overlays.

the size of drawing entities such as text and dimensions. First, determine what you have to draw and what size of paper it will be plotted on. Knowing that, you can determine the scale at which the drawing should be plotted. For example, suppose you decide that a drawing of a mechanical part must be plotted at a scale of ¼" = 1". This means that when you use the plot routine in AutoCAD and enter the proper scale, all entities in the drawing will be reduced to one quarter of their original size. What would happen to text in this case if it were drawn in AutoCAD .125" high? It would be reduced by one quarter, which means the plotted text would be .03125" high! This would be entirely too small. That is why you must plan your drawing carefully, and take the final plotted text height into consideration.

Knowing that text and dimensions are directly affected by the plotting scale, you can compensate for this early in the planning of your drawing. First use the appropriate dimension variables in AutoCAD, to set the proper sizes and offsets. Refer to your specific AutoCAD textbook or the AutoCAD Reference Manual for information on dimension variables. Then set the DIMSCALE dimension variable to the *reciprocal* of the intended plotted scale. You can easily find the reciprocal of any scale by the following math:

$$¼" = 1"$$
$$.25" = 1"$$
$$1/.25 = 4$$

The number 4 is the reciprocal of the scale ¼" = 1". We know that the part is reduced four times when it is plotted. Therefore, you must set DIMSCALE to the value of 4. Now, whenever you place a dimension on your drawing, all of its size components such as text, arrowheads, and offsets are multiplied by the DIMSCALE value of 4. This makes them four times their actual size, but they will look correct on your monitor when you draw the features of the part at their actual size.

The second important aspect of using the reciprocal of the drawing scale is the text size. Always multiply the intended plotted text height by the reciprocal to find the AutoCAD text height. For example, if you wish to have text displayed .125" high on a drawing with a scale of ¼" = 1", first multiply this by the reciprocal, in this case 4.

$$.125 \times 4 = .5$$

The value of .5 is the text height. Therefore, you should set your text height to .5 when drawing in AutoCAD. It will appear correct on the screen. When you plot the drawing at the scale of ¼" = 1", the text height is multiplied by the scale to produce the plotted text height. Thus .5 ÷ ¼ = **.125,** which is the proper text height.

This same technique is used for architectural, civil, and any other scaled drawings. As you can see, planning your drawing regarding scales and scale factors is a

PROFESSIONAL PERSPECTIVE

Working with a CADD program can be exciting, frustrating, and fun. The frustration may come in the early stages of learning on a CADD system. It results from looking at the screen too long, trying to remember new commands and techniques, and working at the computer too many hours because it is so fascinating. If you have never worked with a computer before to create drawings, you are in for a lot of fun.

Be prepared to be challenged by the complexity of CADD programs, the speed of the computer, and the program's desire for information. You will be communicating with the computer. It will ask you questions and prompt you for information. You will need to think quickly and know your next move. Thorough planning should be a part of drawing with the computer. This means even drawing a freehand sketch before you sit down at the computer. You can meet the challenge by organizing your work and using school or company standards.

Keep in mind that when you work with a CADD system, you are working with a machine that is highly predictable, but occasionally frustrating. The computer will talk back to you and demand information. It can occasionally do things you don't expect, but seldom does it do anything that you did not tell it to do. It allows you to be extremely productive and creative. And it can eat your drawing in a split second. So save your work regularly!

There are abundant jobs for qualified CADD operators. The best operators are those who have good organizational and planning skills, good drawing skills, and know the capabilities of the CADD system with which they were trained. The best jobs will be obtained by those students who can use the computer's commands effectively to solve problems, and who know where and how to find the information they need quickly. They must develop an intimate knowledge of the reference manuals and on-screen help available with the CADD program. In addition, those students who possess the skills needed to customize the CADD software to suit their specific needs will be in high demand by employers. Your portfolio should contain diskettes of your best drawings in all specialty fields of drafting and engineering. An excellent presentation for these drawings is in the form of a continuously running "slide show" that automatically displays your drawings.

critically important task and one that should become an integral part of your drawing habits.

▼ STORING A DRAWING

The storage file for the computer drafting system is the disk, or flexible disk. For permanent storage the magnetic tape is the storage file. The disk drive does the actual storing of your drawing on the disk. It can also read the data from the disk and send it to the computer.

The act of storing a drawing that is on the screen is handled by menu commands such as SAVE, and END. After selecting the proper command you may be given a chance to change the file name of the drawing. After

that, a press of the RETURN, or ENTER, key sends the drawing on its way to the disk. Some systems eliminate the need for the operator to store the drawing, because they are constantly sending the drawing data to the disk. This is a great feature because it removes the consequences of a computer crash, or malfunction. Should the computer go down (malfunction) during the course of work on a drawing, nothing would be lost except the command that was in progress.

How often should a CADD operator store a drawing? This could be a matter of personal preference or company policy. Practice varies, but it is a good habit to save your work every 10 to 15 minutes, just in case something happens to the computer.

MATH APPLICATION

One of the most important aspects of learning to operate a CADD system is the rectangular coordinate system. The polar system is one form of rectangular coordinates that uses a distance and angle for measurement. Understanding angles and how they relate to each other is useful in the solution of a variety of problems. For example, you may be given a flat sheet metal part or a property plat to draw, and the sketch contains only included angles. It is up to you to convert these included angles into polar coordinates in order to draw the object. (See the figure below.)

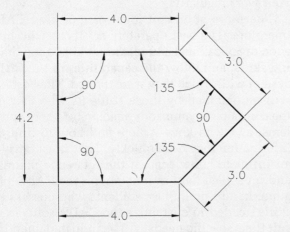

The first line can be drawn easily by entering a length, 4", and an angle of 0°. The angle of the second line is determined by subtracting 135° from 180° (the angle of a straight line). This gives the angle of 45° measured counterclockwise from horizontal. Thus,

to draw to point C, enter a length of 3" and an angle of 45°. (See the figure below.)

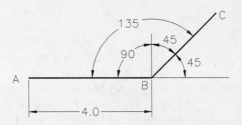

In order to draw to point D you must remember the rule that intersecting lines create equal opposite angles. By drawing a horizontal line through point C and projecting the previous line past the object, you see that one half of the 90° angle is above the horizontal. Therefore, the angle of the line to point D is the result of subtracting 45° from 180°. To draw a line from point C to point D, enter a length of 3" and an angle of 135°.

Try to complete the remainder of the object using the previous examples as a guide.

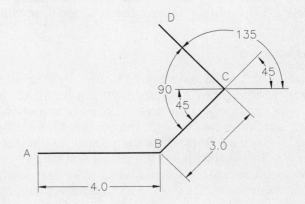

INTRODUCTION TO CADD TEST

MULTIPLE CHOICE

1. CADD systems allow data entry with a _____.
   - a. stylus
   - b. plotter
   - c. monitor
   - d. printer

2. The combination of the entire design, material handling, manufacturing, and packaging process is referred to as _____.
   - a. CIM
   - b. CAM
   - c. CAE
   - d. CIS

3. The process that fills in a 3-D model's surfaces with colors and shades is _____.
   - a. hatching
   - b. morphing
   - c. rendering
   - d. sectioning

4. The principal display device of a CADD system is the _____.
   - a. plotter
   - b. digitizer
   - c. printer
   - d. monitor

5. A single dot on a monitor is called a _____.
   - a. pixel
   - b. byte
   - c. matrix
   - d. RAM

6. The memory of a computer that is cleaned out when the computer is turned off is called _____.
   - a. ROM
   - b. REM
   - c. RAM
   - d. CPU

7. Graphic instructions can be entered from the keyboard as _____.
   - a. XYZ coordinates
   - b. polar coordinates
   - c. relative coordinates
   - d. all of the above

8. Which of the following is *not* a pointing device?
   - a. function key
   - b. stylus
   - c. mouse
   - d. puck

9. The keys along the top of a keyboard numbered from 1 to 12 are called _____ keys.
   - a. cursor
   - b. command
   - c. option
   - d. function

10. What peripheral device can use pens to create a paper drawing copy?
    - a. printer
    - b. digitizer
    - c. plotter
    - d. scanner

11. The principal permanent storage media of most microcomputer systems is _____.
    - a. magnetic disk
    - b. magnetic tape
    - c. RAM
    - d. ROM

12. A disk that holds 1,200,000 bytes of data is referred to as a _____ disk.
    - a. 1.2KB
    - b. 1.2CPU
    - c. 1.2MB
    - d. 1.2BM

13. The fastest-growing digital storage technology is the _____.
    - a. CD RAM
    - b. CD ROM
    - c. hard disk
    - d. CPU

14. A menu overlay is used on a _____.
    - a. display device
    - b. function board
    - c. digitizer table
    - d. keyboard

15. A point located below the origin of a rectangular coordinate system has a _____ value.
    - a. +X
    - b. +Y
    - c. –X
    - d. –Y

16. A relative coordinate is located from the _____.
    - a. previous point
    - b. absolute origin
    - c. polar region
    - d. Cartesian origin

17. This coordinate system locates points with a distance and angle from the previous point.
    - a. incremental
    - b. polar
    - c. absolute
    - d. relative

18. Entering X, Y, and Z coordinates allows you to construct _____.
    - a. wire forms
    - b. surface models
    - c. 3-D objects
    - d. all of the above

19. Invisible dots that are helpful in locating points on the screen are called a _____.
    - a. coordinate system
    - b. snap grid
    - c. grid system
    - d. orthographic grid

20. Symbols drawn once and used repeatedly are called _____.
    - a. menus
    - b. entities
    - c. blocks
    - d. polygons

CHAPTER 2

INTRODUCTION TO CADD PROBLEMS

Problem 2–1 This problem involves writing, not drawing. Choose one of the topics listed below and write a report on that subject. Should you choose to write a report on the topic 1. CADD system, select one manufacturer and focus your report on that specific brand of system. Be specific, especially if the manufacturer makes more than one model of CADD system. Check with your instructor about any specifics relating to your report. Reference materials can be trade magazines and journals, textbooks, company brochures and instruction manuals, telephone interviews, personal interviews, field trips and attendance at seminars, workshops, conferences and conventions.

1. CADD system (Choose a specific brand and model)
2. Computer: mainframe, mini, or micro
3. Input devices (hand-held)
4. Digitizer, graphics tablet
5. Output devices: plotter, printer, display screen (monitor or terminal)
6. Storage devices: disk drive, tape drive
7. Drawing media; paper, film, pens
8. Storage media: disks, tape

Problem 2–2 Write a short report that discusses the differences among pen plotters, electrostatic plotters, dot-matrix printers, and laser printers. You may use any sources that can provide the needed information, such as textbooks, magazines, advertising brochures, demonstrations, interviews, and personal experiences. Be sure to discuss the following points in your report:

• Type of media required (paper, pens, etc.)
• Media size
• Price
• Quality of final product
• Usefulness in your application
• Warranties and service contracts
• Durability
• Reputation
• Length of time on the market

Problem 2–3 Create a specification sheet for purchasing a single CADD workstation and the software to run it. You can do this as a report, a form, or a list. The assignment should include the following pieces of equipment and software:

• IBM compatible microcomputer—486 or 586 (Pentium)—33Mhz—200Mb hard disk—2 1.44Mb flexible disks—SVGA color graphics card—2 serial ports
• 17" color monitor
• 11" × 11" digitizer tablet with a 12-button cursor
• B-size, 8-pen plotter
• MS-DOS software—latest version
• CADD software (AutoCAD, VersaCAD, etc.)—latest version

Keep the following things in mind when creating your specification list.

• Contact at least three computer hardware vendors for prices.
• Provide each vendor with the same specifications.
• Do not specify brand names of equipment unless instructed to do so.
• Determine the brand of CADD software you want before obtaining prices.
• You may have to go to different vendors for the CADD software.

Problem 2–4 Using existing equipment, or vendor's catalogs, measure the equipment required for a microcomputer CADD workstation (computer, monitor, keyboard, and pointing device), and design a workstation table. Sketch the table first, then draw it manually or on your CADD system.

After drawing the table, create a computer lab arrangement of from 12 to 20 CADD workstations using your table design. Consider the following when designing the lab:

• Chalkboard or marker board location
• Glare from windows
• Aisles and access
• Power connections
• Location of pen plotters or printers
• Storage for paper, plotter pens, and supplies
• Instructor podium or workstation
• Projection system for instructor workstation (use overhead projector)

Problem 2–5 Construct a comparison sheet on three pen plotters, each made by a different manufacturer, that are all in the same price range. The plotters should have the following features:

• Hold up to D-size paper
• Hold 8 pens, either felt-tip or wet-ink. Interchangeable pen carousels are acceptable.
• Portable (mounted on rollers)
• Allow manual pen control
• Allow for viewing the drawing in the middle of a plot (pause)
• Cost less than $6000

Check with local computer vendors for brochures and specifications. Read computer magazines and journals for plotter advertisements and articles that compare and rate plotters. Arrange for demonstrations by vendors or companies that use pen plotters. Plotters can have features additional to those mentioned above, or you can delete and add features to the list as desired or instructed.

MATH PROBLEMS

1. Draw the following sheet metal part using CADD software. If you are using AutoCAD, the best point entry method to use is polar coordinates. Space is provided in which you can write the coordinate values needed to draw a line to each point.

A to B _____
B to C _____
C to D _____
D to E _____
E to F _____
F to G _____
G to A _____

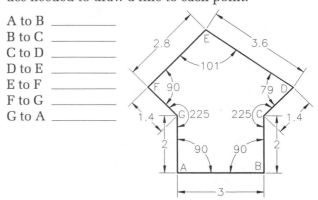

2. The following property plat has been given to you in the form of a sketch with included angles. You are to draw the property plat full size using the dimensions give. Do not use decimal values for angles or distances. Space is provided in which you can write the coordinate values needed to draw a line to each point. Write your answers using a polar coordinate notation. For example, a line 87'–3" long, at an angle of 65° would be written @87'3<65.

A to B _____
B to C _____
C to D _____
D to E _____
E to F _____
F to A _____

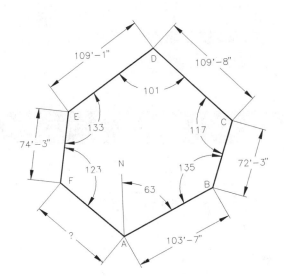

SOLUTION TO CADD DRAWING (from page 71)

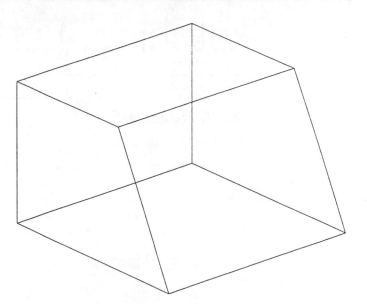

CHAPTER 3

Sketching, Lettering, and Lines

LEARNING OBJECTIVES

After completing this chapter you will:

- ○ Sketch lines, circles, arcs and multiviews.
- ○ Do freehand lettering, and use mechanical lettering equipment.
- ○ Use a CADD system to prepare text.
- ○ Draw ANSI standard lines using manual drafting and computer-aided drafting.
- ○ Solve an engineering problem using manual and computer-aided drafting.

THE ENGINEERING DESIGN PROCESS

The engineering drafter often works from sketches or written information provided by the engineer. When you receive an engineer's sketch and are asked to prepare a formal drawing, you should follow these steps:

Step 1. Whether you are using manual or CADD drafting, prepare your own sketch the way you think it should look on the final drawing, taking into account correct drafting standards.

Step 2. Evaluate the size of the object so you can determine the scale and sheet size for your final manual drawing or the screen limits for your final computer-aided drawing.

Step 3. Lay out the drawing either very lightly using construction lines or light blue lead. The construction lines may be easily erased if you make a mistake. If you are using a computer, begin the layout on the screen and if you make an error, just erase it and try again. Editing your CADD work is fast and easy.

Step 4. Complete the final manual drawing by darkening all construction lines to proper ANSI standard line weights and run a copy for checking. After completing the computer drawing, a check plot may be made on the plotter or printer.

The example in Figure 3–1 shows a comparison between the engineer's rough sketch and the finished drawing.

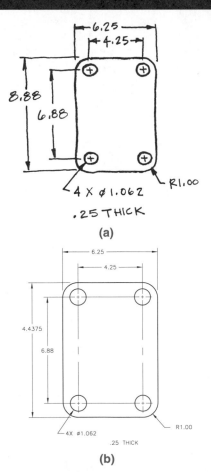

Figure 3–1 A comparison between **(a)** an engineer's rough sketch and **(b)** the finished drawing.

(Continued)

ENGINEERING DESIGN PROCESS *(Continued)*

A CADD drafter can easily work from points established on a Cartesian coordinate system as discussed in Chapter 2. When this is required, you need to first determine the type of coordinate system being used: absolute, incremental, or polar. When a situation of this kind occurs you may be given the X and Y values for each point in a chart similar to Figure 3–2.

Drawing lines between all of the listed points in this demonstration produces the drawing shown in Figure 3–3.

POINT	X	Y
1.	2.8	2.3
2.	7.7	2.3
3.	7.7	5.6
4.	5.2	5.6
5.	4.1	4.1
6.	2.8	4.1
7.	2.8	2.3

Figure 3–2 Absolute values for X and Y point coordinates.

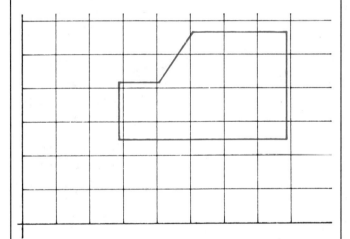

Figure 3–3 The solution to a Cartesian coordinate system drawing problem.

▼ SKETCHING

Sketching is freehand drawing, that is, drawing without the aid of drafting equipment. Sketching is convenient since all that is needed is paper, pencil, and an eraser. There are a number of advantages and uses for freehand sketching. Sketching is fast visual communication. The ability to make an accurate sketch quickly can often be an asset when communicating with people at work or at home. Especially when technical concepts are the topic of discussion, a sketch may be the best form of communication. Most drafters prepare a preliminary sketch to help organize thoughts and minimize errors on the final drawing. The computer operator usually prepares a sketch on graph paper to help establish the coordinates for drawing components. Some drafters use sketches to help record the stages of progress when designing until a final design is ready for implementation into formal drawings. A sketch can be a useful form of illustration in technical reports. Sketching is also used in job shops where one-of-a-kind products are made. In the job shop, the sketch is often used as a formal production drawing. When the drafter's assignment is to prepare working drawings for existing parts or products, the best method to gather shape and size description about the project is to make a sketch. The sketch can be used to quickly lay out dimensions of features for later transfer to a formal drawing.

▼ TOOLS AND MATERIALS

Sketching equipment is not very elaborate. As mentioned, all you need is paper, pencil, and an eraser. The pencil should have a soft lead; a common number 2 pencil works fine or an automatic 0.7 or 0.9 mm pencil with F or HB lead is good also. The pencil lead should not be sharp. A dull, slightly rounded pencil point is best. Different thicknesses of line, if needed, can be drawn by changing the amount of pressure you apply to the pencil. The quality of the paper is not too critical either. A good sketching paper is newsprint, although almost any kind works. Actually, paper with a surface that is not too smooth is best. Many engineering designs have been created on a napkin around a lunch table. Sketching paper should not be taped down to the table. The best sketches are made when you are able to move the paper to the most comfortable drawing position. Some people make horizontal lines better than vertical lines. If this is your situation, then move the paper so that vertical lines become horizontal. Such movement of the paper may not always be possible, so it does not hurt to keep practicing all forms of lines for best results.

▼ STRAIGHT LINES

Lines should be sketched in short, light, connected segments as shown in Figure 3–4. If you sketch one long stroke in one continuous movement, your arm tends to make the line curved rather than straight, as shown in Figure 3–5. Also, if you make a dark line, you may

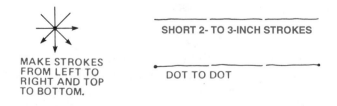

Figure 3–4 Sketching short line segments.

Figure 3–5   Long movements tend to cause a line to curve.

Figure 3–6   Step 1, use dots to identify both ends of a line.

Figure 3–7   Step 3, use short light strokes.

A ●————————————————————————● B

Figure 3–8   Step 4, darken to finish the line.

have to erase if you make an error, whereas if you draw a light line there often is no need to erase.

Following is the procedure used to sketch a horizontal straight line with the dot-to-dot method:

Step 1.   Mark the starting and ending positions, as in Figure 3–6. The letters A and B are only for instruction. All you need are the points.

Step 2.   Without actually touching the paper with the pencil point, make a few trail motions between the marked points to adjust the eye and hand to the anticipated line.

Step 3.   Sketch very light lines between the points by stroking in short light strokes (2- to 3-in. long). Keep one eye directed toward the end point while keeping the other eye directed on the pencil point. With each stroke, an attempt should be made to correct the most obvious defects of the preceding stroke so the finished light lines are relatively straight. (See Figure 3–7.)

Step 4.   Darken the finished line with a dark, distinct, uniform line directly on top of the light line. Usually the darkness can be obtained by pressing on the pencil. (See Figure 3–8.)

▼ CIRCULAR LINES

Figure 3–9 shows the parts of a circle. There are several sketching techniques to use when making a circle; this text explains the trammel and hand-compass methods.

Trammel Method

Step 1.   Make a trammel to sketch a 6-in. diameter circle. Cut or tear a strip of paper approximately 1 in. wide and longer than the radius, 3 in. On the strip of paper, mark an approximate 3-in. radius with tick marks such as A and B in Figure 3–10.

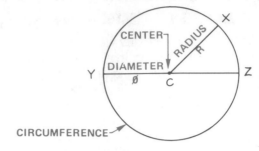

Figure 3–9   The parts of a circle.

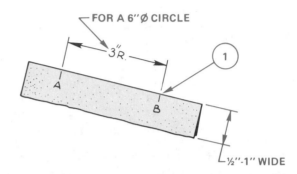

Figure 3–10   Step 1, make a trammel.

Step 2.   Sketch a straight line representing the circle radius at the place where the circle is to be located. On the sketched line, locate with a dot the center of the circle to be sketched. Use the marks on the trammel to mark the other end of the radius line as shown in Figure 3–11. With the trammel next to the sketched line, be sure point B on the trammel is aligned with the center of the circle you are about to sketch.

Step 3.   Pivot the trammel at point B, making tick marks at point A as you go, as shown in Figure 3–12, until the circle is complete.

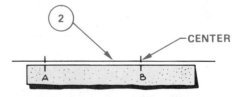

Figure 3–11   Step 2, locate the center of the circle.

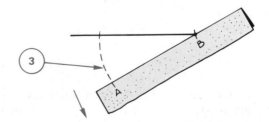

Figure 3–12   Step 3, begin the circle construction.

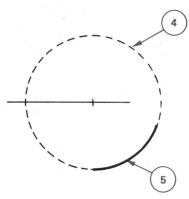

Figure 3–13 Steps 4 and 5, darken the circle.

Step 4. Lightly sketch the circumference over the tick marks to complete the circle, then darken it (step 5) as shown in Figure 3–13.

Hand-compass Method

The hand-compass method is a quick and fairly accurate method of sketching circles, although it is a method that takes some practice.

Step 1. Be sure that your paper is free to rotate completely around 360°. Remove anything from the table that might stop such a rotation.

Step 2. To use your hand and a pencil as a compass, place the pencil in your hand between your thumb and the upper part of your index finger so your index finger becomes the *compass point* and the pencil becomes the *compass lead.* The other end of the pencil rests in your palm as shown in Figure 3–14.

Step 3. Determine the circle radius by adjusting the distance between your index finger and the pencil point. Now, with the desired approximate radius established, place your index finger on the paper at the proposed center of the circle.

Figure 3–14 Step 2, holding the pencil in the hand compass.

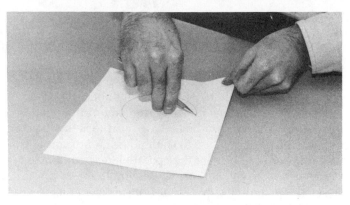

Figure 3–15 Step 4, rotate the paper under your finger center point.

Step 4. With the desired radius established, keep your hand and pencil point in one place while rotating the paper with your other hand. Try to keep the radius steady as you rotate the paper. (See Figure 3–15.)

Step 5. You can perform step 4 very lightly and then go back and darken the circle or, if you have had a lot of practice, you may be able to draw a dark circle as you go.

Another method, generally used to sketch large circles, is to tie a string between a pencil and a pin. The distance between the pencil and pin is the radius of the circle. Use this method when a large circle is to be sketched, since the other methods may not work as well.

▼ MEASUREMENT LINES AND PROPORTIONS

When sketching objects, all the lines that make up the object are related to each other by size and direction. In order for a sketch to communicate accurately and completely, it must be drawn in the same proportion as the object. The actual size of the sketch depends on the paper size and how large you want the sketch to be. The sketch should be large enough to be clear, but the proportions of the features are more important than the size of the sketch.

Look at the lines in Figure 3–16. How long is line 1? How long is line 2? Answer these questions without measuring either line, but instead relate each line to the other. For example, line 1 could be stated as being half as long as line 2, or line 2 called twice as long as line 1. Now you know how long each line is in relationship to the other (proportion), but we do not know how long either line is in relationship to a measured scale. No scale

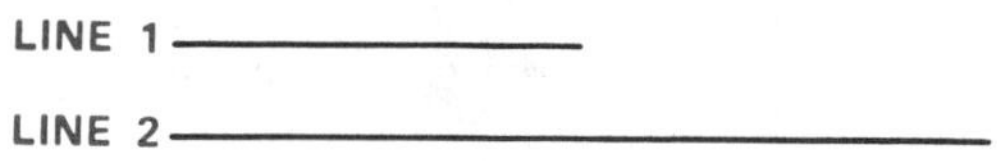

LINE 1 ——————————

LINE 2 ————————————————

Figure 3–16 Measurement lines.

is used for sketching, so this is not a concern. Whatever line you decide to sketch first determines the scale of the drawing. This first line sketched is called the *measurement line*. Relate all the other lines in the sketch to that first line. This is one of the secrets of making a sketch look like the object being sketched.

The second thing you must know about the relationship of the two lines in the above example is their direction and position relative to each other. For example, do they touch each other, are they parallel, perpendicular, or at some angle to each other? When you look at a line ask yourself the following questions (for this example use the two lines given in Figure 3–17):

1. How long is the second line?
   a. same length as the first line?
   b. shorter than the first line? How much shorter?
   c. longer than the first line? How much longer?
2. In what direction and position is the second line related to the first line?

A typical answer to these questions for the lines in Figure 3–17 would be as follows:

1. The second line is about three times as long as the first line.
2. Line two touches the lower end of the first line with about a 90° angle between them.

Carrying this concept a step further, a third line can relate to the first line or the second line and so forth. Again, the first line drawn (measurement line) sets the scale for the whole sketch.

This idea of relationship can also apply to spaces. In Figure 3–18, the location of the hole can be determined by space proportions. A typical verbal location for the hole in this block might be as follows: the hole is located about one-half hole width from the top of the object or about two hole widths from the bottom, and about one hole width from the right side or about three hole widths from the left side of the object. All the parts must be related to the whole object.

Figure 3–17   Measurement line.

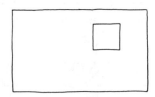

Figure 3–18   Space proportions.

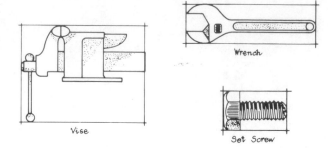

Figure 3–19   Block technique.

Block Technique

Any illustration of an object can be surrounded with some sort of a rectangle overall, as shown in Figure 3–19. Before starting a sketch, see the object to be sketched inside a rectangle in your mind's eye. Then use the measurement-line technique with the rectangle, or block, to help you determine the shape and proportion of your sketch.

▼ PROCEDURES IN SKETCHING

Step 1. When starting to sketch an object, visualize the object surrounded with a rectangle overall. Sketch this rectangle first with very light lines. Sketch the proper proportion with the measurement-line technique, as shown in Figure 3–20.

Step 2. Cut sections out or away using proper proportions as measured by eye, using light lines, as in Figure 3–21.

Step 3. Finish the sketch by darkening in the desired outlines for the finished sketch. (See Figure 3–22.)

Figure 3–20   Step 1, outline the drawing area with a block.

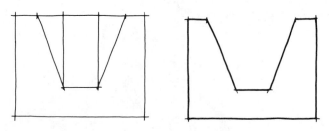

Figure 3–21   Step 2, draw features to proper proportions.   **Figure 3–22**   Step 3, darken the object lines.

Irregular Shapes

By using a frame of reference or an extension of the block method, irregular shapes can be sketched easily to their correct proportions. Follow these steps to sketch the cam shown in Figure 3–23.

Step 1. Place the object in a lightly constructed box. (See Figure 3–24.)

Step 2. Draw several equally spaced horizontal and vertical lines as shown in Figure 3–25. If you are sketching an object already drawn, just draw your reference lines on top of the object's lines to establish a frame of reference. If you are sketching an object directly, you have to visualize these reference lines on the object you sketch.

Step 3. On your sketch, correctly locate a proportioned box similar to the one established on the original drawing or object, as shown in Figure 3–26.

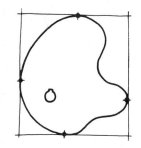

Figure 3–23 Cam.

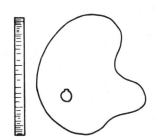

Figure 3–24 Step 1, box the object.

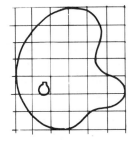

Figure 3–25 Step 2, evenly spaced grid.

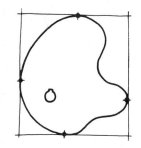

Figure 3–26 Step 3, proportioned box.

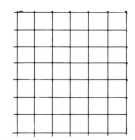

Figure 3–27 Step 4, regular grid.

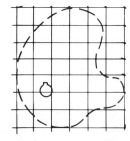

Figure 3–28 Step 5, sketched shape using the regular grid.

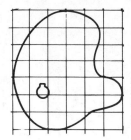

Figure 3–29 Step 6, completely darken the outline of the object.

Step 4. Using the drawn box as a frame of reference, include the grid lines in correct proportion, as seen in Figure 3–27.

Step 5. Then, using the grid, sketch the small irregular arcs and lines that match the lines of the original, as in Figure 3–28.

Step 6. Darken the outline for a complete proportioned sketch, as shown in Figure 3–29.

▼ MULTIVIEW SKETCHES

Multiview projection is also known as orthographic projection. Multiviews are two-dimensional views of an object that are established by a line of sight that is perpendicular (90°) to the surface of the object. When making multiview sketches, a systematic order should be followed. Most drawings are in the multiview form. Learning to sketch multiview drawings will save you time when making a formal drawing. The pictoral view shows the object in a 3-D picture.

Multiview Alignment

To keep your drawing in a common form, sketch the front view in the lower left portion of the paper, the top view directly above the front view, and the right-side view to the right side of the front view. (See Figure 3–30.) The views needed may differ depending on the object. Multiview arrangement is explained in detail in Chapter 5.

Multiview Sketching Technique

Steps in sketching:

Step 1. Sketch and align the proportional rectangles for the front, top, and right side of the object given in Figure 3–30. Sketch a 45° line to help transfer width dimensions. The 45° line is established by projecting the width from the top view across and the width from the right-side view up until the lines intersect as shown in Figure 3–31.

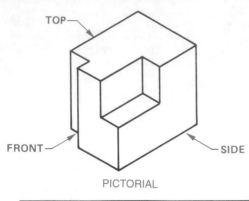

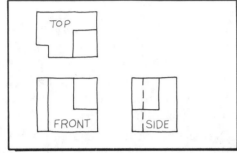

Figure 3–30 Views of object shown in pictorial and multi-view.

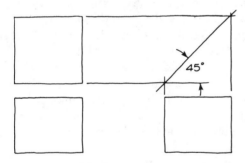

Figure 3–31 Step 1, block out views.

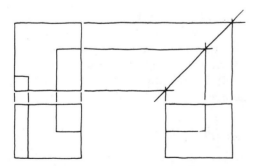

Figure 3–32 Step 2, block out shapes.

Step 2. Complete the shapes by cutting out the rectangles, as shown in Figure 3–32.

Step 3. Darken the lines of the object as in Figure 3–33. Remember, keep the views aligned for ease of sketching and understanding.

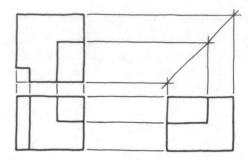

Figure 3–33 Step 3, darken all object lines.

Figure 3–34 Step 4, draw hidden features.

Step 4. In the views where some of the features are hidden, show those features with hidden lines, which are dashed lines as shown in Figure 3–34.

▼ ISOMETRIC SKETCHES

Isometric drawings provide a three-dimensional pictorial representation of an object. Isometric drawings are easy to draw and make a very realistic exhibit of the object. The surface features or the axes of the objects are drawn at equal angles from horizontal. Isometric sketches tend to represent the objects as they appear to the eye. Isometric sketches help in the visualization of an object because three sides of the object are sketched in a single three-dimensional view. Chapter 18 covers isometric drawings in detail.

Establishing Isometric Axes

In setting up an isometric axis, you need four beginning lines: a horizontal reference line, two 30° angular lines, and one vertical line. Draw them as very light construction lines. (See Figure 3–35.)

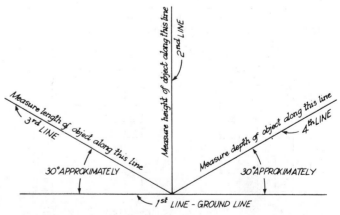

Figure 3–35 Isometric axis.

Step 1. Sketch a horizontal reference line (consider this the ground-level line.)

Step 2. Sketch a vertical line perpendicular to the ground line and somewhere near its center. The vertical line is used to measure height.

Step 3. Sketch two 30° angular lines, each starting at the intersection of the first two lines as shown in Figure 3–35.

Making an Isometric Sketch

The steps in making an isometric sketch are as follows:

Step 1. Select an appropriate view of the object.

Step 2. Determine the best position in which to show the object.

Step 3. Begin your sketch by setting up the isometric axes. (See Figure 3–36.)

Step 4. By using the measurement-line technique, draw, using correct proportion, a rectangular box which could surround the object to be drawn. Use the object shown in Figure 3–37 for this explanation. Imagine the rectangular box in your mind. Begin to sketch the box by marking off the width at any convenient length (measurement line), as in Figure 3–38. Next estimate and mark the length and height as related to the measurement line. (See Figure 3–39.) Sketch the three-dimensional box by using lines parallel to the original axis lines. (Figure 3–40.) Sketching the box is the most critical part of the construction. It must be done correctly; otherwise your sketch will be out of proportion. All lines drawn in the same direction must be parallel.

Step 5. Lightly sketch in the slots, holes, insets, and other features, that define the details of the object. By estimating distances on the rectangular box, the features of the object are easier to sketch in correct proportion than trying to draw them without the box. (See Figure 3–41.)

Step 6. To finish the sketch, darken all the object lines (outlines), as in Figure 3–42. For clarity, do not show any hidden lines.

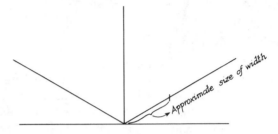

Figure 3–38   Step 4, lay out the width.

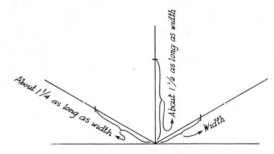

Figure 3–39   Step 4, lay out the length and height.

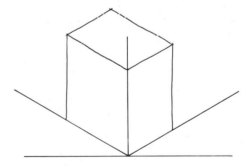

Figure 3–40   Step 4, sketch the 3-D box.

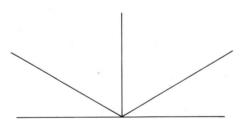

Figure 3–36   Step 3, sketch the isometric axis.

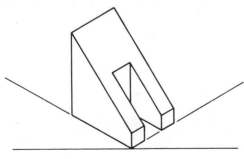

Figure 3–37   Given object.

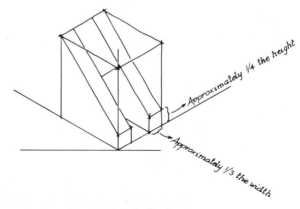

Figure 3–41   Step 5, sketch the features.

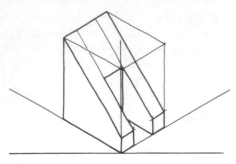

Figure 3–42 Step 6, darken the outline.

Nonisometric Lines

Isometric lines are lines that are on or parallel to the three original isometric axes lines. All other lines are nonisometric lines. Isometric lines can be measured in true length. *Nonisometric lines* appear either longer or shorter than they actually are. (See Figure 3–43.) You can measure and draw nonisometric lines by connecting their end points. You can find the end points of the nonisometric lines by measuring along isometric lines. To locate where nonisometric lines should be placed, you have to relate to an isometric line. Follow through these steps, using the object in Figure 3–44 as an example.

Step 1. Develop a proportional box, as in Figure 3–45.

Step 2. Sketch in all isometric lines, as shown in Figure 3–46.

Step 3. Locate the starting and end points for the nonisometric lines. (See Figure 3–47.)

Step 4. Sketch the nonisometric lines, as shown in Figure 3–48, by connecting the points established in step 3. Also darken all outlines.

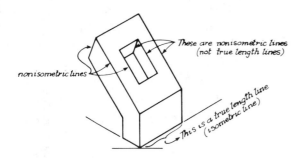

Figure 3–43 Nonisometric lines.

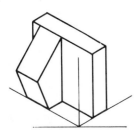

Figure 3–44 Guide.

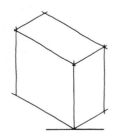

Figure 3–45 Step 1, sketch the box.

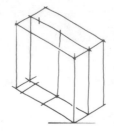 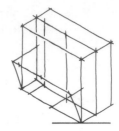

Figure 3–46 Step 2, sketch isometric lines.   **Figure 3–47** Step 3, locate nonisometric line end points.

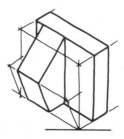

Figure 3–48 Step 4, complete the sketch and darken all outlines.

Sketching Isometric Circles

Circles and arcs appear as ellipses in isometric views. To sketch isometric circles and arcs correctly, you need to know the relationship between circles and the faces, or planes, of an isometric cube. Depending on which face the circle is to appear, isometric circles look like one of the ellipses shown in Figure 3–49. The angle the ellipse (isometric circle) slants is determined by the surface on which the circle is to be sketched.

To practice sketching for isometric circles, you need an isometric surface to put them on. The surfaces can be found by first sketching a cube in isometric. A *cube* is a box with six equal sides. Notice, as shown in Figure 3–50, that only three of the sides can be seen in an isometric drawing.

Four-center Method. The four-center method of sketching an isometric ellipse is easier to perform, but care must be taken to form the ellipse arcs properly so the ellipse does not look distorted.

Step 1. Draw an isometric cube similar to Figure 3–50.

Step 2. On each surface of the cube, draw line segments that connect the 120° corners to the centers of the opposite sides. (See Figure 3–51.)

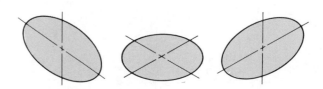

Figure 3–49 Isometric circles.

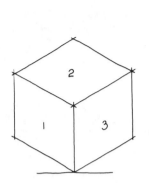

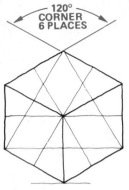

Figure 3–50　Step 1, draw an isometric cube.

Figure 3–51　Step 2, four-center isometric ellipse construction.

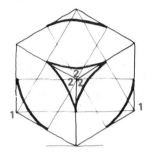

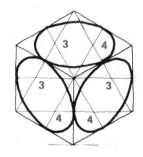

Figure 3–52　Step 3, sketch arcs from points 1 and 2 as centers.

Figure 3–53　Step 4, sketch arcs from points 3 and 4 as centers.

Step 3.　With points 1 and 2 as the centers, sketch arcs that begin and end at the centers of the opposite sides on each isometric surface. (See Figure 3–52.)

Step 4.　On each isometric surface, with points 3 and 4 as the centers, complete the isometric ellipses by sketching arcs that meet the arcs sketched in step 3. (See Figure 3–53.)

Sketching Isometric Arcs

Sketching isometric arcs is similar to sketching isometric circles. First, block out the overall configuration of the object, then establish the centers of the arcs. Finally, sketch the arc shapes as shown in Figure 3–54. Remember that isometric arcs, just as isometric circles, must lie in the proper plane and have the correct shape.

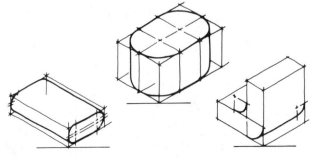

Figure 3–54　Sketching isometric arcs.

▼ LETTERING

Information on drawings that cannot be represented graphically by lines may be presented by lettered dimensions, notes, and titles. It is extremely important that these lettered items be exact, reliable, and entirely legible in order for the reader to have confidence in them and never have any hesitation as to their meaning. This is especially important when using reproduction techniques that require a drawing to be reduced in size such as with photocopy or microfilm, or when secondary originals are made. The quality of each generation of secondary original is reduced, thus requiring the highest quality from the original drawing. Poor lettering ruins an otherwise good drawing.

▼ SINGLE-STROKE GOTHIC LETTERING

ANSI　The standard for lettering was established in 1935 by the American National Standards Institute (ANSI). This standard is now conveyed by the document ANSI Y14.2M, *Line Conventions and Lettering.*

The standardized lettering format was developed as a modified form of the Gothic letter font. The term *font* refers to a complete assortment of any one size and style of letters. The simplification of the Gothic letters resulted in elements for each letter that became known as single-stroke Gothic lettering. The name sounds complex but it is not. The term *single stroke* comes from the fact that each letter is made up of a single straight or curved line element that makes it easy to draw and clear to read. There are upper- and lower-case, vertical, and inclined Gothic letters, but industry has become accustomed to using vertical upper-case letters as standard. (See Figure 3–55.)

▼ MICROFONT LETTERING

When adequately sized standard single-stroke Gothic lettering is created with clear open features there is generally no problem with microfilm or photocopy reductions, and many companies that use microfilm continue to use standard lettering. However, an alternate lettering design has been developed by the National Micrographics Association

ABCDEFGHIJKLMNOP
QRSTUVWXYZ&
1234567890

Figure 3–55　Vertical upper case single-stroke Gothic letters and numbers.

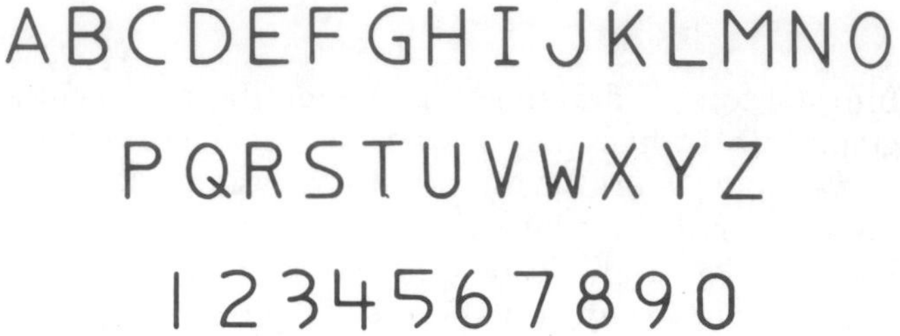
ABCDEFGHIJKLMNO
PQRSTUVWXYZ
1234567890

Figure 3–56 Microfont alphabet and numerals.

R abcdefghijklmnopqrstuvwxyz
H
2/3 x H
M abcdefghijklmnopqrstuvwxyz
H
2/3 x H

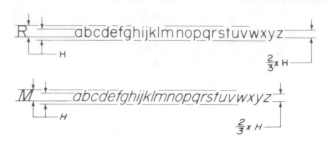

Figure 3–58 Lower-case lettering.

that is intended to provide greater legibility when reduced for microfilm applications. This lettering, known as *Microfont,* is a style that provides a more open font for better reproduction capabilities. These letters may be made freehand, with lettering templates and guides, on typewriters, or with computer graphics. (See Figure 3–56.)

▼ OTHER LETTERING STYLES

Inclined Lettering

Some companies prefer inclined lettering. The general slant of inclined letters is 68°. One edge of the Ames Lettering Guide has a 68° slant, which may be used to help maintain the proper angle. Structural drafting is one field where slanted lettering is commonly found. Figure 3–57 shows slanted upper-case letters.

ABCDEFGHIJKLMNOP
QRSTUVWXYZ&
1234567890
5
2

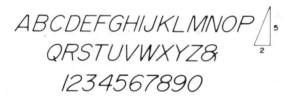

Figure 3–57 Upper-case inclined letters and numbers.

Lower-case Lettering. Occasionally, lower-case letters are used; however, they are very uncommon in mechanical drafting. Civil or map drafters use lower-case lettering for some practices. Figure 3–58 shows lower-case lettering styles.

Architectural Styles

Architectural lettering is much more varied in style than mechanical lettering; however, neatness and readability are essential. (See Figure 3–59.)

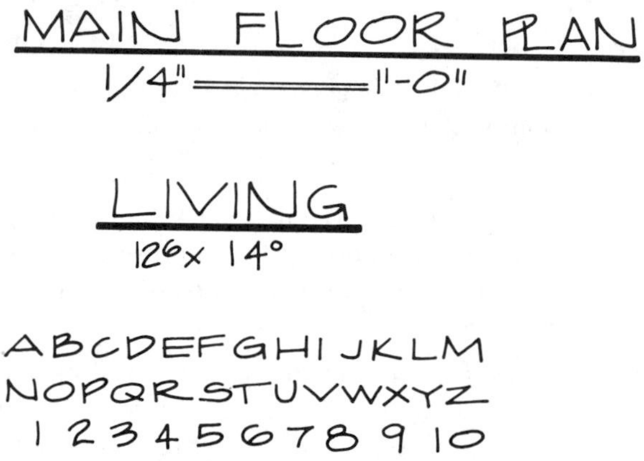
MAIN FLOOR PLAN
1/4" = 1'-0"
LIVING
12⁶ x 14⁰
ABCDEFGHIJKLM
NOPQRSTUVWXYZ
1 2 3 4 5 6 7 8 9 10

Figure 3–59 Architectural lettering.

CADD Applications

LETTERING

Lettering with a CADD system is one of the easiest tasks associated with computer-aided drafting. It is just a matter of deciding on the style or font of lettering to use, and then locating the text where it is needed. Lettering is called text when using CADD.

The CADD drafter often rejoices when the time comes to place text and notes on the drawing because no freehand lettering is involved. The computer places text of a consistent shape and size on a drawing in any number of styles, or fonts. The FONT or TEXT command is one of several that can be found in a section of the menu labeled *text*, or *text attributes.* The drafter is also able to specify the height, width, and slant angle of characters (letters and numbers). Most systems maintain a certain size of text called the *default* size that is used if the operator does not specify one. The term *default* refers to any value that is maintained by the computer for a command or function that has variable parameters. The default text height may be ⅛ in., but the user can change it if need be.

Lettering Styles

Many CADD systems possess a variety of lettering styles, or fonts. The drafter can select the style to use simply by picking a menu command, symbol, or by typing a command at the keyboard. Figure 3–60 shows some of the styles and sizes of characters that can be used in CADD. The size and style of characters used is dictated by the nature of the drawing.

Locating Text

The process of locating text has not changed. The drafter still needs to decide where to locate dimensions, notes, and parts lists, but with CADD the process of placing notes is just a bit more technical. Most CADD systems provide for keyboard entry of location and size coordinates, thus allowing the drafter to accurately locate text. But using the keyboard to input text location coordinates is a tedious process. Often key strokes are combined with digitizer stylus commands to generate text.

Text is located by pointing to one of several places on the lettering. Figure 3–61 shows an example of several points on the text that can be used for location purposes. Not all CADD systems use all the points shown, but most systems have a command known as *text* that is used for placing written information on

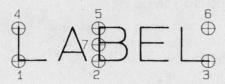

Figure 3–61 Text can be placed on a drawing by choosing one of several location points shown here.

the drawing. Some systems may allow the operator to locate text between two points. The computer calculates the size of each letter so the text fits in the desired space.

The first decision regarding text is to determine its height, width, and slant angle. Most CADD systems maintain a default text size that is used by the computer if the operator decides not to change it. The angle of rotation, direction, or text path is also determined by the drafter. The text path is the angle from horizontal that the text will lie on. An example of text path is shown in Figure 3–62. Text located on a horizontal line has a direction of 0°, and text that reads from the bottom up vertically has a direction of 90°.

With the size and slant angle decided, you then need to locate it on the drawing and type the text at the keyboard. This process can occur in a couple of ways. You could be asked to first digitize the text location using a pointing input device and then type the text. The second method is the reverse of the first. Type the text and it appears on the screen with the crosshairs at the point of location that you previously specified. Then *drag* the text to the proper location, press a button on the puck or keyboard, and it is in place. The nice thing about locating notes and labels on a drawing with a CADD system is the ability to move them around instantaneously, as often as you want, without erasing holes in your drawing.

ABCDEFGHIJKLMNOPQRSTUVWXYZ
acbcdefghijklmnopqrstuvwxyz
1234567890

ABCDEFGHIJKLMNOPQRSTUVWXYZ
acbcdefghijklmnopqrstuvwxyz
1234567890

ABCDEFGHIJKLMNOPQRSTUVWXYZ
acbcdefghijklmnopqrstuvwxyz
1234567890

Figure 3–60 Samples of CADD character font styles.

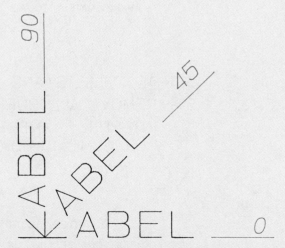

Figure 3–62 Examples of text rotation angles.

▼ LETTERING LEGIBILITY

ANSI The minimum recommended lettering size on engineering drawings is .125 in. (3 mm). All dimension numerals, notes, and other lettered information should be the same height except for titles, drawing numbers, section and view letters, and other captions which are .25 in. (6 mm) high.

Either vertical or inclined lettering may be used on a drawing depending on company preference. However, only one style of lettering should be used on a drawing. Lettering must be dark, crisp, and opaque for the best possible reproducibility.

The composition or spacing of letters in words and between words in sentences should be such that the individual letters are uniformly spaced with approximately equal background areas. This usually requires that letters such as I, N, or S be spaced slightly farther from their adjacent letters than L, A, or W. A minimum recommended space between letters in words is approximately .06 in. (1.5 mm). The space between words in a note or sentence should be about the same as the height of the letters. The space between two numerals with a decimal point between them is a minimum of two-thirds of the letter height.

Notes should be lettered horizontally on the sheet. When lettering notes, sentences, or dimensions requires more than one line, the vertical space between the lines should be a minimum of one-half the height of letters. The maximum recommended space between lines of lettering is equal to the height of the letters. Some companies prefer to use the minimum space to help conserve space while other companies prefer the maximum space for clarity. (See Figure 3–63.)

▼ VERTICAL FREEHAND LETTERING

Vertical freehand lettering is the standard for mechanical drafting. The ability to perform good-quality lettering quickly is important. A common comment among employers hiring entry-level drafters is the ability to do quality lettering and line work. Although standard, not all companies require freehand lettering. Some companies allow drafters the flexibility of freehand lettering or using a template. Many companies are now changing to computer-aided drafting and traditional lettering skills may become obsolete.

Always use lightly drawn horizontal guidelines that are spaced equal to the height of the letters. Some people need vertical guidelines to help keep their letters vertical. The ability to perform quality freehand lettering requires a great deal of practice for most people.

Vertical Capital Letters

Straight Elements. Use a 0.5 mm automatic pencil for lettering. This kind of pencil does not need sharpening and H, F, or HB leads are usually easy to control for effective lettering. Remember, you need to experiment with leads to determine which gives you the best results. Letters should be dark and crisp. Fuzzy letters print as fuzzy lines.

Letters composed of straight lines are shown in Figure 3–64. You should become familiar with these letter forms and the strokes needed to make them. The arrows in Figure 3–64 indicate the direction of the stroke used to form the letters. However, if these strokes are uncomfortable, you should develop your own procedure. Try the recommended strokes first. Because these letters are made of single-stroke elements, try not to let the strokes

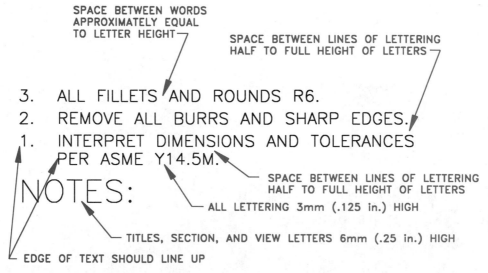

Figure 3–63 Spacing of letters, words, and notes.

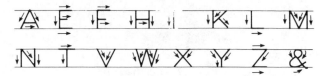

Figure 3–64 Recommended strokes for vertical uppercase Gothic letters with straight elements.

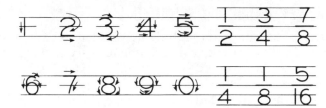

Figure 3–67 Vertical numerals and fractions.

combine as the result may be curves where straight elements should be.

Horizontal guidelines should be used for all lettering at all times. Use vertical guidelines if you have difficulty keeping your letters vertical. Guidelines should be very light 6H or 4H pencil lines. Some drafters prefer to use a light-blue lead rather than a graphite lead for guidelines. Light blue will not reproduce in the diazo or photocopy processes. When lettering, protect your drawing by resting your hand on a clean protective sheet placed over your drawing. This prevents smearing and smudging, as shown in Figure 3–65.

Curved Elements.   Letters that contain arcs are shown in Figure 3–66. Notice the difference in the sizes of arcs and circles used for different letters. The recommended letter elements, as with straight elements, are made up of a series of suggested strokes. These strokes when used as shown provide the best lettering results.

Vertical guidelines are often an asset for the best lettering results.

Vertical Numerals and Fractions

Vertical numerals, as seen in Figure 3–67, are also made up of recommended strokes. Numerals are the same height as capital letters.

Fractions.   Fractions are not as commonly used on engineering drawings as decimal inches or millimeters, but are common on architectural and structural drawings. When fractions are used on a drawing, the fraction numerals should be the same size as other numerals on the drawing. The fraction bar should be drawn in line with the direction of the dimension. For example, if all dimensions read horizontally, all fraction bars are horizontal. The fraction's numerals should not touch the fraction bar. A space of 1.5 mm or .06 in. between the numeral and bar is suggested. The fraction bar may be diagonal in certain situations such as when used in a general note, in a drawing title, or when using CADD. (See Figure 3–68.)

Decimal Points.   The placement of the decimal point in a decimal dimension is critical. If the decimal point is crowded or drawn too lightly, it may be lost and the result is an unclear dimension. Always make the decimal point dark and bold. Also space the numerals far enough to clearly provide room for the decimal point. Two-thirds the height of letters is recommended. (See Figure 3–69.)

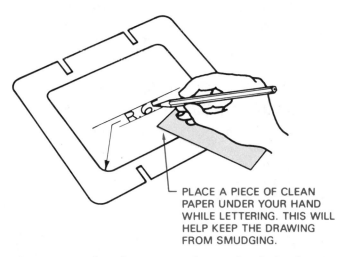

Figure 3–65 Place clean paper under your hand when lettering.

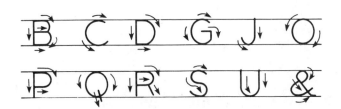

Figure 3–66 Recommended strokes for vertical uppercase Gothic letters with curved strokes.

Figure 3–68 Fractions.

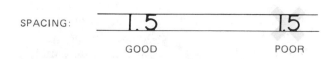

Figure 3–69 Spacing of decimal point in numerals.

▼ LETTERING TECHNIQUES

Always use guidelines. Straight, even letters of consistent height look better than letters of varying heights. Even when using guidelines be sure to extend each letter directly to the guidelines. Letters that periodically extend beyond or fall short of the guidelines tend to make the words or notes irregular.

Try an H or F pencil for lettering if you have a light touch or try a 2H pencil if your touch is heavy. Use a 0.5 mm automatic pencil. Lines should be black, crisp, and sharp. All vertical lines are made perpendicular, starting at the top for each stroke; all horizontal lines are made from left to right. Balance angles for the letters A, V, W, X, and Y about a vertical guideline. Use a round form for curved letters. Be careful to not allow any space to show between the letter and the guideline.

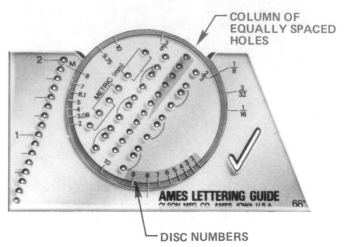

Figure 3–70   Ames lettering guide. *Courtesy Olson Manufacturing Company, Inc.*

▼ COMPOSITION

As a rule of thumb, curved letters can be placed close together; straight letters should be placed further apart. Good lettering composition is evident when all letters in a word look as if they have the same amount of space between them. To achieve this appearance it is often necessary to shorten the horizontal strokes of open letters such as L and J. When strokes are parallel and next to each other as the WA in WALL and both Ns and I in PLANNING should be placed a little farther apart.

If your letters are wiggly or if you are nervous, try pressing hard to make your lines straighter. If you are pressing too hard, try to relax the pressure. Also try making each letter as rapidly as possible. This tends to eliminate wiggly letters. If your lead is too hard, wiggly letters could result; try a softer lead.

▼ MAKING GUIDELINES

Guidelines are very lightly drawn lines equal to the height of letters in distance apart. As previously mentioned, some drafters prefer to use a light-blue lead so guidelines will not reproduce.

Ames Lettering Guide

Probably the most commonly used device for making guidelines is the versatile Ames Lettering Guide shown in Figure 3–70. With an Ames Lettering Guide it is possible to draw guidelines and sloped lines for lettering from ¹⁄₁₆ to 2 in. in height. Instructions for use are normally found with the lettering guide when purchased. The lettering guide may also be used for parallel lines, needed for such purposes as section lines, including brick, tile, and concrete block, or a music staff.

▼ OTHER LETTERING AIDS AND GUIDELINE METHODS

Other guideline lettering aids for equidistant spacing of lines have parallel slots ranging in width from ¹⁄₁₆ to ¼ in. These lettering guideline aids are not as complex as the Ames Lettering Guide, but they are also not as flexible.

Another method of making guidelines used by a few drafters is to place an ⅛-in. grid paper under the drawing. The lines show through the drawing sheet and guidelines need not be drawn. Be careful with this method, because lines of lettering may not be as straight as with conventional guidelines.

▼ BASIC LETTERING CONSIDERATIONS

Always use two guidelines. Three are helpful for beginners. The center guideline is for the letter crossbars.

To adjust eye and hand coordination, first letter lightly, then darken your lettering. Doing so allows you to correct mistakes. This technique should not be necessary after you get some lettering experience.

Make all letters and numbers on the drawing at least ⅛ in. high. All letters should be the same height. Be consistent.

Make all lettering dark. You may have to press on the pencil to get dark letters.

When practicing lettering, practice no more than 15 minutes per day. Otherwise your hand may cramp and any further practice may be of less value. Practice every day to gain speed and neatness as you letter.

▼ LETTERING GUIDE TEMPLATES

Some companies prefer that drafters use lettering guides so uniformity is maintained. Standard lettering guide

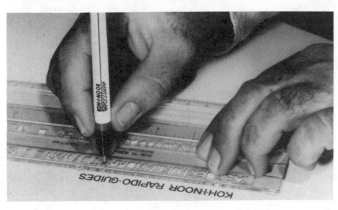

Figure 3–71 Lettering guide template in use. *Courtesy Koh-I-Noor, Inc.*

templates are available with vertical Gothic letters and numerals ranging in height from ³⁄₃₂ to ³⁄₈ in. (See Figure 3–71.) Lettering guides are also available in many other lettering styles, including slanted Gothic, and Microfont letters in either upper- or lower-case.

▼ MECHANICAL LETTERING EQUIPMENT

Mechanical lettering equipment has typically been used on government projects and in civil drafting. Mechanical lettering equipment is available in kits with templates for letters and numerals in a wide range of sizes. A complete lettering equipment kit includes a scriber, templates, tracing pins, and lettering pens. Figure 3–72 shows the component parts of a lettering equipment set. Complete instructions are normally found with the lettering equipment when purchased.

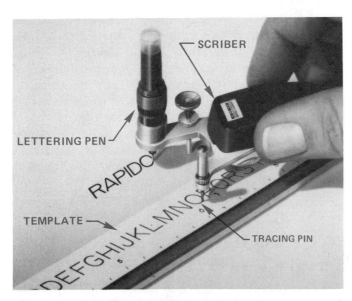

Figure 3–72 Components of a lettering equipment set, and using the mechanical lettering equipment. *Courtesy Koh-I-Noor, Inc.*

Figure 3–73 Lettering machine. *Courtesy AED Corporation.*

▼ MACHINE LETTERING

Lettering machines are available in a variety of fonts, styles, and letter sizes to prepare drawing titles, labels, or special headings. These types of lettering features are especially useful for making display letters or cover sheets, or rendering drawing titles. Figure 3–73 shows a lettering machine that has over 300 different lettering styles and sizes available on disks. This particular machine uses a typewriter keyboard for quick preparation of lettering. A computer may also interface with the lettering machine to increase speed and provide additional flexibility. Lettering machines prepare strips of lettering on clear tape with an adhesive back for placement on drawing originals. The tape is also available in a variety of colors for special displays and presentation drawings.

▼ TYPEWRITERS

Specially designed typewriters are used to type certain information on engineering drawings. Some companies use such a machine to help save time on items that would require many hours of freehand or template lettering. The information that may be typed on drawings includes general notes, parts lists, or other lengthy documentation. (See Figure 3–74.)

▼ TRANSFER LETTERING

A large variety of transfer lettering fonts, styles, and sizes are available on sheets. These transfer letters may be used in any combination to prepare drawing titles, labels, or special headings. They may be used to improve the quality of a presentation drawing, or simply for all drawing titles. Figure 3–75 shows a few of the variety of letters available.

Figure 3–74 Typewriter used in industry. *Courtesy Hyster Company.*

Figure 3–75 Variety of transfer lettering fonts. *Courtesy Chartpak.*

Transfer letters may be purchased as vinyl letter sheets where individual letters are removed from a sheet and placed on the drawing in the desired location, or on sheets where letters are placed on the drawing by rubbing with a burnishing tool. Rub-on letters, as they are often called, offer a high-quality lettering. Drawing aids, in the form of transfer tapes, may be used to prepare borders, line drawings, or special symbols as shown in Figure 3–76.

▼ CADD REPLACES MANUAL LETTERING METHODS

CADD is rapidly replacing most manual lettering methods for drafting applications and graphic arts. While some of the applications previously discussed, such as the typewriter and lettering machine, are automated, they require a certain amount of manual preparation. A CADD system has the ability to perform all of the same tasks with more speed and productivity.

▼ LINES

Drafting is a graphic language using lines, symbols, and notes to describe objects to be manufactured or built.

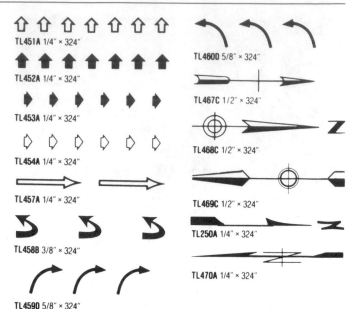

Figure 3–76 Drafting aids; borders, lines, and special symbols. *Courtesy Chartpak.*

Lines on drawings must be of a quality that will easily reproduce. All lines are dark, crisp, sharp, and of the correct thickness when properly drawn. There is no variation in darkness, only a variation in thickness, known as *line contrast.* Certain lines are drawn thick so they stand out clearly from other information on the drawing. Other lines are drawn thin. Thin lines are not necessarily less important than thick lines, but they are subordinate for identification purposes.

ANSI The American National Standards Institute (ANSI) recommends two line thicknesses with bold lines twice as thick as thin lines. This line standard relates to both manual and computer-aided drafting. Standard line thicknesses are 0.6 mm for thick lines, and 0.3 mm for thin lines. The actual width of lines may be more or less than the recommended thickness depending on the size of the drawing or the size of the final reproduction. Drawings that meet military documentation standards require three thicknesses of lines: thick, medium, and thin. Figure 3–77 shows widths and types of lines as taken from ANSI standard, *Line Conventions and Lettering.* ANSI Y14.2M–1992. Figure 3–78 shows a sample drawing using the various kinds of lines.

▼ TYPES OF LINES

Construction and Guidelines

Construction lines are used for laying out a drawing. Construction lines are drawn very lightly so they will not reproduce and will not be mistaken for any other line on the drawing. Construction lines are drawn with a 4H to

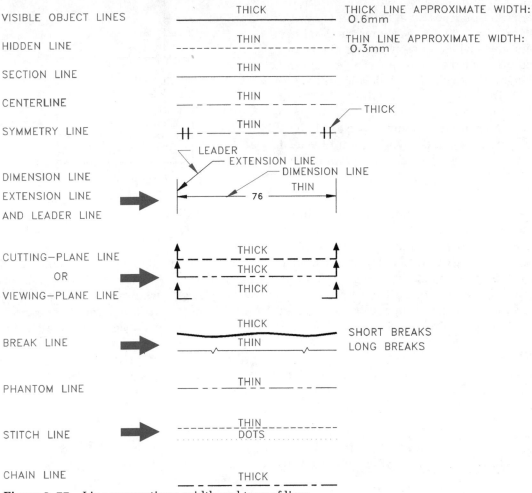

Figure 3–77 Line conventions, width and type of lines.

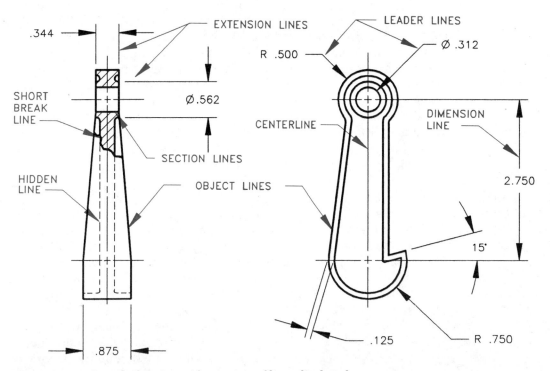

Figure 3–78 Sample drawing with a variety of lines displayed.

6H pencil and, if drawn properly, will not need to be erased. Use construction lines for all preliminary work.

Guidelines are drawn the same as construction lines and will not reproduce when properly drawn. Guidelines are used to keep lines of lettering in perfect alignment and of a constant height. For example, if lettering on a drawing is to be .125 in. high, then guidelines are drawn very lightly .125 in. apart. Guidelines must always be used for lettering. Some drafters prefer to use a light-blue lead for all guidelines and layout work. Light-blue lead will not reproduce and may be cleaner than graphite lead.

Object Lines

Object lines, also called visible lines, or outlines, describe the visible surface or edge of the object. They are drawn as thick lines as shown in Figure 3–79. Thick lines, remember, are usually drawn at 0.6 mm or .032 in. wide. Drafters usually use a soft lead, H or F, to draw object lines properly. You need to experiment to see which lead gives you the best line quality. For inked drawings, object lines can be made with a number 2 (0.6 mm) pen.

Object Line Quality.   To determine if your object lines are dark and crisp enough, turn the paper over and

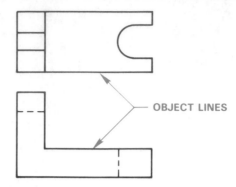

Figure 3–79   Object lines.

hold it up to the light. The line should have a dark, consistent density. A light table, if available, also works very well to check darkness.

You can also make a diazo or photocopy reproduction of your drawing to observe the quality. Properly drawn lines and lettering make quality prints. If the print lines are fuzzy, your original probably needs more work.

Most manual drafters use automatic pencils with a 0.7 mm lead for object lines.

CADD Applications

DRAWING LINES

The basics of drawing lines with a CADD system are applicable to all systems, whether they use a puck, stylus, light pen, thumbwheels, joystick, or mouse. The term *pointers* refers to the pointing device used to select commands and digitize points. The term *digitize* means to press the stylus to the tablet or press the proper puck or keyboard button to digitize, or draw, a point. Several different commands will be given for one function in the following discussions, to give the reader an idea of the terms used in different systems.

Once you are in the digitize mode, your location on the screen is indicated by crosshairs. When you move the pointer, the crosshairs move. Unless you press a button or click the pointer, nothing is drawn on the screen. Some systems may continually display the coordinates of the crosshair location on the screen. These values change as you have the pointer. When constructing your drawing it is best to look at the video display as you draw. You will then be able to see if your lines are not straight, or if the drawing does not look right.

Begin drawing a line by first selecting the proper command such as POINT, INSERT POINT, PLACE LINE, or LINE. This command is used to locate corners of the object and connect them with straight lines. Digitize a point on the tablet and the coordinates of that point are displayed on the screen. Select another point and a straight line is drawn between the two. This shows as line 1 in Figure 3–80a. Each selection of the pointer after that draws a line. The steps required to draw an object are shown in Figure 3–80a–d. There is no need to select the INSERT POINT COMMAND between each digitized point. If you wish to break the line and move to another feature, select the INSERT POINT command again. This allows you to locate a new coordinate point without drawing an unwanted line.

Some systems use what is called a *rubber band line.* Once you have digitized one point, a line is attached to that point and to the crosshairs and moves with the crosshairs. Digitizing another point freezes the first line and another is now attached to the crosshairs.

Different types of lines can be indicated either before or after digitizing the points. Specifying the type of line before digitizing points is done by selecting a

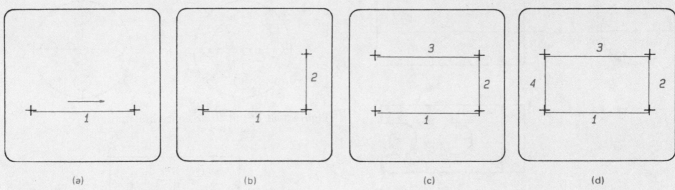

Figure 3–80   Digitizing an object one point at a time.

command such as LINE TYPE, LINE, or DASH. This command may then display several prompts asking for dash length, spacing, frequency, and the like. Any line drawn after you have entered a type will have the characteristics that you have choosen, and the type of line will change only after you select a new type of line. Should you wish to change an existing line to a different type, you can select the LINE TYPE command, set the new parameters, then digitize a point on the line. Some systems may require you to first activate an edit command such as MODIFY because you are changing something that has already been drawn.

Hidden Lines

A *hidden line* represents an invisible edge on an object. Hidden lines are thin lines, half as thick as object lines for contrast. Figure 3–81 shows hidden lines properly drawn with .125 in. dashes spaced .06 in. apart. Hidden lines, as all thin lines, can be drawn effectively with a 0.3 mm to 0.5 mm automatic drafting pencil, or a number 00 (0.3 mm) technical pen tip when inking. The best way to draw hidden lines is to draw dash lengths and spaces by eye. This takes some practice, but is the fastest way.

Hidden Line Rules

The drawings in Figure 3–82 show situations where hidden lines meet or cross object lines and other hidden lines. These situations represent rules that should be followed when possible.

Centerlines

Centerlines are used to show and locate the centers of circles and arcs, and are used to represent the center

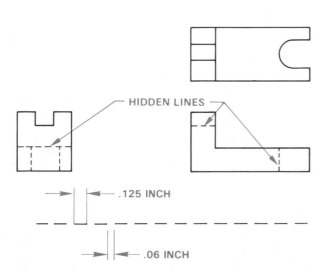

Figure 3–81   Hidden line representation.

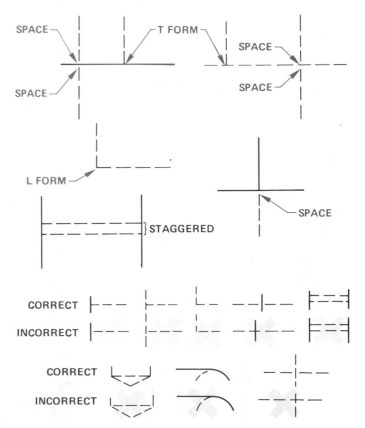

Figure 3–82   Hidden line rules.

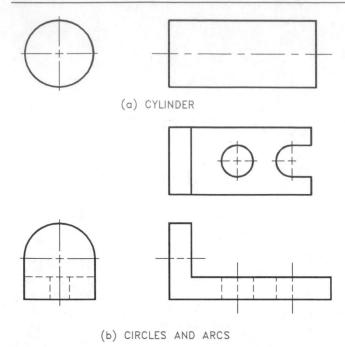

(a) CYLINDER

(b) CIRCLES AND ARCS

Figure 3–83 Centerline representation and examples.

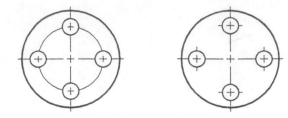

Figure 3–86 Bolt circle centerline options.

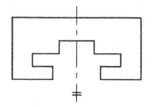

Figure 3–87 Symmetry symbol.

axis of a circular or symmetrical form. Centerlines are thin lines on a drawing. They should be about half as thick as an object line. The long dash is about .75 to 1.50 in. The spaces between dashes are about .062 in. and the short dash about .125 in. long. The length of long lines varies for the situation and size of the drawing. Try to keep the long lengths uniform throughout the centerline. (See Figure 3–83.) Small centerline dashes should cross only at the center of a circle or arc. (See Figure 3–84.) Small circles should have centerlines as shown in Figure 3–85.

Centerlines for holes in a bolt circle may be drawn either of two ways, depending on how the holes are located, as shown in Figure 3–86. A *bolt circle* is a pattern of holes arranged in a circle.

When a centerline represents symmetry, as in the centerplane of an object, the symmetry symbol, shown in Figure 3–87, may be used if needed for clarity.

The centerline is commonly drawn with a 0.3 to 0.5 mm automatic pencil. A 0.3 mm pencil may provide best results with a soft lead and a 0.5 mm pencil with a harder lead. Remember the results should be dark, crisp, and sharp lines that are half as thick as object lines. When inking, a number 00 (0.3 mm) technical pen gives the best results.

Extension Lines

Extension lines are thin lines used to establish the extent of a dimension. Extension lines begin with a short space from the object and extend to about .125 in. beyond the last dimension, as shown in Figure 3–88. Extension lines may cross object lines, centerlines, hidden lines, and other extension lines, but they may not cross dimension lines. Circular features, such as holes, are located by their centers in the view where they appear as circles. In this practice, centerlines become extension lines as shown in Figure 3–89.

Dimension Lines and Leader Lines

Dimension lines are thin lines capped on the ends with arrowheads and broken along their length to provide a

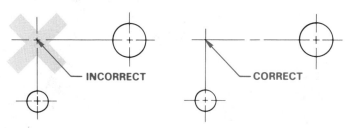

Figure 3–84 Centerline rules.

Figure 3–85 Centerlines for small circles.

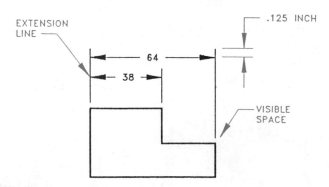

Figure 3–88 Extension lines.

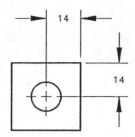

Figure 3–89 The centerline becomes an extension line when used for dimensioning.

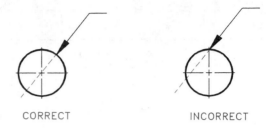

Figure 3–92 Circle to leader line relationship.

space for the dimension numeral. *Dimension lines* indicate the length of the dimension. (See Figure 3–90.)

Leaders, or leader lines, are thin lines used to connect a specific note to a feature as shown in Figure 3–91. Leaders may be drawn at any angle, but 45°, 30°, or 60° lines are most common. Slopes greater than 75° or less than 15° from horizontal should be avoided. The leader has a .25 in.

shoulder at one end that begins at the center of the vertical height of the lettering and an arrowhead at the other end pointing to the feature. If the leader were to continue from the point where the arrowhead touches the circle, it would intersect the center. (See Figure 3–92.)

Arrowheads

Arrowheads are used to terminate dimension lines and leaders. Properly drawn arrowheads should be three times as long as they are wide. All arrowheads on a drawing should be the same size. Do not use small arrowheads in small spaces. (Limited space dimensioning is covered in Chapter 10.) Some companies require that arrowheads be drawn with an arrow template, while others accept properly drawn freehand arrowheads. (See Figure 3–93.) Individual company preference dictates whether arrowheads are filled in or left often as shown. Microfilming requires that arrowheads be filled in for clarity.

Cutting-plane and Viewing-plane Lines

Cutting-plane lines are thick lines used to identify where a sectional view is taken. *Viewing-plane lines* are also thick and are used to identify where a view is taken for view enlargements or for partial views. Cutting-plane and viewing-plane lines are properly drawn in either of the two ways. The approximate dash and space

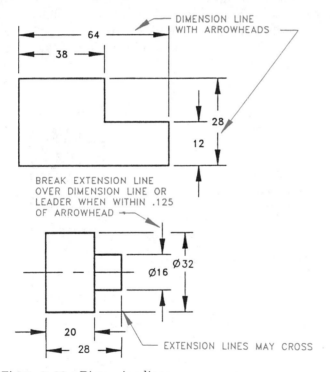

Figure 3–90 Dimension lines.

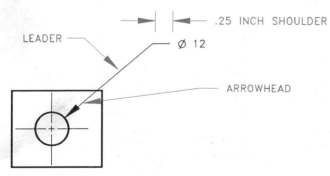

Figure 3–91 Leader line.

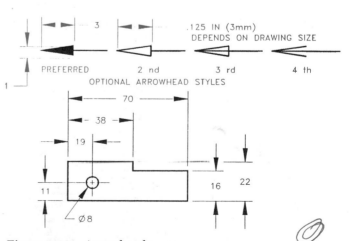

Figure 3–93 Arrowheads.

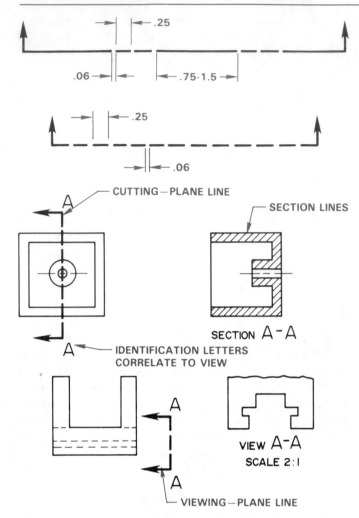

Figure 3–94 Cutting and viewing plane lines.

sizes are shown in Figure 3–94. The cutting-plane line takes precedence over the centerline when used in the place of a centerline.

The scale of the view may be increased or remain the same as the view from the viewing plane, depending on the clarity of information presented. When the location of the cutting plane or viewing plane is easily understood or if the clarity of the drawing is improved, the portion of the line between the arrowheads may be omitted as shown in Figure 3–95.

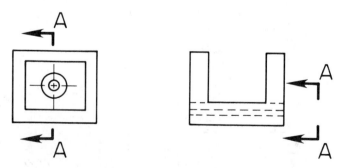

Figure 3–95 Simplified cutting and viewing plane lines.

Section Lines

Section lines are thin lines used in the view of a section to show where the cutting-plane line has cut through material. (See Figure 3–96.) Section lines are drawn equally spaced at 45°, but may not be parallel or perpendicular to any line of the object. Any convenient angle may be used to avoid placing section lines parallel or perpendicular to other lines of the object; 30° and 60° are common. Section lines that are more than 75° or less than 15° from horizontal should be avoided. Section lines should be drawn in opposite directions on adjacent parts. (See Figure 3–102.) For additional adjacent parts any suitable angle may be used to make the parts appear clearly separate. The space between section lines may vary depending on the size of the object. (See Figure 3–97.) Figure 3–98 shows correct and incorrect applications of section lines. When a very large area requires section lining, you may elect to use outline section lining, as shown in Figure 3–99.

The section lines shown in Figures 3–96 through 3–99 were all drawn as general section-line symbols. General section lines can be used for any material and are specifically used for cast or malleable iron. Coded sec-

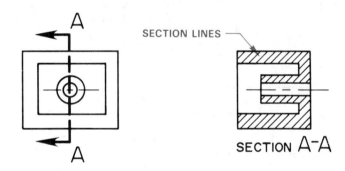

Figure 3–96 Section lines.

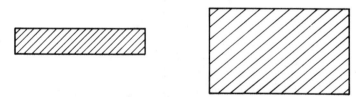

Figure 3–97 Space between section lines.

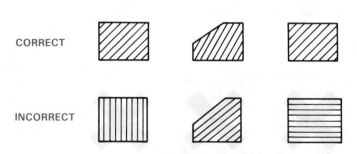

Figure 3–98 Correct and incorrect section lines.

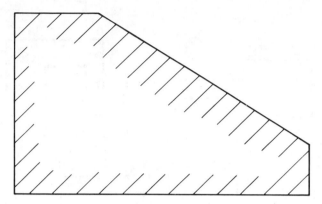

Figure 3–99   Outline section lines.

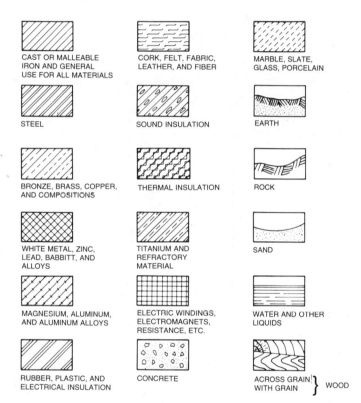

Figure 3–100   Coded section lines.

CAST OR MALLEABLE IRON AND GENERAL USE FOR ALL MATERIALS

STEEL

BRONZE, BRASS, COPPER, AND COMPOSITIONS

WHITE METAL, ZINC, LEAD, BABBITT, AND ALLOYS

MAGNESIUM, ALUMINUM, AND ALUMINUM ALLOYS

RUBBER, PLASTIC, AND ELECTRICAL INSULATION

CORK, FELT, FABRIC, LEATHER, AND FIBER

SOUND INSULATION

THERMAL INSULATION

TITANIUM AND REFRACTORY MATERIAL

ELECTRIC WINDINGS, ELECTROMAGNETS, RESISTANCE, ETC.

CONCRETE

MARBLE, SLATE, GLASS, PORCELAIN

EARTH

ROCK

SAND

WATER AND OTHER LIQUIDS

ACROSS GRAIN / WITH GRAIN } WOOD

tion-line symbols, as shown in Figure 3–100, are not commonly used on detail drawings, as they are more difficult to draw and the drawing title block usually identifies the type of material in the part. Coded section lines may be used when the material must be clearly represented, as in the section through an assembly of parts made of different materials. (See Figure 3–101.) Very thin parts less than 4 mm thick may be shown without section lining; only the outline is shown. This option is often used for a gasket as shown in Figure 3–102.

Break Lines

There are two types of *break lines:* the short break and long break line. The thick, short break is very common

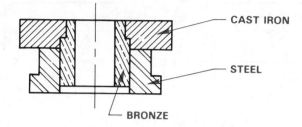

Figure 3–101   Coded section lines in assembly.

CAST IRON

STEEL

BRONZE

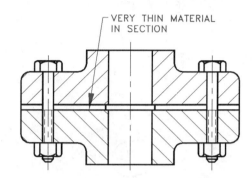

Figure 3–102   Very thin material in section.

VERY THIN MATERIAL IN SECTION

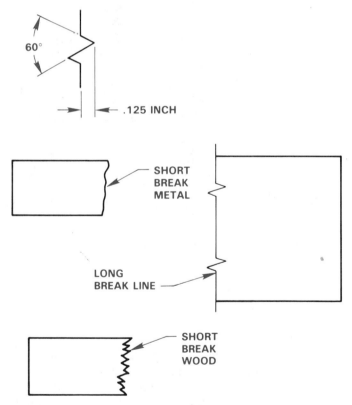

60°

.125 INCH

SHORT BREAK METAL

LONG BREAK LINE

SHORT BREAK WOOD

Figure 3–103   Long and short break lines.

on detail drawings, although the thin, long break may be used for breaks of long distances at the drafter's discretion. (See Figure 3–103.) Other conventional breaks may be used for cylindrical features as in Figure 3–104.

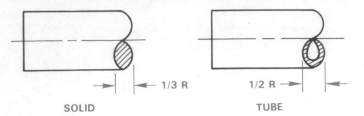

Figure 3–104   Cylindrical conventional breaks.

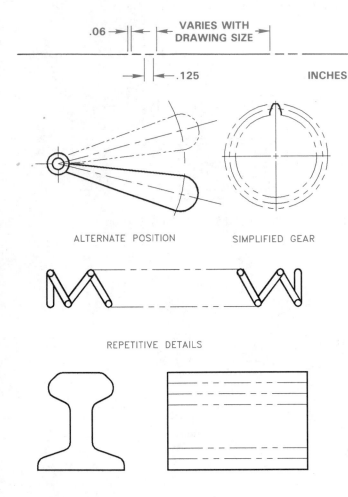

Figure 3–105   Phantom line representation and examples.

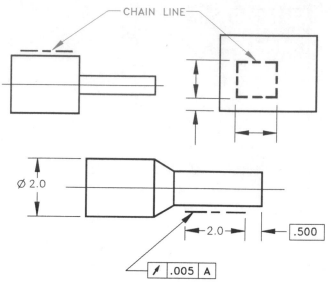

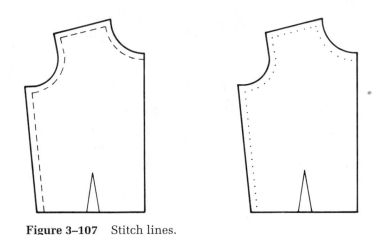

Figure 3–106   Chain lines.

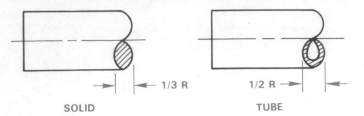

Figure 3–107   Stitch lines.

Phantom Lines

Phantom lines are thin lines made of one long and two short dashes alternately spaced. Phantom lines are used to identify alternate positions of moving parts, adjacent positions of related parts, repetitive details, or the contour of filleted and rounded corners. (See Figure 3–105.)

Chain Lines

Chain lines are thick lines of alternately spaced long and short dashes used to indicate that the portion of the surface next to the chain line receives some specified treatment. (See Figure 3–106.)

Stitch Lines

There are two types of acceptable *stitch lines.* One is drawn as thin, short dashes, the other as .016-in. diameter dots spaced .12 in. apart. They are used to indicate the location of a stitching or sewing process as shown in Figure 3–107.

▼ PENCIL AND INK LINE TECHNIQUES
Pencil Techniques

There are four basic properties of correctly drawn lines: uniformity, contrast, darkness, and sharpness. *Line uniformity* means that all lines are drawn their proper thickness without variation. For example, all thick lines (object, cutting-plane) are the same degree of thickness. All thin lines (center, dimension, extension) are the same thinness. *Line contrast* is the variation that exists between different types of lines; thick and thin.

ANSI/MIL According to ANSI standards, thick and thin are the two line thicknesses used. Military (MIL) standards recommend three line thicknesses; thick (cutting and viewing plane, short break, and object), medium (hidden and phantom), and thin (center, dimension, extension, leader, long break, and section).

All lines should be the same darkness with no variation. Lines must be drawn opaque so light will not pass through. Lines are sharp when edges are clear and crisp. Fuzzy edges on lines result from the texture of the paper, or too-soft pencil lead. When using an automatic pencil, do not let the lead out too far.

Wash your hands and equipment frequently to keep the drawing clean. Try not to handle your pencil leads.

Lay out your entire drawing with construction lines before darkening any lines. A combination of pressure and pencil lead hardness makes the best lines. If you are right handed, it is generally best to draw horizontal lines from left to right and vertical lines from bottom to top. The opposite may be true for left-handed drafters. Do not use excessive pressure. You do not have to dig trenches in the drawing board. The more pressure you use, the more difficult it is to erase. If your hand cramps while drawing or lettering, you are forcing yourself into an unnatural stroking technique. Relax! Try different methods until you find a suitable system of strokes.

Make thick lines with a 0.7 or 0.9 mm automatic pencil held vertically to establish full contact between the lead and the paper. In order to obtain lines that are dark, crisp, and sharp, you may need to go over each line several times. If light passes through your lines, they are drawn too lightly. Hold your drawing up to a light or place it on a light table to see if light passes through.

Always draw radii first, then connect the straight lines. This is also a good technique for ink drawings. (See Figure 3–108.) If you make the straight lines first, you may have trouble aligning the arcs.

Line Layout

First, draw thin horizontal lines beginning from the top of the sheet and work downward. Draw vertical thin lines next, starting on the left side of the sheet and working toward the right side for right-handed drafters. Next draw all circles and arcs. Last, draw all horizontal and vertical thick lines in the same direction as thin lines. Do all lettering when the line work is complete. Try to avoid going back over lines or lettering that has already been drawn in order to help avoid smudging.

MAKE ARCS FIRST   CONNECT ARCS

Figure 3–108 Draw circles and arcs first.

Inking Techniques

Ink or graphite only adheres cleanly to one side of vellum. If the ink beads or skips while making a line, you are working on the wrong side of the paper. Paper with preprinted borders and title blocks have the correct side predetermined. Some manufacturers have a water mark that can be seen when the paper is held up to the light. When the water mark is readable, the paper is right side up. For Mylar®, there is either a single or double matte surface. Matte refers to the textured surface as opposed to a glossy surface. Double matte surface may be drawn on either side. Single matte surface requires you to draw on the matte side, not the shiny, or glossy, side.

Use template lifters or templates and other equipment designed for inking that keep the edge of the tool slightly off the paper. Template lifters are plastic shapes that can be attached to the back of a template or other instrument to help keep the instrument off the drafting surface, as seen in Figure 3–109. Another option is to place a template under the one being used. (See Figure 3–110.) There are also template risers available that are long plastic strips that fit on the edges of the template to help keep it off of the drawing surface. Figure 3–111 shows how the template riser functions when attached to a template. Drafting machine scales have long been manufactured with edge relief for inking. Now, templates, tri-

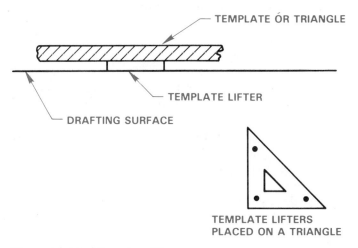

TEMPLATE OR TRIANGLE

TEMPLATE LIFTER

DRAFTING SURFACE

TEMPLATE LIFTERS
PLACED ON A TRIANGLE

Figure 3–109 Template lifters.

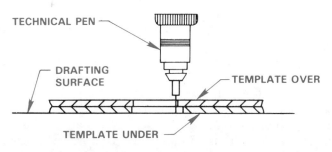

TECHNICAL PEN

DRAFTING SURFACE

TEMPLATE OVER

TEMPLATE UNDER

Figure 3–110 Second template as a spacer.

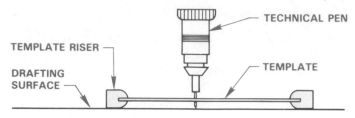

Figure 3–111   Template risers.

Figure 3–112   Built-in ink risers. *Courtesy Chartpak.*

angles, and other equipment are being made with ink risers built in, as shown in Figure 3–112. If you draw with ink without taking these precautions, the ink could easily flow under the template and cause a mess.

Periodically check your technical fountain pens for leaks around the tip to prevent your hands from smearing your drawing. Also check the tip for a drop of ink before you begin a line. Have a piece of cloth or a tissue available to help keep the tip free of ink drops. Keep your pens clean and loaded with fresh ink to help keep proper ink flow for trouble free use. Also, keep a piece of paper handy for scratching your pen to start the ink flowing.

Ink your lines beginning with thin horizontal lines and work from top to bottom and from left to right if right handed. The technique of drawing horizontal lines top to bottom and vertical lines left to right allows inked lines to dry as you go along and saves valuable drafting time. Inked lines are dry when they do not appear glossy. Glossy lines are wet; keep equipment away from them. Next, ink all circles and arcs. Ink all thick horizontal and vertical lines in the same manner as thin horizontal and vertical lines. Finally, do all lettering.

When using technical pens, hold the pen perpendicular (90°) to the paper for the best results. Do not apply any pressure to the pen. Allow the pen to flow easily over the vellum or Mylar. Figure 3–113 shows proper technical pen use. If the technical pen is not held 90° to the surface, the resulting line may be fuzzy or rough on the edge. Move your technical pen at a constant speed. Do not go too fast and do not slow at the ends of lines. Following these hints will help keep your line consistent in width and the pen point from skipping.

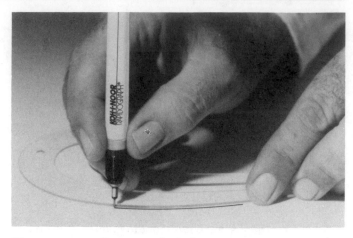

Figure 3–113   Proper technical pen use. *Courtesy Koh-I-Noor, Inc.*

Erasing Ink

Erasing ink from vellum is possible in small areas but doing so may destroy its surface. A smooth surface cannot be inked over again satisfactorily. An electric eraser would be best used here, but overenthusiasm will burn a hole through the paper. Electric erasers are excellent tools, but be careful to touch the paper surface lightly.

To erase on polyester film, it is best to apply a little water with a felt tip or a clean cotton swab. Remove any excess water with a blotter. Allow the area to dry thoroughly. The ink remover recommended by the ink manufacturer can also be used. A polyester eraser also removes ink. Use an eraser with care so you do not destroy the mat surace, which makes further drafting in that area difficult.

Polyester Lead on Film

When pencils are used on Mylar, the recommended lead is made of polyester. The quality of a line on polyester film depends on how well the mat finish is maintained. If you use single matte film, do not try to draw on the glossy side; use polyester lead on the matte side.

Pencil Skills with Polyester Lead

○ Draw with a single line, in one direction. Tracing a line in both directions deposits a double line, and causes smearing and damage to the matte.

○ Draft with a light touch. Drafting films require up to 40 percent less pressure than other media. Smearing and embossing can be reduced with less pressure. It will take practice to use polyester lead after drawing with graphite lead.

○ The surface of your drafting board is also a factor in line quality. Use a recommended backing material on your table. Check with your local vendor.

○ Erase with a vinyl eraser. If an electric eraser is used, be very careful not to destroy the matte.

PROFESSIONAL PERSPECTIVE

Sketching

Freehand sketching is an important skill if you are a manual or a CADD drafter. Preparing a sketch before you begin a formal drawing may save many hours of work. The sketch assists you in the layout process in that it allows you to:

○ Decide how the drawing should appear when finished.
○ Decide how big to make the drawing.
○ Determine the sheet size for manual drafting or the screen limits for CADD.
○ Establish the coordinate points for the computer drawing.

A little time spent sketching and planning your work saves a lot of time in the final drafting process. Sketches are also a quick form of communication in any professional environment. You can often get your point across or communicate more effectively with a sketch.

Lettering

The appearance of manual drawings may be greatly enhanced by quality freehand lettering. An otherwise good drawing may look unprofessional with poor lettering. However, good freehand lettering does not come easily for some people; it takes a lot of practice. The only substitute for practice is an inherent talent. For some people, lettering comes naturally.

CADD lettering is a different story. Lettering on a computer is as easy as typing, and the lettering comes out fast and is perfect every time. Another exciting aspect of CADD lettering is the many styles available. There are even lettering styles available that duplicate the artistic appearance of the best freehand architectural lettering. Freehand lettering may not be important when the computer age totally takes over the drafting industry. For now, however, lettering on manual drawings is very important.

Lines

Quality line work is dependent on the opaqueness of the lines. The diazo reproduction process creates an excellent print if the light is unable to pass through the lines. If lines appear fuzzy, take a magnifying glass and look at the line density. If the line does not make an opaque image, you will not get a good print. Do not blame it on the print machine. You cannot make poor lines look good if they are not opaque. The job is difficult when drawing with pencil on vellum. You need the right combination of pressure, lead hardness, and skill to produce properly executed lines. This combination is different for each individual. Some people draw acceptable lines with an F lead while others are successful with a 2H lead. Also, one lead may not work well for all lines. You may need to have a few automatic pencils with different lead grades. If you are inking on polyester film, you will get good-quality lines, but you may have trouble getting used to working with the pitfalls of ink, such as smearing and error cleanup. Either manual technique takes a lot of practice.

Drafting with the computer is a different story, because all you need to do is make lines on the screen and then it is up to the plotting or printing process to reproduce a quality drawing on paper. You can choose a check print on a dot matrix printer for a quick copy where quality is unimportant, or you can ink plot on vellum or polyester film for excellent plots. You can even plot different line thicknesses by using thick or thin pens as appropriate. You can also plot in different colors if you want to emphasize a part of the drawing.

MATH APPLICATION

FRACTIONAL ARITHMETIC

The United States is one of the few countries of the world that does not commonly use a metric system of measurement. It is important for a drafter to be able to handle the arithmetic of both common fractions and decimal fractions. Determine the overall height of Figure 3–114 and the overall width of fifteen of these pieces laid side to side.

The solution to overall height involves an addition of mixed numbers:

$$2\frac{1}{16} + 1\frac{1}{4} + 2\frac{3}{8} + 2\frac{1}{16} + 1\frac{4}{16} + 2\frac{6}{16} = 5\frac{11}{16}$$

The solution to the width of 15 pieces requires multiplication:

$$5\frac{7}{8} \times 15 = \frac{47}{8} \times \frac{15}{1} = \frac{705}{8} = 88\frac{1}{8}" = 7'\ 4\frac{1}{8}"$$

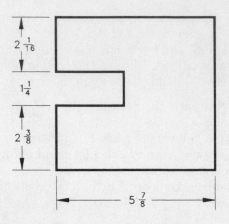

Figure 3–114 Fractionally dimensioned part.

3 SKETCHING, LETTERING, AND LINES TEST

DIRECTIONS

Answer the questions with short, complete statements or drawings as needed.

PART 1. SKETCHING

1. Define sketching.
2. How are sketches useful as related to computer graphics?
3. Describe the proper sketching tools.
4. When sketching, should the paper be taped to the drafting board or table? Why?
5. What kind of problem can occur if a long straight line is drawn without moving the hand?
6. What type of paper should be used for sketching?
7. Describe a method that can be used to sketch irregular shapes.
8. Define an isometric sketch.
9. What is the difference between an isometric line and a nonisometric line? An example may be used.
10. What do the use of proportions have to do with sketching techniques?

PART 2. LETTERING

1. What type of lettering characters are recommended by ANSI for mechanical drafting?
2. According to ANSI standards, what are the minimum recommended lettering heights?
3. How should the letters within words be spaced?
4. What is the recommended minimum space between two numerals having a decimal point between them?
5. What is the recommended horizontal space between sentences?
6. Discuss the general rule that applies to the vertical spacing between lines of lettering and between individual notes on a drawing.
7. Identify the ANSI document that provides the recommended lettering standards.
8. Describe guidelines.
9. Why are guidelines necessary for freehand lettering?
10. Describe the recommended leads and points used in mechanical and automatic pencils for freehand lettering.
11. Identify a method to use when lettering to help avoid smudging the drawing.

12. When lettering fractions, what is the recommended relationship of the fraction division line to the fraction numerals?
13. List two manual methods that can be used to make guidelines rapidly.
14. Why do some companies prefer the use of lettering guide templates?
15. When should the mechanical lettering template be placed along a straightedge when lettering?
16. Describe an advantage of machine lettering.
17. Discuss a use for typewriters on engineering drawings.
18. What is meant by text direction, and how is it measured?
19. Identify a use for transfer letters.
20. What is a lettering font?
21. What aspects of a label must first be set before the text is located on the drawing?
22. What is meant by text path?
23. Describe two methods for locating and entering text.

PART 3. LINES

1. Identify the American National Standards Institute document that governs line standards.
2. What are construction lines and guidelines used for, and how should they be drawn?
3. Discuss line uniformity and line contrast.
4. Should there be any difference in line darkness?
5. What is the recommended thickness of object lines?

6. What do hidden lines represent on a drawing?
7. Descibe two functions that centerlines serve on a drawing.
8. Extension lines are thin lines that are used for what purpose?
9. Where should the extension lines begin in relationship to the object and end in relation to the last dimension line?
10. Describe leaders.
11. What is the correct length-to-width ratio of properly drawn arrowheads?
12. Should arrowheads on a drawing all be the same size? Why?
13. Describe the difference between a cutting-plane and a viewing-plane line.
14. Should cutting-plane lines be drawn thick or thin?
15. Discuss the recommended spacing and angle of section lines.
16. List two uses for phantom lines.
17. What type of line will be used to denote that a portion of a surface or feature will receive a specific treatment?
18. Discuss the proper recommended technique when using a mechanical pencil with graphite lead to draw lines on vellum.
19. How will the line technique differ when using a mechanical pencil as compared to using an automatic pencil?
20. Describe the proper technique to use when drawing lines with a technical fountain pen.

CHAPTER 3 — SKETCHING, LETTERING, AND LINES PROBLEMS

PART 1. SKETCHING PROBLEMS

DIRECTIONS

Use proper sketching materials and techniques to solve the following sketching problems, on 8½ × 11 in. bond paper or newsprint. Use very lightly sketched construction lines for all layout work. Darken the finished lines, but do not erase the layout lines.

Problem 3–1   List on a separate sheet of paper the length, direction, and position of each line shown in the drawing. Remember, do not measure the lines with a scale. Example: Line 2 is the same length as line 1, and touches the top of line 1 at a 90° angle.

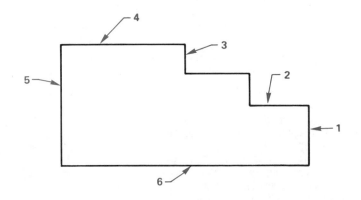

Problem 3–2 Use the trammel and hand compass methods to sketch a circle with approximately a 4-in. diameter.

Problem 3–3 Find a stapler, tape dispenser, or coffee cup and sketch a two-dimensional frontal view using the block technique. Do not measure the object. Use the measurement-line method to approximate proper proportions.

Problem 3–4 Find an object with an irregular shape such as a French curve and sketch a two-dimensional view using the regular grid method. Sketch the correct proportions of the object without measuring.

Problem 3–5 Given the three objects in the figure below, sketch, without measuring, the front, top, and side views of each. Use the multiview-alignment technique discussed in this chapter.

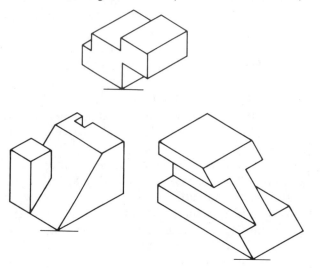

Problem 3–6 Using the same object selected for problem 3–3, sketch an isometric representation. Do not measure the object but use the measurement-line technique to approximate proportions.

PART 2. LETTERING PROBLEMS

DIRECTIONS

Use vertical freehand Gothic lettering, or computer-aided drafting as required by the specific instructions for each assignment. Do all freehand lettering on an A-size drawing sheet with .125-in. letters. Space lines of lettering .125 in. apart. Use guidelines for all freehand lettering.

Problem 3–7 Use vertical Gothic freehand lettering to letter the following statement:

THE QUALITY OF THE FREEHAND LETTERING GREATLY AFFECTS THE APPEARANCE OF THE ENTIRE DRAWING. DRAFTERS LETTER IN PENCIL OR INK. PROPER FREEHAND LETTERING IS DONE WITH A SOFT, SLIGHTLY ROUNDED POINT IN A MECHANICAL DRAFTING PENCIL, OR A 0.5 MM AUTOMATIC PENCIL WITH 2H, H, OR F GRADE LEAD DEPENDING UPON INDIVIDUAL PRESSURE. LETTERING IS DONE BETWEEN VERY LIGHTLY DRAWN GUIDE-LINES. THESE GUIDELINES ARE DRAWN PARALLEL AND SPACED EQUAL TO THE HEIGHTS OF THE LETTERS. GUIDELINES HELP TO KEEP YOUR LETTERING UNIFORM IN HEIGHT. LETTERING STYLES MAY VARY BETWEEN COMPANIES. SOME COMPANIES REQUIRE THE USE OF LETTERING DEVICES.

Problem 3–8 Use vertical Gothic freehand lettering to letter the following statement:

MOST MECHANICAL DRAFTING THAT DOES NOT USE CAD DOES USE VERTICAL FREEHAND LETTERING. THE QUALITY OF THE FREEHAND LETTERING GREATLY AFFECTS THE APPEARANCE OF THE ENTIRE DRAWING. MANY MECHANICAL DRAFTING TECHNICIANS USE FREEHAND LETTERING WITH PENCIL ON VELLUM OR WITH POLYESTER LEAD ON MYLAR. LETTERING IS COMMONLY DONE WITH A SOFT, SLIGHTLY ROUNDED LEAD IN A MECHANICAL PENCIL OR A 0.5 MM LEAD IN AN AUTOMATIC PENCIL. LETTERING IS DONE BETWEEN VERY LIGHTLY DRAWN GUIDELINES. GUIDELINES ARE SPACED PARALLEL AT A DISTANCE EQUAL TO THE HEIGHT OF THE LETTERS. GUIDELINES ARE REQUIRED TO HELP KEEP ALL LETTERS THE SAME UNIFORM HEIGHT.

Problem 3–9 Use vertical Gothic freehand lettering and a CADD system to letter the following notes:

1. INTERPRET DIMENSIONS AND TOLERANCES PER ASME Y14.5M-1994.
2. UNLESS OTHERWISE SPECIFIED, ALL DIMENSIONS ARE IN MILLIMETERS.
3. REMOVE ALL BURRS AND SHARP EDGES.
4. ALL FILLETS AND ROUNDS R6.
5. CASEHARDEN 62 ROCKWELL C SCALE.
6. AREAS WHERE MATERIAL HAS BEEN REMOVED SHALL HAVE SMOOTH TRANSITIONS AND BE FREE OF SCRATCHES, GRIND MARKS, AND BURNS.
7. FINISH BLACK OXIDE.
8. PART TO BE CLEAN AND FREE OF FOREIGN DEBRIS. COMPARE THE DIFFERENCE BETWEEN FREEHAND AND CADD LETTERING AS TO SPEED AND APPEARANCE.

PART 3. LINE PROBLEMS

DIRECTIONS

1. Using the selected engineer's layout as a guide only, make an original drawing using manual or CADD as required by your course objectives. Select an appropriate scale. Draw *only* the object lines, centerlines, hidden lines, and phantom lines as appropriate for each problem. Do not dimension. Keep in mind that the engineer's sketches are rough and not meant for tracing.
2. Use construction lines (very lightly drawn) to prepare the entire drawing. When satisfied with the product, darken the drawing using proper line technique.
3. Complete the title block.
   a. The title of the drawing is given.
   b. The material the part is made of is given.
   c. The drawing number is the same as the problem number.
   d. Use FULL scale.

Problem 3–10 Object lines (in.)

Part Name: Plate
Material: .25-in.-Thick Mild Steel

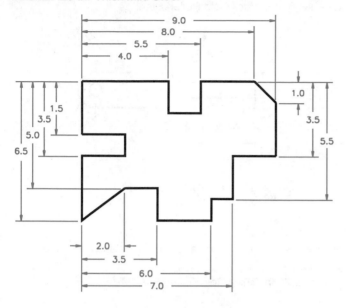

Problem 3–13 Arcs, object lines, and centerlines (in.)

Part Name: Latch
Material: 25-in.-Thick Mild Steel

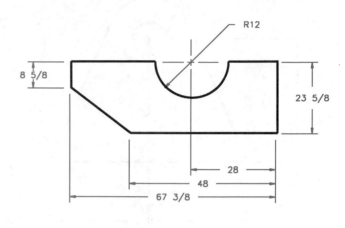

Problem 3–11 Straight object lines only (in.)

Part Name: Milk Stencil
Material: .015-in.-thick wax-coated cardboard
Used as a stencil to spray paint identification on crates of milk.

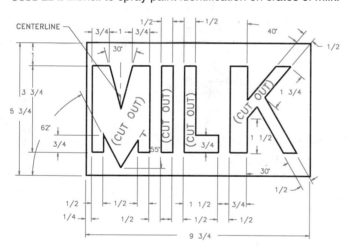

Problem 3–14 Straight line and arc object lines and centerlines (metric)

Part Name: Plate
Material: 10-mm-Thick HC–112
Used as a spacer to separate electronic components in a computer chasis.

Problem 3–12 Circle and arc object lines and centerlines (in.)

Part Name: Connector
Material: .25-in.-Thick Steel

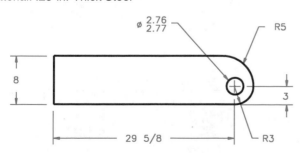

Problem 3–15 Circle and arc object lines and centerlines (in.)

Part Name: Bogie Lock
Material: .25-in.-Thick Mild Steel

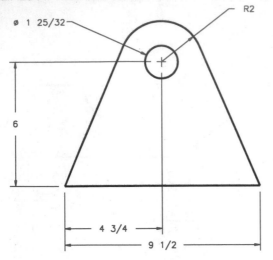

Problem 3–16 Circle, object lines, and centerlines (in.)

Part Name: Stove Back
Material: .25-in.-Thick Mild Steel

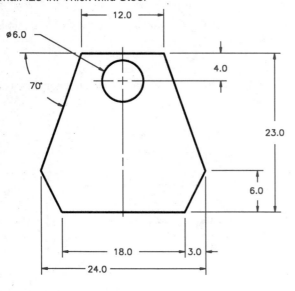

Problem 3–17 Circle and arc object lines and centerlines (in.)

Part Name: Bogie Lock
Material: .25-in.-Thick Mild Steel

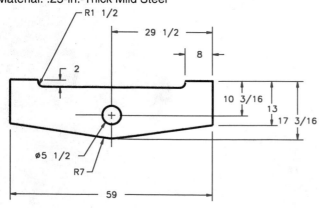

Problem 3–18 Circle and arc object lines and centerlines (metric)

Part Name: T-Slot Cleaner
Material: 6-mm-Thick Cold Rolled Steel

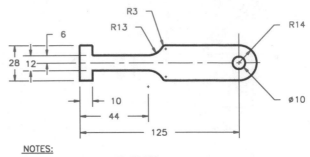

NOTES:
1. BREAK ALL SHARP EDGES.
2. STAMP NAME ON HANDLE.

Problem 3–19 Circle and arc object lines and centerlines (in.)

Part Name: T-Slot Cleaner (inch)
Material: .25-in.-Thick Cold Rolled Steel

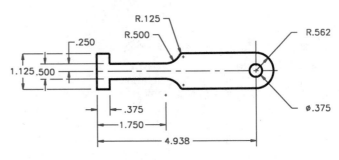

NOTES:
1. BREAK ALL SHARP EDGES.
2. STAMP NAME ON HANDLE.

Problem 3–20 Circle and arc object lines and centerlines (in.)

Part Name: Plate
Material: .125-in.-Thick Aluminum

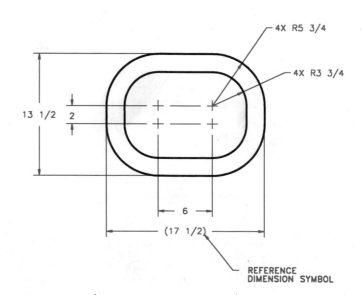

Problem 3–21 Circle and arc object lines and centerlines (in.)

Part Name: Gasket
Material: .062-in.-Thick Neoprene

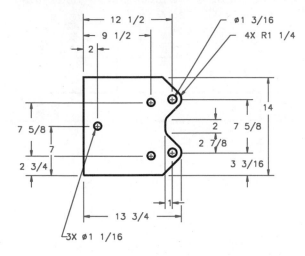

Problem 3–22 Circle and arc object lines and centerlines (in.)

Part Name: Gasket
Material: .062-in.-Thick Cork
Gasket for hydraulic pump.

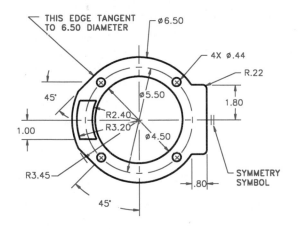

Problem 3–23 Arc object lines, centerlines, phantom lines, and leader lines (in.)

Part Name: Bogie Lock
Material: .25-in.-Thick Mild Steel
Connect the leader lines and place the notes on the drawing.

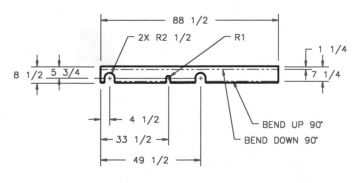

Problem 3–24 Object lines and hidden lines (in.)

Part Name: V-block
Material: 4.00-in.-Thick Mild Steel

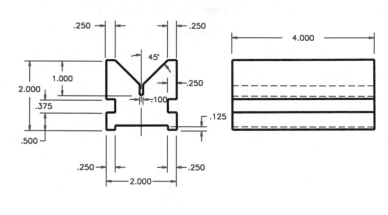

PART 4. DRAWING WITH THE COMPUTER PROBLEMS

DIRECTIONS

The problems in this section are representative of the skills required when using a computer graphics system. The use of the Cartesian coordinate system is stressed because of its universal use in computer-aided drafting. Problems 3–25, 3–26, and 3–27 test this skill of locating points.

Problem 3–25 You will need graph paper with 10 squares per inch for this problem. The 11 points given below each have an X and Y coordinate. Plot each of the points on the grid paper using Cartesian coordinates and connect them with a straight line. The origin of your coordinate system does not have to be the extreme lower left corner of the paper.

POINT	X	Y
1	0	1.5
2	3	1.5
3	3	0
4	7.12	0
5	7.12	2.25
6	6.18	2.25
7	6.18	.75
8	3.93	.75
9	3.93	2.25
10	0	2.25
11	0	1.5

Problem 3–26 The two objects given below (objects A and B) are similar to the previous problem, but each has two views, front and right side. Use 10-squares-to-the-inch graph paper and follow the same procedure you used in problem 3–25. The lower left corner of each view should be the origin point for that view. Each view should have its own separate Cartesian coordinate system.

		FRONT		RIGHT SIDE	
	POINT	**X**	**Y**	**X**	**Y**
OBJECT A	1	0	0	0	0
	2	3	0	2.75	0
	3	3	3.6	2.75	3.6
	4	0	3.6	0	3.6
	5	0	0	0	0
OBJECT B	1	0	0	0	0
	2	4	0	3.8	0
	3	4	2.1	3.8	4.2
	4	3	2.1	0	4.2
	5	3	4.2	0	2.1
	6	1	4.2	0	0
				Move to:	
	7	1	2.1	0	2.1
	8	0	2.1	3.8	2.1
	9	0	0		

Problem 3–27 The following problem contains drawing data given in both incremental and polar coordinates. The points should be plotted and drawn on 10-squares-to-the-inch grid paper.

A. INCREMENTAL COORDINATES			B. POLAR COORDINATES		
Point	**X**	**Y**	**Point**	**Angle**	**Radius**
1	0	0	1	0	0
2	3	0	2	0	1.25
3	0	2	3	90	1
4	–1	0	4	0	1.25
5	0	1	5	270	1
6	–1	0	6	0	1.50
7	0	–2	7	90	2
8	–1	0	8	180	2.75
9	0	–1	9	90	1
			10	180	1.75
			11	270	3

Problem 3–28 Arcs, circles, and centerlines (inch)

Part Name: Bracket
Material: Mild Steel (MS)

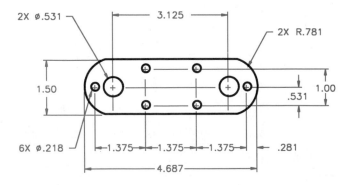

MATH PROBLEMS

For the illustration of the block shown, determine:

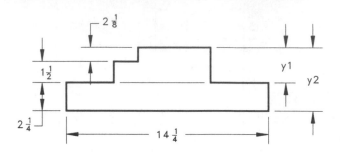

1. dimension y1
2. dimension y2
3. The overall width of 12 of the blocks laid side to side.
4. Convert each of the four given dimensions of the block to decimal fractions.
5. Twenty-five pieces of metal are needed for a job. Each piece is to be 3³⁄₃₂" in length. Disregarding the cutting loss, what length of stock is needed?
6. A piece of stock 25⅞" long is to be cut into five equal lengths. Allowing ¹⁄₁₆" lost per cut, what will be the length of each of the five pieces?
7. A drawing shows a dimension of 4.1875". Convert this decimal fraction to a common fraction.

Geometric Construction

After completing this chapter you will use manual or computer-aided drafting to:

○ Draw parallel and perpendicular lines.
○ Construct bisectors and divide lines and spaces into equal parts.
○ Draw polygons.
○ Draw tangencies.
○ Draw ellipses.
○ Solve an engineering problem by making a formal drawing with geometric constructions from an engineer's sketch or layout.

THE ENGINEERING DESIGN PROCESS

You should always approach an engineering drafting problem in a systematic manner. Plan out how you propose to solve the problem. As an entry-level drafter you need to do this planning with sketches and written notes. These sketches and notes help you decide:

○ The scale to use to effectively display the drawing. If the scale is too small, complex detail may be lost, but if the scale is too large, the drawing may take up too much space and cause you to spend too much time.
○ What paper size or CADD drawing limits should be established. This depends on the final plotted scale of the drawing, the amount of detail, and the specifications required.

After these decisions are made, you need to proceed with the problem in a logical manner. Refer to the engineer's sketch shown in Figure 4–1 as you follow these layout steps for either manual drafting or CADD:

Step 1. Do all preliminary manual work using construction lines so if you make an error, it is easy to erase. Begin by establishing the centers of the Ø57 and Ø25.5 circles and then draw the circles.

Step 2. Draw the concentric circles of Ø71.5 and Ø40.

Step 3. Locate and draw the 6×R7.5 arcs and then using tangencies blend the R3 radii with the outside arcs.

Step 4. Draw the 2×R7 arcs tangent to the large inside arcs.

Step 5. Draw the centerlines for the 6×Ø7 circles and then draw the circles in place.

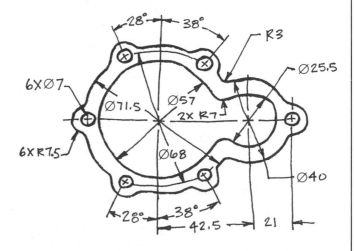

Figure 4–1 Engineer's sketch. **(Continued)**

ENGINEERING DESIGN PROCESS *(Continued)*

Step 6. Darken all thin lines and then darken all object lines. If there are arcs and straight lines, draw the arcs first and then blend the straight lines into the tangencies. The finished drawing is shown in Figure 4–2.

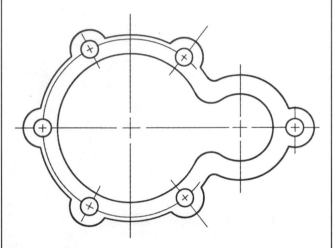

Figure 4–2 The completed drawing (without dimensions) for the engineer's sketch shown in Figure 4–1.

Machine parts and other manufactured products are made up of different geometric shapes ranging from squares and cylinders to complex irregular curves. Geometric constructions are methods that may be used to draw various geometric shapes or to perform drafting tasks related to the geometry of product representation and design. The proper use of geometric constructions requires a fundamental understanding of plane geometry. When using geometric construction techniques extreme accuracy and the proper use of drafting instruments are important. Pencil or compass leads should be sharp and instruments should be in good shape. Always use very lightly drawn construction lines for all preliminary work.

When computers are used in drafting, the task of creating most geometric construction related drawings becomes much easier, although the theory behind the layout of geometric shapes and related constructions remains the same.

▼ CHARACTERISTICS OF LINES

A straight line segment is a line of any given length, such as lines A-B shown in Figure 4–4.

A curved line may be in the form of an arc with a given center and radius or an irregular curve without a defined radius as shown in Figure 4–5.

CADD Applications

A LAYOUT APPROACH FOR CADD APPLICATIONS

Make a form like the example given in Figure 4–3 to list all of your drawing steps. Write the command in the left column. Choose from the following list of commands and provide the information indicated.

POINT—Cartesian coordinate location
SCALE—numerical size
ORIGIN—coordinates
CIRCLE—center, diameter
ARC—center, beginning and ending angles (degrees)
POLYGON—center, radius flat to flat (RF), or radius point to point (RP)
DASHED LINE—end points
TEXT (NOTES)—1) height, width, slant angle; 2) angle direction; 3) lower left or lower right coordinates.

Indicate the type of line in column 2, the coordinates in column 3, and additional parameters (diameter, radius, angles, etc.) in column 4.

SAMPLE FORM

Student Name_____ File Name_____
Scale_____ Date_____

① COMMAND	② LINE TYPE	③ X	Y	④ PARAMETERS
Point	Solid	0	0	
''	''	2	0	
''	''	2	1	
''	''	0	1	
Circle	Solid	1	.5	Dia. = .25
Arc	''	0	.5	R=.5, 90° - 180°

Figure 4–3 CADD planning form.

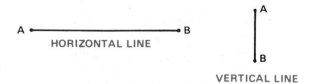

Figure 4–4 Horizontal and vertical lines.

Figure 4–5 Arc and irregular curve.

Two or more lines may intersect at a point as in Figure 4–6. The opposite angles of two intersecting lines are equal; a = a, b = b.

Parallel lines are lines equidistant throughout their length, and if they were to extend indefinitely they would never cross. (See Figure 4–7.)

Perpendicular lines intersect at a 90° angle as shown in Figure 4–8.

▼ GEOMETRIC SHAPES
Angles

Angles are formed by the intersection of two lines. Angles are sized in degrees (°). Components of a degree are minutes and seconds. There are 60 minutes (') in one degree and there are 60 seconds (") in one minute. $1° = 60'$, $1' = 60"$. The four basic types of angles are shown in Figure 4–9. A straight angle equals 180° while a right angle equals 90°. An acute angle contains more than 0° but less than 90° and an obtuse angle contains more than 90° but less than 180°.

The parts of an angle are shown in Figure 4–10. An angle is labeled by giving the letters defining the line

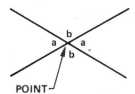

Figure 4–6 Intersecting lines.

Figure 4–7 Parallel lines.

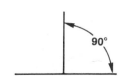

Figure 4–8 Perpendicular lines.

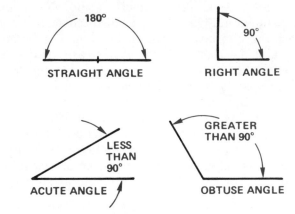

Figure 4–9 Types of angles.

ends, and vertex, with the vertex always between the ends, such as XYZ where Y is the vertex.

Triangles

A triangle is a geometric figure formed by three intersecting lines creating three angles. The sum of the interior angles always equals 180°. The parts of a triangle are shown in Figure 4–11. Triangles are labeled by lettering the vertex of each angle such as ABC, or by labeling the sides abc as shown in Figure 4–12.

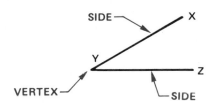

Figure 4–10 Parts of an angle.

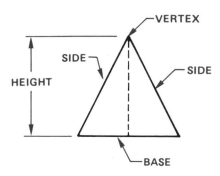

Figure 4–11 Parts of a triangle.

Figure 4–12 Labeling triangles.

There are three kinds of triangles: acute, obtuse, and right as shown in Figure 4–13. Two special acute triangles are the equilateral, which has equal sides and angles, and the isosceles, which has two equal sides and angles. An isosceles triangle may also be obtuse. In a scalene triangle, no sides or angles are equal.

Right triangles have certain unique geometric characteristics. Two internal angles equal 90° when added. The side opposite the 90° angle is called the hypotenuse, as seen in Figure 4–14. A semicircle (half circle) is always formed when an arc is drawn through the vertices of a right triangle as shown in Figure 4–15.

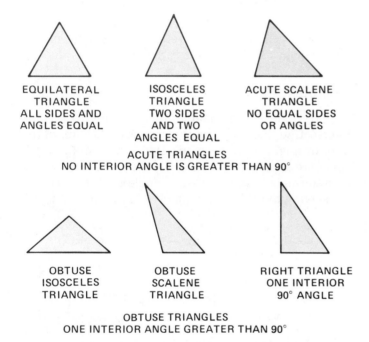

Figure 4–13 Types of triangles.

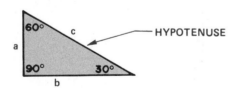

Figure 4–14 Right triangle.

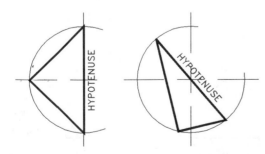

Figure 4–15 A semicircle is formed when an arc is drawn through the vertices of a right triangle.

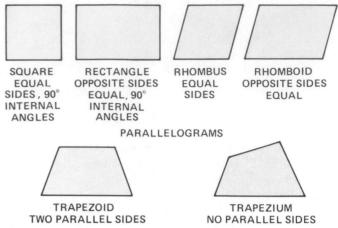

Figure 4–16 Quadrilaterals.

Quadrilaterals

Quadrilaterals are four-sided polygons that may have equal or unequal sides or interior angles. The sum of the interior angles is 360°. Quadrilaterals with parallel sides are called parallelograms. (See Figure 4–16.)

Regular Polygons

Some of the most commonly drawn geometric shapes are regular polygons. Regular polygons have equal sides and equal internal angles. Polygons are closed figures with any number of sides, but no less than three. The relationship of a circle to a regular polygon is that a circle may be drawn to touch the corners (circumscribed) or the sides, known as flats, (inscribed) of a regular polygon. (See Figure 4–17.) This relationship is an advantage in constructing regular polygons. Some common regular polygons are shown in Figure 4–18.

Regular Solids

Solid objects constructed of regular polygon surfaces are called regular polyhedrons. (See Figure 4–19.) A *polyhedron* is a solid formed by plane surfaces. The surfaces are referred to as *faces*.

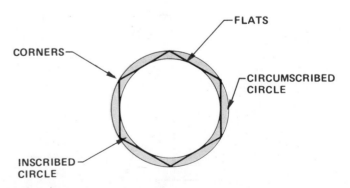

Figure 4–17 Regular polygon.

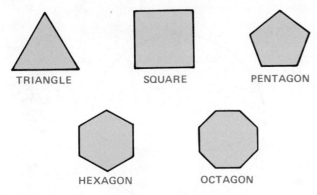

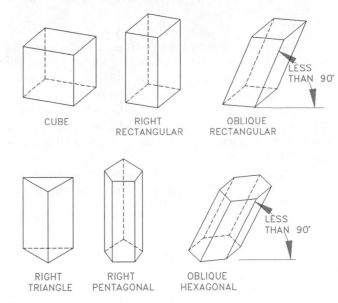

Figure 4–18   Common regular polygons.

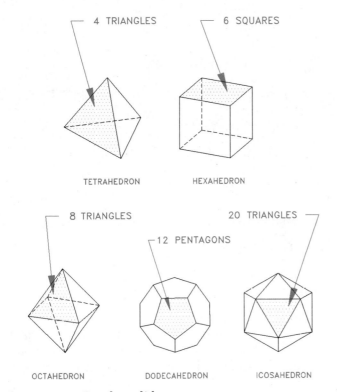

Figure 4–19   Regular solids.

Figure 4–20   Common prisms.

Prisms

A *prism* is a geometric solid object with ends that are the same size and shaped polygons and sides that connect the same corresponding corners of the ends. Figure 4–20 shows a few common examples. A *right prism* has sides that meet 90° with the ends. An *oblique prism* has sides that are at an angle to the ends.

Pyramids

A *pyramid* has a regular polygon-shaped base and sides that meet at one point called a *vertex* as shown in Figure 4–21. A geometric solid is *truncated* when a portion is removed and a plane surface is exposed. The *axis* is an imaginary line that connects the vertex to the midpoint of the base. It may also be referred to as an imaginary line around which parts are regularly arranged.

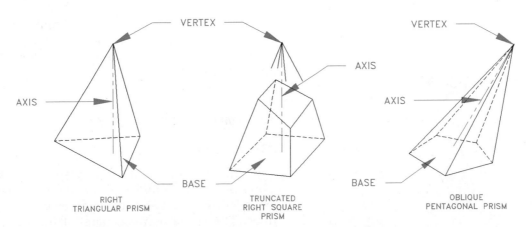

Figure 4–21   Common pyramids.

Circles

A circle is a closed curve with all points along the curve at an equal distance from a point called the center. The circle has a total of 360°. The circumference is the distance around the circle. The radius is the distance from the center to the circumference. The diameter is the distance across the circle through the center. Figure 4–22 shows the parts of a circle and circle relationships.

An arc is part of the circumference of a circle. The arc may be identified by a radius, an angle, or a length (Figure 4–23).

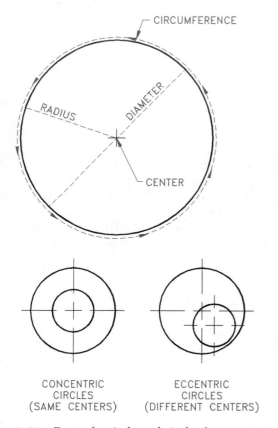

Figure 4–22   Parts of a circle and circle characteristics.

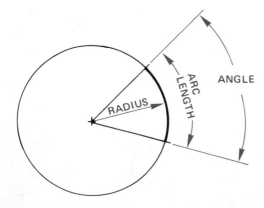

Figure 4–23   Arc.

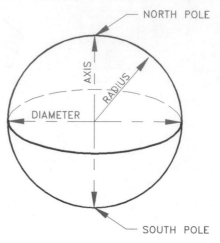

Figure 4–24   Sphere.

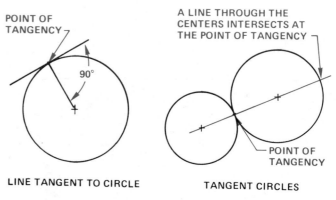

Figure 4–25   Tangency.

Sphere

A sphere is the shape of a ball. Every point on the surface of a sphere is equidistant from the center. If you think of the earth as a sphere, the North and South poles are at the end of its axis as shown in Figure 4–24.

Tangents

Straight or curved lines are tangent to a circle or arc when the line touches the circle or arc at only one point. If a line were to connect the center of the circle or arc to the point of tangency, the tangent line and the line from the center would form a 90° angle. (See Figure 4–25.)

▼ COMMON GEOMETRIC CONSTRUCTIONS
Parallel Lines

Parallel lines are lines evenly spaced at all points along their length and will not intersect even when extended. Figure 4–26 shows an example of parallel lines. The space between parallel lines may be any distance.

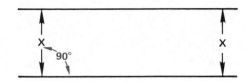

Figure 4–26   Parallel lines.

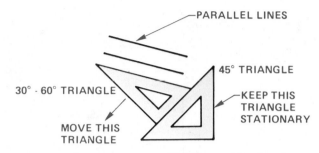

Figure 4–27   Constructing parallel lines with triangles.

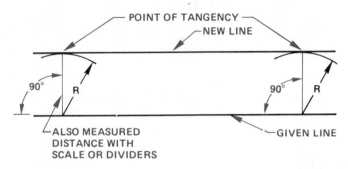

Figure 4–28   Constructing parallel lines with a straightedge and compass.

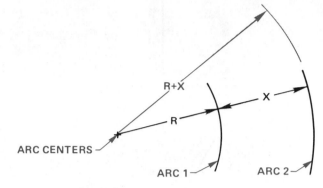

Figure 4–29   Parallel arcs.

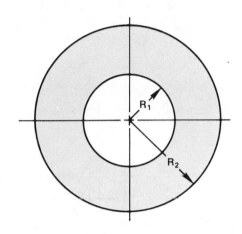

Figure 4–30   Concentric circles.

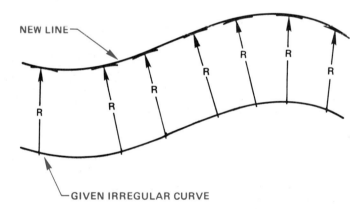

Figure 4–31   Constructing parallel irregular curves.

Parallel lines may be drawn horizontally, vertically, or at any angle using a drafting machine. When a drafting machine is not available, parallel lines may be drawn with the aid of standard triangles as shown in Figure 4–27. Parallel lines may be drawn with a straightedge and compass when the distance between lines is established. Use the compass radius to draw arcs near the ends of the given line. The parallel line is then drawn at the points of tangency of the arcs as shown in Figure 4–28. Draw very light construction lines for all construction layout.

Parallel or concentric arcs may be drawn where the distance between arcs is equal. Establish parallel arcs by adding the radius of Arc 1 (R_1) to the distance between the arcs (X), as shown in Figure 4–29.

Concentric circles are parallel circles drawn from the same center. The distance between circles is equal, as shown in Figure 4–30.

Parallel irregular curves are drawn any given distance (R) apart by using a compass set at the given distance. A series of arcs is lightly drawn all along the given curve at the required distance. The line to be drawn parallel is then constructed with an irregular curve at the points of tangency of the construction arcs. (See Figure 4–31.)

Perpendicular Lines

Perpendicular lines intersect at 90°, or a right angle, as seen in Figure 4–32. Perpendicular lines are drawn with the horizontal and vertical scales of a drafting machine, even if the scales are set at an angle away from horizontal, as seen in Figure 4–33. Perpendicular lines can also be drawn with a straightedge and triangle or two triangles as shown in Figure 4–34.

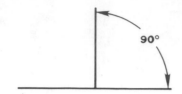

Figure 4–32 Perpendicular lines.

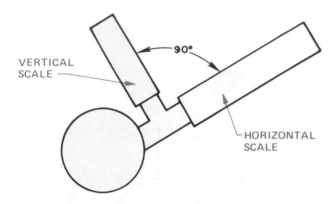

Figure 4–33 Perpendicular drafting machine scales.

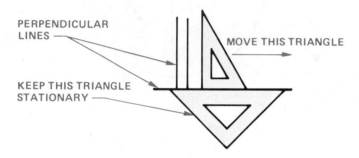

Figure 4–34 Perpendicular lines with triangles.

Perpendicular Bisector

The perpendicular bisector of a line may be obtained using a straightedge and a compass.

Step 1. Set the compass radius more than halfway across the given line segment and draw two intersecting arcs from the line ends.

Step 2. Connect the points where the arcs intersect, as in Figure 4–35.

Bisecting an Angle

Any given angle may be divided into two equal angles by using a compass and straightedge.

Step 1. Adjust the compass to any radius and draw an arc as shown in Figure 4–36.

Step 2. At the points where the arc intersects the sides of the angle, draw two arcs of equal radius.

Step 3. Connect a straight line from the vertex of the angle to the point of intersection of the two arcs.

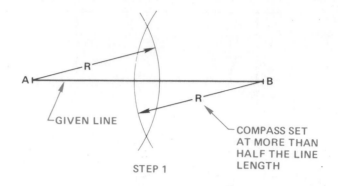

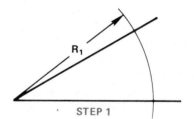

Figure 4–35 Constructing a perpendicular bisector.

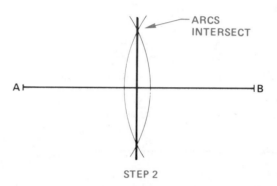

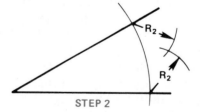

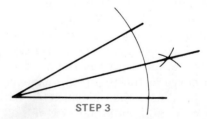

Figure 4–36 Bisecting an angle.

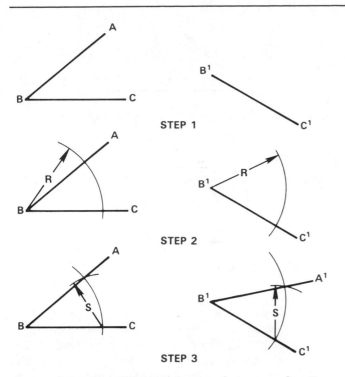

STEP 1

STEP 2

STEP 3

Figure 4–37 Transferring a given angle to a new location.

Transferring an Angle

A given angle may be transferred to a new location as shown in Figure 4–37. This method also works well for transferring a triangle.

Step 1. Given angle ABC, draw line B¹C¹ in the new location and position.

Step 2. Draw any arc, R, with its center at B on the given angle. Transfer and draw arc R with its center at point B¹.

Step 3. Using the intersection of one side of the given angle and arc R as center, set a compass distance to the opposite intersection of arc R. Then transfer the new radius, S, to the new position using the intersection of B¹C¹ and arc R as center. Connect a line from B¹ to the intersection of the two arcs, R and S, as shown in Figure 4–37.

Dividing a Line Into Equal Parts

Any given line can be divided into any number of equal parts. Divide given line AB, shown in Figure 4–38, into eight equal parts.

Step 1. Draw a construction line at any angle from either end of line AB as shown in Figure 4–39.

A ├────────────────────┤ B

Figure 4–38 Given line AB to divide into equal parts.

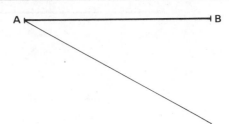

Figure 4–39 Step 1, angled construction line.

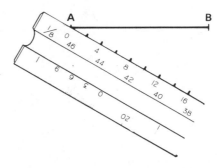

Figure 4–40 Step 2, divide angled line into required number of equal parts.

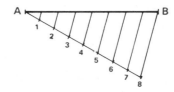

Figure 4–41 Step 3, connect parallel line segments.

Step 2. Use a scale or dividers to divide the angled construction line into eight equal parts. Use any size division that will extend down the angled line to approximately the length of the given line. (See Figure 4–40.)

Step 3. Connect the last point (8) of the construction line to point B of the given line. Then draw lines parallel to line 8B from each of the numbered points on the construction line to the given line AB. (See Figure 4–41.) You now have the given line AB, divided into eight equal parts. This same process may be used for any number of equal parts.

Dividing a Space Into Equal Parts

You can divide any given space into any number of equal parts. Given the space shown in Figure 4–42, divide it into 12 equal parts.

Step 1. Between the lines that establish the given space, place a scale with the required number of increments (12) so that the zero on the scale is on either line and the 12 is on the other line. Mark each increment. (See Figure 4–43.)

Figure 4–42   Dividing a given space into equal parts.

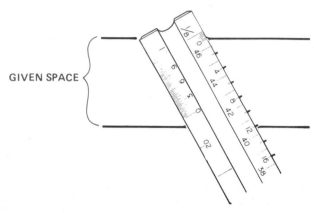

Figure 4–43   Step 1, place a scale with required number of increments between the lines.

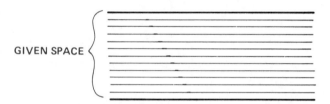

Figure 4–44   Step 2, given space divided into equal parts.

Step 2.   Remove the scale. Each increment you have marked has divided the given space into 12 equal spaces. Example 4–44 shows parallel lines drawn at each mark to complete the space division.

▼ CONSTRUCTING POLYGONS

Triangles

The following technique is known as triangulation. Using given triangle sides x, y, and z, in Figure 4–45, draw a triangle as shown in the following steps:

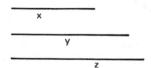

Figure 4–45   Construct a triangle given three sides.

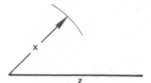

Figure 4–46   Step 1, layout line Z and swing arc X.

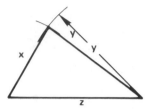

Figure 4–47   Step 2, swing arc Y.

Step 1.   Lay out one of the given sides. Select the side that will be the base: z has been chosen as shown in Figure 4–46. From one end of line z strike an arc equal in length to one of the other lines, x, for example, as in Figure 4–46.

Step 2.   From the other end of line z strike an arc with a radius equal to the remaining line, y. Allow this arc to intersect the previous arc. Where the two arcs cross, draw a line to the end of the base line, as in Figure 4–47 to complete the triangle.

Any given polygon may be constructed or transferred to a new location using this triangulation method. Given the triangle, abc, in Figure 4–48, transfer it to a new location as shown in the following steps:

Step 1.   Transfer one of the sides of the triangle to the new location by measuring its length or using dividers. (See Figure 4–49.)

Step 2.   From one end of line a in the new location, draw an arc equal in length to line b. From the other end of line a, draw an intersecting arc equal in length to line c. (See Figure 4–50a.)

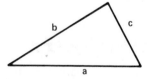

Figure 4–48   Transfer a given triangle to a new location.

Figure 4–49   Step 1, establish new location.

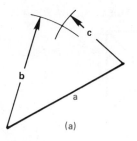

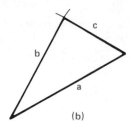

Figure 4–50 **(a)** Step 2, swing arcs b and c. **(b)** Step 3, connect the ends of line a to intersection of arcs.

Step 3. Connect both ends of line a to the intersection of the arcs to form the triangle in its new position as shown in Figure 4–50b.

A right angle may be drawn when the length of the two sides adjacent to the 90° angle are given as shown in Figure 4–51.

Step 1. Draw line a perpendicular to line b. (See Figure 4–52.)

Step 2. Connect a line from the end of line a to the end of line b to establish the right triangle. (See Figure 4–53.)

Draw a right triangle given the length of the hypotenuse and one side as in Figure 4–54.

Step 1. Draw line c and establish its center. You can use the perpendicular bisector method to find the center. (See Figure 4–55.)

Step 2. From the center of line c, draw a 180° arc with the compass point at the line center and the compass lead at the line end. (See Figure 4–56.)

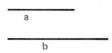

Figure 4–51 Construct a right triangle given two sides.

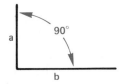

Figure 4–52 Step 1, draw line a perpendicular to line b.

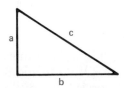

Figure 4–53 Step 2, connect hypotenuse.

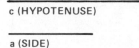

Figure 4–54 Construct a right triangle given a side and hypotenuse.

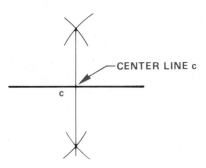

Figure 4–55 Step 1, draw the hypotenuse and establish its center.

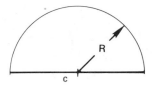

Figure 4–56 Step 2, draw a 180° arc from the center.

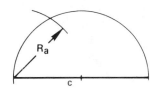

Figure 4–57 Step 3, use radius Ra to establish the end of line a.

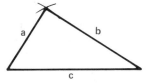

Figure 4–58 Step 4, complete the right triangle.

Step 3. Set the compass with a radius equal in length to the other given line, a. From one end of line c draw an arc intersecting the previous arc as shown in Figure 4–57.

Step 4. From the intersection of the two arcs in step 3, draw two lines connecting the ends of line c to complete the required right triangle. (See Figure 4–58.)

An equilateral triangle may be drawn, given the length of one side, as demonstrated in the following steps.

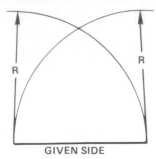

Figure 4–59 Step 1, to construct an equilateral triangle given one side, swing equal arcs r.

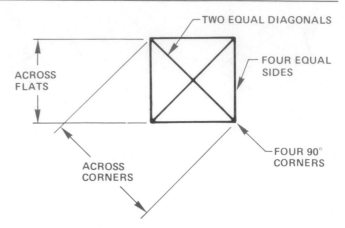

Figure 4–61 Elements of a square.

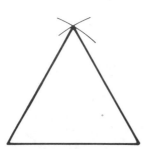

Figure 4–60 Step 2, complete the equilateral triangle.

Step 1. Draw the given length. Then set a compass with a radius equal to the length of the given line. Use this compass setting to draw intersecting lines from the ends of the given line as in Figure 4–59.

Step 2. Connect the point of intersection of the two arcs to the ends of the given side to complete the equilateral triangle as shown in Figure 4–60.

Squares

Square-head bolts or square nuts are often drawn in conjunction with manufactured parts. Use geometric constructions when necessary; however, when practical use square templates or special square-head bolt and nut templates. There are several methods that may be used to draw a square, each of which relates to the characteristics of the square shown in Figure 4–61.

Draw a square with the length of one side given as follows:

Step 1. Draw the length of the given side. Then with the drafting machine, or other appropriate method of obtaining a 90° angle, draw lines from each end equal in length to the given side. (See Figure 4–62.)

Step 2. Connect the end of sides 1 and 2 drawn in step 1 to complete the square with side 3, shown in Figure 4–63.

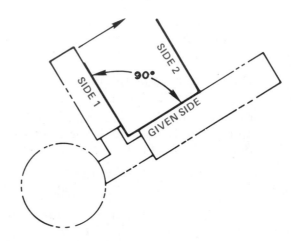

Figure 4–62 Step 1, use the drafting machine to establish 90° angles and sides.

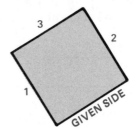

Figure 4–63 Step 2, complete the square by drawing side 3.

Draw a square with the distance across the flats or across the corners given.

Step 1. Draw a circle equal in diameter to the distance across the flats, Figure 4–64a, and a circle with a diameter equal to the distance across the corners, Figure 4–64b. Notice that the position of the centers dictates the position of the squares.

Step 2. Draw 45° lines tangent to the circle and extending to the centerlines as seen in Figure 4–65a. Draw lines inside the circle connecting the intersections of the centerlines and the circle as seen in Figure 4–65b.

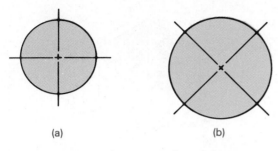

Figure 4–64 Step 1, draw a circle to construct a square with flats or corners given.

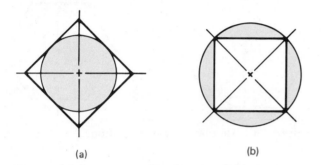

Figure 4–65 Step 2, a circumscribed square at a and an inscribed square at b.

Regular Hexagon

Hexagons are six-sided polygons. Each side of a regular hexagon is equal and its interior angles are 120° as shown in Figure 4–66. Hexagons are commonly used as the shape for the heads of bolts and for nuts. Generally the dimension given for the size of a hexagon is the distance across the flats. The distances across the flats and across the corners are shown in Figure 4–66.

There are two easy geometric construction methods to draw a hexagon. One gives the distance across the flats; the other gives the distance across the corners. When practical use a hexagon template or a special hex-head bolt, screw, or hex nut template.

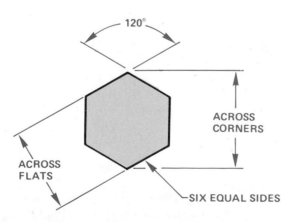

Figure 4–66 Elements of a regular hexagon.

Across the Flats. Given the distance across the flats, do the following to construct a hexagon:

Step 1. Lightly draw a circle with the given distance as the diameter. (See Figure 4–67.)

Step 2. Set your drafting machine at 30° from horizontal or use a triangle. Then lightly draw lines 1 and 2 tangent to the circle as seen in Figure 4–67.

Step 3. Now set the angle at 30° the other way from horizontal and draw lines 3 and 4 tangent to the circle as shown in Figure 4–68.

Step 4. Draw lines 5 and 6 vertical and tangent to the circle and darken the six lines to complete the required hexagon as in Figure 4–69.

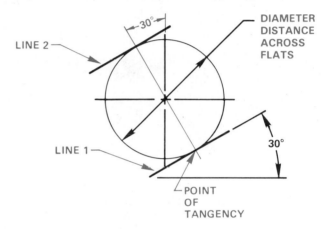

Figure 4–67 Steps 1 and 2, constructing a hexagon.

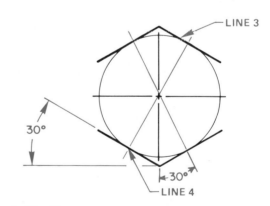

Figure 4–68 Step 3.

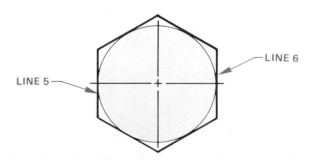

Figure 4–69 Step 4, complete the hexagon.

Across the Corners. Given the distance across the corners of a hexagon, use the following construction steps:

Step 1. Lightly draw centerlines in the desired location of the hexagon center. Then lightly draw a circle with the diameter equal to the given distance across the corners. Use construction lines as shown in Figure 4–70.

Step 2. From the location where the vertical centerline touches the circle, draw two lines 30° from horizontal, inside the circle as shown in Figure 4–71.

Step 3. Draw lines 3 and 4 at 30° from horizontal in the other direction, as seen in Figure 4–72.

Step 4. Draw vertical lines 5 and 6 and darken the object lines to create the required hexagon shown in Figure 4–73.

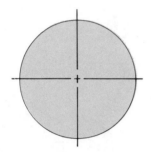

Figure 4–70 Step 1, construct a hexagon given the distance across the corners.

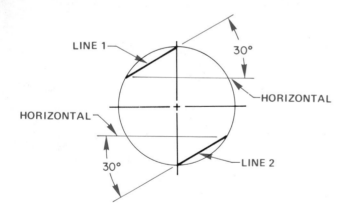

Figure 4–71 Step 2.

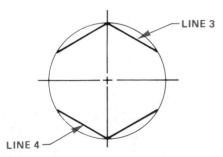

Figure 4–72 Step 3.

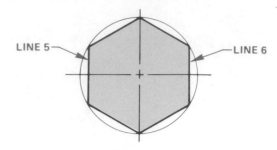

Figure 4–73 Step 4, complete the hexagon.

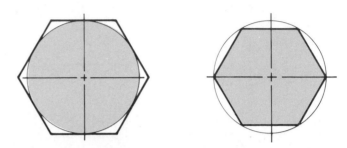

Figure 4–74 Alternate hexagon positions.

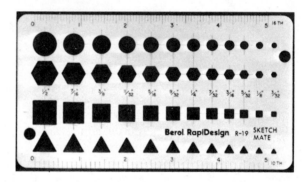

Figure 4–75 General purpose template. *Courtesy Berol Corporation.*

The position of a hexagon may be rotated by establishing the six sides in an alternate relationship to the circle centerlines, as shown in Figure 4–74. Be sure that the included angles are 120° and each side is equal in length. The template shown in Figure 4–75 is a handy aid in drawing a variety of shapes, including the hexagon.

Regular Octagon

The octagon may be drawn using the same principle applied to the construction of a hexagon. A regular octagon has eight equal sides. If you use a 45° line as shown in Figure 4–76 you can easily draw an octagon about the given circle with a diameter the distance across the flats.

Polygons

A polygon is any closed plane geometric figure with three or more sides or angles. Hexagons and octagons

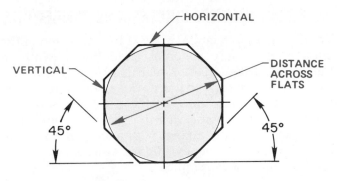

Figure 4–76   Elements of an octagon.

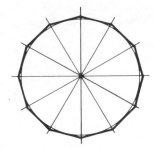

Figure 4–77   Step 1, divide the circle into 12 parts; Step 2, connect the 12 radial lines.

are polygons. You can draw any polygon (if you know the distance across the flats and the number of sides) by the same method used to draw the hexagon and octagon across the flats.

There are 360° in a circle; so, if you need to draw a 12-sided polygon divide 360° by 12 (360° ÷ 12 = 30°) to determine the central angle of each side. Since there are 30° between each side of a 12-sided polygon, divide a circle with a diameter equal to the distance across the flats into equal 30° parts as shown in Figure 4–77.

Now, connect the 12 radial lines with line segments that are tangent to each arc segment, as shown in Figure 4–77.

CADD Applications

DRAWING GEOMETRIC SHAPES

The process used for drawing geometric shapes involves three steps:

1. select the command
2. locate the feature
3. input the parameters, that is, draw the feature

Circle. A circle, for example, may be drawn several different ways. After selecting the CIRCLE command, the information that you have about the circle will determine the method by which you draw it. The methods are shown in Figure 4–78, (a) through (d). The first method (a) requires that two points, the center (1) and a point on the circle (2), be digitized before the circle can be drawn. The second, example (b), requires that two opposite points (1 and 2) on the circumference (the diameter) be digitized. The third, example (c), requires that three points be digitized on the circumference. The final method (d) requires that the center point be digitized and the radius be entered as a number selected from the keypad on the menu tablet, or typed from the keyboard.

Ellipse. Ellipses are quick and easy to draw on a CADD system. Most programs have a command such as ELLIPSE. In this command you are asked for the

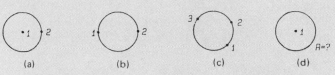

Figure 4–78   Four ways to digitize a circle.

major and minor diameters and the computer automatically draws an ellipse with the information.

Arc. Arcs are constructed in the same manner as circles. But arcs are only portions of circles and therefore the two end points of the arc must be digitized. The methods of constructing an arc are shown in Figure 4–79, (a) through (d). The fourth method (d) has two variations, each requiring numerical input. The first variation requires point number 1 (beginning point) to be digitized, as well as point number 2, a location on the arc. Then a numerical value for the radius and length of the arc must be entered from the menu keypad or from the keyboard.

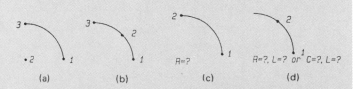

Figure 4–79   Four ways to digitize an arc.

(Continued)

CADD APPLICATIONS (Continued)

The second variation requires the same two points, but instead of entering the radius, the center point location of the arc is entered at the keypad.

Tangencies. Tangencies are easy to draw with the computer. Most CADD programs have a command or option such as TANGENT, which allows you to pick a circle or arc and automatically draw a line or another circle or arc tangent to the selected object.

Polygon. A polygon can be drawn using the same methods as the CIRCLE command, but some additional information is needed. Since polygons have corners, or vertices, the computer needs to know where the first vertex should be located. A prompt asking for the location will appear on the screen and the answer should be in degrees. Figure 4–80 shows two examples of polygons that have been given different starting angles. The final bit of information the computer needs is the number of sides. After this number is entered from the menu or keyboard, the polygon is drawn.

Rectangle. A rectangle is one of the easiest shapes to draw. Only two points are needed. The first is one corner of the rectangle, which may be any of the four corners. Once you digitize this corner, the direction location of the second point has already been determined. The second point is the corner of the rectangle opposite the first corner that was digitized. The computer then connects these two points with the straight lines needed to form the rectangle. Figure 4–81 illustrates this process. Of course rectangles and boxes can be drawn using the LINE command, but it takes longer.

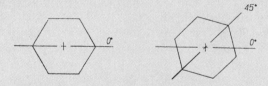

Figure 4–80 Polygon starting angles.

Figure 4–81 Rectangle construction using opposite corners.

▼ CONSTRUCTING CIRCLES AND TANGENCIES

To be tangent to a circle or arc, a line must touch the circle or arc at only one point, and a line drawn from the center of the circle or arc must be perpendicular to the tangent line at the point of tangency. (See Figure 4–82.) Lines drawn tangent to circles or arcs are common in engineering drafting. It is important that tangent lines be drawn with a smooth transition at the point of tangency. Figure 4–83 shows some common tangency examples.

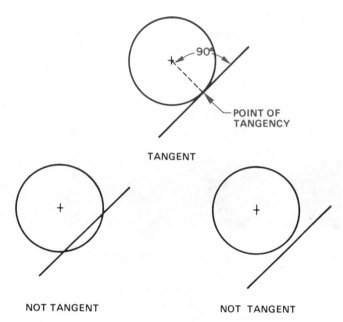

Figure 4–82 Tangency.

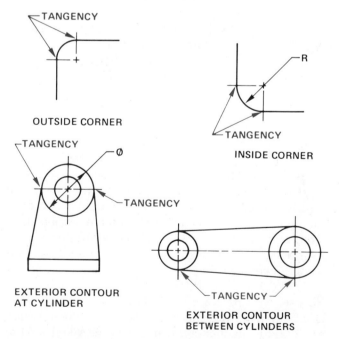

Figure 4–83 Common tangencies.

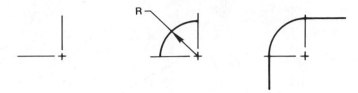

Figure 4–84   Drawing lines tangent to an arc and an arc at a 90° corner.

GOOD   POOR

Figure 4–85   Good and poor tangency examples.

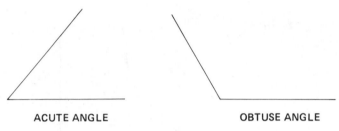

ACUTE ANGLE   OBTUSE ANGLE

Figure 4–86   Construct an arc tangent to given angles.

When drafting tangent features it is usually best to lightly draw the centerlines, then draw the arcs or circles, and finally draw the tangent lines. It is easier to draw a line tangent to a circle or arc than it is to draw a circle or arc tangent to a line. (See Figure 4–84.) When possible, use a template to draw the arcs and circles. Keep in mind that template circles are calibrated in diameters while arcs have radius dimensions. So, for example, a .50-in. radius requires using a 1.00-in. diameter circle. The combination of templates and a drafting machine provides the easiest and fastest results; use a compass only if necessary. Figure 4–85 shows the relationship between a well-drawn and a poorly drawn tangency.

Draw an arc tangent to a given acute or obtuse angle in much the same way as previously shown for lines and circles. Given the acute and obtuse angles in Figure 4–86, draw an arc with a 12 mm radius tangent to the sides of each angle.

Step 1.   Select a circle template with a 24 mm diameter circle. Move the template into position at each angle, leaving a small space for the lead or pen point thickness, and draw the arc as shown in Figure 4–87.

Step 2.   Remove the template and draw the sides of each angle to the points of tangency of the arcs. (See Figure 4–88.)

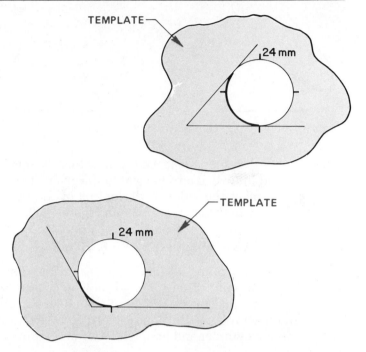

Figure 4–87   Step 1, draw a tangent arc with a template.

Figure 4–88   Step 2, completed tangent arcs.

If the arc center is required for dimensioning purposes, use the following procedure:

Step 1.   Establish the arc centers by lightly drawing lines parallel to the given sides at a distance from the sides equal to the radius of the arc. In this example the radius is 12 mm. (See Figure 4–89.)

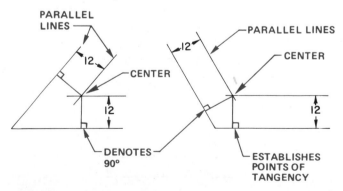

Figure 4–89   Step 1, drawing tangent arcs with a compass. This procedure also works for arcs at 90° corners.

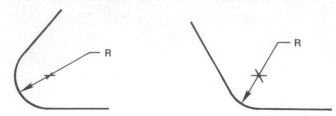

Figure 4–90   Step 2, completed tangent arcs.

Step 2.   Set the compass at the given radius, 12 mm, and draw two arcs tangent to the given sides. Connect the sides to the points of the tangency of the arcs as shown in Figure 4–90.

Arc Tangent to a Given Line and a Circle

The important point to remember is that the distance from the point of tangency to the center of the tangent arc is equal to the radius of the tangent arc. Given the machine part in Figure 4–91, draw an arc tangent to the cylinder and base with a 24 mm radius.

Step 1.   Draw a line parallel to the base at a distance equal to the radius of the arc, 24 mm. Then draw an arc that intersects the line drawn with a radius equal to the given arc plus the radius of the cylinder (24 + 11 = 35 mm). (See Figure 4–92.)

Step 2.   Use a compass or a template to draw the required radius from the center point established in step 1. (See Figure 4–93.)

The previous example demonstrated an internal arc tangent to a circle or arc and a line. A similar procedure is used for an external arc tangent to a given circle and line as shown in Figure 4–94. In this case the radius of the cylinder (circle) is subtracted from the arc radius to obtain the center point.

The following examples, Figures 4–95 through 4–97, are situations that commonly occur on machine parts where an arc may be tangent to given cylindrical or arc shapes. Keep in mind that the methods used to draw the tangents are the same as just described. The key is

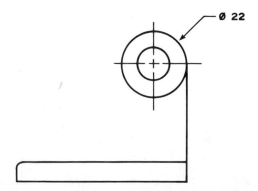

Figure 4–91   Construct an arc tangent to a given line and circle.

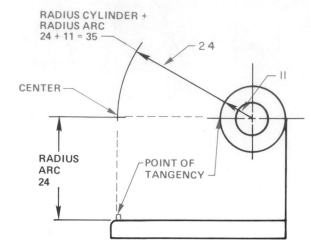

Figure 4–92   Step 1.

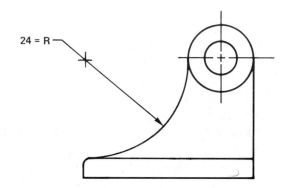

Figure 4–93   Step 2.

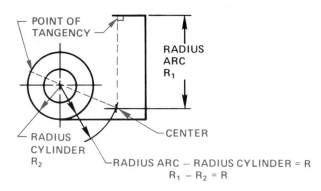

STEP 1

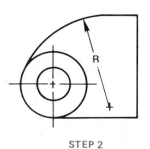

STEP 2

Figure 4–94   Constructing an external arc tangent to a line and cylinder.

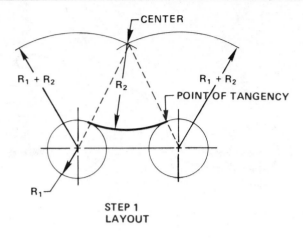

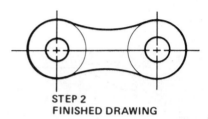

Figure 4–95   Chain link.

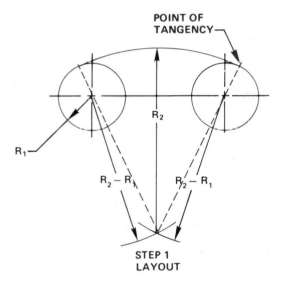

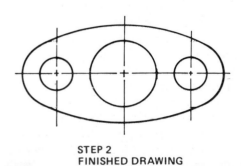

Figure 4–96   Gasket.

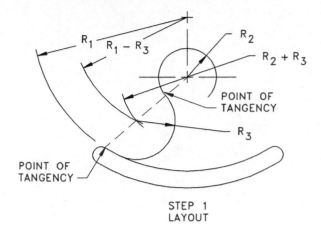

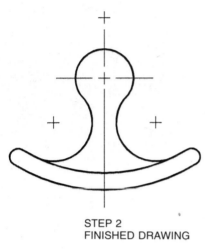

Figure 4–97   Hammer head.

that the center of the required arc is always placed at a distance equal to its radius from the points of tangencies. To achieve this means that you will either have to add or subtract the given radius and the required radius, depending upon the situation. Use a template to draw circles and arcs when possible.

Ogee Curve

An S curve, commonly called an *ogee* curve, occurs in situations where a smooth contour is needed between two offset features. The following steps show how to obtain an S curve with equal radii.

Step 1. Given the offset points A and B in Figure 4–98, draw a line between points A and B and bisect that line to find point C.

Step 2. Draw the perpendicular bisector of lines AC and CB. From point A draw a line perpendicular to the lower line that will intersect the AC bisector at X. From point B draw a line perpendicular to the upper line that will intersect the CB bisector at Y. (See Figure 4–99.)

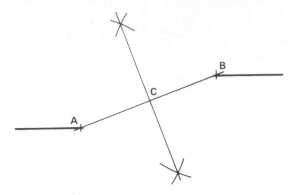

Figure 4–98   Step 1, constructing an S curve.

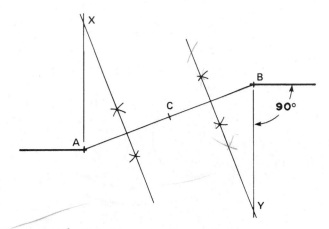

Figure 4–99   Step 2, constructing an S curve.

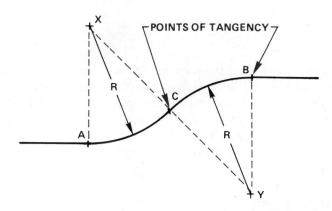

Figure 4–100   Step 3, constructing an S curve.

Step 3.   With X and Y as the centers, draw a radius from A to C and a radius from B to C as shown in Figure 4–100.

Step 4.   If the S curve line, AB, represents the centerline of the part, then develop the width of the part parallel to the centerline using concentric arcs to complete the drawing. (See Figure 4–101.)

When an S curve has unequal radii the procedure is similar to the previous example, as shown in Figure 4–102.

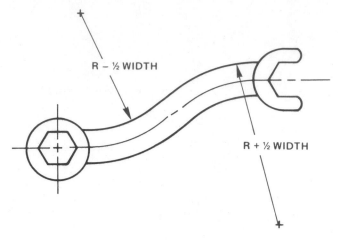

Figure 4–101   Step 4, a wrench with a complete S curve.

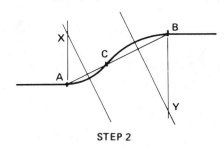

STEP 1

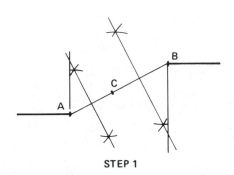

STEP 2

Figure 4–102   Constructing an S curve with unequal radii.

▼ CONSTRUCTING AN ELLIPSE

When a circle is viewed at an angle, an ellipse is observed. The angular relationship of an ellipse to a circle is shown in Figure 4–103. If a surface with a through hole is inclined 45°, the representation would be a 45° ellipse, as shown in Figure 4–104. An ellipse has a major diameter and a minor diameter. In Figure 4–104 the minor diameter is established by projecting the top and bottom edge from the slanted surface. The major diameter, in this case, is equal to the diameter of the circle.

If possible, always use a template to draw elliptical shapes. Ellipse templates are available from 10° to 85°. The ellipse shown in Figure 4–104 could have been drawn by projecting the centerlines and then lining up the center marks of a 45° ellipse template. (See Figure 4–105.)

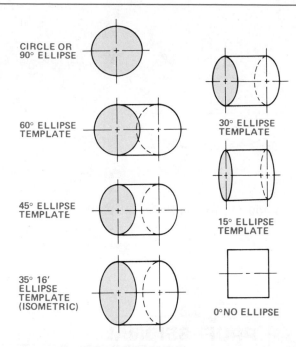

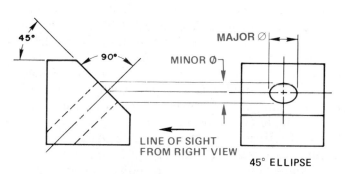

Figure 4–103 Ellipses are established by their relationship to a circle turned at various angles.

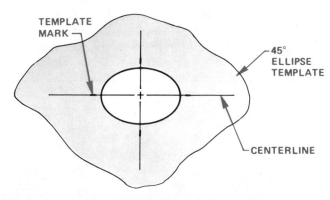

Figure 4–104 Elliptical view.

Figure 4–105 Drawing an ellipse with a template.

CADD Applications

Drawing objects with tangencies using manual drafting techniques is often tricky. The feature must touch at exactly the point of tangency. While the methods discussed in this chapter allow you to do this with consistent success, the use of CADD systems makes this task automatic. For example, AutoCAD has a TTR (Tangent, Tangent, Radius) option in the CIRCLE command. This allows you to automatically draw a circle tangent to lines, circles, or arcs simply by picking the objects and specifying the desired radius of the circle as shown in Figure 4–106.

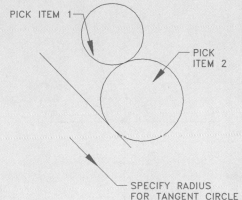

Figure 4–106 Automatically drawing tangencies with CADD.

Approximating the Ellipse

If the elliptical shape does not exactly fit the available template calibrations, then use one that is close. How close depends on your company standards. Most companies would rather have the drafters draw a close representation than take the time to lay out an ellipse geometrically.

Using the Concentric Circle Construction Method. When templates are not available or are not close enough for the particular ellipse that must be drawn, several methods can be used to draw elliptical shapes. One of the most practical is the concentric circle method described in the following steps:

Step 1.   Use construction lines to draw two concentric circles, one with a diameter equal to the minor ellipse diameter and the other with a diameter equal to the major ellipse diameter. (See Figure 4–107.)

Step 2.   Use the drafting machine or triangles to divide the circles into at least 12 equal parts; 30° and

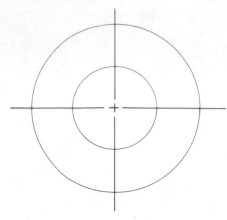

Figure 4–107   Step 1, constructing an ellipse, concentric circle method.

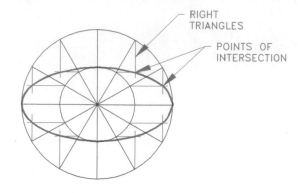

Figure 4–109   Steps 3 and 4, complete ellipse.

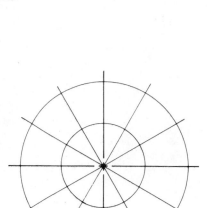

Figure 4–108   Step 2, concentric circle method.

60° each way from horizontal. More parts give better accuracy, but also consume more time. (See Figure 4–108.)

Step 3.   From the points of intersection of each line at the circumference of the circles, draw horizontal lines from the minor diameter and vertical intersecting lines from the major diameter, essentially creating a series of right triangles, although the intersection points only are needed. (See Figure 4–109.)

Step 4.   Using an irregular curve carefully connect the points created in Step 3 to complete the ellipse as shown in Figure 4–109.

PROFESSIONAL PERSPECTIVE

Geometric constructions are found in nearly every engineering drafting assignment. Some geometric constructions are as simple as a radius corner while others are very complex. As a professional drafter, you should be able to quickly identify the type of geometric construction involved in the problem and solve it with one of the techniques you have learned. When you solve these problems, accuracy and carefully drawn construction lines are critical. Use the best tools available. For example, a professional drafter hardly ever draws an ellipse using construction methods; he or she uses the large variety of ellipse templates available. Always use a template if you can, although, if all else fails, you do know how to construct the ellipse. Even though the geometric construction methods are presented for the manual drafter, the principles are the same for the CADD drafter. You should always refer to these principles and techniques to insure that your CADD application has performed the task correctly. You will find with CADD that some of the very time-consuming constructions, such as dividing a space into equal parts or drawing a pentagon, are not only extremely fast but they are almost perfectly accurate.

MATH APPLICATION

INTERIOR ANGLES OF REGULAR POLYGONS

For a regular polygon with n sides, the formula for the interior angles is $\Theta = \left(\frac{n-2}{n}\right)180°$. (See Figure 4–110.)

The Greek letter "theta" (Θ) is commonly used in geometry to stand for the measurement of an angle. As an example of using this formula, the interior angle of the pictured hexagon is:

$$\Theta = \left(\frac{6-2}{6}\right)180° = \left(\frac{4}{6}\right)180° = 120°$$

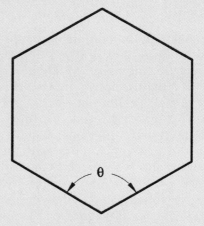

Figure 4–110 Regular hexagon.

GEOMETRIC CONSTRUCTION TEST

DIRECTIONS

Answer the questions with short complete statements or drawings as needed.

QUESTIONS

1. What is known to be true about the opposite angles created by intersecting lines?
2. Define *parallel lines*.
3. Define *perpendicular*.
4. How many degrees are there in a circle?
5. How are angles sized?
6. Name four basic types of angles.
7. Define *triangle*.
8. Show how an angle is named.
9. Name and show an example of three types of triangles.
10. Show two methods of naming triangles.
11. Name four types of quadrilaterals.
12. What name is given to the distance from the center to the circumference of a circle?
13. What is the name of the distance that goes across a circle through its center?
14. What is a circle called that touches the flats of a polygon?
15. What is a circle called that touches the corners of a polygon?
16. Show and label an example of concentric and eccentric circles.
17. Show an example of a line tangent to a circle and two circles tangent to each other.
18. When should templates be used to make circles and arcs?
19. When should templates be used to make elliptical shapes?
20. Which of the following methods could be used to most easily produce circles and arcs: compass, templates, computer graphics system?

4 GEOMETRIC CONSTRUCTION PROBLEMS

DIRECTIONS

1. Problems may be completed using manual or computer-aided drafting depending on your course guidelines.

2. Use very lightly drawn lines (construction lines) for all preliminary work and darken in only the completed product. *Do not erase the construction lines when you complete the problems. This will help your instructor observe your drawing technique.*

3. Geometric construction problems are presented as written instructions, drawings, or a combination of both. If the problems are presented as drawings without dimensions, transfer the drawing from the text, using dividers and scales, to your drawing sheet. Draw dimensioned problems full scale (1:1) unless otherwise specified. Draw all object, hidden, and centerlines. Do not draw dimensions.

Problem 4–1   Draw two tangent circles with their centers on a horizontal line. Circle 1 has a 64 mm diameter and circle 2 has a 50 mm diameter.

Problem 4–2   Make a perpendicular bisector of a horizontal line that is 79 mm long.

Problem 4–3   Make an angle 48° with one side vertical. Bisect the angle.

Problem 4–4   Divide a 96 mm line into 7 equal spaces.

Problem 4–5   Draw two parallel horizontal lines each 50 mm long with one 44 mm above the other. Divide the space between the lines into 8 equal spaces.

Problem 4–6   Transfer triangle a,b,c to a new position with side c 45° from horizontal.

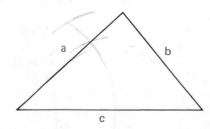

Problem 4–7   Make a right triangle with one side 35 mm long and the other side 50 mm long.

Problem 4–8   Make a right triangle with the hypotenuse 75 mm long and one side 50 mm long.

Problem 4–9   Make an equilateral triangle with 65 mm sides.

Problem 4–10   Make a square with a distance of 50 mm across the flats.

Problem 4–11   Draw a hexagon with a distance of 2.5 in. across the flats.

Problem 4–12   Draw a hexagon with a distance of 2.5 in. across the corners.

Problem 4–13   Draw an octagon with a distance of 2.5 in. across the flats.

Problem 4–14   Draw a rectangle 50 mm × 75 mm with 12 mm radius tangent corners.

Problem 4–15   Draw two separate angles, one a 30° acute angle and the other a 120° obtuse angle. Then draw a .5-in. radius arc tangent to the sides of each angle.

Problem 4–16   Given the following incomplete part, draw an inside arc with a 2.5 in. radius tangent to ØA and line B.

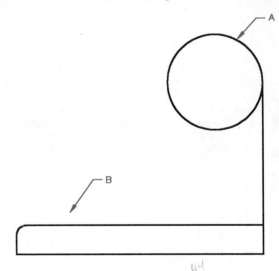

Problem 4–17   Draw two circles with 25 mm diameters and 38 mm between centers on horizontal. On the upper side between the two circles draw an inside radius 44 mm. On the lower side draw an outside radius 76 mm.

Problem 4–18   Draw an S curve with equal radii between points A and B given below:

B

A

Problem 4–19   Draw two lines, one on each side of the ogee curve drawn in problem 4-18 and .5 in. away.

Problem 4–20   Make an ellipse with a 38 mm minor diameter and a 50 mm major diameter.

Problem 4–21 Corner arcs (in.)

Part Name: Plate Spacer
Material: .250-Thick Aluminum

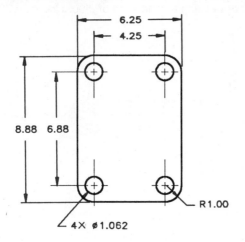

Problem 4–22 Circles (in.)

Part Name: Flange
Material: .50-Thick Cast Iron (CI)

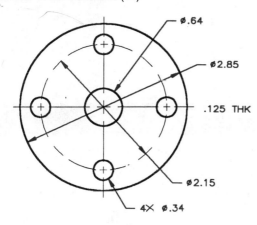

Problem 4–23 Hexagon (in.)

Part Name: Sleeve
Material: Bronze
Draw the hexagon with circles view only.

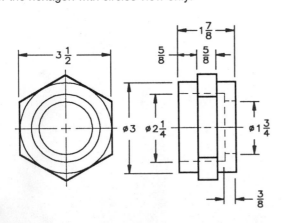

Note: This engineer's sketch shows dimensions to hidden features, which is not a standard practice.

Problem 4–24 Tangencies (metric)

Part Name: Coupler
Material: SAE 1040

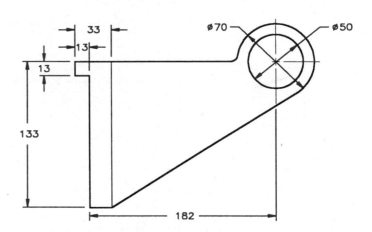

Problem 4–25 Tangencies (in.)

Part Name: Hanger
Material: Mild Steel (MS)

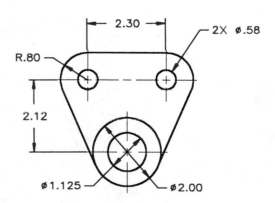

Problem 4–26 Tangencies (in.)

Part Name: Bracket
Material: Aluminum

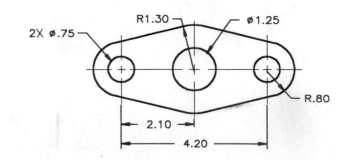

Problem 4–27   Arcs and tangencies (in.)

Part Name: Gusset
Material: .250-Thick Aluminum
Specific instructions: sizes are given as limit dimensions. Limit

dimensioning is discussed in Chapter 8. For now, when a dimension is shown as .531/.468, for example, you produce the drawing using an even numeral between the two. In this case you draw using a dimension of .500. *Courtesy TEMCO*

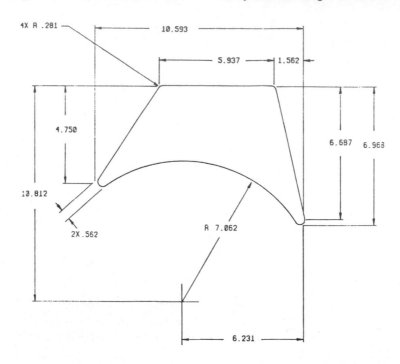

Problem 4–28   Circle and arc object lines and centerlines (metric)

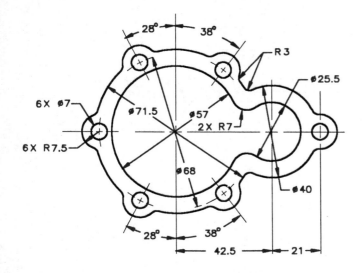

MATH PROBLEMS

For problems 1 through 5, find the interior angles of these types of regular polygons:

1. Octagon.
2. Equilateral triangle.
3. Pentagon (5 sides).
4. Decagon (10 sides).
5. Square.
6. How many sides does a regular polygon have if its interior angle is 140°?

Multiviews

THE ENGINEERING DESIGN PROCESS

The engineer has just handed you a sketch of a new part design. (See Figure 5–1.) The engineer explains that a multiview drawing is needed on the shop floor as soon as possible so a prototype can be manufactured and tested. As the engineering drafter, it is your responsibility to create a drawing that shows all necessary manufacturing data. Based on drafting guidelines, you first decide which view should be the front view. Having established the front view, you must now determine what other views are required in order to depict all of the features of the part. Using visualization techniques based on the glass box helps you to decide which views are needed. (See Figure 5–2.) Unfolding the box puts all of the views in their proper positions, so sketch the layout and complete the drawing. (See Figure 5–3.)

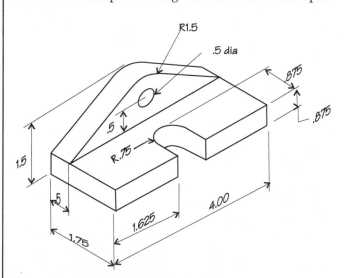

Figure 5–1 Engineer's rough sketch.

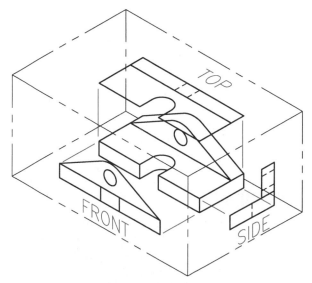

Figure 5–2 Using the glass box principle to visualize the needed views.

(Continued)

ENGINEERING DESIGN PROCESS *(Continued)*

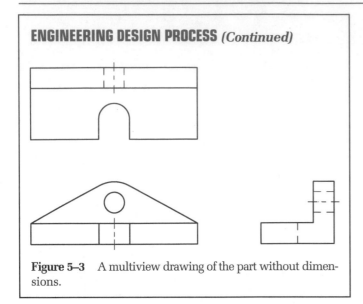

Figure 5–3 A multiview drawing of the part without dimensions.

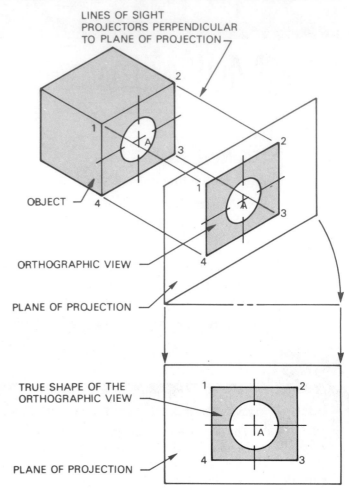

Figure 5–4 Orthographic projection to form orthographic view.

ANSI This chapter is developed in accordance with the ANSI standard for multiview presentation, titled *Multi and Sectional View Drawings,* ANSI Y14.3. This standard is available from the American National Standards Institute, 1430 Broadway, New York, NY 10018. The content of this discussion provides an in-depth analysis of the techniques and methods of multiview presentation.

Orthographic projection is any projection of the features of an object onto an imaginary plane called a plane of projection. The projection of the features of the object is made by lines of sight that are perpendicular to the plane of projection. When a surface of the object is parallel to the plane of projection, the surface appears in its true size and shape on the plane of projection. In Figure 5–4, the plane of projection is parallel to the surface of the object. The line of sight (projection from the object) is perpendicular to the plane of projection. Notice also that the object appears three-dimensional (width, height, and depth) while the view on the plane of projection has only two dimensions (width and height). In situations where the plane of projection is not parallel to the surface of the object, the resulting orthographic view is foreshortened, or shorter than true length. (See Figure 5–5.)

▼ MULTIVIEWS

Multiview projection establishes views of an object projected upon two or more planes of projection by using orthographic projection techniques. The result of multiview projection is a multiview drawing. A multiview drawing represents the shape of an object using two or more views. Consideration should be given to the choice and number of views used so, when possible, the surfaces of the object are shown in their true size and shape.

It is easier for an individual to visualize a three-dimensional picture of an object than it is to visualize a two-dimensional drawing. In mechanical drafting, however, the common practice is to prepare completely dimensioned detail drawings using two-dimensional views, known as *multiviews.* Figure 5–6 shows an object represented by a three-dimensional drawing, also called a *pictorial,* and three two-dimensional views, or multiviews, also known as orthographic projection. The multivew method of drafting represents the *shape* description of the object.

Glass Box

If we place the object in Figure 5–6 in a glass box so the sides of the glass box are parallel to the major surfaces of the object, we can project those surfaces onto the sides of the glass box and create multiviews. Imagine the sides of the glass box are the planes of projection that were previously discussed. (See Figure 5–7.) If we look at all sides of the glass box, we have six total views: FRONT, TOP, RIGHT-SIDE, LEFT-SIDE, BOTTOM, and REAR. Now unfold the glass box as if the corners were hinged about the front (except the rear view) as demon-

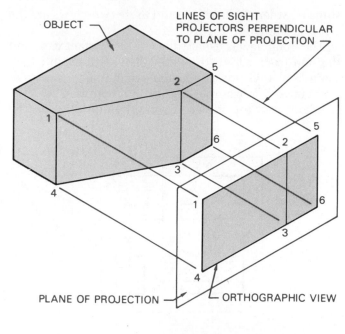

LINES OF SIGHT
PROJECTORS PERPENDICULAR
TO PLANE OF PROJECTION

OBJECT

PLANE OF PROJECTION — ORTHOGRAPHIC VIEW

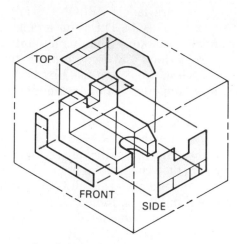

Figure 5–7 The glass box principle.

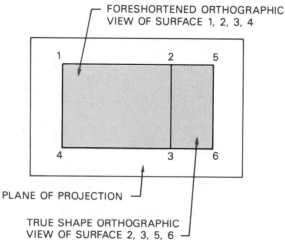

FORESHORTENED ORTHOGRAPHIC
VIEW OF SURFACE 1, 2, 3, 4

PLANE OF PROJECTION —

TRUE SHAPE ORTHOGRAPHIC
VIEW OF SURFACE 2, 3, 5, 6 —

Figure 5–5 Projection of a foreshortened orthographic surface.

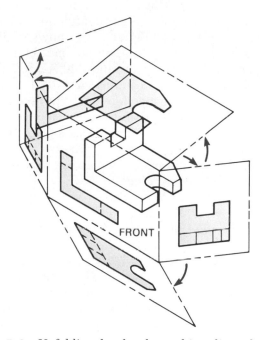

Figure 5–8 Unfolding the glass box at hinge lines also called fold lines.

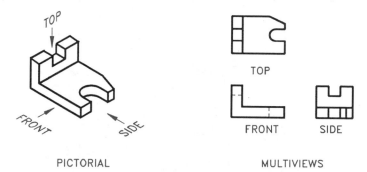

PICTORIAL   MULTIVIEWS

Figure 5–6 Pictorial vs. multiview.

strated in Figure 5–8. These hinge lines are commonly called *fold lines.*

Completely unfold the glass box onto a flat surface and you have the six views of an object represented in multiview. Figure 5–9 shows the glass box unfolded.

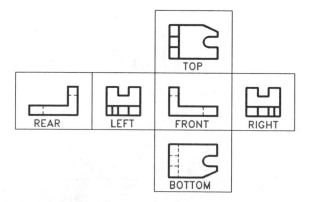

Figure 5–9 Glass box unfolded.

Notice the views are labeled FRONT, TOP, RIGHT, LEFT, REAR, and BOTTOM. This is the arrangement that the views are always found when using multiviews in third-angle projection. *Third-angle projection* is the principal multiview projection used by United States industries. *First-angle projection* is commonly used in European countries.

Look at Figure 5–9 in more detail so you can see the items that are common between views. Knowing how to identify features of the object that are common between views aids you in the visualization of multiviews. Notice in Figure 5–10 that the views are aligned. The top view is directly above and the bottom view is directly below the front view. The left-side view is directly to the left, while the right-side view is directly to the right of the front view. This format allows the drafter to project points directly from one view to the next to help establish related features on each view.

Now, look closely at the relationship among the front, top, and right-side views. (A similar relationship exists with the left-side view.) Figure 5–11 shows a 45° line projected from the corner of the fold, or reference line (hinge), between the front, top, and side views. This 45° line is used as an aid in projecting views. All of the features established on the top view can be projected to the 45° line and then down onto the side view. This projection works because the depth dimension is the same between the top and side views. The reverse is also true. Features from the side view may be projected to the 45° line and then over to the top view.

The same concept of projection achieved in Figure 5–11 using the 45° line also works by using a compass at the intersection of the horizontal and vertical fold lines. The compass establishes the common relationship between the top and side views as shown in Figure 5–12. Another method commonly used to transfer the size of features from one view to the next is to use dividers to transfer distances from the fold line of the top view to the fold line of the side view. The relationships between the fold line and these two views are the same, as shown in Figure 5–13.

The front view is usually the most important view and the one from which the other views are established. There is always one dimension common between adjacent views. For example, the width is common between the front and top views and the height between the

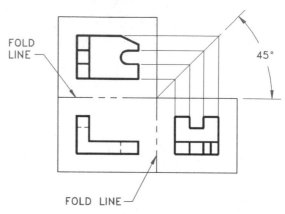

Figure 5–11 45° projection line.

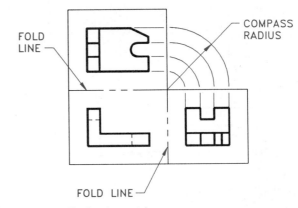

Figure 5–12 Projection with a compass.

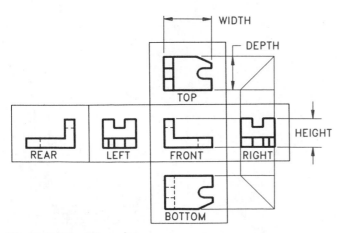

Figure 5–10 View alignment.

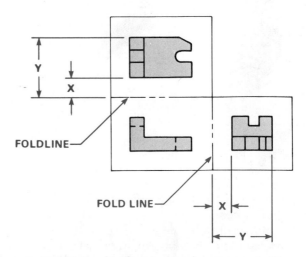

Figure 5–13 Using dividers to transfer view projections.

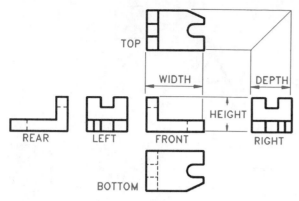

Figure 5–14 Multiview orientation.

front and side views. Knowing this allows you to relate information from one view to another. Take one more look at the relationship among the six views shown in Figure 5–14.

▼ VIEW SELECTION

Although there are six primary views that you can select to completely describe an object, it is seldom necessary to use all six. As a drafter, you must decide how many views are needed to properly represent the object. If you draw too many views you are wasting time, which costs your employer money. If you draw too few views, you may not have completely described the object. The manufacturing department then has to waste time trying to determine the complete description of the object, which again costs your employer money.

Selecting the Front View

Usually, you should select the front view first. The front view is generally the most important view and, as you learned from the glass box description, the front view is the origin of all other views. There is no exact way for everyone to always select the same front view, but there are some guidelines to follow. The front view should:

○ represent the most natural position of use.
○ provide the best shape description or most charac- teristic contours.
○ have the longest dimension.
○ have the fewest hidden features.
○ be the most stable and natural position.

Take a look at the pictorial drawing in Figure 5–15. Notice the front view selection. This front view selection violates the best shape description and fewest hidden features guidelines. However, the selection of any other view as the front would violate other rules, so in this case there is possibly no absolutely correct answer. Given the pictorial drawings in Figure 5–16, identify the view that you believe would be the best front view for each.

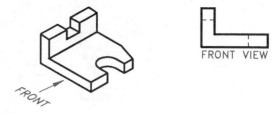

Figure 5–15 Front view selection.

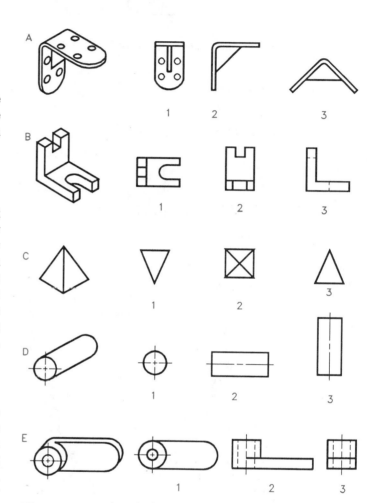

Figure 5–16 Select the best front views that correspond to the pictorial drawings at the left. You may make a first and second choice, if you wish.

A. 2,1. B. 3,2. C. 3. D. 2,1. E. 2,1.

Other View Selection

Use the same rules when selecting other views needed as you do when selecting the front view:

○ most contours.
○ longest side.
○ least hidden features.
○ best balance or position.

Given the six views of the object in Figure 5–17, which views would you select to completely describe the

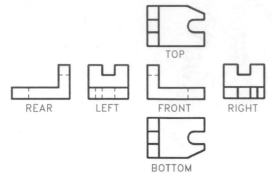

Figure 5-17 Select the necessary views to describe the object.

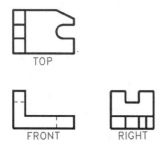

Figure 5-18 The selected views.

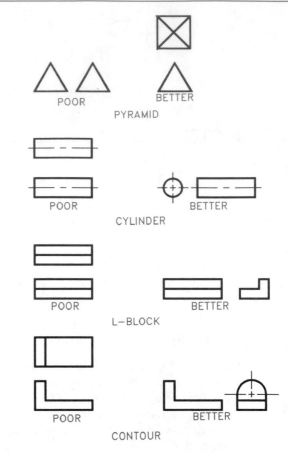

Figure 5-19 Selecting two views.

object? If your selection is the front, top, and right side, then you are correct. Let's take a closer look. Figure 5-18 shows the selected three views. The front view shows the best position and the longest side, the top view clearly represents the angle and the arc, and the right-side view shows the notch. Any of the other views have many more hidden features. You should always avoid hidden features if you can.

It is not always necessary to select three views. Some objects can be completely described with two views or even one view. When selecting fewer than three views to describe an object, you must be careful which view other than the front view you select. (See Figure 5-19.)

One-view drawings are also often practical. When an object has shape and a uniform thickness, more than one view is unnecessary. Figure 5-20 shows a gasket drawing where the thickness of the part is identified in the materials specifications of the title block. The types of parts that may fit into this category include gaskets, washers, spacers, or similar features.

Although two-view drawings are generally considered the industry's minimum recommended views for a part, other objects that are clearly identified by dimensional shape may be drawn with one view, as in Figure 5-21.

In Figure 5-21, the shape of the pin is clearly identified by the 25 mm diameter. In this case, the second view would be a circle and would not necessarily add any more valuable information to the drawing. The primary question to keep in mind is, can the part be easily manufactured, without confusion, from the drawing? If there is any doubt, then the adjacent view should probably be drawn. As an entry-level drafter, you should ask your

Figure 5-20 One-view drawing.

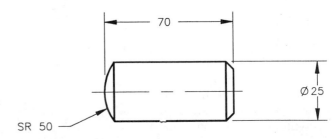

Figure 5-21 One-view drawing.

drafting supervisor to clarify the company policy regarding the number of views to be drawn.

Contour Visualization

Some views do not clearly identify the shape of certain contours. You must then draw the adjacent views to visualize the contour. (See Figure 5-22.)

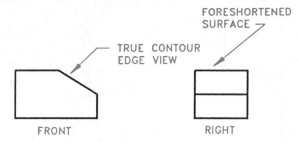

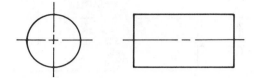

Figure 5–22 Contour representation.

Figure 5–25 Cylindrical shape representation.

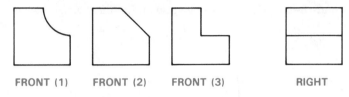

Figure 5–23 Select the front view that properly goes with the given right-side view.

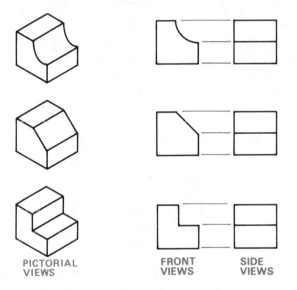

Figure 5–24 The importance of a view that clearly shows the contour of a surface.

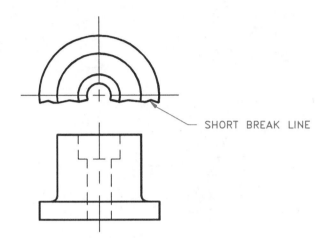

Figure 5–26 Partial view.

The edge view of a surface shows that surface as a line. The true contour of the slanted surface is seen as an edge in the front view of Figure 5–22 and describes the surface as an angle, while the surface is foreshortened, slanting away from your line of sight, in the right-side view. In Figure 5–23, select the front view that properly describes the given right-side view. All three front views in Figure 5–23 could be correct. The side view does not help the shape description of the front view contour. (See Figure 5–24.)

Cylindrical shapes appear round in one view and rectangular in another view, as seen in Figure 5–25. Both views in Figure 5–25 may be necessary as one shows the diameter shape and the other shows the length. The ability to visualize from one view to the next is a critical skill for a drafter. You may have to train your-

self to look at two-dimensional objects and picture three-dimensional shapes. You can also use some of the techniques discussed here to visualize features from one view to another.

Partial Views

When symmetrical objects are drawn in limited space or when there is a desire to save valuable drafting time, partial views may be used. The top view in Figure 5–26 is a partial view. Notice how the short break line is used to show that a portion of the view is omitted. Caution should be exercised when using partial views, as confusion could result in some situations. If the partial view reduces clarity, then draw the entire view.

View Enlargement

When part of a view has detail that cannot be clearly dimensioned due to the drawing scale or complexity, a view enlargement may be used. To establish a view enlargement, a thick phantom line circle is placed around the area to be enlarged. This circle is broken at a convenient location and an identification letter (.18 to .25 in. in height) is centered in the break. Arrowheads are then placed on the line next to the identification letter, as shown in Figure 5–27a. An enlarged view of the detailed area is then shown in any convenient location in the field of the drawing. The view identification and scale is then placed below the enlarged view. (See VIEW A, Figure 5–27b.)

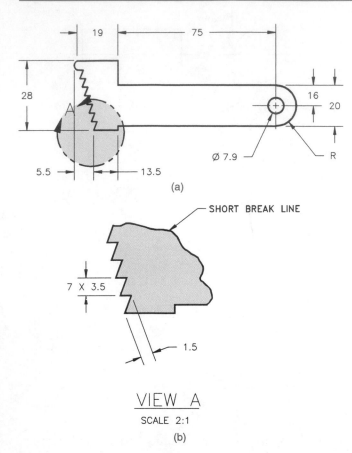

(a)

(b)

Figure 5–27 View enlargement example.

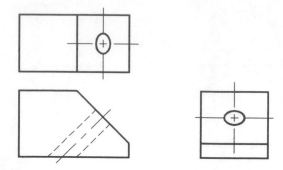

Figure 5–29 Hole represented on an inclined surface as an ellipse.

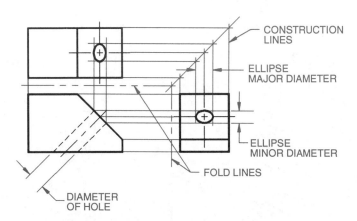

Figure 5–30 Establish an ellipse in the inclined surface.

▼ PROJECTION OF CIRCLES AND ARCS

Circles on Inclined Planes

When the line of sight in multiviews is perpendicular to a circular feature, such as a hole, the feature appears round, as shown in Figure 5–28. When a circle is projected onto an inclined surface, its view is elliptical in shape, as shown in Figure 5–29. The ellipse shown in the top and right-side views of Figure 5–29 was established by projecting the major diameter from the top to the side view and the minor diameter to both views from the front view, as shown in Figure 5–30. The major diameter in this example is the hole diameter.

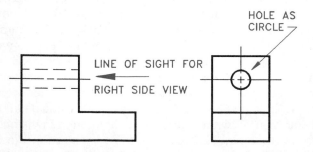

Figure 5–28 View of a hole projected as a circle and its hidden view through the part.

The rectangular areas projected from the two diameters in the top and right-side views of Figure 5–30 provide the parameters of the ellipse to be drawn in these views. The easiest method of drawing the ellipse is to find an ellipse template that has a shape that fits within the area.

Ellipse templates, as discussed in Chapters 1 and 3, designate elliptical shapes by the angle of the circle in relationship to the line of sight. If your slanted surface is 30°, then a 30° ellipse template would be used to draw the elliptical shape. Keep in mind that the other view would require a 60° ellipse. When the angle is 45°, then both views would be the same. The ellipse template selected should have a shape that fits into the rectangle constructed, as shown in Figure 5–31. If you cannot find a representative ellipse that fits exactly, then use the closest size possible, or verify the drawing procedure with your supervisor. Establishing an elliptical shape with geometric construction is much more time-consuming but may be necessary when templates are not available. The elliptical shapes are a representation and may not have to be perfect, depending on company policy. Also, notice that the hidden lines for the hole have been omitted in the top and side views of Figure 5–30. These lines would require a great amount of extra work without adding significantly to the function of the drawing. Such a practice of omitting lines for an accurate representation may be

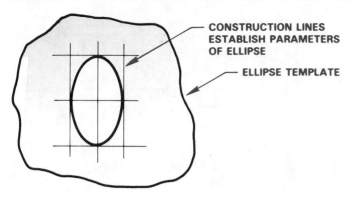

Figure 5–31 Ellipse template.

used with caution. Remember that the drawing must easily communicate the manufacture of the part.

Arcs on Inclined Planes

When a curved surface from an inclined plane must be drawn in multiview, a series of points on the curve establishes the contour. Begin by selecting a series of points on the curved contour as shown in the right-side view of Figure 5–32. Project these points from the right side to the inclined front view. From the point of intersection on the inclined surface in the front view, project lines to the top view. Then project corresponding points from the right side to the 45° projection line and onto the top view. Corresponding lines create a pattern of points, as shown in the top view of Figure 5–32. You may also use a compass or dividers to transfer measurements from the fold lines, as previously discussed. After the series of points is located in relationship to the front and top views, connect the points with an irregular curve. See the completed curve in Figure 5–33.

Fillets and Rounds

Fillets are slightly rounded inside curves at corners, generally used to ease the machining of inside corners or to allow patterns to release more easily from casting and forgings. Fillets may also be designed into a part to

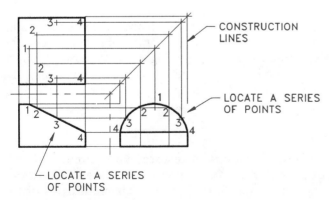

Figure 5–32 Locating an inclined curve in multiview.

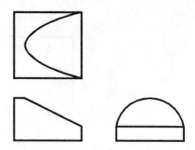

Figure 5–33 Completed curve.

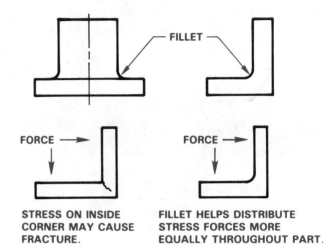

Figure 5–34 Fillets.

allow additional material on inside corners for stress relief. (See Figure 5–34.)

Other than the concern of stress factors on parts, certain casting methods require that inside corners have fillets. The size of the fillet often depends on the precision of the casting method. For example, very precise casting methods may have smaller fillets than green-sand casting, where the exactness of the pattern requires large inside corners.

Rounds are rounded outside corners that are used to relieve sharp exterior edges. Rounds are also necessary in the casting and forging process for the same reasons as fillets. Figure 5–35 shows rounds represented in views.

A machined edge causes sharp corners, which may be desired in some situations. However, if these sharp cor-

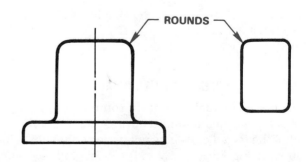

Figure 5–35 Rounds.

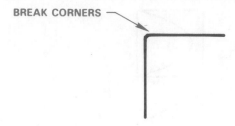

Figure 5–36 Break corner.

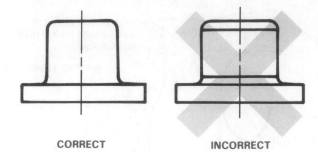

CORRECT   **INCORRECT**

Figure 5–38 Rounds and fillets in multiview.

ners are to be rounded, the extent of roundness depends on the function of the part. When a sharp corner has only a slight relief, it is referred to as a break corner, as shown in Figure 5–36.

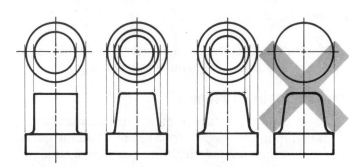

Figure 5–39 Rounded curves and cylindrical shapes in multiview.

CADD Applications

FILLETS AND ROUNDS

CADD programs allow you to draw a fillet or round between any two lines or a line and an arc simply by picking the two lines and specifying the fillet radius. The AutoCAD command for this operation is called FILLET, but it draws any radius exterior or interior corner. If the lines intersect, they are automatically trimmed and the fillet is drawn. If the lines do not meet, they are automatically extended and the fillet is drawn. If all of the lines of an object are connected as one entity (referred to as a polyline in AutoCAD), picking anywhere on the line automatically fillets all corners. Figure 5–37 shows some before and after examples of using the FILLET command.

BEFORE   AFTER   BEFORE   AFTER

Figure 5–37 Automatically drawing fillets and rounds using CADD.

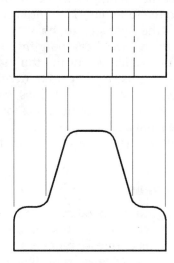

Figure 5–40 Contour in multiview.

identifies the height and the top shows the diameter. Figure 5–39 shows how these cylindrical shapes should be represented in multiview. Figure 5–40 shows the representation of the contour of an object as typically displayed in multiview using phantom lines. This is done to clearly accent the rounded feature.

Rounded Corners in Multiview

An outside or inside slightly rounded corner of an object is represented in multiview as a contour only. The extent of the round or fillet is not projected into the view as shown in Figure 5–38. Cylindrical shapes may be represented with a front and top view, where the front

Runouts

The intersections of features with circular objects are projected in multiview to the extent where one shape runs into the other. The characteristics of the intersecting features are known as runouts. The runout of features intersecting cylindrical shapes is projected from the point of tangency of the intersecting feature, as shown in Figure 5–41. Notice also that the shape of the runout varies when drawn at the cylinder depending on the shape of the intersecting feature. Rectangular-shaped features have a fillet at the runout, while curved (elliptical or round) features contour toward the centerline at the runout. Runouts may also exist when a feature such as a web intersects another feature, as shown in Figure 5–42.

▼ LINE PRECEDENCE

When drawing multiviews, it is common for one type of line to fall in line with a different line type. As a drafter, you need to decide which line to draw. This is known as *line precedence*. You draw the line that is most important based on these rules:

○ Object lines take precedence over hidden lines and centerlines.
○ Hidden lines take precedence over centerlines.

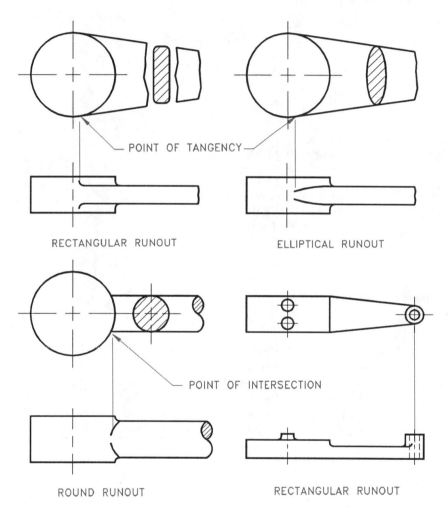

Figure 5–41   Runouts.

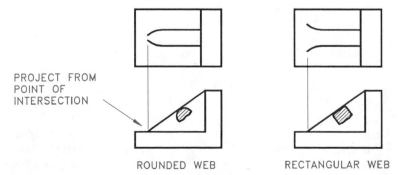

Figure 5–42   Other types of runouts.

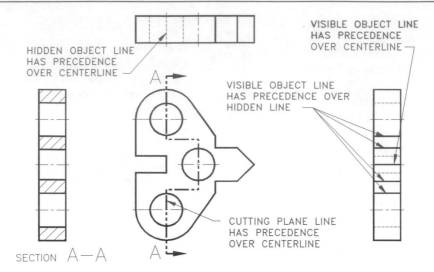

Figure 5–43 Line precedence.

○ In sectioning (covered in Chapter 12), cutting plane lines take precedence over centerlines.

When an object line is drawn over a centerline, the ends or tails of the centerline may be drawn slightly beyond the outside of the view. Figure 5–43 shows examples of line precedence.

▼ THIRD-ANGLE PROJECTION

The method of multiview projection described in this chapter is also known as third-angle projection. This is the method of view arrangement that is commonly used in the United States. In the previous discussion on multiview projection, the object was placed in a glass box so the sides of the glass box were parallel to the major surfaces of the object. Next, the object surfaces were projected onto the adjacent surfaces of the glass box. This achieved the same effect as if the viewer's line of sight were perpendicular to the surface of the box and looking directly at the object, as shown in Figure 5–44. With the multiview concept in mind, assume an area of space is divided into four quadrants, as shown in Figure 5–45.

If the object were placed in any of these quadrants, the surfaces of the object would be projected onto the adjacent planes. When placed in the first quadrant, the method of projection is known as first-angle projection. Projections in the other quadrants are termed second-, third-, and fourth-angle projections. Second- and fourth-angle projection are not used, though first- and third-angle projection are very common.

Third-angle projection, as commonly used in the United States, is achieved when we take the glass box from Figure 5–44 and place it in quadrant three from Figure 5–45. Figure 5–46 shows the relationship of the glass box to the projection planes in the third-angle

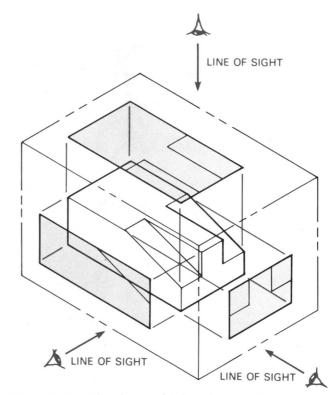

Figure 5–44 Glass box in third-angle projection.

projection. In this quadrant, the projection plane is between the viewer's line of sight and the object. When the glass box in the third-angle projection quadrant is unfolded, the result is the multiview arrangement previously discussed and shown in Figure 5–47.

A third-angle projection drawing may be accompanied by a symbol on or next to the drawing title block. The standard third-angle projection symbol is shown in Figure 5–48.

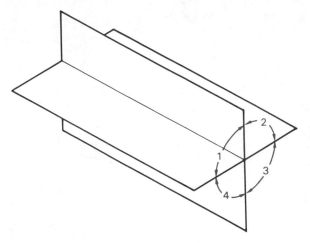

Figure 5–45 Quadrants of spacial visualization.

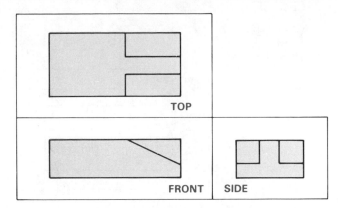

Figure 5–47 Third-angle projection.

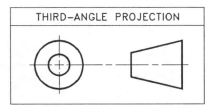

Figure 5–48 Third-angle projection symbol.

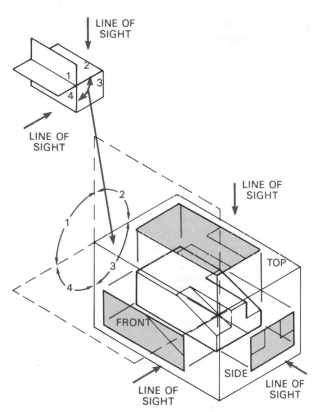

Figure 5–46 Glass box placed in the third quadrant for third-angle projection.

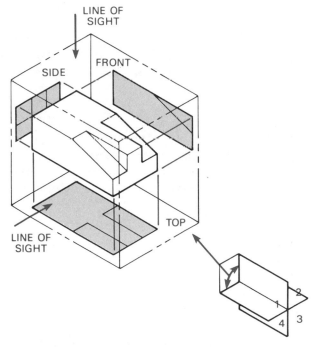

Figure 5–49 Glass box in first-angle projection.

▼ FIRST-ANGLE PROJECTION

First-angle projection is commonly used in Europe and other countries of the world. This method of projection places the glass box in the first quadrant. Views are established by projecting surfaces of the object onto the surface of the glass box. In this projection arrangement, however, the object is between the viewer's line of sight and the projection plane, as you can see in Figure 5–49. When the glass box in the first-angle projection quadrant is unfolded, the result is the multiview arrangement shown in Figure 5–50.

A first-angle projection drawing may be accompanied by a symbol on or adjacent to the drawing title block. The standard first-angle projection symbol is shown in Figure 5–51. Figure 5–52 shows a comparison of the same object in first- and third-angle projections.

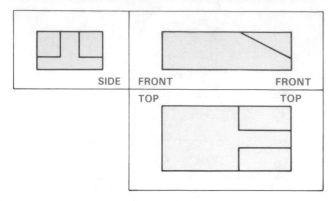

Figure 5–50   First-angle projection.

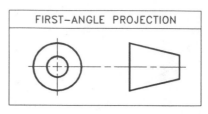

Figure 5–51   First-angle projection symbol.

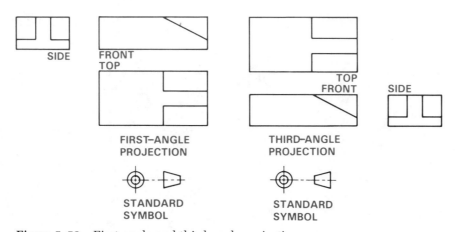

Figure 5–52   First-angle and third-angle projection.

▼ RECOMMENDED REVIEW

It is recommended that you review the following parts of Chapter 3 before you begin working on multiview drawings. This will refresh your memory about how related lines are properly drawn:

○ Object Lines.
○ Viewing-Planes.
○ Hidden Lines.
○ Break Lines.
○ Centerlines.
○ Phantom Lines.

▼ LAYOUT

Many factors influence the drawing layout. Your prime goal should be a clear and easy-to-read interpretation of selected views with related information. Although this chapter deals with multiview presentation, it is not totally realistic to consider view layout without thinking about the effects of dimension placement on the total drawing. Chapter 10 correlates the multiview drawings of *shape description* with dimensioning, known as *size description*.

The initial steps in view layout should be performed using rough sketches. By using rough sketches, you can analyze which views you need before you begin working on vellum or Mylar.

Sketching the Layout

Consider the engineering sketch in Figure 5–53 as you evaluate the proper view layout.

Step 1.   Select the front view using the rules discussed in this chapter. Sketch the front view that you

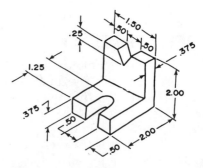

Figure 5–53   Engineering sketch.

Figure 5-54 Sketch the front view.

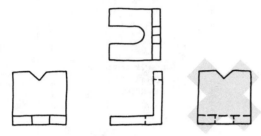

Figure 5-55 Sketch the required views.

have picked. Try to keep your sketch proportional to the actual object, as in Figure 5–54.

Step 2. Select the other views needed to completely describe the shape of the V-block mount, as shown in Figure 5–55.

The front, top, and left-side views clearly define the shape of the V-block mount. Now lay out the formal drawing using the sketch as a guide. Several factors must be considered before you begin:

1. Size of drawing sheet
2. Scale of the drawing
3. Number and size of views
4. Amount of blank space required for future revisions
5. Dimensions and notes (not drawn at this time)

Drawing the Layout

Step 1. Use an A3-size (297 × 420 mm) formerly B-size (11 × 17 in.) drawing sheet. The recommended working area is shown in Figure 5–56. The amount of blank area on a drawing depends on company standards. Some companies want the drawing to be easy to read with no crowd-

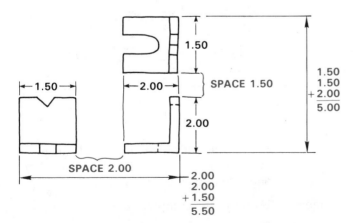

Figure 5–56 Rough sketch with overall dimensions and selected space between views.

ing: others may want as much information as possible on a sheet. Generally, .50 in. should be the minumum space between the drawing and border line. An area for future revision generally (but not always) should be left between the title block and upper right corner. An area for general notes should be available to the left of the title block for ANSI standard layout or in the upper left corner for military standard layout. The remaining area is the space available for drawing. On Figure 5–56, this area is about 10 × 6 in.

Step 2. After determining the approximate working area, use your rough sketch as a guide to establish the actual size of the drawing by adding the overall dimensions and the space between views. (See Figure 5–56.) The amount of space selected will not crowd the views. The amount to select is an arbitrary decision and one where the drafter must use good judgment. Keep in mind that, for now, we will not consider dimensions when evaluating space requirements. The effect of dimensions on drawing layouts is discussed in Chapter 10.

Step 3. Now, in each direction subtract the total drawing size from the total space available and divide by two. This gives you the boundaries of the drawing area. Use construction lines to block out the total drawing areas that you have selected. (See Figure 5–57.)

Calculations:

Total space height	6.00	
Total drawing height	−5.00	
	1.00	÷ 2 = .50
Total space width	10.00	
Total drawing width	−5.50	
	4.50	÷ 2 = 2.25

Step 4. Within the area established, use construction lines to block out the views that you selected in the rough sketch. Use multiview projection as discussed in this chapter, beginning with the front view. (See Figure 5–58.)

The construction lines, if properly drawn, will not have to be erased. Remember that construction lines are drawn very lightly with a 6H, 4H, or light-blue lead. In any case, the construction lines should not reproduce in the diazo or photocopy process, if properly drawn.

Step 5. Complete your drawing by using proper techniques to draw the finish lines. To help keep the final drawing neat, remember to:

a. Work from top to bottom.
b. Work from left to right if right handed or from right to left if left handed.

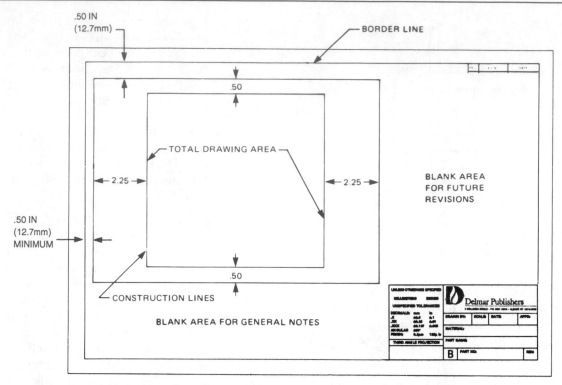

Figure 5–57 The recommended working area and layout of the total drawing area.

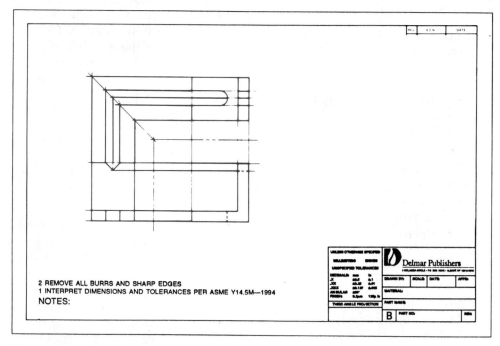

Figure 5–58 Lay out the views using construction lines.

c. Draw thin lines first.
d. Draw circles and arcs next.
e. Draw object lines.
f. Do all lettering and place a clean paper under your hand while lettering.
g. Avoid moving equipment or your hands over completed areas.

h. Keep your equipment and hands clean.
i. Cover large completed areas with blank paper to help avoid smudging.

Figure 5–59 shows the completed drawing.

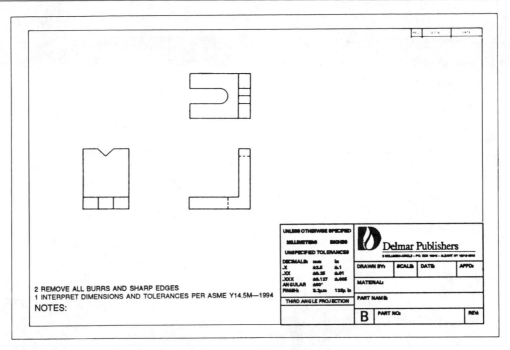

Figure 5–59 Completed multiview drawing without dimensions.

CADD Applications

The previous discussion explained and showed you how to set up multiviews by projecting features from one view to another. The lines that you use to project between views are construction lines. They are used for layout purposes only and may either be erased or left on the drawing if they are very light and do not reproduce. When using a CADD system, construction lines may also be used. Construction lines in a CADD system are generally placed on their own layer and in their own color. After they have been used to serve their purpose, such as establish views, their layer may be frozen or turned off. The construction lines are only used for layout and construction and are not part of the drawing file when they are turned off. The AutoCAD program has construction lines that are drawn using the XLINE command. The XLINE command creates construction lines of infinite length. They may be drawn horizontally, vertically, at an angle, or offset a given distance from another line or object. The XLINE may also be used to bisect an angle. Figure 5-60 shoes how construction lines may be used in CADD to project features between views.

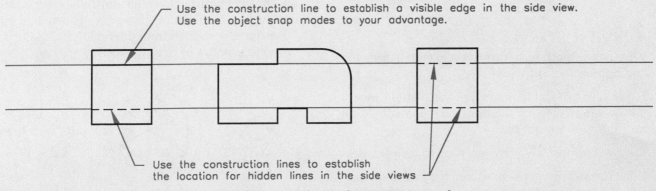

Use the construction line to establish a visible edge in the side view. Use the object snap modes to your advantage.

Use the construction lines to establish the location for hidden lines in the side views

Figure 5–60 CADD drawing with construction lines created using the XLINE command.

MATH APPLICATION

WEIGHT OF AN ELLIPTICAL PLATE

Problem: A large steel plate in the shape of an ellipse is to be moved. The plate has the dimensions shown in Figure 5–61.

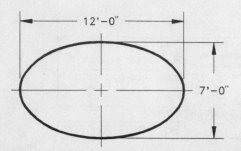

Figure 5–61 Elliptical steel plate.

The steel is known by its thickness to weigh 19 pounds per square foot. What is the weight of this plate?

Solution: The area of an ellipse, as well as other figures, is found in Appendix N. It is given by the formula, Area = $\pi \, x \, y$, as defined by Figure 5–62:

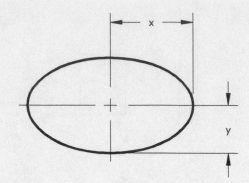

Figure 5–62 Ellipse.

This formula says to multiply the constant π (which is about 3.14159) by dimension x, then to multiply that result by dimension y. For the steel plate, $x = 6'$ and $y = 3.5'$, so the area is (3.14159) (6) (3.5) or 65.98 square feet (ft^2). Now, the weight may be found:

$$65.98 \text{ ft}^2 \times 19 \, \frac{\text{lb}}{\text{ft}^2} = 1253 \text{ lb}$$

CHAPTER

5   MULTIVIEWS TEST

DIRECTIONS

Answer the questions with short, complete statements or drawings as needed.

QUESTIONS

1. Why is it important to prepare a sketch before you begin the actual drawing?

2. Explain in brief statements five basic guidelines to follow when selecting the front view of an object for multiview presentation.

3. List two reasons why views are aligned on a drawing.

4. What method do we use to space views on a drawing?

5. Show an example of third-angle projection.

6. Show an example of first-angle projection.

7. Given the pictorial drawings below, draw the six basic views of each using multiview projection. Show all construction lines. Use your dividers to transfer dimensions from the pictorials. Make your drawings 2× as large as the given.

8. Given the objects below, complete the unfinished views using proper runouts. Transfer the drawings to your answer sheet. Make your drawings 2× as large as the given.

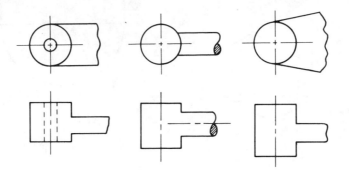

MULTIVIEWS PROBLEMS

PART 1

DIRECTIONS

Missing lines. Determine the missing lines in problems 5–1 through 5–8 and redraw the following objects in multiview. Do not draw the pictorial; it is only given to aid visualization. Use a scale or dividers to transfer dimensions to your drawing from the given sketches.

Problem 5–2 Angle Gage

Problem 5–1 Pocket Block

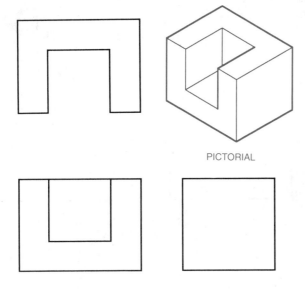

PICTORIAL

Problem 5–3 Base

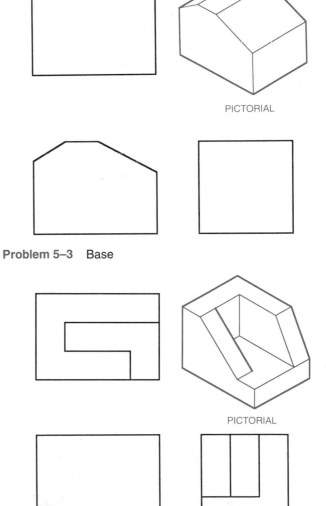

PICTORIAL

Problem 5–4 Corner Block

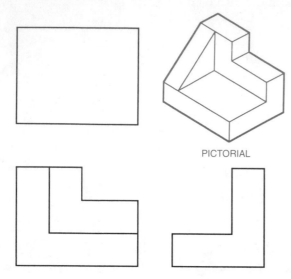

PICTORIAL

Problem 5–7 Gib

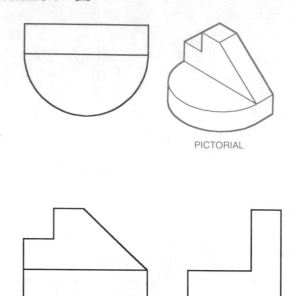

PICTORIAL

Problem 5–5 Cylinder Block

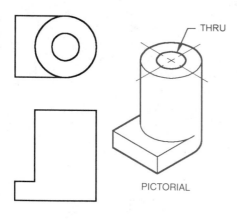

THRU

PICTORIAL

Problem 5–8 Eccentric

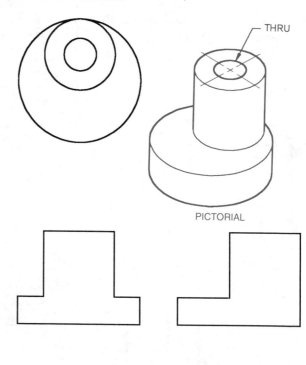

THRU

PICTORIAL

Problem 5–6 Shaft Block

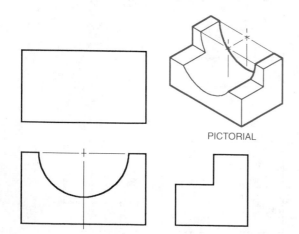

PICTORIAL

DIRECTIONS

Missing views. Determine the missing views for problems 5–9 through 5–12 and redraw the objects in multiview. Do not draw the pictorial; it is only given to aid visualization. Use a scale or dividers to transfer dimensions to your drawing from the given sketches.

Problem 5–9   Guide Book

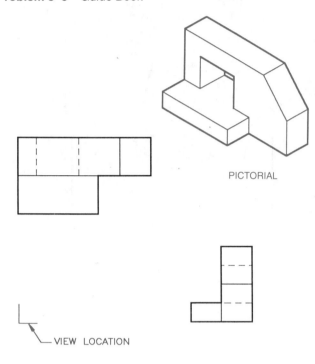

PICTORIAL

VIEW LOCATION

Problem 5–10   Key Slide

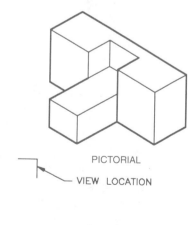

PICTORIAL
VIEW LOCATION

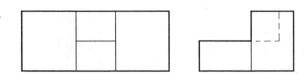

Problem 5–11   Angle Bracket

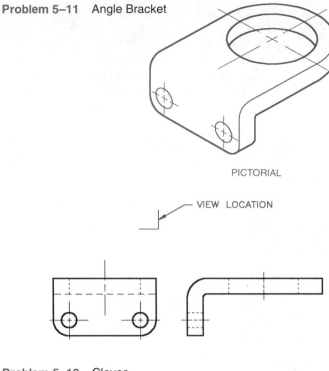

PICTORIAL

VIEW LOCATION

Problem 5–12   Cleves

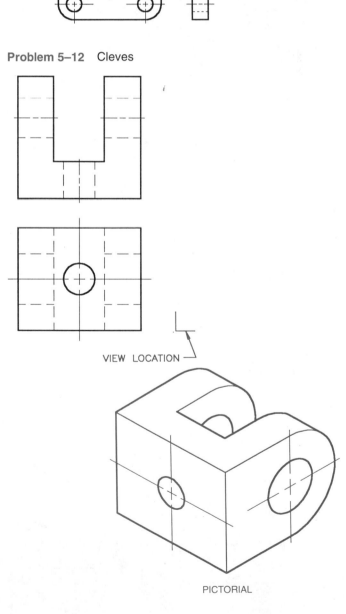

VIEW LOCATION

PICTORIAL

DIRECTIONS

Multiviews. Redraw the objects in problems 5–13 and 5–14 in multiview. The proposed view locations are given. Use dimensions given on pictorial sketches. Do not dimension views. Do not draw the pictorial view.

Problem 5–13 V-block (metric)

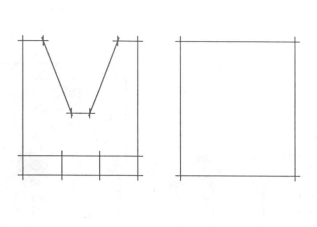

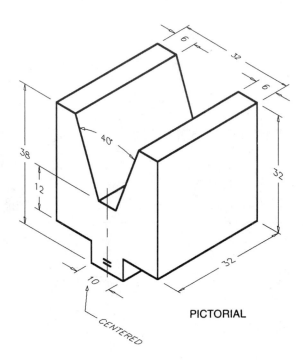

PICTORIAL

Problem 5–14 Guide Base (in.)

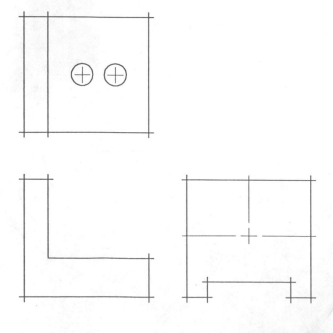

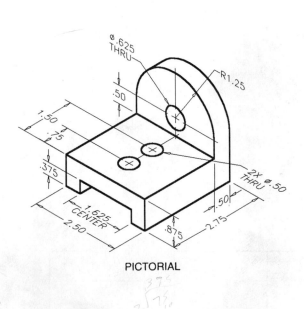

PICTORIAL

PART 2

DIRECTIONS

Problems 5–15 through 5–26 contain one or two views. Draw the needed views shown without dimensions. Place notes, if given, in lower left corner of your drawing.

Problem 5–15 Cylindrical object (in.)

Part Name: Sleeve Bearing
Material: Phosphor Bronze
Problem based on original art courtesy Production Plastics.

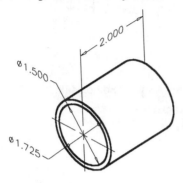

Problem 5–16 Arc and hole (in.)

Part Name: Door Lock
Material: Mild Steel

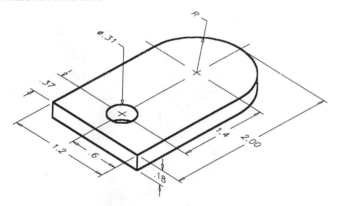

Problem 5–17 Circles and spherical radius (in.).

Part Name: Spherical Insert
Material: Bronze
Problem based on original art courtesy Production Plastics.

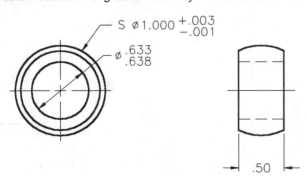

Problem 5–18 Planes and slot (in.)

Part Name: Step Block
Material: Mild Steel

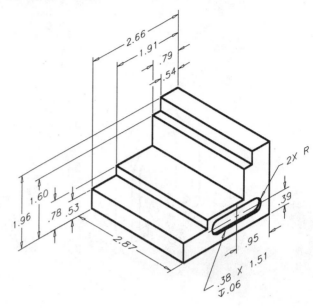

Problem 5–19 Cylinders and circles (in.)

Part Name: Roll End Bearing
Material: Phosphor Bronze
Problem based on original art courtesy Production Plastics.

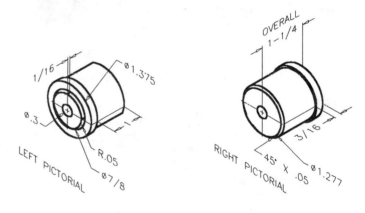

Problem 5–20 Cylinders and circles (in.)

Part Name: Roll End Bearing
Material: Phosphor Bronze

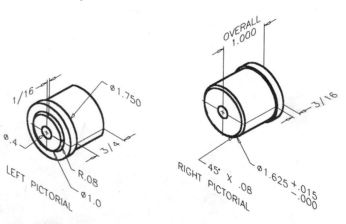

Problem 5–21   Circle arcs and planes (in.)

Part Name: V-block Clamp
Material: SAE 1080

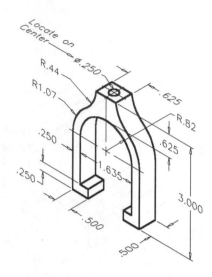

Problem 5–22   View enlargement (in.)

Part Name: Drill Gauge
Material: 16-GA Mild Steel

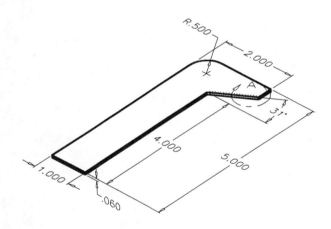

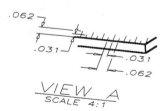

VIEW A
SCALE 4:1

Problem 5–23   Multiple features (in.)

Part Name: Lock Ring
Material: SAE 1020

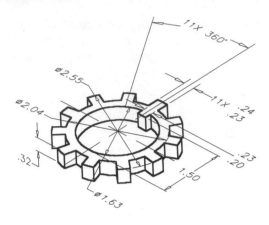

Problem 5–24   Angles and holes (in.)

Part Name: Bracket
Material: SAE 1020

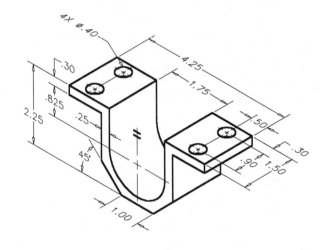

Problem 5–25   Single view (metric)

Part Name: Symmetrical Chart Plate
Material: 10-mm-thick Stainless Steel

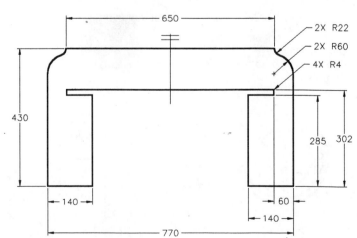

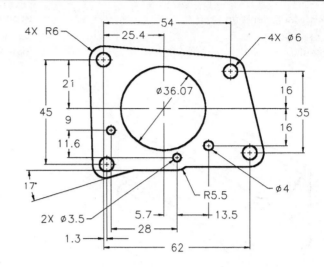

Problem 5–26 Single view (metric)

Part Name: Body Flange Gasket
Material: 0.4 mm Nitrile Rubber
Unspecified Radii: 3 mm
Problem based on original art courtesy Vellumoid, Inc.

PART 3

DIRECTIONS

1. Determine the front view, then establish the other views, if any, of Problems 5–27 through 5–38. Use manual drafting or CADD to draw complete multiviews for each object as specified by your course objectives.
2. Make a multiview sketch of the selected problems as close to correct proportions as possible.

This will help you determine sheet size and view spacing for your final layout.
3. Use your sketch as a guide to draw an original multiview drawing on an adequately sized sheet at an appropriate scale.
4. Do not draw dimensions.

Problem 5–27 Multiviews (metric)

Part Name: Guide Rail
Material: SAE 4310

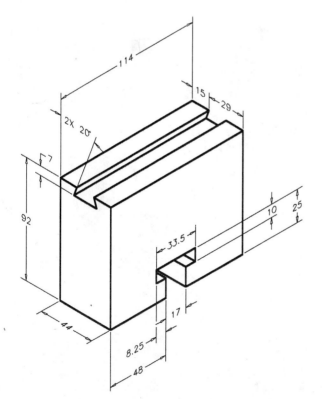

Problem 5–28 Multiviews (in.)

Part Name: Support Bracket
Material: Plastic

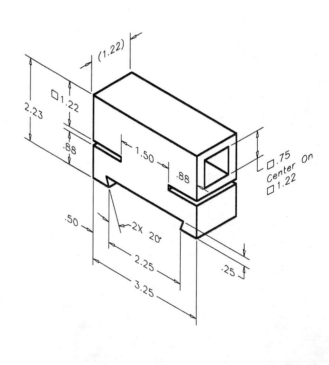

Problem 5–29   Multiviews (in.)

Part Name: V-block
Material: A-Steel

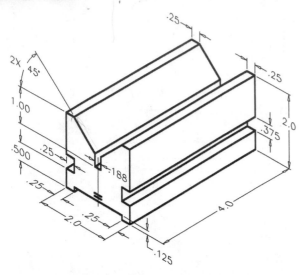

Problem 5–30   Multiviews, angles, and holes (metric)

Part Name: Angle Bracket
Material: Mild Steel

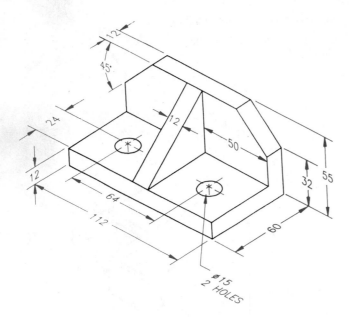

Problem 5–31   Multiviews (in.)

Part Name: Support
Material: Mild Steel

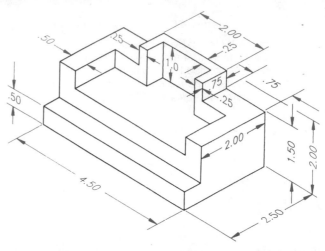

Problem 5–32   Multiviews, circles, and arcs (in.)

Part Name: Chain Link
Material: SAE 4320

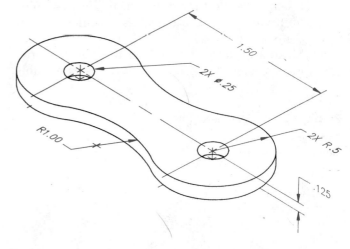

Problem 5–33   Multiviews, circles, and arcs (in.)

Part Name: Pivot Bracket
Material: Cold-Rolled Steel

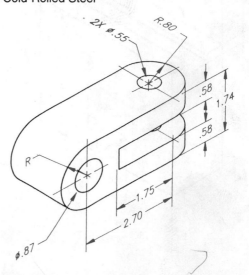

Problem 5–34 Multiview arcs and circles (metric)

Part Name: Hinge Bracket
Material: Cast Aluminum

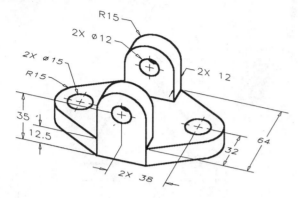

Problem 5–35 Multiview circles and arcs (in.)

Part Name: Bearing Support
Material: SAE 1040

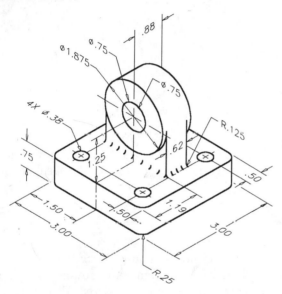

Problem 5–36 Multiviews (metric)

Part Name: Gate Latch Base
Material: Aluminum

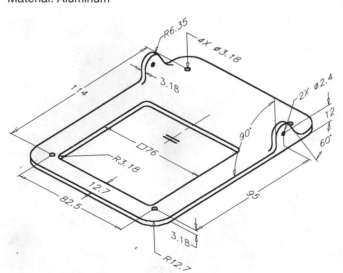

Problem 5–37 Multiviews (in.)

Part Name: Mounting Bracket
Material: SAE 1020
All Fillets R.13

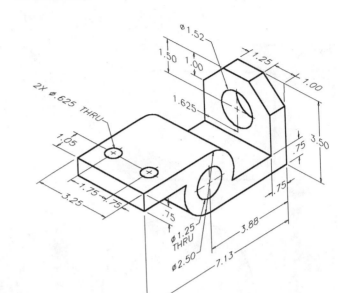

Problem 5–38 Multiviews (in.)

Part Name: Support Base
Material: SAE 1040

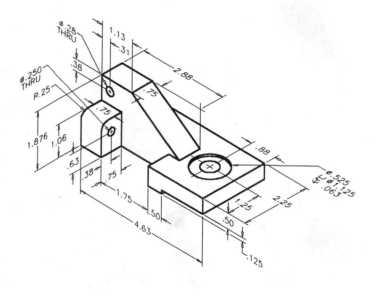

MATH PROBLEMS

Find the areas of these figures:

1.

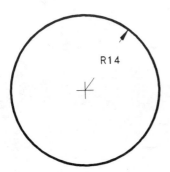

2.

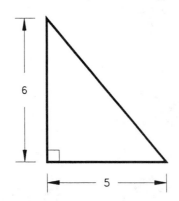

3.

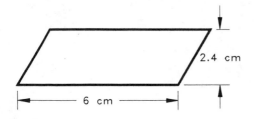

4.

5.

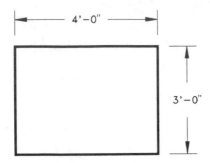

6.

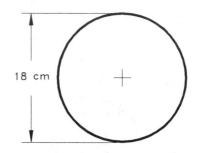

7. What is the weight of a steel plate in the shape of a circle with a diameter of 10'? The steel weighs 14 lb per ft².

Find the volume of:

8. A spherical holding tank 15' in diameter.
9. A cylinder with a diameter of 10' and a height of 4'.
10. A 7'-high cone having a base with a radius of 3'.

Auxiliary Views

THE ENGINEERING DESIGN PROCESS

You have been asked to create a multi-view detail drawing for a guide bracket, and the engineer has provided a sketch of the part. (See Figure 6–1.) As you study the sketch, you discover that one of the surfaces on the bracket is not parallel to any of the six principal viewing planes. You determine that your multiview drawing will need an auxiliary view in order to show the angled face in its true size and shape. Through sketching your layout ideas, you conclude that a complete auxiliary view would not add clarity to the drawing and decide instead on a partial auxiliary view. The final layout is shown in Figure 6–2.

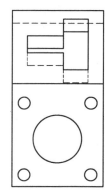
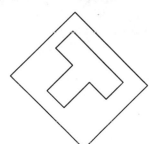
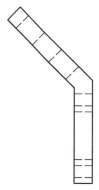

Figure 6–2 Final layout with partial auxiliary view (dimensions not included).

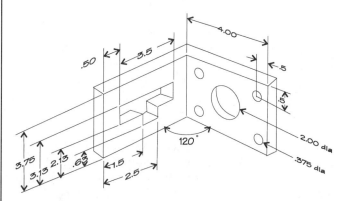

Figure 6–1 Engineer's rough sketch.

▼ AUXILIARY VIEWS

ANSI The standard for auxiliary view presentation is found in ANSI Y14.3 titled *Multi and Sectional View Drawings*.

Auxiliary views are used to show the true size and shape of a surface that is not parallel to any of the six principal views. When a surface feature is not perpendicular to the line of sight, the feature is said to

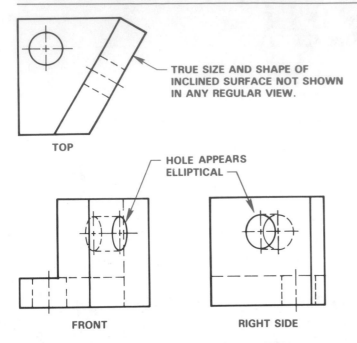

TRUE SIZE AND SHAPE OF INCLINED SURFACE NOT SHOWN IN ANY REGULAR VIEW.

TOP

HOLE APPEARS ELLIPTICAL

FRONT

RIGHT SIDE

Figure 6–3   Foreshortened surface auxiliary view needed.

be foreshortened, or shorter than true length. These foreshortened views do not give a clear or accurate representation of the feature. It is not proper to place dimensions on foreshortened views of objects. Figure 6–3 shows three views of an object with an inclined foreshortened surface.

An auxiliary view allows you to look directly at the inclined surface in Figure 6–3 so you can view the surface and locate the hole in its true size and shape. An auxiliary view is projected from the inclined surface in the view where that surface appears as a line. The projection is at a 90° angle. (See Figure 6–4.) The height dimension, *H,* is taken from the view that shows the height in its true length.

Notice in Figure 6–4 that the auxiliary view shows only the true size and shape of the inclined surface. This is known as a *partial auxiliary view.* A full auxiliary view, Figure 6–5, also shows all the other features of the

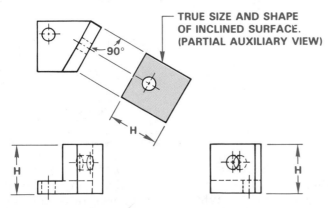

TRUE SIZE AND SHAPE OF INCLINED SURFACE. (PARTIAL AUXILIARY VIEW)

90°

H

H

H

Figure 6–4   Partial auxiliary view.

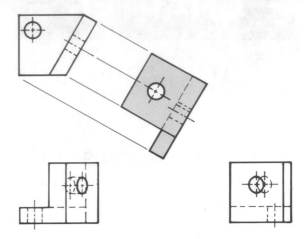

Figure 6–5   Complete auxiliary view.

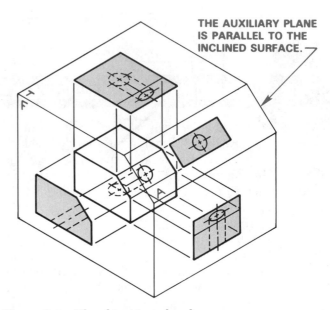

THE AUXILIARY PLANE IS PARALLEL TO THE INCLINED SURFACE.

Figure 6–6   The object in a glass box.

object projected onto the auxiliary plane. Normally the information needed from the auxiliary view is the inclined surface only, and the other areas do not usually add clarity to the view. However, each object must be considered separately. Usually, you do not need to draw those areas that do not add clarity to the drawing.

Look at the glass box principle that was discussed earlier as it applies to auxiliary views, as shown in Figure 6–6. Figure 6–7 shows the glass box unfolded. Notice that the fold line between the front view and the auxiliary view is parallel to the edge view of the slanted surface.

The auxiliary view is projected from the edge view of the inclined surface, which establishes true length. The true width of the inclined surface may be transferred from the fold lines of the top or side views as shown in Figure 6–8. Projection lines are drawn as very light construction lines and should not reproduce. Hidden lines are generally not shown on auxiliary views unless

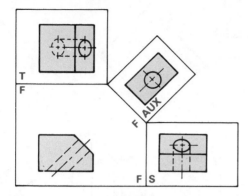

Figure 6–7 The glass box unfolded.

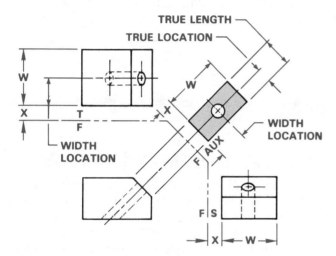

Figure 6–8 Establishing auxiliary view with fold lines.

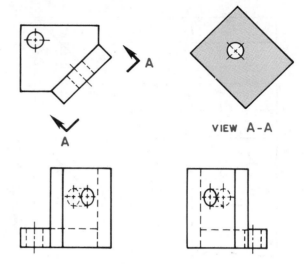

VIEW A-A

Figure 6–9 Establishing auxiliary view with viewing-plane.

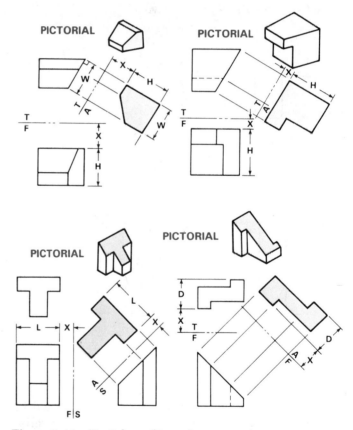

Figure 6–10 Partial auxiliary view examples.

the use of hidden lines helps clarify certain features. The auxiliary view may be projected directly from the inclined surface, as in Figure 6–4. When this is done, a centerline or projection line may continue between the views to indicate alignment and view relationship. This centerline or projection line is often used as an extension line for dimensioning purposes. If view alignment is clearly obvious, the projection line may be omitted. When it is not possible to align the auxiliary view directly from the inclined surface, then viewing-plane lines may be used and the view placed in a convenient location on the drawing (Figure 6–9).

The viewing-plane lines are labeled with letters so the view can be clearly identified. This is especially necessary when several viewing-plane lines are used to label different auxiliary views. Place views in the same relationship as the viewing-plane lines indicate. Do not rotate the auxiliary but leave it in the same relationship as if it were projected from the slanted surface. Where multiple views are used, orient views from left to right and from top to bottom.

Figure 6–10 shows some other examples of partial auxiliary views in use. It is important to visualize the relationship of the slanted surfaces, edge view, and

auxiliary view. If you have trouble with visualization, it is possible to establish the auxiliary view through the mechanics of view projection. Use the following steps:

Step 1. Number each corner of the inclined view so the numbers coincide from one view to the other, as shown in Figure 6–11. Carefully project one point at a time from view to view. Some points have two numbers depending on the view.

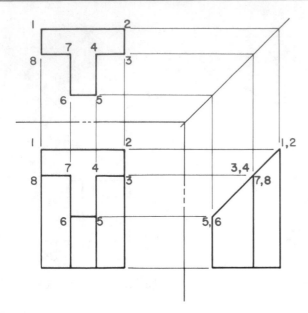

Figure 6–11 Step 1, multiview layout.

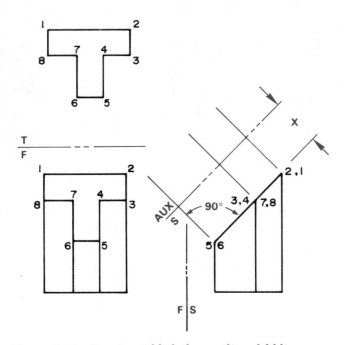

Figure 6–12 Step 2, establish the auxiliary fold line.

Step 2. With the corresponding points numbered in each view, draw an auxiliary fold line parallel to the edge view of the slanted surface. The auxiliary fold line may be any convenient distance from the edge view. Project the points on the edge view perpendicular (90°) across the auxiliary fold line. (See Figure 6–12.)

Step 3. Establish the distance to each point from the adjacent view and transfer the distance to the auxiliary view as shown in Figure 6–13a. Sometimes it is helpful to sketch a small pictorial to assist in visualization. (See Figure 6–13b.)

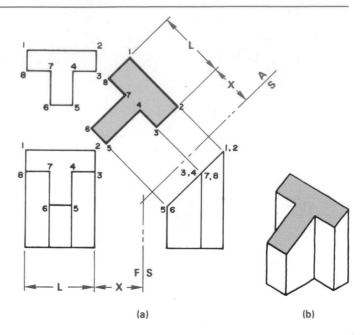

(a)   (b)

Figure 6–13 **(a)** Step 3, layout auxiliary view. **(b)** Pictorial to help visualize object.

▼ PLOTTING CURVES IN AUXILIARY VIEWS

Curves are plotted in auxiliary views in the same manner as those shapes described previously. When corners exist, they are used to lay out the extent of the auxiliary surface. An auxiliary surface with irregular curved contours requires that the curve be divided into elements so the element points can be transferred from one view to the next, as shown in Figure 6–14.

The contour of elliptical shapes may be plotted, or if the shape coincides with an available elliptical template, a template should always be used to save time. (See Figure 6–15.)

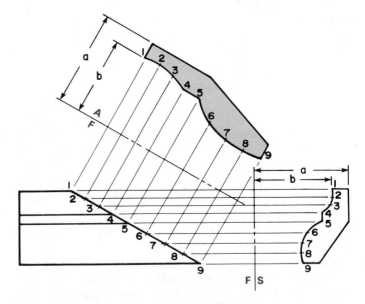

Figure 6–14 Plotting curves in auxiliary view.

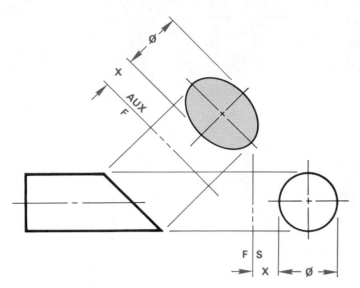

Figure 6–15 Elliptical auxiliary view.

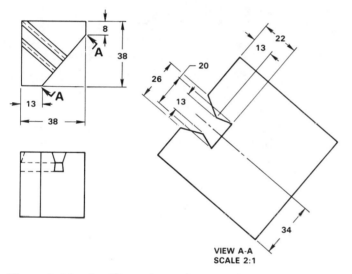

VIEW A-A
SCALE 2:1

Figure 6–16 Auxiliary view enlargement.

▼ ENLARGEMENTS

In some situations you may need to enlarge an auxiliary view so a small detail can be shown more clearly. Figure 6–16 shows an object with a foreshortened surface. The two principal views are clearly dimensioned at a 1:1 scale. The 1:1 scale is too small to clarify the shape and size of the slot through the part. A viewing-plane line is placed to show the relationship of the auxiliary view. The auxiliary view can then be drawn in any convenient location at any desired scale; in this case a 2:1 scale is used.

▼ SECONDARY AUXILIARY VIEWS

There are some situations when a feature of an object is in an oblique position in relationship to the principal planes of projection. These inclined, or slanting, surfaces

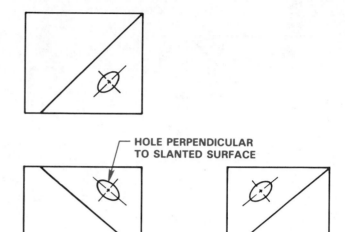

HOLE PERPENDICULAR
TO SLANTED SURFACE

Figure 6–17 Oblique surface. There is no edge view in the principal views.

do not provide an edge view in any of the six possible multiviews. The inclined surface in Figure 6–17 is foreshortened in each view, and an edge view also does not exist. From the discussion on primary auxiliary views you realize that projection must be from an edge view to establish the auxiliary.

In order to obtain the true size and shape of the inclined surface in Figure 6–17, a secondary auxiliary view is needed. The following steps may be used to prepare a secondary auxiliary.

Step 1. Only two principal views are necessary to work from, as the third view does not add additional information. Establish an element in one view that is true in length, as shown in Figure 6–18. Label the corners of the inclined surface and draw a fold line perpendicular to one of the true-length elements.

Step 2. The purpose of this step is to establish a primary auxiliary view that displays the slanted surface as an edge. Project the slanted surface onto the primary auxiliary plane, as shown in Figure 6–19. Doing so results in the inclined surface appearing as an edge view or line.

Step 3. Now, with an edge view established, the next step is the same as the normal auxiliary view procedure. Draw a fold line parallel to the edge view. Project points from the edge view perpendicular to the secondary fold line to establish points for the secondary auxiliary, as shown in Figure 6–20.

In Figure 6–20, the primary auxiliary view established all corners of the inclined surface in a line, or edge view. This edge view is necessary so the perpendicular line of projection (sight) for the secondary auxiliary

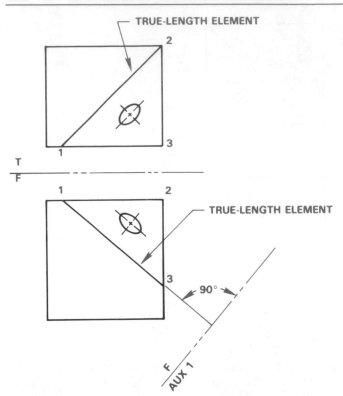

Figure 6–18 Step 1, draw a fold line perpendicular to a true length element.

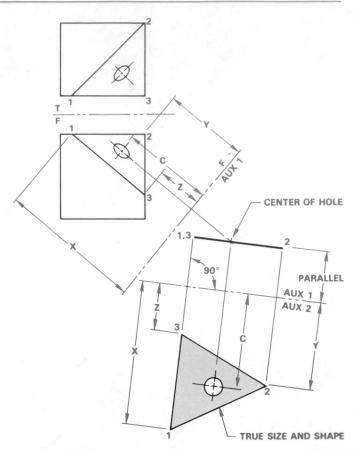

Figure 6–20 Step 3, secondary auxiliary projected from the edge view to get the true size and shape of the oblique surface.

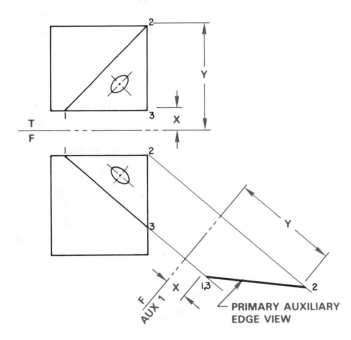

Figure 6–19 Step 2, primary auxiliary edge view of oblique surface.

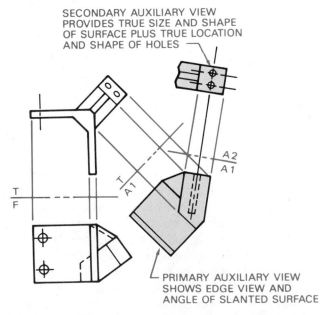

Figure 6–21 The primary and secondary auxiliary views.

assists in establishing the true size and shape of the surface. In many situations both the primary and secondary auxiliary views are used to establish the relationship between features of the object. (See Figure 6–21.) The pri-

mary auxiliary view shows the relationship of the inclined feature to the balance of the part, and the secondary auxiliary view shows the true size and shape of the inclined features plus the true location of the holes.

CADD Applications

MULTIVIEW AND AUXILIARY VIEW DRAWING

So far in your study you have learned that multiview and auxiliary view drawings are the orthographic projection of two or more planes at right angles to each other. One important aspect of this technology is accuracy. As a manual drafter, the accuracy depends on your sharp pencil and a keen eye for technique. Even solutions in the best manual have a degree of error due to line widths or the slight tilt of the pencil in relation to the paper. Multiview and auxiliary view drawings produced with CADD are, on the other hand, nearly perfect. The reason for this perfection has to do with the capabilities of the hardware and software, and the fact that the computer is actually displaying the mathematical counterpart of the graphic analysis. In other words, the views are a representation of the mathematical coordinates of the problem being presented. Therefore, the drawing accuracy is only controlled by the accuracy potential of the computer. Figure 6–22 shows a multiview drawing on a CADD monitor.

Most CADD systems provide the drafter with "drawing aids" such as a screen grid similar to the light-blue lines preprinted on vellum to assist the manual drafter. The screen grid is displayed as dots. These grid dot patterns may be displayed in any convenient incre-
ment to assist in accurately placing the views. The drafter may also establish designated increments for the screen cursor crosshairs to snap. This allows the drafter to accurately determine distances on the drawing without guesswork. Another asset of the CADD system is the display of drawing distance and coordinates. For example, when you draw a line on the screen, the computer displays the exact line length. Drawing multiviews with these kinds of drawing aids makes the job easy and accurate. When drawing auxiliary views, the CADD drawing aids are flexible and may be rotated at the angle of the auxiliary view to assist in accurately laying out the view as shown in Figure 6–23.

A CADD system also allows you to work in layers and colors for a clearer picture of the problem and solution. For example, you can set a layer for projection lines shown in any color, such as red. This helps you visualize the entire drawing and allows you to accurately project a feature from one view to the next. The layers may be turned on or off at your convenience. When you complete the drawing and you are ready to make a plot, all you have to do is turn off the layer with projection lines and the drawing is ready to plot.

(Continued)

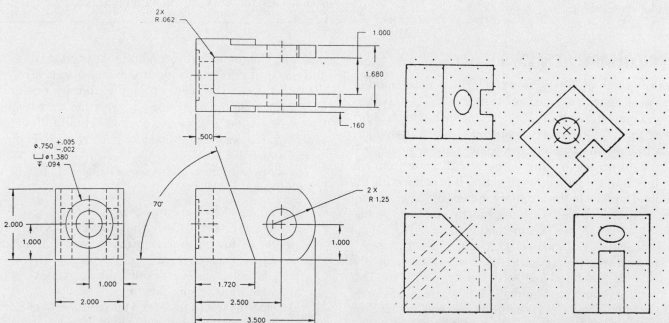

Figure 6–22 Using CADD to draw multiviews.

Figure 6–23 Using the rotated grid points in CADD to help draw the auxiliary view.

CADD APPLICATIONS *(Continued)*

The lines that you use to project between views are construction lines. The construction lines are only used for layout and construction and are not part of the drawing file when they are turned off. The AutoCAD program has construction lines that are drawn using the XLINE command. The XLINE command creates construction lines of infinite length. They may be drawn horizontally, vertically, at an angle, or offset a given distance from another line or object. The XLINE may also be used to bisect an angle. When laying out auxiliary views, the XLINE command Angle option may be used. This allows you to project from an inclined surface to establish the correct true size and shape, and location of the auxiliary view as shown in Figure 6-24.

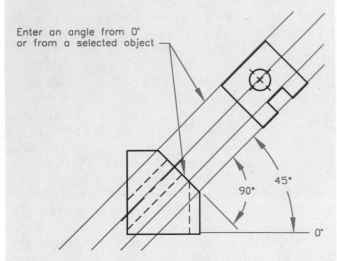

Figure 6–24 Laying out an auxiliary view using the XLINE command.

PROFESSIONAL PERSPECTIVE

For many people, one of the most difficult aspects of engineering drafting is the need to visualize two-dimensional views. This "spacial" ability is natural for some, while others must carefully analyze every part of a multiview drawing in order to fully visualize the product. Ideally, you should be able to look at the multiviews of an object and readily formulate a pictorial representation in your mind. The principal reason that multiviews are aligned is to assist in the visualization and interpretation process. The front view is the most important view, and that is why it contains the most significant features. As you look at the front view, you should be able to gain a lot of information about the object, then visually project from the front view to other views to gain an understanding of the entire product.

It is up to the drafter to select the best views to completely describe the object. Too few views make the object difficult or impossible to interpret, while too many views make the drawing too complex and waste valuable drafting time. If you follow the guidelines set up in this chapter, you should be able to successfully complete the task. Remember, always begin with a sketch. With the sketch you can quickly determine view arrangement and spacing. Without the preliminary sketch, you might use a lot of drafting time and end up discovering that the layout does not work as you expected.

▼ AUXILIARY VIEW LAYOUT

Working with auxiliary views is a little more complex than dealing with the normal multiviews. The auxiliary view is often difficult to place in relationship to the other views. Space requirements many times cause the auxiliary view to interfere with other views. Whenever possible, it is best to project the auxiliary view directly from the inclined surface with one projection line connecting the inclined surface to the auxiliary view. This arrangement makes it easier for the reader to correctly interpret the relationship of the views. This method is shown in the layout steps provided in Figures 6–11, 6–12, and 6–13.

When drawing space is limited, it is possible to use the viewing-plane method to display the auxiliary view. This technique is shown in Figure 6–10. The advantage of the viewing-plane method is that it allows you to place the auxiliary view in any convenient place on the drawing. Multiple auxiliary views should be placed in a group arranged from right to left and from top to bottom on the drawing. One unorthodox method that is often used by entry-level drafters is to place viewing planes all over the drawing and shift views from their normal position. This practice should be avoided, because the normal multiview projection is always best. You should follow the same layout procedure outlined for multiviews in Figures 5–53 through 5–57.

The steps summarized here work the same for manual drafting or CADD:

○ Select and sketch the front view.
○ Select and sketch the other principal views in proper relationship to the front view, eliminating any unnecessary views.
○ An additional step is to sketch the auxiliary view or views in their proper positions.
○ Establish the working area and the sheet size to be used.
○ Determine the total drawing area.
○ Lay out the views using construction lines.
○ Complete the formal drawing.

MATH APPLICATION

PROJECTED AREA

Problem: Suppose the steel plate in the shape of an ellipse with an area of 65.98 ft² (from the Math Application in Chapter 5) is set at a 60° angle on the ground. With the sun overhead, what is the area of the *shadow* of the plate on the ground? (See Figure 6–25.)

Solution: The shadow's area is called the *projected area* in math. It is found by multiplying the true area by the *cosine* of the inclined angle. Using a calculator, cos 60° = .5, so the area of the shadow is 65.98 × .5 = 32.99 or 33 ft². The principle of multiplying true area by cosine to obtain projected area applies to all 2-D shapes. Math Applications in later chapters explore right triangles and trigonometry functions (like cosine) further.

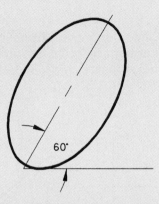

Figure 6–25 Tilted ellipse.

CHAPTER

6 AUXILIARY VIEWS TEST

DIRECTIONS

Answer the questions with short, complete statements or drawings as needed.

1. Describe the purpose of auxiliary views.

2. When projecting an auxiliary view from the edge view of the surface, what is the angle of projection in relationship to the edge view?

3. Discuss how viewing-plane lines may be used to establish an auxiliary view.

4. When would it be necessary to enlarge an auxiliary view?

5. Describe the procedure for drafting an enlarged auxiliary view.

6. Describe a situation when a secondary auxiliary view will be necessary.

7. Identify the ANSI document that governs auxiliary views.

8. What method is used to show the alignment between the auxiliary view and the direct projection from the inclined surface?

9. If the axis of a hole is perpendicular to an inclined surface, what is the shape of the hole in an auxiliary view of the inclined surface?

10. If the axis of a hole is at 45° to an inclined surface, what is the shape of the hole in an auxiliary view of the inclined surface?

AUXILIARY VIEWS PROBLEMS

PART 1

DIRECTIONS

Missing lines in auxiliary views. Given the sketches for Problems 6–1 through 6–9, determine the missing lines in the partial auxiliary views. Redraw the objects in multiview, including the correct auxiliary views. Use a scale or dividers to transfer dimensions from the given sketches to your drawing. Do not dimension your drawings.

Problem 6–1

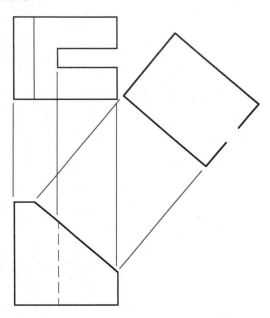

Problem 6–3

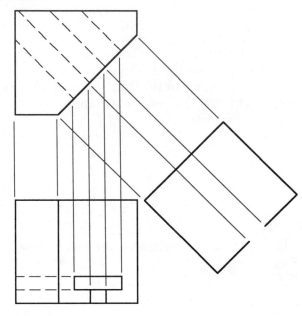

Problem 6–4

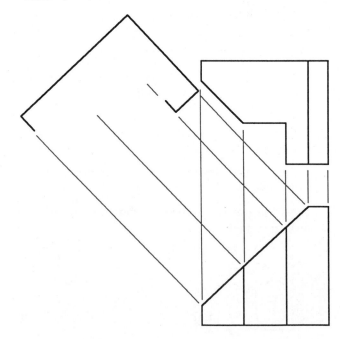

Problem 6–2

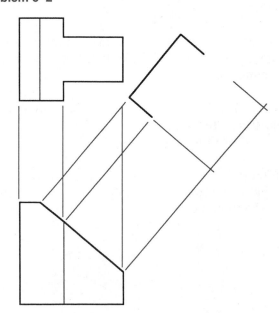

Problem 6–5

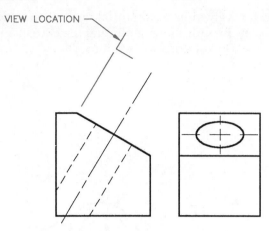

Problem 6–6

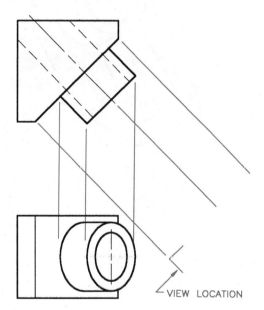

Problem 6–7

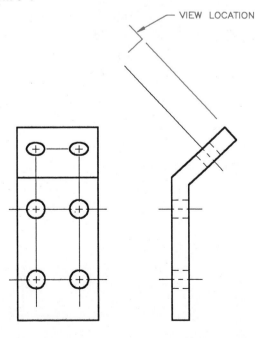

Problem 6–8

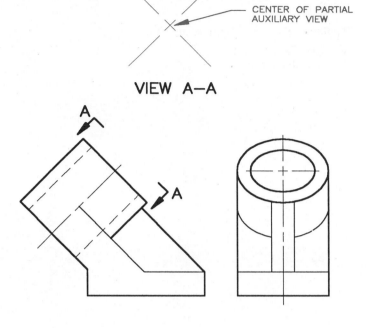

VIEW A–A

Problem 6–9

Title: Angle V-block (in.)
Material: SAE 4320

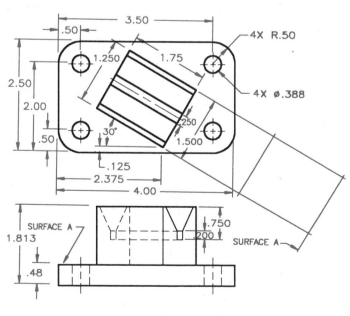

DIRECTIONS

Given the sketches for Problems 6–10 through 6–18, determine the missing view based on the object and dimensions given in the pictorial view; then, redraw the objects in multiview, including the correct auxiliary view. Do not draw dimensions.

Problem 6–10   Primary auxiliary view (in.)

Title: Cylinder Support
Material: Cast Iron

TOP

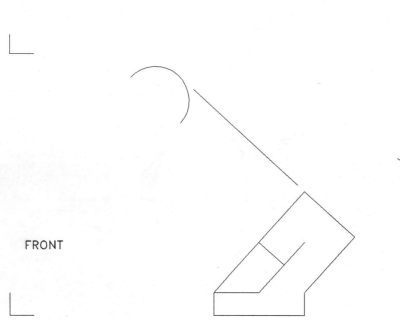

FRONT

SUGGESTED VIEW LAYOUT

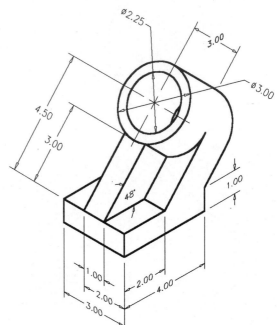

Problem 6–11   Primary auxiliary view (in.)

Title: 135 Bracket
Material: Aluminum

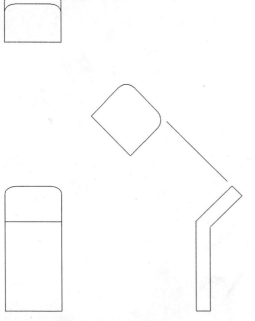

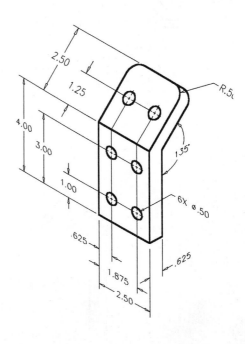

SUGGESTED VIEW LAYOUT

Problem 6–12   Primary auxiliary view (in.)

135-1 Bracket
Material: Aluminum

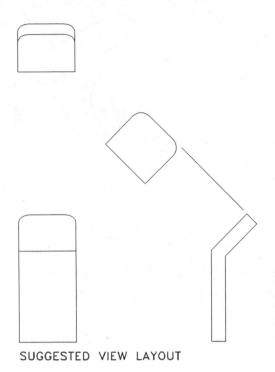

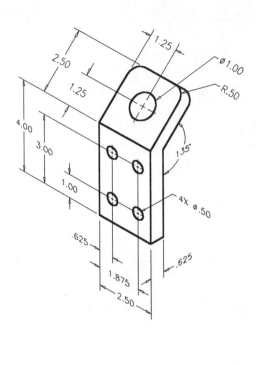

SUGGESTED VIEW LAYOUT

Problem 6–13   Primary auxiliary view (in.)

Title: Support Base
Material: Cast Aluminum

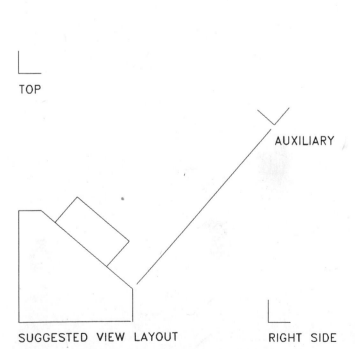

TOP

AUXILIARY

SUGGESTED VIEW LAYOUT   RIGHT SIDE

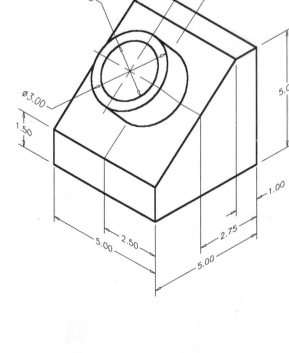

Problem 6–14   Primary auxiliary view (in.)

Title: Shaft Support
Material: SAE 1020

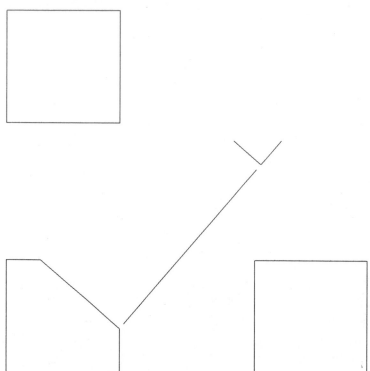

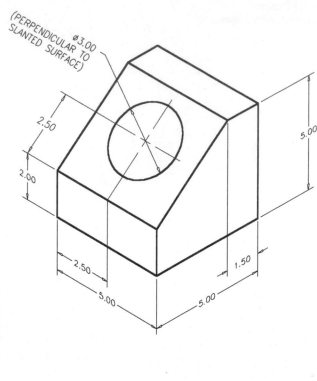

SUGGESTED VIEW LAYOUT

Problem 6–15   Primary auxiliary view (in.)

Title: Spacer
Material: Cast Iron

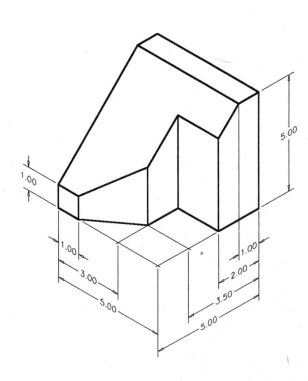

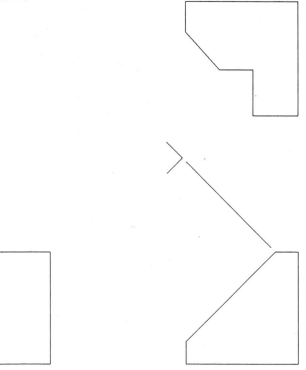

SUGGESTED VIEW LAYOUT

Problem 6–16 Primary auxiliary view (in.)

Title: T-Block
Material: SAE 4320

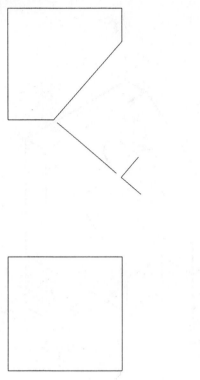

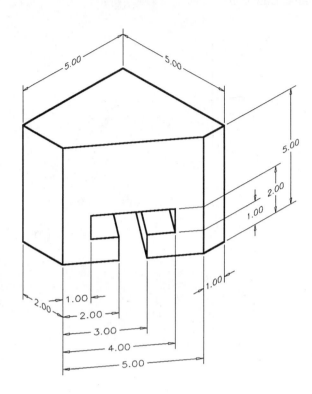

SUGGESTED VIEW LAYOUT

Problem 6–17 Primary auxiliary view (in.)

Title: T-Wedge
Material: Mild Steel

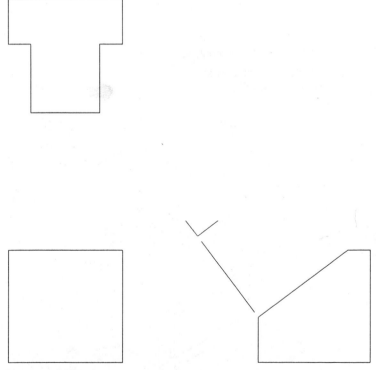

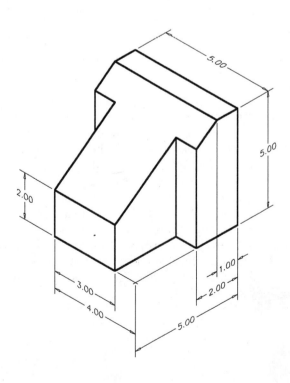

SUGGESTED VIEW LAYOUT

Problem 6–18   Primary auxiliary view (in.)

Title: Brace
Material: Cast Iron

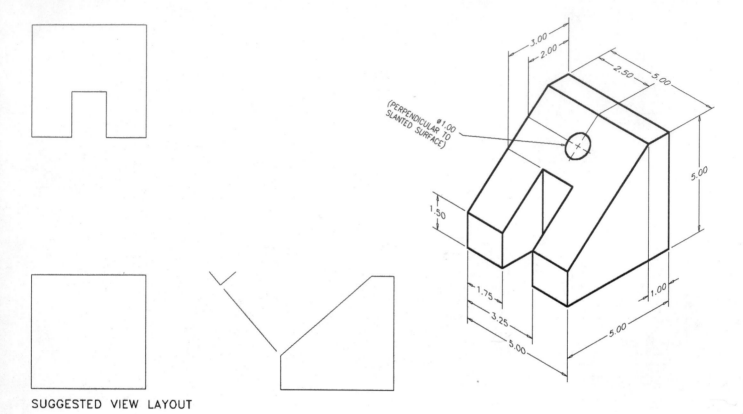

SUGGESTED VIEW LAYOUT

PART 2

DIRECTIONS

1. Make a preliminary sketch of the objects in Problems 6–19 through 6–26. Try to determine which view would be the best front view. Sketch at least one view showing the true shape of the slanted surface(s). Be sure to relate that view correctly to the standard views.

2. Using your sketch as a guide, make a multiview drawing that includes the necessary standard views and the full or partial auxiliary view(s) needed. Use a proper scale on the appropriate size drawing sheet. Use computer-aided drafting if required by your course. Do not dimension your drawings.

Problem 6–19   Primary auxiliary view (in.)

Part Name: Angle Base
Material: Mild Steel

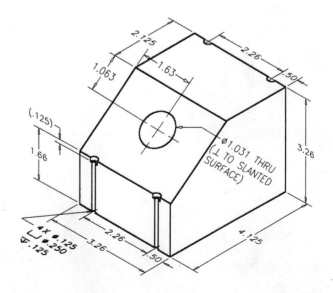

Problem 6–20 Primary auxiliary view (metric)

Part Name: Belt Guide
Material: Aluminum

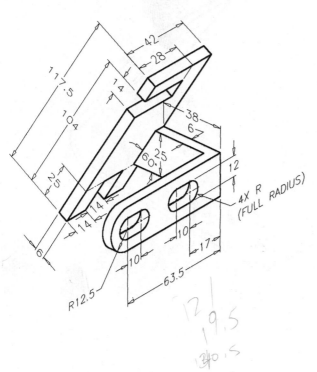

Problem 6–22 Primary auxiliary views (metric)

Part Name: Pivot Link
Material: Aluminum Number 195

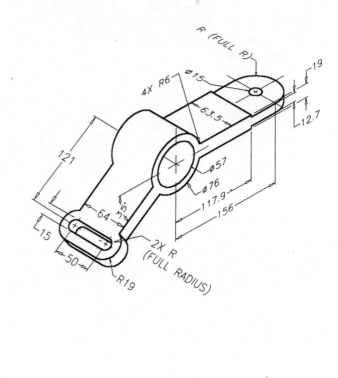

Problem 6–21 Primary auxiliary view

Part Name: Angle V-Block
Material: Mild Steel

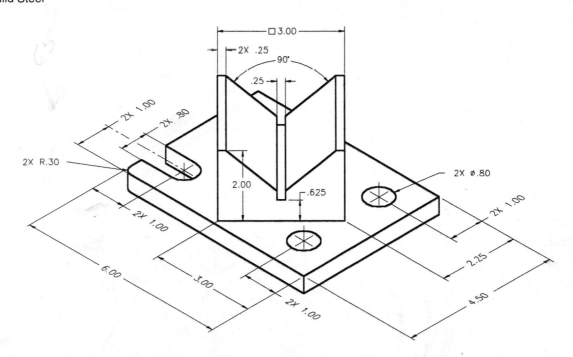

Problem 6–23   Primary auxiliary views

Part Name: Mounting Bracket
Material: Mild Steel

Problem 6–24   Secondary auxiliary view (metric)

Part Name: Skewed Face Plate
Material: SAE 1040

Problem 6–25   Secondary auxiliary views

Part Name: Angle Bracket
Material: Cast Iron

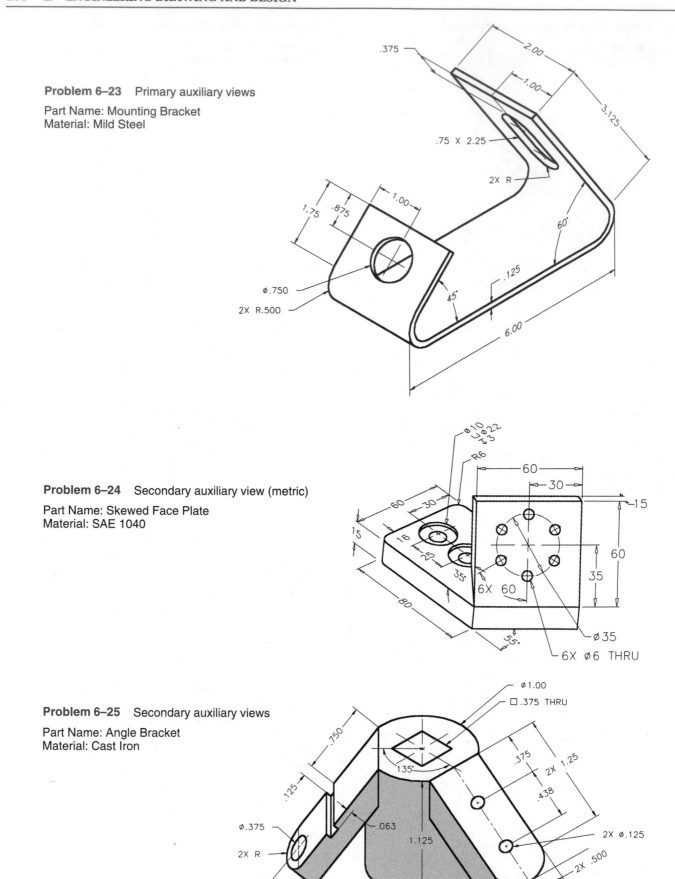

Problem 6–26   Secondary auxiliary views (metric)

Part Name: Chassis Switch Plate
Material: 4-mm-thick .416 Stainless Steel

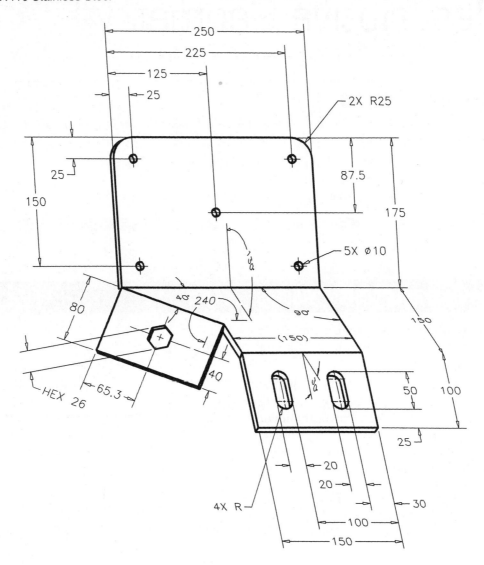

MATH PROBLEMS

A solar panel is in the shape of a 4' × 8' rectangle. With the sun overhead, what is the area of the shadow it casts on the ground if it is inclined at the following angles?

1. 10°
2. 20°
3. 30°
4. 80°
5. 90°
6. What should be the area of a solar panel inclined at 45° if the projected area from an overhead sun is to be 100 ft^2?

Descriptive Geometry I

THE ENGINEERING DESIGN PROCESS

Following is a sample descriptive geometry problem solved by the engineering design process.

1. *Identify and define the problem.* Find the total length of the ridge and hips of the roof on the house in Figure 7–1 in order to cover them with the proper length of sheet metal flashing. Restated, find the true length of the ridge and hips shown. Understand that the hips and ridge are lengths to be determined. Know what flashing is.
2. *Research and generate solutions.* To find the true lengths, many methods can be used.
   a. Fold line technique. (As explained in this chapter.)
   b. The revolution method. (As explained in this chapter.)
   c. Mathematics: Apply trigonometry concepts.
   d. Approximation: Compare the lengths to known distances and guess.
   e. Climb up on the roof and measure the actual distance.

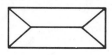

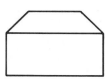

Figure 7–2   Scaled drawing of a house.

3. *Evaluate the possible solutions.* Assume there is an accurately scaled two-view drawing to work from as in Figure 7–2. The quickest solutions would be 2(a) and 2(b) or a combination of the two, listed above. The choice depends on which one is easier for you to do.
4. *Use the best solution.* The ridge is already shown in true length in the top projection plane. The hip lengths can be revolved, or use the fold line technique to determine the true lengths. (See Figure 7–3.)
5. *Evaluate the best solution.* Did it work? The answer is yes, and it was quick and simple.
6. *Finalize the solution.* In this case, the solution worked. Therefore, the problem is solved.

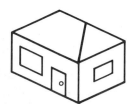

Figure 7–1   Problem set-up.

(Continued)

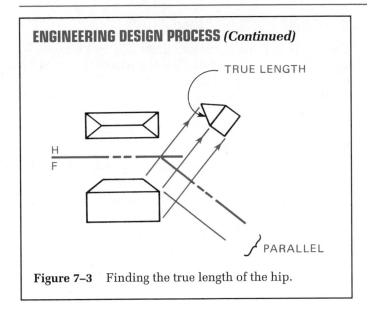

ENGINEERING DESIGN PROCESS *(Continued)*

Figure 7–3   Finding the true length of the hip.

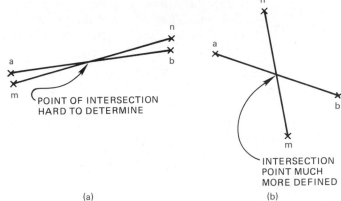

(a)   (b)

Figure 7–4   Accuracy of intersecting lines.

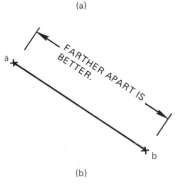

TOO CLOSE

MANY LINES CAN BE DRAWN
THROUGH POINTS a AND b WITH
APPARENT ACCURACY. IT IS
DIFFICULT TO LOCATE PLACEMENT
OF THE CORRECT LINE.

(a)

FARTHER APART IS BETTER.

(b)

Figure 7–5   Locating a line by two points.

In descriptive geometry, problems are solved *graphically* by using the relationships between points, lines, planes, and curved surfaces. These relationships can be found by making projections from projection planes of these points, lines, and surfaces. This chapter will use problem-solving principles similar to those explained in previous chapters and will also discuss some additional concepts. Descriptive geometry principles are most valuable for determining the angle between two lines, two planes, or a line and a plane, or for locating the intersection between two planes, a cone and a plane, or two cylinders.

The solution of descriptive geometry problems is entirely graphical. When using CADD, the solutions are nearly perfect. The reason for this perfection has to do with the capabilities of the hardware and software, and the fact that you are actually displaying the mathematical counterpart of the graphic analysis. In other words, the points, lines, or planes are a representation of the mathematical coordinates of the problem being presented. Therefore, the solution's accuracy is only controlled by the accuracy potential of the computer. A CADD system also allows you to work in layers and colors for a clearer picture of the problem and solution. For example, you can set a layer for projection lines where these lines show in red. This may help you visualize the entire problem and keep your solution clearer. However, when you are using the board method for solutions, the accuracy of the answer depends on the accuracy used in creating the solution and, sometimes, the method of solution. For example, the degree of accuracy in locating the intersection of two lines depends on the angle between the lines. The degree of accuracy increases as the angle between the lines approaches 90°, as shown in Figure 7–4. When the lines are nearly parallel, you should find the point of intersection by another method for better accuracy. It should be noted that when the

location of a line is to be determined by finding two points on it, the points should be located as far apart as possible, as shown in Figure 7–5. Always use the method that produces the most accuracy.

▼ DESCRIPTIVE GEOMETRY FUNDAMENTALS

You should be able to solve the descriptive geometry problems in these two chapters if you thoroughly understand the following two concepts and know how to apply them:

Concept 1:   You must know how to *locate a point* on additional projection planes when you are *given* that point location on two *adjacent* projection planes. This procedure is deceivingly simple, but it is very important that you do it properly.

Concept 2:   When creating a new fold line to establish another projection plane, you must understand how to *position* that *fold line* to the *given line* or to the *true length of the line* to acquire the information you want.

Applications appear throughout Chapters 7 and 8 on descriptive geometry when these concepts apply.

One good way to learn the concepts of descriptive geometry is to understand the process of projecting a point onto various projection planes. Add to that the projection of a line onto various projection planes (two points can define a line), the projection of plane onto various projection planes (three points not in a straight line can define a plane), and you have mastered the basics of descriptive geometry.

In addition to the practical applications of descriptive geometry, these concepts provide valuable training in the skills of visualizing points, lines, and planes in space. If visualization continues to be difficult, then memorizing the basic steps in solving descriptive geometry problems will be necessary.

There are usually several good ways to solve descriptive geometry problems. An attempt has been made in this chapter to include at least one and in some cases, two understandable solutions for each situation.

Whether solving descriptive geometry on the board or by CADD, you need to define the problem. The best way to understand the problem is to develop a sketch of the given situation. Be sure to include all known information. Secondly, sketch the processes of your possible solutions while not being too concerned about accuracy at this point. When you feel you have determined the correct technique to solve the problem, lay out the solution on the board or with CADD. It will then be a matter of drawing the process carefully to create an accurate solution.

▼ STANDARDS OF PROJECTION

The following elements are illustrated in Figure 7–6, which is a pictorial drawing of some basic descriptive geometry terminology. Figure 7–7 illustrates the same multiview box in third-angle projection of the projection planes unfolded.

To understand the explanations in this chapter, the following definitions are assumed.

1. *Line of sight*—An imaginary straight line from the eye of the observer to a point on the object. All lines of sight for a particular projection plane are assumed to be parallel, which means that the observer is either an infinite distance away or the

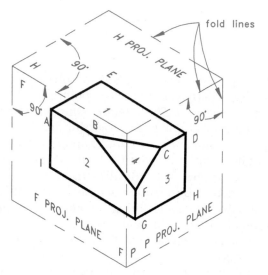

Figure 7–6 Basic descriptive geometry terminology illustrated on a standard third-angle projection multi-projection box system.

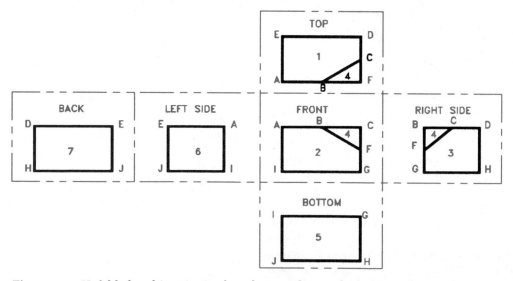

Figure 7–7 Unfolded multi-projection box showing the standard planes of projection in flat projection layout.

observer occupies a slightly different position when looking at each point on the object.

2. *Projection plane*—An imaginary surface on which the view of the object is projected and drawn. This surface is imagined to exist between the object and the observer. The observer's line of sight is always perpendicular to the projection plane.

3. *Fold line*—The line of intersection between two projection planes. A fold line appears as an edge view of a projection plane when that projection plane is folded back 90° from the plane being viewed. It is the line on which one plane is folded to bring it into the plane of the adjacent (touching) projection plane. The fold line represents a 90° change of direction. Fold lines are called out in the text as H/F, F/P, and F/A.

4. *Projection lines* or *projectors*—Straight lines drawn perpendicular (at 90°) to the fold lines, which connect the projection of a point in a projection plane to the projection of the same point in an adjacent projection plane. These lines are required to obtain the point locations in the adjacent projection planes.

5. *Front projection planes*—Shows the projected size of the front of the object. The lines of sight are horizontal and the projection plane is vertical in an elevation view.

6. *Profile projection plane*—Shows the projected size of the vertical side of an object. The lines of sight are horizontal and the projection plane is vertical in an elevation view.

7. *Horizontal* or *top projection plane*—Shows the projected size of the vertical side of an object. The lines of sight are vertical, and the projection plane is level with the horizon in an elevation view.

8. *Adjacent projection planes*—Perpendicular (90°) projection planes that are next to each other in space. They have a common fold line between them. An object projected on any two adjacent projection planes will contain all the dimensions of that object.

▼ PROJECTION OF A POINT

Figure 7–8 shows a point that has been projected onto three mutually perpendicular projection planes. The actual point itself is shown by a small six-pointed cross. Any method can be used to identify a point; the idea is to be consistent. In this explanation, the real point is identified by a capital letter. The letter A was chosen arbitrarily in Figure 7–8.

Everywhere point A is projected, a cross and a lower-case letter with an attached subscript lower-case letter, a, is used. The subscript lower-case letter corresponds to the projection plane onto which the point was projected. For example, a_f is the projection of point A on the front projection plane. In Figure 7–9a, the fold lines

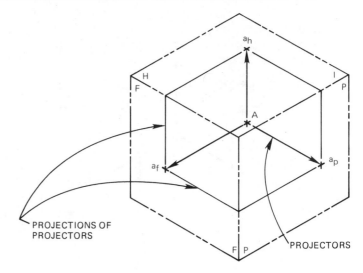

Figure 7–8   Projection of a point on multi-projection representation.

are identified by two upper-case letters corresponding to the two adjacent projection planes. Right angle symbols have been used in this example to remind you that the projectors of point A project to the projection planes at 90° angles to the plane—*always!*

Given the horizontal and front projection planes of point A, Figure 7–9b shows the sequence used in locating the correct position in the P projection plane.

Step 1.   Draw a projection line from a_h, to a_f, then draw a projector from a_f perpendicular through the fold line F/P onto the projection plane, P, at any distance at this step. Then, measure the distance, D, from point a_h to fold line H/F.

Step 2.   Transfer distance D from fold line F/P along the projector of a_f onto projection plane P. Mark and identify this spot with a cross and the letters a_p.

In Figure 7–10, the same method is used to locate a point in an auxiliary projection plane. In this example, the location of fold line H/A was chosen arbitrarily, but later on, the exact placement of the fold line will be important when solving advanced descriptive geometry problems. To project a point through successive auxiliary projection planes, use the fold lines for reference measurements as shown in Figure 7–11. Distance D is determined first again, then distance X is established. Notice that in each situation, a projection plane has been *skipped* to determine the correct distance to locate the point in the correct projection plane.

▼ PROJECTION OF A LINE

A line is represented by two points that are connected. We have already learned how to project a point onto dif-

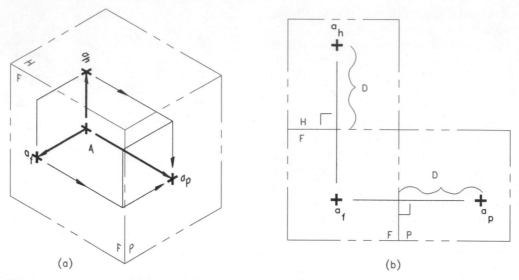

Figure 7–9 Locating a point on the profile projection plane: **(a)** multi-projection representation; **(b)** descriptive geometry flat projection layout.

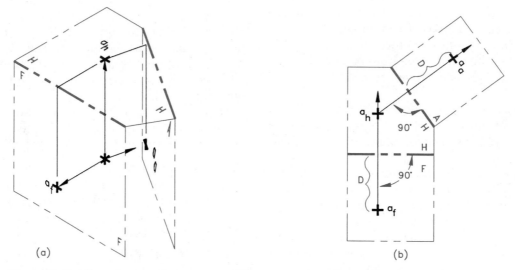

Figure 7–10 Locating a point on an auxiliary projection plane: **(a)** multi-projection representation; **(b)** descriptive geometry flat projection layout.

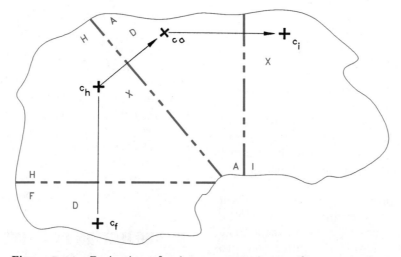

Figure 7–11 Projection of point on successive auxiliary projection planes using flat projection layout system.

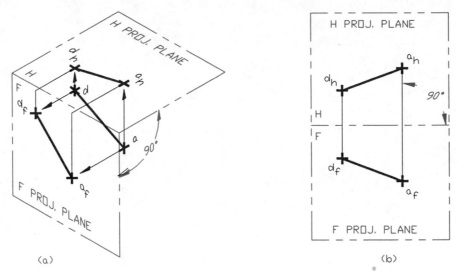

Figure 7–12   A line shown projected onto two adjacent projection planes: **(a)** multi-projection representation; **(b)** descriptive geometry flat projection layout.

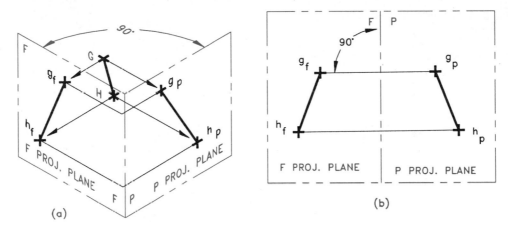

Figure 7–13   A line shown projected onto two adjacent projection planes: **(a)** multi-projection representation; **(b)** descriptive geometry flat projection layout.

ferent projection planes. Project a line onto different projection planes in exactly the same manner as you do one point, except project two points. Figures 7–12 and 7–13 show two lines projected onto two adjacent projection planes.

Placement of the Fold Line

When solving real problems, the new fold line must always be placed parallel or perpendicular to the given line depending on what you are trying to find. This is explained in more detail further in this chapter. Observe Figures 7–14 through 7–24. The fold line is placed parallel to the given line to obtain the true length. Notice in Figure 7–27 the fold line is placed perpendicular to the line where it is shown in true length in order to see the line as a point.

▼ CLASSIFICATION OF LINES IN SPACE

In general, lines are classified or named according to the projection plane to which they are parallel.

○ *Front* or *frontal line*—A line that is parallel to the front projection plane and appears in true length in the front projection plane. (See Figure 7–14.)

○ *Horizontal line*—A line that is parallel to the horizontal projection plane and appears true length in the horizontal projection plane. (See Figure 7–15.)

○ *Profile line*—A line that is parallel to the profile projection plane and appears true length in the profile projection plane. (See Figure 7–16.)

○ *Vertical line*—A line that is both parallel to the front projection plane and the profile projection plane.

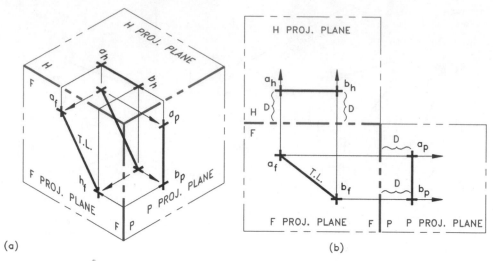

Figure 7–14 A frontal line: **(a)** multi-projection representation; **(b)** descriptive geometry flat projection layout.

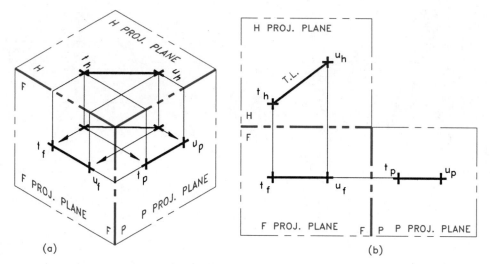

Figure 7–15 A horizontal line: **(a)** multi-projection representation; **(b)** descriptive geometry flat projection layout.

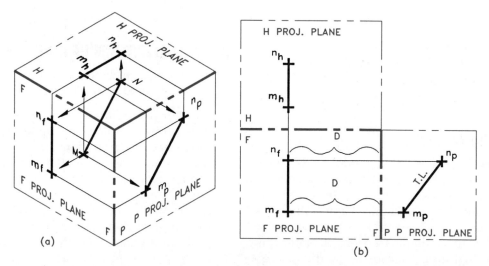

Figure 7–16 A profile line: **(a)** multi-projection representation; **(b)** descriptive geometry flat projection layout.

○ *Inclined line*—A line that appears inclined in one principal projection plane and is parallel to the other two principal projection planes.
○ *Oblique line*—A line that is parallel to none of the front, profile, or horizontal projection planes. (See Figure 7–17.)
○ *Parallel lines*—Lines that are equal distance from each other throughout their lengths. (See Figure 7–18.) Parallel lines will appear parallel in all projection planes with only two exceptions. They will not appear parallel when one line is projected behind the other as shown in the front projection plane in Figure 7–18 where they appear as *one line*. And, they will not appear parallel when they are projected onto a projection plane as ends of the lines. In Figure 7–18, they appear as *points* in the profile projection line.

○ *Foreshortened line*—If a line appears shorter than it really is, it is called foreshortened.
○ *True length line*—The true length of a line is the actual straight-line distance between its two ends.

Finding the True Length of Frontal, Profile, or Horizontal Lines

The *true length* of any line is found only upon a projection plane parallel to that line. In Figure 7–19, the *frontal line* A-B is parallel to the front projection plane; therefore, its true length is shown in the front projection plane. In Figure 7–20, the *profile line* C-D is parallel to the profile projection plane and shows its true length in the profile projection plane. Likewise, in Figure 7–21 the *horizontal line* R-S is parallel to the horizontal projection plane and shows its true length in the horizontal pro-

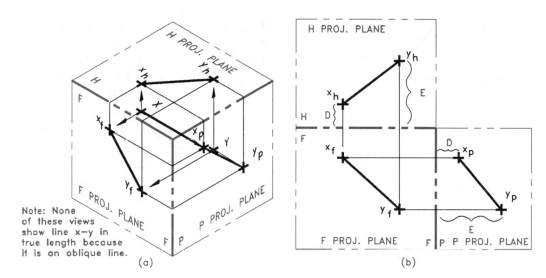

Figure 7–17   An oblique line: **(a)** multi-projection representation; **(b)** descriptive geometry flat projection layout.

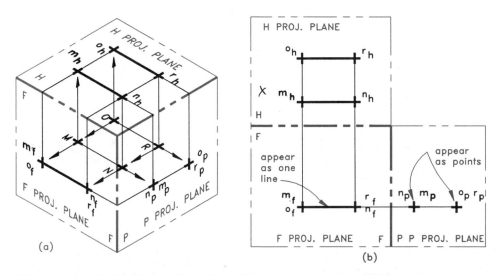

Figure 7–18   Parallel lines: **(a)** multi-projection representation; **(b)** descriptive geometry flat projection layout.

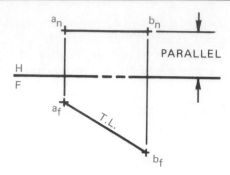

Figure 7–19 True length shown on the frontal projection plane.

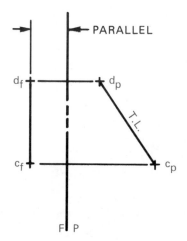

Figure 7–20 True length shown on the profile projection plane.

jection plane. Note the *T.L.* identifying the line where it shows as *true length*.

Finding the True Length of an Oblique Line

To find the true length of any line, oblique or otherwise, you must be given or be able to construct two

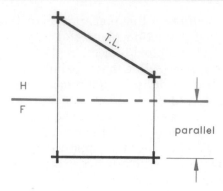

Figure 7–21 True length shown on the horizontal projection plane.

adjacent projection planes showing the location of that line on each projection plane. Given this information, two good methods are used to find the true length.

Fold Line Method. The fold line method represents the concept of the viewer *moving around the line* in such a position that the line is parallel to the line of sight. In this position, the viewer will see the line in true length. To do this graphically, follow these steps:

Step 1. Refer to Figure 7–22a and b. Locate new fold line H/A parallel to line C-D creating a new projection plane A in part (a) or locate new fold line F/R parallel to line C-D creating new projection plane R in part (b).

Step 2. Refer to Figure 7–23a and b. Extend projectors from points c_h and d_h perpendicular to fold line A/H and onto projection plane A at any convenient lengths in part (a) or extend projectors from points C_f and d_f perpendicular to fold line F/R and onto projection plane R at any convenient lengths in part (b) as shown.

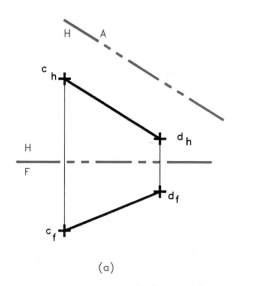

or

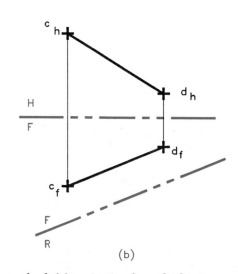

(a)   (b)

Figure 7–22 Step 1, finding true length by fold line method: **(a)** projecting from the horizontal projection plane; **(b)** projecting from the frontal plane.

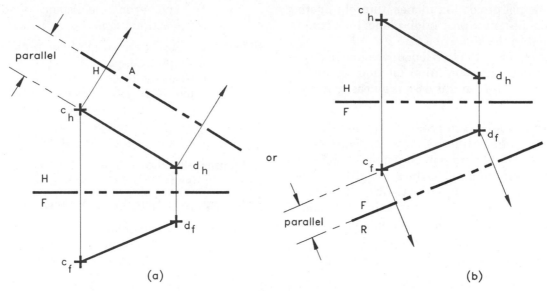

Figure 7–23   Step 2, finding true length by fold line method: **(a)** projecting from the horizontal projection plane; **(b)** projecting from the frontal plane.

Step 3.   Refer to Figure 7–24a. Transfer distance X, the distance from fold line H/F to point c_f onto projection plane A along the projector from point c_h from fold line A/H creating point c_a. Then, transfer distance Y, the distance from fold line H/F to point d_f onto projection plane A along the projector from point d_h from fold line H/F creating point d_a. Connect point c_a to point d_a with a straight line and you will have created the true length of line C-D on the A projection plane.

Refer to Figure 7–24b. Transfer distance M, the distance from fold line H/F to point c_h onto projection plane R along the projector from point c_f from fold line F/R creating point c_r. Then, transfer distance N, the distance from fold line H/F to point d_n onto projection plane R along the projector from point d_f from fold line F/R creating point d_r. Connect point c_r to point d_r with a straight line and you will have created the true length of line C-D on the R projection plane.

Revolution Method.   When using the revolution method to find the true length of a line, *we graphically move the line*, not the viewer. This can usually be done

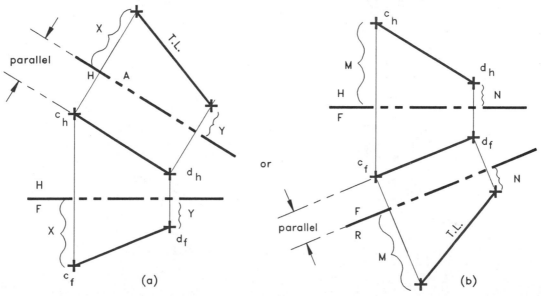

Figure 7–24   Step 3, finding true length by fold line method: **(a)** projecting from the horizontal projection plane; **(b)** projecting from the frontal plane.

in the two original projection planes, thus eliminating the need to create a new projection plane; for example, if we need to find the slope of a line.

We will need to find the true length of a line in an elevation projection. The revolution method for finding the true length of a line in this case is an easier method to use.

Step 1.   Given two adjacent projection planes showing the location of line C-D, refer to Figure 7–25b. Revolve d_h around point c_h on the H projection plane until point d_h is the same distance from fold line H/F as point c_h. Create and label the new point as d_{h1}. This makes line C-D parallel to fold line H/F in the H projection plane.

Step 2.   Project a projector from point d_{h1} in the H projection plane perpendicular to fold line

H/F and beyond the distance of point d_f on the F projection plane.

Step 3.   On the F projection plane, project a projector from point d_f parallel to fold line H/F until it crosses the projector of point d_{h1}. Identify the intersection created as point d_{f1}. Connect point c_f and point d_{f1} with a straight line and you will have developed the true length of line C-D on the F projection plane.

Figure 7–26 shows how to find the true length by revolving line C-D about *either* end point.

End View of a True Length Line

When viewing the end of a true length line, it will appear as a *point*. There are several problems in descriptive geometry, such as finding the edge view of a plane

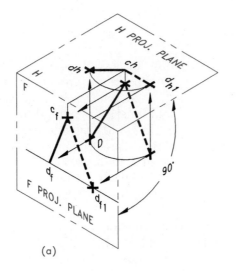

(a)

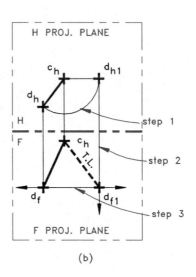

(b)

Figure 7–25   Find the true length of a line by the revolution method: **(a)** multi-projection representation; **(b)** descriptive geometry flat projection layout.

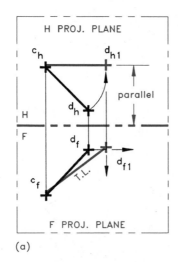

(a)

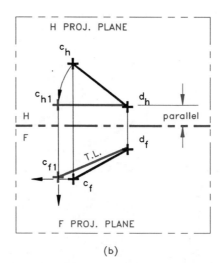

(b)

Figure 7–26   Finding the true length of a line by revolution: **(a)** revolving line C-D around point C; **(b)** revolving line C-D around point D.

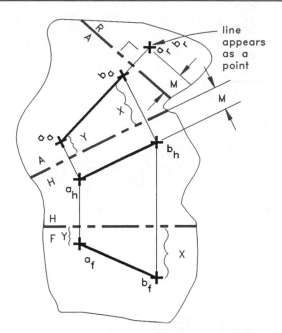

Figure 7–27 Showing the end view of a true length line A-B as a point on a descriptive geometry flat projection layout.

where this concept will be necessary to know. Refer to Figure 7–27. After finding the true length of line A-B, place new fold line A/R perpendicular to the true length line A-B located on the A projection plane. Find the point a_r and point d_r on the R projection plane as previously explained in locating a point. Notice when you do this that both points are located in the *same* position. In other words, line A-B appears as a *point*. Notice also that the point is identified by the letters b_r-a_r where b_r comes first because this end is seen first on the R projection plane. The dash is included to further clarify that this is the end of a *line,* not just a *point.*

▼ PLANES

Without question, problems involving planes and plane surfaces are some of the most common found in indus-

try. The basic principles involving planes are applicable in most industrial fields. A *plane* is a flat surface that is not curved or warped, and in which a straight line will lay. Figure 7–28 shows four different ways a plane may be represented in graphic form: by three points not in a straight line, by one straight line and one point not on the line, by two parallel lines, and by two intersecting lines.

▼ CLASSIFICATION OF PLANE SURFACES

○ *Normal planes*—Plane surfaces are classified similar to lines in that they are called normal planes if the planes are parallel to any of the primary projection planes.
   a. *Front* or *frontal plane*—A plane parallel to the front projection plane, such as plane 2 in Figure 7–29. It appears in true shape on the front projection plane.
   b. *Horizontal plane*—A plane parallel to the horizontal projection plane, such as plane 1 in Figure 7–29. It appears in true shape on the horizontal projection plane.
   c. *Profile plane*—A plane parallel to the profile projection plane, such as plane 3 in Figure 7–29. It appears in true shape on the profile projection plane.
○ *Inclined plane*—When a plane appears distorted on two projection planes and as a line or edge on the other projection plane.
○ *Oblique plane*—A plane that is not parallel to any of the primary projection planes, such as plane 4 in Figure 7–29. Some or none of the dimensions will be true in an oblique plane.
○ *True shape of the plane*—The actual or real size and shape of the plane. In Figure 7–30, plane A-B-C is parallel to the horizontal projection plane and will appear in true shape and size on this projection plane.

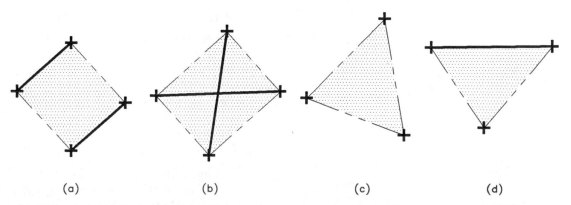

(a)   (b)   (c)   (d)

Figure 7–28 Different ways of representing planes: **(a)** three points not in a staight line; **(b)** one straight line and one point not on the line; **(c)** two intersecting straight lines; **(d)** two parallel straight lines.

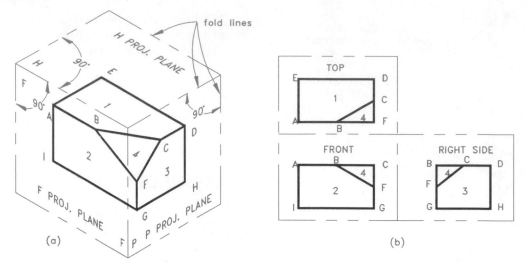

Figure 7–29 Planes: **(a)** planes on a multi-projection box; **(b)** planes shown on multi-projection box unfolded.

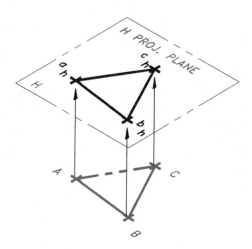

Figure 7–30 True shape of a horizontal plane.

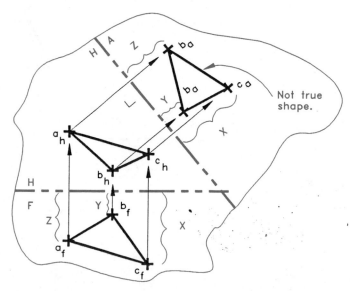

Figure 7–31 Projecting a plane on an auxiliary projection plane.

Projection of a Plane

A plane can be represented by *three points not in a straight line* that are connected with straight lines. To project a plane into different planes do it in exactly the same manner as you do one point, except project three points. (See Figure 7–31.)

Finding the Edge View of a Plane

Knowing how to find the *edge view* of a plane is an essential concept to understand in order to be able to find the true shape of a plane. The edge view of a plane appears as a *straight line.* The edge view of a plane (a straight line) can be seen on a projection plane that shows the *end view* of any line (a point) contained in that plane. A *normal plane* shown on the standard projection planes will *already* appear on an *edge view* on its adjacent projection planes.

A plane can consist of a series of parallel lines. The end view of a true length line appears as a point. Therefore, when you look at the end view of a true length line included in the plane, you see the plane as an *edge view* (straight line).

Finding the Edge View of an Oblique Plane

To find the edge view of an oblique plane, we must first find a true length line in that plane. When we find the projection plane where the true length line on the oblique plane appears as a point, the oblique plane will also appear as an edge view.

Given: An oblique plane shown on two adjacent projection planes.

Find: An edge view of the oblique plane.

Step 1. Find a line in true length on the oblique plane in one of the two given adjacent projection planes. Figure 7–32 shows one method to find a true length line on the oblique plane. Line

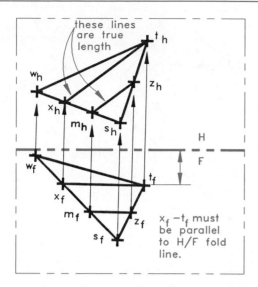

Figure 7–32 Finding a true length line on an oblique plane.

M-N is drawn parallel to fold line H/F in the F projection plane. *Note:* We could have drawn line M-N in the H projection plane parallel to the H/F fold line and the solution would be the same. Line M-N can be drawn anywhere on the oblique plane as long as it is parallel to fold line H/F. However, when solving by the board, better accuracy will result when the line is located where the end points of the line have the most distance between them. In this example, the other line, X-T, makes use of the longer distance as well as using the existing point t_f, requiring the need to find only one new point, not two. Find point X on the H projection plane, and identify it as point x_h. When point t_h and point x_h are connected with a straight line, you have found the true length of line X-T.

Step 2. To find the projection plane where line x_h-t_h will appear as a point, place new fold line H/A perpendicular to true length line x_h-t_h as shown in Figure 7–33. Then, find all the rest of the points of the oblique plane, W-S-T, on the A projection plane just as you have done in previous examples of projecting points. The points w_a, s_a, t_a, and x_a, when located accurately, will form into a straight line representing the edge view of the oblique plane W-S-T.

Finding the True Shape of a Plane by the Fold Line Method

In order to find the true shape of any plane, the edge view of the plane must first exist or be found as in the previous example. Use plane W-S-T as shown in Figure 7–33 to work from the edge view found. Figure 7–34 shows how to find the true shape of plane W-S-T. The approach here uses the fold line method (walking around in order to view the plane in true shape). Place new fold

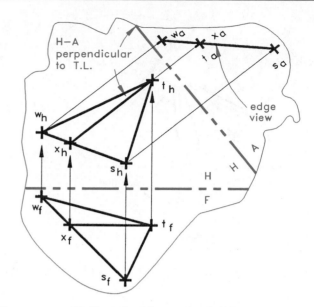

Figure 7–33 Finding the edge view of a plane.

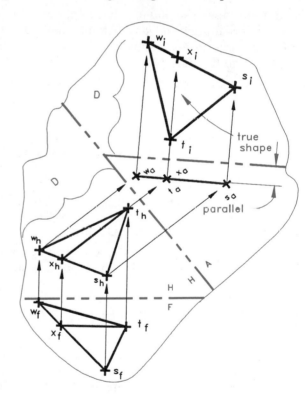

Figure 7–34 Finding the true shape of a plane by the fold line method.

line A/I parallel to the edge view of plane W-S-T. Then, project points w_a, s_a, and t_a correctly onto the I projection plane. When the points are connected with straight lines, they will form the true size and shape of plane X-Y-Z.

Finding the True Shape of a Plane by the Revolution Method

In this method you must also have an edge view or find the edge view of the plane to be able to show the

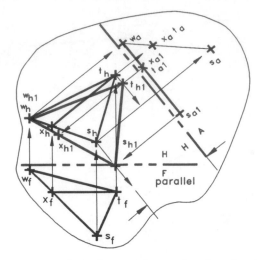

Figure 7–35 Finding the true shape of a plane by the revolution method.

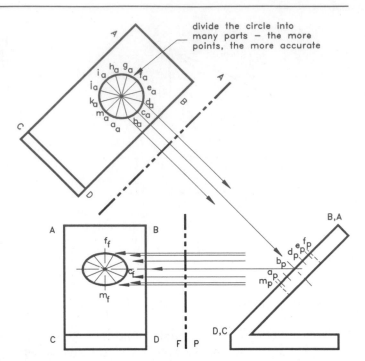

Figure 7–36 Finding the true shapes of irregular features.

plane in true shape. Use plane W-S-T as shown in Figure 7–31 to work from the edge view found on the A projection plane. Figure 7–35 shows how the plane was revolved around the point w_a.

Step 1. Revolve points s_a, t_a, and x_a parallel to fold line H/A and rename them s_{a1}, t_{a1}, and x_{a1}. Project these points perpendicular onto the H projection plane until they pass the distances from fold line H/A as points s_h, t_h, and x_h are from fold line H/A. Then, draw a projector from point s_h parallel to fold line H/A until it crosses the projector from point s_{a1}. Label this intersection as point s_{h1}. Revolve points t_{a1} and x_{a1} in the same manner as point s_{a1}.

Step 2. Connect points w_h, s_{h1}, x_{h1}, and t_{h1} with straight lines to form the true shape and size of plane W-S-T.

Projecting Irregular Shaped Planes

To determine the shapes of irregular or circular features on the inclined surfaces on the original projection planes you first need to find the true outer shape of the inclined surface as explained in the previous examples of finding the true shape of a plane.

Step 1. As shown in Figure 7–36, find the true shape of the inclined surface A-B-C-D of the object as found in the A projection plane. Locate the feature(s) in their correct position(s) on the true shape of the surface of the object.

Step 2. Create and project various points on the irregular shape(s) back onto the projection plane that shows the edge view of the inclined surface. The more points used, the more accurate looking the feature(s) will be projected back onto the original projection planes.

Step 3. Project the points from the edge view back onto the original projection planes. When the points are connected, they will establish what the features will look like as viewed from those projection planes.

PROFESSIONAL PERSPECTIVE

Finding the correct solutions to descriptive geometry problems is mainly dependent on two things: The *method* used to solve the problems and the *accuracy* in constructing the drawing. There are different techniques that can be used to solve each problem graphically, and the challenge is to select the best method. There also must be extreme accuracy taken when drawing and constructing the various lines on the board. For people who tend to visualize objects easily, descriptive geometry is not too difficult to solve. Others who do not visualize easily will need to follow a step-by-step procedure—a cookbook approach—and memorize specific procedures to obtain the correct answers. Either way, solving descriptive geometry problems can and should be fun and intriguing.

MATH APPLICATION

It is important to be able to convert from one type of scale measurement to another when setting up descriptive geometry problems in order to have accuracy in your answers. The process for solution may be correct, but the answers will not be. Some examples of conversions follow.

1 in. = 2.54 cm

1 cm = .394 in. = $\frac{3}{8}$ + in.

1 ft = 30.48 cm

1 m = 100 cm = 39.37 in.

You may also need to calculate areas of planes in descriptive geometry problems. The area formulas are included for review. (See Figures 7–37 through 7–41.)

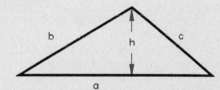

Figure 7–37   Area of a right triangle = ab/2

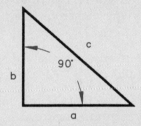

Figure 7–38   Area of a general triangle = ah/2

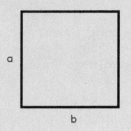

Figure 7–39   Area of a square = a^2

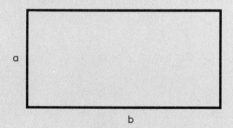

Figure 7–40   Area of a rectangular feature = ab

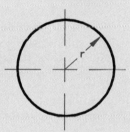

Figure 7–41   Area of a circle = πr^2

When calculating areas, most planes can be separated into at least one or more of these basic shapes. For example, see Figure 7–42.

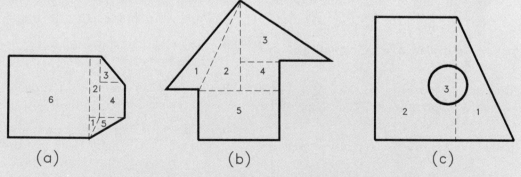

(a)   (b)   (c)

Figure 7–42   Calculating areas of odd shaped planes: **(a)** area separated into a rectangle and plane; **(b)** area separated into a triangle and a rectangle; **(c)** area separated into a rectangle and triangle minus area of a circle.

CHAPTER

7

DESCRIPTIVE GEOMETRY I TEST

DIRECTIONS

Answer the following questions with short, complete statements and/or show proportional thumbnail sketches to explain your answer.

1. Give two reasons why drawing accuracy is important when solving descriptive geometry problems on the board.

2. Describe two methods that can be used when finding the true length of a line or true shape of a plane.

3. Show by thumbnail sketches examples of the concepts:

   a. Parallelism
   b. Perpendicularity
   c. Fold line
   d. True length line
   e. Plane

4. In a standard three flat multi-projection layout (a) which two projection planes show the height of the object, and (b) which two projection planes will show the overall length of the object?

5. If two lines are parallel in space, they will appear parallel in all the projection planes except two. List them.

6. How must parallel lines in space appear in order to see the true distance between them?

7. How are normal lines classified?

8. How ar normal planes classified?

9. What does the end view of a point look like?

10. List four ways a plane can be described.

11. What does the end view of a true length line appear as?

12. A fold line must be parallel to the _____ of a plane to be able to determine the true shape of the plane.

13. Any surface seen in true size can also be seen in _____ .

14. To see a true length line as a point, you must rotate the line _____ degrees.

15. If three lines appear parallel in two projection planes, they will appear parallel in _____ other projection planes.

16. An edge view of a plane is represented by _____ .

17. The projection of a line onto an auxiliary projection plane placed parallel to the projection of the line will show the line in _____ .

18. How high a point in space is from the ground may be seen in any _____ projection plane.

19. A vertical line will appear as a _____ in the front projection plane.

20. Distances measured off in auxiliary projection planes are obtained from the _____ projection planes.

21. Projectors are always projected at _____ degrees to the fold line.

22. Horizontal distances can be found in the _____ projection plane.

23. A line in space can only be defined by _____ .

24. There is/are _____ method(s) used to solve descriptive geometry problems.

25. A line is oblique when it appears _____ in three of the normal projection planes.

26. A plane is a _____ surface.

27. When finding the distances to project a point onto an auxiliary projection plane, you must _____ one projection plane to get the dimension distances.

28. The fold line represents a _____ degree change of direction.

29. An inclined line would appear _____ in two of the normal projection planes.

30. If descriptive geometry problems are hard for you to solve, you may have to use _____ to solve descriptive geometry problems.

CHAPTER
7

DESCRIPTIVE GEOMETRY I PROBLEMS

DIRECTIONS

Please read the problems carefully before you begin working. Solve problems 7–1 through 7–25, showing all the projectors used, identifying brackets of the distances used, and correctly identifying the point(s) in the required view(s). Use a separate sheet of graph paper for each problem and the graph lines included with each problem for proper set-up. Use a scale of at least three squares to one to obtain accuracy in the problems' solutions.

Problem 7–1 Point Projection. Locate point A in the P projection plane.

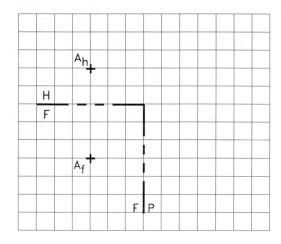

Problem 7–2 Point projection. Locate point R in the A projection plane.

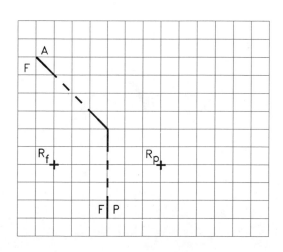

Problem 7–3 Point projection. Locate point T in the A projection plane.

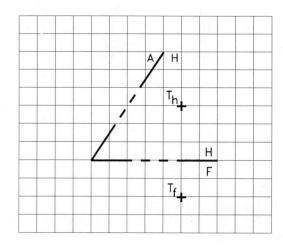

Problem 7–4 Point projection. Locate point F in projection planes A, G, and S.

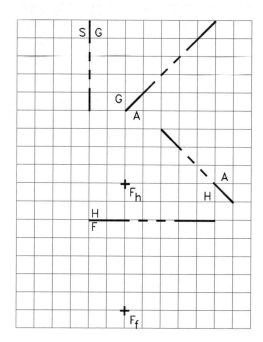

Problem 7–5 Point projection. Locate point W in projection planes B, C, F, G, R, and S.

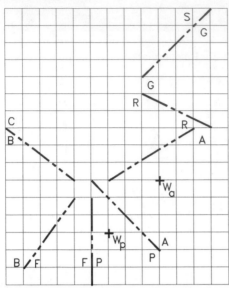

Problem 7–8 Line projection. Locate line M-N in the P projection plane. Identify the classification of line M-N.

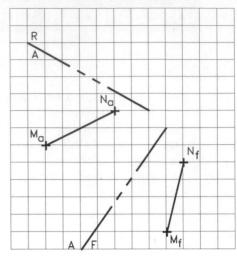

Problem 7–6 Line projection. Locate line A-B in the H projection plane. Identify the classification of line A-B.

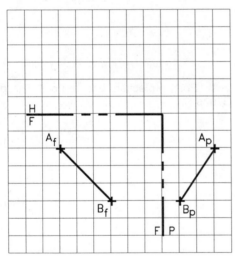

Problem 7–9 Line projection. Locate line D-E in the F projection plane. Identify the classification of line D-E.

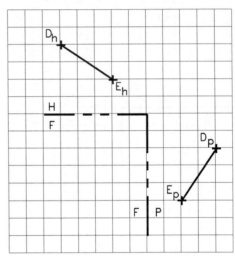

Problem 7–7 Line projection. Locate line S-T in the H projection. Identify the classification of line S-T.

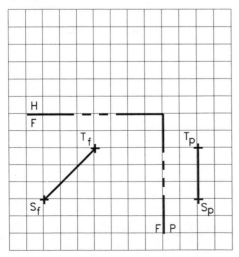

Problem 7–10 True length. Find the true length of line J-K by the (a) fold line method and (b) revolution method.

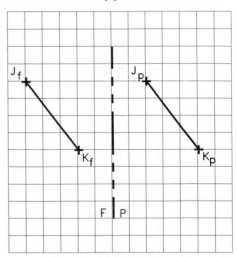

Problem 7–11 True length. Find the true length of line G-H by the (a) fold line method and (b) revolution method.

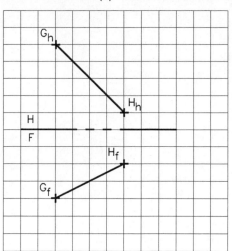

Problem 7–12 True length. Find the true length of line R-S by the (a) fold line method and (b) revolution method.

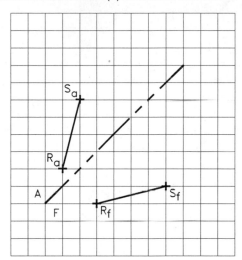

Problem 7–13 True length. Find the true length of line C-D by the (a) fold line method and (b) revolution method.

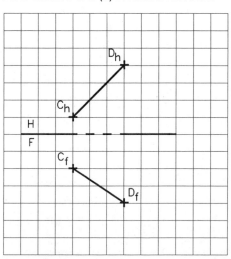

Problem 7–14 End view. Locate line A-B in a projection plane where it appears as a point.

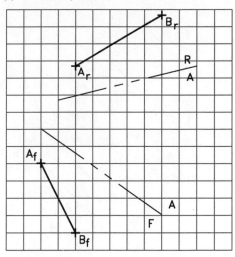

Problem 7–15 End view. Locate line G-H in a projection plane where it appears as a point.

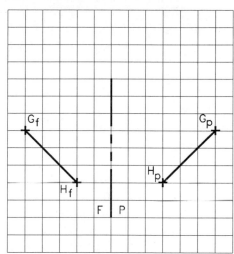

Problem 7–16 End view. Locate line M-N in a projection plane where it appears as a point.

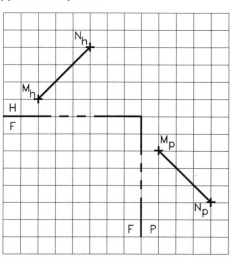

Problem 7–17 End view. Locate lines M-N and A-B in a projection plane where they appear as points.

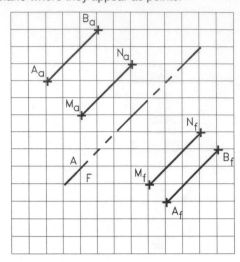

Problem 7–18 Projection of a plane. Complete the missing P projection plane of plane A-B-C.

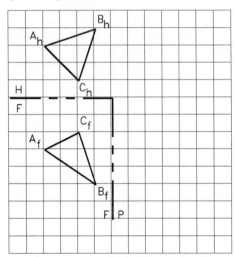

Problem 7–19 Projection of a plane. Complete the missing H projection plane of plane R-S-T-U.

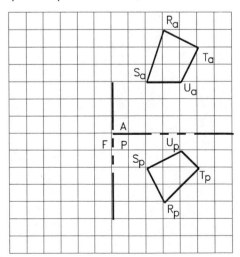

Problem 7–20 Projection of a plane. Complete the missing R projection plane of plane X-Y-Z.

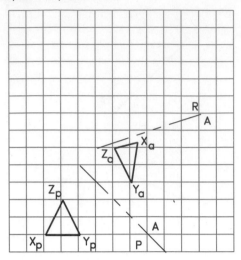

Problem 7–21 Projection of a plane. Complete the missing two projection planes of plane M-N-O-P-R-S.

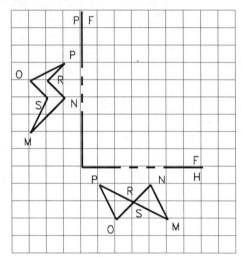

Problem 7–22 Edge view of a plane. Locate plane A-B-C in a projection plane where it appears as a line.

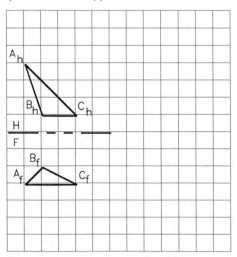

Problem 7–23   Edge view of a plane. Locate plane W-X-Y-Z in a projection plane where it appears as a line.

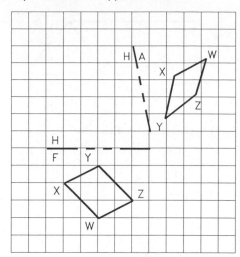

Problem 7–24   True shape of a plane. Locate plane R-S-T in a projection plane where you can see it in true shape.

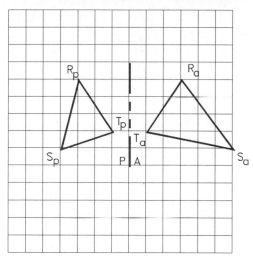

Problem 7–25   True shape of a plane. Locate plane A-B-C-D-E-F in a projection plane where you can see it in true shape.

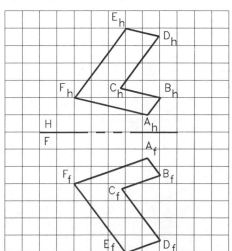

For problems 7–26 through 7–32, draw required multiviews needed to solve problems. Do not draw dimensions.

Problem 7–26   Find the true shape of surface A by the fold line method.

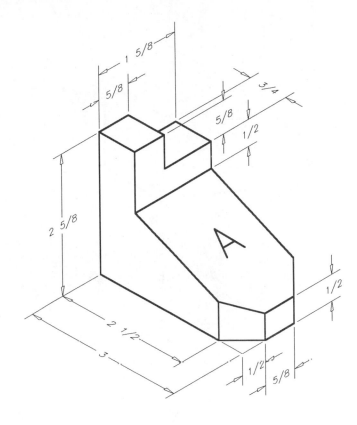

Problem 7–27   Find the true shape of surface B by the fold line method.

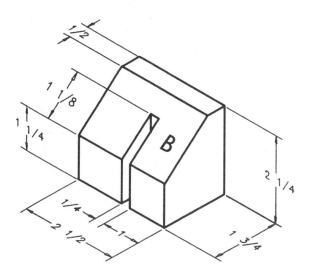

Problem 7–28 Find the true shapes of surfaces A and B by the fold line method.

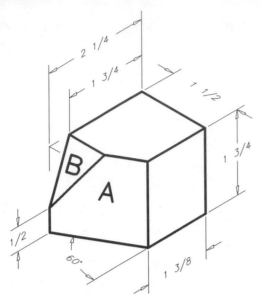

Problem 7–29 Find the true shape of surface D.

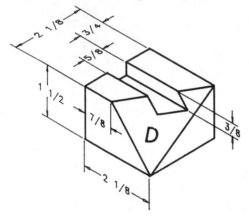

Problem 7–30 True shape of a plane. Find the true shapes of the surfaces A, B, C, D, E, and F of the concrete pier block.

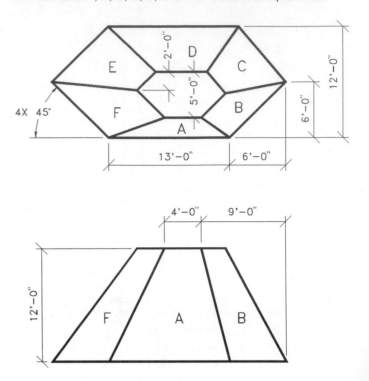

Problem 7–31 From the given engineer's sketch, draw the top, front, and appropriate projection planes of the sloped surface. Use the fold line method to create the true shape of the sloped surface.

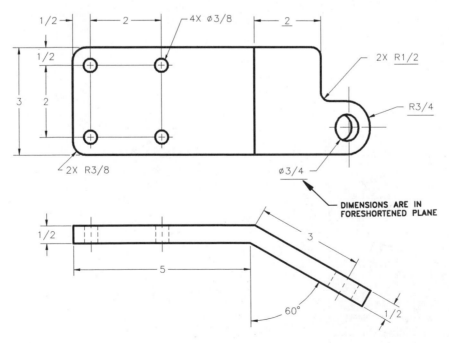

DIMENSIONS ARE IN FORESHORTENED PLANE

Problem 7–32 From the given engineer's sketch, draw the top, front, and appropriate projection plane of the sloped surface. Use the fold line method to create the true shape of the sloped surface.

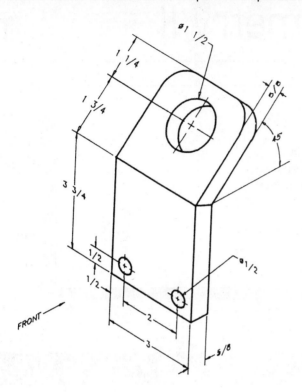

MATH PROBLEMS

Convert the following inch and feet dimensions to metric dimensions.

1. 3 in.
2. 100 ft
3. 5'–6"

Convert the following metric dimensions to inch dimensions.

4. 50 cm
5. 2.54 cm
6. 150 mm

Convert the following given measurement to the required measurements.

7. 32 feet equal how many meters?
8. 104 inches equals how many millimeters?
9. .3125 decimal inches equals how many mm?
10. 66 inches equals how many mm?
11. 5 miles equals how many km?
12. Determine the area of the feature shown.

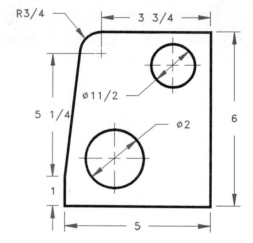

Descriptive Geometry II

THE ENGINEERING DESIGN PROCESS

A sample descriptive geometry problem solved by the design process follows.

1. *Identify and define the problem.* Given an isometric and partial detail drawing of a building with a chip blower and its discharge pipe location, determine the location and the shape of the roof opening shown in Figure 8–1. Also, the angle between the pipe and the roof must be determined to complete the design of the pipe support bracket. Finally, determine the slope of the discharge pipe. The partial plane view shows the location of a chip blower in a 40' wide × 100' long metal building. The building has 16' high walls

and a 4/12 slope on the roof. The blower point 1 is 4' above the floor and the discharge point 2 is 28' above the floor.

2. *Research and generate solutions.* To see how much can be understood from the problem statement, take it word by word. "Determine the location of the roof opening" means there will be a pipe intersecting or going through the roof. This is the concept of a line intersecting a plane. "Determine the shape of the roof opening" explains itself. The true shape of the opening is asked for here. You are finding the true shape of a plane to determine the angle between the pipe and the

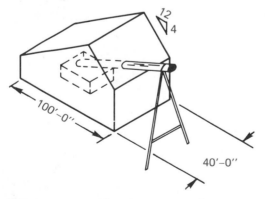

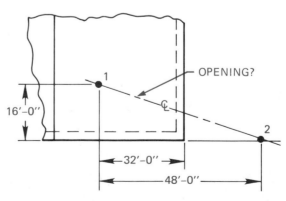

Figure 8–1   Problem layout with partial detail.

(Continued)

ENGINEERING DESIGN PROCESS *(Continued)*

roof. The concept of the angle between a line and a plane is used to determine the slope of the discharge pipe. Therefore, the problem requires you to find:

a. The intersection of a line and a plane.
b. The true shape of a plane.
c. The angle between a line and a plane.
d. The slope of a line.

3. *Evaluate the possible solutions.* Draw the top, front, and side projection planes or, as in architectural drawing, the top, front elevation, and side elevation of the given problem. This will help in the evaluation if there is doubt about which part of the information is useful. Only one solution for each part has been determined, so this step is limited in its usefulness.

4. *Use the best solution.* Decide which views or projection planes will give the necessary information. The front, top, and side views or projec-

tion planes do not give the true lengths, shapes or angles needed. Auxiliary views or projection planes must be found and constructed, as in Figure 8–2. The roof plane must be seen as a true shape in order to see the true shape of the hole in it. You must see the edge of the roof plane and the true length of the pipe in the same view or projection plane to determine where they intersect and the angle between them. To find the slope of the discharge pipe, you must see the true length of the centerline of the pipe in an elevation view or projection plane.

5. *Evaluate the best solution.* Check the answers by other methods if possible. Does everything in the answer make sense?

6. *Finalize the solution.* Present the solution to your employer or the instructor with the completed notations.

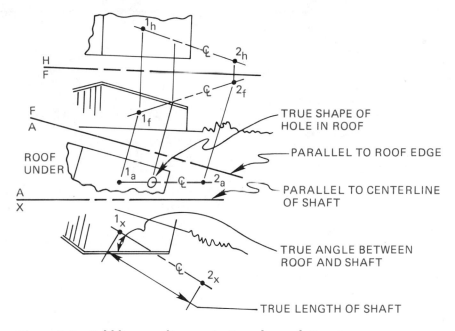

Figure 8–2   Fold line auxiliary projection plane solution.

▼ DIRECTIONS OF LINES AND PLANES

The procedure to locate lines and planes in space was discussed in Chapter 7. To solve many engineering problems, the direction of lines and planes must also be known. The direction of lines and planes is identified in space by a variety of ways, depending on their uses.

Horizontal Directions

Bearings are generally used for map directions and can only be determined in a *horizontal projection plane*

or a top view. Knowing the bearing angle of a line or plane is especially useful in mining and civil engineering. The direction of north is always understood to be toward the top of a drawing layout unless otherwise indicated by a north-pointing arrow pointing in another direction. South is always opposite north, while west is to the left and east is to the right forming the word *WE* as shown in Figure 8–3. A bearing angle is always 90° or less and will be identified either from the north or south. Lines pointing directly north are identified as due north. Also, lines pointing directly east, west, or south

Figure 8–3 Compass directions.

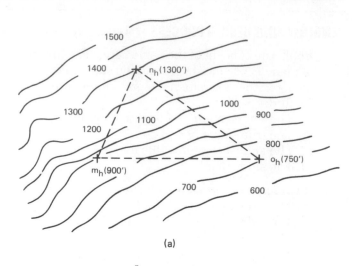

(a)

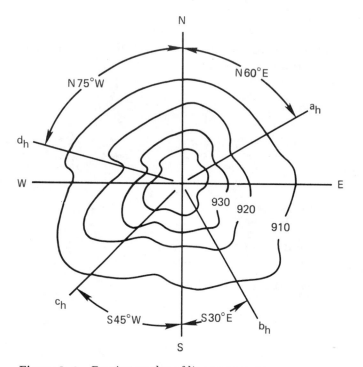

Figure 8–4 Bearing angles of lines on a contour map.

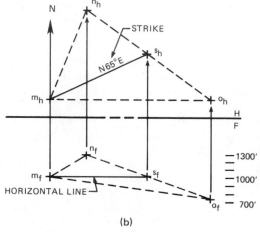

(b)

Figure 8–5 Stratums: **(a)** horizontal view of a stratum; **(b)** finding the strike of the stratum.

are identified due east, due west, and due south, respectively. The bearings of several angles are shown in Figure 8–4. Notice that the bearing is first identified by a north or south direction indicator. The acute angle formed from the vertical direction and east or west follows next. Lastly, the corresponding side of the acute angle direction east or west is noted. Examples include: N 60° E, N 75° W, S 30° E, and S 45° W.

Bearing is Found in the Horizontal Plane. The bearing of a line can only be found in the horizontal projection plane. The line does **not** have to be in true length to represent the bearing.

There is really no such thing as the bearing of a plane; but in mining applications, it is common practice

to locate a *stratum* by using the bearing of a horizontal line in the plane of a stratum. Stratums in their simplest form are horizontal layers of rock. Within a specific area it is reasonable to assume that a stratum is uniform in thickness and that it lies between two parallel planes. For the discussion in this text, these conditions will be assumed. The bearing of a horizontal line in a stratum is called the *strike*. To find the strike of plane M-N-O in Figure 8–5, you can use the same technique we used in Chapter 7 by locating a horizontal line in the front projection plane and finding that same line in the horizontal projection plane where is appears as true length.

Step 1. Draw horizontal line m_f-s_f in the F projection plane.

Step 2. Find line M-S in the H projection plane. (Line m_h-s_h being shown in true length is immaterial, because this is where the bearing of the line is shown whether it is in true length or not.)

Step 3. Measure the angle of line m_h-s_h from the north or south direction. This bearing N 65° E of line M-N as seen in the horizontal projection plane is the strike of the plane.

This discussion of the application of descriptive geometry to mining problems is brief for the purposes of this chapter. It will be necessary to study in more detail the concepts of contour lines, saddles, ridges, ravines, stratas, veins, faults, outcrops, cuts, and fills to adequately utilize descriptive geometry for civil engineering applications. Many of the calculations to determine cross sections for highways, and the designs and drawings from many civil engineering projects are currently being made using computers.

Vertical Directions

The vertical direction of a line can only be found when the following two conditions exist: *the line must be in an elevation projection plane* and *the line must appear in true length.* The vertical direction of a line is identified by the acute angle formed between the vertical line and the horizon (ground). This angle will vary from true horizontal (0°) to vertical (+90°). If the angle is between the horizon and +90°, or the line goes upwards from its origin, the angle is called *inclination* and is assigned a positive value. If the angle is between the horizon and −90°, or the line goes downward from its origin, it is called *declination* and is assigned a negative value. (See Figure 8–6.) The terminology used to identify the inclination of a line is specified differently for various engineering fields. Each of the terms that follow are those ways that express the relationship of a line to a horizontal plane. In all cases, the same two conditions must exist to be able to identify the true direction of a vertical line. The line must be in an elevation plane where it appears in true length.

Rules for Finding the Slope of a Line. Always start from the horizontal projection plane when solving problems that require you to find the slope of a line. For example, when using the fold line technique, place the fold line parallel to the line where it exits on the horizontal projection plane. When using the revolution method, revolve the line where it exists in the horizontal projection plane parallel to the existing fold line.

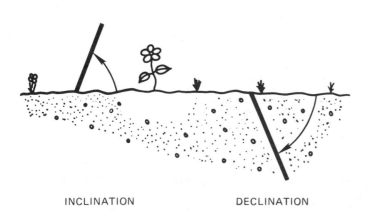

Figure 8–6   Inclination and declination.

Either method will result in showing the line in true length in an elevation projection plane. This is the only place you can measure the true slope of a line.

The *slope angle* or *slope* of a line is the angle in degrees that the line makes with the horizontal (ground or level) plane. The slope angle can be expressed by the following formula:

Slope angle (s) is the angle where tangent (arctan s) $= \dfrac{\text{Rise}}{\text{Run}}$

Again, the true slope of a line can only be seen in an elevation projection plane where the line appears in true length. (See Figure 8–7.)

The *grade* of a line is expressed in percent form and is found by using the following formula:

$$\text{Percent grade} = \frac{\text{Rise}}{\text{Run}} \times 100$$

This is illustrated in Figure 8–8a. Observe also that the grade is the tangent of the slope angle multiplied by 100. From trigonometry, we know that the tangent of an angle is defined as dividing the length of the opposite side of a triangle of the included angle by the adjacent side of that angle as seen in Figure 8–8b. One of the most

Figure 8–7   Slope angle.

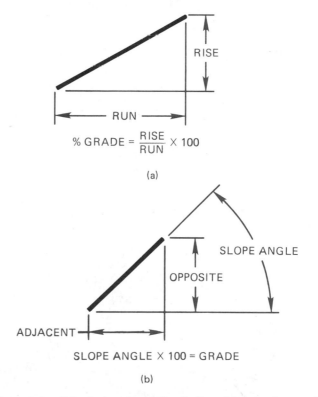

Figure 8–8   Calculating the grade of a line: **(a)** formula one; **(b)** formula two.

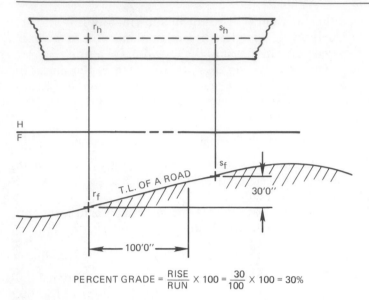

$$\text{PERCENT GRADE} = \frac{\text{RISE}}{\text{RUN}} \times 100 = \frac{30}{100} \times 100 = 30\%$$

Figure 8–9   Grade of a road.

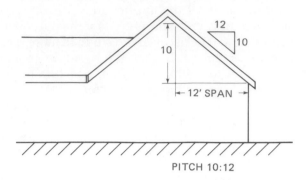

PITCH 10:12

Figure 8–10   Pitch of a roof.

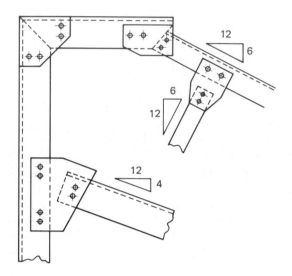

Figure 8–11   Bevel of a beam.

common ways of measuring run and rise is to use an engineer's scale, which has divisions in multiples of ten, or set the computer scale in CAD to the appropriate scale. Grade is simply another way to express slope; therefore, the conditions for finding grade of a line are the same as for finding the true slope of a line. The term *grade* is usually used when referring to a line that has an angle with less than a 45° inclination. Grade is usually associated with civil engineering but not always. Slope is usually applied to a line that has an angle with greater than 45° inclination, such as the slope of a roof on a house. Slope is used more with architectural engineering but not always.

In architectural engineering, the slope may also be referred to as the *pitch*. When describing the angle the roof plane has with the horizon, it is expressed as the ratio of the vertical rise to 12' of span. The graphical method of showing pitch on a drawing is illustrated in Figure 8–10.

In structural engineering, slope is usually referred to as the *bevel* of a line. Figure 8–11 shows the bevel of several beams. In civil engineering, the terms *grade* and *slope* are frequently used as shown in Figure 8–12a. In addition, *batter* is applied when dealing with the slope

of concrete footings as shown in Figure 8–12b. *Dip* is used when finding the slope of a stratum (plane). The dip is the slope of the strike. Figure 8–13 shows how to find the dip of a stratum.

In review, *slope, grade, pitch, bevel, batter,* and *dip* are all representations of the angle between a true length line and the horizon. Finding this angle graphically is done with the same procedure in each situation.

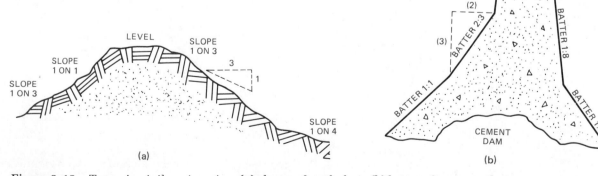

Figure 8–12   Terms in civil engineering: **(a)** slopes of earth dam; **(b)** batter of concrete dam.

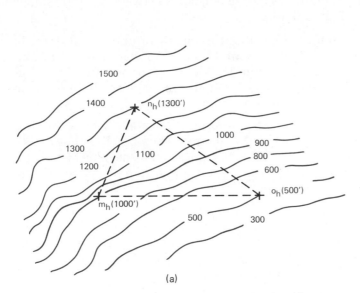

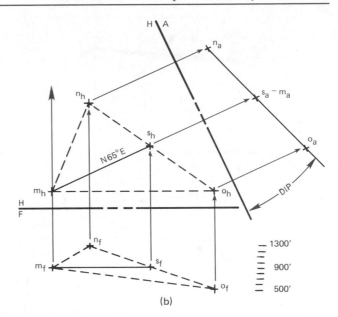

Figure 8–13 Fold line method of finding the dip of a stratum.

▼ TRUE DISTANCE BETWEEN A POINT AND A LINE

To find the true distance between a point and a line, you must create a projection plane that will show the line and the point appearing as *two points*. This occurs when you are viewing the end of the line and the point on one projection plane. As with most problems, there are several ways to solve them. Figure 8–14 shows how to find this distance by the fold line method.

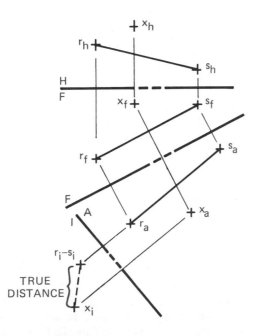

Figure 8–14 True distance between a line and a point.

Finding the Distance Between Two Points with the Fold Line Method

The first step in the solution of this type of problem is always to place a fold line parallel to the given line. The line will be in true length when it is projected onto the new projection plane. The second step is to place another fold line perpendicular to the true length line. The line will then appear as a point when it is projected onto the next new projection plane. After projecting the point onto the two new projection planes, it is just a matter of measuring the distance between two points.

Step 1. Find the true length of line R-S and the location of point X on the A projection plane.

Step 2. Find the end view of the true length line R-S and the location of point X on the I projection plane. The distance measured between point X and the end view of line R-S on the I projection plane is the true distance between point X and line R-S.

▼ TRUE DISTANCE BETWEEN TWO SKEW LINES

An example of two *skew lines* (two nonparallel, nonintersecting lines) is shown in Figure 8–15. The true distance between two skew lines is the perpendicular between them. To be able to determine the true distance between the two skew lines you must create a projection plane where one line appears as a point. (*Note:* While the one line appears as a point, the other line will not necessarily appear in true length.) The perpendicular distance between the line appearing as a point and the

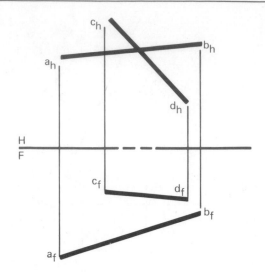

Figure 8–15 Skew lines.

other line is the true distance between both lines. (See Figure 8–16.)

Step 1. Find the true length of one of the lines. In this example, line R-S was selected to find the true length. The true length of line R-S and the location of line M-N has been found on the A projection plane. (See Figure 8–16a.)

Step 2. Create the projection plane where line R-S appears as a point. Then line R-S appears as a point in the I projection plane. Line M-N is also carried onto this projection plane. (See Figure 8–16b.)

Step 3. On the I projection plane, the true distance can be measured between line R-S and line M-N by constructing a perpendicular line from line R-S (appearing as a point) and line M-N. (See Figure 8–16c.)

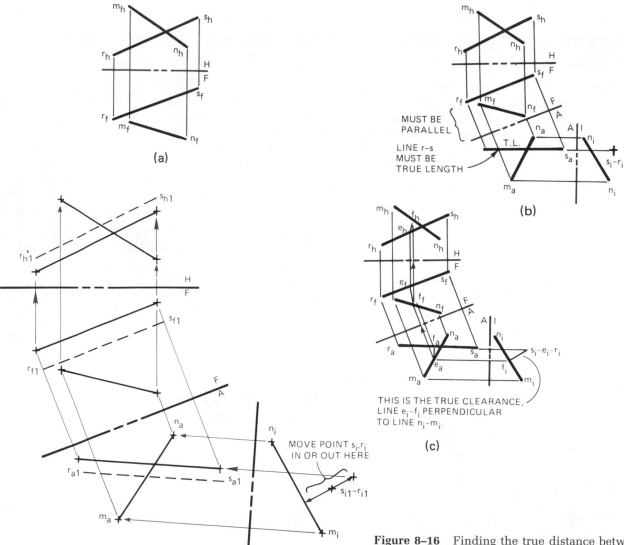

Figure 8–16 Finding the true distance between skew lines: **(a)** Step 1; **(b)** Step 2; **(c)** Step 3; **(d)** adjusting the true distance.

Sometimes a problem may require you to adjust the distance between two skew lines or make sure there is enough clearance between the two lines. An application for determining the true distance or clearance determination might be for the location of a valve in a piping system or the clearance between power lines and a nearby metal structure in an electrical power substation. When you have found the true distance between the two skew lines as in Step 3 in the previous problem, it is just a matter of adjusting the point (line R-S) in or out along the perpendicular distance from line M-N, as shown in Figure 8–16d. You then can relocate the new position of line R-S in all the previous projection planes. If line R-S cannot be moved in the original problem, you will need to solve the problem by starting with line M-N rather than line R-S as in this example.

▼ INTERSECTION OF A LINE AND A PLANE

The location and anchoring of guy wires; the location of holes for shafts, cables, pipes, and wires; and the angles that are created by these features are important considerations for the designer in structural and mining problems.

Finding the Piercing Point of a Line and a Plane

To find the point of intersection, *piercing point* of a line and a plane you must create a projection plane where the *plane appears as an edge.* You then can project this apparent point of intersection back into the original projection planes.

Use the following steps if you have a problem that requires finding the intersection of the roof and the circular shaft as shown in Figure 8–17a (piercing point problem.)

Step 1.   Find a projection plane that shows the roof plane as an edge. Figure 8–17b shows that the P projection plane contains the roof plane as an edge. When the centerline of the shaft is projected onto the P projection plane, the apparent point of intersection, w_p of the centerline of the shaft and the roof plane can be seen.

Step 2.   Project this apparent point of intersection, w_p back onto the previous projection planes. As Figure 8–17c shows, a projector is projected from the A projection plane back to the F projection plane intersecting the centerline of the shaft. This is the actual point of intersection,

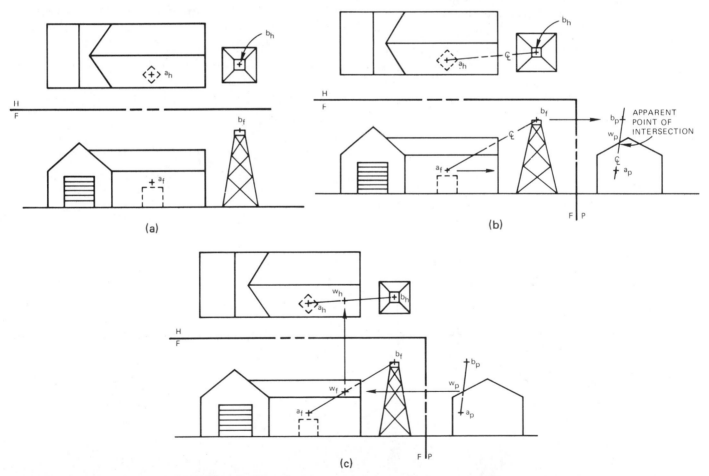

Figure 8–17   Piercing point problem: **(a)** set-up; **(b)** Step 1, connecting two points and Step 2, finding the apparent point of intersection; **(c)** Step 3, finding the actual piercing point.

point w_f of the roof plane shown on the F projection plane. You can also show the proper location of point W in the H projection plane by the normal procedure of locating a point in space.

▼ VISIBILITY

When two lines are shown in space, sometimes it is difficult to know which line is on top of which or if they intersect. Problems dealing with the intersection of planes require you to know which line is in front of the other in order to correctly solve and complete the problem as you will see later in this chapter. For example, look at Figure 8–18a. Looking at the F projection plane you cannot tell if lines A-B and R-S intersect or if line A-B is in front of or behind line R-S. There is a simple way to determine which situation exists.

Step 1. On the F projection plane shown in Figure 8–18b, project a projector from the apparent point of intersection of lines A-B and R-S perpendicular to fold line F/H. Continue projecting until it crosses one or both of the lines on the H projection plane.

Step 2. If the projector passes through the apparent point of intersection on the H projection plane, these would be intersecting lines and this would be the real point of intersection. But this is *not* the case in our example. The projector crosses line R-S in the H projection plane first and verifies that line R-S is closest to us in the F projection plane.

Step 3. We can apply the same procedure when determining the visibility of lines R-S and A-B on the H projection plane. Figure 8–18c shows a projector being projected from the apparent point of intersection on the H projection plane perpendicular to the H/F fold line. The projector is then continued on until it crosses one or both of lines A-B and R-S in the F projection plane. The projector crosses line A-B in the F projection plane first and verifies that line A-B is closest to us in the H projection plane.

▼ INTERSECTION OF PLANES

Structural skins, houses, sheet metal forms, and many other things have intersecting surfaces. Determining the lines of intersection is essential in completing each view of a drawing and eventually determining the shape and size of each surface for the construction of each part in the fabrication shop. If two planes are not parallel, they will intersect forming a straight line that will be common to both planes. Only two points that lay on both planes will need to be found to fix the position of this line of intersection that is common to both planes. To find the line of intersection of two planes, a projection plane must be created that shows both planes where one appears as an edge view. Figure 8–19a shows two projection planes in which the two planes appear to intersect.

Step 1. Create a projection plane where one of the planes appears as an edge view. Figure 8–19b shows finding plane A-B-C as an edge view in

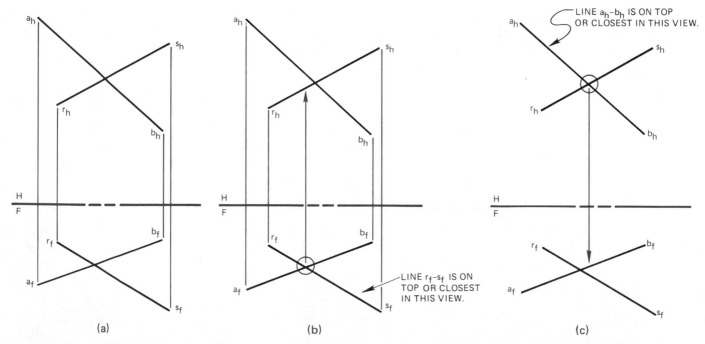

Figure 8–18 Visibility: **(a)** two lines in space; **(b)** finding visibility of lines in the F projection plane; **(c)** finding the visibility of lines in the H projection plane.

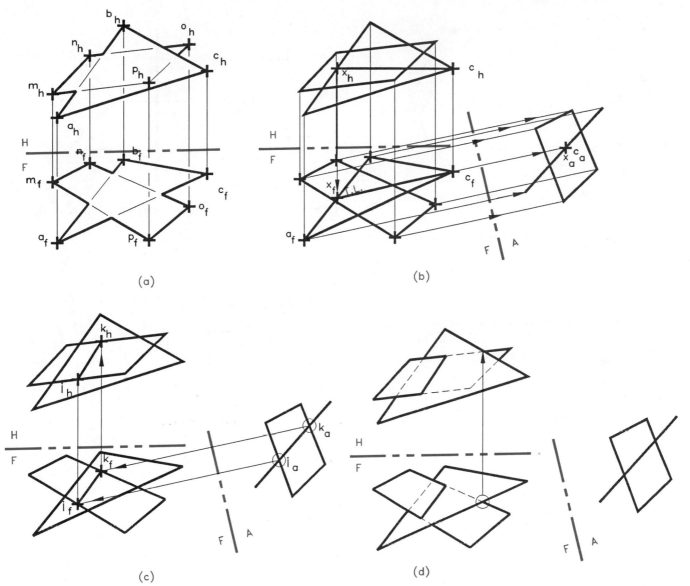

Figure 8–19 Two planes that appear to intersect: **(a)** problem set-up; **(b)** Step 1, finding the edge view of a plane; **(c)** Step 2, finding the line of intersection of two planes; **(d)** Step 3, determining visibility and completing the visible and hidden lines of two planes.

the A projection plane. You already know how to do this from Chapter 7.

Step 2. Figure 8–19c shows the common points J and K of intersection of the two planes. These points can now be projected back onto the F and H projection planes and completed as the line of intersection.

Step 3. Identify the visibility of the intersecting lines of each plane on the F and H projection planes by the process previously explained. You will need to do this in order to show which lines are visible and which lines are hidden on the two planes. Figure 8–19d shows this process completed.

▼ DIHEDRAL ANGLES

Working with structural skins or various sheet metal projects requires knowing the exact angle between those surfaces or planes. The true angle between two planes, or *dihedral angle,* is found in a projection plane in which both of the intersecting planes appear as an edge. (See Figures 8–20 and 8–21.) Figure 8–22a shows the top and front views of a sheet metal funnel on the H and F projection planes. The problem here is to determine the dihedral angle between the front side and the right side of the funnel.

Steps in Finding the True Dihedral Angle

To find the true dihedral angle between two planes, you need to find the end view of the line that is created by

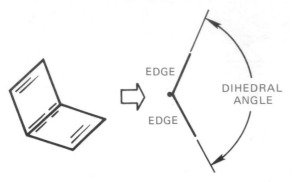

Figure 8–20 Pictorial representation of a dihedral angle.

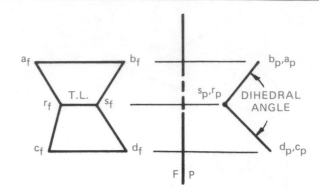

Figure 8–21 Descriptive geometry flat layout showing the dihedral angle of two planes.

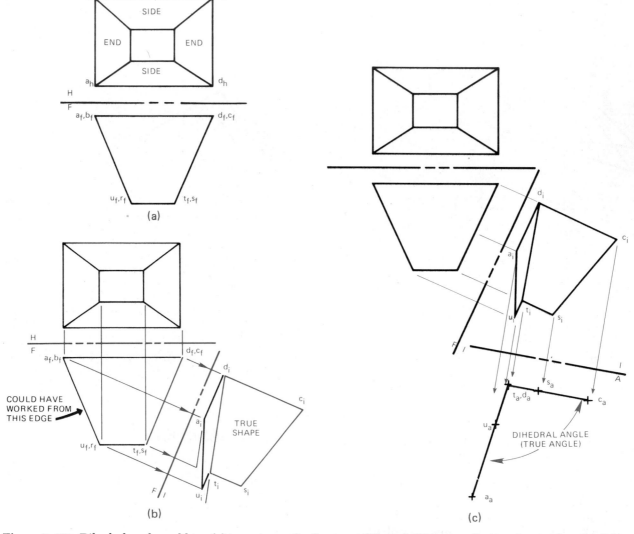

Figure 8–22 Dihedral angle problem: **(a)** two views of a sheet metal funnel; **(b)** Step 1, finding the true length of the line of intersection; **(c)** Step 2, finding the end view of a true length line and creating the dihedral angle of two planes.

the intersection of the two planes. The first step is to place a fold line parallel to this line of intersection. Then, place a fold line perpendicular to this true length line.

Step 1.   Create a projection plane in which you show the true length of the line of intersection of the two

planes you are trying to determine the dihedral angle of. In Figure 8–22b the true length of line D-T, the line of intersection, has been found in the I projection plane. You already know how to find the true length of a line from Chapter 7.

Then, project only the points that identify the rest of the two intersecting planes onto the I projection plane as shown.

Step 2. Create a projection plane that shows the intersecting line D-T appearing as a point. In Figure 8–22c, line D-T has been found on the A projection plane. It appears as a point. Notice also that when you project all the other points of the two planes onto the A projection plane, they will form two intersecting straight lines. The angle formed between these two planes is the true or dihedral angle. This angle can be measured with a protractor or calculated on the computer.

▼ AN INTRODUCTION TO VECTOR GEOMETRY

Vector analysis is a branch of mathematics that includes the manipulation of vectors. Vector geometry is included in this chapter because vectors are a vital part of many of the engineering sciences. Graphic solutions are usually sufficiently accurate for engineering analysis of forces and other directional quantities in structural design, mechanics, and other physical sciences. Vectors can be manipulated in some very complex ways, but the discussion here is limited to vector addition and its practical purposes. Since vectors are segments of straight lines, the principles of descriptive geometry are directly applicable to the graphic solution of problems involving vectors.

A *vector quantity* is one that has *magnitude* and *direction,* such as displacement, acceleration, momentum, force, velocity, and torque. The speed of a space shuttle, for example, describes only the magnitude of its velocity, which is a physical or *scalar* quality. The velocity of that space shuttle, on the other hand, is a vector quality that describes not only its speed but its direction of motion in space. A vector quantity is usually represented by a straight line with an attached arrowhead pointed in the same direction in which the quantity is acting. The length of the vector is drawn proportionally to the magnitude of that quantity. An example of a *vector diagram* representation of a force of 10 lb directed N 45° E is shown in Figure 8–23. To solve problems using vector geometry, it is essential to know the difference between a vector diagram and a *space diagram.* Figure 8–24

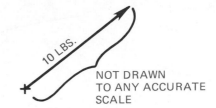

Figure 8–24 Space diagram.

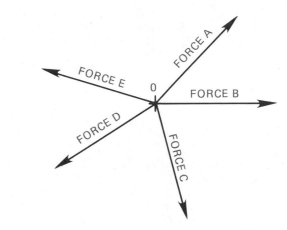

Figure 8–25 Space diagram of concurrent forces.

shows the space diagram of the 10-lb vector. Notice that the space diagram is *not* drawn to scale. It shows sense and the accurate direction. When two or more vectors act on an object, it is called a vector or *force system.* The discussion in this chapter will be limited to forces that are *concurrent,* because they are the most common. In concurrent forces, vectors pass through a common point, as seen in Figure 8–25. Forces with a line of action lying in one plane are called *coplanar* forces. Figure 8–26 shows five coplanar vectors pictorially and with their corresponding vector diagram. When those forces lie in more than one plane but still meet at one point, they are identified as *noncoplanar,* as shown pictorially and in their corresponding vector diagram in Figure 8–27.

Addition of Concurrent Coplanar Vectors

To determine the net effect of a vector system, vector quantities must be combined. The graphic addition of concurrent, coplanar vectors simply requires you to draw a vector diagram locating the vector *head-to-tail* in their correct direction. The sum or the space that is left open is called the *resultant.* The resultant has the same effect as the total of the individual vectors added together. In Figure 8–28a, the two vectors M and N shown in the space diagram are to be added. The magnitude of vector M is 40 lb and vector N is 60 lb. Remember that the space diagram shows the *correct direction,* so in the vector diagram, Figure 8–28b, the direction of the vectors is duplicated and the vector length is scaled to the same

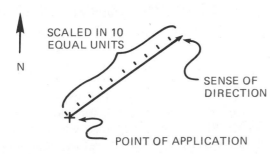

Figure 8–23 Vector representation/vector diagram.

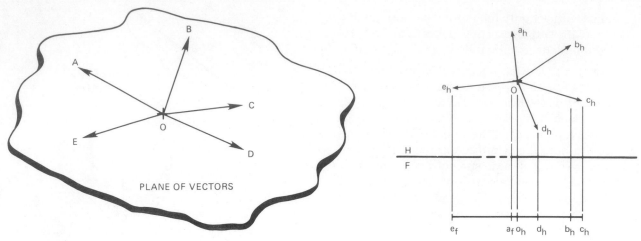

Figure 8–26 Concurrent coplanar vector system.

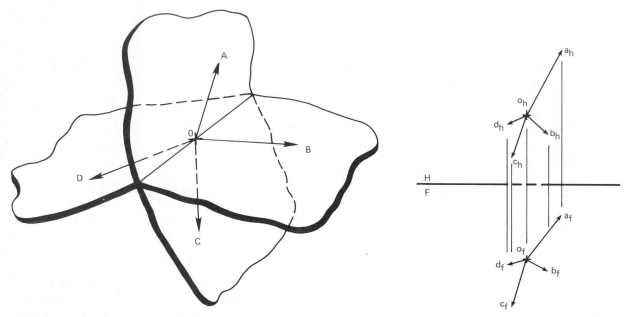

Figure 8–27 Concurrent noncoplanar vector system.

units. After placing each vector head-to-tail, in any order, the space left is connected with a straight line. This line is called the resultant and the direction arrow is in the same direction as the added vectors. Measuring the length of the resultant to scale gives a length of 80 lb. If the direction of the resultant is reversed as shown in Figure 8–28c, it is called the *equilibrant.* An equilibrant is a force that will balance all other forces through a point. An equilibrant is the same as the resultant of zero. If you have just two vectors to be added, the *parallelogram method* can be used. This method requires more construction, in that the sides of the parallelogram are drawn parallel to the given vectors. Then the diagonal of the parallelogram is drawn, which is the resultant of the two vectors, as shown in Figure 8–28d. When you have three or more vectors, similar methods can be used. In Figure 8–29a the vectors are drawn to scale, using the space dia-

gram. In Figure 8–29b the correct direction is obtained as previously explained. The closing side of the vector polygon is the *resultant.* As before, if the equilibrant is required, the direction of the arrowhead of the resultant is reversed, as shown in Figure 8–29c.

Resolution of Concurrent, Coplanar Vectors

Two or more vectors can be added to form a single resultant vector. By reversing that process, a single vector can be resolved into two or more component vectors. In the case of coplanar vectors, only two can be unknown in magnitude or direction. Most commonly, a vector must be resolved into two other vectors having specified directions. In Figure 8–30a, a given vector T is to be resolved into two component vectors having direction of OR and OS. The magnitude of each will need to be

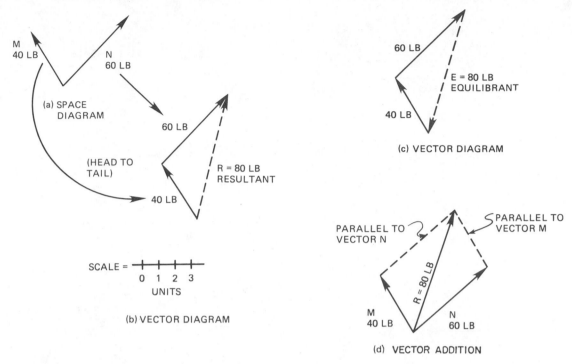

Figure 8–28 Addition of concurrent coplanar vectors: **(a)** space diagram; **(b)** vector diagram showing the resultant; **(c)** vector diagram showing the equilibrant; **(d)** vector addition by parallelogram method.

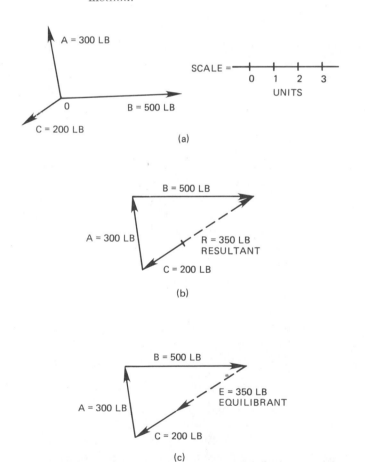

Figure 8–29 Addition of concurrent, coplanar vectors: **(a)** space diagram; **(b)** vector diagram showing the resultant; **(c)** vector diagram showing the equilibrant.

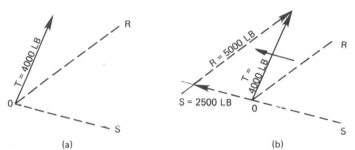

Figure 8–30 Resolution of a single vector into two components: **(a)** space diagram; **(b)** vector diagram.

found. By using the parallelogram method for the addition of two vectors, the solution is obtained in reverse. Vector R is moved parallel until it crosses the arrow end of the given vector. (See Figure 8–30b.) The S vector is extended through the tail end of the given vector. Where the two moved vectors cross will determine the lengths of each. The direction of each of the two vectors will be opposite that of the given vector.

Addition of Concurrent, Noncoplanar Vectors

The addition of noncoplanar vectors is the same as for coplanar vectors, except that three dimensions are used now. Remember that noncoplanar means *in more than one plane,* so the principles of descriptive geometry will be needed to solve these problems. Figure 8–31a shows the space diagram on the F and H projection planes for the given vector quantities. To solve this

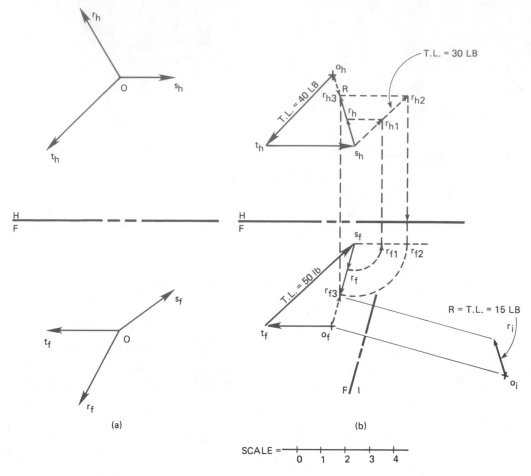

Figure 8–31 Addition of concurrent, noncoplanar vectors: **(a)** space diagram; **(b)** vector diagram.

problem, you must create a projection plane that shows the true length of the resultant force.

Step 1. Set up the space diagram on a frontal projection plane and the adjacent horizontal projection plane as shown in Figure 8–31a.

Step 2. Find the resultant vector on the H and F projection planes just as you would for coplanar vectors with one major exception. You will need to find the true lengths of those vectors, which are *not* shown in true length on either the H projection plane or the F projection plane. To do this you can use the fold line or the revolution method. Then project the true lengths back onto the H and F projection planes correctly. In Figure 8–31b, the S vector is already shown in true length on the F projection plane. Therefore, you can scale the S vector on the F projection plane. The T vector is shown already in true length on the H projection plane. Therefore, you can scale the T vector on the H projection plane. However, the R vector is not shown in true length on either the H projection plane or the F projection plane.

To find the true length of the R vector, use the same procedure as you did when you had to find the slope of a line when given only the bearing of that line as explained in Chapter 7. In this case, line s_h-r_h was established with an arbitrary length after the vector diagram was completed and placed in the correct direction on the H projection plane. Then, the corresponding vector diagram was completed in the F projection plane. Vector R was placed in the correct direction as shown in the space diagram on the F projection plane with the corresponding length as projected from the H projection plane. The true length of line s_f-r_f was found in the H projection plane. Notice that the line needed to be lengthened to equal the scaled 30 lb. Line s_h-r_{h2} now represents the true length of vector R on the H projection plane. This length was then projected into the vector diagrams as shown in Figure 8–31b.

Step 3. Find the true length of the resultant force, the space left in the vector diagrams. The resultant force is *not* shown in true length on either the H or F projection planes, so the fold line method was used this time to find the true length of the resultant force on the I projection plane. The resultant force is measured at 15 lb as shown on the I projection plane. (See Figure 8–31b.)

▼ FORCES IN EQUILIBRIUM

Equilibrium is a condition in which the series of forces acting upon a body or structure equals zero and it will remain at rest. Graphically, the vector polygon closes, with vectors in a continuous head-to-tail arrangement. (See Figure 8–32.) When a body is known to be in equilibrium under the action of certain known and unknown forces, these graphic conditions can be used to evaluate the unknown forces. However, the number of unknown forces is strictly limited. Coplanar vector systems can have up to three unknown forces. Vector diagrams can more easily be constructed if you isolate the structure or part to which the known forces are applied. This isolation simply means free from all adjacent bodies and is referred to as *free-body*. Construction of a free-body diagram should be the first step in the analysis of every equilibrium problem.

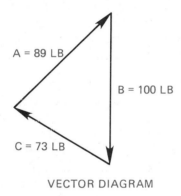

A = 89 LB
B = 100 LB
C = 73 LB

VECTOR DIAGRAM

Figure 8–32   Vector diagram showing forces in equilibrium.

Finding the Equilibrium of Two Unknowns in a Coplanar System

The following steps may be used to find the equilibrium of two unknowns in a coplanar system when you are given a weight that is supported by two ropes as shown in Figure 8–33a.

Step 1.   Imagine the ropes are cut, making the knot an isolated free body, as shown in Figure 8–33b. Draw a free-body diagram with the tension forces in the free-body diagram being represented with force vectors pointing away from the knot.

Step 2.   Make a vector diagram. Start with the known vector force of 100 lb by drawing it to scale with the proper direction. The direction of forces A and B are known but not their magnitudes. It was previously explained that for forces to be in equilibrium, the vector diagram must close. As shown in Figure 8–33c, the vectors A and B are drawn parallel to their corresponding vectors in the free-body diagram shown in Figure 8–33b. When the vectors are drawn properly, point L is established in the vector diagram.

Step 3.   Scale the magnitudes of vectors A and B. In this example, both vector A and vector B measure to be 60 lb.

A similar type of problem occurs when two directions are unknown. In this situation, you are given the space diagram in Figure 8–34a which shows a weight of 10 lb supported by two cables that pass over pulleys, which, in turn, support two other weights of 6 and 8 lb, respectively.

Step 1.   Draw the space diagram. The forces acting in the cables are known, but the direction of the vectors are not. As shown in Figure 8–34b,

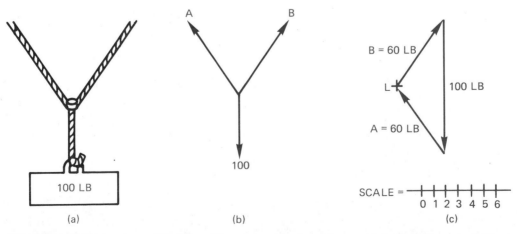

(a)

100 LB

(b)

A

B

100

(c)

B = 60 LB
L
100 LB
A = 60 LB

SCALE = 0 1 2 3 4 5 6

Figure 8–33   Forces in equilibrium with two magnitudes unknown: **(a)** space diagram; **(b)** free-body diagram; **(c)** vector diagram.

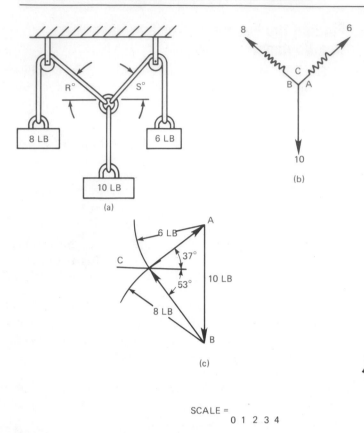

(a)

(b)

(c)

SCALE =
0 1 2 3 4

Figure 8–34 Forces in equilibrium with two directions unknown: **(a)** space diagram; **(b)** free-body diagram; **(c)** vector diagram.

start with the vector you know the force and direction of. The vector that represents 10 lb is drawn in the down position. As shown, draw the other two vector forces at any two different directions similar as shown in Figure 8–34b. The squiggly lines indicate that the real directions are unknown.

Step 2. Draw the free-body diagram. This is solved by using the method of creating a triangle when three sides of the triangle are known. This concept was covered in Chapter 4. Start with the known force of 10 lb. Draw it in its correct direction and magnitude. Next, with a radius equal to 6 lb, an arc is drawn with A as its center. (See Figure 8–34c.) With a radius equal to 8 lb, another arc is drawn with B as its center. These arcs intersect at C completing the triangle.

Step 3. Determine the directions of the 6-lb and 8-lb vectors. The directions are indicated by the angles shown and are determined by placing the arrowheads in a head-to-tail manner in the diagram. (See Figure 8–34c.)

PROFESSIONAL PERSPECTIVE

In this chapter, many new concepts have been introduced. But as you may have noticed, the techniques used to solve each of the unknown angles and locations of the intersections are the same as those used in Chapter 7 with, for the most part, only the terminology changed. Again, accuracy in the construction and drawing of lines is essential. In most cases, the problem situations are more complex; so it is essential to understand how to break each one down into various simple elements. To solve vector problems, the correct terminology must be learned for the different types of vectors and diagrams. This will be especially true if there has been no previous experience with the related physics. Problem solving, while difficult for many, can be made easier if a logical system, such as the one used at the beginning of this chapter, is used. This system is essentially the one most engineers or anyone can best use to solve any type of problem.

MATH APPLICATION

Given algebraic formulas, you may have to be able to manipulate the formulas to find the unknown quantities. For example, in the formula:

$$\text{Slope of a line} = \frac{\text{Rise}}{\text{Run}}$$

$$\text{Slope} = \frac{75}{100} = .75$$

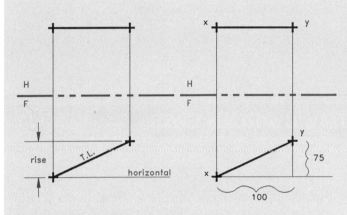

Figure 8–35   Calculations for slope.

In the formula for slope angle:

$$\text{Slope angle of a line} = \text{Tangent of slope angle}$$

$$\text{Tangent of angle a} = \frac{\text{Opposite side}}{\text{Adjacent side}}$$

$$\text{Slope angle a} = \text{Tangent angle a}$$

$$\text{Tangent angle a} = \text{Tangent}\frac{75}{100}$$

$$\text{Tangent angle a} = 37°$$

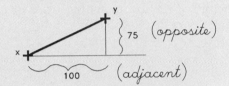

Figure 8–36   Calculations for slope angle.

In the formula for percent of grade:

$$\text{Percent of grade} = \text{Slope of a line} \times 100,\ \text{or}$$

$$= \frac{\text{Rise}}{\text{Run}} \times 100$$

$$= .75 \times 100$$

$$= 75\%$$

CHAPTER

8

DESCRIPTIVE GEOMETRY II TEST

DIRECTIONS

Answer the following questions with short, complete statements and/or sketch thumbnail sketches to explain your answers.

1. If a line has a bearing of N 30° E, what does this mean?
2. What is the relationship beween bearing and slope?
3. In what projection planes can you find the true slope of a line?
4. What two conditions must exist before you can determine the true grade or slope of a line?
5. Explain the difference between the inclination and declination of a line.
6. What are skew lines?
7. The dihedral angle is _____.

8. What two features must a vector have?
9. Explain what concurrent forces are.
10. Explain the difference between coplanar and non-coplanar vectors.
11. Forces acting upon a body equal _____ when it is in equilibrium.
12. Sketch an example of a space diagram.
13. List the six steps in the problem-solving process.
14. The _____ step in the problem-solving process is the most important.
15. Bearings can be determined in a _____ projection plane.
16. Stratums in their simplest form are _____ of rock.
17. The bearing of a horizontal line in a stratum is called the _____.
18. An acute angle is any angle less than _____ degrees.

19. Vertical directions can only be found in _____ projection planes.
20. A slope angle is designated in _____.
21. The grade of a line is designated in _____.
22. In architectural engineering, the pitch is similar to the _____.
23. The dip of a stratum is similar to _____.
24. The true distance between a point and a line is the _____.
25. Visibility refers to the _____.
26. If two planes are parallel, they will _____ along a straight line common to both planes.
27. The true angle between two intersecting planes is called _____.
28. Vector analysis is a branch of _____.
29. Vectors are _____.
30. A vector's direction is shown by a/an _____ symbol.

31. Space diagrams (are/are not) drawn to scale.
32. Forces whose line of action lie in one plane are called _____.
33. In a vector diagram, the vectors are placed _____.
34. A body in equilibrium is the same as a resultant of _____.
35. A free-body diagram is a diagram isolating a few forces from _____.
36. Coplanar systems can have only _____ unknowns.
37. Identifying what the problem is in problem solving is _____ step in problem solving.
38. Problem-solving techniques will work with _____ types of problems.
39. When you see the intersection of a line and a plane appearing as an edge, you can find the _____ point of intersection of the plane and the line.
40. A vector quantity has both _____ and _____.

CHAPTER

8 DESCRIPTIVE GEOMETRY II PROBLEMS

DIRECTIONS

Please read the problems carefully before you begin working. Your instructor may assign one or more of the following problems. Use a separate sheet of drawing paper with graph lines on it so that the proper set-up can be duplicated where necessary for each problem. Use one sheet per problem.

Problem 8–1 Given the adjacent horizontal and front projection planes showing the cable R-S, find its bearing, slope, and grade.

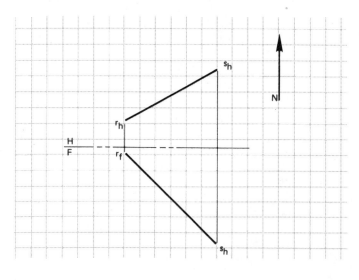

Problem 8–2 Bearings. The property lines of a building site have been found in some old files. The lines are readable, but the bearing notations of each segment are blurred. Redetermine those bearings and list them on an included chart.

LINE	BEARING
AB	_____
BC	_____
CD	_____
DE	_____
EA	_____

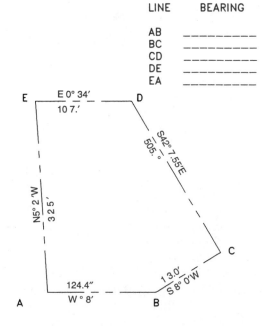

Problem 8–3   Bearing, slope, and true length. From the given data, draw separate adjacent horizontal and vertical projection planes that contain the following lines:

Line	Bearing	Slope	True Length
B-C	N 22° W	+30°	3 in.
R-S	S 75° E	−10°	4 in.
M-N	N 60° E	+15°	5 in.

Problem 8–4   Slope and grade. In this problem, two sections of pipe that are connected together are given. Using only the centerlines of each pipe, draw the necessary projection planes to answer the following questions. Leave all construction lines on the drawing. Identify on the drawing where the answer appears.

    a. Identify and calculate the true length of pipe A-B.
    b. Identify and calculate the slope in degrees of pipe B-C.
    c. Identify and calculate the grade in percent of pipe A-B.
    d. Identify and calculate the true angle between the pipes.

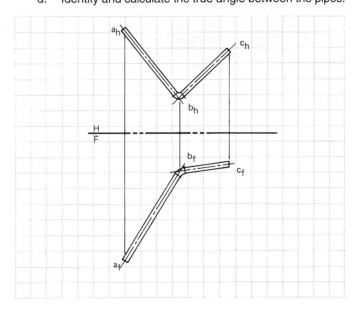

Problem 8–5   True length, true angle. A designer's sketch, on the right, shows the adjacent front and horizontal projection planes of the roof plane A-B-C-D and the location where the antenna is to be mounted on the roof plane. The top end of the antenna is to project 8' above the ridge of the house. Find:

    a. Identify and calculate the true length of the antenna in feet and inches.
    b. The place where the bottom of the antenna touches the roof in the front projection plane.
    c. Create a projection plane that will show the true angle made by the roof plane and then calculate the true angle in degrees.
    d. Show in the horizontal and front projection planes three equally spaced guy wires anchored *on the roof plane.* Two guy wires are attached 5' from the base of the antenna along the roof plane, and the third is attached at the point shown in the horizontal projection plane. All three guy wires are anchored to the antenna at a distance of 6' above the ridge.
    e. Identify and calculate the total length of guy wire used to stabilize the antenna to the nearest foot.

Problem 8–6   True distance between point and a plane. Find the shortest distance between cable R-T and point E.

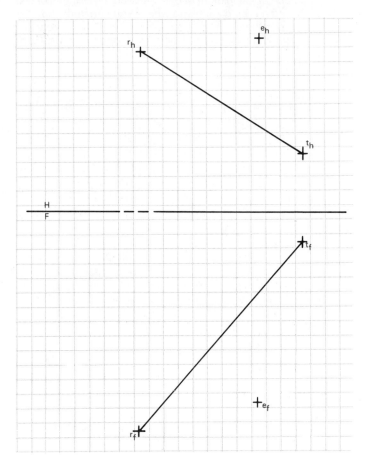

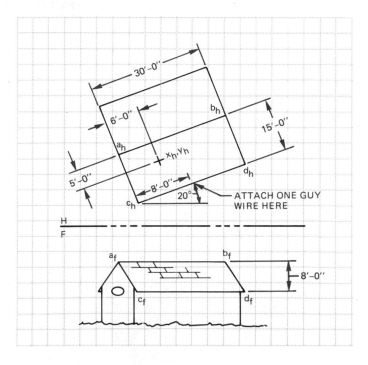

Problem 8–7   Skew lines. It is important to keep the clearance between the two cable segments A-B and M-N of a diameter of ½" each at a minimum of one inch and a maximum of two inches. Create a projection plane where you can verify how far apart they are in the set-up. If they do not fit this criteria, readjust cable A-B in all views so it will be met. (Drawing scale: Each square = ¼")

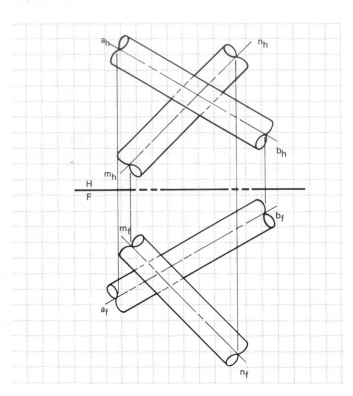

Problem 8–8   True distance between parallel lines. Two parallel cable segments M-N and R-S are shown in adjacent horizontal and front projection planes. Create a projection plane that will show the true distance between the cable segments.

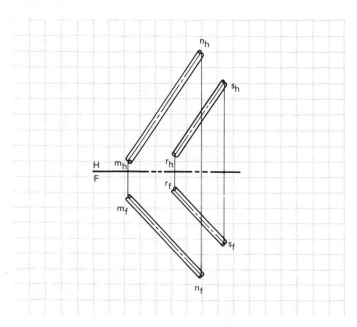

Problem 8–9   True angles. Two parallel pipe corner segments, K-P and R-S, are shown in adjacent horizontal and front projection planes. Create a projection plane that will show the true angle of the pipe segments.

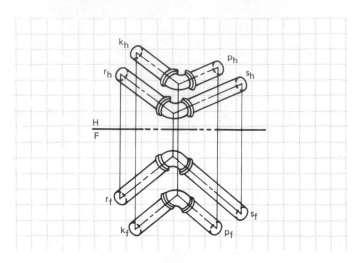

Problem 8–10   Piercing point. A vein of ore has been determined by plane D-H-G. A mine tunnel, R-S, is being dug toward this vein. Create a projection plane that will show the true distance of how much further the tunnel will need to be extended from point R to reach the vein. Calculate the distance to the nearest foot. (Drawing scale: Four squares = 200')

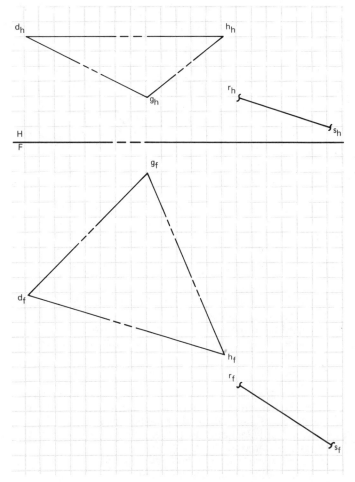

Problem 8–11 Visibility. The adjacent front and horizontal projection planes of three cable segments A-B, M-N, and R-S are shown. Determine which cable segments are in front of which in each projection plane. Then, correctly complete the cable segment drawings by adding the missing object lines in each projection plane. (Drawing scale: One square = ¼")

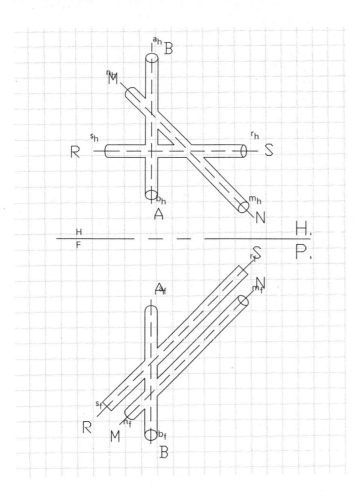

Problem 8–12 True shapes and angles between planes. A designer's sketch of a roof plan of a residential house with specifications for size of the roof and chimney flashing detail is given. It is necessary to determine some information about the roof area and the amount of roof flashing needed to keep the building job on schedule. Use the sketch and detail information to develop the necessary projection planes that will supply the information needed. You need to:

a. Create the projection planes that will show how many squares of roofing will be needed to cover the entire roof area. (100 ft² of roof surface = 1 square). Calculate to the nearest square how much roofing will be needed to cover the roof, with any extra to be left for repairs.

b. Calculate how much 18" wide flashing will be needed to flash all the ridges, valleys, and hips to the nearest foot.

c. Create the projection planes that will show all the pitches of the roof. Calculate each of the pitches in the correct pitch ratio.

d. Create the projection plane that will show the ridge angle needed to bend the chimney flashing correctly. Calculate the ridge angle to the nearest degree.

HEIGHT ABOVE GROUND OF:
RIDGE= 24'–0"
EAVES= 17'–0"

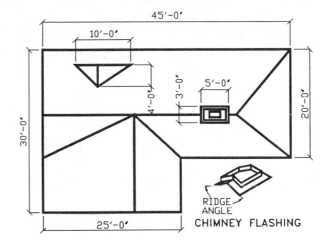

Problem 8–13 Dihedral angle. Given are two adjacent front and horizontal projection planes of a laundry chute. Find the true angle between side A and side B. (Drawing scale: One square = ¼")

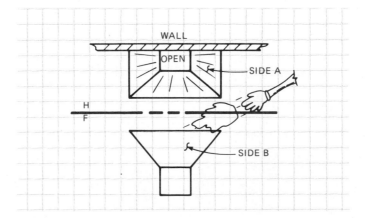

Problem 8–14 True shapes and dihedral angle. This sketch of a sand collection hopper was given to you by an engineer. He asked you to find the following:

a. Create a projection plane that will show the true shape of plane A including the hinged door.

b. Create a projection plane that will show how far the top of the door, which is hinged at the bottom, will be from the bot-

tom of the hopper when the door is hanging down in the open position. Show the measurement of this distance in feet and inches.

c. Create a projection plane that will show the true angle between plane A and plane B. Show the measurement of this angle to the nearest degree.

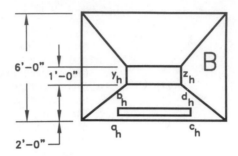

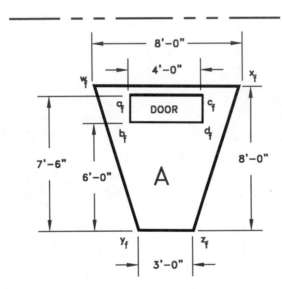

Problem 8–15 Vectors. Draw the correct vector diagram to determine the resultant force of the concurrent, coplanar forces below. Measure the resultant force to the nearest 10 lb.

Problem 8–16 Vectors. Given two adjacent front and horizontal projection planes showing three concurrent noncoplanar forces, find the resultant force of vectors A-B, A-C, and A-D to the nearest pound.

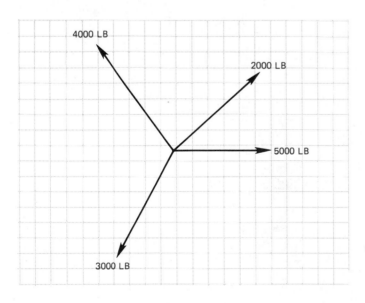

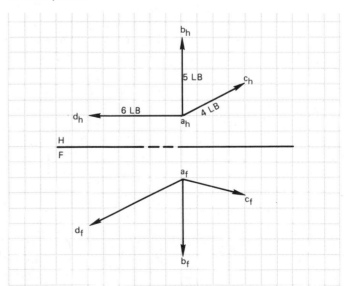

Problem 8–17 Vectors. Given the front projection plane showing the space diagram of three forces, find the tensions in members A-B and A-C acted upon by the given force.

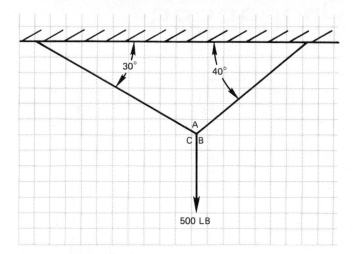

Problem 8–18 Vectors. Given the front projection plan showing a space diagram of three forces in equilibrium, find angles A and B to the nearest degree.

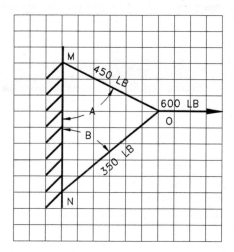

Problem 8–19 Given the following information on sloping lines, determine their slope, slope angle, and percent of grade.
1. Line A-B: run = 50 feet; rise = 34 feet
2. Line C-D: run = 100 feet; rise = 25 feet
3. Line D-E: run = 148 yards; rise = 100 yards
4. Line F-G: run = 65 inches; rise = 0 inches
5. Line HJ: run = 100 meters; rise = 75 meters

CHAPTER 9

Manufacturing Processes

LEARNING OBJECTIVES

After completing this chapter, you will:

○ Define and describe various manufacturing materials, material terminology, numbering systems, and material treatment.
○ Discuss casting processes and terminology.
○ Explain the forging process and terminology.
○ Describe manufacturing processes.
○ Define and draw the representation of various machined features.
○ Explain tool design and drafting practices.
○ Discuss the statistical process quality control assurance system.
○ Evaluate the results of an engineering and manufacturing problem.
○ Explain the use of computer-aided manufacturing (CAM) in today's industry.
○ Discuss robotics in industry.

THE ENGINEERING DESIGN PROCESS

Your company produces die cast aluminum parts. A customer has explained that any delay in parts shipment will require them to shut down a production line. It would be very costly. It is critical to your customer that certain features on their castings fall within specification limits, and the engineering department has decided to develop an early warning system in an attempt to prevent production delays.

You are asked to develop new quality control charts with two levels of control. The maximum allowable deviation is indicated by the upper and lower specification limits as specified by the dimensional tolerances on the part drawing. A mean value is established between the nominal dimension and dimensional limits, and these are set up as upper and lower control limits. The chart instructs the inspector to immediately notify the supervisor if the feature falls outside of control limits and halt production if it is outside of specification limits. This provides a means of addressing problems before they can interfere with production, shifting the emphasis from revision to prevention. A sample chart with two levels of control is shown in Figure 9–1.

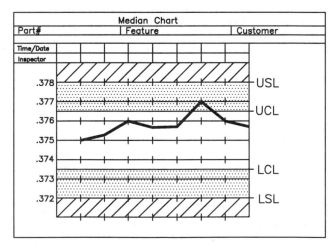

Figure 9–1 Sample quality control chart with both control and specification limits.

A number of factors influence product manufacturing. Beginning with research and development (R & D), a product should be designed to meet a market demand, have good quality, and be economically produced. The sequence of product development begins with an idea and results in a marketable commodity, as shown in Figure 9–2.

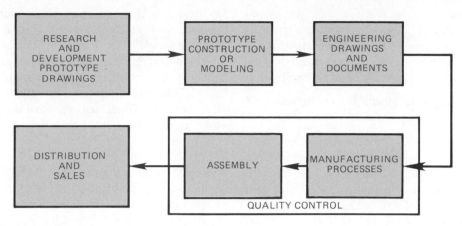

Figure 9–2 Sequence of product development.

▼ MANUFACTURING MATERIALS

There is a wide variety of materials available for product manufacturing that fall into three general categories: metal, plastic, and inorganic materials. Metals are classified as *ferrous, nonferrous,* and *alloys.* Ferrous metals contain iron, such as cast iron and steel. Nonferrous metals do not have iron content; for example, copper and aluminum. Alloys are a mixture of two or more metals.

Plastics or *polymers* have two types of structure: thermoplastic and thermoset. *Thermoplastic* material may be heated and formed by pressure and upon reheating, the shape can be changed. *Thermoset* plastics are formed into a permanent shape by heat and pressure and may not be altered by heating after curing. Plastics are molded into shape and only require machining for tight tolerance situations or when holes or other features are required that would be impractical to produce in a mold. It is common practice to machine some plastics for parts, such as gears and pinions.

Inorganic materials include carbon, ceramics, and composites. Carbon and graphite are classified together and have properties that allow molding by pressure. These materials have low *tensile strength* (ability to be stretched) and high compressive strength with increasing strength at increased temperatures. Ceramics are clay, glass, refractory, and inorganic cements. Ceramics are very hard, brittle materials that are resistant to heat, chemicals, and corrosion. Clay and glass materials have a crystalline structure, while *refractories* must be bonded together by applying temperatures. Due to their great heat resistance, refractories are used for high-temperature applications, such as furnace liners. *Composites* are two or more materials that are bonded together by adhesion. *Adhesion* is a force that holds together the molecules of unlike substances when the surfaces come in contact. These materials generally require carbide cutting tools or special methods for machining.

Cast Iron

There are several classes of cast iron, including gray, white, chilled, alloy, malleable, and nodular cast iron.

Cast iron is primarily an alloy of iron and 1.7% to 4.5% of carbon, with varying amounts of silicon, manganese, phosphorus, and sulfur.

Gray Cast Iron.   Gray iron is a popular casting material for automotive cylinder blocks, machine tools, agricultural implements, and cast iron pipe. Gray cast iron is easily cast and machined. It contains 1.7% to 4.5% carbon and 1% to 3% silicon.

ANSI/ASTM   The American National Standards Institute (ANSI) and *American Society for Testing Materials* (ASTM) specifications A48–76 group gray cast iron into two classes: easy to manufacture (20A, 20B, 20C, 25A, 25B, 25C, 30A, 30B, 30C, 35A, 35B, 35C); and more difficult to manufacture (40B, 40C, 45B, 45C, 50B, 50C, 60B, 60C). The prefix denotes the minimum tensile strength in thousand pounds per square inch.

White Cast Iron.   White cast iron is extremely hard, brittle, and has almost no *ductility* (ability to be stretched, drawn, or hammered thin without breaking). Caution should be exercised when using this material, because thin sections and sharp corners may be weak and the material is less resistant to impact uses. This cast iron is suited for products with more compressive strength requirements than gray cast iron (compare over 200,000 pounds per square inch [psi] with 65,000 to 160,000 psi). White cast iron is used where high wear resistance is required.

Chilled Cast Iron.   When gray iron castings are chilled rapidly, an outer surface of white cast iron results. This material has the internal characteristics of gray cast iron and the surface advantage of white cast iron.

Alloy Cast Iron.   Elements such as nickel, chromium, molybdenum, copper, or manganese may be alloyed with cast iron to increase the properties of strength, wear resistance, corrosion resistance, or heat resistance. Alloy iron castings are commonly used for such items as pistons, crankcases, brake drums, and crushing machinery.

Malleable Cast Iron. The term *malleable* means the ability to be hammered or pressed into shape without breaking. Malleable cast iron is produced by heat-treating white cast iron. The result is a stronger, more ductile, and shock-resistant cast iron that is easily machinable.

ANSI/ASTM The specifications for malleable cast iron are found in ANSI/ASTM A47–77.

Nodular Cast Iron. Special processing procedures, along with the addition of magnesium or cerium bearing alloys, result in a cast iron with spherical-shaped graphite rather than flakes, as in gray cast iron. The results are iron castings with greater strength and ductility. Nodular cast iron may be chilled to form a wear-resistant surface ideal for use in crankshafts, anvils, wrenches, or heavy-use levers.

Steel

Steel is an alloy of iron containing 0.8% to 1.5% carbon. Steel is a readily available material that may be worked in either a heated or cooled state. The properties of steel can be changed by altering the carbon content and heat treating. *Mild steel* (MS) is low in carbon (less than 0.3%), and is commonly used for forged and machined parts, but may not be hardened. *Medium carbon steel* (0.3% to 0.6% carbon) is harder than mild steel yet remains easy to forge and machine. *High carbon steel* (0.6% to 1.50% carbon) may be hardened by heat treating, but is difficult to forge, machine, or weld.

Hot-rolled steel (HRS) characterizes steel that is formed into shape by pressure between rollers or by forging when in a red-hot state. When in this hot condition, the steel is easier to form than when it is cold. An added advantage of hot forming is a consistency in the grain structure of the steel, which results in a stronger, more ductile metal. The surface of hot-rolled steel is rough, with a blue-black oxide buildup. The term *cold-rolled steel* (CRS) implies the additional forming of steel after initial hot rolling. The cold-rolling process is used to clean up hot-formed steel, provide a smooth, clean surface, insure dimensional accuracy, and increase the tensile strength of the finished product.

Steel alloys are used to increase such properties as hardness, strength, corrosion resistance, heat resistance, and wear resistance. Chromium steel is the basis for stainless steel and is used where corrosion and wear resistance is required. Manganese alloyed with steel is a purifying element that adds strength for parts that must be shock and wear resistant. Molybdenum is added to steel when the product must retain strength and wear resistance at high temperatures. When tungsten is added to steel, the result is a material that is very hard and ideal for use in cutting tools. Tool steels are high in carbon and/or alloy content so that the steel will hold an edge when cutting other materials. When the cutting tool requires deep cutting at high speed, then the alloy and hardness characteristics are improved for a classification known as *high-speed steel.* Vanadium alloy is used when a tough, strong, nonbrittle material is required.

Steel castings are used for machine parts where the use requires heavy loads and the ability to withstand shock. These castings are generally stronger and tougher than cast iron. Steel castings have uses as turbine wheels, forging presses, gears, machinery parts, and railroad car frames.

Steel Numbering Systems. **AISI/SAE** The American Iron and Steel Institute (AISI) and Society of Automotive Engineers (SAE) provide similar steel numbering systems. Steels are identified by four numbers, except for some chromium steels, which have five numbers. For a steel with the identification SAE 1020 the first two numbers (10) identify the type of steel and the last two numbers (20) specify the approximate amount of carbon in hundredths of a percent, (0.20% carbon). The letter *L* or *B* may be placed between the first and second pair of numbers. When this is done, the *L* means that lead is added to improve machinability, and the *B* identifies a boron steel. The prefix *E* means that the steel is made using the electric furnace method. The prefix *H* indicates that the steel is produced to hardenability limits. Steel that is degassed and deoxidized before solidification is referred to as *killed steel* and is used for forging, heat treating, and difficult stampings. Steel that is cast with little or no degasification is known as *rimmed steel,* and has applications where sheets, strips, rods, and wires with excellent surface finish or drawing requirements are needed. General applications of SAE steels are shown in Appendix B, Table 25. For a more in-depth analysis of steel and other metals, refer to the *Machinery's Handbook.**

Hardening of Steel. The properties of steel may be altered by *heat treating.* Heat treating is a process of heating and cooling steel using specific, controlled conditions and techniques. When steel is initially formed and allowed to cool naturally, it is fairly soft. *Normalizing* is a process of heating the steel to a specified temperature and then allowing the material to cool slowly by air, which brings the steel to a normal state. In order to harden the steel, the metal is first heated to a specified temperature which varies with different steels. Next the steel is *quenching,* which means to cool suddenly by plunging into water, oil, or other liquid. Steel may also be *case hardened* using a process known as *carburization.* Case hardening refers to the hardening of the surface layer of the metal. Carburization is a process where carbon is introduced into the metal by heating to a specified temperature range while in contact with a

---

*Erick Oberg, Franklin D. Jones, and Holbrook L. Horton, *Machinery's Handbook,* 24th ed. (New York: Industrial Press Inc., 1992).

solid, liquid, or gas material consisting of carbon. This process is often followed by quenching to enhance the hardening process. *Tempering* is a process of reheating a normalized or hardened steel through a controlled process of heating the metal to a specified temperature, followed by cooling at a predetermined rate to achieve certain hardening characteristics. For example, the tip of a tool may be hardened while the balance of the tool remains unchanged.

Under certain heating and cooling conditions and techniques, steel may also be softened using a process known as *annealing.*

Hardness Testing. There are several methods of checking material hardness. The techniques have common characteristics based on the depth of penetration of a measuring device or other mechanical systems that evaluate hardness. The Brinell and Rockwell hardness tests are popular. The Brinell test is performed by placing a known load using a ball of a specified diameter, in contact with the material surface. The diameter of the resulting impression in the material is measured and the Brinell Hardness Number (BHN) is then calculated. The Rockwell hardness test is performed using a machine that measures hardness by determining the depth of penetration of a spherical-shaped device under controlled conditions. There are several Rockwell hardness scales depending on the type of material, the type of penetrator, and the load applied to the device. A general or specific note on a drawing that requires a hardness specification may read: CASE HARDEN 58 PER ROCKWELL "C" SCALE. For additional information, refer to the *Machinery's Handbook.*

Nonferrous Metals

Metals that do not contain iron have properties that are better suited for certain applications where steel may not be appropriate.

Aluminum. Aluminum is corrosion resistant, lightweight, easily cast, conductive of heat and electricity, may be easily *extruded* (shaped by forcing through a die), and is very malleable. Pure aluminum is seldom used, but alloying it with other elements provides materials that have an extensive variety of applications. Some aluminum alloys lose strength at temperatures generally above 121° C (250° F), while they gain strength at cold temperatures. There are a variety of aluminum alloy numerical designations used. A two- or three-digit number is used, and the first digit indicates the alloy type, as follows: 1 = 99% pure, 2 = copper, 3 = manganese, 4 = silicon, 5 = magnesium, 6 = magnesium and silicon, 7 = zinc, 8 = other. For designations above 99%, the last two digits of the code are the amount over 99%. For example, 1030 means 99.30% aluminum. The second digit is any number between 0 and 9 where 0 means no control of

specific impurities and numbers 1 through 9 identify control of individual impurities.

Copper Alloys. Copper is easily rolled and drawn into wire, has excellent corrosion resistance, is a great electrical conductor, and has better ductility than any metal except for silver and gold. Copper is alloyed with many different metals for specific advantages, which include improved hardness, casting ability, machinability, corrosion resistance, elastic properties, and lower cost.

Brass. Brass is a widely used alloy of copper and zinc. Its properties include corrosion resistance, strength, and ductility. For most commercial applications, brass has about a 90% copper and 10% zinc content. Brass may be manufactured by any number of processes including casting, forging, stamping, or drawing. Its uses include valves, plumbing pipe and fittings, and radiator cores. Brass with greater zinc content may be used for applications requiring greater ductility, such as cartridge cases, sheet metal, or tubing.

Bronze. Bronze is an alloy of copper and tin. Tin in small quantities adds hardness and increases wear resistance. Tin content in coins and medallions, for example, ranges from 4% to 8%. Increasing amounts of tin also improve the hardness and wear resistance of the material, but cause brittleness. Phosphorus added to bronze (phosphor bronze) increases its casting ability and aids in the production of more solid castings, which is important for thin shapes. Other materials, such as lead, aluminum, iron, and nickel, may be added to copper for specific applications.

Precious and Other Specialty Metals

Precious metals include gold, silver, and platinum. These metals are valuable because they are rare, costly to produce, and have specific properties that influence use in certain applications.

Gold. Gold for coins and jewelry is commonly hardened by adding copper. Gold coins, for example, are 90% gold and 10% copper. The term *carat* is used to refer to the purity of gold, where $\frac{1}{24}$ gold is one carat. Therefore, $24 \times \frac{1}{24}$ or 24 carats is pure gold. Fourteen-carat gold, for example, is $\frac{14}{24}$ gold and $\frac{10}{24}$ copper. Gold is extremely malleable, corrosion resistant, and is the best conductor of electricity. In addition to use in jewelry and coins, gold is used as a conductor in some electronic circuitry applications. Gold is also sometimes used in applications where resistance to chemical corrosion is required.

Silver. Silver is alloyed with 8% to 10% copper for use in jewelry and coins. Sterling silver for use in such items as eating utensils and other household items is

$^{925}/_{1000}$ silver. Silver is easy to shape, cast, or form, and the finished product may be polished to a high-luster finish. Silver has uses similar to gold because of corrosion resistance and ability to conduct electricity.

Platinum. Platinum is rarer and more expensive than gold. Industrial uses include applications where corrosion resistance and a high melting point are required. Platinum is used in catalytic converters because it has the unique ability to react with and reduce carbon monoxide and other harmful exhaust emissions in automobiles. The high melting point of platinum makes it desirable in certain aerospace applications.

Columbium. Columbium is used in nuclear reactors because it has a very high melting point, 2403° C (4380° F), and is resistant to radiation.

Titanium. Titanium has many uses in the aerospace and jet aircraft industries because it has the strength of steel, the approximate weight of aluminum, and is resistant to corrosion and temperatures up to 427° C (800° F).

Tungsten. Tungsten has been used extensively as the filament in light bulbs because of its ability to be drawn into very fine wire and its high melting point. Tungsten, carbon, and cobalt are formed together under heat and pressure to create tungsten carbide, the hardest manmade material. Tungsten carbide is used to make cutting tools for any type of manufacturing application. Tungsten carbide saw blade inserts are used in saws for carpentry so the cutting edge will last longer. Such blades make a finer and faster cut than plain steel saw blades.

▼ MANUFACTURING PROCESSES

Casting, forging, and machining processes are used extensively in the manufacturing industry. It is a good idea for the entry-level drafter or pre-engineer to be generally familiar with types of casting, forging, and machining processes, and to know how to prepare related drawings. A number of methods may be used by industry to prepare casting, forging, and machining drawings. For this reason, it is best for the beginning drafter to remain flexible and adapt to the standards and techniques used by the specific company. As the drafter gains knowledge of company products, processes, and design goals, he or she may begin to produce designs. It is not uncommon for a drafter to become a designer after three years of practical experience.

Castings

Castings are the end result of a process called founding. *Founding,* or *casting,* as the process is commonly called, is the pouring of molten metal into a hollow or wax-filled mold. The mold is made in the shape of the desired casting. There are several casting methods used in industry. The results of some of the processes are castings that are made to very close tolerances and with smooth finished surfaces. In the simplest terms, castings are made in three separate steps: (1) a pattern that is the same shape as the desired finished product is constructed; (2) using the pattern as a guide, a mold is made by packing sand or other material around the pattern; (3) when the pattern is removed from the mold, molten metal is poured into the hollow cavity. After the molten metal solidifies, the surrounding material is removed and the casting is ready for cleanup or machining operations.

Sand Casting. Sand casting is the most commonly used method of making castings. There are two general types of sand castings: green sand and dry sand molding. *Green sand* is a specially refined sand that is mixed with specific moisture, clay, and resin, which work as binding agents during the molding and pouring procedures. New sand is light brown in color; the term green sand refers to the moisture content. In the *dry sand* molding process, the sand does not have any moisture content. The sand is bonded together with specially formulated resins. The end result of the green sand or the dry sand molds is the same.

Sand castings are made by pounding or pressing the sand around a split pattern. The first or lower half of the pattern is placed upside down on a molding board, then sand is pounded or compressed around the pattern in a box called a *drag.* The drag is then turned over, and the second or upper half of the pattern is formed when another box, called a *cope,* is packed with sand and joined to the drag. A fine powder is used as a parting agent between the cope and drag at the parting line. The parting line is the separating joint between the two parts of the pattern or mold. The entire box, made up of the cope and drag, is referred to as a *flask.* (See Figure 9–3.) Before the molten metal can be poured into the cavity, a passageway for the metal must be made. The passageway is called a *runner* and *sprue.* The location and design of the sprue and runner are important to allow for a rapid and continuous flow of metal. Additionally, vent holes are established to allow for gases, impurities, and metal to escape from the cavity. Finally, a riser (or group of risers) is used, depending on the size of the casting, to allow for the excess metal to evacuate from the mold and, more importantly, to help reduce shrinking and incomplete filling of the casting. (See Figure 9–4.) After the casting has solidified and cooled, the filled risers, vent holes, and runners are removed.

Cores. In many situations a hole or cavity is desired in the casting to help reduce the amount of material removal later or to establish a wall thickness. When this is necessary, a core is used. Cores are made from either clean sand mixed with binders, such as resin, and baked in an

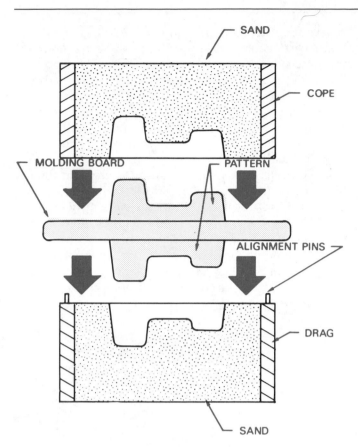

Figure 9-3 Components of sand casting process.

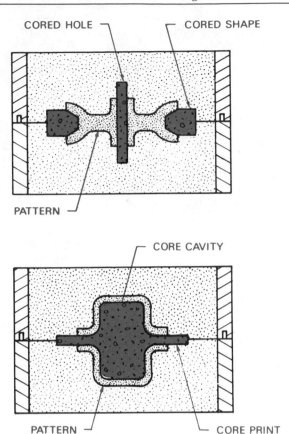

Figure 9-5 Cores in place.

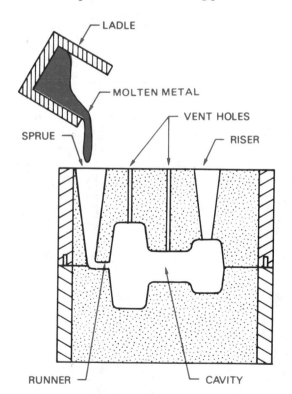

Figure 9-4 Pouring molten metal into a sand casting mold.

oven for hardening, or ceramic products (when a more refined surface finish is required). When the pattern is made, a place for positioning the core in the mold is

established; this is referred to as the core print. After the mold is made, the core is then placed in position in the mold at the core print. The molten metal, when poured into the mold, flows around the core. After the metal has cooled, the casting is removed from the flask and the core is cleaned out, usually by shaking or tumbling. Figure 9-5 shows cores in place. Cored features help reduce casting weight and save on machining costs. Cores used in sand casting should generally be more than one inch (25.4 mm) in cross section. Cores used in precision casting methods may have much closer tolerances and fine detail. Certain considerations must be taken for supporting very large or long cores when placed in the mold. Usually in sand casting, cores require extra support when they are three times longer than the cross-sectional dimension. Depending on the casting method, the material, the machining required, and the quality, the core holes should be a specified dimension smaller than the desired end product if the hole is to be machined to its final dimension. Cores in sand castings should be between .125 to .5 in. (3.2 to 12.7 mm) smaller than the finished size.

Centrifugal Casting. Objects with circular or cylindrical shapes lend themselves to *centrifugal casting*. In this casting process, a mold is revolved very rapidly while molten metal is poured into the cavity. The molten metal is forced outward into the mold cavity by cen-

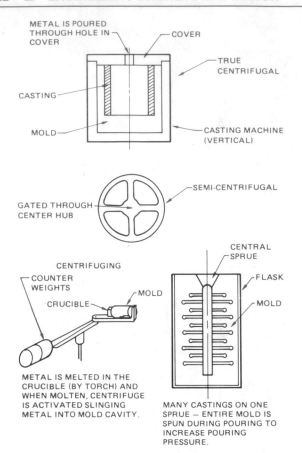

Figure 9–6 Centrifugal casting.

trifugal forces. No cores are needed because the fast revolution holds the metal against the surface of the mold. This casting method is especially useful for casting cylindrical shapes such as tubing, pipes, or wheels. (See Figure 9–6.)

Die Casting. Some nonferrous metal castings are made using the *die casting* process. Zinc alloy metals are the most common, although brass, bronze, aluminum, and other nonferrous products are also made using this process. Die casting is the injection of molten metal into a steel or cast iron die under high pressure. The advantage of die casting over other methods, such as sand casting, is that castings can be produced quickly and economically on automated production equipment. When multiple dies are used, a number of parts can be cast in one operation. Another advantage of die casting is that high-quality precision parts can be cast with fine detail and a very smooth finish.

Permanent Casting. *Permanent casting* refers to a process in which the mold can be used many times. This type of casting is similar to sand casting in that molten metal is poured into a mold. It is also similar to die casting because the mold is made of cast iron or steel. The result of permanent casting is a product that has better finished qualities than can be gained by sand casting.

Investment Casting. *Investment casting* is one of the oldest casting methods. It was originally used in France for the production of ornamental figures. The process used today is a result of the *ciré perdue,* or *lost-wax,* casting technique that was originally used. The reason that investment casting is called lost-wax casting is that the pattern is made of wax. This wax pattern allows for the development of very close tolerances, fine detail, and precision castings. The wax pattern is coated with a ceramic paste. The shell is allowed to dry and then is baked in an oven to allow the wax to melt and flow out; thus "lost" as the name implies. The empty ceramic mold has a cavity that is the same shape as the precision wax pattern. This cavity is then filled with molten metal. When the metal solidifies, the shell is removed and the casting is complete. Generally very little cleanup or finishing is required on investment castings. (See Figure 9–7.)

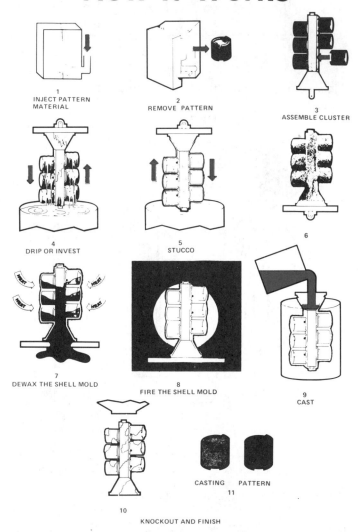

Figure 9–7 Investment casting. *Courtesy Precision Castparts Corporation.*

Forgings

Forging is a process of shaping malleable metals by hammering or pressing between dies that duplicate the desired shape. The forging process is shown in Figure 9–8. Forging may be accomplished on hot or cold materials. Cold forging is possible on certain materials or material thicknesses where hole punching or bending is the required result. Some soft, nonferrous materials may be forged into shape while cold. Ferrous materials such as iron and steel must be heated to a temperature that results in an orange-red or yellow color. This color is usually achieved between 982° to 1066°C. Forging is used for a large variety of products and purposes. The advantage of forging over casting or machining operations is that the material is not only shaped into the desired form, but in the process it retains its original grain structure. Forged metal is generally stronger and more ductile than cast metal, and exhibits a greater resistance to fatigue and shock than machined parts. Notice in Figure 9–9 that the grain structure of the forged material remains parallel to the contour of the part, while the machined part cuts through the cross section of the material grain.

Hand Forging. Hand forging is an ancient method of forming metals into desired shapes. The method of heating metal to a red color and then beating it into shape is called *smithing* or, more commonly, *blacksmithing*. Blacksmithing is used only in industry for finish work, but is still used for horseshoeing and the manufacture of specialty ornamental products.

Machine Forging. Types of machine forging include upset, swaging, bending, punching, cutting, and welding. *Upset* forging is a process of forming metal by pressing along the longitudinal dimension to decrease the length while increasing the width. For example, bar stock is upset forged by pressing dies together from

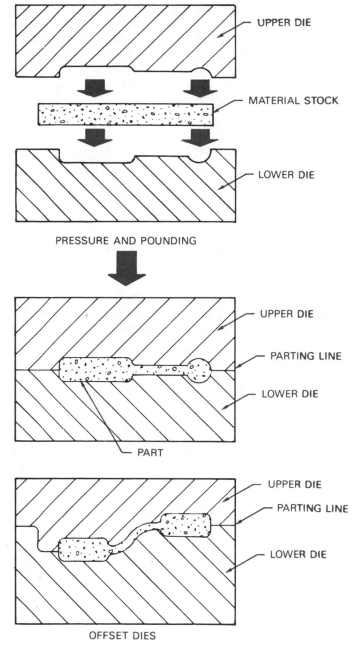

PRESSURE AND POUNDING

OFFSET DIES

Figure 9–8 The forging process.

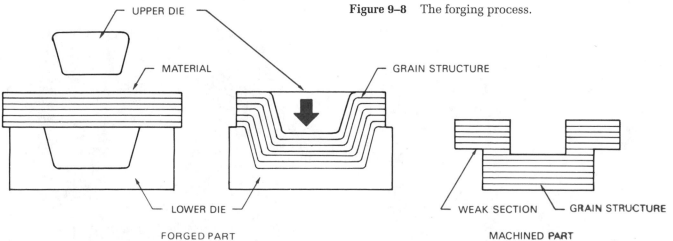

FORGED PART

MACHINED PART

Figure 9–9 Forging compared to machining.

the ends of the stock to establish the desired shape. *Swaging* is the forming of metal by using concave tools or dies that result in a reduction in material thickness. *Bending* is accomplished by forming metal between dies, changing it from flat stock to a desired contour. Bending sheet metal is a cold forging process in which the metal is bent in a machine called a break. *Punching* and *cutting* are performed when the die penetrates the material to create a hole of any desired shape and depth or to remove material by cutting away. In forge *welding,* metals are joined together under extreme pressure. Material that is welded in this manner is very strong. The resulting weld takes on the same characteristics as the metal before joining.

Mass production forging methods allow for the rapid production of the high-quality products shown in Figure 9–10. In machine forging, the dies are arranged in sequence so that the finished forging is done in a series of steps. Complete shaping may take place after the material has been moved through several stages. Additional advantages of machine forging include:

1. The part is formed uniformly throughout the length and width.
2. The greater the pressure exerted on the material, the greater the improvement of the metallic properties.
3. Fine grain structure is maintained to help increase the part's resistance to shock.
4. A group of dies may be placed in the same press.

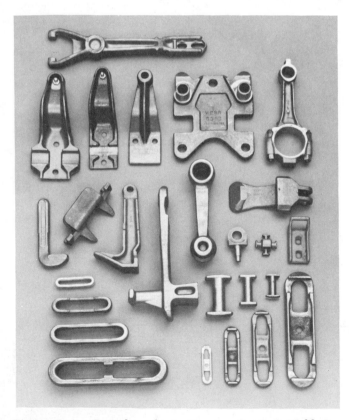

Figure 9–10 Forged products. *Courtesy Jarvis B. Webb Co.*

CADD Applications

COMPUTER NUMERICAL CONTROL MACHINE TOOLS (CNC)

Computer-aided design and drafting (CADD) has a direct link to computer-aided manufacturing (CAM) in the form of *computer numerical control* (CNC) machine tools. A flow chart for the computer numerical control process is shown in Figure 9–11. Figure 9–12 shows a CNC machine. In most cases, the drawing is generated in a computer and this information is sent directly to the machine tool for production. In some applications, information in the computer is transferred to a punched numerical control tape. The numerical control tape may be used as a backup system to the CNC, or the tape may be used as a copy for the numerical control program. Nearly all types of machine tools have been designed to operate from numerical control tape. However, this method is becoming obsolete.

Many CNC systems offer a microcomputer base that incorporates a monitor display and full alphanumeric keyboard, making programming and on-line editing easy and fast. For example, data input for certain types of machining may result in the program-

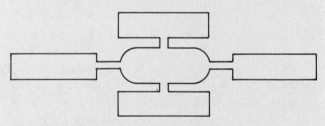

Figure 9–11 The computer numerical control (CNC) process.

Figure 9–12 CNC machines. *Courtesy Boston Digital Corporation.*

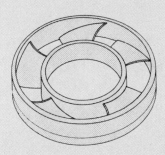

Figure 9–13   CNC programming of a part with five equally spaced blades. This figure demonstrates the CNC programming of one of the five equally spaced blades and the automatic orientation and programming of the other four blades. *Courtesy Boston Digital Corporation.*

ming of one of several identical features while the other features are oriented and programmed automatically, such as the five equally spaced blades of the centrifugal fan shown in Figure 9–13. Among the advantages of CNC machining are increased productivity, reduction of production costs, and manufacturing versatility.

The drafter has a special challenge when preparing drawings for CNC machining. The drawing method must coordinate with a system of controlling a machine tool by instructions in the form of numbers. The types of dimensioning systems that relate directly to CNC applications are tabular, arrowless, datum, and related coordinate systems. The main emphasis is to coordinate the dimensioning system with the movement of the machine tools. This can be best accomplished with datum dimensioning systems in which each dimension originates from a common beginning point. Programming the CNC system requires that cutter compensation be provided for contouring. This task is automatically computer-calculated in some CNC machines, as shown in Figure 9–14. Definitions and examples of the dimensioning systems are discussed in Chapter 10, Dimensioning.

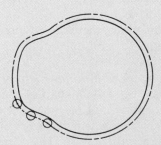

Figure 9–14   Automatic cutter compensation for profile machining. *Courtesy Boston Digital Corporation.*

▼ MACHINE PROCESSES

The concepts covered in this chapter serve as a basis for many of the dimensioning practices presented in Chapter 10. A general understanding of machining processes and the drawing representations of these processes is a necessary prerequisite to dimensioning practices. Problem applications are provided in the following chapters because a knowledge of dimensioning practice is important and necessary to be able to complete manufacturing drawings.

Machine Tools

Drilling Machine.   The *drilling machine,* often referred to as a *drill press* (Figure 9–15), is commonly used to machine drilled holes. Drilling machines are also used to perform other operations, such as reaming, boring, countersinking, counterboring, and tapping. During the drilling procedure, the material is held on a table while the drill or other tool is held in a revolving spindle over the material. When drilling begins, a power or hand-feed mechanism is used to bring the rotating drill in contact with the material. Mass production drilling machines are designed with multiple spindles. Automatic drilling procedures are available on turret drills. These turret drills allow for automatic tool selection and spindle speed. Several operations can be performed in one

Figure 9–15   Drilling machine. *Courtesy Delta International Machining Corporation.*

setup—for example, drilling a hole to a given depth and tapping the hole with a specified thread.

Grinding Machine. A *grinding machine* uses a rotating abrasive wheel, rather than a cutting tool, for the purpose of removing material. (See Figure 9–16.) The grinding process is generally used when a smooth, accurate surface finish is required. Extremely smooth surface finishes can be achieved by honing or lapping. *Honing* is a fine abrasive process often used to establish a smooth finish inside cylinders. *Lapping* is the process of creating a very smooth surface finish using a soft metal impregnated with fine abrasives, or fine abrasives mixed in a coolant that floods over the part during the lapping process.

Lathe. One of the earliest machine tools, the *lathe* (Figure 9–17) is used to cut material by turning cylindrically shaped objects. The material to be turned is held

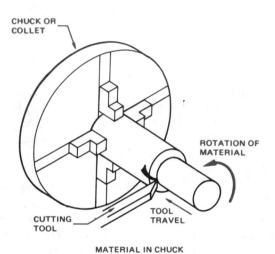

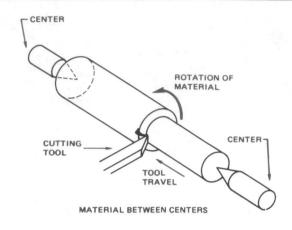

Figure 9–18 Holding material in a lathe.

Figure 9–16 Grinding machine. *Courtesy Litton Industrial Automation.*

between two rigid supports, called *centers,* or in a holding device called a *chuck* or *collet,* as shown in Figure 9–18. The material is rotated on a spindle while a cutting tool is brought into contact with the material. The cutting tool is supported by a tool holder on a carriage that slides along a bed as the lathe operation continues. The turret is used in mass production manufacturing where one machine setup must perform several operations. A turret lathe is designed to carry several cutting tools in place of the lathe tailstock or on the lathe carriage. The operation of the turret provides the operator with an automatic selection of cutting tools at preestablished fabrication stages. Figure 9–19 shows an example of eight turret stations and the tooling used.

Milling Machine. The *milling machine* (Figure 9–20) is one of the most versatile machine tools. The milling machine uses a rotary cutting tool to remove material from the work. The two general types of milling machines are *horizontal* and *vertical* mills. The difference is in the position of the cutting tool, which may be mounted on either a horizontal or vertical spindle. In the operation, the work is fastened to a table that is mechanically fed into the cutting tool, as shown in Figure 9–21. There are

Figure 9–17 Lathe. *Courtesy Hardinge Brothers, Inc.*

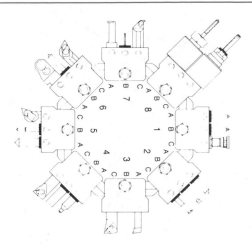

STATION	Model	TOOLING Description
1A		Drill with Bushing
1B		Center Drill with Bushing
1C	T20-⅝	Adjustable Revolving Stock Stop
2A		Drill with Bushing
2B		Boring Bar with Bushing
2C		Threading Tool with Bushing
3A		Grooving Tool
3C		Insert Turning Tool
4A		Center Drill with Bushing
4B		Flat Bottom Drill with Bushing
4C		Insert Turning Tool
5A		Drill with Bushing
5B		Step Drill with Bushing
5C		Insert Turning Tool
6A	T8-⅝	Knurling Tool
6B		Drill with Bushing
6C		Insert Turning Tool
7A		Grooving Tool
7B		Drill with Bushing
7C		Insert Threading Tool
8A	TT-⅝	"Collet Type" Releasing Tap Holder
		Tap Collet
		Tap
8C	TE-⅝	Tool Holder Extension
	T19-⅝	Floating Reamer Holder
		Reamer With Bushing

Figure 9–19 A turret with eight tooling stations and the tools used at each station. *Courtesy Toyoda Machinery USA, Inc.*

Figure 9–20 Close-up of a horizontal milling cutter.

a large variety of milling cutters available that influence the flexibility of operations and shapes that can be performed using the milling machine. Figure 9–22 shows a few of the milling cutters available. Figure 9–23 shows a series of milling cutters grouped together to perform a

VERTICAL MILL

Figure 9–21 Material removal with a vertical machine cutter. *Courtesy The Cleveland Twist Drill Company.*

SIDE CUTTER HEAVY DUTY CUTTER

SQUARE END

BALL END

CORNER ROUNDING ANGLE CONVEX CONCAVE

Figure 9–22 Milling cutters. *Courtesy The Cleveland Twist Drill Company.*

milling operation. End milling cutters, as shown in Figure 9–24, are designed to cut on the end and the sides of the cutting tool. Milling machines that are commonly used in high-production manufacturing often have two or more cutting heads that are available to perform multiple operations. The machine tables of standard horizontal or vertical milling machines move from left to right (x-axis), forward and backward (y-axis), and up and down (z-axis), as shown in Figure 9–25.

Figure 9–23 Grouping horizontal milling cutters for a specific machining operation. *Courtesy The Cleveland Twist Drill Company.*

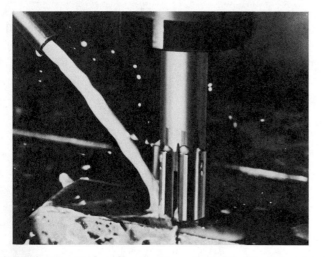

Figure 9–24 End mill operation. *Courtesy The Cleveland Twist Drill Company.*

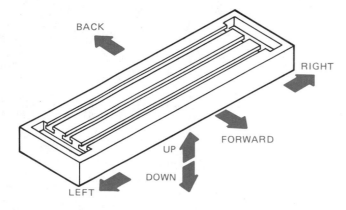

TABLE MOVEMENTS OF THE PLAIN MILLING MACHINE

Figure 9–25 Table movements on a standard milling machine.

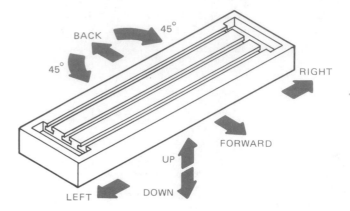

TABLE MOVEMENTS OF THE UNIVERSAL MILLING MACHINE

Figure 9–26 Table movements on a universal milling machine.

The Universal Milling Machine. Another type of milling machine, known as the *universal milling machine,* has table action includes x-, y-, and z-axis movement plus angular rotation. The universal milling machine looks much the same as other milling machines, but has the advantage of additional angular table movement as shown in Figure 9–26. This additional table movement allows the universal milling machine to produce machined features, such as spirals, not possible on conventional machines.

Saw Machines. Saw machines may be used as cutoff tools to establish the length of material for further machining, or saw cutters can be used to perform certain machining operations such as cutting a narrow slot *(kerf).*

Two types of machines that function as cutoff units only are the *power hacksaw* and the *band saw.* These saws are used to cut a wide variety of materials. The hacksaw, as shown in Figure 9–27, operates using a back-and-forth motion. The fixed blade in the power hacksaw cuts material on the forward motion. The metal cutting band saw is available in a vertical or horizontal design, as shown in Figure 9–28. This type of cutoff saw

Figure 9–27 Power hacksaw. *Courtesy JET Equipment and Tool.*

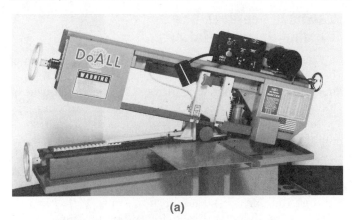

(a)

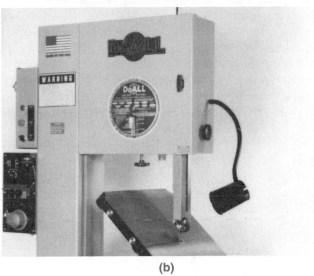

(b)

Figure 9–28 **(a)** Horizontal band saw, **(b)** vertical band saw. *Courtesy DoAll Company*

has a continuous band that runs either vertically or horizontally around turning wheels. Vertical band saws may also be used to cut out irregular shapes.

Saw machines are also made with circular abrasive or metal cutting wheels. The *abrasive saw* may be used for high-speed cutting where a narrow saw kerf is desirable or when very hard materials must be cut. One advantage of the abrasive saw is its ability to cut a variety of materials—from soft aluminum to case-hardened steels. (Cutting a variety of metals on the band or the power hacksaw requires blade and speed changes.) A disadvantage of the abrasive saw is the expense of abrasive discs. Many companies use this saw only when versatility is needed. The abrasive saw is usually found in the grinding room where abrasive particles can be contained, but may also be used in the shop for general purpose cutting. Metal cutting saws with teeth, also known as *cold* saws, are used for precision cutoff operations, cutting saw kerfs, slitting metal, and other manufacturing uses. Figure 9–29 shows a circular saw blade.

Water jet cutting is used on composite materials and thin metal with a computer-controlled 55,000 pounds

Figure 9–29 Circular blade saw. *Courtesy The Cleveland Twist Drill Company.*

per square inch (psi) water jet. Cuts are made with holding tolerances of .0008 in. (0.020 mm) and without generating heat.

Shaper.   The *shaper* is used primarily for production of horizontal, vertical, or angular flat surfaces. Shapers are generally becoming out of date and are rapidly being replaced by milling machines. A big problem with the shaper in mass production industry is that it is very slow and cuts only in one direction. One of the main advantages of the shaper is its ability to cut irregular shapes that cannot be conveniently reproduced on a milling machine or other machine tools. However, other more advanced multiaxis machine tools are now available that quickly and accurately cut irregular contours.

Chemical Machining

Chemical machining uses chemicals to remove material accurately. The chemicals are placed on the material to be removed while other areas are protected. The amount of time the chemical remains on the surface determines the extent of material removal. This process, also known as chemical milling, is generally used in situations where conventional machining operations are difficult. A similar method, referred to as chemical blanking, is used on thin material to remove unwanted thickness in certain areas while maintaining "foil" thin material at the machined area. Material may be machined to within .00008 in. (0.002 mm) using this technique.

Electrochemical Machining.   *Electrochemical machining* (ECM) is a process in which a direct current is passed through an electrolyte solution between an electrode and the workpiece. Chemical reaction, caused by the current in the electrolyte, dissolves the metal, as shown in Figure 9–30.

Electrodischarge Machining (EDM).   In *electrodischarge machining,* the material to be machined and an electrode are submerged in a dielectric fluid which is a

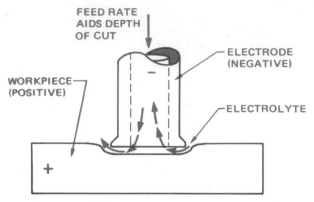

Figure 9–30 Electrochemical machining (ECM).

nonconductor, forming a barrier between the part and the electrode. A very small gap of about .001 inch is maintained between the electrode and the material. An arc occurs when the voltage across the gap causes the dielectric to break down. These arcs occur about 25,000 times per second, removing material with each arc. The compatibility of the material and the electrode is important for proper material removal. The advantages of EDM over conventional machining methods include its success in machining intricate parts and shapes that otherwise cannot be economically machined, and its use on materials that are difficult or impossible to work with, such as stainless steel, hardened steels, carbides, and titanium.

Electron Beam (EB) Cutting and Machining. In this type of chemical machining, an electron beam generated by a heated tungsten filament is used to cut or "machine" very accurate features into a part. This process may be used to machine holes as small as .0002 in. (0.005 mm) or contour irregular shapes with tolerances of .0005 in. (0.013 mm). *Electron beam cutting* techniques are versatile and may be used to cut or machine any metal or nonmetal.

Ultrasonic Machining. *Ultrasonic machining,* also known as impact grinding, is a process in which a high-frequency mechanical vibration is maintained in a tool designed to a specific shape. The tool and material to be machined are suspended in an abrasive fluid. The combination of the vibration and the abrasive causes the material removal.

Laser Machining. The laser is a device that amplifies focused light waves and concentrates them in a narrow, very intense beam. The term LASER comes from the first letters of the words "*Light Amplification by Stimulated Emission of Radiation.*" Using this process, materials are cut or machined by instant temperatures up to 75,000° F (41,649° C). *Laser machining* may be used on any type of material and produces smooth surfaces without burrs or rough edges.

Machined Features and Drawing Representations

The following discussion provides a brief definition of the common manufacturing-related terms. The figures that accompany each definition show an example of the tool, a pictorial of the feature, and the drawing representation. The terms are organized in categories of related features, rather than alphabetical order.

Drill. A *drill* is used to machine new holes or enlarge existing holes in material. The drilled hole may go through the part, in which case the note THRU can be added to the diameter dimension. When the views of the hole clearly show that the hole goes through the part, then the note THRU may be omitted. When the hole does not go through, the depth must be specified. This is referred to as a *blind hole.* The drill depth is the total usable depth to where the drill point begins to taper. A drill is a conical-shaped tool with cutting edges, normally used in a drill press. The drawing representation of a drill point is a 120° total angle. (See Figure 9–31.)

Ream. The tool is called a reamer. The *reamer* is used to enlarge or finish a hole that has been drilled, bored, or cored. A cored hole is cast in place, as previously discussed. A reamer removes only a small amount of material; for example, .005 to .016 in. depending on the size

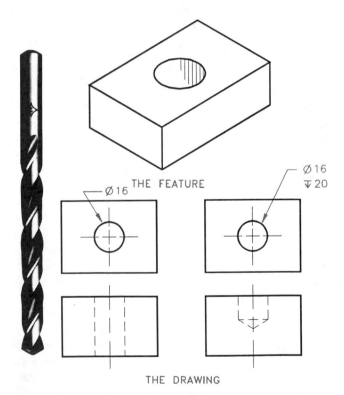

Figure 9–31 Drill. *Tool photo courtesy The Cleveland Twist Drill Company.*

of a hole. The intent of a reamed hole is to provide a smooth surface finish and a closer tolerance than available with the existing hole. A reamer is a conical-shaped tool with cutting edges similar to a drill; however, a reamer will not create a hole as with a drill. Reamers may be used on a drill press, lathe, or mill. (See Figure 9–32.)

Bore.   *Boring* is the process of enlarging an existing hole. The purpose may be to make a drilled or cored hole in a cylinder or part concentric with or perpendicular to other features of the part. A boring tool is used on a lathe for removing internal material. (See Figure 9–33.)

Counterbore.   The *counterbore* is used to enlarge the end(s) of a machined hole to a specified diameter and depth. The machined hole is made first, and then the counterbore is aligned during the machining process by means of a pilot shaft at the end of the tool. Counterbores are usually made to recess the head of a fastener below the surface of the object. The drafter should be sure that the diameter and depth of the counterbore are adequate to accommodate the fastener head and fastening tools. (See Figure 9–34.)

Countersink.   A *countersink* is a conical feature in the end of a machined hole. Countersinks are used to recess

the conically shaped head of a fastener, such as a flathead machine screw. The drafter should specify the countersink note so the fastener head is recessed slightly below the surface. (See Figure 9–35.)

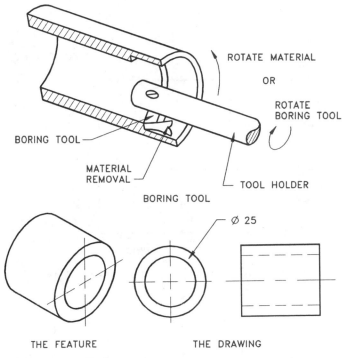

Figure 9–33   Boring.

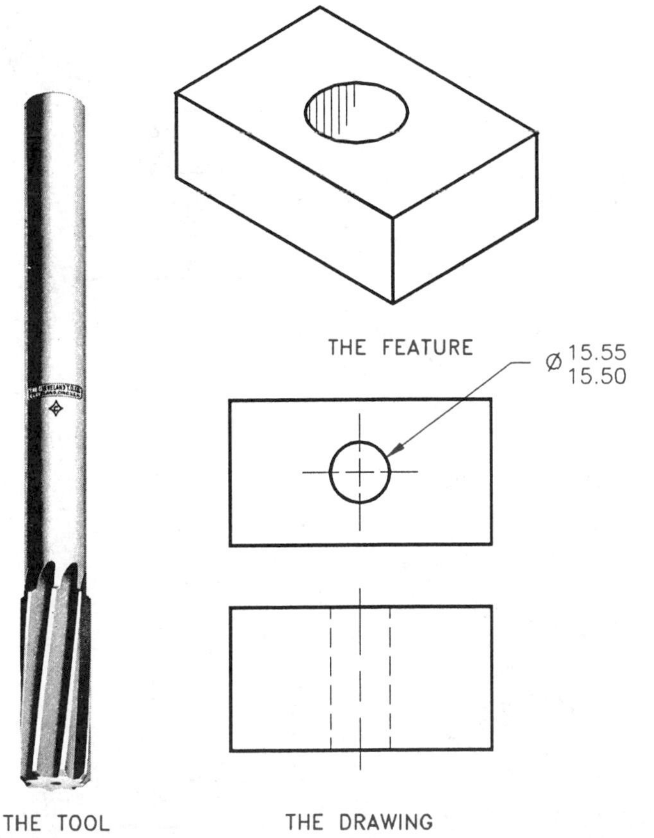

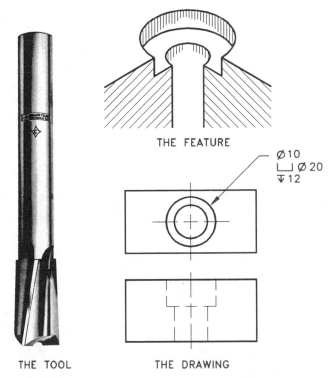

Figure 9–32   Ream. *Tool photo courtesy The Cleveland Twist Drill Company.*

Figure 9–34   Counterbore. *Tool photo courtesy The Cleveland Twist Drill Company.*

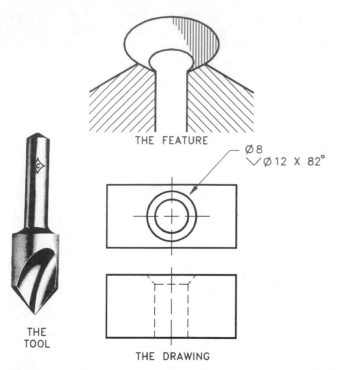

Figure 9–35 Countersink. *Tool photo courtesy The Cleveland Twist Drill Company.*

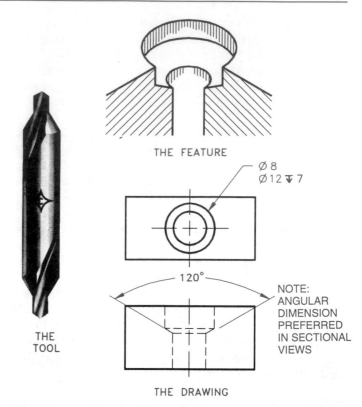

Figure 9–36 Counterdrill. *Tool photo courtesy The Cleveland Twist Drill Company.*

Counterdrill. A *counterdrill* is a combination of two drilled features. The first machined feature may go through the part, while the second feature is drilled, to a given depth, into one end of the first. The result is a machined hole that looks similar to a countersink-counterbore combination. The angle at the bottom of the counterdrill is a total of 120°, as shown in Figure 9–36.

Spotface. A *spotface* is a machined, round surface on a casting, forging, or machined part on which a bolt head or washer can be seated. Spotfaces are similar in characteristics to counterbores, except that a spotface is generally only about 2 mm or less in depth. Rather than a depth specification, the dimension from the spotface surface to the opposite side of the part may be given. This is also true for counterbores; however, the depth dimension is commonly provided in the note. When no spotface depth is given, the machinist will spotface to a depth that establishes a smooth cylindrical surface. (See Figure 9–37.)

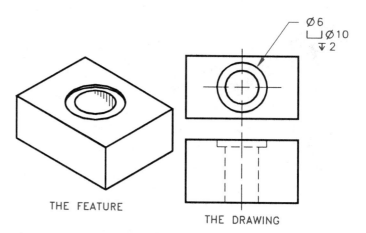

Figure 9–37 Spotface.

Boss. A *boss* is a circular pad on forgings or castings that projects out from the body of the part. While more closely related to castings and forgings, the surface of the boss is often machined smooth for a bolt head or washer surface to seat on. Also, the boss commonly has a hole machined through it to accommodate the fastener's shank. (See Figure 9–38.)

Lug. Generally cast or forged into place, a *lug* is a feature projecting out from the body of a part, usually rec-

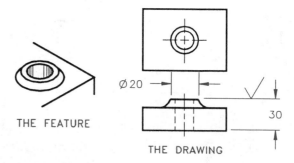

Figure 9–38 Boss.

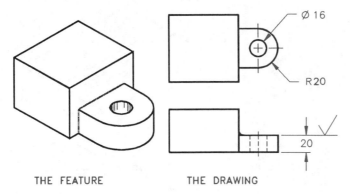

Figure 9–39 Lug.

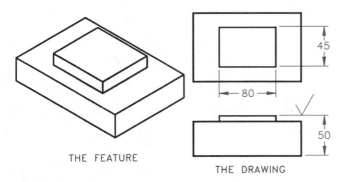

Figure 9–40 Pad.

tangular in cross section. Lugs are used as mounting brackets or function as holding devices for machining operations. Lugs are commonly machined with a drilled hole and a spotface to accommodate a bolt or other fastener. (See Figure 9–39.)

Pad. A *pad* is a slightly raised surface projecting out from the body of a part. The pad surface can be any size or shape. The pad may be cast, forged, or machined into place. The surface is often machined to accommodate the mounting of an adjacent part. A boss is a type of pad, although the boss is always cylindrical in shape. (See Figure 9–40.)

Chamfer. A *chamfer* is the cutting away of the sharp external or internal corner of an edge. Chamfers may be used as a slight angle to relieve a sharp edge, or to assist the entry of a pin or thread into the mating feature. (See Figure 9–41.) Verify alternate methods of dimensioning chamfers in Chapter 10.

Fillet. A *fillet* is a small radius formed between the inside angle of two surfaces. Fillets are often used to help reduce stress and strengthen an inside corner. Fillets are common on the inside corners of castings and forgings to strengthen corners. Fillets are also used to help a casting or forging release a mold or die. Fillets are arcs given as radius dimensions. The fillet size depends on the function of the part and the manufacturing process used to make the fillet. (See Figure 9–42.)

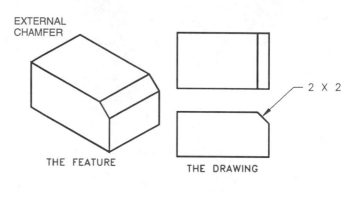

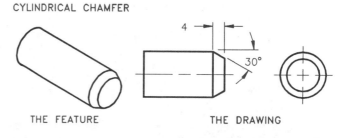

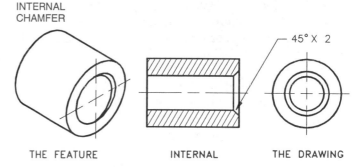

Figure 9–41 Chamfers.

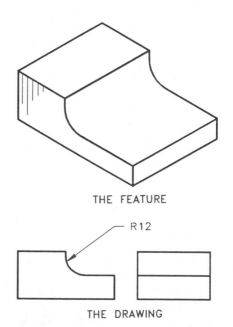

Figure 9–42 Fillet.

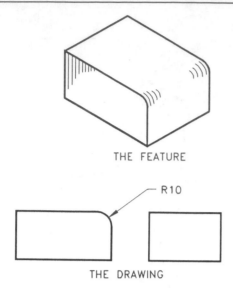

Figure 9–43   Round.

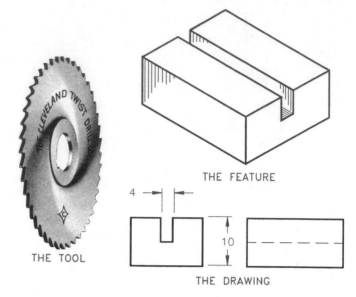

Figure 9–45   Kerf. *Tool photo courtesy The Cleveland Twist Drill Company.*

Round.   A *round* is a small-radius outside corner formed between two surfaces. Rounds are used to refine sharp corners, as shown in Figure 9–43. In some situations where a sharp corner must be relieved and a round is not required, a slight corner relief may be used, referred to as a *break corner*. The note BREAK CORNER may be used on the drawing. Another option is to provide a note that specifies REMOVE ALL BURRS AND SHARP EDGES. *Burrs* are machining fragments that are often left on a part after machining.

Dovetail.   A *dovetail* is a slot with angled sides that may be machined at any depth and width. Dovetails are commonly used as a sliding mechanism between two mating parts. (See Figure 9–44.)

Kerf.   A *kerf* is a narrow slot formed by removing material while sawing or using some other machining operation. (See Figure 9–45.)

Key, Keyseat, Keyway.   A *key* is a machine part that is used as a positive connection for transmitting torque between a shaft and a hub, pulley, or wheel. The key is placed in position in a *keyseat,* which is a groove or chan-

nel cut in a shaft. The shaft and key are then inserted into a hub, wheel, or pulley where the key mates with a groove, called a *keyway.* There are several different types of keys. The key size is often determined by the shaft size. (See Figure 9–46.) Types of keys and key sizes are discussed in Chapter 11, Fasteners and Springs.

Neck.   A *neck* is the result of a machining operation that establishes a narrow groove on a cylindrical part or object. There are several different types of neck grooves, as shown in Figure 9–47. Dimensioning necks is clearly explained in Chapter 10.

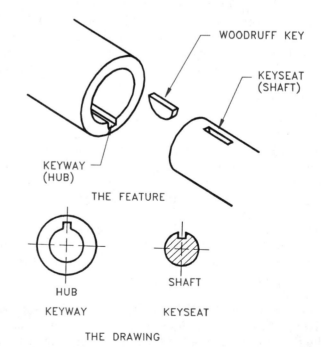

Figure 9–46   Key, keyseat, keyway.

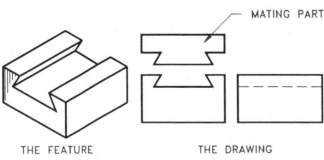

THE FEATURE   THE DRAWING

Figure 9–44   Dovetail.

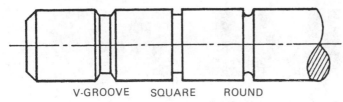

Figure 9–47 Neck.

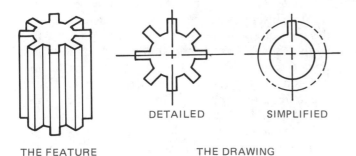

Figure 9–48 Spline.

Spline. A *spline* is a gear-like, serrated surface on a shaft and in a mating hub. Splines are used to transmit torque and allow for lateral sliding or movement between two shafts or mating parts. A spline can be used to take the place of a key when more torque strength is required or when the parts must have lateral movement. (See Figure 9–48.)

Threads. There are many different forms of threads available that are used as fasteners to hold parts together, to adjust parts in alignment with each other, or to transmit power. Threads that are used as fasteners are commonly referred to as *screw threads. External threads* are thread forms on an external feature, such as a bolt or shaft. The machine tool used to make external threads is commonly called a *die.* Threads may be machined on a lathe using a thread-cutting tool. (See Figure 9–49.) *Internal threads* are threaded features on the inside of a hole. The machine tool that is commonly used to cut internal threads is called a *tap.* (See Figure 9–50.)

T-slot. A *T-slot* is a slot of any dimension that is cut to resemble a "T." The T-slot may be used as a sliding mechanism between two mating parts. (See Figure 9–51.)

Knurl. *Knurling* is a cold forming process used to uniformly roughen a cylindrical or flat surface with a diamond or straight pattern. Knurls are often used on handles or other gripping surfaces. Knurls may also be used to establish an interference (press) fit between two mating parts. The actual knurl texture is not displayed on the drawing. (See Figure 9–52.)

Surface Texture. *Surface texture,* or *surface finish,* is the intended condition of the material surface after manufacturing processes have been implemented. Surface texture includes such characteristics as roughness, waviness, lay, and flaws. Surface roughness is one of the most

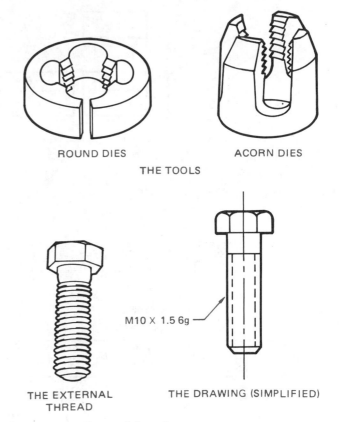

Figure 9–49 External thread.

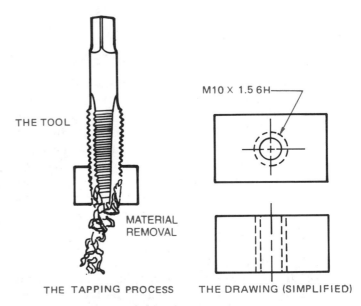

Figure 9–50 Internal thread.

common characteristics of surface finish, which consists of the finer irregularities of the surface texture due, in part, to the manufacturing process. The surface roughness is measured in micrometers (μm) or microinches (μin.) Micro (μ) means millionth. The roughness averages for different manufacturing processes are shown in Appendix B, Table 26. Surface finish is considered a drafting specification and therefore is discussed in detail in Chapter 10.

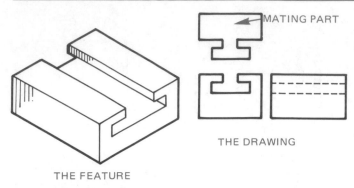

THE DRAWING

THE FEATURE

Figure 9–51   T-slot.

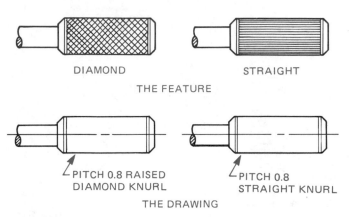

DIAMOND   STRAIGHT

THE FEATURE

PITCH 0.8 RAISED
DIAMOND KNURL

PITCH 0.8
STRAIGHT KNURL

THE DRAWING

Figure 9–52   Knurl.

Design and Drafting of Machined Features

Drafters should gain a working knowledge of the machining processes and capabilities of their companies. Drawings should be prepared that allow machining within the capabilities of the machinery available. If the local machinery will not produce parts that have certain functional requirements, then the machining may need to be performed elsewhere, or the company may have to purchase the necessary equipment. However, the first consideration should be looking for the least-expensive method to get the desired result. Avoid overmachining. Machining processes are expensive, so drawing requirements that are not necessary for the function of the part should be avoided. For example, surface finishes become more expensive to machine as the roughness height decreases. So if a surface roughness of 125 microinches is adequate, then do not use a 32-microinch specification just because you like smooth surfaces. For example, note the difference between 63- and 32-microinch finishes. A 63-microinch finish is a good machine finish that may be performed using sharp tools at high speeds with extra fine feeds and cuts. The 32-microinch callout, on the other hand, requires extremely fine feeds and cuts on a lathe or milling machine and in many cases requires grinding. The 32-microinch finish is more expensive to perform.

CADD Applications

JIG AND FIXTURE DESIGN

Computer-aided design and drafting (CADD) jig and fixture design programs make it possible for the designer or engineer to electronically construct the tooling for a specific application directly on the computer terminal. As with other CADD programs, tooling component libraries, consisting of fully detailed and accurate jig and fixture components, are available. The speed of tooling design sharply increases as compared to manual methods. Items such as clamps, locator pins, rests, or fixture bases can be retrieved from the library and inserted into your design quickly by using the specific functions provided by the software.

In a manufacturing environment where cost and competition are critical considerations, a great deal of thought must be given to meeting the functional and appearance requirements of a product for the least possible cost. It generally does not take very long for an entry-level drafter to pick up these design considerations by communicating with the engineering and manufacturing departments. Many drafters become designers, checkers, or engineers within a company by learning the product and being able to implement designs based on the company's manufacturing capabilities.

▼ TOOL DESIGN

In most production machining operations, special tools are required to either hold the workpiece or guide the machine tool. Tool design involves knowledge of kinematics (study of mechanisms), machining operations, machine tool function, material handling, and material characteristics. *Tool design* is also known as *jig and fixture design*. In mass production industries, jigs and fixtures are essential to insure that each part is produced quickly and accurately within the dimensional specifications. These tools are used to hold the workpiece so that machining operations are performed in the required positions. Application examples are shown in Figure 9–53. Jigs are either fixed or moving devices that are used to hold the workpiece in position and guide the cutting tool. Fixtures do not guide the cutting tool, but are used in a fixed position to hold the workpiece. Fixtures are often used in the inspection of parts to insure that the part is held in the same position each time a dimensional or other type of inspection is made.

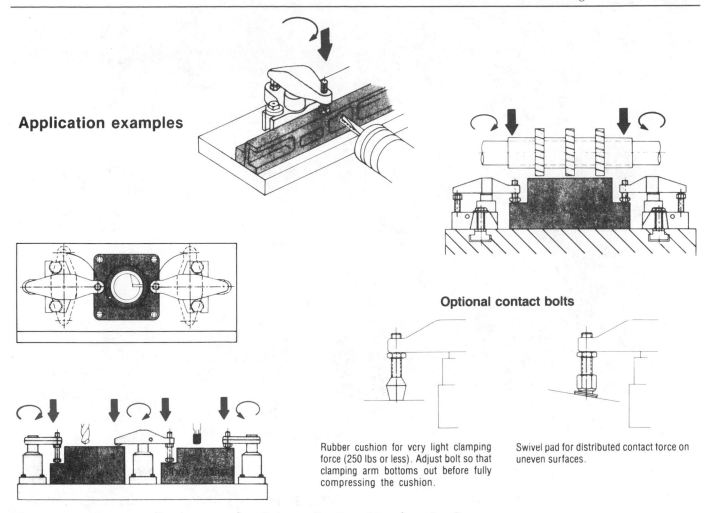

Application examples

Optional contact bolts

Rubber cushion for very light clamping force (250 lbs or less). Adjust bolt so that clamping arm bottoms out before fully compressing the cushion.

Swivel pad for distributed contact force on uneven surfaces.

Figure 9–53   Fixture application examples. *Courtesy Carr Lane Manufacturing Co.*

Jig and fixture drawings are prepared as an assembly drawing where all of the components of the tool are shown as if they were assembled and ready for use, as shown in Figure 9–54. Components of a jig or fixture often include such items as fast-acting clamps, spring-loaded positioners, clamp straps, quick-release locating pins, handles, and knobs, or screw clamps, as shown in Figure 9–55. Normally the part or workpiece is drawn in position using phantom lines, in a color such as red, or a combination of phantom lines and color.

▼ COMPUTER-INTEGRATED MANUFACTURING (CIM)

Completely automated manufacturing systems combine computer-aided design and drafting (CADD), computer-aided engineering (CAE), and computer-aided manufacturing (CAM) into a controlled system known as *computer-integrated manufacturing* (CIM). In Figure 9–56, CIM brings together all the technologies in a management system, coordinating CADD, CAM, CNC, robotics, and material handling from the beginning of the design process through

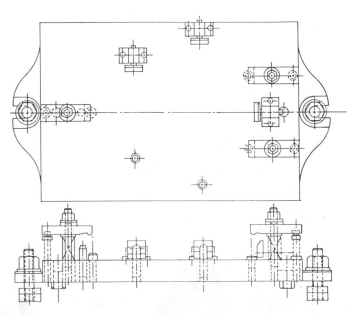

Figure 9–54   Fixture assembly drawing. *Courtesy Carr Lane Manufacturing Co.*

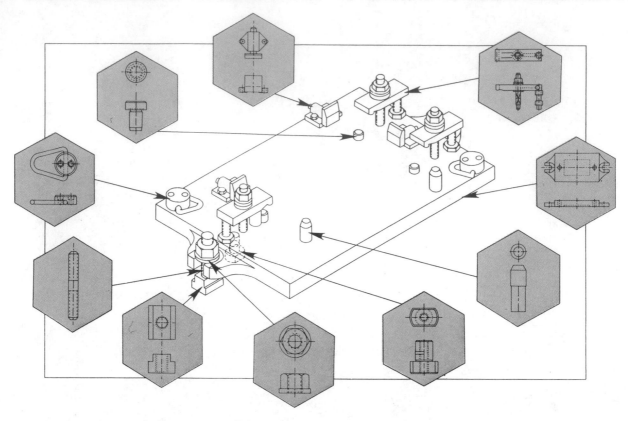

Figure 9–55 Fixture components. *Courtesy Carr Lane Manufacturing Co.*

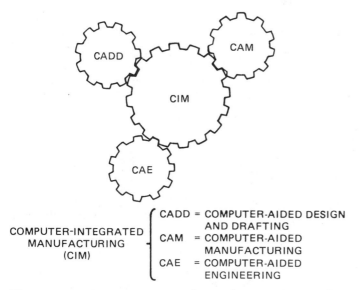

Figure 9–56 Computer-integrated manufacturing (CIM) is the bringing together of the technologies in a management system.

the packaging and shipment of the product. The computer system is used to control and monitor all the elements of the manufacturing system. Before a more complete discussion of the CIM systems, it is a good idea to define some of the individual elements.

Computer-Aided Design and Drafting (CADD)

Engineers and designers use the computer as a flexible design tool. Designs can be created graphically using 3-D models. The effects of changes can be quickly seen and analyzed. This allows the engineer to perform experimentation, perform stress analysis, and make calculations right on the computer. In this manner, engineers can be more creative and improve the quality of the product at less cost. Computer-aided drafting is a partner to the design process. Accurate quality 2-D drawings are created from the designs. Drafting with the computer has increased productivity from two to twenty times over manual techniques depending on the project and skill of the drafter. The computer has also revolutionized the storage of drawings. There is no longer a need for rooms full of drawing file cabinets.

Computer-Aided Engineering (CAE)

Three-dimensional models of the product are used for *finite element analysis.* This is done in a program in which the computer breaks the model up into finite elements, which are small rectangular or triangular shapes. Then the computer is able to analyze each element and determine how it will act under given conditions. It also evaluates how each element acts with the other elements and with the entire model. CAE also allows the engineer to simulate function and motion of the product without the need to build a real prototype. In this manner, the product can be tested to see if it works as it should. This process has also been referred to as predictive engineering, in which a computer software prototype, rather than a physical prototype, is made to test the function and performance of the product. In this manner, design changes may be made right on the computer screen. Key dimensions are placed on the computer model, and by changing a dimension, you can automatically change the design or you can change the values of variables in mathematical engineering equations to automatically alter the design.

Computer Numerical Control (CNC)

Computer numerical control is the use of a computer to write, store, and edit numerical control programs to operate a machine tool. Numerical control (NC) is a method of controlling a machine tool using coded computer language.

Computer-Aided Manufacturing (CAM)

CAM is a concept that surrounds any use of computers to aid in any manufacturing process. CAD and CAM work best when product programming is automatically performed from the model geometry created during the CAD design process. Complex 3-D geometry can be quickly and easily programmed for machining.

Computer-Aided Quality Control (CAQC)

Information on the manufacturing process and quality control is collected by automatic means while parts are being manufactured. This information is fed back into the system and compared to the design specifications or model tolerances. With this type of monitoring of the mass production process, it can be ensured that the highest product quality is maintained.

Concurrent Engineering

Concurrent means happening at the same time. In concurrent engineering, also known as collaborative engineering, product design, manufacturing, marketing, and sales coordinate through every aspect of the company. Traditionally, a product has been sent from one department to the next without substantial teamwork. For example, the engineering department creates the design, which is then passed off to manufacturing. The product then goes to marketing and is sold. Each group essentially works independently of each other. This is not the case with the concurrent engineering concept. Concurrent engineering consists of a team of people who work well together. The team consists of product design people who work directly with marketing, manufacturing, and sales all at the same time. A key element of this system is computer

networking and the exchange of electronic information. Everyone on the team has access to the information while work is in progress. The International Organization for Standardization (ISO) has established a system called ISO 9000, which is designed to ensure quality management between all phases of product development, marketing, engineering, inspection, testing, packaging, storage, sales, installation, technical support, and maintenance. Proper use of this system helps bring the product to market faster and allows the development team the opportunity to create several design options in an effort to improve what goes to market. This is important in keeping a competitive edge.

Parametrics

Parametrics play an important role in the successful implementation of a CIM system. Computerized parametric design refers to the ability to change a variable and automatically have all of the other values change to fit the revised conditions. In the complete computer-integrated manufacturing process, all phases of the system must work together or the system does not work as intended. Parametrics can work to keep the entire system in balance and working together. For example, if one element of the system, such as the production of engineering drawings, has scheduling problems, then the entire program from design concept to product shipping is automatically revised to accommodate the situation. Production planning is an important part of the CIM system. Each machine is tooled and programmed to perform specific jobs. Keeping this machinery operating uniformly ensures the best production throughout the system.

Rapid Prototyping

Rapid prototyping is being used to shorten the time for the design concept stage to the start of manufacturing the product. Rapid prototyping is the ability to turn CADD information into a prototype model made out of paper, resins, powders, or polymers in a few hours. Conventional prototyping using machine tools takes a number of days to months depending on the complexity of the parts. The rapid prototype is just like the real thing in every detail except material. With the prototype in hand, design changes can be made without causing manufacturing or delivery delays. The technologies where CADD designs are quickly transformed into 3-D objects are also referred to as rapid manufacturing, desktop manufacturing, solid free-form fabrication, or conceptual modeling. A technology that produces 3-D prototypes is called the selective laser sintering (SLS) process. This method may use a wide range of powdered materials that sinter (soften and bond when heated). During the SLS process, the 3-D part is created in layers. This process characteristic allows for the production of prototypes and parts featuring virtually any shape or geometry. The process

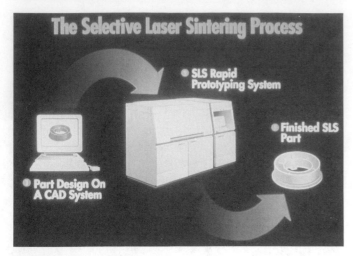

Figure 9–57 A rapid prototyping process. *Courtesy DTM Corporation.*

shown in Figure 9–57 can speed up a company's development cycle. This can help move the final product to market faster, or it can allow more time to refine a complex design.

The SLS process begins with a thin layer of heat-fusible powder deposited into a workspace container. The layer of powder is heated to just below its melting point. The first cross-section of the object under fabrication is traced on the layer of powder by a heat-generated CO_2 laser. The temperature of the powder impacted by the laser is raised to the point of sintering, fusing the powder particles and forming the beginning of the desired shape. Successive layers of powder are deposited and sintered until the part is complete. The basic components of the SLS process are shown in Figure 9–58.

Robotics

According to the Society of Manufacturing Engineers (SME) Robot Institute of America, a *robot* is a reprogrammable multifunctional manipulator designed to move material, parts, tools, or specialized devices through variable programmed motions for the performance of a variety of tasks. *Reprogrammable* means the robot's operating program may be changed to alter the motion of the arm or tooling. Multifunctional means the robot is able to perform a variety of operations based on the program and tooling it uses. A typical robot is shown in Figure 9–59.

To make a complete *manufacturing cell,* the robot is added to all of the other elements of the manufacturing process. The other elements of the cell include the computer controller, robotic program, tooling, associated machine tools, and material handling equipment.

Closed-loop (servo) and *open-loop (non-servo)* are the two types of control used to position the robot tooling. The robot is constantly monitored by position sensors in the closed-loop system. The movement of the robot arm must always conform to the desired path and speed. Open-loop robotic systems do not constantly monitor the position of the tool while the robot arm is moving. The control happens at the end of travel where limit controls position the accuracy of the tool at the desired place.

The Human Factor

The properly operating elements of the CIM system provide the best manufacturing automation available. What has been described in this discussion might lead you to believe that the system can be set up, turned on, and the people can go home. Not so. An important part of CIM is the human element. People are needed to

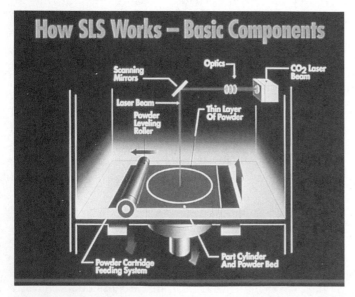

Figure 9–58 The basic components of the selective laser sintering (SLS) process. *Courtesy DTM Corporation.*

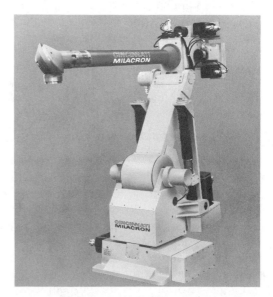

Figure 9–59 A typical robot. *Courtesy of Cincinnati Milacron Marketing Company.*

handle the many situations that happen in the manufacturing cell. The operators constantly observe the machine operation to ensure that material is feeding properly and the quality is maintained. In addition to monitoring the system, people load raw material into the system, change machine tools when needed, maintain and repair the system as needed, and handle the completed work when the product is finished.

▼ STATISTICAL PROCESS CONTROL (SPC)

A system of quality improvement is helpful to anyone who turns out a product, or is engaged in a service and wishes to improve the quality of work, and at the same time increase the output; all with less labor, and at reduced cost. Competition is here to stay, regardless of the nature of the business. Improved quality means less waste and less rework, resulting in increased profits and an improved market position. Customarily, many managers see quality as a drag on profits. Quality is often placed after cost and delivery, because some managers believe that high quality can only be achieved through costly and slow inspection processes. Many managers are now seeing the influence of quality on sales because high quality has become an important criterion in their customers' purchase decisions. In addition, poor quality is expensive. It is estimated that between 15% and 40% of the American manufacturer's product cost is a result of unacceptable output. Regardless of the goods or service produced, it is always less costly to do it right the first time. Improved quality improves productivity, increases sales, reduces cost, and improves profitability. The net result is continued business success.

Traditionally, a type of quality control/detection system has been used in most organizations in the United States. This system comprises customer demand for a product, which is then manufactured in a process made up of a series of steps or procedures. Input to the process includes machines, materials, work force, methods, and environment, as shown in Figure 9–60. Once the product or service is produced, it goes on to an inspection operation where decisions are made to ship, scrap, rework, or otherwise correct any defects discovered (if they are discovered). In actuality, if nonconforming products are being produced, then some are being shipped. Even the best inspection process screens out only a portion of the defective goods. Problems inherent in this system are that it doesn't work very well and it is costly. American businesses have become accustomed to accepting these limitations as the "costs of doing business."

The most effective way to improve quality is to alter the production process, the system, rather than the inspection process. This entails a major shift in the entire organization from the detection system to a prevention mode of operation. In this system, the elements (inputs, process, product or service, customer) remain the same, but the inspection method is significantly altered or eliminated. A primary difference between the two systems is that in the prevention system, statistical techniques and problem-solving tools are used to monitor, evaluate, and provide guidance for adjusting the process to improve quality. *Statistical process control* (SPC) is a method of monitoring a process quantitatively and using statistical signals to either leave the process alone, or change it. (See Figure 9–61.) It involves several fundamental elements:

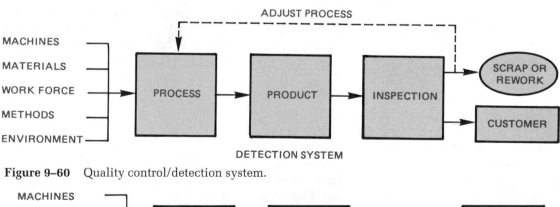

Figure 9–60   Quality control/detection system.

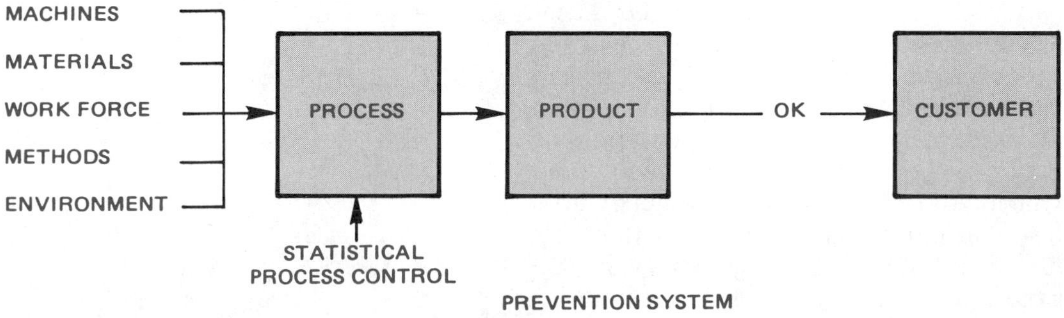

Figure 9–61   Quality control/prevention system.

1. The process, product, or service must be measured. It can be measured using either variables (a value that varies), or attributes (a property or characteristic) from the data collected. The data should be collected as close to the process as possible. If you are collecting data on a particular dimension of a manufactured part, it should be collected by the machinist who is responsible for holding that dimension.

2. The data can be analyzed using control charting techniques. Control charting techniques use the natural variation of a process, determining how much the process can be expected to vary if the process is operationally consistent. The control charts are used to evaluate whether the process is operating as designed, or if something has changed.

3. Action is taken based on signals from the control chart. If the chart indicates that the process is *in control* (operating consistently), then the process is left alone at this point. On the other hand, if the process is found to be *out of control* (changing more than its normal variability allows), then action is taken to bring it back into control. It is also important to determine how well the process meets specifications and how well it accomplishes the task. If a process is not in control, then its ability to meet specifications is constantly changing. Process capability cannot be evaluated unless the process is in control. Process improvement generally involves changes to the process system that will improve quality or productivity or both. Unless the process is consistent over time, any actions to improve it may be ineffective.

Manufacturing quality control often uses computerized monitoring of dimensional inspections. When this is done, a chart is developed that shows feature dimensions obtained at inspection intervals. The chart shows the expected limits of sample averages as two dashed parallel horizontal lines, as shown in Figure 9–62. It is important not to confuse control limits with tolerances—they are not related to each other. The control limits come from the manufacturing process as it is operating. For example, if you set out to make a part 1.000 ± .005 in., and you start the run and periodically take five samples and plot the averages (x) of the samples, the sample averages will vary less than the individual parts. The control limits represent the expected variation of the sample averages if the process is stable. If the process shifts or a problem occurs, the control limits signal that change. Notice in Figure 9–62 that the x values represent the aver-

PROFESSIONAL PERSPECTIVE

Mechanical drafting and manufacturing are very closely allied. The mechanical drafter should have a general knowledge of manufacturing methods, including machining processes. It is common for the drafter to consult with engineers and machinists regarding the best methods to implement a drawing from the design to the manufacturing processes. Part of design problem solving is to create a design that is not only functional, but one that can also be manufactured using available technology and at a cost that justifies the end product. The solutions to these types of concerns depend on how familiar the drafter and designer are with the manufacturing capabilities of the company. For example, if the designer over-tolerances a part or feature, the result could be a rejection of the part by manufacturing or an expensive machining operation. The drafter should also know something about how machining processes operate so that drawing specifications do not call out something that is not feasible to manufacture. The drafter should be familiar with the processes that are used in machine operations so the notes that are placed on a drawing conform to the proper machining techniques. Notes for machining processes are given on a drawing in the same manner that they are performed in the shop. For example, the note for a counterbore is given as the diameter and depth of the hole (first process) and the diameter and depth of the counterbore (second process). Also, the drafter should know that the specification given will, in fact, yield the desired result. One interesting aspect of the design drafter's job is the opportunity to communicate with people in manufacturing and come up with a design that enables the product to be easily manufactured.

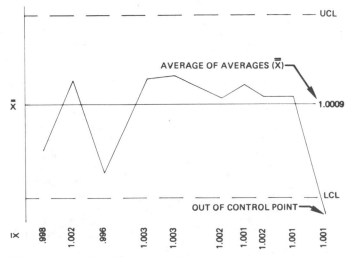

Figure 9–62 Quality control chart.

age of each five samples; $\bar{x}$ is the average of averages over a period of sample taking. The *upper control limit* (UCL) and the *lower control limit* (LCL) represent the expected variation of the sample averages. A sample average may be "out of control" yet remain within tolerance. During this part of the monitoring process, the out-of-control point represents an individual situation that may not be a problem; however, if samples continue to be measured out of control limits, then the process is out of control (no longer predictable) and therefore may be producing parts outside the specification. Action must then be taken to bring the process back into statistical control or 100% inspection must be resumed. The SPC process only works when a minimum of twenty-five sample means are in control. When this process is used in manufacturing, part dimensions remain within tolerance limits and parts are guaranteed to have the quality designed.

 **MATH APPLICATION**

FINDING MEANS FOR SPC

Here is a set of data consisting of weekly samplings of a 2-in. bolt making machine. The process is considered "out of control" if sample averages exceed a UCL of 2.020 in. or are less than an LCL of 1.980 in. In preparation of a chart, find x for each week, and find $\bar{x}$ for the set of 4 weeks. Was the process "out of control" during any week?

Week 1	Week 2	Week 3	Week 4
1.970	2.022	1.980	2.003
2.013	2.014	2.001	1.989
2.051	2.011	1.898	1.993
1.993	2.001	1.888	2.019
1.992	2.009	1.979	2.003

Solution: In the world of math statistics, a bar over a letter denotes an average (mean). The formula for finding the average of x values is usually written as $x = \Sigma x / n$ where x is an individual data value, and n is the total number of data values. The Σ is the capital Greek letter sigma, which stands for repeated addition. The formula is the mathematical way of saying to add up all the data and then divide by the total number of data. Applying this idea to each week separately:

$$x \text{ for Week 1} = (1.970 + 2.013 + 2.051 + 1.993 + 1.992) \div 5$$
$$= (10.019) \div 5 = 2.0038 \text{ or } \mathbf{2.004}$$

Similarly,

$$x \text{ for Week 2} = 2.0114 \text{ or } \mathbf{2.011}$$
$$x \text{ for Week 3} = 1.9492 \text{ or } \mathbf{1.949}$$
$$x \text{ for Week 4} = 2.0014 \text{ or } \mathbf{2.001}$$

$\bar{x}$ is the mean *of these means*. It is found by the same method:

$$\bar{x} = (2.0038 + 2.0114 + 1.9492 + 2.0014) \div 4$$
$$= 1.99145 \text{ or } \mathbf{1.991}$$

The mean for Week 3 is lower than the LCL, so the process was "out of control" then, and corrective measures should have been taken at the end of that week.

CHAPTER

9 MANUFACTURING PROCESSES TEST

DIRECTIONS

Answer the questions with short, complete statements or drawings as needed.

PART 1. MANUFACTURING MATERIALS

1. Define *ferrous metals*.
2. Define *nonferrous metals*.
3. What is another name for plastics?
4. Define *thermoplastic*.
5. Define *thermoset*.
6. Define *tensile strength*.
7. Why are refractories used for such applications as furnace liners?
8. What are composites?
9. Describe the characteristics of gray cast iron.
10. Given the gray cast iron material specification 30A, what does the prefix 30 denote?
11. Define *ductility*.
12. What are the properties of white cast iron?

13. Which cast iron has the internal characteristics of gray cast iron and the exterior properties of white cast iron?
14. Define *malleable*.
15. Identify at least two uses for nodular cast iron.
16. How is it possible to alter the properties of steel?
17. Name a steel that is low in carbon and is commonly used for forged and machined parts.
18. Describe high carbon steel.
19. Describe the difference between hot- and cold-rolled steel.
20. Describe the properties of the following steel alloying elements: manganese, molybdenum, and tungsten.
21. Given the steel identification number SAE 1020, describe the components: SAE, 10, and 20.
22. Identify the steel recommended for the following general applications: agricultural steel, bolts and screws, car and truck gears, transmission shafts.
23. Define *heat treating*.
24. Define *normalizing*.
25. Define *case hardening*.
26. Define *carburization*.
27. Define *tempering*.
28. How is the Rockwell hardness test performed?
29. Describe the properties of aluminum.
30. Define *extruded*.
31. Identify the alloying elements in brass.
32. Identify the alloying elements in bronze.
33. What is the advantage of adding phosphorus to bronze?
34. Describe at least one industrial use for gold.
35. Identify the metal that has the weight advantage of aluminum and the strength of steel.
36. What are the elements of and process for making tungsten carbide?

PART 2. MANUFACTURING PROCESSES

37. What is another name for casting?
38. Define *casting*.
39. Name the most commonly used method of making castings.
40. Discuss the function of cores in the casting process.
41. List at least two advantages of using cores.
42. Describe centrifugal casting.
43. List at least two advantages of the die casting process.
44. Describe the permanent casting process.
45. Name the casting technique that is referred to as "lost wax."
46. Why is shrinkage allowance required in casting design?
47. What is the estimated shrinkage for most irons?
48. Define *draft*.

49. In casting design, is draft added to the minimum or maximum design sizes of the part?
50. Identify at least two reasons why fillets are used in casting design.
51. List at least three considerations that influence the amount of extra material that must be left on a casting for machining allowance.
52. What is the recommended standard finish allowance for iron or steel?
53. Describe hot spots in casting.
54. Define *forging*.
55. What is the grain structure advantage of using the forging process to manufacture a part rather than making a machined part?
56. Describe upset forging.
57. Describe swaging.
58. Describe bending.
59. Describe punching.
60. List at least two advantages of machine forging.

PART 3. MACHINE PROCESSES

61. List at least two reasons why the mechanical drafter should be familiar with machining processes.
62. List four types of machining operations that can be performed on a drilling machine.
63. Identify one of the primary functions of the grinding machine.
64. Describe the main function of a lathe.
65. Describe a feature of milling machines that influences the ability of operations and shapes that can be performed.
66. Identify two types of saw tools that can be used for cut-off and machining operations.
67. Describe chemical machining.
68. Describe electro-chemical machining (ECM).
69. Describe electro-discharge machining (EDM).
70. Describe electron beam (EB) machining.
71. What is another name for ultrasonic machining?
72. Describe the basic function of a laser device.
73. Laser machining may be used on what materials?
74. What does the abbreviation *CNC* mean?
75. Briefly explain the CNC process. Describe Computer-integrated manufacturing.
76. What is the purpose of boring?
77. Identify at least one function of a counterbore.
78. Describe drill depth.
79. Describe and give one application of a knurl.
80. Describe the primary function of a key.
81. Define *keyseat*.
82. What is the function of a reamer?
83. What action will the machinist take when no spot-face depth is given?
84. Define *surface texture*.
85. In what units is surface roughness height measured?

86. Discuss the results of designing a part with specifications that require overmachining.
87. Describe the difference between jigs and fixtures.
88. Normally a jig or fixture is drawn as an assembly of the unit ready for use and the workpiece or part to be held is drawn in position. How is the workpiece drawn in relationship to the jig or fixture?

89. Name the quality control/detection system that uses statistical techniques and problem solving to monitor, evaluate, and provide guidance for adjusting the process to improve quality.
90. Describe the basic function of control charting in quality control.

CHAPTER

MANUFACTURING PROCESSES PROBLEMS

This chapter is intended as a reference for manufacturing processes. The concepts discussed serve as a basis for further study in the following chapters. A thorough understanding of dimensioning practices is necessary before complete manufactured products can be drawn. Problem assignments ranging from basic to complex manufacturing drawings are assigned in the following chapters.

Problems 9–1 through 9–14 Make a two-view drawing of each of the following machined features using manual or computer-aided drafting as required by your instructor. Prepare the drawings on 8½" × 11" vellum unless otherwise specified by your instructor. Dimensioning may be omitted; however, all line representations should be properly drawn using correct techniques. Machined features to be drawn:

9–1. Counterbore
9–2. Chamfer
9–3. Countersink
9–4. Counterdrill
9–5. Drill (not through the material)
9–6. Fillet
9–7. Round
9–8. Spotface
9–9. Dovetail
9–10. Kerf
9–11. Keyseat
9–12. Keyway
9–13. T-slot
9–14. Knurl

Problems 9–15 through 9–42 The following topics require research and/or industrial visitations. It is recommended that you research current professional magazines, visit local industries, or interview professionals in the related field. Your reports should emphasize the following:

• Product
• The process
• Special manufacturing considerations
• The link between manufacturing and engineering
• Current technological advances

Select one or more of the topics listed below or as assigned by your instructor and write a 500-word report for each.

9–15. Casting
9–16. Forging

9–17. Conventional machine shop
9–18. Computer numerical control (CNC) machining
9–19. Surface roughness
9–20. Tool design
9–21. Chemical machining
9–22. Electrochemical machining (ECM)
9–23. Electrodischarge machining (EDM)
9–24. Electron beam (EB) machining
9–25. Ultrasonic machining
9–26. Laser machining
9–27. Statistical process control (SPC)
9–28. Plastic
9–29. Cast iron
9–30. Steel
9–31. Aluminum
9–32. Copper alloys
9–33. Precious and other specialty metals
9–34. Computer-aided manufacturing (CAM)
9–35. Computer-aided engineering (CAE)
9–36. Computer-integrated manufacturing (CIM)
9–37. Robotics
9–38. Computer-aided design
9–39. Computer-aided drafting
9–40. Manufacturing cell
9–41. Rapid prototyping
9–42. Concurrent engineering

MATH PROBLEMS

Here is a set of six monthly samples from a process. The UCL is 5.05 and the LCL is 4.95.

Jan	Feb	Mar	Apr	May	Jun
5.06	6.03	5.04	5.04	4.95	4.95
4.94	5.04	5.04	5.07	4.95	4.97
4.99	5.05	5.05	5.05	4.93	4.98
4.99	4.99	5.06	5.02	4.92	5.01
5.00	4.98	5.06	5.00	4.95	5.00

1. Find x for each month.
2. Find $\bar{\bar{x}}$ for the six month period.
3. Conclude whether the process was "out of control" for any of the months.

CHAPTER 10

Dimensioning

LEARNING OBJECTIVES

After completing this chapter you will:

- Identify and use common dimensioning systems.
- Explain and apply dimensioning standards as interpreted by ASME Y14.5M—1994.
- Apply proper specific notes for manufacturing features.
- Place proper general notes on a drawing.
- Interpret and use correct tolerancing techniques.
- Prepare completely dimensioned multiview drawings from engineering sketches and industrial drawings.
- Dimension CAD/CAM machine tool drawings.
- Prepare casting and forging drawings.
- Provide surface finish symbols on drawings.
- Solve tolerance problems including limits and fits.
- Use an engineering problem as the basis for your layout techniques.

THE ENGINEERING DESIGN PROCESS

As mentioned later in this chapter, a complete detail drawing is made up of multiviews and dimensions. The layout techniques of a detail drawing must include an analysis of how the views and dimensions go together to create the finished drawing. The most effective way to form preliminary layout ideas is through the use of rough sketches as described in Chapter 3.

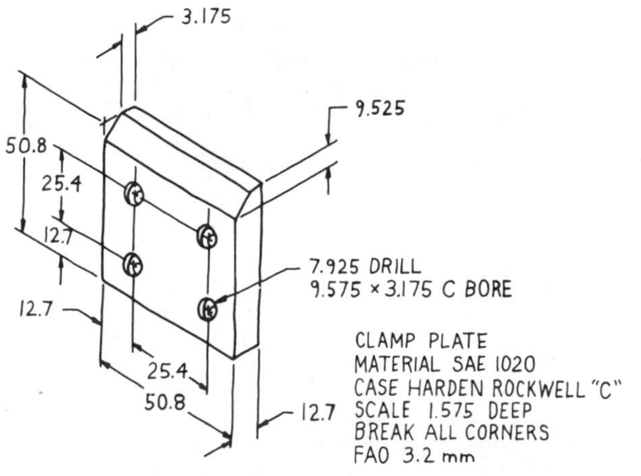

Figure 10–1 Engineering sketch.

Consider the engineering sketch in Figure 10–1 as you evaluate how to prepare a complete detail drawing. The procedure is the same with CADD or manual drafting.

Step 1. Select and make a rough sketch of the proper multiviews. Leave plenty of space between views so dimensions may be added. The engineer's sketch is not always accurate. It is the drafter's responsibility to convert the engineering ideas from the rough stage to a formal drawing. The information may be correct; however, the organization of information and proper layout are the drafter's duties. (See Figure 10–2.)

Step 2. Place dimensions and notes on your multiview sketch. (See Figure 10–3.) Keep the following dimensioning rules in mind as you select and place the dimensions:

- Begin with smallest dimension closest to the object and place dimensions that progressively increase in size further away from the object.
- Do not crowd dimensions.

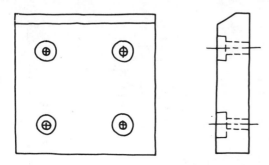

Figure 10–2   Rough sketch of selected multiviews.

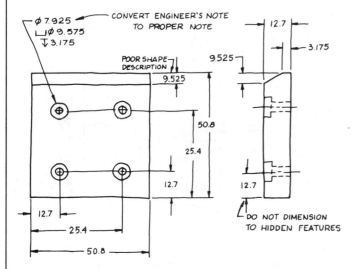

Figure 10–3   Rough sketch with dimensions and notes placed.

○ Dimension between views where possible.

○ Dimension to views that show the best shape of features.

○ Group dimensions when possible.

○ Do not dimension size or location dimensions to hidden features.

○ Stagger adjacent dimension numerals.

○ Use leaders to properly label specific notes.

○ Convert all information to proper drafting standards.

Evaluate these basic rules as you decide where dimensions should be placed.

Step 3.   Determine the scale to use based on the amount of fine detail to be shown and the total size of the object.

The clamp plate can easily be drawn full scale.

Step 4.   Determine the sheet size to be used based on the drawing scale to be used, the total drawing area (including views and dimensions) and the amount of clear area needed for

general notes and future revisions. Refer to Chapter 5.

Look at Figure 10–3 as you follow the calculations to determine the length and height of the drawing area needed for the clamp plate.

Length of drawing area (in inches)
Front view width	2.00
Space between views:	
Front view to .50 dimension	.75
.50 to 1.00 dimension	.50
1.00 to 2.00 dimension	.50
2.00 to .375 dimension	.50
.375 dimension to right side view	.75
Right side view depth	+.50
Drawing area total length	5.50

The space from the view to the first dimension and the spaces between dimensions are drafting decisions. The first space is often, but not always, larger than distances to additional dimension lines. Each dimension line after the first is equally spaced so that dimensions are not crowded.

Height of drawing area (in inches)
Height of views	2.00
Front view to .50 dimension	.75
.50 to 1.00 dimension	.50
1.00 to 2.00 dimension	.50
Right side view to .125 dimension	.75
.125 to .50 dimension	+.50

There should be enough space for the CBORE note	
Drawing area total height	5.00

The total length and height of the drawing area is 5.50 × 5.00. Keep in mind that these calculations are approximate. There should be adequate space on your drawing for some flexibility. Select a B (A3 metric) size drafting sheet for this drawing. Figure 10–4 shows the 5.50 × 5.00 drawing area blocked out on a B-size sheet. Before you proceed, be sure that your equipment, table, and hands are clean.

Step 5.   Use construction lines to draw the selected multiviews.

Step 6.   Use construction lines to place hidden, extension, center, dimension, and leader lines, using selected spacing and placement from rough sketch. Also, use guidelines to prepare for the placement of lettering. (See Figure 10–5.)

(Continued)

ENGINEERING DESIGN PROCESS *(Continued)*

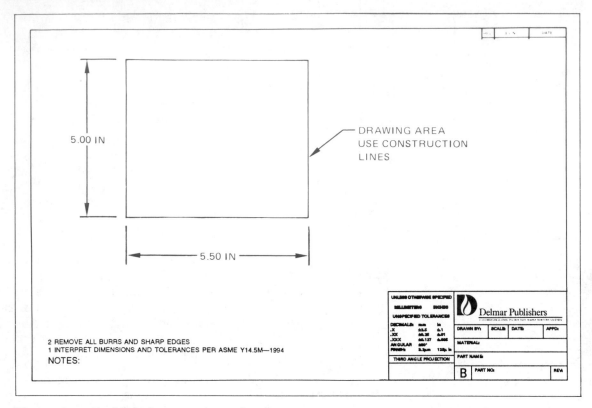

Figure 10–4 Establish the approximate drawing area.

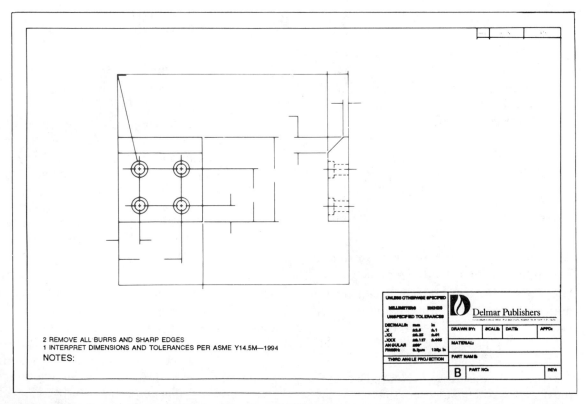

Figure 10–5 Lay out the multiview's object lines, hidden lines, centerlines, extension and dimension lines, leader lines; establish guidelines for lettering.

Step 7. Complete the detail drawing using proper drafting technique in the following sequence:

1. Darken all horizontal thin lines from top to bottom and all vertical thin lines from left to right if right-handed or right to left if left-handed.

2. Draw all circles, arcs, and object lines.

3. Draw all straight object lines in the same order as thin lines.

4. Do all lettering from the top to the bottom of the sheet. Letter all specific and general notes and fill in the title block. To help keep your drawing clean, place a blank sheet of paper under your hand while lettering.

5. Draw all arrowheads.

6. Check your line quality by placing the original over a light table. If light passes through the lines or lettering, the work is not dark enough. Another way to check your work is to run a test-strip diazo copy. If the quality of the print is not adequate, you may need to work more on the drawing. A print will not improve the quality of a drawing.

Figure 10–6 shows the completed detail drawing of the clamp plate.

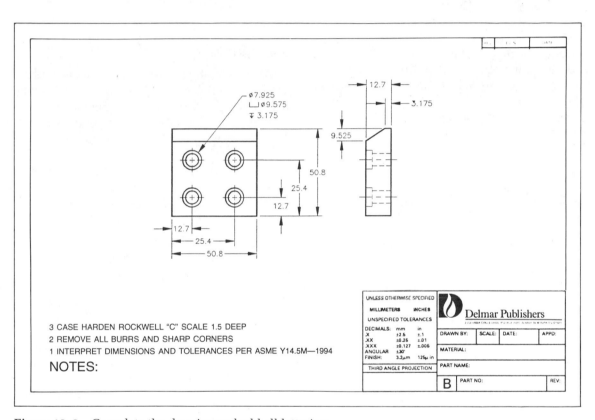

Figure 10–6 Complete the drawing and add all lettering.

ANSI/ASME The standard adopted by the American National Standards Institute (ANSI) and published by The American Society of Mechanical Engineers (ASME) is titled *Dimensioning and Tolerancing, ASME Y14.5M—1994.* This standard is available through the ASME at 345 East 47th Street, New York, NY 10017. The standard that controls general dimensional tolerances found in the title block or in general notes is *ASME Y14.1M—1992, Metric Drawing Sheet Size and Format.*

A complete detail drawing includes multiviews and dimensions, which provide both shape and size description. There are two classifications of dimensions: size and location. *Size dimensions* are placed directly on a feature to identify a specific size or may be connected to a feature in the form of a note. The relationship of features of an object is defined with *location dimensions.*

Notes are a type of dimension that generally identify the size of a feature or features with more than a numerical specification. For example, a note for a counterbore will give size and identification of the machine process used in manufacturing. There are basically two types of notes: local (or specific) notes and general notes. *Local* notes are connected to specific features on the views of the drawing. *General* notes are placed separate from the views and relate to the entire drawing.

It is important for the drafter to effectively combine shape and size descriptions so that the drawing is easy to read and understand. There are many techniques that will help implement this goal. The drafter should carefully evaluate the dimensioning rules while preparing detail drawings and should keep in mind at all times never to crowd information on a drawing. It is better to use larger paper than to crowd the drawing.

▼ DIMENSIONING SYSTEMS

Unidirectional

Unidirectional dimensioning is commonly used in mechanical drafting. It requires that all numerals, figures, and notes be lettered horizontally and be read from the bottom of the drawing sheet. Figure 10–7 shows unidirectional dimensioning in use.

Aligned

Aligned dimensioning requires that all numerals, figures, and notes be aligned with the dimension lines so that they may be read from the bottom (for horizontal dimensions) and from the right side (for vertical dimensions). This method of dimensioning is commonly used in architectural and structural drafting. (See Figure 10–8.)

Tabular

Tabular dimensioning is a system in which size and location dimensions from datums or coordinates (x, y, z axis) are given in a table identifying features on the drawing. Figure 10–9 shows a method of tabular dimensioning.

Arrowless

Also known as dimensioning without dimension lines, *arrowless* dimensioning is similar to tabular dimensioning in that features are identified with letters and keyed to a table. Location dimensions are established with extension lines as coordinates from determined datums. (See Figure 10–10.)

Chart Drawing

Chart drawings are used when a particular part or assembly has one or more dimensions that change depending on the specific application. For example, the diameter of a part may remain constant with several alternate lengths required for different purposes.

The variable dimension is usually labeled on the drawing with a letter in the place of the dimension. The letter is then placed in a chart where the changing values are identified. Figure 10–11 is a chart drawing that shows two

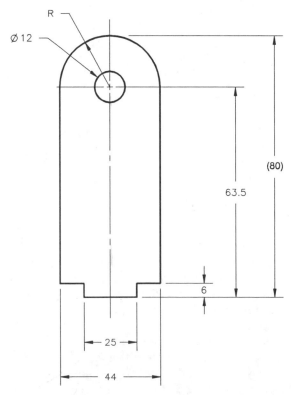

Figure 10–7   Unidirectional dimensioning.

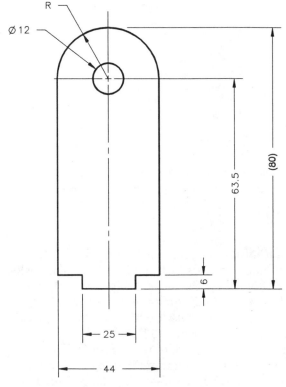

Figure 10–8   Aligned dimensioning.

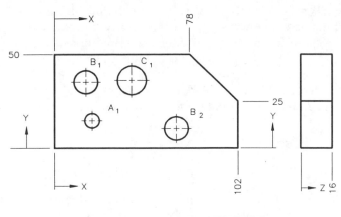

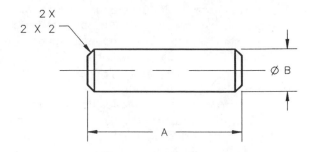

Figure 10–9   Tabular dimensioning.

HOLE SYMBOL	HOLE DIA	LOCATION		DEPTH Z
		X	Y	
A 1	6	12	14	THRU
B 1	9	12	38	9
B 2	9	57	7	12
C 1	12	43	38	THRU

Figure 10–11   Chart drawing.

LENGTH A	B=20.3	B=38.1	B=50.8	B=57.2
	PART NO	PART NO	PART NO	PART NO
76	DP20.3–76.2	DP38.1–76.2	DP50.8–76.2	DP57.2–76.2
101	DP20.3–101.6	DP38.1–101.6	DP50.8–101.6	DP57.2–101.6
127	DP20.3–127	DP38.1–127	DP50.8–127	DP57.2–127
152	DP20.3–152.4	DP38.1–152.4	DP50.8–152.4	DP57.2–152.4

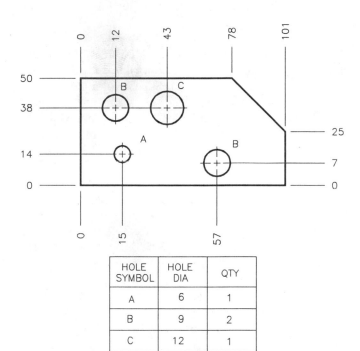

HOLE SYMBOL	HOLE DIA	QTY
A	6	1
B	9	2
C	12	1

Figure 10–10   Arrowless dimensioning.

▼ DIMENSIONING RULES

ANSI   The ANSI/ASME standard document for dimensioning is titled *Dimensioning and Tolerancing* ASME Y14.5M. This standard is available from the American National Standards Institute, 1430 Broadway, New York, NY 10018. This standard is all metric as denoted by the *M* following the number. The following fundamental rules for dimensioning are adapted from ASME Y14.5M.

1. Each dimension shall have a tolerance, except for those dimensions specifically identified as reference, maximum, minimum, or stock. The tolerance may be applied directly to the dimension or indicated by a general note located in the title block of the drawing.

2. Dimensioning and tolerancing shall be complete to the extent that there is full understanding of the characteristics of each feature. Neither measuring the drawing nor assumption of a distance or size is permitted, except drawings such as loft, printed wiring, templates, and master layouts prepared on stable material, provided the necessary control dimensions are given.

3. Each necessary dimension of an end product shall be shown. No more dimensions than those necessary for complete definition shall be given. The use of reference dimensions on a drawing should be minimized.

4. Dimensions shall be selected and arranged to suit the function and mating relationship of a part and shall not be subject to more than one interpretation.

5. The drawing should define a part without specifying manufacturing methods. For example, give only the diameter symbol for a hole without a process note such as drill, ream, or punch. However, in

dimensions that have alternate sizes. The view drawn represents a typical part, and the dimensions are labeled A and B. The correlated chart identifies the various (A) lengths available at given (B) diameters. The chart in this example also shows purchase part numbers for each specific item. This method of dimensioning is commonly used in vendor or specification catalogues for alternate part identification.

those cases in which manufacturing, processing, quality assurance, or environmental information is essential to the definition of engineering requirements, it shall be specified on the drawing or in document references on the drawing.

6. It is permissible to identify as nonmandatory certain processing dimensions that provide for finish allowance, shrink allowance, and other requirements, provided the final dimensions are given on the drawing. Nonmandatory processing dimensions shall be identified by an approprite note, such as NONMANDATORY (MFG DATA).

7. Dimensions should be arranged to provide required information for optimum readability. Dimensions should be shown in true profile views and refer to visible outlines.

8. Wires, cables, sheets, rods, and other materials manufactured to gage or code numbers shall be specified by linear dimensions indicating the diameter or thickness. Gage or code numbers may be shown in parentheses following the dimension.

9. A 90° angle is implied where centerlines and lines depicting features are shown on a drawing at right angles and no angle is specified. The tolerance for these 90° angles is the same as the general angular tolerance specified in the title block or general notes.

10. A 90° basic angle applies where centerlines of features in a pattern or surfaces shown at right angles on the drawing are located or defined by basic dimensions and no angle is specified.

11. Unless otherwise specified, all dimensions are applicable to 20° C (68° F). Compensation may be made for measurements made at other temperatures.

12. All dimensions and tolerances apply in a *free state condition* except nonrigid parts. Free state condition describes distortion of a part after removal of forces applied during manufacturing. *Nonrigid* parts are those that may have dimensional fluctuation due to thin wall characteristics.

13. Unless otherwise specified, all geometric tolerances apply for full depth, length, and width of the feature.

14. Dimensions apply on the drawing where specified.

Dimensioning Definitions

Actual Size. The part size as measured after production is known as actual size, also known as produced size.

Allowance. The tightest possible fit between two mating parts. MMC external feature – MMC internal feature = allowance.

Basic Dimension. A basic dimension is considered to be a theoretically exact size, location, profile, or orientation of a feature or point. The basic dimension provides a basis for the application of tolerance from other dimensions or notes. Basic dimensions are drawn with a rectangle around the numerical value. For example: $\boxed{.625}$ or $\boxed{30°}$. This is discussed more in Chapter 13.

Bilateral Tolerance. A bilateral tolerance is allowed to vary in two directions from the specified dimension.

Inch examples are $.250^{+.002}_{-.005}$ and $.500\pm.005$,

metric examples are $12^{+0.1}_{-0.2}$ and 12 ± 0.2.

Datum. A datum is considered to be a theoretically exact surface, plane, axis, center plane, or point from which dimensions for related features are established.

Datum Feature. The datum feature is the actual feature of the part that is used to establish a datum.

Dimension. A dimension is a numerical value used on a drawing to describe size, shape, location, geometric characteristic, or surface texture.

Feature. A feature is any physical portion of an object, such as a surface or hole.

Geometric Tolerance. The general term applied to the category of tolerances used to control form, profile, orientation, location, and runout. See Chapter 13.

Least Material Condition (LMC). Least material condition is the opposite of maximum material condition. The LMC is the lower limit for an external feature and the upper limit for an internal feature.

Limits of Dimension. The limits of a dimension are the largest and smallest possible boundary to which a feature may be made as related to the tolerance of the dimension. Consider the following inch dimension and tolerance: .750±.005. The limits of this dimension are calculated as follows: .750 + .005 = .755 upper limit and .750 – .005 = .745 lower limit. For the metric dimension 19.00±0.15, 19.00 + 0.15 = 19.15 is the upper limit, and 19.00 – 0.15 = 18.85 is the lower limit.

Maximum Material Condition (MMC). The maximum material condition, given the limits of the dimension, is the situation where a feature contains the most material possible. MMC is the largest limit for an external feature and the smallest limit for an internal feature.

Nominal Size. A dimension used for general identification such as stock size or thread diameter.

Reference Dimension. A dimension usually without tolerance, used for information purposes only. A reference is a repeat of a given dimension or established from other values shown on the drawing. It does not govern production or inspection.

Specified Dimension. The specified dimension is the part of the dimension from which the limits are calculated. For example, the specified dimension of .625±.001 in. is .625. In the following metric example, 15±0.1 mm, 15 is the specified dimension.

Tolerance. The tolerance of a dimension is the total permissible variation in size. Tolerance is the difference of the lower limit from the upper limit. For example, the limits of .500±.005 in. are .505 and .495 making the tolerance equal to .010 in. The tolerance for the following metric example, 12±0.1 mm, is 0.2 mm.

Unilateral Tolerance. A unilateral tolerance is a tolerance that has a variation in only one direction from the specified dimension, as in $.875^{+.000}_{-.002}$ in. or $22^{0}_{-0.2}$, $22^{+0.2}_{0}$ mm.

Dimensioning Units

The metric *International System of Units* (SI) is featured predominately in this text because SI units (millimeters) supersede United States (US) customary units specified on engineering drawings.

Metric units expressed in millimeters or U.S. customary units expressed in decimal inches are considered the standard units of linear measurement on engineering documents and drawings. The selection of millimeters or inches depends on the needs of the individual company. When all dimensions are either in millimeters or inches, the general note, UNLESS OTHERWISE SPECIFIED, ALL DIMENSIONS ARE IN MILLIMETERS (or INCHES), should be lettered on the drawing. Inch dimensions should be followed by *IN.* on predominantly millimeter drawings and *mm* should follow millimeters on predominantly inch-dimensioned drawings.

While **not** an ANSI standard, some companies use a method of placing inch and metric equivalents together on each dimension. This is known as *dual dimensioning.* The inch dimension may be followed by millimeters in brackets or separated by a slash. For example, 1.00 [25.4] or 1.00/25.4. A general note should accompany the drawing to identify this practice: DIMENSIONS IN [ ] ARE MILLIMETERS (or INCHES), or INCH/MILLIMETERS or MILLIMETERS/INCHES.

Decimal Points

Dimension numerals that contain decimal points should be allowed adequate space at the decimal so there is no crowding with the numerals. A minimum of two-thirds the height of the letters is recommended. The decimal should be clear and bold and in line with the bottom of the numerals. For example, 1.750.

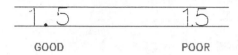

GOOD   POOR

Figure 10–12 Spacing of numerals with decimal.

A specified dimension in inches is expressed to the same number of decimal places as its tolerance, and zeros are added to the right of the decimal point if needed. For example, .500±.002.

Where millimeter dimensions are less than one, a zero shall precede the decimal, as in 0.8. Decimal inch dimensions do not have a zero before the decimal, as in .75. Proper spacing of numerals with a decimal is shown in Figure 10–12.

Neither the decimal point nor a zero is shown when the metric dimension is a whole number. For example, 24. Also, when the metric dimension is greater than a whole number by a fraction of a millimeter, the last digit to the right of the decimal point is not followed by a zero as in 24.5. This is true unless tolerance numerals are involved. (Refer to Figure 10–16.) Both the plus and minus values of a metric or inch tolerance have the same number of decimal places and zeros are added to fill in where needed.

Fractions

Fractions are used on engineering drawings, but they are not as common as decimal inches or millimeters. Fraction dimensions generally mean a larger tolerance than decimal numerals. When fractions are used on a drawing the fraction numerals should be the same size as other numerals on the drawing. The fraction bar should be drawn in line with the direction that the dimension reads. For unidirectional dimensioning, the fraction bars are all horizontal. For aligned dimensioning, the fraction bars are horizontal for dimensions that read from the bottom of the sheet and vertical for dimensions that read from the right. The fraction numerals should not be allowed to touch the fraction bar. A space of 1.5 mm or .06 in. is recommended between the bar and the fraction numerals. In a few situations—for example when a fraction is part of a general note, material specification, or title—the fraction bar may be placed diagonally, as shown in Figure 10–13. This is also a common practice using CADD.

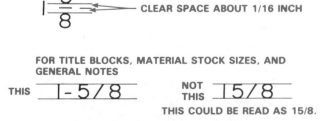

Figure 10–13 Numerals in fractions.

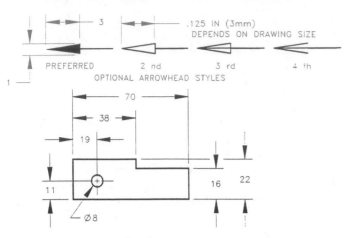

Figure 10–14   Arrowheads.

Arrowheads

Arrowheads are used to terminate dimension lines and leaders. Properly drawn arrowheads should be drawn three times as long as they are high. All arrowheads on a drawing should be the same size. Do not use small arrowheads in small spaces. Limited-space dimensioning practice is covered in this chapter (see Figure 10–16).

Some companies require that arrowheads be drawn with an arrow template, while others accept properly drawn freehand arrowheads. (See Figure 10–14.) Individual company preference dictates if arrowheads are filled in solid or left open as shown. CADD users often prefer the open style, which helps increase regeneration and plotting speed. However, others prefer the appearance of the filled-in arrowhead.

▼ DIMENSIONING FUNDAMENTALS
Dimension Line Spacing

Dimension lines should be placed at a uniform distance from the object, and all succeeding dimension lines should be equally spaced. Figure 10–15 shows the

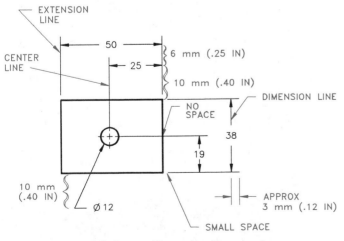

Figure 10–15   Minimum dimension line spacing.

minimum acceptable distances for spacing dimension lines. In actual practice the minimum distance is too crowded. Judgment should be used based on space available and information presented. Never crowd dimensions, if at all possible. Always place the smallest dimensions closest to the object and progressively larger dimensions outward from the object. Group dimensions and dimension between views when possible.

Relationship of Dimension Lines to Numerals

Dimension numerals are centered on the dimension line. Numerals are commonly all the same height and are lettered horizontally (unidirectional). A space equal to at least half the height of lettering should be provided between numerals in a tolerance. The numeral, dimension line, and arrowheads should be placed between extension lines when space allows. When space is limited, other options should be used. Figure 10–16 shows several dimensioning options. Evaluate each of the examples carefully as you dimension your own drawing assignments. Figure 10–17 shows some correct and incorrect dimensioning practices. Keep in mind that some computer-aided drafting programs do not necessarily acknowledge all of the rules or accepted examples. Some flexibility on your part may be needed to become accustomed to some of the potential differences.

Chain Dimensioning

Chain dimensioning, also known as point-to-point dimensioning, is a method of dimensioning from one feature to the next. Each dimension is dependent on the previous dimension or dimensions. This is a common practice, although caution should be used as the tolerance of each dimension builds on the next, which is known as *tolerance buildup,* or *stacking.* Figure 10–18 also shows the common mechanical drafting practice of providing an overall dimension while leaving one of the intermediate dimensions blank. The overall dimension is often a critical dimension which should stand independent in relationship to the other dimensions. Also, if all dimensions are given, then the actual size may not equal the given overall dimension due to tolerance buildup. An example of tolerance buildup is when three chain dimensions have individual tolerances of ±.15 and each feature is manufactured at or toward the +.15 limit: the potential tolerance buildup is three times +.15 for a total of .45. The overall dimension would have to carry a tolerance of ±0.45 to accommodate this buildup. If the overall dimension is critical, such an amount may not be possible. Thus, either one intermediate dimension should be omitted or the overall dimension omitted. The exception to this rule is when a dimension is given only as reference. A reference dimension is enclosed in parentheses, as in Figure 10–19a. Figure 10–19b shows the overall dimension of an object as a reference.

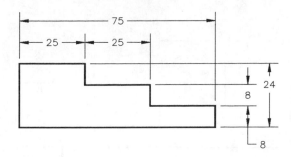

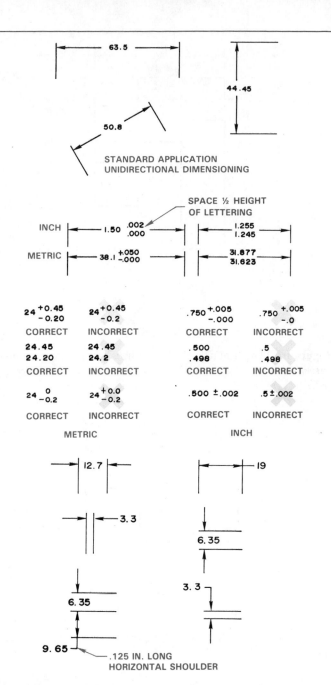

Figure 10–16 Dimensioning applications to limited spaces.

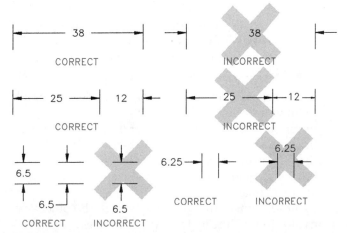

Figure 10–17 Correct and incorrect dimensioning practice.

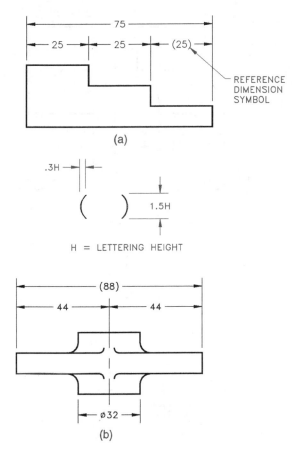

Figure 10–18 Chain dimensioning.

Figure 10–19 Reference dimension examples, and reference dimension symbol.

Datum Dimensioning (Stagger Adjacent Dimensions)

Datum dimensioning is a common method of dimensioning machine parts whereby each feature dimension originates from a common surface, axis, or center plane. (See Figure 10–20.) Each dimension in datum dimensioning is independent so there is no tolerance buildup. Figure 10–21 shows how dimensions can be symmetrical about a center plane used as a datum. Also notice in Figure 10–21 how dimension numerals are staggered rather than being stacked directly above one another. Always **stagger adjacent dimensions** when possible. Doing so helps clarity and reduces crowding. Figure 10–21 also shows use of the symmetrical symbol.

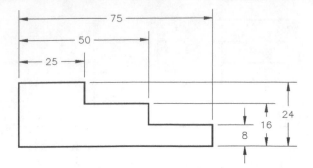

Figure 10–20 Datum dimensioning from a common surface.

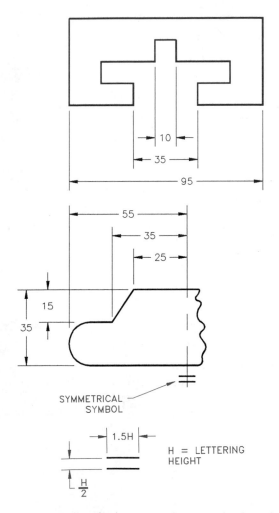

Figure 10–21 Datum dimensioning from center plane datums and symmetrical symbol.

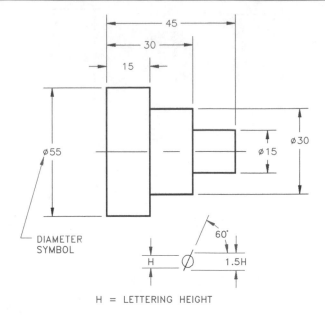

Figure 10–22 Dimensioning cylindrical shapes, and diameter symbol.

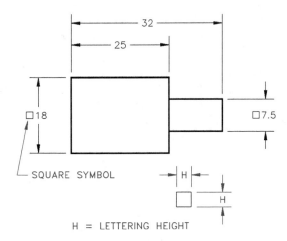

Figure 10–23 Dimensioning square features, and square symbol.

Both halves of the object are the same. Part of the right side is removed, and a short break line is used to save space. Use this practice only if necessary, because it can cause confusion if improperly read.

Dimension cylindrical shapes in the view where the cylinders appear rectangular. The diameters are identified by the diameter symbol, and the circular view may be omitted. (See Figure 10–22.) Square features may be dimensioned in a similar manner using the square symbol shown in Figure 10–23.

▼ PREFERRED DIMENSIONING PRACTICES

The drawings in Figure 10–24 show both correct and incorrect dimensioning techniques. These are mostly suggested options for correct and incorrect dimensioning. Good judgment should be used with each case.

Always try, when possible, to:

❍ avoid crossing extension lines, but **do not** break them when they do cross.

❍ **never** cross extension lines over dimension lines, or break the extension line over the dimension line if there is **absolutely no other solution**.

❍ break extension lines when they cross **over or near** an arrowhead.

❍ **avoid** dimensioning over or through the object.

❍ **avoid** dimensioning to hidden features.

❍ **avoid** unnecessary long extension lines.

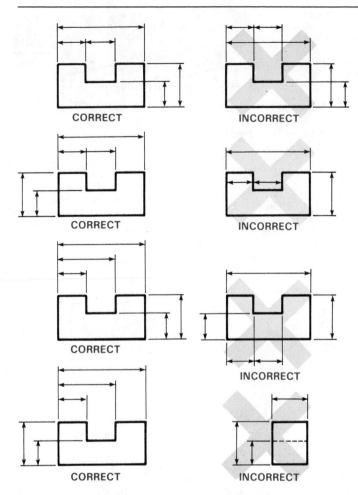

Figure 10–24   Correct and incorrect dimensioning examples.

○   **avoid** using any line of the object as an extension line.
○   dimension **between views** when possible.
○   **group** adjacent dimensions.
○   dimension to views that provide the **best shape description**.

Dimensioning Angles

Angular surfaces may be dimensioned as coordinates, as angles in degrees, or as a flat taper. (See Figure 10–25.) Angles are calibrated in degrees, symbol °. (There are 360° in a circle. Each degree contains 60 minutes, symbol '. Each minute has 60 seconds, symbol ". 1° = 60'; 1' = 60".) Notice in the angular method in Figure 10–25, the dimension line for the 45° angle is drawn as an arc. The radius of this arc is centered at the vertex of the angle.

Dimensioning Chamfers

A chamfer is a slight surface angle used to relieve a sharp corner. Chamfers of 45° are dimensioned with a note, while other chamfers require an angle and size dimension,

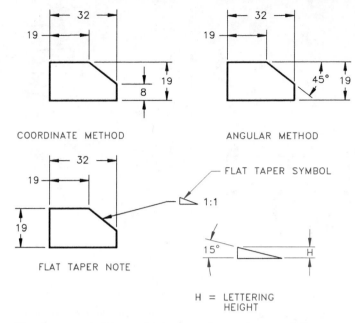

Figure 10–25   Dimensioning angular surfaces and flat taper symbol.

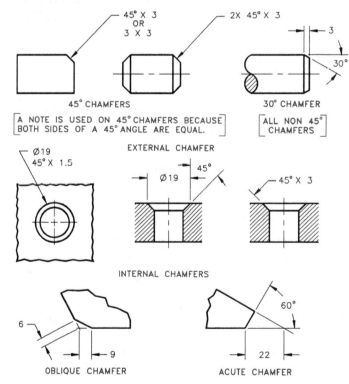

Figure 10–26   Dimensioning chamfers.

as seen in Figure 10–26. A note is used on 45° chamfers because both sides of a 45° angle are equal.

Dimensioning Conical Shapes

Conical shapes should be dimensioned when possible in the view where the cone appears as a triangle, as in Figure 10–27. A conical taper may be treated in one of three possible ways, as shown in Figure 10–28.

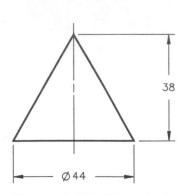

Figure 10–27 Dimensioning conical shapes.

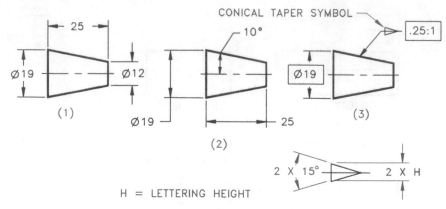

Figure 10–28 Dimensioning conical tapers, and the conical taper symbol.

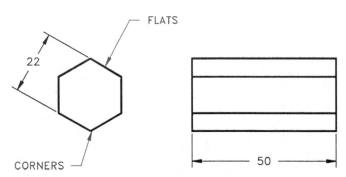

Figure 10–29 Dimensioning hexagons.

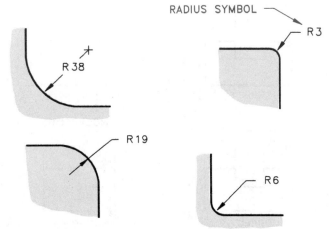

Figure 10–30 Dimensioning arcs—no centers located, and the radius symbol.

Dimensioning Hexagons and Other Polygons

Dimension hexagons and other polygons across the flats in the views where the true shape is shown. Provide a length dimension in the adjacent view, as shown in Figure 10–29.

Dimensioning Arcs

Arcs are dimensioned with leaders and radius dimensions in the views where they are shown as arcs. The leader may extend from the center to the arc or may point to the arc, as shown in Figure 10–30 and 10–31. The letter *R* precedes all radius dimensions. Depending on the situation, arcs may be dimensioned with or without their centers located. Figure 10–31 shows a very large arc with the center moved closer to the object. To save space, a break line is used in the leader and the shortened locating dimension. The length of an arc may also be dimensioned one of three ways, as shown in Figure 10–32.

In a situation where an arc lies on an inclined plane and the true representation is not shown, the note TRUE R may be used to specify the actual radius. However, dimensioning the arc in an auxiliary view of the inclined surface would be better if possible. (See Figure 10–33.)

The symbol *CR* refers to controlled radius. *Controlled radius* means that the limits of the radius tolerance zone must be tangent to the adjacent surfaces, and there can be no reversals in the contour. The *CR* control is more

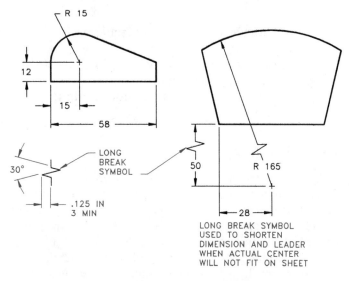

Figure 10–31 Dimensioning arcs—with centers located, and the long break symbol.

restrictive than use of the *R* radius symbol where reversals in the contour of the radius are permitted.

A spherical radius may be dimensioned with the abbreviation *SR* preceding the numerical value, as shown in Figure 10–34.

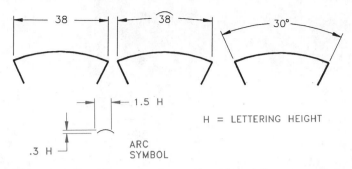

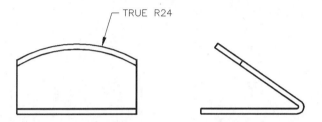

Figure 10–32   Dimensioning arc length and the arc symbol.

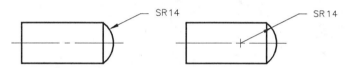

Figure 10–33   Dimensioning a true radius in an inclined plane.

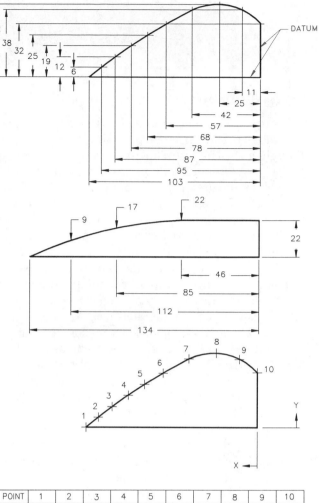

Figure 10–34   Dimensioning a spherical radius.

Dimensioning Contours Not Defined as Arcs

Coordinates or points along the contour are located from common surfaces or datums as shown in Figure 10–35. Figure 10–36 shows a curved contour dimensioned using oblique extension lines. While this technique may be used, it is not as common as the previous method.

POINT	1	2	3	4	5	6	7	8	9	10
X	−103	−95	−87	−78	−68	−57	−42	−25	−11	0
Y	0	6	12	19	25	32	38	41	38	32

Figure 10–35   Dimensioning contours not defined as arcs.

Locating a Point Established by Extension Lines

If the sides of the object in Figure 10–37 were extended beyond the bend they would meet at the intersection of the extension lines. This imaginary point is where the dimension often originates in this type of situation.

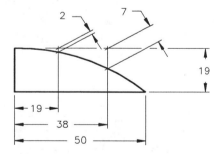

Figure 10–36   Dimensioning a curved contour using oblique extension lines.

Figure 10–37   Locating a point established by extension lines.

CADD Applications

DIMENSIONING

The proper placement and use of dimensions is one of the most difficult aspects of drafting. It requires thought and planning. The drafter must first determine the type of dimension that suits the application, then place it on the drawing. The dimensioning options for the CADD drafter are many, and the proper usage of these options should be learned early and practiced whenever possible. A variety of dimensioning possibilities may be located on the CADD menu under the heading dimension parameters, automatic dimensioning, or simply, dimensions. The menu may show each of the possibilities, or they may be listed on the screen after selecting the DIMENSION or DIM command.

Two basic types of dimensions are horizontal and vertical. These can be labeled on the menu as HORIZ and VERT, or DMH (dimension horizontal) and DMV (dimension vertical). AutoCAD, for example, has the DIMLINEAR (DIMLIN) command that allows you to automatically place any straight dimension whether it be horizontal, vertical, or at an angle. All you have to do is pick the extension line origins and the dimension line location. To place a horizontal dimension, first select the DIMLIN command. You may then be prompted to locate one end of the dimension by placing the screen cursor so that the vertical crosshair is aligned with one edge of the feature to be dimensioned, as shown in Figure 10–38a. Pick this point and then locate the crosshairs on the opposite end of the feature and pick that point, as shown in Figure 10–38b. You have now established the length of the feature. The dimension is calculated by the computer according to the scale and grid spacing that was initially established for the drawing. Next you need to pick the distance away from the feature the dimension line should be located. After this step is done, the dimension may be drawn on the screen. (See Figure 10–38c.) The user also may be prompted for additional information such as location of text, number of decimals, location of arrowheads, and length of extension lines. To give you an idea of the amount of information required by the computer to calculate and place a single dimension, here's a list of the possibilities:

1. Beginning point.
2. End point.
3. Distance from feature.
4. Extension lines (yes or no, on or off)? Figure 10–38d shows the complete dimension with extension lines drawn.

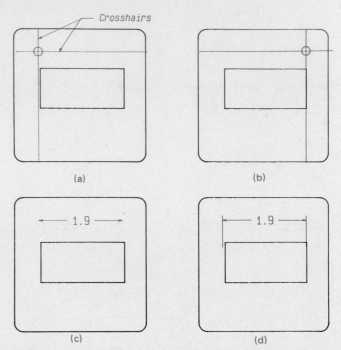

Figure 10–38 Locating a dimension using the screen cursor.

5. Arrowhead location (inside or outside the extension lines).
6. Text size.
7. Text direction.
8. Text location (inside, above, or below the dimension line).
9. Dimension type (linear, datum, radial, or angular dimensioning).
10. Number of decimal places.
11. Fractions or decimals.
12. Grid spacing.
13. Scale.
14. Dimension number placement (automatic, semiautomatic, or manual).

Of course, some of these items are established as initializing parameters when you begin your drawing, and some are only required when you select a specific style of dimensioning, but they all affect the dimensions. Therefore, it is important that you learn and practice all the aspects of dimensioning that are a part of your specific CADD system.

Another way to place dimensions on a drawing is to pick the extension line origins on the object. The CADD dimensioning then automatically calculates the extension line offset and places the extension lines, dimension lines, text, and arrowheads. This can be done in either a datum or chain dimensioning format, as shown in Figure 10–39. When you are dimensioning circles, the dimension line and text numeral are automatically placed inside the circle by picking the circle, or you have the option of dimensioning with a leader. Figure 10–40 illustrates some of the options. Dimensioning arcs is also easy. All you have to do is pick the arc to be

dimensioned and enter the dimension text; the leader with text numeral is placed on the drawing automatically, as shown in Figure 10–41.

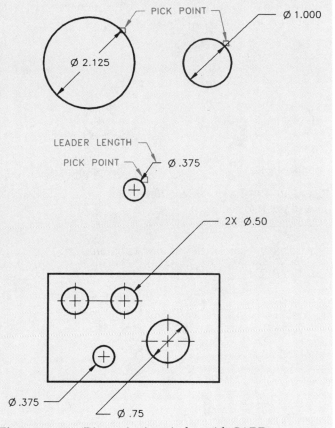

Figure 10–40 Dimensioning circles with CADD.

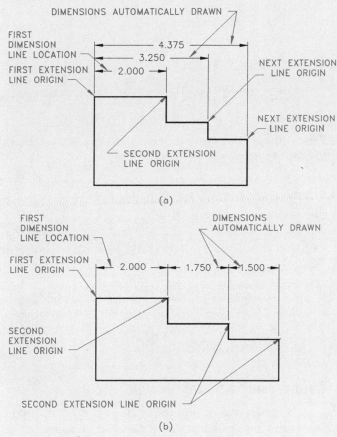

Figure 10–39 Locating dimensions at extension line origins for **(a)** datums and **(b)** chain dimensioning.

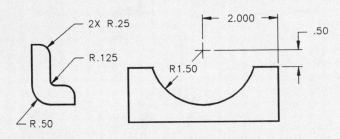

Figure 10–41 Dimensioning arcs with CADD.

▼ NOTES FOR SIZE FEATURES

Holes

Hole sizes are dimensioned with leaders to the view where they appear as circles, or dimensioned in a sectional view. When leaders are used to establish notes for holes, the shoulder should be centered on the beginning or the end of the note. When a leader begins at the left side of a note, it should originate at the beginning of the

note. When a leader begins at the right side of a note, it should originate at the end of the note. (See Figure 10–42.)

Figure 10–43 shows the leader should touch the side of the circle and, if it were to continue, would intersect the center of the circle. Leaders may be drawn at any angle, preferably between 15° and 75° from horizontal. Do not draw horizontal or vertical leaders.

A hole through a part may be noted if not obvious. The diameter symbol precedes the diameter numeral, as di-

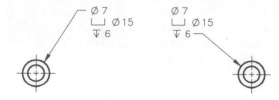

Figure 10–42 Leader orientation to the note. Center leader shoulder at beginning or end of note.

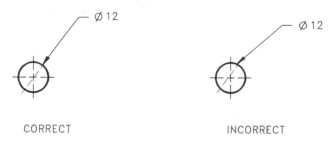

CORRECT INCORRECT

Figure 10–43 Leader orientation to the circle. Leader arrow should *point* to center.

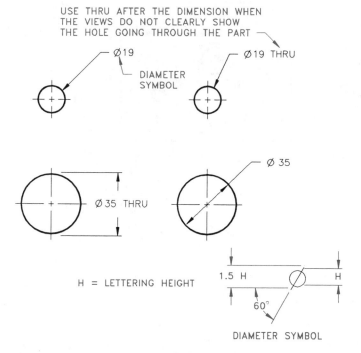

Figure 10–44 Dimensioning hole diameters, and the diameter symbol.

mensioned in Figure 10–44. If a hole does not go through the part, the depth must be noted in the circular view or in section, as shown in Figure 10–45.

Counterbore

A counterbore is often used to machine a diameter below the surface of a part so a bolt head or other fastener may be recessed. Counterbore and other similar notes are given in the order of machine operations with a leader in the view where they appear as circles. (See Figure 10–46a.)

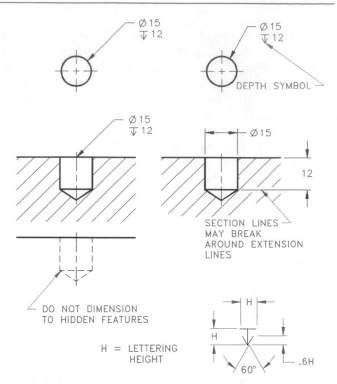

Figure 10–45 Dimensioning hole diameters and depths, and the depth symbol.

Multiple counterbores are dimensioned in a similar manner as shown in Figure 10–46d.

ANSI The ANSI standard recommends that the elements of each note shown in Figures 10–46 through 10–49 be aligned as shown. However, due to individual preference or drawing space constraints the elements of the note may be confined to fewer lines, as shown in Figure 10–46b. **Never** separate individual note components, as shown in Figure 10–46c.

Countersink or Counterdrill

A countersink, or counterdrill, is also often used to recess the head of a fastener below the surface of a part. (See Figure 10–47.)

Spotface

A spotface is used to provide a flat bearing surface for a washer face or bolthead. (See Figure 10–48.) Follow the counterbore guidelines when lettering the spotface note.

Multiple Features

When a part has more than one feature of the same size, they may be dimensioned with a note that specifies the number of like features, as shown in Figure 10–49. If a part contains several features all the same size, the methods shown in Figure 10–50 may be used.

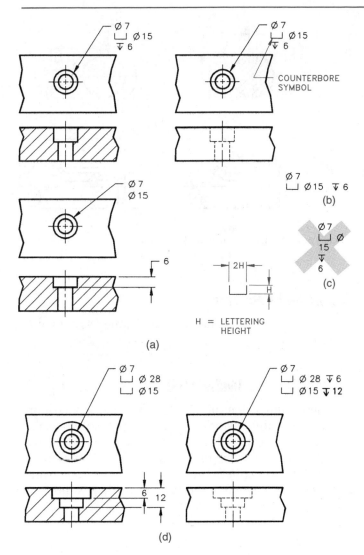

Figure 10–46 **(a)** Counterbore note; **(b)** alternate note with elements of note grouped on one line; **(c)** never split individual note elements. The proper counterbore/spotface symbol is also displayed.

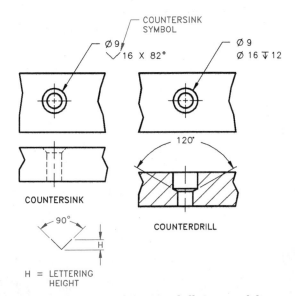

Figure 10–47 Countersink, counterdrill notes, and dimensions. The properly drawn countersink symbol is also shown.

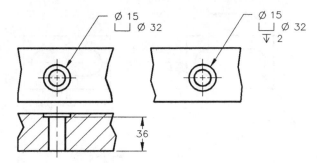

Figure 10–48 Spotface note and dimensions.

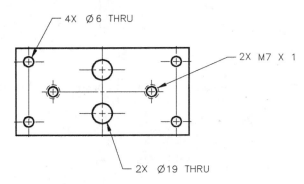

Figure 10–49 Dimension notes for multiple features.

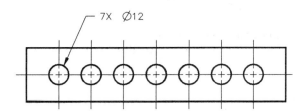

Figure 10–50 Dimension notes for multiple features all of common size.

Slots

Slotted holes may be dimensioned in one of three ways, as shown in Figure 10–51. These methods are used only when the ends are fully rounded and tangent to the sides. When the ends of a slot or external feature have a radius greater than the width of the feature, then the size of the radius must be given, as shown in Figure 10–52.

Keyseats

Keyseats are dimensioned in the view that clearly shows their shape by width, depth, length, and location, as in Figure 10–53.

Knurls

Knurls are dimensioned with notes and leaders that point to the knurl in the rectangular view, as shown in Figure 10–54a. ANSI **does not** recommend showing a knurl representation on the view as in Figure 10–54b.

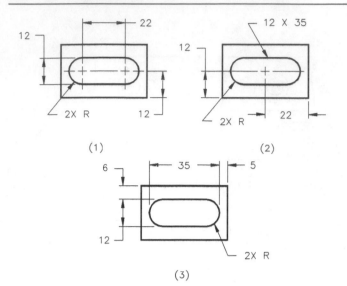

(1)   (2)

(3)

Figure 10–51   Dimensioning slotted holes with full radius ends.

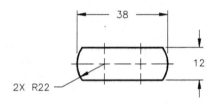

Figure 10–52   Dimensioning slot or external feature with end radius larger than feature width.

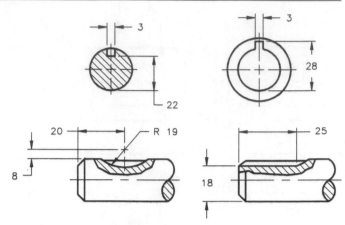

Figure 10–53   Dimensioning keyseats.

Necks and Grooves

Necks and grooves may be dimensioned as shown in Figure 10–55.

▼ LOCATION DIMENSIONS

In general, dimensions either identify a size or a location. The previous dimensioning discussions focused on size dimensions. Location dimensions to cylindrical features, such as holes, are given to the center of the feature in the view where they appear as a circle, or in a sectional view. Rectangular shapes may be located to their sides, and

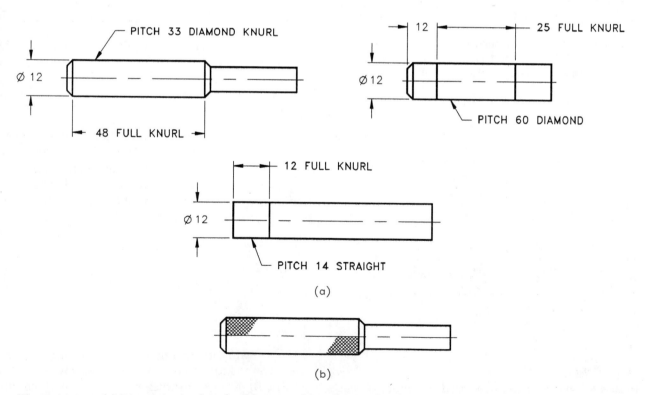

Figure 10–54   **(a)** Dimensioning knurls, **(b)** optional knurl representation.

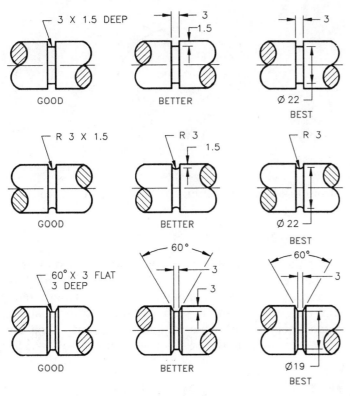

Figure 10–55　Dimensions for necks and grooves.

symmetrical features may be located to their centerline or center plane. Some location dimensions also control size.

Locating Holes

Locate a hole center in the view where the hole appears as a circle, as shown in Figure 10–56.

Rectangular Coordinates

Linear dimensions are used to locate features from planes or centerlines, as shown in Figure 10–57.

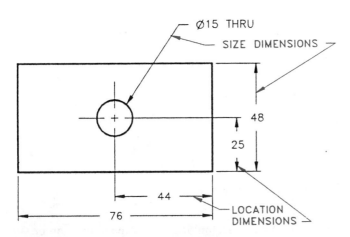

Figure 10–56　Size and location dimensions.

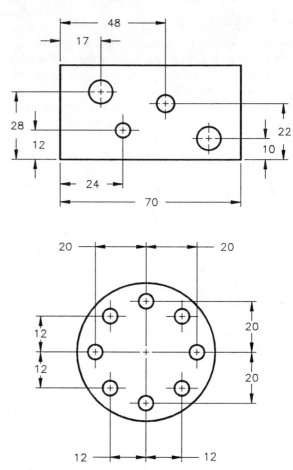

Figure 10–57　Rectangular coordinate location dimensions.

Polar Coordinates

Angular dimensions locate features from planes or centerlines, as shown in Figure 10–58.

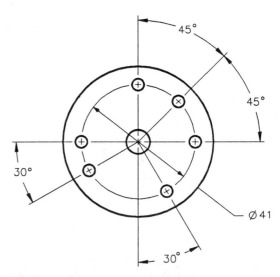

Figure 10–58　Polar coordinate dimensions. If all features are equally spaced, a note such as 6× 60° may be used in one location.

Repetitive Features

Repetitive features may be located by noting the number of times a dimension is repeated, and giving one typical dimension and the total length as reference. This method is acceptable for chain dimensioning. (See Figure 10–59.) Locating multiple tabs is shown in Figure 10–60. This method also works for slots. When repetitive features on an object are nearly the same size, they may be shown with an identification letter, such as *Y*. (See Figure 10–61.)

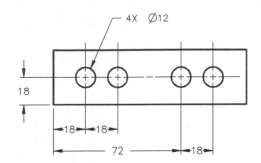

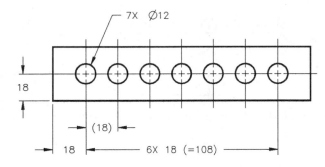

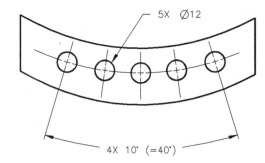

Figure 10–59 Dimensioning repetitive features.

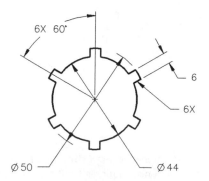

Figure 10–60 Locating multiple tabs.

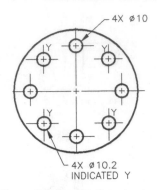

Figure 10–61 Dimensioning similar sized multiple features.

▼ DIMENSION ORIGIN

When the dimension between two features must clearly denote from which feature the dimension originates, the dimension origin symbol may be used. This method of dimensioning means that the origin feature must be established first and the related feature may then be dimensioned from the origin. (See Figure 10–62.)

▼ DIMENSIONING AUXILIARY VIEWS

Dimensions should be placed on views that provide the best size and shape description of an object. In many instances the surfaces of a part are foreshortened and require auxiliary views to completely describe true size, shape, and the location of features. When foreshortened views occur, dimensions should be placed on the auxiliary view for clarity. In unidirectional dimensioning the dimension numerals are placed horizontally so they read from the bottom of the sheet. When aligned di-

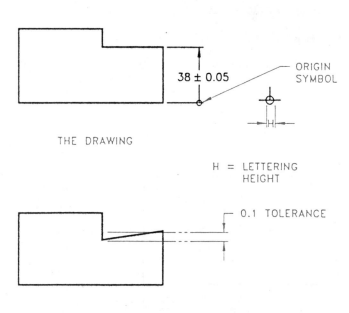

Figure 10–62 Dimension origin, and the dimension origin symbol.

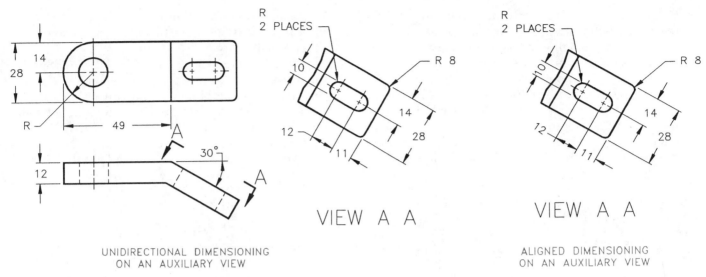

Figure 10–63   Dimensioning auxiliary views.

mensioning is used, the dimension numerals are placed in alignment with the dimension lines. (See Figure 10–63.)

▼ GENERAL NOTES

The notes that were discussed previously are classified as specific notes because they referred to specific features of an object. *General notes*, on the other hand, relate to the entire drawing. Each drawing contains a certain number of general notes either in or near the drawing title block. (See Figure 10–64.) *General notes* are concerned with such items as the following:

- ○ Material specifications.
- ○ Dimensions: inches or millimeters.

- ○ General tolerances.
- ○ Confidential note, copyrights, or patents.
- ○ Drawn by.
- ○ Scale.
- ○ Date.
- ○ Part name.
- ○ Drawing size.
- ○ Part number.
- ○ Number of revisions.
- ○ First-angle or third-angle projection symbol.
- ○ ANSI/ASME, MIL (military), or other standard reference.
- ○ General machining, finish, or paint specifications.
- ○ Identification of features that are the same throughout the drawing.

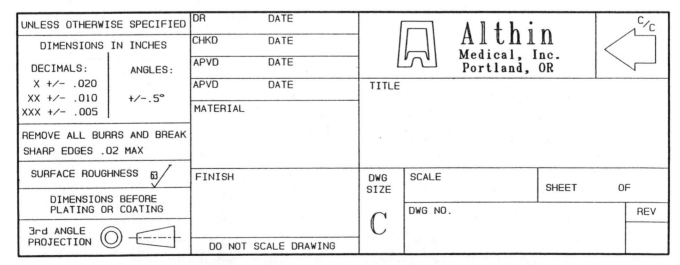

Figure 10–64   General title block information. *Courtesy Althin Medical, Inc.*

ANSI General notes are located next to the title block for ANSI drawings. The exact location of the general notes depends on specific company standards. Military standards specify general notes be located near the upper-left corner of the drawing. A common location for general notes is the lower-left corner of the drawing, usually .5 in. each way from the border line. Figure 10–66 shows some general notes that could commonly be included on ANSI standard format drawings. Other common locations for general notes are the lower right corner of the drawing directly above the title block or just to the left of the title block.

Notice in Figure 10–66 that the word NOTES is lettered first, followed by the first, second, and additional notes. Depending on company standards, the word NOTES is lettered .18 to .25 in. high, but some companies prefer to omit the word NOTES as an unnecessary preface to obvious general notes. The space between notes is from one-half to full height of

CADD Applications

USING LAYERS FOR DIMENSIONING

A time-tested method of creating several drawings containing different information on the same outline is known as *overlay drafting.* A base drawing is generated, then an overlay is placed over it and a specific type of information is drawn on the overlay. For example, an architectural floor plan is drawn as the base, and then overlays are drawn for the plumbing plan, electrical plan, and so on. The base can be combined with any individual overlay or combinations thereof to create a print of just the information required. Also, several drafters can be given copies of the base and each can create an overlay containing specific information. This enables several drafters to essentially work on the same drawing at the same time; a great time-saver. The final drawing is then composed of several layers or levels.

The concept of layering is a prime component of CADD systems, some having the ability to display over 200 layers. This not only allows different drafters to work on the same drawing at the same time, but also addresses a shortcoming of the computer itself: the speed at which a drawing can be redrawn on the screen. The less information that is on the screen at one time, the less time it takes to be redrawn, and the less clutter there is to get in the way of the drafter. Figure 10–65 shows a simple example of the layering concept.

Several commands exist to allow the operator to work with the different layers. The LAYER command itself lets the operator assign a layer number to a drawing. RELAYER gives the drafter the chance to change the numbers of the drawing layers, and DISPLAY LAYER, or LAYER CONTROL, provides for the display of any combination of layers.

Color is a valuable tool when working with layers, and more companies are realizing this fact. Many color systems enable the user to assign any color desired to any layer. Some systems have a limited number of colors, while others possess the ability to create thousands of hues. Color gives even more meaning to the layering concept, because several layers displayed on the screen at the same time can be distinguished from each other. The colors on the screen can be transferred to the plotter by use of the PEN command. This command allows specific colors and widths of pens in the pen plotter to be selected.

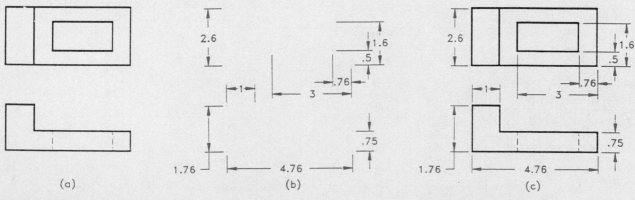

Figure 10–65 Layering enables the drafter to focus on one aspect of the drawing at a time.

ADDITIONAL NOTES

NOTES:

3. ALL FILLETS AND ROUNDS R .125 UNLESS OTHERWISE SPECIFIED.

2. REMOVE ALL BURRS AND SHARP EDGES.

1. INTERPRET DIMENSIONS AND TOLERANCES PER ASME Y14.5M–1994.

Figure 10–66   General notes located in the lower left corner of the sheet.

the lettering, and the notes are often lettered .125 in. in height. The first note, INTERPRET DIMENSIONS AND TOLERANCES PER ASME Y14.5M—1994, should be included on all new drawings. Additional notes depend on the information required to support the drawing.

▼ TOLERANCING

Definitions

As previously mentioned, tolerance is the total permissible variation in a size or location dimension.

A specified dimension is that part of the dimension from which the limits are calculated. For example, 15.8 is the specified dimension of 15.8±0.2.

A bilateral tolerance is allowed to vary in two directions from the specified dimension, as in $6.5^{+0.1}_{-0.3}$ or 6.5±0.2.

A unilateral tolerance varies in only one direction from the specified dimension: $22^{0}_{-0.2}$ or $19.5^{+.05}_{0}$.

Limits are the largest and smallest possible sizes a feature may be made as related to the tolerance of the dimension.

Example 1: 19.0±0.1

Upper limit: 19.0 + 0.1 = 19.1

Lower limit: 19.0 – 0.1 = 18.9

Example 2: $9.5^{0}_{-0.5}$

Upper limit: 9.5 + 0 = 9.5

Lower limit: 9.5 – 0.5 = 9.0

The tolerance, being the total permissible variation in the dimension, is easily calculated by subtracting the lower limit from the upper limit.

Example 3: 22.0 ± 0.1

Upper limit:	22.1
Lower limit:	−21.9
Tolerance	0.2

Example 4: $31.75^{+0.10}_{0}$

Upper limit:	31.85
Lower limit:	−31.75
Tolerance	0.10

All dimensions on a drawing have a tolerance except reference, maximum, minimum, or stock size dimensions. Dimensions on a drawing may read as in Figure 10–67 with general tolerances specified in the title block of the drawing. General tolerance specifications, as given in a typical industry title block, are shown in Figure 10–68. Using this title block as an example, the tolerances of the dimensions in Figure 10–67 are as follows (given in inches):

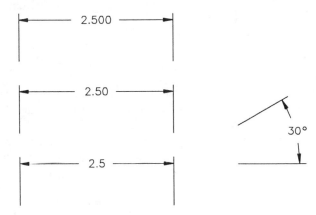

Figure 10–67   Typical drawing dimensions.

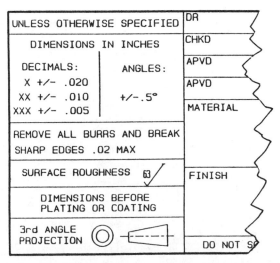

Figure 10–68   General tolerances from a company title block. *Courtesy Althin Medical, Inc.*

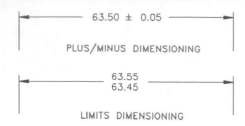

Figure 10–69 Specific tolerance dimensions.

2.500 has .xxx±.005 applied; 2.500±.005, tolerance equals .100.

2.50 has .xx±.010 applied; 2.50±.010, tolerance equals .020.

2.5 has .x±.020 applied; 2.5±.020, tolerance equals .040.

30° has Angles±0.5° applied; 30°±0.5°, tolerance equals 1°.

Dimensions that require tolerances different from the general tolerances given in the title block must be specified in the dimension. These are referred to as *specific tolerance* dimensions and may be represented with *plus/minus* or *limits* dimensioning as in Figure 10–69.

Statistical Tolerancing

Statistical tolerancing is the assigning of tolerances to related dimensions in an assembly based on the requirements of statistical process control (SPC). SPC is discussed in Chapters 9 and 25. Statistical tolerancing is displayed in dimensioning as shown in Figure 10–70a. When the feature could be manufactured using SPC or conventional means, it is necessary to show both the statistical tolerance and the conventional tolerance as in

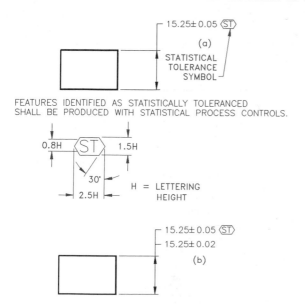

Figure 10–70 Statistical tolerancing application and notes, and the statistical tolerancing symbol.

Figure 10–70b. The appropriate general note should also accompany the drawing as shown in Figure 10–70.

Maximum and Least Material Conditions

Maximum material condition, abbreviated MMC, is the condition of a part or feature when it contains the most amount of material. The key is *most material.* The MMC of an external feature is the upper limit or largest size. (See Figure 10–71.) The MMC of an internal feature is the lower limit or smallest size. (See Figure 10–72.)

The *least material condition* (LMC) is the opposite of MMC. LMC is the least amount of material possible in the size of a feature. The LMC of an external feature is its lower limit. The LMC of an internal feature is its upper limit.

Clearance Fit

A *clearance fit* is a condition when, due to the limits of dimensions, there is always a clearance between mating parts. The features in Figure 10–73 have a clearance fit. Notice that the largest size of the shaft is smaller than the smallest hole size.

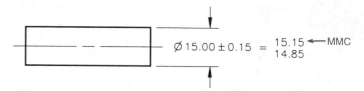

Figure 10–71 Maximum material condition (MMC) of an external feature. Plus/minus dimension shown.

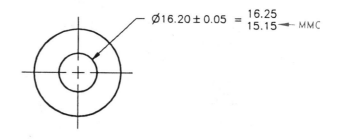

Figure 10–72 Maximum material condition (MMC) of an internal feature. Plus/minus dimension shown.

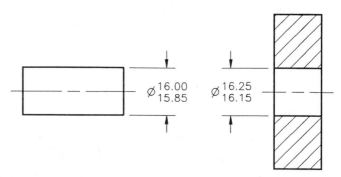

Figure 10–73 A clearance fit between two mating parts. Limits dimensions shown.

Allowance

The *allowance* of a clearance fit between mating parts is the tightest possible fit between the parts. The allowance is calculated with the formula:

MMC Internal Feature

– MMC External Feature

Allowance

The allowance of the parts in Figure 10–73 is:

MMC Internal Feature	16.15
MMC External Feature	−16.00
Allowance	0.15

Interference Fit

An *interference fit,* also known as force or shrink fit, is the condition that exists when, due to the limits of the dimensions, mating parts must be pressed together. Interference fits are used, for example, when a bushing must be pressed onto a housing or when a pin is pressed into a hole. (See Figure 10–74.)

Types of Fits

Selection of Fits. In selecting the limits of size for any application, the type of fit is determined first based on the use or service required from the equipment being designed; then the limits of size of the mating parts are established to insure that the desired fit will be produced. The number of standard fits described here cover most applications.

Designation of Standard ANSI Fits. Standard fits are designated by means of the following symbols, which facilitate reference to classes of fit for educational purposes. The symbols are not intended to be shown on manufacturing drawings; instead, sizes should be specified on drawings. The letter symbols used are as follows:

RC   Running, or Sliding, Clearance Fit
LC   Locational Clearance Fit
LT   Transition Clearance, or Interference Fit
LN   Locational Interference Fit
FN   Force, or Shink, Fit

These letter symbols are used in conjunction with numbers representing the class of fit: thus FN 4 represents a Class 4 force fit.

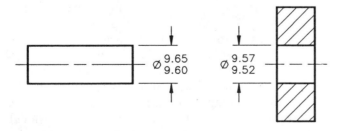

Figure 10–74 An interference fit between two mating parts. Limits dimensions shown.

Description of Standard ANSI Fits. The classes of fits are arranged in three general groups known as running and sliding fits, locational fits, and force fits. Standard Fit Tables are given in Appendix B.

Running and Sliding Fits (RFC). Running and sliding fits are intended to provide a similar running performance with suitable lubrication allowance, throughout their range of sizes. The clearances for the first two classes, used chiefly as sliding fits, increase more slowly with the diameter than do the clearances for the other classes, so that accurate location is maintained even at the expense of free relative motion. These fits may be described as follows:

○ RC1—Close sliding fits are intended for the accurate location of parts that must assemble without perceptible play.

○ RC2—Sliding fits are intended for accurate location, but with greater maximum clearance than class RC1. Parts made to this fit move and turn easily but are not intended to run freely, and in the larger sizes may seize with small temperature changes.

○ RC3—Precision running fits are about the closest fits that can be expected to run freely, and are intended for precision work at slow speeds and light journal pressures, but are not suitable where appreciable temperature differences are likely to be encountered.

○ RC4—Close running fits are intended chiefly for running fits on accurate machinery with moderate surface speeds and journal pressures, where accurate location and minimum play is desired.

○ RC5 and RC6—Medium running fits are intended for high running speeds, or heavy journal pressures, or both.

○ RC7—Free running fits are intended for use where accuracy is not essential, or where large temperature variations are likely to be encountered, or under both these conditions.

○ RC8 and RC9—Loose running fits are intended for use where wide commercial tolerances may be necessary, together with an allowance, on the external member.

Locational Fits (LC, LT, and LN). Locational fits are fits intended to determine only the location of the mating parts; they may provide rigid or accurate location, as with interference fits, or provide some freedom of location, as with clearance fits. Accordingly, they are divided into three groups: clearance fits (LC), transition fits (LT), and interference fits (LN). These fits are described as follows:

○ LC—Locational clearance fits are intended for parts which are normally stationary, but which can be freely assembled or disassembled. They range from snug fits for parts requiring accuracy of location, through medium clearance fits for parts such as spigots, to looser fastener fits where freedom of assembly is of prime importance.

○ LT—Locational transition fits are a compromise between clearance and interference fits. They are for applications where accuracy of location is important but either a small amount of clearance or interference is permissible.

○ LN—Locational interference fits are used where accuracy of location is of prime importance, and for parts requiring rigidity and alignment with no special requirements for bore pressure. Such fits are not intended for parts designed to transmit frictional loads from one part to another by virtue of the tightness of fit. Such conditions are covered by force fits.

Force Fits (FN). Force, or shrink, fits constitute a special type of interference fit normally characterized by maintenance of constant bore pressures throughout its range of sizes. The interference, therefore, varies almost directly with diameter, and the difference between its minimum and maximum values is small so as to maintain the resulting pressures within reasonable limits. These fits are described as follows:

○ FN1—Light drive fits are those requiring light assembly pressures and producing more or less permanent assemblies. They are suitable for thin sections or long fits, or in external cast-iron members.

○ FN2—Medium drive fits are suitable for ordinary steel parts, or for shrink fits on light sections. They are about the tightest fits that can be used with high-grade cast-iron external members.

○ FN3—Heavy drive fits are suitable for heavy steel parts or for shrink fits in medium sections.

○ FN4 and FN5—Force fits are suitable for parts that can be highly stressed, or for shrink fits where the heavy pressing forces required are impractical.

Establishing Dimensions for Standard ANSI Fits. The fit used in a specific situation is determined by the operating requirements of the machine. When the type of fit has been established, the engineering drafter refers to tables that show the standard hole and shaft tolerances for the specified fit. One source of these tables is the *Machinery's Handbook.* Tolerances are based on the type of fit and nominal size ranges, such as 0–.12, .12–.24, .24–.40, .40–.71, .71–1.19, and 1.19–1.97 in. So, if you have a 1-in. nominal shaft diameter and an RC4 fit, refer to Figure 10–75 to determine the shaft and hole limits. The hole and shaft limits for a 1-in. nominal diameter are:

Upper hole limit = 1.000 + .0012 = 1.0012
Lower hole limit = 1.000 + 0 = 1.000
Upper shaft limit = 1.000 − .0008 = .9992
Lower shaft limit = 1.000 − .0016 = .9984

NOMINAL SIZE RANGE IN INCHES	RC4 STANDARD TOLERANCE LIMITS	
	Hole	Shaft
0–.12	+.0006	−.0003
	0	−.0007
.12–.24	+.0007	−.0004
	0	−.0009
.24–.40	+.0009	−.0005
	0	−.0011
.40–.71	+.0010	−.0006
	0	−.0013
.71–1.19	+.0012	−.0008
	0	−.0016
1.19–1.97	+.0016	−.0010
	0	−.0020

Figure 10–75 Standard RC4 fits for nominal sizes ranging from 0 to 1.97 inches. Standard fit tables are given in Appendix B, Table 28.

You then dimension the hole as ∅1.0012–1.0000, and the shaft as ∅.9992–.9984.

Standard ANSI/ISO Metric Limits and Fits. The standard for the control of metric limits and fits is governed by the document ANSI B4.2 *Preferred Metric Limits and Fits.* The system is based on symbols and numbers that relate to the internal or external application and the type of fit. The specifications and terminology for fits are slightly different from the ANSI standard fits previously described. The metric limits and fits are divided into three general categories: clearance fits, transition fits, and interference fits. *Clearance fits* are generally the same as the running and sliding fits explained earlier. With clearance fits, a clearance always occurs between the mating parts under all tolerance conditions. With *transition fits,* a clearance or interference may result due to the range of limits of the mating parts. When interference fits are specified, a press or force situation exists under all tolerance conditions. Refer to Figure 10–76 for the ISO symbol and descriptions of the different types of metric fits.

The metric limits and fits may be designated in a dimension one of three ways. The method used depends on individual company standards and the extent of use of the ISO system. When most companies begin using this system, the tolerance limits are calculated and shown on the drawing followed by the tolerance symbol in parentheses; for example, 25.000–24.979 (25 h7). The symbol in parentheses represents the basic size, 25, and the shaft tolerance code, h7. The term *basic size* denotes the dimension from which the limits are calculated. When companies become accustomed to using the system, they represent dimensions with the code fol-

| TYPE OF FIT | ISO SYMBOL | | DESCRIPTION OF FIT |
	Hole	Shaft	
Clearance Fit	H11/c11	C11/h11	*Loose running*
	H9/d9	D9/h9	*Free running*
	H8/f7	F8/h7	*Close running*
	H7/g6	G7/h6	*Sliding*
	H7/h6	H7/h6	*Locational clearance*
Transition Fit	H7/k6	K7/h6	*Locational transition*
	H7/n6	N7/h6	*Locational transition*
Interference Fit	H7/p6[1]	P7/h6	*Locational interference*
	H7/s6	S7/h6	*Medium drive*
	H7/u6	U7/h6	*Force*

Figure 10–76 Description of metric fits. Standard fit tables are given in Appendix B, Table 28.

| BASIC SIZE | CLOSE RUNNING FIT | |
	Hole (H8)	Shaft (f7)
20	20.033	19.993
	20.000	19.959
25	25.033	24.980
	25.000	24.959
30	30.033	29.980
	30.000	29.959
40	40.039	39.975
	40.000	39.950
50	50.039	49.975
	50.000	49.950

Figure 10–77 Tolerances of close running fits for basic sizes ranging from 20 to 50 mm. Standard fit tables are given in Appendix B, Table 28.

lowed by the limits in parentheses, as follows: 25 h7 (25.000–24.979). Finally, when a company has used the system long enough for interpreters to understand the designations, the code is placed alone on the drawing, like this: 25 h7. When it is necessary to determine the dimension limits from code dimensions, use the charts in ANSI B4.2 or the *Machinery's Handbook*. For example, if you want to determine the limits of the mating parts with a basic size of 30 and a close running fit, refer to the chart shown in Figure 10–77. The hole limits for the 30 mm basic size are ⌀30.033–30.000, and the shaft dimension limits are ⌀29.980–29.959.

▼ DIMENSIONS APPLIED TO PLATINGS AND COATINGS

When platings such as chromium, copper, and brass, or coatings such as galvanizing, polyurethane, and sili-

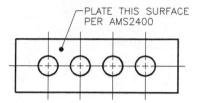

Figure 10–78 A dot replaces the arrowhead on a leader connecting a specific note to a surface.

cone are applied to a part or feature, the specified dimensions should be defined in relationship to the coating or plating process. A general note that indicates that the dimensions apply before or after plating or coating is commonly used and specifies the desired variables; for example, DIMENSIONAL LIMITS APPLY BEFORE (AFTER) PLATING (COATING). A leader connecting a specific note to a surface may also be used. Notice the dot replaces the arrowhead when the leader points to the surface in Figure 10–78.

▼ MAXIMUM AND MINIMUM DIMENSIONS

In some situations, a dimension with an unspecified tolerance may require that the general tolerance be applied in one direction only from the specified dimension. For example, when a 12 mm radius shall not exceed 12 mm, the dimension may read R12 MAX. Therefore, when it is desirable to establish a maximum or minimum dimension the abbreviations MAX or MIN are applied to the dimension. A dimension with a specified tolerance reads as previously discussed:

for example, $R12_{-0.05}^{0}$, or $R12_{0}^{+0.05}$.

▼ CASTING DRAWING AND DESIGN

The end result of a casting drawing is the fabrication of a pattern. The preparation of casting drawings depends on the casting process used, the material to be cast, and the design or shape of the part. The drafter makes the drawing of the part the same as the desired end result after the part has been cast. The drafter may need to take certain casting characteristics into consideration, and the pattern maker needs to adjust the size and shape of the pattern to take into account characteristics that the drafter does not intentionally apply on the drawing.

Shrinkage Allowance

When metals are heated and then cooled, they shrink until the final temperature is reached. The amount of shrinkage depends on the material used. The shrinkage for most iron is about .125 in. per ft, .250 in. per ft for steel, .125 to .156 in. per ft for aluminum, .22 in. per ft for brass, and .156 in. per ft for bronze. Values for shrinkage allowance are approximate since the exact allowance depends on the size and shape of the casting and the contraction of the casting during cooling. The drafter normally does not need to take shrinkage into consideration because the pattern maker uses shrink rules that use expanded scales to take into account the shrinkage of various materials.

Draft

Draft is the taper allowance on all vertical surfaces of a pattern, which is necessary to facilitate the removal of the pattern from the mold. Draft is not necessary on horizontal surfaces because the pattern easily separates from these surfaces without sticking. Draft angles begin at the parting line and taper away from the molding material. (See Figure 10–79.) The draft is added to the minimum design sizes of the product. Draft varies with different materials, size and shape of the part, and casting methods. Little if any draft is necessary in investment casting. The factors that influence the amount of draft are the height of vertical surfaces, the quality of the pattern, and the ease with which the pattern must be drawn from the mold. A typical draft angle for cast iron and steel is .125 in. per ft. Whether the drafter takes draft into consideration on a drawing depends on company standards. Some companies leave draft angles to the pattern maker, while others require drafters to place draft angles on the drawing.

Fillets and Rounds in Casting

One of the purposes of fillets and rounds on a pattern is the same as that of draft angles: to allow the pattern to eject freely from the mold. Also, the use of fillets on inside corners helps reduce the tendency of cracks to develop during shrinkage. (See Figure 10–80.) The radius for fillets and rounds depends on the material to be cast, the casting method, and the thickness of the part. The recommended radii for fillets and rounds used in sand cast-

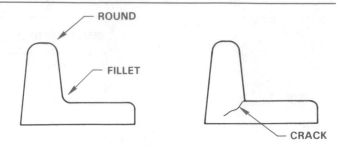

Figure 10–80 Fillets and rounds for castings.

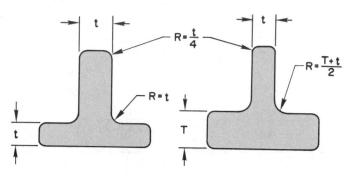

Figure 10–81 Recommended fillet and round radii for sand castings.

ing is determined by part thickness as shown in Figure 10–81.

Machining Allowance

Extra material must be left on the casting for any surface that will be machined. As with other casting design characteristics, the machining allowance depends on the casting process, material, size and shape of the casting, and the machining process to be used for finishing. The standard finish allowance for iron and steel is .125 in., and for nonferrous metals such as brass, bronze, and aluminum .062 in. In some situations the finish allowance may be as much as .5 or .75 in. for castings that are very large or have a tendency to warp.

Other machining allowances may be the addition of lugs, hubs, or bosses on castings that are otherwise hard to hold. The drafter can add these items to the drawing or the pattern maker may add them to the pattern. These features may not be added for product function but they serve as aids for chucking or clamping the casting in a machine. (See Figure 10–82.)

There are several methods that can be used to prepare drawings for casting and machining operations. The method used depends on company standards. A commonly used technique is to prepare two drawings, one a casting drawing and the other a machining drawing. The casting drawing, as shown in Figure 10–83, shows the part as a casting. Only dimensions necessary to make the casting are shown in this drawing. The other drawing is of the same part, but this time only machining in-

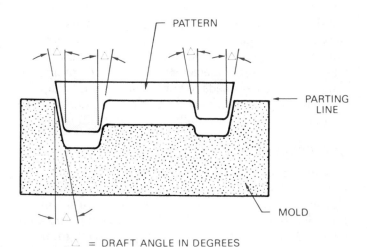

△ = DRAFT ANGLE IN DEGREES

Figure 10–79 Draft angles for castings.

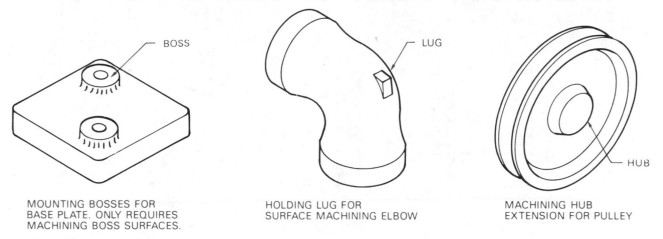

BOSS

LUG

HUB

MOUNTING BOSSES FOR
BASE PLATE. ONLY REQUIRES
MACHINING BOSS SURFACES.

HOLDING LUG FOR
SURFACE MACHINING ELBOW

MACHINING HUB
EXTENSION FOR PULLEY

Figure 10–82   Cast features added for machining.

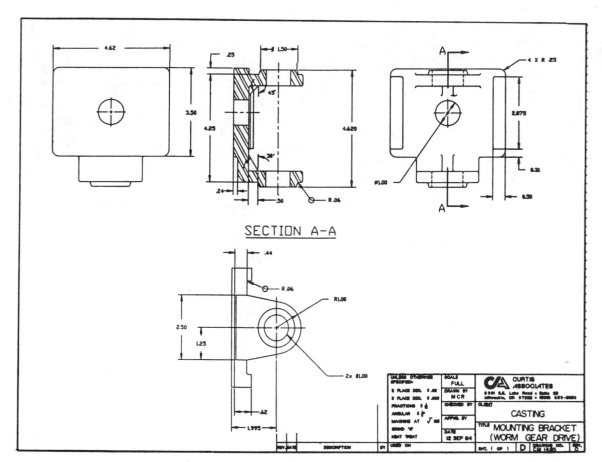

Figure 10–83   Casting drawing. *Courtesy Curtis Associates.*

formation and dimensions are given. The actual casting goes to the machine shop along with the machining drawing so the features can be machined as specified. The machining drawing is shown in Figure 10–84.

Another method of preparing casting and machining drawings is to show both casting and machining information together on one drawing. This technique requires the pattern maker to add machining allowances. The drawing may have draft angles specified in the form of a note. The pattern maker must add the draft angles to

the finished sizes given. With draft angles and finish allowances omitted, the designer or drafter may need to consult with the pattern maker to insure that the casting is properly made. A combination casting/machining drawing is shown in Figure 10–85.

Another technique is to draw the part as a machining drawing and then use phantom lines to show the extra material for machining allowance and draft angles, as shown in Figure 10–86.

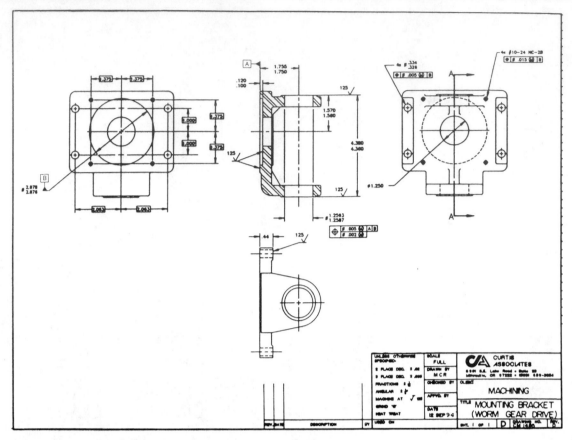

Figure 10–84 Machining drawing. *Courtesy Curtis Associates.*

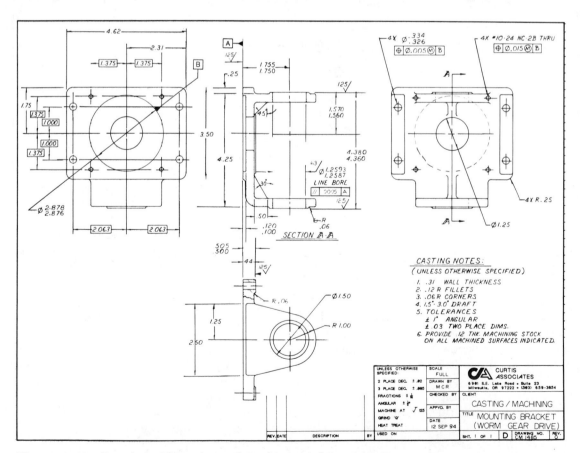

Figure 10–85 Drawing with casting and machining information. *Courtesy Curtis Associates.*

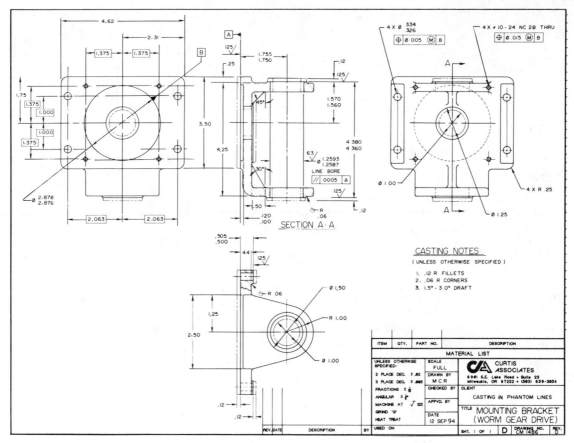

Figure 10–86 Phantom lines used to show machining allowances. *Courtesy Curtis Associates.*

Forging, Design, and Drawing

Draft for forgings serves much the same purpose as draft for castings. The draft associated with forging is found in the dies. The sides of the dies must be angled to facilitate the release of the metal during the forging process. If the vertical sides of the dies did not have draft angle, then the metal would become stuck in the die. Internal and external draft angles may be specified differently because the internal drafts in some materials may have to be greater to help reduce the tendency of the part to stick in the die. While draft angles may change slightly with different materials, the common exterior draft angle recommended is 7°. The internal draft angles for most soft materials is also 7°, but the recommended interior draft for iron and steel is 10°.

The application of fillets and rounds to forging dies is to improve the ejection of the metal from the die. Another reason, similar to that for casting, is increased inside corner strength. One factor that applies to forgings as different from castings is that fillets and rounds that are too small in forging dies may substantially reduce the life of the dies. Recommended fillet and round radius dimensions are shown in Figure 10–87.

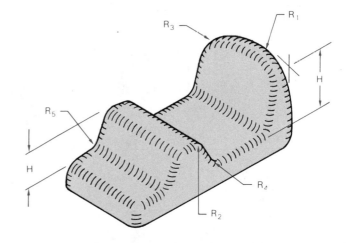

DIMENSIONS IN MILLIMETERS					
H	R_1	R_2	R_3	R_4	R_5
6	1.5	1.5	4.5	3	3
13	1.5	1.5	4.5	3	3
25	3	3	9	6	9
50	4.5	6	13	13	15
75	6	7.5	16	16	25
100	7.5	10.5	25	25	35
125	9	13	23	32	44
150	10.5	15	32	38	50

Figure 10–87 Recommended fillets and rounds for forgings.

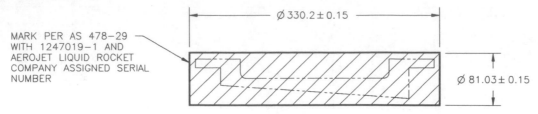

Figure 10–88 Blank material for forging process. *Courtesy Aerojet Propulsion Division.*

Forging Drawings

A number of methods may be used in the preparation of forging drawings. One technique used in forging drawings that is clearly different from the preparation of casting drawings is the addition of draft angles. Casting drawings usually do not show draft angles; forging detail drawings usually do show draft.

Before a forging can be made, the dimensions of the stock material to be used for the forging must be determined. Some companies leave this information to the forging shop to determine; other companies have their engineering department make these calculations. After the stock size is determined, a drawing showing size and shape of the stock material is prepared. The blank material is dimensioned, and the outline of the end product is drawn inside the stock view using phantom lines. (See Figure 10–88.)

Forgings are made with extra material added to surfaces that must be machined. Forging detail drawings may be made to show the desired end product with the outline of the forging shown in phantom lines at areas that require machining. (See Figure 10–89.) Notice the double line around the perimeter showing draft angle. Another option used by some companies is to make two separate drawings, one a forging drawing and the other a machining drawing. The forging drawing shows all of the views, dimensions, and specifications that relate only to the production of the forging, as shown in Figure 10–90. The machining drawing gives views, dimensions, and specifications related to the machining processes, as shown in Figure 10–91.

▼ MACHINED SURFACES
Surface Finish Definitions

Surface Finish. *Surface finish* refers to the roughness, waviness, lay, and flaws of a surface. Surface finish is the specified smoothness required on the finished surface of a part that is obtained by machining, grinding, honing, or lapping. The drawing symbol associated with surface finish is shown in Figure 10–92.

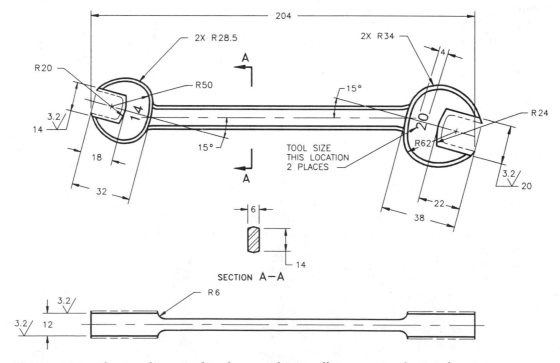

Figure 10–89 Phantom lines used to show machining allowance on a forging drawing.

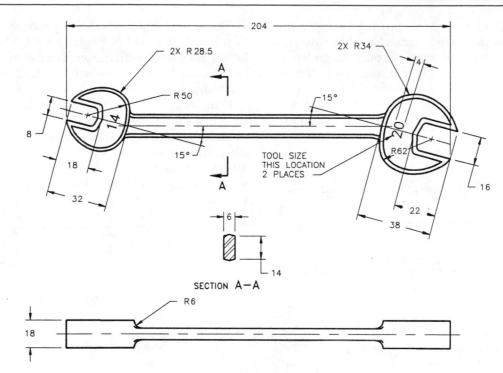

Figure 10–90   Forging drawing with forging dimensions only.

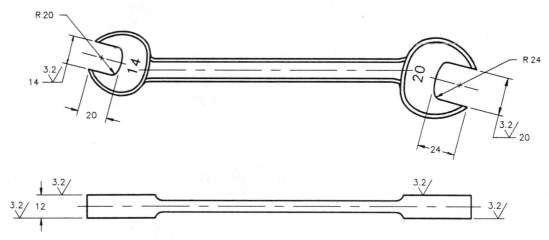

Figure 10–91   Machining drawing with machining dimensions only.

EDGE VIEW OF PART

Figure 10–92   Surface finish symbol.

Surface Roughness.   Surface roughness refers to fine irregularities in the surface finish and is a result of the manufacturing process used. Roughness height is measured in micrometers, μm (millionths of a meter), or in microinches, μin (millionths of an inch).

Surface Waviness.   Surface waviness is the often widely spaced condition of surface texture usually caused by such factors as machine chatter, vibrations, work deflection, warpage, or heat treatment. Waviness is rated in millimeters or inches.

Lay.   Lay is the term used to describe the direction or configuration of the predominant surface pattern. The lay symbol is used if considered essential to a particular surface finish. The characteristic lay symbol may be attached to the surface finish symbol, as shown in Figure 10–93.

Surface Finish Symbol

Some of the surfaces of an object may be machined to certain specifications. When this is done, a surface finish symbol is placed on the view where the surface or surfaces appear as lines (edge view). (See Figure 10–94.)

The finish symbol on a machine drawing alerts the machinist that the surface must be machined to the given specification. The finish symbol also tells the pattern or die maker that extra material is required in a casting or forging.

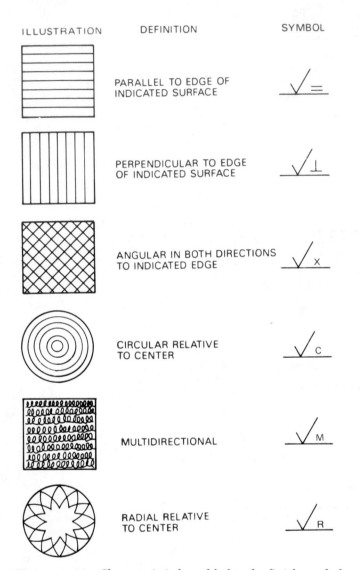

Figure 10–93 Characteristic lay added to the finish symbol.

The surface finish symbol is properly drawn using a thin line as detailed in Figure 10–95. The numerals or letters associated with the surface finish symbol should be the same height as the lettering used on the drawing dimensions and notes.

Often only the surface roughness height is used with the surface finish symbol, which means 3.2 micrometers. When other characteristics of a surface texture are specified, they are shown in the format represented in Figure 10–96. For example, roughness-width cutoff is a numerical value that establishes the maximum width of surface irregularities to be included in the roughness height measurement. Standard roughness-width cutoff values for inch specifications are .003, .010, .030, .100, .300, and 1.000; .030 is implied when no specification is given.

Figure 10–97 is a magnified pictorial representation of the characteristics of a surface finish symbol. Figure 10–98 shows some common roughness height values in

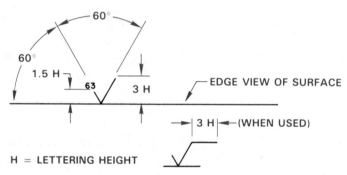

Figure 10–95 Properly drawn surface finish symbol.

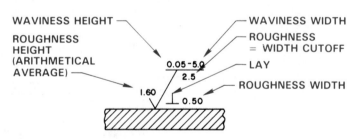

Figure 10–96 Elements of a complete surface finish symbol.

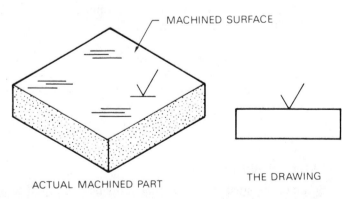

Figure 10–94 Standard surface finish symbol placed on the edge view. The symbol should always be placed horizontally when unidirectional dimensioning is used.

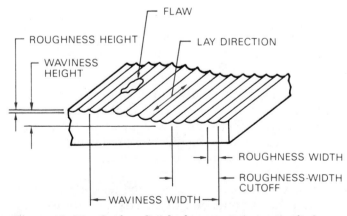

Figure 10–97 Surface finish characteristics magnified.

MICROMETERS	ROUGHNESS HEIGHT RATING MICRO INCHES	SURFACE DESCRIPTION	PROCESS
25	1000	VERY ROUGH	SAW AND TORCH CUTTING, FORGING OR SAND CASTING.
12.5	500	ROUGH MACHINING	HEAVY CUTS AND COARSE FEEDS IN TURNING, MILLING AND BORING.
6.3	250	COARSE	VERY COARSE SURFACE GRIND, RAPID FEEDS IN TURNING, PLANNING, MILLING, BORING AND FILLING.
3.2	125	MEDIUM	MACHINING OPERATIONS WITH SHARP TOOLS, HIGH SPEED, FINE FEEDS AND LIGHT CUTS.
1.6	63	GOOD MACHINE FINISH	SHARP TOOLS, HIGH SPEEDS, EXTRA-FINE FEEDS AND CUTS.
0.80	32	HIGH GRADE MACHINE FINISH	EXTREMELY FINE FEEDS AND CUTS ON LATHE, MILL AND SHAPERS REQUIRED. EASILY PRODUCED BY CENTERLESS, CYLINDRICAL AND SURFACE GRINDING.
0.40	16		
0.20	8	VERY FINE MACHINE FINISH	FINE HONING AND LAPPING OF SURFACE.
0.050 0.100	2-4	EXTREMELY SMOOTH MACHINE FINISH	EXTRA-FINE HONING AND LAPPING OF SURFACE. MIRROR FINISH.
0.025	1	SUPER FINISH	DIAMOND ABRASIVES.

Figure 10–98 Common roughness height values with a surface description and associated machining process.

micrometers and microinches, a description of the resulting surface, and the process by which the surface may be produced. When a maximum and minimum limit is specified, the average roughness height must lie within the two limits.

When a standard or general surface finish is specified in the drawing title block or in a general note, then a surface finish symbol without roughness height specified is used on all surfaces that arc the same as the general specification. When a part is completely finished to a given specification, then the general note, FINISH ALL OVER, or abbreviation FAO, or FAO 125 µIN, may be used. The placement of surface finish symbols on a drawing can be accomplished a number of ways, as shown in Figure 10–99. Additional elements may be applied to the surface finish symbol, as shown in Figure 10–100.

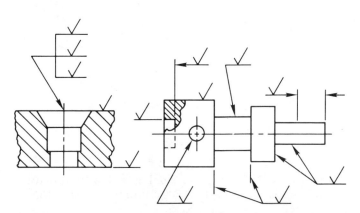

Figure 10–99 Proper placement of surface finish symbols.

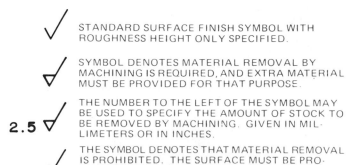

STANDARD SURFACE FINISH SYMBOL WITH ROUGHNESS HEIGHT ONLY SPECIFIED.

SYMBOL DENOTES MATERIAL REMOVAL BY MACHINING IS REQUIRED, AND EXTRA MATERIAL MUST BE PROVIDED FOR THAT PURPOSE.

2.5 THE NUMBER TO THE LEFT OF THE SYMBOL MAY BE USED TO SPECIFY THE AMOUNT OF STOCK TO BE REMOVED BY MACHINING. GIVEN IN MILLIMETERS OR IN INCHES.

THE SYMBOL DENOTES THAT MATERIAL REMOVAL IS PROHIBITED. THE SURFACE MUST BE PRODUCED BY PROCESSES SUCH AS CASTING OR FORGING.

Figure 10–100 Material removal elements added to the surface finish symbol.

CADD Applications

DIMENSIONING SOFTWARE PACKAGES

Most of the major CADD software packages provide for a variety of dimensioning applications depending on the user's need. For example, you may use unidirectional or aligned dimensioning, and datum or chain dimensioning based on your company's standards and procedures. You can also alter the way dimensions are presented by changing any one or more of these variables:

○ Space beween dimension lines.
○ Style and height of text.
○ Size and shape of arrowheads.
○ Extension line length beyond the last dimension line.
○ Space between the object and the start of the extension line.

○ Location of the text in relation to the dimension line.

Dimensioning *default* values are normally set to ASME standards, but the user has the flexibility to change these values for specific applications. A default value refers to a standard preset value. Figure 10–101 shows some of the flexibility available in changing predefined default values.

In addition to the major CADD software packages, there are hundreds of third-party software packages. The term *third-party* refers to programs that support or may be used in conjunction with the main package. In third-party software for dimensioning, the principal emphasis seems to be on datum, tabular, arrowless, and geometric tolerancing packages. These are the types of software that help increase speed and productivity and simplify the dimensioning process. For example, some of the programs have automatic dimensioning, which allows the drafter to pick all of the items for dimensioning and, when finished, press the RETURN key or select AUTOMATIC, depending on the program, and watch everything be dimensioned. Some packages also automatically generate a tabular chart for hole reference. A drawing showing arrowless datum dimensioning is shown in Figure 10–102.

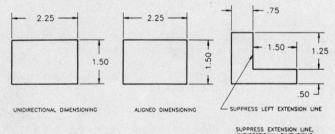

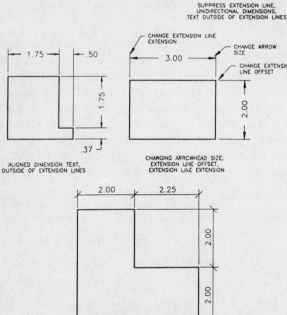

Figure 10–101 Changing dimension default values for specific applications.

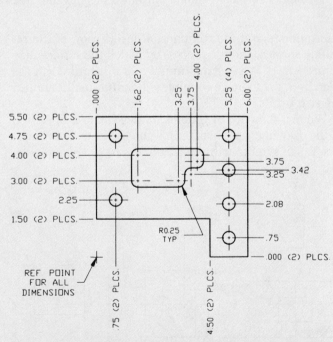

Figure 10–102 Datum dimensioning with a third-party software package, Auto-DATUM. *Courtesy CADMASTER, Inc.*

▼ DESIGN AND DRAFTING OF MACHINED FEATURES

Drafters should gain a working knowledge of machining processes and the machining capabilities of the company for which they work. Drawings should be prepared that allow machining within the capabilities of the machinery available. The first consideration should be the least-expensive method to get the desired result. Avoid overmachining. Machining processes are expensive, so do not call for requirements on a drawing that are not necessary for the function of the part. For example, surface finishes become more expensive to machine as the roughness height decreases. So, if a surface roughness of 125 microinches is adequate, then do not use a 32-microinch specification just because you like smooth surfaces. Another example is demonstrated by the difference between the 63- and 32-microinch finish. A 63-microinch finish is a good machine finish that may be performed using sharp tools at high speeds with extra-fine feeds and cuts. The 32-microinch callout, on the other hand, requires extremely fine feeds and cuts on a lathe or milling machine and in many cases requires grinding. The 32 finish is more expensive to perform.

In a manufacturing environment where cost and competition are critical considerations, a great deal of thought must be given to meeting the functional and appearance requirements of a product at the least possible cost. It generally does not take very long for an entry-level drafter to pick up these design considerations by communicating with engineering and manufacturing department personnel. Many drafters become designers, checkers, or engineers within a company by learning the product and being able to implement designs based on manufacturing capabilities.

▼ SYMBOLS

ANSI standards specify symbols to be used on drawings. With the influence of computer graphics, the use of symbols is easy. When symbols are drawn manually, a template should be used because many of the symbols are difficult to draw properly free-hand, and the template drawing is generally neater and more uniform. Any attempt to draw symbols using drafting tools other than specially designed templates takes too much time. Individual symbols have been represented and detailed throughout this chapter each time they are introduced.

PROFESSIONAL PERSPECTIVE

The proper placement and use of dimensions is one of the most difficult aspects of drafting. It requires careful thought and planning. First, you must determine the type of dimension that fits the application, then place it on the drawing. While the CADD system makes the actual placement of dimensions quick and easy, it does not make the planning process any easier. Making preliminary sketches is very important before beginning a drawing. The preliminary sketch allows you to select and place the views and then place the dimensions. When you were drawing only the views the space requirements were not as critical, but a drawing without dimensions is unrealistic.

The problem with dimensions is that they take up a lot of space. One of the big issues an entry-level drafter faces is determining the space requirements for dimensions. In many cases these requirements are underestimated. As a rule of thumb, if you think the drawing is crowded on a particular size sheet, then play it safe and use the next larger size. Most companies want the drawing information spread out and easy to read, but be careful, because some companies want the

drawing crowded with as much information as you can get on a sheet. This text advocates a clean and easy-to-read drawing that is not crowded. Out on the job, however, you must do what is required by your employer. If you are drawing an uncrowded drawing, you should consider leaving about one-quarter of the drawing space clear of view and dimensional information. Usually this space is above and to the left of the title block. This space provides adequate room for general notes and engineering changes.

Follow the dimensioning rules, guidelines, and examples discussed in this chapter and use proper dimensioning standards. Try as hard as you can to avoid breaking dimensioning standards. Following are some of the pitfalls to watch out for as a beginning drafter:

○ Do not crowd dimensions. Keep your dimension line spacing equal and far enough apart to clearly separate the dimension numerals.

○ Do not dimension to hidden features. This also means do not dimension to the centerline of a

(Continued)

PROFESSIONAL PERSPECTIVE *(Continued)*

hole in the hidden view. Always dimension to the view where the feature appears as an object line.

❍ Dimension to the view that shows the most characteristic shape of an object or feature. For example, dimension shapes where you see the contour, dimension holes to the circular view, and dimension cylindrical shapes in the rectangular view.

❍ Do not stack adjacent dimension numerals; stagger them so they are easier to read.

❍ Group dimensions as much as you can. It is better to keep dimensions concentrated to one location or one side of a feature rather than spread around the drawing. This makes the drawing easier to read.

❍ If you are using manual drafting, use a template to draw symbols. If you use computer-aided draft-

ing, make a standard symbol library or use software with symbols available. Symbols not only speed up the drafting process, they clearly identify the feature. For example, if you draw a single view of a cylinder, the diameter (∅) dimension is identified in the rectangular view and drawing the circular view may not be necessary.

❍ Look carefully at the figures in the text and use them as examples as you prepare your drawings.

Try to put yourself in the place of the person who has to read and interpret your drawing. Make the drawing as easy to understand as possible. Keep the drawing as uncluttered and as simple as possible, yet still complete.

CADD Applications

DIMENSIONING FOR CADD/CAM

The implementation of computer-aided design and drafting (CADD) and computer-aided manufacturing (CAM) in industry is best accomplished when common control of the computer exists between engineering and manufacturing. The success of this automation is, in part, relative to the standardization of operating and documentation procedures. CADD can be accomplished through the same coordinate dimensioning systems previously described in this chapter. Standard dimensioning systems are used to establish a geometric model of the part, which in turn is displayed at a graphics workstation. The data retrieved from this model is the mathematical description of the part to be produced. The drafter must dimension the part completely and accurately so that each contour or geometric shape of the part is continuous. The dimensioning systems that locate features or points on a feature in relation to x, y, and z axes derived from a common origin are most effective, such as datum, tabular, arrowless, and polar coordinate dimensioning. The x, y, and z axes originate from three mutually perpendicular planes which are generally the geometric counterpart of the sides of the part when the surfaces are at right angles as seen in Figure 10–103. If the part is cylindrical, two of the planes intersect at right angles to establish the axis of the cylinder and the third is perpendicular to the intersecting planes as shown in Figure 10–104. The x, y, and z coordinates that are used to establish features on a drawing are con-

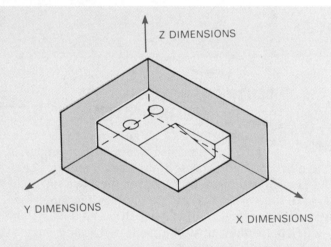

Figure 10–103   Features of a part dimensioned in relation to X, Y, and Z axes.

verted to x, y, and z axes that correspond to the linear and rotary motions that occur in CAM.

The position of the part in relation to mathematical quadrants determines whether the x, y, and z values are positive or negative. The preferred position is the mathematical quadrant that allows for programming positive commands for the machine tool. Notice in Figure 10–105 that the positive x and y values occur in quadrant 1.

In CAD/CAM and computer-integrated manufacturing (CIM) programs, the drawing is made on the computer screen and sent directly to the computer numerical control machine tool without generating a

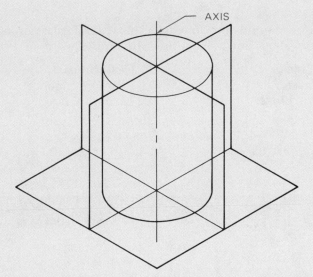

Figure 10–104 The planes related to the axis of a cylindrical feature.

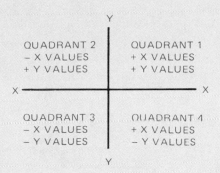

Figure 10–105 Determination of X and Y values in related quadrants.

hard copy of the drawing. In this situation it becomes important for the drafter to understand the machine tool operation. Figure 10–106a shows a drawing created for CAD/CAM, and Figure 10–106b shows the same drawing as represented on the computer screen prior to generating the machine tool program. Notice, the dimensions are not displayed on the drawing in Figure 10–106b. The data used to input the information from the original design drawing is used, by the computer, to establish the machine tool paths. The CAD/CAM program also allows the operator to show the tool path, as shown in Figure 10–107. The tool path display allows the operator to determine if the machine tool will perform the assigned machining operation.

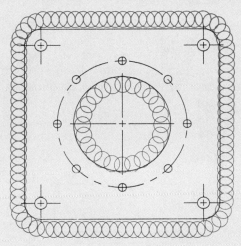

Figure 10–107 The tool path display, shown in color, allows the operator to determine if the machine tool will perform the assigned machining operation.

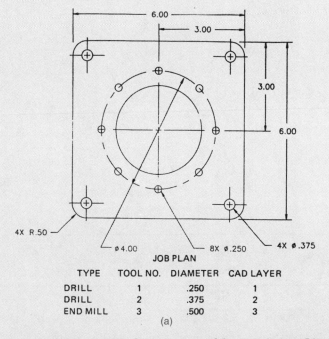

JOB PLAN

TYPE	TOOL NO.	DIAMETER	CAD LAYER
DRILL	1	.250	1
DRILL	2	.375	2
END MILL	3	.500	3

(a)

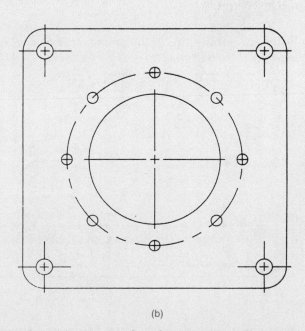

(b)

Figure 10–106 **(a)** A drawing created for CAD/CAM. **(b)** Drawing from (a) represented on the computer screen prior to generating the machine tool program.

▼ RECOMMENDED REVIEW

It is recommended that you review the following parts of Chapter 3 before you begin working on fully dimen-sioned multiview drawings. This will refresh your memory about how related lines are properly drawn.

- ○ Object Lines.
- ○ Viewing Planes.
- ○ Extension Lines.
- ○ Hidden Lines.
- ○ Break Lines.
- ○ Dimension Lines.
- ○ Centerlines.
- ○ Phantom Lines.
- ○ Leader Lines.

Also review Chapters 5 and 6 covering multiview and auxiliary view drawings.

MATH APPLICATION

FINDING DIAGONALS

Suppose you need the distance from one point on a drawing to another, such as between points A and B in Figure 10–108.

A 3-D diagonal may be found by the following formula.

$$r = \sqrt{(\text{difference of x})^2 + (\text{difference of y})^2 + (\text{difference of z})^2}$$

Find the distance from C to D in Figure 10–109.

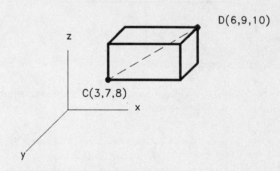

Figure 10–108

You can employ the Pythagorean Theorem. (See Appendix N on page 927.)

$$r = \sqrt{(\text{difference of x})^2 + (\text{difference of y})^2}$$

$$r = \sqrt{(10 - -3)^2 + (7 - 2)^2}$$

$$r = \sqrt{10^2 + 5^2}$$

$$r = \sqrt{100 + 25} = \sqrt{125} = 11.2"$$

Figure 10–109

$$r = \sqrt{(6 - 3)^2 + (9 - 7)^2 + (10 - 8)^2}$$

$$r = \sqrt{3^2 + 2^2 + 2^2} = \sqrt{9 + 4 + 4} = \sqrt{17} = 4.12"$$

CHAPTER 10 DIMENSIONING TEST

DIRECTIONS

Answer the questions with short complete statements or drawings as needed.

PART I. GENERAL

1. Name two classifications of dimensions.
2. Identify and describe two types of notes.
3. Define *unidirectional dimensioning*.

4. Specify the ANSI document that governs the standard for dimensioning and tolerancing.
5. Define the following terms; examples may be used if appropriate: *actual size; bilateral tolerance; dimension; feature; limits of dimension; specified dimension; tolerance; unilateral tolerance.*
6. What are the recommended standard units of linear measurements on engineering documents?
7. When all dimensions are metric, what is the general note that should accompany the drawing?
8. Define *dual dimensioning,* and show an example of a dual dimensioning general note.
9. How should the decimal in numerals be treated?
10. Should all of the dimension arrowheads on a drawing be the same size?
11. What is the recommended length-to-height ratio of arrowheads?
12. Discuss proper dimension line spacing.
13. Identify a possible disadvantage of chain dimensioning.
14. Describe datum dimensioning.
15. How are notes for holes dimensioned on an engineering document?
16. Describe how holes are located.
17. Define *maximum material condition* (MMC).
18. Define *least material condition* (LMC).
19. Describe a clearance fit.
20. Describe running and sliding fits (RC), locational fits (LC, LT, LN), and force fits (FN).
21. List three factors that influence sheet size selection.
22. Identify two factors that influence drawing scale selection.
23. Define *casting.*
24. Define *core.*
25. Define *forging.*
26. Explain the purpose of draft angle on a casting or forging.
27. Define *surface finish.*
28. Identify the units used to measure surface roughness height.

29. Show three examples of the recommended placement of surface finish symbols.
30. Describe the surface condition and process used to establish the following surface roughness heights given in micrometers: 12.5, 6.3, 3.2, 1.6, 0.80, 0.20, 0.050.

PART 2. GENERAL TOLERANCING

31. Identify the tolerance and limits of each of the following dimensions:
Metric: 25.5±0.127; 19±0.25 Inch: .375±.003; 1.6250±.0005.
32. Given the following CAD drawing, calculate the allowance:

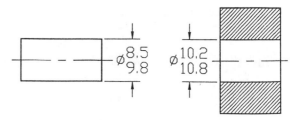

33. From the following list of given conditions, calculate the limits of the shaft and the limits of the hole. (Review the allowance calculation formula.)
a. Metric dimensions.
b. A clearance fit.
c. Allowance = 0.05.
d. Specified dimension of shaft = 12.
e. Shaft tolerance = 0.26 BILATERAL. (Remember the tolerance is the total permissible variation.)
f. Hole tolerance = 0.18.
34. Name five items that a CADD system needs in order to place a dimension.
35. Define *layering.*
36. List at least two reasons why CADD layering can assist the dimensioning process.

CHAPTER

10 DIMENSIONING PROBLEMS

DIRECTIONS

1. From the selected sketch, determine which view should be the front view. Then determine which other views, if any, you need to draw to fully display the part in a multiview drawing.
2. Make a multiview sketch of the selected problem as close to correct proportions as possible. Be sure to indicate where you intend to place the dimension lines, extension lines, arrowheads, and hidden features, to help you determine the spacing for your final drawing.
3. Using the sketch you have just developed as a guide, make an original multiview drawing on an adequate size drawing sheet and at an appropriate scale. Include all dimensions needed using uni-

(Continued)

DIRECTIONS *(Continued)*

directional dimensioning. Use computer-aided drafting if specified by your course outline. Include the following general notes at the lower-left corner, .5 in. each direction from the borderline using ⅛-in.-high lettering. Letter the word NOTES with ³⁄₁₆-in.-high lettering.

2. REMOVE ALL BURRS AND SHARP EDGES.
1. INTERPRET DIMENSIONS AND TOLERANCES PER ASME Y14.5M—1994.

   NOTES:

Additional notes may be required depending on the specifications of each individual assignment. Remember that each problem assignment is given as an engineer's layout to help simulate actual drafting conditions.

Dimensions and views on engineer's layouts may not be placed in accordance with acceptable standards. You need to carefully review the chapter material when preparing the layout sketch. In some problems, the engineer's layout may "assume" certain information, such as the symmetry of a part or the alignment of holes. You need to place enough dimensions or draw lines between features to "fully" dimension the part.

   UNSPECIFIED TOLERANCES:

DECIMALS	mm	IN.
X	±2.5	±.1
XX	±0.25	±.01
XXX	±0.127	±.005
ANGULAR ±30'		
FINISH	3.2µm	125µIN.

Problem 10–1   Basic practice (metric)

Part Name: Step Block
Material: SAE 1020

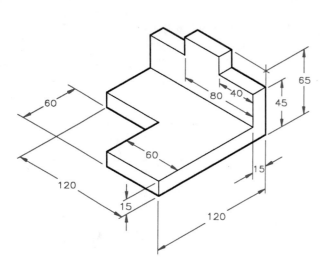

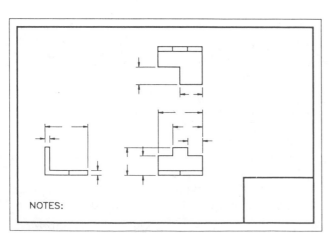

LAYOUT SKETCH STARTED

Problem 10–2   Basic practice (metric)

Part Name: Machine Tool Wedge Plate
Material: SAE 4320

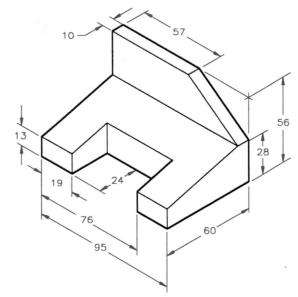

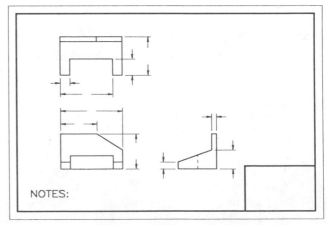

LAYOUT SKETCH STARTED

Problem 10–3 Dimensioning basic practice (in.)

Part Name: V-guide
Material: SAE 4320

Fractions: ±1/32

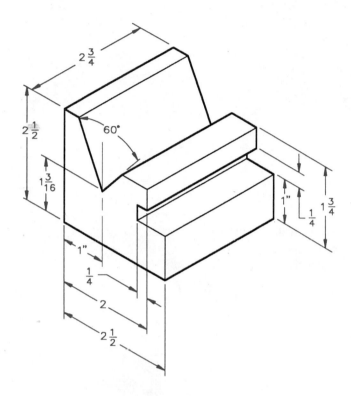

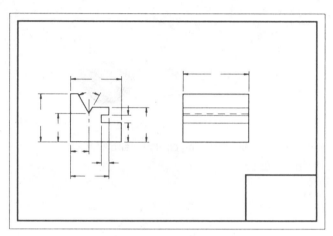

LAYOUT SKETCH
OPTIONAL SINGLE VIEW WITH LENGTH
GIVEN IN GENERAL NOTE OR TITLE BLOCK

Problem 10–4 Circles, arcs, and counterbores (in.)

Part Name: Rest Pad
Material: SAE 1040
Fillets: R.125

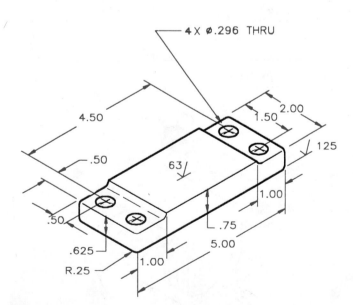

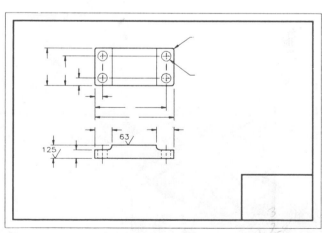

LAYOUT SKETCH

Problem 10–5   Limited space (metric)

Part Name: Angle Bracket
Material: Mild Steel (MS)

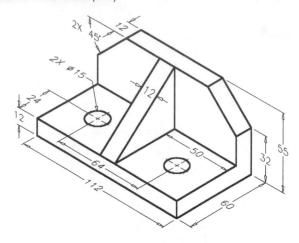

Problem 10–6   Dimensioning contours and limited spaces (in.)

Part Name: Support
Material: Aluminum

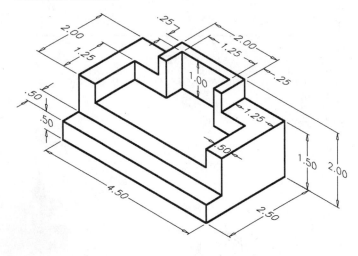

Problem 10–7   Limited spaces (metric)

Part Name: Selector Slide Kicker
Material: Al 1510

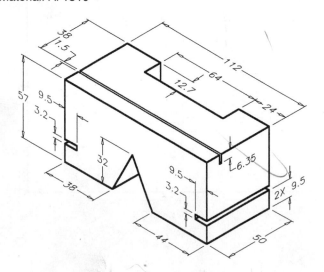

Problem 10–8   Circles and arcs (in.)

Part Name: Chain Link
Material: SAE 4320

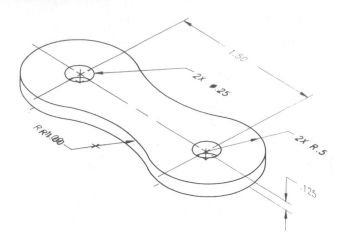

Problem 10–9   Holes and limited space (in.)

Part Name: Pivot Bracket
Material: SAE 1040

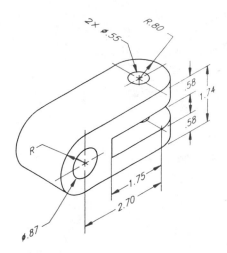

Problem 10–10   Holes, angles, and arcs (in.)

Part Name: Journal Bracket
Material: Cast Iron (CI)

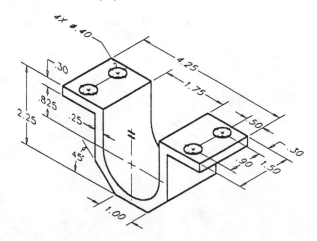

Problem 10–11   Dimensioning multiple features (in.)

Part Name: Lock Ring
Material: SAE 1020

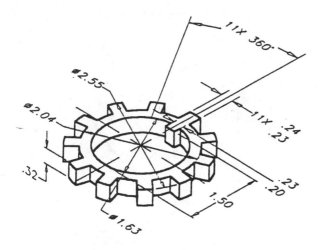

Problem 10–13   Single view (in.)

Part Name: Idler Gear Shaft
Material: MIL-S-7720
Problem based on original art courtesy Aerojet TechSystems Co.

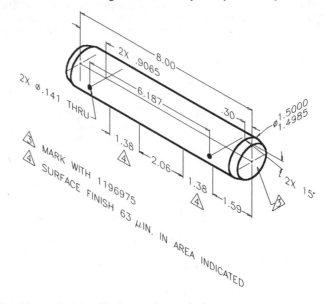

Problem 10–12   Circles and arcs, single view (in.)

Part Name: Thermostat Gauge Standard Motor
Material: Stainless Steel .067/.057 in. Thick
Problem based on original art courtesy Vellumoid Inc.

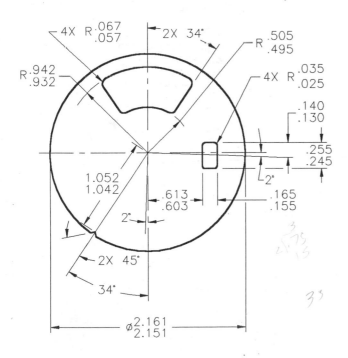

Problem 10–14   Circles and arcs (in.)

Part Name: Bearing Support
Material: SAE 1040

Fillets: R6

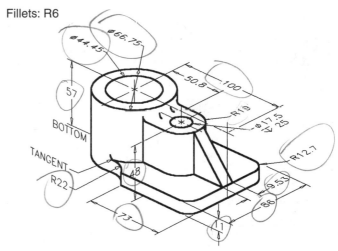

Problem 10–15   Circles and arcs (metric)

Part Name: Hinge Bracket
Material: Cast Aluminum

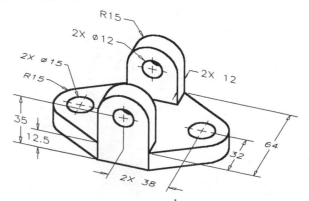

Problem 10–16 Missing view (in.)

Part Name: Bearing Pillow Block
Material: SAE 1040
SPECIFIC INSTRUCTIONS:
Determine the missing view so that holes can be shown and dimensioned as circles. Convert all fractional location dimensions to three-place decimals, and size fractional dimensions to two-place decimals. Verify and add any missing dimensions.
Courtesy Production Plastics, Inc.

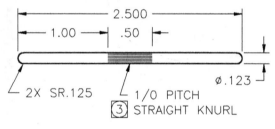

UNLESS OTHERWISE SPECIFIED:
ALL FRACTIONAL DIMENSIONS ±1/32   ALL THREE PLACE DECIMALS ±.005
ALL TWO PLACE DECIMALS ±.010

Problem 10–17 Knurling (in.)

Part Name: Screw Handle
Material: SAE 1020

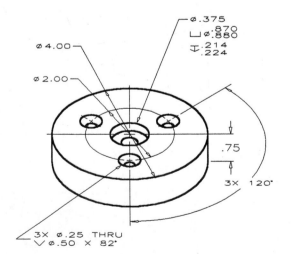

2X SR.125   1/0 PITCH
③ STRAIGHT KNURL

③ KNURL AFTER SIZING, KNURL DIA.
TO BE GREATER THAN ∅.125

Problem 10–18 Machine features (in.)

Part Name: Spacer
Material: SAE 1030

Problem 10–19 Polar dimensioning (in.)

Part Name: Rotating Bracket
Material: .25 THK Stainless Steel

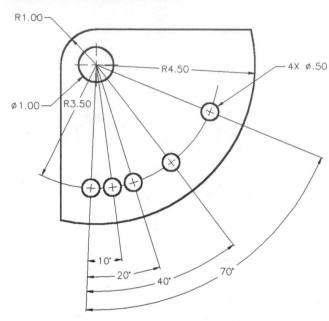

Problem 10–20 Polar coordinate dimensioning (in.)

Part Name: Spacer
Material: Plastic

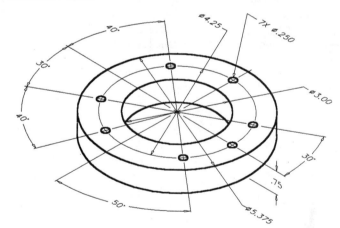

Problem 10–21 Repetitive features (in.)

Part Name: Slot Plate
Material: Aluminum

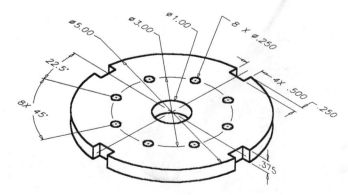

Problem 10–22   Tabular dimensioning (metric)

Part Name: Mounting Base
Material: Stainless Steel

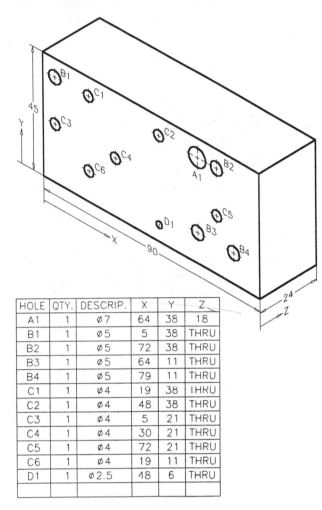

HOLE	QTY.	DESCRIP.	X	Y	Z
A1	1	⌀7	64	38	18
B1	1	⌀5	5	38	THRU
B2	1	⌀5	72	38	THRU
B3	1	⌀5	64	11	THRU
B4	1	⌀5	79	11	THRU
C1	1	⌀4	19	38	THRU
C2	1	⌀4	48	38	THRU
C3	1	⌀4	5	21	THRU
C4	1	⌀4	30	21	THRU
C5	1	⌀4	72	21	THRU
C6	1	⌀4	19	11	THRU
D1	1	⌀2.5	48	6	THRU

Problem 10–23   Dimensioning circles, arcs, and slots (in.)

Part Name: Top Pipe Support Bracket
Material: SAE 1020
Note: All Fillets R.12

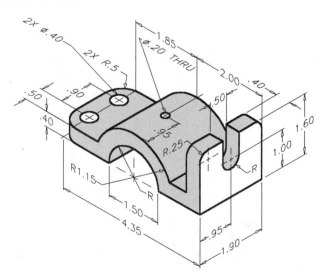

Problem 10–24   Chain dimensioning (metric)

Part Name: Control Housing Cover
Material: Cast Iron

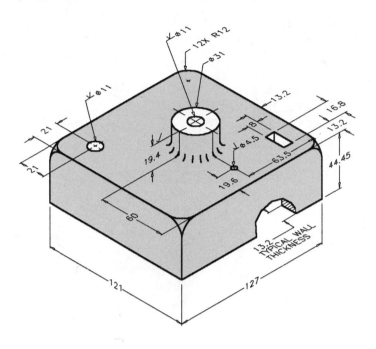

Problem 10–25   Dimensioning compound circles and arcs

Part Name: Multiple Shaft Support
Material: SAE 1030
Note: Fillets and Rounds R.08

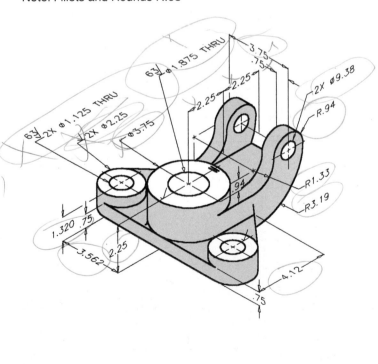

Problem 10–26 Dimensioning auxiliary views (in.)

Part Name: Connector
Material: SAE 4320
Finish All Over 63 μin.

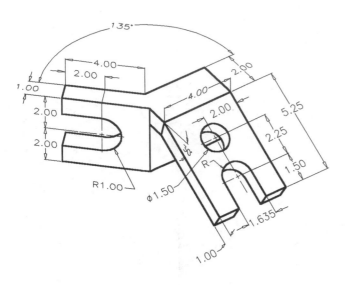

Problem 10–27 Dimensioning small spaces (in.)

Part Name: Guide Bracket
Material: SAE 4020
Fillets and Rounds: R.125

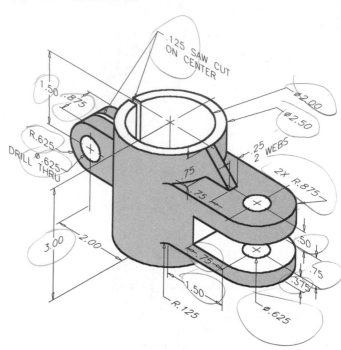

Problem 10–28 Chart drawing (metric)

Part Name: Chart Plate
Material: 10 mm Brass

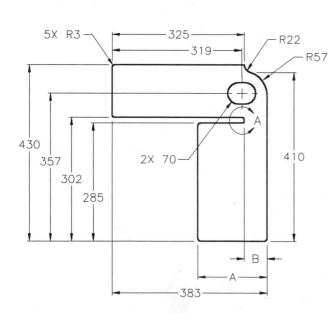

VIEW A

PART NO.	DIM A	DIM B
342142	172	75
342143	140	60
342144	108	45
342145	76	30
342146	44	15

Problem 10–29 Chart drawing (in.)

Part Name: Tank Bracket
Material: 11 GA A570-30
Problem based on original art courtesy TEMCO

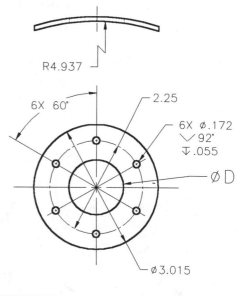

1.574	1.448	1.322	D DIA

Problem 10–30 Arrowless dimensioning (in.)

Part Name: Metering Box Gasket
Material: Neoprene
Notes: Dimensions have been corrected to allow for steel rule blade.
SPECIFIC INSTRUCTIONS:
Convert the engineering sketch to arrowless tabular dimension-

ing. Set up a table that identifies hole diameters, arc radii, and location dimensions from x and y coordinates. Dimensions above Y are + dimensions and below Y are – dimensions. Dimensions to the right of X are + dimensions and to the left of X are – dimensions.
Problem based on original art courtesy Vellumoid Inc.

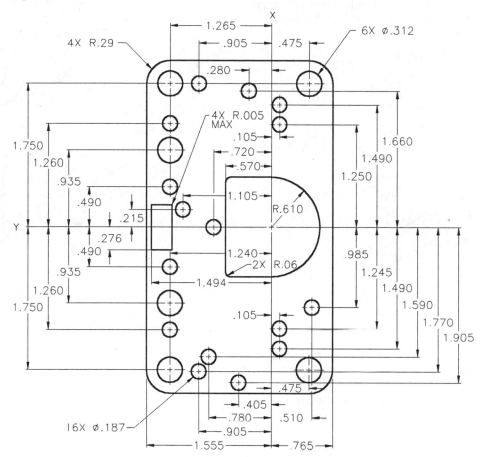

Problem 10–31 Casting drawing (metric)

Part Name: Slider
Material: ASTM 60 CI
Hardness: Brinell 180-220
SPECIFIC INSTRUCTIONS:
Draw part as finished. Include all machining and casting dimensions on one detail drawing. The pattern maker will apply unspecified allowances during production.

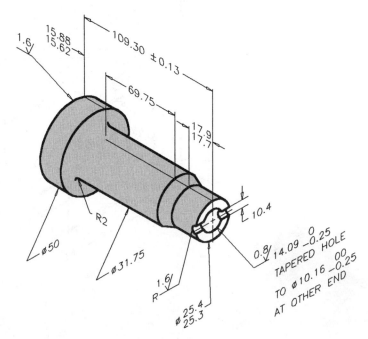

Problem 10–32 Forging and machining drawings (metric)

Part Name: Pump Pivot Support
Material: HRMS (Hot-rolled Mild Steel)
SPECIFIC INSTRUCTIONS:
Prepare two drawings: one showing only forging-related views, dimensions, and notes; and the other showing only machining-related views, dimensions, and notes. Provide the draft angles recommended for steel (refer to chapter). Add 3 mm to forging where finish surface identified. Do not draw thread symbol or thread note (M12 × 1.75).

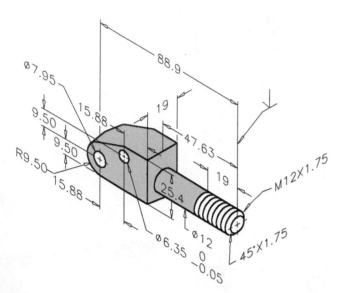

Problem 10–33 Angles, holes, and arcs (in.)

Part Name: Mounting Bracket
Material: SAE 1020
Fillets: R.13

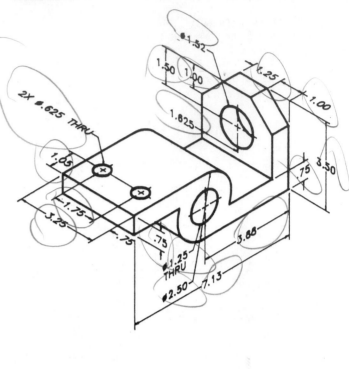

Problem 10–34 Contours, machine features, and limited space (in.)

Part Name: Support Base
Material: SAE 1040

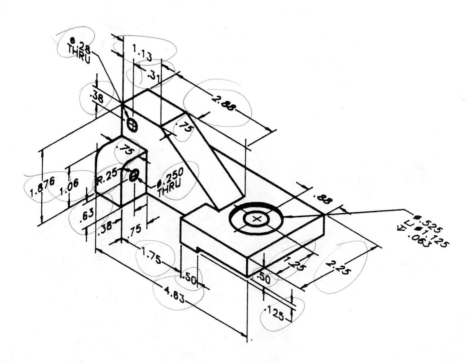

Problem 10–35 CAD/CAM (in.)

Part Name: Spacer
Material: SAE 1030

JOB PLAN			
Tool Type	Tool No.	Diameter	CAD Layer
Drill	1	.250	1
Drill	2	.375	2
End Mill	3	.500	3

Problem 10–36 CAD/CAM (in.)

Part Name: Cover Plate
Material: .500 THK Aluminum

JOB PLAN				
Tool Type	Tool No.	Diameter	Length	CAD Layer
End Mill	1	.750	1.000	1
End Mill	2	.500	1.000	2
End Mill	3	.250	1.000	3
Spot Drill	4	.125	1.000	4
Drill	5	.500	1.000	5
Drill	6	1.000	1.000	6
Clamps				7

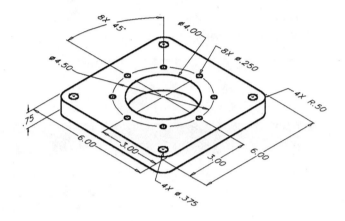

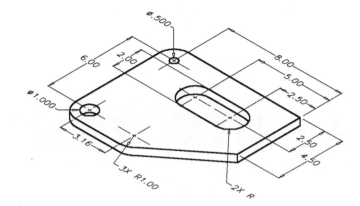

MATH PROBLEMS

Find the distance between these points on a 2-D drawing:

1. (5, 7) and (8, 9)
2. (3, –4) and (7, 10)
3. (–2, –5) and (8, 3)
4. (6, 4) and (6, 12)

Find the distance between these points on a 3-D drawing:

5. (0, 0, 0) and (5, 7, 9)
6. (1, 2, 3) and (12, 18, 20)
7. (–1, 0, 6) and (1, 3, 9)
8. (6, 12, 3) and (–2, 12, 8)

CHAPTER 11

Fasteners and Springs

LEARNING OBJECTIVES

After completing this chapter you will:

○ Draw screw thread representations and provide correct thread notes.
○ Prepare drawings for fastening devices.
○ Draw completely dimensioned spring representations.
○ Prepare formal drawings from engineer's sketches and actual industrial layouts.

THE ENGINEERING DESIGN PROCESS

You are asked to make a drawing from engineer's notes for a fastening device that is needed for one of the products the engineer is designing. The engineer's notes look like Figure 11–1. You may be thinking:

○ A 10-gage shaft threaded with 32–UNF–2A. Maybe you can use a 10–32UNF–2A threaded rod and cut to .375 lengths.
○ Then you need to chamfer both ends with .015 × 35° and provide a slot in one end that is .030 wide and .047 deep.

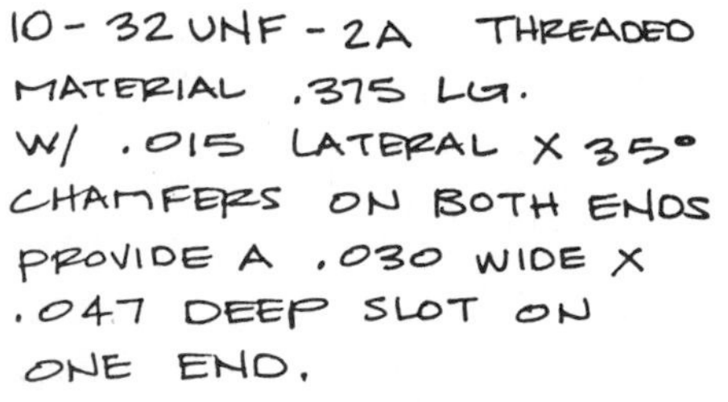

> 10 - 32 UNF - 2A   THREADED
> MATERIAL .375 LG.
> W/ .015 LATERAL X 35°
> CHAMFERS ON BOTH ENDS
> PROVIDE A .030 WIDE X
> .047 DEEP SLOT ON
> ONE END.

Figure 11–1   Engineer's notes.

So, you go to work preparing a drawing like the one shown in Figure 11–2. The point here is that the engineer provided a lot of information in the thread note. In fact, standard screws and other fasteners may be completely described without a drawing. A written specification can be used to completely describe an object. For example, 1/2–13UNC–2 × 1.5 LG FILLISTER HEAD MACHINE SCREW.

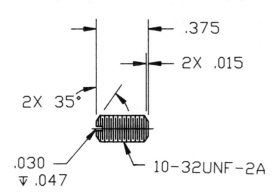

Figure 11–2   CADD drawing from engineer's notes.

ANSI   This chapter introduces you to the methods of calling for and drafting fasteners and springs. Fasteners include screw threads, keys, pins, rivets, and weldments. There are two types of springs, helical and flat. The American National Standards Institute documents that govern the standards for fasteners and springs are *Screw Thread Representation* ANSI Y14.6, *Screw Thread Representation (Metric Supplement)* ANSI Y14.6aM, *Symbols for Welding and Nonde-* *structive Testing Including Brazing* ANSI/AWS A2.4, and *Mechanical Spring Representation* ANSI Y14.13M.

▼ SCREW-THREAD FASTENERS

The standardization of screw threads was achieved among the United States, United Kingdom, and Canada in 1949. A need for interchangeability of screw-thread fasteners was the purpose of this standardization and resulted in the Unified Thread Series. The Unified Thread Series is now the American standard for screw threads. Prior to 1949 the United States standard was the American National screw threads. The unification standard occurred as a result of combining some of the characteristics of the American National screw threads with the United Kingdom's long accepted Whitworth screw threads. Screw thread systems were revised again in 1974 for metric application. The modifications were minor and based primarily on metric translation. In order to emphasize that the Unified screw threads evolve from inch calibrations, the term Unified Inch Screw Threads is used while the term Unified Screw Threads Metric Translation is used for the metric conversion.

Screw threads are a helix or conical spiral form on the external surface of a shaft or internal surface of a cylin-drical hole, as shown in Figures 11–3 and 11–4. Screw threads are used for an unlimited number of services, such as for holding parts together as fasteners, for leveling and adjusting objects, and for transmitting power from one object or feature to another.

Screw-Thread Terminology

Refer to Figure 11–3 and Figure 11–4 as a reference for the following definitions related to external and internal threads.

Axis. The thread axis is the centerline of the cylindrical thread shape.

Body. That portion of a screw shaft that is left un-threaded.

Chamfer. An angular relief at the last thread to help allow the thread to more easily engage with a mating part.

Classes of Threads. A designation of the amount of tolerance and allowance specified for a thread.

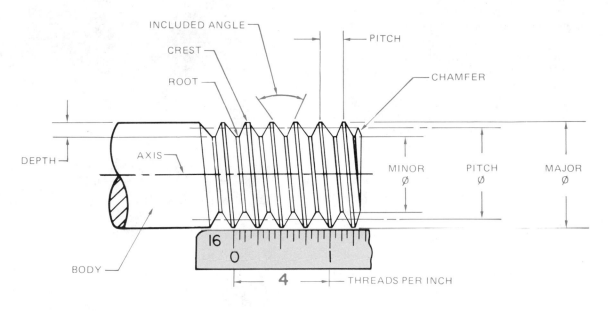

Figure 11–3 External screw thread components.

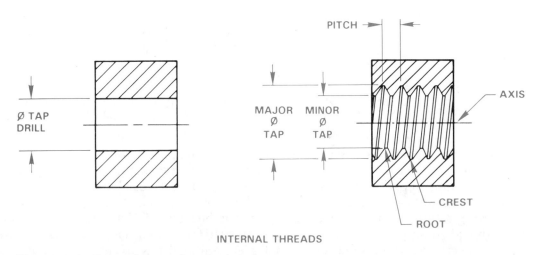

Figure 11–4 Internal screw thread components.

Crest. The top of external and the bottom of internal threads.

Depth of Thread. Depth is the distance between the crest and the root of a thread, measured perpendicular to the axis.

Die. A machine tool used for cutting external threads.

Fit. Identifies a range of thread tightness or looseness.

Included Angle. The angle between the flanks (sides) of the thread.

Lead. The lateral distance a thread travels during one complete rotation.

Left-hand Thread. A thread that engages with a mating thread by rotating counterclockwise, or with a turn to the left when viewed toward the mating thread.

Major Diameter. The distance on an external thread from crest to crest through the axis. For an internal thread the major diameter is measured from root to root across the axis.

Minor Diameter. The dimension from root to root through the axis on an external thread and measured across the crests through the center for an internal thread.

Pitch. The distance measured parallel to the axis from a point on one thread to the corresponding point on the adjacent thread.

Pitch Diameter. A diameter measured from a point halfway between the major and minor diameter through the axis to a corresponding point on the opposite side.

Right-hand Thread. A thread that engages with a mating thread by rotating clockwise, or with a turn to the right when viewed toward the mating thread.

Root. The bottom of external and the top of internal threads.

Tap. A tap is the machine tool used to form an interior thread. Tapping is the process of making an internal thread.

Tap Drill. A tap drill is used to make a hole in metal before tapping.

Thread. The part of a screw thread represented by one pitch.

Thread Form. The design of a thread determined by its profile.

Thread Series. Groups of common major diameter and pitch characteristics determined by the number of threads per inch.

Threads per Inch. The number of threads measured in one inch. The reciprocal of the pitch in inches.

▼ THREAD-CUTTING TOOLS

A tap is a machine tool used to form an internal thread as shown in Figure 11–5. A die is a machine tool used to form external threads. (See Figure 11–6.)

A tap set is made up of a taper tap, a plug tap, and a bottoming tap as shown in Figure 11–7. The taper tap is generally used for starting a thread. The threads are tapered to within ten threads from the end. The tap is tapered so the tool more evenly distributes the cutting

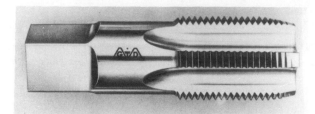

Figure 11–5 Tap. *Courtesy Greenfield Tap & Die, Division of TRW, Inc.*

Figure 11–6 Die. *Courtesy The Cleveland Twist Drill, an Acme-Cleveland Company.*

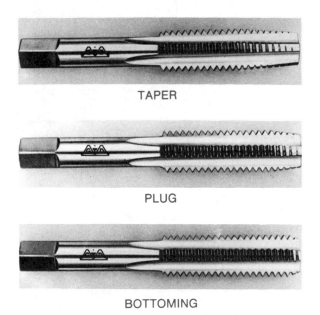

TAPER

PLUG

BOTTOMING

Figure 11–7 Tap set includes taper, plug, and bottoming taps. *Courtesy Greenfield Tap & Die, Division of TRW, Inc.*

edges through the depth of the hole. The plug tap has the threads tapered to within five threads from the end. The plug tap can be used to completely thread through material or thread a *blind hole* (a hole that does not go through the material) if full threads are not required all the way to the bottom. The bottoming tap is used when threads are needed to the bottom of a blind hole.

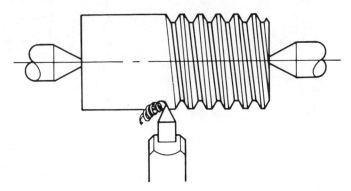

Figure 11–8   Thread cutting on a lathe.

The die is a machine tool used to cut external threads. Thread cutting dies are available for standard thread sizes and designations.

External and internal threads may also be cut on a lathe. A lathe is a machine that holds a piece of material between two centers or in a chucking device. The material is rotated as a cutting tool removes material while traversing along a carriage which slides along a bed. Figure 11–8 shows how a cutting tool can make an external thread.

▼ THREAD FORMS

Unified threads are the most common threads used on threaded fasteners. Figure 11–9 shows the profile of a Unified thread.

American National threads, shown in profile in Figure 11–10, are similar to the Unified thread but have a flat root. Still in use today, the American National thread generally replaced the sharp-V thread form.

The *sharp-V thread,* although not commonly used, is a thread that will fit and seal tightly. It is difficult to manufacture since the sharp crests and roots of the threads are easily damaged. (See Figure 11–11.) The sharp-V thread was the original United States standard thread form.

Metric thread forms vary slightly from one European country to the next. The International Organization for Standardization (ISO) was established to standardize metric screw threads. The ISO thread specifications are similar to the Unified thread form. (See Figure 11–12.)

Whitworth threads are the original British standard thread forms developed in 1841. These threads have been referred to as parallel screw threads. The Whitworth thread forms are primarily being used for replacement parts. (See Figure 11–13.)

Square thread forms, shown in Figure 11–14, have a longer pitch than Unified threads. Square threads were developed as threads that would effectively transmit power; however, they are difficult to manufacture because of their perpendicular sides. These are modified square threads with ten-degree sides. The square thread is generally replaced by Acme threads.

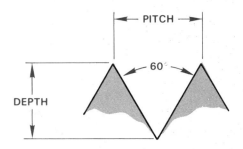

Figure 11–11   Sharp-V thread form.

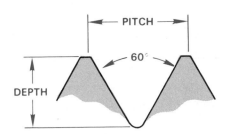

Figure 11–9   Unified thread form.

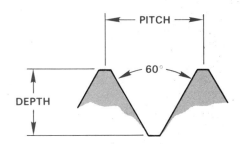

Figure 11–12   Metric thread form.

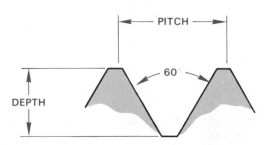

Figure 11–10   American National thread form.

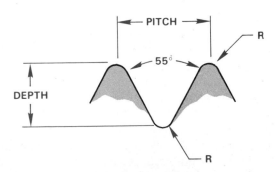

Figure 11–13   Whitworth thread form.

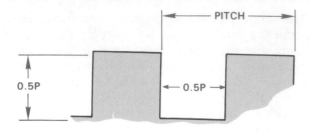

Figure 11–14   Square thread form.

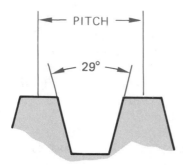

Figure 11–15   Acme thread form.

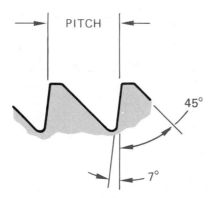

Figure 11–16   Buttress thread form.

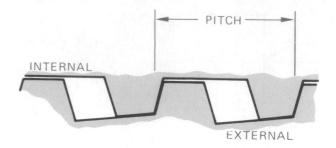

Figure 11–17   Dardelet self-locking thread form.

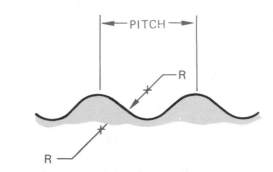

Figure 11–18   Rolled thread form.

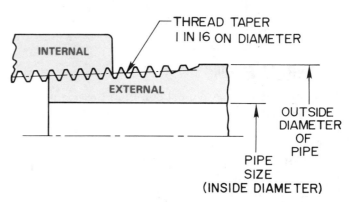

Figure 11–19   American National Standard taper pipe thread form.

Acme thread forms are commonly used when rapid traversing movement is a design requirement. Acme threads are popular on such designs as screw jacks, vice screws, and other equipment and machinery that requires rapid screw action. A profile of the Acme thread form is shown in Figure 11–15.

Buttress threads are designed for applications where high stress occurs in one direction along the thread axis. The thread flank or side that distributes the thrust or force is within 7° of perpendicularity to the axis. This helps reduce the radial component of the thrust. The buttress thread is commonly used in situations where tubular features are screwed together and lateral forces are exerted in one direction. (See Figure 11–16.)

Dardelet thread forms are primarily used in situations where a self-locking thread is required. These threads resist vibrations and remain tight without auxiliary locking devices. (See Figure 11–17.)

Rolled thread forms are used for screw shells of electric sockets and lamp bases. (See Figure 11–18.)

American National Standard taper pipe threads are the standard threads used on pipes and pipe fittings. These threads are designed to provide pressure-tight joints or not, depending on the intended function and materials used. American pipe threads are measured by the *nominal pipe size,* which is the inside pipe diameter. For example, a ½-in. pipe size has an outside pipe diameter of .840 in. (See Figure 11–19.)

▼ THREAD REPRESENTATIONS

There are three methods of thread representation in use: detailed, schematic, and simplified, as shown in Figure 11–20. The *detailed* representation may be used in special situations that require a pictorial display of threads such as in a sales catalog or a display drawing. Detailed threads are not common on most manufactur-

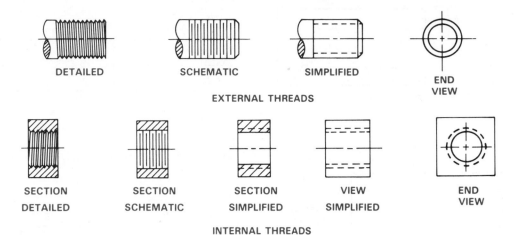

Figure 11–20 Thread representations.

ing drawings since they are much too time-consuming to draw. *Schematic* representations are also not commonly used in industry. Although they do not take the time of detailed symbols, they do require extra time to draw. Some companies, however, may continue to use the schematic thread representation.

The actual use and purpose of the drawing helps determine which thread symbol to use. It may even be possible to mix representations on a particular drawing if clarity is improved. The simplified representation is the most common method of drawing thread symbols. *Simplified* representations clearly describe threads and they are easy and quick to draw. They are also very versatile as they can be used in all situations, while the other representations cannot be used in all situations. Figure 11–21 shows simplified threads in different applications.

Also, notice how the use of a thread chamfer slightly changes the appearance of the thread. Chamfers are commonly applied to the first thread to help start a thread in its mating part.

When an internal screw thread does not go through the part, it is common to drill deeper than the depth of the required thread when possible. This process saves time and reduces the chance of breaking a tap. The thread may go to the bottom of a hole but to produce it requires an extra process using a bottoming tap. Figure 11–22 shows a simplified representation of a thread that does not go through. The bolt should be shorter than the depth of thread so the bolt does not hit bottom. Notice in Figure 11–22 the hidden lines representing the major and minor thread diameters are spaced far enough apart to be clearly separate. This spacing is important because on some threads the difference between the major and minor diameters is very small and, if drawn as they actually appear, the lines would run together. The hidden-line dashes are also drawn staggered for clarity. Figure 11–23 shows a bolt fastener as it would appear drawn in assembly with two parts using simplified thread representation.

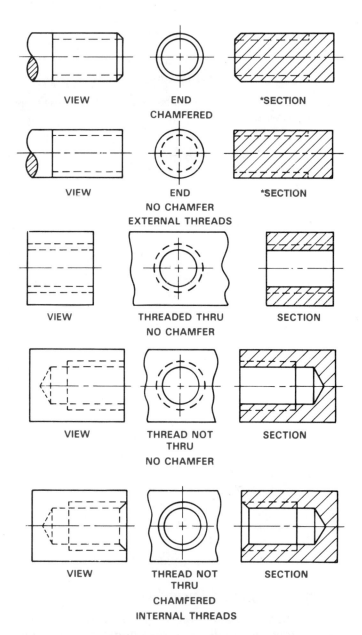

Figure 11–21 Simplified thread representations. *Threaded shafts are not sectioned unless there is a need to expose an internal feature.

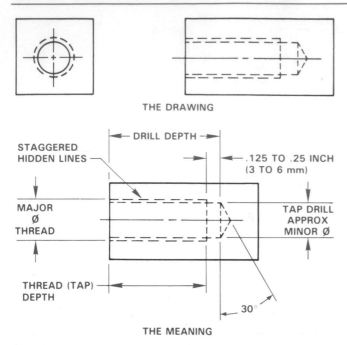

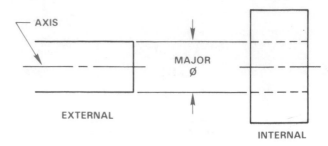

Figure 11–24 Step 1, the simplified major thread diameter.

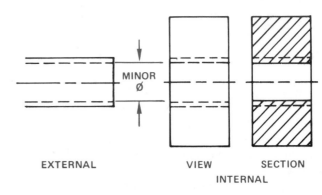

Figure 11–25 Step 2, the simplified minor diameter.

Figure 11–22 The simplified internal thread that does not go through the part.

Step 1. Draw the major thread diameter as an object line for external threads and hidden line for internal threads as in Figure 11–24.

Step 2. Draw the minor thread diameter, which is about equal to the tap drill size (found in a tap drill chart). If the minor diameter and major diameter are too close together, then slightly exaggerate the space. The minor diameter is a hidden line for the external thread and a hidden line staggered with the major diameter lines for the internal thread. The minor diameter is an object line for the internal thread in section. (See Figure 11–25.)

Drawing Schematic Threads

Schematic thread representations are drawn to approximate the appearance of threads by spacing lines equal to the pitch of the thread. When there are too many threads per inch to draw easily, then the distance can be exaggerated for clarity. The following steps show how to draw schematic thread representations.

Step 1. Draw the major diameter of the thread. Schematic symbols can only be drawn in section for internal threads. Then lay out the number of threads per inch (if convenient and space is available) using a thin line at each space. Figure 11–26 uses eight threads per inch.

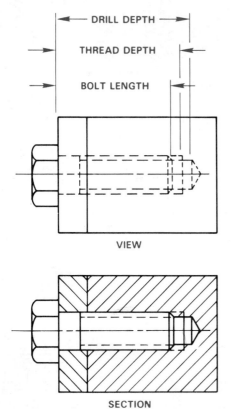

Figure 11–23 The simplified external thread in assembly.

Drawing Simplified Threads

Simplified representations are the easiest thread symbols to draw and are the most commonly used in industry. The following steps show how to draw simplified threads.

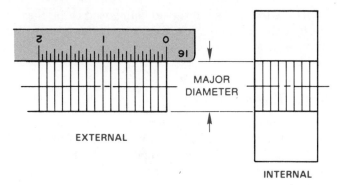

Figure 11–26 Step 1, the schematic thread.

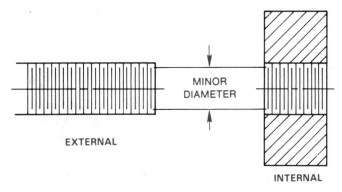

Figure 11–27 Step 2, the schematic thread.

Step 2. Draw a thick line equal in length to the minor diameter between each pair of thin lines drawn in Step 1. (See Figure 11–27.)

Drawing Detailed Threads

Detailed thread representations are the most difficult and time consuming thread symbols to draw. They may be necessary for some applications as they most closely approximate the actual thread. Detailed external and sectioned threads may be drawn but detailed internal threads may not be drawn in multiview. Detailed internal threads may only be drawn in section. The following steps show how to draw detailed thread representations for external and internal threads in section.

Step 1. Use construction lines to lightly draw the major and minor diameters of the thread as shown in Figure 11–28.

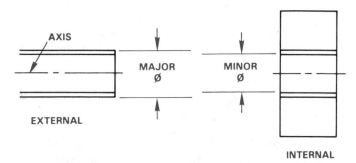

Figure 11–28 Step 1, the detailed thread.

Step 2. Divide one edge of the thread into equal parts. In this case, eight threads per inch so the pitch is .125 in., as shown in Figure 11–29. If the pitch is too small (much less than .125 in.) then exaggerate the distance. Remember, these symbols are representations and while they should have an appearance close to the actual thread, they do not have to be exact if the result is too difficult and time consuming to draw.

Step 3. Stagger the opposite side one-half pitch and draw parallel thin lines equal to the spaces established in Step 2. (See Figure 11–30.)

Step 4. Draw Vs at 60° to form the root and crest of each thread. (See Figure 11–31.)

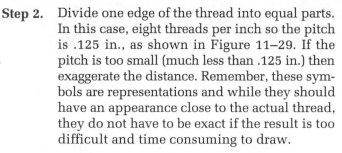

Figure 11–29 Step 2, the detailed thread.

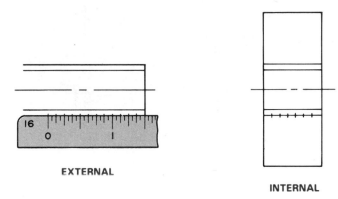

Figure 11–30 Step 3, the detailed thread.

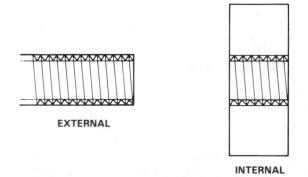

Figure 11–31 Step 4, the detailed thread.

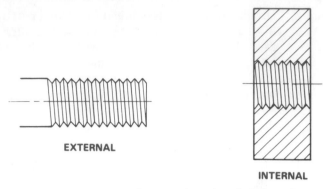

EXTERNAL

INTERNAL

Figure 11–32 Step 5, the complete detailed thread representation.

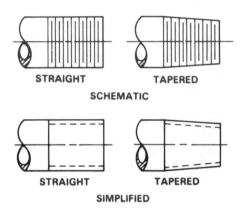

STRAIGHT   TAPERED

SCHEMATIC

STRAIGHT   TAPERED

SIMPLIFIED

Figure 11–33 Straight and tapered pipe thread representations.

Step 5. Complete the detailed thread representation by connecting the roots of opposite threads by drawing parallel lines as shown in Figure 11–32. Accuracy is very important if detailed threads are to turn out satisfactory.

Detailed thread representations may be used to draw any thread form using the same steps previously shown. The difference occurs in drawing the profile of the thread. For example, Vs are used to draw Unified, sharp-V, American Standard, or metric threads. The profile changes for other threads. The detailed thread drawings for American National Standard taper pipe threads are the same as Unified threads except that the major and minor diameters taper at a rate of .0625 in. per inch. Figure 11–33 shows that pipe threads may be drawn tapered or straight depending on company preference. The thread note clearly defines the type of thread.

▼ THREAD NOTES

Simplified, schematic, and detailed thread representations clearly show where threads are displayed on a drawing. However, the representations alone do not give the full information about the thread. As the term representation implies, the symbols are not meant to be exact but they are meant to describe the location of a thread when used. The information that clearly and completely identifies the thread being used is the *thread note.* The thread note must always be in the same order and be accurate, otherwise the thread may be manufactured incorrectly.

Metric Threads

ISO The metric thread notes shown below is the recommended standard as specified by the International Organization for Standardization (ISO). The note components are described as follows:

M   10 X 1.5—6H
(A)   (B)   (C)   (D)   (E) (F)

(A)   M is the symbol for ISO metric threads.
(B)   The nominal major diameter in milllimeters, followed by the symbol X meaning, *by.*
(C)   The thread pitch in millimeters, followed by a dash (—).
(D)   The number may be a 3, 4, 5, 6, 7, 8, or 9, which identifies the grade of tolerance from fine to coarse. The larger the number, the larger the tolerance. Grades 3 through 5 are fine, and 7 through 9 are coarse. Grade 3 is very fine and grade 9 is very coarse. Grade 6 is the most commonly used and is the medium tolerance metric thread. The grade 6 metric thread is comparable to the class 2 Unified Screw Thread. A letter placed after the number gives the thread tolerance class of the internal or external thread. Internal threads are designated by upper-case letters, such as *G* or *H,* where *G* means a tight allowance and *H* identifies an internal thread with no allowance. The term *allowance* refers to the tightness of fit between the mating parts. External threads are defined with lower-case letters such as *e, g,* or *h.* For external threads, *e* denotes a large allowance, *g* is a tight allowance, and *h* establishes no allowance. Grades and tolerances below 5 are intended for tight fits with mating parts; those above 7 are a free class of fit intended for quick and easy assembly. When the grade and allowance are the same for both the major diameter and the pitch diameter of the metric thread, then the designation is given as shown, 6H. In some situations where precise tolerances and allowances are critical between the major and pitch diameters, separate specifications could be used, for example 4H 5H, or 4g 5g, where the first group (4g) refers to the grade and allowance of the pitch diameter, and the second group (5g) refers to the grade and tolerance of the major diameter. A fit between a pair of threads is indicated in the same thread note by specifying the internal thread followed by the external thread specification separated by a slash, for example, 6H/6g.

(E) A blank space at (E) denotes a right-hand thread, a thread that engages when turned to the right. A right-hand thread is assumed unless an LH is lettered in this space. LH, which describes a left-hand thread, must be specified for a thread that engages when rotated to the left.

(F) The depth of internal threads or the length of external threads in millimeters is provided at the end of the note. When the thread goes through the part, this space is left blank although some companies prefer to letter the description, THRU.

Unified and American National Threads

The thread note shall always be drawn in the order shown. The components of the note are described as follows:

.50 — 13 UNC — 2 A
(A)   (B) (C)   (D)(E)(F)(G) (H)

(A) The major diameter of the thread in inches followed by a dash (—).

(B) Number of threads per inch.

(C) Series of threads are classified by the number of threads per inch as applied to specific diameters and thread forms, such as coarse or fine threads. UNC (in the example) means Unified National Coarse. Others include UNF for Unified National Fine, UNEF for Unified National Extra Fine, or UNS for Unified National Special. The UNEF and UNS thread designations are for special combinations of diameter, pitch, and length of engagement. American National screw threads are identified with UN for external and internal threads, or UNR, a thread designed to improve fatigue strength of external threads only. The series designation is followed by a dash (—).

(D) Class of fit is the amount of tolerance. 1 means a large tolerance, 2 is a general-purpose moderate tolerance, and 3 is for applications requiring a close tolerance.

(E) A means an external thread (shown in the example) while B means an internal thread. (B replaces A in this location.) The A or B may be omitted if the thread is clearly external or internal, as shown on the drawing.

(F) A blank space at (F) means a right-hand thread. A right-hand thread is assumed. LH in this space identifies a left-hand thread.

(G) A blank space at (G) identifies a thread with a single lead, that is, a thread that engages one pitch when rotated 360°. If a double or triple lead is required, then the word DOUBLE or TRIPLE must be lettered here.

(H) This location is for internal thread depth or external thread length in inches. When the drawing clearly shows that the thread goes through, this space is left blank. If clarification is needed, then the word THRU may be lettered here.

Other Thread Forms

Other thread forms, such as Acme, are noted on a drawing using the same format. For example, 5/8 — 8 ACME — 2G describes an Acme thread with a ⅝-in. major diameter, 8 threads per inch, and a general purpose (G) class 2 thread fit. For a complete analysis of threads and thread forms refer to the *Machinery's Handbook* published by Industrial Press, Inc.

American National Standard taper pipe threads are noted in the same manner with the letters NPT used to designate the thread form. A typical note may read 3/4—14 NPT.

Thread Notes on a Drawing

The thread note is usually applied to a drawing with a leader in the view where the thread appears as a circle for internal threads as shown in Figure 11–34. External threads may be dimensioned with a leader as shown in Figure 11–35, with the thread length given as a dimension or at the end of the note. An internal thread that does not go through the part may be dimensioned as in Figure 11–36. Some companies may require the drafter to indicate the complete process required to machine a thread including noting the tap drill size, tap drill depth if not through, the thread note, and thread depth if not

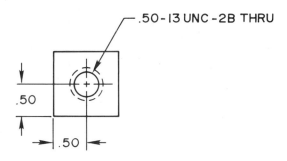

UNIFIED SCREW THREAD

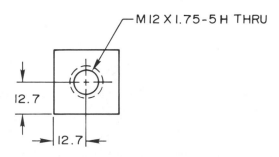

METRIC SCREW THREAD

Figure 11–34 Drawing and noting internal screw threads (simplified representation).

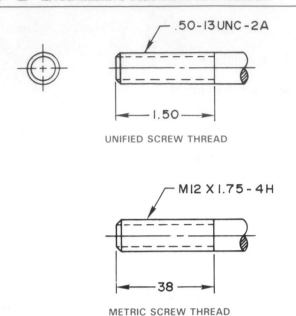

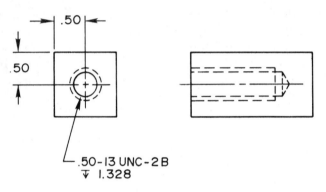

Figure 11–35   Drawing and noting external screw threads.

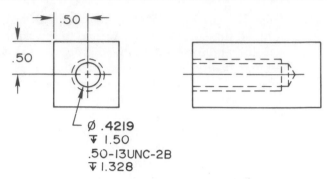

Figure 11–37   Showing tap drill depth and thread depth.

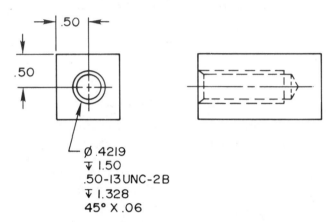

Figure 11–38   Showing tap drill and thread depth with a chamfer.

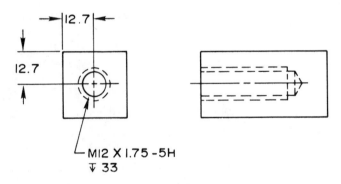

Figure 11–36   Drawing and noting internal screw threads with a given depth.

through. (See Figure 11–37.) A thread chamfer may also be specified in the note as shown in Figure 11–38.

Many companies require only the thread note and depth. The complete process is determined in manufacturing.

▼ MEASURING SCREW THREADS

When measuring features from prototypes or existing parts, the screw thread size can be determined on a fastener or threaded part by measurement. Measure the major diameter with a vernier caliper or micrometer. Determine the number of threads per inch when a rule or scale is the only available tool by counting the number of threads between inch graduations. The quickest and easiest way to determine the thread specification is with a screw pitch gage, which is a set of thin leaves with teeth on the edge of each leaf that correspond to standard thread sections. Each leaf is stamped to show the number of threads per inch. Therefore, if the major diameter measures .625 in. and the number of teeth per inch is 18, then by looking at a thread variation chart you find that you have a 5/8—18 UNF thread.

▼ THREADED FASTENERS
Bolts and Nuts

A *bolt* is a threaded fastener with a head on one end and is designed to hold two or more parts together with a nut or threaded feature. The *nut* is tightened upon the bolt or the bolt head may be tightened into a threaded feature. Bolts can be tightened or released by torque

CADD Applications

DRAWING FASTENERS

The use of simplified thread representations in manual drafting is mostly due to ease and time savings, while schematic thread representations are used to add a little realism to drawings. Detailed thread representations, on the other hand, are usually avoided unless a specific reason requires this time-consuming drawing method. With CADD the drafter has any type of thread representation available in an instant. With this flexibility CADD drafters often select the detailed thread because it makes a drawing look more realistic and artistic. However, choosing simplified or schematic thread representations still make sense from an economic standpoint, because screen regeneration and plotting time are less. CADD fastener software allows the user to select the following variables:

○ Type of screw, such as machine, cap, or self-tapping.
○ Type of nut, if an internal threaded feature.
○ Type of head, if any.
○ Type of thread representation: simplified, schematic, or detailed.
○ Threading of holes or studs is also an option.

After selecting these variables, the drafter gives the thread specifications. The software uses parametrics based on these variables to draw and label the exact thread required in a fraction of the time it would take to manually draw the same fastener. A sample CADD fastener overlay template is shown in Figure 11–39.

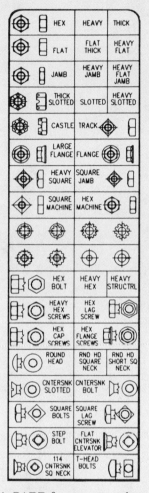

Figure 11–39 A CADD fastener template overlay symbol library. *Courtesy Chase Systems.*

applied to the head or to the nut. Bolts are identified by a thread note, length, and head type. For example, 5/8—11 UNC—2 X 1 1/2 LONG HEXAGON HEAD BOLT. Figure 11–40 shows various types of bolt heads. Figure 11–41 shows common types of nuts. Nuts are classified by thread specifications and type. Nuts are available with a flat base or a washer face.

Machine Screws

Machine screws are a thread fastener used for general assembly of machine parts. Machine screws are available in coarse (UNC) and fine (UNF) threads, in diameters ranging from .060 in. to .5 in., and in lengths from ⅛ in. to 3 in. Machine screws are specified by thread, length, and head type. Machine screws have no chamfer. There are several types of heads available for machine design flexibility. (See Figure 11–42.)

Cap Screws

Cap screws are fine-finished machine screws that are generally used without a nut. Mating parts are fastened where one feature is threaded. Cap screws have a variety of head types and range in diameter from .060 in. to 4 in., with a large range of lengths. Lengths vary with diameter, for example, lengths increase in ¹⁄₁₆ in. increments for diameters up to 1 in. For diameters larger than 1 in., lengths increase in increments of ⅛ in. or ¼ in.; the other extreme is a 2-in. increment for lengths over 10 in. Cap screws have a chamfer to the depth of the first thread. Standard cap screw head types are shown in Figure 11–43.

Set Screws

Set screws are used to help prevent rotary motion and to transmit power between two parts such as a pul-

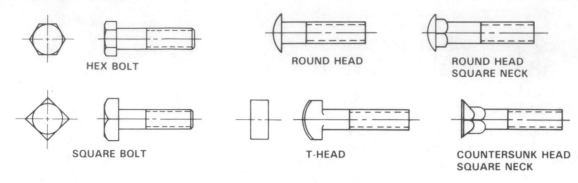

HEX BOLT

ROUND HEAD

ROUND HEAD
SQUARE NECK

SQUARE BOLT

T-HEAD

COUNTERSUNK HEAD
SQUARE NECK

Figure 11–40   Bolt head types.

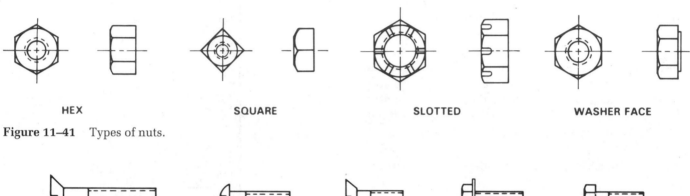

HEX

SQUARE

SLOTTED

WASHER FACE

Figure 11–41   Types of nuts.

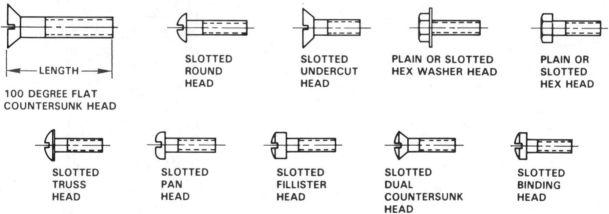

LENGTH

100 DEGREE FLAT
COUNTERSUNK HEAD

SLOTTED
ROUND
HEAD

SLOTTED
UNDERCUT
HEAD

PLAIN OR SLOTTED
HEX WASHER HEAD

PLAIN OR
SLOTTED
HEX HEAD

SLOTTED
TRUSS
HEAD

SLOTTED
PAN
HEAD

SLOTTED
FILLISTER
HEAD

SLOTTED
DUAL
COUNTERSUNK
HEAD

SLOTTED
BINDING
HEAD

Figure 11–42   Types of machine screw heads.

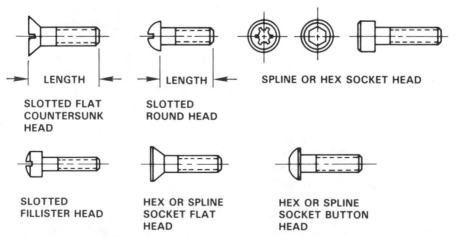

LENGTH

LENGTH

SPLINE OR HEX SOCKET HEAD

SLOTTED FLAT
COUNTERSUNK
HEAD

SLOTTED
ROUND HEAD

SLOTTED
FILLISTER HEAD

HEX OR SPLINE
SOCKET FLAT
HEAD

HEX OR SPLINE
SOCKET BUTTON
HEAD

Figure 11–43   Cap screw head styles.

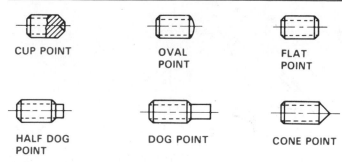

Figure 11–44 Set screw point styles.

ley and shaft. Purchased with or without a head, set screws are ordered by specifying thread, length, head or headless, and type of point. Headless set screws are available in slotted and with hex or spline sockets. The shape of a set screw head is usually square. Standard square-head set screws have cup points, although other points are available. Figure 11–44 shows optional types of set screw point styles.

Lag Screws and Wood Screws

Lag screws are designed to attach metal to wood or wood to wood. Before assembly with a lag screw, a pilot hole is cut into the wood. The threads of the lag screw then form their own mating thread in the wood. Lag screws are sized by diameter and length. Wood screws are similar in function to lag screws and are available in a wide variety of sizes, head styles, and materials.

Self-tapping Screws

Self-tapping screws are designed for use in situations where the mating thread is created by the fastener. These screws are used to hold two or more mating parts when one of the parts becomes a fastening device. A clearance fit is required through the first series of features or parts while the last feature receives a pilot hole similar to a tap drill for unified threads. The self-tapping screw then forms its own threads by cutting or displacing material as it enters the pilot hole. There are several different types of self-tapping screws with head variations similar to cap screws. The specific function of the screw is important as these screws may be designed for applications ranging from sheet metal to hard metal fastening.

Thread Inserts

Screw thread inserts are helically formed coils of diamond-shaped wire made of stainless steel or phosphor bronze. The inserts are used by being screwed into a threaded hole to form a mating internal thread for a threaded fastener. Inserts are used to repair worn or damaged internal threads and to provide a strong thread surface in soft materials. Some screw thread inserts are

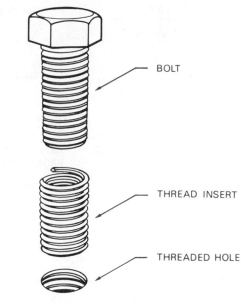

Figure 11–45 Thread insert.

designed to provide a secure mating of fasteners in situations where vibration or movement could cause parts to loosen. Figure 11–45 shows the relationship among the fastener, thread insert, and tapped hole.

How to Draw Various Types of Screw Heads

As you have found from the previous discussion, screw fasteners are classified in part by head type, and there is a large variety of head styles. A valuable drafting reference is the *Machinery's Handbook,* which clearly lists specifications for all common head types.

Hexagon head fasteners are generally drawn with the hexagon positioned across the corners vertically in the front view. When projected across the flats the hex head looks like a square head. The following steps show how to draw a hex head bolt. Hex nuts are drawn in the same manner.

Step 1. Use construction lines to draw the end view of the hexagon with the distance across the flats. A ¾-in. nominal thread measures 1⅛ in. across the flats. Position the bolt head so a front view projection is across the corners as shown in Figure 11–46.

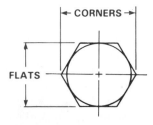

Figure 11–46 Step 1, the flats and corners of a hexagon.

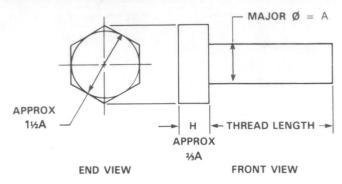

Figure 11–47 Step 2, layout with construction lines.

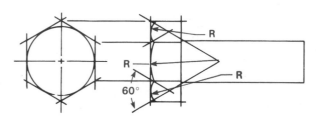

Figure 11–48 Step 3, hex head layout.

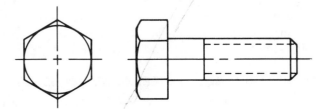

Figure 11–49 Step 4, completed drawing.

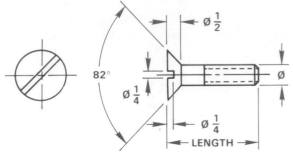

FLAT HEAD CAP SCREW

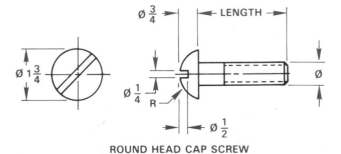

ROUND HEAD CAP SCREW

HEX SOCKET HEAD CAP SCREW

Figure 11–50 Layout specifications for common cap screws.

Step 2. Block out the major diameter and length, and bolt head height. The same ¾-in. hex bolt has a head height, H, of ¹¹⁄₃₂ in. (See Figure 11–47.)

Step 3. Project the hexagon corners to the front view. Then establish radii, *R*, centers with 60° angles as shown in Figure 11–48. Draw the radii.

Step 4. Draw a 30° chamfer tangent on each side to the small radius arcs. Use object lines to complete both views and draw the thread representation. (See Figure 11–49.)

Square-head bolts are drawn in the same manner as hex-head bolts.

A few common cap screw heads are shown with approximate layout dimensions in Figure 11–50.

Templates

When bolt or screw heads are drawn only occasionally, then it may be satisfactory to use the layout techniques. However, whether you draw bolt or screw heads a few times or many times each day, it is always better to use templates. There is a large variety of templates designed to allow you to quickly and easily draw bolt or screw heads and detailed screw threads.

Nuts

A nut is used as a fastening device in combination with a bolt to hold two or more pieces of material together. The nut thread must match the bolt thread for acceptable mating. Figure 11–51 shows the nut and bolt relationship. The hole in the parts must be drilled larger than the bolt for clearance.

There are a variety of nuts in hexagon or square shapes. Nuts are also designed slotted to allow them to

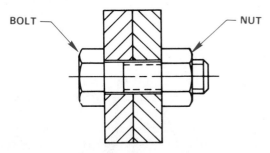

Figure 11–51 Nut and bolt relationship known as a floating fastener.

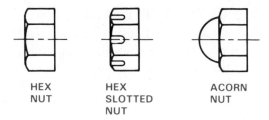

Figure 11–52 Common nuts.

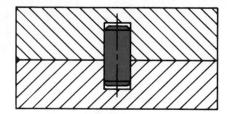

Figure 11–54 Dowel pin in place, sectional view.

be secured with a pin or key. Acorn nuts are capped for appearance. Self-locking nuts are available with neoprene gaskets that help keep the nut tight when movement or vibration is a factor. Figure 11–52 shows some common nuts.

▼ WASHERS

Washers are flat, disk-shaped objects with a center hole to allow a fastener to pass through. Washers are made of metal, plastic, or other materials for use under a nut or bolt head, or at other machinery wear points, to serve as a cushion, or a bearing surface, to prevent leakage or to relieve friction or locking device. (See Figure 11–53.) Washer thickness varies from .016 in. to .633 in.

▼ DOWEL PINS

Dowel pins used in machine fabrication are metal cylindrical fasteners that retain parts in a fixed position or keep parts aligned. Generally, depending on the function of the parts, one or two dowel pins are sufficient for holding adjacent parts. Dowel pins must generally be pressed into a hole with an interference tolerance of between .0002 in. to .001 in. depending on the material and the function of the parts. Figure 11–54 shows the section of two adjacent parts and a dowel pin. Figure 11–55 is a chart drawing of some standard dowel pins.

▼ TAPER AND OTHER PINS

For applications that require perfect alignment of accurately constructed parts, *tapered dowel pins* may be better than straight dowel pins. Taper pins are also

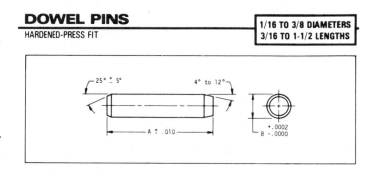

DOWEL PINS
HARDENED-PRESS FIT

1/16 TO 3/8 DIAMETERS
3/16 TO 1-1/2 LENGTHS

Material: 416 Stainless Steel (Clear Passivate) Hardened to: Rockwell C 36-40
Conforms to specification MS-16555
For 303 stainless pins, please see Cat. No. EPS-A1 and EPS-B1

	1/16 DIA. PIN B = .0626		3/32 DIA. PIN B = .0938		1/8 DIA. PIN B = .1251		5/32 DIA. PIN B = .1563	
A	Cat. No.	Price	Cat. No.	Price	Cat. No.	Price	Cat. No.	Price
3/16	EPS-D1-1	$.09	*EPS-D2-1	$.14	-		-	
1/4	EPS-D1-2	.09	*EPS-D2-2	.15	*EPS-D3-2	$.14	-	
5/16	EPS-D1-3	.10	EPS-D2-3	.11	*EPS-D3-3	.15	-	
3/8	EPS-D1-4	.10	EPS-D2-4	.11	EPS-D3-4	.13	*EPS-D4-4	$.17
7/16	EPS-D1-5	.10	EPS-D2-5	.11	EPS-D3-5	.13	*EPS-D4-5	.17
1/2	EPS-D1-6	.11	EPS-D2-6	.11	EPS-D3-6	.13	EPS-D4-6	.17

Figure 11–55 Dowel pin chart drawing. *Courtesy NORDEX, Inc.*

used for parts that have to be taken apart frequently or where removal of straight dowel pins may cause excess hole wear. Figure 11–56 shows an example of a taper pin assembly. Taper pins, shown in Figure 11–57, range in diameter, *D*, from 7/0, which is .0625 in., to .875 in., and lengths, *L*, vary from .375 in. to 8 in.

Other types of pins serve functions similar to taper pins, such as holding parts together, aligning parts, locking parts, and transmitting power from one feature to another. Other common pins are shown in Figure 11–58.

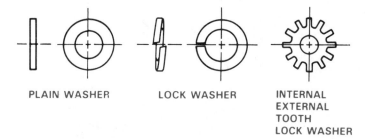

Figure 11–53 Types of washers.

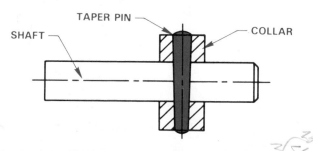

Figure 11–56 Taper pin in assembly, sectional view.

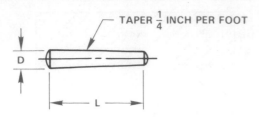

Figure 11–57 Taper pin.

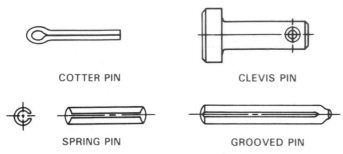

Figure 11–58 Other common pins.

▼ RETAINING RINGS

Internal and external retaining rings are available as fasteners to provide a stop or shoulder for holding bearings or other parts on a shaft. They are also used internally to hold a cylindrical feature in a housing. Common retaining rings require a groove in the shaft or housing for mounting with a special plier tool. Also available are self-locking retaining rings for certain applications. (See Figure 11–59.)

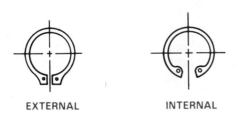

Figure 11–59 Retaining rings.

▼ KEYS, KEYWAYS, AND KEYSEATS

Standards for keys were established to control the relationship among key sizes, shaft sizes, and tolerances for key applications. A key is an important machine element, which is employed to provide a positive connection for transmitting torque between a shaft and hub, pulley, or wheels. The key is placed in position in a keyseat, which is a groove or channel cut in a shaft. The shaft and key are then inserted into the hub, wheel, or pulley, where the key mates with a groove called a keyway. Figure 11–60 shows the relationship among the key, keyseat, shaft, and hub.

Standard key sizes are determined by shaft diameter. For example, shaft diameters ranging from $7/8$ in. to $1\frac{1}{4}$ in. would require a $\frac{1}{4}$-in. nominal width key. Keyseat depth dimensions are established in relationship to the shaft diameter. Figure 11–61 shows the standard dimensions for the related features. For a shaft with a $\frac{3}{4}$-in. diameter the recommended shaft dimension, *S,* would be .676 in. and the hub dimension, *T,* would be .806 in. when using a rectangular key. Types of keys are shown in Figure 11–62.

▼ RIVETS

A *rivet* is a metal pin with a head used to fasten two or more materials together. The rivet is placed through holes in mating parts and the end without a head extends through the parts to be headed-over (formed into a

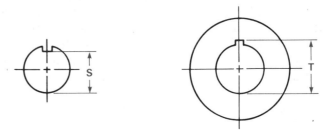

Figure 11–61 Keyseats and keyways are generally sized as related to shaft size.

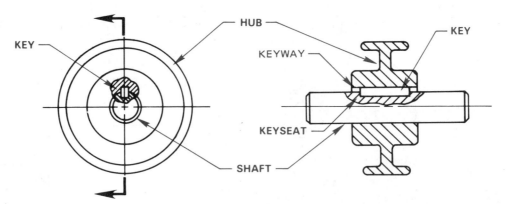

Figure 11–60 Relationship between the key, keyseat, keyway, shaft, and hub.

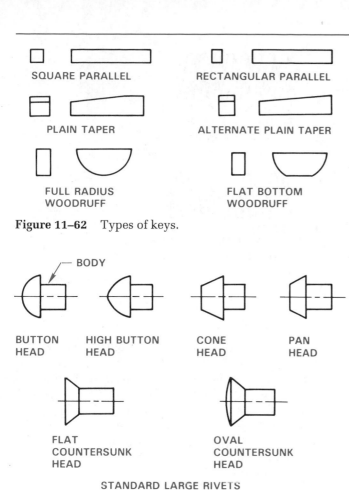

Figure 11–62　Types of keys.

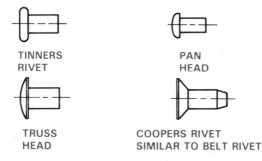

Figure 11–63　Common rivets.

head) by hammering, pressing, or forging. The end with the head is held in place with a solid steel bar known as a dolly while a head is formed on the other end. Rivets are classified by body diameter, length, and head type. (See Figure 11–63.)

▼ SPRINGS

A *spring* is a mechanical device, often in the form of a helical coil, that yields by expansion or contraction due to pressure, force, or stress applied. Springs are made to return to their normal form when the force or stress is removed. Springs are designed to store energy for the purpose of pushing or pulling machine parts by reflex action

into certain desired positions. Improved spring technology provides springs with the ability to function for a long time under high stresses. The effective use of springs in machine design depends on five basic criteria including material, application, functional stresses, use, and tolerances.

Continued research and development of spring materials have helped to improve spring technology. The spring materials most commonly used include high-carbon spring steels, alloy spring steels, stainless spring steels, music wire, oil-tempered steel, copper-based alloys, and nickel-based alloys. Spring materials, depending on use, may have to withstand high operating temperatures and high stresses under repeated loading.

Spring design criteria are generally based on material gage, kind of material, spring index, direction of the helix, type of ends, and function. Spring wire gages are available from several different sources ranging in diameter from number 7/0 (.490 in.) to number 80 (.013 in.). The most commonly used spring gages range from 4/0 to 40. There are a variety of spring materials available in round or square stock for use depending on spring function and design stresses. The spring index is a ratio of the average coil diameter to the wire diameter. The index is a factor in determining spring stress, deflection, and the evaluation of the number of coils needed and the spring diameter. Recommended index ratios range between 7 and 9 although other ratios commonly used range from 4 to 16. The direction of the helix is a design factor when springs must operate in conjunction with threads or with one spring inside of another. In such situations the helix of one feature should be in the opposite direction of the helix for the other feature. Compression springs are available with ground or unground ends. Unground, or rough, ends are less expensive than ground ends. If the spring is required to rest flat on its end, then ground ends should be used. Spring function depends on one of two basic factors, compression or extension. Compression springs release their energy and return to their normal form when compressed. Extension springs release their energy and return to the normal form when extended. (See Figure 11–64.)

Spring Terminology

The springs shown in Figure 11–65 show common characteristics.

Ends.　Compression springs have four general types of ends: open or closed ground ends and open or closed unground ends, shown in Figure 11–66. Extension

COMPRESSION SPRING　　EXTENSION SPRING

Figure 11–64　Compression and extension spring.

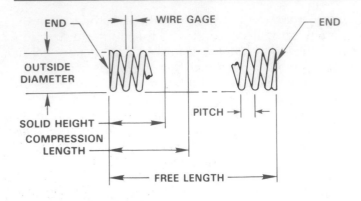

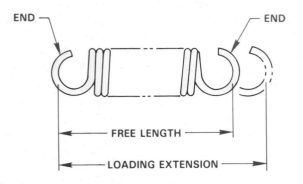

COMPRESSION SPRING

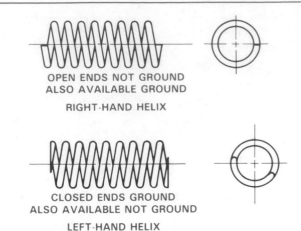

OPEN ENDS NOT GROUND
ALSO AVAILABLE GROUND

RIGHT-HAND HELIX

CLOSED ENDS GROUND
ALSO AVAILABLE NOT GROUND

LEFT-HAND HELIX

Figure 11–66 Helix direction and compression spring end types.

EXTENSION SPRING

Figure 11–65 Spring characteristics.

springs have a large variety of optional ends, a few of which are shown in Figure 11–67.

Helix Direction. The helix direction may be specified as right hand or left hand. (See Figure 11–66.)

Free Length. The length of the spring when there is no pressure or stress to affect compression or extension.

Compression Length. The compression length is the maximum recommended design length for the spring when compressed.

Solid Height. The solid height is the maximum compression possible. The design function of the spring

IN LINE MACHINE LOOP AND HOOK
ALSO AVAILABLE AT RIGHT ANGLES

FULL LOOP ON SIDE WITH SMALL
EYE ON CENTER
ALSO AVAILABLE WITH FULL LOOP
CENTERED

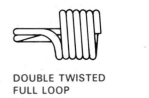

DOUBLE TWISTED
FULL LOOP

SMALL OFFSET HOOK

LONG ROUNDED END

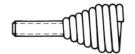

CONED END WITH
SHORT SWIVEL EYE

CONED END WITH
SWIVEL BOLT

MANY OTHER COMBINATIONS ARE AVAILABLE

Figure 11–67 Extension spring end types.

should not allow the spring to reach solid height when in operation unless this factor is a function of the machinery.

Loading Extension. The extended distance to which an extension spring is designed to operate.

Pitch. The pitch is one complete helical revolution, or the distance from a point on one coil to the same corresponding point on the next coil.

Torsion Springs

Torsion springs are designed to transmit energy by a turning or twisting action. *Torsion* is defined as a twisting action that tends to turn one part or end around a longitudinal axis while the other part or end remains fixed. Torsion springs are often designed as antibacklash devices or as self-closing or self-reversing units. (See Figure 11–68.)

Flat Springs

Flat springs are arched or bent flat-metal shapes designed so when placed in machinery they cause tension on adjacent parts. The tension may be used to level parts, provide a cushion, or position the relative movement of one part to another. One of the most common examples of flat springs is leaf springs on an automobile.

Spring Representations

There are three types of spring representations, detailed, schematic, and simplified as seen in Figure 11–69. Detailed spring drawings are used in situations that require a realistic representation, such as vendors' catalogs, assembly instructions, or detailed assemblies. Schematic spring representations are commonly used on drawings. The single-line schematic symbols are easy to draw and clearly represent springs without taking the additional time required to draw a detailed spring. The use of simplified spring drawings is limited to situations where the clear resemblance of a spring is not necessary.

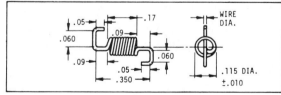

Material: 302 Stainless Steel (Spring Temper) Passivated

Figure 11–68 Torsion spring, also called antibacklash spring. *Courtesy NORDEX, Inc.*

DETAILED   SCHEMATIC   SIMPLIFIED

Figure 11–69 Spring representations.

SPRING DATA:
MATERIAL: 15 MOEN GAGE
NUMBER OF COILS: 12
TYPE OF ENDS: GROUND
HELIX DIRECTION: OPTIONAL

Figure 11–70 Detailed spring drawing with spring data chart.

While very easy to draw, the simplified spring symbol must be accompanied by clearly written spring specifications. The simplified spring representation is not very useful in assembly drawings or other situations that require a visual comparison of features.

Spring Specifications

No matter which representation is used several important specifications must accompany the spring symbol. Spring information is generally lettered in the form of a specific or general note.

Spring specifications include outside or inside diameter, wire gage, kind of material, type of ends, surface finish, free and compressed length, and number of coils. Other information, when required, may include spring design criteria, and heat treatment specifications. The information is often provided on a drawing as shown in Figure 11–70. The material note is usually found in the title block.

Drawing Detailed Spring Representations

Detailed spring drawing requires a great degree of accuracy. First determine the outside diameter, number of coils, wire diameter, and free length or compressed length.

The following steps show how to draw a detailed spring representation with the following specifications:

Material: 2.5 mm diameter high-carbon spring steel
Outside Diameter: 16 mm
Free Length: 50 mm
Number of Coils: 6

Step 1. Using construction lines, draw a rectangle equal to the outside diameter (16 mm) wide and the free length (50 mm) long. (See Figure 11–71.)

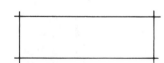

Figure 11–71 Step 1, preliminary spring layout.

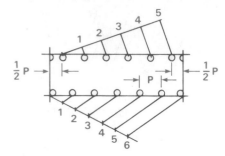

Figure 11–72 Step 2, spacing the coils.

Figure 11–73 Step 3, connect the coils to complete the detailed spring representation.

Figure 11–74 Detailed spring representation in section.

Step 2. Along one inside edge of the length of the rectangle, lay out seven equally-spaced full circles with a 2.5 mm diameter (wire size). The layout may be done by dividing the length into six equal spaces. (See Figure 11–72.) On the other inside edge, lay out a half circle at each end. Beginning at a distance of ½ P away from one end, draw the first of six full (2.5 mm) circles with equal spaces between them.

Step 3. Connect the circles drawn in Step 2 to make the coils. Draw lines from a point of tangency on one circle to a corresponding point on a circle on the other side. Draw the last element on each side down the edge of the rectangle for ground ends. To draw unground ends, make the last element terminate at the axis of the spring. (See Figure 11–73.)

For detailed coils in a longitudinal sectional view leave the circles from Step 2 and draw that part of the spring that appears as if the front half were removed as shown in Figure 11–74. Then fill in or section-line the circles.

Drawing Schematic Spring Representations

Schematic spring symbols are much easier to draw than detailed representations while clearly resembling a spring. The following steps show how to draw a schematic spring symbol for the same spring previously drawn.

CADD Applications

DRAWING SPRINGS

Parametric capabilities are the key to drafting and design flexibility with CADD. *Parametric* means that you have the ability to change a value or values in a situation and automatically have the entire design changed to fit the circumstances. Many CADD packages have this feature available, and CADD programs for spring design are no exception. For instance, if you are designing a compression spring, all you have to do is make the selection from a menu and then enter the following information as prompted by the computer:

○ Wire diameter.
○ Number of coils.
○ Outside diameter.
○ Free length.
○ Type of ends.

When you enter this data, the computer calculates the rest of the information, such as free length and compression length, and then asks if you want an external or sectional view. After all the information is complete, the SPRING program automatically draws and dimensions the spring on the screen, followed by prompts for you to rotate or move the spring to the desired location.

Step 1. Use construction lines to draw a rectangle equal in size to the outside diameter by the free length as seen in Figure 11–71.

Step 2. Establish six equal spaces at P distance along one edge of the rectangle. Along the opposite edge begin and end with a space equal to ½ P. Establish five equal spaces between the ½ P ends. (See Figure 11–75.)

Step 3. Beginning on one side, draw the elements of each spring coil as shown in Figure 11–76.

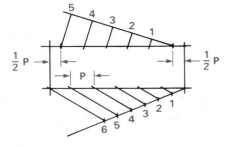

Figure 11–75 Step 2, spacing the coils.

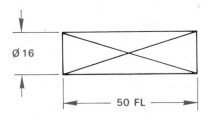

Figure 11–76 Step 3, complete the schematic representation.

Ø 16

50 FL

Figure 11–77 Simplified spring representation.

Drawing a Simplified Spring Representation

Simplified spring representations are drawn by making a rectangle equal in size to the major diameter and free length and then drawing diagonal lines as shown in Figure 11–77.

▼ RECOMMENDED REVIEW

It is recommended that you review the following parts of Chapter 3 before you being working on fully dimensioned fastener and spring drawings. This will refresh your memory about how related lines are properly drawn.

○ Object Lines.
○ Viewing Planes.
○ Extension Lines.
○ Hidden Lines.
○ Break Lines.
○ Dimension Lines.
○ Centerlines.
○ Phantom Lines.
○ Leader Lines.

Also review Chapters 5 and 6 covering multiview and auxiliary view drawings, and Chapter 10 covering dimensioning practices.

PROFESSIONAL PERSPECTIVE

There are three basic methods for drawing screw thread and spring representations: using construction techniques, using a template, or using CADD. The manual construction techniques should be avoided in most cases because they are very time-consuming. However, in some special applications these methods may be used effectively. The manual drafter should use a template when at all possible to draw screw threads, fastener heads, and spring representations. For the CADD user, there are many software programs available that allow drawing of screw threads in simplified, schematic, or detailed representations. A complete variety of head types can be drawn in addition to springs in simplified, schematic, or detailed representations. This type of CADD software makes an otherwise complex task very simple and saves a lot of drafting time. The entry-level drafter often spends unnecessary time trying to make a screw thread representation look exactly like the real thing. It is important to have an accurate drawing, but with regard to screw threads, the thread note tells the reader the exact thread specifications. This is why threads are shown on a drawing as a representation rather than a duplication of the real thing. The simplified representation is the most commonly used technique, because it is fast and easy to draw.

Your goal as a professional drafter is to make the drawing communicate so that there is no question about what is intended. In simple terms, make the drawing clear and accurate using the easiest method, and make sure the thread note and specifications are complete and correct.

MATH APPLICATION

RATIO AND PROPORTION OF A TAPER PIN

Problem: The taper of a pin is .25:12. What will the taper be for the 3-in. long pin shown in Figure 11–78?

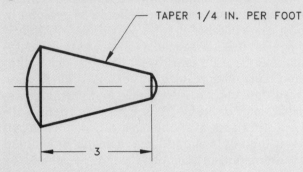

TAPER 1/4 IN. PER FOOT

3

Figure 11–78 Taper pin.

Solution: Set up a proportional equation. (See Appendix N.)

$$\frac{.25 \text{ taper}}{12 \text{ length}} = \frac{x \text{ taper}}{3 \text{ length}}$$

Cross-multiply and solve:

$$(12)(x) = (.25)(3)$$
$$12x = .75$$
$$x = .75 \div 12 = \mathbf{.0625 \text{ in.}}$$

CHAPTER

11 FASTENERS AND SPRINGS TEST

DIRECTIONS

Answer the questions with short complete statements or drawings as needed.

QUESTIONS

1. Define the following screw thread terms: axis, classes of threads, fit, die, major diameter, minor diameter, pitch, tap, tap drill, thread, and threads per inch.

2. Name and describe three types of taps.

3. Name the three methods of thread representation.

4. What thread representation is most commonly used and why?

5. Why is it common, when possible, to drill deeper than the intended depth of the tap?

6. Why should an internal tap be deeper than the fastener thread?

7. Identify the components of the following metric thread note.

    M 2.5 X 0.45—6g

8. Identify the components of the following Unified screw thread note.

    ⅞—14 UNF—2 B

9. When is the abbreviation LH used at the end of a thread note?

10. What assumption is made about the thread if LH is not placed at the end of the thread note?

11. Identify two functions of thread inserts.

12. What are set screws used for?

13. What type of information is required to completely identify a bolt or screw?

14. Identify two applications for taper pins.

15. Identify two applications for washers.

16. What are retaining rings used for?

17. What is the principal application for keys?

18. How are standard key sizes determined?

19. Define *keyseat.*

20. What are rivets used for, and how are they applied?

DIRECTIONS

1. From the selected problems determine which views and dimensions should be used to completely detail the part. Use simplified representation unless otherwise specified.

2. Make a multiview sketch, to proper proportions, including dimensions and notes.

3. Using the sketch as a guide, draw an original multiview drawing on an adequately sized drawing sheet. Add all necessary dimensions and notes using unidirectional dimensioning. Use manual or computer-aided drafting as required by your course guidelines.

4. Include the following general notes at the lower left corner of the sheet .5 in. each way from the corner border lines:

2. REMOVE ALL BURRS AND SHARP EDGES.
1. INTERPRET DIMENSIONS AND TOLERANCES PER ASME Y14.5M—1994.

NOTES:
Additional general notes may be required depending on the specifications of each individual assignment. The following should be part of your title block unless otherwise specified by your instructor:

UNSPECIFIED TOLERANCES:

DECIMALS	mm	IN.
X	±2.5	±.1
XX	±0.25	±.01
XXX	±0.127	±.005
ANGULAR ±30'		
FINISH	3.2μm	125μIN.

Problem 11–1   (in.)

Part Name: Full Dog Point Gib Screw
Material: 10-32 UNF-2A1 × .75 Long

SR.075

.25

.125
.750

.200

SIDE VIEW

Problem 11–2   (in.)

Part Name: Thumb Screw
Material: SAE 1315 Steel

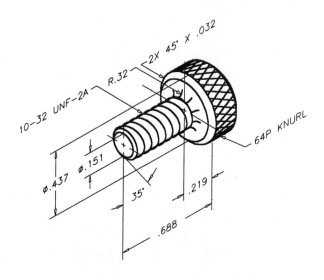

2X 45° X .032
R.32
10-32 UNF-2A
64P KNURL
ø.437   ø.151
35°
.219
.688

Problem 11–3 (in.)

Part Name: H-Step Threading Screw
Material: SAE 3130

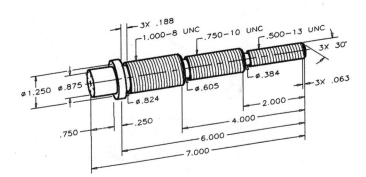

.875–9 UNC–2A
(SHOW IN DETAIL
REPRESENTATION)

.625–11 UNC–2A
(SHOW IN SCHEMATIC)

.50–20 UNF–2A
(SHOW IN
SIMPLIFIED)

ø.72
ø.51
ø.43

MEDIUM
DIAMOND
KNURL

ø1.00

4X 45° X .06

1.00
.25
1.00
.25
1.00
.25
.25
4.75

Problem 11–4 (metric)

Part Name: Knurled Hex Soc Hd Step Screw
Material: SAE 1040
Case Harden: 1.6 mm-deep per Rockwell C Scale
Finish: 2 µm; Black Oxide

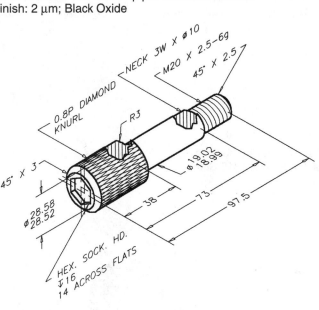

NECK 3W X ø10
M20 X 2.5–6g
45° X 2.5
0.8P DIAMOND
KNURL
R3
45° X 3
ø19.02
18.99
ø28.58
28.52
38
73
97.5
HEX. SOCK. HD.
⌴16
14 ACROSS FLATS

Problem 11–6 (metric)

Part Name: Lathe Dog
Material: Cast Iron

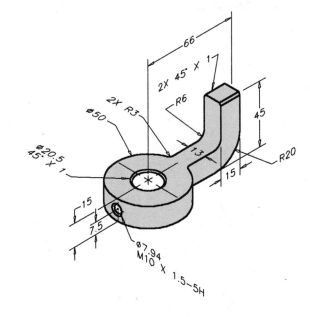

66
2X 45° X 1
R6
2X R3
45
ø50
1.3
R20
ø20.5
45° X 1
15
15
7.5
ø7.94
M10 X 1.5–5H

Problem 11–5 (in.)

Part Name: Machine Screw
Material: Stainless Steel
Finish All Over: 2 µm

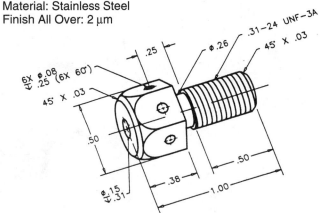

.31–24 UNF–3A
45° X .03
.25
ø.26
6X ø.08
.25 (6X 60°)
45° X .03
.50
.38
1.00
ø.15
⌴.31
.50

Problem 11–7 (in.)

Part Name: Threaded Step Shaft
Material: SAE 1030

3X .188
1.000–8 UNC
.750–10 UNC
.500–13 UNC
3X 30°
ø1.250
ø.875
ø.824
ø.605
ø.384
3X .063
.750
.250
2.000
4.000
6.000
7.000

Problem 11–8 (in.)

Part Name: Washer Face Nut
Material: SAE 1330 Steel

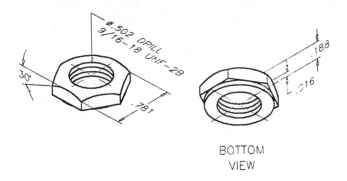

BOTTOM
VIEW

Problem 11–9 (in.)

Part Name: Shoulder Screw
Material: SAE 4320 Steel

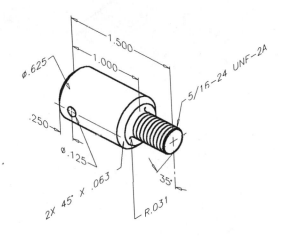

Problem 11–10 (in.)

Part Name: Stop Screw
Material: SAE 4320
Hex Depth: .175
NOTE: Medium Diamond Knurl at head.

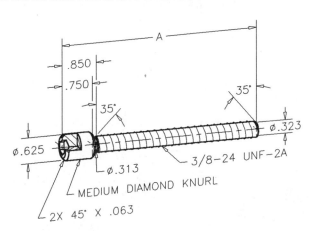

5/16 HEX

CHART A	
PART #	LENGTH
1DT–1011	4.438
1DT–1012	3.688

Problem 11–11 (in.)

Part Name: Drain Fitting
Material: Ø1.25 6061-T6 Aluminum
Problem based on original art courtesy TEMCO.

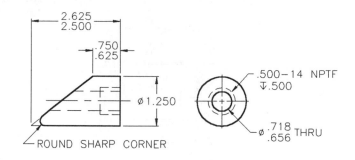

OUTSIDE SURFACE OF COUPLING MUST
BE FREE OF OXIDIZATION AND INK.

Problem 11–12 (metric)

Part Name: Vice Base
Material: Cast Iron

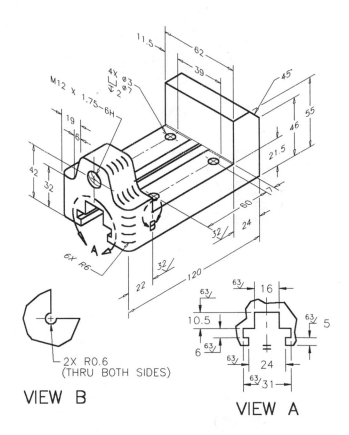

VIEW B

VIEW A

Problem 11–13 (in.)

Part Name: Valve Spring
Material: ∅.024 Type 302 ASTM A313 Spring Steel
Courtesy TEMCO.

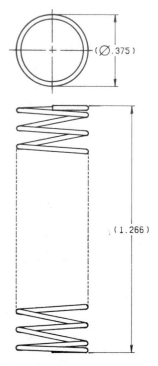

NOTE:
1. MFG SPRING TO .375 DIA X 1.266 FL; .281 LB @
.391 DEF; LH COIL; CLOSED ENDS

Problem 11–14 (metric)

Part Name: Compression Spring (use schematic representation)
Material: 2.5 mm Steel Spring Wire
Ends: Plain Ground
Outside Diameter: 25
Free Length: 75
Number of Coils: 16
Finish: Chrome Plate

Problem 11–15 (metric)

Part Name: Antibacklash Spring (use detailed representation)
Material: 302 Stainless Steel Spring Tempered 0.38 Diameter
Problem based on original art courtesy NORDEX, Inc.

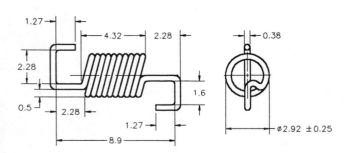

Problem 11–16 (metric)

Part Name: Flat Spring
Material: 3.5 mm Spring Steel
Finish: Black Oxide
Heat Treat: 1 mm-deep Rockwell C Scale

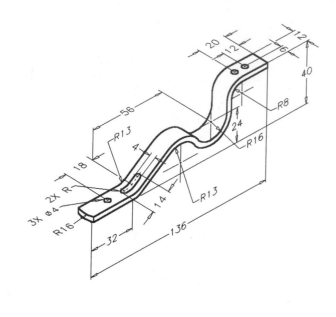

Problem 11–17 (in.)

Part Name: Retaining Ring
Material: Stainless Steel

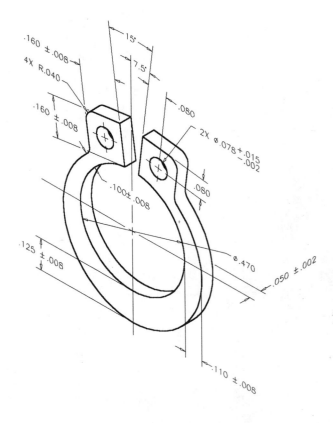

Problem 11–18 (in.)

Part Name: Half Coupling
Material: ∅1.250 C1215 Steel
Problem based on original art courtesy TEMCO.

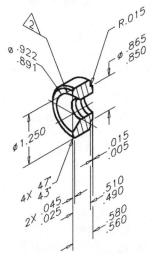

HALF OF PART SHOWN FOR CLARITY.

OUTSIDE SURFACE OF COUPLING MUST BE
FREE OF OXIDIZATION AND INK.

△2
.500–14 NPTF: L–1 GAGE, PLUS 1/MINUS 1
TURNS FROM NOMINAL. TAP FROM THIS END.

Problem 11–19 (in.)

Part Name: Collar
Material: SAE 1020

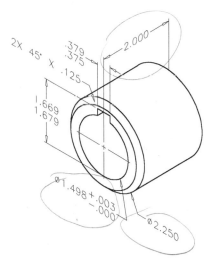

Problem 11–20 (in.)

Part Name: Bearing Nut
Material: SAE 1040

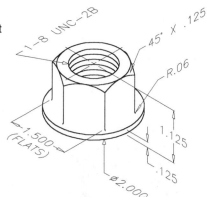

Problem 11–21 (in.)

Part Name: Adjustment Screw
Material: SAE 2010 Steel

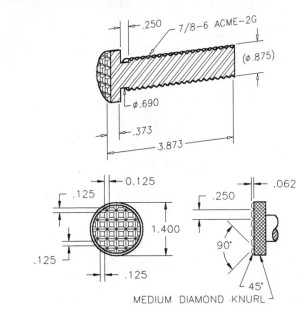

MEDIUM DIAMOND KNURL

Problem 11–22 (in.)

Part Name: Screw Shaft
Material: ¾ Hex × 4⅛ Stock Mild Steel

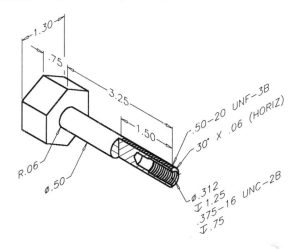

Problem 11–23 (in.)

Part Name: Packing Nut
Material: Bronze
A: Spanner slots .250 wide × .063 deep.

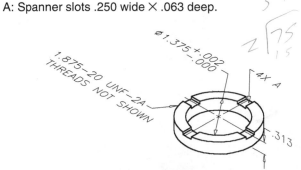

A: SLOT WIDTH .250
   SLOT DEPTH .063

Problem 11–24 (in.)

Part Name: Pump Pivot Support
Material: Cold-rolled Mild Steel

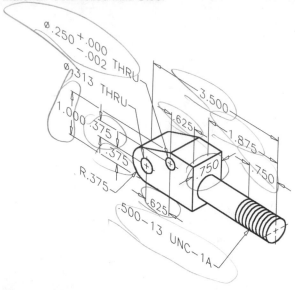

Problem 11–25 (in.)

Part Name: Set Screw
Material: Steel
SPECIFIC INSTRUCTIONS:
Prepare a detailed drawing from the written instructions below.

10 - 32 UNF - 2A THREADED
MATERIAL .375 LG.
W/ .015 LATERAL X 35°
CHAMFERS ON BOTH ENDS
PROVIDE A .030 WIDE X
.047 DEEP SLOT ON
ONE END.

MATH PROBLEMS

1. The taper of a pin is .5:12. What would be the taper for a 5 in. long pin?
2. A lathe operator turns out 9 brass bushings in 2 hours 15 minutes. At the same rate, how many bushings can be turned in 8 hours?
3. Seventeen drills cost $27.50. What would seven drills cost?
4. The scale on a landscape map is 1":50'. If two points on the map are 3.5" apart, what is the actual distance between the two points?
5. A crew of twelve people assembles 45 units a day. If production is to be increased to 80 units per day, how many people will be needed?
6. If 5 dozen shop cloths are enough for six mechanics, how many cloths will be needed for eleven mechanics?
7. A shop manual chain hoist requires a 22-lb pull to lift one ton (2000 lb). How much pull would it take to lift a 1500-lb load?

CHAPTER 12

Sections, Revolutions, and Conventional Breaks

THE ENGINEERING DESIGN PROCESS

You are working with an engineer on a special project and she gives you a rough sketch of a part from which you are to make a formal drawing. At first glance, you think it is an easy part to draw, but with further study you realize that it is more difficult than it looks. The rough sketch, shown in Figure 12–1, has a lot of hidden features. So your first thought is, "What do I use for the front view?" You decide that a front exterior view would show only the diameter and length. So instead of drawing an outside view, you decide to use a full section to expose all of the interior features and, at the same time, give the overall length and diameter. "This is great!" you say. Then you realize there are 6 holes spaced 60° apart that remain as hidden features in the left side view. This is undesirable, because you do

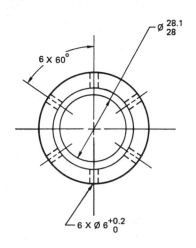
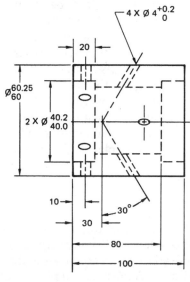

Figure 12–1   Engineer's layout drawing.

(Continued)

ENGINEERING DESIGN PROCESS *(Continued)*

not want to section all of the holes. So you decide to use a broken-out section to expose only two holes. This allows you to dimension the 6×60° and the $6\times \varnothing 6^{+0.2}_{0}$ holes all at the same time. Now with all

this thinking and planning out of the way, it only takes you three hours to completely draw and dimension the part, and you are ready to give the formal drawing, shown in Figure 12–2, to the checker.

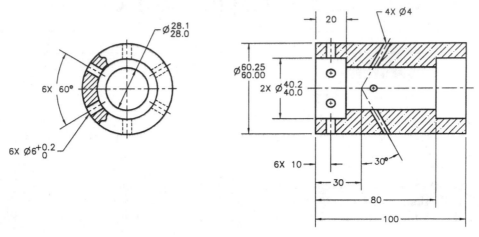

Figure 12–2   CADD solution to engineering problem.

ANSI   The American National Standards Institute document that governs sectioning techniques is titled *Multi and Sectional View Drawings* ANSI Y14.3. The engineering standard *Line Conventions and Lettering* ASME Y14.2M covers the principles of drawing recommended section lines. The content of this chapter is based upon the ANSI standard and provides an indepth analysis of the techniques and methods of sectional view presentation.

▼ SECTIONING

Sections, or *sectional views,* are used to describe the interior portions of an object that are otherwise difficult to visualize. Interior features that are described using hidden lines may not appear as clear as if they were exposed for viewing as visible features. It is also a poor practice to dimension to hidden features, but the sectional view allows you to expose the hidden features for dimensioning. Figure 12–3 shows an object in conventional multiview representation and using a sectional view. Notice how the hidden features are clarified in the sectional view.

▼ CUTTING-PLANE LINES

The sectional view is created by placing an imaginary cutting plane through the object as if you were to cut away the area to be exposed. The adjacent view then becomes the sectional view by removing the portion of the object between the viewer and the cutting plane. (See Figure 12–4).

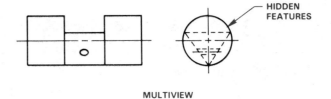

MULTIVIEW

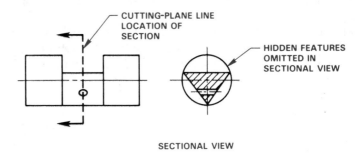

SECTIONAL VIEW

Figure 12–3   Conventional multiview compared to a sectional view.

The sectional view should be projected from the view that has the cutting plane as you would normally project a view in multiview. The *cutting-plane line* is a thick line that represents the cutting plane as shown in Figure 12–4. The cutting-plane line is capped on the ends with arrowheads that show the direction of sight of the sectional view. When the extent of the cutting plane is obvious, only the ends of the cutting-plane line may be used as shown in Figure 12–5. Such treatment of the cutting plane also helps keep the view clear of excess lines.

If lack of space restricts the normal placement of a sectional view, the view may be placed in an alternate

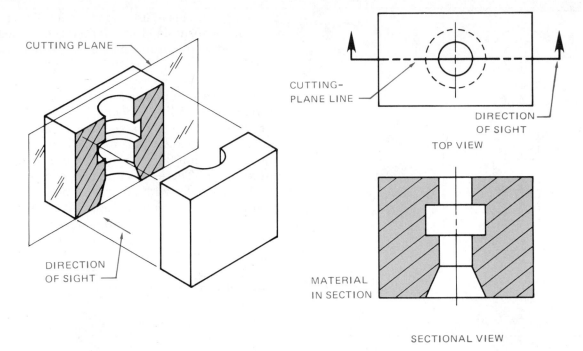

Figure 12–4   Cutting-plane line.

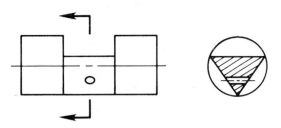

Figure 12–5   Simplified cutting-plane line.

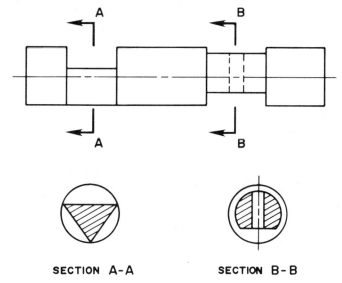

SECTION A-A   **SECTION B-B**

Figure 12–6   Labeled cutting-plane lines and related sectional views.

location. When this is done, the sectional view should not be rotated but should remain in the same orientation as if it were a direct projection from the cutting plane. The cutting planes and related sectional views should be labeled with letters beginning with A, as shown in Figure 12–6.

The cutting-plane line may be completely omitted when the location of the cutting plane is clearly obvious. (See Figure 12–7.) When in doubt, use the cutting-plane line.

▼ SECTION LINES

Section lines are thin lines used in the view of the section to show where the cutting-plane line has cut through material. (See Figure 12–8.) Also refer to Chapter 3. Section lines are usually drawn equally spaced at 45° but may not be parallel or perpendicular to any line of the object. Any convenient angle may be used to avoid placing section lines parallel or perpendicular to other lines of the object. Angles of 30° and 60° are common. Section lines that are more than 75° or less than 15° from hori-

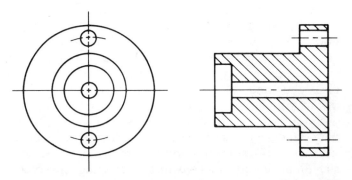

Figure 12–7   An obvious cutting-plane line may be omitted.

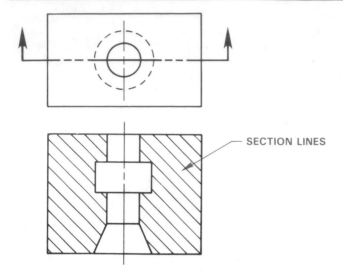

Figure 12–8 Section lines.

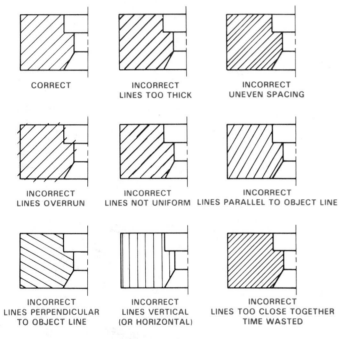

Figure 12–9 Common section line errors.

The other option is to use coded section lining, which is more time-consuming to draw than general section lining. Coded section lining may be used effectively when a section is taken through an assembly of adjacent parts of different materials as seen in Figure 12–10.

General section lines are evenly spaced. The amount of space between lines depends on the size of the part. Very large parts have larger spacing than very small parts. Use your own judgment. The key is to clearly represent section lines without an unnecessary expenditure of time. There are a number of ways to space section lines equally without measuring each space. One good way is to calibrate the space using the same point or mark on the drafting machine scale as you draw each line. This method works with clear plastic scales as shown in Figure 12–11. Another technique is to use the Ames Lettering Guide for drawing equally-spaced section lines. Refer to Chapter 3 to review the use of this tool.

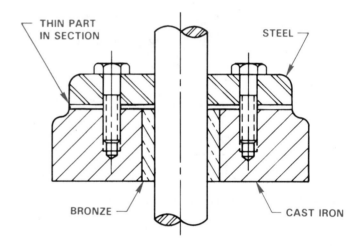

Figure 12–10 Assembly section, coded section lines.

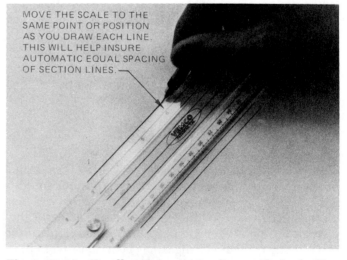

Figure 12–11 Equally spacing section lines with the drafting machine scale.

zontal should be avoided. Section lines must never be drawn either horizontally or vertically. Figure 12–9 shows some common errors in drawing section lines. Section lines should be drawn in opposite directions on adjacent parts, and when several parts are adjacent any suitable angle may be used to make the parts appear clearly separate. When a very large area requires section lining, you may elect to use outline section lining.

Equally spaced section lines specify either a general material designation or cast iron. This method of drawing section lines is quick and easy with the actual material identification located in the drawing title block.

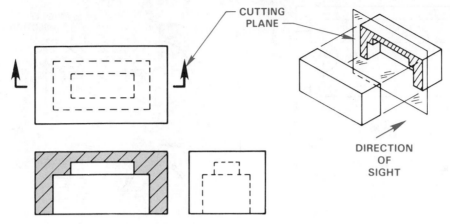

Figure 12–12 Full section.

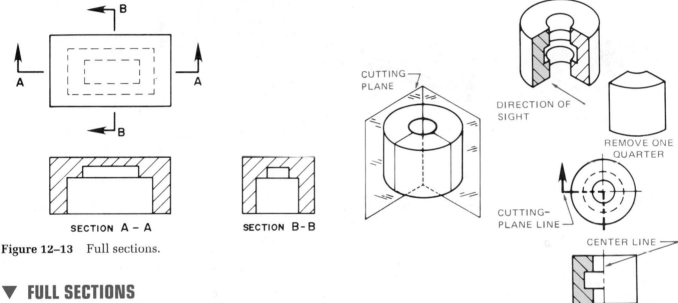

Figure 12–13 Full sections.

Figure 12–14 Half section.

▼ FULL SECTIONS

A *full section* is drawn when the cutting plane extends completely through the object, usually along a center plane as shown in Figure 12–12. The object shown in Figure 12–12 could have used two full sections to further clarify hidden features. In such a case, the cutting planes and related views are labeled. (See Figure 12–13.)

▼ HALF SECTIONS

A *half section* may be used when a symmetrical object requires sectioning. The cutting-plane line of a half section actually removes one-quarter of the object. The advantage of a half section is that the sectional view shows half of the object in section and the other half of the object as it normally appears. Thus the name half section. (See Figure 12–14.) Notice that a centerline is used in the sectional view to separate the sectioned portion from the unsectioned portion. Hidden lines are generally omitted from sectional views unless their use improves clarity.

▼ OFFSET SECTIONS

Staggered interior features of an object may be sectioned by allowing the cutting-plane line to *offset* through the features as shown in Figure 12–15. Notice in Figure 12–15 that there is no line in the sectional view indicating a change in direction of the cutting-plane line. Normally the cutting-plane line in an offset section extends completely through the object to clearly display the location of the section.

▼ ALIGNED SECTIONS

Similar to the offset section, the *aligned section* cutting-plane line also staggers to pass through offset features of

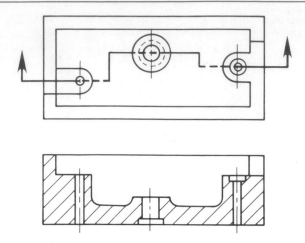

Figure 12–15 Offset section.

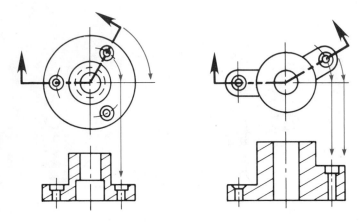

Figure 12–16 Aligned section.

an object. Normally the change in direction of the cutting-plane line is less than 90° in an aligned section. When this section is taken, the sectional view is drawn as if the cutting plane is rotated to a plane perpendicular to the line of sight as shown in Figure 12–16.

▼ UNSECTIONED FEATURES

Specific features of an object are commonly left unsectioned in a sectional view if the cutting-plane line passes through the feature and parallel to it. The types of features that are left unsectioned for clarity are bolts, nuts, rivets, screws, shafts, ribs, webs, spokes, bearings, gear teeth, pins, and keys. (See Figure 12–17.) When the cutting-plane line passes through the previously described features perpendicular to their axes, then section lines are shown as seen in Figure 12–18.

While it is most common to draw the outline of the previously discussed features and display them without section lines, a less used method called alternate section lines may be used. The normally unsectioned feature is drawn using hidden lines and every other section line is drawn through the feature, as shown in Figure 12–19.

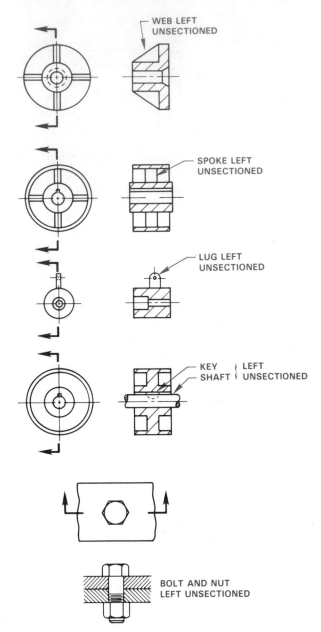

Figure 12–17 Certain features are not sectioned when a cutting-plane line passes parallel to their axis.

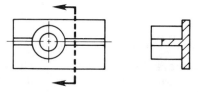

Figure 12–18 Cutting-plane perpendicular to normally unsectioned features.

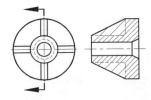

Figure 12–19 Alternate section line method.

CADD Applications

SECTIONING

Everyone should have the chance to spend what often seems like hours drawing section lines in a complex sectional view to really appreciate the speed and accuracy of drawing section lines with a computer. One important aspect of drawing section lines manually is getting the lines uniform and equally spaced. This task is automatic with CADD. Additionally, most CADD drafting packages have more section-line symbols available than you would ever have the need to use. CADD section-line commands such as HATCH or BHATCH are used to select one of a variety of patterns and you have the opportunity to change the section-line scale if you wish. Figure 12–20 shows three different section-line scale factors available.

When using CADD to section-line an area you need to define the boundaries of the area. In other words, tell the computer where the section lines are to be placed so you do not end up with lines in an unwanted area. If you want to section-line inside a circle or square, all you need to do is pick the circle or square. (See Figure 12–21.)

Some CADD programs allow you to draw section lines in an enclosed area simply by picking a point within the area. In AutoCAD, for example, the enclosed area is called a boundary. You get an error message, "BOUNDARY AREA NOT CLOSED," if there is a gap in the boundary. A common boundary gap might be a corner where lines do not meet. You may have to "ZOOM" in on the corner to actually see a very small gap. If this happens, correct the problem and try to section the area again as shown in Figure 12–22.

The computer makes drawing section lines for any drafting field quick and easy. There are standard material section lines for the mechanical drafter, and brick and other patterns for the architect. A large variety of section-line symbols and graphic patterns are available in most CADD packages, or as tablet menu overlays as shown in Figure 12–23.

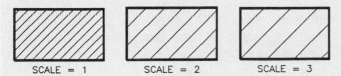

Figure 12–20 Three different section line scale factors for CADD applications.

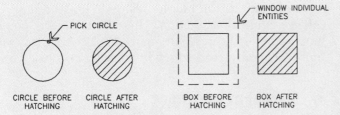

Figure 12–21 Selecting objects for sectioning with CADD.

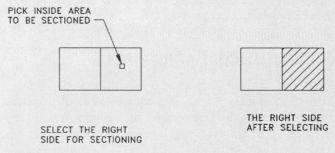

Figure 12–22 Recommended drawing sequence for section lining adjacent areas with CADD.

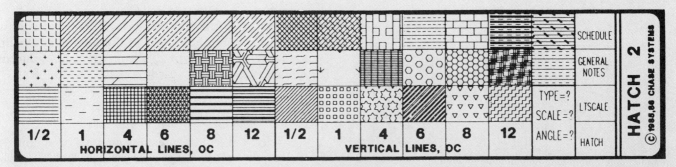

Figure 12–23 CADD tablet menu overlay for section lines and graphic patterns. *Courtesy Chase Systems.*

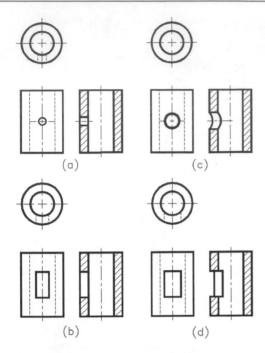

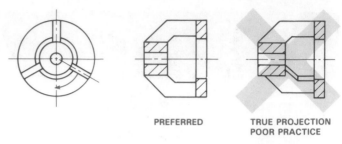

Figure 12–26 Conventional revolution in section.

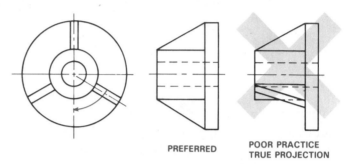

Figure 12–27 Conventional revolution in multiview.

▼ INTERSECTIONS IN SECTION

When a section is drawn through a small intersecting shape, the true projection is ignored. The detail is too complex to represent. (See Figure 12–24a and b.) However, larger intersecting features are drawn as their true representation. (See Figure 12–24c and d.) The professional decision is up to you.

▼ CONVENTIONAL REVOLUTIONS

When the true projection of a feature results in foreshortening, the feature should be revolved onto a plane perpendicular to the line of sight as in Figure 12–25. The revolved spoke shown gives a clear representation with a minimum of drafting time. Figure 12–26 shows another illustration of conventional revolution compared to true projection. Notice how the true projection results in

a distorted and foreshortened representation of the spoke. The revolved spoke in the preferred view is clear and easy to draw. The practice illustrated here also applies to features of unsectioned objects in multiview as shown in Figure 12–27.

▼ BROKEN-OUT SECTIONS

Often a small portion of a part may be broken away to expose and clarify an interior feature. This technique is called a *broken-out section.* There is no cutting-plane line used as you can see in Figure 12–28. A short break line is generally used with a broken-out section.

▼ AUXILIARY SECTIONS

A section that appears in an auxiliary view is known as an *auxiliary section.* Auxiliary sections are generally projected directly from the view of the cutting plane. If these sections must be moved to other locations on the drawing sheet, they should remain in the same relationship (not rotated) as if taken directly from the view of the cutting plane. (See Figure 12–29.)

▼ CONVENTIONAL BREAKS

When a long object of constant shape throughout its length requires shortening, *conventional breaks* may be used. These breaks may be used effectively to save time, paper, or space or to increase the scale of an otherwise very long part. Figure 12–30 shows typical conventional breaks. Used on metal shapes, the short break line is

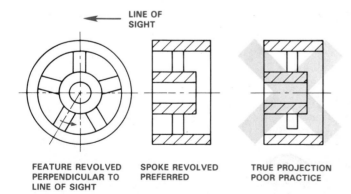

Figure 12–24 Intersections in section.

Figure 12–25 Conventional revolution.

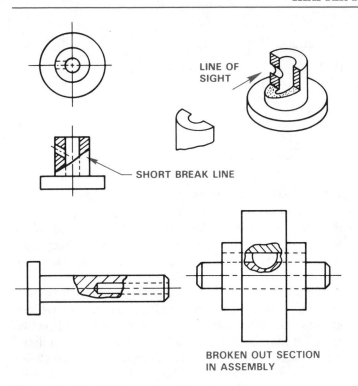

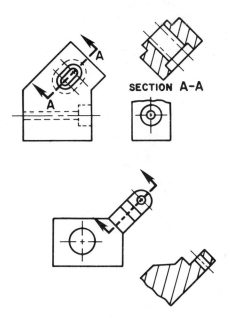

Figure 12–28　Broken-out sections.

Figure 12–29　Auxiliary sections.

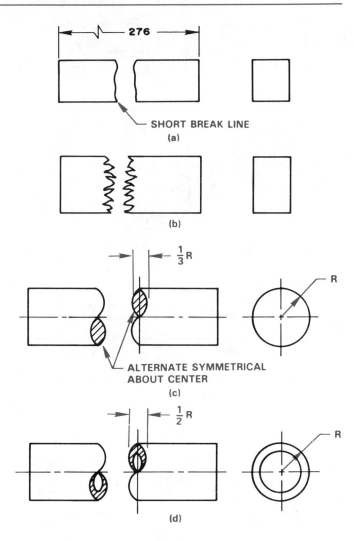

Figure 12–30　Conventional breaks for various shapes are **(a)** metal shapes, **(b)** wood shapes, **(c)** cylindrical solid shapes, **(d)** cylindrical tubular shapes.

The break lines for tubular round shapes may be drawn freehand or with an irregular curve or template. The total shape widths should be approximately ½ radius and should be symmetrical about the horizontal centerline and vertical guidelines as shown.

▼ REVOLVED SECTIONS

When a feature has a constant shape throughout the length that cannot be shown in an external view, a *revolved section* may be used. The desired section is revolved 90° onto a plane perpendicular to the line of sight as shown in Figure 12–31. Revolved sections may be represented on a drawing one of two ways as shown in Figure 12–32. In Figure 12–32a the revolved section is drawn on the part. The revolved section also may be broken away as seen in Figure 12–32b. The surrounding space may be used for dimensions as shown in Figure 12–33.

drawn thick, freehand, and slightly irregular. Notice that the actual length can be given with a long break line used as a dimension line. For wood shapes, the short break line is drawn freehand as a thick, very irregular line.

The break line for solid round shapes may be drawn freehand or with an irregular curve or template. The shape widths should be approximately ⅓ radius and should be symmetrical about the horizontal centerline and the vertical guidelines as shown.

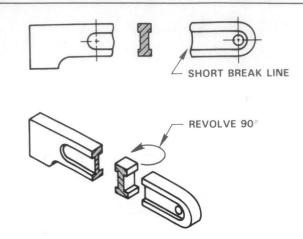

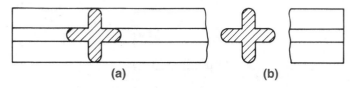

Figure 12–31 Revolved section.

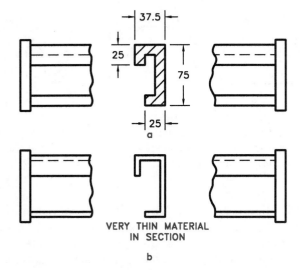

Figure 12–32 **(a)** Revolved section not broken away; **(b)** revolved section broken away.

VERY THIN MATERIAL
IN SECTION

b

Figure 12–33 **(a)** Dimensioning a broken-away revolved section, **(b)** a revolved section through thin material. Section lines are omitted when thin material less than 4 mm thick is sectioned.

Notice in Figure 12–33 that very thin parts, less than 4 mm thick, in sections may be left unsectioned. This practice is also common when sectioning a gasket or similar feature.

▼ REMOVED SECTIONS

Removed sections are similar to revolved sections except they are removed from the view. A cutting-plane line is

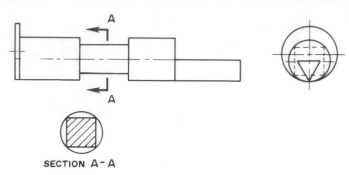

SECTION A-A

Figure 12–34 Removed section.

placed through the object where the section is taken. Removed sections are not generally placed in direct alignment with the cutting-plane line but are placed in a surrounding area as shown in Figure 12–34.

Removed sections may be preferred when a great deal of detail makes it difficult to effectively use a revolved section. An additional advantage of the removed section is that it may be drawn to a larger scale so close detail may be more clearly identified as shown in Figure 12–35. The sectional view should be labeled, as shown, with SECTION A-A and the revised scale which in this example is 2:1. The predominant scale of the principal views is shown in the title block.

Multiple removed sections are generally arranged on the sheet in alphabetical order from left to right and top to bottom. (See Figure 12–36.) Notice that hidden

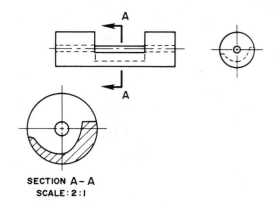

SECTION A-A
SCALE: 2:1

Figure 12–35 Enlarged removed section.

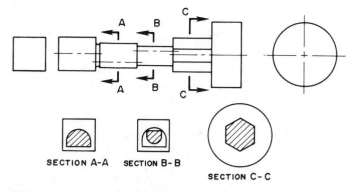

SECTION A-A   SECTION B-B

SECTION C-C

Figure 12–36 Multiple removed sections.

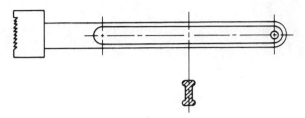

Figure 12–37 Alternate removed section method.

lines have been omitted for clarity. The cutting planes and related sections are labeled alphabetically excluding the letters *I, O,* and *Q,* as they may be mistaken for numbers. When the entire alphabet has been used, label your sections with double letters beginning with *AA, BB,* and so on.

Another method of drawing a removed section is to extend a centerline adjacent to a symmetrical feature and revolve the section of the centerline as shown in Figure 12–37. The removed section may be drawn at the same scale or enlarged as necessary to clarify detail.

▼ RECOMMENDED REVIEW

It is recommended that you review the following parts of Chapter 3 before you begin working on fully dimensioned sectioning drawings. This will refresh your memory about how related lines are properly drawn.

- ❍ Object lines.
- ❍ Viewing-planes.
- ❍ Extension Lines.
- ❍ Hidden Lines.
- ❍ Break Lines.
- ❍ Dimension Lines.
- ❍ Centerlines.
- ❍ Phantom Lines.
- ❍ Leader Lines.

Study these topics more carefully as related directly to sectioning practices.

- ❍ Cutting-planes.
- ❍ Section Lines.
- ❍ Coded Section Lines.

Also review Chapters 5 and 6 covering multiview and auxiliary view drawings and Chapter 10 for recommended dimensioning practices.

PROFESSIONAL PERSPECTIVE

Your role as an engineering drafter is communication. You are the link between the engineer and manufacturing, fabrication, or construction. Up to this point you have learned a lot about how to perform this communication task, but the hard part comes when you have to apply what you have learned. Your goal is to make every drawing clear, complete, and easy to interpret. There are a lot of factors to consider, including:

- ❍ Selection of the front view and related views.
- ❍ Deciding if sections are to be used.
- ❍ Placement of dimensions in a logical format so that the size description and located features are complete.

This is a difficult job, but you are motivated to do it. And just when you have mastered how to select and dimension multiviews, you are faced with preparing sectional drawings, which means a dozen main options and endless applications. (But don't worry because you have made it this far!) Here are some key points to consider:

- ❍ You need to completely show and describe the outside of the part first.
- ❍ If there are a lot of hidden lines for internal features, you know immediately that a section is needed, because you cannot dimension to hidden lines.
- ❍ Analyze the hidden lines and decide on the type of section that will work best.
- ❍ Review each type of sectioning technique until you find one that looks best for the application.
- ❍ Be sure you clearly label the cutting plane-lines and related multiple sections so they correlate.
- ❍ Half sections are good for showing both the outside and inside of the object at the same time, but be cautious about their use, because sometimes they confuse the reader.

You're not alone in this new venture. The problem assignments in this chapter recommend a specific sectioning technique during this learning process.

MATH APPLICATION

DISTANCE BETWEEN HOLES ON A BOLT CIRCLE

Problem: Find the center-to-center distance between adjacent holes on the bolt circle shown in Figure 12–38.

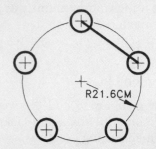

Figure 12–38 Bolt circle.

Solution: Problems like this involve finding the side of a right triangle. The first step is to construct a right triangle onto the drawing, as shown in Figure 12–39.

From the definition of *sine*. (See Appendix N.)

$$\sin A = \frac{y}{r}$$

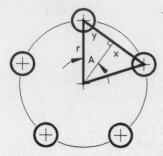

Figure 12–39 Bolt circle with a right triangle drawn.

Because there are five holes on the circle, angle A must be (half of 360°) ÷ 5, or 36°. Hypotenuse r is equal to the radius of 21.6 cm, so you now have $\sin 36° = \frac{y}{21.6}$. Using a calculator to find the sine of 36° gives $.5878 = \frac{y}{21.6}$.

Multiplying both sides of this equation by 21.6 gives the length of side y of the triangle: y = (21.6) (.5878) = 12.696. Finally, side y is half of the required center-to-center distance, so d = 2(12.696) = **25.4 cm.**

CHAPTER

12 SECTIONING TEST

DIRECTIONS

Answer the questions with short, complete statements or drawings as needed on an A-size drawing sheet.

QUESTIONS

1. Describe the potential advantage of a sectional view over a multiview.
2. Explain and use a sketch to show the relationship of the cutting-plane line arrowheads to the sectional view.
3. How much of the object does a full section actually remove?
4. At what angles may section lines be drawn?
5. What are coded section lines?
6. Describe the purpose and advantage of a half section.
7. What type of line is used in a half-sectional view to separate the sectioned area from the nonsectioned area?

8. List four items that would remain unsectioned in a sectional view when the cutting-plane line cuts through parallel to the axis of the items.
9. Describe the error found in drawings (a), (b), (c), (d), and (e) on page 359.
10. Describe a situation when a cutting-plane line may be omitted in sectioning drawings.
11. Discuss the advantage of using conventional revolutions.
12. Discuss the cutting-plane line associated with a broken-out section.
13. What type of break line is generally associated with a broken-out section?
14. In what situations are conventional breaks commonly used?
15. Discuss the difference between a revolved and removed section. An example may be used.
16. Show an example of conventional breaks for the following shapes: (A) Solid cylindrical shape Ø.5 in., (B) Tube Ø.5 in. with .06 in. wall thickness, (C) Solid steel □ .5 in.

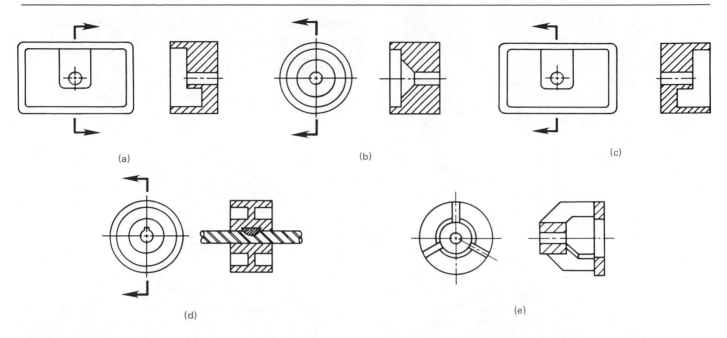

(a)

(b)

(c)

(d)

(e)

12 SECTIONING PROBLEMS

DIRECTIONS

1. From the selected engineer's sketch or layout, determine the views and sections that are needed.
2. Make a sketch of the selected views and sections as close to correct proportions as possible. Do not spend a lot of time as the sketch is only a guide. Indicate where the cutting-plane lines and dimensions are to be placed.
3. Using the sketch you have developed as a guide, draw an original sectioned multiview drawing on an adequately sized drawing sheet. Select a scale that properly details the part on the selected sheet size. Use unidirectional dimensioning. Use manual or computer-aided drafting as required by your course guidelines.
4. Include the following general notes at the lower left corner of the sheet .5 in. each way from the corner border lines:

   2. REMOVE ALL BURRS AND SHARP EDGES.
   1. INTERPRET DIMENSIONS AND TOLER-ANCES PER ASME Y14.5M—1994.

   NOTES:

Additional notes may be required depending on the specifications of each individual assignment. A tolerance block is recommended as shown in problems for Chapter 10 unless otherwise specified.

UNSPECIFIED TOLERANCES:

DECIMALS	mm	IN.
X	±2.5	±.1
XX	±0.25	±.01
XXX	±0.127	±.005
ANGULAR ±30'		
FINISH	3.2μm	125μIN.

5. The engineering layouts may not be dimensioned properly. Verify the correct practice before placing dimensions; for example, the diameter symbol should precede the diameter dimension, and leaders should not cross over dimension lines. Check other line and dimensioning techniques for proper standards. Actual industrial drawings are provided as advanced problems throughout.

Problem 12–1 Full section (in.)

Part Name: Fitting
Material: Bronze
Finish All Over: 63 μin.
NOTE: Refer to Chapter 10 for proper dimensioning practices. Engineering sketches may not display correct practices. This problem may require a front view, side view showing the hexagon, and a full section to expose the interior features for dimensioning.

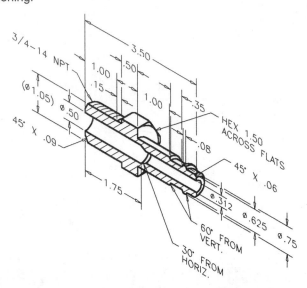

Problem 12–2 Full section from actual industry drawing (in.)

Part Name: Weld Washer Slot
Material: SAE 1040
SPECIFIC INSTRUCTIONS:
Set up dimensions L, T, and W as a chart drawing where the dimensions change for each of the following parts:
Courtesy Production Plastics, Inc.

	L	T	W
PART A	2.125	.500	1.875
PART B	2.250	.625	2.000
PART C	2.500	.750	2.125
PART D	2.625	.875	2.250

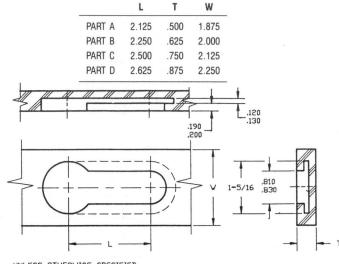

UNLESS OTHERWISE SPECIFIED:

ALL FRACTIONAL DIMENSIONS ±1/32
ALL TWO PLACE DECIMALS ±.010
ALL THREE PLACE DECIMALS ±.005

SPECIFIC INSTRUCTIONS:
Add diameter symbol to all diameter dimensions.
Courtesy TEMCO.

Problem 12–3 Full section from actual industry drawing (in.)

Part Name: Weld Coupling
Material: AISI 1010, Killed

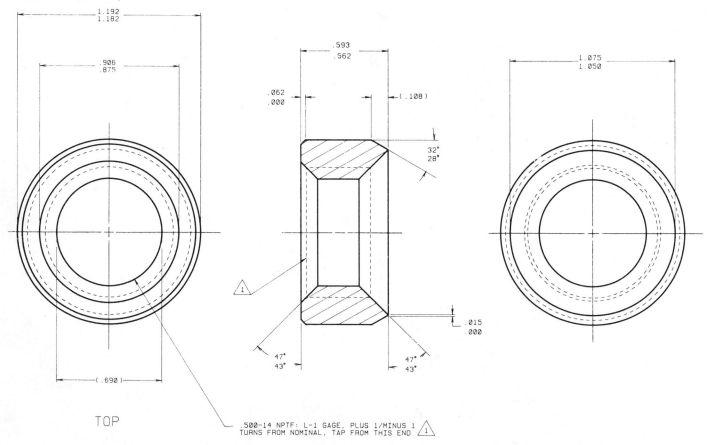

Problem 12–4   Full section from actual industry drawing (in.)

Part Name: Bolt Guard
Material: Aluminum
Courtesy Stanley Hydraulic Tools, a Division of the Stanley Works.

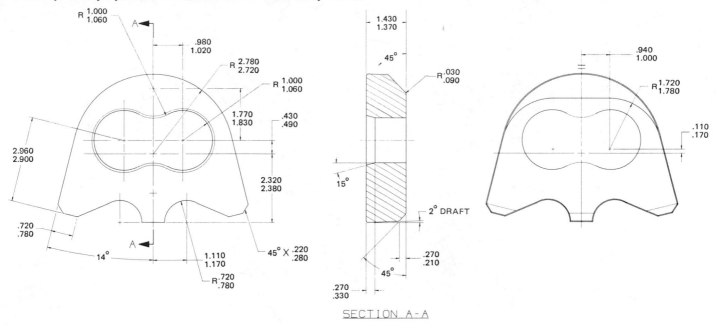

SECTION A-A

Problem 12–5   Full section and view enlargement (in.)

Part Name: Spring
Material: SAE 1060
Problem based on original art courtesy Stanley Hydraulic Tools, a Division of the Stanley Works.
Heat Treat:

1.   Austenitize at 1475° F.
2.   Direct quench in agitated oil.
3.   Temper to R_cC 44-46.

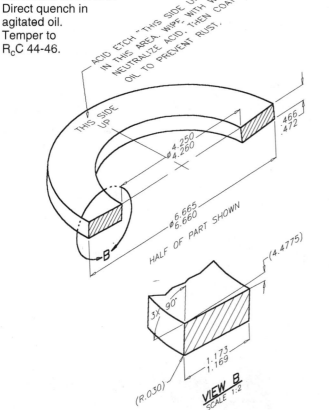

Problem 12–6   Full section (metric)

Part Name: Hydraulic Valve Cylinder
Material: Phosphor Bronze
All Fillets and Rounds R.1

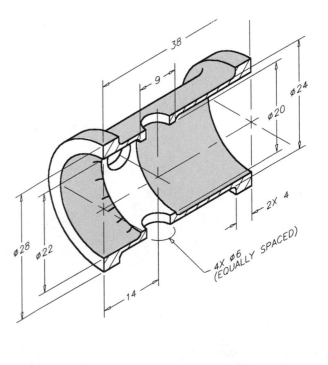

Problem 12–7   Full section (in.)

Part Name: Machine Plate
Material: 6160 T6 Steel

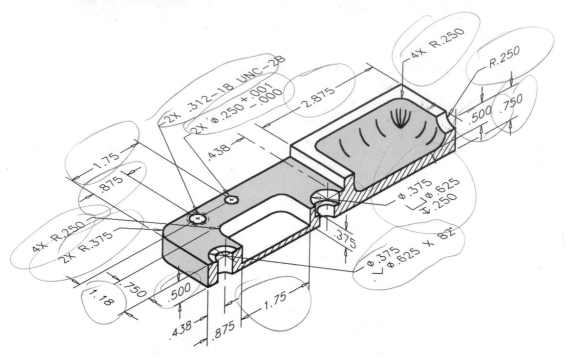

Problem 12–8   Advanced full section and partial auxiliary section from actual industry drawing (in.)

Part Name: Accumulator Plug
Material: Phosphor Bronze
Courtesy Stanley Hydraulic Tools, a Division of the Stanley Works.

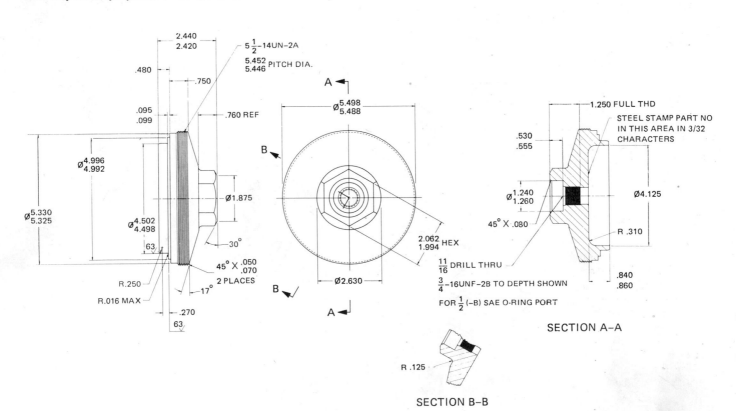

Problem 12–9 Full section (in.)

Part Name: Face Plate
Material: 6160 T6 Steel

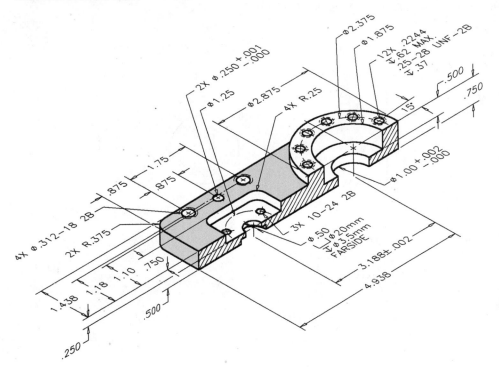

Problem 12–10 Full section, dimensioning cylindrical shapes (in.)

Part Name: Plug
Material: Phosphor Bronze

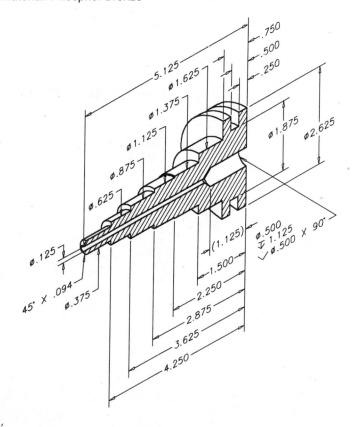

Problem 12–11   Advanced full section and view enlargement from actual industry drawing (in.)

Part Name: Swivel
Material: SAE 5150
Courtesy Stanley Hydraulic Tools, a Division of the Stanley Works.

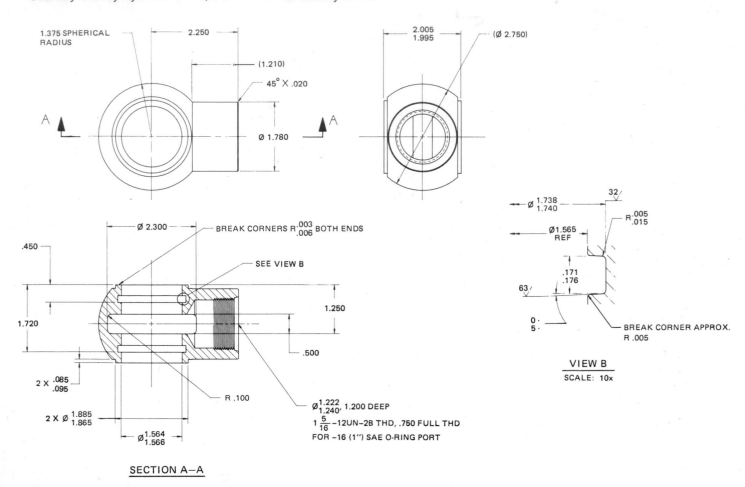

SECTION A–A

VIEW B
SCALE: 10x

Problem 12–12   Full section or half section (in.)

Part Name: Hub
Material: Cast Iron
SPECIFIC INSTRUCTIONS:
Convert the broken-out section in the given drawing to a full section.
NOTE: Type of section used affects view selection.

Problem 12–13   Full and broken-out section (metric)

Part Name: Hydraulic Valve Cylinder
Material: Phosphor Bronze
NOTE: Proposed sections and section lines are not given in engineer's layout.

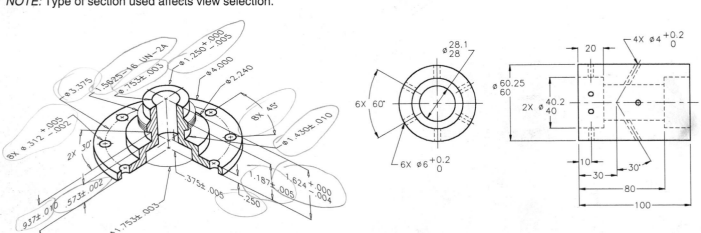

Problem 12–14　Advanced full section, auxiliary section, and view enlargement from actual industry drawing (in.)

Part Name: Bulk Head
Material: Cast Iron
Courtesy Stanley Hydraulic Tools, a Division of the Stanley Works.

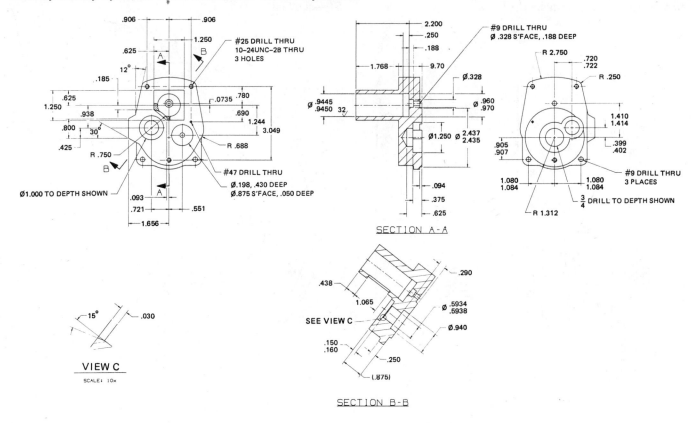

Problem 12–15　Full section (in.)

Part Name: Hanger
Material: SAE 1030
Fillets and Rounds: R.062

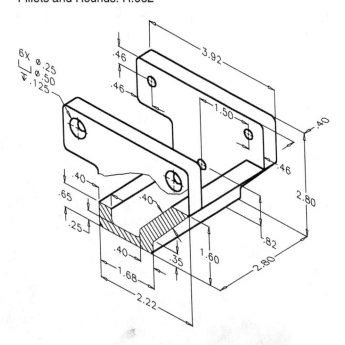

Problem 12–16   Advanced full section from engineer's sketch (metric)

Part Name: Bearing Housing
Material: SAE 1015
Courtesy Aerojet TechSystems Co.

NOTE: Use dimensioning symbols based on the ANSI standards discussed in Chapter 10.

Note the following abbreviations:
CBORE—counterbore
SF—spotface

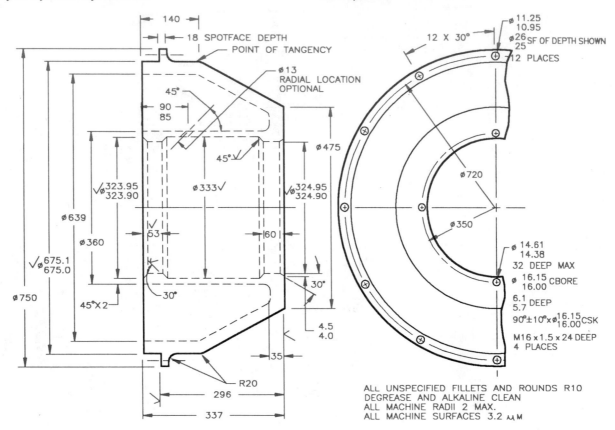

ALL UNSPECIFIED FILLETS AND ROUNDS R10
DEGREASE AND ALKALINE CLEAN
ALL MACHINE RADII 2 MAX.
ALL MACHINE SURFACES 3.2 μM

Problem 12–17   Half section (in.)

Part Name: Dial
Material: Bronze

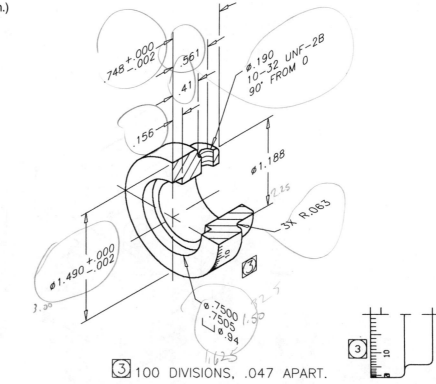

③ 100 DIVISIONS, .047 APART.

Problem 12–18 Half section and auxiliary view (metric)

Part Name: Nozzle Base
Material: Titanium, ASTM-B367 Grade C3
NOTES:
Clean casting by mechanical blasting prior to alpha case removal. Chemically remove alpha case prior to inspection of castings using procedure approved by procurement activity. Cast surfaces shall be visually inspected and be free from cracks, tears, laps, shrinkage, and porosity.
Radiographic inspect castings per MIL-STD-271 acceptance criteria for porosity and inclusions per ASTM-E-446 and for cavity shrinkage per ASTM-E-192 (plates for .75 wall thickness). Severity levels of casting defects shall be no greater than tabulated below for the indicated casting areas.
NOTE: Provide adequate views to avoid crossing extension lines over dimension lines.

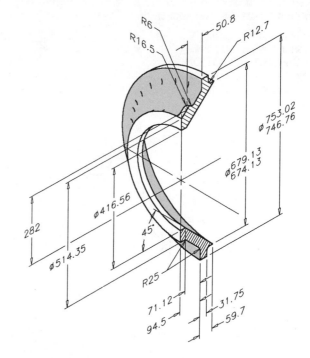

Problem 12–19 Half section and broken-out section (in.)

Part Name: Bench Block
Material: AISI 1018
Case Harden: .020 Deep 59-60 Rockwell C Scale or AISI 4140 Oil Quench 40-45C.
REAM TOLERANCE: $+.0002$ $-.0000$

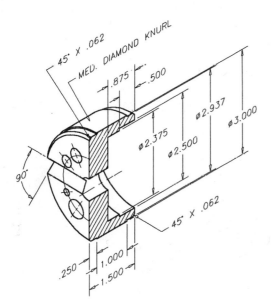

Problem 12–20 Half section (metric)

Part Name: Idler Pully
Material: SAE 4310
Fillets and Rounds: R3.2 mm
NOTE: Many diameter dimensions should be placed on left and right side views.

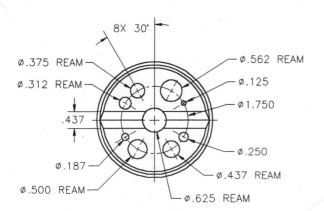

FRONT VIEW

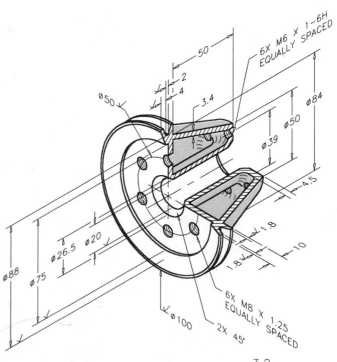

ALL MACHINED SURFACES 3.2

Centerline ⅛

Space 1/32

Problem 12–21 Offset section (in.)

Part Name: Die Casting
Material: SAE 6150
SPECIFIC INSTRUCTIONS:
Convert all dimensions to ANSI Y14.5M standards as shown and discussed in this text.
Courtesy of Kris Altmiller.

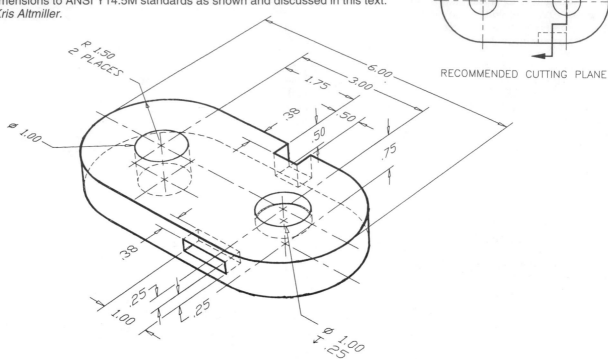

RECOMMENDED CUTTING PLANE

Problem 12–22 Offset section (in.)

Part Name: Drill Plate
Material: SAE 1020
Case Harden: 55 Rockwell C Scale
Fillets and Rounds R.12
FAO 63 μin.

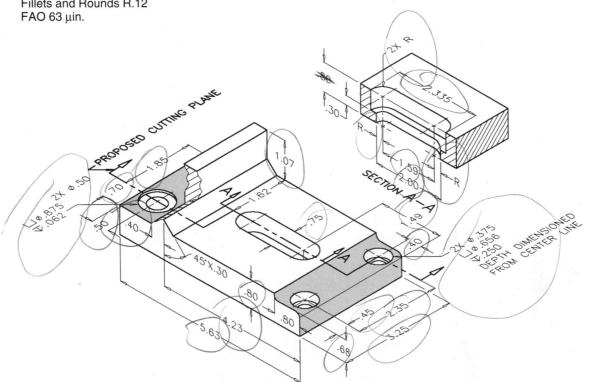

Problem 12–23 Aligned section (in.)

Part Name: Hub
Material: SAE 3145
Fillets and Rounds: R.125

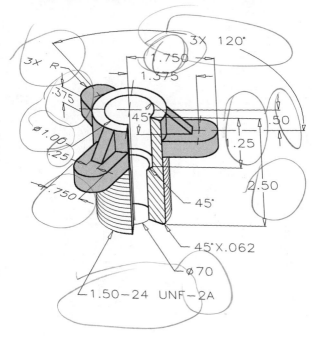

Problem 12–24 Broken-out section (in.)

Part Name: Taper Shaft
Material: SAE 4320
FAO 16 μin.

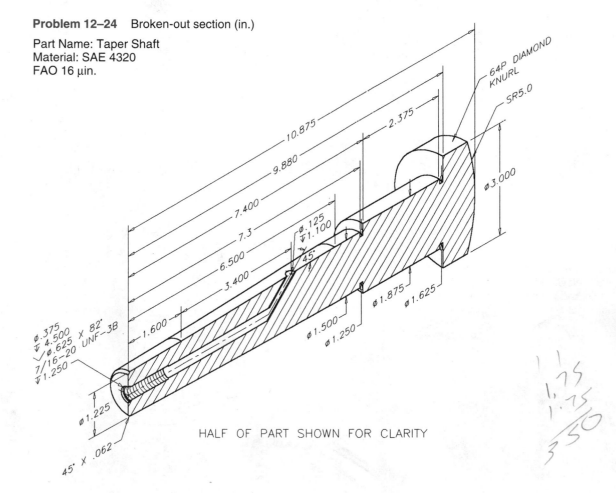

HALF OF PART SHOWN FOR CLARITY

Problem 12–25   Auxiliary section (metric)

Part Name: Gear Base
Material: SAE 2340
Finish All Over: 0.8 μm

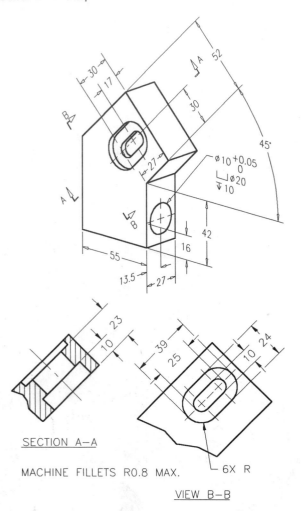

SECTION A–A

MACHINE FILLETS R0.8 MAX.

VIEW B–B

Problem 12–27   Broken-out section (metric)

Part Name: 25 mm 45° Elbow
Material: Cast Iron
Fillets and Rounds: R4 mm
Consider bottom view or auxiliary view for hole pattern dimensions.

Problem 12–26   Broken-out section, view enlargement (in.)

Part Name: Clamp Cap
Material: Cast Aluminum
Fillets and Rounds: R.06

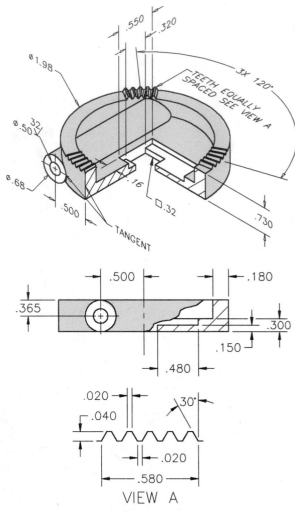

VIEW A

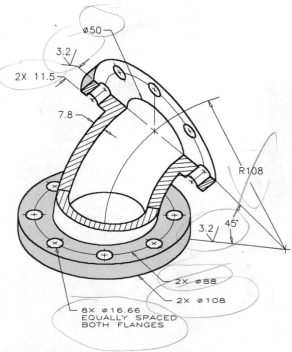

Problem 12–28 Revolved section (in.)

Part Name: End Loading Arm
Material: SAE 2310
Fillets and Rounds: R.25

Problem 12–29 Revolved section (in.)

Part Name: Offset Handwheel
Material: Bronze
All Fillets and Rounds: R.12
Finishes: 125 μin.

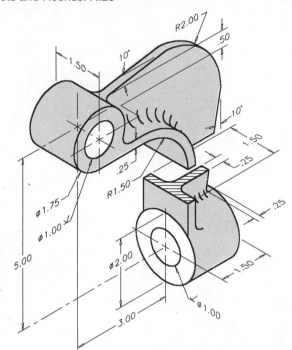

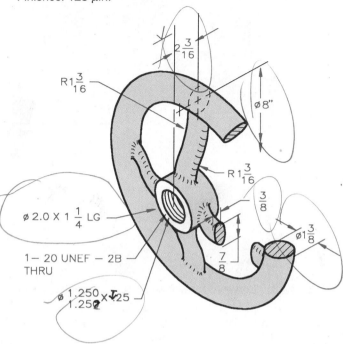

Problem 12–30 Advanced removed sections and view enlargements from actual industry drawing (in.)

Part Name: Swivel Stem
Material: SAE 4340
Courtesy Stanley Hydraulic Tools, a Division of the Stanley Works.

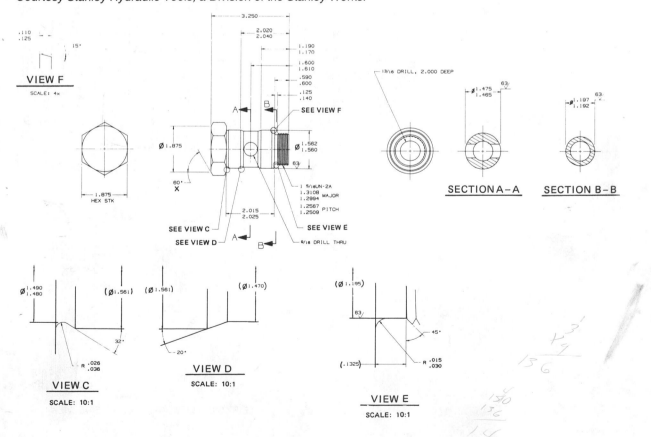

Problem 12–31 Removed section view enlargement, keyways (in.)

Part Name: Taper Shaft
Material: SAE 3130

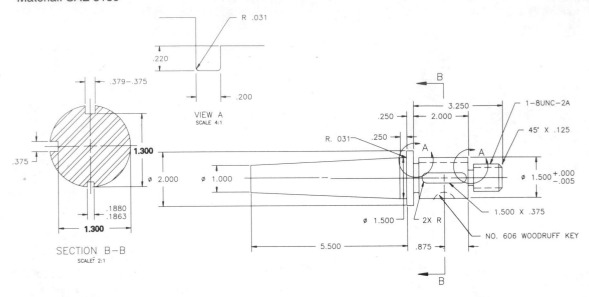

Problem 12–32 Broken-out, revolved section, view enlargement (in.)

Part Name: Pipe Wrench Handle
Material: SAE 5120
Fillets and Rounds: R.06
SPECIFIC INSTRUCTIONS:
Engineering sketch is given in fractional inches, convert all size dimensions to two-place decimals and location dimensions to three-place decimals for final drawing.

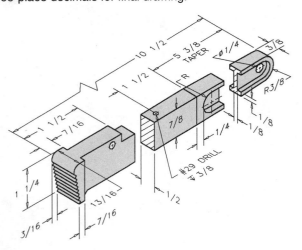

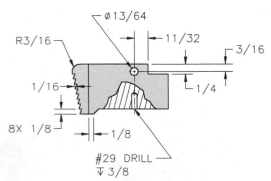

Problem 12–33 Removed sections (metric)

Part Name: Valve Stem
Material: Phosphor Bronze
Finish All Over: 1 μm

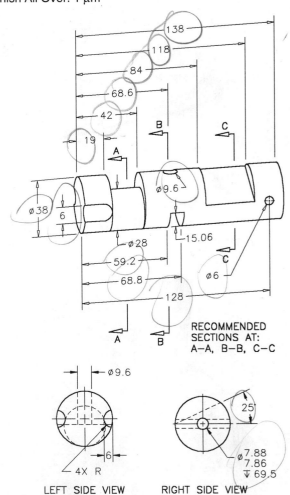

Problem 12–34 Offset section (in.)

Part Name: Die Plate Casting
Material: SAE 3120
Courtesy of Kris Altmiller.

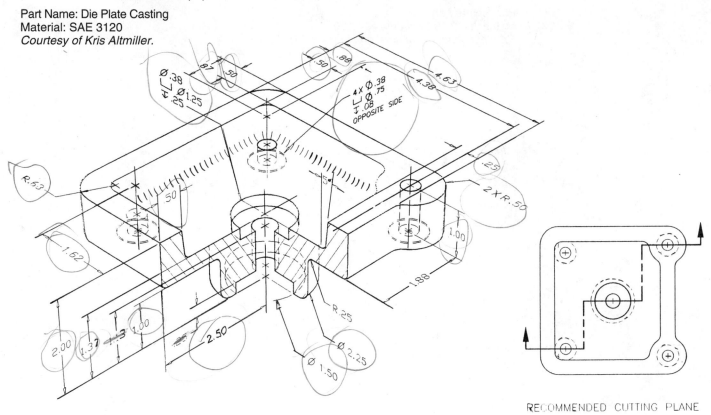

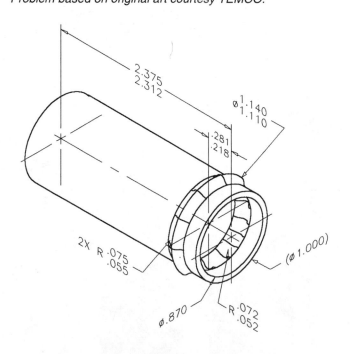

RECOMMENDED CUTTING PLANE

ALL FILLETS AND ROUNDS R.25 UNLESS OTHERWISE SPECIFIED

Problem 12–35 Broken-out section (in.)

Part Name: Slide Bar Connector
Material: SAE 4120
SPECIFIC INSTRUCTIONS:
Convert point-to-point dimensioning to datum dimensioning.

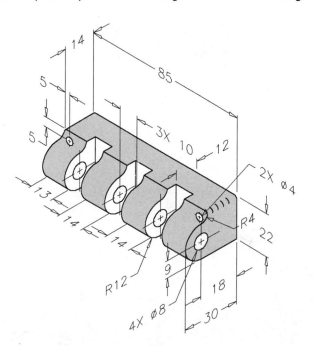

Problem 12–36 Broken-out section (in.)

Part Name: Drain Tube
Material: ∅1.00 × .065
Problem based on original art courtesy TEMCO.

Problem 12–37   Conventional break (in.)

Part Name: Leg
Material: Ø2.00 Schedule 40 A120
Problem based on original art courtesy TEMCO.

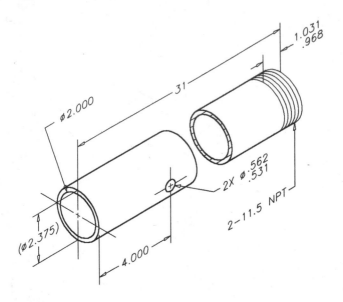

Problem 12–38   Revolved section, conventional revolution (in.)

Part Name: Crank Arm
Material: Cast Steel 80,000
All Fillets and Rounds: R.12

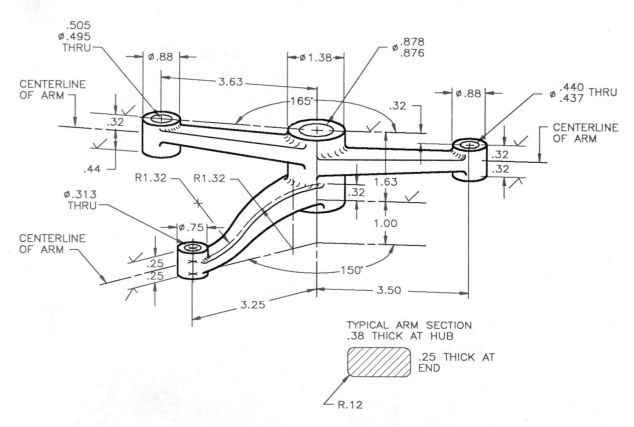

Problem 12–39 You select the sectioning technique (in.)

Part Name: Base Support
Material: Mild Steel

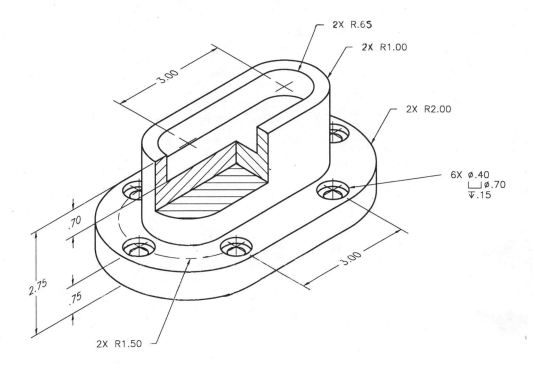

Problem 12–40 You select the sectioning technique (in.)

Part Name: Mounting Base
Material: Cast Aluminum

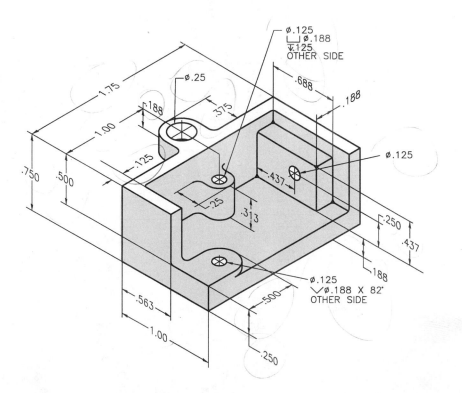

Problem 12–41 Advanced half and full sections from actual industry drawing (in.)

Part Name: Diffuser Casting
Material: Titanium

SPECIFIC INSTRUCTIONS:
Convert all dimensions to ASME Y14.5M standards as shown and discussed in this text. Display the casting material using phantom lines as shown in the engineer's layout. Use ANSI standard cutting-plane line.

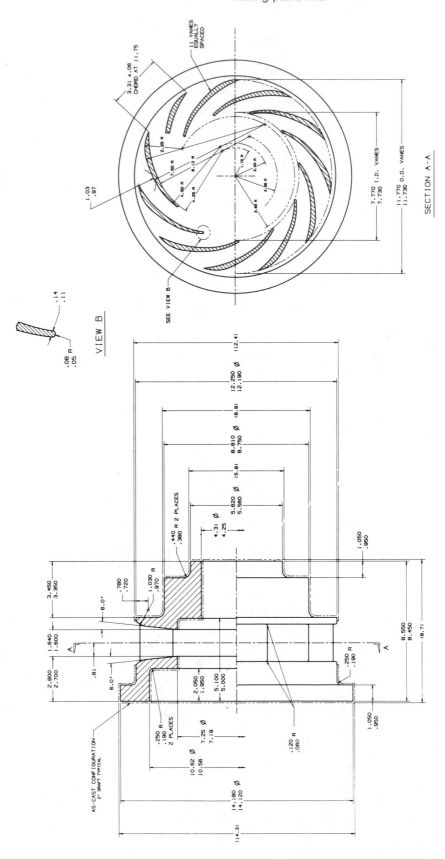

Problem 12–42   Advanced aligned section from actual industry drawing (in.)

Part Name: Crankshaft Adapter
Material: CI

SPECIFIC INSTRUCTIONS:
1.   Convert given dimensions to ASME standards.
2.   Use ANSI standard cutting-plane line.
Courtesy American Hoist and Derrick Company.

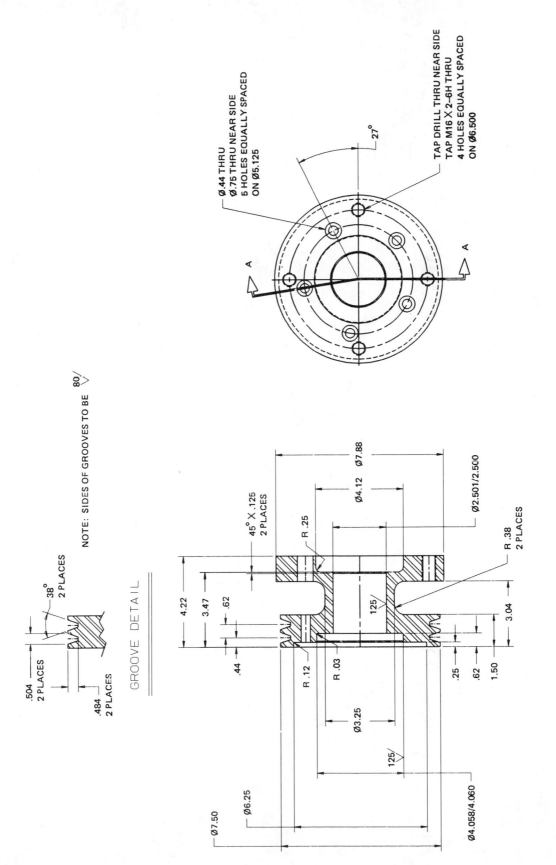

Problem 12–43   Advanced aligned sections from actual industry drawing (metric)

Part Name: Steering Pump Adapter
Material: HC-80
Courtesy Hyster Company.

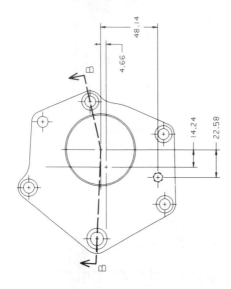

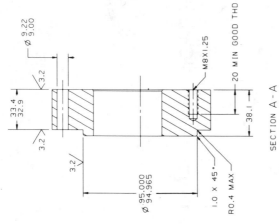

SECTION A - A

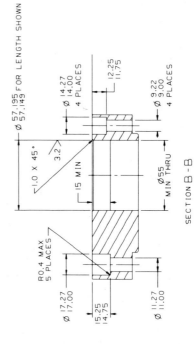

SECTION B - B

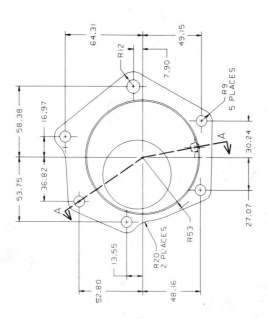

Problem 12–44 Advanced offset section, enlarged views, from actual industry drawing (metric)

Part Name: Pad-Monotrol
Material: Hi-mount Pro-fax 8623
SPECIAL INSTRUCTIONS:
Avoid leaders crossing dimension lines.
Courtesy Hyster Company.

Problem 12–45 Advanced half and full sections (in.)

Refer to the drawing for Problem 12–41 (page 376).

Part Name: Diffuser Casting
Material: Titanium
SPECIFIC INSTRUCTIONS:
Convert all dimensions to ASME Y14.5M standards as shown and discussed in this text. Casting shown in phantom lines. Omit phantom line technique for showing casting and make two drawings: one a *casting* drawing with only casting information, and the other a *machining* drawing with only machining dimensions and information. Refer to Chapters 9 and 10 for a review of this method.

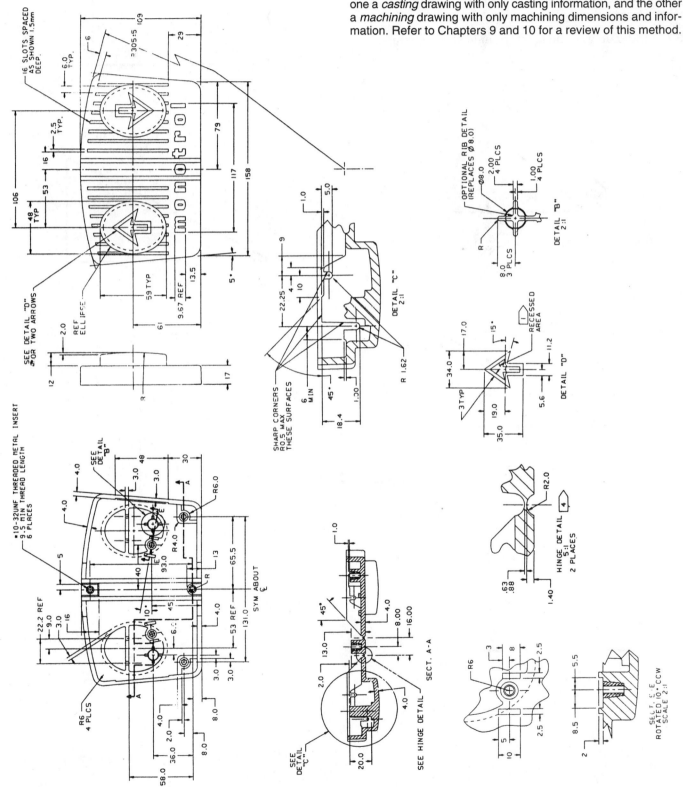

MATH PROBLEMS

1. A bolt circle with radius 14 in. has six equally spaced holes. Find the center-to-center distance between the adjacent holes.

2. A bolt circle with radius 16 in. has eight equally spaced holes. Find the center-to-center distance between the adjacent holes.

3. Find side y of the right triangle shown below.

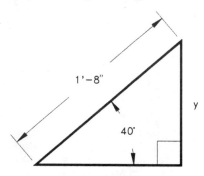

4. Find side x of the right triangle shown below. (Hint: Use cosine.)

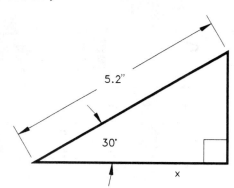

5. Find angle A of the right triangle shown below.

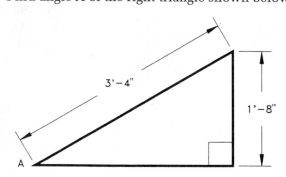

6. Find angle Θ of the right triangle shown below.

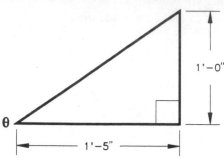

7. Find side y of the right triangle shown below.

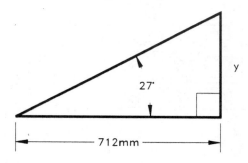

8. Find side r of the right triangle shown below.

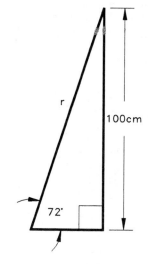

CHAPTER 13

Geometric Tolerancing

LEARNING OBJECTIVES

After completing this chapter, you will:

○ Label datum features on a drawing.
○ Establish basic dimensions where appropriate.
○ Place proper feature control frames on drawings establishing form, profile, orientation, runout, and location geometric tolerances.
○ Use and interpret material condition symbols.
○ Determine the virtual condition of features.
○ Use geometric tolerancing to completely dimension objects from an engineer's sketch and actual industrial layouts.
○ Solve an engineering problem from engineer's notes.

THE ENGINEERING DESIGN PROCESS

There have been some problems in manufacturing one of the parts due to the extreme tolerance variations possible. The engineer requests that you revise the drawing shown in Figure 13–1 and gives you these written instructions:

○ Establish datum A with three equally spaced datum target points at each end of the Ø28.1–28.0 cylinder.

○ Establish datum B at the left end surface.
○ Make the bottom surfaces of the 2X Ø40.2–40.0 features perpendicular to datum A by 0.06.
○ Provide a cylindricity tolerance of 0.3 to the outside of the part.
○ Make the 2X Ø40.2–40.0 features concentric to datum A by 0.1.
○ Locate the 6X Ø6 + 0.2 holes with reference to

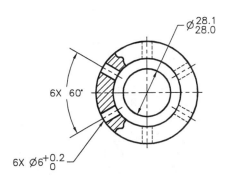

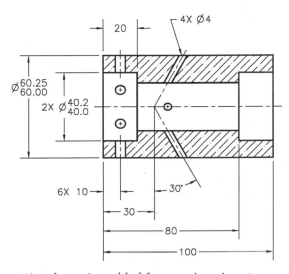

Figure 13–1   The original drawing to be revised with geometric tolerancing added from engineer's notes.

(Continued)

datum A at MMC and datum B with a position tolerance of 0.05 at MMC.

○ Locate 4X Ø4 holes with reference to datum A at MMC and datum B with a position tolerance of 0.04 at MMC.

This is an easy job because you did the original drawing on CADD and you have a custom geometric tolerancing package. You go back to the workstation and within one hour you have the check plot shown in Figure 13–2 ready for evaluation.

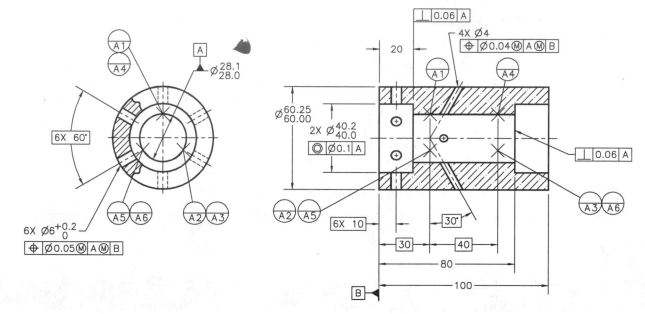

Figure 13–2   The revised drawing with geometric tolerancing added.

Geometric tolerancing (GT) is the dimensioning and tolerancing of individual features of a part where the permissible variations relate to characteristics of form, profile, orientation, runout, or the relationship between features. This subject is commonly referred to as geometric dimensioning and tolerancing (GD & T).

ANSI   The standards for geometric dimensioning and tolerancing (GD & T) are governed by the American National Standards document ASME Y14.5M—1994 titled *Dimensioning and Tolerancing,* published by the American Society of Mechanical Engineers. The proper representation of all dimension and tolerancing symbols is displayed in Appendix F.

▼ GENERAL TOLERANCING

Tolerancing was introduced in Chapter 10 with definitions and examples of how dimensions are presented on a drawing. Review Chapter 10 to establish a solid understanding of dimensioning and tolerancing terminology.

The term *general tolerancing* as used here implies dimensioning without the use of geometric tolerancing. General tolerancing applies a degree of form and location control by increasing or decreasing the tolerance. The *limits of size* of a feature control the amount

of variation in size and geometric form. This is the boundary between maximum material condition (MMC) and least material condition (LMC). The form of the feature may vary between the upper limit and lower limit of a size dimension. This is known as *extreme form variation.* (See Figure 13–3.)

▼ SYMBOLOGY

Symbols on drawings represent specific information that would otherwise be difficult and time consuming to duplicate in note form. Symbols must be clearly drawn to required size and shape so they communicate the desired meaning uniformly. The exact drafting specifications are given when each symbol is first introduced. Use these specifications to prepare your own CADD blocks or symbols library. Geometric tolerancing symbols are divided into the following types:

1.   Datum symbols.
2.   Geometric characteristic symbols.
3.   Material condition (modifying) symbols.
4.   Feature control frame.
5.   Supplementary symbols.

A variety of professional templates and CADD programs are available to help save time when preparing GT

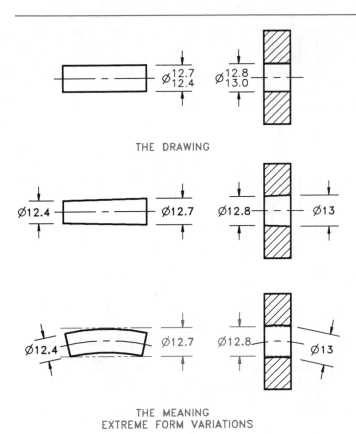

THE DRAWING

THE MEANING
EXTREME FORM VARIATIONS

Figure 13–3 Extreme form variations of given tolerances.

symbols. It is recommended that all dimensioning symbols be drawn with a template or CADD program for good representation, accuracy, and speed.

▼ DATUM FEATURE SYMBOLS

Datums are considered to be theoretically perfect surfaces, planes, points, or axes that are established from the true geometric counterpart of the datum feature. The datum feature, as shown in Figure 13–4, is the actual feature of the part that is used to establish the datum.

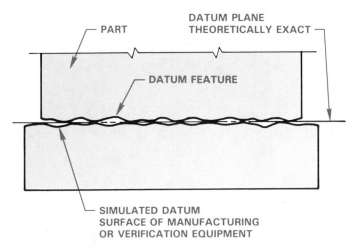

Figure 13–4 Magnified representation of a datum feature.

Datums are used to originate size and location dimensions. Without the use of GT, datums are often implied. When dimensions originated at a common surface, for example, that surface was assumed to be the datum. The only problem with this method was that the manufacturing operation may not have read the datum in the same way as the engineering department, or the implied datum may have been ignored all together. With the use of defined datums each part is made with dimensions that begin from the same origin; there is no variation.

Datum feature symbols are commonly drawn using thin lines with the symbol size related to the size of the lettering on a drawing, as shown in Figure 13–5. The triangular base may be filled or unfilled. However, the filled triangle helps the symbol show up better on a drawing.

Datum feature symbols are placed in the view that shows the edge of a surface or are attached to a diameter or symmetrical dimension when associated with a centerline or center plane. (See Figure 13–6.)

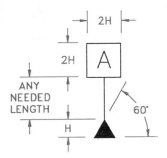

H = LETTERING HEIGHT

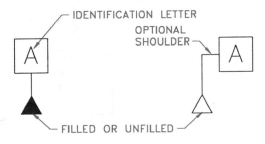

Figure 13–5 Datum feature symbol.

▼ DATUM REFERENCE FRAME

Datum referencing is used to relate features of a part to an appropriate datum or datum reference frame. A datum indicates the origin of a dimensional relationship between a toleranced feature and a designated feature or features on a part. The datum identification symbol on a drawing relates to an actual feature on the part known as the *datum feature*. The part is then placed on the surface of an inspection table or manufacturing verification equipment. This inspection table or equipment is referred to as the *simulated datum*. The *datum* is the theoretically

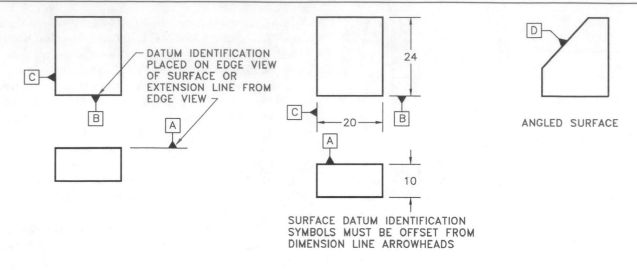

SURFACE DATUMS

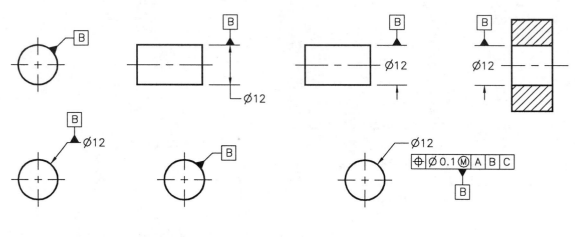

AXIS DATUMS

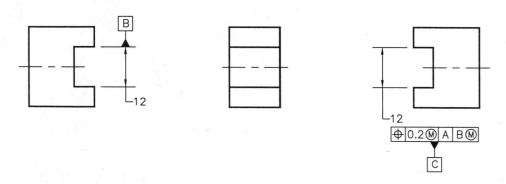

AXIS AND CENTER PLANE DATUM IDENTIFICATION
SYMBOLS MUST ALIGN WITH OR REPLACE THE
DIMENSION LINE ARROWHEAD OR BE PLACED ON
THE FEATURE, LEADER SHOULDER, OR FEATURE
CONTROL FRAME.

CENTER PLANE DATUMS

Figure 13–6   Datum identification.

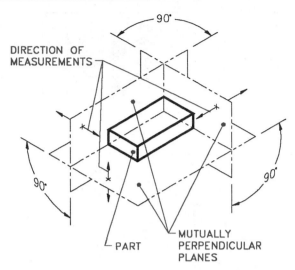

Figure 13–7 Datum reference frame.

exact plane, axis, or point that is established by the true geometric counterpart of the specified datum feature. Measurements cannot be made from the true geometric counterpart, because it is theoretical or assumed to exist. The machine tables, surface plates, or inspection tables are of such high quality that they are used to simulate the datums from which measurements are taken and dimensions are verified. In this way, each dimension always originates from the same reliable location. Dimensions are never taken or verified from one surface of the part to another.

Sufficient datum features, those most important to the design of a part, are chosen to position the part in relation to a set of three mutually perpendicular planes, jointly called a *datum reference frame.* This reference frame exists in theory only and not on the part. Therefore, it is necessary to establish a method for simulating the theoretical reference frame from the actual features of the part. This simulation is accomplished by positioning the part on appropriate datum features to adequately relate the part to the reference frame and to restrict the motion of the part in relation to it. (See Figure 13–7.) These planes are simulated in a mutually perpendicular relationship to provide direction as well as the origin for related dimensions and measurements. Thus when a part is positioned on the datum reference frame, dimensions related to the datum reference frame by a feature control frame or note are thereby mutually perpendicular. This theoretical reference frame constitutes the three plane dimensioning system used for datum referencing.

▼ DATUM FEATURES

A datum feature is selected on the basis of its geometric relationship to the toleranced feature and the requirements of the design. To ensure proper part interface and assembly, corresponding features of mating parts are also selected as a datum feature where practical. Datum features must be readily recognizable on the part. Therefore, in the case of symmetrical parts or parts with identical features, physical identification of the datum feature on the part may be necessary. A datum feature should be accessible on the part and be of sufficient size to permit processing operations.

The three reference planes are mutually perpendicular unless otherwise specified. Planes in the datum reference frame are placed in order of importance. For example, the plane that originates most size and location dimensions, or has the most functional importance, is the *primary,* or first, datum plane. The next datum is the *secondary,* or second, datum plane and the third element of the frame is called the *tertiary,* or third, datum plane. The desired order of precedence is indicated by entering the appropriate datum reference letters from left to right in the feature control frame. For instructional purposes it may be convenient to label datums as A, B, and C although in industry other letters are also used to identify datums such as D, E, and F, or X, Y, and Z. The letters that should be avoided are O, Q, and I. Figure 13–8a shows a part and the planes that are chosen as datum features. Notice in Figure 13–8b how the order of precedence of datum features relates the part to the datum reference frame. The datum features are identified as surfaces D, E, and F. These surfaces are most important to the design and function of the part. Surfaces D, E, and F are the primary, secondary, and tertiary datum features respectively since they appear in that order in the feature control frame.

Multiple datum frames may be established for some parts depending on the complexity and function of the part. The relationship between datum frames is often controlled by a representative angle. (See Figure 13–9.)

Partial Surface Datums

In some situations it may be more realistic to apply a datum feature symbol to a portion of a surface rather than the entire surface. For example, when a long part has related features located in one or more concentrated places, then the datum may be located next to the important features. When this is done, a chain line is used to identify the extent of the datum feature. The location and length of the chain line must be dimensioned. (See Figure 13–10.)

Coplanar Surface Datums

Coplanar surfaces are two or more surfaces that are in the same plane. The relationship of coplanar surface datums may characterize the surfaces as one plane or datum in correlated geometric callouts as shown in Figure 13–11.

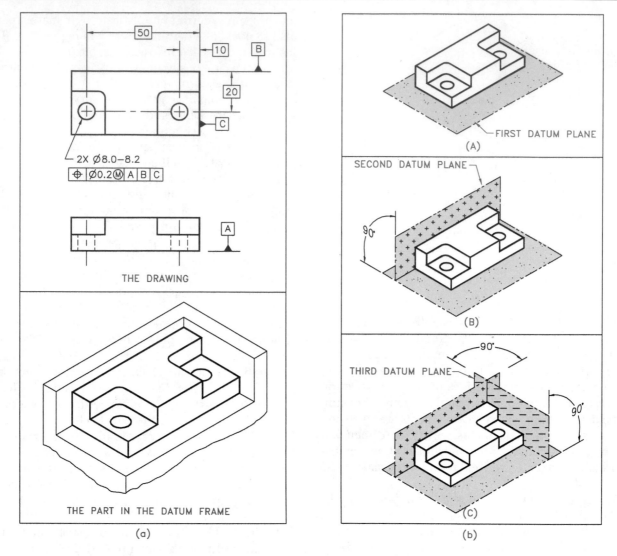

Figure 13–8 **(a)** Part in which datum features are plane surfaces; **(b)** sequence of datum features relates part to datum reference frame.

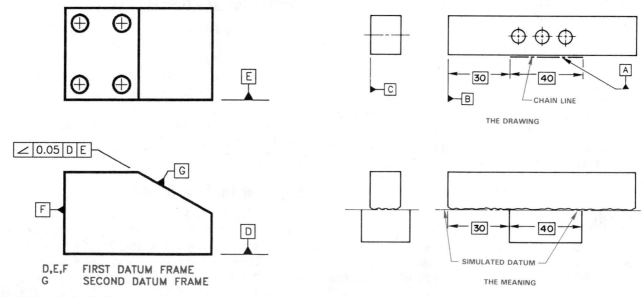

Figure 13–9 Multiple datum frames.

Figure 13–10 Partial surface datums.

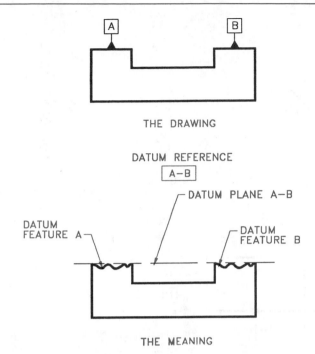

Figure 13–11   Coplanar surface datums.

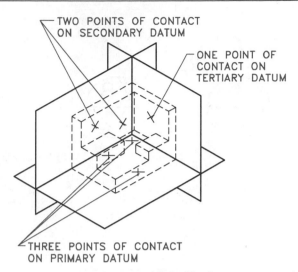

Figure 13–12   Datum frame established by datum target symbols.

▼ DATUM TARGET SYMBOLS

In many situations it is not possible to establish an entire surface or surfaces as datums, or it may not be practical to coordinate a partial datum surface with the part features. When this happens due to the size or shape of the part, then datum targets may be used to establish datum planes. Datum targets are especially useful on parts with surface or contour irregularities such as sand castings or forgings, on some sheet metal parts subject to bowing or warpage, or on weldments where heat can induce warpage. Datum targets are designated points, lines, or surface areas that are used to establish the datum reference frame. The primary datum is established by three points or contact locations. The three locations are placed on the part so they form a stable, well-spaced pattern and are not in a line. The secondary datum is created by two points or locations. The tertiary datum is established by one point or location. Remember the purpose of the datum frame is to prepare a stable position for the part dimensions to be established in relationship to corresponding datums. For this reason, the three points on the primary datum are used to provide stability similar to a three legged stool. The two points on the secondary datum provide the needed amount of stability when the object is placed against the secondary datum. Finally, the tertiary datum requires only one point of contact to complete the stability between the three datum frame elements. (See Figure 13–12.)

The datum target symbol is drawn as a circle using thin lines and is connected with a leader that points to a target point, line, or surface area. The datum target symbol is divided into two halves by a horizontal line.

The top half of the symbol is reserved for identification of the datum target area size when used. The bottom half of the symbol is used for datum target identification. For example, if there are three datum target points on a datum, the first point may be labeled A1, the second A2, and the third A3. The datum target point is located on the surface or edge view from adjacent datums with basic dimensions. (See Figure 13–13.) Chain dimensioning is commonly used to locate datum target points and target areas. The location dimensions must originate from datums. Datum target areas are located to their centers. (See Figure 13–14.) When the surface of a part has an irregular contour or different levels, the contact points, lines, or areas may lie on the same plane and different lengths of locating pins may be used. The pins used to make point contact are usually rounded. The pins for target area contact have a flat surface equal in diameter to the target area.

▼ DATUM AXIS

The datum frame established by a cylindrical object is the base of the part and the two theoretical planes, represented in Figure 13–15 by the X and Y center planes, which cross to establish the datum axis. The actual secondary datum is the cylindrical surface. The center planes located at X and Y are used to indicate the direction of dimensions that originate from the datum axis. (See Figure 13–16.)

Datum target points, lines, or surface areas may also be used to establish a datum axis. A primary datum axis may be established by two sets of three equally spaced targets: a set at one end of the cylinder and the other set near the other end as shown in Figure 13–17. When two cylindrical features of different diameter are used to establish a datum axis then the datum target points are identified in correspondence to the adjacent cylindrical datum feature as shown in Figure 13–18.

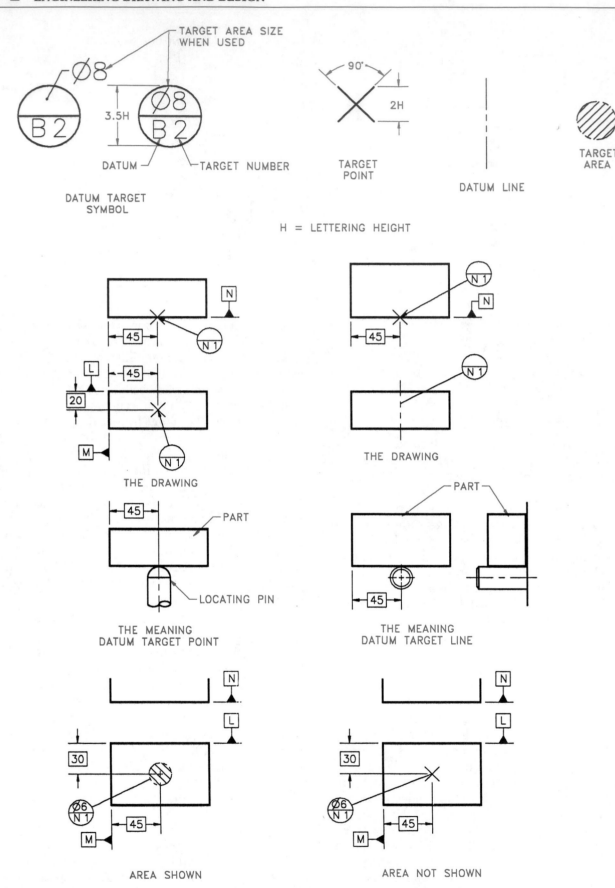

Figure 13–13 Datum target symbol, target point, datum line, and target area.

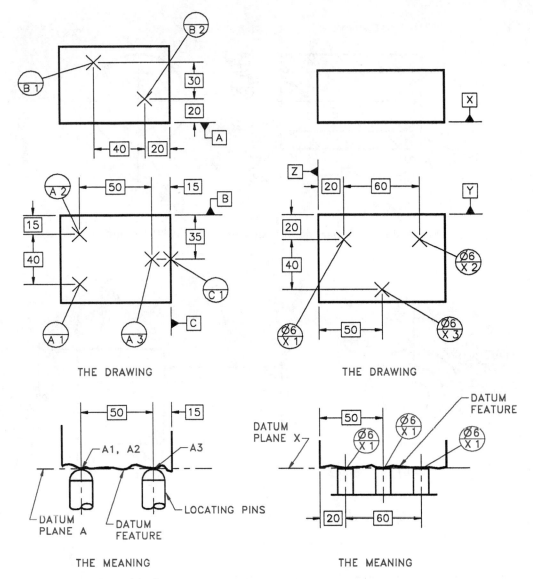

THE DRAWING   THE DRAWING

THE MEANING   THE MEANING

Figure 13–14   Locating datum targets.

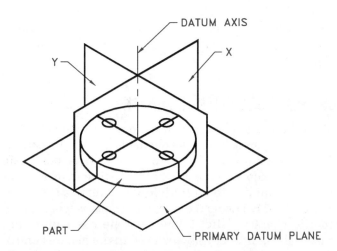

Figure 13–15   Datum frame for cylindrical object.

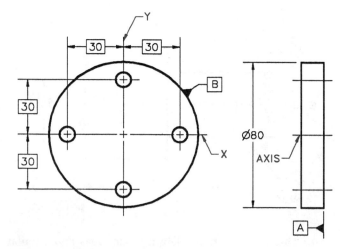

Figure 13–16   Datum axis.

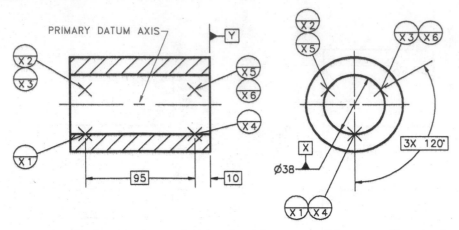

Figure 13–17 Establishing a primary datum axis with two sets of three equally spaced targets.

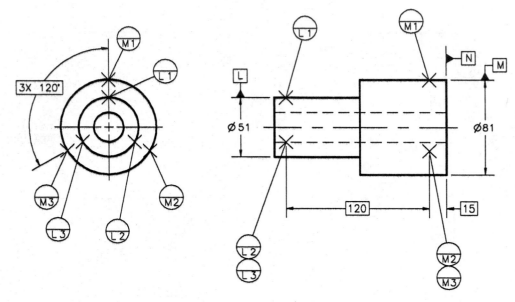

Figure 13–18 Datum targets identified on adjacent cylindrical features.

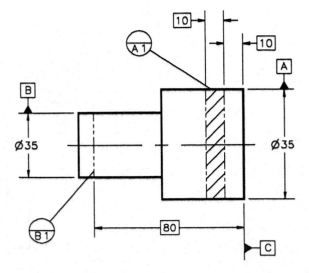

Figure 13–19 Cylindrical datum target areas and circular datum target lines.

Cylindrical datum target areas and circular datum target lines may also be used to establish the datum axis of cylindrically shaped parts as shown in Figure 13–19. A secondary datum axis is established by placing three equally spaced targets on the cylindrical surface. (See Figure 13–20.)

▼ FEATURE CONTROL FRAME

The feature control frame is used to relate a geometric tolerance to a part feature. The elements in a feature control frame must always be in the same order. The most basic format is when a geometric characteristic and related tolerance is applied to an individual feature as shown in Figure 13–21. The next expanded format is when a geometric characteristic, tolerance zone descriptor, and material condition symbol is used in the feature control frame. (See Figure 13–22.) One, two, or three datum ref-

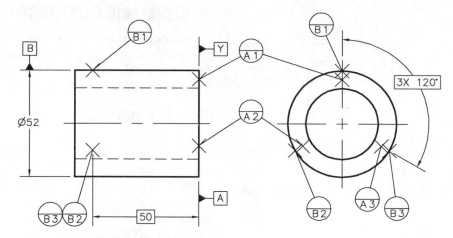

Figure 13–20 Establishing a secondary datum on a cylindrical object with three equally spaced targets.

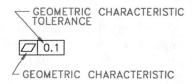

Figure 13–21 Feature control frame with geometric characteristic and related tolerance.

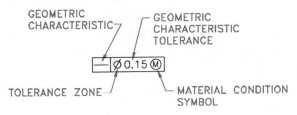

Figure 13–22 Feature control frame with geometric characteristic, tolerance zone descriptor, and material condition symbol.

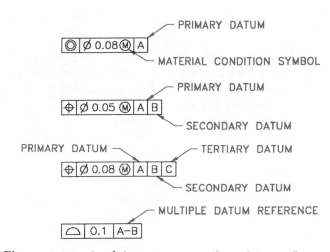

Figure 13–23 Applying one, two, or three datum references to the feature control frame.

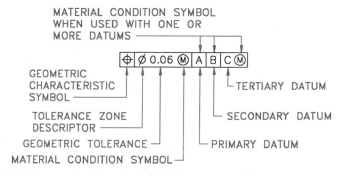

Figure 13–24 Elements of a feature control frame.

erences may be included in a feature control frame as shown in Figure 13–23. The material condition symbol that applies to the feature tolerance is always placed after the tolerance. When a material condition symbol is applied to a datum reference, it is placed after the datum reference and in the same compartment as shown in Figure 13–24.

The feature control frame is drawn with thin lines and connected to the feature with a leader or extension line. (See Figure 13–25.) The feature control frame may be combined with a datum feature symbol when the feature controlled by the geometric tolerance also serves as a datum. The datum feature symbol and the datum reference in the feature control frame are considered separately. When a datum feature symbol and a feature control frame are combined, the feature control frame should be shown first. The datum feature symbol is centered on the base of the feature control frame. (See Figure 13–26.)

▼ BASIC DIMENSIONS

A basic dimension is defined as any size or location dimension that is used to identify the theoretically exact size, profile, orientation, or location of a feature or datum target. Basic dimensions are the basis from which permissible variations are established by tolerances on

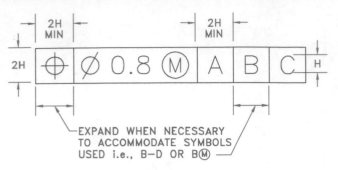

H = LETTERING HEIGHT

Figure 13–25   Detailed feature control frame and application.

Figure 13–26   Combination datum feature symbol and feature control frame.

other dimensions in notes or in feature control frames. A basic dimension is described on a drawing by placing a thin-line box around the dimension. Metric and inch basic dimension numerals follow the same rules applied to other dimension numerals. Refer to Chapter 10. Metric basic dimensions contain only the number of decimal places needed to display the intended control. While the same is true with inch basic dimensions, the basic dimension value is expressed in the same number of decimal places as the related tolerance. (See Figure 13–27.)

▼ GEOMETRIC TOLERANCES

Geometric tolerances are divided into five types: form, profile, orientation, runout, and location. These tolerances are subdivided into thirteen characteristics plus modifying terms, all of which are discussed in this section.

Form Tolerance

A tolerance of form is commonly applied to individual features or elements of single features and is not related to datums. The amount of given form variation must fall within the specified size tolerance zone.

Straightness.   The straightness symbol is detailed in Figure 13–28. Perfect straightness exists when a surface element or the axis of a part is a straight line. A straightness tolerance allows for a specified amount of variation from a straight line. The straightness feature control frame may be attached to the surface of the object with a leader or combined with the diameter dimension of the part. When straightness is connected to the surface of the feature with a leader, *surface straightness* is implied as

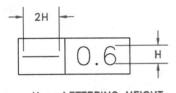

H = LETTERING HEIGHT

Figure 13–28   Straightness feature control frame.

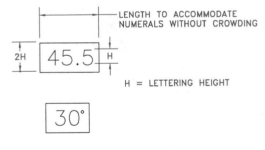

H = LETTERING HEIGHT

BASIC DIMENSION SYMBOLS

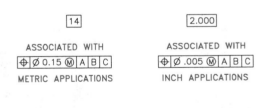

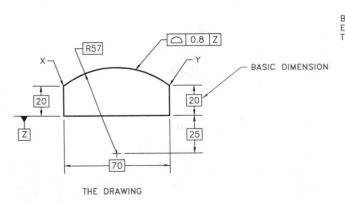

THE DRAWING

Figure 13–27   Basic dimensions.

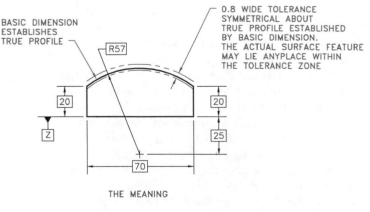

THE MEANING

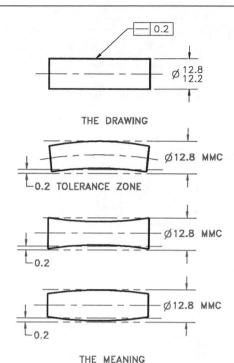

Figure 13–29 Surface straightness.

shown in Figure 13–29. When the straightness feature control frame is combined with the diameter dimension of the part, then *axis straightness* is specified. (See Figure 13–30.) While a straightness tolerance is common on cylindrical parts, straightness may also be applied to the surface of noncylindrical parts. When specifying center feature straightness, a center plane is implied and the diameter tolerance zone descriptor is omitted. The straightness geometric tolerance must always be less than the size tolerance.

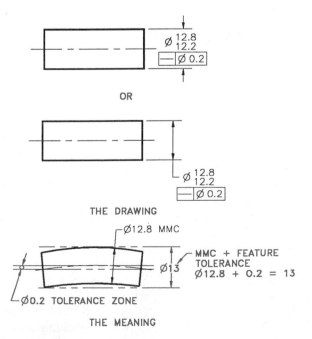

Figure 13–30 Axis straightness.

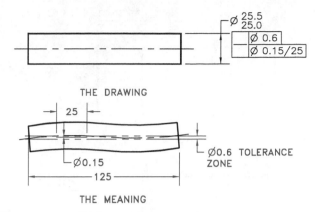

Figure 13–31 Unit straightness.

Unit straightness is a situation where a specified tolerance is given per unit of length, for example 25 mm units, and a greater amount of tolerance is provided over the total length of the part. (See Figure 13–31.) This is intended to control excessive waviness in a long part.

Flatness. The flatness tolerance symbol is detailed in relationship to the feature control frame in Figure 13–32. A surface is considered flat when all of the surface elements lie in one plane. A flatness geometric tolerance callout allows for a specified amount of surface variation from a flat plane. The flatness tolerance zone establishes two parallel planes. The actual surface of the object may not extend beyond the boundary of the size tolerance zone and, when associated with the size dimension, the flatness tolerance must be smaller than the size tolerance. The flatness feature control frame may be connected to the edge view of the surface with a leader or with an extension line. (See Figure 13–33.)

Unit flatness may be specified when it is desirable to control the flatness of surface units. The unit size may be 25 × 25 mm or 1.00 × 1.00 in. This is used to prevent abrupt surface variation. A unit flatness callout may be presented with or without a separate tolerance for the total area. Unit flatness, just as unit straightness, often have a total tolerance in order to avoid a situation where the unit tolerance gets out of control. The size of the unit area is given after the unit tolerance specification. The feature control frame is doubled in height and the total area flatness is given first as shown in Figure 13–34.

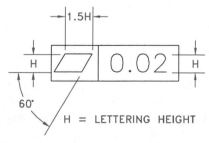

Figure 13–32 Flatness feature control frame.

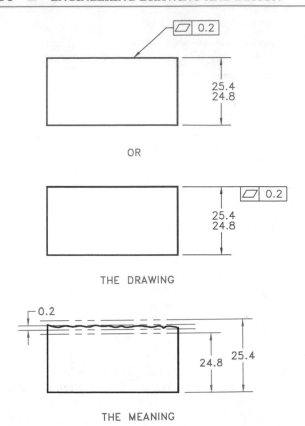

OR

THE DRAWING

THE MEANING

Figure 13–33 Flatness representation.

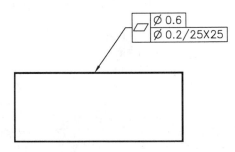

Figure 13–34 Unit flatness.

Circularity. The circularity geometric characteristic symbol is detailed in a feature control frame in Figure 13–35. Circularity may be applied to cylindrical, conical, or spherical shapes. Circularity exists when all of the elements of a circle are the same distance from the center. Circularity is a cross-sectional evaluation of the feature to determine if the circular surface lies between a tolerance zone that is made up of two concentric circles. The cross-sectional tolerance zone is established perpendicular to the axis of the feature. The term *circular* or *line element* is used to refer to a cross-sectional or single line tolerance zone as opposed to a blanket or entire surface tolerance zone. The circularity geometric tolerance zone is a radius dimension. The circularity feature control frame may be connected to the part with a leader in the circular or rectangular view as shown in Figure 13–36. The circularity tolerance must always be less than the size

H = LETTERING HEIGHT

Figure 13–35 Circularity feature control frame.

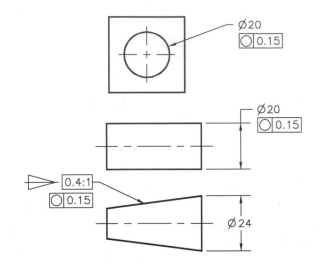

THE DRAWING

0.15 TOLERANCE ZONE

THE MEANING

Figure 13–36 Circularity representation.

H = LETTERING HEIGHT

Figure 13–37 Using the free state symbol.

tolerance except for parts subject to free state variations. *Free state variation* is a term used to describe distortion of a part after removal of forces applied during manufacturing. The part may have to meet the tolerance specifications while in free state, or it may be necessary to hold features in a simulated mating part to verify dimensions. The free state symbol shown in Figure 13–37 is placed in the feature control frame after the geometric tolerance and material condition symbol, if any.

Cylindricity. The cylindricity geometric characteristic symbol is detailed in Figure 13–38. Cylindricity is sim-

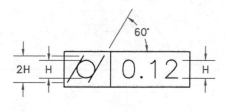

H = LETTERING HEIGHT

Figure 13–38 Cylindricity feature control frame.

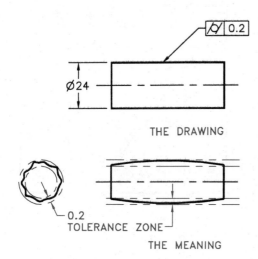

THE DRAWING

THE MEANING

Figure 13–39 Cylindricity representation.

ilar to circularity in that both have a radius tolerance zone. The difference is that circularity is a cross-sectional tolerance which results in a feature that must lie between two concentric circles, while cylindricity is a blanket tolerance that results in a feature lying between two concentric cylinders. The cylindricity feature control frame may be connected with a leader to either the circular or rectangular view. (See Figure 13–39.) The cylindricity tolerance must be less than the size tolerance.

Profile Tolerance

Profile is used to control form or a combination of size, form, and orientation. Profile callouts are commonly used on arcs, curves, or irregular shaped features but can also be applied to plane surfaces. The shape and size of the profile may be defined with basic dimensions or tolerance dimensions. When tolerance dimensions are used to establish profile, the profile tolerance zone must be within the size tolerance. The profile feature control frame is connected to the longitudinal view with a leader line. The two types of profile are profile of a line and profile of a surface. (See Figure 13–40.)

The profile of a line is a cross-sectional or single-line tolerance that extends through the specified feature. The profile of a line may be controlled in relation to datums or without datums. The profile of a line may be all around the part by using the all around symbol on the leader, as seen in Figure 13–41, or between two specified points on

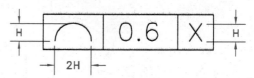

PROFILE OF A LINE

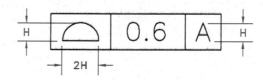

PROFILE OF A SURFACE

Figure 13–40 Profile feature control frames.

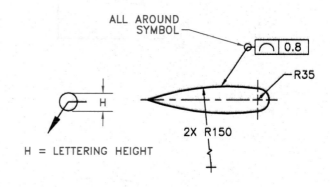

THE DRAWING

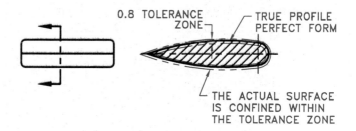

THE MEANING

Figure 13–41 Profile of a line all around, and the all around symbol.

the part by using the between symbol and identifying the points, as shown in Figure 13–42.

The profile of a surface is a blanket tolerance zone that affects all the surface elements of a feature equally. Profile of a surface is generally referenced to one or more datums so proper orientation of the profile boundary can be maintained. Profile of a surface may be applied all around a part as in Figure 13–43, or between two specified points as shown in Figure 13–44.

The profile of a line or the profile of a surface is a bilateral tolerance zone when the leader line points to the surface of the part without any additional symbology

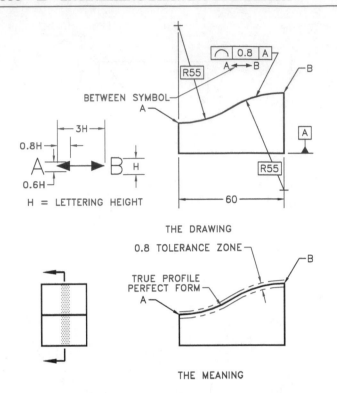

THE DRAWING

THE MEANING

Figure 13–42 Profile of a line between two given points, and the between symbol.

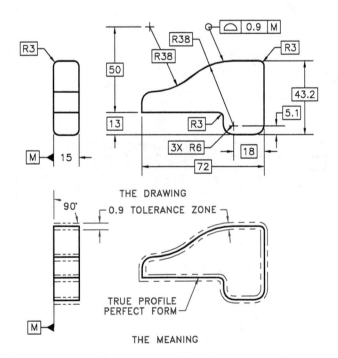

THE DRAWING

THE MEANING

Figure 13–43 Profile of a surface all around.

as was shown in Figures 13–40 through 13–44. A *bilateral profile tolerance* means that the tolerance zone is split equally on each side of the specified perfect form. Either type of profile callout may also have a unilateral tolerance zone specified. A unilateral tolerance zone places the entire zone on only one side of the true profile or perfect form. This is accomplished on a drawing

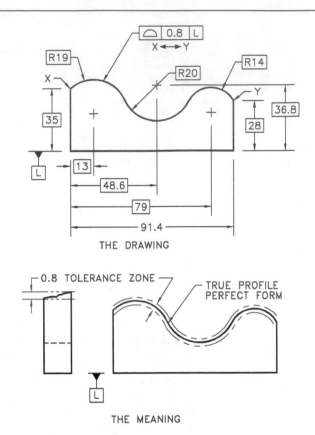

THE DRAWING

THE MEANING

Figure 13–44 Profile of a surface between two given points.

by placing a phantom line parallel to the surface where the leader arrowhead touches the part. The phantom line is placed any clear distance from the part surface and is either inside or outside depending on the specified direction of the unilateral tolerance from the true profile. The actual feature is confined between the true profile and the given tolerance reference. (See Figure 13–45.)

The profile of a coplanar surfaces may also be specified by placing a phantom line between the surfaces in the view where they appear as edges. The profile feature control frame is then connected to the phantom line with a leader. (See Figure 13–46.) Profile may also be used to control the angle of an inclined surface in relationship to a datum. (See Figure 13–47.)

Orientation Tolerance

Orientation tolerances refer to a specific group of geometric characteristics that establish a relationship between the features of an object. The tolerances that orient one feature to another are parallelism, perpendicularity, angularity and, in some applications, profile. Orientation tolerances require that one or more datums be used to establish the relationship between features. Parallelism, perpendicularity, and angularity also controls flatness. The geometric tolerance must fall within the size tolerance of the part.

Parallelism. The parallelism geometric characteristic symbol is detailed in Figure 13–48. A parallelism tol-

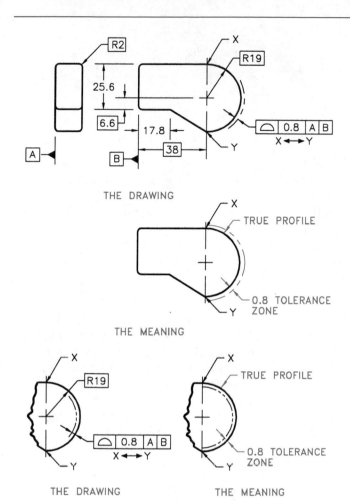

THE DRAWING

THE MEANING

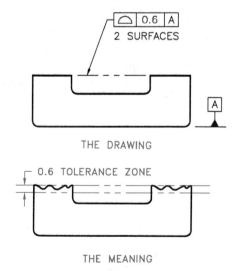

THE DRAWING · THE MEANING

Figure 13–45 A unilateral profile representation.

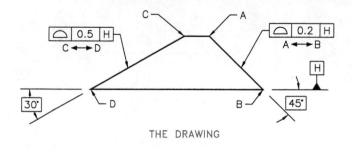

THE DRAWING

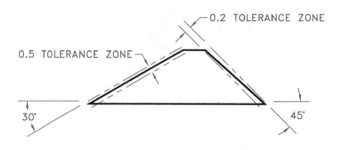

THE MEANING

Figure 13–47 Profile of an inclined surface.

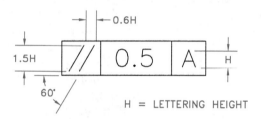

H = LETTERING HEIGHT

Figure 13–48 Parallelism feature control frame.

THE DRAWING

THE MEANING

Figure 13–46 Profile of coplanar surfaces.

erance zone requires that the actual feature is between two parallel planes or lines that are parallel to a datum. The parallelism feature control frame may be attached to the feature with a leader or on an extension line of the surface. (See Figure 13–49.) Unless otherwise specified a parallelism tolerance zone is a total tolerance that

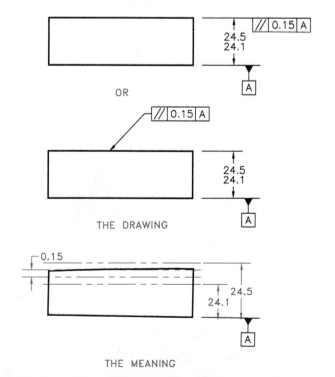

Figure 13–49 Parallelism representation.

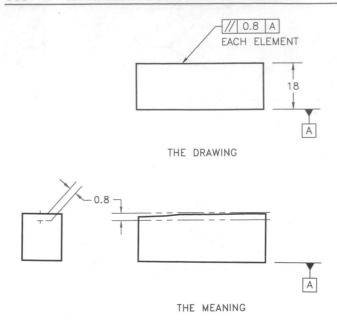

THE DRAWING

THE MEANING

Figure 13–50 Single line element parallelism.

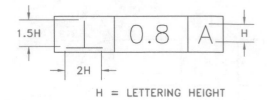

H = LETTERING HEIGHT

Figure 13–52 Perpendicularity feature control frame.

covers the entire surface. If a single line element is to be specified rather than the surface, then the note, EACH ELEMENT, or EACH RADIAL ELEMENT for an arc shaped surface, must be added below the feature control frame. (See Figure 13–50.) This technique also applies to perpendicularity and angularity.

In certain instances parallelism may also be applied to a cylindrical feature. The tolerance zone is a cylindrical shape that is parallel to a datum axis reference. The feature control frame that relates to an axis specification is attached to the diameter dimension, and a diameter zone descriptor should precede the geometric tolerance. (See Figure 13–51.)

Perpendicularity. The *perpendicularity* geometric characteristic symbol is detailed in Figure 13–52. A per-

pendicularity geometric tolerance requires that a given feature be located between two parallel planes or lines, or within a cylindrical tolerance zone that is a basic 90° to a datum. The perpendicularity feature control frame may be connected to the feature surface with a leader or an extension line, or attached to the diameter dimension for axis perpendicularity. (See Figure 13–53.)

Angularity. The *angularity* geometric characteristic symbol is represented in Figure 13–54. An angularity tolerance zone places a given feature between two parallel planes that are at a specified basic angular dimension from a datum. The basic angle from the datum may be any amount except 90°. The angularity feature control frame may be connected to the surface with a leader or from an extension line, or attached to the diameter dimension for axis angularity. (See Figure 13–55.)

Runout Tolerance

Runout is used to control the relationship of radial features to a datum axis and features that are 90° to a datum axis. These types of features include cylindrical, tapered, and curved shapes as well as plane surfaces that are at right angles to a datum axis. A runout tolerance is determined when a dial indicator is placed on the surface to be inspected and the part is rotated 360°. The *full indicator movement* (FIM) on the dial indicator must not exceed the amount of specified tolerance. The runout fea-

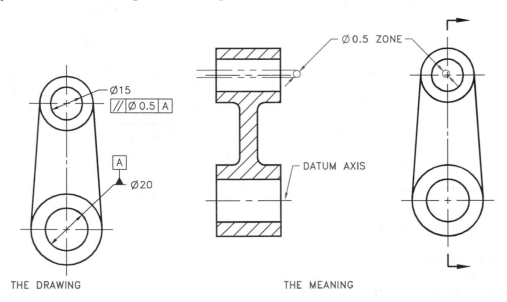

THE DRAWING   THE MEANING

Figure 13–51 Axis parallelism.

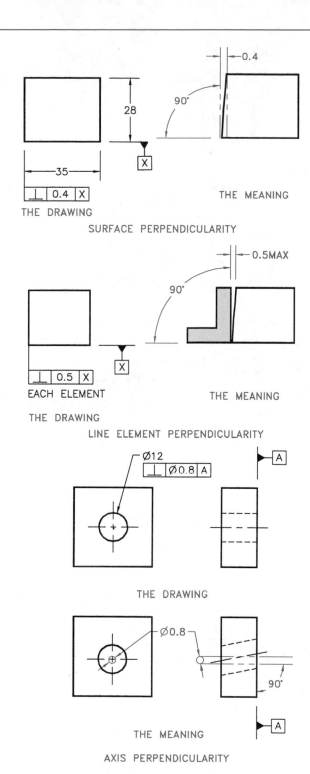

SURFACE PERPENDICULARITY

LINE ELEMENT PERPENDICULARITY

AXIS PERPENDICULARITY

Figure 13–53 Perpendicularity representations.

H = LETTERING HEIGHT

Figure 13–54 Angularity feature control frame.

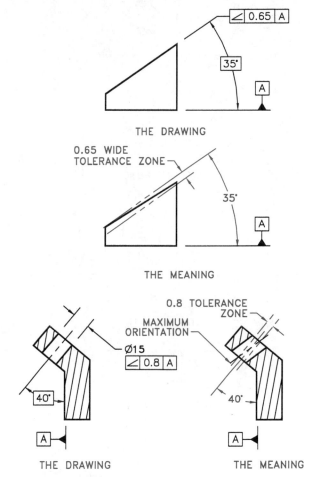

Figure 13–55 Angularity representations.

ture control frame is attached to the required surface with a leader. The runout tolerance applies to the length of the intended surface or until there is a break, or change in shape or diameter of the surface. There are two types of runout, circular and total. (See Figure 13–56.)

Circular Runout. Circular runout provides control of single circular elements of a surface. Circular runout controls circularity and the relationship of the common axes (coaxial) of parts. This tolerance, established by the FIM of a dial indicator, is placed at one location on the feature as the part is rotated 360°. (See Figure 13–57.)

Total Runout. Total runout provides composite control of all surface elements. The tolerance is applied simultaneously to all circular and profile measuring positions as the part is rotated 360°. (See Figure 13–58.) Where applied to surfaces constructed around a datum axis, total runout is used to control cumulative variations of circularity, straightness, coaxiality, angularity, taper, and profile of a surface. Where applied to surfaces constructed at right angles to a datum axis, total runout controls cumulative variations of perpendicularity to detect wobble and flatness, and to detect concavity or convexity.

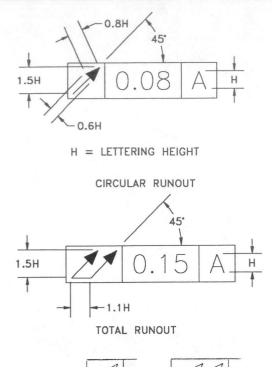

H = LETTERING HEIGHT

CIRCULAR RUNOUT

TOTAL RUNOUT

OPEN ARROWS OPTIONAL

Figure 13–56   Runout feature control frames.

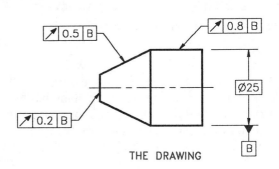

THE DRAWING

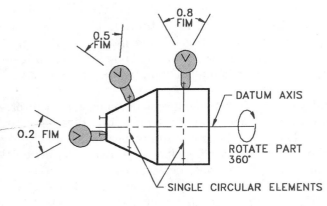

THE MEANING

Figure 13–57   Circular runout.

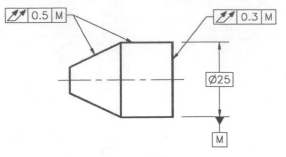

THE DRAWING

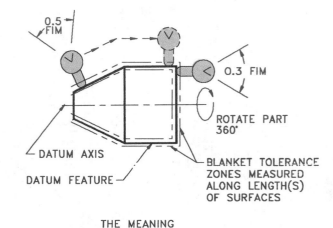

THE MEANING

Figure 13–58   Total runout.

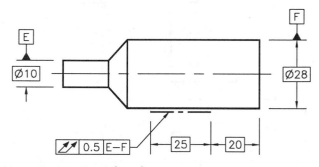

Figure 13–59   Partial surface runout.

A portion of a surface may have a specified runout tolerance if it is not desired to control the entire surface. This is done by placing a chain line in the linear view adjacent to the desired location. The chain line is located with basic dimensions as shown in Figure 13–59. When a part has compound datum features, that is, where more than one datum feature is used to establish a common datum, then the combined datum features are shown separated by a dash in the feature control frame. The datums are of equal importance. (See Figure 13–59.)

▼ MATERIAL CONDITION SYMBOLS

Material condition symbols are used in conjunction with the feature tolerance or datum reference in the

CADD Applications

GEOMETRIC TOLERANCING

The implementation of geometric tolerancing and dimensioning into a mechanical drafting CADD program is quite practical. The GT symbology makes this application a bonus to the mechanical drafting system.

Some dimensioning and tolerancing guidelines for use in conjunction with CADD/CAM are outlined, in part, from ASME Y14.5M as follows:

1. Major features of the part should be used to establish the basic coordinate system but are not necessarily defined as datums.
2. Subcoordinated systems that are related to the major coordinates are used to locate and orient features on a part.
3. Define part features in relation to three mutually perpendicular reference planes and along features that are parallel to the motion of CAM equipment.
4. Establish datums related to the function of the part and relate datum features in order of precedence as a basis for CAM usage.
5. Completely and accurately dimension geometric shapes. Regular geometric shapes may be defined by mathematical formulas although a profile feature that is defined with mathematical formulas should not have coordinate dimensions unless required for inspection or reference.
6. Coordinate or tabular dimensions should be used to identify approximate dimensions on an arbitrary profile.
7. Use the same type of coordinate dimensioning system on the entire drawing.
8. Continuity of profile is necessary for CADD. Clearly define contour changes at the change or point of tangency. Define at least four points along an irregular profile.
9. Circular hole patterns may be defined with polar coordinate dimensioning.
10. When possible dimension angles in degrees and decimal parts of degrees; for example, 45° 30' = 45.5°.
11. Base dimensions at the mean of a tolerance because the numerical control (NC) programmer will normally split a tolerance and work to the mean. Establish dimensions without limits that conform to the NC machine capabilities and part function where possible. Bilateral profile tolerances are also recommended for the same reason.
12. Geometric tolerancing is necessary to control specific geometric form and location.

The standard ASME Y14.5.1 *Mathematical Definition of Y14.5,* is the mathematical model of all specifications found in ASME Y14.5M. This allows for convenient digitizing of a part so all gaging is in a CADD system.

feature control frame. The material condition symbols are required to establish the relationship between the size or location of the feature and the geometric tolerance. The use of different material condition symbols alter the effect of this relationship. The material condition modifying elements are maximum material condition, MMC; regardless of feature size, RFS; and least material condition, LMC. There is no symbol for RFS because it is assumed for all geometric tolerances and datum references unless another material condition symbol is specified. The standard material condition symbols are shown in Figure 13–60.

Perfect Form Envelope

The form of a feature is controlled by the size tolerance limits. The envelope, or boundary, of these size limits is established at MMC. Remember from the discussion in Chapter 10 that MMC is the largest limit for an exter-

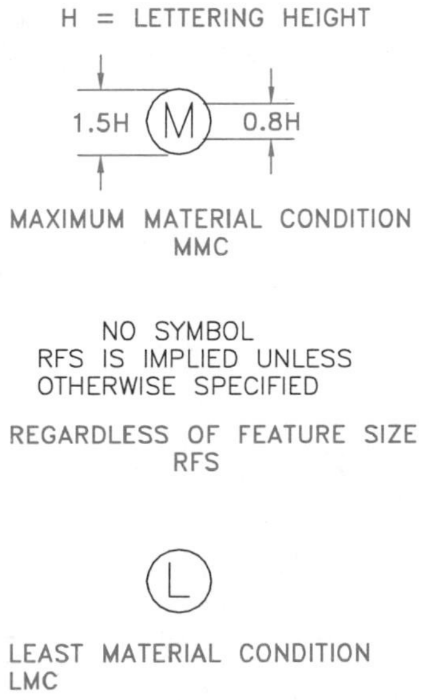

H = LETTERING HEIGHT

1.5H (M) 0.8H

MAXIMUM MATERIAL CONDITION
MMC

NO SYMBOL
RFS IS IMPLIED UNLESS
OTHERWISE SPECIFIED

REGARDLESS OF FEATURE SIZE
RFS

(L)

LEAST MATERIAL CONDITION
LMC

Figure 13–60 Material condition symbols.

nal feature and the smallest limit for an internal feature. The key is *most material*. The true geometric form of the feature is at MMC. This is known as the *perfect form boundary*. This is also referred to as the perfect form envelope. If the part feature is produced at MMC, it is considered to be at perfect form. When it is desired to permit a surface or surfaces of a feature to exceed the boundary of perfect form at MMC, a note such as PERFECT FORM AT MMC NOT REQUIRED is specified, exempting the pertinent size dimension. If a feature is produced at LMC, the opposite of MMC, the form is allowed to vary within the geometric tolerance zone or to the extent of the MMC envelope.

Regardless of Feature Size

Regardless of feature size means that the geometric tolerance applies at any produced size. The tolerance remains as the specified value regardless of the actual size of the feature. Regardless of feature size is implied (assumed) for all geometric characteristics and related datums unless otherwise specified.

Effect of RFS on Surface Straightness

RFS is implied for the straightness geometric characteristic. When a surface straightness tolerance is used by connecting a leader from the feature control frame to the surface of the part, then the geometric tolerance remains the same regardless of the feature size, and the actual size may not exceed the perfect form envelope at MMC. An acceptable part may be produced between the given size tolerance, and any straightness irregularity may not be greater than the specified geometric tolerance. Perfect form is required at MMC. (See Figure 13–61.)

Effect of RFS on Axis Straightness

When the straightness tolerance is applied to the axis of the feature by a relationship with the diameter dimension, RFS is implied unless otherwise specified, and the actual feature size plus the geometric tolerance may exceed the MMC perfect form envelope. (See Figure 13–62.)

Datum Feature RFS

Datum features that are influenced by size variations, such as diameters and widths, are also subject to variations in form. RFS is implied unless otherwise specified. When a datum feature has a size dimension and a geometric form tolerance, the size of the simulated datum is the MMC size limit. This rule applies except for axis straightness where the envelope is allowed to exceed MMC. Figure 13–63 shows the effect of RFS on the primary datum feature with axis and center plane datums. When the datum features are secondary or tertiary then the axis or center plane shall also have an angular relationship to the primary datum. (See Figure 13–64.)

Maximum Material Condition

The use of MMC in conjunction with the geometric tolerance in a feature control frame means that the given tolerance is held at the MMC produced size, and as the feature dimension departs from MMC the geometric tolerance is allowed to increase equal to the change. The maximum amount of change is at the LMC

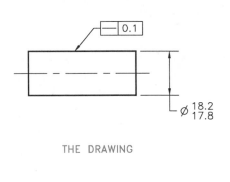

THE DRAWING

ACTUAL SIZE	GEOMETRIC TOLERANCE
18.2 MMC	0–/PERFECT FORM
18.1	0.1
18.0	0.1
17.9	0.1
17.8 LMC	0.1

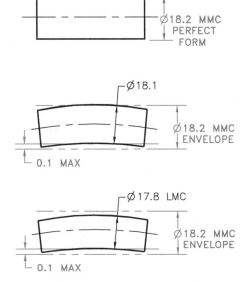

THE MEANING

Figure 13–61   Effect of regardless of feature size, RFS, on surface straightness.

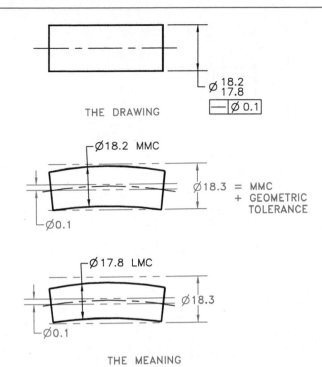

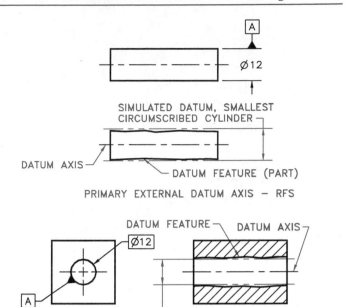

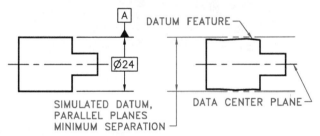

Figure 13–62   Effect of RFS on axis straightness.

Figure 13–63   Effect of RFS on the primary datum feature with axis and center plane datums.

ACTUAL SIZE	GEOMETRIC TOLERANCE
18.2 MMC	0.1
18.1	0.1
18.0	0.1
17.9	0.1
17.8 LMC	0.1

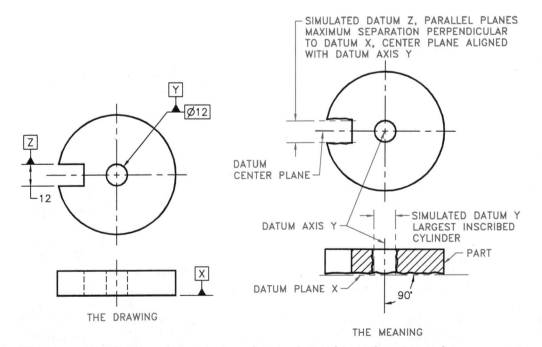

Figure 13–64   The secondary or tertiary datum relationship to the primary datum.

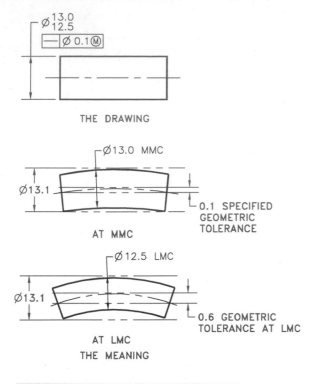

Figure 13–65   Effect of maximum material condition, MMC, on form tolerance, axis straightness.

ACTUAL SIZE	GEOMETRIC TOLERANCE
13.0 MMC	0.1
12.9	0.2
12.8	0.3
12.7	0.4
12.6	0.5
12.5 LMC	0.6

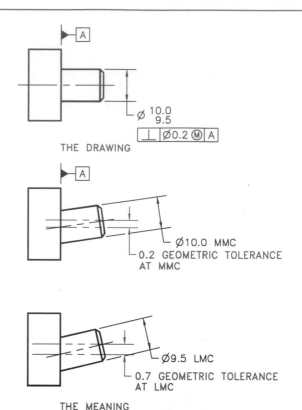

Figure 13–66   Effect of MMC on form tolerance, axis perpendicularity.

ACTUAL SIZE	GEOMETRIC TOLERANCE
10.0 MMC	0.2
9.9	0.3
9.8	0.4
9.7	0.5
9.6	0.6
9.5 LMC	0.7

produced size. MMC must be specified for any geometric characteristic where its application is desired.

Effect of MMC on Form Tolerance

When MMC is specified in conjunction with an axis straightness callout, the MMC feature size envelope is exceeded by the given geometric tolerance. The given geometric tolerance is held at the MMC produced size, then as the actual produced size departs from MMC the geometric tolerance is allowed to increase equal to the change from MMC to a maximum amount of departure at LMC. (See Figure 13–65.) The same situation may also be demonstrated with a perpendicularity tolerance where a datum axis is perpendicular to a given datum as in Figure 13–66.

Least Material Condition

The application of LMC is not as prevalent as MMC. LMC may be used when it is desirable to control minimum wall thickness. When an LMC material condition symbol is used in conjunction with a geometric tolerance, the specified tolerance is held at the LMC pro-

duced size. Unlike MMC condition, an LMC value is never permitted to be less than LMC. As the actual produced size deviates from LMC toward MMC, the tolerance is allowed an increase equal to the amount of change from LMC. (See Figure 13–67.)

▼ LOCATION TOLERANCE

Location tolerances include concentricity, symmetry, and position. *True position* is the theoretically exact location of the axis or center plane of a feature. The true position is located using basic dimensions. The locational tolerance specifies the amount that the axis of the feature is allowed to deviate from true position. All geometric tolerances imply RFS. MMC or LMC must be specified if desired. Reference to true position dimensions must be provided as basic dimensions at each required location on the drawing or a general note may be used to specify that UNTOLERANCED DIMENSIONS LOCATING TRUE POSITION ARE BASIC. Basic dimensions are theoretically perfect, so that datum or chain

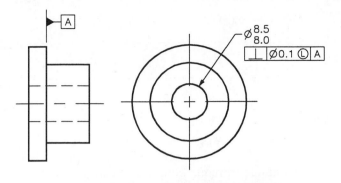

ACTUAL SIZE	GEOMETRIC TOLERANCE
8.0 MMC	0.6
8.1	0.5
8.2	0.4
8.3	0.5
8.4	0.2
8.5 LMC	0.1

Figure 13–67 Application of least material condition, LMC.

dimensioning may be used equally to locate true position because there is no tolerance build-up. The location feature control frame is generally added to the note or dimension of the related feature.

Concentricity

The concentricity geometric characteristic symbol is detailed in Figure 13–68. Concentricity is the relationship of the axes of cylindrical shapes. Perfect concentricity exists when the axes of two or more cylindrical features are in perfect alignment. (See Figure 13–69.) A concentricity geometric tolerance allows for a specified amount of deviation of the axes of concentric cylinders as shown in Figure 13–70. The geometric tolerance and related datums imply RFS. It is difficult to control the axis relationship specified by concentricity, so runout is often used. Where the balance of a shaft is critical, concentricity may be used.

Figure 13–68 Concentricity feature control frame.

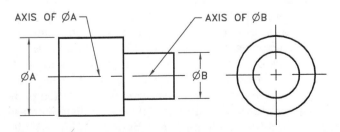

Figure 13–69 Perfect concentricity when both axes coincide.

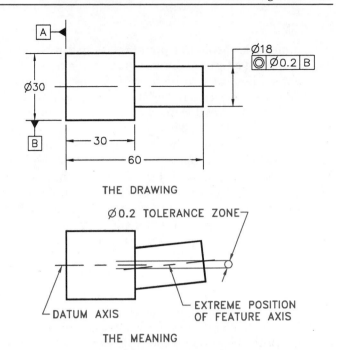

Figure 13–70 Concentricity represented.

Position Tolerance of Symmetrical Features

Position tolerancing may be used to locate the center plane of one or more features relative to a datum center plane. Figure 13–71 shows optional feature control frames where the position geometric tolerance and related datum reference is controlled on an RFS or MMC basis. When the position tolerance is at MMC, the related datum feature may be RFS, MMC, or LMC depending on design requirements. If the position tolerance is held RFS, then the related datum reference is RFS. RFS is implied unless otherwise specified.

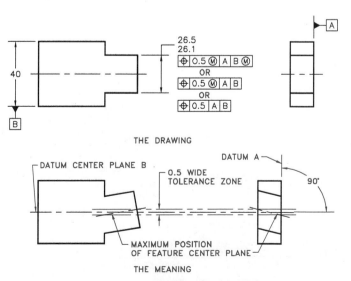

Figure 13–71 Position geometric characteristic specifying symmetry.

Symmetry

Symmetry is the center plane relationship between two or more features. Perfect symmetry exists when the center plane of two or more features is in alignment. The geometric characteristic symbol used to specify symmetry is detailed in the feature control frame shown in Figure 13–72. The symmetry geometric tolerance is a zone in which the median points of opposite symmetrical surfaces align with the datum center plane. The symmetry geometric tolerance and related datum reference are applied only on an RFS basis. (See Figure 13–73.) A zero position tolerance at MMC is used when it is required to control the symmetrical relationship of features within their size limits. An explanation of zero position tolerance at MMC follows.

Position

The position geometric characteristic symbol, also used for symmetry, is detailed in Figure 13–74. RFS is implied for the position geometric tolerance and related datum reference unless otherwise specified. MMC or LMC is applied to the position tolerance and datum reference as needed for design. The maximum material condition application is common although some cases require the use of regardless of feature size or least mate-

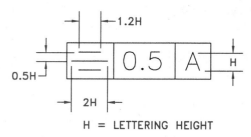

Figure 13–72   The symmetry symbol detailed in a feature control frame.

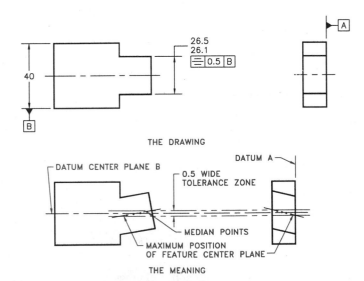

Figure 13–73   Symmetry application.

Figure 13–74   Position feature control frame.

rial condition. The feature control frame is applied to the note or diameter dimension of the feature. A positional tolerance defines a zone within which the center, axis, or center plane of a feature or size is permitted to vary from the true, theoretically exact, position. Basic dimensions establish the true position from specified datum features and between interrelated features. A positional tolerance is indicated by the position symbol, a tolerance, and appropriate datum references placed in a feature control frame. The location of each feature is given by basic dimensions. Dimensions locating true position must be excluded from the drawing's general tolerances by applying the basic dimension symbol to each basic dimension or by specifying on the drawing or drawing reference the general note, UNTOLERANCED DIMENSIONS LOCATING TRUE POSITION ARE BASIC.

Positional Tolerance at Maximum Material Condition (MMC).   Positional tolerance at MMC means that the given tolerance is held at the MMC produced size. Then, as the feature size departs from MMC the position tolerance increases equal to the amount of change from MMC to the maximum change at LMC. Positional tolerance at MMC may be defined by the feature axis or surface. The datum references commonly establish true position perpendicular to the primary datum and the coordinate location dimensions to the secondary and tertiary datums. The position tolerance zone is a cylinder equal in diameter to the given position tolerance. The cylindrical tolerance zone extends through the thickness of the part unless otherwise specified. The actual centerline of the feature may be anywhere within the cylindrical tolerance zone. (See Figure 13–75.) Another explanation of position may be related to the surface of a hole. The hole surface may not be inside a cylindrical tolerance zone established by the MMC diameter of the hole less the position tolerance. (See Figure 13–76.)

Zero Positional Tolerance at MMC.   When the positional tolerance is associated with the MMC symbol, the tolerance is allowed to exceed the specified amount when the actual feature size departs from MMC. When the actual sizes of features are manufactured very close to MMC, it is critical that the feature axis or surface not exceed the boundaries discussed previously. When parts are rejected due to this situation, it is possible to increase the acceptability of mating parts by reducing the MMC size of the feature to minimum allowance with the mat-

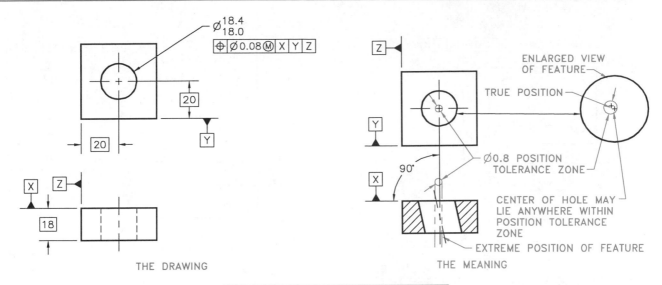

ACTUAL SIZE	POSITIONAL TOLERANCE
18.0 MMC	0.08
18.1	0.18
18.2	0.28
18.3	0.38
18.4 LMC	0.48

Figure 13–75 Positional tolerance at maximum material condition, MMC.

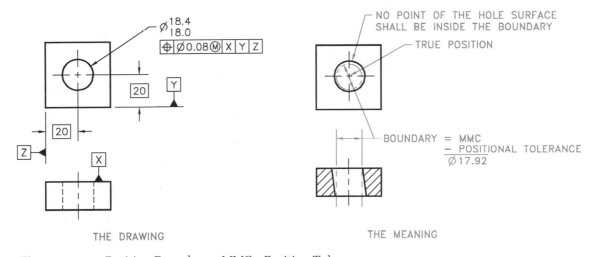

Figure 13–76 Position Boundary = MMC – Position Tolerance

ing part (virtual condition) and providing a zero positional tolerance at MMC. The positional tolerance is dependent on the feature size. When zero positional tolerance is used, no positional tolerance is allowed when the part is produced at MMC. True position is required at MMC. As the actual size departs from MMC, the positional tolerance increases equal to the amount of change to the maximum tolerance at LMC. (See Figure 13–77.)

Positional Tolerance at RFS. RFS may be applied to the positional tolerance when it is desirable to maintain the given tolerance at any produced size. This application results in close positional control. (See Figure 13–78.)

Positional Tolerance at LMC. Positional tolerance at LMC is used to control the relationship of the feature surface and the true position at largest hole size. The function of the LMC specification is generally for the control of minimum-edge distances. When LMC is used the given positional tolerance is held at the LMC produced size where perfect form is required. As the actual size departs from LMC toward MMC, the positional tolerance zone is allowed to increase equal to the amount of change. The maximum positional tolerance is at the MMC produced size. (See Figure 13–79.)

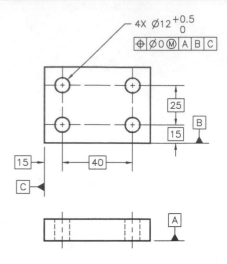

ACTUAL SIZE	POSITIONAL TOLERANCE
12.0 MMC ZERO ALLOWANCE 12mm FASTENER AT MMC	0
12.1	0.1
12.2	0.2
12.3	0.3
12.4	0.4
12.5	0.5

Figure 13–77   Zero positional tolerances at MMC.

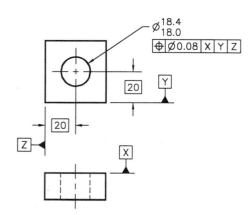

ACTUAL SIZE	POSITIONAL TOLERANCE
18.0 MMC	0.08
18.1	0.08
18.2	0.08
18.3	0.08
18.4 LMC	0.08

Figure 13–78   Positional tolerance at RFS.

▼ DATUM PRECEDENCE AND MATERIAL CONDITION

The effect of material condition on the datum and related feature may be altered by changing the datum precedence and the applied material condition symbol. The datum precedence is established by the order of

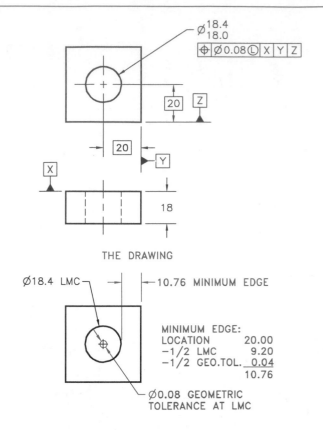

ACTUAL SIZE	POSITIONAL TOLERANCE
18.0 MMC	0.48
18.1	0.38
18.2	0.28
18.3	0.18
18.4 LMC	0.08

Figure 13–79   Positional tolerance at LMC.

placement in the feature control frame. The first datum listed is the primary datum; subsequent datums are secondary and tertiary. Figure 13–80 shows the effect of altering datum precedence and material condition.

▼ POSITION OF MULTIPLE FEATURES

The location of multiple features is handled in a manner similar to the location of a single feature. The true positions of the features are located with basic dimensions using rectangular or polar coordinates. The features are then identified by quantity, size, and position. (See Figure 13–81.) When two or more separate patterns are referenced to the same datums and with the same datum precedence, then the patterns are functionally the same. Verification of location and size is performed together. If this situation occurs and the interrelationship between patterns is not desired, then the specific note SEP REQT, meaning separate requirement, shall be placed beneath each affected feature control frame.

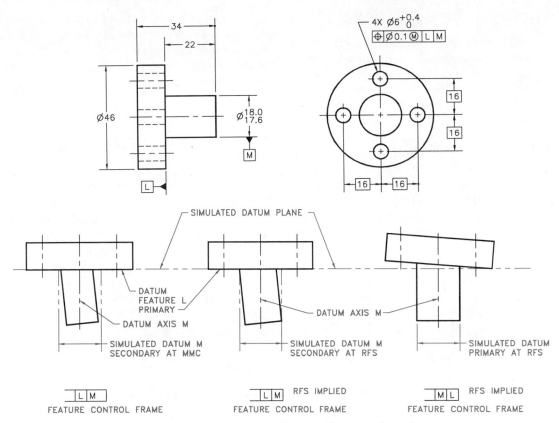

Figure 13–80   The effect of altering datum precedence and material condition.

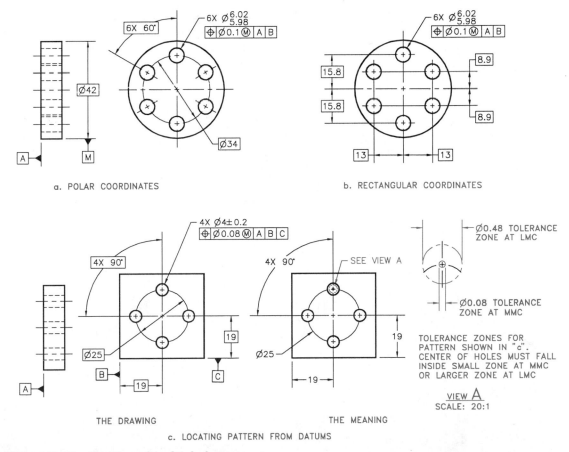

Figure 13–81   Position of multiple features.

▼ COMPOSITE POSITIONAL TOLERANCING

In some situations it may be permissible to allow the location of individual features in a pattern to differ from the tolerance related to the items as a group. When this is done, the group of features is given a positional tolerance that is greater in diameter than the zone specified for the individual features. The tolerance zone of the individual features must fall within the group zone. The positional tolerance of the individual elements controls the perpendicularity of the features. An individual tolerance zone may extend partly beyond the group zone only if the feature axis does not fall outside the confines of both zones. When such is the case, the feature control frame is expanded in height and divided into two parts. The upper part, known as the *pattern locating control,* specifies the larger positional tolerance for the pattern of features as a group. The lower entry, commonly called the feature relating control, specifies the smaller positional tolerance for the indi-

vidual features within the pattern. The pattern locating control is located first with basic dimensions. The *feature relating control* is established at the actual position of the feature center. Only the primary datum is represented in the feature relating control. (See Figure 13–82.)

▼ TWO SINGLE-SEGMENT FEATURE CONTROL FRAMES

The composite position tolerance is specified by a feature control frame doubled in height with one position symbol shown in the first compartment and only a single datum reference for orientation given for the feature relating control. The *two single-segment feature control frame* is similar, except there are two position symbols, each displayed in a separate compartment, and a two datum reference in the lower feature control frame. Look at Figure 13–83. The top feature control frame is

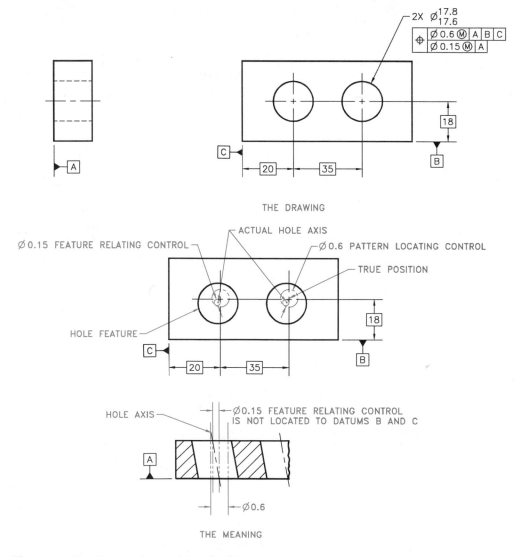

Figure 13–82 Composite positional tolerance.

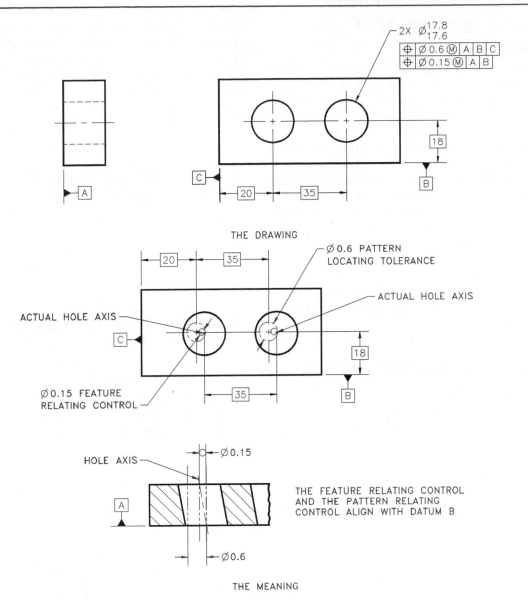

Figure 13–83 Two single-segment feature control frame.

the pattern locating control and works as previously discussed. The lower feature control frame is the feature-relating control in which two datums control the orientation and alignment with the pattern locating control. This type of position tolerance provides a tighter relationship of the holes within the pattern than the composite position tolerance.

▼ POSITIONAL TOLERANCE OF TABS

Positional tolerancing of tabs may be accomplished by identifying related datums, dimensioning the relationship between the tabs, and providing the number of units followed by the size and feature control frame. (See Figure 13–84.)

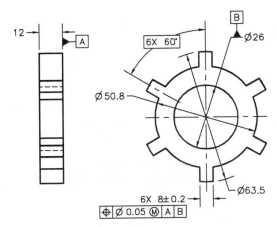

Figure 13–84 Positional tolerance of tabs, similiar practice for slots.

▼ POSITIONAL TOLERANCE OF SLOTTED HOLES

The positional tolerance of slotted holes may be accomplished by locating the slot centers with basic dimensions and providing a feature control frame to both the length and width of the slotted holes. (See Figure 13–85.) The word BOUNDARY is placed below each feature control frame. This means that each slotted feature is controlled by a theoretical boundary of identical shape that is located at true position. The size of each slot must remain within their size limits and no portion of the surface may enter the theoretical boundary. The size of the boundary calculated by MMC (each dimension, length and width) – Position Tolerance. Normally there is a greater position tolerance for the length than the width. When the position tolerance is the same for the length and the width, the feature control frame is separated from the size dimensions and connected to one of the slots with a leader.

▼ POSITIONAL TOLERANCE OF THREADED FEATURES AND GEARS

The orientation or positional tolerance of screw threads applies to the datum axis. The datum feature is established by a pitch diameter cylinder. If a different datum feature is required, the note should be lettered below the feature control frame, for example, MINOR DIA or MAJOR DIA. Similarly, the intended datum feature for gears or splines should be identified below the feature control frame. The options include MAJOR DIA, PITCH DIA, or MINOR DIA.

▼ POSITIONAL TOLERANCE OF COUNTERBORED HOLES

Counterbored holes are a type of coaxial feature. *Coaxial* means that one or more features have the same axis. There are three different ways to provide positional tolerancing to counterbored holes as shown in Figure 13–86.

▼ POSITIONAL TOLERANCING OF COAXIAL HOLES

Coaxial holes that are aligned through different features may be controlled with positional tolerancing using a feature relating zone for each hole and a pattern locating zone for the holes as a group. Some possible options are shown in Figure 13–87.

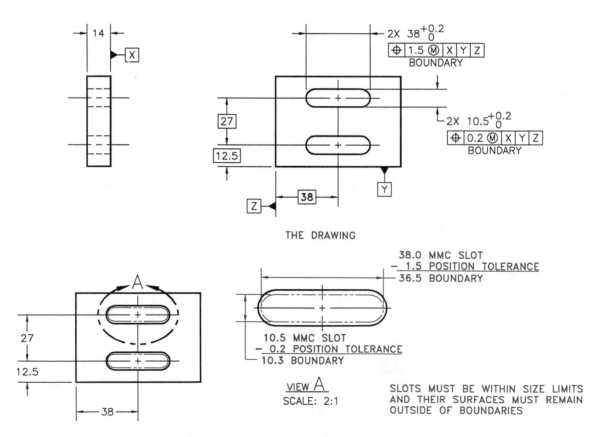

Figure 13–85 Positional tolerance for slotted holes.

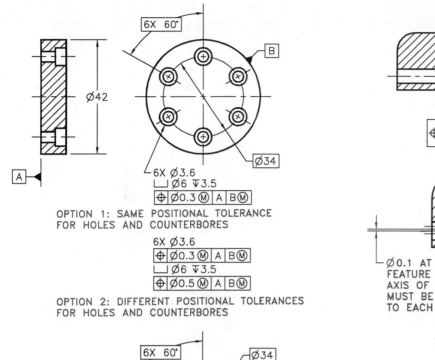

OPTION 1: SAME POSITIONAL TOLERANCE
FOR HOLES AND COUNTERBORES

OPTION 2: DIFFERENT POSITIONAL TOLERANCES
FOR HOLES AND COUNTERBORES

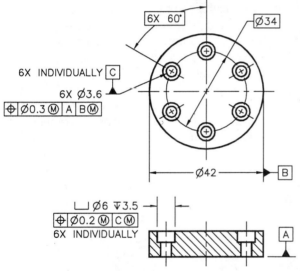

OPTION 3: POSITIONAL TOLERANCE FOR HOLES
AND COUNTERBORES WITH DIFFERENT DATUM
REFERENCE AND EACH CONTROLLED INDIVIDUALLY

Figure 13–86   Positional tolerance of counterbored holes.

▼ POSITIONAL TOLERANCING OF NONPARALLEL HOLES

Positional toleracing may be applied to holes that are not parallel to each other. The axis may not be perpendicular to a surface, as shown in Figure 13–88.

▼ PROJECTED TOLERANCE ZONE

The standard application of a positional tolerance implies that the cylindrical zone extends through the thickness of the part or feature. In some situations where there is the possibility of interference with mating parts,

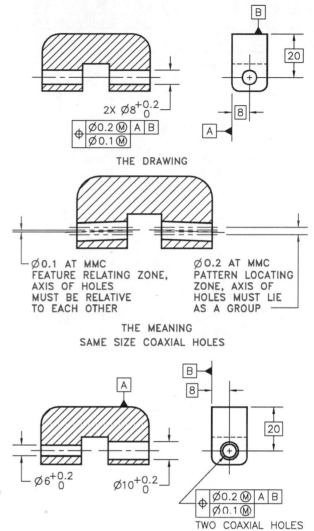

THE DRAWING

Ø0.1 AT MMC
FEATURE RELATING ZONE,
AXIS OF HOLES
MUST BE RELATIVE
TO EACH OTHER

Ø0.2 AT MMC
PATTERN LOCATING
ZONE, AXIS OF
HOLES MUST LIE
AS A GROUP

THE MEANING
SAME SIZE COAXIAL HOLES

COAXIAL HOLES OF DIFFERENT SIZE

Figure 13–87   Positional tolerance for coaxial holes.

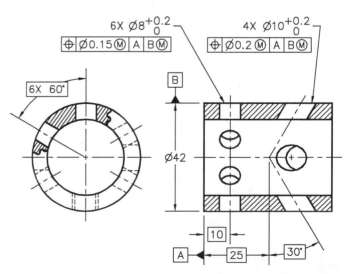

Figure 13–88   Positional tolerancing of nonparallel holes.

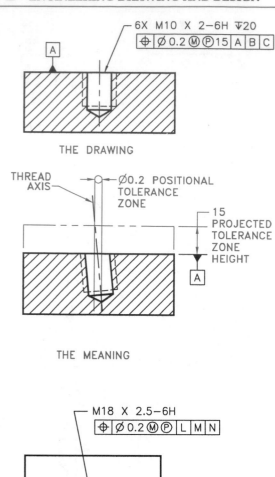

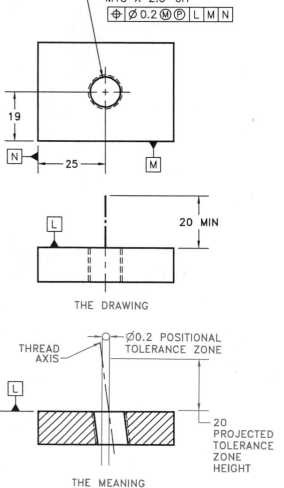

Figure 13–89 Projected tolerance zone.

the tolerance zone could be extended or projected away from the primary datum controlling the axis of the related feature. The amount of projection is equal to the thickness of the mating part. This works because the axis of the threaded or press fit feature is limited by the exact location and angle of the thread or hole within which they are assembled. This type of application is especially useful when the mating features are screws, pins, or studs. These types of conditions are referrred to as *fixed fasteners* because the fastener is fixed in the mating part and there is no clearance allowance. The projected tolerance zone may be handled in one of two ways using the projected tolerance zone symbol. (See Figure 13–89.) When calculating the positional tolerance zones that are applied to the parts of a fixed fastener, the following formula may be used: MMC Hole – MMC Fastener (nominal thread size) ÷ 2 = Positional Tolerance Zone of Each Part. In some situations it may be desirable to provide more tolerance to one part than another, for example 60 percent to the threaded part and 40 percent to the unthreaded part.

When parts are assembled with fasteners, such as bolts and nuts or rivets, and where all the parts have clearance holes to accommodate the fasteners, the application is referred to as a *floating fastener*. Floating fasteners require that the fastening device be secured on each side of the part, such as with a bolt and nut. With a fixed fastener, one of the parts is a fastening device. Greater tolerance flexibility with floating fasteners is due to the fastener clearance at each part. When calculating the positional tolerance zone for floating fastener parts, the following formula may be used: MMC Hole – MMC Fastener (nominal thread size) = Positional Tolerance Zone of Each Part.

CADD Applications

GEOMETRIC TOLERANCING SYMBOLS

Make a sketch of the desired symbol and decide how it will be placed on a drawing. Name the symbol; this is also the CADD file name. Select a point on each symbol that is a convenient point of origin when placing the symbol on a drawing. Figure 13–90 shows the point of origin determined for a sample of geometric tolerancing symbols.

After you have drawn a specific symbol, store it as a symbol by using such commands as SYMBOL, BLOCK, or WBLOCK. When you have created a group of symbols, it is time to organize them into a symbol library using a command such as CREATE

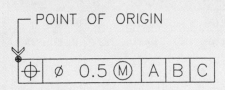

Figure 13–90 The insertion point for geometric tolerancing symbols may be located in any convenient position. This example shows the upper-left corner as a convenient insertion point.

MENU or TEMPLATE. A common computer prompt is TYPE THE SYMBOL NAME. Then digitize the box on the template where you want the symbol located. Follow this process until all of the symbols that you created are located on the template. When all of the symbols have been placed on the template, assign a name such as GEOMETIC TOLERANCING TEMPLATE, or LIBRARY. Print out or plot the symbols on a sheet so they correspond to the locations in which they were placed on the template. This new symbol library may then be placed as an overlay on the menu tablet.

Symbol templates may be selected for use by entering a command such as SELECT or ACTIVATE MENU, or by using the pointing device to select the template name from the menu tablet. When the symbol template is ready to use, individual symbols may be entered on a drawing by pressing the pointing device inside the specific box on the symbol library and then pressing the desired location on the drawing. Some systems automatically display the symbol at the cross-hairs location on the screen. Then as the point device is moved, the symbol also moves on the screen. When the symbol is positioned in the desired location, a button is pushed that places the symbol. Some symbols may require informational prompts such as ZONE DESCRIPTOR, GEOMETRIC TOLERANCE, MATE-

RIAL CONDITION, and/or DATUM REFERENCE after placement.

If you choose not to design your own custom geometric tolerancing symbols, there are predesigned GT symbol libraries available for most major software packages. Refer to the third-party applications catalog with your software to find any software available. Figure 13–91 shows an example of a tablet menu symbol library overlay designed for GT applications.

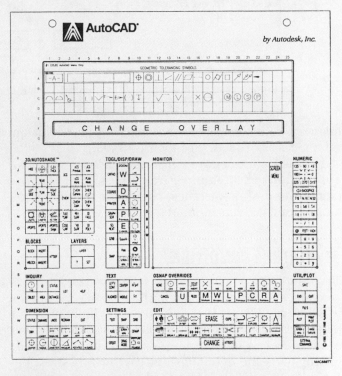

Figure 13–91 A customized tablet menu symbol library overlay for geometric tolerancing symbols. *Tablet menu courtesy Autodesk, Inc. GD&T symbols library courtesy Drafting Technology Services. Inc. Bartlesville, OK.*

▼ GEOMETRIC DIMENSIONING AND TOLERANCING WITH AUTOCAD

AutoCAD Release 13 initiated the ability to add GD&T symbols to your drawings. The feature control frame and related GD&T symbols may be created using the TOLERANCE and LEADER commands. These commands may be used to access the Symbol dialogue which contains the geometric characteristic symbols. (See Figure 13–92.)

The symbol that you pick in the Symbol dialogue box is displayed in the Sym text box of the Geometric Tolerance dialogue box. The Geometric Tolerance dialogue box is divided into compartments that relate to the compartments found in the feature control frame. These compartments allow you to create the desired feature control frame by entering diameter symbols, geometric

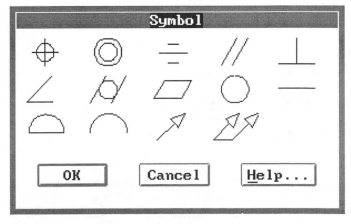

Figure 13–92 The AutoCAD Symbol dialogue box.

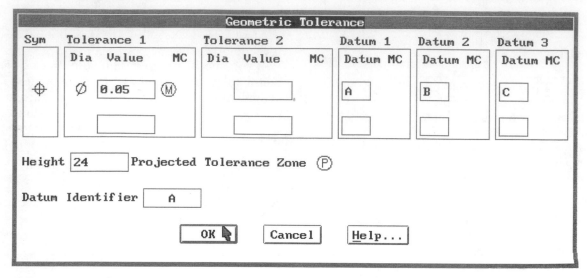

Figure 13–93 The AutoCAD Geometric Tolerance dialogue box.

tolerance values, material condition symbols, and datum references as needed. (See Figure 13–93.)

Double feature control frames may also be created for applications such as unit straightness, unit flatness, composite profile tolerance, composite positional tolerance, or coaxial positional tolerance.

▼ VIRTUAL CONDITION

The tolerances of a feature that relate to size, form, orientation, and location, including the possible application of MMC or RFS, are determined by the function of the part. Consideration must be given to the collective effect of these factors in determining the clearance between mating parts and in establishing gage feature sizes. *Virtual condition* is the resultant boundary of such considerations. Virtual condition is the sole condition where a feature size may be outside of MMC. Controlling the clearance between mating parts is critical to the design process. When features are dimensioned using a combination of size and geometric tolerances, the resulting effects of the specifications should be considered to insure that parts will always fit together. The boundary created by the combined effects of size, MMC, and the geometric tolerance is known as the virtual condition.

When a positional tolerance is applied to an internal feature, the Virtual Condition = MMC Hole − Positional Tolerance. This calculation determines the maximum feature size that should be allowed to fit within the hole. (See Figure 13–94.)

When perpendicularity is applied to an external diameter, such as a pin, the Virtual Condition = MMC Feature + Perpendicularity Geometric Tolerance. The virtual condition determines the smallest acceptable mating feature that fits over the given part while maintaining a positive connection between the surfaces at datum A as shown in Figure 13–95.

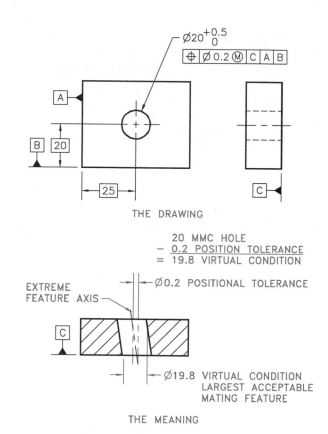

Figure 13–94 Virtual Condition of Hole = MMC − Positional Tolerance

▼ STATISTICAL TOLERANCING WITH GEOMETRIC CONTROLS

Methods of tolerancing for statistical process control (SPC) were introduced in Chapter 10. Statistical tolerances may also be used with geometric controls by placing the statistical tolerancing symbol in the feature control frame as shown in Figure 13–96.

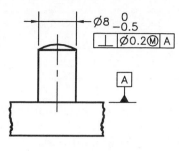

THE DRAWING

8 MMC FEATURE + 0.2 GEOMETRIC TOLERANCE
= 8.2 VIRTUAL CONDITION

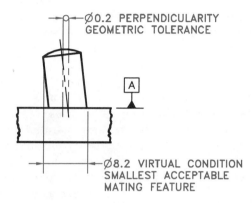

THE MEANING

Figure 13–95   Virtual Condition Perpendicular Pin = MMC
Feature + Geometric Tolerance

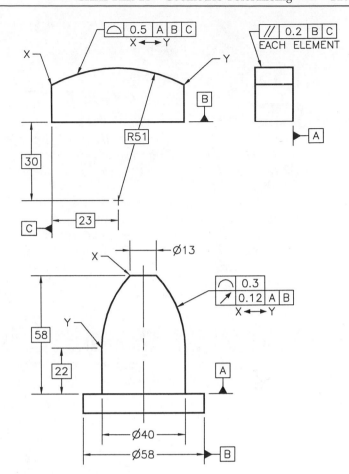

Figure 13–97   Combination controls (profile and runout).

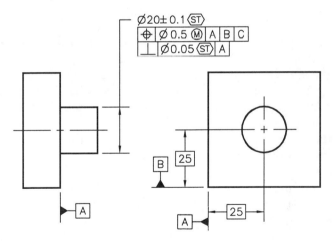

Figure 13–96   Statistical tolerancing with geometric controls.

▼ COMBINATION CONTROLS

In some situations, compatible geometric characteristics
may be combined in one feature control frame or separate frames associated with the same surface. This is normally done when the combined effect of two different
geometric characteristics and tolerance zones is desired.
The profile tolerance may be used to illustrate the combination of geometric characteristics. Profile and paral-

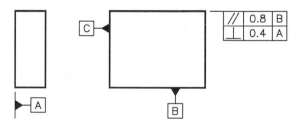

Figure 13–98   Combination control (parallelism and perpendicularity).

lelism may be combined to control the profile of a surface plus the parallelism of each element to a datum. Profile and runout may be combined to control the line
elements within the profile specification and circular elements within the runout tolerance as shown in Figure
13–97. The combined control of parallelism and perpendicularity may be represented on the same feature as
shown in Figure 13–98. Other combined geometric
characteristics may be used when the callouts are compatible and the design function of the part requires
such specific controls.

PROFESSIONAL PERSPECTIVE

Written By Michael A. Courtier, Supervisor/Instructor, CAD/CAM/CAE Employee Training and Development, Boeing Aerospace

In working with engineers and drafters for years on the subject of geometric tolerancing, I find that questions keep popping up in several key areas—datums; limits of size; geometric tolerances as a refinement of size; the tolerance zones (2D or 3D, cylindrical or total wide, etc.); and the apparent overlap of controls (circularity, cylindricity, circular runout, total runout, position). It is in these "weak" spots that you should dig a little deeper and fill in your knowledge.

New and old users alike tend to miss the key rules of thumb in selecting *datums,* and most textbooks on the subject don't address this area.

❍ Make sure you select datums that are real, physical surfaces of the part. (I've seen many poor drawings with datums on centerlines, at the centers of spheres, or even at the center of gravity.)

❍ Be sure to select large enough datum surfaces. (It's hard to chuck up on a cylinder 1/10" long, or get three non-colinear points of contact on a plane of similar dimensions.)

❍ Choose datums that are accessible. (If manufacturing or quality assurance has to build a special jig to reach the datum, extra costs are incurred for the ultimate part.)

❍ Given the above constraints, select functional surfaces and corresponding features of mating parts where possible.

Remember, datums are the *basis* for subsequent dimensions and tolerances, so they should be selected carefully. (You might also be well advised to assign a little form, orientation, or runout control to the datum surface to assure an accurate basis.)

In the area of *limits of size,* there are three possible interpretations of a size dimension. Every class I've taught has been divided on what the proper extent of a size dimension's control should be—the overall size (or geometric form), the size at various cross-sectional checks, or both. The correct answer, per definition in the standard, is **BOTH!**

The words *refinement of size* are a source of general confusion to new users of geometric tolerancing. Let's say a part is dimensioned on the drawing as .500 ±.005" thick, with parallelism control to .003" on the upper surface. If a production part has a convex upper surface with low points at .504, then the high point can only be .505, *not* .507. In other words, only .001" of the .003" of parallelism control can be used since parallelism is *a refinement of size within the overall size limits.*

Another one of the often misunderstood aspects of geometric tolerancing is the *tolerance zone.* Many engineers, technicians, drafters, machinists, and inspectors misinterpret the *shape* of the tolerance zone. I recommend that all new users make a chart showing the tolerance zones for each type of tolerance and a sampling of its various applications. This chart should include—at a minimum—the 2D zones between two lines or concentric circles (e.g., straightness, circularity) versus the 3D zones between parallel planes (e.g., flatness, perpendicularity); the cylindrical versus the total wide zones (e.g., straightness to an axis versus surface straightness, or position of a hole versus position of a slot); the cross-sectional zones versus the total surface zones (e.g., circularity versus cylindricity); and many, many others. If you can properly visualize the tolerance zone, you're well on your way to the proper interpretation of the geometric tolerance.

One more thing to watch out for is the *apparent overlap of controls.* While it may appear that there is overlap, there are subtle yet distinct differences between all the geometric tolerance controls. Consider cylindrical shapes, for instance. You can use a wide variety of controls for them, including straightness, circularity, cylindricity, concentricity, circular runout, total runout, and position, to name a few. Remember that the first three are form controls, so no datum relationship is controlled. Straightness can control the axis or the surface merely by placing the feature control frame in a different location on the drawing. Straightness is a longitudinal control, whereas circularity is a circular element control (both are 2D); cylindricity is a 3D control of the surface relative to itself (no datum). The last four controls are with respect to a datum (or datums), but they have their differences also. Concentricity and the runouts are always RFS, whereas positional tolerancing can be LMC or MMC and get bonus tolerances. Runout is a rotational consideration, while positional tolerancing does not imply rotation.

In summary, ASME Y14.5M is a very powerful tool for engineering drawings. Digging deep into the subtleties will separate the amateur from the professional, who knows it will save time and money in every discipline from design to manufacturing to quality assurance testing. Used properly, the engineering design team can convey more information about the overall design to downstream areas or subcontracting concerns that may have no knowledge of the final assembly. The proper assessment of the symbology requires no interpreter for the reader who is well versed in this international sign language called *geometric tolerancing.*

MATH APPLICATION

ROUNDING NUMBERS

Calculators and computers usually indicate more precision in their answers than is warranted by the original data. In addition to the information on rounding in Appendix N (page 920), here is the mathematical convention for rounding calculations:

When Adding or Subtracting:
1. Round each figure to the coarsest (least "precise") piece of data.
2. Do the arithmetic.

For example, to add this group 4.39 + 7.9 + 6.42, first round each number to the nearest tenth (because of the 7.9) giving 4.4 + 7.9 + 6.4. Then, add for the answer of **18.7.**

When Multiplying or Dividing:
1. Do the arithmetic.

2. Round the answer to the least number of significant digits in the original data.

Digits are *significant* when they give us information other than holding a decimal point in place; 13.09 has four significant digits, but 13,000,000 has only two.

For example, to calculate 13.78 × 6.1 × 4.385 first multiply, giving 368.59433. Because the 6.1 has only two significant digits, we round the answer to **370**, which has only two significant figures (the zero of 370 being merely a place-keeper).

When working with complicated formulas or when running computer programs, do not round off until the final answer is obtained. Constants in formulas are not considered data. When using a geometrical formula with π in it, it's best to push the calculator's π button instead of manually entering 3.14.

CHAPTER
13 GEOMETRIC TOLERANCING TEST

DIRECTIONS

Answer the questions with short complete statements or drawings as needed.

PART 1. QUESTIONS

1. Give a brief definition of geometric tolerancing (GT).
2. Give an example of how general tolerancing has some control on form.
3. Define *datum*.
4. Describe or show an example of how a datum feature symbol is shown on a plane surface and a centerline axis.

5. What is another name for the three-plane dimensioning system?
6. Name the elements of the three-plane dimensioning system.
7. Discuss the order of precedence of datum features.
8. When is it practical to use a partial surface datum?
9. Describe how coplanar surface datums are drawn.
10. Show an example of how datum target points are located and identified on a primary datum surface.
11. Show an example of how a datum target line may be shown on a drawing.
12. Show an example of how datum target areas may be shown on a drawing.
13. What is a feature control frame used for?
14. Show and label the complete order of elements in a feature control frame.

15. Define basic and show an example of basic dimension.
16. List the geometric characteristics for form, profile, orientation, location, and runout.
17. Show examples of how the following geometric characteristics are represented on a drawing: straightness, flatness, circularity, cylindricity, profile of a surface, parallelism, perpendicularity, angularity, and runout.
18. Which geometric tolerance is more confining, circularity or cylindricity?
19. Show an example of how unit straightness may be shown on a drawing.
20. Which profile tolerance is most confining?
21. Must the parallelism geometric tolerance zone be within the related size tolerance?
22. What does the note, EACH ELEMENT, denote when placed below a feature control frame?
23. Is a basic angle required to establish an angular surface for an angularity geometric tolerance?
24. Name the runout tolerance that is most confining.
25. How are the specified limits of total runout determined?
26. Describe or show an example of how a surface portion can have a specified runout tolerance.
27. Define *perfect form envelope*.
28. Clearly define regardless of feature size (RFS).
29. When is RFS implied?
30. Clearly define maximum material condition (MMC) as related to the effect of MMC on the geometric tolerance.
31. Clearly define least material condition (LMC) as related to the effect of LMC on the geometric tolerance.
32. True or False: position tolerances must have a correlated material condition symbol (MMC or LMC) applied to the tolerance and related datums.
33. Define the concentricity geometric characteristic.
34. Define the symmetry geometric characteristic.
35. Describe the position tolerance zone and how it is located.
36. Define *true position*.
37. Describe a projected tolerance zone.
38. Show an example of a combination control.

PART 2. CALCULATIONS

1. Given:
   a. Shaft ∅ 24.00/23.92.
   b. Straightness geometric tolerance 0.02.

   What is the geometric tolerance at the actual sizes specified below for the type of straightness and material condition shown?

ACTUAL SIZE	SURFACE STRAIGHTNESS	AXIS STRAIGHTNESS	
	RFS	RFS	MMC
24.00			
23.99			
23.98			
23.96			
23.94			
23.92			

2. Given:
   a. Positional tolerance ∅0.02 at true position in reference to datums L, M, N.
   b. Hole size ∅8.50/8.40.

   What is the positional tolerance using different material condition symbols at the actual sizes shown below?

ACTUAL SIZES	MATERIAL CONDITION APPLIED TO TOLERANCE		
	MMC	RFS	LMC
8.50			
8.49			
8.48			
8.46			
8.44			
8.42			
8.40			

3. If the positional tolerance of the hole in problem 2 above is zero at MMC, then what would the positional tolerance be at the actual produced sizes given below?

ACTUAL SIZES	MMC
8.50	
8.48	
8.46	
8.44	
8.42	
8.40	

4. What is the virtual condition of a ∅12.2/12.0 hole that is located with a positional tolerance of ∅0.05 at MMC?
5. What is the virtual condition of a ∅6.0/5.9 pin established with perpendicularity to a datum A by ∅0.02 at MMC?
6. Calculate the positional tolerance for the location of holes with the following specifications:
   a. Floating fastener.
   b. Fastener: M20 × 2.5.
   c. Hole through two parts: ∅21.2/20.8.

   Positional tolerance for holes in part 1 equals _____, part 2 equals _____.

7. Calculate the positional tolerance for the location of holes with the following specifications:

    a. Fixed fastener.
    b. Part 1 hole: ⌀9.0/8.6.

c. Part 2 hole: M8 × 1.25.
d. Equal positional tolerance for each part.

Positional tolerance for holes in part 1 equals _____, part 2 equals _____.

CHAPTER

13 GEOMETRIC TOLERANCING PROBLEMS

DIRECTIONS

1. From the selected engineering layouts, determine which views and dimensions should be used to completely detail the part.

2. Make a multiview sketch to proper proportions including dimensions and notes.

3. Using the sketch as a guide, draw an original multiview drawing on an adequately sized drawing sheet. Add all necessary dimensions and notes using unidirectional dimensioning. Use manual or computer-aided drafting as required by your course guidelines.

4. Include the following general notes at the lower left corner of the sheet .5 in. each way from the corner border lines:

2. REMOVE ALL BURRS AND SHARP EDGES.
1. INTERPRET DIMENSIONS AND TOLERANCES PER ASME Y14.5M—1994.

NOTES:
Additional general notes may be required depending on the specifications of each individual assignment. A tolerance title block is recommended.

UNSPECIFIED TOLERANCES:

DECIMALS	mm	IN.
X	±2.5	±.1
XX	±0.25	±.01
XXX	±0.127	±.005
ANGULAR ±30'		
FINISH	3.2μm	125μIN.

Problem 13–1 Geometric tolerancing (metric)

Part Name: Flow Pin
Material: Bronze
Finish: Finish All Over 0.20 μm.

Problem 13–2 Geometric tolerancing (metric)

Part Name: LN2 Test Pump Lock Nut
Material: AMS 5732.
Additional General Notes:
1. ⚠3: Mark per AS478 Class D with 1193125 and applicable dash number.
2. Finish All Over 1.6 μm.

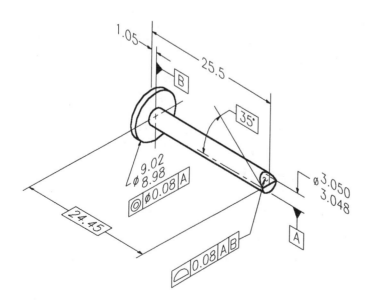

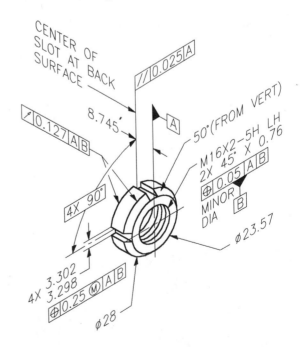

Problem 13–3 (inch)

Part Name: Half Coupling
Material: Ø1.250 6061-T6 Aluminum
SPECIFIC INSTRUCTIONS:
Provide MMC material condition after position tolerance except for RFS at threads.
Problem based on original art courtesy TEMCO.

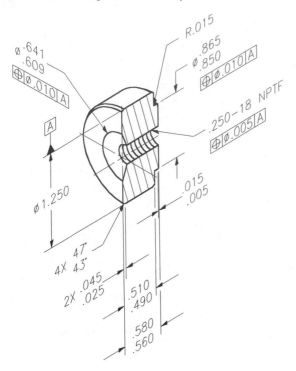

Problem 13–5 (inch)

Part Name: Half Coupling
Material: Ø1.625 6061-T6511
Problem based on original art courtesy TEMCO.

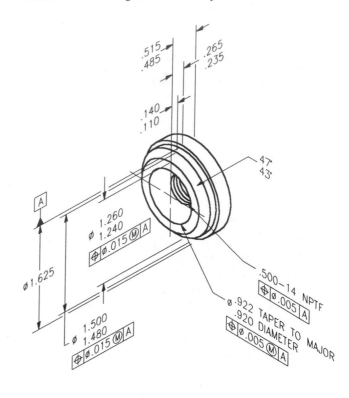

Problem 13–4 (inch)

Part Name: Coupling
Material: AISI 1010, Killed
SPECIFIC INSTRUCTIONS:
Provide MMC material condition after position tolerance except for RFS at threads.
Problem based on original art courtesy TEMCO.

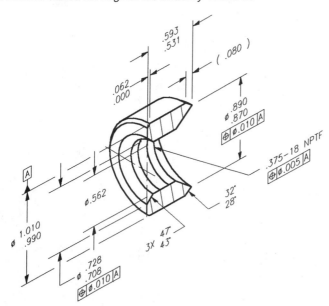

Problem 13–6 (metric)

Part Name: Spline Plate
Material: SAE 3135

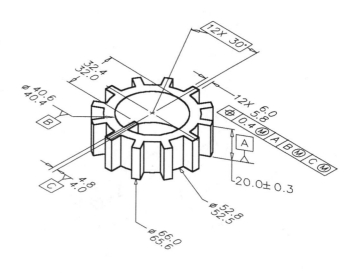

Problem 13–7 (inch)

Part Name: Nut
Material: No. 10 Bronze

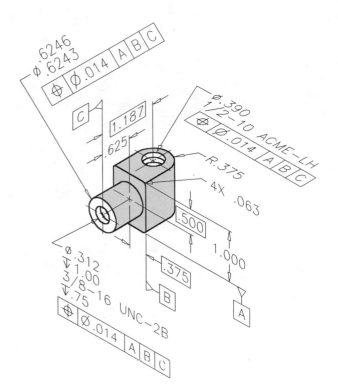

Problem 13–9 (inch)

Part Name: Thrust Washer
Material: SAE 5150

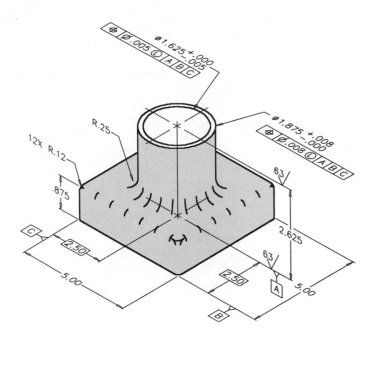

Problem 13–8 (metric)

Part Name: Coupling Bracket
Material: SAE 4310 Steel

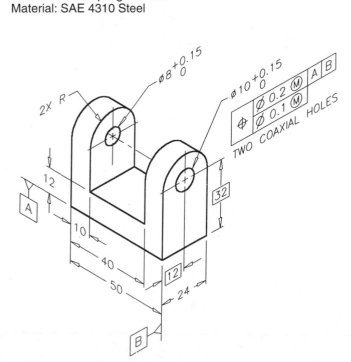

Problem 13–10 (mctric)

Part Name: Spacer
Material: SAE 4310

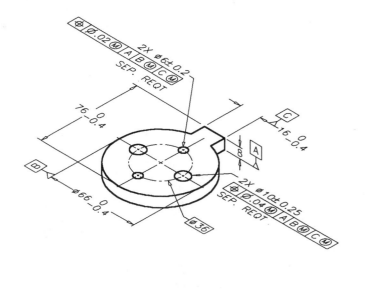

Problem 13–11 (metric)

Part Name: Bearing Support
Material: SAE 1040

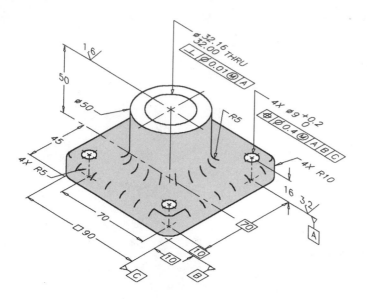

Problem 13–13 (metric)

Part Name: Lock Nut
Material: SAE 3130

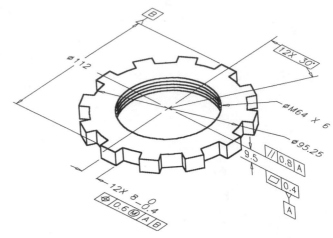

Problem 13–12 (inch)

Part Name: Slide Screw
Material: SAE 4320

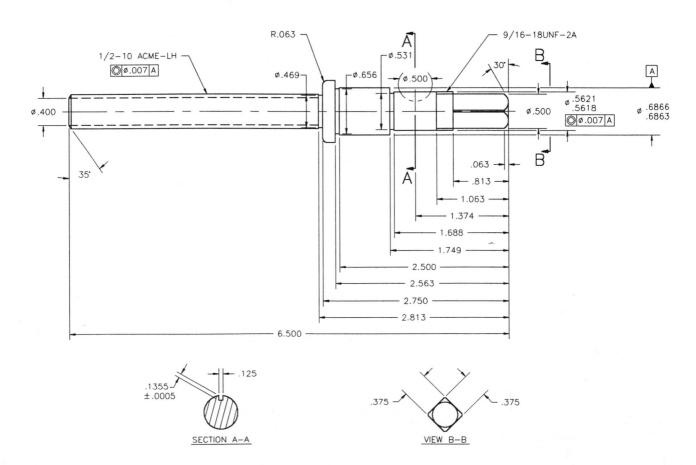

SECTION A–A

VIEW B–B

Problem 13–14   (inch)

Part Name: Cover Plate
Material: Phosphor Bronze

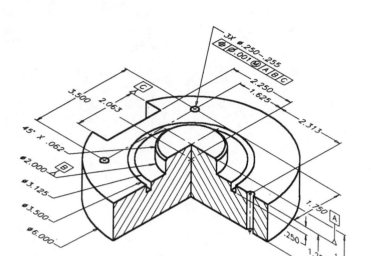

Problem 13–16   (metric)

Part Name: Hub
Material: SAE 3310

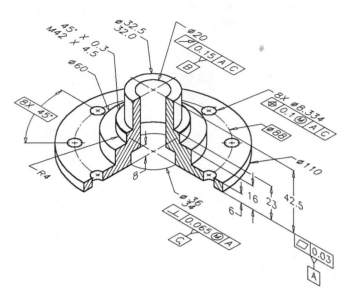

Problem 13–15   (inch)

Part Name: Angle Support Mounting
Material: SAE 3110

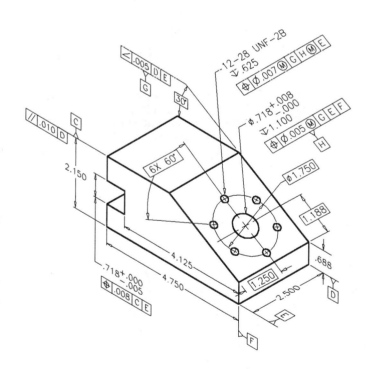

Problem 13–17   Geometric tolerancing (metric)

Part Name: Fixture MIBRDA—1265
Material: SAE 4320
Harden: Brinell 200-240
Additional General Notes:
1.   Finish All Over 0.80 μm.

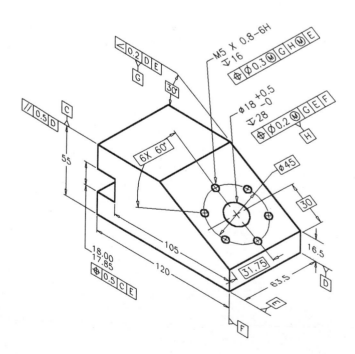

Problem 13–18 Geometric tolerancing (metric)

Part Name: Mounting Bracket
Material: Stainless Steel
Additional General Notes:
1. All Fillets and Rounds R24.
2. Finish All Over 1.6 μm.

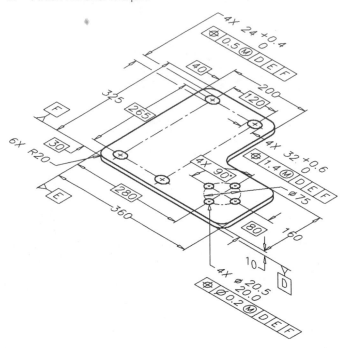

Problem 13–20 (inch)

Part Name: Hub
Material: SAE 4320

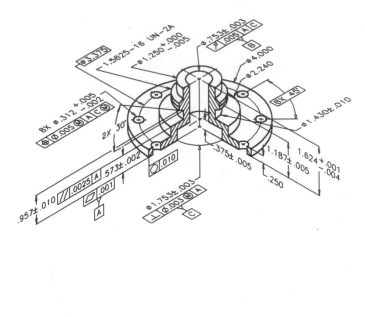

Problem 13–19 Geometric tolerancing (metric)

Part Name: Oscillator Housing
Material: Phosphor Bronze
Additional General Notes:
1. Finish All Over 0.80 μm.
SPECIFIC INSTRUCTIONS:
Use the following engineer's notes to complete the geometric tolerancing of the Oscillator Housing:

• Establish datum A with three equally spaced datum target points at each end of the ∅28.1-28.0 cylinder.

• Establish datum B at the left end surface.
• Make the bottom surfaces of the 2✕ ∅40.2-40.0 features perpendicular to datum A by 0.06.
• Provide a cylindricity tolerance of 0.3 to the outside of the part.
• Make the 2✕ ∅40.2-40.0 features concentric to datum A by 0.1.
• Locate the 6✕ ∅6 holes with reference to datum A at MMC and datum B with a position tolerance of 0.05 at MMC.
• Locate the 4✕ ∅4 holes with reference to datum A at MMC and datum B with a position tolerance of 0.04 at MMC.

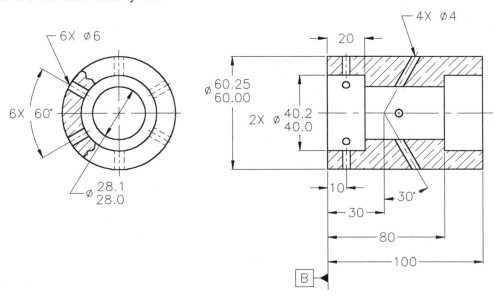

Problem 13–21   (inch)

Part Name: Lock Spacer
Material: SAE 1030

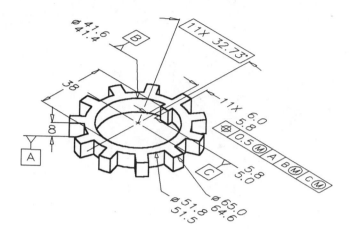

Problem 13–22   (metric)

Part Name: Mounting Plate
Material: SAE 4140

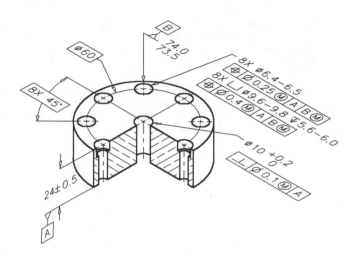

Problem 13–23   Geometric tolerancing (metric)

Part Name: Side Panel Mounting Plate
Material: SAE 30308

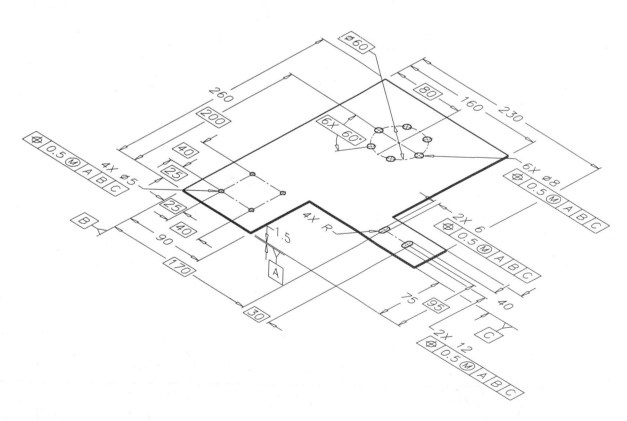

Problem 13–24 Advanced geometric tolerancing from actual industry drawing (metric)

Part Name: Pinion Gear Shaft
Material: CRES 15-5PH ASTM A564
Additional General Notes:
1. Finish All Over 1.6 μm.
2. Heat Treat Per Mil-H-6875 to H1100 Condition.
3. Penetrant Inspect Finished Part Per Mil-Std-271, Group III. No evidence of Linear Indications Permitted.
4. Part to be Clean and Free of Foreign Debris.
Problem based on original art courtesy Aerojet Tech-Systems Co.

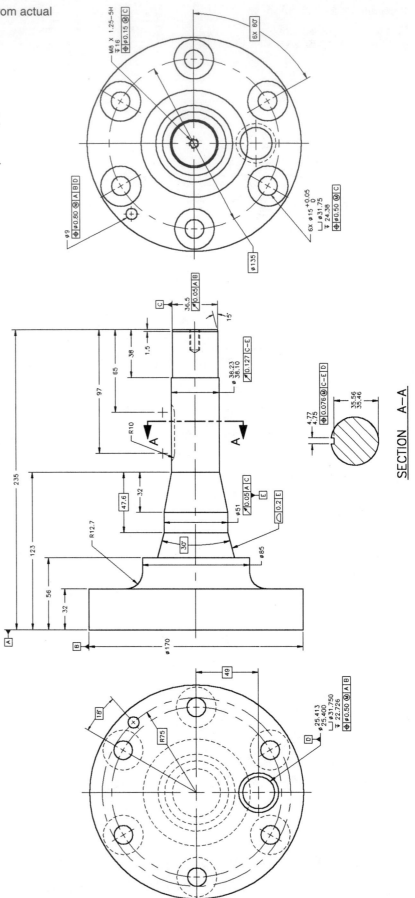

SECTION A–A

MATH PROBLEMS

Round to the nearest tenth.

1. 4.849
2. 3.650
3. .275
4. 5.249

Round to the nearest hundredth.

5. 4.849
6. 7.0574
7. .27499

Perform the following calculations, rounding appropriately.

8. 3.7 + 4.19 + 8.00004
9. 7.8 × 6.3 × 5.29
10. Find C using the formula C = πD for D = 6542 cm.

Mechanisms: Linkages, Cams, Gears, and Bearings

LEARNING OBJECTIVES

After completing this chapter, you will:

○ Draw linkage diagrams.
○ Create cam displacement diagrams.
○ Design cam profile drawings from previously drawn cam displacement diagrams.
○ Make detail gear drawings using simplified representations and gear data charts.
○ Establish unknown data for gear trains.
○ Calculate bearing information from specifications.
○ Design a complete gear reducer from engineering data and sketches.

THE ENGINEERING DESIGN PROCESS

Your latest assignment is to produce a drawing of an in-line follower plate cam profile for the manufacturing department. The engineer has provided the specifications in a cam displacement diagram. (See Figure 14–1.) Your manufacturing department is using a sophisticated computer-aided manufacturing (CAM) system and needs a CADD drawing from which the required tooling paths will be established.

Using a CADD system, the process is fairly simple and completely accurate. Using appropriate techniques, you find and draw the base circle and the prime circle. Next, you create an array of lines extending outwards from the center of the circle at 30° increments. (See Figure 14–2.) Finally, by utilizing offsets of the base circle as indicated by the engineer's diagram, you find all the specified control points. Using a spline curve that intersects each of these points, you create a profile drawing to generate tooling paths for the CAM system. (See Figure 14–3.)

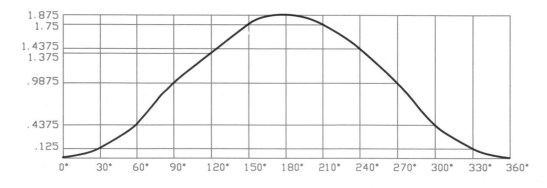

BASE CIRCLE: 2.50
PRIME CIRCLE: 2.50+.50=3.00

Figure 14–1   Engineer's cam displacement diagram.

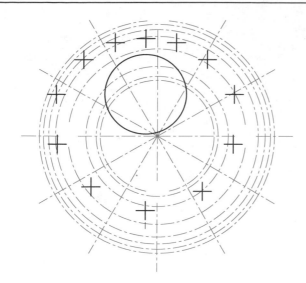

Figure 14–2   Construction of a plate cam profile for computer-aided manufacturing (CAM).

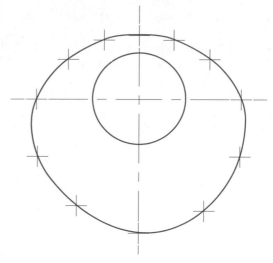

Figure 14–3   The plate cam profile used to generate the CAM tool path.

CADD Applications

MECHANISM DESIGN

Computer-aided design and drafting is a natural in mechanism design and drafting. Mechanism design drawings are created using symbols and techniques that are easily adapted to CADD systems. You will see later that linkage mechanisms, for example, are often designed by displaying the mechanism in several different positions. Using CADD to do this allows you to place each different position of movement on a separate layer and in a different color. Any one or more layers may be turned on or off at the designer's convenience to evaluate the function of the mechanism. In addition, the CADD system is much faster and more accurate than manual design and drafting methods.

There are computer programs available to make the design and drafting of mechanisms easy. For example, there are programs that may be used to simulate the movement of linkage mechanisms. Most of the illustrations in this chapter were created using a CADD system.

Using CADD in linkage design allows you to quickly and accurately establish a series of positions for the mechanism. After the first position is determined, you may easily use commands such as COPY and ROTATE to place the linkages in alternate positions. You can even set up a continuous slide show with each position of the mechanism as a screen display in the slide show. The term *slide show* refers to screen displays that are sequenced in a specific order and for a given period of time. If you allow the slide show to run fairly fast, the viewer gets a good understanding and display of the mechanism in action. This type of a CADD display can be established in 2-D, or 3-D as shown in Figure 14–4.

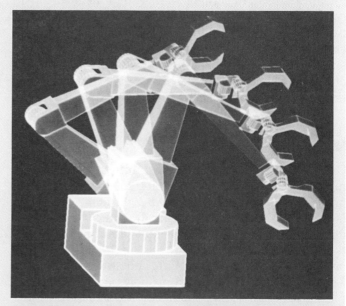

Figure 14–4   Linkage mechanism designed with 3-D CADD showing the movement at several positions. *Courtesy Spectragraphics.*

▼ MECHANISMS

A *mechanism* is an arrangement of parts in a mechanical device or machine. This chapter deals with the design and drafting of elements of a mechanism, including linkages, cams, gears, and bearings. The study of mechanisms is part of the physical sciences known as *mechanics.* Mechanics includes statics and dynamics. *Statics* is the study of physics dealing with nonmoving objects acting as weight. *Dynamics* is the branch of physics that studies the motion of objects and the effects of the forces that cause motion. Dynamics is divided into two categories, kinetics and kinematics. *Kinetics* is an element of physics that deals with the effects of forces that cause motion in mechanisms. The linkages, cams, and gears discussed in this chapter relate to the branch of physics known as kinematics. *Kinematics* is the study of mechanisms without reference to the forces that cause the movement.

Mechanisms in Our Daily Lives

Mechanisms play an important role in our daily activities. Every modern convenience, from the toaster used to make part of your breakfast to the automobile that you drive to work, is made up of one or more mechanisms. The car, for example, is a complex combination of mechanisms that includes all of the types of mechanisms to be discussed in this chapter.

▼ LINKAGES

The elements of any mechanism are referred to as *links.* Links or linkages may be defined as any rigid element of the mechanism. In actual practice all of the mechanism components are links, including levers, bars, sliders, cams, or gears. The frame of the device is even considered a fixed link and is an important part of the mechanism. The first part of this chapter separates the types of linkages that deal primarily with motion caused by levers, rockers, cranks, and sliders.

▼ LINKAGE SYMBOLS

One of the nice things about designing linkage mechanisms is that the drawings are in the form of schematic or single-line representations. After the complete design is created using these schematic drawings, the designer may go to work creating the actual components in relation to the schematic design. Schematic drawings have only a few basic components as shown in Figure 14–5. The symbols may be drawn proportional to the examples. The actual scale depends on the size of the drawing. Typical sizes are shown for most applications in this chapter. You can see here how easy it would be to develop a CADD symbols library for use when drawing these mechanisms.

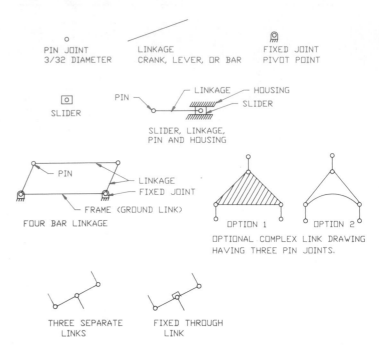

Figure 14–5 Linkage diagram symbols.

▼ TYPES OF LINKAGES

There are many different combinations of linkage mechanisms. These devices are broken into a few basic elements. The illustrations in this discussion show you the linkage mechanism in a simple pictorial drawing of the actual device and in the schematic representation.

Crank Mechanism

A *crank* is a link that makes a complete revolution around a fixed point. When working with the crank, keep in mind that the crank link is a fixed distance equal to the radius of the movement as shown in Figure 14–6.

Lever, or Rocker, Mechanism

A *lever,* or *rocker,* is a link that moves back and forth or oscillates through a given angle as illustrated in Figure 14–7.

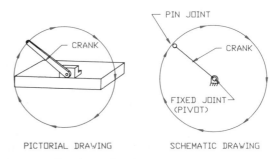

Figure 14–6 Crank mechanism.

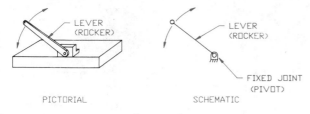

Figure 14–7   Lever or rocker mechanism.

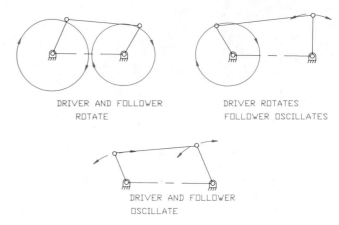

Figure 14–10   Four-bar linkage movements.

Rocker Arm Mechanism

A *rocker arm* is different from the rocker previously discussed because it has a pivot point near the center and oscillates through a given angle as shown in Figure 14–8. Notice the symbol for a fixed link. This symbol is used to indicate that the link between points A and B remains rigid.

Bell Crank Mechanism

Another form of a rocker arm is the *bell crank.* The bell crank is a more complex form of the linkage mechanism. The bell crank drawing shown in Figure 14–9 is a three-joint mechanism where the distances between points A, B, and C are fixed. Other bell crank designs may contain more than three pivot points depending on the design requirements. Notice that the schematic representation in Figure 14–9 shows two alternatives. Consult with your instructor or employer on the technique to use.

Four-bar Linkage

The most commonly used linkage mechanism is called a four-bar linkage. There are many alternate designs of the four-bar linkage, but the basic form has four links, one of which is the ground link or machine

frame. One of the rotating links is called the driver or crank and the other is called the follower or rocker. The link connected between the crank and rocker is called the connecting rod or coupler. The two pivoted links both rotate through 360°, or one rotates while the other oscillates, or they both oscillate, depending on the lengths and arrangement of the links. (See Figure 14–10.)

Normally when determining the function of a four-bar linkage, the designer must show the extreme right and extreme left positions. If the crank rotates and the rocker oscillates, then the angle of oscillation may be established with this technique. For example, determine the angle of oscillation for rocker link CD in the four-bar linkage shown in Figure 14–11. If you are using manual drafting, it is best to use a sharp pencil and make careful measurements. Color pencils are often helpful in showing the different positions. If you are using CADD to draw the diagrams, then the different positions may be shown on different layers and in different colors for a clear analysis of the operation. Follow these steps:

Step 1. and Step 2.   Draw a circle through point B with the center at A and a circle through C with the center at D. Determine the extreme right position of link CD by adding the length of links AB and BC (1.00 + 2.50 = 3.50). This puts AB and BC in a straight line as shown in Figure 14–12.

Step 3.   Establish the extreme left position of link CD by subtracting the length of link AB from BC (2.50 – 1.00 = 1.50). This places link AB and

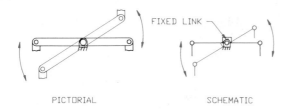

PICTORIAL   SCHEMATIC

Figure 14–8   Rocker arm and fixed link symbol.

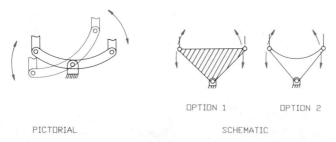

PICTORIAL   SCHEMATIC

Figure 14–9   Belcrank mechanism.

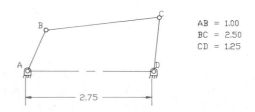

AB = 1.00
BC = 2.50
CD = 1.25

Figure 14–11   Four-bar linkage example.

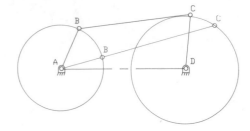

Figure 14–12 Steps 1 and 2, four-bar linkage solution.

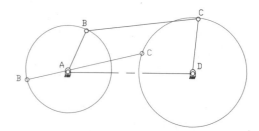

Figure 14–13 Step 3, four-bar linkage solution, extreme left position.

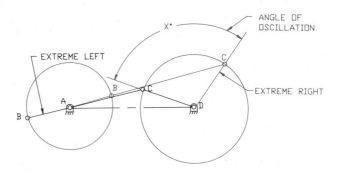

Figure 14–14 Step 4, four-bar linkage solution, extreme right position.

BC in a straight line to the left as shown in Figure 14–13.

Step 4. Determine the angle between the extreme right and extreme left positions of CD as shown in Figure 14–14.

Slider Crank Mechanism

A *slider crank* is a linkage mechanism that is commonly used in machines such as engine pistons, pumps, and clamping devices where a straight line motion is required. Figure 14–15 shows the type of slider design used to move a piston back and forth in a straight line. Notice that the distance the slider travels from the extreme left to the extreme right position is called the *stroke*. When working with the design of an engine, it is important to establish the piston stroke and diameter. The distance the piston travels is the stroke. A combination of the stroke and piston diameter determines the piston displacement. Therefore, a four-cylinder

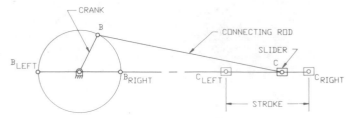

Figure 14–15 Slider mechanism.

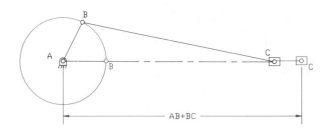

Figure 14–16 Steps 1 and 2, slider mechanism extreme right position.

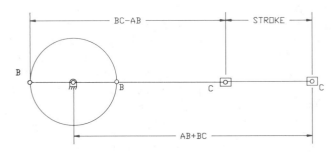

Figure 14–17 Step 3, slider mechanism extreme left position.

2000 cc engine has a displacement of 500 cc per piston cylinder. You can graphically show the stroke of a piston if you have an engine with a crank length of 1.25 in. and a connecting rod length of 3.25 in. Determine the length of the piston stroke as follows:

Step 1. and Step 2. Draw an arc with point A as its center and AB = 1.25 in. as the radius. Show the connecting rod attached to the piston as link BC. Add AB to BC (1.25 + 3.25 = 4.50). Measure this distance from point A along the centerline through A and C. This establishes the extreme right position as shown in Figure 14–16.

Step 3. Subtract AB from BC (3.25 − 1.25 = 2.00). Measure from point A along the centerline through A and C. This establishes the extreme left position of point C. Measure the distance between the extreme left and right positions to determine the stroke as shown in Figure 14–17.

Combination Four-bar Linkage and Slider Mechanism

The design of linkage mechanisms is only limited by the imagination of the designer. There may be any vari-

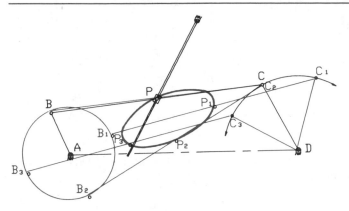

Figure 14–18　Solution to four-bar linkage/slider mechanism example.

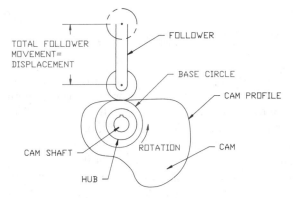

Figure 14–19　Elements of a cam mechanism.

ety of combinations. In the previous examples, the extreme right and left positions were established to determine the function of the mechanism. However, in many situations the designer must establish many positions to completely analyze the movement. One such example is shown in Figure 14–18. This mechanism is a combination four-bar linkage and slider. The objective is to determine the path of point P as the link AB rotates 360°. In order to effectively solve a problem of this type, it is necessary to plot the path of point P by moving link AB a minimum of 30° increments. This is best accomplished if you use a different color or line type for each position. It can become confusing; therefore, careful labeling of each position is important.

Step 1.　Draw a circle through point B with A as the center. Point B must remain on this circle through each movement. Draw an arc through point C with D as the center. Point C must remain on this arc through each movement of the mechanism.

Step 2.　While 30° increments are recommended, this example uses 90° increments for convenience and clarity. Move point B clockwise to B_1. (See Figure 14–18.) From B_1 draw an arc with a radius equal to BC until it intersects arc CD, establishing point C_1. Draw link B_1C_1. Measure distance BP and transfer it to link B_1C_1 thus establishing point P_1.

Step 3.　Follow the same procedure for positions B_2 and B_3. This determines the path of coupler point P at positions P_2 and P_3.

▼ CAMS

A *cam* is a rotating mechanism that is used to convert rotary motion into a corresponding straight motion. The timing involved in the rotary motion is often the main design element of the cam. For example, a cam may be designed to make a follower rise a given amount in a given degree of rotation, then remain constant for an additional period of rotation, and finally fall back to the beginning in the last degree of rotation. The total movement of the cam follower happens in one 360° rotation of the cam. This movement is referred to as the displacement. Cams are generally in the shape of irregular plates, grooved plates, or grooved cylinders. The basic components of the cam mechanism are shown in Figure 14–19.

▼ CADD Applications

CAMS AND GEARS

Designing cams and gears is easy with CADD if a parametric program is used. In a system of this type all you have to do is change the variables and automatically create a new cam or gear. For cam design the variables are:

○ Type of cam motion.
○ Follower displacement.
○ The specific rise, dwell, and fall configuration.
○ Prime circle and base circle diameters.
○ In-line or offset follower.
○ Hub diameter, hub projection, face width, shaft and keyway specifications.

In gear design the variables are:

○ Type of gear.
○ Pitch diameter, base circle diameter, pressure angle.
○ Diametral pitch.
○ Hub diameter, hub projection, face width, shaft and keyway specifications.

The program automatically calculates the rest of the data and draws the cam profile or a detailed drawing of the gear in simplified or detailed representation.

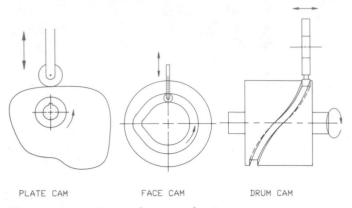

PLATE CAM   FACE CAM   DRUM CAM

Figure 14–20   Types of cam mechanisms.

Cam Types

There are basically three different types of cams: the plate cam, face cam, and drum cam. (See Figure 14–20.) The plate cam is the most commonly used type of cam.

Cam Followers

There are several types of cam followers. The type used depends on the application. The most common type of follower is the roller follower. The roller follower works well at high speeds, reduces friction and heat, and keeps wear to a minimum. The arrangement of the follower in relation to the cam shaft may differ depending on the application. The roller followers shown in Figure 14–21 include: the in-line follower where the axis of the follower is in line with the cam shaft; the offset roller follower; and the pivoted follower. The pivoted follower requires spring tension to keep the follower in contact with the cam profile.

Another type of cam follower is the knife-edged follower shown in Figure 14–22a. This follower is used for only low-speed and low-force applications. The knife-edged follower has a low resistance to wear, but is very responsive and may be effectively used in situations that require abrupt changes in the cam profile.

The flat-faced follower shown in Figure 14–22b is used in situations where the cam profile has a steep rise or fall. Designers often offset the axis of the follower. This practice causes the follower to rotate while in

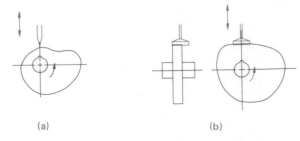

(a)   (b)

Figure 14–22   **(a)** Knife-edged cam follower. **(b)** Flat-faced cam follower.

operation. This rotating action allows the follower surface to wear evenly and last longer.

▼ CAM DISPLACEMENT DIAGRAMS

Cams are generally designed to achieve some type or sequence of a timing cycle in the movement of the follower. There are several predetermined types of motion from which cams are designed. These forms of motion may be used alone, in combination, or custom designed to suit specific applications. The following discussion shows you how to set up a cam displacement diagram given a specific type of cam motion. The cam displacement diagram is similar to a graph representing the cam profile in a flat pattern of one complete 360° revolution of the cam. The terms associated with the displacement diagram include *cycle, period, rise, fall, dwell,* and *displacement.* A complete cam cycle has taken place when the cam rotates 360°. A period of the cam cycle is a segment of follower operation such as rise, dwell, or fall. Rise exists when the cam is rotating and the follower is moving upward. Fall is when the follower is moving downward. Dwell exists when the follower is constant, not moving either up or down. A dwell is shown in the displacement diagram as a horizontal line for a given increment of degrees. When developing the cam displacement diagram, the height of the diagram is drawn to scale and is equal to the total follower displacement. (See Figure 14–23.) The horizontal scale is equal to one cam revolution or 360°. The horizontal

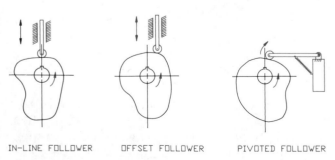

IN–LINE FOLLOWER   OFFSET FOLLOWER   PIVOTED FOLLOWER

Figure 14–21   Types of cam roller followers.

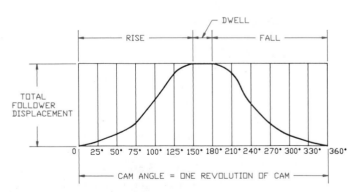

Figure 14–23   Cam displacement diagram.

scale may be drawn without scale. Some engineering drafters prefer to make this scale equal to the circumference of the base circle, or any convenient length. The base circle is an imaginary circle with its center at the center of the cam shaft and its radius tangent to the cam follower at zero position. The horizontal scale is then divided into increments of degrees. Each rise and fall is divided into six increments. So, if the follower rises 150°, then each increment is 150°/6 = 25°. If the cam falls between 180° and 360°, this represents a total fall of 180°. The fall increments are 180°/6 = 30° as shown in Figure 14–23.

Simple Harmonic Motion

Simple harmonic motion may be used for high-speed applications if the rise and fall are equal at 180°. Moderate speeds are recommended if the rise and fall are unequal or if there is a dwell in the cycle. This application causes the follower to jump if the speeds are too high.

Draw a cam displacement diagram using simple harmonic motion when the total displacement is 2.00 in. and the cam follower rises the length of the total displacement in 180° and falls back to 0° in 180°. Use the following procedure to set up the displacement diagram:

Step 1.   Draw a rectangle equal in height (vertical scale) to the total displacement of 2.00 in. and equal in length (horizontal scale) to 360°. The horizontal scale should have 6°–30° increments for the rise from 0° to 180° and 6°–30° increments for the fall from 180° to 360°. The horizontal scale may be any convenient length. Draw a thin vertical line from each horizontal increment as shown in Figure 14–24.

Step 2.   Draw a half circle at one end of the displacement diagram equal in diameter to the rise of the cam. Divide the half circle into six equal parts as shown in Figure 14–25.

Step 3.   The cam follower begins its rise at 0°. The rise continues by projecting point 1 on the half circle over to the first (30°) increment on the horizontal scale. Continue this process for points 2, 3, 4, 5, and 6 on the half circle, each

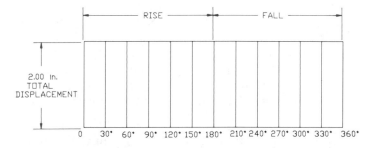

Figure 14–24   Step 1, layout for simple harmonic motion cam displacement diagram.

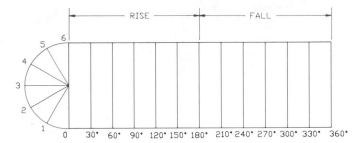

Figure 14–25   Step 2, layout for simple harmonic motion cam displacement diagram.

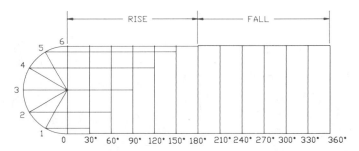

Figure 14–26   Steps 3, 4, and 5, layout for simple harmonic motion cam displacement diagram.

intersecting the next increment on the horizontal scale.

Step 4.   Notice the pattern of points created in the preceding step. Use your irregular curve to carefully connect these points. If you are using CADD, use your curve-fitting command to draw the cam profile.

Step 5.   Develop the fall profile by projecting the points from the half circle in the reverse order discussed in Step 3. (See Figure 14–26.)

Constant Velocity Motion

Constant velocity motion is also known as straight-line motion. This is used for the feed control of some machine tools, when it is required for the follower to rise and fall at a uniform rate. Constant velocity motion is only used at slow speeds because of the abrupt change at the beginning and end of the motion period. The displacement diagram is easy to draw. All you have to do is draw a straight line from the beginning of the rise or fall to the end as shown in Figure 14–27.

Modified Constant Velocity Motion

Modified constant velocity motion was designed to help reduce the abrupt change at the beginning and end of the motion period. This type of motion may be adjusted to accomplish specific results by altering the degree of modification. This is done by placing a curve at the beginning and end of the rise and fall. The radius

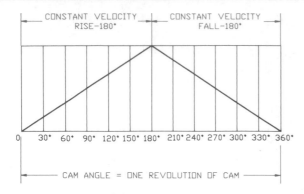

Figure 14–27 Cam displacement diagram for constant velocity motion.

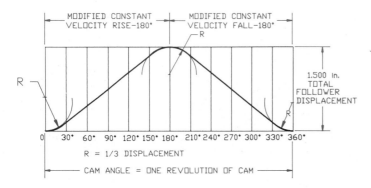

Figure 14–28 Cam displacement diagram for modified constant velosity motion.

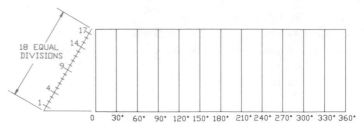

Figure 14–29 Steps 1 and 2, layout for uniform accelerated motion cam displacement diagram.

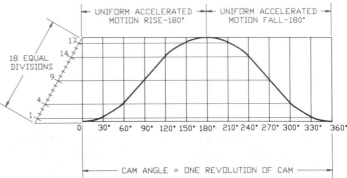

Figure 14–30 Step 3, layout for uniform accelerated motion cam displacement diagram.

of this arc depends on the amount of smoothing required, but the radius normally ranges from one-third to full displacement. If the motion were modified to one-third the displacement, then the cam displacement diagram is drawn as shown in Figure 14–28.

Uniform Accelerated Motion

Uniform accelerated motion is designed to reduce the abrupt change at the beginning and end of a period. It is recommended for moderate speeds, especially when associated with a dwell. The advantage of this motion is its use when constant acceleration for the first half of the rise and constant deceleration for the second half of the rise are required.

Use the following technique to draw a uniform accelerated motion displacement diagram where the follower rises a total of 2.00 in. in 180° and falls back to 0° in 180°:

Step 1.
and
Step 2.
Set up the displacement diagram with the height equal to 2.00 in. total rise and the horizontal scale divided into 30° increments. Keep in mind that the horizontal scale is divided into 30° increments if the rise and fall is in 180° each. If the rise, for example, were 120°, then the increments would be 120°/6 = 20° each. Set up a scale with eighteen

equal divisions at one end of the displacement diagram and mark off the first, fourth, ninth, fourteenth, and seventeenth divisions as shown in Figure 14–29.

Step 3. Establish the rise by projecting from the first division on the scale to the 30° increment on the diagram, then continue with the fourth division to 60°, and so on until each division is used. Continue this same procedure in reverse order to establish the profile of the fall. Connect all of the points to complete the displacement diagram as shown in Figure 14–30.

Cycloidal Motion

Cycloidal motion is the most popular cam profile development for smooth running cams at high speeds. The term *cycloidal* comes from the word *cycloid*. A cycloid is a curved line generated by a point on the circumference of a circle as the circle rolls along a straight line. The cycloidal cam motion is developed in this same manner and the result is the smoothest possible cam profile. Cycloidal motion is a little more complex to set up than other types of motion. Use the following procedure to develop a cam displacement diagram for cycloidal motion with a total rise of 2.500 in. in 180°:

Step 1. Begin the displacement diagram with a total rise of 2.500 in. in 180°. Only half of the diagram is shown for this example.

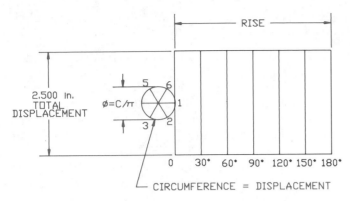

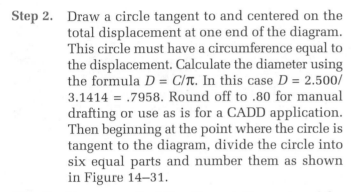

Figure 14–31 Steps 1 and 2, layout for cycloidal motion cam displacement diagram.

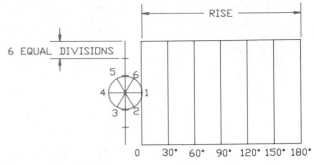

Figure 14–32 Step 3, layout for cycloidal motion cam displacement diagram.

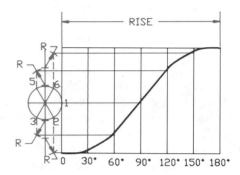

Figure 14–33 Step 4, layout for cycloidal motion cam displacement diagram.

Step 2. Draw a circle tangent to and centered on the total displacement at one end of the diagram. This circle must have a circumference equal to the displacement. Calculate the diameter using the formula $D = C/\pi$. In this case $D = 2.500/3.1414 = .7958$. Round off to .80 for manual drafting or use as is for a CADD application. Then beginning at the point where the circle is tangent to the diagram, divide the circle into six equal parts and number them as shown in Figure 14–31.

Step 3. Draw a vertical line through the center of the circle equal in length to the displacement. Divide this line into six equal parts as shown in Figure 14–32.

Step 4. Use a radius equal to the radius of the circle to draw arcs from the divisions on the vertical line as shown in Figure 14–33. These arcs should intersect the dashed lines drawn from points 2, 3, 5, and 6 on the circle. Where the dashed lines and the arcs intersect, draw horizontal lines into the displacement diagram, intersecting the

appropriate increment from the horizontal scale. Connect the points of intersection with a smooth curve as shown in Figure 14–34.

Developing a Cam Displacement Diagram with Different Cam Motions

Most cam profiles are not as simple as the preceding examples. Many designs require more than one type of cam motion and may also incorporate dwell. Construct

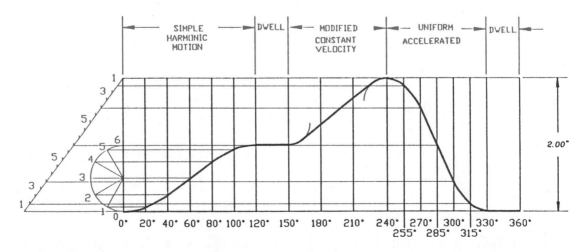

Figure 14–34 The development of a cam displacement diagram with different cam motions.

a cam displacement diagram from the following information:

❍ Total displacement equals 2.00 in.
❍ Rise 1.00 in. simple harmonic motion in 120°.
❍ Dwell for 30°.
❍ Rise 1.00 in. modified constant velocity motion in 90°.
❍ Fall 2.00 in. uniform accelerated motion in 90°.
❍ Dwell for 30° through the balance of the cycle.

Look at Figure 14–34 as you review the method of development for each type of cam motion displayed.

▼ CONSTRUCTION OF AN IN-LINE FOLLOWER PLATE CAM PROFILE

Each of the cam displacement diagrams may be used to construct the related cam profiles. The *cam profile* is the actual contour of the cam. In operation, the cam follower is stationary and the cam rotates on the cam shaft. In cam profile construction, however, the cam is drawn in one position and the cam follower is moved to a series of positions around the cam in relationship to the cam displacement diagram. The following technique is used to draw the cam profile for the cam displacement diagram given in Figure 14–35 and for a cam with a

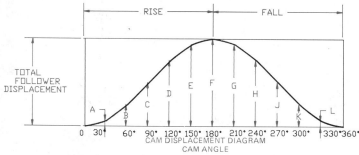

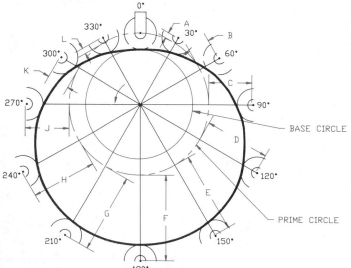

Figure 14–35 Construction of an in-line follower plate cam profile.

2.50 in. base circle, a .75 in. cam follower, and counter-clockwise rotation:

Step 1. Refer to Figure 14–35. Use construction lines for all preliminary work. Draw the cam follower in place near the top of the sheet with phantom lines. Draw the base circle (∅2.50 in.). The base circle is tangent to the follower at the 0° position. Draw the prime circle 2.50 + .75 = ∅3.25 in. The *prime circle* passes through the center of the follower at the 0° position. The cam displacement diagram is placed on Figure 14–35 for easy reference. In actual practice, you may have the displacement diagram next to the cam profile drawing or on a CADD layer to be inserted on the screen for reference.

Step 2. Begin working in a direction opposite from the rotation of the cam. Since this cam rotates counterclockwise, work clockwise. Starting at 0° draw an angle equal to each of the horizontal scale angles on the displacement diagram. In this case the angles are 30° increments. This is not true with all cam displacement diagrams. Some have varying increments.

Step 3. Notice the measurements labeled A through L on the displacement diagram in Figure 14–35. Begin by transferring the distance A along the 30° element (in the profile construction drawing) by measuring from the prime circle along this line. This establishes the center of the cam follower at this position. Do the same with each of the measurements B through L located on each corresponding increment.

Step 4. Lightly draw the cam followers in position with centers located at each of the points found in Step 3. (See Figure 14–35.)

Step 5. Draw the cam profile by connecting a smooth curve tangent to the cam followers at each position. This is one place where CADD drafting plays an important role regarding accuracy and speed.

Preparing the Formal Plate Cam Drawing

You are ready to prepare the formal plate cam drawing after you have constructed the plate cam profile using the previously described techniques. The previous drawing may be used because all lines were drawn as construction lines or on a CADD layer that may be turned off when the formal drawing is complete. The information needed on the plate cam drawing includes:

❍ Cam profile.
❍ Hub dimensions, including cam shaft, outside diameter, width, keyway dimensions.

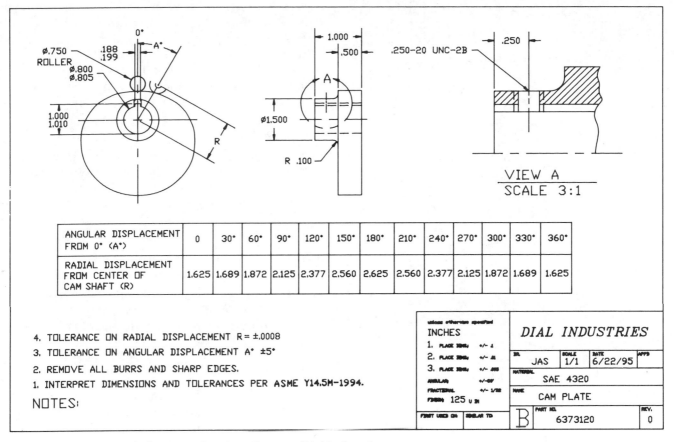

ANGULAR DISPLACEMENT FROM 0° (A°)	0	30°	60°	90°	120°	150°	180°	210°	240°	270°	300°	330°	360°
RADIAL DISPLACEMENT FROM CENTER OF CAM SHAFT (R)	1.625	1.689	1.872	2.125	2.377	2.560	2.625	2.560	2.377	2.125	1.872	1.689	1.625

4. TOLERANCE ON RADIAL DISPLACEMENT R = ±.0008

3. TOLERANCE ON ANGULAR DISPLACEMENT A° ±5°

2. REMOVE ALL BURRS AND SHARP EDGES.

1. INTERPRET DIMENSIONS AND TOLERANCES PER ASME Y14.5M-1994.

NOTES:

Figure 14–36   Formal plate cam drawing. *Courtesy Dial Industries.*

○ Roller follower placed in one convenient location, such as 60°, using phantom lines.

○ The drawing is set up as a chart drawing where $A°$ equals the angle of the follower at each position, and R equals the radius from the center of the cam shaft to the center of the follower at each position.

○ A chart giving the values of the angles A and the radii R at each follower position.

○ Side view showing the cam plate thickness and set screw location with thread specification, if used.

○ Tolerances, unless otherwise specified.

Establish all measurements for dimensions A and R at each of the follower positions. This may be done graphically by measuring from the profile construction, or mathematically using trigonometry. If a CADD system is used, the measurements may be taken directly from the layout. Figure 14–36 shows a formal plate cam drawing.

▼ CONSTRUCTION OF AN OFFSET FOLLOWER PLATE CAM PROFILE

When an offset cam follower is used, the method of construction is a little more complex than the technique used for the in-line follower. For this example, the follower is offset .75 in. as shown in Figure 14–37. Prepare the cam profile drawing as follows:

Step 1. Refer to Figure 14–37. Use construction lines or a construction CADD layer for all preliminary work. Draw the cam follower in place near the top of the sheet with phantom lines. Draw the base circle (Ø2.50 in.). The base circle is tangent to the follower at 0° position. Draw the prime circle (2.50 + .75 = Ø3.25 in.). The prime circle passes through the center of the follower at 0° position. Draw the offset circle (Ø1.50 in.). The offset circle is drawn with a radius equal to the follower offset distance. The cam displacement diagram is placed on Figure 14–37 for easy reference. In actual practice, you may have the displacement diagram next to the cam profile drawing or on a CADD layer to be inserted on the screen for reference.

Step 2. Begin working in a direction opposite from the rotation of the cam. Since this cam rotates counterclockwise, work clockwise. Starting at 0° on the offset circle, draw an angle equal to each of the horizontal scale angles on the displacement diagram. In this case the angles are 30° increments. This is not true with all cam displacement diagrams. Some have varying increments.

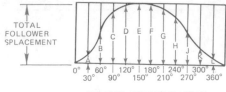

CAM DISPLACEMENT DIAGRAM

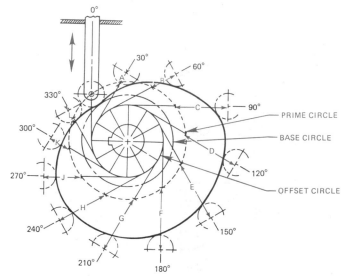

Figure 14–37 Construction of an offset follower plate cam profile.

Step 3. Notice where each of the angle increments drawn in Step 2 intersects the offset circle. At each of these points, draw another line tangent to the offset circle. Make these lines long enough to pass beyond the reference circle.

Step 4. Notice the measurements labeled A through L on the displacement diagram in Figure 14–37. Begin by transferring the distance A along the line tangent to the 30° element by measuring from the prime circle along this line. This establishes the center of the cam follower at this position. Do the same with each of the measurements B through L located on the tangent line from each corresponding increment.

Step 5. Lightly draw the cam followers in position with centers located at each of the points found in Step 4. (Refer to Figure 14–37.)

Step 6. Draw the cam profile by connecting a smooth curve tangent to the cam followers at each position. This is one place where CADD drafting plays an important role regarding accuracy and speed.

▼ DRUM CAM DRAWING

Drum cams are used when it is necessary for the follower to move in a path parallel to the axis of the cam. The drum cam is a cylinder with a groove machined in the surface the shape of the cam profile. The cam follower moves along the path of the groove as the drum is rotated. The displacement diagram for a drum cam is actually the pattern of the drum surface as if it were rolled out flat. The height of the displacement diagram is equal to the height of the drum. The length of the displacement diagram is equal to the circumference of the drum. Refer to the drum cam drawing in Figure 14–38 as you follow the construction steps:

Step 1. Draw the top view showing the diameter of the drum, the cam shaft and keyway, and the roller follower in place at 0°. Draw the front view as shown in Figure 14–38. Draw the outline of the cam displacement diagram equal to the height and circumference of the drum.

Step 2. Draw the roller follower on the displacement diagram at each angular interval.

Step 3. Draw curves tangent to the top and bottom of each roller follower position. These curves represent the development of the groove in the surface of the cam.

Step 4. In the top view, draw radial lines from the cam shaft center equal to the angle increments shown on the displacement diagram. Be sure to lay out the angles in a direction opposite the cam rotation. The points where these lines intersect the depth and the outside circumference of the groove are labeled A, A₁, B, B₁, respectively. Notice that the same corresponding points are labeled on the displacement diagram.

Step 5. From points A, A₁ on the displacement diagram, project horizontally until each point intersects a vertical line from the same corresponding point in the top view. This establishes the points in the front view along the outer and inner edges of the groove, both top and bottom. Continue this process for each pair of points on the drum cam displacement diagram.

▼ GEARS

Gears are toothed wheels used to transmit motion and power from one shaft to another. Gears are rugged and durable and can transmit power with up to 98 percent efficiency with long service life. Gear design involves a combination of material, strength, and wear characteristics. Most gears are made of cast iron or steel, but brass and bronze alloys, and plastic are also used for some applications. Gear selection and design is often done through vendors' catalogs or the use of standard formulas. A gear train exists when two or more gears are in combination for the purpose of transmitting power. Generally, two gears in mesh are used to increase or reduce speed, or change the direction of motion from one

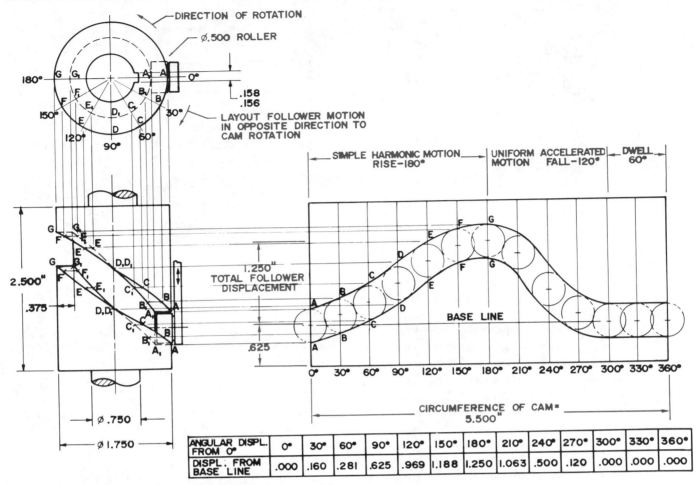

The table within the figure:

ANGULAR DISPL. FROM 0°	0°	30°	60°	90°	120°	150°	180°	210°	240°	270°	300°	330°	360°
DISPL. FROM BASE LINE	.000	.160	.281	.625	.969	1.188	1.250	1.063	.500	.120	.000	.000	.000

2. TOLERANCE ON ANGULAR DISPLACEMENT ±.5°.
1. TOLERANCE ON DISPLACEMENT FROM BASELINE ±.0008.

NOTES:

Figure 14–38 Construction of a drum cam drawing.

CADD Applications

CAM DISPLACEMENT DIAGRAMS AND PROFILES

CADD systems make drawing cam displacement diagrams and cam profiles easy and accurate without the need for mathematical calculations. Constructing the cam displacement diagram and converting the information to the cam profile is the same as with manual drafting except the accuracy is increased substantially. Drawing the cam profile using irregular curves is difficult and time consuming. Most CADD systems have a curve-fitting command that makes it possible to automatically draw the cam profile through the points of tangency at the cam follower positions.

CAM MANUFACTURING

Modern cams are manufactured using computer numerical control (CNC) machining. In many instances the drawing is prepared on a CADD system and transferred to a computer-aided manufacturing (CAM) program for immediate coding for the CNC machine. Some CAM programs allow the designer to create the cam profile and transfer the data to the CNC machine tool without ever generating a drawing. Refer to Chapter 9 on Manufacturing Processes for more information.

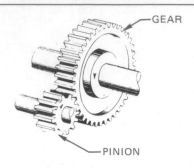

Figure 14–39   The gear and pinion.

shaft to another. When two gears are in mesh, the larger is called the gear and the smaller is called the pinion. (See Figure 14–39.)

AGMA/ANSI   Gear selection generally follows the guidelines of the American Gear Manufacturers Association (AGMA) or the American National Standards Institute (ANSI).

It is important for engineering drafters to fully understand gear terminology and formulas. However, many drafters will not draw gears, because gears are commonly supplied as purchase parts. When this happens, the drafter may be required to make gear selections for specific applications or draw gears on assembly drawings. Details of gears are often drawn using simplified techniques as described in this chapter. In some situations, the gears are drawn as they actually exist for display on assembly drawings or in catalogs. When this is necessary, CADD can make the job easy and increase productivity. If manual drafting is used, a gear tooth template is needed.

▼ GEAR STRUCTURE

Gears are made in a variety of structures depending on the design requirements, but there is some basic termi-

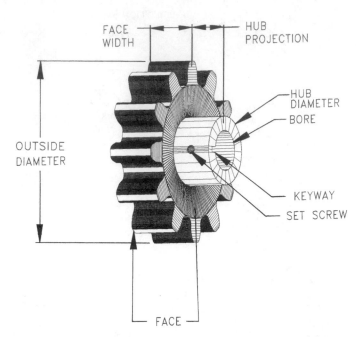

Figure 14–40   Elements of the gear structure.

nology that can typically be associated with gear structure. The elements of the gear structure shown in Figure 14–40 include the outside diameter, face, hub, bore, and keyway.

Types of Hubs

The hubs are the lugs or shoulders projecting from one or both faces of some gears. These may be referred to as A, B, or C hubs. An A hub is also called a flush hub because there is no projection from the gear face. B hubs have a projection on one side of the gear while C hubs have projections on both sides of the gear face. (See Figure 14–41.)

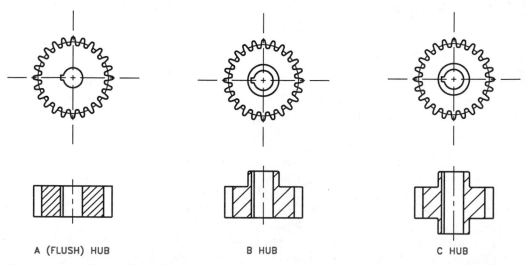

A (FLUSH) HUB   B HUB   C HUB

Figure 14–41   Types of gear hubs.

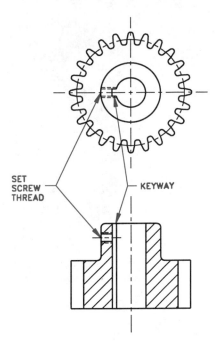

Figure 14–42 Gear, keyways, and set screw.

Keyways, Keys, and Set Screws

Gears are ususally held on the shaft with a key, keyway, and set screw. Refer to Chapter 9 for information on the manufacture of keyways, Chapter 10 for proper dimensioning, and Chapter 11 for screw thread specifications. One or more set screws are usually used to keep the key secure in the keyway. (See Figure 14–42.)

▼ SPLINES

Splines are teeth cut in a shaft and a gear or pulley bore and are used to prevent the gear or pulley from spinning on the shaft. Splines are often used when it is necessary for the gear or pulley to easily slide on the shaft. Splines may also be nonsliding and in all cases are stronger than keyways and keys. The standardization of splines is established by the Society of Automotive Engineers (SAE) so that any two parts with the same spline specifications should fit together. The following is an example of an SAE spline specification:

SAE 2 1/2–10 B SPLINE
 (A)   (B)   (C)(D)

The note components are described below:

(A)   Society of Automotive Engineers
(B)   The outside diameter of the spline.
(C)   The number of teeth.
(D)   A = this is a fixed non-sliding spline.
      B = this spline slides under no load conditions.
      C = this spline slides under load conditions.

Involute Spline

Another spline standard is the *involute spline*. The teeth on the involute spline are similar to the curved teeth found on spur gears. The spline teeth generally have a shorter whole depth than standard spur gears, and the pressure angle is normally 30°.

▼ GEAR TYPES

The most common and simplest form of gear is the spur gear. Bevel gears and worm gears are also used. Gear types are designed based on one or more of the following elements:

○   The relationship of the shafts; either parallel, intersecting, nonintersecting shafts, or rack and pinion.
○   Manufacturing cost.
○   Ease of maintenance in service.
○   Smooth and quiet operation.
○   Load-carrying ability.
○   Speed reduction capabilities.
○   Space requirements.

Parallel Shafting Gears

Many different types of mating gears are designed with parallel shafts. These include spur and helical gears.

Spur Gears.   There are two basic types of spur gears; external and internal spur gears. (See Figure 14–43.) When two or more spur gears are cut on a single shaft, they are referred to as cluster gears. External spur gears are designed with the teeth of the gear on the outside of a cylinder. External spur gears are the most common type of gear used in manufacturing. Internal spur gears have the teeth on the inside of the cylindrical gear. The advantages of spur gears over other types is their low

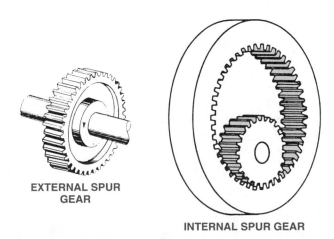

EXTERNAL SPUR GEAR

INTERNAL SPUR GEAR

Figure 14–43 Spur gears. *Reprinted from Baril, MODERN MACHINING TECHNOLOGY, ©1987, Delmar Publishers Inc.*

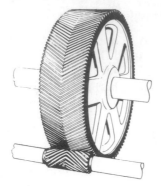
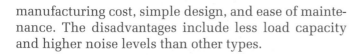

Figure 14–44 Helical gear.

Figure 14–45 Herringbone gear.

Figure 14–46 Bevel gears.

Figure 14–47 Crossed helical gears. *Courtesy Browning Mfg. Division of Emerson Electric Co.*

manufacturing cost, simple design, and ease of maintenance. The disadvantages include less load capacity and higher noise levels than other types.

Helical Gears. Helical gears have their teeth cut at an angle that allows more than one tooth to be in contact. (See Figure 14–44.) Helical gears carry more load than equivalent-sized spur gears and operate more quietly and smoothly. The disadvantage of helical gears is that they develop *end thrust*. End thrust is a lateral force exerted on the end of the gear shaft. Thrust bearings are required to reduce the effect of this end thrust. Double helical gears are designed to eliminate the end thrust and provide long life under heavy loads. However, they are more difficult and costly to manufacture. The herringbone gear shown in Figure 14–45 is a double helical gear without space between the two opposing sets of teeth.

Intersecting Shafting Gears

Intersecting shafting gears allow for the change in direction of motion from the gear to the pinion. Different types of intersecting shafting gears include bevel and face gears.

Bevel Gears. Bevel gears are conical in shape, allowing the shafts of the gear and pinion to intersect at 90° or any desired angle. The teeth on the bevel gear have the same shape as the teeth on spur gears except they taper toward the apex of the cone. Bevel gears provide for a speed change between the gear and pinion. (See Figure 14–46.) Miter gears are the same as bevel gears except both the gear and pinion are the same size and are used when shafts must intersect at 90° without speed reduction. Spiral bevel gears have the teeth cut at an angle, which provides the same advantages as helical gears over spur gears.

Face Gears. The face gear is a combination of bevel gear and spur pinion, or bevel gear and helical pinion. This combination is used when the mounting accuracy is not as critical as with bevel gears. The load-carrying capabilities of face gears is not as good as that of bevel gears.

Nonintersecting Shafting Gears

Gears with shafts that are at right angles but not intersecting are referred to as nonintersecting shafts. Gears that fall into this category are crossed helical, hypoid, and worm gears.

Crossed Helical Gears. Also known as right angle helical or spiral gears, crossed helical gears provide for nonintersecting right angle shafts with low load-carrying capabilities. (See Figure 14–47.)

Hypoid Gears. Hypoid gears have the same design as bevel gears except the gear shaft axes are offset and do not intersect. (See Figure 14–48.) The gear and pinion are often designed with bearings mounted on both sides for improved rigidity over standard bevel gears. Hypoid gears are very smooth, strong, and quiet in operation.

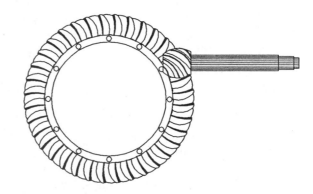

Figure 14–48 Hypoid gears.

Figure 14–49 Worm and worm gear.

Figure 14–50 Rack and pinion.

Worm Gears. A worm and worm gear are shown in Figure 14–49. This type of gear is commonly used when a large speed reduction is required in a small space. The worm may be driven in either direction. When the gear is not in operation, the worm automatically locks in place. This is a particular advantage when it is important for the gears to have no movement of free travel when the equipment is shut off.

Rack and Pinion

A rack and pinion is a spur pinion operating on a flat straight bar rack. (See Figure 14–50.) The rack and pinion is used to convert rotary motion into straight line motion.

▼ SPUR GEAR DESIGN

Spur gear teeth are straight and parallel to the gear shaft axis. The tooth profile is designed to transmit power at a constant rate, and with a minimum of vibration and noise. To achieve these requirements, an *involute curve* is used to establish the gear tooth profile. An involute curve is a spiral curve generated by a point on a chord as it unwinds from the circle. The contour of a gear tooth, based on the involute curve, is determined by a base circle, the diameter of which is controlled by a pressure angle. The *pressure angle* is the direction of push transmitted from a tooth on one gear to a tooth on the mating gear or pinion. (See Figure 14–51.) Two standard pressure angles, 14.5° and 20°, are used in spur gear design. The most commonly used pressure angle is 20° because it provides a stronger tooth for quieter running and heavier load-carrying characteristics. One of the basic rules of spur gear design is to have no fewer than 13 teeth on the running gear and 26 teeth on the mating gear.

Standard terminology and formulas control the drawing requirements for spur gear design and specifications. Figure 14–52 shows a pictorial representation of the spur gear teeth with the components labeled. As an engineering drafter, it is important that you become familiar with the terminology and associated mathematical formulas used to calculate values.

Diametral Pitch

The diametral pitch actually refers to the tooth size and has become the standard for tooth size specifications. As you look at Figure 14–53, notice how the tooth size increases as the diametral pitch decreases. One of the most elementary rules of gear tooth design is that mating teeth must have the same diametral pitch.

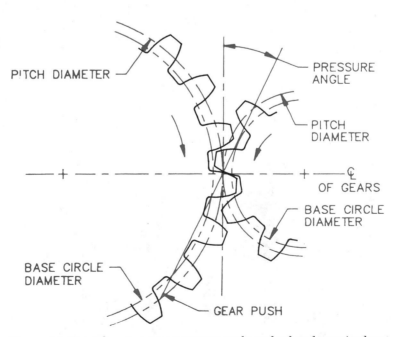

Figure 14–51 The spur gear pressure angle and related terminology.

TERM	DESCRIPTION	FORMULA
Pitch Diameter (D)	The diameter of an imaginary pitch circle on which a gear tooth is designed. Pitch circles of two spur gears are tangent.	$D = N/P$
Diametral Pitch (P)	A ratio equal to the number of teeth on a gear per inch of pitch diameter.	$P = N/D$
Number of Teeth (N)	Number of teeth on a gear.	$N = D \times P$
Circular Pitch (p)	The distance from a point on one tooth to the corresponding point on the adjacent tooth, measured on the pitch circle.	$p = 3.1416 \times D/N$ $p = 3.1416/P$
Center Distance (C)	The distance between the axis of two mating gears.	$C = \text{sum of pitch DIA}/2$
Addendum (a)	The radial distance from the pitch circle to the top of the tooth.	$a = 1/P$
Dedendum (b)	The radial distance from the pitch circle to the bottom of the tooth. (This formula is for 20° teeth only.)	$b = 1.250/P$
Whole Depth (h_t)	The full height of the tooth. It is equal to the sum of the addendum and the dedendum.	$h_t = a + b$ $h_t = 2.250/P$
Working Depth (h_k)	The distance that a tooth occupies in the mating space. A distance equal to two times the addendum.	$h_k = 2a$ $h_k = 2.000/P$
Clearance (c)	The radial distance between the top of a tooth and the bottom of the mating tooth space. It is also the difference between the addendum and dedendum.	$c = b - a$ $c = .250/P$
Outside Diameter (D_o)	The overall diameter of the gear. It is equal to the pitch diameter plus two dedendums.	$D_o = D + 2a$
Root Diameter (D_r)	The diameter of a circle coinciding with the bottom of the tooth spaces.	$D_r = D - 2b$
Circular Thickness (t)	The length of an arc between the two sides of a gear tooth measured on the pitch circle.	$t = 1.5708/P$
Chordal Thickness (t_c)	The straight line thickness of a gear tooth measured on the pitch circle.	$t_c = D \sin (90°/N)$
Chordal Addendum (a_c)	The height from the top of the tooth to the line of the chordal thickness.	$a_c = a + t^2/4D$
Pressure Angle (ϕ)	The angle of direction of pressure between contacting teeth. It determines the size of the base circle and the shape of the involute teeth.	
Base Circle Dia. (D_B)	The diameter of a circle from which the involute tooth form is generated.	$D_B = D \cos \phi$

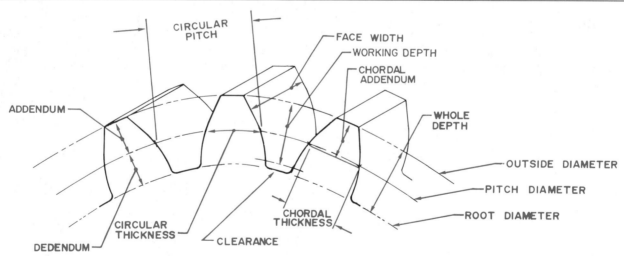

Figure 14–52   Gear terminology and formulas.

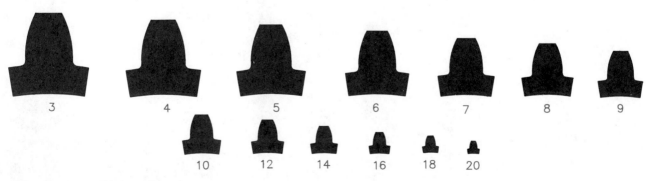

Figure 14–53   Diametral pitch.

◣ CADD Applications

GEAR TEETH

There are custom CADD programs that allow the gear teeth to be drawn automatically. To do this, the computer prompts the drafter for variables such as pitch diameter, diametral pitch, pressure angle, and number of teeth. The program then automatically notifies the

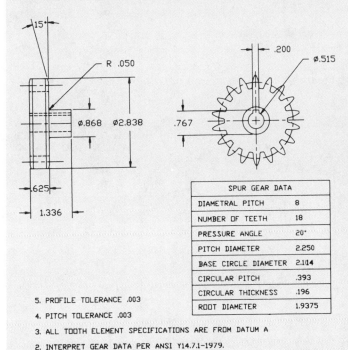

SPUR GEAR DATA	
DIAMETRAL PITCH	8
NUMBER OF TEETH	18
PRESSURE ANGLE	20°
PITCH DIAMETER	2.250
BASE CIRCLE DIAMETER	2.1114
CIRCULAR PITCH	.393
CIRCULAR THICKNESS	.196
ROOT DIAMETER	1.9375

NOTES:
1. INTERPRET DIMENSIONS AND TOLERANCES PER ASME Y14.5M-1994.
2. INTERPRET GEAR DATA PER ANSI Y14.7.1-1979.
3. ALL TOOTH ELEMENT SPECIFICATIONS ARE FROM DATUM A.
4. PITCH TOLERANCE .003
5. PROFILE TOLERANCE .003

Figure 14–54 Gear detailed drawing using CADD to automatically draw teeth and gear data chart from given specifications.

drafter whether the information is accurate, provides all additional data, and creates a detail drawing of the gear. CADD was used to draw the teeth in the gear displayed in Figure 14–54. The CADD drafter may easily draw gears displayed in detailed representation, or use the simplified technique to save regeneration and plotting time.

Some software programs do more than assist in the design process. For example, objects such as gear teeth can be subjected to simulated tests and stress analysis on the computer screen as shown in Figure 14–55.

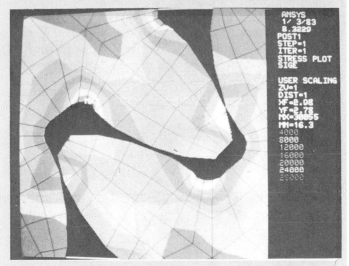

Figure 14–55 Objects such as these gear teeth can be subjected to simulated tests and stress analysis on the computer screen. *Courtesy Swanson Analysis Systems, Inc.*

▼ GEAR ACCURACY

AGMA

A system has been established by the American Gear Manufacturers Association (AGMA) for the classification of gears based on the accuracy of the maximum tooth-to-tooth and total composite tolerances allowed. This number is called the AGMA *quality number*. The AGMA quality numbers and corresponding maximum tolerances are established by diametral pitch and pitch diameter. The AGMA quality numbers are listed in *AGMA Gear Handbook, 2000–A88*. The higher the AGMA quality number, the tighter the tolerance. For example, an AGMA Q6 allows approximately .004 total composite error, Q10 is

less than .001, and Q13 is less than .0004. The *AGMA Gear Handbook* also displays a list of gear applications and the quality number suggested for each application.

▼ DRAWING SPECIFICATIONS AND TOLERANCES

The final step in designing a gear is the presentation of a drawing that displays the dimensioned gear using multiviews and gear data charts. While companies do not always provide the complete gear data charts on their drawings, errors can happen when complete data is not provided. Gear data formulas and sample gear data charts have been shown with previous examples in this chapter; however, the AGMA recommended information for spur and helical gears is also in Appendix M.

▼ DESIGNING AND DRAWING SPUR GEARS

ANSI The drafting standard that governs gear drawings is the American National Standards document ANSI Y14.7.1 *Gear Drawing Standards—Part I.*

Because gear teeth are complex and time consuming to draw, simplified representations are used to make the practice easier. (See Figure 14–56.) The simplified method shows the outside diameter and the root diameters as phantom lines and the pitch diameter as a centerline in the circular view. In addition, a side view is often required to show width dimensions and related features. (See Figure 14–56a.) If the gear construction has webs, spokes, or other items that require further clarification, then a full section is normally used. (See Figure 14–56b.) Notice in the cross section that the gear tooth is left unsectioned and the pitch diameter is shown as a centerline.

When cluster gears are drawn, the circular view may show both sets of gear tooth representations in simplified form, or two circular views may be drawn. (See Figure 14–57.) When cluster gears are more complex than those shown here, multiple views and removed sections may be required.

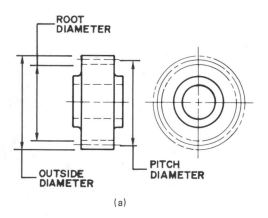

(a)

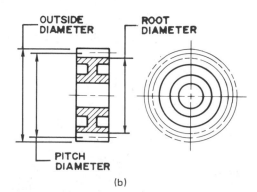

(b)

Figure 14–56 Typical spur gear drawings using simplified gear teeth representation.

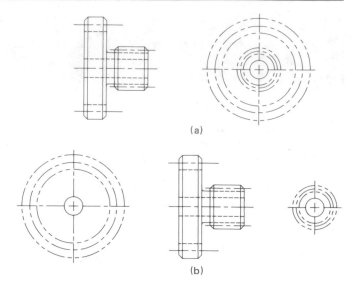

Figure 14–57 Cluster gear drawings.

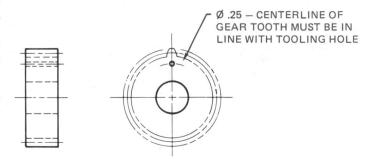

Ø .25 – CENTERLINE OF GEAR TOOTH MUST BE IN LINE WITH TOOLING HOLE

Figure 14–58 Showing the relationship of one gear tooth to another feature on the gear.

One or more teeth may be drawn for specific applications. For example, when a tooth must be in alignment with another feature of the gear, the tooth may be drawn as shown in Figure 14–58.

Gear drawings typically have a chart that shows the manufacturing information associated with the teeth, and with related part detail dimensions placed on the specific views.

▼ DESIGNING SPUR GEAR TRAINS

A gear train is an arrangement of two or more gears connecting driving and driven parts of a machine. Gear reducers and transmissions are examples of gear trains. The function of a gear train is to:

○ transmit motion between shafts.
○ decrease or increase the speed between shafts.
○ change the direction of motion.

It is important for you to understand the relationship between two mating gears in order to design gear trains. When gears are designed, the end result is often a specific *gear ratio*. Any two gears in mesh have a gear

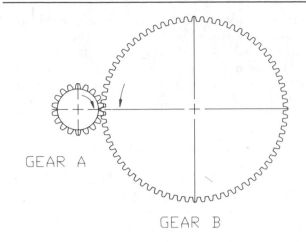

Figure 14–59 Calculating gear data.

	(D) PITCH DIAMETER	(N) NUMBER OF TEETH	(P) DIAMETRAL PITCH	RPM	DIRECTION
GEAR A	6	18	3	1200	C. WISE
GEAR B	18	72	3	400	C.C.WISE

ratio. The gear ratio is expressed as a proportion, such as 2:1 or 4:1, between two similar values. The gear ratio between two gears is the relationship between the following characteristics:

○ Number of teeth.
○ Pitch diameters.
○ Revolutions per minute (rpm).

If you have gear A (pinion) mating with gear B, as shown in Figure 14–59, the gear ratio is calculated by dividing like values of the smaller gear into the larger gear as follows:

$$\frac{\text{Number Teeth}_{\text{Gear B}}}{\text{Number Teeth}_{\text{Gear A}}} = \text{Gear Ratio}$$

$$\frac{\text{Pitch Diameter}_{\text{Gear B}}}{\text{Pitch Diameter}_{\text{Gear A}}} = \text{Gear Ratio}$$

$$\frac{\text{rpm}_{\text{Gear A}}}{\text{rpm}_{\text{Gear B}}} = \text{Gear Ratio}$$

Now, calculate the gear ratio for the two mating gears in Figure 14–56 if gear A has 18 teeth, 6 in. pitch diameter, and operates at 1200 rpm, and gear B has 54 teeth, 18 in. pitch diameter, and operates at 400 rpm:

Number of Teeth	=	54/18	=	3:1
Pitch Diameter	=	18/6	=	3:1
rpm	=	1200/400	=	3:1

You can solve for unknown values in the gear train if you know the gear ratio you want to achieve, the number of teeth and pitch diameter of one gear, and the input speed. For example, gear A has 18 teeth, a pitch diameter of 6 in., and an input speed of 1200 rpm, and the ratio between gear A and gear B is 3:1. In order to keep this information well organized, it is recommended that you set up a chart similar to the one shown in Fig-

ure 14–59. The unknown values are shown in color for your reference. Determine the number of teeth for gear B as follows:

$$\text{Teeth}_{\text{Gear A}} \times \text{Gear Ratio} = 8 \times 3 = \text{Teeth}_{\text{Gear B}} = 54$$

Or, if you know that gear B has 54 teeth and the gear ratio is 3:1, then:

$$\frac{\text{Teeth}_{\text{Gear B}}}{\text{Gear Ratio}} = \frac{54}{3} = \text{Teeth}_{\text{Gear A}} = 18$$

Determine the rpm of gear B:

$$\frac{\text{rpm}_{\text{Gear A}}}{\text{Gear Ratio}} = \frac{1200}{3} = \text{rpm}_{\text{Gear B}} = 400$$

Determine the pitch diameter of gear B:

$$\text{Pitch Diameter}_{\text{Gear A}} \times \text{Gear Ratio} = 6 \text{ in.} \times 3 = \text{Pitch Diameter Gear B} = 18 \text{ in.}$$

In some situations it is necessary for you to refer to the formulas given in Figure 14–52 to determine some unknown values. In this case it is necessary to calculate the diametral pitch using the formula $P = N/D$, where P = diametral pitch, N = number of teeth, and D = pitch diameter:

$$P_{\text{Gear A}} = \frac{N}{D} = \frac{18}{6} = 3$$

Because the diametral pitch is the tooth size and the teeth for mating gears must be the same size, the diametral pitch of gear B is also 3.

Keep in mind that the preceding example presents only one set of design criteria. Other situations may be different. Always solve for unknown values based on information that you have and work with the standard formulas presented in this chapter. Following are some important points to keep in mind as you work with the design of gear trains:

○ The rpm of the larger gear is always slower than the rpm of the smaller gear.
○ Mating gears always turn in opposite directions.
○ Gears on the same shaft (cluster gears) always turn in the same direction and at the same speed (rpm).
○ Mating gears have the same size teeth (diametral pitch).
○ The gear ratio between mating gears is a ratio between the number of teeth, the pitch diameters, and the rpms.
○ The distance between the shafts of mating gears is equal to $1/2D_{\text{Gear A}} + 1/2D_{\text{Gear B}}$. This distance is 3 + 9 = 12 in. between shafts in Figure 14–59.

▼ DESIGNING AND DRAWING THE RACK AND PINION

A rack is a straight bar with spur teeth used to convert rotary motion to reciprocating motion. Figure 14–60

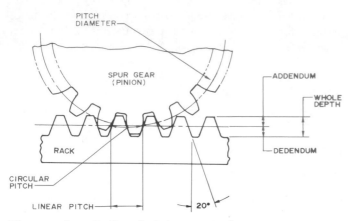

Figure 14–60 Rack and pinion terminology.

shows a spur gear pinion mating with a rack. Notice the circular dimensions of the pinion become linear dimensions on the rack. (Linear Pitch = Circular Pitch)

When preparing a detailed drawing of the rack, the front view is normally shown with the profile of the first and last tooth drawn. Phantom lines are drawn to represent the top and root, and a centerline is used for the pitch line. (See Figure 14–61.) Related tooth and length dimensions are also placed on the front view and a depth dimension is placed on the side view. A chart is then placed in the field of the drawing to identify specific gear-cutting information.

▼ DESIGNING AND DRAWING BEVEL GEARS

Bevel gears are usually designed to transmit power between intersecting shafts at 90°, although they may be designed for any angle. Some gear design terminology

and formulas relate specifically to the construction of bevel gears. These formulas and a drawing of a bevel gear and pinion are shown in Figure 14–62. Most of the gear terms discussed for spur gears apply to bevel gears. In many cases information must be calculated using the formulas provided in the spur gear discussion shown in Figure 14–52.

The drawing for a bevel gear is similar to the drawing for a spur gear, but in many cases only one view is needed. This view is often a full section. Another view may be necessary to show the dimensions for a keyway. As with other gear drawings a chart is used to specify gear-cutting data. (See Figure 14–63.) Notice in this example that only a partial side view is used to specify the bore and keyway.

▼ DESIGNING AND DRAWING WORM GEARS

Worm gears are used to transmit power between non-intersecting shafts, and they are used for large speed reductions in a small space as compared to other types of gears. Worm gears are also strong, move in either direction, and lock in place when the machine is not in operation. A single lead worm advances one pitch with every revolution. A double lead worm advances two pitches with each revolution. As with bevel gears, the worm gear and worm have specific terminology and formulas that apply to their design. The representative worm gear and worm technology and design formulas are shown in Figure 14–64.

When drawing the worm gear, the same techniques are used as discussed earlier. Figure 14–65 shows a worm gear drawing using a half-section method. A side

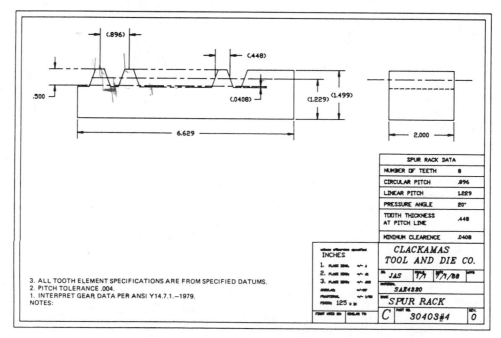

Figure 14–61 A detailed drawing of a rack.

TERM	DESCRIPTION	FORMULA
Pitch Diameter (D)	The diameter of the base of the pitch cone.	$D = N/P$
Pitch Cone	In Figure 14–64, the pitch cone is identified as XYZ.	
Pitch Angle (⌀)	The angle between an element of a pitch cone and its axis. Pitch angles of mating gears depend on their relative diameters (gear a and gear b).	$\tan ⌀_a = D_a/D_b$ $\tan ⌀_b = D_b/D_a$
Cone Distance (A)	Slant height of pitch cone.	$A = D/2 \sin ⌀$
Addendum Angle (δ)	The angle subtended by the addendum. It is the same for mating gears.	$\tan δ = a/A$
Dedendum Angle (Ω)	The angle subtended by the dedendum. It is the same for mating gears.	$\tan Ω = b/A$
Face Angle (⌀ₒ)	The angle between the top of the teeth and the gear axis.	$⌀_o = ⌀ - δ$
Root Angle (⌀ᵣ)	The angle between the bottom of the tooth space and the gear axis.	$⌀_r = ⌀ - Ω$
Outside Diameter (Dₒ)	The diameter of the outside circle of gear.	$D_o = D + 2a(\cos ⌀)$
Crown Height (X)	The distance between the cone apex and the outer tip of the gear teeth.	$X = .5\,(D_o)/\tan ⌀_o$
Crown Backing (Y)	The distance from the rear of the hub to the outer tip of the gear tooth, measured parallel to the axis of the gear.	
Face Width	A distance that should not exceed one-third of the cone distance (A).	

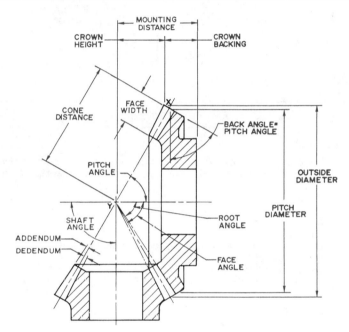

Figure 14–62   Bevel gear terminology and formulas.

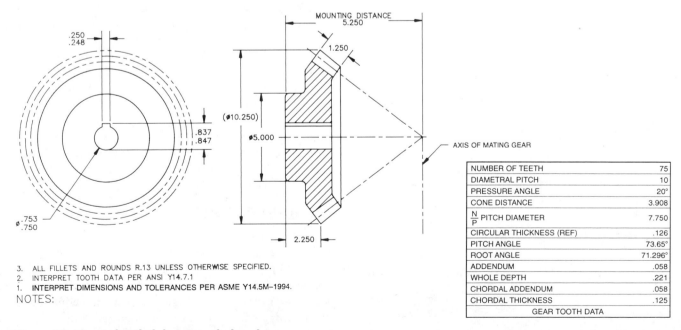

NUMBER OF TEETH	75
DIAMETRAL PITCH	10
PRESSURE ANGLE	20°
CONE DISTANCE	3.908
$\frac{N}{P}$ PITCH DIAMETER	7.750
CIRCULAR THICKNESS (REF)	.126
PITCH ANGLE	73.65°
ROOT ANGLE	71.296°
ADDENDUM	.058
WHOLE DEPTH	.221
CHORDAL ADDENDUM	.058
CHORDAL THICKNESS	.125
GEAR TOOTH DATA	

3. ALL FILLETS AND ROUNDS R.13 UNLESS OTHERWISE SPECIFIED.
2. INTERPRET TOOTH DATA PER ANSI Y14.7.1
1. **INTERPRET DIMENSIONS AND TOLERANCES PER ASME Y14.5M–1994.**
NOTES:

Figure 14–63   A detailed drawing of a bevel gear.

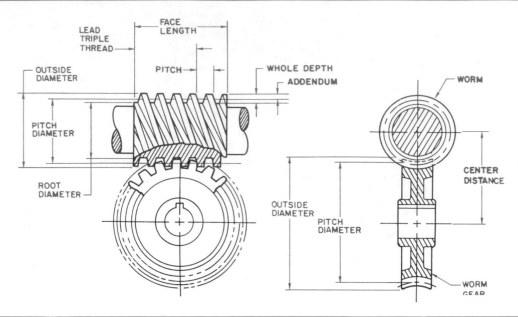

TERM	DESCRIPTION	FORMULA
.Pitch Diameter (worm) (D$_w$)		D$_w$ = 2C − D$_g$
Pitch Diameter (gear) (D$_g$)		D$_g$ = 2C − D$_w$
Pitch (P)	The distance from one tooth to the corresponding point on the next tooth measured parallel to the worm axis. It is equal to the circular pitch on the worm gear.	P = L/T
Lead (L)	The distance the thread advances axially in one revolution of the worm.	L = D$_g$/R L = P × T
Threads (T)	Number of threads or starts on worm.	T = L/P
Gear Teeth (N)	Number of teeth on worm gear.	N = πD$_g$/P
Ratio (R)	Divide number of gear teeth by the number of worm threads.	R = N/T
Addendum (a)	For single and double threads.	a = .318P
Whole Depth (WD)	For single and double threads.	WD = .686P

NOTE: REFER TO THE FORMULAS ON PAGE 448 FOR ADDITIONAL CALCULATIONS.

Figure 14–64 Worm and worm gear terminology and formulas.

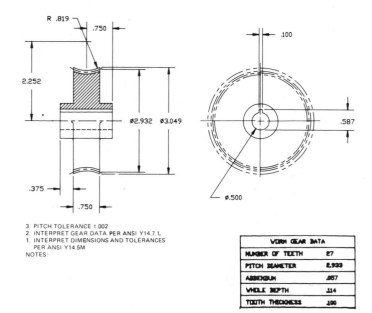

3. PITCH TOLERANCE ±.002
2. INTERPRET GEAR DATA PER ANSI Y14.7.1.
1. INTERPRET DIMENSIONS AND TOLERANCES
   PER ANSI Y14.5M
NOTES:

WORM GEAR DATA	
NUMBER OF TEETH	27
PITCH DIAMETER	2.932
ADDENDUM	.057
WHOLE DEPTH	.114
TOOTH THICKNESS	.100

Figure 14–65 A detailed drawing of a worm gear.

view is provided to dimension the bore and keyway, and a chart is given for gear-cutting data.

A detailed drawing of the worm is prepared using two views. The front view shows the first gear tooth on each end with phantom lines between for a simplified representation. The side view is the same as a spur gear drawing with the keyway specifications. The gear-cutting chart is then placed on the field of the drawing or over the title block. Figure 14–66 shows the worm drawing using CADD to provide a detailed representation of the worm teeth in the front view rather than phantom lines as in the simplified representation.

▼ PLASTIC GEARS

The following is taken in part from *Plastic Gearing* by William McKinlay and Samuel D. Pierson, published by ABA/PGT Inc., Manchester, CT. Gears can be molded of many engineering plastics in various grades and in filled varieties. Filled plastics are those in which a material has been added to improve the mechanical properties. The additives normally used in gear plastics are glass, poly-

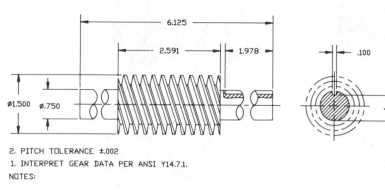

WORM DATA	
PITCH DIAMETER	1.251
LEAD RIGHT OR LEFT	RIGHT
CENTER DISTANCE	1.858
WORKING DEPTH	.218
CLEARANCE	.0312
PRESSURE ANGLE	20°
ADDENDUM	.125
WHOLE DEPTH	.25
CHORDAL THICKNESS	.125

2. PITCH TOLERANCE ±.002
1. INTERPRET GEAR DATA PER ANSI Y14.7.1.
NOTES:

Figure 14–66 A detailed drawing of a worm.

tetrafluoroethylene (PTFE), silicones, and molybdenum disulphide. Glass fiber reinforcement can double the tensile strength and reduce the thermal expansion. Carbon fiber is often used to increase strength. Silicones, PTFE, and molybdenum disulphide are used to act as built-in lubricants and provide increased wear resistance. Plastic gears are designed in the same manner as gears made from other materials. However, the physical characteristics of plastics make it necessary to follow gear design practices more closely than when designing gears that are machined from metals.

Advantages of Molded Plastic Gears

Gears molded of plastic are replacing stamped and cut metal gears in a wide range of mechanisms. Designers are turning to molded plastic gears for one or more of the following reasons:

◯ reduced cost.
◯ increased efficiency.
◯ self-lubrication.
◯ increased tooth strength with nonstandard pressure angles.
◯ reduced weight.
◯ corrosion resistance.
◯ less noise.
◯ available in colors.

Disadvantages of Molded Plastic Gears

Plastic gearing has the following limitations when compared with metal gearing:

◯ lower strength.
◯ greater thermal expansion and contraction.
◯ limited heat resistance.
◯ size change with moisture absorption.

Accuracy of Molded Plastic Gears

Today's technology permits a very high degree of precision. In general, tooth-to-tooth composite tolerances can be economically held to .0005 or less for fine pitch gears. Total composite tolerance varies depending

on configuration, evenness of product cross-section, and the selection of the molding material.

▼ BEARINGS

Bearings are mechanical devices used to reduce friction between two surfaces. They are divided into two large groups known as plain and rolling element bearings. Bearings are designed to accommodate either rotational or linear motion. Rotational bearings are used for radial loads, and linear bearings are designed for thrust loads. Radial loads are loads that are distributed around the shaft. Thrust loads are lateral. Thrust loads apply force to the end of the shaft. Figure 14–67 shows the relationship between rotational and linear motion.

Plain Bearings

Plain bearings are often referred to as sleeve, journal bearings, or bushings. Their operation is based on a sliding action between mating parts. A clearance fit between the inside diameter of the bearing and the shaft is critical to ensure proper operation. Refer to fits between mating parts in Chapter 10 for more information. The bearing has an interference fit between the outside of the bearing and the housing or mounting device as shown in Figure 14–68.

The material from which plain bearings are made is important. Most plain bearings are made from bronze or

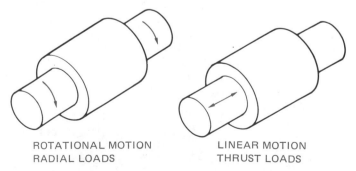

ROTATIONAL MOTION
RADIAL LOADS

LINEAR MOTION
THRUST LOADS

Figure 14–67 Radial and thrust loads.

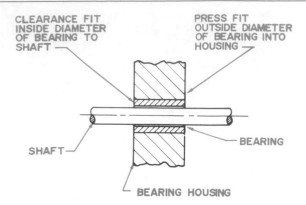

Figure 14–68 Plain bearing terminology and fits.

Figure 14–70 Typical ball bearings.

phosphor bronze. Bronze bearings are normally lubricated, while phosphor bronze bearings are commonly impregnated with oil and require no additional lubrication. Phosphor bronze is an excellent choice when antifriction qualities are important and where resistance to wear and scuffing are needed.

Rolling Element Bearings

Ball and roller bearings are the two classes of rolling element bearings. Ball bearings are the most commonly used rolling element bearings. In most cases, ball bearings have higher speed and lower load capabilities than roller bearings. Even so, ball bearings are manufactured for most uses. Ball bearings are constructed with two grooved rings, a set of balls placed radially around the rings, and separator that keeps the balls spaced apart and aligned as shown in Figure 14–69.

Single-row ball bearings are designed primarily for radial loads, but they can accept some thrust loads. Double-row ball bearings may be used where shaft alignment is important. Angular contact ball bearings support a heavy thrust load and a moderate radial load. Thrust bearings are designed for use in thrust load situations only. When both thrust and radial loads are necessary, both radial and thrust ball bearings are used together. Some typical ball bearings are shown in Figure 14–70. Ball bearings are available with shields and seals. A

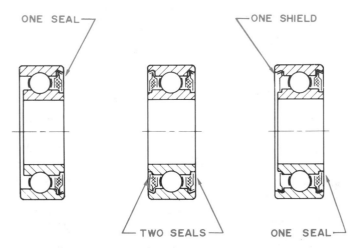

Figure 14–71 Bearing seals and shields.

shield is a metal plate on one or both sides of the bearing. The shields act to keep the bearing clean and retain the lubricant. A sealed bearing has seals made of rubber, felt, or plastic placed on the outer and inner rings of the bearing. The sealed bearings are filled with special lubricant by the manufacturer. They require little or no maintenance in service. Figure 14–71 shows the shields and seals used on ball bearings.

Roller bearings are more effective than ball bearings for heavy loads. Cylindrical roller bearings have a high radial capacity and assist in shaft alignment. Needle roller bearings have small rollers and are designed for the highest load-carrying capacity of all rolling element bearings with shaft sizes under 10 in. Tapered roller bearings are used in gear reducers, steering mechanisms, and machine tool spindles. Spherical roller bearings offer the best combination of high load capacity, tolerance to shock, and alignment, and are used on conveyors, transmissions, and heavy machinery. Some common roller bearings are displayed in Figure 14–72.

▼ DRAWING BEARING SYMBOLS

Only engineering drafters who work for a bearing manufacturer normally make detailed drawings of roller or ball bearings. In most other industries, bearings are not drawn because they are purchase parts. When bearings are drawn as a representation on assembly drawings or product catalogs, they are displayed using symbols.

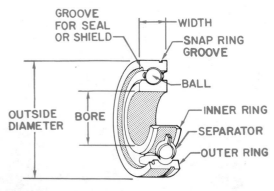

Figure 14–69 Ball bearing components.

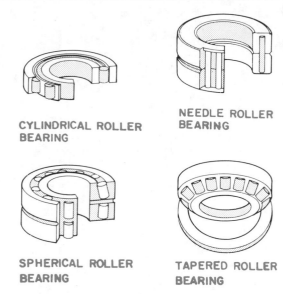

CYLINDRICAL ROLLER
BEARING

NEEDLE ROLLER
BEARING

SPHERICAL ROLLER
BEARING

TAPERED ROLLER
BEARING

Figure 14–72 Typical roller bearings.

CADD Applications

BEARING SYMBOLS

When CADD is used, the bearing symbols may be drawn and saved in a symbols library for immediate use at any time. This use of CADD helps increase productivity and accuracy over other drafting techniques. (See Figure 14–73.)

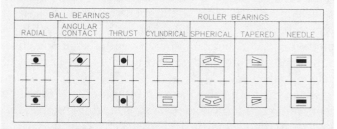

Figure 14–73 Bearing symbols drawn using CADD.

▼ BEARING CODES

Bearing manufacturers use similar coding systems for the identification and ordering of different bearing products. The bearing codes generally contain the following type of information:

○ Material.
○ Bearing type.
○ Bore size.
○ Lubricant.
○ Type of seals or shields.

A sample bearing numbering system is shown in Figure 14–74.

▼ BEARING SELECTION

A variety of bearing types are available from manufacturers. Bearing design may differ depending on the use requirements. For example, a supplier may have light, medium, and heavy bearings available. Bearings may have specially designed outer and inner rings. Bearings are available open (without seals or shields), or with one or two shields or seals. Light bearings are generally designed to accommodate a wide range of applications involving light to medium loads combined with relatively high speeds. Medium bearings have heavier con-

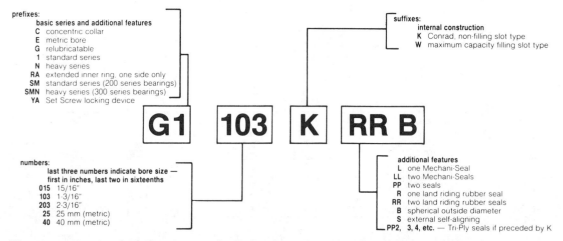

Figure 14–74 A sample bearing numbering system. *Courtesy The Torrington Company.*

struction than light bearings and provide a greater radial and thrust capacity. They are also able to withstand greater shock than light bearings. Heavy bearings are often designed for special service where extra heavy shock loads are required. Bearings may also be designed to accommodate radial loads, thrust loads, or a combination of loading requirements.

Bearing Bore, Outside Diameters, and Width

Bearings are dimensioned in relation to the bore diameter, outside diameter, and width. These dimensions are shown in Figure 14–75. After the loading requirements have been established, the bearing is selected in relationship to the shaft size. For example, if an approximate ∅1.5 inch shaft size is required for a medium service bearing, then a vendor's catalog chart similar to the one shown in Figure 14–76 is used to select the bearing.

Referring to the chart shown in Figure 14–76, notice that the first column is the vendor's bearing number, followed by the bore size (B). To select a bearing for the approximate 1.5 inch shaft, go to the chart and pick the bore diameter of 1.5748, which is close to 1.5. This is the 308K bearing. The tolerance for this bore is specified in the chart as 1.5748 + .0000 and − .0005. Therefore, the limits dimension of the bore in this example is 1.5748 − 1.5743. The outside diameter is 3.5433 + .0000 − .0006 (3.5433 − 3.5427). The width of this bearing is .906 + .000 − .005 (.906 −.901). The fillet radius is the maximum shaft or housing fillet radius in which the bearing corners will clear. The fillet radius for the 308K bearing is R.059. Notice that the dimensions are also given in millimeters.

Shaft and Housing Fits

Shaft and housing fits are important, because tight fits may cause failure of the balls, rollers, or lubricant, or

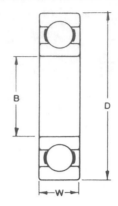

Figure 14–75 Bearing dimensions. *Courtesy The Torrington Company.*

overheating. Loose fits can cause slippage of the bearing in the housing, resulting in overheating, vibration, or excessive wear.

Shaft Fits. In general, for precision bearings, it is recommended that the shaft diameter and tolerance be the same as the bearing bore diameter and tolerance. The shaft diameter used with the 308K bearing is dimensioned ∅1.5748 − 1.5743.

Housing Fits. In most applications with rotating shafts, the outer ring is stationary and should be mounted with a push of the hand or light tapping. In general, the minimum housing diameter is .0001 larger than the maximum bearing outside diameter and the maximum housing diameter is .0003 larger than the minimum housing diameter. With this in mind, the housing diameter for the 308K bearing is 3.5433 + .0001 = 3.5434 and 3.5434 + .0003 = 3.5437. The housing diameter limits are 3.5437 − 3.5434.

DIMENSIONS — TOLERANCES

Bearing Number	Bore B				Outside Diameter D				Width W		Fillet Radius[1]		Wt.		Static Load Rating C_O		Extended Dynamic Load Rating C_E	
			tolerance + .0000″ + 000 mm to minus				tolerance + .0000″ + 000 mm to minus		+ .000″, − .005″ + .00 mm. − 13 mm									
	in.	mm	in.	mm	in.	mm	in.	mm	in.	mm	in.	mm	lbs.	kg	lbs.	N	lbs.	N
300K	.3937	10	.0003	.008	1.3780	35	.0005	.013	.433	11	.024	6	.12	.054	850	3750	2000	9000
301K	.4724	12	.0003	.008	1.4567	37	.0005	.013	.472	12	.039	1.0	.14	.064	850	3750	2080	9150
302K	.5906	15	.0003	.008	1.6535	42	.0005	.013	.512	13	.039	1.0	.18	.082	1270	5600	2900	13200
303K	.6693	17	.0003	.008	1.8504	47	.0005	.013	.551	14	.039	1.0	.24	.109	1460	6550	3350	15000
304K	.7874	20	.0004	.010	2.0472	52	.0005	.013	.591	15	.039	1.0	.31	.141	1760	7800	4000	17600
305K	.9843	25	.0004	.010	2.4409	62	.0005	.013	.669	17	.039	1.0	.52	.236	2750	12200	5850	26000
306K	1.1811	30	.0004	.010	2.8346	72	.0005	.013	.748	19	.039	1.0	.78	.354	3550	15600	7500	33500
307K	1.3780	35	.0005	.013	3.1496	80	.0005	.013	.827	21	.059	1.5	1.04	.472	4500	20000	9150	40500
308K	1.5748	40	.0005	.013	3.5433	90	.0006	.015	.906	23	.059	1.5	1.42	.644	5600	24500	11000	49000
309K	1.7717	45	.0005	.013	3.9370	100	.0006	.015	.984	25	.059	1.5	1.90	.862	6700	30000	13200	58500
310K	1.9685	50	.0005	.013	4.3307	110	.0006	.015	1.063	27	.079	2.0	2.48	1.125	8000	35500	15300	68000
311K	2.1654	55	.0006	.015	4.7244	120	.0006	.015	1.142	29	.079	2.0	3.14	1.424	9500	41500	18000	80000
312K	2.3622	60	.0006	.015	5.1181	130	.0008	.020	1.220	31	.079	2.0	3.89	1.765	10800	48000	20400	90000

[1] Maximum shaft or housing fillet radius which bearing corners will clear.

Figure 14–76 Bearing selection chart. *Courtesy The Torrington Company.*

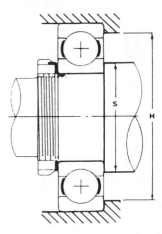

Figure 14–77 Shaft shoulder and housing shoulder dimensions. *Courtesy The Torrington Company.*

The Shaft Shoulder and Housing Shoulder Dimensions

Next, you should size the shaft shoulder and housing shoulder diameters. The shaft shoulder and housing shoulder diameter dimensions are represented in Figure 14–77 as S and H. The shoulders should be large enough to rest flat on the face of the bearing and small enough to allow bearing removal. Refer to the chart in Figure 14–78 to determine the shaft shoulder and housing shoulder diameters for the 308K bearing selected in the preceding discussion. Find the basic bearing number 308 and determine the limits of the shaft shoulder and the housing shoulder. The shaft shoulder diameter is 2.00 – 1.93 and the housing shoulder diameter is 3.19 – 3.06. Now you are ready to detail the bearing location on the housing drawing and on the shaft drawing. A partial detailed drawing of the shaft and housing for the 308K bearing is shown in Figure 14–79.

Surface Finish of Shaft and Housing

The recommended surface finish for precision bearing applications is 32 microinches (0.80 micrometer) for

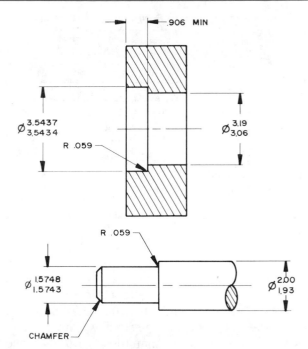

Figure 14–79 A partial detail drawing of the shaft and housing for the 308K bearing.

the shaft finish on shafts under 2 inches in diameter. For shafts over 2 inches in diameter, a 63 microinch (1.6 micrometer) finish is suggested. The housing diameter may have a 125 microinch (3.2 micrometer) finish for all applications.

Bearing Lubrication

It is necessary to maintain a film of lubrication between the bearing surfaces. The factors to consider when selecting lubrication requirements include the:

- type of operation, such as continuous or intermittent.
- service speed in rpm (revolutions per minute).
- bearing load, such as light, medium, or heavy.

Bearings may also be overlubricated, which may cause increased operating temperatures and early failure.

Extra-Light • 9100 Series								Light • 200, 7200WN Series								Medium • 300, 7300WN Series							
Basic Bearing Number	Shoulder Diameters							Basic Bearing Number	Shoulder Diameters							Basic Bearing Number	Shoulder Diameters						
	shaft, S				housing, H				shaft, S				housing, H				shaft, S				housing, H		
	max.		min.		max.		min.		max.		min.		max.		min.		max.		min.		max.		min.
	in.	mm	in.	mm	in.	mm	in.	mm		in.	mm	in.	mm	in.	mm	in.	mm		in.	mm	in.	mm	in.	mm	in.	mm
9100	.52	13.2	.47	11.9	.95	24.1	.91	23.1	200	.56	14.2	.50	12.7	.98	24.9	.97	24.6	300	.59	15.0	.50	12.7	1.18	30.0	1.15	29.2
9101	.71	18.0	.55	14.0	1.02	25.9	.97	24.6	201	.64	16.3	.58	14.7	1.06	26.9	1.05	26.7	301	.69	17.5	.63	16.0	1.22	31.0	1.21	30.7
9102	.75	19.0	.67	17.0	1.18	30.0	1.13	28.7	202	.75	19.0	.69	17.5	1.18	30.0	1.15	29.2	302	.81	20.6	.75	19.0	1.42	36.1	1.40	35.6
9103	.81	20.6	.75	19.0	1.30	33.0	1.25	31.8	203	.84	21.3	.77	19.6	1.34	34.0	1.31	33.3	303	.91	23.1	.83	21.1	1.61	40.9	1.60	40.6
9104	.98	24.9	.89	22.6	1.46	37.1	1.41	35.8	204	1.00	25.4	.94	23.9	1.61	40.9	1.58	40.1	304	1.06	26.9	.94	23.9	1.77	45.0	1.75	44.4
9105	1.18	30.0	1.08	27.4	1.65	41.9	1.60	40.6	205	1.22	31.0	1.14	29.0	1.81	46.0	1.78	45.2	305	1.31	33.3	1.14	29.0	2.17	55.1	2.09	53.1
9106	1.38	35.1	1.34	34.0	1.93	49.0	1.88	47.8	206	1.47	37.3	1.34	34.0	2.21	56.1	2.16	54.9	306	1.56	39.6	1.34	34.0	2.56	65.0	2.44	62.0
9107	1.63	41.4	1.53	38.9	2.21	56.1	2.15	54.6	207	1.72	43.7	1.53	38.9	2.56	65.0	2.47	62.7	307	1.78	45.2	1.69	42.9	2.80	71.1	2.72	69.1
9108	1.81	46.0	1.73	43.9	2.44	62.0	2.39	60.7	208	1.94	49.3	1.73	43.9	2.87	72.9	2.78	70.6	308	2.00	50.8	1.93	49.0	3.19	81.0	3.06	77.7
9109	2.03	51.6	1.94	49.3	2.72	69.1	2.67	67.8	209	2.13	54.1	1.94	49.3	3.07	78.0	2.97	75.4	309	2.28	57.9	2.13	54.1	3.58	90.9	3.41	86.6
9110	2.22	56.4	2.13	54.1	2.91	73.9	2.86	72.6	210	2.34	59.4	2.13	54.1	3.27	83.1	3.17	80.5	310	2.50	63.5	2.36	59.9	3.94	100.1	3.75	95.2
9111	2.48	63.0	2.33	59.2	3.27	83.1	3.22	81.8	211	2.54	64.5	2.41	61.2	3.68	93.5	3.56	90.4	311	2.75	69.8	2.56	65.0	4.33	110.0	4.13	104.9
9112	2.67	67.8	2.53	64.3	3.47	88.1	3.42	86.9	212	2.81	71.4	2.67	67.8	3.98	101.1	3.87	98.3	312	2.94	74.7	2.84	72.1	4.65	118.1	4.44	112.8

Figure 14–78 Shaft shoulder and housing shoulder dimension selection chart. *Courtesy The Torrington Company.*

Oil Viscosities and Temperature Ranges for Ball Bearing Lubrication

Maximum Temperature Range Degrees F	Optimum Temperature Range Degrees F	Speed Factor, S_i (inner race bore diameter (inches) × RPM)	
		Under 1000	Over 1000
		Viscosity	
–40 to +100	–40 to –10	80 to 90 SSU (at 100 deg. F)	70 to 80 SSU (At 100 deg. F)
–10 to +100	–10 to +30	100 to 115 SSU (at 100 deg. F)	80 to 100 SSU (At 100 deg. F)
+30 to +150	+30 to +150	SAE 20	SAE 10
+30 to +200	+150 to +200	SAE 40	SAE 30
+50 to +300	+200 to +300	SAE 70	SAE 60

Figure 14–80 Selection of oil viscosity based on temperature ranges and speed factors.

Selection of the proper lubrication for the application should be determined by the manufacturer's recommendations. The ability of the lubricant is due, in part, to *viscosity*. Viscosity is the internal friction of a fluid, which makes it resist a tendency to flow. Fluids with low viscosity flow more freely than those with high viscosity. The chart in Figure 14–80 shows the selection of oil viscosity based on temperature ranges and speed factors.

Oil Grooving of Bearings

In situations where bearings or bushings do not receive proper lubrication, it may be necessary to provide grooves for the proper flow of lubrication to the bearing surface. The bearing grooves help provide the proper lubricant between the bearing surfaces, and maintain adequate cooling. There are several methods of designing paths for the lubrication to the bearing surfaces as shown in Figure 14–81.

Sealing Methods

Machine designs normally include means for stopping leakage and keeping out dirt and other contaminants when lubricants are involved in the machine operation. This is accomplished using static or dynamic sealing devices. Static sealing refers to stationary devices that are held in place and stop leakage by applied pressure. Static seals such as gaskets do not come in contact with the moving parts of the mechanism. Dynamic seals are those that contact the moving parts of the machinery, such as packings.

Gaskets are made from materials that prevent leakage or access of dust contaminants into the machine cavity. Silicone rubber gasket materials are used in applications such as water pumps, engine filter housings, and oil pans. Gasket tapes, ropes, and strips provide good cushioning properties for dampening vibration, and the adhesive sticks well to most materials. Nonstick gasket materials such as paper, cork, and rubber are available for certain applications. Figure 14–82 shows a typical gasket mounting.

Dynamic seals include packings and seals that fit tightly between the bearing or seal seat and the shaft. The

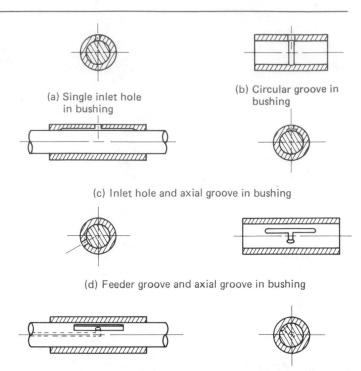

(a) Single inlet hole in bushing

(b) Circular groove in bushing

(c) Inlet hole and axial groove in bushing

(d) Feeder groove and axial groove in bushing

(e) Feeder groove and straight axial groove in the shaft

Figure 14–81 Methods of designing paths for lubrication to bearing surfaces.

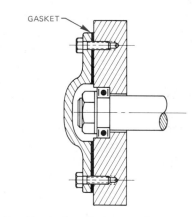

GASKET

Figure 14–82 Typical gasket mounting.

pressure applied by the seal seat or the pressure of the fluid causes the sealing effect. Molded lip packings that provide sealing as a result of the pressure generated by the machine fluid are available. Figure 14–83 shows examples of molded lip packings. Molded ring seals are placed in a groove and provide a positive seal between the shaft and bearing or bushing. Types of molded ring seals include labyrinth, O-ring, lobed ring, and others. Labyrinth, which means maze, refers to a seal that is made of a series of spaced strips that are connected to the seal seat, making it difficult for the lubrication to pass. Labyrinth seals are used in heavy machinery where some leakage is permissible. (See Figure 14–84.) The O-ring seal is the most commonly used seal because of

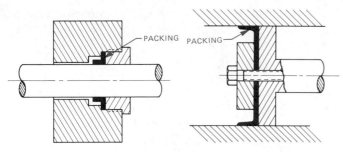

Figure 14–83   Molded lip packings.

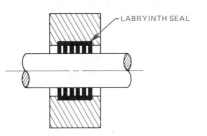

Figure 14–84   Labryinth seal.

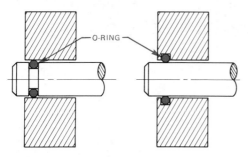

Figure 14–85   O-ring seals.

its low cost, ease of application, and flexibility. The O-ring may be used for most situations involving rotating or oscillating motion. The O-ring is placed in a groove that is machined in either the shaft or the housing as shown in Figure 14–85. The lobed ring has rounded lobes that provide additional sealing forces over the standard O-ring seal. A typical lobed ring seal is shown in Figure 14–86.

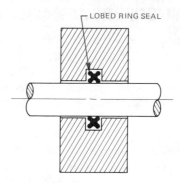

Figure 14–86   Lobed ring seal.

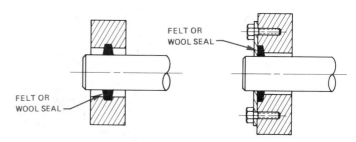

Figure 14–87   Felt and wool seals.

Felt and wool seals are used where economical cost, lubricant absorption, filtration, low friction, and a polishing action are required. However, the ability to completely seal the machinery is not as positive as with the seals described earlier. (See Figure 14–87.)

Bearing Mountings

There are a number of methods used for holding the bearing in place. Common techniques include a nut and lock washer, a nut and lock nut, or a retaining ring. Other methods may be designed to fit the specific application or requirements, such as a shoulder plate. Figure 14–88 shows some examples of mountings.

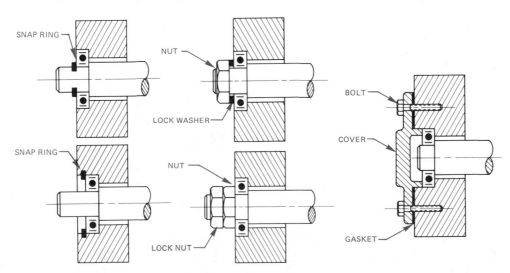

Figure 14–88   Typical bearing mountings.

▼ GEAR AND BEARING ASSEMBLIES

Gear and bearing assemblies show the parts of the complete mechanism as they appear assembled. (See Figure 14–89.) When drawing assemblies, you need to use as few views as possible, but enough to adequately display how all the parts fit together. In some situations all that is needed is a full sectional view that displays all of the internal components. An exterior view such as a front or top view plus a section sometimes works. Dimensions are normally omitted from the assembly unless the dimensions are needed for assembly purposes. For example, when a specific dimension regarding the relationship of one part to another is required to properly assemble the parts, each part is identified with a number in a circle. This circle is referred to as a balloon. The balloons are connected to the part being identified with a leader. The balloons are about one-half inch in diameter. The identification number is quarter-inch-high lettering or text. The balloon numbers correlate with a parts list. The parts list is normally placed on the drawing above or adjacent to the title block, or on a separate sheet as shown in Figure 14–90.

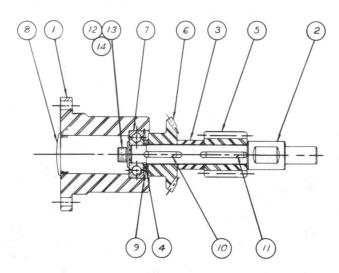

Figure 14–89 Assembly drawing. *Courtesy Curtis Associates.*

ASSEMBLY _Cross Shaft Assembly_   USED ON _____

NUMBER OF UNITS _____   DATE _____

HAVE	NEED	P/Ø NO. W/Ø NO.	DET. NO.	PART NO.	DWG.	QTY.	PART NAME	DESCRIPTION	VENDOR
			1		B	1	Bearing Retainer	Ø 3" C.D. Bar	
			2		B	1	Cross Shaft	Ø 3/4" C.D. Bar	
			3		A	1	Spacer	Ø 3/4 O.D. x 11GA Wall Tube x .738 Thick	
			4		A	1	Spacer	Ø 3/4 O.D. x 11GA Wall Tube x .125 Thick	
			5		–	1	Steel Worm 12 D.P. Single Thread	Boston Gear #H 1056 R.H.	
			6		–	1	Bevel Gear 20° P.A. 2" P.D.	Boston Gear #HL 149 Y-G	
			7		–	1	Ball Bearing .4724 Ø Bore	T.R.W. or Equivalent #MRC 201-S22	
			8		–	1	End Plug		
			9		–	1	Snap Ring	Waldes-Truarc #N 5000-125	
			10		–	1	Key Stock	1/8 Sq. x 3/4 Lg.	
			11		–	1	Key Stock	1/8 Sq. x 1 Lg.	
			12		–	1	Socket Head Cap Screw	1/4 UNC x 3/4 Lg.	
			13		–	1	Lockwasher	1/4 Nominal	
			14		–	1	Flat Washer	1/4 Nominal	

Figure 14–90 Parts list. *Courtesy Curtis Associates.*

PROFESSIONAL PERSPECTIVE

In every situation regarding linkage, cam, and gear design there is a need to investigate all of the manufacturing alternatives and provide a solution the customer can afford. Competition is so tough in the manufacturing industry that you should evaluate each design to find the best way to produce the product. The best way to understand this concept as an entry-level engineering drafter is to talk to experienced designers, engineers, and machinists. Go to the shop to see how things are done and determine the drawing requirements directly from the people who know. According to one design engineer, "If you don't know how it is going to be made, you're not a good drafter. Know the manufacturing capabilities of each piece of equipment." In addition to your drafting courses it is a good idea to take some manufacturing technology classes. Math is also an important part of your program. Drafters in this field use a lot of geometry and trigonometry. After you have a strong educational background, the experienced engineer says, "Keep an open mind and look at all the alternatives."

For example, you work as an engineering drafter for a foundry. For years the flasks have been handled either by hand or with a lift truck. Both of these methods are time consuming and dangerous, and the company has some contracts that require founding some castings that are too heavy to handle. So, the engineering department plans the design for a hydraulic flask handler. Your team is responsible for the handling mechanism that must be designed with this criteria:

❍ Use a hydraulic piston with a 6-in. stroke.
❍ Handle up to 44-in. wide flasks.
❍ Take up no more than 28 in. in overall height.

Your team began the problem-solving process and came up with these steps to solve the problem:

❍ Develop a single-line schematic kinematic diagram representing the movement of the mechanism based on the design information.
❍ Identify and locate the available materials needed to build the flask handler.

The preliminary design is shown in Figure 14–91. All your team needs to do now is ask the other design teams for input, and after revisions are made, prepare a complete set of working drawings for fabrication.

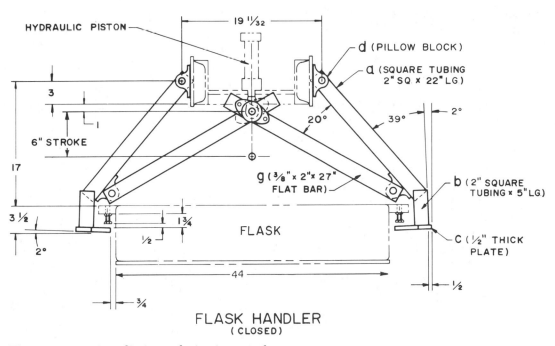

Figure 14–91 A preliminary design is created.

MATH APPLICATION

LENGTH OF A CONNECTING ROD

Problem: Find the length of the connecting rod for the slider mechanism of Figure 14–92.

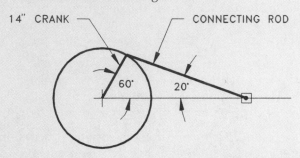

14" CRANK — CONNECTING ROD

60° 20°

Figure 14–92 Slider mechanism.

Solution: Using the Law of Sines (from section N4e of Appendix N)

$$\frac{a}{\sin A} = \frac{b}{\sin B}$$

$$\frac{14"}{\sin 20°} = \frac{b}{\sin 60°}$$

Solving this proportional equation for b gives b = **35.4"**.

CHAPTER

14 MECHANISMS TEST

DIRECTIONS

Answer the questions with short complete statements or drawings as needed.

QUESTIONS

1. Draw a four-bar linkage and label the four links.
2. List at least two uses for a slider crank mechanism.
3. Explain why schematic symbols are used for linkage drawings.
4. The total movement or displacement of a slider mechanism is called _____.
5. Describe the function of a cam mechanism.
6. List three types of cam mechanisms.
7. Explain the relationship of the motion of the cam follower to the cam shaft for the three types of cam described in Question 6.
8. Name the most commonly used cam follower.
9. Describe the purpose of the cam displacement diagram and how it relates to the cam profile construction.
10. List at least four types of cam motion.
11. What type of cam motion is best for smooth-running high-speed applications?
12. Which type of cam motion is suited for only very slow speed applications?

13. Explain why constant velocity motion is modified with a curve at the bottom and top of the period.
14. The length of the displacement diagram for a drum cam is equal to _____.
15. Define *gear.*
16. List at least two functions of a gear.
17. Identify the parts of the spur gear shown in the following drawing:

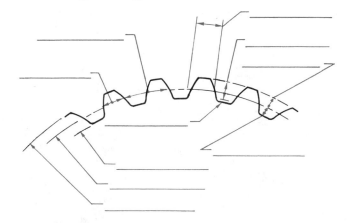

18. List three types of parallel shafting gears.
19. Describe the function of a bevel gear.
20. Describe the function of a worm gear and worm.

21. Given a spur gear with 54 teeth, diametral pitch of 6, and a pitch diameter of 9.000 in., calculate the following:
    a. Addendum
    b. Dedendum
    c. Circular pitch
    d. Outside diameter
    e. Root diameter
    f. Circular thickness
    g. Chordal thickness
    h. Base circle diameter

22. What is the gear ratio if the gear in question 21 mates with a pinion having 18 teeth?

23. Explain the difference between plain and roller element bearings.

24. What is the best material for use in the manufacture of plain bearings?

25. Describe the difference between shielded and sealed bearings.

CHAPTER 14 MECHANISMS PROBLEMS

DIRECTIONS

Please read problems carefully before you being working. Complete each problem on an appropriately sized sheet. Precision work is important for accurate solutions.

LINKAGE PROBLEMS

Problem 14–1 Following is a pictorial drawing and a linkage schematic of a vise grip in the closed position. Reproduce the schematic exactly as shown and also show the open position in a second color.

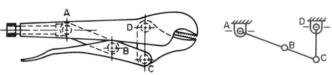

Problem 14–2 Following is a pictorial drawing and a linkage schematic of a toggle clamp in the closed position. Reproduce the schematic exactly as shown and also show the open position in a second color.

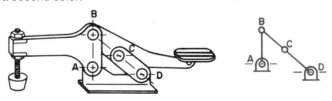

Problem 14–3 Determine the extreme right and left positions of link CD in the figure shown below. Determine and label the angle through which CD oscillates.

AB = 1.625 in.   CD = 1.125 in.
BC = 2.125 in.   AD = 3.125 in.

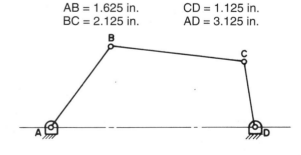

Problem 14–4 Draw the mechanism below in the position shown. Using different colors, draw the mechanism in the extreme right and left positions. Dimension the stroke.

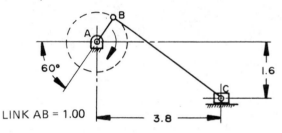

LINK AB = 1.00

Problem 14–5 Draw the combination bellcrank slider mechanism in the position shown below. Determine the stroke of the slider if A moves to position A^1.

AB = 1.500"
BC = 2.125"
CD = 3.375"

ANGLE ABC = 80°

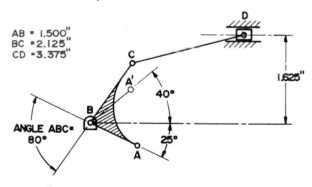

Problem 14–6 The following drawing is a linkage schematic of an oscillating law sprinkler. The spray tube, shown in section, is part of link CD. Link AB moves through 360°, while points A and D are stationary. Determine and dimension the angle of oscillation through which the spray moves.

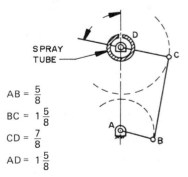

SPRAY TUBE

$AB = \frac{5}{8}$

$BC = 1\frac{5}{8}$

$CD = \frac{7}{8}$

$AD = 1\frac{5}{8}$

Problem 14–7 Given the windshield wiper mechanism shown below determine and dimension the angle of oscillation of the wiper blades. The electric motor rotates link ED continuously through 360°. ABC is one link with a 90° angle at B. A tension spring is located in the center of link BG.

AB = FG = ED = 1.25 in.	BC = .875 in.
AF = BG = 16 in.	CD = 7.625 in.

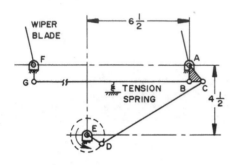

Problem 14–9 Draw the mechanism in the position shown below. Draw the path of point P as linkage AB moves every 30° through a total of 360°. Use a different color and/or line type for each position.

Dimension the angle through which CD oscillates.
BCP is a through link.
Point P slides on EF.
EF = 6.5 in.

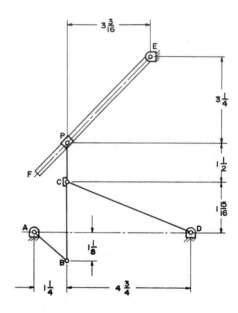

Problem 14–8 Draw the mechanism shown below. Using different colors, show the path of point D in a total of five equally spaced positions, including the extreme right and left positions.

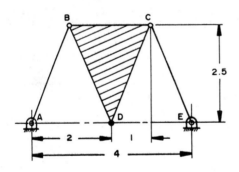

Problem 14–10 Given the mechanism shown to the right, rotate link BC at 60° intervals clockwise through 360°. Plot and draw the paths of points D, E, and F. Dimension the angle of oscillation of link AF.

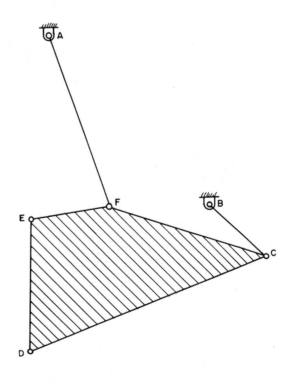

Problem 14–11 Given the assembly drawing of the foundry flask handler shown below, draw a mechanism schematic showing the two extreme positions of movement. The handler is operated by a hydraulic piston with a 6-in. stroke.

a is welded to b.
b is welded to c.

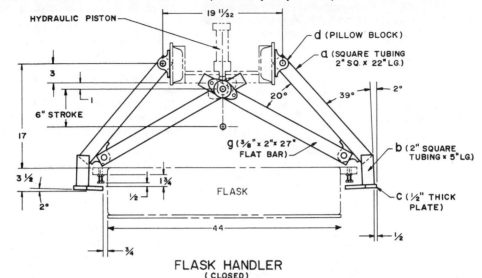

FLASK HANDLER
(CLOSED)

Problem 14–12 Given the pivot hoist shown below, draw and scale exactly as shown. Rotate ADE clockwise so link AE is horizontal. Determine and dimension the extended length of the spring between C and E. Determine and dimension the angle between AE and CE when E is in the new position.

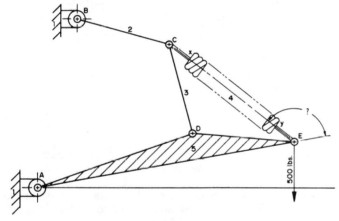

Problem 14–13 Given the figure shown below, determine and dimension the stroke of point E moving in a straight line as link AB rotates 360°. Also dimension the angle of oscillation of link CD.

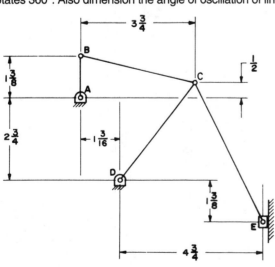

Problem 14–14 Find two examples of linkage mechanisms at home or school. Explain in a short complete statement the function of each mechanism. Using schematic representations, show and dimension the extreme positions of each mechanism.

CAM DISPLACEMENT DIAGRAMS

Problem 14–15 Construct a cam displacement diagram for a cam follower that rises in simple harmonic motion a total of 2.00 in. in 150°, dwells for 30°, falls 2.00 in. simple harmonic motion in 120°, and dwells for 60°. Draw the horizontal scale 6.00 in.

Problem 14–16 Construct a cam displacement diagram for a cam follower that rises in uniform accelerated motion a total of 2.00 in. in 180°, dwells for 30°, falls 2.00 in. uniform accelerated motion in 120°, and dwells for 30°. Draw the horizontal 6.00 in.

Problem 14–17 Construct a cam displacement diagram for a cam follower that rises in modified constant velocity for 3.00 in. in 180°, falls 3.00 in. modified constant velocity motion in 120°, and dwells for 60°. Use a modified constant velocity motion designed with one-third of the displacement.

Problem 14–18 Construct a cam displacement diagram for a cam follower that rises 2.000 in. cycloidal motion in 120°, dwells for 60°, and falls 2.000 in. in cycloidal motion in 180°.

Problem 14–19 Construct a cam displacement diagram for a cam follower that rises in simple harmonic motion a total of 1.250 in. in 90°, dwells for 60°, rises .750 in. in 45° simple harmonic motion, falls 2.00 in. with cycloidal motion in 120°. Draw the horizontal scale 12 in.

Problem 14–20 Construct a cam displacement diagram for a cam follower that rises in modified constant velocity motion (modified to one-third the displacement) for 3.000 in. in 180°, dwells for 30°, and falls 3.000 in. simple harmonic motion in 30°, and dwells to the end of the cycle. Draw the horizontal scale 12 in.

Problem 14–21 Construct a cam displacement diagram for a cam follower that rises 3.500 in. in 90° cycloidal motion, dwells for 45°, falls 2.500 in. cyclloidal motion in 135°, dwells for 30°, falls 1.000 in. simple harmonic motion in 30°, and dwells to the end of the cycle. Draw the horizontal scale 12 in.

Problem 14–22 Construct a cam displacement diagram for a cam follower that rises in cycloidal motion for 3.000 in. in 90°, dwells for 30°, falls 1.000 in. in simple harmonic motion in 90°, dwells for 30°, and falls the remaining 2.000 in. in uniform accelerated motion in 120°. Draw the horizontal scale 12 in.

Problem 14–23 Construct a cam displacement diagram for a cam follower that rises in cycloidal motion a total of 2.1875 in. in 150°, dwells for 30°, falls back to the original level in simple harmonic motion in 150°, then dwells through the remainder of the cycle. Use a 12-in. horizontal scale.

Problem 14–24 Construct a cam displacement diagram for a cam follower that raises a .375 in. diameter in-line roller follower 1.500 in. in uniform accelerated motion in 150°, dwells 45°, falls with modified constant velocity (one-third displacement) in 120°, and dwells the remainder of the cycle.

Problem 14–25 Construct a cam displacement diagram for a cam follower that rises 3.500 in. 90° cycloidal motion, dwells for 30°, falls 2.250 in. cycloidal motion in 150°, falls 1.250 in. simple harmonic motion in 60°, and dwells for 30°. Draw the horizontal scale equal in circumference to a 3.000-in. diameter circle.

CAM PROFILE DRAWINGS

Problem 14–26 Use the displacement diagram constructed in Problem 14–15 and the information given in the illustration below to lay out the plate cam profile drawing. The cam rotates counterclockwise. Make a two-view detailed drawing of the cam and dimension it as shown in this chapter.

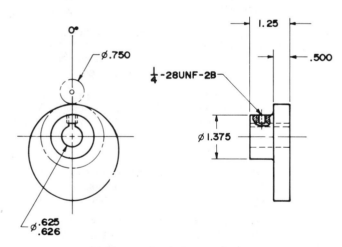

IN-LINE ROLLER FOLLOWER	= ∅ .750
BASE CIRCLE	= ∅ 2.000
KEY SIZE USED	= 1/4 x 1/4 SQ. KEY
PLATE THICKNESS	= .500
HUB THICKNESS	= .750
HUB DIAMETER	= ∅ 1.375
SHAFT DIAMETER	= ∅ .625/.626

Problem 14–27 Use the displacement diagram constructed in Problem 14–18 and the information given in the illustration for Problem 14–26 to lay out the plate cam profile. The cam is rotating clockwise. Make a two-view detailed drawing of the cam, properly toleranced and dimensioned.

Problem 14–28 Make a two-view detailed drawing of a plate cam using the displacement diagram from Problem 14–24. Completely dimension the drawing using the following information:

The cam rotates counterclockwise.

In-line roller follower = ∅.375	Hub projection = .500
Base circle = ∅2.750	Plate thickness = .375
Shaft = ∅1.000	All dimensions in inches.
Hub diameter = ∅1.250	

Problem 14–29 Make a two-view detailed drawing of a plate cam using the displacement diagram from Problem 14–20. Completely dimension the drawing using the following information:

The cam rotates counterclockwise.

In-line roller follower = ∅.750	Hub projection = .500
Base circle = ∅2.000	Plate thickness = .500
Shaft = ∅.625	All dimensions in inches.
Hub diameter = ∅1.250	

Problem 14–30 Make a two-view detailed drawing of a plate cam using the displacement diagram from Problem 14–18. Completely dimension the drawing using the information shown below:

The cam rotates clockwise.

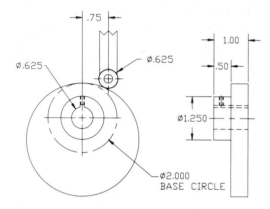

Problem 14–31 Use the displacement diagram constructed in Problem 14–25 and the following illustration to lay out the profile of the groove in the drum cam. The cam rotates clockwise. Make a two-view detailed drawing of the drum cam, with tolerances and dimensions as discussed and shown in this chapter.

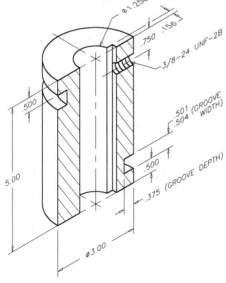

HALF OF OBJECT SHOWN

GEAR PROBLEMS

Problem 14–32 Given the gear train and chart shown below, calculate or determine the missing information to complete the chart.

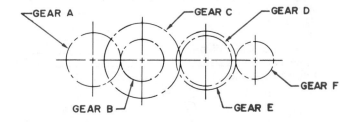

GEAR	DIAMETRAL PITCH (D)	NUMBER OF TEETH (N)	PITCH DIAMETER (P)	RPM	DIRECTION	CENTER DISTANCE	GEAR RATIO
A	4		7.5"	240	CLOCKWISE		
B		18					
C			10.0"	400			
D	5	40					
E	7						
F		14		1500			

Problem 14–33 Given the ten-gear power transmission and chart shown below, calculate or determine the missing information to complete the chart.

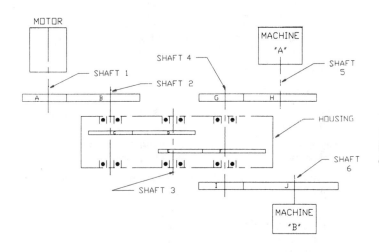

Gear Data Block						
Gear	Pitch Diameter	No. of Teeth	Diametral Pitch	RPM	Ctr. Distance Between Mating Gears	Direction
A	3.00			3600	4.00	Counter-clockwise
B			5			
C		48				
D			12	1080		
E	4.00	40				
F		100				
G	5.00				6.00	
H			6			
I			4			
J		40		108		

Problem 14–34 Given the following information, use ANSI standards to make a detailed drawing of the spur gear shown below:

20 teeth	Face width = 2.500 in.
Diametral pitch = 5	Shaft diameter = ∅1.125
20° pressure angle	Keyway for a .25 in. square key

Place the centerline of the keyway in line with a radial line through the center of one tooth (one tooth profile needed to show alignment). Include the necessary spur gear data in a chart placed over the title block. Use the formulas given in this chapter to solve for unknown values.

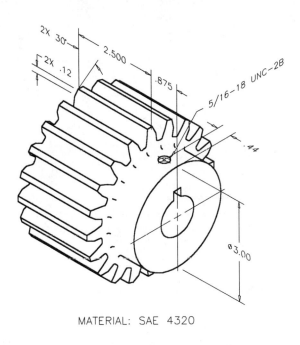

MATERIAL: SAE 4320

Problem 14–35 Use ANSI standards to make a detailed drawing of a rack that mates with the spur gear in Problem 14–34. The overall length is 24 in.

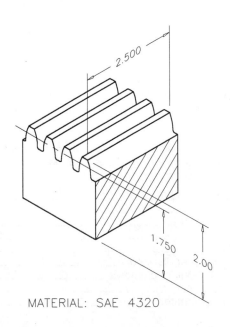

MATERIAL: SAE 4320

Problem 14–36 Use ANSI standards to make a detailed drawing of a straight bevel gear given the following information and the illustration shown below:

Pitch diameter = ⌀8.000 in.
Pressure angle = 20°
32 teeth
Diametral pitch = 4
Face width = 1.400
Shaft diameter = ⌀1.125 in.
Use a .250 square key.

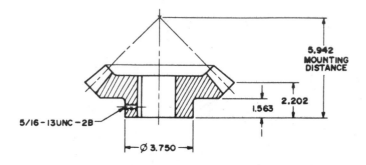

BEARING PROBLEMS

Problem 14–37 Use the charts shown in this chapter to establish the following medium-service bearing dimensions for an approximate ⌀1.25-in. shaft.

Bearing catalog number _____
Bore _____
Outside diameter _____
Width _____
Fillet radius _____
Shaft shoulder diameter _____
Housing shoulder diameter _____
Shaft diameter _____
Housing diameter _____

Problem 14–38 Use the charts shown in this chapter to establish the following medium-service bearing dimensions for an approximate ⌀3.5-in. shaft:

Bearing catalog number _____
Bore _____
Outside diameter _____
Width _____
Fillet radius _____
Shaft shoulder diameter _____
Housing shoulder diameter _____
Shaft diameter _____
Housing diameter _____

Problem 14–39 Use the charts shown in this chapter to establish the following medium-service bearing dimensions for an approximate ⌀20-mm shaft.

Bearing catalog number _____
Bore _____
Outside diameter _____
Width _____
Fillet radius _____
Shaft shoulder diameter _____
Housing shoulder diameter _____
Shaft diameter _____
Housing diameter _____

Problem 14–40 Use the charts shown in this chapter to establish the following medium-service bearing dimensions for an approximate ⌀60-mm shaft.

Bearing catalog number _____
Bore _____
Outside diameter _____
Width _____
Fillet radius _____
Shaft shoulder diameter _____
Housing shoulder diameter _____
Shaft diameter _____
Housing diameter _____

GEAR DESIGN, BEARING SELECTION, SHAFT DESIGN PROBLEM

Problem 14–41 Design a two-speed gear reducer that will operate eight to ten hours per day and receive moderate shock while in operation. Use the following information:

A 5-HP 1750 electric motor supplies the input power.
There are six gears arranged approximately as shown on page 471.
Gear C-D is a cluster gear sliding on the countershaft.
The output speed is 625 rpm when gear C is engaged with gear E.
The output speed is 437.5 rpm when gear D is engaged with gear F.
Gear A has 32 teeth, gear B has 64 teeth, gear C has 25 teeth, and gear D has 24 teeth.

Using the gear and bearing information from this chapter to design the gear reducer and do the following:

1. Determine the diametral pitches for all six gears.

2. Determine the number of teeth for gears E and F.

3. Use tolerances, surface finishes, and fit as discussed in this chapter.

4. Use manufacturer's catalogs shown in this chapter or supplied by your instructor to select standard parts.

5. Make a detailed drawing of the cluster gear, including spur gear data charts for both gears on one sheet.

6. Make detailed drawings of the three shafts—input, output, and countershaft—each on a separate sheet. The shafts are approximately ⌀1.250 in. Design the shafts based on bearing specifications and fits. Design the keyseats based on the shaft diameter given and as specified in the *Machinery's Handbook* or other source.

7. Make an assembly drawing of the entire product. Prepare a complete parts list placed over the title block. All parts should be ballooned on the drawing to coordinate with the parts list. The dimensions given in the illustration on page 471 are for reference only. Detailed dimensions are not required on the assembly drawing.

Problem 14–41 (Continued)

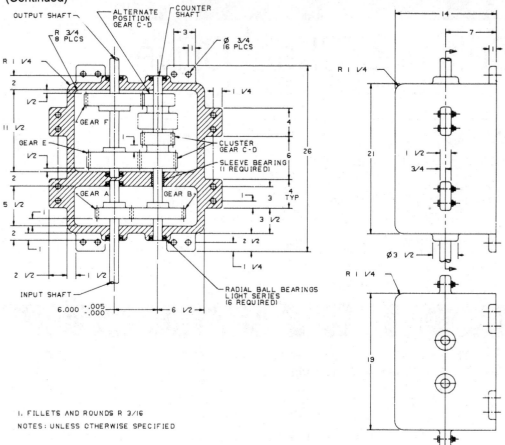

I. FILLETS AND ROUNDS R 3/16

NOTES: UNLESS OTHERWISE SPECIFIED

Problem 14–42 Prepare detail drawing of the pinion and gear from the following gear data:

Dimensions for 20° straight bevel gear
90° shaft angle

	PINION	GEAR
NUMBER OF TEETH	22	75
DIAMETRAL PITCH	10	10
FACE WIDTH	1.25	1.25
PRESSURE ANGLE	20°	20°
SHAFT ANGLE	90°	90°
WORKING DEPTH	0.200	0.200
WHOLE DEPTH	0.221	0.221
PITCH DIAMETER	2.200	7.500
PITCH ANGLE	16.348°	73.652°
CONE DISTANCE	3.908	3.908
CIRCULAR PITCH	0.314	0.314
ADDENDUM	0.142	0.058
DEDENDUM	0.077	0.161
CLEARANCE	0.021	0.021
DEDENDUM ANGLE	1.126°	2.356°
FACE ANGLE OF BLANK	18.704°	74.778°
ROOT ANGLE	15.222°	71.296°
OUTSIDE DIAMETER	2.473	7.533
PITCH APEX TO CROWN	3.710	1.044
CIRCULAR THICKNESS	0.188	0.126
BACKLASH	0.002	0.002
CHORDAL THICKNESS	0.186	0.125
CHORDAL ADDENDUM	0.146	0.058
TOOTH ANGLE	107.149 min	107.149 min
LIMIT POINT WIDTH	0.046	0.046
TOOL ADVANCE	0.002	0.002

MATH PROBLEMS

For the oblique triangle shown below (which is not drawn to scale):

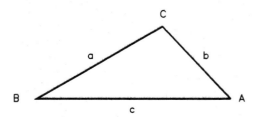

1. Find side b given a = 125, A = 54.7°, B = 65.2°.

2. Find side c given b = 321, A = 75.3°, C = 38.5°. (Hint: A + B + C = 180°)

3. Find angle C given b = 50.4, C = 33.3, B = 118.5°.

4. Find angle A given b = 51.5, a = 62.5, B = 40.7°, and given that angle A is greater than 90°.

5. Find side b given a = 320, c = 475, A = 35.3°, and given that angle C is less than 90°.

6. Find side a given b = 50.4, c = 33.3, B = 118.5°.

CHAPTER 15

Belt and Chain Drives

LEARNING OBJECTIVES

After completing this chapter, you will:

○ Design belt drive systems using vendor specifications.
○ Design chain drive systems from vendor specifications.
○ Draw sprocket and chain designs from given engineering layouts.

THE ENGINEERING DESIGN PROCESS

The problem is to design a two-belt sheave for Class D conventional industrial belts. The ID should be 5.00 and the OD 6.375. Provide 1.126 between belt centers. Keep the weight down as much as possible; try for .375 web thickness. Design for a ∅.875 shaft. Refer to the *Machinery's Handbook* for the bore diameter and the recommended keyway for a parallel rectangular key. Your research concludes that you need the following specifications:

○ 38° V in sheave for belts.

○ ∅.874–.875 bore.

○ .203 keyway × 1.032 (measured from bottom of bore).

○ You decide on a ∅2.00 hub for good material strength.

The problem solution is shown in Figure 15–1.

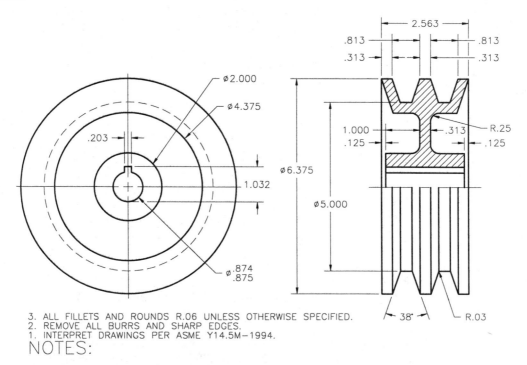

3. ALL FILLETS AND ROUNDS R.06 UNLESS OTHERWISE SPECIFIED.
2. REMOVE ALL BURRS AND SHARP EDGES.
1. INTERPRET DRAWINGS PER ASME Y14.5M–1994.
NOTES:

Figure 15–1   Solution to engineering problem.

The function of belt, chain, and gear drives is to transmit power between rotating shafts. It is important to evaluate the advantages and disadvantages of each type of mechanism for the proper design and implementation of the final product. Gears were discussed in Chapter 14 because they fall into a class of linkage mechanism where two or more elements of the linkage system are in direct contact with each other. Gears are also the most rugged and durable of the mechanisms, transmitting motion with little or no slippage. They can satisfy high power demands at efficiencies of up to 98 percent.

▼ ADVANTAGES OF GEAR DRIVES

Gear drives are commonly used in some applications rather than belt or chain drives for the following reasons:

○ They are more rugged and durable than most belt and chain drives.
○ They are more efficient than belt or chain drives.
○ They are more practical when space limitations require the shortest distance between centers.
○ They have great maximum speed ratios.
○ They work better when high horsepower and load capabilities are important.
○ They are preferable when nonparallel shafting is the design requirement.
○ They may be used where timing and synchronization are required. Chain drives may work effectively in these applications, but belt drives are not recommended because of slippage.

▼ ADVANTAGES OF BELT DRIVES

Belt drives are often the choice of the designer when some of the following characteristics are important:

○ They are up to 95 percent efficient, especially at high speeds.
○ They are designed to slip when an overload occurs.
○ They resist abrasion.
○ They require no lubrication, because there is no metal-to-metal contact, except at shaft bearings.
○ They are normally smooth running and less noisy.
○ Belt drives, especially those using flat belts, may be used more effectively when very long shaft center distances are required.
○ They can operate effectively at high speed ranges.
○ They have flexible shaft center distances, where gear drives are restricted.
○ They are less expensive than gear or chain drives.
○ They are easy to assemble and install and have flexible tolerances.
○ They absorb shock well.
○ They are easy and inexpensive to maintain.

▼ ADVANTAGES OF CHAIN DRIVES

Chain drive mechanisms have certain design advantages over both gear and belt drives:

○ They have flexible shaft center distances, where gear drives are restricted.
○ They are less expensive than gear drives, in most cases.
○ They have simpler installation and assembly tolerances than gear drives.
○ They provide better shock absorbing qualities than gears.
○ They have no slippage as compared to belt drives, resulting in more efficient operation.
○ They have lower loads on shaft bearings because tension is not required as with belt drives.
○ They are easy to install.
○ They are not affected by sun, heat, or oil and do not deteriorate with age as do belts.
○ They are more effective at lower speeds than belts.
○ They require little adjustment, while belt drives require frequent adjustment.

▼ BELTS AND BELT DRIVES

Belts are used to transmit power from one shaft to another smoothly, quietly, and inexpensively. Belts are made of continuous construction from materials such as rubberized cord, leather strands, fabric, or, more commonly, reinforced nylon, rayon, steel, or glass fiber. Belts operate on shaft-mounted pulleys and sheaves. *Pulleys* are wheels constructed with a groove in the circumference to match the shape of the belt and transmit power to the belt. *Sheaves* are the same as pulleys with the differentiation being that sheaves are generally the drive and pulleys are the driven. Sheaves and pulleys both guide the belt in its operation, so the terms are normally used interchangeably.

▼ BELT TYPES

Different types of belts are designed for individual applications. The type of belt you are probably most familiar with is the V belt, which is used for most automobile and household applications. Other belts, such as flat belts, are used for long center distances and high-speed applications. Positive drive belts fit notched pulleys and are designed for more positive power transmission than other belt designs.

V Belts

V belts are, as just mentioned, the most commonly used belts. These belts have a wide range of applications and operating conditions. They operate efficiently at speeds between 1500 and 6500 fpm (feet per minute) and a temperature range of −30° F to 185° F.

SAE V-belt manufacturers code the belts according to shape to help ensure interchangeability between suppliers. V belts fall into three main categories: automotive, agricultural, and industrial. Automotive belts are manufactured in accordance with six Society of Automotive Engineers (SAE) approved sectional shapes. Agricultural V belts are similar in shape to automotive V belts. They are identified with a letter *H*. Industrial V belts fall into the categories of conventional, narrow, light duty, and double V cross section. A comparison of the standard V-belt cross sections is shown in Figure 15–2.

The load-carrying capability of a V belt is provided by reinforcing cords in the belt. The reinforcement is made of rayon, nylon, steel, or glass fibers. Figure 15–3 shows the construction of a V

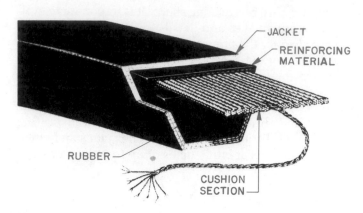

Figure 15–3   V-belt with steel reinforcing cables. *Courtesy Browning Mfg. Emerson Power Transmission.*

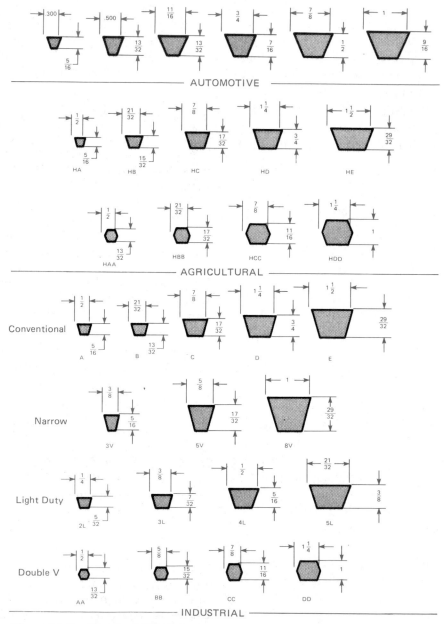

Figure 15–2   Standard V-belt types.

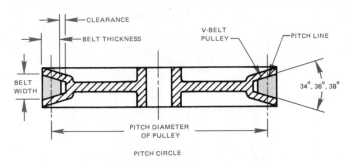

Figure 15–4 V-belt and pulley (sheave) cross section.

Figure 15–5 Positive drive belt. *Courtesy Browning Mfg. Emerson Power Transmission.*

belt with steel reinforcing cables. The reinforcing cords are usually embedded in a soft rubber material called a cushion section. The cushion section is surrounded by a hard rubber or other material and the internal structure is covered with an abrasion-resistant material.

The *pitch line* is the only portion of the belt that does not change length as the belt bends around the pulley. The pitch line is used to determine the pitch diameter of the pulley. The *pitch diameter* is the effective diameter of the pulley, which is used to establish the speed ratio. Most pulleys are made of cast iron or steel. The construction and terminology of a pulley and mating belt are shown in Figure 15–4.

Flat Belts

Flat belts are typically used where high-speed applications are more important than power transmission, and when long center distances are necessary. Flat belts are less efficient at moderate speeds, because they tend to slip under load. Flat belts are also used where drives with nonparallel shafts are required, because the belt may be twisted to accommodate the relationship of the shafts.

Positive Drive Belts

Positive drive belts have a notched underside that contacts a pulley with the same design on the circumference. These belts have some of the same advantages as chain and gear drives due to their positive contact between the belt and the pulley. These belts are more suitable for operations requiring high efficiency, timing, or constant velocity. They may also be used to decrease the pulley size and give the same operating performance of a larger sized V-belt pulley. A sample positive drive belt is shown in Figure 15–5.

▼ TYPICAL BELT DRIVE ARRANGEMENTS

Belt drives are normally arranged with or without an idler. The *idler* is often used to help maintain constant tension on the belt. (See Figure 15–6.)

When belts are in operation, they stretch. Machines that are operated by belt drives must have an idler, an

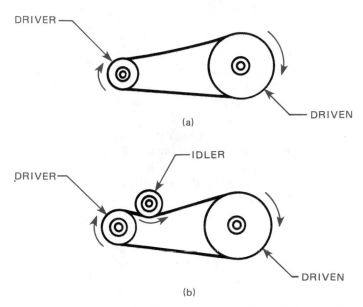

Figure 15–6 **(a)** V-belt drive arrangement. **(b)** Belt drive with an idler.

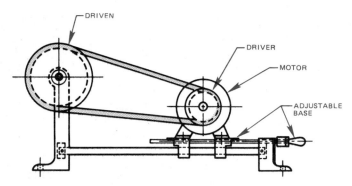

Figure 15–7 V-belt drive arrangement with adjustable motor base.

adjustable motor base, or both to provide for belt adjustment. Maintenance of the tension between the shafts is important in controlling the efficiency of the belt drive. Figure 15–7 shows an assembly drawing of an adjustable motor base. There are many different types of adjustable motor bases, including the sliding and tilting types.

In some cases adjustable motor bases are not practical. When this occurs, idler pulleys are used to keep belts tight. A V-belt and flat-belt idler are displayed in Figure

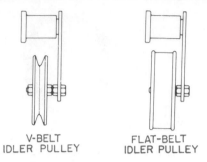

V-BELT
IDLER PULLEY

FLAT-BELT
IDLER PULLEY

Figure 15–8 Idler pulleys. *Courtesy Lovejoy, Inc., Downers Grove, IL.*

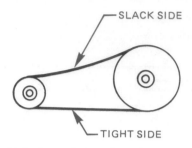

SLACK SIDE

TIGHT SIDE

Figure 15–9 Belt drive slack side and tight side.

15–8. When idler pulleys are used, their size and location in relation to the belt drive are important. When referring to the placement of the idler pulley, the terms *slack side* and *tight side* are used. When the belt is in operation, the slack side is the belt on the top of the drive and the tight side is normally on the bottom of the drive when the driver and driven shafts are in horizontal alignment as shown in Figure 15–9. There are several typical design positions for the location of the idler pulley in relation to the driver pulley. An inside idler pulley is placed near the driver on the slack side of the belt. It should be the same diameter or larger than the driver. An inside idler pulley may also be placed on the tight side near the driven pulley. An outside idler pulley is placed near the driver on the outside of the belt on the slack side. The best design for this application is to have the idler pulley slightly larger in diameter than the driver pulley. When an outside idler pulley is placed on the tight side, it is normally placed near the driven pulley. (See Figure 15–10.)

▼ BELT DRIVE SELECTION

Belts are selected for a specific application based on the function of the machinery. The factors influencing the selection and design of belt drives include:

○ Horsepower of driving shaft.
○ Speed of driving shaft.
○ Service conditions.
○ Type of machine.
○ Center distance between shafts.
○ Belt length.
○ Number of belts required.

Designing V-belt Drives

Belt manufacturers provide formulas and catalogs of the design and selection of belt drives based on the criteria described previously. The following discussion is based on engineering data provided in the Browning Power Transmission catalog, produced by the Browning Manufacturing Division of Emerson Power Transmission Company. In order to select a drive using standard

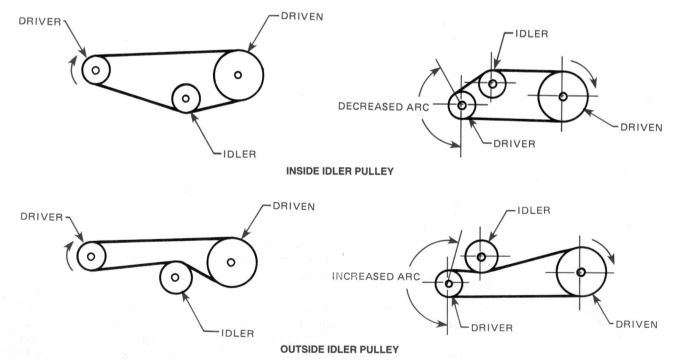

DRIVER

DRIVEN

IDLER

INSIDE IDLER PULLEY

IDLER

DECREASED ARC

DRIVER

DRIVEN

DRIVER

DRIVEN

IDLER

OUTSIDE IDLER PULLEY

IDLER

INCREASED ARC

DRIVER

DRIVEN

Figure 15–10 Idler pulley arrangements.

drive selection tables, this sample problem is used, courtesy of Browning Manufacturing Division of Emerson Power Transmission Company.

A drive is required for a 20-HP (horsepower) motor driving a fan 16 hours per day. The motor speed is 3600 rpm and the shaft size is 1.625 in. The fan speed is approximately 2250 rpm and the fan shaft is 1.438 in. The center distance is 38 in. minimum and 41 in. maximum.

Step 1. Determine the design horsepower. The design HP of a machine takes into account the type of service and the efficiency of a particular machine. Design horsepower is calculated using the formula:

Design HP = Rated HP × Service Factor

The rated horsepower is the horsepower specified on the driver motor. For example, the label on the motor may read 2 HP. The service factor is determined by the type of machinery, the number of hours of daily operation, and the type of driven unit. Service factors may be selected from the chart shown in Figure 15–11. Based on the example problem, a fan operating 16 hours per day has a service factor of 1.3. Calculate the design HP:

20 HP × 1.3 = 26 Design HP

Step 2. Determine the driven speed. To establish the required driven speed refer to Figure 15–12. Find the "Nominal Driven Speeds and Horsepower per Belt" column. Under the column labeled "3500-rpm Motor," find the "Nominal Driven rpm" column. Select the RPMs equal to or less than the given fan speed of 2250 rpm. Fifteen combinations that provide 2244 rpm for the driven pulley are highlighted in Figure 15–12.

Step 3. Determine the available selections. Refer again to Figure 15–12 and look under the "HP per Belt" column to determine which sheave combination results in the smallest sheave diameters, the least number of belts, and the lightest belt selection for the sample problem. To begin this process, assume your selection will be found in the "Gripnotch" column under HP per Belt. If you use a "Super" belt, another selection would be made. Read down the "Gripnotch" column until you find the largest HP per belt (12.94) without exceeding the Design HP (26) when multiplied by 1, 2, or 3 belts. Your goal is to use the least number of belts. Therefore, the best solution in this case is *2 belts × 12.94 = 25.88.* 25.88 is less than the 26-Design HP. These items must be equal to or less than the given data. Of the fifteen selections from Step 2, the proposed selections are highlighted in Figure 15–12. They are:

TYPES OF DRIVEN MACHINES	TYPES OF DRIVING UNITS		
	AC Motors; Normal Torque, Squirrel Cage, Synchronous and Split Phase. DC Motors; Shunt Wound. Multiple Cylinder Internal Combustion Engines.		
	Intermittent Service (3-5 Hours Daily or Seasonal)	Normal Service (8-10 Hours Daily)	Continuous Service (16-24 Hours Daily)
Agitators for Liquids Blowers and Exhausters Centrifugal Pumps and Compressors Fans up to 10 HP Light Duty Conveyors	1.0	1.1	1.2
Belt Conveyors For Sand, Grain, etc. Dough Mixers Fans Over 10 HP Generators Line Shafts Laundry Machinery Machine Tools Punches-Presses-Shears Printing Machinery Positive Displacement Rotary Pumps Revolving and Vibrating Screens Speed Reducers, All Types	1.1	1.2	1.3
Brick Machinery Bucket Elevators Exciters Piston Compressors Conveyors (Drag-Pan-Screw) Hammer Mills Paper Mill Beaters Piston Pumps Positive Displacement Blowers Pulverizers Saw Mill and Woodworking Machinery Textile Machinery	1.2	1.3	1.4
Crushers (Gyratory-Jaw-Roll) Mills (Ball-Rod-Tube) Hoists Rubber Calenders-Extruders-Mills	1.3	1.4	1.5

A minimum Service Factor of 2.0 is suggested for equipment subject to choking.*

*
> **CAUTION—Drives requiring high Overload Service Factors, such as crushing machinery, certain reciprocating compressors, etc. subjected to heavy shock load without suitable fly wheels, may need heavy duty web type sheaves rather than standard arm type. For any such application, consult the Browning Engineers.**

Figure 15–11 Service factors for belt drives. *Courtesy Browning Mfg. Emerson Power Transmission.*

Pitch Diameter (PD) Driver = 5.4 in.
PD Driven = 8.4 in.
Belt Section = B
Number of Belts = 1 minimum and 3 maximum (exact number to be determined later)
HP per Belt = 12.94

Step 4. Determine the belt part number for the approximate center distance and the F factor for the belt. The selected center distance, as stated in the problem, must be between 38 and 41 in. The F factor is a correction factor for the loss of arc of contact. The standard arc contact for a belt to sheave is 180°. When the belt does not contact 180°, there is loss or gain or arc contact. When there is loss or gain or arc contact, a correction factor must be used to establish the corrected horsepower and number of belts required. Look at Figure 15–13, which is an extension of the chart in Figure 15–12. Find the center distance (CD) in the horizontal column that is closest to the required center distance. If you cannot find an acceptable CD between

3500 and 1750 RPM Motors

Ratio	Driver P.D.	Driven P.D.	Belt Section	Grooves Min.	Grooves Max.	3500 Nom. Driven RPM	3500 H.P. Super	3500 H.P. Gripnotch	1750 Nom. Driven RPM	1750 H.P. Super	1750 H.P. Gripnotch	Belt No.	C.D.	F.	Belt No.	C.D.	F.	Belt No.	C.D.	F.
1.54	10.4	16.0	B	1	3	–	–	–	1136	14.97	17.10	B67	13.4	0.87	B74	16.9	0.90	B81	20.5	0.94
1.54	13.0	20.0	C	1	12	–	–	–	1136	27.81	33.05	C85	17.7	0.85	C100	25.3	0.89	C112	31.3	0.91
1.55	3.8	5.9	B	1	3	2258	4.33	8.03	1129	3.32	5.33	B28	7.2	0.73	B45	15.7	0.84	B56	21.3	0.89
1.55	4.0	6.2	A	1	6	2258	5.19	6.35	1129	3.31	4.17	A26	5.5	0.76	A36	10.6	0.84	A46	15.6	0.89
1.55	4.0	6.2	B	1	6	2258	5.00	8.71	1129	3.75	5.76	B28	6.8	0.71	B45	15.3	0.84	B56	20.9	0.87
1.55	9.6	14.9	5V	2	10	–	–	–	1129	27.32	32.36	5V670	14.0	0.85	5V800	20.6	0.89	5V950	28.1	0.93
1.56	3.2	5.0	A	1	6	2244	3.62	4.35	1122	2.32	2.92	A24	6.1	0.78	A35	11.7	0.88	A45	16.7	0.88
1.56	3.2	5.0	B	1	2	2244	2.18	5.87	1122	1.99	4.00	B28	8.4	0.73	B45	16.9	0.84	B56	22.4	0.89
1.56	3.6	5.6	A	1	6	2244	4.43	5.38	1122	2.82	3.56	A24	5.3	0.76	A35	10.9	0.84	A45	15.9	0.88
1.56	3.6	5.6	B	1	6	2244	3.63	7.33	1122	2.88	4.89	B28	7.6	0.73	B45	16.1	0.84	B56	21.6	0.89
1.56	3.9	6.1	B	1	3	2244	4.67	8.38	1122	3.54	5.55	B28	7.0	0.71	B45	15.5	0.84	B56	21.0	0.87
1.56	4.1	6.4	B	1	3	2244	5.33	9.05	1122	3.97	5.97	B28	6.6	0.71	B45	15.1	0.84	B56	20.6	0.87
1.56	4.1	6.4	3V	1	4	2244	7.72	8.76	1122	4.47	4.90	3V280	5.6	0.79	3V335	8.4	0.84	3V400	11.7	0.89
1.56	4.5	7.0	A	1	3	2244	6.07	7.50	1122	3.90	4.90	A29	6.0	0.78	A39	11.0	0.84	A49	16.1	0.90
1.56	4.5	7.0	5V	2	5	2244	–	17.45	1122	–	10.28	5V500	15.9	0.82	5V630	22.4	0.86	5V800	30.9	0.92
1.56	4.8	7.5	A	1	1	2244	6.56	8.15	1122	4.26	5.36	A31	6.3	0.79	A40	10.9	0.85	A49	15.4	0.90
1.56	5.4	8.4	B	1	3	2244	9.06	12.94	1122	6.66	8.65	B34	6.9	0.74	B47	13.5	0.84	B58	19.0	0.88
1.56	5.4	8.4	5V	2	3	2244	–	24.53	1122	–	14.42	5V500	14.1	0.82	5V630	20.6	0.86	5V800	29.1	0.90
1.56	5.5	8.6	A	1	3	2244	7.57	9.53	1122	5.05	6.36	A36	7.4	0.81	A45	12.0	0.87	A54	16.5	0.92
1.56	6.4	10.0	A	1	3	2244	8.58	11.03	1122	6.02	7.58	A42	8.6	0.84	A50	12.6	0.89	A58	16.7	0.93
1.57	2.8	4.4	A	1	2	2229	2.76	3.26	1115	1.80	2.28	A24	6.9	0.78	A35	12.5	0.84	A45	17.5	0.90
1.57	3.0	4.7	A	1	3	2229	3.19	3.81	1115	2.06	2.60	A24	6.5	0.78	A35	12.1	0.84	A45	17.1	0.90
1.57	3.5	5.5	A	1	3	2229	4.23	5.13	1115	2.69	3.40	A24	5.5	0.76	A35	11.0	0.84	A45	16.0	0.88
1.57	3.7	5.8	A	1	3	2229	4.62	5.63	1115	2.94	3.71	A24	5.1	0.75	A35	10.6	0.84	A45	15.7	0.88
1.57	4.2	6.6	A	1	6	2229	5.56	6.83	1115	3.55	4.48	A27	5.5	0.76	A37	10.6	0.84	A47	15.6	0.89
1.58	4.8	7.6	A	1	6	2215	6.56	8.16	1108	4.26	5.36	A31	6.3	0.79	A40	10.8	0.85	A49	15.3	0.90
1.58	5.0	7.9	B	1	1	2215	8.04	11.85	1108	5.86	7.86	B32	6.6	0.73	B46	13.7	0.84	B57	19.2	0.87
1.58	5.2	8.2	A	1	10	2215	7.16	8.97	1108	4.71	5.94	A34	7.0	0.80	A43	11.5	0.86	A52	16.1	0.92
1.58	5.7	9.0	A	1	3	2215	7.83	9.90	1108	5.27	6.64	A37	7.4	0.82	A46	12.0	0.88	A55	16.5	0.93
1.58	6.0	9.5	A	1	1	2215	8.18	10.41	1108	5.60	7.05	A39	7.8	0.82	A47	11.8	0.88	A55	15.9	0.92
1.58	6.6	10.4	B	1	3	2215	11.51	15.74	1108	8.97	10.95	B43	8.8	0.80	B53	13.9	0.86	B63	19.0	0.89
1.58	6.7	10.6	A	1	1	2215	8.85	11.46	1108	6.33	7.98	A44	8.9	0.85	A52	12.9	0.84	A60	16.9	0.94
1.58	6.9	10.9	B	1	1	2215	11.95	16.30	1108	9.52	11.50	B45	9.2	0.81	B55	14.3	0.86	B65	19.3	0.90
1.58	7.2	11.4	C	2	6	–	–	–	1108	12.92	19.41	C51	12.2	0.76	C75	24.2	0.84	C96	34.8	0.89
1.58	7.4	11.7	5V	2	8	–	–	–	1108	18.97	23.39	5V500	9.8	0.79	5V630	16.4	0.85	5V800	24.9	0.90

Figure 15–12 Motor speed and drive selection table. *Courtesy Browning Mfg. Emerson Power Transmission.*

3500 and 1750 RPM Motors

Nominal Center Distance (C.D.) and Arc-Length Factor (F) using Browning Belts

Belt No.	C.D.	F.	Belt No.	C.D.	F.	Belt No.	C.D.	F.	Belt No.	C.D.	F.	Belt No.	C.D.	F.	Belt No.	C.D.	F.	Driver P.D.	Driven P.D.	Ratio
5V1120	36.7	0.95	5V1320	46.7	0.98	5V1600	60.7	1.03	5V1900	75.7	1.06	5V2240	92.7	1.08	5V3550	158.2	1.17	9.6	14.9	1.55
A55	21.7	0.95	A65	26.7	0.98	A75	31.7	1.01	A85	36.7	1.05	A95	41.7	1.06	A180	84.2	1.19	3.2	5.0	1.56
B67	27.9	0.93	B78	33.4	0.97	B88	38.4	1.00	B99	43.9	1.02	B136	62.5	1.09	B360	173.7	1.31	3.2	5.0	1.56
A55	20.9	0.95	A65	25.9	0.98	A75	30.9	1.01	A85	35.9	1.04	A95	40.9	1.06	A180	83.4	1.19	3.6	5.6	1.56
B67	27.2	0.93	B78	32.7	0.97	B88	37.7	0.99	B99	43.2	1.02	B136	61.7	1.09	B360	172.9	1.31	3.6	5.6	1.56
B67	26.5	0.93	B78	32.0	0.97	B88	37.0	0.99	B99	42.5	1.01	B136	61.0	1.09	B360	172.3	1.31	3.9	6.1	1.56
B67	26.1	0.93	B78	31.6	0.97	B88	36.6	0.99	B99	42.1	1.01	B136	60.6	1.09	B360	171.9	1.31	4.1	6.4	1.56
3V475	15.5	0.92	3V560	19.7	0.95	3V670	25.2	1.00	3V800	31.7	1.03	3V950	39.2	1.07	3V1400	61.7	1.15	4.1	6.4	1.56
A59	21.1	0.94	A69	26.1	0.99	A79	31.1	1.02	A89	36.1	1.04	A100	41.6	1.07	A180	81.6	1.19	4.5	7.0	1.56
5V1000	40.9	0.95	5V1250	53.5	1.00	5V1600	71.0	1.04	5V2000	91.0	1.08	5V2500	116.0	1.11	5V3550	168.5	1.17	4.5	7.0	1.56
A58	19.9	0.94	A67	24.5	0.97	A76	29.0	1.01	A85	33.5	1.04	A94	38.0	1.05	A180	81.0	1.19	4.8	7.5	1.56
B69	24.5	0.92	B80	30.0	0.97	B90	35.0	0.99	B103	41.5	1.02	B144	62.0	1.11	B360	169.3	1.31	5.4	8.4	1.56
5V1000	39.1	0.95	5V1250	51.6	0.99	5V1600	69.1	1.04	5V2000	89.2	1.08	5V2500	114.2	1.11	5V3550	166.7	1.17	5.4	8.4	1.56
A63	21.0	0.96	A72	25.5	0.98	A81	30.0	1.01	A90	34.5	1.05	A100	39.5	1.07	A180	79.6	1.19	5.5	8.6	1.56
A66	20.7	0.97	A74	24.7	0.99	A82	28.7	1.01	A90	32.7	1.03	A98	36.7	1.07	A180	77.7	1.19	6.4	10.0	1.56
A55	22.5	0.95	A65	27.5	0.98	A75	32.5	1.02	A85	37.5	1.05	A95	42.5	1.06	A180	85.0	1.19	2.8	4.4	1.57
A55	22.1	0.95	A65	27.1	0.98	A75	32.1	1.01	A85	37.1	1.05	A95	42.1	1.06	A180	84.6	1.19	3.0	4.7	1.57
A55	21.1	0.95	A65	26.1	0.98	A75	31.1	1.01	A85	36.1	1.04	A95	41.1	1.06	A180	83.6	1.19	3.5	5.5	1.57
A55	20.7	0.93	A65	25.7	0.98	A75	30.7	1.01	A85	35.7	1.04	A95	40.7	1.05	A180	83.2	1.19	3.7	5.8	1.57
A57	20.6	0.94	A67	25.6	0.99	A77	30.6	1.01	A87	35.6	1.04	A97	40.6	1.07	A180	82.2	1.19	4.2	6.6	1.57
A58	19.9	0.94	A67	24.4	0.97	A76	28.9	1.01	A85	33.4	1.04	A94	37.9	1.05	A180	80.9	1.19	4.8	7.6	1.58
B68	24.7	0.92	B79	30.2	0.97	B89	35.2	0.99	B100	40.7	1.02	B140	60.7	1.10	B360	170.0	1.31	5.0	7.9	1.58
A61	20.6	0.95	A70	25.1	0.98	A79	29.6	1.00	A88	34.1	1.04	A97	38.6	1.07	A180	80.1	1.19	5.2	8.2	1.58
A64	21.0	0.96	A73	25.5	0.98	A82	30.1	1.01	A91	34.6	1.05	A103	40.6	1.07	A180	79.1	1.19	5.7	9.0	1.58
A63	19.9	0.96	A71	23.9	0.98	A79	27.9	1.00	A87	31.9	1.02	A95	35.9	1.05	A180	78.5	1.19	6.0	9.5	1.58
B73	24.0	0.93	B82	29.0	0.94	B92	33.5	0.97	B105	40.0	1.03	B144	59.5	1.10	B360	166.8	1.31	6.6	10.4	1.58
A68	21.0	0.97	A76	25.0	0.99	A84	29.0	1.01	A92	33.0	1.03	A103	38.5	1.05	A180	77.0	1.18	6.7	10.6	1.58
B75	24.3	0.94	B84	28.8	0.96	B96	33.9	0.98	B112	42.9	1.04	B154	63.9	1.11	B360	166.2	1.31	6.9	10.9	1.58
C112	42.8	0.94	C136	54.8	0.98	C162	67.8	1.02	C210	91.8	1.08	C270	120.8	1.14	C420	195.8	1.24	7.2	11.4	1.58
5V1000	34.9	0.93	5V1250	47.4	0.99	5V1600	65.0	1.03	5V2000	85.0	1.07	5V2500	110.0	1.11	5V3550	162.5	1.17	7.4	11.7	1.58

Figure 15–13 Nominal center distances, number of belts, arc length factor chart. *Courtesy Browning Mfg. Emerson Power Transmission.*

the given requirements, find a pair of CDs as highlighted in Figure 15–13. In this case, you have located:

> Belt B90 with a CD of 35.0 in. and F of 0.99
> Belt B103 with a CD of 41.5 and F of 1.02

To establish the exact belt for this application, you must *interpolate.* Interpolate means to make an approximation between existing known values. Interpolate to find the correct belt for the application using this method:

> For every inch of belt length difference between belts, there is about 1/2 in. of center distance change. The belt numbers represent a relationship. For example, the B90 belt is 13 in. shorter than the B103 belt (B103 − B90 = 13). For this example, if you subtract 3 in. from a B103 belt, you get a B100 belt. If there is 1/2 in. of CD change for every inch of belt length, then 3 in. of length change equals 1.5 in. of CD change. Therefore, the B100 belt has a CD of 40 in.

Now determine the F factor of the B100 belt by comparing the values of the known belts:

> B90   CD = 35 in.   F = .99
> B100   CD = 40 in.   F = ?
> B103   CD = 41.5 in.   F = 1.02

Establish the difference in F factor between the B90 and B103 belts.

> 1.02 − .99 = .03 (difference in F factor)

Establish the difference in CD between the B90 and B103 belts:

> 41.5 − 35 = 6.5 in. (difference in CD)

Now divide the difference in F factor by the difference in CD to determine the F factor per inch of CD:

> .03 divided by 6.5 = .0046
> (F factor per inch of CD)

The B100 belt has a 40-in. CD, which is 1.5 in. less than the B103 belt; so, the B100 belt has an F factor of 1.5 × .0046 less than the B103 belt:

> 1.5 × .0046 = .0069
> F 1.02 (B103 belt) − .0069 = 1.01
> (rounded from 1.031)
> F factor for the B100 belt.

Therefore, the B100 belt has an F factor of 1.01.

Step 5. Determine the corrected horsepower and number of belts for the drive. The corrected horsepower is an adjustment of the design horsepower

in relation to the F factor. It is calculated with this formula:

> F Factor × HP per Belt = Corrected Horsepower

The corrected horsepower for the sample problem is:

> 1.01 (F) × 12.94 (HP per belt) = 13.06
> Corrected HP per Belt

Now determine the number of belts required for the drive using this formula:

> Design HP/Corrected HP = Number of Belts

If the number calculated in this formula is one or less, then one belt is required; or two belts for two or less; three belts for three or less; and so on. The number of belts for the sample problem is:

> 26 (Design HP)/13.06 (Corrected HP) = 1.99
> = 2 Belts Required

In conclusion, the design solution to the sample problem is as follows:

> 26 Design HP
> 12.94 Design HP per Belt
> 13.06 Corrected HP per Belt
> B100 Belt Number
> 40 in. Center Distance
> 2 Belts Required
> ∅5.4 in. Driver Pitch Diameter (PD)
> ∅8.4 in. Driven PD
> B Belt Section

Determine the Belt Drive Ratio

The belt drive ratio is the relationship between the drive and driven pulleys. The ratio may be determined by dividing the rpm of the driver by the rpm of the driven pulley. If you use the sample problem presented earlier, for example, the motor rpm from Figure 15–12 is 3500 rpm, and the nominal driven rpm from the same chart is 2244 rpm. Therefore, the Ratio = 3500 rpm/2244 rpm = 1.5597 = 1.56. Now, refer back to Figure 15–12 and notice the 1.56 highlighted in the Ratio column at the far left. This 1.56 ratio coincides with the other items selected for the application.

Determine the Belt Velocity

Belt velocity is the speed a belt travels in feet per minutes (fpm) for a given application. In some cases, the belt velocity may be important. In many applications, though, it is not a factor. Belt velocity becomes important at very high speeds. For example, cast iron sheaves should not travel at speeds greater than 6500 fpm. Another factor that influences design at high speeds is vibration, which may require balancing of the sheaves. Belt velocity is calculated using this formula:

Belt Velocity fpm = Pitch Diameter (PD)
Sheave × 0.2618 × Sheave Speed rpm
(0.2618 is a constant to be used for all applications)

The belt velocity for the driver sheave of the sample problem is:

∅5.4 (PD Sheave) × 0.2618 × 3500 rpm = 4948 fpm

▼ CHAIN DRIVES

Chain drives, like gear and belt drives, are used to transmit power from one shaft to another. As discussed earlier, chain drives have certain advantages over gear and belt drives. In general, chain drives are less efficient and durable than gear drives, but they offer greater power capacity, more position power transmission, durability, and longer service life than belts, especially at high temperatures.

In chain-drive applications, toothed wheels called *sprockets* mate with a chain to transmit power from one shaft to another. (See Figure 15–14.)

▼ CHAIN DRIVE SPROCKETS

Sprockets are designed for various applications. Depending on the shaft mounting requirements, sprockets are available with a hub on one or both sides or without hubs. Sprockets are designed with a solid web in most applications, although weight is reduced by designing recessed webs or spokes.

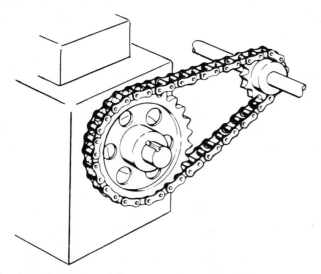

Figure 15–14 Chain drive sprockets and chain. *Courtesy Morse Industrial.*

Sprockets are machined from cast iron, steel, aluminum alloy, powdered metal, or plastic. Split sprockets are used when mounting between bearings requires easy installation. Some applications have sprockets mounted to a removable steel or cast-iron hub when frequent replacement is required.

▼ CHAIN CLASSIFICATION AND TYPES

Chains are classified in relation to the accuracy of construction between the sprocket and the chain links. There are three broad categories: precision, nonprecision, and light-duty chains. Precision chains are designed for smooth, free-running operation at high speed and high power. Common precision chains include the roller, offset sidebar, silent inverted tooth, and double pitch chains. Nonprecision chains do not have as high a degree of precision between the sprocket and chain links. These lower cost chains are generally used for low-speed applications on machinery rated below 50 HP. Nonprecision chains include detachable, pintle, and welded chains. Light-duty chains are designed for application in low-power situations such as equipment control mechanisms such as computers, typewriters, or appliance controls.

▼ PRECISION CHAINS
Roller Chain

Roller chain is the most commonly used power transmission chain. Roller chain is rated up to 630 horsepower for single chain drives referred to as single strand chain. Roller chains may be added in multiple units, known as multiple strand chains, to substantially increase the horsepower under which they may operate. Roller chains provide efficient, quiet operation, and should be lubricated for the best service life. Figure 15–15 shows the parts of a typical roller chain.

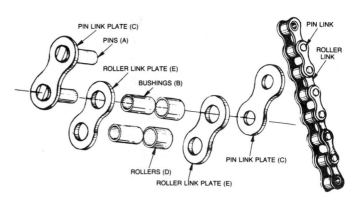

Figure 15–15 Roller chain assembly. *Courtesy Rainbow Industrial Products Corporation.*

CADD Applications

SPROCKETS

The same simplified representation used in gear CADD drawings may also be used for sprocket CADD drawings, or custom software is available to automatically generate detailed drawings from given data, Figure 15–16.

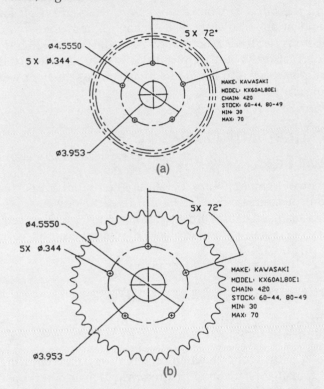

Figure 15–16 Sprocket drawing: **(a)** simplified representation, **(b)** detailed representation.

The *chain pitch* is important when designing a chain drive. The chain pitch is the distance from the center of one pin to the center of the other pin in one link. Figure 15–17 shows a single strand chain with the pitch dimensioned and a multiple strand chain. Roller chains are available with pitch lengths ranging from 1/4 to 3 in.

Double Pitch Roller Chain

Double pitch roller chain is designed mostly for situations with long center distances such as conveyors. While the efficiency of the double pitch chain is as good as the roller chain, it is intended for lighter duty operation than regular pitch roller chain. (See Figure 15–18.)

Offset Sidebar Roller Chain

The offset sidebar, shown in Figure 15–19, is the least expensive precision chain. It is designed to carry heavier loads than nonprecision chains. Offset sidebar chains are designed to handle loads up to 425 HP and speeds up to 36 feet per second. One of the advantages

Figure 15–18 Double pitch roller chain. *Photo courtesy Morse Industrial.*

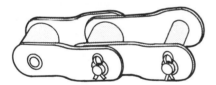

Figure 15–19 Offset sidebar roller chain.

Figure 15–17 Single-strand and multiple-strand roller chain. *Photo courtesy Morse Industrial.*

Figure 15–20 Inverted tooth silent chain. *Photo courtesy Morse Industrial.*

of the offset sidebar chain over the other chain types is its open construction, which allows it to withstand dirt and contaminants that might cause other precision chains to bind and wear out rapidly. For this reason and because it is rugged and durable, offset sidebar chains are often used to drive construction machinery.

Inverted Tooth Silent Chain

The most expensive precision chain to manufacture is the inverted tooth silent chain shown in Figure 15–20. This chain is used where high speed and smooth, quiet operation are required in rigorous applications. Lubrication is required to keep these chains running trouble free. Applications include machine tools, pumps, and power drive units.

▼ NONPRECISION CHAINS

Detachable Chain

The detachable chain shown in Figure 15–21 is the lightest, simplest, and least expensive of all chains. The detachable chain is capable of transmitting power up to 25 HP at low speeds. One common application is farm machinery requiring speeds of up to 350 fpm. Detachable chains do not require lubrication.

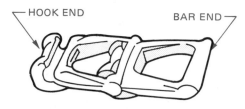

Figure 15–21 Detachable chains.

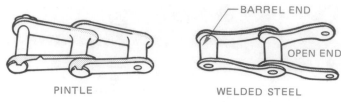

Figure 15–22 Pintle and welded steel chains.

Pintle and Welded Steel Chain

The pintle and welded steel chains shown in Figure 15–22 are a combination design between the detachable and the offset sidebar roller chain. They are used for applications similar to the detachable chain where more rigorous service is required for up to 40 HP and 425 fpm. Lubrication is not required, so it may also be used where lubrication is not effective, such as dusty or wet conditions.

▼ LIGHT-DUTY CHAIN
Bead Chain

Bead chains, Figure 15–23, are commonly used for control mechanisms for light-duty, low-power applications. Bead chains are rated from 15 to 75 pounds and are available in standard bead diameters of 3/32, 1/8, 3/16, and 1/4 in.

▼ CHAIN DRIVE ARRANGEMENTS

Chain drives have some of the same characteristics as belt drives in the following ways:

❍ Chains elongate when in operation, thus requiring shaft adjustment or link removal to maintain proper tightness.
❍ Chains have a tight side and a slack side when in operation.

The arrangement and direction of rotation of the driver sprocket and the driven sprocket are important to the proper function of the chain drive mechanism. For chain drives with long center distances, the slack side should be on the bottom, because upper side slack could cause the chain on the top to meet the bottom as the chain elongates. Figure 15–24 shows the recommended and not recommended arrangements.

Figure 15–23 Bead chain.

On short center drive arrangements, as shown in Figure 15–25, the slack side should be on the bottom, because slack on the top could cause the chain to jump out of the sprocket. The relationship of the shaft centers is also important. The shaft centers should be designed in a horizontal position or inclined no more than 45°. When the shafts are inclined up to 45°, the slack side may be on either the top or bottom. (See Figure 15–26.)

Shaft arrangements between 45° and vertical should be avoided. However, when they are required due to the machine design, the chain should be kept tighter than normal applications. This type of arrangement requires that the chain be tightened frequently. Properly designed situations with the shafts inclined greater than 45° provide for the slack to be on the side closest to horizontal. (See Figure 15–27.)

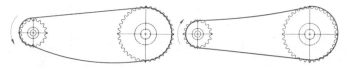

Figure 15–24 Chain drives with long center distances should have slack on bottom side.

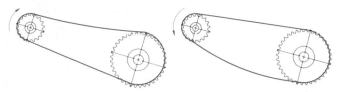

Figure 15–25 Chain drives with short center distances should have slack on the lower side.

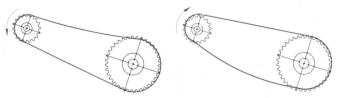

Figure 15–26 Chain drive shaft centers should be on a horizontal line or on a line inclined not more than 45°, in which case the slack may be on the upper or lower side.

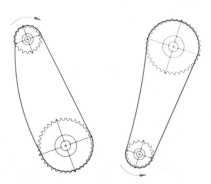

Figure 15–27 When the shaft centers are aligned more than 45° from horizontal, the slack side should be in the side closest to horizontal.

▼ ROLLER CHAIN DRIVE SELECTION

Designing roller chain drives includes selecting the sprocket sizes, the chain pitch, the chain length, the distance between shaft centers, and the lubrication requirements. Factors that contribute to selecting the proper chain drive variables are rpm of the sprockets, the horsepower (HP) of the motor, and the load conditions.

Chain manufacturers provide formulas and catalogs for the design and selection of chain drives. The following discussion is based on engineering data provided in the Morse Power Transmission Products catalog, provided by the Morse Industrial Products Division, Emerson Power Transmission Corporation.

A chain drive is required for a 2-HP electric motor operating laundry machinery. The drive spocket turns at 1800 rpm, which produces 900 rpm at the driven sprocket. The approximate center distance is 12 in.

Step 1. Determine the design horsepower. The design horsepower (HP) of a machine takes into account the type of service and the efficiency of the machine. A service factor is applied to the horsepower rating of the drive motor to determine the design horsepower of a chain drive. The service factors are established by the power source, the nature of the load, and the strain or shock on the drive. Three basic operating characteristics are used to establish the service factors. (See Figure 15–28.)

Now, determine the service factor for the given problem. You have a 2-HP electric motor operating laundry machinery. Laundry machinery operates under moderate shock. Refer to the table in Figure 15–29 to determine the service factor. An electric motor operating under moderate shock has a service factor of 1.3.

Establish the design horsepower using the formula:

Motor HP × Service Factor = Design HP

For the sample problem, the design HP is:

2 HP × 1.3 = 2.6 Design HP

Step 2. Determine the number of teeth of the drive sprocket. Now that you know the design HP

Smooth: Running load is fairly uniform. Starting and peak loads may be somewhat greater than running load, but occur infrequently.

Moderate Shock: Running load is variable. Starting and peak loads are considerably greater than running load and occur frequently.

Heavy Shock: Starting loads are extremely heavy. Peak loads and overloads occur continuously and are of maximum fluctuation.

Figure 15–28 Chain drive operating classifications. *Courtesy Morse Industrial.*

service factors

Service Factors are selected below for various applications after first determining the prime mover or power source type.

Prime Mover	TYPE
Internal Combustion Engine with Hydraulic Coupling or Torque Converter Electric Motor Turbine Hydraulic Motor	A
Internal Combustion Engine with Mechanical Drive	B

service factor table

APPLICATION	Type of Prime Mover		APPLICATION	Type of Prime Mover		APPLICATION	Type of Prime Mover	
	A	B		A	B		A	B
AGITATORS (paddle or propeller)			CRUSHING MACHINERY			PAPER INDUSTRY MACHINERY		
Pure liquid	1.1	1.3	Ball mills, crushing rolls, Jaw crushers	1.6	1.8	Agitators, bleachers	1.1	1.3
Liquids—variable density	1.2	1.4	DREDGES			Barker—mechanical	1.6	1.8
BAKER MACHINERY			Conveyors, cable reels	1.4	1.6	Beater, Yankee Dryer	1.3	1.5
Dough Mixer	1.2	—	Jigs & screens	1.6	1.8	Calendars, Dryer & Paper Machines	1.2	1.4
BLOWERS	See Fans			Consult Morse		Chippers & winder drums	1.5	1.7
BREWING & DISTILLING EQUIPMENT			Cutter head drives			PRINTING MACHINERY		
Bottling Machinery	1.0	—	Dredge pumps	See Pumps		Embossing & flat bed presses, folders	1.2	—
Brew Kettles, cookers, mash tubs	1.0	—	FANS & BLOWERS			Paper cutter, rotary press & linotype machine	1.1	—
Scale Hopper—Frequent starts	1.2	—	Centrifugal, propeller, vane	1.3	1.5	Magazine & newspaper presses	1.5	—
BRICK & CLAY EQUIPMENT			Positive blowers (lobe)	1.5	1.7	PUMPS		
Auger machines, cutting table	1.3	1.5	GRAIN MILL MACHINERY			Centrifugal, gear, lobe & vane	1.2	1.4
Brick machines, dry press, & granulator	1.4	1.6	Sifters, purifiers, separators	1.1	1.3	Dredge	1.6	1.8
Mixer, pug mill, & rolls	1.4	1.6	Grinders and hammer mills	1.2	1.4	Pipe line	1.4	1.6
CENTRIFUGES	1.4	1.6	Roller mills	1.3	1.5	Reciprocating		
COMPRESSORS			GENERATORS & EXCITERS	1.2	1.4	3 or more cyl.	1.3	1.5
Centrifugal & rotary (lobe)	1.1	1.3	MACHINE TOOLS			1 or 2 cyl.	1.6	1.8
Reciprocating			Grinders, lathes, drill press	1.0	—	RUBBER & PLASTICS INDUSTRY EQUIPMENT		
1 or 2 cyl.	1.6	1.8	Boring mills, milling machines	1.1	—	Calendars, rolls, tubers		
3 or more	1.3	1.5	MARINE DRIVES	Consult Morse		Tire-building and Banbury Mills	1.5	1.7
CONSTRUCTION EQUIPMENT OR OFF-HIGHWAY VEHICLES			MILLS			Mixers and sheeters	1.6	1.8
Drive line duty, power take-off, accessory drives	Consult Morse		Rotary type:			Extruders	1.5	1.7
			Ball, Pebble, Rod, Tube, Roller	1.5	1.7	SCREENS		
			Dryers, Kilns, & tumbling barrels	1.6	1.8	Conical & revolving	1.2	1.4
CONVEYOR			Metal type:			Rotary, gravel, stone & vibrating	1.5	1.7
Apron, bucket, pan & elevator	1.4	1.6	Draw bench carriage & main drive	1.5	—	STOKERS	1.1	—
Belt (ore, coal, sand, salt)	1.2	1.4	Forming Machines	Consult Morse		TEST STANDS & DYNAMOMETERS	Consult Morse	
Belt—light package, oven	1.0	1.2	MIXERS			TEXTILE INDUSTRY		
Screw & flight (heavy duty)	1.6	1.8	Concrete	1.6	1.8	Spinning frames, twisters, wrappers & reels	1.0	—
CRANES & HOISTS			Liquid & Semi-liquid	1.1	1.3	Batchers, calendars & looms	1.1	—
Main hoist—medium duty	1.2	1.4	OIL INDUSTRY MACHINERY					
Main hoist—heavy duty, skip hoist	1.4	1.6	Compounding Units	1.1	1.3			
			Pipe line pumps	1.4	1.6			
			Slush pumps	1.5	1.7			
			Draw works	1.8	2.0			
			Chillers, Paraffin filter presses, Kilns	1.5	1.7			

Figure 15–29 Chain drive service factors. *Courtesy Morse Industrial.*

(2.6 HP) and the rpm of the drive sprocket (1800 rpm), you can determine the number of teeth of the drive sprocket. The drive sprocket is normally the "Small" sprocket as shown in the left column of the chart in Figure 15–30. Look at Figure 15–30 and find the rpm—Small Sprocket highlighted along the top horizontal column. Move down the 1800 rpm column until you find a design HP equal to or greater than the design HP of 2.6 for the sample problem. Find 2.93 highlighted in Figure 15–30. Now move horizontally to the left

No. 25—¼" pitch   standard single strand roller chain

No. of Teeth Small Spkt.	50	100	300	500	700	900	1200	1500	1800	2100	2500	3000	3500	4000	4500	5000	5500	6000	6500	7000	7500	8000	8500	9000	10000
										Revolutions per Minute—Small Sprocket															
12	0.03	0.06	0.16	0.25	0.34	0.43	0.55	0.68	0.80	0.92	1.07	1.26	1.45	1.57	1.32	1.12	0.97	0.86	0.76	0.68	0.61	0.56	0.51	0.47	0.40
13	0.04	0.06	0.17	0.27	0.37	0.47	0.60	0.74	0.87	1.00	1.17	1.38	1.58	1.77	1.49	1.27	1.10	0.96	0.86	0.77	0.69	0.63	0.57	0.53	0.45
14	0.04	0.07	0.19	0.30	0.40	0.50	0.65	0.80	0.94	1.08	1.27	1.49	1.71	1.93	1.66	1.42	1.23	1.08	0.96	0.86	0.77	0.70	0.64	0.59	0.50
15	0.04	0.07	0.20	0.32	0.43	0.54	0.70	0.86	1.01	1.17	1.36	1.61	1.85	2.08	1.84	1.57	1.36	1.20	1.06	0.95	0.86	0.78	0.71	0.65	0.56
16	0.04	0.08	0.22	0.34	0.47	0.58	0.76	0.92	1.09	1.25	1.46	1.72	1.98	2.23	2.03	1.73	1.50	1.32	1.17	1.05	0.94	0.86	0.78	0.72	0.61
17	0.05	0.08	0.23	0.37	0.50	0.62	0.81	0.99	1.16	1.33	1.56	1.84	2.11	2.38	2.22	1.90	1.64	1.44	1.28	1.14	1.03	0.94	0.86	0.79	0.67
18	0.05	0.09	0.25	0.39	0.53	0.66	0.86	1.05	1.24	1.42	1.66	1.96	2.25	2.53	2.42	2.07	1.79	1.57	1.39	1.25	1.12	1.02	0.93	0.86	0.73
19	0.05	0.09	0.26	0.41	0.56	0.70	0.91	1.11	1.31	1.50	1.76	2.07	2.38	2.69	2.62	2.24	1.94	1.70	1.51	1.35	1.22	1.11	1.01	0.93	0.79
20	0.06	0.10	0.28	0.44	0.59	0.74	0.96	1.75	1.38	1.59	1.86	2.19	2.52	2.84	2.83	2.42	2.10	1.84	1.63	1.46	1.32	1.20	1.09	1.00	0.86
21	0.06	0.11	0.29	0.46	0.62	0.78	1.01	1.24	1.46	1.68	1.96	2.31	2.66	2.99	3.05	2.60	2.26	1.98	1.76	1.57	1.42	1.29	1.17	1.08	0.92
22	0.06	0.11	0.31	0.48	0.66	0.82	1.07	1.30	1.53	1.76	2.06	2.43	2.79	3.15	3.27	2.79	2.42	2.12	1.88	1.69	1.52	1.38	1.26	1.16	0.99
23	0.06	0.12	0.32	0.51	0.69	0.86	1.12	1.37	1.61	1.85	2.16	2.55	2.93	3.30	3.50	2.98	2.59	2.27	2.01	1.80	1.62	1.47	1.35	1.24	1.06
24	0.07	0.13	0.34	0.53	0.72	0.90	1.17	1.43	1.69	1.94	2.27	2.67	3.07	3.46	3.73	3.18	2.76	2.42	2.15	1.92	1.73	1.57	1.44	1.32	1.12
25	0.07	0.13	0.35	0.56	0.75	0.94	1.22	1.50	1.76	2.02	2.37	2.79	3.21	3.61	3.96	3.38	2.93	2.57	2.28	2.04	1.84	1.67	1.53	1.40	1.20
26	0.07	0.14	0.37	0.58	0.79	0.98	1.28	1.56	1.84	2.11	2.47	2.91	3.34	3.77	4.19	3.59	3.11	2.73	2.42	2.17	1.95	1.77	1.62	1.49	1.27
28	0.08	0.15	0.40	0.63	0.85	1.07	1.38	1.69	1.99	2.29	2.68	3.15	3.62	4.09	4.54	4.01	3.47	3.05	2.70	2.42	2.18	1.98	1.81	1.66	1.42
30	0.08	0.16	0.43	0.68	0.92	1.15	1.49	1.82	2.15	2.46	2.88	3.40	3.90	4.40	4.89	4.45	3.85	3.38	3.00	2.68	2.42	2.20	2.01	1.84	1.57
32	0.09	0.17	0.46	0.73	0.98	1.23	1.60	1.95	2.30	2.64	3.09	3.64	4.18	4.72	5.25	4.90	4.25	3.73	3.30	2.96	2.67	2.42	2.21	2.03	1.73
35	0.10	0.19	0.51	0.80	1.08	1.36	1.76	2.15	2.53	2.91	3.41	4.01	4.61	5.20	5.78	5.60	4.86	4.26	3.78	3.38	3.05	2.77	2.53	2.32	1.98
40	0.12	0.22	0.58	0.92	1.25	1.57	2.03	2.48	2.93	3.36	3.93	4.64	5.32	6.00	6.68	6.85	5.93	5.21	4.62	4.13	3.73	3.38	3.09	2.83	2.42
45	0.13	0.25	0.66	1.05	1.42	1.78	2.31	2.82	3.32	3.82	4.47	5.26	6.05	6.82	7.58	8.17	7.08	6.21	5.51	4.93	4.45	4.04	3.69	3.38	2.89
	TYPE A				TYPE B									TYPE C											

TYPE A: Manual or Drip Lubrication (500 fpm max.)
TYPE B: Bath or Disc Lubrication (3500 fpm max.)
TYPE C: Oil Stream Lubrication (Up to max. speed shown).
The limiting RPM for each lubrication type is read from the column to the left of the boundary line shown.
The ratings on this page are in accordance with the standards of the American Chain Association. Copyright 1974.

For Multi. Strand Chain use

No. of Strands	Strand Factor
2	1.7
3	2.5
4	3.3

Figure 15–30 Horsepower ratings for single-strand roller chain. *Courtesy Morse Industrial.*

from 2.93 to find the number of teeth of the small sprocket in the left column. Notice that 40 teeth is highlighted in Figure 15–30.

Step 3. Determine the chain pitch. The *chain pitch* is the center distance between pins of one chain link. (See Figure 15–31.) The number of teeth of the small sprocket is taken from the horsepower rating table for a chain with a specific pitch. If you look back at Figure 15–30, you can see the pitch identified in the heading as "No. 25—1/4" Pitch." Chains are manufactured with standard size pitches available in increments from 1/4–3 in. Chains are also classified by

number; for example, a 1/4-in. pitch is No. 25, a 3/8-in. pitch is No. 35, a 1/2-in. pitch is No. 40, and a 5/8-in. pitch is No. 50.

Step 4. Determine the ratio and the number of teeth for the driven sprocket. The chain drive ratio is the relationship between the drive and driven sprockets. The ratio may be determined by dividing the rpm of the driver by the rpm of the driven sprocket. If you use the sample problem, the driver is 1800 rpm and the driven is 900 rpm. The ratio is 1800 rpm/900 rpm = 2:1. Now, multiply the number of teeth of the small sprocket by the ratio to get the number of teeth for the large sprocket: 40 (teeth small sprocket) × 2 (ratio) = 80 teeth large sprocket.

Step 5. Determine the center distance in chain pitches. As a general rule, the preferred center distance between shafts is between 30 and 50 chain pitches. The maximum recommended spacing between centers is 80 pitches. This is to help insure that there is clearance between the two sprockets and to allow for a minimum of 120° of chain contact around the small sprocket. (See Figure 15–32.) The center dis-

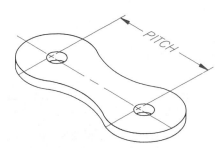

Figure 15–31 Chain pitch.

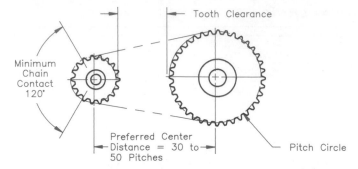

Figure 15–32 The center distance between shafts and a 120° minimum recommended chain contact.

D	K	D	K	D	K	D	K
32	25.94	63	100.54	94	223.82	125	395.79
33	27.58	64	103.75	95	228.61	126	402.14
34	29.28	65	107.02	96	233.44	127	408.55
35	31.03	66	110.34	97	238.33	128	415.01
36	32.83	67	113.71	98	243.27	129	421.52
37	34.68	68	117.13	99	248.26	130	428.08
38	36.58	69	120.60	100	253.30	131	434.69
39	38.53	70	124.12	101	258.39	132	441.36
40	40.53	71	127.69	102	263.54	133	448.07
41	42.58	72	131.31	103	268.73	134	454.83
42	44.68	73	134.99	104	273.97	135	461.64
43	46.84	74	138.71	105	279.27	136	468.51
44	49.04	75	142.48	106	284.67	137	475.42
45	51.29	76	146.31	107	290.01	138	482.39
46	53.60	77	150.18	108	295.45	139	489.41
47	55.95	78	154.11	109	300.95	140	496.47
48	58.36	79	158.09	110	306.50	141	503.59
49	60.82	80	162.11	111	312.09	142	510.76
50	63.33	81	166.19	112	317.74	143	517.98
51	65.88	82	170.32	113	323.44	144	525.25
52	68.49	83	174.50	114	329.19	145	532.57
53	71.15	84	178.73	115	334.99	146	539.94
54	73.86	85	183.01	116	340.84	147	547.36
55	76.62	86	187.34	117	346.75	148	554.83
56	79.44	87	191.73	118	352.70	149	562.36
57	82.30	88	196.16	119	358.70	150	569.93
58	85.21	89	200.64	120	364.76	151	577.56
59	88.17	90	205.18	121	370.86	152	585.23
60	91.19	91	209.76	122	377.02	153	592.96
61	94.25	92	214.40	123	383.22	154	600.73
62	97.37	93	219.08	124	389.48	155	608.56

Figure 15–33 Sprocket teeth factors "K." *Courtesy Morse Industrial.*

tance between shafts should be no less than the difference between sprocket diameters for ratios greater than 3:1. To determine the center distance in chain pitches, divide the center distance by the chain pitch. For the sample problem, this is:

$$\frac{12 \text{ in. (Center Distance)}}{0.25 \text{ in. (Pitch)}} = 48 \text{ Chain Pitches}$$

Step 6. Determine the chain length. First, determine the chain length in pitches using the formula:

$$2C + \frac{S}{2} + \frac{K}{C}, \text{ where:}$$

C = Center distances between sprockets in chain pitches.

S = Number of teeth in small sprocket plus number of teeth in large sprocket.

K = Number of teeth in large sprocket minus number of teeth in small sprocket equals D. Look at the table in Figure 15–33 to find the value K corresponding to D.

For the same problem:

C = 48 (from Step 5)
S = 40 + 80 = 120
K = 80 – 40 = 40
D = 40

Look at Figure 15–33 to find K = 40.53.

$$2C + \frac{S}{2} + \frac{K}{C} = 2(48) + \frac{120}{2} + \frac{40.53}{48} =$$

$$96 + 60 + .84 = 156.84$$

The chain length in pitches must be the next higher whole number, because you cannot have a partial chain pitch. So, 156.84 = 157 chain length in pitches. Whenever possible, use an even number of chain pitches. Using an odd number of chain pitches requires an offset link, which is not preferred. For this prob-

lem, therefore, use 158 chain length in pitches rather than 157. Now, multiply the chain length in pitches by the chain pitch to find the chain length in inches:

$$158 \times .25 = 39.5 \text{ inches of chain length}$$

Step 7. Determine chain lubrication. Correct lubrication is important in maintaining long life for chain drives. Lubrication methods are governed by the speed and function of the machine. Periodic manual lubrication works well for slow-speed chain drives. To do this, use a brush to apply a medium-consistency mineral oil while a machine is stopped. For moderate-speed chain drives, a drip system is often designed to keep the chain lubricated. For high-speed chain drives, an oil bath or an oil spray is often provided. To determine the type of oil application recommended for the sample problem, refer back to the chart in Figure 15–30. Notice the notes Type A, Type B, and Type C at the bottom of the chart. These refer to the recommended lubrication as follows:

Type A = Manual or drip lubrication.
Type B = Bath or disc lubrication.
Type C = Oil stream lubrication.

Notice the HP rating of 2.93 for the sample problem falls in the Type B lubrication category.

PROFESSIONAL PERSPECTIVE

This chapter covers the use of vendors' catalogs to find specific belt and chain drives for a given application. As a drafter in any engineering field, you will find that using vendors' catalogs and specifications is a very important part of your job. Most companies have a complete library of catalogs of purchase parts. Purchase parts, often referred to as standard parts, are products that can be purchased already made. Common purchase parts are bolts, nuts, belts, chains, and sprockets. You should become familiar with the types of purchase parts your company uses and how to find them in the vendors' catalogs. Also determine the suppliers that have the best prices, highest quality, and fastest availability.

MATH APPLICATION

BELT LENGTH

Problem: Find the length of a driving belt running around two pulleys of radii 14" and 8" if the distance between the centers of the two pulleys is 25", as shown in Figure 15–34.

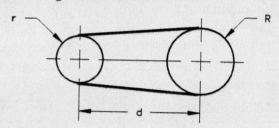

Figure 15–34   Drive belt around pulleys.

Solution: The formula for belt length between two pulleys is

$$L = 2(d)(\cos \Theta) + \pi(R + r) + 2\Theta(R - r)$$

$$\text{where } \Theta = \text{Inv sin}\left(\frac{R - r}{d}\right)$$

"Inv sin" stands for "inverse sine" and means "angle whose sine is." For example, Inv sin .7 is equal to 44.4° (or .775 radians) because sin 44.4° = .7. It is important to note that this formula requires angle Θ to be in *radian* measure. The first step in this rather complicated formula to find L is to determine Θ. It is easiest to put your calculator in the "radian mode" to work this problem:

$$\Theta = \text{Inv sin}\left(\frac{14 - 8}{25}\right) = \text{Inv sin } .24 = .2424 \text{ radians}$$

Then substituting into the formula for L:

$$L = 2(25)(\cos .2424) + \pi(14 + 8) + 2(.2424)(14 - 8)$$
$$L = 2(25)(.9708) + \pi(22) + 2(.2424)(6)$$
$$L = 48.54 + 69.12 + 2.91 = 174.03 \text{ or } \mathbf{174"}$$

15 BELT AND CHAIN DRIVES TEST

DIRECTIONS

Answer the questions with short complete statements or drawings as needed on an A-size drawing sheet as follows:

QUESTIONS

1. List at least five advantages of gear drives.
2. List at least five advantages of belt drives.
3. What would be three disadvantages of belt drives?
4. List at least five advantages of chain drives.
5. Most V belts function most effectively at what speed and temperature?
6. Are the terms *pulley* and *sheave* used interchangeably?
7. Define *pitch line*.
8. What is the function of an idler in belt drive design?
9. Name the three main categories of V belts.
10. Why are adjustable motor bases used in belt drives?
11. Describe the typical use for flat belts.
12. Describe positive drive belts.
13. List at least five factors that are involved in the selection of a correct belt drive.

14. What is the ratio of the driver pulley to the driven pulley if the driver operates at 2400 rpm and the driven operates at 600 rpm?
15. What is the belt velocity for the driver described in Question 14 if the pitch diameter is 4.5 in.?
16. What is the most commonly used power chain in industry?
17. Define *sprocket*.
18. The preferred position of chain drive shafts is in a horizontal line or on a line inclined not more than how many degrees?
19. List at least four of the items that must be determined when designing roller chain drives.
20. If a small sprocket has 12 teeth and travels at 1600 rpm, and the large sprocket travels at 400 rpm, how many teeth does the large sprocket have?
21. What is the center distance between shafts in chain pitches if the center distance measures 16 in. and a No. 25 chain is used?
22. Why is correct lubrication of chain drives important?
23. Identify at least three methods of lubricating chain drives.
24. When establishing chain length in chain pitches, why is it recommended that the chain length in chain pitches be rounded up to an even number of chain pitches?

15 BELT AND CHAIN DRIVES PROBLEMS

PART 1

DIRECTIONS

Solve Problems 15–1 through 15–7 using the information and tables given in this chapter and in Figure 15–35. All motors are AC motors with normal torque, squirrel cage, synchronous, and split phase, or shunt wound DC motors, or multiple cylinder internal combustion engines. Include the following information with each solution:

○ Service factor.
○ Design horsepower.
○ Driven speed.
○ Pitch diameter driver and driven.
○ Belt section.
○ Number of belts.
○ Belt number and center distance.
○ Corrected horsepower.
○ Belt drive ratio.
○ Belt velocity of driver sheave.

Problem 15–1 A 3-HP, 1750-rpm motor is to operate a furnace blower having a shaft speed of approximately 1115 rpm under normal service. The center distance between the motor and blower shafts is about 16 in.

Problem 15–2 A 3-HP, 1750-rpm motor is used to operate a drill press speed reducer under intermittent service. The spindle speed is about 1136 rpm. The center distance between the motor and spindle shafts is about 20.5 in.

1.50	12.0	18.0	A	1	10	—	—	—	1167	10.57	13.63	A77	15.3	0.96	A81	17.3	0.98	A85	19.4	0.99
1.50	12.0	18.0	C	2	12	—	—	—	1167	25.92	31.20	C75	15.1	0.82	C96	25.7	0.88	C112	33.8	0.92
1.50	13.0	19.5	A	1	3	—	—	—	1167	11.07	14.40	A83	16.3	0.98	A86	17.8	0.99	A89	19.4	0.99
1.51	3.7	5.6	A	1	3	2318	4.61	5.62	1159	2.94	3.71	A24	5.3	0.76	A35	10.8	0.84	A45	15.8	0.88
1.51	3.9	5.9	B	1	3	2318	4.64	8.35	1159	3.53	5.53	B28	7.1	0.73	B45	15.7	0.84	B56	21.2	0.89
1.53	3.4	5.2	B	1	6	2288	2.90	6.59	1144	2.43	4.44	B28	8.1	0.73	B45	16.6	0.84	B56	22.1	0.89
1.53	3.6	5.5	A	1	3	2288	4.42	5.37	1144	2.81	3.55	A24	5.4	0.76	A35	11.0	0.84	A45	16.0	0.88
1.53	3.6	5.5	3V	1	4	2288	6.28	7.30	1144	3.63	4.07	3V250	5.3	0.78	3V300	7.8	0.83	3V355	10.6	0.85
1.53	3.8	5.8	A	1	6	2288	4.81	5.87	1144	3.06	3.86	A24	5.0	0.76	A35	10.6	0.84	A45	15.6	0.88
1.53	3.8	5.8	B	1	6	2288	4.31	8.01	1144	3.31	5.32	B28	7.3	0.73	B45	15.8	0.84	B56	21.3	0.89
1.54	2.4	3.7	A	1	1	2273	1.85	2.12	1136	1.27	1.61	A24	7.8	0.79	A35	13.3	0.86	A45	18.3	0.90
1.54	2.6	4.0	A	1	2	2273	2.31	2.70	1136	1.54	1.94	A24	7.4	0.79	A35	12.9	0.84	A45	18.0	0.90
1.54	3.5	5.4	A	1	3	2273	4.22	5.12	1136	2.69	3.40	A24	5.6	0.76	A35	11.1	0.84	A45	16.1	0.88
1.54	3.7	5.7	A	1	3	2273	4.62	5.62	1136	2.94	3.71	A24	5.2	0.76	A35	10.7	0.84	A45	15.7	0.88
1.54	3.9	6.0	B	1	3	2273	4.66	8.36	1136	3.53	5.54	B28	7.0	0.73	B45	15.6	0.84	B56	21.1	0.89

Figure 15–35 Nominal center distances, number of belts, arc length factor chart.

Problem 15–3 A 1 1/2-HP, 1750-rpm electric motor is used to operate a woodworking band saw with the blade turning at 1144 rpm, intermittent service. The center-to-center distance is about 16 in.

Problem 15–4 A 2-HP electric motor with a shaft speed of 1750 rpm operates a printing machine at normal service. The shaft on the printing machine is to operate at 1167 rpm. The center-to-center distance is about 18 in.

Problem 15–5 A 2-HP electric motor with a shaft speed of 1750 rpm operates a punch machine at continuous service. The shaft on the punch machine is to operate at 1108 rpm. The center-to-center distance is about 17 in.

Problem 15–6 A 1.5-HP motor with a shaft speed of 1750 rpm operates a compressor at normal service. The shaft on the compressor is to operate at 1167 rpm. The center-to-center distance is about 18 in.

Problem 15–7 A 2-HP electric motor with a shaft speed of 1750 rpm operates a printing machine at normal service. The shaft on the printing machine is to operate at 1115 rpm. The center-to-center distance is about 17.5 in.

Problem 15–8 Given the following layout, prepare a detailed drawing of the pulley on appropriately sized vellum or using the CADD system, unless otherwise specified by your instructor.

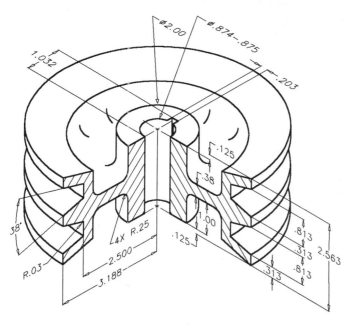

DIRECTIONS

Solve Problems 15–9 through 15–15 using the information and tables given in this chapter. Include the following information with each solution:

- Service factor.
- Design horsepower.
- Number of teeth drive sprocket and driven sprocket.
- Chain pitch.
- Ratio.
- Center distance in chain pitches.
- Chain length.
- Lubrication.

Problem 15–9 A chain drive is required for a 2-HP electric motor operating a paper machine. The drive sprocket turns 1200 rpm and the driven sprocket turns 400 rpm. The shaft center distance is about 22 in.

Problem 15–10 A chain drive is required for a 1/10-HP electric motor operating a speed reducer for a belt light package conveyor. The drive sprocket turns 50 rpm and the driven sprocket turns 17 rpm. The shaft center distance is about 40 in.

Problem 15–11 A chain drive is required for a 1-HP electric motor operating a centrifuge. The drive sprocket turns 6000 rpm and the driven sprocket turns 4000 rpm. The shaft center distance is about 18 in.

Problem 15–12 A chain drive is required for a 2-HP electric motor operating a centrifugal pump. The drive sprocket turns 2500 rpm and the driven sprocket turns 1562.5 rpm. The shaft center distance is about 28 in.

Problem 15–13 A chain drive is required for a 5-HP electric motor operating a flour grinder. The drive sprocket turns 3500 rpm and the driven sprocket turns 1500 rpm. The shaft center distance is about 18 in.

Problem 15–14 A chain drive is required for a 2.5-HP internal combustion tractor engine operating a generator. The drive sprocket turns 3000 rpm and the driven sprocket turns 2000 rpm. The shaft center distance is about 24 in.

Problem 15–15 A chain drive is required for a 2-HP electric motor operating a pure liquid agitator. The drive sprocket turns 2100 rpm and the driven sprocket turns 700 rpm. The shaft center distance is about 30 in.

PART 2

DIRECTIONS

Given the following engineering layouts, prepare detailed drawings of the sprockets. These problems may be done using manual drafting; however, the problems may be completed much faster if computer-aided drafting is available. Problems courtesy of Production Plastics, Inc.

NOTE: Dimensions given on engineering problems may not meet ASME standards. Convert all dimensions to comply with ASME Y14.5M—1994 as specified in Chapters 10 and 13 of this text.

UNSPECIFIED TOLERANCES:

DECIMALS	mm	IN.
X	±2.5	±.1
XX	±0.25	±.01
XXX	±0.127	±.005
ANGULAR ±30'		
FINISH	3.2μm	125μIN.

Problem 15–16   Chain saver

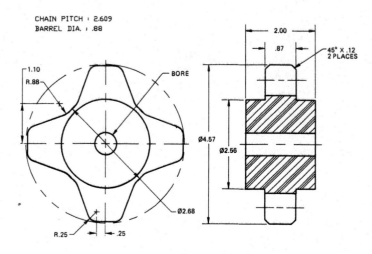

CHAIN PITCH : 2.609
BARREL DIA. : .88

Problem 15–17   FWB SD 18 sprocket

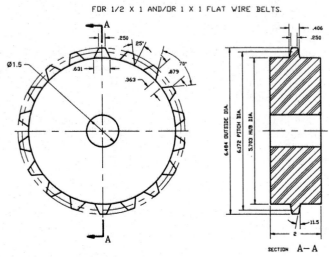

FOR 1/2 X 1 AND/OR 1 X 1 FLAT WIRE BELTS.

UNLESS OTHERWISE SPECIFIED:
ALL FRACTIONAL DIMENSIONS ±1/32
ALL TWO-PLACE DECIMALS ±.010
ALL THREE-PLACE DECIMALS ±.005

Problem 15–18   Chain saver sprocket

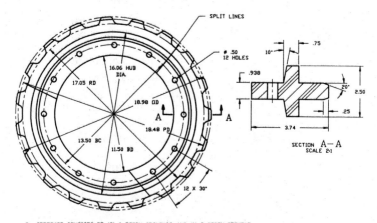

2. SPROCKET CONSISTS OF (2) 6 TOOTH SEGMENTS, AND (1) 7-TOOTH SEGMENT.
1. TOLERANCES TO BE +.003/-.000 ON BD, ALL OTHERS ±.005.
NOTES:

PART 3

DIRECTIONS

Draw the following chain chart drawings. These are excellent computer-aided drafting problems, if available. *Courtesy HHK Chain Corporation.*

Problem 15–19   Heavy series roller chain

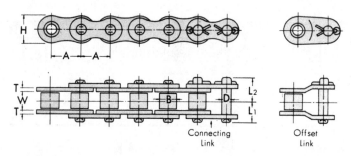

● HEAVY SERIES ROLLER CHAIN

DIMENSIONS—INCHES

| HKK Chain No. | Pitch | Roller | | Pin | Plate | | Overall Dimension | | Average Ultimate Strength (lbs.) | Average Weight Per Foot (lbs.) |
| | | Dia. | Width | Dia. | Height | Thickness | | | | |
	A	B	W	D	H	T	L₁	L₂		
HKK 60H	3/4	0.469	1/2	0.234	0.689	0.125	0.565	0.644	11,300	1.17
HKK 80H	1	0.625	5/8	0.312	0.921	0.156	0.699	0.817	20,000	2.05
HKK 100H	1-1/4	0.750	3/4	0.375	1.154	0.187	0.831	0.969	28,900	3.01

Problem 15–20   Conveyor-type standard roller chain

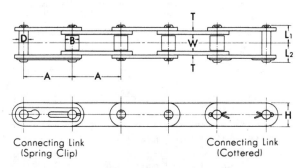

Standard Roller Type

DIMENSIONS—INCHES

| HKK Chain No. (ANSI No.) | Pitch | Roller | | Pin | Plate | | Overall Dimension | | Average Ultimate Strength (lbs.) | Average Weight Per Foot (lbs.) | |
| | | Dia. | Width | Dia. | Height | Thickness | | | | Riveted | Cottered |
	A	B	W	D	H	T	L₁	L₂			
STANDARD ROLLER TYPE											
C 2040	1	0.312	5/16	0.156	0.450	0.060	0.323	0.378	4,300	0.32	—
C 2050	1-1/4	0.400	3/8	0.200	0.591	0.080	0.400	0.467	7,200	0.55	—

Problem 15–21 Offset sidebar chain

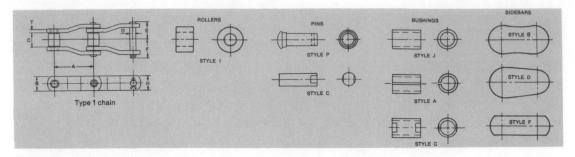

Type 1 chain

(Units in Inches or Pounds)

Chain number	Chain type	Ave. pitch	Roller			Pin				Sidebar			Bush	Average ultimate strength	Average working load	No. of links/ approx. 10 ft.	Weight /ft.
			dia.	width	Style	dia.	to rivet	to end	Style	height	thick.	Style	Style				
		A	B	C		D	E	F		H	T						
ED 501	1	1.500	.813	.906	1	.313	.9	1.0	C	1.000	.156	F	C	10.000	1.350	80	2.6
ED 1534	1	1.500	.875	1.000	1	.438	1.0	1.2	"	1.375	.188	8	A	34.000	2.100	80	4.2
ED 620	1	1.654	.875	1.000	1	.375	.9	1.0	"	1.125	.125	A	C	8.000	1.350	73	2.9
ED 622	1	1.654	.875	1.000	1	.438	1.1	1.2	"	1.125	1.88	"	A	20.000	2.100	73	3.1

Problem 15–22 Sticker chain for the forest industries

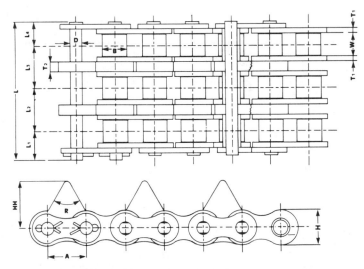

HKK No.	Pitch	Roller		Pin	Plate		Overall Dimension					Sticker Height	Sticker Thickness	Sticker Angle	Average Ultimate Strength (LBS.)	Average Weight per ft. (LBS.)
		Dia.	Width	Dia.	Height	Thickness										
	A	B	W	D	H	T₁	L	L₁	L₂	L₃	L₄	HH	T₂	R		
108110	1.000	0.625	0.625	0.312	0.914	0.125	3.678	0.745	1.153	1.153	0.627	1.250	0.125	60°	53,000	5.54
108139	1.000	0.625	0.625	0.312	0.914	0.125	3.678	0.745	1.153	1.153	0.627	1.250	0.250	60°	53,100	5.54
110131	1.250	0.750	0.750	0.375	1.154	1.585	4.503	0.913	1.409	1.409	0.772	1.535	0.312	80°	78,600	8.05

MATH PROBLEMS

Find the belt length for these pulley radii and center-to-center distances:

1. r = 8", R = 10", d = 27"
2. r = 12", R = 12", d = 100"
3. r = 5", R = 30", d = 40"
4. r = 7", R = 15", d = 30"
5. r = 4", R = 8", d = 35"
6. r = 101.6 mm, R = 203.2 mm, d = 889 mm

CHAPTER 16

Working Drawings

LEARNING OBJECTIVES

After completing this chapter, you will:

❍ Draw complete sets of working drawings, including details, assemblies, and parts lists.
❍ Prepare written specifications of purchase parts for the parts list.
❍ Properly group information on the assembly drawing with identification numbering systems.
❍ Explain the engineering change process and prepare engineering changes.

THE ENGINEERING DESIGN PROCESS

The engineering department has been asked to develop a new product line consisting of an adjustment knob replacement kit. As the engineering drafter, you have been supplied with appropriate engineering sketches and asked to develop a complete drawing package.

In the new product development process, several departments within your company will need specific information. The manufacturing department will need the dimensional data as well as material and finish specifications. Since one of the items in the kit is a purchase part, the purchasing department must be provided with the necessary information. Packaging information will be required to get the product ready for shipping, and the sales department will need some presentation drawings to show potential clients. Additionally, customer assembly drawings are to be included in an instruction sheet to be supplied with the kit.

It is up to you to develop a complete set of working drawings. Figures 16–1 through 16–6 show all of the drawings and information used to complete this project.

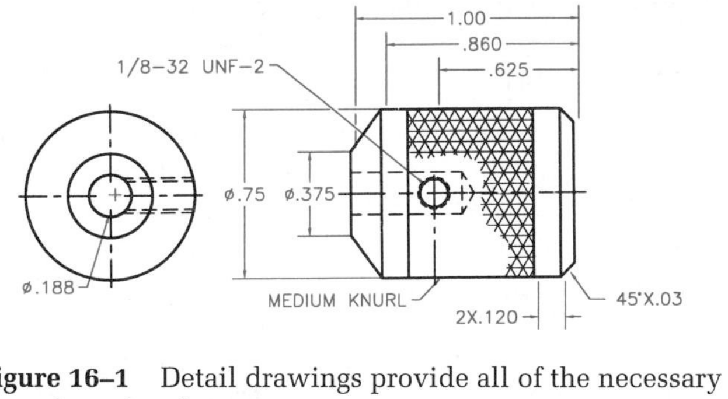

Figure 16–1 Detail drawings provide all of the necessary manufacturing data.

B	ADDED KNURL	3/7	RJ
A	⌀.375 WAS .438	2/9	RJ
REV	DESCRIPTION	DATE	BY

Figure 16–2 Engineering changes must be recorded.

2	1	RAK–002	1/8–32UNF SET SCREW	ACME PART#15–125–32A
1	1	RAK–001	ADJUSTMENT KNOB	A360
ITEM	QTY	PART #	DESCRIPTION	MATERIAL

Figure 16–3 Purchasing information is included in the bill of materials or parts list.

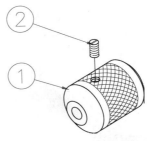

Figure 16–4 Isometric assembly drawings show pre-assembly data and packaging information.

(Continued)

ENGINEERING DESIGN PROCESS *(Continued)*

Figure 16–5 Pictorial illustrations can be a valuable sales tool.

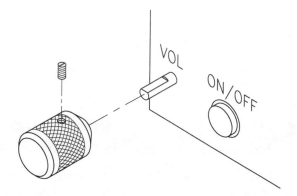

Figure 16–6 Technical illustrations are commonly used to provide customers with instructions on the assembly and use of a product.

The drawings that have been shown as examples or assigned as problems in this text are called *detail drawings.* Detail drawings are the kind of drawings that most entry-level mechanical drafters prepare. When a product is designed and drawings are made for manufacturing, each part of the product must have a drawing. These drawings of individual parts are referred to as detail drawings. Component parts are assembled to create a final product, and the drawing that shows how the parts go together is called the assembly drawing. Associated with the assembly drawing and coordinated to the detail drawings is the parts list. When the detail drawings, assembly drawing, and parts list are combined, they are referred to as a complete set of working drawings. Working drawings, then, are a set of drawings that supply all the information necessary to manufacture any given product. A set of working drawings includes all the information and instructions needed for the purchase or construction of parts and the assembly of those parts into a product.

▼ DETAIL DRAWINGS

Detail drawings, used by workers in manufacturing, are drawings of each part contained in the assembly of a product. The only parts that may not have to be drawn

1/2 - 13UNC-2 X 1.5 LG, SOCKET HEAD CAP SCREW

Figure 16–7 Written description of standard or purchase part.

are standard parts. *Standard parts* are items that can be purchased from an outside supplier more economically than they can be manufactured. Examples of standard, or purchased, parts are common bolts, screws, pins, keys, and any other product that can be purchased from a vendor. Standard parts do not have to be drawn because a written description will clearly identify the part as shown in Figure 16–7. Detail drawings contain some or all of the following items:

1. Necessary multiviews.
2. Dimensional information.
3. Identity of the part, project name, and part number.
4. General notes and specific manufacturing information.
5. The material of which the part is made.
6. The assembly that the part fits (could be keyed to the part number).
7. Number of parts required per assembly.
8. Name(s) of person(s) who worked on or with the drawing.
9. Engineering changes and related information.

In general, detail drawings have information that is classified into three groups:

1. Shape description, which shows or describes the shape of the part.
2. Size description, which shows the size and location of features on the part.
3. Specifications regarding items such as material, finish, and heat treatment.

Figure 16–8 shows an example of a detail drawing.

Sheet Layout

Detail drawings may be prepared with one part per sheet, referred to as a *monodetail* drawing, as in Figure 16–8, or with several parts grouped on one sheet, which is called a *multidetail* drawing. The method of presentation depends on the choice of the individual company. The drawing assignments in this text have been done with one detail drawing per sheet, which is common industry practice. Some companies, however, draw many details per sheet. The advantage of one detail per sheet is that each part stands alone so the drawing of the part can be distributed to manufacturing without several other parts attached. Drawing sheet sizes vary depending on the part size, scale used, and information presented. This procedure requires that drawings be filed with numbers that allow the parts to be located in relation to the assembly. The advantage of drawing several details per sheet is one of economics. The drafter may be able to draw several parts on one sheet depend-

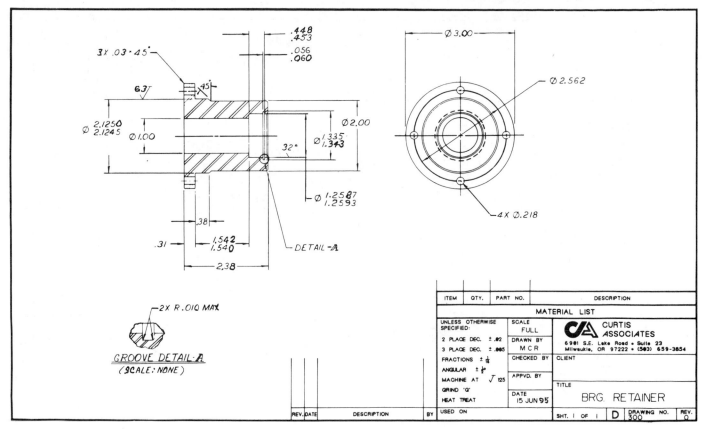

Figure 16–8   Monodetail drawing (one part per sheet). *Courtesy Curtis Associates.*

ing on the size of the parts, the scale used, and the information associated with the parts. If there are six parts on one sheet, then the number of title blocks that must be completed is reduced by five. The company may use one standard sheet size and encourage drafters to place as many parts as possible on one sheet. When this practice is used, there may be a group of sheets with parts detailed for one assembly. The sheet numbers correlate the sheets to the assembly, and each part is keyed to the assembly. The sheets are given page numbers identifying the page number and total number of pages in the set. For example, if there are three pages in a set, then the first page is identified as 1 of 3, the second as 2 of 3, and the third as 3 of 3. Figure 16–9 shows an example of a *multidetail* drawing with several detail drawings on one sheet. Some companies may use both methods at different times depending on the purpose of the drawings and the type of product. For example, it is more common for the parts of a weldment to be drawn grouped on sheets as opposed to one per sheet because the parts may be fabricated at one location in the shop.

Steps in Making a Detail Drawing

Review Chapter 5, covering the layout of multiviews, and Chapter 10, which explains in detail how to lay out a dimensioned drawing. Review Chapter 12 for sectioning techniques.

Detail Drawing Manufacturing Information

Detail drawings may be drawn to suit the needs of the manufacturing processes. A detail drawing may have all of the information necessary to completely manufacture the part—for example, casting and machining information on one drawing. In some situations a completely dimensioned machining drawing may be sent to the pattern or die maker. The pattern or die is then made to allow for extra material where machined surfaces are specified. When company standards require, two detail drawings may be prepared for each part. One detail gives views and dimensions that are necessary only for the casting or forging process. Another detail is drawn that does not give the previous casting or forging dimensions but provides only the dimensions needed to perform the machining operations on the part. Examples of these drawings are given in Chapter 10.

▼ ASSEMBLY DRAWINGS

Most products are composed of several parts. A drawing showing how all of the parts fit together is called an *assembly drawing.* Assembly drawings may differ in the amount of information provided, and this decision often depends on the nature or complexity of the product. Assembly drawings are generally multiview drawings. The drafter's goal in the preparation of assembly

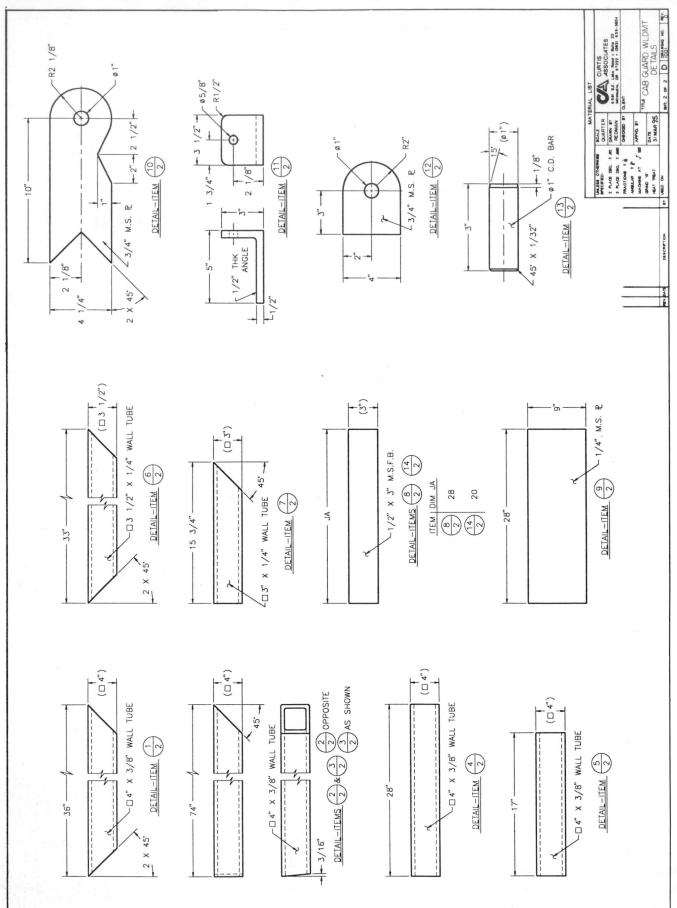

Figure 16-9 Multidetail drawing (several detail drawings on one sheet). *Courtesy Curtis Associates.*

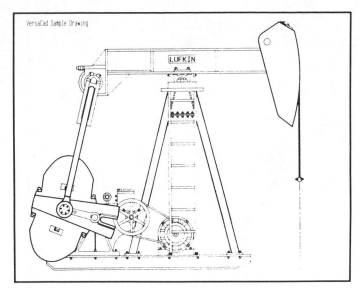

Figure 16–10 CADD drawing of assembly with single front view. *Courtesy T&W Systems, Inc.*

drawings is to use as few views as possible to completely describe how each part goes together. In many cases a single front view does the job. (See Figure 16–10.) Full sections are commonly associated with assembly drawings because the full section may expose the assembly of most or all of the internal features, as shown in Figure 16–11. If one section or view is not enough to show how the parts fit together, then a number of views or sections may be necessary. In some situations a front view or group of views with broken-out sections is the best method of showing the external features while exposing some of the internal features. (See Figure 16–12.) The drafter must make the assembly drawing clear enough for the assembly department to put the product together. Other elements of assembly drawings that make them different from detail drawings is that they usually contain few or no hidden lines or dimensions. Hidden lines should be avoided on assembly drawings unless absolutely necessary for clarity. The common practice is to draw an exterior view to clarify outside features and a sectional view to expose interior features. Dimensions serve no purpose on an assembly drawing unless the dimensions are used to show the assembly relationship of one part to another. These *assembly dimensions* are only necessary when a certain distance between parts must exist before proper assembly can take place. Machining processes and other specifications are generally not given on an assembly drawing unless a machining operation must take place after two or more parts are assembled. Other assembly notes may include bolt tightening specifications, assembly welds, or cleaning, painting, or decal placement that must take place after assembly. Figure 16–13 shows a process

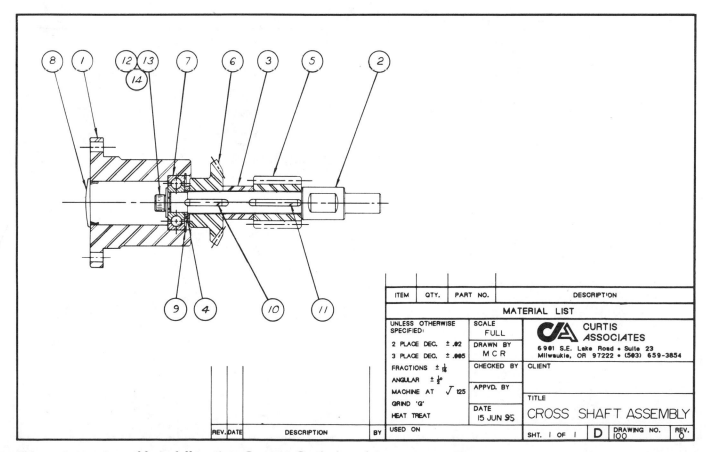

Figure 16–11 Assembly in full section. *Courtesy Curtis Associates.*

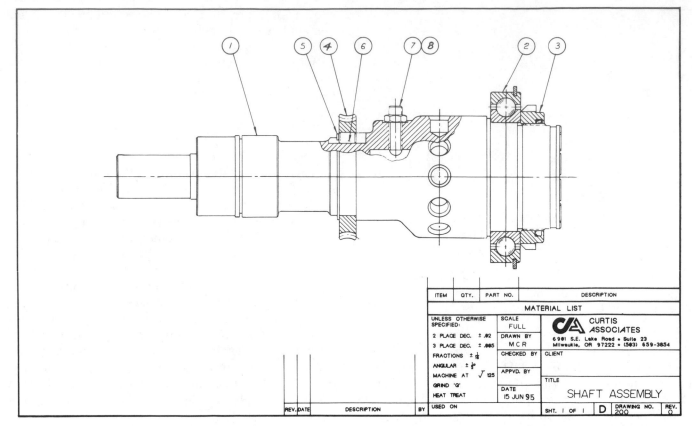

Figure 16–12 Assembly with broken-out sections. *Courtesy Curtis Associates.*

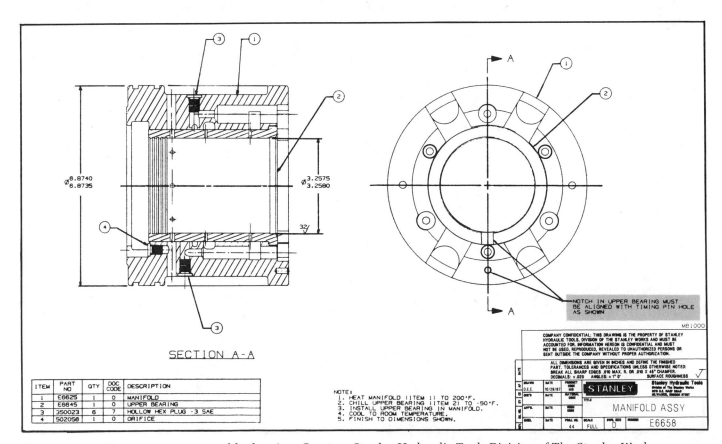

Figure 16–13 Process note on assembly drawing. *Courtesy Stanley Hydraulic Tools Division of The Stanley Works.*

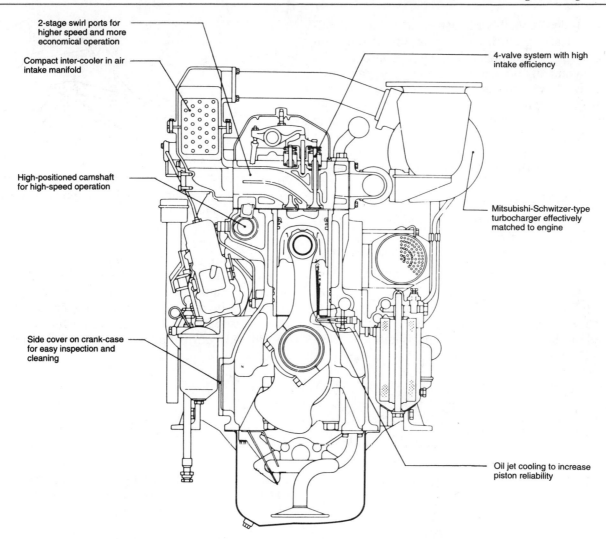

2-stage swirl ports for
higher speed and more
economical operation

Compact inter-cooler in air
intake manifold

High-positioned camshaft
for high-speed operation

Side cover on crank-case
for easy inspection and
cleaning

4-valve system with high
intake efficiency

Mitsubishi-Schwitzer-type
turbocharger effectively
matched to engine

Oil jet cooling to increase
piston reliability

Figure 16–14 Full-section assembly with section lines omitted. *Courtesy Mitsubishi Heavy Industries America, Inc.*

note applied to an assembly drawing. Some company standards or drawing presentations prefer that assemblies be drawn with sectioned parts shown without section lines. Figure 16–14 shows a full-section assembly with parts left unsectioned.

Assembly drawings may contain some or all of the following information:

1. One or more views.
2. Sections necessary to show internal features, function, and assembly.
3. Any enlarged views necesssary to show adequate detail.
4. Arrangement of parts.
5. Overall size and specific dimensions necessary for assembly.
6. Manufacturing processes necessary for or during assembly.
7. Reference or item numbers that key the assembly to a parts list and to the details.
8. Parts list or bill of materials.

▼ TYPES OF ASSEMBLY DRAWINGS

There are several different types of assembly drawings used in industry:

1. Layout, or design, assembly.
2. General assembly.
3. Working-drawing or detail assembly.
4. Erection assembly.
5. Subassembly.
6. Pictorial assembly.

Layout Assembly

Engineers and designers may prepare a design layout in the form of a sketch or as an informal drawing. These engineering design drawings are used to establish the relationship of parts in a product assembly. From the layout the engineer prepares sketches or informal detail drawings for prototype construction. This research and development (R & D) is the first step in the process of taking a design from an idea to a manufactured product.

Layout, or design, assemblies may take any form depending on the drafting ability of the engineer, the time frame for product implementation, the complexity of the product, or company procedures. In many companies the engineers work with drafters who help prepare formal drawings from engineering sketches or informal drawings. The R & D department is one of the most exciting places for a drafter to be. Figure 16–15 shows a simple layout assembly of a product in the development stage. The limits of operation are shown in phantom lines.

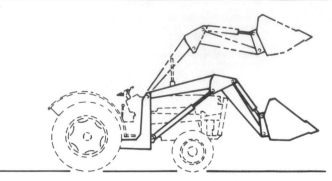

Figure 16–15 Layout assembly. The limits of operation are shown in phantom lines.

General Assembly

General assemblies are the most common types of assemblies that are used in a complete set of working drawings. A set of working drawings contains three parts: detail drawings, an assembly drawing, and a parts list.

Working-drawing, or Detail, Assembly

When a drawing is created where details of parts are combined on the same sheet with an assembly of those parts, a detail assembly is the result. While this practice is not as common as general assemblies, it is a practice at some companies. The use of working-drawing assemblies may be a company standard, or this technique may be used in a specific situation even when it is not considered a normal procedure at a particular company.

CADD Applications

WORKING DRAWINGS

Using CADD to prepare working drawings may increase your productivity while you are drawing details. Some drafters indicate a 1:1 productivity ratio while others boast up to a 10:1 ratio for CADD versus manual drafting. One engineer explained that it normally takes a drafter 100 hours of training and daily use of the CADD system before he or she is on a 1:1 level of productivity with the manual drafter. After the initial 100 hours, the productivity of the CADD drafter increases. This all depends on some contributing factors such as:

❍ The complexity of the part.
❍ The experience of the drafter with the CADD system.
❍ The ability of the individual or group to customize the CADD system for specific applications.
❍ The number of common symbols, notes, or details that can be used on a variety of drawings.

After all of the detail drawings are made, the drafter can easily and quickly prepare the assembly drawing. It used to take the manual drafter hours or even days to complete the assembly drawing, depending on the complexity of the product. With CADD even the most complex drawings can be done in a short period of time. The assembly should be drawn after the detail drawings of the individual parts are complete because the drafter can use the details to complete the assembly. To do this, follow these helpful techniques.

❍ Do all detail drawings first.
❍ Have the views on a layer separate from the dimensions.
❍ Make the principal view of each detail drawing a BLOCK or SYMBOL that can be sent between drawings.
❍ Start the assembly drawing by bringing the main detail view into the layout and position it where the other detail views can be conveniently added.
❍ Bring each detail view into position to begin building the assembly. Use a command such as SCALE to increase or reduce the size of each detail view to fit the scale of the assembly.
❍ After all of the detail views are assembled, you can use the editing functions to erase, trim, or otherwise clean up the assembly.
❍ Add balloons, notes, and dimensions (if any).
❍ Add the parts list. It is best to have a standard parts list format that can be called up and added to any drawing. Then all you have to do is add the text and the drawing is done.

This method saves a lot of time because you do not have to start over again drawing each part in the assembly.

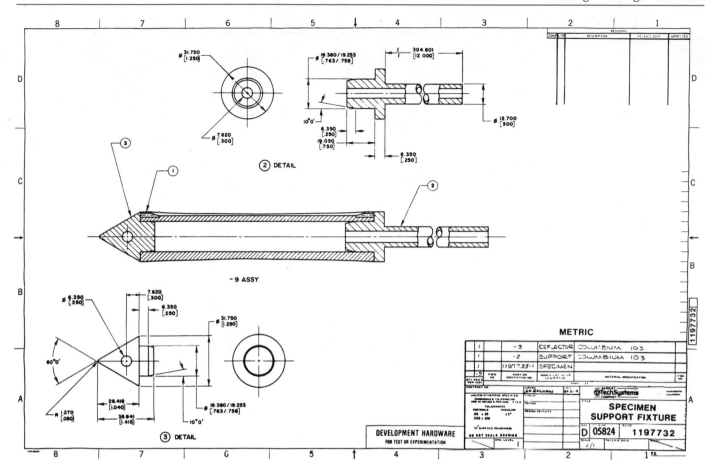

Figure 16–16   Working-drawing, or detail, assembly. *Courtesy Aerojet Propulsion Division.*

The detail assembly may be used when the end result dictates that the details and assembly be combined on as few sheets as possible. An example may be a product with few parts that is produced only once for a specific purpose. (See Figure 16–16.)

Erection Assembly

Erection assemblies usually differ from general assemblies in that dimensions and fabrication specifications are commonly included. Typically associated with products that are made of structural steel, or cabinetry, erection assemblies are used for both fabrication and assembly. Figure 16–17 shows a CADD-produced erection assembly with multiviews, fabrication dimensions, and an isometric drawing that also helps display how the parts fit together.

Subassembly

The complete assembly of a product may be made up of several component assemblies. These individual unit assemblies are called *subassemblies*. A complete set of working drawings may have several subassemblies, each with its own detail drawings, and the general assembly. The general assembly of an automobile, for

example, includes the subassemblies of the drive components, the engine components, and the steering column, just to name a few. A subassembly, such as an engine, may have other subassemblies, such as the carburetor or the generator. Figure 16–18 (see page 503) shows a subassembly with a parts list.

Pictorial Assembly

Pictorial assemblies are used to display a pictorial rather than multiview representation of the product, which may be used in other types of assembly drawings. Pictorial assemblies may be made from photographs or artistic renderings. The pictorial assembly may be as simple as the isometric drawing in Figure 16–17, which was used to more clearly assist workers in the assembly of the product. Pictorial assemblies are commonly used in product catalogs or brochures. The purpose of these pictorial representations may be for sales promotion, customer self-assembly, or maintenance procedures. (See Figure 16–19, page 504.) Pictorial assemblies may also take the form of exploded technical illustrations, also commonly known as illustrated parts breakdowns. These exploded multiview or isometric pictorials are used in vendors' catalogs and instruction manuals, for maintenance or assembly. (See Figure 16–20, page 505.)

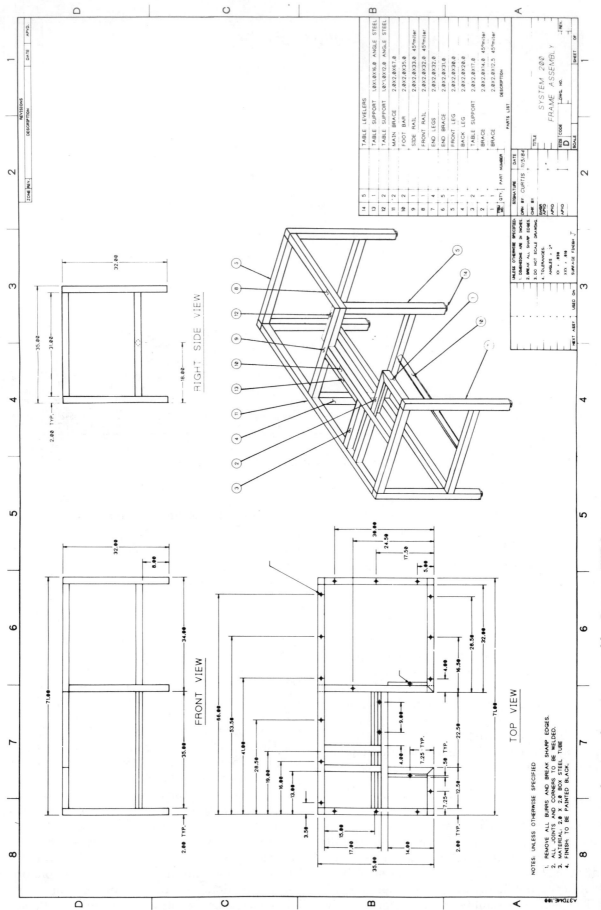

Figure 16-17 CADD drawing of erection assembly. *Courtesy EFT Systems.*

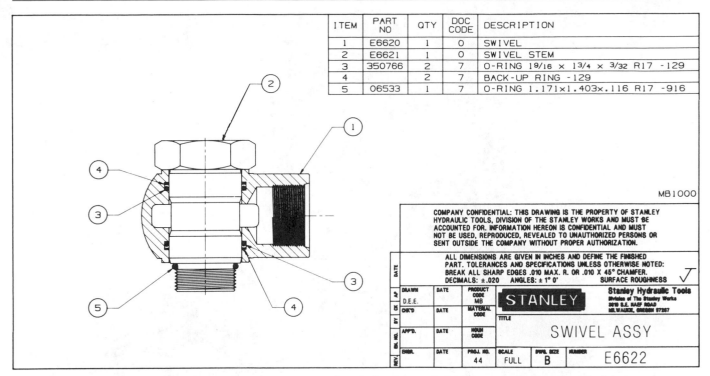

ITEM	PART NO	QTY	DOC CODE	DESCRIPTION
1	E6620	1	O	SWIVEL
2	E6621	1	O	SWIVEL STEM
3	350766	2	7	O-RING 19/16 × 13/4 × 3/32 R17 -129
4		2	7	BACK-UP RING -129
5	06533	1	7	O-RING 1.171×1.403×.116 R17 -916

MB1000

COMPANY CONFIDENTIAL: THIS DRAWING IS THE PROPERTY OF STANLEY HYDRAULIC TOOLS, DIVISION OF THE STANLEY WORKS AND MUST BE ACCOUNTED FOR. INFORMATION HEREON IS CONFIDENTIAL AND MUST NOT BE USED, REPRODUCED, REVEALED TO UNAUTHORIZED PERSONS OR SENT OUTSIDE THE COMPANY WITHOUT PROPER AUTHORIZATION.

ALL DIMENSIONS ARE GIVEN IN INCHES AND DEFINE THE FINISHED PART. TOLERANCES AND SPECIFICATIONS UNLESS OTHERWISE NOTED: BREAK ALL SHARP EDGES .010 MAX. R. OR .010 X 45° CHAMFER. DECIMALS: ±.020   ANGLES: ± 1° 0'   SURFACE ROUGHNESS

DRAWN D.E.E. | PRODUCT CODE MB
CHK'D | MATERIAL CODE
APP'D | HOUR CODE

STANLEY

Stanley Hydraulic Tools
Division of The Stanley Works
2010 S.E. HAEF ROAD
MILWAUKIE, OREGON 97267

TITLE
SWIVEL ASSY

ENGR. | PROJ. NO. 44 | SCALE FULL | DWG. SIZE B | NUMBER E6622

Figure 16–18 Subassembly with parts list. *Courtesy Stanley Hydraulic Tools Division of The Stanley Works.*

▼ IDENTIFICATION NUMBERS

Identification or item numbers are used to key the parts from the assembly drawing to the parts list. Identification numbers are generally placed in balloons. *Balloons* are circles that are connected to the related part with a leader line. Several of the assembly drawings in this chapter show examples of identification numbers and balloons. Numbers in balloons are common, although some companies prefer to use identification letters. Balloons are drawn between .375 and 1 in. in diameter depending on the size of the drawing and the amount of information that must be placed in the balloon. The leaders that connect the balloons to the parts are thin lines that may be presented in any one of several formats depending on company standards. Figure 16–21 (see page 505) shows the common methods that are used to connect balloon leaders. Notice that the leaders may terminate with arrowheads or dots. Whichever method of connecting balloons is used, the same technique should be used throughout the entire drawing.

The item numbers in balloons should be grouped so they are in an easy-to-read pattern. This is referred to as *information grouping.* Good information grouping occurs when balloons are in alignment in a pattern that is easy to follow, as opposed to scattered about the drawing. (See Figure 16–22, page 505.) In some situations when particular part groups are so closely related that individual identification is difficult, the identification balloons may be grouped next to each other. For example, a cap screw, lock washer, and flat washer may require

that the balloons be placed in a cluster or side-by-side, as shown as items 12, 13, and 14 in Figure 16–11.

In some cases the balloons key the detail drawings to the assembly and parts list, and also key the assembly drawing and parts list to the page on which the detail drawing is found. (See Figure 16–23, page 505.) Figure 16–24 (see page 506) is an assembly drawing and parts list that is located on page 1 of a two-page set of working drawings. Notice how the balloons key the parts from the assembly and parts list to the detail drawings found on page 2. The page 2 details are located in Figure 16–9 (see page 496).

▼ PARTS LISTS

The parts list is usually combined with the assembly drawing yet remains one of the individual components of a complete set of working drawings. The information that is associated with the parts list generally includes:

1. Item number—from balloons.
2. Quantity—the amount of that particular part needed for this assembly.
3. Part or drawing number, which is a reference back to the detail drawing.
4. Description, which is usually a part name or complete description of a purchase part or stock specification including sizes or dimensions.
5. Material identification—the material that the part is made of.
6. Information about vendors for purchase parts.

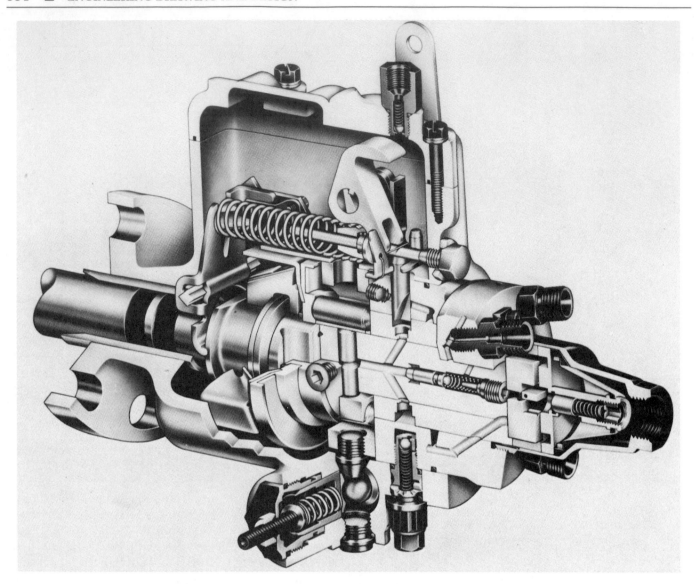

Figure 16–19 Pictorial assembly. *Courtesy Stanadyne Diesel Systems.*

Parts lists may or may not contain all of the previous information, depending on company standards. The elements listed 1 through 4 are the most common items. When all six elements are provided, the parts list may also be called a *list of materials.*

When drawn on the assembly drawing, the parts list may be located above the title block, in the upper right or left corner, or in a convenient location on the drawing field. The location depends on company standards, although the position over the title block is most common. The information on the parts list is usually presented with the first item number followed by consecutive item numbers. When the parts list is so extensive that the columns fill the page, a new group of columns is added next to the first. If additional parts are added to the assembly, space on the parts list is available. This is the reason that parts list data is provided from the bottom of the sheet upward or the top of the sheet downward.

Some companies prepare parts lists on a computer so information can be retrieved and edited more easily. When this is done, the parts list may be plotted or typed directly on the drawing original or on an adhesive Mylar® sheet that is added to the drawing. If the entire drawing is prepared on a CADD system, then the parts list can easily be plotted on the original.

The parts list is not always placed on the assembly drawing. Some companies prefer to prepare parts lists on separate sheets, which allows for convenient filing. This method also allows for the parts list to be computer-generated or typed separately from the drawings. Separate parts lists are usually prepared on a computer so information may be edited conveniently. Another option is to have parts lists typed on a standard parts list form, Figure 16–25 (see page 507).

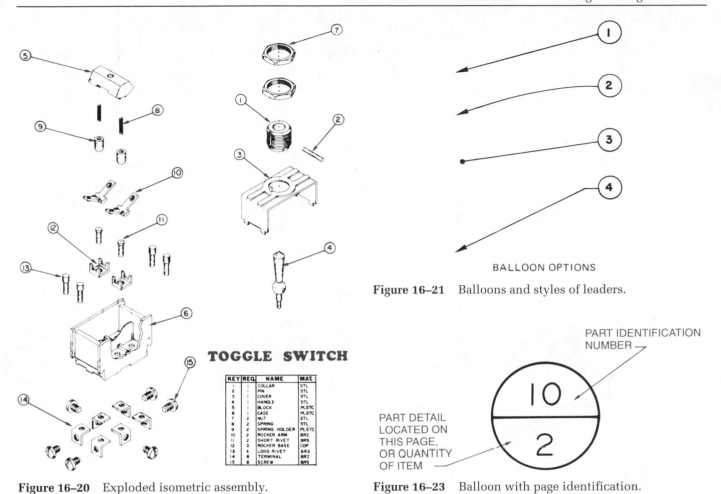

Figure 16–20 Exploded isometric assembly.

TOGGLE SWITCH

KEY	REQ	NAME	MAT.
1	1	COLLAR	STL
2	1	PIN	STL
3	1	COVER	STL
4	1	HANDLE	STL
5	1	BLOCK	PLSTC
6	1	CASE	PLSTC
7	2	NUT	STL
8	2	SPRING	STL
9	2	SPRING HOLDER	PLSTC
10	2	ROCKER ARM	BRZ
11	2	SHORT RIVET	BRS
12	2	ROCKER BASE	COP
13	4	LONG RIVET	BRS
14	6	TERMINAL	BRZ
15	6	SCREW	BRS

Figure 16–21 Balloons and styles of leaders.

BALLOON OPTIONS

PART IDENTIFICATION NUMBER

PART DETAIL LOCATED ON THIS PAGE, OR QUANTITY OF ITEM

Figure 16–23 Balloon with page identification.

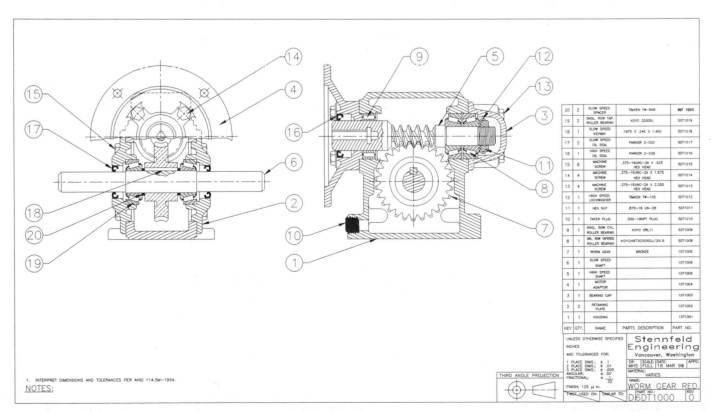

KEY	QTY.	NAME	PARTS DESCRIPTION	PART NO.
20	2	SLOW SPEED SPACER	TIMKEN TW-506	50T 1020
19	2	SNGL. ROW TAP. ROLLER BEARING	KOYO 32005J	50T1019
18	1	SLOW SPEED KEYWAY	.1875 X .245 X 1.450	50T1018
17	2	SLOW SPEED OIL SEAL	PARKER 2-020	50T1017
16	1	HIGH SPEED OIL SEAL	PARKER 2-028	50T1016
15	8	MACHINE SCREW	.375-16UNC-2A X .625 HEX HEAD	50T1015
14	4	MACHINE SCREW	.375-16UNC-2A X 1.875 HEX HEAD	50T1014
13	4	MACHINE SCREW	.375-16UNC-2A X 2.250 HEX HEAD	50T1013
12	1	HIGH SPEED LOCKWASHER	TIMKEN TW-105	50T1012
11	1	HEX NUT	.875-16 UN-2B	50T1011
10	1	TAPER PLUG	.500-18NPT PLUG	50T1010
9	1	SNGL. ROW CYL. ROLLER BEARING	KOYO CRL11	50T1009
8	1	DBL. ROW TAPERED ROLLER BEARING	KOYO46T30305DJ/29.5	50T1008
7	1	WORM GEAR	BRONZE	10T1005
6	1	SLOW SPEED SHAFT		10T1006
5	1	HIGH SPEED SHAFT		10T1005
4	1	MOTOR ADAPTOR		10T1004
3	1	BEARING CAP		10T1003
2	2	RETAINING PLATE		10T1002
1	1	HOUSING		10T1001

UNLESS OTHERWISE SPECIFIED
INCHES
AND TOLERANCES FOR:
1 PLACE DIMS. ± .1
2 PLACE DIMS. ± .01
3 PLACE DIMS. ± .005
ANGULAR; ± 30′
FRACTIONAL; ± 1/32
FINISH; 125 μ in.

THIRD ANGLE PROJECTION

Stennfeld Engineering
Vancouver, Washington

DR: MHS	SCALE: FULL	DATE: 18 MAR 96	APPD:
MATERIAL: VARIES			
NAME: WORM GEAR RED.			
FIRST USED ON:	SIMILAR TO:	PART NO.: D 6DT1000	REV: 0

1. INTERPRET DIMENSIONS AND TOLERANCES PER ANSI Y14.5M—1994.
NOTES:

Figure 16–22 Assembly drawing and parts list. *Courtesy Mark Stennfeld.*

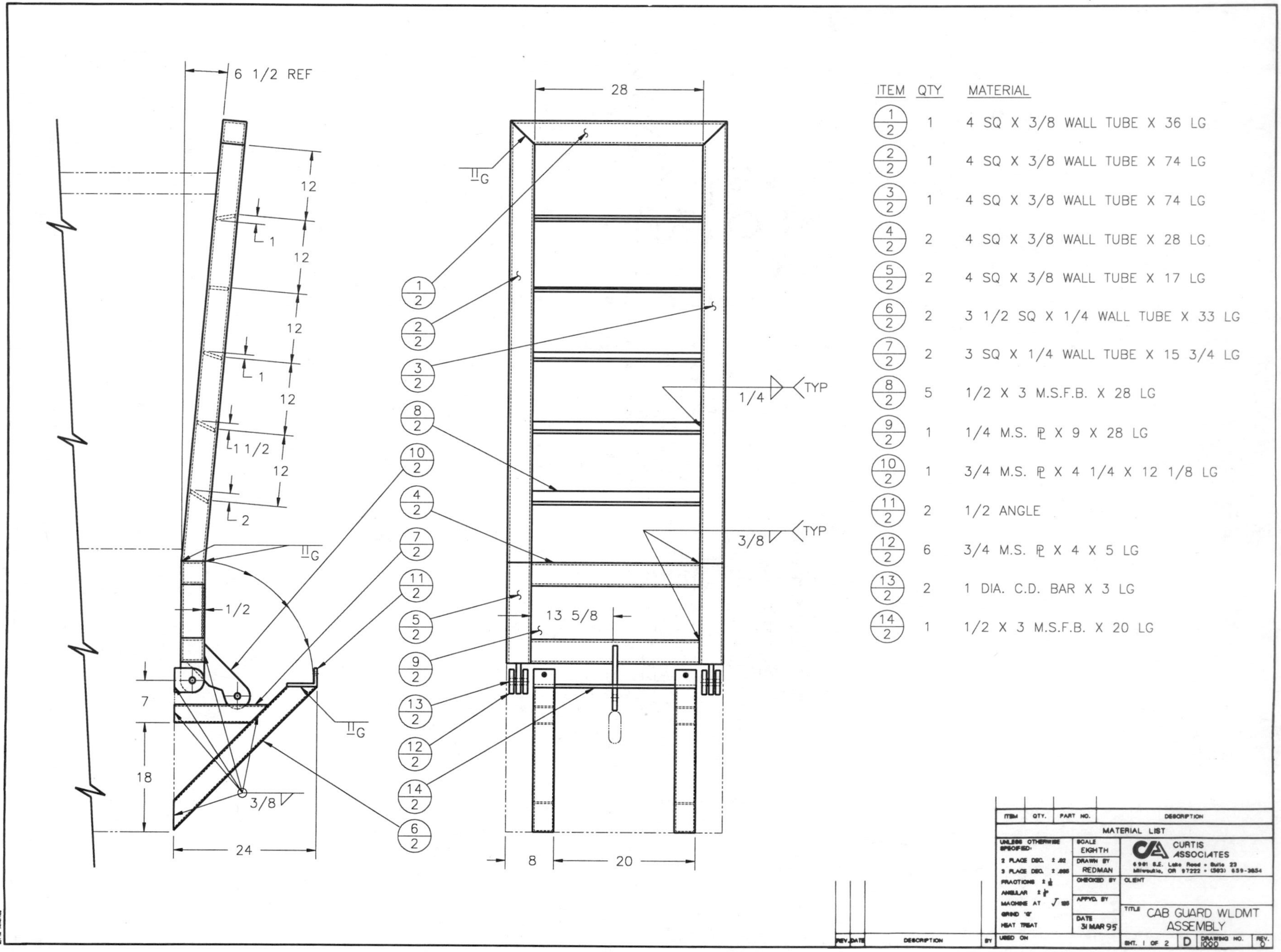

ITEM	QTY	MATERIAL
1/2	1	4 SQ X 3/8 WALL TUBE X 36 LG
2/2	1	4 SQ X 3/8 WALL TUBE X 74 LG
3/2	1	4 SQ X 3/8 WALL TUBE X 74 LG
4/2	2	4 SQ X 3/8 WALL TUBE X 28 LG
5/2	2	4 SQ X 3/8 WALL TUBE X 17 LG
6/2	2	3 1/2 SQ X 1/4 WALL TUBE X 33 LG
7/2	2	3 SQ X 1/4 WALL TUBE X 15 3/4 LG
8/2	5	1/2 X 3 M.S.F.B. X 28 LG
9/2	1	1/4 M.S. PL X 9 X 28 LG
10/2	1	3/4 M.S. PL X 4 1/4 X 12 1/8 LG
11/2	2	1/2 ANGLE
12/2	6	3/4 M.S. PL X 4 X 5 LG
13/2	2	1 DIA. C.D. BAR X 3 LG
14/2	1	1/2 X 3 M.S.F.B. X 20 LG

Figure 16–24 Assembly and parts list with page identification balloons. *Courtesy Curtis Associates.*

HAVE	NEED	P/Ø NO. W/Ø NO.	DET. NO.	PART NO.	DWG.	QTY.	PART NAME	DESCRIPTION	VENDOR
			1		B	1	Bearing Retainer	Ø 3" C.D. Bar	
			2		B	1	Cross Shaft	Ø 3/4" C.D. Bar	
			3		A	1	Spacer	Ø 3/4 O.D. x 11GA Wall Tube x .738 Thick	
			4		A	1	Spacer	Ø 3/4 O.D. x 11GA Wall Tube x .125 Thick	

ASSEMBLY Cross Shaft Assembly   USED ON

NUMBER OF UNITS   DATE

Figure 16–25 Parts list separate from assembly drawing. *Courtesy Curtis Associates.*

▼ PURCHASE PARTS

Purchase or standard parts, as previously mentioned, are parts that are manufactured and available for purchase from an outside vendor. It is generally more economical for a company to buy such items than to make them. Parts that are available from suppliers can be found in the *Machinery's Handbook, Fastener's Handbook,* or vendors' catalogs. Purchase parts do not require a detail drawing, because a written description completely describes the part. For this reason, the purchase parts found in a given assembly must be described clearly and completely in the parts list. Some companies have a purchase parts book that is used to record all purchase parts used in their product line. The standard parts book gives a reference number for each part, which is placed on the parts list for convenient identification.

▼ ENGINEERING CHANGES

ASME The standards controlling the use of engineering changes are specified in ASME Y14.35M—1992.

Engineering change documents are used to initiate and record changes to products in the manufacturing industry. Changes to engineering drawings can be requested from any branch of a company that deals directly with production and distribution of the product. For example, the engineering department may implement product changes as research and development results show a need for upgrading. Manufacturing requests changes as problems arise in product fabrication or assembly. The sales staff may also initiate change proposals that stem from customer complaints. Engineering changes are treated in the same way as original drawings; they are initiated, approved, sent to the drafting department, and filed when completed. It is a good idea to leave the area below the revision block for future changes.

Engineering Change Request (ECR)

Before a drawing can be changed by the drafting department, an engineering change request (ECR) is needed. This ECR is the document used to initiate a change in a part or assembly. The ECR is usually attached to a print of the part affected. The print and the ECR show by sketches and written descriptions what changes will be made. The ECR also contains a number that becomes the reference record of the change to be made. Figure 16–26 shows a sample ECR form. Generally when a drafter receives an ECR, the following procedure is used:

1. Get the drawing original, sepia, or computer file.
2. Make the requested change using the same media as that used for the original.
3. Fill out the engineering documents using freehand lettering or a word processor.
4. Distribute a changed copy of the drawing and the completed engineering change forms for release approval.
5. Refile the drawing original.

Engineering Change Notice (ECN)

Records of changes are kept so reference can be made between the existing and proposed product. To make sure that records of these changes are kept, special notations are made on the drawing, and engineering change records are kept. These records are commonly known as engineering change notices (ECN) or engineering change orders (ECO).

When a change is to be made, an engineering change request initiates the change to the drafting department. The drafter then alters the original drawing or computer file to reflect the change request. When the drawing of a part is changed, a revision letter is placed next to the change. For example, the first change is lettered *A*, the second change is *B*, and so on. The letters *I, O, S, X,* and *Z* are not used. When all of the available letters *A*

Aerojet Liquid Rocket Company
ENGINEERING CHANGE REQUEST & ANALYSIS

| DATE | ECR/ANS | PAGE |
| | | OF |

| ORIGINATOR NAME | | DEPT | EXT | DATE | DOCUMENT NEED DATE | PROPOSED EFFECTIVITY |

| PART DOCUMENT NO | CURRENT REV LTR | PART DOCUMENT NAME |

| USED ON NEXT ASSY NO | PROGRAM S AFFECTED | DEPT S AFFECTED |

		QTSS ITEM	YES _____ NO _____
		REQUAL REQD	YES _____ NO _____
		REVISE QTSS FORM	YES _____ NO _____

DESCRIPTION OF CHANGE

JUSTIFICATION OF CHANGE

| PROJECT ENGINEER SIGNATURE | DEPT | EXT | DATE | CCB REP SIGNATURE | DATE |

DESIGN ENGINEERING TECHNICAL EVALUATION

| DESIGN ENGINEER SIGNATURE | DEPT | EXT | DATE | CAUSING CONT OR WORK ORDER NO |

ANY AFFECT ON	YES	NO		YES	NO		YES	NO
1 PERFORMANCE			5 WEIGHT			9 OPERATIONAL COMPUTER PROGRAMS		
2 INTERCHANGEABILITY			6 COST SCHEDULE			10 RETROFIT		
3 RELIABILITY			7 OTHER END ITEMS			11 END ITEM IDENT		
4 INTERFACE			8 SAFETY EMI			12 VENDOR CHANGE CRITICAL ITEMS ONLY		

CCB DECISION		SIGNATURES	DEPT	CON CUR	DIS SENT
		QUALITY ASSURANCE			
		MANUFACTURING			
		ENGINEERING			
		PRODUCT SUPPORT			
		MATERIAL			
		TEST OPERATION			

| CCB CHAIRMAN SIGNATURE | DATE | CLASS I ☐ CLASS II ☐ |
| CUSTOMER SIGNATURE | DATE | EFFECTIVITY |

Figure 16–26 Sample engineering change request (ECR) form. *Courtesy Aerojet Propulsion Division.*

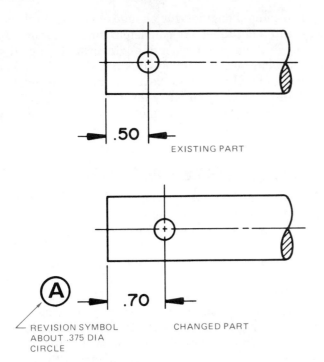

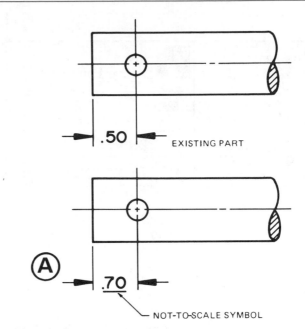

Figure 16–27 An existing part and the same part after a change has been made.

Figure 16–28 An existing part and the same part changed using the not-to-scale symbol.

through *Y* have been used, double letters such as *AA* and *AB,* or *BA* and *BB* are used next. A circle drawn around the revision number helps the identification stand out clearly from the other drawing numerals. Figure 16–27 shows a part as it exists and also after a change has been made.

When the drafter chooses to make the change by not altering the drawing of the part but only changing the dimension, that dimension is labeled not-to-scale. The method of making the change will be the decision of the drafting department based on the extent of the change and the time required to make it. The not-to-scale symbol may be used to save time. Figure 16–28 shows a change made to a part with the new dimension identified as not to scale with a thick straight line placed under the new dimension. With CADD, it is generally easy and preferred to make the change to scale and avoid using the not-to-scale method.

After the part has been changed on the drawing and the proper revision letter is placed next to the change, the drafter records the change in the ECN column of the drawing. The location of the ECN column varies with different companies. This ECN identification may be found next to the title block or in a corner of the drawing.

ANSI The ANSI document, *Drawing Sheet Sizes and Format,* ANSI Y14.1, recommends the ECN column be placed in the upper right corner of a drawing.

The drafter records the revision number (number of times revised), ECN number (usually given on the ECR), and the date of the change. Some companies also have

a column for a brief description of the change (as recommended by ANSI) and an approval. Figure 16–29 shows an expanded and condensed ECN column format. Notice that changes are added in alphabetic order from top to bottom.

The condensed ECN column in Figure 16–29b shows that the drawing has been changed twice. The first

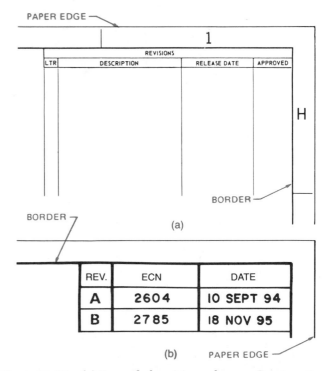

Figure 16–29 **(a)** Expanded revision columns. *Courtesy Aerojet Propulsion Division.* **(b)** Condensed revision columns.

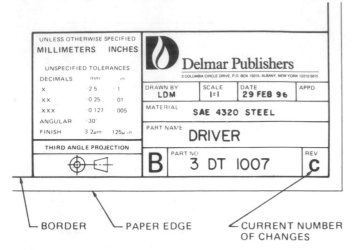

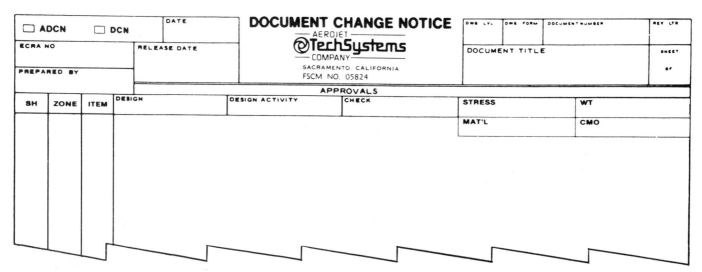

Figure 16–30 Title block displaying the current number of times the drawing has been changed.

change was initiated by ECN number 2604 on September 10, 1994, and the second by ECN 2785 on November 18, 1994. Some companies may also identify the number of times a drawing has been changed by providing the letter of the current change in the title block, as shown in Figure 16–30. The drafter alters this letter to reflect each change.

When the drafter has made all of the drawing changes as specified on the ECR, than an engineering change notice (ECN) is filled out. The ECN completes the process and is filed for future reference. Usually the ECN completely describes the part as it existed before the change and the change that was made as presented on the ECR. Figure 16–31 shows a typical ECN form. With a changed drawing and a filed ECN, anyone can verify what the part was before the change and the reason for the change. The ECN number is a reference for reviewing the change.

The general engineering change elements, terminology, and techniques are consistent among companies, although the actual format of engineering change documents may be considerably different. Some formats are very simple while others are much more detailed. One of the first tasks of an entry-level drafter is to become familiar with the specific method of preparing engineering changes.

▼ DRAWING FROM A PROTOTYPE

Occasionally the engineering drafter has to take an existing product or structure and draw a set of working drawings. This may be a prototype. A *prototype* is a model or original design that has not been released for production. The prototype is often used for testing and performance evaluations. In most cases there are prototype drawings from which the prototype was built, but during the modification process changes may have been made. Your job, then, is to complete the production drawings, which are established by measuring the prototype, redrawing the prototype drawings with changes, or a combination of these methods. These revised drawings are often referred to as *as-built drawings.* In this type of situation you need to be well versed in the use of measuring equipment such as calipers, micrometers, and surface gages. You need to consult with the engineer, the testing department, and the shop personnel who made any modifications to the prototype. With this research you can fully understand how the new drawings should be created. You need to be careful to make the drawings accurate and take into account the manufacturing capabilities of your company. Then the set of working drawings that you complete is checked by the checker and the engineer before the product is released for production. Once the product is released for production, changes are often costly because

Figure 16–31 Sample engineering change notice. *Courtesy Aerojet Propulsion Division.*

tooling and patterns may have to be altered. These types of changes are usually submitted as engineering change requests (ECR). They must be approved by the engineering and manufacturing departments.

▼ ANALYSIS OF A SET OF WORKING DRAWINGS
The Design Sketches

The design of a product generally begins in the research and development (R & D) department of a company. The engineer or designer may prepare engineering sketches or layout drawings. Figure 16–32 shows the engineer's sketches for a screwdriver to be manufactured by the company. Notice in this example that the engineer's sketch may be rough. It is your responsibility, as a professional drafter, to prepare the final drawings in accordance with ASME/ANSI, MIL, or the appropriate standards adopted by the company.

The Detail Drawings

The engineer's sketches go to the drafting department for the preparation of formal drawings. Depending on the complexity of the project, the engineer and drafter may work together, or there may be a team of engineers and drafters working on various components of the project. The drafter may first complete the detail drawings as shown in Figure 16–33 and Figure 16–34.

PROFESSIONAL PERSPECTIVE

Many entry-level drafters are often involved in preparing detail drawings or making changes to detail drawings. When a drafter has gained valuable experience with drafting practices, standards, and company products, there is usually an opportunity to advance to a design drafter position. Drafting jobs at these levels can be exciting, as the drafters work closely with engineers to create new and updated product designs. An individual drafter in a small company may be teamed with an engineer to implement designs. In larger companies a team of drafters, designers, and engineers may work together to design new products. These are the types of situations where a drafter may have the opportunity to prepare some or all of the drawings for a complete set of working drawings. Generally in these research and development departments of companies the preliminary product drawings are used to build prototypes of the designs for testing. After sufficient tests have been performed on the product, the drawings are revised and released for production. The new product will now become reality.

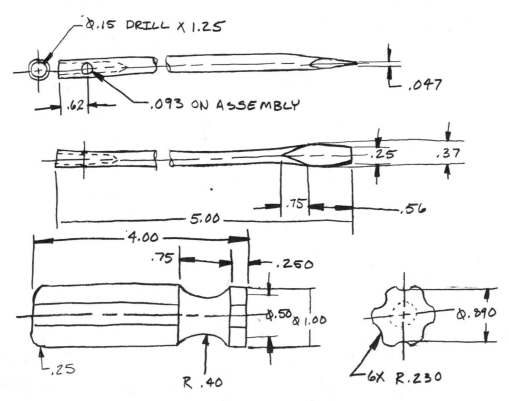

Figure 16–32 Very rough engineer's sketch. Engineering sketches may not be as good as the one shown here. The engineering drafter must convert the sketch to proper ASME, MIL, or ISO standards as appropriate.

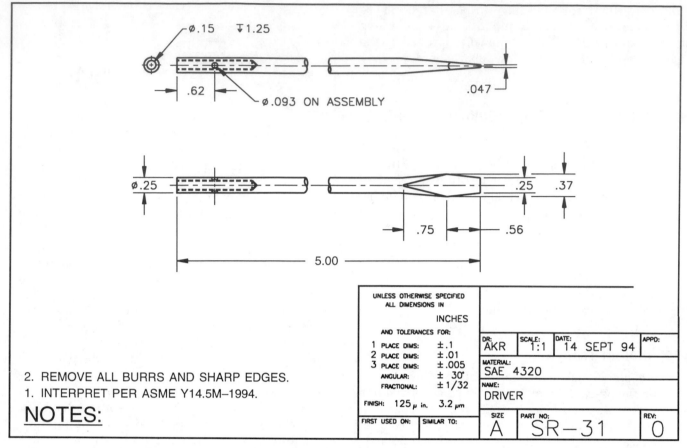

2. REMOVE ALL BURRS AND SHARP EDGES.
1. INTERPRET PER ASME Y14.5M–1994.

NOTES:

Figure 16–33 The DRIVER detail drawing before an ECR is issued. Notice the "O" in the lower right corner. This indicates an original unchanged drawing.

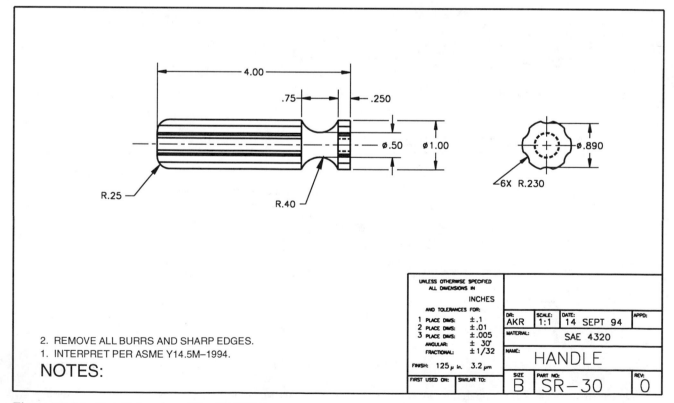

Figure 16–34 Detail drawing of the HANDLE.

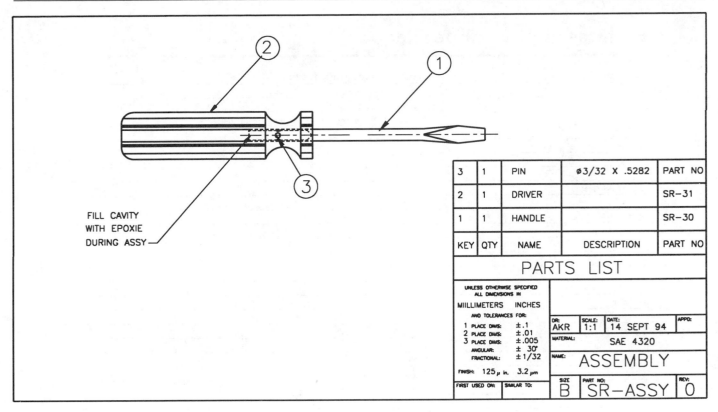

3	1	PIN	⌀3/32 X .5282	PART NO
2	1	DRIVER		SR–31
1	1	HANDLE		SR–30
KEY	QTY	NAME	DESCRIPTION	PART NO

PARTS LIST

UNLESS OTHERWISE SPECIFIED ALL DIMENSIONS IN		
MIILLIMETERS INCHES		
AND TOLERANCES FOR:		
1 PLACE DIMS: ±.1		
2 PLACE DIMS: ±.01		
3 PLACE DIMS: ±.005		
ANGULAR: ± 30°		
FRACTIONAL: ± 1/32		
FINISH: 125 μ in. 3.2 μm		

FILL CAVITY
WITH EPOXIE
DURING ASSY

DR: AKR	SCALE: 1:1	DATE: 14 SEPT 94	APPD:
MATERIAL:		SAE 4320	
NAME:		ASSEMBLY	
FIRST USED ON:	SIMILAR TO:		
SIZE B	PART NO: SR–ASSY		REV: 0

Figure 16–35   Assembly drawing.

The Assembly Drawing

As discussed earlier in this chapter, the assembly drawing is used to show the assembly department how the parts fit together and if there are any special instructions required for the assembly process. When CAD is used, the drafter often freezes the dimensions and isolates the individual parts, which are brought together in a separate drawing. Using CAD allows the drafter to scale the parts into one common size and then move them together in the assembled position. Assembly notes and dimensions, if any, are added. Each part receives a balloon and a correlating parts list is created as shown for the screwdriver design in Figure 16–35. When the complete set of working drawings has been drawn and approved, the product is released to production.

The Product Catalog

The power of 3-D CAD applications makes it possible for companies to display their products using pictorial assemblies or exploded pictorial assemblies. Figure 16–36 shows a pictorial assembly of the screwdriver that was designed in the previous discussion.

Engineering Changes

When a product is in production, a change can be requested from any department including manufacturing, engineering, or sales. These changes may be based on new ways to manufacture, redesign, customer feed-

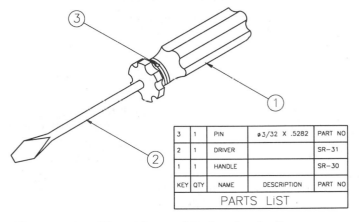

3	1	PIN	⌀3/32 X .5282	PART NO
2	1	DRIVER		SR–31
1	1	HANDLE		SR–30
KEY	QTY	NAME	DESCRIPTION	PART NO

PARTS LIST

Figure 16–36   Pictorial assembly drawing for the company catalog.

back, or a variety of issues. Before a drawing can be changed by the drafting department, it usually requires an engineering change request (ECR). The ECR is also referred to as an engineering change order (ECO). The ECR is the document that is used to initiate a change in a part or assembly. The ECR is often attached to a redlined print of the part or drawing affected. The print and the ECR normally contain a change number that is used later by the drafter to properly document the change. Every company has an ECR form that looks different. The ECR form shown in Figure 16–37 displays a change requested for the DRIVER shown in Figure 16–33.

Engineering Change Request

Request to be completed
Through ECN# | 1171

Engineering:

Engineer

Date: 14 Sept 94

Manufacturing:

Date:

Sales:

Date:

Description of Request:

Refer to part number SR−31 (Driver)

Change 5" length to 6"

Change Hole location dimension from .62 to .69

Reason for Request:

Longer tool dimension will make part more stable when locked in handle.

Hole location was design error.

| Castings & Forgings Affected?(yes or no) NO | | Disposition of Production Stock: | | ☐ Use in production | ☐ |
| Approved: | Date: 14 Sept 94 | Drawing by: Student | Scrap ☒ | Transfer to service stock | ☐ |

Figure 16–37 The engineering change request (ECR).

Companies use various methods to change drawings depending on the complexity of the change, the company standards, and the time allowed. If dimensions are changed and the feature is not drawn to its proper scale, the not-to-scale symbol should be used. Changes made to CAD drawings are usually easier to accurately scale than those created using manual drawing. When CAD is used, the company often desires to have the changed drawing accurately reflect the product.

After a change is made on the drawing, a balloon is placed next to the drawing revision and a letter identifying the change is placed in the balloon. Balloons are usually 3/8 to 1/2 inch (8 to 12 mm) in diameter depending on the drawing size. Some companies use a square, a hexagon, or another symbol for change balloons. The letter in the balloon designates the change. An *A* is used for the first change, a *B* for the second change, and so on. Some companies use R1, R2, and R3. However, letters are common to differentiate from balloon numbers on assembly drawings. Next, the drafter records the change in the revision block of the drawing. The revision block is usually in the upper right corner in accordance with ANSI standards. The current change is also recorded in the title block as "A," "B," or "C" as appropriate. Figure 16–38 shows the DRIVER in Figure 16–33

changed in accordance with the ECR in Figure 16–37.

The next procedure in the change process is for the drafter to fill out the engineering change notice (ECN). The change documents are a formal part of the drafting process and must be professionally prepared, recorded, and filed for future reference. The ECN should be hand lettered, typed, or electronically prepared if the document is available in a word processor. The ECN form used in companies and the process for filling out the forms vary. The ECN form shown in Figure 16–39 has been created from a combination of formats found in industry. The large numbers on the ECN refer to the changes requested for the DRIVER in Figure 16–33 and the ECR in Figure 16–37. The following steps are used when filling out this ECN form. (NOTE: Blank ECR and ECN forms are available in the *Instructor's Guide* to be copied for use in engineering change projects.)

1. Put the ECN number in at the top and bottom right side of page, on this example. (Get this number from the engineering change request.) In this case it is 1171.

2. List quantity of the part affected. (On this drawing, a detail-only part can be affected.) NOTE: Quantity = 1.

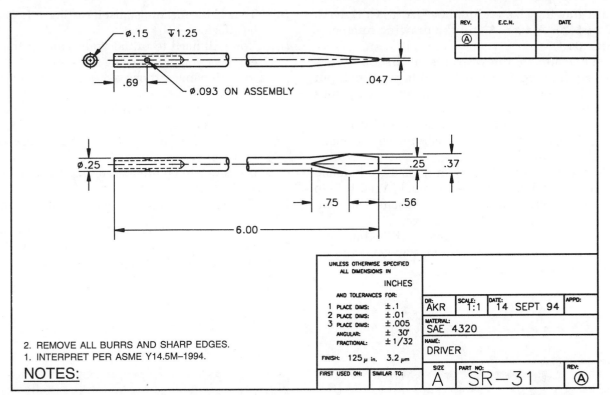

Figure 16–38 Changes to the DRIVER based on the ECR shown in Figure 16–37. Notice the balloons next to each change, the identification of the change in the upper right corner, and the current change "A" in the lower right corner.

Engineering Change Notice

ECN NO. 1171 ①

Disposion of production stock:
A =Alter or rework U=Use in production
T=Transfer to service stock   S=Scrap

Qty.		Drawing Size Part No.	R/N	Description	Change	Other Usage in Production	D/S
01	1	A SR–31	A	DRIVER	6" WAS 5"		S
02	②	③	④	⑤	.69 Hole Location Dimension		⑦
03					was .62		
04					⑥		
05							
06							
07							
08							
09							
10							
11							
12							
13							
14							
15							
16							
17							
18							

Reason: ⑧

Line 1–This dimension needed to be longer to help make the part more stable when locked in handle.

Line 2–3 This was a design error.

⑫

Castings & forgings affected? ☐ Yes ☒ No ⑨	Design engineer: ⑩	Supervisor approval: ⑪	Release date: ⑬	Page 1/1

Figure 16–39 The completed engineering change notice (ECN) form. The circled numbers refer to the steps for completing the form given in the text. The ECN form must be professionally typed or hand lettered.

3. List drawing size and number (A, B, C, D, etc.). This is an A size. A and the drawing number = SR – 31.

4. List revision. (In this case, it is the first one. If it is a new drawing, just put N or 0 here.) If not, put revision letter (A, B, C, D, etc). Some companies use numbers.

5. List title for drawing or part here (DRIVER).

6. List changes you made. Describe the change made in short terms.

7. List what is to be done with parts already made under the old design. Use a symbol (A, U, T, S) to show this. The definitions of these letters are conveniently located on the ECN form.

8. List your reasons for the change in brief but complete statements so in the future someone can find out why the change was made.

9. To help the machinist, list if casting or forgings will be affected.

10. You will need to initial your name or that of the engineer you are working for, so if questions arise to the changes those asking will know who to query.

11. Your supervisor should approve the changes you have made.

12. Pages are listed as 1/1 or 1/3, if three pages are used to complete one change. If there are three pages, the pages would be numbered 1/3, 2/3, 3/3, or 1 of 3, 2 of 3, 3 of 3.

13. The date you made the change must be recorded.

The part or product is released again for continued manufacturing after the engineering change documents have been properly prepared and approved.

MATH APPLICATION

DISTANCE BETWEEN TANGENT CIRCLES

Problem: It is desired to find dimension z, the distance between the centers of the two smaller circles of the illustration shown in Figure 16–40.

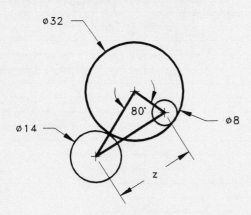

Figure 16–40 Tangent circles.

Solution: It is necessary to visualize the triangle connecting the centers of the circles and make use of the known radii to calculate two of the sides. The side of the triangle connecting the 32" and 14" diameter circle centers must be the sum of their radii, 16 + 7 = 23", and the side connecting the 32" and 8" diameter circle centers must be the difference of their radii, 16 – 4 = 12". (See Figure 16–41.)

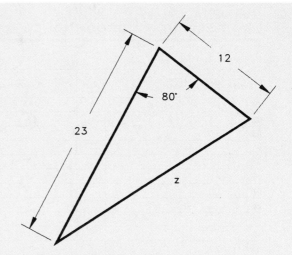

Figure 16–41 Oblique triangles for the tangent circles.

Since the triangle is not a right triangle, we cannot use a single trig function, and since we do not know a side *and* the angle opposite the side, we cannot use the Law of Sines. Instead, we must use the Law of Cosines (See Appendix N):

$$z^2 = (23)^2 + (12)^2 - 2(23)(12)\cos 80°$$
$$z^2 = 529 + 144 - 2(23)(12)(.1736)$$
$$z^2 = 529 + 144 - 95.854$$
$$z^2 = 577.146$$
$$z = \sqrt{577.146} = 24.024 \text{ or } \mathbf{24.0"}$$

CHAPTER
16 WORKING DRAWINGS TEST

DIRECTIONS

Answer the questions with short complete statements or drawings as needed.

QUESTIONS

1. Name three elements that make up a complete set of working drawings.
2. Define *detail drawing*.
3. Identify six items of information that may be found on a detail drawing.
4. List the three general groups of information that are generally found on detail drawings.
5. Describe one advantage of having one detail per sheet.
6. Descrive one advantage of drawing several details per sheet.
7. List five elements to consider when selecting paper size for detail drawings.
8. Define *assembly drawing*.
9. List six items of information that may be found on an assembly drawing.
10. List the six types of assembly drawings.
11. Define *balloons*.
12. What is used to connect balloons to their related parts?
13. Describe proper information grouping.
14. List five items that are normally found on the parts list.
15. Where is the parts list normally located?
16. When are parts lists prepared on separate sheets?
17. What is the advantage of computerized parts lists?
18. What are purchase parts?
19. Provide an example of a purchase part.
20. What does the abbreviation ECR denote?
21. What does the abbreviation ECN denote?
22. How is a revision identified on a drawing?
23. How are revision numbers or letters placed on a drawing so they clearly stand out from other drawing numerals or letters?
24. Give an example of how a dimension is shown not to scale.

CHAPTER
16 WORKING DRAWINGS PROBLEMS

DIRECTIONS

1. From the selected engineering sketches or layouts prepare a complete set of working drawings, including details assembly and parts list. Determine which views and dimensions should be used to completely detail each part. Also, determine the views, parts list, dimensions, and notes, if any, for the assembly drawing. Use ANSI standards. Use manual drafting or CADD as required by your course guidelines and objectives.
2. The complete set of working drawings will be prepared with one detail drawing per sheet using multiview projection and with the assembly drawing and parts list on one sheet unless otherwise specified. All purchase (standard) parts will be completely identified in the parts list. Using the sketches as a guide, draw original multiview drawings on an adequately sized sheet. Add all necessary dimensions and notes using unidirec-

tional dimensioning. Use computer-aided drafting if required by your course guidelines. Problems 16–1 through 16–10 are designed to be manufactured as projects in the manufacturing (machine) technology lab.
3. Some problems in this chapter may contain errors, missing information, or slight inaccuracies. This is intentional and is meant to encourage you to apply appropriate problem solving methods, engineering, and drafting standards in order to solve the problems. This is meant to force you to think about each part and how parts fit together in the assembly. As in "real-world" projects, the engineering problem should be considered as a basis for your preliminary layouts. Always question inaccuracies in project designs and consult with the proper standards and other sources. In some cases, an error might be the source of engineering

(Continued)

DIRECTIONS *(Continued)*

changes provided by the instructor; however, this is determined by your specific course objectives. Other situations may require that corrections be made during the development of the original design drawings. This is not intended as a source of frustration, but is considered to be part of the engineering drafter's daily responsibility in project development.

4. Include the following general notes at the lower left corner of the sheet .5 in. each way from the corner border lines for each detail drawing: (NOTE: Number 2 does not apply to the assembly drawings.)

2. REMOVE ALL BURRS AND SHARP EDGES.
1. INTERPRET DIMENSIONS AND TOLERANCES PER ASME Y14.5M—1994.

NOTES:

Additional general notes may be required depending on the specifications of each individual assignment. The following unspecified tolerances should be part of your title block:

UNSPECIFIED TOLERANCES:

DECIMALS	mm	IN.
X	±2.5	±.1
XX	±0.25	±.01
XXX	±0.127	±.005
ANGULAR ±30'		
FINISH	3.2µm	125µIN.

Problem 16–1 Working-drawing assembly (metric)

Assembly Name: Plumb Bob
SPECIFIC INSTRUCTIONS:
Prepare a working-drawing assembly that has a detail drawing of each part, an assembly drawing, and a parts list on one sheet.

PARTS LIST:

ITEM	QTY	NAME	MATERIAL
1	1	PLUMB BOB	BRONZE
2	1	CAP	BRONZE

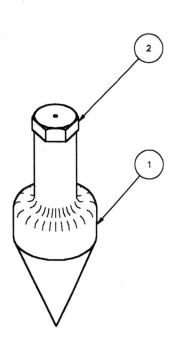

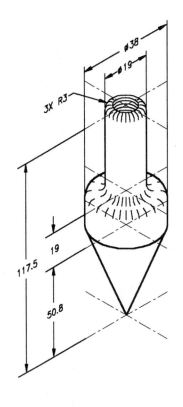

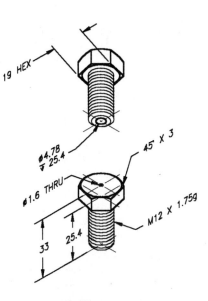

TWO PICTORIAL VIEWS OF THE CAD ARE PROVIDED FOR CLARITY

Problem 16–2 Working drawing (in.)

Assembly Name: Hammer
SPECIFIC INSTRUCTIONS:
Prepare a detail drawing for the hammer head and two optional
hammer handles on one sheet. Make the assembly drawing
and parts list on another sheet.

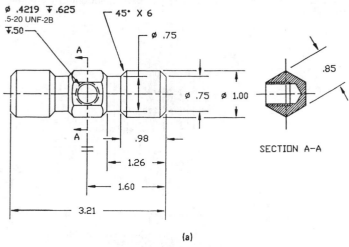

(a)

MATERIAL: BRASS
HAMMERHEAD

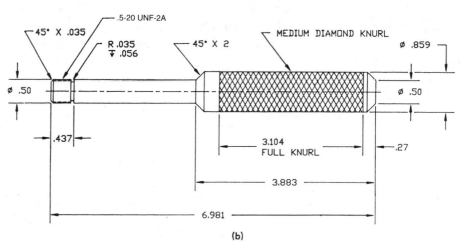

(b)

MATERIAL: SAE 6061 T6
HAMMER HANDLE (OPTION ONE)

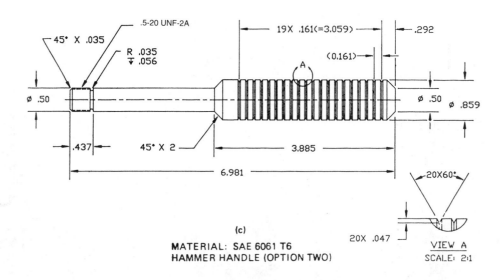

(c)

MATERIAL: SAE 6061 T6
HAMMER HANDLE (OPTION TWO)

Problem 16–3   Working drawing (in.)

Assembly Name: Key Holder
SPECIFIC INSTRUCTIONS:
Prepare a detail drawing for each part, an assembly drawing, and
a parts list on one sheet, unless otherwise specified by your
instructor.

No. 30 KEY RING, 2 REQD

ASSEMBLY

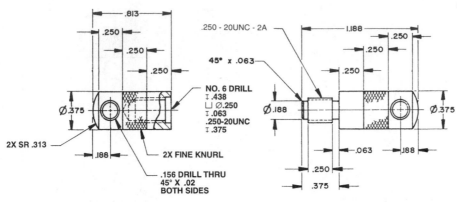

BODY PART A

BODY PART B

Problem 16–4   Working drawing (metric)

Assembly Name: C-clamp
SPECIFIC INSTRUCTIONS:
Prepare a complete set of working drawings with all of the detail
drawings on one sheet and the assembly drawings and parts list
on another sheet. Use multiview projection for view layout.

PARTS LIST:

ITEM	QTY	NAME	MATERIAL
1	1	PIN	CRMS
2	1	BODY	SAE 4320
3	1	SCREW	SAE 4320
4	1	SWIVEL	SAE 1020

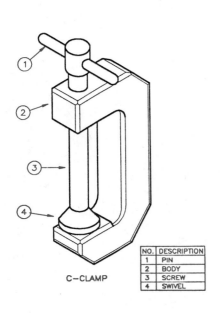

C–CLAMP

NO.	DESCRIPTION
1	PIN
2	BODY
3	SCREW
4	SWIVEL

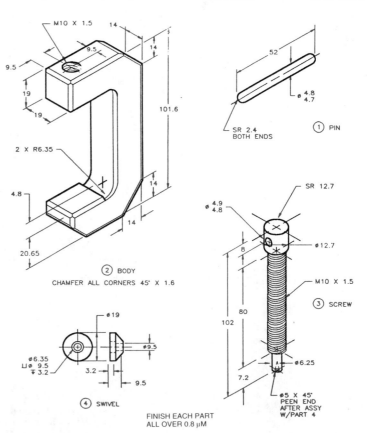

FINISH EACH PART
ALL OVER 0.8 μM

Problem 16–5 Working drawing (in.)

Assembly Name: Mill Work Stop
SPECIFIC INSTRUCTIONS:
Prepare a complete set of working drawings with one detail drawing per sheet and the assembly drawing with parts list on one sheet. Notice: part number 10, TEE STUD, is cut from standard 3/8-16UNC ALL THREAD. Parts 12 and 13, BLANK KNOBS, are purchased as knobs without the threads machined. You will draw the knobs as a "representation" of the purchase part and give the thread note as identified in the parts list. The actual dimensions of the knob to be purchased are not important; however, accurate specifications for the threads to be machined into the purchase part are important.
NOTE: Some design is required throughout the completion of the project.

PARTS LIST:

ITEM	QTY	NAME	DESCRIPTION	MATERIAL
1	1	VERTICAL MEMBER	3/4 × 2 × 4	SAE 6061
2	1	BASE MEMBER	1 × 2 × 3.5	SAE 6061
3	2	RIB	1/4 PLATE	SAE 6061
4	1	ARM	3/8 × 1 × 5-1/2	SAE 1018
5	1	CLAMP	1/2 × 1 × 1-5/16	SAE 1018
6	1	TEE PLATE	5/16 × 1 × 1-1/4	SAE 1018
		DRILL AND TAP AT CENTER FOR 5/16-18UNC-2 THREAD THRU		
7	1	STOP ROD	∅5/16 × 6	SAE 303 SS
8	1	WING NUT	1/4-28 × 1/2	
9	1	BUSHING	∅1/2 × .31	SAE 1018
10	1	TEE STUD	3/8-16 × 2 LG ALL THREAD	
11	1	CARRIAGE BOLT	5/16-18UNC-2	
12	21	BLANK KNOB	VLIER HK-3 (DRILL AND TAP FOR 3/8-16UNC-2B)	
13	1	BLANK KNOB	VLIER HK-3 (DRILL AND TAP FOR 5/16-18UNC-2B)	

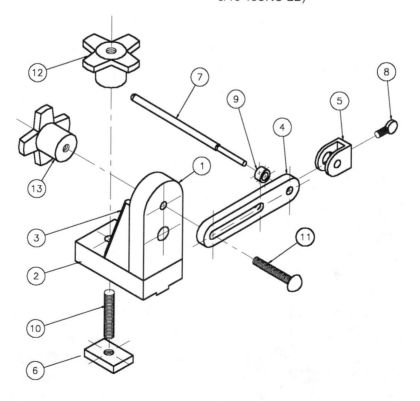

Problem 16–5 (Continued)

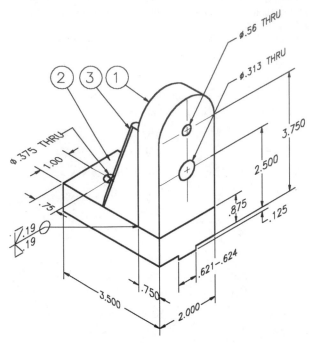

DEBUR ALL EDGES .020

BODY ASSEMBLY WELDMENT
FINISH: OOLOR ANODIZED

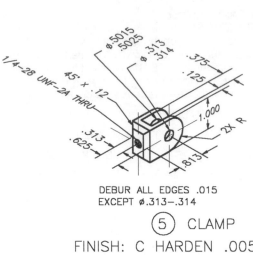

DEBUR ALL EDGES .015
EXCEPT ∅.313–.314

⑤ CLAMP
FINISH: C HARDEN .005/.010
BLACK OXIDE

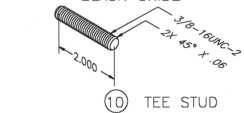

⑩ TEE STUD

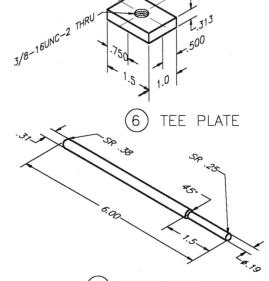

⑥ TEE PLATE

⑦ STOP ROD

FINISH: POLISH

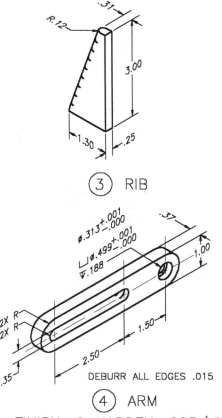

③ RIB

DEBURR ALL EDGES .015

④ ARM
FINISH: C HARDEN .005/.010
BLACK OXIDE

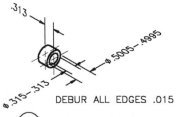

DEBUR ALL EDGES .015

⑨ BUSHING
FINISH: C HARDEN .005/.010
BLACK OXIDE

Problem 16–6 Working drawing (metric)

Assembly Name: Tool Holder
SPECIFIC INSTRUCTIONS:
Prepare a complete set of working drawings with one detail drawing per sheet, and the assembly and parts list on another sheet. NOTE TO STUDENT: When preparing the assembly drawings, use separate balloons for each part in each view, or use only one balloon in the view that most clearly identifies the part.

PARTS LIST:

ITEM	QTY	NAME	MATERIAL
1	1	PARTING TOOL, 3/32 in. × ½ in. PURCHASE PART	TOOL STEEL
2	1	TOOL HOLDER BODY	06 STEEL
3	1	ADJUSTMENT SCREW	SAE 1035 STEEL
4	1	SHIM	SAE 4320 STEEL
5	1	KNURL NUT	SAE 3130 STEEL
6	1	WASHER	SAE 1060 STEEL
7	1	STUD	SAE 1035 STEEL
8	1	M 10 × 1.5 HEX NUT PURCHASE PART	

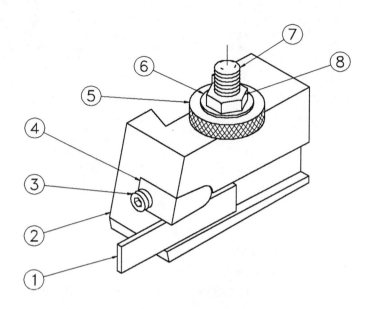

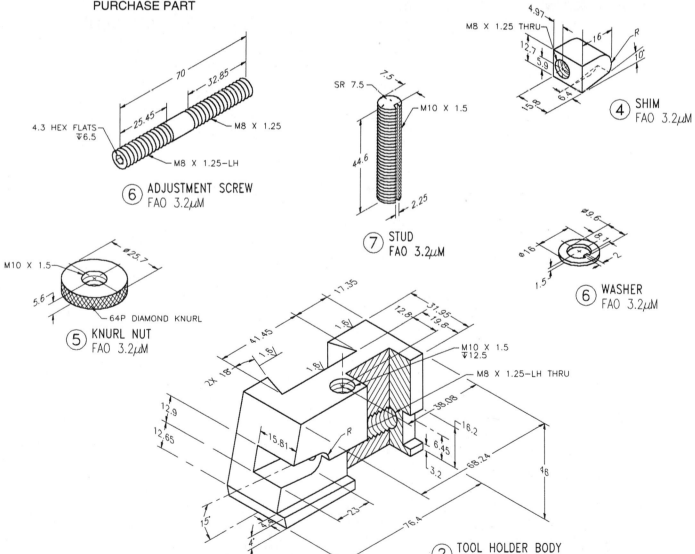

Problem 16–7 Working drawing (in.)

Assembly Name: Adjustable Attachment
SPECIFIC INSTRUCTIONS:
Prepare a complete set of working drawings with detail drawings
of individual parts combined on one or more sheets depending
on the size of sheet selected. The assembly drawing and parts
list will be combined on one sheet.

PARTS LIST:

ITEM	QTY	NAME	DESCRIPTION	MATERIAL
1	1	CAP SCREW	.25-20UNC-2 × .75 HEX SOC HEAD	STL
2	1	FRAME		SAE 4340
3	1	ADJUSTING SCREW		SAE 1045
4	1	SET SCREW	8-32UNC-2 HEX SOC FLAT POINT	STL
5	1	TAPER PIN	0 × .625	STL
6	1	KNURL KNOB		SAE 1024
7	1	CLAMP SCREW		SAE 2330
8	1	PILOT SCREW		SAE 2330
9	1	ADJUSTING NUT		SAE 3130

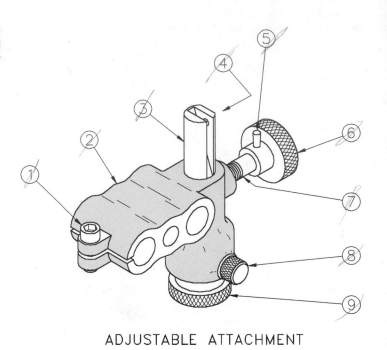

ADJUSTABLE ATTACHMENT

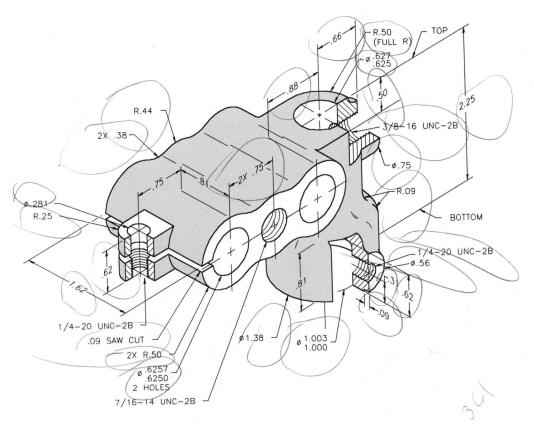

② FRAME
SAE 4340
REMOVE ALL BURRS & SHARP EDGES
FAO
FILLETS AD ROUNDS R.125
1 REQ'D

Problem 16–7 (Continued)

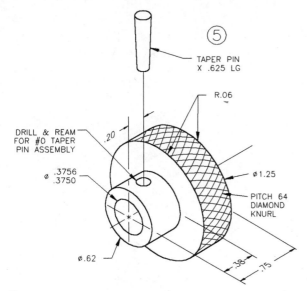

TAPER PIN
X .625 LG

R.06

DRILL & REAM
FOR #0 TAPER
PIN ASSEMBLY

.20

ø .3756
.3750

ø1.25

PITCH 64
DIAMOND
KNURL

ø.62

.38 .75

⑥ **KNURL KNOB**
SAE 1024
REMOVE ALL BURRS & SHARP EDGES
FAO
1 REQD

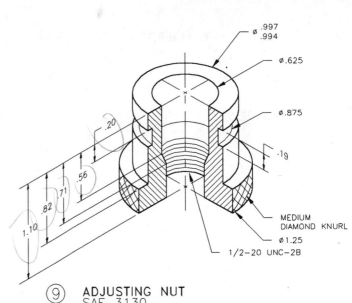

ø .997
.994

ø.625

ø.875

.19

.20

.56

.71

.82

1.10

MEDIUM
DIAMOND KNURL

ø1.25

1/2–20 UNC–2B

⑨ **ADJUSTING NUT**
SAE 3130
REMOVE ALL BURRS & SHARP EDGES
1 REQD

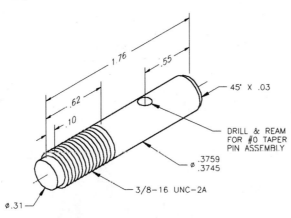

1.76 .55

.62

.10

45° X .03

DRILL & REAM
FOR #0 TAPER
PIN ASSEMBLY

ø .3759
.3745

3/8–16 UNC–2A

ø.31

⑦ **CLAMP SCREW**
SAE 2330
REMOVE ALL BURRS & SHARP EDGES
1 REQD

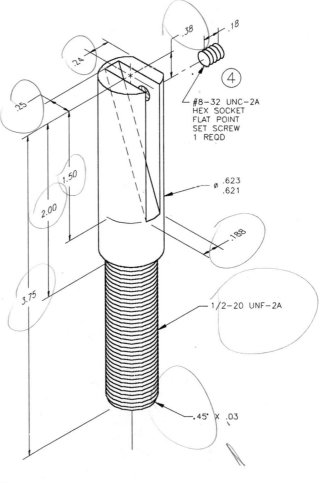

.38 .18

.24

.25

④

#8–32 UNC–2A
HEX SOCKET
FLAT POINT
SET SCREW
1 REQD

1.50

2.00

ø .623
.621

.188

3.75

1/2–20 UNF–2A

.45° X .03

③ **ADJUSTING SCREW**
SAE 1045
REMOVE ALL BURRS & SHARP EDGES
FAO
1 REQD

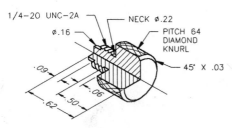

1/4–20 UNC–2A

NECK ø.22

ø.16

PITCH 64
DIAMOND
KNURL

45° X .03

.09

.06

.62 .50

⑧ **PILOT SCREW**
SAE 2330
REMOVE ALL BURRS & SHARP EDGES
FAO
1 REQD

Problem 16–8 Working drawing (in.)

Assembly Name: Precision Vice

PARTS LIST:

ITEM	QTY	NAME	DESCRIPTION	MATERIAL
1	1	BASE		SAE 1040
2	1	BALL WASHER		SAE 1040
3	1	NUT		SAE 1040
4	1	CAP SCREW	⁵⁄₁₆-24NF × 1½ HEX SOCKET HEAD	STL
5	1	JAW		SAE 1040
6	1	JAW INSERT		SAE 4330
7	2	MACH SCREW	¼-28NF × ½ FLAT HD	STL
8	2	CAP SCREW	¼-28NF × ¾ HEX SOCKET HEAD	STL
9	1	JAW INSERT		SAE 4330

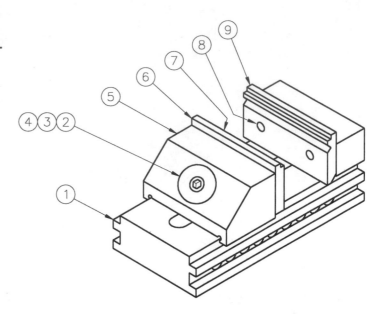

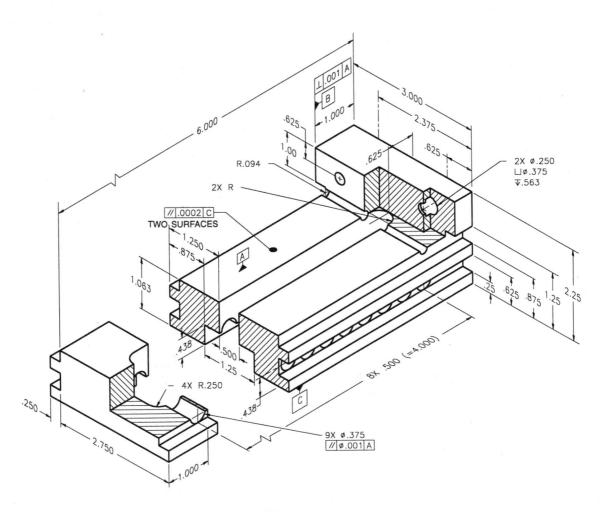

① BASE
FAO 32μIN

Problem 16–8 (Continued)

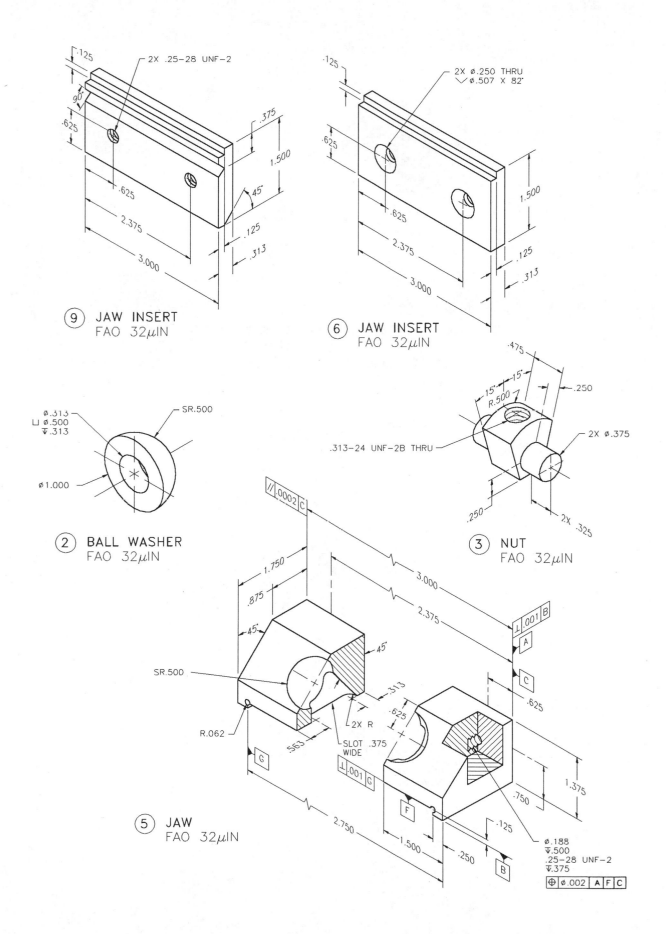

.125

2X .25–28 UNF–2

90°

.625

.375

1.500

.625

45°

2.375

.125

3.000

.313

⑨ JAW INSERT
FAO 32μIN

.125

2X ∅.250 THRU
∨∅.507 X 82°

.625

.625

1.500

2.375

.125

3.000

.313

⑥ JAW INSERT
FAO 32μIN

∅.313
⌴ ∅.500
▽.313

SR.500

∅1.000

② BALL WASHER
FAO 32μIN

.475

15° 15°

R.500

.250

.313–24 UNF–2B THRU

2X ∅.375

.250

2X .325

③ NUT
FAO 32μIN

// .0002 C

1.750

.875

3.000

2.375

45°

⊥ .001 B

A

SR.500

45°

C

.625

.313

R.062

.625

2X R

⊥ .001 G

G

SLOT .375
WIDE

F

.563

1.375

.750

.125

⑤ JAW
FAO 32μIN

2.750

1.500

.250

B

∅.188
▽.500
.25–28 UNF–2
▽.375

⊕ ∅.002 A F C

Problem 16–9   Working drawing (in.)

Assembly Name: Machine Vice
SPECIFIC INSTRUCTIONS:
When preparing the assembly drawings, use separate balloons
for each part in each view, or use only one balloon in the view that
most clearly identifies the part.

PARTS LIST:

ITEM	QTY	NAME	DESCRIPTION	MATERIAL
1	1	SCREW		SAE 4320
2	1	HANDLE		MS
3	2	HANDLE CAP		MS
4	2	MACHINE SCREW	(10).190-32UNF-2 × 6 SLOT FIL HD	STL
5	1	SET SCREW	¼-20UNC-2 × .250 FULL DOG POINT	STL
6	1	MOVABLE JAW		SAE 1020
7	1	MOVABLE JAW PLATE		SAE 4320
8	2	MACHINE SCREW	(10).190-32UNF-2 × .875 SLOT FIL HD	STL
9	1	FIXED JAW PLATE		SAE 4320
10	1	BODY		SAE 4320
11	1	GUIDE		SAE 1020
12	2	MACHINE SCREW	¼-20UNC-2 × .500 SLOT FIL HD	STL

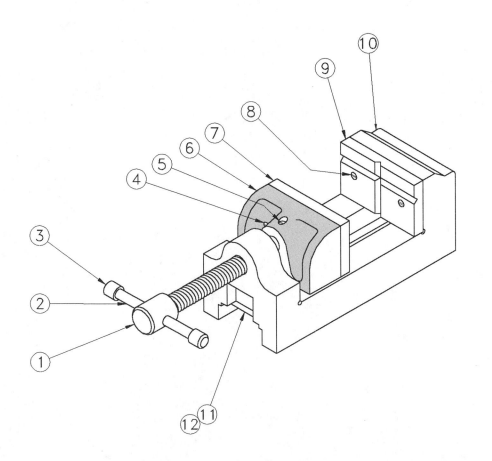

Problem 16–9 (Continued)

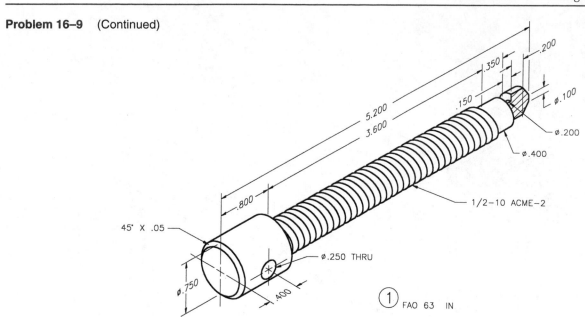

45° X .05

5.200
3.600
.800
.150
.350
.200
ø.100
ø.200
ø.400
1/2–10 ACME–2
ø.250 THRU
ø.750
.400

① FAO 63 IN

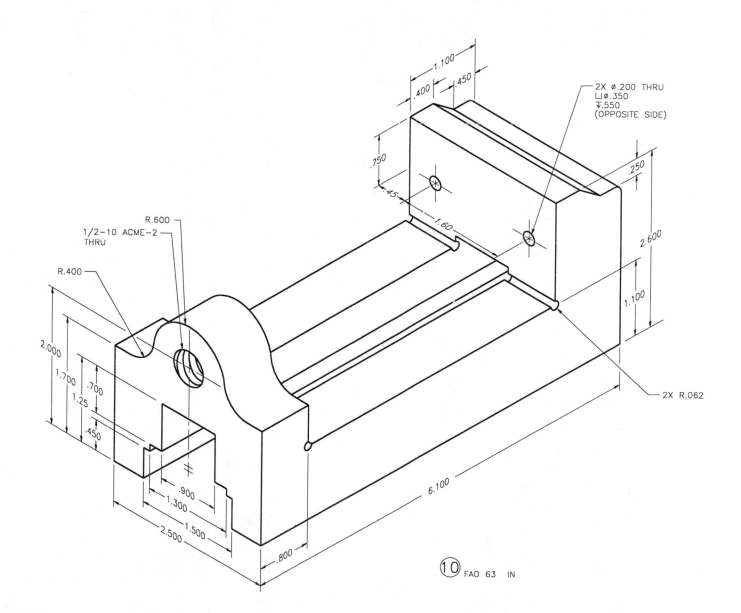

1.100
.400
.450
2X ø.200 THRU
⊔ø.350
▼.550
(OPPOSITE SIDE)
.750
.45
.250
1.60
2.600
1.100
R.600
1/2–10 ACME–2
THRU
R.400
2.000
1.700
.700
1.25
.450
.900
1.300
1.500
2.500
.800
6.100
2X R.062

⑩ FAO 63 IN

Problem 16–9 (Continued)

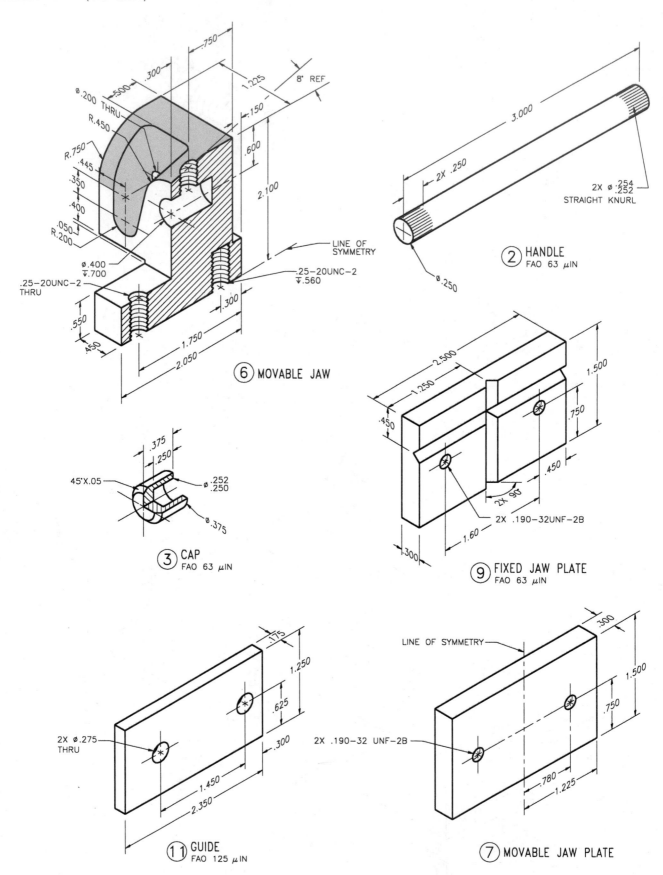

② HANDLE
FAO 63 μIN

⑥ MOVABLE JAW

③ CAP
FAO 63 μIN

⑨ FIXED JAW PLATE
FAO 63 μIN

⑪ GUIDE
FAO 125 μIN

⑦ MOVABLE JAW PLATE

Problem 16–10   Working drawing (in.)

Assembly Name: Arbor Press

PARTS LIST:

ITEM	QTY	NAME	DESCRIPTION	MATERIAL
1	1	BASE		SAE 1020
2	1	TABLE PIN		SAE 1020
3	1	TABLE		SAE 1020
4	2	BALL END		SAE 1020
5	1	SLEEVE		SAE 1020
6	1	HANDLE		SAE 1020
7	1	GEAR		SAE 4320
8	1	RACK PAD		SAE 4320
9	1	COVER PLATE		SAE 1020
10	4	CAP SCREWS	8-32UNC-2 × .50 HEX SOC	STL
11	1	RACK		SAE 4320
12	1	SCREW		SAE 1040
13	1	COLUMN		SAE 1020
14	1	MACHINE SCREW	⅜-16UNC-2 × 1.00 HEX HEAD	STL

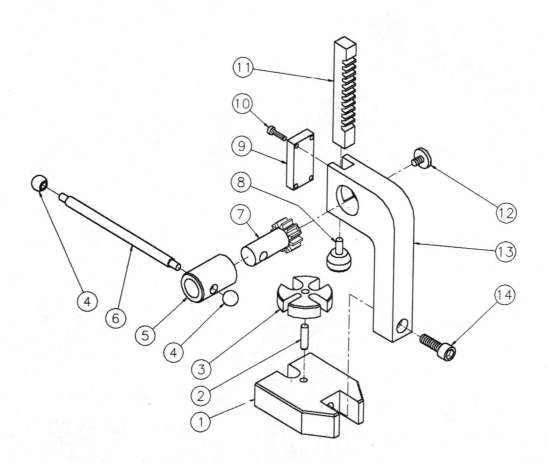

Problem 16–10 (Continued)

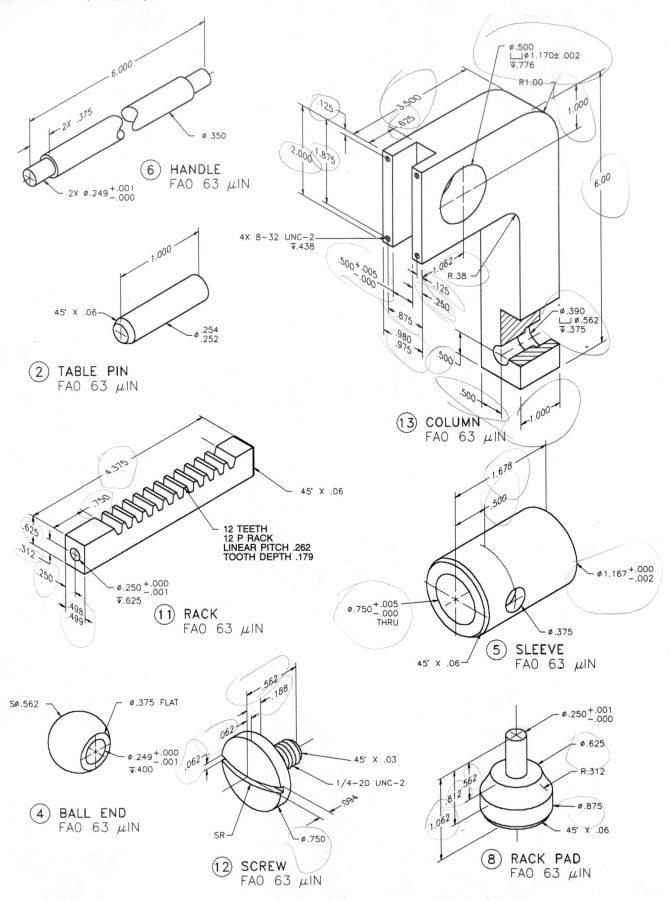

(6) HANDLE
FAO 63 μIN

(2) TABLE PIN
FAO 63 μIN

(11) RACK
FAO 63 μIN

12 TEETH
12 P RACK
LINEAR PITCH .262
TOOTH DEPTH .179

(13) COLUMN
FAO 63 μIN

(5) SLEEVE
FAO 63 μIN

(4) BALL END
FAO 63 μIN

(12) SCREW
FAO 63 μIN

(8) RACK PAD
FAO 63 μIN

Problem 16–10 (Continued)

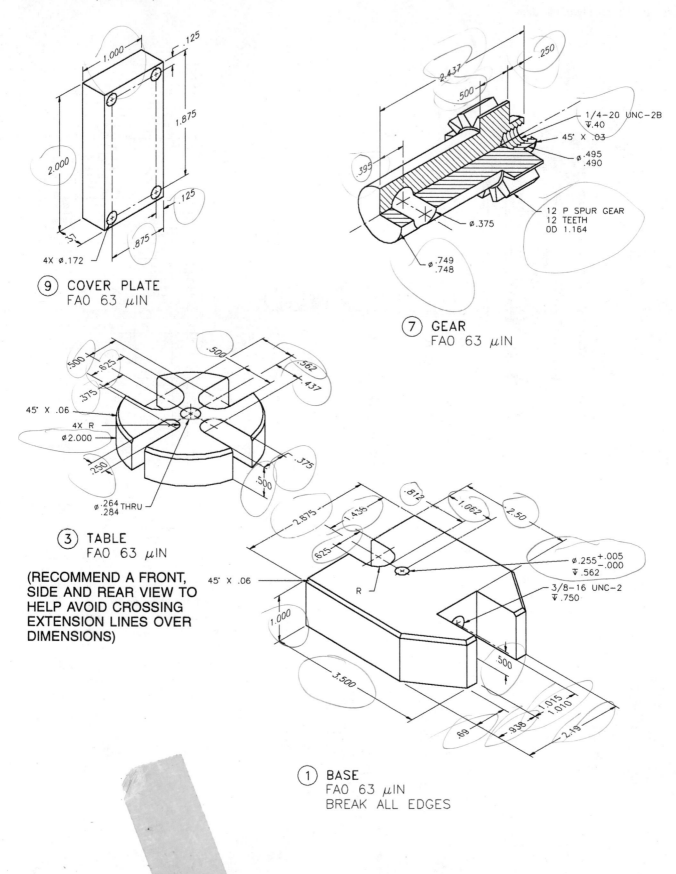

⑨ COVER PLATE
FAO 63 μIN
4X ⌀.172

⑦ GEAR
FAO 63 μIN
1/4–20 UNC–2B
▼.40
45° X .03
⌀ .495 .490
12 P SPUR GEAR
12 TEETH
OD 1.164
⌀.375
⌀ .749 .748
.395

③ TABLE
FAO 63 μIN
45° X .06
4X R
⌀2.000
⌀ .264 THRU .284

(RECOMMEND A FRONT,
SIDE AND REAR VIEW TO
HELP AVOID CROSSING
EXTENSION LINES OVER
DIMENSIONS)

① BASE
FAO 63 μIN
BREAK ALL EDGES
45° X .06
⌀.255 +.005 −.000
▼.562
3/8–16 UNC–2
▼.750
R

Problem 16–11   Working drawing (in.)

Assembly Name: Hydraulic Jack

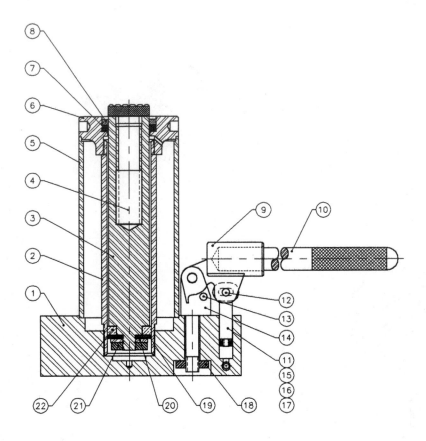

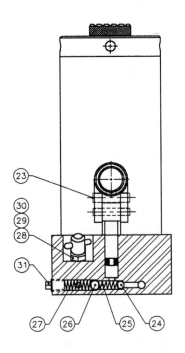

KEY	QTY	NAME	DESCRIPTION	PART NO.
31	5	PLUG	1/8–27 NPT	5DT1031
30	1	HANDLE	NEEDLE VALVE	1DT1030
29	1	VALVE	NEEDLE	1DT1029
28	1	NUT	NEEDLE VALVE	1DT1028
27	1	SPRING	⌀.343 X 1.250 CLOSED ENDS, 9 COILS	5DT1027
26	1	BALL	⌀.3125	5DT1026
25	1	SPRING	⌀.281 X 1.000 CLOSED ENDS, 6 COILS	5DT1025
24	1	BALL	⌀.250	5DT1024
23	1	PIN	PIVOT	1DT1023
22	1	GUIDE	BRONZE	1DT1022
21	1	NUT	.437–20 UNF–2B	5DT1021
20	1	WASHER	PISTON	1DT1020
19	1	CUP	⌀1.500 LEATHER	5DT1019
18	1	NUT	.437–14 UNC–2B	5DT1018
17	1	NUT	10–32 UNF–2B	5DT1017
16	1	WASHER	ANS TYPE B PLAIN NO. 10 N SERIES	5DT1016
15	1	CUP SEAL	⌀.445 NEOPRENE	5DT1015
14	1	SUPPORT	PUMP PIVOT	1DT1014
13	1	PIN	STOP	1DT1013
12	1	PIN	DRIVE	1DT1012
11	1	PLUNGER	PUMP	1DT1011
10	1	HANDLE	PUMP	1DT1010
9	1	SOCKET	PUMP HANDLE	1DT1009
8	1	NUT	PACKING	1DT1008
7	1	PACKING	⌀1.500 LEATHER	5DT1007
6	1	CAP	TOP	1DT1006
5	1	TUBE	RESERVOIR	1DT1005
4	1	SCREW	JACK	1DT1004
3	1	PISTON	JACK	1DT1003
2	1	TUBE	CYLINDER	1DT1002
1	1	BASE	JACK	1DT1001
KEY	QTY	NAME	DESCRIPTION	PART NO.
PARTS LIST				

Part number code 1

1 = Machine part
5 = Purchase part
DT = Drafting Technology
1001 thru 1031 part number
code use 6DT1000 for
assembly drawing number

Problem 16–11   (Continued)

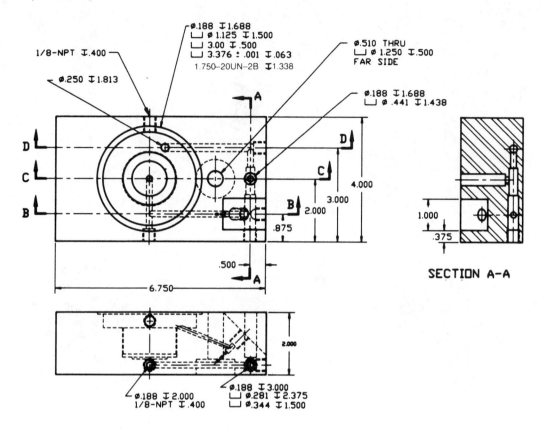

1/8-NPT ⊤.400

⌀.250 ⊤1.813

⌀.188 ⊤1.688
⊔ ⌀ 1.125 ⊤1.500
⊔ 3.00 ⊤.500
⊔ 3.376 ± .001 ⊤.063
1.750–20UN–2B ⊤1.338

⌀.510 THRU
⊔ ⌀ 1.250 ⊤.500
FAR SIDE

⌀.188 ⊤1.688
⊔ ⌀ .441 ⊤1.438

4.000
3.000
2.000
.875
.500
6.750

1.000
.375

SECTION A-A

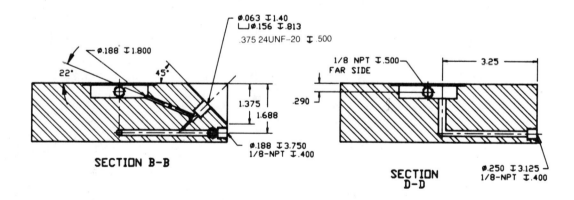

⌀.188 ⊤2.000
1/8-NPT ⊤.400

⌀.188 ⊤3.000
⊔ ⌀.281 ⊤2.375
⊔ ⌀.344 ⊤1.500

2.000

⌀.063 ⊤1.40
⊔⌀.156 ⊤.813
.375 24UNF–20 ⊤.500

⌀.188 ⊤1.800

22°   45°

1.375
1.688

⌀.188 ⊤3.750
1/8-NPT ⊤.400

SECTION B-B

1/8 NPT ⊤.500
FAR SIDE

3.25
.290

**SECTION
D-D**

⌀.250 ⊤3.125
1/8-NPT ⊤.400

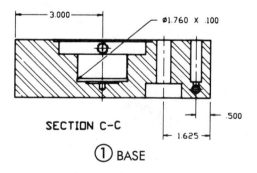

3.000
⌀1.760 X .100

.500
1.625

SECTION C-C

① BASE

Problem 16–11 (Continued)

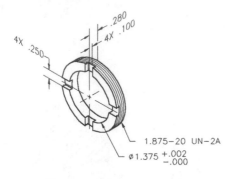

.280
4X .100
4X .250

1.875–20 UN–2A

⌀1.375 +.002 −.000

⑧ PACKING NUT

.125

⌀.447

⌀1.250

⑳ PISTON WASHER

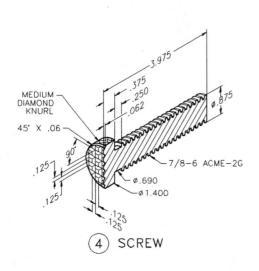

3.975

.375
.250
.062

⌀.875

MEDIUM
DIAMOND
KNURL

45° X .06

90°

.125

.125

.125
.125

7/8–6 ACME–2G

⌀.690

⌀1.400

④ SCREW

5.345

.150

⌀3.200 +.002 −.000

⌀3.375 +.000 −.002

⌀3.068

⑤ RESERVOIR TUBE

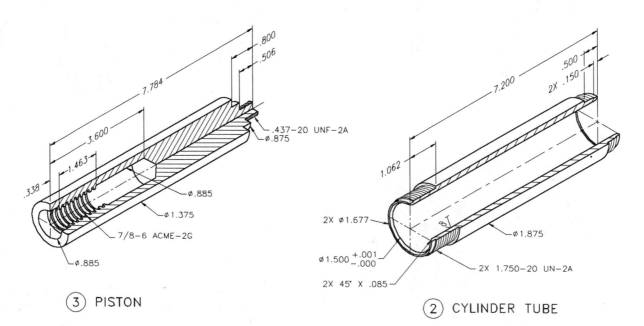

.800

.506

7.784

3.600

.437–20 UNF–2A
⌀.875

1.463

.338

⌀.885

⌀1.375

7/8–6 ACME–2G

⌀.885

③ PISTON

7.200

.500

2X .150

1.062

2X ⌀1.677

⌀1.875

⌀1.500 +.001 −.000

2X 1.750–20 UN–2A

2X 45° X .085

② CYLINDER TUBE

Problem 16–11 (Continued)

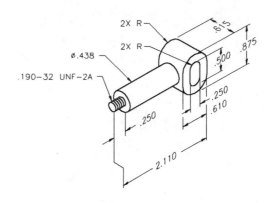

(11) PUMP PLUNGER
HARDEN TO 55 RC

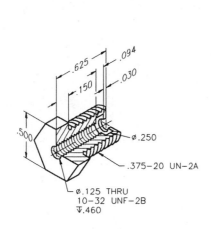

(28) NEEDLE VALVE NUT

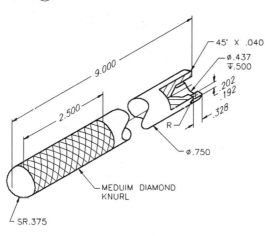

(10) PUMP HANDLE

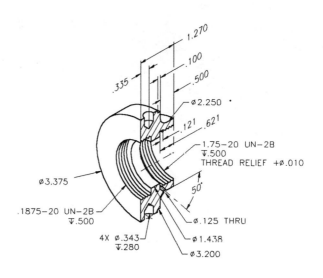

(6) TOP CAP

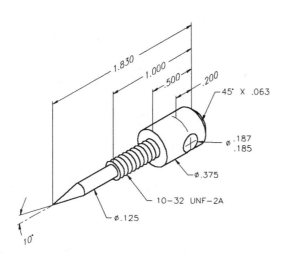

(29) NEEDLE VALVE

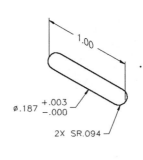

(30) NEEDLE VALVE HANDLE

Problem 16–11   (Continued)

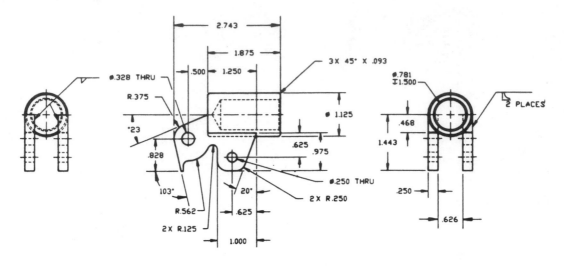

(9) PUMP HANDLE SOCKET

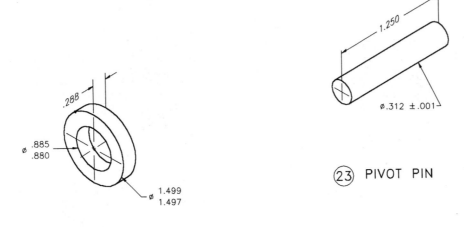

(22) BRONZE GUIDE

(23) PIVOT PIN

(13) STOP PIN

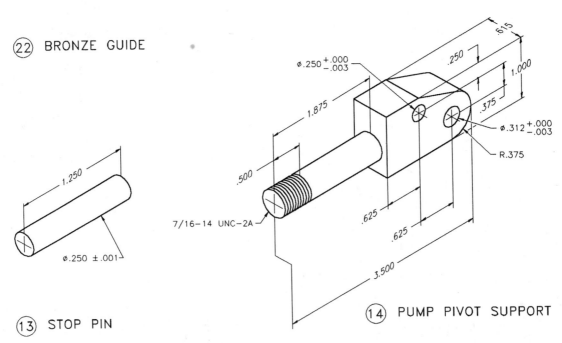

(14) PUMP PIVOT SUPPORT

Problem 16–12   Working drawing (in.)

Assembly Name: Worm Gear Reducer

Advanced project: design changes may be required.

PARTS LIST:

ITEM	QTY	NAME	DESCRIPTION	ITEM	QTY	NAME	DESCRIPTION
1	1	HOUSING		12	4	HIGH SPEED LOCKWASHER	TIMKEN TW-105
2	2	RETAINING PLATE		13	4	MACHINE SCREW	.375-16UNC-2A × 2.250 HEX HEAD
3	1	BEARING CAP		14	4	MACHINE SCREW	.375-16UNC-2A × 1.875 HEX HEAD
4	1	MOTOR ADAPTOR					
5	1	HIGH SPEED SHAFT		15	8	MACHINE SCREW	.375-16UNC-2A × .625 HEX HEAD
6	1	SLOW SPEED SHAFT					
7	1	WORM GEAR	BRONZE	16	1	HIGH SPEED OIL SEAL	PARKER 2-028
8	1	DBL. ROW TAPERED ROLLER BEARING	KOYO46T30305DJ/29.5	17	2	SLOW SPEED OIL SEAL	PARKER 2-020
9	1	SNGL. ROLE CYL. ROLLER BEARING	KOYO CRL11	18	1	SLOW SPEED KEYWAY	1875 × .245 × 1.450
				19	1	SNGL. ROW TAP ROLLER BEARING	KOYO 32005J
10	1	TAPER PLUG	500-16NPT PLUG				
11	1	HEX NUT	.875-16 UN-28	20	2	SLOW SPEED SPACER	TIMKEN TW-506

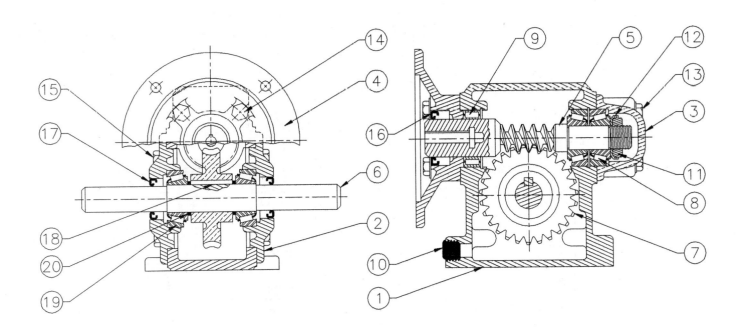

Problem 16–12 (Continued)

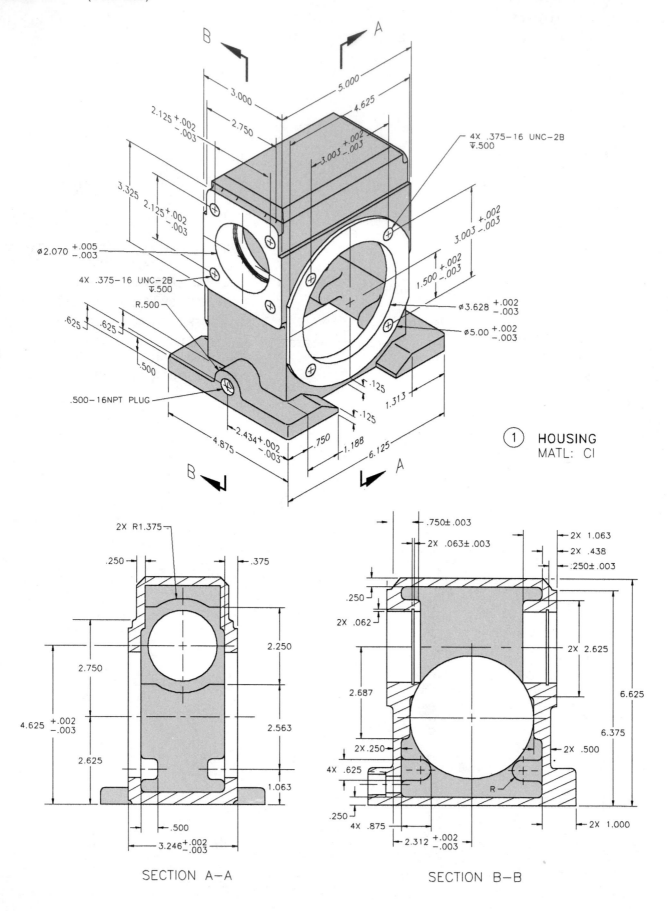

4X .375–16 UNC–2B
▽.500

2.125 +.002 −.003

2.750

5.000

4.625

3.003 +.002 −.003

3.000

3.325

2.125 +.002 −.003

ø2.070 +.005 −.003

4X .375–16 UNC–2B
▽.500

R.500

.625   .625

.500

.500–16NPT PLUG

2.434 +.002 −.003

4.875

.750

1.188

3.003 +.002 −.003

1.500 +.002 −.003

ø3.628 +.002 −.003

ø5.00 +.002 −.003

.125

.125

1.313

6.125

① HOUSING
MATL: CI

SECTION A–A

2X R1.375

.250

.375

2.750

4.625 +.002 −.003

2.625

.500

3.246 +.002 −.003

2.250

2.563

1.063

SECTION B–B

.750±.003

2X .063±.003

2X 1.063

2X .438

.250±.003

.250

2X .062

2.687

2X .250

4X .625

.250

4X .875

2.312 +.002 −.003

2X 2.625

2X .500

R

6.625

6.375

2X 1.000

Problem 16–12 (Continued)

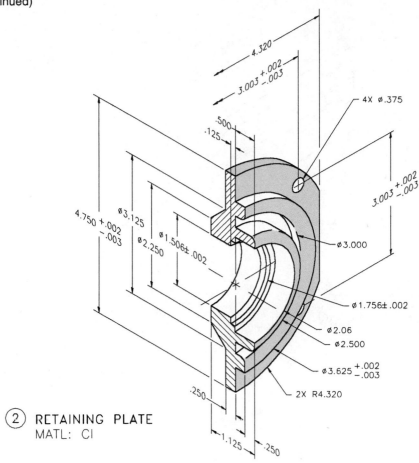

4.320

$3.003 \, ^{+.002}_{-.003}$

.500

.125

4X ∅.375

$3.003 \, ^{+.002}_{-.003}$

∅3.125

∅2.250

∅1.506±.002

∅3.000

$4.750 \, ^{+.002}_{-.003}$

∅1.756±.002

∅2.06

∅2.500

$∅3.625 \, ^{+.002}_{-.003}$

.250

2X R4.320

1.125

.250

② RETAINING PLATE
MATL: CI

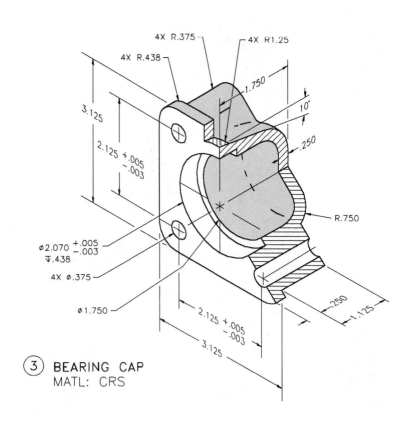

4X R.375

4X R1.25

4X R.438

1.750

3.125

10°

.250

$2.125 \, ^{+.005}_{-.003}$

R.750

$∅2.070 \, ^{+.005}_{-.003}$
▽.438

4X ∅.375

∅1.750

$2.125 \, ^{+.005}_{-.003}$

.250

1.125

3.125

③ BEARING CAP
MATL: CRS

Problem 16–12 (Continued)

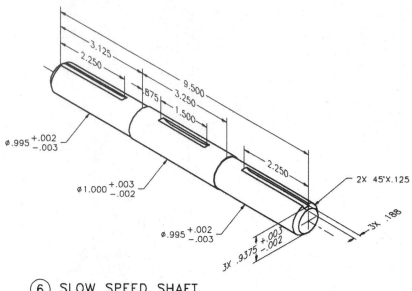

⑥ SLOW SPEED SHAFT
MATL: SAE 4320

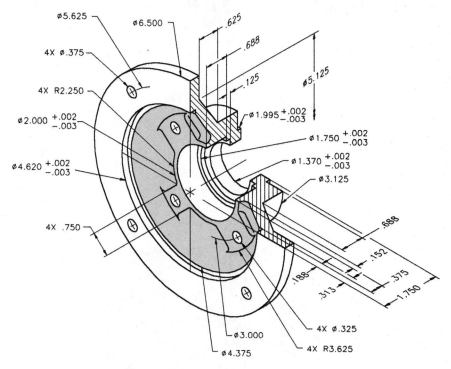

④ MOTOR ADAPTER
MATL: CI

Problem 16–12 (Continued)

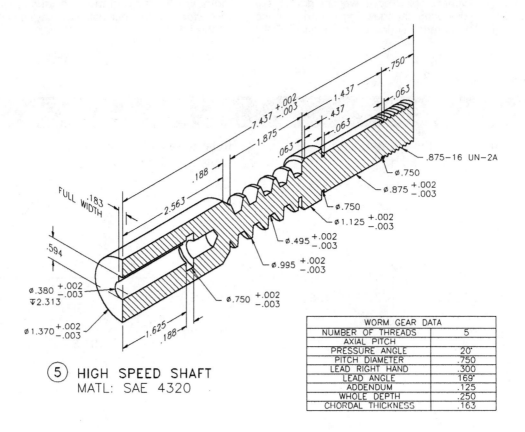

⑤ HIGH SPEED SHAFT
MATL: SAE 4320

WORM GEAR DATA	
NUMBER OF THREADS	5
AXIAL PITCH	
PRESSURE ANGLE	20°
PITCH DIAMETER	.750
LEAD RIGHT HAND	.300
LEAD ANGLE	169°
ADDENDUM	.125
WHOLE DEPTH	.250
CHORDAL THICKNESS	.163

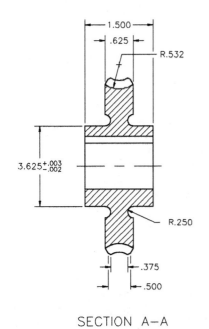

SECTION A–A

⑦ WORM GEAR
MATL: PHOSPHOR BRONZE

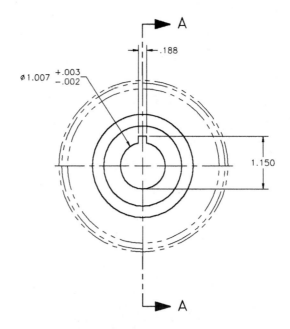

OPTIONAL
DETAILED
REPRESENTATION

SPUR GEAR DATA	
DIAMETRAL PITCH	8
NUMBER OF TEETH	27
PRESSURE ANGLE	20°
PITCH DIAMETER	3.375
BASE CIRCLE DIAMETER	3.6187
CIRCULAR PITCH	.3927
CIRCULAR THICKNESS	.1964
ROOT DIAMETER	3.125

Problem 16–13 Engineering Changes

Your instructor will use your complete set of working drawings to prepare a series of engineering changes. Typical engineering change request (ECR) and engineering change notice (ECN) forms are available in the *Engineering Drawing and Design Instructor's Guide*. Your instructor may copy the forms as needed for this problem.

MATH PROBLEMS

1. Find dimension L in the following figure.

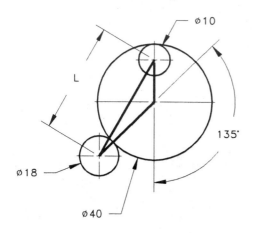

Refer to the following drawing of a general oblique triangle (not drawn to scale).

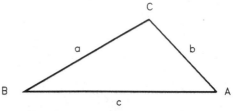

3. Find angle A if a = 15.4, b = 13.1, and c = 12.6.
4. Find side c if C = 61.5°, a = 7.56, and b = 8.94.
5. Find angle B if A = 37.6°, b = 7.63, and c = 8.72.
6. Find angle C if a = 10.1, b = 15.3, and c = 12.7.

2. Find angle Θ in the following drawing.

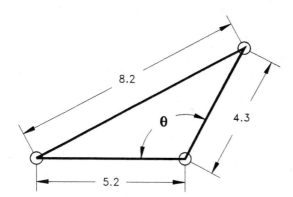

CHAPTER 17

Pictorial Drawings

LEARNING OBJECTIVES

After completing this chapter, you will:

- ○ Draw three-dimensional objects using 3-D coordinates.
- ○ Construct objects using isometric, dimetric, or trimetric methods.
- ○ Construct objects using oblique drawing methods.
- ○ Draw objects using one-, two-, or three-point perspective.
- ○ Apply a variety of shading techniques to pictorial drawings.
- ○ Given an orthographic engineering sketch of a part or assembly, draw it in pictorial form using proper line contrasts and shading techniques.

THE ENGINEERING DESIGN PROCESS

The design and manufacturing department has proposed a new part using 2-D orthographic sketches. The sketches are extremely crude, and it is your task to construct an isometric drawing that can be used for visualization purposes and to construct a prototype. Your first task is to create a readable 2-D sketch from the sketch provided. Check with the designers to verify dimensions and sizes. The following engineering sketch is used to draw an isometric view of the object.

After the 2-D sketch is completed and checked by the designers, begin construction of the isometric view. Use the following steps to construct the part.

1. Choose the view of the object that best shows most of the features of the part. Orient this view facing to the left or right.
2. Use the centerline layout method to locate the axis lines of the circular features.
3. Lay out additional thicknesses and features using the coordinate, or box-in, method.
4. Use proper ellipses to draw circles and arcs.
5. If inking, use different pen widths for line contrast.
6. Apply shading as required.

The completed object is shown below.

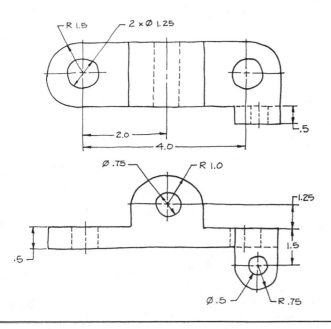

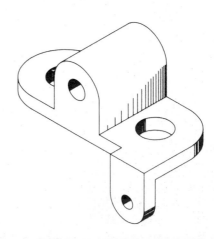

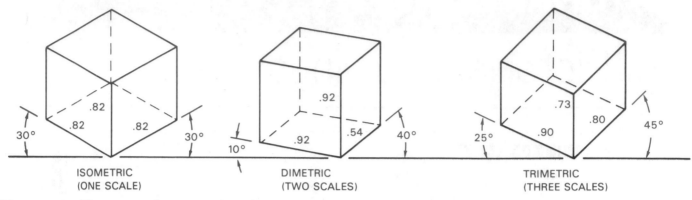

Figure 17–1 Three types of axonometric projections.

▼ PICTORIAL DRAWINGS

Most products are made from orthographic drawings that allow you to view an object with the line of sight perpendicular to the surface you are looking at. The one major shortcoming of this form of drawing is the lack of depth. Certain situations demand a single view of the object that provides a more realistic representation. This realistic single view is achieved with pictorial drawing.

The most common forms of pictorial drawing used in engineering drawing are isometric and oblique. These two basic forms of pictorial drawing are easy to master, if you can visualize objects in orthographic projection and three dimensions.

Isometric drawing belongs to a family of pictorial representation known as *axonometric* projection. Two other similar forms of drawing occupy this group. Dimetric projection involves the use of two different scales as opposed to the single scale used in isometric. Trimetric projection is the most involved of the three and uses three different scales for measurement. Figure 17–1 illustrates the differences in scale between isometric and a common dimetric and trimetric drawing.

The terms drawing and projection should be clarified. A *projection* is an exact representation of an object projected onto a plane from a specific position. The observer's line of sight to the various points on the object passes through a projection plane. The representation on a drawing sheet of the points of the object on the projection plane becomes the *drawing*. Exact projections of objects are time-consuming to make and often involve the use of odd angles and scales. Therefore most drafters and illustrators work with axonometric drawing techniques rather than true projection techniques. Creating an axonometric drawing involves the use of approximate scales and angles that are close enough to the projection scales and angles to be acceptable.

The most realistic type of pictorial illustration is perspective drawing. The use of vanishing points in the projection of these drawings gives them the depth and distortion that you see with your eyes. You will examine each of these types of pictorial drawings in this chapter and discuss step-by-step construction methods for them. In addition you will see how pictorial drawings can be drawn with a computer drafting system.

Technical Illustration

Pictorial drawing is a term that is often interchangeably with technical illustration. But pictorial drawing includes only line drawings done in one of several three-dimensional methods, whereas technical illustration involves the use of a variety of artistic and graphic arts skills and a wide range of media in addition to pictorial drawing techniques. Figure 17–2 is an example of a pictorial drawing, and Figure 17–3 shows a technical illustration. Pictorial drawings are most often the basis for technical illustrations.

Uses of Pictorial Drawings

Pictorial drawings are excellent aids in the design process, for they allow designers and engineers to view the objects at various stages in their development. Pictorial drawings are used in instruction manuals, parts

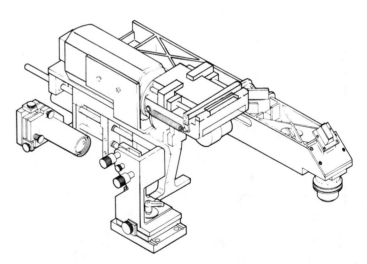

Figure 17–2 Pictorial drawing. *Courtesy Industrial Illustrators, Inc.*

Figure 17–3 Cutaway technical illustration. *Courtesy Industrial Illustrators, Inc.*

catalogs, advertising literature, technical reports and presentations, and as aids in the assembly and construction of products.

▼ ISOMETRIC PROJECTION

The word *isometric* means equal (iso) measure (metric). The three principal planes and edges make equal angles with the plane of projection. An isometric projection is achieved by first revolving the object, in this case a 1-in. cube, 45° in a multiview drawing, Figure 17–4a, then tilting it forward until the diagonal line AE is perpendicular to the projection plane, as seen in the side view of Figure 17–4b. This creates an angle of 35° 16' between the vertical axis AD and the plane of projection. When viewed in the isometric or front view, this axis appears vertical. The remaining two principal axes, AB and AC, are at 30° to a horizontal line.

The three principal axes are called isometric lines, and any line parallel to them is also an isometric line. These lines can all be measured. Any lines not parallel to these three axes are nonisometric lines and cannot be measured. The angles between each of these three isometric axes are 120°. The three planes between the isometric axes, and any plane parallel to them, are called isometric planes. (See Figure 17–5.)

Isometric Scale

The isometric projection is a true representation of an object rotated and tilted in the manner just described. An isometric projection must be drawn using an isometric scale. An isometric scale is created by first laying a regular scale at 45° and projecting the increments of that scale vertically down to a blank scale drawn at an angle of 30°. The resulting isometric scale is seen in Figure 17–6. A 1-in. measurement on the regular scale now measures .816 in. on the isometric scale.

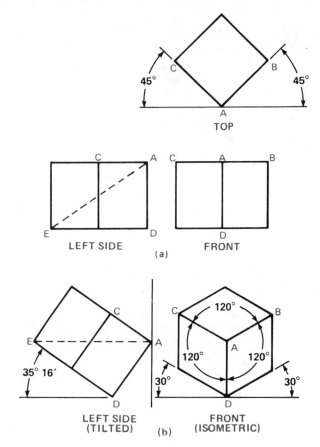

Figure 17–4 Construction of an isometric projection.

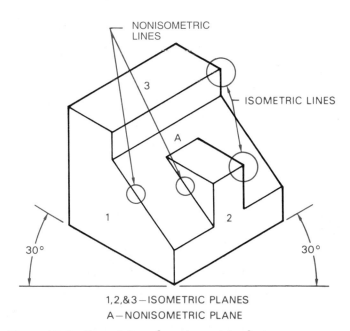

1,2,&3—ISOMETRIC PLANES
A—NONISOMETRIC PLANE

Figure 17–5 Isometric and nonisometric planes.

An isometric drawing is done using a regular scale. This is most common in industry because it does not involve the creation of special scales. The only difference between an isometric drawing and an isometric projection is the size. The drawing appears slightly larger

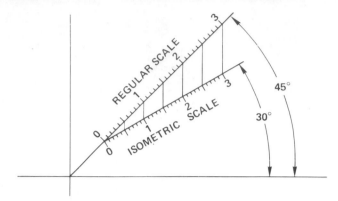

Figure 17–6 Projection of regular scale to isometric scale.

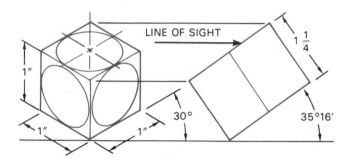

ISOMETRIC DRAWING
BASED ON TRUE MEASUREMENT IN ISOMETRIC VIEW
(PREFERRED METHOD)

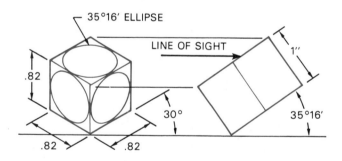

ISOMETRIC PROJECTION
BASED ON TRUE MEASUREMENT IN ORTHOGRAPHIC
(SELDOM USED)

Figure 17–7 The differences between isometric drawing and isometric projection.

than the projection. Figure 17–7 illustrates the differences between the drawing and the projection.

▼ TYPES OF ISOMETRIC DRAWINGS

Isometric drawing is a form of pictorial drawing in which the receding axes are drawn at 30° from the horizontal, as shown in Figure 17–4. There are three basic forms of isometric drawing: these are known as regular, reverse, and long-axis isometric.

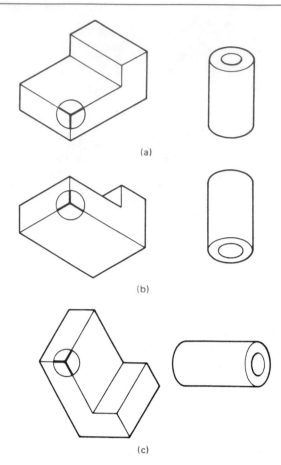

Figure 17–8 Isometric axis variations: **(a)** regular; **(b)** reverse; **(c)** long-axis.

Regular Isometric

The top of an object can be seen in the regular isometric form of drawing. An example can be seen in Figure 17–8a. This is the most common form of isometric drawing, and when using it, the illustrator can choose to view the object from either side.

Reverse Isometric

The only difference between reverse and regular isometric is that you can view the bottom of the part instead of the top. The 30° axis lines are drawn downward from the horizontal line instead of upward. Figure 17–8b shows an example of reverse isometric.

Long-axis Isometric

The long-axis isometric drawing is normally used for objects that are long, such as shafts. Figure 17–8c shows an example of the long-axis form.

The designer or illustrator should choose the view that will give the most realistic presentation of the object. For example, if the object is normally seen from below, then the reverse isometric would be the proper form to use.

▼ ISOMETRIC CONSTRUCTION TECHNIQUES

Just as objects differ in their geometric makeup, so do the construction methods used to draw the object. Different techniques exist to assist the drafter in constructing the various shapes.

Box, or Coordinate, Method

The most common form of isometric construction is the box, or coordinate, method and is used on objects that have angular or radial features. The orthographic views of the object to be drawn are shown in Figure 17–9a.

First an isometric box the size of the overall dimensions of the object (X, Y, and Z) must be drawn. (See Figure 17–9b.) Then the measurements of the features of the object can be transferred to the isometric box. To locate points D and C, just measure dimension U from points F and G. Point E is located at the lower left corner of the box. Draw a line from E to D. Next locate point H by measuring up distance W from the bottom right corner of the box. Draw a light construction line at a 30° angle

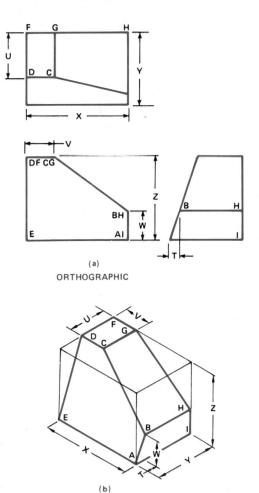

(a)
ORTHOGRAPHIC

(b)
ISOMETRIC

Figure 17–9   Box method of isometric construction: **(a)** orthographic view; **(b)** isometric view.

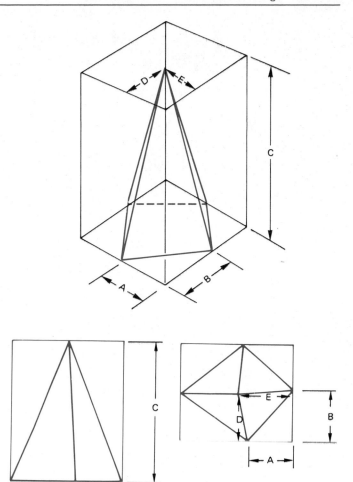

Figure 17–10   Isometric box method for an irregular object.

toward the front vertical axis. Now draw a line from A parallel to line ED until it intersects the line from H. This intersection will be point B. The location of point B can also be found by measuring the horizontal dimension T from point A, then the vertical dimension W.

It may be necessary to draw construction lines on the orthographic views and transfer measurements directly from these views to the isometric view. This method may work well with irregularly shaped objects such as the one shown in Figure 17–10.

Centerline Layout Method

The centerline method begins with the skeleton of the object, the centerlines. This technique is used on objects with many circles and arcs. The use of this method is seen in Figure 17–11a through d. The center points of all of the circles and arcs should be located first. Begin with a good reference point from which you can work, such as the bottom of the object in Figure 17–11a. Center points A and B are first established, then vertical axis lines are drawn up from these. Now the vertical locations of center points C, D, and E can be measured as seen in Figure 17–11b. Then determine the various sizes of the

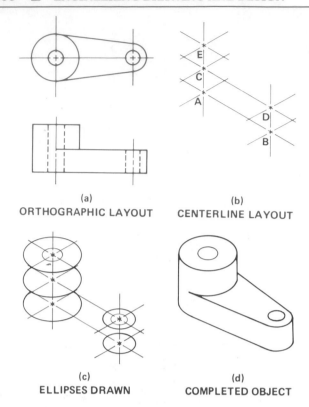

(a)
ORTHOGRAPHIC LAYOUT

(b)
CENTERLINE LAYOUT

(c)
ELLIPSES DRAWN

(d)
COMPLETED OBJECT

Figure 17–11 Isometric centerline layout method: **(a)** orthographic layout; **(b)** centerline layout; **(c)** ellipses drawn; **(d)** completed object.

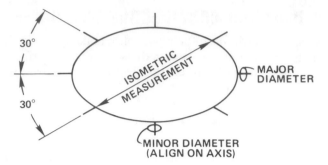

Figure 17–13 Meaning of isometric ellipse template markings.

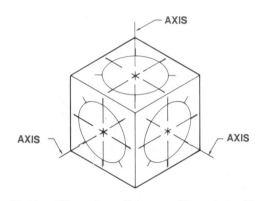

Figure 17–14 Align minor diameter of isometric ellipse template on axis of hole.

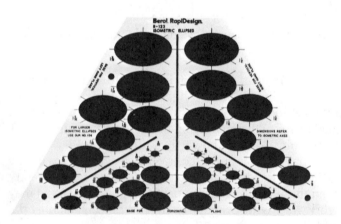

Figure 17–12 Isometric ellipse template. *Courtesy Berol USA/RapiDesign.*

ellipses and draw them at their proper locations as shown in Figure 17–11c. The finished drawing is shown in Figure 17–11d.

Circle and Arc Construction

After the center points of circles and arcs are located, you need to select the correct size of circle from an isometric ellipse template, such as the one shown in Figure 17–12. Note that each ellipse on the template has several tick marks around it. A description of these tick marks is shown in Figure 17–13. An isometric ellipse is mea-

sured on the marks at 30° angles from the large (horizontal) diameter of the ellipse. When aligning an isometric ellipse on the centerlines of your layout, always align the tick marks on the *minor* diameter of the ellipse with the *axis* of the feature, as shown in Figure 17–14. When you do this, two other sets of tick marks on the template will align with the ellipse centerlines.

Arc locations are found in the same way as circles, and the use of the isometric ellipse template is also the same, except just a portion of the circle is drawn. First, find the center point of the arc, then the tangent points. (See Figure 17–15.) Draw the axis line of the arc lightly and again align the minor diameter tick marks with the axis line. Now draw only the portion of the ellipse that is needed for the arc.

Constructing isometric arcs is easy with CADD software such as AutoCAD. Simply use the isocircle option of the ELLIPSE command. The isometric crosshairs can be toggled while the ellipse is attached to the crosshairs, and you can see the ellipse in the left, top, and right side orientations.

Fillets and Rounds

The key to drawing isometric circles and arcs is getting the ellipse aligned properly. Study the object in Figure 17–16. Determine the axis of the circular feature, then align the minor diameter tick marks of the template with the axis of the fillet or round.

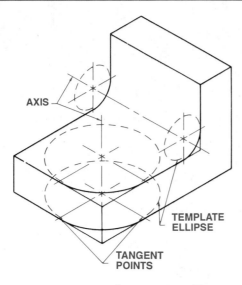

Figure 17–15 Isometric arc layout using ellipse template.

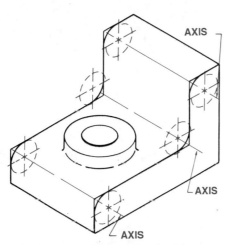

Figure 17–16 Isometric fillet and round layout.

Intersections

A common type of intersection is that of a cylindrical hole passing through an oblique plane. First draw an isometric ellipse at the top of the boxed surface of the object. (See Figure 17–17.) Then draw a series of construction lines parallel to one side of the box that pass through the ellipse. Project each of these lines down the sides of the box and across the oblique surface. Each line forms a trapezoid or parallelogram depending on the shape of the object. The points at which each line intersects the ellipse (A and B, for example) should be projected straight down to the opposite side of the parallelogram. These new points, A_1 and B_1, are now points on the perimeter of the hole in the oblique surface.

Another common form of intersection occurs when two cylinders meet. This construction is similar to the one just described. In Figure 17–18, the center point location of the branch cylinder is determined, and the ellipse is drawn. Next draw as many construction lines

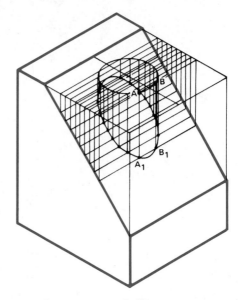

Figure 17–17 Construction of ellipse on isometric oblique plane.

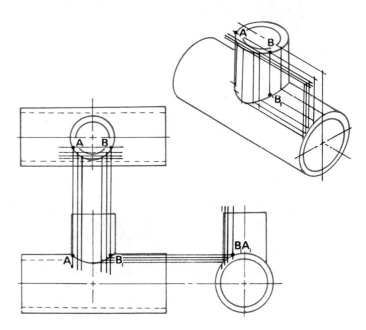

Figure 17–18 Isometric construction of two intersecting cylinders.

passing through the ellipse as needed to produce a smooth curve on the intersection. Project these lines to the end of the main cylinder, down to the edge, then back along the cylinder. Now project points on the ellipse down to the corresponding construction line on the cylinder, A to A_1 and B to B_1, for example. Then connect the points using an irregular curve or polyline in AutoCAD to produce the intersection.

Sections

Full and half sections are common in technical illustration and should be drawn along the isometric axes.

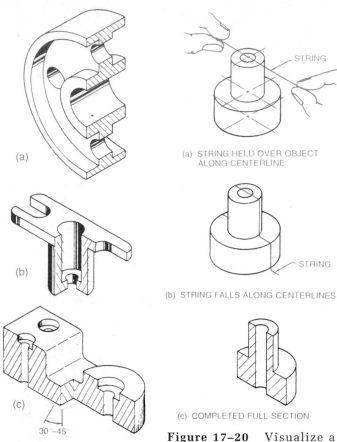

(a)

(b)

(c)

30°–45°

Figure 17–19 Isometric sections: **(a)** full; **(b)** half; **(c)** offset.

(a) STRING HELD OVER OBJECT ALONG CENTERLINE

STRING

(b) STRING FALLS ALONG CENTERLINES

STRING

(c) COMPLETED FULL SECTION

Figure 17–20 Visualize a string dropped along the cutting plane to create an isometric full section.

AXIS

MARK THREAD SPACING WITH TICKS AND CENTERLINES ON AXIS.

Figure 17–21 Isometric thread representation.

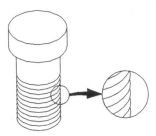

Figure 17–22 Detailed isometric thread representation.

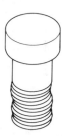

Figure 17–23 Off-center ellipses produce "wandering" threads.

The section lines in full sections should all be drawn in the same direction, while those on a half section should appear in opposite directions. See the section lines illustrated in Figure 17–19a and b. The section lines in offset sections should change directions with each jog in the part, as seen in Figure 17–19c.

There is no preferred way to draw an isometric section. One technique that you might try is to imagine that the cutting-plane line is a long string, and you are stretching it out just above the axis along which you plan to cut. If you let the string fall, it would come to rest along the axis where the cut is to be made, as illustrated in Figure 17–20.

Threads

ANSI Screw threads can be drawn in isometric by first measuring equal spaces along the shaft or hole to be threaded. Then, using the same size ellipse as the diameter of the shaft or hole, draw a series of parallel ellipses. (See Figure 17–21.) These ellipses represent the crests of the threads. This method achieves a simple isometric representation of threads.

A more detailed thread appearance can be achieved by giving each ellipse rounded ends instead of butting the ellipse into the straight sides of the shaft. (See Figure 17–22.) When using this technique, begin drawing the threads from the head of the shaft. With manual drawing, be sure to lay out guidelines for the shaft diameter so the ends of the ellipse do not fall short of the edge of the shaft or go beyond it, as seen in Figure 17–23. This can create a phenomenon known as wandering threads.

Drawing threads, like those shown in Figure 17–22, is quick using software such as AutoCAD. Draw the uppermost thread using an arc or ellipse and edit as needed by trimming or breaking the ellipse so it fits properly. Then use a command like ARRAY, and create a rectangular array of the required number of threads.

Spheres

A sphere drawn isometrically is nothing more than a true circle. Figure 17–24 illustrates the construction of a sphere using three isometric ellipses drawn to represent the perpendicular axes of the circle. If you need to draw

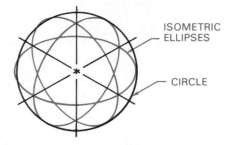

Figure 17–24   Isometric sphere construction.

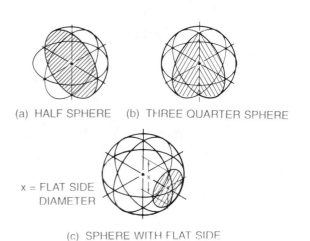

(a) HALF SPHERE   (b) THREE QUARTER SPHERE

x = FLAT SIDE
DIAMETER

(c) SPHERE WITH FLAT SIDE

Figure 17–25   Portions of isometric spheres: **(a)** half sphere; **(b)** three-quarter sphere; **(c)** sphere with flat side.

a sphere, remember to choose a circle that is 1¼ times larger than the actual sphere, because isometric drawings are 1¼ times larger than the actual representation.

Should you need to draw spheres that have been cut in some manner, refer to Figure 17–25. They illustrate a half sphere, three-quarters of a sphere, and a sphere with a flat side.

Isometric Dimensioning

It is not common for isometric drawings to be dimensioned; however, some isometric piping drawings rely heavily on dimensioning to get their message across. One technique that can be used when dimensioning is to letter vertical strokes parallel to extension lines, as shown in Figure 17–26. This gives the appearance that the dimension is lying in the plane of the extension lines. A second simple technique is to draw the heel of the arrowhead parallel with the extension line. This further emphasizes the plane that the dimension is lying in. (See Figure 17–26a.) Examples of unidirectional, aligned, and one-plane (horizontal) isometric dimensioning are shown in Figure 17–26b.

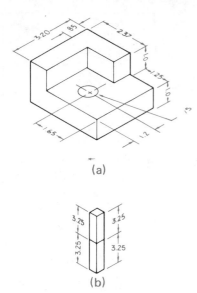

Figure 17–26   Isometric dimensioning: **(a)** dimensions parallel to extension lines; **(b)** various styles of dimensioning.

▼ DIMETRIC PICTORIAL REPRESENTATION

Dimetric projection is similar to isometric projection, but instead of all three axes forming equal angles with the plane of projection, only two axes form equal angles. These can be greater than 90° and less than 180°, but cannot have an angle of 120° because that is the isometric angle. The third axis may have an angle less or greater than the two equal axes, depending on the angles chosen. The two equal angles are said to be foreshortened because they are not measured at full scale. They are foreshortened equally. Since the third axis is projected at a different angle, it is foreshortened at a different scale. Full-size and foreshortened approximate scales are used on the dimetric axes. Some common approximate dimetric angles and scales are shown in Figure 17–27.

▼ TRIMETRIC PICTORIAL REPRESENTATION

The term trimetric refers to three measurements. It is a type of pictorial drawing in which none of the three principal axes makes an equal angle with the plane of projection. Since all three angles are unequal, the scales used to measure on the three axes are also unequal. Trimetric projection provides an infinite number of projections.

Trimetric projection is similar to dimetric projection. Figure 17–28 shows some common trimetric angles for the width and depth axes, plus the scales to use on each of the three axes. The size of angle ellipse to use on each principal plane is also indicated.

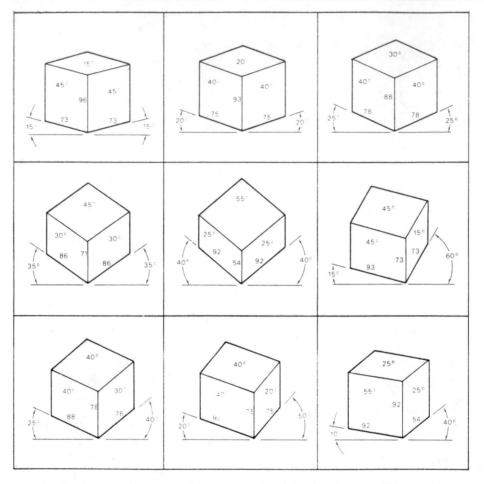

Figure 17–27   Common approximate dimetric angles, scales, and ellipse angles.

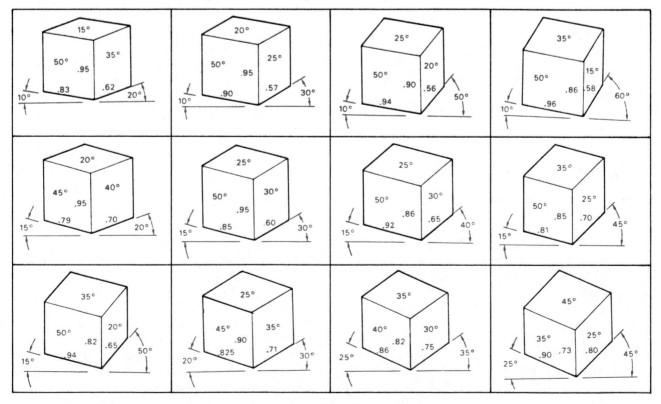

Figure 17–28   Common approximate trimetric angles and scales.

CADD Applications

3-D CAPABILITIES

Computer-generated pictorial line drawings have given drafters, designers, and engineers the ability to draw an object once and then create as many different displays of that object as they can think of. Three-dimensional wire-form capabilities, seen in Figure 17–29, are available with most commercial CADD systems. They allow the operator to view the object as if it were constructed of wire. All edges can be seen at the same time. This is somewhat limiting because complex parts soon become a maze of lines. The solid form, shown in Figure 17–30, is a step closer to reality because it shows only the surfaces that would actually be seen.

The greatest realism in pictorial presentation can be achieved by CADD systems that possess color solids-modeling capabilities. These systems allow the operator to draw the object, shade or color its different parts or surfaces, and rotate it in any direction desired. The operator can also cut into the object at any location to view internal features and then rotate it to achieve the best view. This screen display can then be sent to a color hardcopy unit, a pen plotter, an electrostatic plotter, a 35-mm slide production unit, or videotape.

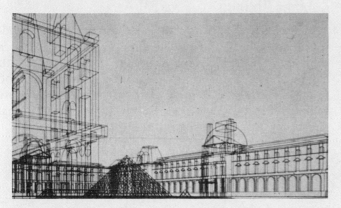

Figure 17–29 The Louvre art museum in Paris, France, drawn as a wire form. *Courtesy Computervision Corporation.*

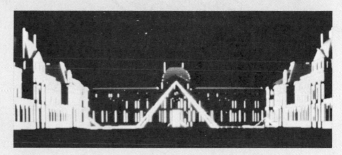

Figure 17–30 Another view of the Louvre, drawn as a solid model. *Courtesy Computervision Corporation.*

▼ EXPLODED PICTORIAL DRAWING

Complicated parts and mechanisms are often illustrated as an exploded assembly in order to show the relationship of the parts in the most realistic manner. (See Figure 17–31a.)

An exploded assembly is a collection of parts, each drawn in the same axonometric method. Any of the pictorial drawing methods mentioned in this chapter can be used to create exploded assemblies. The most important aspect of this type of drawing is to select the viewing direction that illustrates as much of the assembly as possible.

Centerlines are used in exploded views to represent lines of explosion. These lines aid the eye in following a part to its position in the assembly. Centerlines should avoid crossing other centerlines and parts. Centerlines should always be drawn parallel to the axis lines of the drawing regardless of the drawing method selected. The neatest presentation is achieved by leaving gaps where the centerline intersects the part to which it applies and the feature it is mating with. This is shown in Figure 17–31b.

When drawing an exploded assembly with CADD, it is not necessary to draw all the parts in their final positions. Draw each part as needed, placing construction lines for the centerlines, or "lines of explosion." When all parts are completed, move them around as needed to construct the final layout.

▼ OBLIQUE DRAWING

Oblique drawing is a form of pictorial drawing in which the plane of projection is parallel to the front surface of the object. The lines of sight are at an angle to the plane of projection and are parallel to each other. This allows the viewer to see three faces of the object. The front face, and any surface parallel to it, is shown in true shape and size, while the other two faces are distorted in relation to the angle and scale used. Oblique drawing is useful if one face of an object needs to be shown without distortion.

There are three methods of oblique drawing: cavalier, cabinet, and general.

Cavalier Oblique

The *cavalier* projection is one in which the receding lines are drawn true size, or full scale. This form of oblique drawing is usually drawn at an angle to a horizontal of 45°, which approximates a viewing angle of 45°. Because of the scale used on the receding axis, it is not a good idea to draw long objects with the long axis perpendicular to the front face, or projection plane. Objects that have a depth that is smaller than the width can be drawn in the cavalier form without too much distortion. Figure 17–32 shows an object drawn in cavalier oblique.

Cabinet Oblique

A *cabinet* oblique drawing is also drawn with a receding angle of 45°, but the scale along the receding axis is half size. Objects having a greater depth than width can be drawn in this form without the appearance of too much distortion. Cabinet makers often manually drew their cabinet designs using this type of oblique drawing. A cabinet drawing is shown in Figure 17–33.

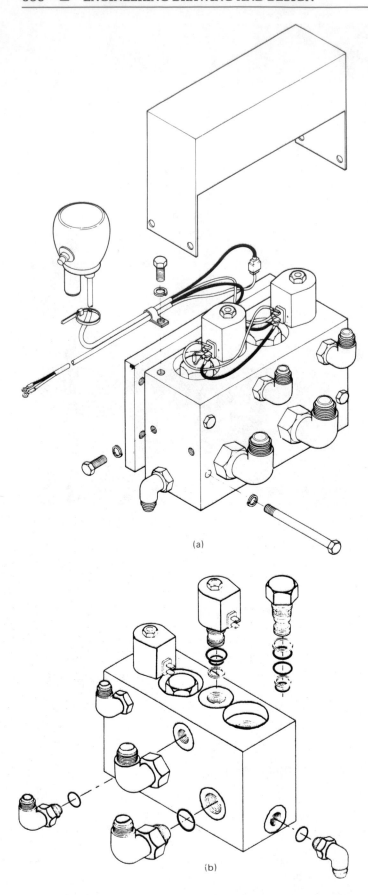

(a)

(b)

Figure 17–31 Exploded assemblies. *Courtesy Industrial Illustrators, Inc.*

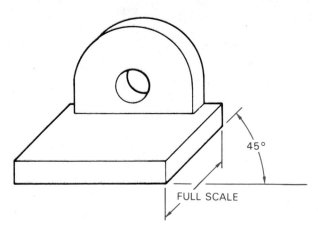

Figure 17–32 Cavalier oblique.

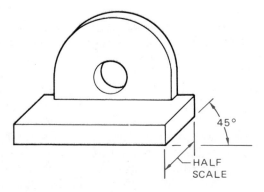

Figure 17–33 Cabinet oblique.

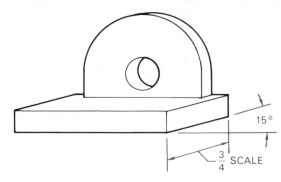

Figure 17–34   General oblique.

General Oblique

The *general* oblique drawing is normally drawn at an angle other than 45°, and the scale on the receding axis is also different from those used in cavalier and cabinet. The most common angles for a general oblique drawing are 30°, 45°, and 60°, but any angle can be used. Any scale from half to full size can be used. A general oblique drawing is shown in Figure 17–34.

▼ PERSPECTIVE DRAWING

Perspective drawing is the most realistic form of pictorial illustration. It reflects the phenomenon of objects appearing smaller the farther away they are until they vanish at a point on the horizon. Lines that are drawn parallel to the principal orthographic planes appear to converge on a vanishing point in perspective drawing.

Computer-aided drafting enables drafters and designers to create pictorial and perspective drawings without much knowledge of the construction techniques involved in either kind of drawing. Architects and drafters have long been using perspective grids to aid in the creation of perspective drawings. But the knowledge of how perspective drawings are made is valuable even for those who use drafting aids and computer systems.

The three types of perspective drawing techniques take their names from the number of vanishing points used in each. *One-point*, or *parallel*, perspective has one vanishing point and is used most often when drawing interiors of rooms. *Two-point*, or *angular*, perspective is the most popular and is used to illustrate exteriors of houses, small buildings, civil engineering projects, and, occasionally, machine parts. The third type, *three-point* perspective, has three vanishing points and is used to illustrate objects having great vertical measurements, such as tall buildings. It is a lengthy process to draw in three-point perspective; therefore it is not used as often as two-point perspective.

The following discussions cover the construction of perspective drawings using manual drawing methods, or CADD 2-D drawing methods. Keep in mind that once an object is drawn in 3-D using some CADD systems, it can then be displayed in a perspective format.

General Concepts

Two principal components of a perspective drawing are the eye of the person viewing the object, and the location of the person in relation to the object. The eye level of the observer is the *horizon line* (HL). This line is established in the elevation (front) view. (See Figure 17–35.) The position of the observer in relation to the object is the *station point* (SP) and is established in relation to the plan (top) view. The location of the station point determines how close the observer is to the object, and the angle at which the observer is viewing the object. The *ground line* (GL) is the line on which the object rests. The *picture plane* (PP), or plane of projection, is the surface (drawing sheet) on which the object is projected. The picture plane can be situated anywhere between the observer and the object. The picture plane can also be located beyond the object. The observer's lines of sight, or visual rays, determine what will show on the picture plane and where. Finally, the three vanishing points discussed in the following sections are referred to as vanishing point right (VPR), vanishing point left (VPL), and vanishing point vertical (VPV).

▼ ONE-POINT PERSPECTIVE

This form of perspective, also known as parallel perspective, has only one vanishing point. The plan view is oriented so that the front surface of the object is parallel to the picture plane. The elevation view is situated below and to the right or left of the plan and rests on the ground line. The following steps will enable you to construct a one-point perspective. Refer to Figure 17–36 as you read the instructions.

Step 1.   Locate the station point between the picture plane and the elevation view of the object. The station point can be anywhere, depending on the part of the object you wish to view. The visual rays from the station point to the extreme corners of the object should form an included angle of approximately 30° to provide the most realistic perspective.

Step 2.   Determine the eye level of the observer in the elevation view. If you wish to look over the object to see the top surface, the horizon line should be drawn above the elevation view. The horizon line can be drawn at any eye level.

Step 3.   The vanishing point is located on the horizon line directly in line with the station point.

Step 4.   Project all points in the plan view that touch the picture plane to the corresponding points

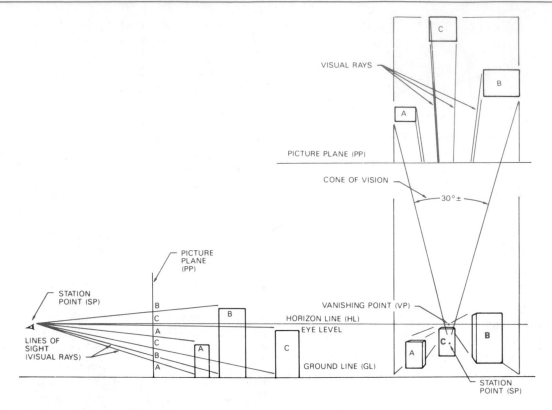

Figure 17–35 Principal components of perspective drawings.

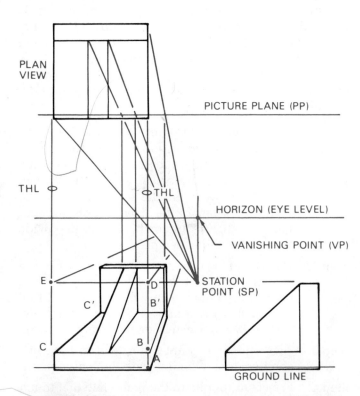

re 17–36 Constructing a one-point perspective drawing.

in the elevation view. These lines are true scale.

Step 5. Draw visual rays from the station point to the rear corners of the object in the plan view.

Step 6. The points where the visual rays intersect the picture plane are projected vertically down to the perspective view. When drawing complex objects, work with only a portion of these points at one time to avoid confusion.

Step 7. Project points A, B, and C toward the vanishing point to intersect the corresponding projectors from the picture plane. The horizontal line B'C' can then be drawn. Remember, any line that is parallel to the PP in the plan view must be parallel to the GL in the perspective view.

Step 8. Project the height of the object from the elevation view to points D and E on the two *true-height lines* (THL). True-height lines are projected from points that touch the PP.

Step 9. Project the height at points D and E back toward the vanishing point to intersect the projectors from the rear portion of the object in the plan view.

Step 10. Complete the object by connecting the ends of the sloping portion and projecting the base of the sloped feature toward the vanishing point.

▼ TWO-POINT PERSPECTIVE

The two-point perspective method is also termed angular perspective because it is turned so that two of its principal planes are at an angle to the picture plane. This is the most popular form of perspective drawing. Its two vanishing points allow parallel lines on the two principal planes to be projected in two directions, thus giving another dimension to the depth of the perspective.

We will use the same object drawn previously in the discussion of one-point perspective. The object has been positioned at an angle in the plan view, with one corner touching the PP. The elevation view, SP, HL, and GL have all been established. Refer to Figure 17–37 as you read the instructions.

Step 1. Draw a line from the station point, parallel to each side of the object in the plan view, to intersect the PP.

Step 2. Project the points on the PP down to the HL. This establishes vanishing point right (VPR) and vanishing point left (VPL) on the horizon line.

Step 3. Project point A on the PP down to the GL. This becomes the true-height line AB in the perspective view.

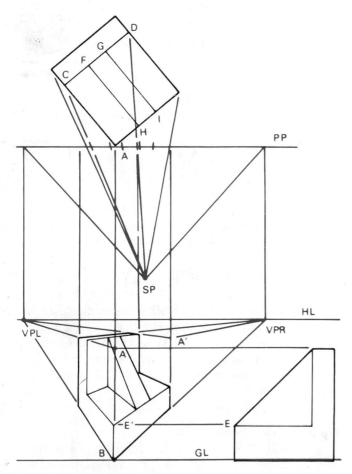

Figure 17–37 Constructing a two-point perspective drawing.

Step 4. Begin blocking in the object by projecting points A and B to the two vanishing points. This establishes the two angular, or perspective, sides of the object.

Step 5. Project visual rays from the SP to the extreme corners of the object in the plan view. Remember that this cone of vision formed by the visual rays should be approximately 30°.

Step 6. Project the intersection of these visual rays with the PP down to intersect with the projectors from points A and B to the two vanishing points. This blocks in the two sides of the object.

Step 7. Draw lines from the SP to points C and D. Where these lines intersect the PP, project vertical lines down into the perspective view. The height of the object at points A and A' can now be projected toward the VPL to intersect with the projectors from points C and D.

Step 8. The height at E can be projected to E' on the THL. Project E' toward both vanishing points to create the basic shape of the part.

Step 9. Project points F, G, H, and I to the station point. Where these lines intersect the PP, drop vertical projectors down to the perspective view. These lines will intersect corresponding points on the perspective drawing. Connect the points as shown in Figure 17–37 to complete the perspective view.

▼ THREE-POINT PERSPECTIVE

Three vanishing points are used in three-point perspective. These drawings require more time to construct than do two-point perspectives, and often occupy a considerable area on the drawing sheet. Three-point perspective drawings are used when certain effects are needed for visual stimulation. You will examine the method of constructing a three-point perspective that requires the least amount of drawing space. Refer to Figure 17–38 as you read the instructions.

Step 1. Draw an equilateral triangle to occupy as much of the sheet as possible. Label the three corners as vanishing points VPR, VPL, and VPV.

Step 2. Construct perpendicular bisectors of each of the three sides. Their intersection at the center of the triangle should be labeled SP (station point). (See Figure 17–38a.)

Step 3. Draw a horizontal line through the SP. Measure and mark the length of the object to the left of the SP. Measure and mark the width of the object to the right of the SP. This will allow you to draw an object in perspective as if you

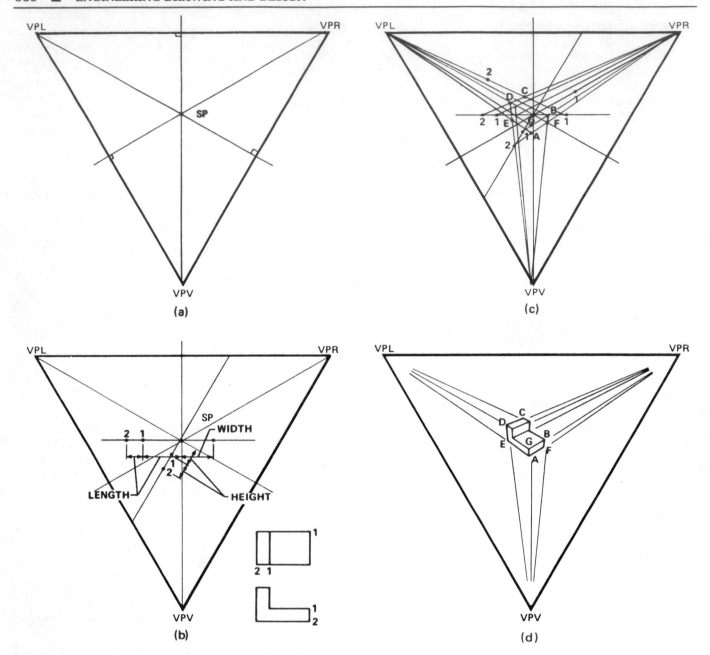

Figure 17–38 Constructing a three-point perspective drawing: **(a)** establishing vanishing points and the station point; **(b)** establishing the length, width, and height of the object; **(c)** establishing the outline of the object; **(d)** the completed object.

had turned the plan view 45° to the picture-plane line. (See Figure 17–38b.)

Step 4. Draw a line parallel to line VPV, VPR. This becomes the height measuring line. Measure the height of the object from the SP down to the left along this line, as shown in Figure 17–38b. Note that numbers have been placed at intervals along the measuring lines. These indicate measurements of features on the object.

Step 5. Project lines from points 1 and 2 to the left of the SP to VPR. Next, draw a line from point 1 at the right of the SP to VPL. Draw lines from

points 1 and 2 on the height measuring line to VPR. (See Figure 17–38c.) You have now blocked in the overall measurements of the object. The following points have been located: lower front corner (A), upper right corner (B), upper far corner (C), and upper left corner (D). Note in Figure 17–38c that the station point (SP) is the upper front corner of a box that encloses the object.

Step 6. Project points B and D to VPV, then project A to VPR and VPL. The intersections of these lines are points E and F, the lower left and right corners respectively. Point G is the height of the front corner of the object as shown in Figure 17–38c. Keep in mind that any surface of the object that is parallel to one of the principal planes of the orthographic view must project to one of the vanishing points as you draw the remainder of the object's features. The completed object is shown in Figure 17–38d.

▼ CIRCLES AND CURVES IN PERSPECTIVE

In most cases, circles in perspective will appear as ellipses. But if a surface of an object is parallel to the picture plane, any circle on that surface will appear as a circle. (See Figure 17–39.) Circles located in planes that are at an angle to the picture plane will appear as ellipses and can be drawn by a method of intersecting lines projected from the elevation and plan view, which is referred to as the coordinate method.

The object having the circles should first be drawn in plan and elevation, and all of the necessary lines and points for your perspective drawing should be determined and placed on the drawing. (See Figure 17–40.) Next divide the circle in the elevation view into a convenient number of pie-shaped sections. Project the intersections of these section lines with the circle to the top and side of the object. The points along the side of the object can then be projected onto the perspective view. Now transfer the distances formed by the intersection of the top of the object and the lines projected from the pie-shaped sections in the elevation view to the plan view. Next draw visual rays from the SP to the points in the plan view. Where the visual rays intersect the picture

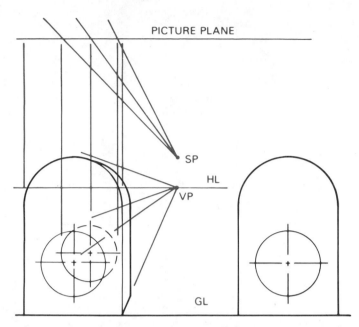

Figure 17–39 Circles in surfaces parallel to the picture plane.

plane line, project the points straight to the perspective view. These projectors will intersect the ones drawn from the elevation view. Connect these intersection points with an irregular curve or an ellipse template.

The same method can be used to draw irregular curves. Establish a grid, or coordinates, on the curve in the elevation view and place the same divisions on a plan view. Project these two sets of coordinates to the perspective view and then use an irregular curve to connect the points. (See Figure 17–41.)

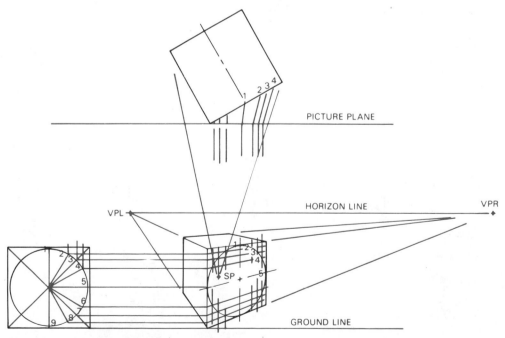

Figure 17–40 Circles plotted in perspective by the coordinate method.

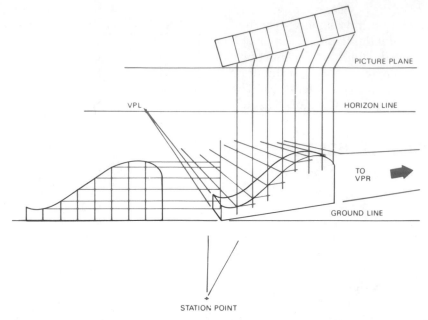

PICTURE PLANE

HORIZON LINE

VPL

TO VPR

GROUND LINE

STATION POINT

Figure 17–41   Irregular curve plotted in perspective by the coordinate method.

▼ BASIC SHADING TECHNIQUES

Line Shading

Most pictorial drawings are created to illustrate shape description and to portray the relationship of parts in an assembly clearly. Highly artistic renderings are not normal for most industrial purposes; therefore any shading techniques used should be simple while conveying the desired effect.

The objects shown in Figure 17–42 illustrate the most basic form of shading, called line-contrast shading. Vertical lines opposite the light source and bottom edges of the part are drawn with the thick lines. Some illustrators will outline the entire object in a heavy line to make it stand out as shown in Figure 17–42b.

Straight-line shading is a series of thin straight lines that can be varied to achieve any desired shading. They can be used on flat and curved surfaces. Note in Figure 17–43 the two ways that straight lines can be used on curved surfaces.

Additional emphasis can be given to curved surfaces with the use of block shading. Figure 17–44 shows how block shading on one or both sides of a curved surface can produce a highlight effect. One to three lines of shading are normally used to achieve the block effect. The total amount of block shading should be approximately one-third the width of the object.

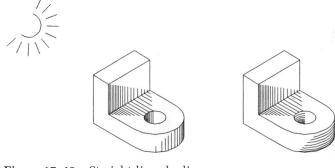

Figure 17–43   Straight-line shading.

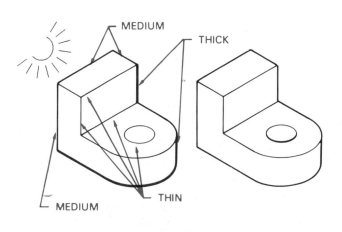

(a) LINE CONTRAST   (b) OUTLINE

Figure 17–42   Line-contrast shading.

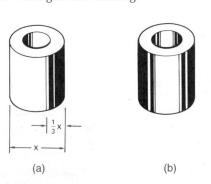

(a)   (b)

Figure 17–44   Block shading.

Figure 17–45 Stipple shading.

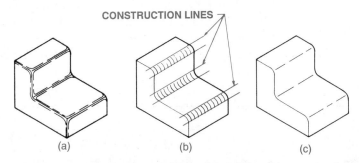

Figure 17–46 Depicting fillets and rounds.

Stipple shading is just dots. (See Figure 17–45.) The closer they are, the darker the shading. Although stippling takes longer than line shading, the results can be quite pleasing.

Fillets and Rounds

Fillets and rounds can be depicted in three ways, which are shown in Figure 17–46. Example (a) uses three lines to indicate the two outside edges (tangent points) of the radius and the centerline of the radius. Note the occasional gaps in the outside line.

With the technique in example (b), the same size ellipse is moved along the axis of the radius and repeated at regular intervals. It is important to draw construction lines first for the outside edges of the fillet or round. This ensures that all of the ellipses are aligned. Draw these ellipses with a thin pen or pencil point. Using a CADD system, you can copy or array the arc or ellipse and do not need to draw the construction lines.

The example at (c) is the simplest and requires the least amount of time to draw. A broken line is drawn along the axis of the radius to indicate the curved surface.

▼ LAYOUT TECHNIQUES

The best way to begin any pictorial drawing, especially if you are working from an orthographic drawing, is to make a quick freehand sketch. This will enable you to see the part in a 3-D form before you try to create a drawing. Always know what the final drawing is going to look like before you begin.

Objects that have boxy or angular shapes should be drawn by first laying out the overall sizes of the part. Box in the shape. Then begin to cut away from the box using measurements from the part or drawing. That is called the coordinate, or "box-in," method. Use construction lines to assist in the layout, be it a manual or a CADD drawing. If the object has circular features, use the following layout techniques.

Step 1. Lay out the centerlines and axis lines of these features first. (See Figure 17–47a.) They act as the skeleton of the part.

Step 2. Draw the circular features using ellipses. (See Figure 17–47b.)

Step 3. Join ellipses with straight lines. (See Figure 17–47c.)

Step 4. Add additional features such as shading, notes, and item tags. If you are drawing manually, trace the object lines in pencil or ink and add shading as required. Any notes or item tags that are needed can be added last. If you are working with a CADD drawing, you can edit the drawing so only the object lines show. If you have created a special "construction" layer for your CADD drawing, turn this layer off when the drawing is completed. The finished drawing is shown in Figure 17–47d.

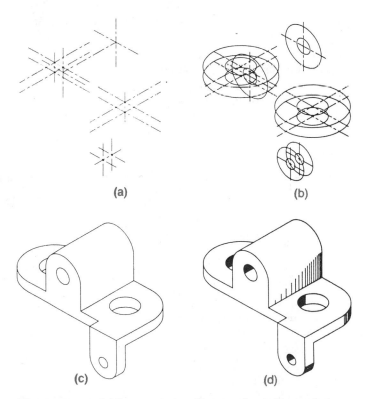

Figure 17–47 **(a)** Lay out centerlines and axis lines of circular features; **(b)** draw circular features using ellipses; **(c)** join ellipses with straight lines; **(d)** complete the drawing with shading.

PROFESSIONAL PERSPECTIVE

Pictorial drawing requires a good ability to visualize objects in three-dimensional form. Companies that produce parts catalogs, instruction manuals, and presentation drawings require the services of a technical illustrator or someone skilled in pictorial drawing. A part of your professional portfolio should be 8" × 10" photo reductions or laser prints of your best pictorial drawings. Reductions of large drawings can look good, and small mistakes tend to fade away when reduced.

The field of technical illustration is much more limited than engineering and drafting but is wide open to the free-lancer or student looking for quick jobs. Always keep examples of a variety of pictorial drawing types in your portfolio. Your portfolio should be a three-ring binder with tabbed dividers for the different types of designs and drawings you have created. Manila pocket folders that fit into the binder are excellent for holding folded blueprints and reduced pictorial drawings. Remember that often a small free-lance job can lead to bigger jobs and even to owning your own company.

MATH APPLICATION

An isometric drawing is constructed using the actual measurements of the part. But because isometric drawing does not take into account "foreshortening," the object appears larger in the drawing than it is in reality. To compensate for this, isometric projection can be used to show the object in its foreshortened appearance. If a part measured one inch on a side, a true isometric projection would create an isometric line that measured .816 in.

In order to construct an isometric view of an object that is a true isometric projection, it is necessary to multiply the actual measurements by .816 to obtain the drawing measurements. The following measurements are converted as follows:

1.45 in. × .816 = **1.183 in.**
3.67 in. × .816 = **2.995 in.**
.89 in. × .816 = **.726 in.**

CHAPTER

17 PICTORIAL DRAWINGS TEST

MULTIPLE CHOICE

Select the best choice a, b, c, or d that answers or completes each question or statement.

1. Isometric drawing belongs to a family of pictorial representation known as:
   a. orthometric
   b. axonometric
   c. dimetric
   d. trimetric

2. What is the angle in degrees of a true isometric ellipse?
   a. 34° 16'
   b. 36° 15'
   c. 35° 15'
   d. 35° 16'

3. What is the angle between each of the three isometric axes?
   a. 130°
   b. 125°
   c. 120°
   d. 140°

4. Which type of isometric drawing is best for representing cylindrical parts?
   a. long-axis
   b. short-axis
   c. regular
   d. reverse

5. Which layout method is best for objects with many circles and arcs?
   a. box-in
   b. long-axis
   c. dimetric
   d. centerline

6. The diameter of an isometric circle is measured along its:
   a. major diameter
   b. minor diameter
   c. 35° 16' from horizontal diameter
   d. 30° from horizontal diameter

7. What is the most realistic form of pictorial drawing?
   a. perspective   c. isometric
   b. dimetric   d. trimetric

8. Which of the following is not a form of oblique drawing?
   a. cabinet   c. long-axis
   b. general   d. cavalier

9. Where are the vanishing points located in perspective drawing?
   a. ground line   c. horizon line
   b. picture plane   d. eye level

10. What is the best angle for the cone of vision in perspective drawing?
    a. 45°   c. 40°
    b. 30°   d. 35°

11. Points of the object in the plan view of a perspective drawing that touch the _____ line, are the only lines that are true scale in the perspective view.
    a. ground   c. picture plane
    b. horizon   d. eye level

12. The vanishing point is located directly above or below the _____ in one-point perspective drawing.
    a. ground line   c. picture plane
    b. horizon line   d. station point

13. Which of the following is *not* a commonly used form of shading?
    a. stipple   c. fillet
    b. block   d. line contrast

14. Block shading normally covers _____ of the object surface.
    a. 1/2   c. 1/8
    b. 1/4   d. 1/3

CHAPTER

17 PICTORIAL DRAWINGS PROBLEMS

DIRECTIONS

Choose the best axis to show as many features of the object as possible in your axonometric or oblique drawing.

For axonometric problems—problems numbered 17–1 through 17–11—draw isometric, dimetric, or trimetric as assigned. For oblique problems—problems numbered 17–12 through 17–17—draw cavalier, cabinet, or general oblique as assigned. Remember that circular features are best shown in the front plane of the oblique view.

For perspective problems—problems numbered 17–18 through 17–25—make a one-, two-, or three-point perspective drawing as assigned except for problem 17–18, which should be done as a one-point perspective view. All objects may be turned at any angle on the picture-plane line for viewing from the station point except problem 17–18, which should be drawn in the direction indicated.

1. Make a freehand sketch of the object to assist in visualization and layout of axonometric and oblique problems.

2. For axonometric and oblique problems, select a scale to fit the drawing comfortably on an A- or B-size drawing sheet. Use a C- or D-size drawing sheet for drawing an initial layout of the perspective problems on sketch or butcher paper.

3. Dimension axonometric or oblique problems only if assigned by your instructor. Do not place dimensions on a perspective view.

4. Perspective objects without dimensions can be measured directly and scaled up as indicated or assigned.

5. Trace the perspective view in pencil or ink on vellum or Mylar®.

6. Make your drawings using a CADD system if appropriate with course guidelines.

Problem 17–1   Axonometric projection

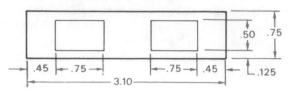

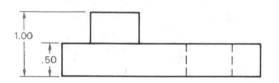

Problem 17–4   Axonometric projection (metric)

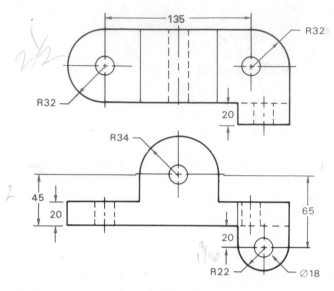

Problem 17–2   Axonometric projection

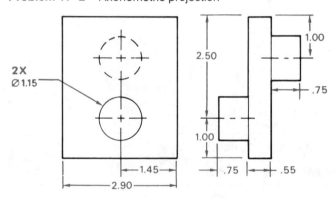

Problem 17–5   Axonometric projection

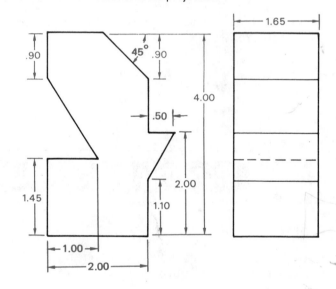

Problem 17–3   Axonometric projection

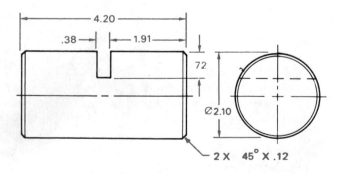

Problem 17–6   Axonometric projection (metric)

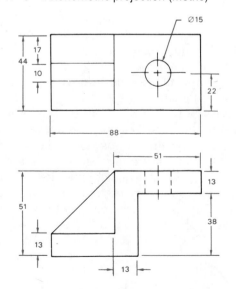

Problem 17–7   Axonometric projection

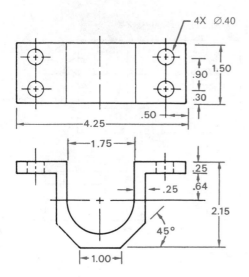

Problem 17–10   Axonometric projection

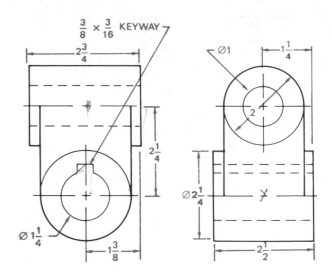

Problem 17–8   Axonometric projection (metric)

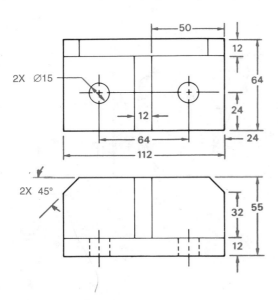

Problem 17–11   Oblique projection

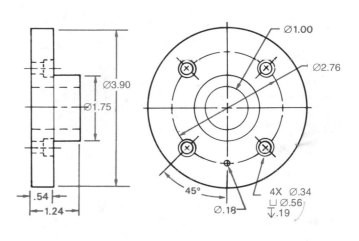

Problem 17–9   Axonometric projection (metric)

Problem 17–12   Oblique projection

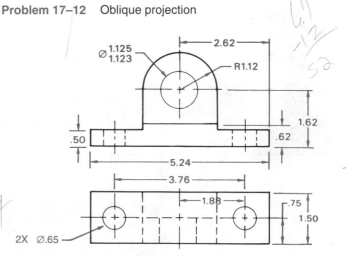

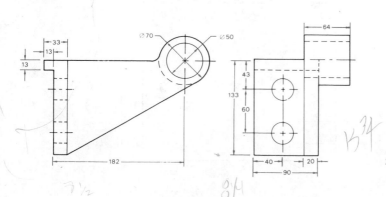

Problem 17–13   Oblique projection

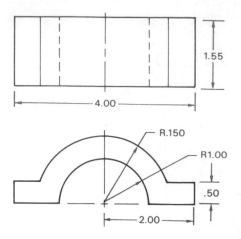

Problem 17–16   Oblique projection

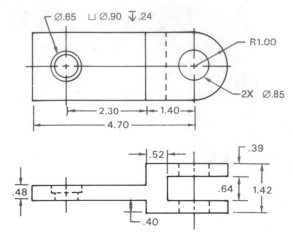

Problem 17–14   Oblique projection (metric)

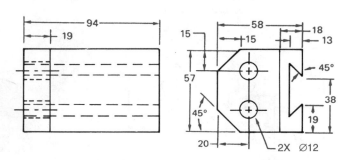

Problem 17–17   Oblique projection

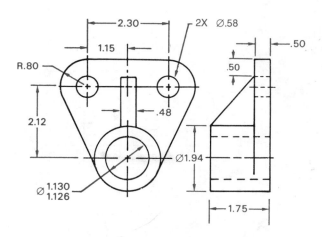

Problem 17–15   Oblique projection

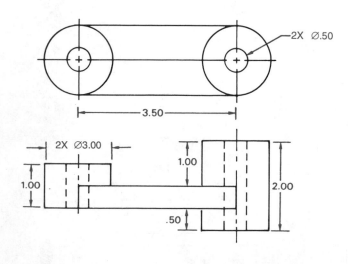

Problem 17–18   Perspective

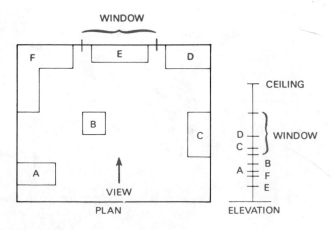

Problem 17–19 Perspective

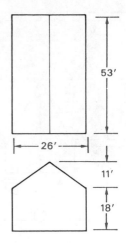

Problem 17–20 Perspective

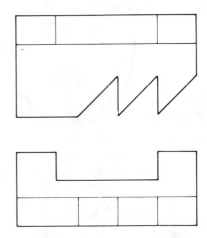

MEASURE AND INCREASE 2X

Problem 17–21 Perspective

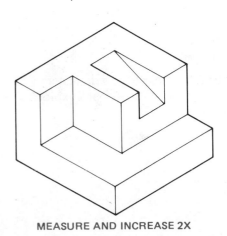

MEASURE AND INCREASE 2X

Problem 17–22 Perspective

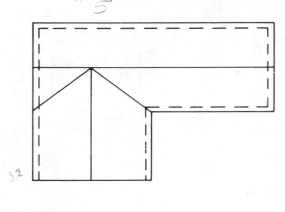

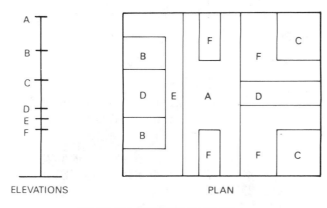

MEASURE AND INCREASE 2X

Problem 17–23 Perspective

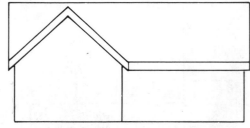

MEASURE AND INCREASE 3X

Problem 17–24 Perspective

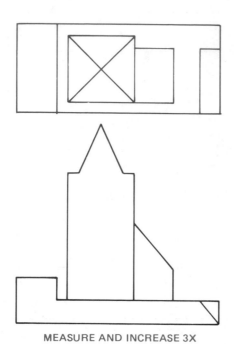

MEASURE AND INCREASE 3X

Problem 17–25 Perspective

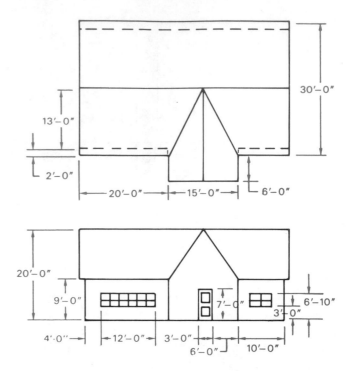

MATH PROBLEMS

1. Provide the measurements required to construct isometric, dimetric, and trimetric projections using the following actual dimensions of the parts. The dimetric and trimetric angles are given.

DIMENSION		ISOMETRIC	DIMETRIC	TRIMETRIC
Left axis	1.55	_____	(10°)____	(25°)____
Right axis	2.95	_____	(40°)____	(45°)____
Vertical axis	2.07	_____	_____	_____

2. Provide the dimetric projection measurements for an object with the following actual dimensions. The left and right axis angles are given.

	DIMENSION	DRAWING MEASUREMENT
Left axis 25°	2.87	_____
Right axis 40°	4.46	_____
Vertical	3.22	_____

3. Provide the trimetric projection measurements for an object with the following actual dimensions. The left and right axis angles are given.

	DIMENSION	DRAWING MEASUREMENT
Left axis 10°	4.95	_____
Right axis 60°	3.18	_____
Vertical	2.86	_____

4. If a 3/4 scale is used to draw the receding axis of a general oblique object, how long would you draw the following dimensions on the receding axis?

3.23 _____
6.54 _____
2.48 _____
4.29 _____

5. If a cylindrical object has a diameter of 3.75 in., what is the approximate width of the block shading that should be applied to it?

CHAPTER 18

Welding Processes and Representations

THE ENGINEERING DESIGN PROCESS

The engineer has just handed you a sketch showing the details for a couple of weldments. The welding details are not shown as standard welding symbols but rather as written instructions, and it is up to you to design the appropriate weld symbols for the weldment drawing.

The first weldment sketch shows an assembly of two aluminum bars connected in a lap joint, the instructions are as follows: *Use 3/8" plug weld, 7/16" deep with a 45° included angle. Space the welds at 3" on center, for the length of the material.* Since aluminum welding requires special considerations, you examine the thickness of the material and decide on a *gas tungsten arc welding* process. Using this information, you design the weld symbol shown in Figure 18–1.

The second weldment sketch shows two low carbon steel plates joined in a T-joint. The instructions are: *Use 1/4" intermittent fillet weld on both sides. The welds should be 2" long with a pitch of 10" and staggered. Field weld at installation site.* The low carbon steel material and T-joint suggest a *shielded metal arc welding* process, and you design an appropriate symbol (Figure 18–2).

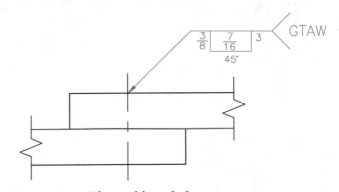

Figure 18–1 Plug weld symbol.

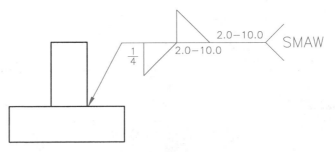

Figure 18–2 Fillet weld symbol.

Welding is a process of joining two or more pieces of like metals by heating the material to a temperature high enough to cause softening or melting. The location of the weld is where the materials actually combine the grain structure from one piece to the other. The parts that are welded become one and the properly welded joint is as strong or stronger than the original material. Welding may be performed with or without pressure applied to the materials. Some materials may actually be welded together by pressure alone. Most welding operations, however, are performed by filling a heated joint between pieces with molten metal.

Welding is actually a method of fastening adjacent parts. Welding was not discussed with fasteners in Chapter 11 because the weld is a more permanent fastening application than screw threads or pins, for example. Welding is a common fastening method used in many manufacturing applications and industries from automobile to aircraft manufacturing, and from computers to ship building. Some of the advantages of welding over other fastening methods include better strength, better weight distribution and reduction, a possible decrease in the size of castings or forgings needed in an assembly, and a potential savings of time and manufacturing costs.

▼ WELDING PROCESSES

There are a large number of welding processes available for use in industry, as shown in Figure 18–3. The most common welding processes include oxygen gas welding, shielded metal arc welding, gas tungsten arc welding, and gas metal arc welding.

Oxygen Gas Welding

Oxygen gas welding, commonly known as *oxyfuel* or *oxyacetylene welding,* may also be performed with such fuels as natural gas, propane, or propylene. Oxyfuel welding is most typically used to fabricate thin materials, such as sheet metal and thin-wall pipe or tubing. Oxyfuel processes are also used for repair work and metal cutting. One advantage of oxyfuel welding is that the equipment and operating costs are less than with other methods. But other welding methods have advanced over the oxyfuel process because they are faster, cleaner, and cause less material distortion. Common oxyfuel welding and cutting equipment is shown in Figure 18–4.

Also associated with oxyfuel applications are soldering, brazing, and braze welding. These methods are

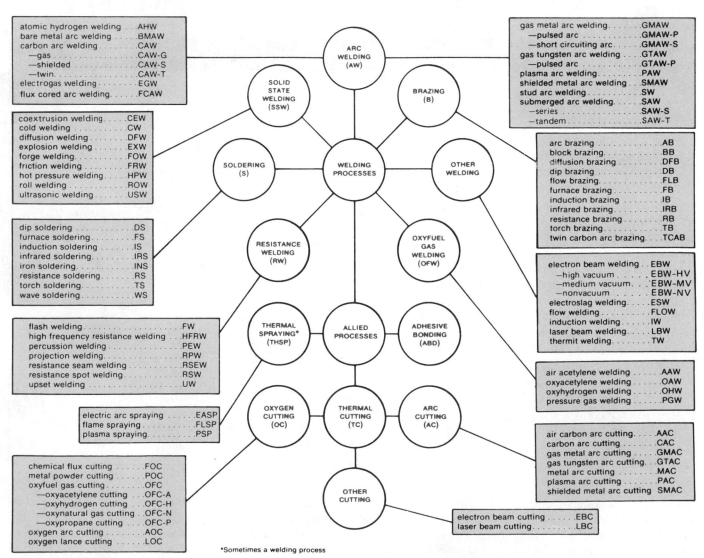

Figure 18–3   Master chart of welding and allied processes. *Courtesy American Welding Society.*

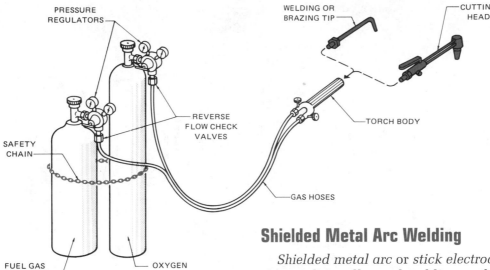

Figure 18–4   Oxyfuel welding and cutting equipment.

more of a bonding process than welding, as the base material remains solid while a filler metal is melted into a joint. Soldering and brazing differ in application temperature. *Soldering* is done below 450° C and *brazing* above 450° C. Like alloys may be used depending upon their melting temperatures. The filler generally associated with soldering is solder. Solder is an alloy of tin and lead. The filler metal associated with brazing is an alloy of copper and zinc. Brazing is a process of joining two very closely fitting metals by heating the pieces, causing the filler metal to be drawn into the joint by capillary action. *Braze welding* is more of a joint filling process that does not rely on capillary action. Another process that uses an oxyfuel mixture is *flame cutting.* This process uses a high-temperature gas flame to preheat the metal to a kindling temperature, at which time a stream of pure oxygen is injected to cause the cutting action.

Shielded Metal Arc Welding

Shielded metal arc or *stick electrode* welding is the most traditionally used welding method. High-quality welds on a variety of metals and thicknesses can be made rapidly with excellent uniformity. This method uses a flux-covered metal electrode to carry an electrical current forming an arc that melts the work and the electrode. The molten metal from the electrode mixes with the melting base material, forming the weld. Shielded metal arc welding is popular because of low-cost equipment and supplies, flexibility, portability, and versatility. Figure 18–5 shows a shielded metal arc welding set-up.

Gas Tungsten Arc Welding

The *gas tungsten arc welding* process is sometimes referred to as *TIG (tungsten inert gas)* welding, or as *Heliarc®,* which is a trademark of the Union Carbide Corporation. Gas tungsten arc welding can be performed on a wider variety of materials than shielded metal arc welding, and produces clean, high-quality welds. This welding process is useful for certain materials and applications. Gas tungsten arc welding is generally limited to thin materials, high-integrity joints, or small

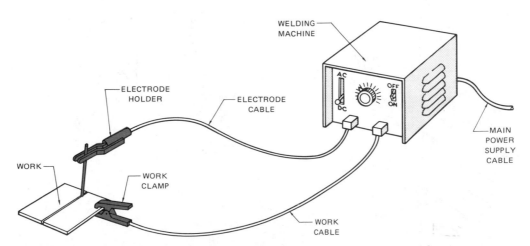

Figure 18–5   Shielded metal arc welding equipment.

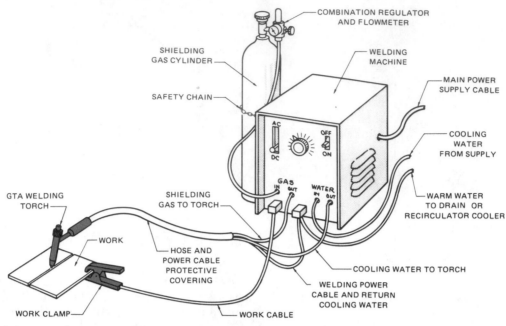

Figure 18–6 Gas tungsten arc welding equipment.

parts, because of its slow welding speed and high cost of equipment and materials. (See Figure 18–6.)

Gas Metal Arc Welding

Another welding process that is extremely fast, economical, and produces a very clean weld is *gas metal arc welding.* This process may be used to weld thin material or heavy plate. It was used previously for welding aluminum using a metal inert gas shield, a process which became referred to as MIG. The present application employs a current-carrying wire that is fed into a joint between pieces to form the weld. This welding process is used in industry with automatic or robotic welding machines to produce rapidly made, high-quality welds in any welding position. While the capital expense of the equipment remains high, the cost is declining due to its popularity. Figure 18–7 shows gas metal arc welding equipment.

▼ ELEMENTS OF WELDING DRAWINGS

Welding drawings are made up of several parts to be welded together. These drawings are usually called *weldments,* or *welding assemblies* or *subassemblies.*

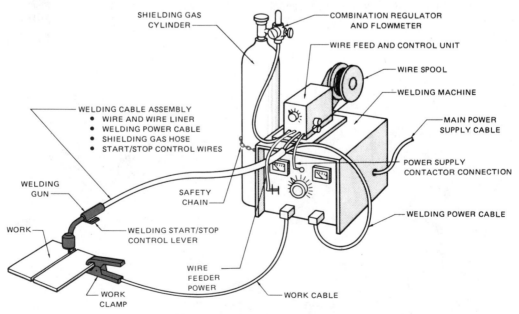

Figure 18–7 Gas metal arc welding equipment.

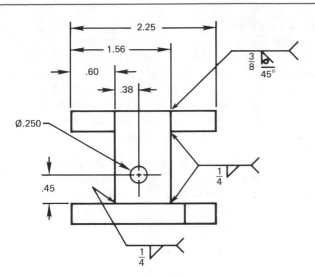

Figure 18–8 Welded subassembly.

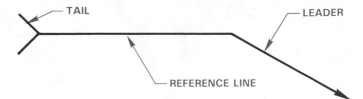

Figure 18–10 Welding symbols: reference line, tail, and leader.

The welding assembly will typically show the parts together in multiview with all of the fabrication dimensions, types of joints, and weld symbols. Welding symbols identify the type of weld, the location of the weld, the welding process, the size and length of the weld, and other weld information. The welding assembly has a list of materials which generally provides a key to the assembly, the number of each part, part size, and material. Figure 18–8 shows a welding subassembly. When additional clarity of component parts must be given, then detailed drawings of each part are prepared, as shown in Figure 18–9.

Welding Symbols

AWS The welding symbol represents complete information about the weld. The welding symbols discussed here are in accordance with the American Welding Society document AWS A2.4.

There are a few basic components of a welding symbol, beginning with the reference line, tail, and leader which are drawn as thin lines, as shown in Figure 18–10. The reference line is usually the first welding symbol element to be located on the drawing. The reference line leader may be drawn using the same rules associated with leaders for notes. The leader may be drawn at any angle, although angles less than 15° or greater than 75° should be avoided. Also, leaders for notes typically run from the shoulder directly to the feature. This practice should be used for welding leaders, although sometimes welding leaders bend to point into difficult-to-reach places, or with more than one leader extending from the same reference line, as shown in Figure 18–11.

After the reference line has been established, additional information is placed on the reference line to continue the weld specification. Figure 18–12 shows the standard location of welding symbol elements as related to the reference line, tail, and leader.

Types of Welds

The next information that is applied to the reference line is the type of weld. The type of weld is associated with the weld shape and/or the type of groove to which the weld is applied. Figure 18–13 shows the information that is associated with the types of welds.

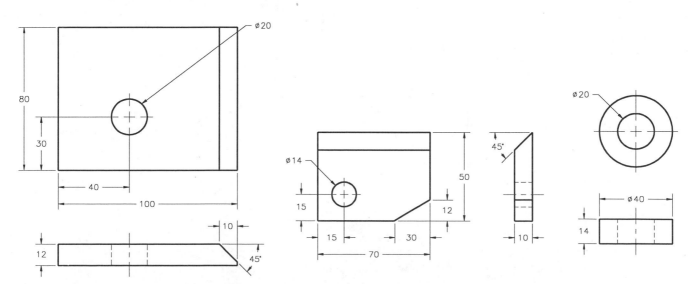

Figure 18–9 Drawings of each part of the welded subassembly.

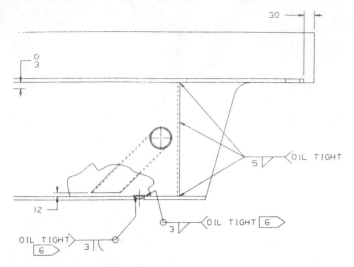

Figure 18–11 Welding symbol leader use. *Courtesy Hyster Company.*

Fillet Weld. A *fillet weld* is formed in the internal corner of the angle formed by two pieces of metal. The size of the fillet weld is shown on the same side of the reference line as the weld symbol and to the left of the symbol. When both legs of the fillet weld are the same, the size is given once, as shown in Figure 18–14. When the leg lengths are different in size, the vertical dimension is followed by the horizontal dimension. (See Figure 18–15.)

Square Groove Weld. A *square groove weld* is applied to a butt joint between two pieces of metal. The two pieces of metal are spaced apart a given distance, known as the root opening. If the root opening distance is a standard in the company then this dimension is assumed. If the root opening is not standard, then the specified dimension is given to the left of the square groove symbol, as shown in Figure 18–16 (see page 578).

V Groove Weld. A *V groove weld* is formed between two adjacent parts when the side of each part is beveled to form a groove between the parts in the shape of a V. The included angle of the V may be given with or without a root opening, as shown in Figure 18–17 (see page 578).

Bevel Groove Weld. The *bevel groove weld* is created when one piece is square and the other piece has a beveled surface. The bevel weld may be given with a bevel angle and a root opening, as shown in Figure 18–18 (see page 578).

U Groove Weld. A *U groove weld* is created when the groove between two parts is in the form of a U. The angle formed by the sides of the U shape, the root, and the weld size are generally given. (See Figure 18–19, page 578.)

J Groove Weld. The *J groove weld* is necessary when one piece is a square cut and the other piece is in a J-shaped groove. The included angle, the root opening, and the weld size are given, as shown in Figure 18–20 (see page 578).

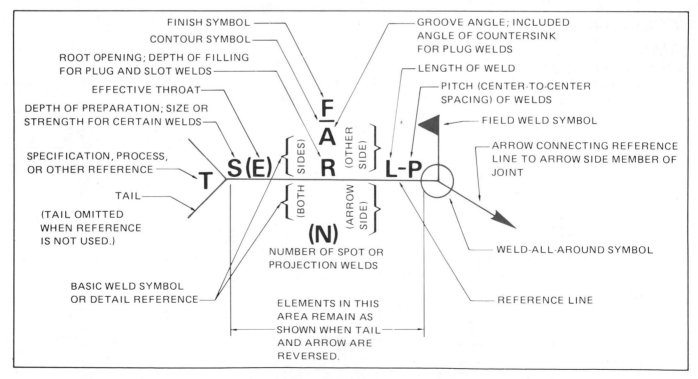

Figure 18–12 Standard location of elements of a welding symbol. *Courtesy American Welding Society.*

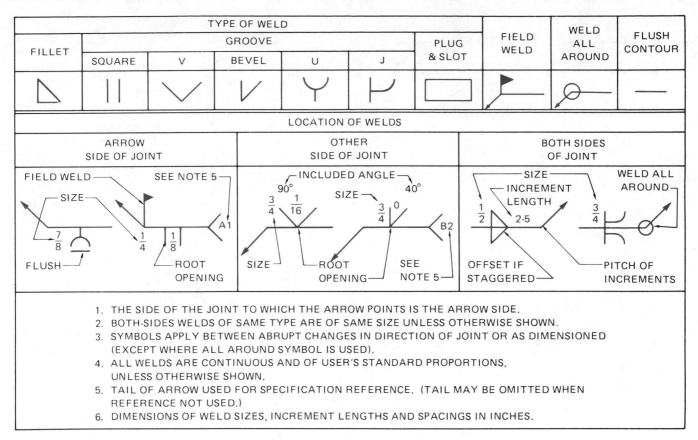

Figure 18–13   Standard welding symbols. *Courtesy American Welding Society.*

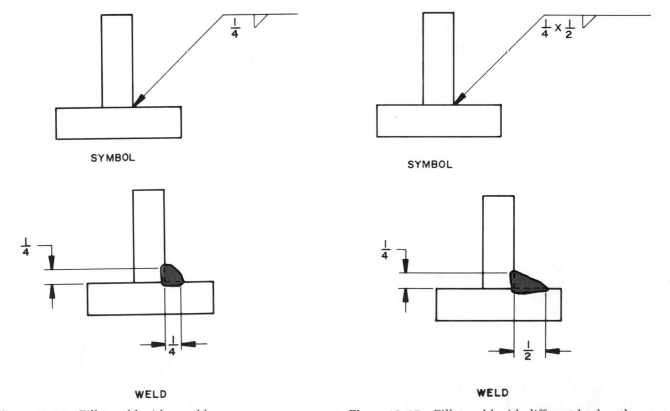

Figure 18–14   Fillet weld with equal legs.

Figure 18–15   Fillet weld with different leg lengths.

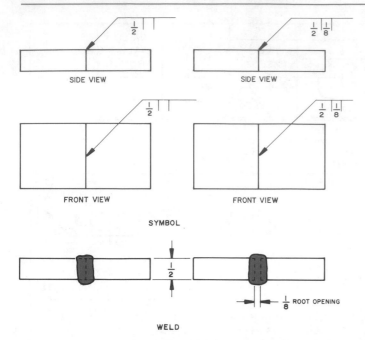

Figure 18–16 Square groove weld showing the root opening.

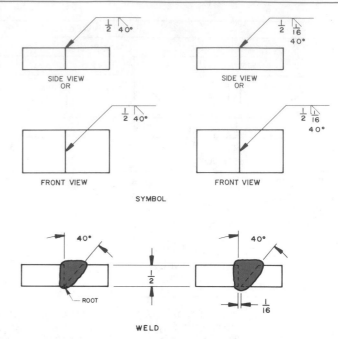

Figure 18–18 Bevel groove weld.

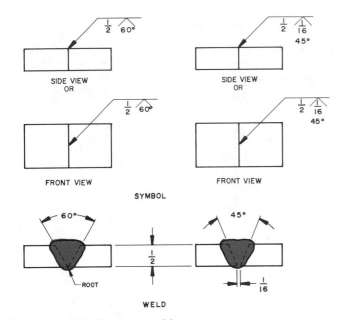

Figure 18–17 V groove weld.

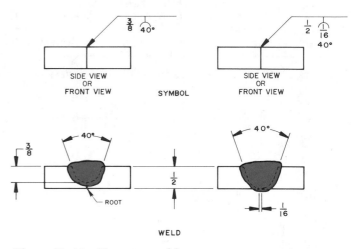

Figure 18–19 U groove weld.

Plug Welds. A *plug weld* is made in a hole in one piece of metal that is lapped over another piece of metal. These welds are specified by giving the weld size, angle, depth, and pitch. (See Figure 18–21a.) The same type of weld may be applied to a slot. This is referred to as a slot weld as shown in Figure 18–21b.

Field Weld. A *field weld* is a weld that is performed in the field as opposed to in a fabrication shop. The reason for this application may be that the individual components may be easier to transport disassembled, or the mounting procedure may require job site installation.

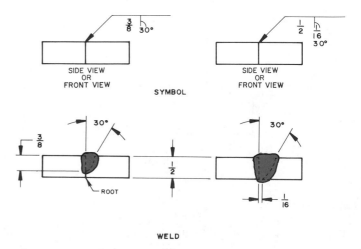

Figure 18–20 J groove weld.

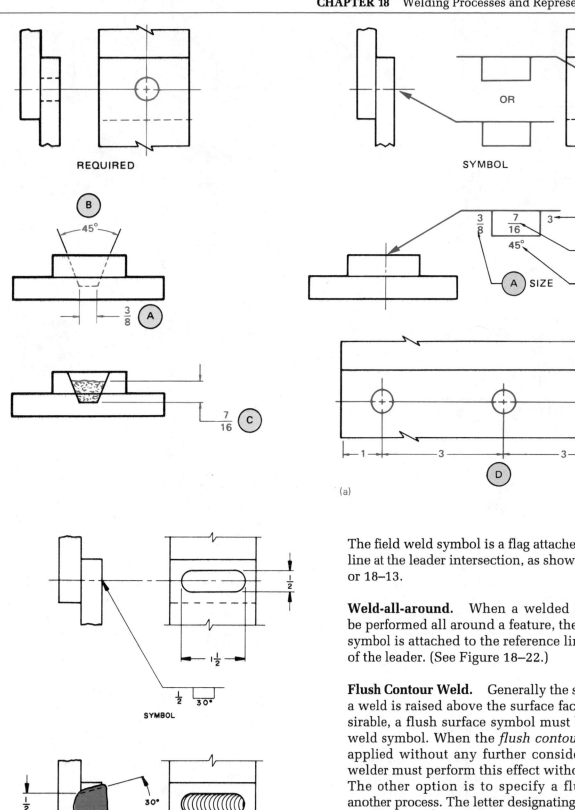

Figure 18–21 (a) Plug welds. (b) Slot weld.

The field weld symbol is a flag attached to the reference line at the leader intersection, as shown in Figure 18–12 or 18–13.

Weld-all-around. When a welded connection must be performed all around a feature, the *weld-all-around* symbol is attached to the reference line at the junction of the leader. (See Figure 18–22.)

Flush Contour Weld. Generally the surface contour of a weld is raised above the surface face. If this is undesirable, a flush surface symbol must be applied to the weld symbol. When the *flush contour weld* symbol is applied without any further consideration, then the welder must perform this effect without any finishing. The other option is to specify a flush finish using another process. The letter designating the other process is placed above the flush contour symbol for another side application, or below the flush contour symbol for an arrow-side application. The options include: C = chipping, G = grinding, M = machining, R = rolling, or H = hammering. (See Figure 18–23.)

Weld Length and Increment. When a weld is not continuous along the length of a part, then the weld length

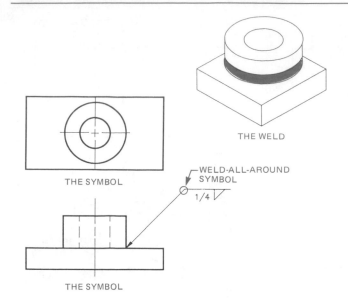

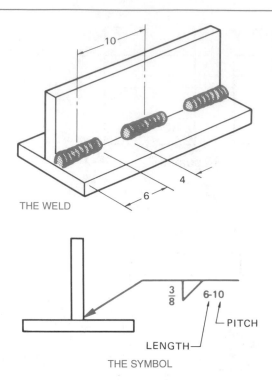

Figure 18–22   Weld-all-around.

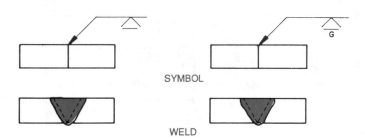

Figure 18–23   Flush contour weld.

Figure 18–24   Intermittent fillet weld.

should be given. In some situations, the weld along the length of a feature is given in lengths spaced a given distance apart. The distance from one point on a weld length to the same corresponding point on the next weld is called the *pitch;* generally from center to center of welds. The weld length and increment are shown to the right of the weld symbol, as shown in Figure 18–24.

Spot Weld.   *Spot welding* is a process of resistance welding where the base materials are clamped between two electrodes and a momentary electric current produces the heat for welding at the contact spot. Spot welding is generally associated with welding sheet metal lap seams. The size of the spot weld is given as a diameter to the left of the symbol. The center-to-center pitch is given to the right of the symbol. (See Figure 18–25.) The strength of the spot welds may be given as minimum shear strength in pounds per spot to the left of the symbol, as shown in Figure 18–26a. When a specific number of spot welds is required in a seam or joint, then the quantity is placed above or below the symbol in parentheses, as shown in Figure 18–26b.

Seam Weld.   A *seam weld* is a continuous weld made between or upon overlapping members. The continuous weld may consist of a single weld bead or a series of overlapping spot welds. The dimensions of seam welds, shown on the same side of the reference line as the weld symbol, relate to either size or strength. The weld size width is shown to the left of the symbol in fractional or decimal inches, or millimeters. (See Figure 18–27a.) The weld length, when specified, is provided to the right of the symbol, as shown in Figure 18–27b. The strength of a seam weld is expressed as minimum acceptable shear strength in pounds per linear inch, placed to the left of the weld symbol, as shown in Figure 18–28 (see page 582).

Weld Symbol Leader Arrow Related to Weld Location

Welding symbols are applied to the joint as the basic reference. All joints have an *arrow side* and an *other side.* When fillet and groove welds are used, the welding symbol leader arrows connect the symbol reference line to one side of the joint known as the arrow side. The side opposite the location of the arrow is called the other side. If the weld is to be deposited on the arrow side of the joint, the proper weld symbol is placed *below* the reference line, as shown in Figure 18–29a (see page 582). If the weld is to be deposited on the side of the joint opposite the arrow, then the weld symbol is placed

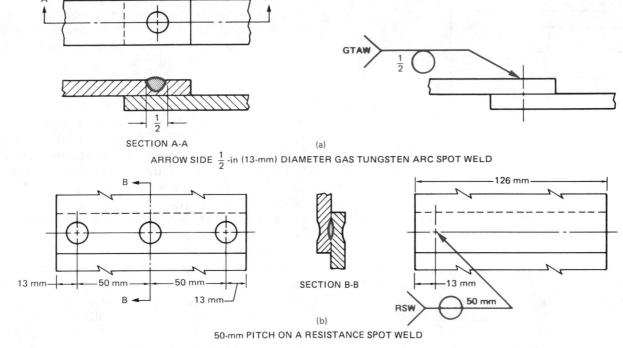

SECTION A-A

(a)

ARROW SIDE $\frac{1}{2}$-in (13-mm) DIAMETER GAS TUNGSTEN ARC SPOT WELD

SECTION B-B

(b)

50-mm PITCH ON A RESISTANCE SPOT WELD

Figure 18–25 Designating spot welds.

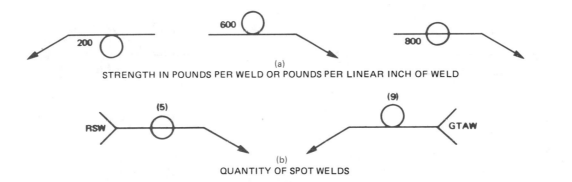

(a)

STRENGTH IN POUNDS PER WELD OR POUNDS PER LINEAR INCH OF WELD

(b)

QUANTITY OF SPOT WELDS

Figure 18–26 Designating strength and number of spot welds.

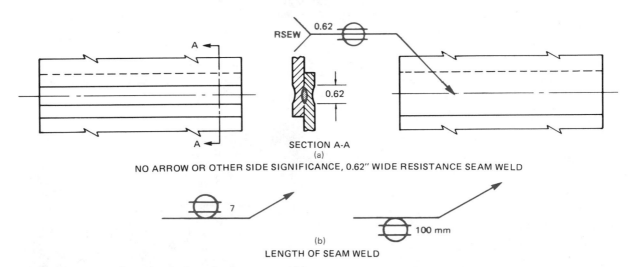

SECTION A-A

(a)

NO ARROW OR OTHER SIDE SIGNIFICANCE, 0.62″ WIDE RESISTANCE SEAM WELD

(b)

LENGTH OF SEAM WELD

Figure 18–27 Indicating the length of a seam weld.

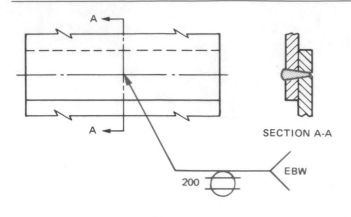

Figure 18–28 Indicating the strength of a seam weld.

above the reference line. (See Figure 18–29b.) When welds are to be deposited on both sides of the joint, then the same weld symbol is shown *above and below* the reference line, as shown in Figure 18–29c and d.

For plug, spot, seam, or resistance welding symbols, the leader arrow connects the welding symbol refer-

ence line to the outer surface of one of the members of the joint at the center line of the desired weld. The member that the arrow points to is considered the arrow side member. The member opposite of the arrow is considered the other side member.

Additional Weld Characteristics

Weld Penetration. Unless otherwise specified, a weld penetrates through the thickness of the parts at the joint. The size of the groove weld remains to the left of the weld symbol. Figure 18–30a shows the size of grooved welds with partial penetration. Notice in Figure 18–30b that a weld with partial penetration may specify the depth of the groove followed by the depth of weld penetration in parentheses, with both items placed to the left of the weld symbol.

When single-groove and symmetrical double-groove welds penetrate completely through the parts being joined, the weld size may be omitted, as shown in Figure 18–31. The depth of penetration of flare-formed groove welds is assumed to extend to the tangent points of the members, as shown in Figure 18–32.

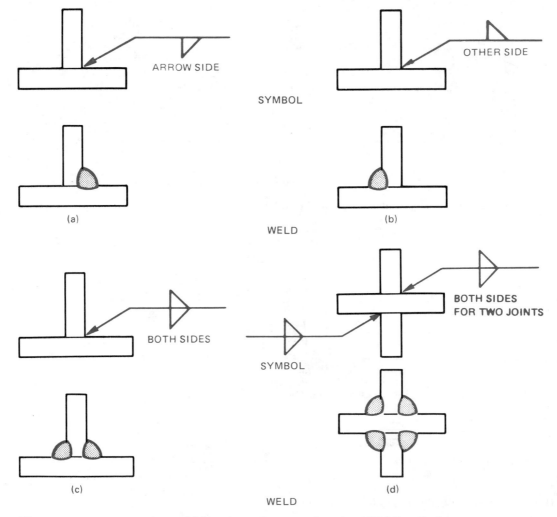

Figure 18–29 Designating weld locations. *Courtesy American Welding Society.*

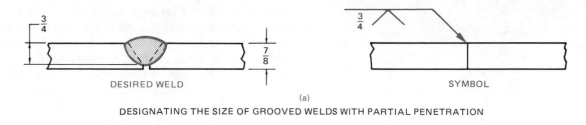

DESIGNATING THE SIZE OF GROOVED WELDS WITH PARTIAL PENETRATION

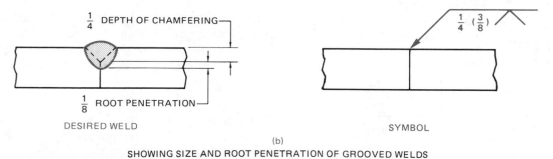

SHOWING SIZE AND ROOT PENETRATION OF GROOVED WELDS

Figure 18–30 **(a)** Designating the size of grooved welds with partial penetration. **(b)** Showing size and penetration of grooved welds. *Courtesy American Welding Society.*

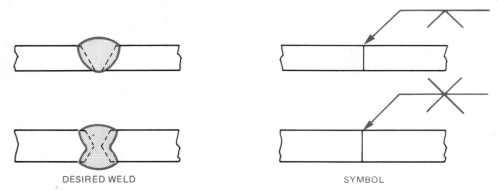

Figure 18–31 Designating single- and double-groove welds with complete penetration. *Courtesy American Welding Society.*

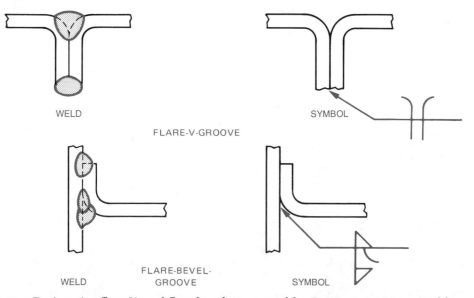

Figure 18–32 Designating flare-V- and flare-bevel-groove welds. *Courtesy American Welding Society.*

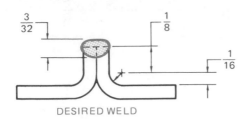

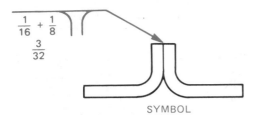

Figure 18–33 Applying dimensions to flange welds. *Courtesy American Welding Society.*

Flange Welds. *Flange welds* are used on light-gauge metal joints where the edges to be joined are flanged or flared. Dimensions of flange welds are placed to the left of the weld symbol. Further, the radius and height of the weld above the point of tangency are indicated by showing both the radius and the height separated by a plus (+) symbol. The size of the flange weld is then placed outward of the flange dimensions. (See Figure 18–33.)

Welding Process Designation. The tail is added to the welding symbol when it is necessary to designate the welding specification, procedures, or other supplementary information needed to fabricate the weld. (See Figure 18–34.)

Weld Joints. The types of weld joints are often closely associated with the types of weld grooves already discussed. The weld grooves may be applied to any of the typical joint types. The weld joints used in most weldments are the butt, lap, tee, outside corner, and edge joints shown in Figure 18–35.

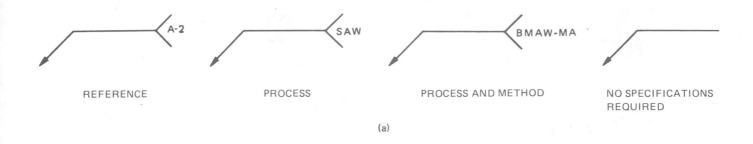

(a)

	Welding Process	Letter Designation
Brazing	Torch brazing	TB
	Induction brazing	IB
	Resistance brazing	RB
Flow Welding	Flow welding	FLOW
Induction Welding	Induction welding	IW
Arc Welding	Bare metal arc welding	BMAW
	Submerged arc welding	SAW
	Shielded metal arc welding	SMAW
	Carbon arc welding	CAW
Gas Welding	Oxyhydrogen welding	OHW
	Oxyacetylene welding	OAW

The following suffixes may be added if desired to indicate the method of applying the above processes:

	Automatic welding	-AU
	Machine welding	-ME
	Manual welding	-MA
	Semiautomatic welding	-SA

(b)

Figure 18–34 **(a)** Locations for weld specifications, processes, and other references on weld symbols. **(b)** Designation of welding processes.

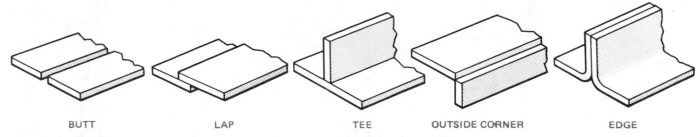

| BUTT | LAP | TEE | OUTSIDE CORNER | EDGE |

Figure 18–35 Types of joints.

CADD Applications

WELDING SYMBOLS AND APPLICATIONS

Computer-aided design and drafting is entering every engineering drafting field including welding processes. Most software packages use a main menu that remains the same for every application. Industry-specific menu overlays are then used to customize the main template for individual applications. The advantage of this is that the drafter becomes familiar with the commands on the main menu, because it is used all of the time. Then as specific needs arise, such as drawing welding symbols, customized symbol libraries are placed over the main menu in locations provided for replaceable tablet menus. This kind of flexibility allows increased efficiency and makes the CADD system as flexible and powerful as possible. A custom tablet overlay featuring a welding symbol library menu is shown in Figure 18–36.

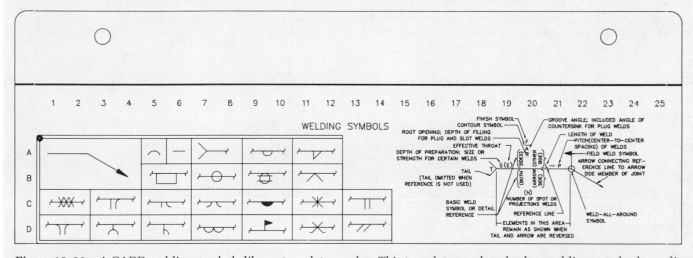

Figure 18–36 A CADD welding symbols library template overlay. This template overlay also has welding standards applications reproduced for your reference. *Courtesy Drafting Technology Services, Inc., Bartlesville, OK.*

▼ WELDING TESTS

There are two types of welding tests—destructive and nondestructive tests.

Destructive Tests (DT)

Destructive tests (DT) use the application of a specific force on the weld until the weld fails. These types of tests may include the analysis of tensile, compression, bending, torsion, or shear strength. Figure 18–37 shows the relationship of forces that may be applied to a weld. The *continuity* of a weld is when the desired characteristics of the weld exist throughout the weld length. *Discontinuity* or lack of continuity exists when a change in the shape or structure exists. The types of problems that alter the desired weld characteristic may include

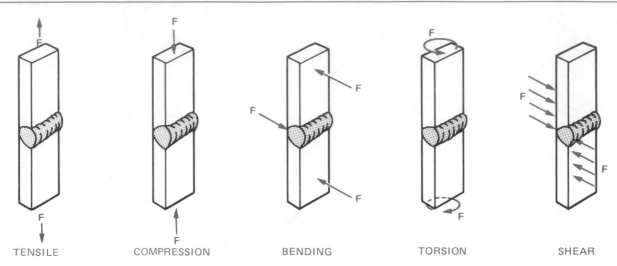

TENSILE   COMPRESSION   BENDING   TORSION   SHEAR

Figure 18–37   Forces on a weld.

cracks, bumps, seams or laps, or changes in density. The intent of destructive testing is to determine how much of a discontinuity can exist in a weld before the weld is considered to be flawed. Parts may be periodically selected for destructive testing. The weld that is tested is unfit for any further use.

Nondestructive Tests (NDS)

Nondestructive tests (NDS) are tests for potential defects in welds that are performed without destroying or damaging the weld or the part. The types of nondestructive tests and the corresponding symbol for each test are shown in Figure 18–38. The testing symbols are used in conjunction with the weld symbol to identify the area to be tested and the type of test to be used. (See Figure 18–39.)

The location of the testing symbol above, below, or placed in a break on the reference line has the same reference to the weld joint as the weld symbol application. Test symbols below the reference line mean arrow side tests. Symbols above the reference line are for other side tests, and a test symbol placed in a break on the line indicate no preference of side to be tested. Test symbols placed on both sides of the reference line require the weld to be tested on both sides of the joint. (See Figure 18–40.)

Two or more different tests may be required on the same section or length of weld. Methods of combining

Type of Nondestructive Test	Symbol
Visual	VT
Penetrant	PT
Dye penetrant	DPT
Fluorescent penetrant	FPT
Magnetic particle	MT
Eddy current	ET
Ultrasonic	UT
Acoustic emission	AET
Leak	LT
Proof	PRT
Radiographic	RT
Neutron radiographic	NRT

Figure 18–38   Standard nondestructive testing symbols.

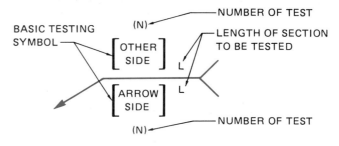

Figure 18–39   Basic nondestructive testing symbol.

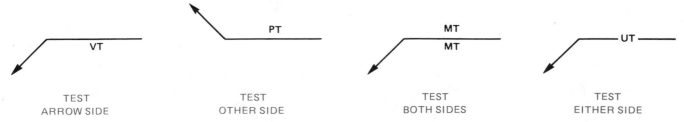

TEST ARROW SIDE   TEST OTHER SIDE   TEST BOTH SIDES   TEST EITHER SIDE

Figure 18–40   Testing symbols used to indicate what side is to be tested.

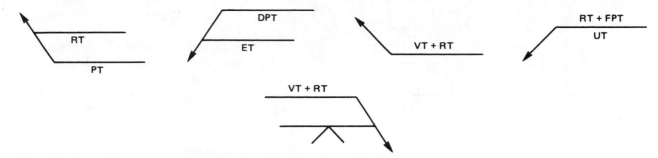

Figure 18–41   Methods of combining testing symbols.

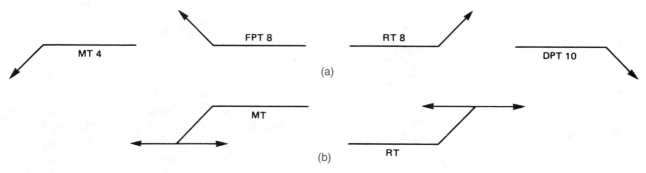

Figure 18–42   Two methods of designating the length of weld to be tested.

welding test symbols to indicate more than one test procedure are shown in Figure 18–41. The length of the weld to be tested may be shown to the right of the test symbol, or may be provided as a dimension line giving the extent of the test length, as shown in Figure 18–42. The number of tests to be made may be identified in parentheses below the test symbol for arrow side tests, or above the symbol for other side tests, as shown in Fig-

ure 18–43. The welding symbols and nondestructive testing symbols can be combined as shown in Figure 18–44. The combination symbol is appropriate to help the welder and inspector identify welds that require special attention. When a radiograph test is needed, a special symbol and the angle of radiation may be specified, as shown in Figure 18–45.

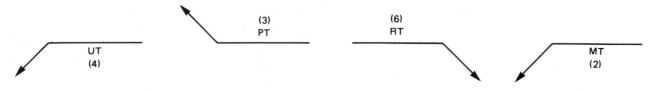

Figure 18–43   Method of specifying number of tests to be made.

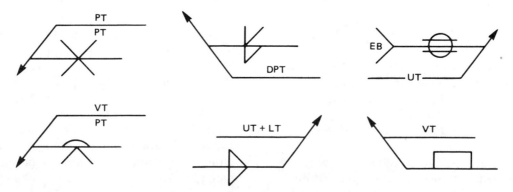

Figure 18–44   Combination welding and nondestructive testing symbols.

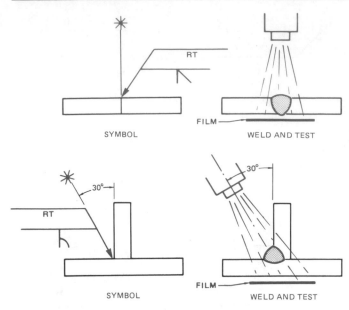

SYMBOL   WELD AND TEST

SYMBOL   WELD AND TEST

Figure 18–45 Combination symbol for welding and radiation location for testing.

▼ WELDING SPECIFICATIONS

A *welding specification* is a detailed statement of the legal requirements for a specific classification or type of product. Products manufactured to code or specification requirements commonly must be inspected and tested to ensure compliance.

There are a number of agencies and organizations that publish welding codes and specifications. The application of a particular code or specification to a weldment can be the result of one or more of the following requirements:

○ Local, state, or federal government regulations.
○ Bonding or insurance company requirements.
○ Customer requirements.
○ Standard industrial practice.

Commonly used codes include:

○ No. 1104, American Petroleum Institute (API). Used for pipeline specifications.
○ Section IX, American Society of Mechanical Engineers (ASME). Used to specify welds for pressure vessels.
○ D1.1, American Welding Society (AWS). Welding specifications for bridges and buildings.
○ AASHT, American Association of State Highway and Transportation Officials.
○ AIAA, Aerospace Industries Association of America.
○ AISC, American Institute of Steel Construction.
○ ANSI, American National Standards Institute.

○ AREA, American Railway Engineering Association.
○ AWWA, American Water Works Association.
○ AAR, Association of American Railroads.
○ MILSTD, Military Standards, Department of Defense.
○ SAE, Society of Automotive Engineers.

Note in the problem assignments that the welding specifications and notes are provided under SPECIFICATIONS and in the general notes.

▼ PREQUALIFIED WELDED JOINTS

AISC/AWS The American Institute of Steel Construction (AISC) and the Structural Welding Code of the American Welding Society (AWS) exempt from tests and qualification most of the common welded joints used in steel construction. These specific common welded joints are referred to as *prequalified*. Work on prequalified welded joints must be in accordance with the Structural Welding Code. Generally, fillet welds are considered prequalified if they conform to the requirements of the AISC and the AWS code.

▼ WELD DESIGN

In most cases, the weld should be about as wide as the thickness of the metal to be welded. Undersized or oversized welds can result in joint failure. Undersized welds may not have enough area to hold the parts together under loading conditions. Oversized welds could result in a joint that is too stiff. The lack of flexibility in the oversized weld could cause the metal near the joint to become over stressed and result in failure adjacent to the weld joint. Welds between materials of different thickness should provide for more weld next to the thickest piece. If this design precaution is not taken, the weld may result in good bonding to the thin material and poor bonding to the thick material because the welding heat is more concentrated at the thinner material. Another option would be to reduce the thickness of the thicker material, or to build up the thickness of the thinner material at the weld joint.

Welding design is generally an engineering decision. When weldments are used in the design the drafter must be able to recognize what the weld means and establish the best location on the drawing for the weld symbol. As entry-level drafters become familiar with company products and the welds used in them, under certain conditions they may be given more design-related tasks.

PROFESSIONAL PERSPECTIVE

Welding processes are used in many manufacturing situations from heavy equipment manufacturing to electronic chassis fabrication to steel building construction. While the specific applications in these fields are different, the display of welding symbols is similar. If there is a strong chance that you will be employed as an engineering drafter in a field that performs a lot of welding operations, then it is a good idea for you to take a class in welding technology. Several classes may even be necessary for you to gain a good understanding of the welding methods and materials. The welding drafting standards and techniques discussed in this chapter are in accordance with the American Welding Society and the American Institute of Steel Construction. If you work in either mechanical engineering or steel construction and fabrication, you should become familiar with the welding applications demonstrated and discussed in these standards.

As an entry-level drafter your drafting assignments will probably be engineer's sketches, or prints marked for revision. As you gain some experience, the engineer may explain verbally or in writing what needs to be done. As you gain experience, you will start to design with welding symbols based on the methods used in the past, material thickness, and welding processes. Assume you are working as an engineering drafter with a mechanical engineer and one of the parts for the project you are working on is a weldment. You have a good idea about what to do, but you ask the engineer for input. The engineer says, "Use a fillet weld-all-around on the diameter 41 side and a

4.5 mm deep bevel weld on the diameter 44 side, but be sure the bevel weld is ground smooth so it doesn't interfere with the mating part." You say, "Okay, now should the process be shielded metal arc welding?" The engineer indicates with a nod, "Yes, you have it!" Now you go back to work and come up with the drawing shown in Figure 18–46.

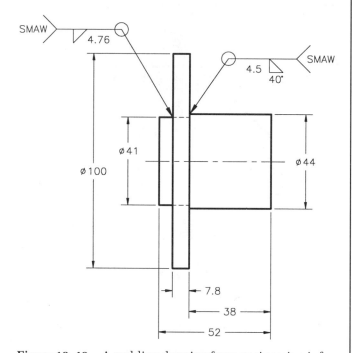

Figure 18–46   A welding drawing from engineering information.

MATH APPLICATION

PERCENTAGE OF WELDED PARTS

Problem: Out of a lot of 1600 welded parts, 2 percent were rejected. How many were accepted?

Solution: This is a problem in percentages. (See Appendix N.) Following percentage problem-solving terminology, the *base* is 1600 and the *rate* is 2 percent. The number of rejected parts is the *part*. To find the part, multiply: 1600 × .02 and get 32. Then the number of accepted parts must be 1600 − 32, or **1568.**

The math problems for this chapter involve finding the base and rate as well as the part.

DIRECTIONS

Answer the questions with short complete statements or drawings as needed.

QUESTIONS

1-10. Given the following welding symbols with letters (a, b, c, etc.) pointing to various components, identify the components related to each letter.

1.
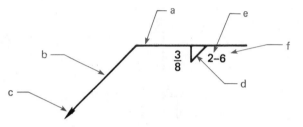

2.
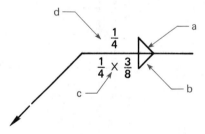

3.
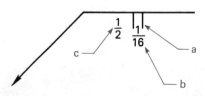

4.
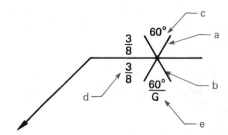

5.
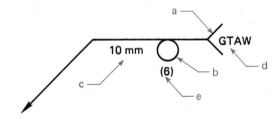

6.
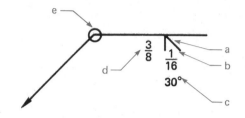

7.
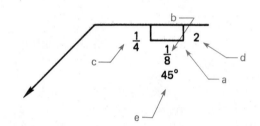

8.
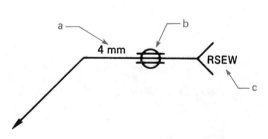

9.
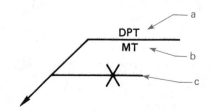

10.

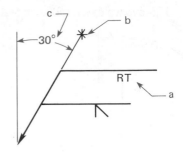

11. Define field weld and show an example of the field weld symbol.
12. List five types of weld joints.
13. Name four categories of the most common welding processes.
14. Describe what is meant by arrow side, other side, and both sides.
15. When is it possible to omit the weld size?
16. What are prequalified welded joints?

CHAPTER 18

WELDING PROCESSES PROBLEMS

DIRECTIONS

1. Given the engineer's sketches and layouts, draw the required number of views and proper welding symbols for each problem. Completely dimension unless otherwise specified.

2. For Problems 18–1 through 18–7: Given the drawings showing actual welds or drawings and specifications, draw the necessary views and proper welding symbols.

Problem 18–2 Fillet and bevel groove (inches)

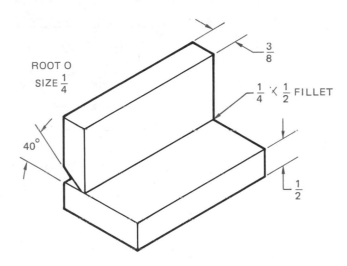

Problem 18–1 Fillet (inches)

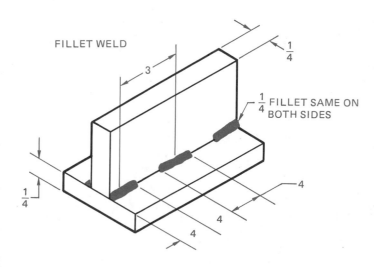

Problem 18–3 V groove (inches)

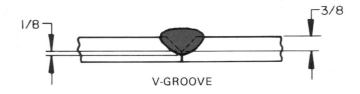

Problem 18–4 Double V groove (inches)

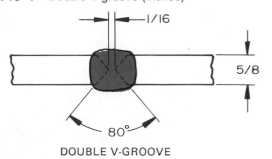

Problem 18–5 Flange (inches)

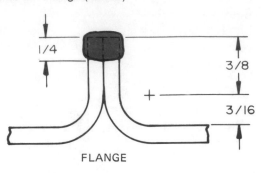

FLANGE

Problem 18–6 J groove weld with radiation test (inches)

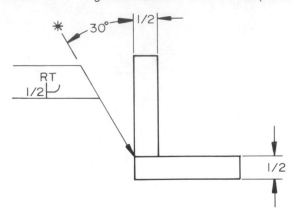

Problem 18–7 Resistance spot weld (metric)

SPOT WELDS (RESISTANCE SPOT WELDS)

(METRIC)

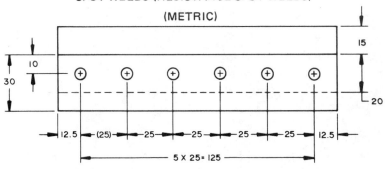

Problem 18–8

Part Name: Spring Housing Weldment
Material: MS

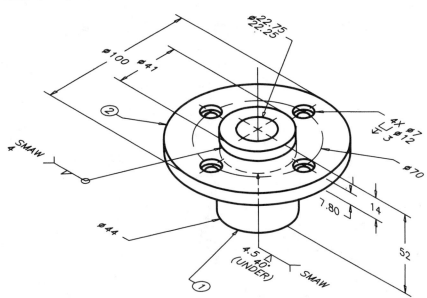

KEY	NAME
BODY	BODY
BODY	BODY

Problem 18–9   (inches)

Drawing Name: Column Base Plate Detail
Material: MS

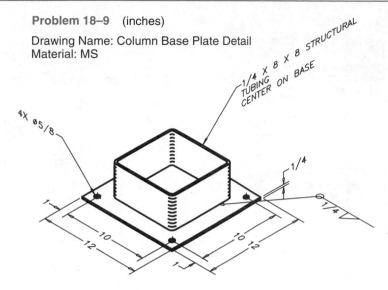

Problem 18–10   (inches)

Drawing Name: Column Base Detail
Material: MS

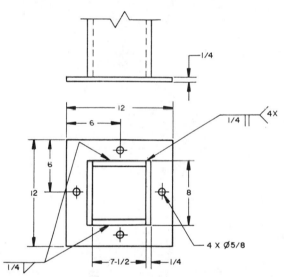

Problem 18–11   (inches)

Drawing Name: Column Intersection Detail
Material: MS

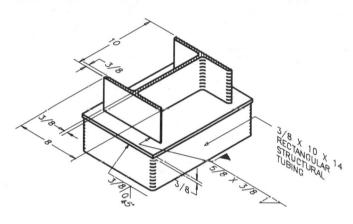

Problem 18–12

Drawing Name: Framed Beam Connection (inches)
Material: Steel
Dimensions of Members:
   W10 × 39: depth = 9-7/8", width = 8", web thickness = 5/16", flange thickness = 1/2".

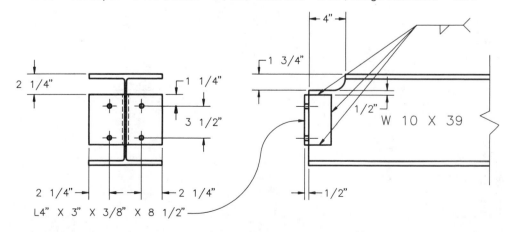

Problem 18–13   (inches)

Drawing Name: Framed Beam Connection
Material: Steel
Dimensions of Members:
   W12 × 40: depth = 12", width = 8", web thickness = 3/16", flange thickness = 1/2".

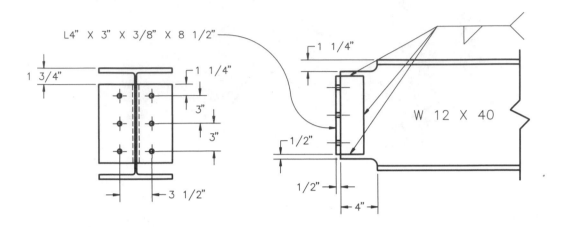

Problem 18–14   (inches)

Drawing Name: Motor Support
Material: See Materials List
Dimensions of Members:
   C4 × 5.4: depth = 4", width = 1-5/8", web thickness = 3/16", flange thickness = 5/16".
Courtesy Production Plastics.

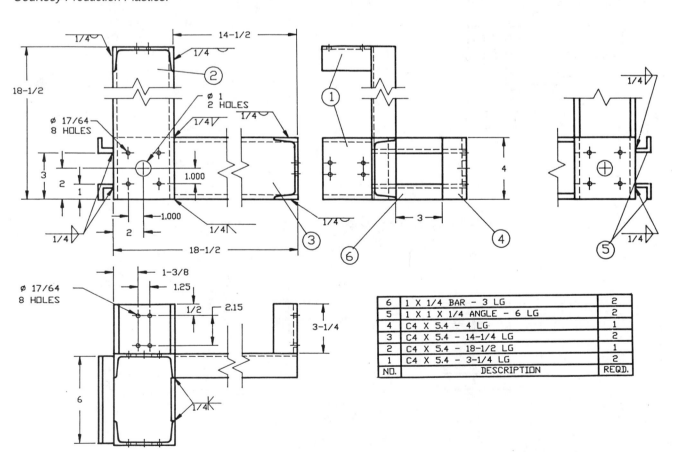

6	1 X 1/4 BAR – 3 LG	2
5	1 X 1 X 1/4 ANGLE – 6 LG	2
4	C4 X 5.4 – 4 LG	1
3	C4 X 5.4 – 14-1/4 LG	2
2	C4 X 5.4 – 18-1/2 LG	1
1	C4 X 5.4 – 3-1/4 LG	2
NO.	DESCRIPTION	REQD.

Problem 18–15

Drawing Name: Mounting Bracket (metric)
Material: See Parts List
Problem based on original art courtesy TEMCO.

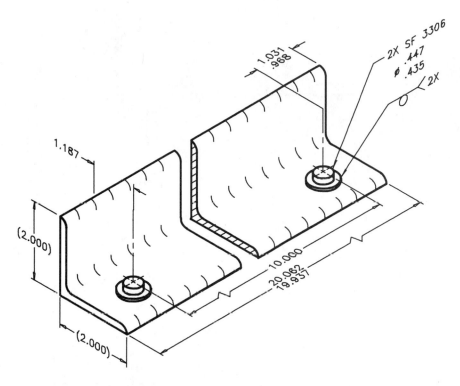

2	WELD NUT	OHIO/ SF 3306 (3/8–16)
1	ANGLE	M821W200L2000 (2X2X3/16 A36)
–01	NAME OF PART	P/N DESCRIPTION MATERIAL

Problem 18–16

Drawing Name: Sideplate Weldment (metric)
SPECIFIC INSTRUCTIONS:
Drawing not to scale; make your own drawing proportional to the problem and reproduce on C size vellum. Some dimensions given for reference. Include all general notes as shown. *Courtesy HYSTER Company.*

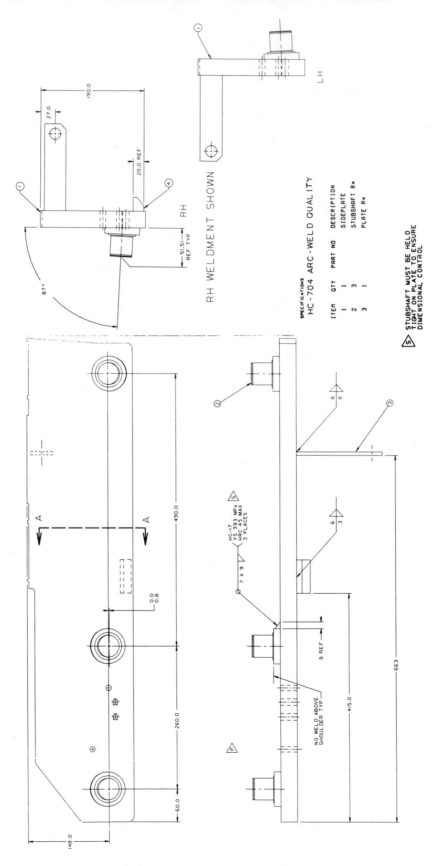

Problem 18–17

Drawing Name: Hydraulic Tank Weldment (metric)
SPECIFIC INSTRUCTIONS:
Drawing not to scale; make your drawing proportional to the problem and reproduce on C size vellum. Some dimensions given for reference. Include all general notes as shown.
Courtesy HYSTER Company.

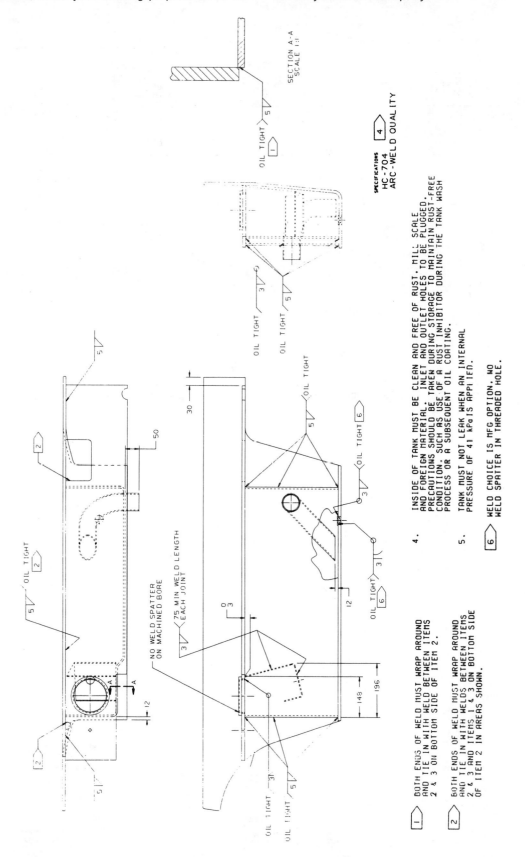

Problem 18–18

DRAWING NAME: Hydraulic Tank Weldment (metric)
SPECIFIC INSTRUCTIONS:
Drawing not to scale; make your drawing proportional to the problem and reproduce on B size vellum. Some dimensions given for reference. Include all general notes as shown.

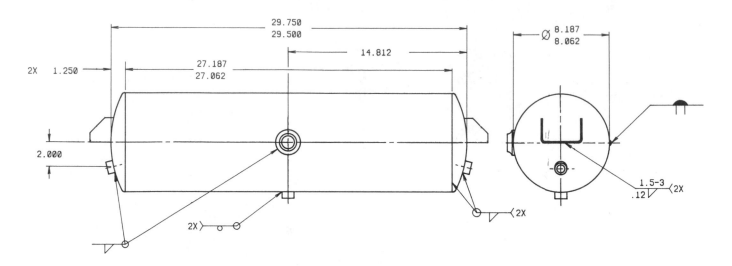

Problem 18–19

Drawing Name: LPG Bracket Weldment (metric)
SPECIFIC INSTRUCTIONS:
Drawing not to scale; make your drawing proportional to the problem and reproduce on C size vellum. Some dimensions given for reference. Include all general notes as shown.
Courtesy HYSTER Company.

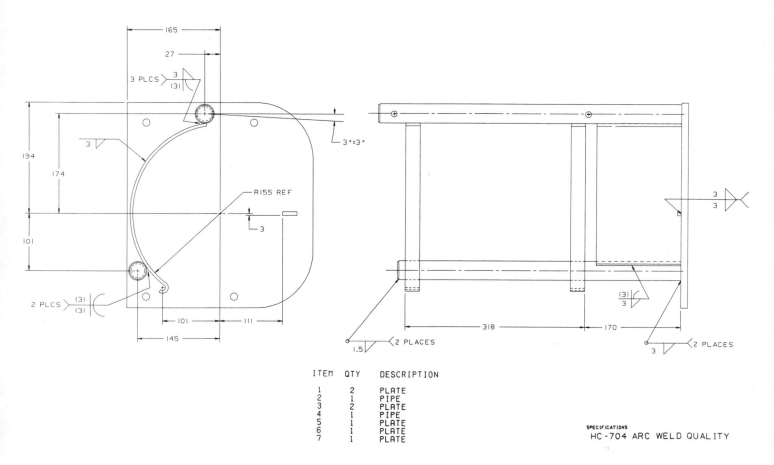

ITEM	QTY	DESCRIPTION
1	2	PLATE
2	1	PIPE
3	2	PLATE
4	1	PIPE
5	1	PLATE
6	1	PLATE
7	1	PLATE

SPECIFICATIONS
HC-704 ARC WELD QUALITY

Problem 18–20

DRAWING NAME: Cab Guard Weldment (inches)
SPECIFIC INSTRUCTIONS:
Drawing not to scale; make your drawing proportional to the problem and reproduce on D size vellum. Some dimensions given for reference. Include all general notes as shown.
Problem based on original art courtesy Curtis Associates.

ITEM	QTY	MATERIAL
1	1	4 SQ X 3/8 WALL TUBE X 36 LG
2	1	4 SQ X 3/8 WALL TUBE X 74 LG
3	1	4 SQ X 3/8 WALL TUBE X 74 LG
4	2	4 SQ X 3/8 WALL TUBE X 28 LG
5	2	4 SQ X 3/8 WALL TUBE X 17 LG
6	2	3 1/2 SQ X 1/4 WALL TUBE X 33 LG
7	2	3 SQ X 1/4 WALL TUBE X 15 3/4 LG
8	5	1/2 X 3 M.S.F.B. X 28 LG
9	1	1/4 M.S. PL X 9 X 28 LG
10	1	3/4 M.S. PL X 4 1/4 X 12 1/8 LG
11	2	1/2 ANGLE
12	6	3/4 M.S. PL X 4 X 5 LG
13	2	1 DIA. C.D. BAR X 3 LG
14	1	1/2 X 3 M.S.F.B. X 20 LG

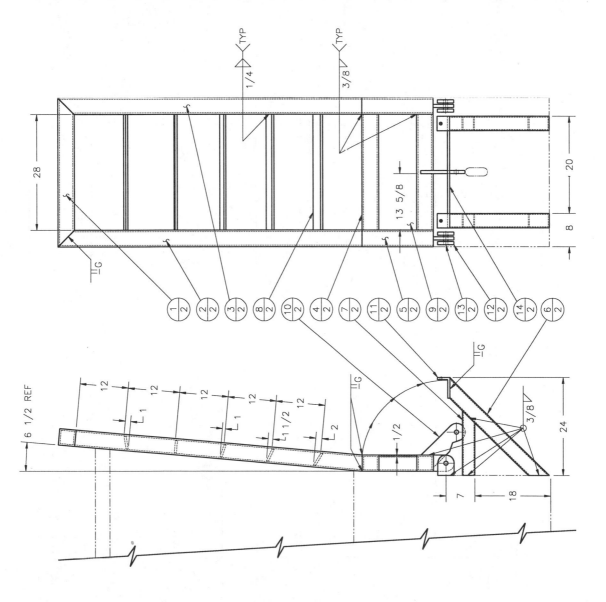

Problem 18–20 Continued

Refer to the Cab Guard Weldment Details below for part dimensions.
Problem based on original art courtesy Curtis Associates.

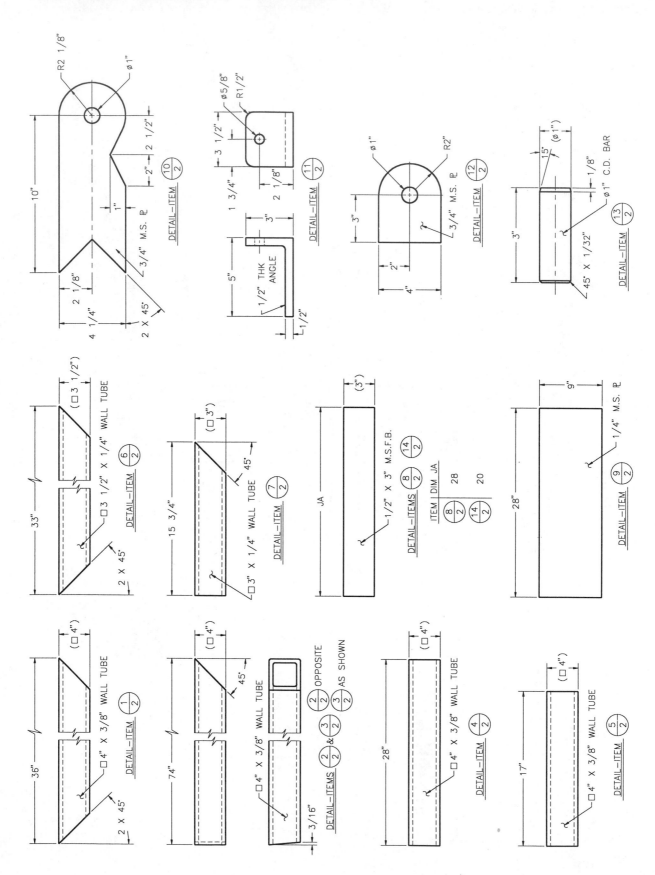

Problem 18–21

Drawing Name: Beam Column
Connector Detail (inches)

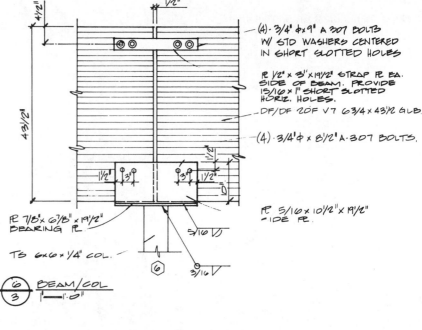

(4) - 3/4" φ x 9" A 307 BOLTS
W/ STD WASHERS CENTERED
IN SHORT SLOTTED HOLES

PL 1/2" x 3' x 19 1/2" STRAP PL EA.
SIDE OF BEAM. PROVIDE
15/16 x 1" SHORT SLOTTED
HORIZ. HOLES.

DF/DF 20F V7 6 3/4 x 43 1/2 GLB.

(4) - 3/4" φ x 8 1/2" A 307 BOLTS.

PL 5/16 x 10 1/2" x 19 1/2"
SIDE PL.

PL 7/8" x 6 7/8" x 19 1/2"
BEARING PL.

TS 6x6 x 1/4" COL.

6 / 3   BEAM/COL
1" — 1'-0"

Problem 18–22

Drawing Name: Beam Wall
Connector Detail (inches)

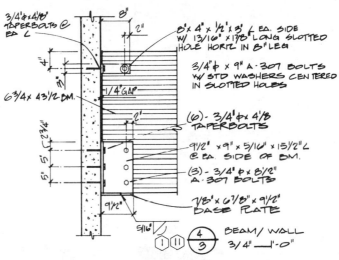

3/4" φ x 4 1/8"
TAPERBOLTS @
EA L

8' x 4" x 1/2" x 3' L EA. SIDE
W/ 13/16" x 1 7/8" LONG SLOTTED
HOLE HORIZ IN 8" LEG

3/4" φ x 9" A 307 BOLTS
W/ STD WASHERS CENTERED
IN SLOTTED HOLES

1/4" GAP

6 3/4 x 43 1/2 BM.

(6) - 3/4" φ x 4 1/8
TAPERBOLTS

9 1/2" x 9" x 5/16" x 15 1/2" L
@ EA. SIDE OF BM.

(3) - 3/4" φ x 8 1/2"
A 307 BOLTS

7/8" x 6 7/8" x 9 1/2"
BASE PLATE

9 1/2"

4 / 3   BEAM/ WALL
3/4" — 1'-0"

Problem 18–23

Drawing Name: Beam-to-Beam
Connector Detail (inches)

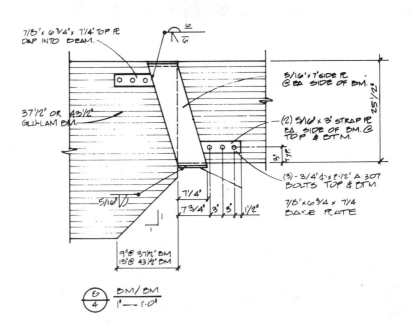

7/8" x 6 3/4 x 7 1/4" TOP PL
DAP INTO BEAM.

5/16 x 7" SIDE PL.
@ EA SIDE OF BM.

37 1/2" OR 43 1/2"
GLU-LAM BM.

(2) 5/16 x 3' STRAP PL
EA. SIDE OF BM. @
TOP & BTM.

(3) - 3/4" φ x 8 1/2" A 307
BOLTS TOP & BTM.

7/8" x 6 3/4 x 7 1/4
BASE PLATE

7 1/4"
7 3/4" 3" 5" 1 1/2"

9" @ 37 1/2" BM
15 @ 43 1/2" BM

8 / 4   BM/ BM
1" — 1'-0"

Problem 18–24

Drawing Name: Mast Brace Weldment
Courtesy American Hoist and Derrick Company.

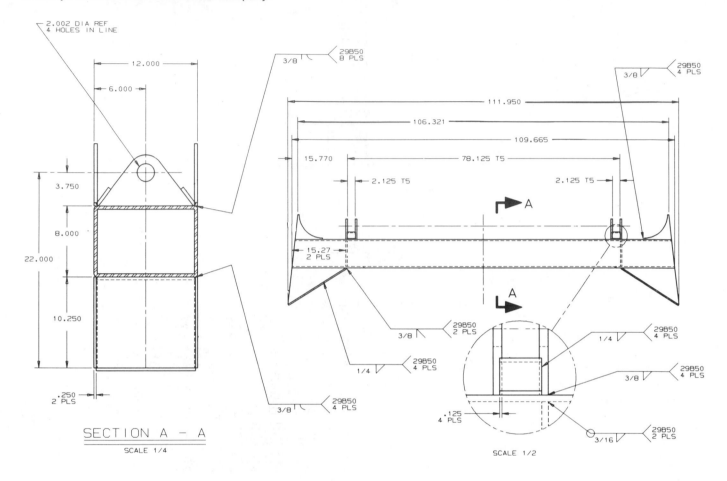

MATH PROBLEMS

1. Twenty-seven percent of the shop output is done on automatic welding machines. If the total output has a value of $57,750.00, what is the value of the work done on the automatic machines?

2. If ⅞ of the total weight of rod used on a job is 67 lb, what is 37½ percent of the total?

3. A battery-operated tool has a useful life of 110 hours. If the battery is used 50 hours, what percent of its useful life is left?

4. What percent is wasted when 2.4 of every 120 sheets of metal are spoiled?

5. The repairs on an electric motor cost 24.5 percent of the original cost of $750.00 What amount was charged for repairs?

6. How much should an article sell for to make a 35 percent profit on a cost of $125.00?

7. A circuit requires 90 volts at a certain point. The true voltage is 72. What is the percent error?

CHAPTER 19

Fluid Power

LEARNING OBJECTIVES

After completing this chapter, you will:

○ Calculate unknown values in given fluid power applications.
○ Draw graphic diagrams of hydraulic circuits.
○ Make graphic diagrams of hydraulic circuits based on sequence of operations functional descriptions.
○ Prepare pictorial diagrams of hydraulic circuits.
○ Make graphic diagrams of pneumatic circuits.
○ Develop fluid power systems from engineering sketches, component lists, or from sketches of revised units.

THE ENGINEERING DESIGN PROCESS

When a draftsperson begins a fluid power drafting project, she or he works from one of several starting points. The drafter may receive an engineer's sketch and a list of components for a new unit, or a sketch for an addition to an existing unit. The drafter may need to show the finished product with a Component List (or Bill of Materials) and/or include a Sequence of Operations List.

In Example 1, the engineer has given the draftsperson a rough sketch and bill of materials for a hydraulic circuit. (See Figure 19–1.) The drafter used a CADD system to produce the final product shown in Figure 19–2. Figure 19–3 shows the actual top-mounted hydraulic power unit described in Figure 19–2. A top-mounted unit means that the pump and motor are mounted on top of the tank. In a skid-mounted unit, they are mounted on the bottom of the tank.

In Example 2, the drafter has received the following sequence of operations and functional description to describe what takes place in the hydraulic circuit. Figure 19–4 shows the resulting hydraulic graphic diagram. Note the use and function of a sequence valve (7) in this example.

Sequence of Operations

1. With motor (3) running, lever on valve (6) is raised manually. Delivery of pump (4) is directed into head end of cylinder (8) for pressing phase.
2. When pressure reaches setting of valve (7), flow is sequenced through (7) to drive motor (9).
3. Manually depressing lever shifts (6) to return cylinder (8) and permit pump (9) to stop.
4. When piston of (8) returns, valve (6) is mechanically centered, unloading delivery of (4) to tank through valve (6), but maintaining sufficient pressure drop to hold work head in raised position.

Functional Description

Valve (5-A) provides overload protection. Valve (7) causes work to be pressed with cylinder (8) before motor (9) starts to rotate, and assures minimum pressure during operation of (9). It also controls maximum thrust of cylinder (8). Valve (5-B) limits maximum torque of motor (9). Valve (6) controls direction of motion of cylinder (8) and the running and stopping of (9).

(Continued)

ENGINEERING DESIGN PROCESS *(Continued)*

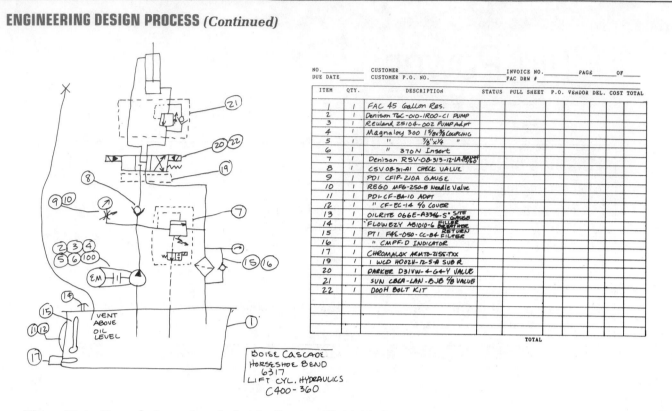

Figure 19–1 Example 1—engineer's sketch. *Courtesy Fluid-Air Components, Inc.*

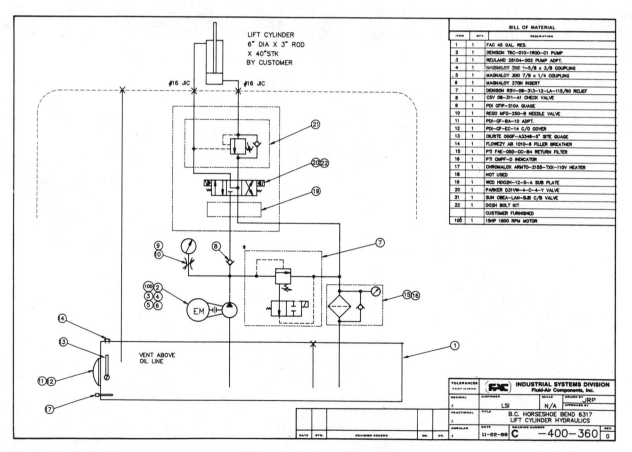

Figure 19–2 Example 1—final product. *Courtesy Fluid-Air Components, Inc.*

Figure 19–3 Top-mounted hydraulic power unit. *Courtesy Fluid-Air Components, Inc.*

COMPONENT LIST		
Key	Quantity	Name, Model Number, and Manufacturer
(1)	1	Reservoir, oil, 30 gal. cap., Co. A
(2)	1	Strainer, ST104, Co. A
(3)	1	Motor, Electric, 5 HP, 1800 rpm, NO-13-53, Co. JKS
(4)	1	Pump, 7 1/2 GPH at 1800 rpm, FE-22-52, Co. VCS
(5)	2	Valve, relief, FE-12-76, Co. EECS
(6)	1	Valve, directional control, DE-04-77, Co. QJS
(7)	1	Valve, sequence, DE-18-49, Co. CS
(8)	1	Cylinder, differential, AU-21-43, Co. CPM
(9)	1	Axial motor, fixed displacement, MO-1313, torque 15 in. lb./100 psi, displacement .96 cu. in./rev., Co. E

In Example 3, the same sequence of operations and functional description are used for a pneumatic circuit. The resulting pneumatic graphic diagram is shown in Figure 9–5.

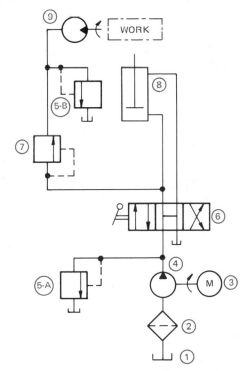

Figure 19–4 Example 2—hydraulic graphic diagram.

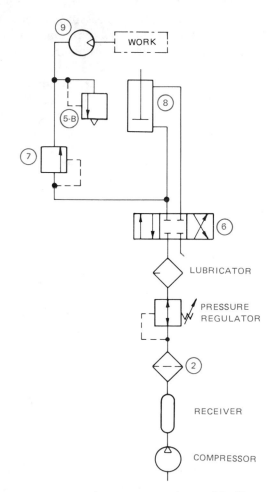

Figure 19–5 Example 3—pneumatic graphic diagram.

For thousands of years, water has been controlled for various uses by utilizing dams, water wheels, and tanks. The practical application of this fluid in motion was the beginning of the science of hydraulics.

Today, hydraulics and pneumatics are referred to as fluid power or fluidics. A *fluid* is defined as something that can flow and is able to move and change shape without separating when under pressure. Fluid power, therefore, includes both liquids and gases. The science of hydraulics refers to liquids and the science of pneumatics refers to gases.

The fluid power technology of using the flow characteristics of a liquid or gas to perform work is used extensively in the operation and control of automobile and aircraft systems, machine tools, earth moving equipment, ships, and spacecraft. To understand the concept of fluid power systems, it is necessary to first look at the basic principles of force, pressure, work, and power.

▼ FORCE, PRESSURE, WORK, POWER

Force can be described as any action that produces motion or alters the position of an object. In the U.S. Customary System, force is usually expressed in pounds, whereas in the International System it is expressed in newtons.

Pressure is the force per unit area exerted on an object. It is shown mathematically as: Pressure = Force/Unit Area. When using force expressed in pounds, the common unit area is a square inch. The pressure is described as pounds per square inch, or psi. When using force expressed in newtons, the unit area is one square meter. However, in this case, instead of the pressure being described as one newton per square meter, it is called a pascal (Pa). Since this measurement is inconveniently small for most engineering work, the kilopascal (kPa), which is 1000 newtons per square meter, is more commonly used. One psi equals 6.895 kPa.

Work is the measurement of force applied to an object multiplied by the distance the object is moved. Therefore, no work is done unless the object is moved or displaced. Work is described mathematically as: Work = Force × Distance. In the U.S. Customary System, distance is expressed in either inches or feet. Work, therefore, is measured in in.-lb or ft-lb. For example, if a force of 800 pounds displaces an object 3 feet, the amount of work done is 2400 ft-lb. In the International System the distance is expressed in meters. If a force of 300 newtons displaces an object 40 meters, the amount of work done is 12,000 newtonmeters, or joules (J). The *joule* is the amount of work done when one newton is displaced a distance of one meter. A kilojoule (kJ), which is 1000 joules, is more commonly used than a joule. Therefore, 12,000 joules would be shown as 12kJ. A joule equals .7377 ft-lb.

Power is described as the work accomplished per unit of time. In other words, an equal amount of work can be done by a high-powered motor in a short time or by a low-powered motor in a long time. The mathematical expression is Power = Work/Time. Units of power are expressed in foot-pounds per minute, or in joules per second. The more common expressions are horsepower, which equals 33,000 ft-lb per minute; and watt, which equals one joule of work per second.

For example, imagine an assembly line motor pushing an object from one conveyor belt to another belt 8 feet away with a force of 1000 lb in 5 seconds. The work being done is 8000 ft-lb. The amount of time is 1/12th of a minute, or .08333 minute. In one minute, the amount of work done is 8000 ft-lb ÷ 1/12 = 96,000 ft-lb/minute. The motor, therefore, has a horsepower of 96,000/33,000 or 2.9.

▼ HYDRAULICS

The science of hydraulics had its beginning in about 1650 when a French mathematician and physicist named Blaise Pascal first observed the law that became known as Pascal's principle. It states that if a pressure is exerted at one portion of fluid that is at rest in a closed container, then that pressure is transmitted equally in all directions without loss through the rest of the fluid and to the walls of the container. (See Figure 19–6.)

What this means practically in a hydraulic circuit is that pressure applied to one part of the system (a piston, for example) will affect another part of the circuit (another piston) with the same pressure. The amount of force produced on the second piston depends, of course, on the area of that piston. Similarly, the amount of work done depends on the distance that the second piston was moved.

Figure 19–7 shows an example of this principle. In this simple hydraulic circuit, the surface of piston A is 10 square inches. When a force of 50 lb is applied, the pressure exerted in the fluid, on all walls, and against the surface of piston B is 5 lb per square inch (psi). The force of piston B is, therefore, 5 psi × 100 square inches, or 500 lb. If piston A moves a distance of 20 feet, the amount of

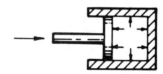

Figure 19–6 Pascal's principle.

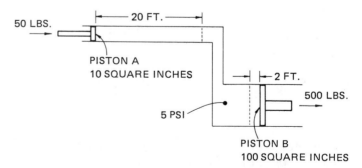

Figure 19–7 Pascal's principle in a hydraulic circuit.

work done is the force × distance, or 50 lb × 20 feet, or 1000 ft-lb. The same work will be done at piston B. Since the force at piston B is 500 lb, the work will be 1000 ft-lb divided by 500 lb (work divided by force equals distance), and the piston will move 2 feet. Another way to conceptualize this is to imagine the fluid at piston A being displaced to piston B. In other words, since the surface area of piston A is 10 square inches and piston A moves 20 feet (240 inches), the total amount of fluid displaced to piston B is 2400 cubic inches. Since piston B has a surface area of 100 square inches, the piston will move 24 inches, or 2 feet.

▼ HYDRAULIC SYSTEMS AND EQUIPMENT

Hydraulic systems perform work by transmitting energy from a power source through pressurized fluid to actuators (in the previous example, the actuator was piston B). In most cases the pressurized fluid is a water-soluble oil or water-glycol mixture, with oil being the fluid used most frequently.

In all hydraulic circuits, there are five basic elements, regardless of the work performed or the complexity of the system. These five elements are: a reservoir, a driver, a pump, valves, and an actuator.

Reservoir

The reservoir, similar to the drawing in Figure 19–8, is the holding tank for the hydraulic fluid. It can also help in separating air and contaminants from the fluid, as well as dissipating some of the heat that is produced within the system.

Driver

The driver may be an electric motor or an internal combustion engine which drives the pump.

Hydraulic Pump

The hydraulic pump is used to pressurize the liquid in the hydraulic system. The pump brings in air at its inlet by creating a partial vacuum, thereby creating the atmospheric pressure that forces the hydraulic liquid through the rest of the system. Pumps such as this, in which the liquid is displaced mechanically, are called positive displacement pumps. Most pumps used in hydraulic systems are of this type. These pumps are divided into two types: reciprocating and rotary. A reciprocating pump pressurizes the liquid by using a back and forth, straight-line motion such as that produced by a piston, plunger, or diaphragm. A rotary pump uses a circular motion such as that produced by gears, vanes, or cams. (See Figure 19–9.) Remember that a hydraulic pump only pressurizes the liquid, thereby producing the flow. It does not pump pressure. A piston pump can be seen in Figure 19–10.

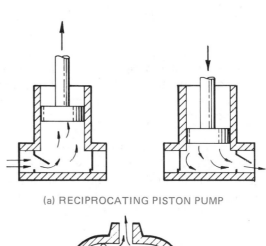

(a) RECIPROCATING PISTON PUMP

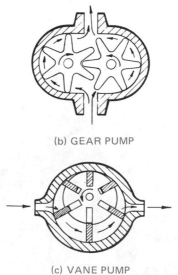

(b) GEAR PUMP

(c) VANE PUMP

Figure 19–9 Positive displacement pumps.

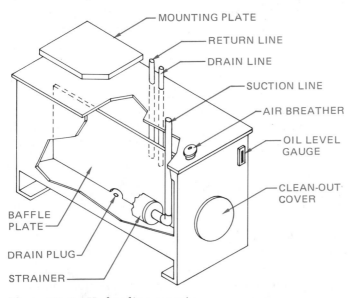

Figure 19–8 Hydraulic reservoir.

Figure 19–10 Piston pump. *Courtesy Parker Hannifin Corporation.*

Valves

Valves are devices that control the pressure, direction, and flow of liquids in the hydraulic system. They accomplish this by opening, closing, or partially obstructing passageways throughout the system. A variety of hydraulic valves is shown in Figure 19–11. The discussion of valves in the following section is divided into three categories: pressure control valves, directional control valves, and flow control valves.

Pressure Control Valves. These are used to maintain a particular pressure within the system. The relief valve is the most common of this type. It remains closed until a predetermined pressure is reached, at which time it

Figure 19–11 Hydraulic valves. *Courtesy Parker Hannifin Corporation.*

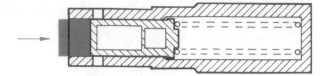

Figure 19–12 Closed relief valve.

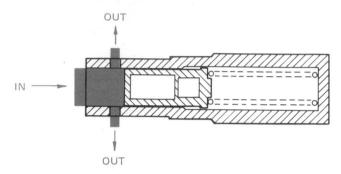

Figure 19–13 Open relief valve.

opens automatically, allowing the fluid to pass through the valve to return to the reservoir. Figures 19–12 and 19–13 show how this is accomplished. (The type of valve illustrated is a spool valve, so named because of the movable portion inside the casting.) In Figure 19–12, we see the valve in the closed position. The spring, which forces the spool to the far left, has a particular pressure setting. When the inlet pressure of the hydraulic fluid exceeds the spring setting, as in Figure 19–13, the spool is forced against the spring, thereby allowing the fluid to pass through the outlet holes to the reservoir or tank. (Both the inlet and outlet holes are called "ports.") These valves provide protection to other parts of the system from the damage that can be caused by pressure that is too high.

A sequence valve operates on basically the same principle as that of the relief valve. The difference is that instead of the fluid being returned to the reservoir, it is routed to another part of the system to perform more work. This is necessary in systems that must provide work in the proper sequence.

Pressure reducing valves, unlike relief valves and sequence valves, are normally open. One of their functions in hydraulic systems is to allow a secondary circuit to operate at a lower pressure than the primary circuit. See Figure 19–14 for an example of a pressure control valve.

Directional Control Valves. These are used to control the direction that the fluid flows in the system. The simplest of directional control valves is the check valve. (See Figure 19–15.) This ball check valve allows the fluid to flow in only one direction. As long as the inlet pressure is greater than the pressure of the internal spring, the fluid will flow through the valve and to the rest of the system. If the flow begins to reverse or if the pressure drops below the pressure of the spring, the

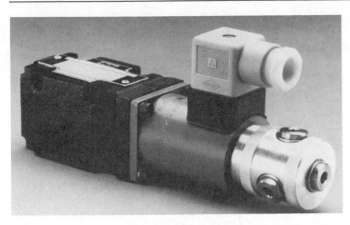

Figure 19–14   Pressure control valve. *Courtesy Parker Hannifin Corporation.*

Figure 19–15   Check valve.

spring pressure will seat the ball and the flow will be stopped.

Multiple way valves provide for the opening or closing of different flow paths. They usually contain a spool. These valves are classified by both the number of ports they contain and the number of spool positions. For example, two different two-way (referring to two ports), two-position valves are illustrated in Figure 19–16. Valve A in position 1 is normally closed, or in its unactuated position. When the push button is pressed and the valve is actuated (position 2), the spool slides to the left and the fluid is allowed to flow through the valve

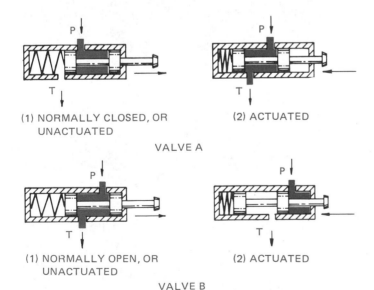

(1) NORMALLY CLOSED, OR
   UNACTUATED

(2) ACTUATED

VALVE A

(1) NORMALLY OPEN, OR
   UNACTUATED

(2) ACTUATED

VALVE B

Figure 19–16   Two-way, two-position directional control valves.

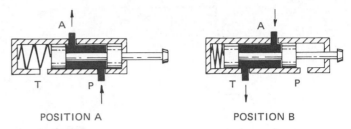

POSITION A   POSITION B

Figure 19–17   Three-way, two-position valve.

Figure 19–18   Directional control valves. *Courtesy Parker Hannifin Corporation.*

from port P (pressure) to port T (tank). With valve B, the unactuated position is open (position 1). When the push button is pressed and the valve is actuated, it then becomes closed as in position 2.

Figure 19–17 shows a three-way, two-position valve. In position A, the fluid flows from port P through the valve and out port A. Port T is blocked. In position B, port P is blocked and the fluid flows from port A to port T.

Two types of directional control valves are shown in Figure 19–18.

Flow Control Valves.   These valves control the rate of flow through the hydraulic system. One type of flow control valve, the throttle valve, is illustrated in Figure 19–19. Figure 19–20 shows a flow control valve with a fixed output. In this type of valve, the rate of flow is not affected by variations in the inlet pressure. Two types of flow valves are shown in Figure 19–21.

Actuators

An actuator in a hydraulic system is the device that converts the fluid power to mechanical energy for the purpose of performing work. Actuators are either linear or rotary.

Linear actuators are most often a cylinder or ram. The single-acting cylinder is the simplest of this type. (See

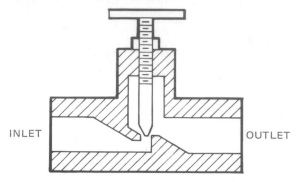

Figure 19–19   Throttle valve.

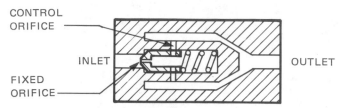

Figure 19–20   Flow control valve.

Figure 19–21   Flow control valves. *Courtesy Parker Hannifin Corporation.*

Figure 19–22.) In this cylinder, the fluid force is applied to only one surface of the piston, which is the head end of the cylinder. The piston is retracted by an external force, such as a spring or the force of gravity.

In a double-acting cylinder, such as the one illustrated in Figure 19–23, the fluid force can be applied to either surface of the piston. This allows the movement of the piston to be controlled hydraulically in two directions. This double-acting cylinder with a single piston rod is a differential type because there is a difference in the piston surface area between the right and left. Since the area at the left is larger, the force applied to that surface is greater, and the work stroke will be slower and more powerful than the opposite work stroke. The nondifferential type of double-acting cylinder shown in Figure 19–24 has a double-ended piston rod that extends through both ends of the cylinder. The surface areas of

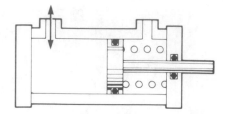

Figure 19–22   Single-acting cylinder. *Courtesy International Standards Organization (ISO).*

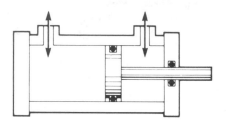

Figure 19–23   Double-acting cylinder. *Courtesy International Standards Organization (ISO).*

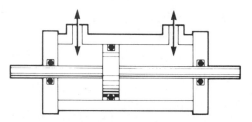

Figure 19–24   Double-acting, nondifferential cylinder. *Courtesy International Standards Organization (ISO).*

Figure 19–25   Cylinders. *Courtesy Parker Hannifin Corporation.*

both sides of the piston are equal, so the forces in both directions will also be equal. Several types of cylinders are shown in Figure 19–25.

Rotary actuators can be of the gear, vane, or piston type (refer to Figure 19–9).

Filters and strainers are also a necessary part of the hydraulic system to ensure long life of the components.

Figure 19–26 Filters. *Courtesy Parker Hannifin Corporation.*

They keep the hydraulic fluid clean by removing foreign particles. See Figure 19–26 for a wide variety of filters.

▼ FLUID POWER DIAGRAMS

Types of Diagrams

Four types of diagrams are used when representing fluid power systems. They are: graphic, pictorial, cutaway, and combination diagrams. Each type emphasizes a different aspect of the system.

A graphic diagram emphasizes the function of the circuit and of each component. The components consist of simple geometric shapes that are linked together with interconnecting lines. (See Figure 19–27.) This type of diagram is most frequently used for designing and troubleshooting fluid power circuits.

A pictorial diagram, Figure 19–28, is used to show the piping between components. The drawings of the com-

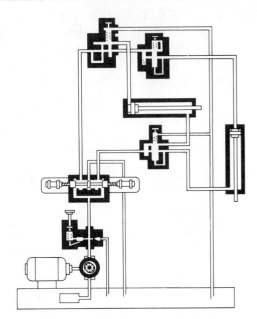

Figure 19–29 Cutaway diagram.

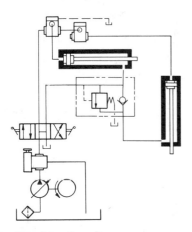

Figure 19–30 Combination diagram.

ponents themselves are pictorial and do not attempt to show the function or method of operation.

The purpose of a cutaway diagram is to show the principal internal working parts and the function of each component. Sometimes several cutaway drawings are used to show the different flow paths that are possible depending upon the position of the various moving parts. (See Figure 19–29.)

A combination diagram uses graphic, pictorial, and cutaway symbols with interconnecting lines. (See Figure 19–30.) This type of diagram provides a way to emphasize function, piping, or flow paths for each component as needed.

Symbols

ANSI Symbols are arranged in the diagram to facilitate the use of direct and straight interconnecting lines. Where components have definite mechanical, functional, or otherwise important relationships to one

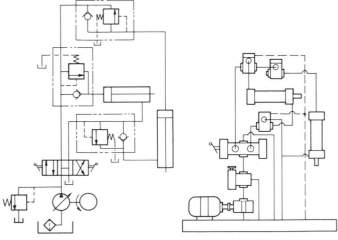

Figure 19–27 Graphic diagram. **Figure 19–28** Pictorial diagram.

another, their symbols are so placed in the diagram. Single lines are used in graphic diagrams. Double lines are used in cutaway diagrams. Pictorial and combination diagrams may use single or double lines or both. ("Fluid Power Diagrams," ANSI/(NFPA)T3.28.9R1—1989)

Graphic diagrams and symbols are best suited to international use and standardization because of their simplicity. The remainder of this section shows graphic symbols that were approved by the International Organization for Standardization (ISO) in 1976. Figure 19–31 shows several graphic symbols commonly used.

Control valves except for nonreturn valves are usually shown in single or multiple squares known as envelopes, with ports shown on the active envelope. (See Figure 19–32.)

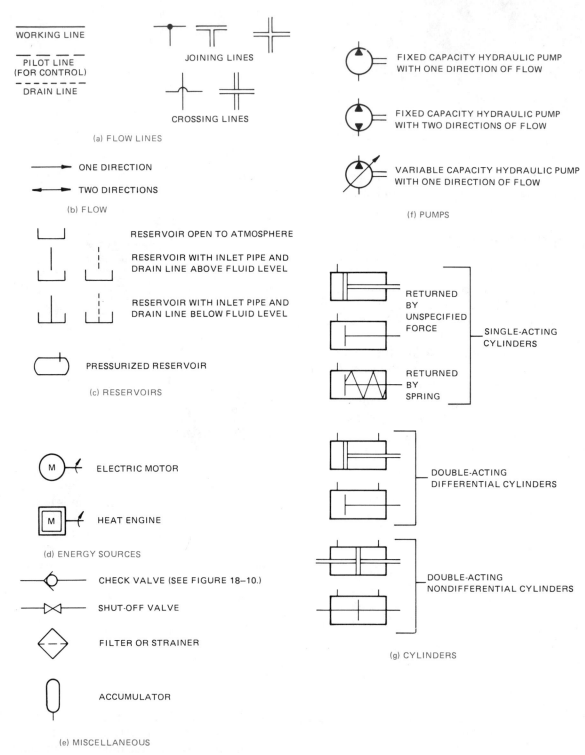

Figure 19–31 Graphic symbols.

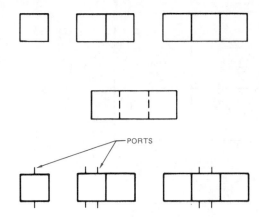

Figure 19–32 Envelopes.

ISO Single envelopes indicate pressure or flow control valves in which there are an infinite number of positions possible. This allows the system to operate at a constant predetermined pressure or flow.

Pressure control valve symbols are shown in Figure 19–33. The symbol for the pressure relief valve is the one that would be used for the valve shown in Figures 19–12 and 19–13.

The sequence valve differs from the pressure relief valve only in that the fluid flows to other parts of the circuit to perform more work instead of returning to the reservoir.

Flow control valve symbols are shown in Figure 19–34. The throttle valve symbol would be used for the valve in Figure 19–19, with the arrow indicating that the valve is adjustable. The symbol for a flow control valve with variable output would be used for the valve in Figure 19–20. Directional control valves are shown in multiple envelopes with each envelope indicating a distinct operating position. Several possible flow paths

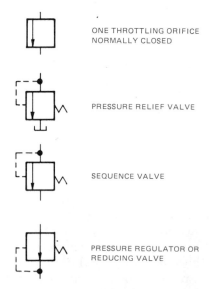

Figure 19–33 Pressure control valve symbols.

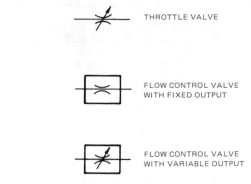

Figure 19–34 Flow control valve symbols.

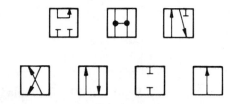

Figure 19–35 Flow paths.

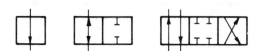

Figure 19–36 Valves with ports open.

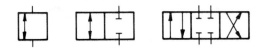

Figure 19–37 Valves with ports closed or blocked.

for these and other valves are shown in Figure 19–35. Figure 19–36 shows examples of valves with the ports open, and Figure 19–37 shows examples of valves with ports closed or blocked. Figure 19–38 shows the symbols used for various methods of actuating (or controlling) valves.

Figure 19–39 shows how these graphic symbols correlate with unactuated and actuated directional control valves. In position 1, this unactuated two-way, two-position valve is normally closed. In the graphic symbol shown above it, the active ports are blocked. In position 2, the valve is actuated and flow through the valve occurs. Its graphic symbol shows the active ports on the envelope with the flow path.

ANSI Note that when possible, the envelope nearest the control symbol (in this case, a push-button control) represents the condition that occurs when the valve is actuated.

A three-way, two-position directional control valve is shown in Figure 19–40 along with its graphic symbol. Since the valve is controlled by an unspecified pressure

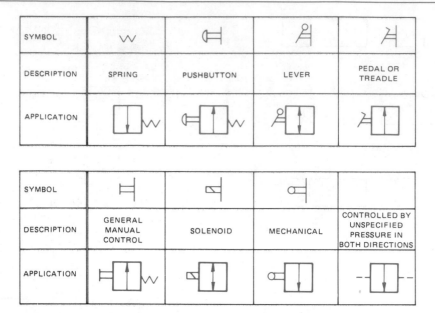

SYMBOL	⌇	⊏H	ㅅH	ㅓH
DESCRIPTION	SPRING	PUSHBUTTON	LEVER	PEDAL OR TREADLE
APPLICATION				

SYMBOL	H⊏H	⊏⊿H	⊏○H	
DESCRIPTION	GENERAL MANUAL CONTROL	SOLENOID	MECHANICAL	CONTROLLED BY UNSPECIFIED PRESSURE IN BOTH DIRECTIONS
APPLICATION				

Figure 19–38 Symbols for methods for actuating valves.

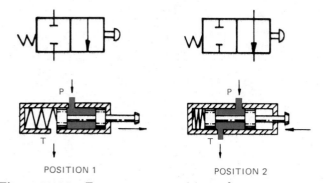

POSITION 1   POSITION 2

Figure 19–39 Two-way, two-position valve.

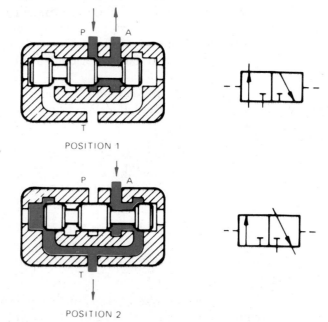

POSITION 1

POSITION 2

Figure 19–40 Three-way, two-position valve.

in both directions, it is not possible to tell from the graphic symbol which position is actuated. In position 1, the flow from the pressure port P exits port A. In position 2, port P is blocked, and the flow from port A goes to the reservoir via port T.

This valve could be used in a hydraulic circuit with a single-acting cylinder as shown in Figure 19–41. When the valve is in position 1, the pressurized flow from the pump flows through the valve, through the adjustable flow control valve, into the cylinder, and forces the piston up. When the valve is in position 2, the fluid pressure from port P is blocked and is diverted back to the reservoir through the pressure relief valve. Gravity from the piston forces the fluid in the cylinder back through the check valves and through port A to the reservoir.

ANSI Note that the graphic symbol for reservoir can be used in one graphic diagram as often as necessary.

A three-way, three-position valve is shown in Figure 19–42 with its corresponding graphic symbol. This is an example of a directional control valve with an intermediate position in which all ports are blocked.

ISO The dashed lines between the envelopes indicate that the center position is not a distinct position; it represents a transitory intermediate condition.

A four-way, three-position valve can be used with a double-acting cylinder as shown in Figure 19–43. In the unactuated position, the only flow that occurs is the flow from the pump through the valve to the reservoir. When the valve is actuated right, the pressurized flow from the pump goes through the directional control valve and through the adjustable flow control valve

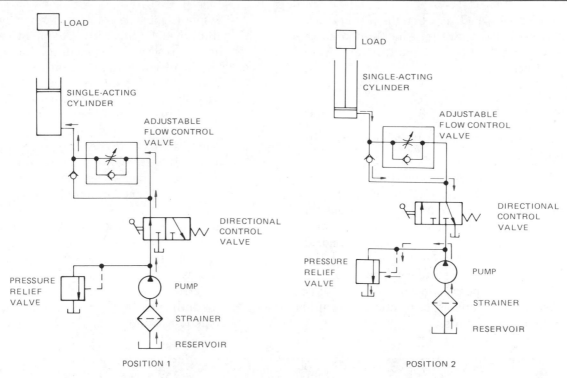

Figure 19–41 Hydraulic circuit.

Figure 19–42 Three-way, three-position valve.

to the lower chamber of the cylinder, pushing the piston up. At the same time, the fluid in the upper chamber flows through the directional control valve to the reservoir. When the valve is actuated left, the pressurized flow is directed to the upper chamber, and the fluid in the lower chamber flows through the check valves and through the directional control valve to the reservoir. If at any time the pressure in the system exceeds

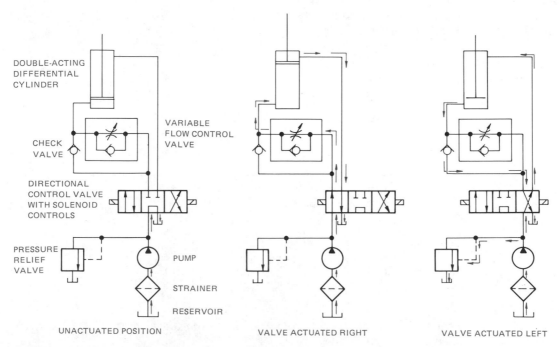

Figure 19–43 Four-way, three-position valve.

a certain preset amount, such as when the piston is actuated all the way to the top, then the fluid from the pump flows through the pressure control valve (relief valve) to the reservoir.

▼ SYMBOL RULES

ANSI These symbol rules apply to both hydraulics and pneumatics.

1. Symbols show connections, flow paths, and functions of components represented. They can indicate conditions occurring during transition from one flow path arrangement to another. Symbols do not indicate construction, nor do they indicate values, such as pressure, flow rate, and other component settings.

2. Symbols do not indicate location of ports, direction of shifting of spools, or positions of controls on actual components.

3. Symbols may be rotated or reversed without altering their meaning except in cases of (a) lines to reservoir, (b) accumulator.

4. Line width does not alter meaning of symbols.

5. Basic symbols may be shown in any suitable size. Size may be varied for emphasis or clarity. Relative sizes should be maintained.

6. Letter combinations used as parts of graphic symbols are not necessarily abbreviations.

7. In multiple envelope symbols, the flow condition shown nearest an actuator symbol takes place when that control is caused or permitted to actuate.

8. Each symbol is drawn to show normal, at-rest, or neutral condition of component unless multiple diagrams are furnished showing various phases of circuit operation.

9. An arrow through a symbol at approximately 45° indicates that the component can be adjusted or varied.

CADD Applications

APPLICATIONS FOR FLUID POWER

Fluid power graphic diagrams lend themselves well to CADD applications because of their simplicity and standardization. Several companies offer a Fluid Power Symbols Library that can be made compatible with most micro-, mini-, and mainframe CADD systems. A symbols library usually includes a full spectrum of graphic symbols from basic flow and pressure control valves to complex hydrostatic transmissions. Figure 19–44a shows a sample page from one company's disk, and Figure 19–44b shows a CADD template overlay menu from another company.

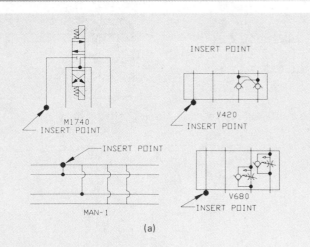

(a)

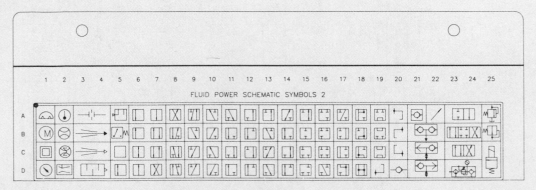

(b)

Figure 19–44 **(a)** CADD system fluid power symbols. *Courtesy Price Engineering Company Inc.*; **(b)** CADD system fluid power symbols. *Courtesy Drafting Technology Services, Inc.*

10. External ports are located where flow lines connect to basic symbols, except where the component enclosure symbol is used.

11. External ports are located at intersections of flow lines and component enclosure symbols when enclosure is used.

▼ PNEUMATICS

In review, a fluid is defined as something that can flow and is able to move and change shape without separating when under pressure. Fluid power includes both liquids and gases. *Pneumatics* is the science that pertains to gaseous pressure and flow.

Pneumatic devices include any tool or instrument that utilizes compressed air, such as riveters, paint sprayers, atomizers, and rock drills. Using compressed-air power is economical and safe. Pneumatic devices have no spark hazard and can be used under wet conditions without electric shock hazard. Other advantages are that pneumatic systems have relatively few moving parts, and devices can be easily exchanged with one another by pipe, tubing, or flexible hose.

Pascal's Principle applies to pneumatics as well as hydraulics. It states that if a pressure is exerted at one portion of fluid that is at rest in a closed container, then that pressure is transmitted equally in all directions without loss through the rest of the fluid and to the walls of the container. (See Figure 19–6.)

Another basic physical law pertaining to pneumatics is Boyle's Law, which states: the absolute pressure of a fixed mass of gas varies inversely to the volume, provided the temperature remains constant. Note that this law is in terms of absolute pressure (psia or pounds per square inch absolute), not gauge pressure (psig or pounds per square inch gauge). At sea level, the weight of the earth's atmosphere is 14.7 psi. This is the pressure that is actually being exerted on the gauge, even though the gauge reading is zero. To find absolute pressure, 14.7 psi must be added to the gauge pressure. Figure 19–45 demonstrates Boyle's Law with gauges reading absolute pressure. As the volume is decreased by one-half, the pressure doubles. Figure 19–46 shows the adjustments that need to be made when the gauge reading at sea level is zero. In position a, 14.7 psi is added to the gauge pressure of 30 psi to get an absolute pressure of 44.7 psia. This figure is then doubled in position b to get 89.4 psia, and 14.7 psi is subtracted to get 74.7 psig. The same procedure applies to position c.

Charles's Law also applies to pneumatics. It states that the volume of a fixed mass of gas varies directly with absolute temperature, provided the pressure remains constant. This law has many implications for pneumatic equipment, as will be seen later.

Another law is that air flow will occur only when there is a difference in pressure. The flow will be from high pressure to low pressure.

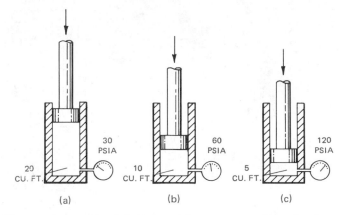

Figure 19–45 Boyle's Law with gauges reading absolute pressure.

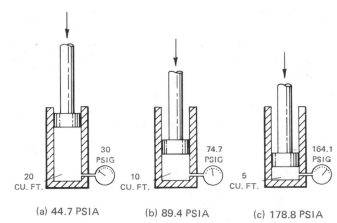

Figure 19–46 Boyle's Law with difference between gauge pressure and absolute pressure.

▼ PNEUMATIC SYSTEMS AND EQUIPMENT

There are eight elements involved in a complete pneumatic circuit: a driver, an air compressor, an air receiver, a filter, a pressure regulator, an air lubricator, valves, and pneumatic devices. A driver can be an electric motor or some other power source that drives the air compressor.

An air compressor is a machine that forces air into a smaller space than it normally occupies. Two things happen with the air at this point: (1) the air pressure increases (Boyle's Law), and (2) the air temperature increases due to the increased pressure (Charles's Law). The amount of pressurized air available for useful work from the compressor is expressed in cubic feet per minute (or standard cu. ft. per min. or SCFM). This measurement of air is known as free air delivered, or FAD. The air compressor is often mounted on the top of the air receiver, as shown in Figure 19–47.

An air receiver is the storage tank for the compressed air. Because the temperature of the compressed air has increased, the amount of water vapor has increased also. This water needs to be removed. A drain is usually

Figure 19–47 Tank-mounted air compressor. *Courtesy Dayton Electric Manufacturing Company.*

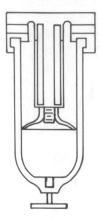

Figure 19–48 Air filter.

Figure 19–50 Pneumatic valves. *Courtesy Parker Hannifin Corporation.*

provided in the air receiver to remove any precipitation that takes place.

An air filter removes water vapor and dirt. (See Figure 19–48.) The air enters the filter and is quickly forced into a rotary motion. The centrifugal force spins out the moisture and dirt, which then collects at the bottom and is drained by an automatic or manual valve.

A pressure regulator is necessary in pneumatic systems to consistently supply the correct pressure to the pneumatic tools. (See Figure 19–49.) The tools usually operate with compressed air at about 90 psig.

An air lubricator (pneumatic lubricator) adds measured amounts of lubricant to the air supply for the purpose of lubricating the equipment receiving the air. In an oil-fog-type lubricator, the oil in the container enters the metering chamber. Because there is a difference in pressure at that point, the oil is then sprayed into the pipeline as fog. An air lubricator should always be downstream of a pressure regulator because some oils can react with the regulator diaphragm and contaminate the air.

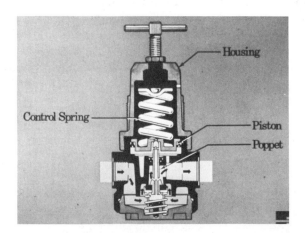

Figure 19–49 Pressure regulator. *Courtesy Parker Hannifin Corporation.*

Valves are devices that control the direction of compressed air in the pneumatic system. They are actuated manually, electrically, or by air. Pneumatic valves operate on the same basic principle as hydraulic directional control valves. For example, a two-way, two-position valve is similar in function to the hydraulic valve shown in Figure 19–16. A variety of pneumatic valves is shown in Figure 19–50.

Pneumatic devices are the elements that utilize compressed air to perform work. One type of element is the cylinder, which can be either single-acting or double-acting. (Refer to Figures 19–22 and 19–23.) Another type is referred to as an air motor. Air motors are divided into two groups based on their type of driving method (either reciprocating piston or rotor). The two main types of air motors with reciprocating pistons are the axial piston (Figure 19–51) and the radial piston (Figure 19–52). Figure 19–53 shows a rotary vane air motor. A tool such as a grinding wheel may use a rotor type of air motor, whereas a riveting hammer may use a reciprocating piston air motor.

In a hydraulic circuit the fluid is returned to the reservoir after going through the system. In a pneumatic circuit the compressed air is returned to the atmosphere after being utilized.

▼ PNEUMATIC DIAGRAMS

Graphic symbols for pneumatic systems are identical to those for hydraulic systems with the exception of those shown in Figure 19–54. Pictorial and cutaway diagrams for pneumatic systems are also the same as for hydraulic systems. Figure 19–55 shows a simple pneumatic circuit with a nondifferential double-acting cylinder.

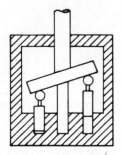

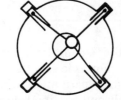

Figure 19–51   Axial piston.   **Figure 19–52**   Radial piston.

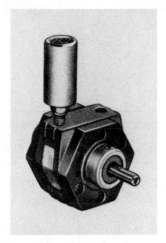

Figure 19–53   Rotary vane air motor. *Courtesy Dayton Electric Manufacturing Company.*

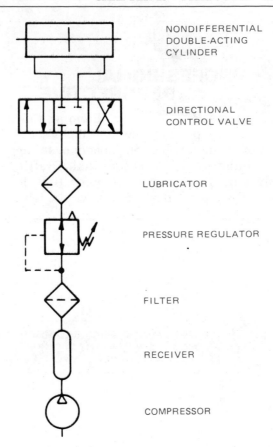

NONDIFFERENTIAL
DOUBLE-ACTING
CYLINDER

DIRECTIONAL
CONTROL VALVE

LUBRICATOR

PRESSURE REGULATOR

FILTER

RECEIVER

COMPRESSOR

Figure 19–55   Simple pneumatic circuit.

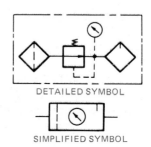

DETAILED SYMBOL

SIMPLIFIED SYMBOL

Figure 19–56   Conditioning unit.

Sometimes a conditioning unit is used in place of a filter, pressure regulator and gauge, and lubricator. A detailed symbol and simplified symbol for this unit is shown in Figure 19–56. Note the style of line used around the detailed symbol. This is sometimes used to represent an enclosure around several components that are combined in one unit.

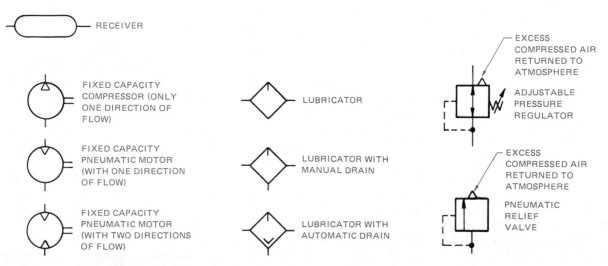

RECEIVER

FIXED CAPACITY
COMPRESSOR (ONLY
ONE DIRECTION OF
FLOW)

FIXED CAPACITY
PNEUMATIC MOTOR
(WITH ONE DIRECTION
OF FLOW)

FIXED CAPACITY
PNEUMATIC MOTOR
(WITH TWO DIRECTIONS
OF FLOW)

LUBRICATOR

LUBRICATOR WITH
MANUAL DRAIN

LUBRICATOR WITH
AUTOMATIC DRAIN

EXCESS
COMPRESSED AIR
RETURNED TO
ATMOSPHERE

ADJUSTABLE
PRESSURE
REGULATOR

EXCESS
COMPRESSED AIR
RETURNED TO
ATMOSPHERE

PNEUMATIC
RELIEF
VALVE

Figure 19–54   Pneumatic graphic symbols.

PROFESSIONAL PERSPECTIVE

From a professional point of view, drafting is not simply a matter of tracing symbols on paper or or shifting symbols around on a CADD terminal. Drafting also involves inquisitiveness and innovation. It sometimes takes a great deal of creative thought to put together a complicated pneumatic circuit or add two extra pumps to an existing hydraulic circuit. Although you may learn many things from experienced draftspeople and qualified salespeople, your ability to think for yourself and question situations that do not make sense to you will remain your greatest assets.

CHAPTER

19 FLUID POWER TEST

DIRECTIONS

Answer the questions with short complete statements or drawings as needed.

QUESTIONS

1. Why does fluidics refer to both hydraulics and pneumatics?
2. What is the difference between hydraulics and pneumatics?
3. What is the common unit of measurement of work in the U.S. Customary System? In the International System?
4. Describe the meaning of ft-lb.
5. What is a "Pa" and what does it describe?
6. Describe the elements in a hydraulic circuit.
7. Define a positive displacement pump and describe the two categories of this kind of pump. Do they pump pressure?
8. A relief valve is the most common pressure control valve. Describe how it works.
9. What is the difference between a relief valve and a sequence valve?
10. Compare a relief valve to a check valve.
11. What is a port in a valve?
12. Describe the purpose of a directional control valve in a fluid power system.
13. What does "two-way, two-position directional control valve" mean?
14. Describe the difference between a differential double-acting cylinder and a nondifferential double-acting cylinder.
15. What is difference between a single-acting cylinder and a double-acting cylinder?
16. What is an actuator?
17. Which type of fluid power diagram would normally be used to show the function of each component? Which would be used for troubleshooting? Which type is internationally standardized?
18. What does it mean when a valve is actuated?
19. In a graphic diagram, what is the name given to the squares used in directional control valves?
20. In a graphic diagram, how would you know which envelope of a two-way, two-position directional control valve is unactuated?
21. If a graphic symbol for a directional control valve is drawn upside down, is the meaning of the circuit changed?
22. What symbol indicates when a component can be adjusted or varied?
23. Can a person build a hydraulic unit from a graphic diagram only? Why or why not?
24. Describe the elements in a pneumatic circuit.
25. With an air compressor, air is compressed from 30 cubic feet to 15 cubic feet, and the pressure increases from 40 psi to 80 psi. Is this an example of Pascal's Principle, Boyle's Law, or Charles's Law?
26. In a container of compressed air, the gauge reading is 20 psi. Assuming that the temperature remains constant, what will the gauge read when the air is compressed into one-half the original space? What will the gauge read when the air is compressed into one-fourth the original space? Which law describes this phenomenon?
27. Should a pressure regulator be upstream or downstream of an air lubricator? Why?

28. What is FAD?
29. Is there any difference in the basic function between a hydraulic directional control valve and a pneumatic directional control valve?
30. If only one pneumatic diagram is shown, are the valves shown in actuated or unactuated phases?

31. What is a pneumatic device? What are the two types?
32. What special considerations are necessary for pneumatics that do not apply to hydraulics?
33. What special considerations are necessary for hydraulics that do not apply to pneumatics?

CHAPTER 19 FLUID POWER PROBLEMS

Problem 19–1 What is the pressure exerted on a 10 in. × 24 in. panel when a force of 1000 pounds is applied to it?

Problem 19–2 What is the pressure exerted on a brick wall that is 4 meters high and 28 meters long when a force of 50,000 newtons is applied to it?

Problem 19–3 A 2350 newton force is applied to an object with a surface area of 72 square meters. What is the pressure? What is the pressure in psi?

Problem 19–4 A hydraulic piston pushes an object from one conveyor belt to another conveyor belt, which is 4'–6" away. The amount of work done is 1800 ft-lb. How much force is being used to push the object? How many kilojoules are needed?

Problem 19–5 The hydraulic piston in the previous example accomplishes its task in five seconds. How much horsepower is the piston performing? How many watts of power is this?

Problem 19–6 One piston in a hydraulic circuit exerts a 28 psi pressure in the hydraulic fluid. A second piston in the circuit is 12 ft away from the first piston and is directly affected by the first piston. It has an area of 15 sq in., which is one-half the area of the first piston. What is the pressure exerted on the second piston? With what force does the second piston move? With what force does the first piston move? If the first piston moves 10 in., how far does the second piston move?

Problem 19–7 The gauge reading of a container of compressed air at sea level is 80 psi. The volume of that air is double. What is the psia? What is the psig?

Whenever possible, draw the following diagrams using a CADD terminal and Fluid Power Symbols Library.

Problem 19–8 Make a graphic diagram of a two-way, two-position directional control valve with a solenoid actuator.

Problem 19–9 Make a graphic diagram of a directional control valve that has four ports and an intermediate position.

Problem 19–10 Make a graphic diagram of a pneumatic circuit that will lift a load with a differential single-acting cylinder. Use a conditioning unit and a two-way, two-position directional control valve.

Problem 19–11 Make a graphic diagram of a hydraulic circuit. Use a nondifferential double-acting cylinder, a check valve, relief valve, and a three-way, two-position valve.

Problem 19–12 Make a graphic diagram of the pictorial hydraulic circuit shown in the following figure. Label each component.

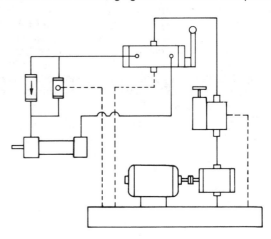

Problem 19–13 Make a pictorial diagram of the graphic hydraulic circuit shown in the following figure. Label each component.

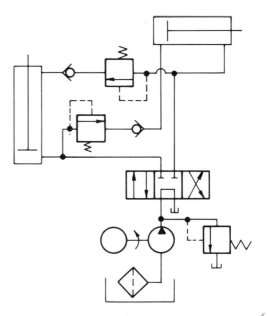

Problem 19–14 Make a graphic diagram of a hydraulic circuit based on the following sequence of operations and functional description. Also make a component list specifying the key number, quantity, and name of component.

Sequence of Operations:

1. With valve (6) in the neutral position, delivery of flow from pump (4) unloads freely through valve (6) to reservoir (1).
2. With motor (3) running, valve (6) is manually actuated right, directing flow from pump (4) to extend clamp cylinder (7-A).
3. When pressure reaches setting of valve (8-A), flow is sequenced through (8-A) to extend nondifferential work cylinder (9) right.
4. When pressure reaches setting of valve (8-B), flow is sequenced through (8-B) to retract clamp cylinder (7-B).
5. Manually actuating valve (6) left, flow is directed from pump (4) to extend clamp cylinder (7-B).
6. When pressure reaches setting of valve (8-C), flow is sequenced through (8-C) to extend work cylinder (9) left.
7. When pressure reaches setting of valve (8-D), flow is sequenced through (8-D) to retract clamp cylinder (7-A).

Functional Description:

Valve (5) provides overload protection. Valve (6) controls direction of motion of (7-A), (7-B), and (9). Valve (8-A) causes work to be clamped by (7-A) before cylinder (9) performs work. Valve (8-B) causes nondifferential work cylinder (9) to be retracted before cylinder (7-B) is retracted. Valve (8-C) causes work to be clamped by (7-B) before cylinder (9) performs work. Valve (8-D) causes nondifferential work cylinder (9) to be retracted before cylinder (7-A) is retracted.

NOTE: Item (2) is the strainer. Work cylinder (9) performs work in two directions, and works in conjunction with cylinders (7-A) and (7-B).

Problem 19–15 Make a graphic diagram of the pneumatic circuit shown.

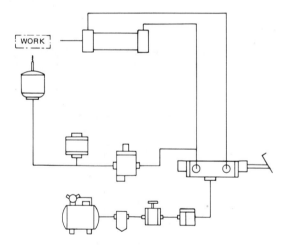

Problem 19–16 The hydraulic sketch in the following figure is received from the engineer. Make an appropriate graphic drawing, label the components, and make a component list.

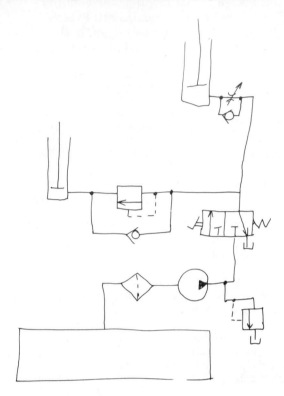

Problem 19–17 Make a pneumatic graphic drawing using the engineer's sketch in the previous example. Make the appropriate changes needed to depict a pneumatic circuit. Do not use a conditioning unit.

CHAPTER 20

Industrial Process Piping

THE ENGINEERING DESIGN PROCESS

This problem requires that the student use an engineering sketch (see Figure 20–1), to create an isometric freehand sketch, then calculate the lengths of straight runs of pipe shown as "1" and "2" in the sketch. If you are not familiar with isometric drawing, see Chapter 17. The following instructions should be used in this problem.

Step 1. Sketch the pipe run in isometric format.

Step 2. Calculate the lengths of the hypotenuse of triangles A and B. (See Figure 20–1, page 624.)

Step 3. Use the Pythagorean Theorem or trigonometric formulae. Each method is shown in the solution.

Step 4. Calculate overall lengths of straight runs of pipe (1 and 2) including fittings.

Step 5. Add the fitting lengths and subtract from the overall length of the pipe run. Use Appendix A to find pipe fitting and flange dimensions. All fittings and flanges are 150# rating.

The term *piping* can refer to any kind of pipe used in a wide range of applications. The word *plumbing* refers to the small diameter pipes within our houses that carry water, gas, and wastes. Pipes of this kind can be copper, steel, cast iron, and plastic. Large underground pipes that transport water and gas to our homes from public utilities, and collect the waste water that is taken to treatment plants, are known as *civil piping* because of their municipal nature. Steel, cast iron, and concrete are the principal materials used for this piping.

Industrial plants involve a process in which raw materials are converted to finished products. Water, air, and steam are used in many industries for processing the raw materials. Piping used to transport fluids between storage tanks and processing equipment is called *process piping*. (See Figure 20–2.)

Large diameter pipes called *pipelines* carry crude oil, water, petroleum products, gases, coal slurries, and a variety of liquids hundreds of miles. This type of pipe is known as *transportation piping*, as shown in Figure 20–3.

▼ WHERE IS INDUSTRIAL PIPING USED?

A web of piping lies beneath the ground in your neighborhood. Miles of pipes for water, sewers, storm drains, natural gas, and electrical power provide vital services. (See Figure 20–4.) Oil and gas pipelines provide the compounds that enable us to drive to work and school daily. The petrochemical plants that process the crude oil into a tremendous variety of products, employ thousands of miles of piping. (See Figure 20–5.)

This paper was once a tree that was chopped into chips, cooked into a pulpy stew and pumped via pipes through a series of processes in a paper mill. The food

ENGINEERING DESIGN PROCESS *(Continued)*

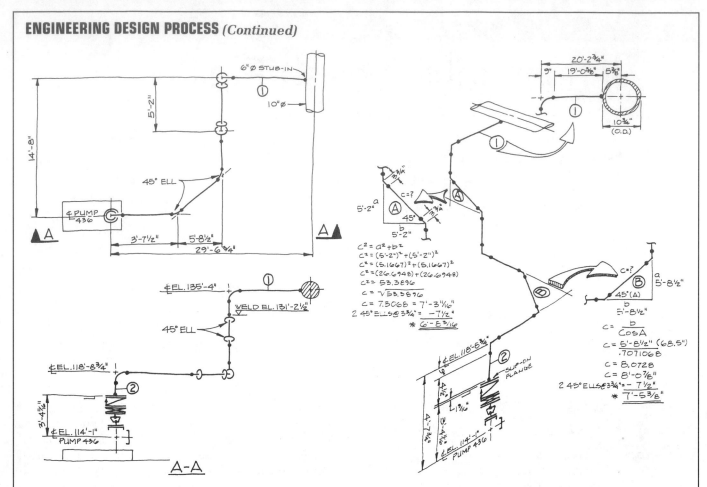

Figure 20–1 This engineering sketch is used to construct an isometric sketch and to calculate the lengths of straight pipe in the angular offsets.

Figure 20–2 Process piping transports fluids between storage tanks and processing equipment at this oil refinery. *Courtesy Amoco Corporation.*

Figure 20–3 This pipeline is a form of transportation piping. *Courtesy Amoco Corporation.*

you consume is processed with the aid of pipes filled with water, chemicals, and liquid mixtures. The beverages you drink were moved in pipes through the various stages of production. The electric wires that provide

power are resting inside pipes (electrical conduit). The electric power inside the wires is produced at steam and nuclear power plants containing intricate webs of piping. (See Figure 20–6.) Even the air you are breathing, if you are in a school or office, was probably brought to you through pipe-shaped ducts.

When you think about all the uses for pipe, and realize how seldom it is seen, it becomes obvious why piping drafting and design has often taken a back seat to other visible manifestations of drafting.

▼ WHAT IS PIPE DRAFTING?

Pipe drafting is a specialized field that calls upon the drafter's skill of visualization and the ability to see pipe and fittings in several planes (depth) in an orthographic view. Pipe fittings must often be turned and rotated at odd angles. Visualization can be confusing for beginning pipe drafters and engineers, therefore ample study should be given to pictorial (isometric and 3-D) drawing techniques. Pipe can be drawn in two forms, double line and single line, as shown in Figure 20–7. The single-line method usually gives beginners the most visualization problems. This method is discussed later in this chapter.

The principal area of learning, other than refining the ability to visualize, is the realm of pipe fittings and joining methods. A major portion of this chapter will be spent on these areas. Those who may have visualization problems should spend additional time studying and drawing pipe fittings.

Pipe drafting involves the creation of a variety of drawing types from maps (site plans) to mechanical drawings (piping details). One of the easiest types of drawings to construct is the *flow diagram* as shown in Figure 20–8. This is a schematic, nonscale diagram that

Figure 20–4 Underground piping supplies natural gas for homes and industries. *Courtesy Pacific Gas and Electric.*

Figure 20–5 Petroleum refineries employ miles of interwoven piping. *Courtesy Amoco Corporation.*

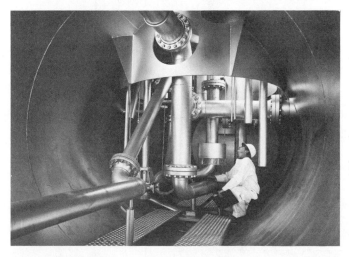

Figure 20–6 Piping is an integral part of steam generation power plants. *Courtesy Washington Public Power Supply System.*

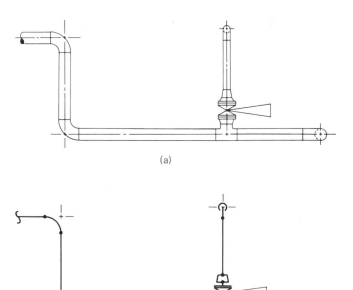

(a)

(b)

Figure 20–7 Pipe can be drawn in **(a)** double-line or **(b)** single-line representation.

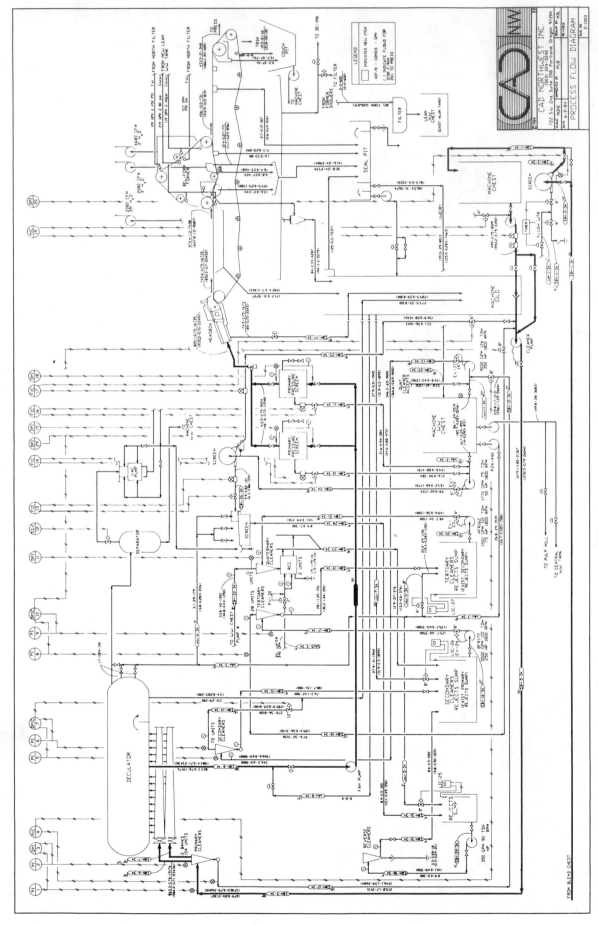

Figure 20–8   This piping flow diagram is a nonscale view of a piping system. *Courtesy Autodesk, Inc.*

illustrates the layout and composition of a system using symbols. The *piping drawing* is the most complex. It is a scale drawing that provides plan, elevation, and section views. All equipment, fittings, dimensions, and notes are shown on this type of drawing. Examples of plans and sections are shown in Figure 20–9a and Figure 20–9b. The drafter must use flow diagrams, structural, mechanical, and instrumentation drawings, and vendor catalogs to construct these drawings.

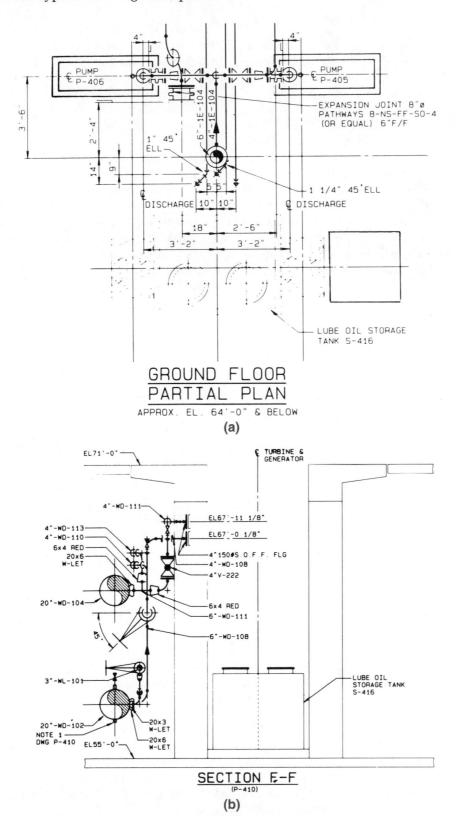

Figure 20–9   **(a)** The piping plan; **(b)** section containing detailed piping information. *Courtesy Schuchart and Associates.*

CADD Applications

PIPING

Most piping drawings involve the use of symbols. Symbols need to be drawn only once, saved in a symbol library, and displayed on a menu attached to a digitizer or menu tablet. When a symbol is needed, it is simply picked from the menu, placed on the drawing, and scaled or rotated as needed. If your CADD system does not have a piping symbol library yet, it will save you time in the future if you begin to create one now.

Software companies create standard and specialty symbols that are assembled in libraries. The tablet menu for the CADD package has a "user" area that was designed for symbol overlays. A productive part of the piping drafter's job is the use and placement of these standard symbols and shapes on new drawings. An example of a tablet menu piping overlay is shown in Figure 20–10.

Many of the symbol programs are "intelligent." That is, they contain pieces of information called *attributes* or *tags* that give meaning to the symbol. For example, a valve symbol may contain hidden attributes pertaining to its diameter, pressure rating, material, operating mechanism, specific type of styling, weight, price, and manufacturer. When the valve symbol is used on a drawing, all of the attributes become a part of the drawing's database.

Drafters using CADD systems should keep an easily accessible, up-to-date record of all piping details. They may be needed again on the same project or on another one. Time savings are involved by just placing an entire detail on a drawing instead of constructing it again. An individual detail may not be exactly the same as a previous one, but even revising an existing detail is much quicker than redrawing it.

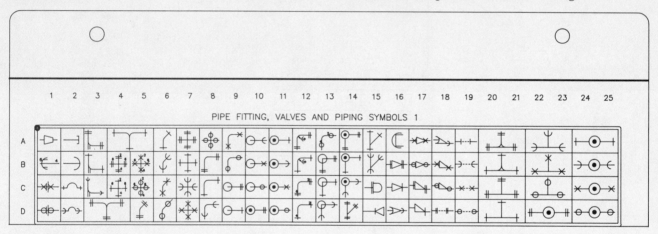

Figure 20–10   Example of a tablet menu overlay that is attached to the user menu area. *Courtesy Drafting Technology Services, Inc., Bartlesville, OK.*

The piping *isometric* is a pictorial drawing that illustrates a pipe run in three-dimensional form. (See Figure 20–11.) Information from this drawing is obtained from the piping drawings. The drafter's ability to view piping runs in three planes is important when working with piping isometrics. Some pipe drawings are done in isometric form because lines can be measured and fits checked. Isometrics are used extensively in chemical process industries.

Subassemblies of pipe and fittings are constructed by pipe fitters and welders who use the piping *spool* drawing. This drawing is usually drawn orthographically and shows all of the pipe and fittings needed to assemble a segment of piping. Spools are often nonscale drawings, but show all dimensions needed for assembly. A typical spool drawing is shown in Figure 20–12.

▼ TYPES OF PIPE

When an engineer or designer decides to use a specific type of pipe, that decision is based on a number of considerations. Principal among those are temperature, pressure, and corrosion. Safety and cost are also important factors. And finally, any decision must comply with the project specifications and local and national codes.

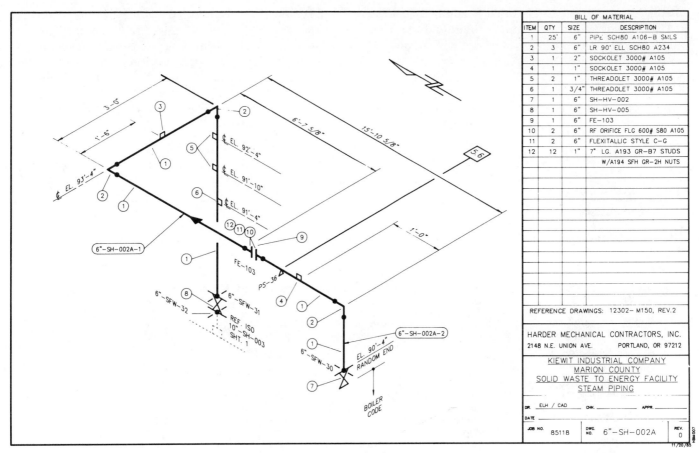

BILL OF MATERIAL

ITEM	QTY	SIZE	DESCRIPTION
1	25'	6"	PIPE SCH80 A106-B SMLS
2	3	6"	LR 90° ELL SCH80 A234
3	1	2"	SOCKOLET 3000# A105
4	1	1"	SOCKOLET 3000# A105
5	2	1"	THREADOLET 3000# A105
6	1	3/4"	THREADOLET 3000# A105
7	1	6"	SH-HV-002
8	1	6"	SH-HV-005
9	1	6"	FE-103
10	2	6"	RF ORIFICE FLG 600# S80 A105
11	2	6"	FLEXITALLIC STYLE C-G
12	12	1"	7" LG. A193 GR-B7 STUDS
			W/A194 SFH GR-2H NUTS

REFERENCE DRAWINGS: 12302-M150, REV.2

HARDER MECHANICAL CONTRACTORS, INC.
2148 N.E. UNION AVE.   PORTLAND, OR 97212

KIEWIT INDUSTRIAL COMPANY
MARION COUNTY
SOLID WASTE TO ENERGY FACILITY
STEAM PIPING

DR. ELH / CAD   CHK.   APPR.
DATE
JOB NO. 85118   DWG. NO. 6"-SH-002A   REV. 0

Figure 20–11   The piping isometric shows a pictorial view of a single run of pipe. *Courtesy Harder Mechanical Contractors, Inc.*

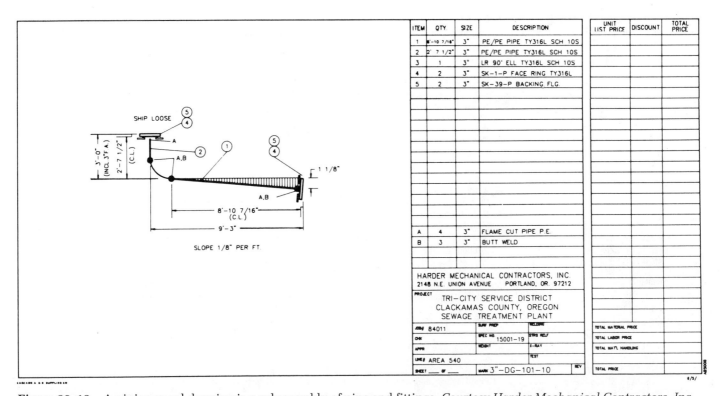

ITEM	QTY	SIZE	DESCRIPTION	UNIT LIST PRICE	DISCOUNT	TOTAL PRICE
1	8'-10 7/16"	3"	PE/PE PIPE TY316L SCH 10S			
2	2' 7 1/2"	3"	PE/PE PIPE TY316L SCH 10S			
3	1	3"	LR 90° ELL TY316L SCH 10S			
4	2	3"	SK-1-P FACE RING TY316L			
5	2	3"	SK-39-P BACKING FLG.			
A	4	3"	FLAME CUT PIPE P.E.			
B	3	3"	BUTT WELD			

HARDER MECHANICAL CONTRACTORS, INC.
2148 N.E. UNION AVENUE   PORTLAND, OR. 97212

PROJECT   TRI-CITY SERVICE DISTRICT
CLACKAMAS COUNTY, OREGON
SEWAGE TREATMENT PLANT

JOB# 84011   SPEC NO. 15001-19
LINE# AREA 540
MARK 3"-DG-101-10

TOTAL MATERIAL PRICE
TOTAL LABOR PRICE
TOTAL MAT'L HANDLING
TOTAL PRICE

Figure 20–12   A piping spool drawing is a subassembly of pipe and fittings. *Courtesy Harder Mechanical Contractors, Inc.*

ANSI The American National Standards Institute (ANSI) specifications that apply to pipes and fittings are listed in Figure 20–13 should you wish to research certain dimensions or properties.

Cast Iron

The most common types of pipe used for commercial and industrial applications are steel and cast iron.

AWWA Cast iron is specified by the American Water Works Association (AWWA) for underground water lines. Cast iron pipe is corrosion resistant and has a heavy wall construction. It is good for use under pavement because its long life obviates the need to dig up the pavement due to frequent leaks. Cast iron is used extensively for water, gas, and sewage piping.

SPECIFICATION NUMBER	DESCRIPTION
Graphic Standards	
ASA Z32.2.3–1949	Graphical Symbols for Pipe Fittings, Valves, and Piping
ANSI Y32.4–1977	Graphic Symbols for Plumbing Fixtures for Diagrams Used in Architecture and Building Construction
ASA Y32.11–1961	Graphical Symbols for Process Flow Diagrams
Pipe, Steel	
ANSI/ASTM A430–79	Austenitic Steel Forged and Bored Pipe for High-Temperature Service
ANSI/ASTM A524–80	Seamless Carbon Steel Pipe for Process Piping
ANSI B36.19–1976	Stainless Steel Pipe
ANSI B36–10–1979	Welded and Seamless Wrought Steel Pipe
ANSI/ASTM A135–79	Electric-Resistance-Welded Steel Pipe
Pipe and Fittings, Plastic	
ANSI/ASTM D1527–77	Acrylonitrile-Butadiene-Styrene (ABS) Plastic Pipe, Schedules 40 and 80
ANSI/ASTM F443–77	Bell-End Chlorinated Poly(Vinyl Chloride) (CPVC) Pipe
ANSI/ASTM D2446–73	Cellulose Acetate Butyrate (CAB) Plastic Pipe (SDR-PR) and Tubing
ANSI/ASTM D1503–73	Cellulose Acetate Butyrate (CAB) Plastic Pipe, Schedule 40
ANSI/ASTM F441–77	Chlorinated Poly(Vinyl Chloride) (CPVC) Plastic Pipe, Schedule 40, 80, and 120
ANSI/ASTM D2104–74	Polyethylene (PE) Plastic Pipe, Schedule 40
ASTM D2469–76	Socket-Type Acrylonitrile-Butadiene-Styrene (ABS) Plastic Pipe Fittings, Schedule 80
ANSI/ASTM D2466–78	Socket-Type Poly(Vinyl Chloride) (CPVC) Plastic Pipe Fittings, Schedule 40
Pipe, Miscellaneous	
ANSI/ASTM C700–78a	Extra Strength and Standard Strength Clay Pipe and Perforated Clay Pipe
ANSI/ASTM B43–80	Seamless Red Brass Pipe, Standard Sizes
ANSI/ASTM C599–70(1977)	Process Glass Pipe and Fittings
ANSI A 40.5–1943	Threaded Cast-Iron Pipe for Drainage, Vent, and Waste Services
ANSI/AWWA C151/A21.51–(1981)	Ductile-Iron Pipe, Centrifugally Cast, in Metal Molds or Sand-Lined Molds for Water or Other Liquids
Fittings and Flanges	
ANSI B16.26–1975	Cast Copper Alloy Fittings for Flared Copper Tubes
ANSI B16.1–1975	Cast Iron Pipe Flanges and Flanged Fittings, Class 25, 125, 250, and 800
ANSI B16.4–1977	Cast-Iron Threaded Fittings, Class 125 and 250
ANSI B16.9–1978	Factory-Made Wrought Steel Buttwelding Fittings
ANSI B16.42–1979	Fittings, Class 150 and 300, Ductile Iron Pipe Flanges and Flanged Fittings
ANSI/AWWA C207–78	Flanges for Water Works Service, 4 in. through 144 in. Steel
ANSI B16.11–1980	Forged Steel Fittings, Socket-Welding and Threaded
ANSI/AWWA C606–81	Joints, Grooved and Shouldered Type
ANSI B16.3–1977	Malleable-Iron Threaded Fittings, Class 150 and 300
ANSI B16.21–1978	Nonmetallic Flat Gaskets for Pipe Flanges
ANSI B16.5–1981	Pipe Flanges and Flanged Fittings, Steel Nickel Alloy and Other Special Alloys
ANSI B16.39–1977	Pipe Unions (Class 150, 250, and 300), Malleable Iron Threaded
ANSI B16.20–1973	Ring-Joint Gaskets and Grooves for Steel Pipe Flanges
ANSI B16.36–1975	Steel Orifice Flanges, Class 300, 600, 900, 1500, and 2500 (includes supplement ANSI B16.36a–1979)
ANSI B16.22–1980	Wrought Copper and Copper Alloy Solder-Joint Pressure Fittings
ANSI B16.28–1978	Wrought Steel Buttwelding Short Radius Elbows and Returns
Valves	
ANSI B16.34–1981	Valves, Flanged and Butt-Welding End-Steel, Nickel Alloy, and Other Special Alloys
ANSI B16.10–1973	Face-to-Face and End-to-End Dimensions of Ferrous Valves
ANSI/FCI 74–1–1979	Valves, Spring Loaded Lift Disc Check
ANSI/AWWA C509–80	Valves, 3 through 12 NPS, for Water and Sewage Systems, Resilient-Seated Gate
ANSI/AWWA C500–80	Valves, 3 through 48 NPS, for Water and Sewage Systems, Gate
Pipe Hangers	
ANSI/MSS SP-58–1979	Pipe Hangers and Supports — Materials, Design, and Manufacture

Figure 20–13 ANSI specifications for pipe, fittings, valves, and pipe hangers.

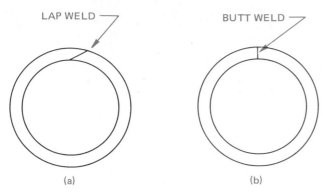

Figure 20–14 **(a)** Lap-welded steel pipe; **(b)** butt-welded steel pipe.

Steel

Carbon steel pipe is the preferred above-ground pipe used in industry today. It is strong, relatively durable, and it can be welded and machined. It is also not as expensive as other materials. In conditions of high temperature it tends to lose strength, so stainless steel and other alloys that withstand high temperatures are then used.

Steel pipe is manufactured by several processes. *Forged* pipe is formed to a special *outside diameter* or O.D. (usually one inch greater than the finished pipe) then bored to the required *inside diameter* (I.D.). Seamless pipe is created by piercing a solid billet and rolling the resulting cylinder to the required diameter. *Welded* pipe is formed from plate steel. The edges are welded together in a *lap* weld or a *butt* weld. (See Figure 20–14.)

Copper and Copper Alloys

Copper pipe and copper alloys can be manufactured by the hot piercing and rolling process or by an extrusion method. Copper pipe is corrosion resistant and has good heat transfer properties, but has a low melting point and is expensive. It is good for instrument lines and food processing, although stainless steel is being used more frequently in these applications. Copper pipe is also used in residential water lines.

Copper tubing is softer and more pliable than rigid copper and brass pipe, and is used for steam, air, and oil piping. It is also used extensively for *steam tracing,* the small diameter pipe that is attached to larger pipes to protect them from freezing or to keep the fluids inside the pipe warm. The outside diameter of copper tubing is the same as its *nominal pipe size* (NPS). For example, a 3/4" tube has an outside diameter of 3/4".

Plastic

Thermoplastic pipe was first developed in Germany in 1921, but was not used to any great extent in the United States until 1951. This first type of plastic pipe was *polyvinyl chloride* (PVC). PVC piping is now used for acids, salt solutions, alcohols, crude oil, and a variety of highly corrosive chemicals.

Polyethylene pipe (PE) was introduced into the United States in 1947. Plastic piping can handle temperatures up to 150° F and is suitable for water and vent piping of corrosive and acidic gases. PE piping is used as conduit for electrical and phone lines, water lines, farm sprinkler systems, salt water disposal, and chemical waste lines. Other forms of plastic piping include *acrylonitrile-butadiene-styrene* (ABS), which is popular for sewage piping, *cellulose-acetate-butyrate* (CAB), and *fiberglass-reinforced pipe* (FRP).

Clay

Clay pipe is formed under high pressure from fire clays or shales, or a combination of the two. It is dried then fired at 2100° F. Under these extreme temperatures, they clay particles actually fuse or "weld" together to form a solid, durable pipe. It is one of the most corrosion-proof pipes available for use in sanitary and industrial sewers, and can carry any known chemical waste except hydrofluoric acid.

Glass

The chemical resistance, transparency, and cleanliness of glass makes it popular for applications in the chemical, food and beverage, and pharmaceutical industries. It can withstand temperatures up to 450° F.

Wood

Continuous stave wood pipe is used in the Pacific Northwest because of the abundance of redwood and douglas fir. Tongue and groove staves are milled to exact radii for the specific diameter of pipe. These staves are fitted together, then strapped with wire, flat steel bands or steel rod threaded on the ends and tightened with bolts. (See Figure 20–15.) The wood used for this pipe is not resistant, especially when water under constant pressure is flowing through it. Wood stave pipe is used almost exclusively for transporting water, and is available in sizes from 10 in. to 16 ft. in diameter.

Steel Tubing

Tubing is small diameter pipe and is often flexible, thus eliminating the need for many common fittings. Tubing is specified by its outside diameter and wall thickness. Its uses include external heating applications, boilers, superheaters, and hydraulic lines in the automotive and aircraft industries.

Pipe Sizes and Wall Thickness

Pipe is available in sizes from 1/8" to 44" in diameter. Sizes greater than this can be ordered but are not normally stocked by suppliers. The typical range of commonly stocked pipe is from 1/2" to 24". These are the

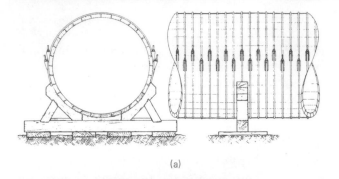

(a)

(b)

Figure 20–15 **(a)** Side and end views of a wood stave pipe resting in its foundation cradle; **(b)** installed wood stave pipe. *Courtesy National Tank and Pipe Co.*

sizes most often used in process piping. Sizes from 1/8" to 1/2" are used for instrument lines and service piping.

Pipe is specified by its *nominal pipe size* (NPS). The NPS for pipe 1/8" to 12" is the inside diameter (I.D.), and pipe 14" and above uses the outside diameter (O.D.) as the NPS. (See Figure 20–16.) If you wish to determine the actual inside diameter of a pipe you should double the wall thickness and subtract that number from the outside diameter.

ANSI The wall thickness of pipe varies in relation to the size and weight of the pipe. The ANSI specifications for wall thickness (B36.10M—1985) classifies pipe as *schedule* (SCH) numbers. These numbers

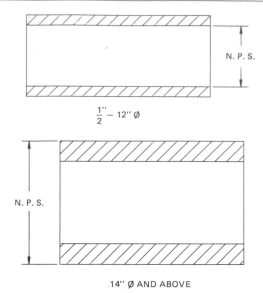

.14″ Ø AND ABOVE

Figure 20–16 How to measure nominal pipe size (NPS).

range from 10 to 160, the higher number representing the thicker wall. The ANSI schedule numbers incorporate the classifications of the ASTM and ASME, which use the designations of standard (STD), extra strong (XS), and double extra strong (XXS). These three classifications are drawn from manufacturer's dimensions. The STD wall thickness compares to SCH 40, XS to SCH 80, and XXS has no comparable schedule number (it is a thicker wall than SCH 160).

▼ PIPE CONNECTION METHODS

Butt-welded Connection

Butt welding is the most common method of joining pipes used in industry today. It is used on steel pipe 2" in diameter and more to create permanent systems.

Butt-welded pipe and fittings provide a uniform wall thickness throughout the system. The smooth inside surface creates gradual direction changes, thus generating little turbulence. One circumferential weld is required to join two pieces of pipe (the number of welds may vary depending on pipe NPS). Figure 20–17 shows a cross section of a butt weld. The weld is strong, leakproof, and relatively maintenance free. Pipe joined in this manner

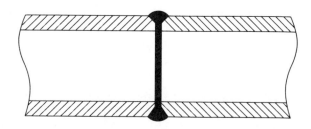

Figure 20–17 Cross-section of a butt-welded pipe connection.

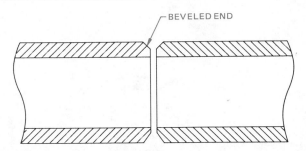

Figure 20-18 Beveled-end pipe is used for butt-welded pipe connections.

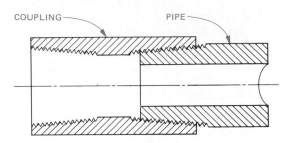

Figure 20-20 Cross section of a screwed connection.

is self-contained, withstands high temperature and pressure, is easy to insulate, and requires less space for construction and hanging than do other methods.

Pipe and fittings joined by butt welding are prepared with a *beveled end* (BE). This type of end preparation provides space for the welding operation. (See Figure 20-18.)

Socket-welded Connection

The *socket-welded* connection shown in Figure 20-19 is used on pipe 2 in. and smaller. It forms a reliable leakproof connection. The pipe has *plain end* (PE) preparation and slips into the fitting. One exterior weld is required, thus no weld material protrudes into the pipe. Since the pipe is slipped inside the fitting, the connection is self-aligning. Socket welding small diameter pipe is less expensive than other welded systems.

Screwed Connection

Screwed connections are used on steel, malleable iron, cast iron, and cast brass pipe less than 2½" in diameter. It is the least leakproof of the pipe joining methods, and is used where the temperature and pressure are low. The end preparation for screwed pipe is termed *threaded & coupled* (T&C) because a coupling is usually supplied with a straight length of pipe. American Standard pipe threads are the most common. The tightest fit is achieved by using fittings with straight threads and pipe with tapered threads. Pipe sealing compounds and teflon tape aid in producing a tighter fit.

A typical coupling and pipe screwed connection is shown in Figure 20-20.

Flanged Connection

The *flanged* method of connecting pipes utilizes a fitting called a flange. The flange has an outside diameter greater than the pipe, and contains several bolt holes. Flanges are welded to the ends of pipe and then sections of pipe can be bolted together, as shown in Figure 20-21. Most flanges are forged steel, cast steel, or iron. Flanged pipe is easily assembled and disassembled, but is considerably heavier than butt-welded pipe. It occupies more space and is also more expensive to support or hang.

Two steel flanges bolted together do not form a tight fit. There must be some sort of sealing material placed between the two flange faces. This sealer is called a *gasket*. It can be soft or semimetallic material. When the bolts are tightened the gasket is squeezed until it is pressed into the machining grooves of the flange faces.

Soft gaskets can be cork, rubber, asbestos, or a combination of materials. Semimetallic gaskets contain both metal and soft material. The soft material provides resilience and gives a tight seal while the metal helps to retain the gasket in place against high pressure and

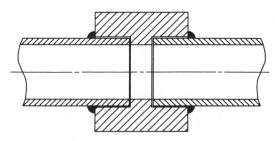

Figure 20-19 The socket-welded connection forms a leakproof joint.

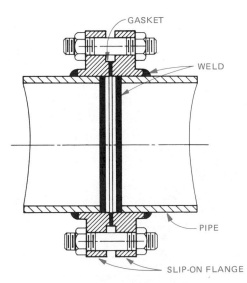

Figure 20-21 Cross section of a typical flanged connection.

temperature. Gaskets are chosen from a wide range of materials that resist deterioration caused by high temperatures, and that are not chemically affected by the fluids in the pipe.

Soldered Connection

Copper and brass water tubes are most often joined by *soldered* connections. Domestic water systems where temperatures and pressures are low are the most common use of this method. Both rigid and soft pipe can be soldered.

Bell (Hub) and Spigot Connection

Underground sewage, water, and gas lines are common applications of the *bell and spigot* connection method. (See Figure 20–22.) Cast iron pipe is used where pressures and loads are low. There are many variations of the bell and spigot method, but they all utilize some sort of sealer. The most common is lead and oakum (a fibrous sealer), although cement is used in certain situations.

Mechanical Unit

The *mechanical joint* is a modification of the bell and spigot connection in which flanges and bolts are used with gaskets, packing rings, or grooved pipe ends providing a seal. Some mechanical joints allow for angular deflection and lateral expansion of the pipe. This type of connection is used in low pressure applications, and where vibration may be excessive. Figure 20–23 illustrates a mechanical joint.

Solvent Welding

This type of pipe connection is also known as *cementing* or *gluing*. It is used on plastic pipe. The solvent is

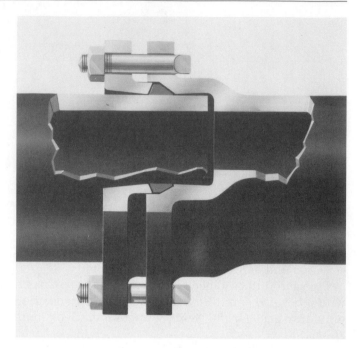

Figure 20–23 Typical mechanical joint connection. *Courtesy Griffin Pipe Products Co.*

applied to the pipe end, which is then inserted in the "socket" opening of a fitting and twisted. The solvent creates a bonding that acts much the same as a weld. Plastic pipe can also be joined by *hot gas welding* in which a torch is used to melt the pipe and fitting together.

Flaring

Soft copper pipe and tubing can be joined by a method called *flaring*. A special tool is clamped on the pipe and the flaring tip is inserted into the end of the pipe. As the tip is rotated it is forced into the pipe, spreading the end of the pipe open, or "flaring" it. The end of the pipe resembles an old blunderbuss rifle. A fitting called a *swage* is then used to connect the flared end to pipe or a fitting.

▼ FITTINGS

Pipe fittings enable pipe to change direction and size, and provide for branches and connections. Each type of pipe and connection method discussed previously uses the same type of fittings. But special fittings may be required by the nature of the connection method used. The following is a general discussion of common fittings used in industry today.

Welded Fittings

Seamless forged steel fittings are prepared with a beveled end to accommodate butt welding. A welding

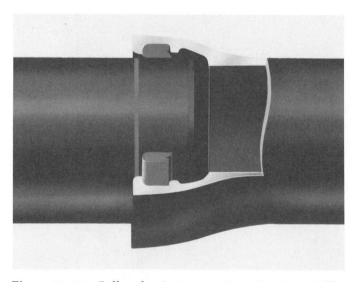

Figure 20–22 Bell and spigot connection. *Courtesy Griffin Pipe Products Co.*

ring (see Figure 20–24a) is placed between pipe and fitting to aid in alignment, provide even spacing, and to prevent weld material from falling into the pipe (see Figure 20–24b). Since fittings have the same schedule numbers and wall thickness as pipe, a smooth inside surface is produced at the joints when fittings and pipe of the same schedule number are welded.

The most familiar fittings are *elbows*. These and other welded fittings are seen in Figure 20–25. Standard shapes are the 90° and 45° elbows. The 90° *reducing elbow* reduces the pipe size in addition to changing direction. A reversal in direction can be achieved by using a 180° elbow. Standard elbows can also be cut to any angle required. A *mitered* elbow is produced by cutting and welding straight pipe. The mitered elbow can be composed of one, two, three or more welded joints. (See Figure 20–26.) It is not used often since it produces considerably more turbulence than standard

LONG NUBS

SHORT NUBS

(a)

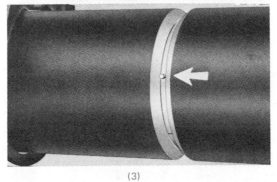

(1)

This photograph shows the ease with which a welder can fit a backing ring into the beveled end of pipe. This is an NPS 6″ schedule 80 pipe. Tack welds are not necessary. The spacer nubs melt and become part of the weld metal.

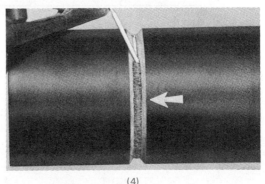

(2)

The adjoining pipe is slipped onto the backing ring until the edge of the bevel touches the spacer nubs. Short pieces of pipe will be self supported by the close tolerances of the ring diameter.

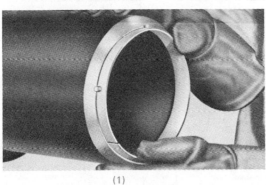

(3)

Note the close fit between the ring and the pipe . . . the minimum gap at the split. The carefully aligned spacer nubs center the backing ring perfectly across the welding groove and provide exact, even spacing between the pipe ends. The metal arrow locates the position of the spacer nubs in X-ray photographs.

(4)

The weld is started at the bottom of the pipe. As this weld is being made in a fixed horizontal position, the split in the ring is on the bottom of the weld to prevent icicle formation inside the pipe. The ring is designed to allow a gap of only 1/16″ at the split, minimizing the operation of filling the gap.

(b)

Figure 20–24   **(a)** Common welding rings with long nubs and short nubs; **(b)** pipe-joining process using a welding ring. *Courtesy ITT Grinnell Corp.*

Figure 20–25 Welded seamless fittings. *Courtesy ITT Grinnell Corp.*

elbows. An installation of the mitered elbow is seen in Figure 20–27.

The diameter of the pipe can be changed by *reducers*. The *concentric reducer* tapers the pipe equally about the axis centerline of the pipe. The *eccentric reducer* has a flat side that allows one side of the pipe—typically the top or bottom—to remain level.

The *straight tee* is a standard fitting that creates a branch or inlet. The straight tee has the same diameter on all three openings. The *reducing tee* has a smaller

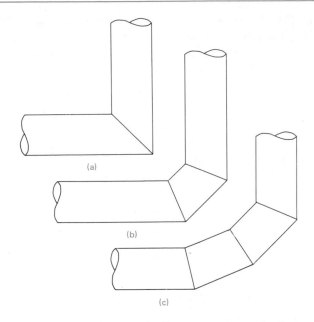

Figure 20–26 The mitered elbow is composed of several cut pipe segments: **(a)** two-piece miter; **(b)** three-piece miter; and **(c)** four-piece miter.

Figure 20–27 Mitered elbows in a vacuum pump installation at a paper mill. *Courtesy Ingersoll-Rand.*

CADD Applications

THREE-DIMENSIONAL SYSTEM MODELS

In addition to increasing productivity in generating drawings and materials lists, some CADD systems enable the drafter or designer to create a 3-D, full-color model of the system. This model can be shown in wire or solid form, colored and shaded in any fashion, and rotated to suit the needs of the user. (See Figure 20–28.) Hardcopy prints of any aspect of the piping system can be generated for use in the design or construction of the system.

A CADD-generated model can be used to create a scale model and/or orthographic and pictorial drawings for the project. This is a major step in the effort to generate a single data base for an entire construction project that is accessible to all engineers, technicians, drafters, and clients. Once all the design information is in the system, any type of data, in the form of plotted drawings, reports, tables, material take-off, estimates, bills, and a variety of screen displays can be requested. As changes are made to the design, it is reflected in all displays, hardcopy prints, and plotted drawings.

Future product and industrial piping design and construction may be handled completely by a computer system. This central system containing an extensive data base, including building or site sizes and construction codes, may be able to automatically route and lay out pipe, valves, fittings, and equipment. The more routine tasks will be performed by the computer and the overall design of the system will be the function of the engineer, technician, or design drafter. Therefore, a knowledge of CADD and a continuing interest in computer and software advances will enable the piping engineer or drafter/designer to learn and change as our technology advances.

Figure 20–28 This 3-D model of a piping system has been shaded and rendered to achieve a realistic presentation. *Courtesy International Software Systems, Inc.*

opening on the outlet side of the fitting. A branch or inlet can also be created with a *lateral*. This fitting provides a 45° branch of the *run* pipe. The run is the straight portion of the fitting. The lateral is available in both straight and reducing forms. The *true* Y is similar to the lateral, but produces branches that are at a 90° angle, thus forming a "Y" with the run pipe. A branching fitting similar to the tee is the *cross*. It has two branches opposite each other. It can be either straight or reducing. It is employed in special situations and tight spaces, but is seldom used because of its expense.

There are several small fittings, less expensive than regular fittings, that are used to create branch connections in new installations or on pipe that is already assembled. They are commonly called *weldolets®*, but are manufactured to accept welded, screwed, socket-welded and brazed pipe. These little fittings can be welded to an elbow *(elbolet®)*, at a 45° angle to the run

pipe *(latrolet®)*, or at a 90° angle to the run pipe *(weldolet®)*. (See Figure 20–29.)

Dimensions for seamless welded fittings can be found in Appendix I.

Screwed Fittings

All of the welded fittings are available in threaded form. In addition to these, there are several special fittings that are used with screwed pipe. Appendix J provides dimensions for galvanized malleable iron fittings.

○ *Union*—Composed of two threaded sleeves and a threaded union ring, this fitting provides a connection point in a straight run of pipe. It enables pipe to be broken apart without tearing down the entire run of pipe. Figure 20–30 illustrates screwed fittings.

○ *Coupling*—Threaded at both ends (TBE) with inter-

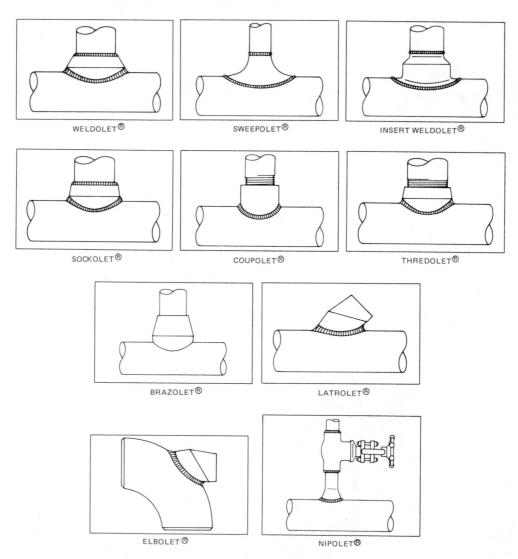

Figure 20–29 Branch connections can create inlets or outlets of varying sizes and angles to the main run of pipe and at less expense than regular fittings. *Courtesy Bonney Forge Division, Gulf and Western.*

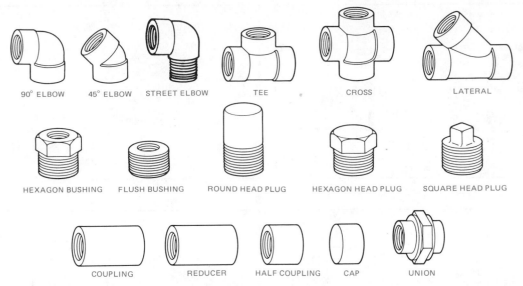

Figure 20–30 Common screwed fittings. *Courtesy Alaskan Copper Works.*

nal threads, it is used to attach two lengths of pipe together.

○ *Half coupling*—Threaded at one end (TOE), this fitting is often welded to pipes and used for instrument connections.

○ *Street elbow*—This 90° elbow is threaded at one end with internal threads and the other end with external threads. It can be attached directly to a fitting, thus eliminating the need of a short piece of threaded pipe *(nipple)*.

○ *Bushing*—A reducing fitting used to connect small pipe to larger fittings.

○ *Plug*—A fitting with external threads on one end that is used to seal the screwed end of a fitting.

▼ FLANGES

The component that creates a bolted connection point in welded pipe is the *flange*. It is a circular piece of steel containing a center bore that matches the pipe I.D. to which it is attached. It has several bolt holes evenly spaced around the center bore. Flanges are used most extensively on welded pipe, but are also available for most other types of pipe. Figure 20–31 shows several types of flanges. Flange dimensions are given in Appendix I.

Slip-on Flange

The *slip-on* flange can be used only on straight pipe because it is bored to slip over the end of the pipe. Two welds are required to attach this flange to pipe. (See Figure 20–31.) There are two types of slip-on flanges.

ASME When the Type 1 slip-on flange is used, the pipe is set back from the face of the flange. It has two-thirds the strength of a weld-neck flange and is limited to 300 lb service by the ASME Pressure Piping Code. The Type 2 slip-on flange allows the pipe to be welded flush with the face of the flange, which is then machined flat. This type of flange is used on lines having 400 lb pressure and above.

Weld-neck Flange

The *weld-neck* flange is forged steel and prepared with a beveled end for butt welding to pipe or fittings. It is always used when a flange must be attached directly to a fitting.

Blind Flange

A pipe can be temporarily sealed with a *blind flange.* It is basically a steel plate with bolt holes in it.

Stub-end or Lap-joint Flange

The *stub-end* or *lap-joint* flange is composed of two parts, the stub end and the flange ring. This flange ring can be carbon steel if expensive pipe such as stainless steel is used. Only the stub end need be stainless.

Reducing and Expander Flange

A change in line size can be achieved with a *reducing flange,* but it should not be used where increased turbulence would be undesirable. A reduction in line size can also be created with an *expander flange.* This fitting is a flange/reducer combination and can be used in place of a weld-neck flange and a reducer.

Orifice Flange

The flow rate inside a pipe can be measured using *orifice flanges* and an *orifice plate.* The two orifice flanges are drilled and tapped to accommodate tubing and a pressure gauge used to measure the flow rate. The orifice plate is a flat disc with a small hole drilled in its center.

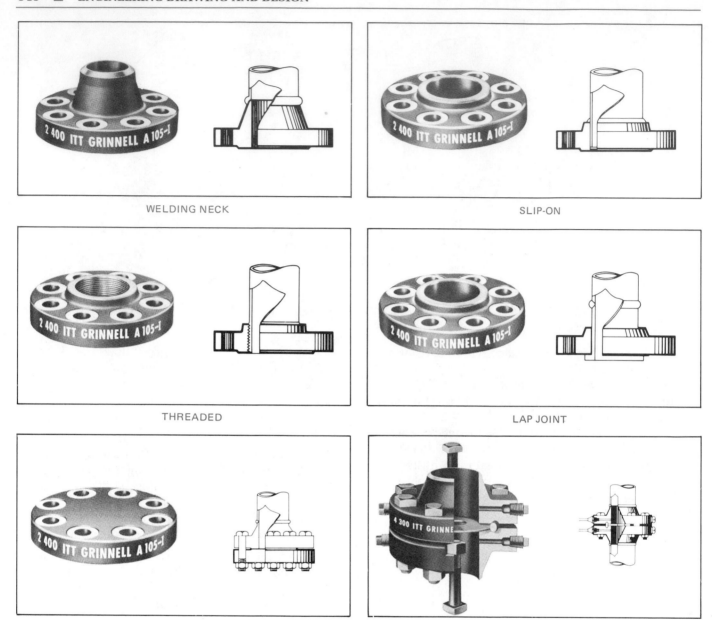

WELDING NECK

SLIP-ON

THREADED

LAP JOINT

BLIND

ORIFICE

Figure 20–31   Common flanges used on butt-welded steel pipe. *Courtesy ITT Grinnell Corp.*

The two flanges are welded to the pipe, the orifice plate and gaskets are placed between them, and the flanges are bolted together. As fluid flows through the hole in the orifice, a pressure differential is created on either side of the plate. This differential can be read on the pressure gauge.

Flange Faces

The *facing* of a flange is the type of machining that is done to the contact surface of the flange. A common flange used in industry is the *raised face* (RF) flange. The face of this flange protrudes a certain distance beyond the flange itself. The *male and female* facings interlock with each other. A recessed face on the female flange accepts both the gasket and the raised face of the male flange. The *tongue-and-groove* facing is interlocking. The male tongue-and-groove flange is 1/4" longer than the female flange to accommodate the height of the tongue. The tongue fits into the groove creating a tighter lock and seal. Figure 20–32 shows cross sections of the flange faces mentioned here.

The *ring-type joint* (RTJ) flange connection requires two flanges that both have a machined groove in the face. This groove is cut to accept an oval or octagonal ring. The grooves on both flanges fit over the ring which creates a seal. Examples of the ring joint flange are seen in the cutaway photo of the orifice flange in Figure 20–31, and the section view in Figure 20–32.

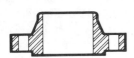

The Raised Face is the most common facing employed with steel flanges; it is $\frac{1}{16}$" high for Class 150 and Class 300 flanges and $\frac{1}{4}$" high for all other pressure classes. The facing is machine-tool finished with spiral or concentric grooves (approximately $\frac{1}{64}$" deep on approximately $\frac{1}{32}$" centers) to bite into and hold the gasket. Because both flanges of a pair are identical, no stocking or assembly problems are involved in its use. Raised face flanges generally are installed with soft flat ring composition gaskets. The width of the gasket is usually less than the width of the raised face. Faces for use with metal gaskets preferably are smooth finished.

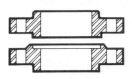

Male-and-Female Facings are standardized in both large and small types. The female face is $\frac{3}{16}$" deep and the male face $\frac{1}{4}$" high and both are usually smooth finished since the outer diameter of the female face acts to locate and retain the gasket. The width of the large male and female gasket contact surface, like the raised face, is excessive for use with metal gaskets. The small male and female overcomes this but provides too narrow a gasket surface for screwed flanges assembled with standard weight pipe.

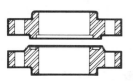

Tongue-and-Groove Facings are also standardized in both large and small types. They differ from male-and-female in that the inside diameters of tongue and groove do not extend to the flange bore, thus retaining the gasket on both its inner and outer diameter; this removes the gasket from corrosive or erosive contact with the line fluid. The small tongue-and-groove construction provides the minimum area of flat gasket it is advisable to use, thus resulting in the minimum bolting load for compressing the gasket and the highest joint efficiency possible with flat gaskets.

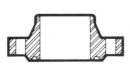

Ring Joint Facing is the most expensive standard facing but also the most efficient, partly because the internal pressure acts on the ring to increase the sealing force. Both flanges of a pair are alike, thus reducing the stocking and assembling problem found with both made-and-female and tongue-and-groove joints. Because the surfaces the gasket contacts are below the flange face, the ring joint facing is least likely of all facings to be damaged in handling or erecting. The flat bottom groove is standard.

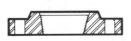

Flat Faces are a variant of raised faces, sometimes formed by machining off the $\frac{1}{64}$" raised face of Class 150 and Class 300 flanges. Their chief use is for mating with Class 125 and Class 250 cast iron valves and fittings. A flat-faced steel flange permits employing a gasket whose outer diameter equals that of the flange or is tangent to the bolt holes. In this manner the danger of cracking the cast iron flange when the bolts are tightened is avoided.

Figure 20–32 Cross section views of common flange faces. *Courtesy ITT Grinnell Corp.*

▼ VALVES

Valves are the components in the piping system that control and regulate fluids. Valves not only provide on/off service in a pipe, but are also used to regulate the flow of fluid in the pipe, maintain a constant pressure, prevent dangerous pressure build-up and prevent backflow in the pipe. A variety of valves are used to perform these tasks.

On/off

The *gate* valve is used exclusively to provide on/off service in a pipe. It is the most common valve used in industry today. (See Figure 20–33.) A gate or disc is moved up and down inside the valve manually or automatically. The design of the gate valve is such that fluid can flow through it with a minimum of friction and pressure loss. The valve *seat* does not interfere with the straight line flow of the fluid. The seat is the material with which the gate makes contact to create a seal. The gate valve is designed specifically for on or off service and infrequent operation. It is unsuitable for throttling or regulating flow. A partially open disc (gate) can cause erosion and wear of the downstream side of the seat and the disc itself.

The sealing mechanism inside a *ball* valve is just that—a ball. The ball has a hole through it that matches

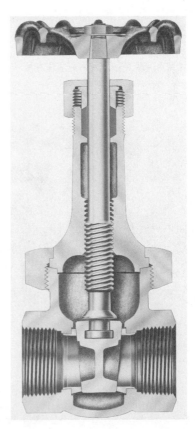

Figure 20–33 The gate valve provides on/off service. *Courtesy Crane Co.*

the I.D. of the pipe. (See Figure 20–34.) It is a quick opening valve, requiring only a one-quarter turn, and is used extensively because of its tight seat. Plastics, nylon, and synthetic rubber are used for the seat material and enable this valve to achieve a tight seal. The ball valve is popular because it has a low profile and low torque requirement for operation. It is also easy and inexpensive to maintain and repair. It is not used in large diameter lines (greater than 12") because the pressure in the line makes it difficult to open and close.

The *plug* valve, also known as *cock* valve, is similar in design to the ball valve and requires only a one-quarter turn to open and close. The opening through the plug can be either rectangular or round. (See Figure 20–35.) When open, there is a pressure drop through the valve, but a high flow efficiency due to the contours of the valve. It has low throttling ability and is best used for on/off service. It can achieve a tighter shutoff than a gate valve, but is normally used on smaller diameter lines.

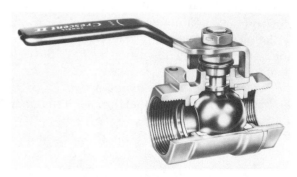

Figure 20–34 The ball valve is quick-opening and is used in pipe less than 12" in diameter. *Courtesy The Wm. Powell Co.*

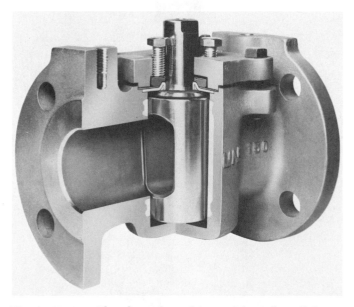

Figure 20–35 The plug valve achieves tighter shutoff than a gate valve but is used on smaller diameter lines. *Courtesy Xomox Corp.*

Regulating

The most common type of regulating valve is the *globe* valve. It is normally used in pipe up to 3" in diameter, but can be used in lines with diameters up to 12". Fluid flowing through the globe valve travels in an "S" pattern, which enables the valve to maintain a close control on the flow and to achieve a tight, positive shutoff. This flow pattern is seen in Figure 20–36. The internal design of the globe valve creates a high flow resistance, leading to a significant pressure drop through the valve. Body pockets in the valve are not drained when the flow is stopped. Situations that require frequent valve operation and maintenance are suited for globe valves because the discs and seats are easily replaced.

A special type of globe valve that creates a 90° direction change in the pipe is an *angle* valve. It is similar in design to a globe valve (see Figure 20–37), and is used in place of a globe valve and a 90° elbow to save money. In high-stress situations, the angle valve is not used.

Low-pressure situations may warrant the use of the *butterfly* valve. The disc is mounted on a stem that is turned a one-quarter turn to open and close. The entire disc moves inside the valve. (See Figure 20–38.) It is a simple operating mechanism that is excellent for regu-

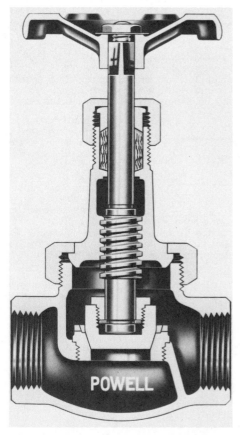

Figure 20–36 The globe valve achieves close flow control but creates high flow resistance. *Courtesy The Wm. Powell Co.*

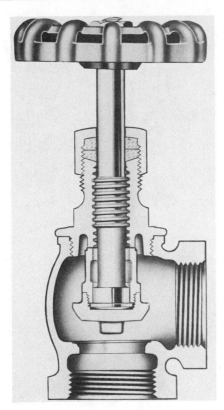

Figure 20–37 The angle valve replaces the globe valve and an elbow. *Courtesy Crane Co.*

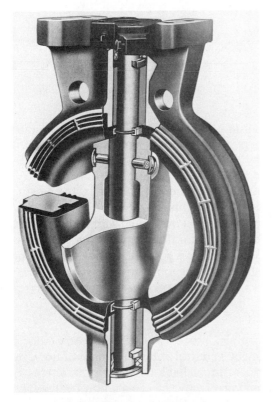

Figure 20–38 The butterfly valve is excellent for regulating flow and creates a minimum pressure drop. *Courtesy Crane Co.*

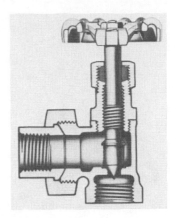

Figure 20–39 The needle valve is good for accurate throttling on instrument and meter lines. *Courtesy Crane Co.*

lating flow. The design creates a minimum pressure drop through the valve. It is light, inexpensive, requires a small installation space and is easy to maintain. All-plastic butterfly valves are available.

The *needle* valve is so named because the end of the stem is needle shaped. (See Figure 20–39.) The stem has fine threads enabling this valve to be adjusted exactly to achieve accurate throttling. The flow through the needle valve changes direction much like the flow through a globe valve. The needle valve is used for high-temperature and high-pressure service in instrument, gauge, and meter lines.

Some types of service require that the working parts of the valve be sealed from the fluid stream. The protection of the valve parts is important when working with fluids that are corrosive, viscous, fibrous, or contain suspended solids and sludges. The protection of the *fluid* is most important when dealing with food and beverages. The *diaphragm* valve suits this need. In place of a disc or other type of sealing mechanism, a diaphragm of rubber, neoprene, butyl, silicone or other flexible material is used. The diaphragm is pushed down by the stem and creates a seal against a seat on the bottom of the valve. (See Figure 20–40.) The diaphragm also serves to protect the working parts of the valve. It has a smooth, stream-lined flow, is easy to maintain, achieves positive flow control and is leaktight. It is suited for both on/off and regulating services up to 400°F.

Checking Backflow

Certain situations require that fluids be prevented from flowing backwards in the pipe should a power failure or pump breakdown occur. *Check* valves prevent backflow, closing when the fluid stops flowing.

The *swing check* valve as shown in Figure 20–41 is similar in construction to the gate valve and is used with it. The disc inside the swing check valve operates by gravity or the weight of the disc. It is best used with low

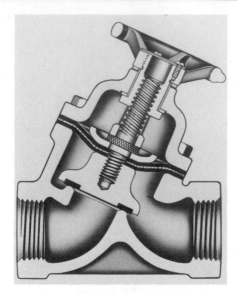

Figure 20–40 The diaphragm valve protects the valve mechanism from the fluid. *Courtesy Crane Co.*

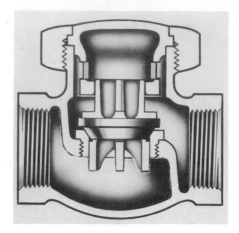

Figure 20–42 The lift check valve is used with the globe valve to check backflow. *Courtesy Crane Co.*

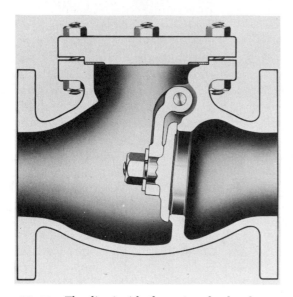

Figure 20–41 The disc inside the swing check valve operates by gravity to check backflow. *Courtesy Crane Co.*

Figure 20–43 The pop safety valve provides momentary release of pressure. *Courtesy Kunkle Valve Co., Inc.*

velocity liquids. The *lift check* valve (see Figure 20–42) is similar to the globe valve and is used with it. It operates by gravity and is available in either horizontal or vertical models.

Safety

The high temperatures and pressures within many industrial processes are potentially dangerous and must be controlled and regulated to prevent serious accidents. The responsibility for keeping pressures at or below a given point is the task of *safety* and *relief* valves.

The *pop safety* valve actually pops wide open when the pressure in a pipe or piece of equipment reaches a set pressure. These valves are used for steam, air, and gas lines only, never liquids. The opening and closing of this valve is instantaneous. (See Figure 20–43.)

The relief valve performs the same function as the pop safety, but is used for liquids only. (See Figure 20–44.) It is also set to open at a specific pressure, but it opens and closes slowly in response to changing pressure. This type of valve is found on most home hot water heaters.

Control

The complex processes used in industry today demand instantaneous control and adjustment of flow, pressure, and temperature. That's where *control* valves enter in. Most any type of valve can be a control valve. The example in Figure 20–45 illustrates one of the types used. The distinguishing component on the control valve is the controller or *actuator*. This is the mechanism that oper-

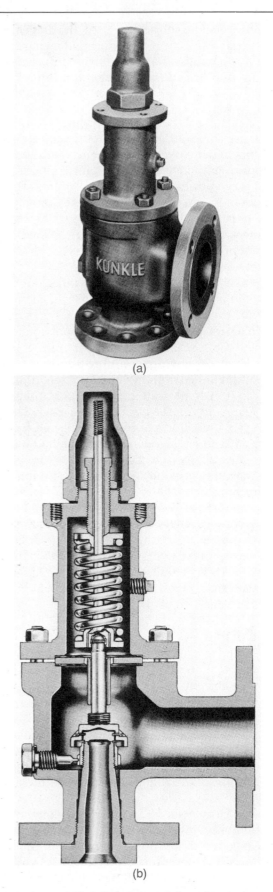

Figure 20–44 **(a)** Relief valve exterior; **(b)** cutaway view; provides slow release of fluids and pressure. *Courtesy Kunkle Valve Co., Inc.*

Figure 20–45 **(a)** Cage guided control valve exterior view; **(b)** cutaway view of same valve. *Courtesy Dezurik Valve Co.*

Figure 20–46 Pressure regulator. *Courtesy Spence Engineering.*

ates the valve. It can be operated by an electric motor, air cylinder, or hydraulic cylinder.

Pressure Regulating

Processes relying on steam require a constant flow at a steady rate. This task falls on *pressure regulators.* This mechanism reduces the incoming pressure of the steam to the required service pressure. The regulator maintains the pressure at the specified level and provides a uniform flow. (See Figure 20–46.)

▼ PIPE DRAFTING

Pipe drafting positions are found with consulting engineering companies, large industrial construction firms that maintain engineering offices, and manufacturers of equipment that utilize piping. The entry-level drafter in these companies may begin by drawing revisions (markups) to existing drawings. These may be flow diagrams (nonscale schematic drawings) or piping drawings. As experience is gained the drafter is assigned more complex drawings that must be constructed from engineering sketches.

Engineers, designers, and technicians design the piping system, select and size the equipment and pipe, and give this information to the drafter. The drafter's job is to construct the required type of drawing or model, often using vendor catalogs, brochures, and charts, or other drawings containing reference information. Once a work-

ing knowledge of the subject is gained, the drafter may be given the responsibility of designing and laying out pipe runs. At this point the drafter begins to utilize his or her knowledge of pipe fittings and reference information.

Plans and Sections (Elevations)

The most common and the most complex type of piping drawings are plans and sections that show building outlines, equipment, pipe, and pipe supports. Examples of a plan and several sections taken from that plan are shown in Figure 20–47. Study these drawings carefully. Look at the plan and sections together to try to determine where specific pipe runs are located. A little time spent now will aid you in working with this type of drawing in the future.

Equipment shown on piping plan and section drawings is seldom drawn in detail, but instead an outline is used to illustrate clearances and access. Tanks, vessels, heat exchangers, pumps, columns, compressors, boilers, dryers, and reactors are just a few examples of some of the equipment with which a piping drafter may work.

In addition to major pieces of equipment, the drafter may be required to indicate pipe supports and hangers. Note in Figure 20–48 (see page 650) the extensive use of pipe hangers in the form of pipe clamps and iron rods. Pipe supports must also come from the bottom up, as seen in Figure 20–49 (see page 650). These are vital to a piping system because they provide stability and anchor points for the pipe. A few common pipe supports and hangers and their application in a sample piping system are shown in Figure 20–50 (see page 650). If the hanger or support is complex or of a special design, it is often illustrated on a "detail" drawing, such as that in Figure 20–51 (see page 651). When standard pipe clamps, anchors, hangers, and supports are used, they may be indicated by a symbol or note such as "P.S."

Single-line Drawings. A time-saving method of drawing pipe is the single-line method. The centerline of the pipe is used to represent the pipe. It may be used to better illustrate new pipe that is added to an existing installation, or to show small diameter pipe. An example of a small diameter, single-line piping drawing is shown in Figure 20–52 (see page 652). Standard pipe symbols are used in single-line drawings (see Appendix L) and the pipe and fittings are often drawn as heavy lines to provide contrast with other lines on the drawing.

Double-line Drawings. The double-line method is the easiest to interpret, for it looks like the actual pipe. But it takes longer to display on a monitor and plot with a pen plotter. This method is used extensively by equipment manufacturers to illustrate standard installation and in presentation drawings that are seldom revised. (See Figure 20–53, page 653.) The double-line method is

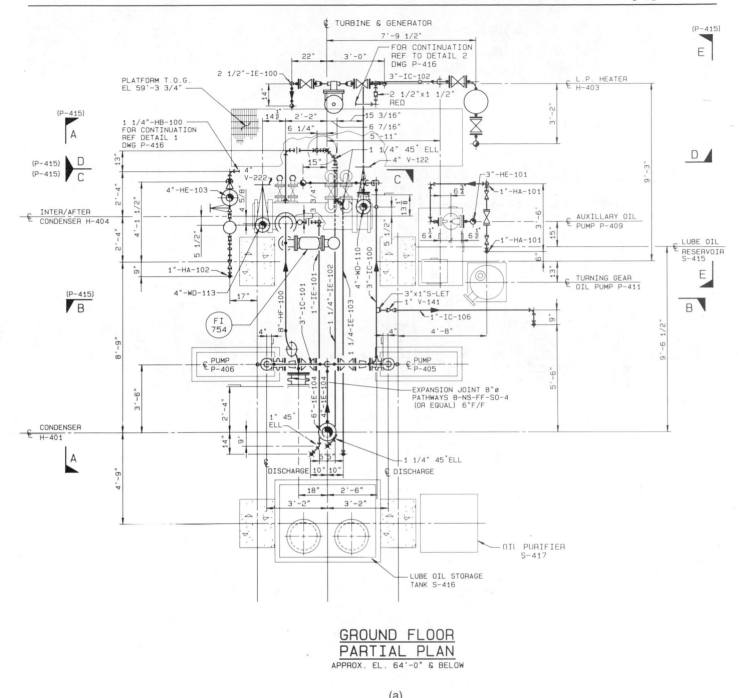

GROUND FLOOR
PARTIAL PLAN
APPROX. EL. 64'-0" & BELOW

(a)

Figure 20–47 **(a)** Piping plan (partial) of a turbine generator installation; **(b–e)** piping sections are taken from plan in (a). *Courtesy Schuchart and Associates.*

also used to portray large diameter pipe, to show existing pipe at an industrial facility, or for small views that may be more easily visualized as double-line drawings (see Figure 20–54, page 653). When new or small pipe is shown in the single-line method on the same drawing, the contrast between the two styles aids in interpretation. The single-line method is seldom used for large diameter pipe because clearances, interferences, spacing, and distances are not as easily seen. The double-line method is preferred where these factors are critical.

When laying out either type of piping drawing, it is best to begin with background information such as building outlines or structural steel column reference points. From these points the equipment locations can be determined. Pumps are often represented by centerlines. Other large equipment must be drawn. Pipe centerline locations must then be determined. CADD drafters can use pipe centerline location lines of a different color layer or as tick marks on the screen, then insert fittings and draw pipe as they go.

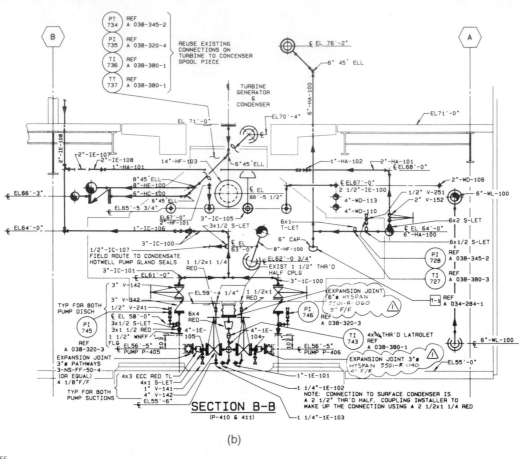

SECTION B-B
(P-410 & 411)

(b)

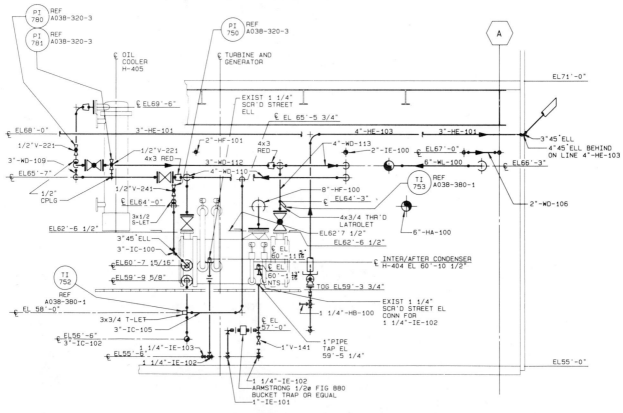

SECTION C-C
(P-410)

(c)

Figure 20-47 (Continued)

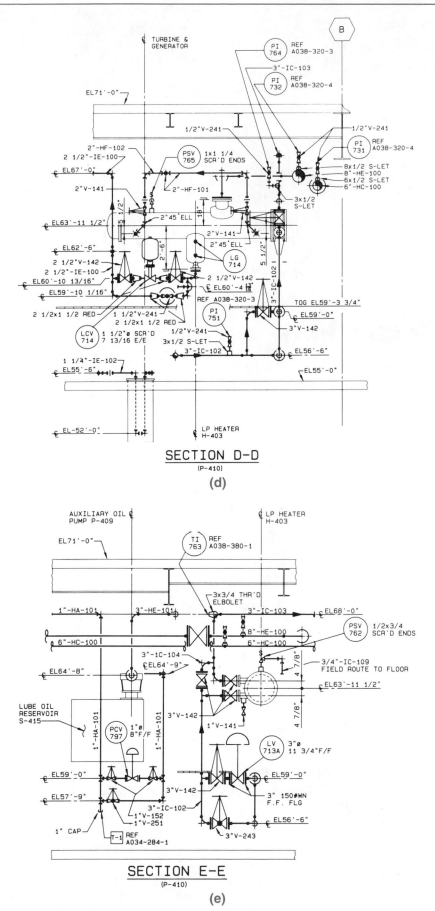

SECTION D-D
(P-410)

(d)

SECTION E-E
(P-410)

(e)

Figure 20–47 (Continued)

Figure 20–48 Pipe hangers are used to suspend pipe and conduit from the ceiling in this paper mill expansion project. *Courtesy Publisher's Paper Co.*

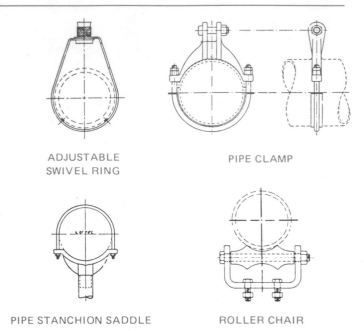

ADJUSTABLE SWIVEL RING   PIPE CLAMP

PIPE STANCHION SADDLE   ROLLER CHAIR

Figure 20–49 Pipe supports anchored to the ground are used on this air compressor piping at a chemical processing plant. *Courtesy Ingersoll-Rand.*

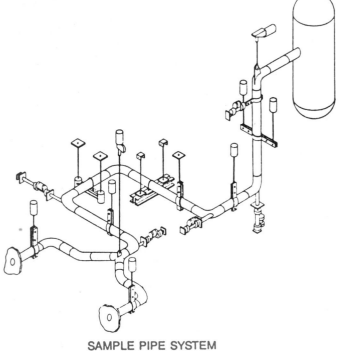

SAMPLE PIPE SYSTEM
WITH HANGERS AND SUPPORTS

Figure 20–50 Examples of common pipe hangers and supports. *Courtesy ITT Grinnell Corp.*

Fitting and Valve Symbols

A major part of the piping drafter's job is drawing fittings and valves. Appendices K and L provide commonly used valve symbols. Manual drafting piping templates, as shown in Figure 20–55 (see page 654), are available in a variety of sizes. Some companies may even have their own special templates made for them.

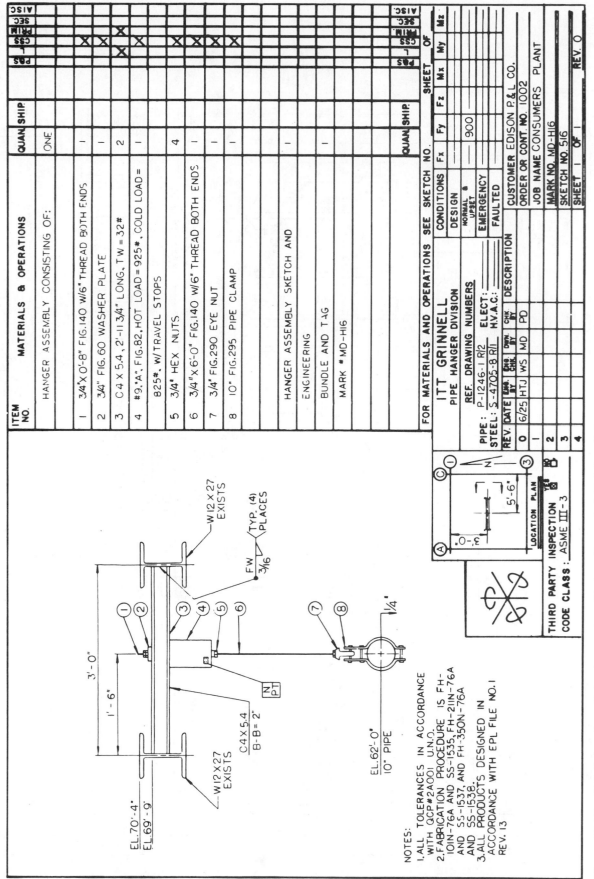

Figure 20–51   Detail drawing of pipe hanger installation. *Courtesy ITT Grinnell Corp.*

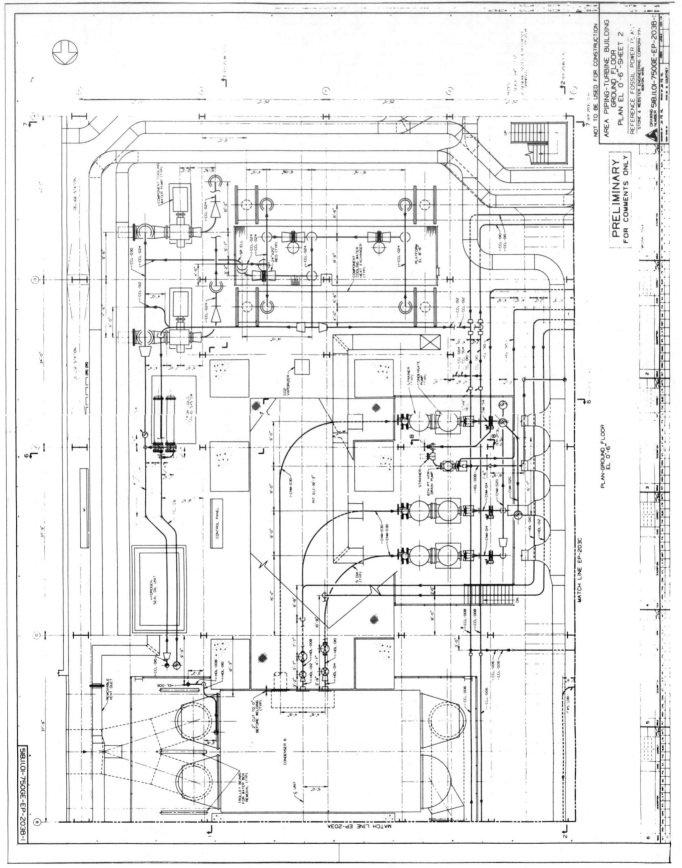

Figure 20-52 Single line piping drawing of a cooling water system. *Courtesy Versatec.*

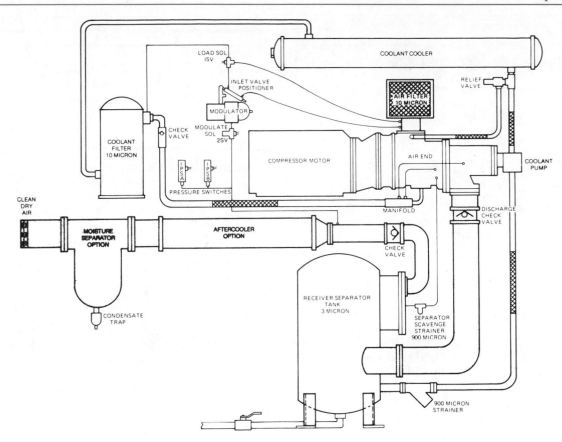

Figure 20–53 Manufacturer's double-line piping drawing of a compressed air system. *Courtesy Ingersoll-Rand.*

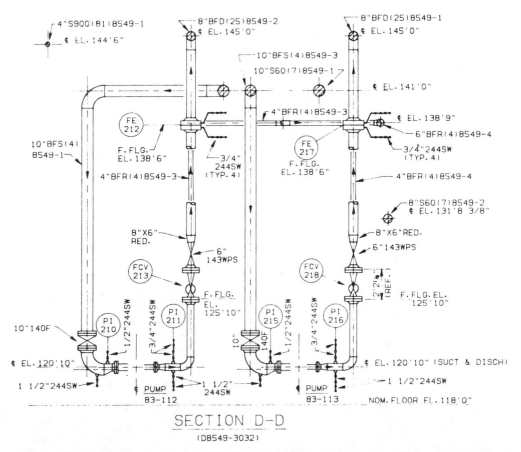

Figure 20–54 Sections from a typical double-line piping drawing. *Courtesy Schuchart and Associates.*

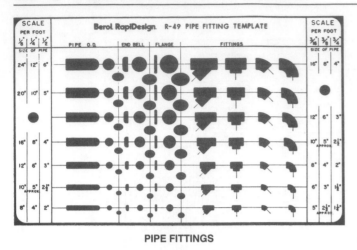

PIPE FITTINGS

Figure 20–55 A common manual drafting piping template. *Courtesy Berol USA Rapi/Design.*

Dimensions and Notes

Piping drawings are normally dimensioned in a style similar to architectural drawings, but some companies may use the mechanical style of broken dimension lines. The former type of dimension lines are unbroken and values are written above horizontal lines, and to the left of and reading up on vertical dimension lines. Values are given in feet and inches with a dash between the two. Refer to the plan in Figure 20–47a for examples of piping plan dimensions.

Piping elevations or sections are not normally dimensioned like plan views. Instead of linear dimensions, elevations are used. Occasionally elevations for the bottom of the pipe (B.O.P.) are required. The sections shown in Figure 20–47b–e illustrate the use of elevations to call out the centerlines of pipes and the tops of concrete (T.O.C.), or steel, for example.

Lay out dimensions before any other written information is added to the drawing. Indicated pipe lengths and dimensions to pipe direction changes should be given from reference points such as equipment and structural steel centerlines. Location dimensions of fittings, flanges, and valves (if needed) should be given to center points or flange faces. Pipe fittings and valves that are attached without pipe between them are often left undimensioned. This is termed "fitting-to-fitting," and their location is simply the result of the combined fitting lengths, and will not vary.

Dimension lines are important and should not be broken if passing through other lines. Pipes, equipment, or extension lines that must pass through dimension lines should be broken and the dimension line should remain intact.

Once dimension lines have been placed on the drawing, local notes and callouts can be given. These, too, should be placed close to where they apply without crowding other entities or creating a confused layout. Try to group local notes together when possible. This enables

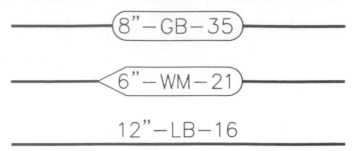

Figure 20–56 Methods of indicating pipe line specifications.

those who must interpret the drawing to find similar information without too much searching.

Pipe specification numbers can be written near the pipe and connected to it with a leader line, or the numbers can reside inside a symbol called a *line balloon*. The line balloon is either rounded or squared on the ends and is 1/4" wide and approximately 1" long. The preferred location for the line balloon is inside the pipe, but may be placed outside the pipe and used with a leader line if space is unavailable. Some examples of pipe specification numbers are shown in Figure 20–56.

When checking your drawing be sure that all of the following items have been given:

1. Pipe length dimensions.
2. Dimension locations of all pipe direction changes.
3. Locations of all fittings and valves if required.
4. Elevation location of all pipe direction changes in section views.
5. Size and type of valves.
6. Size and type of fittings (if not readily identified).
7. Pipe diameter, contents, and identification number.
8. Pipe flow arrows.
9. Equipment names and numbers.

Piping Details

A *piping detail* drawing can be made of any connection, application, or installation that can't be readily distinguished from the other piping drawings. Details can often quickly clarify a point that would have taken valuable time to determine. Common piping detail drawings are used for the following:

❍ Special pipe connections or fittings, or special valve arrangements.
❍ Small diameter pipe and fitting assemblies.
❍ Special pipe support arrangements.
❍ Tank attachment details.
❍ Minor structural alterations (concrete, wood, and steel).
❍ Operating and installation procedures.

The engineers and drafters on the job must determine when piping details are needed and if they are to be drawn orthographically or pictorially. Some common piping details are shown in Figure 20–57.

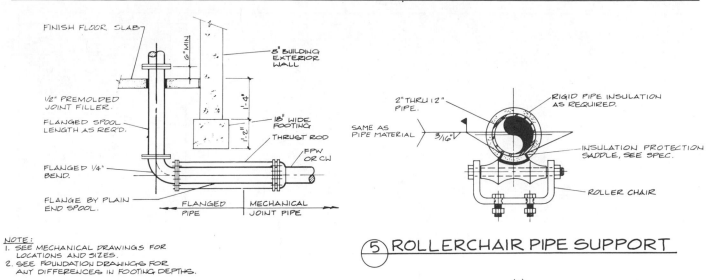

NOTE:
1. SEE MECHANICAL DRAWINGS FOR LOCATIONS AND SIZES.
2. SEE FOUNDATION DRAWINGS FOR ANY DIFFERENCES IN FOOTING DEPTHS.

⑧ THRUST TIE

(a)

⑤ ROLLERCHAIR PIPE SUPPORT

(c)

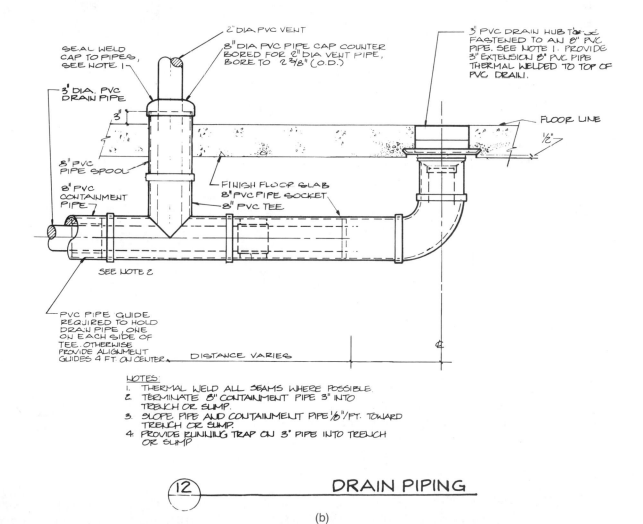

NOTES:
1. THERMAL WELD ALL SEAMS WHERE POSSIBLE.
2. TERMINATE 8" CONTAINMENT PIPE 3" INTO TRENCH OR SUMP.
3. SLOPE PIPE AND CONTAINMENT PIPE ⅛"/FT. TOWARD TRENCH OR SUMP.
4. PROVIDE RUNNING TRAP ON 3" PIPE INTO TRENCH OR SUMP

⑫ DRAIN PIPING

(b)

Figure 20–57 **(a and b)** Piping details of buried pipe; **(c)** pipe hanger. *Courtesy CH₂M Hill—I.D.C.*

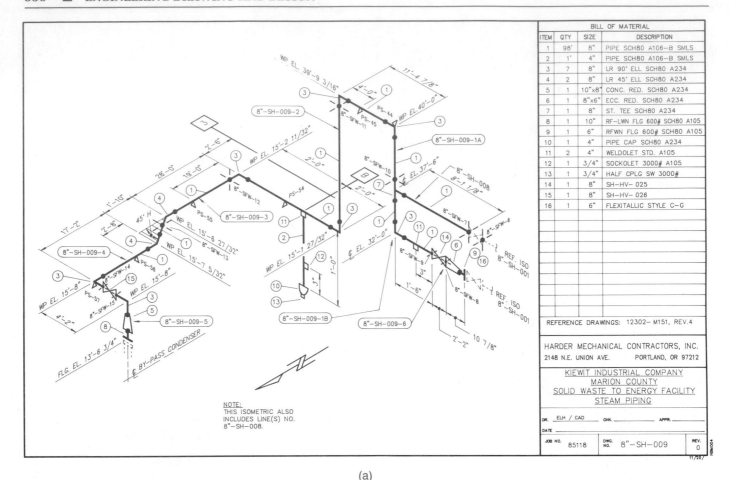

BILL OF MATERIAL

ITEM	QTY	SIZE	DESCRIPTION
1	98'	8"	PIPE SCH80 A106-B SMLS
2	1'	4"	PIPE SCH80 A106-B SMLS
3	7	8"	LR 90° ELL SCH80 A234
4	2	8"	LR 45° ELL SCH80 A234
5	1	10"x8"	CONC. RED. SCH80 A234
6	1	8"x6"	ECC. RED. SCH80 A234
7	1	8"	ST. TEE SCH80 A234
8	1	10"	RF-LWN FLG 600# SCH80 A105
9	1	6"	RFWN FLG 600# SCH80 A105
10	1	4"	PIPE CAP SCH80 A234
11	2	4"	WELDOLET STD. A105
12	1	3/4"	SOCKOLET 3000# A105
13	1	3/4"	HALF CPLG SW 3000#
14	1	8"	SH-HV-025
15	1	8"	SH-HV-026
16	1	6"	FLEXITALLIC STYLE C-G

REFERENCE DRAWINGS: 12302- M151, REV.4

HARDER MECHANICAL CONTRACTORS, INC.
2148 N.E. UNION AVE.   PORTLAND, OR 97212

KIEWIT INDUSTRIAL COMPANY
MARION COUNTY
SOLID WASTE TO ENERGY FACILITY
STEAM PIPING

DR.	ELH / CAD	CHK.		APPR.	
DATE					
JOB NO. 85118		DWG. NO. 8"-SH-009			REV. 0

NOTE:
THIS ISOMETRIC ALSO
INCLUDES LINE(S) NO.
8"-SH-008.

(a)

Piping Isometrics and Spools

A *piping isometric* drawing is a pictorial view of a piping system complete with fittings, valves, dimensions, notes, and possibly instrumentation. This single view of the piping system eliminates the need for additional views. It is often done after the plans and elevations are completed.

The advantages of a piping isometric are many. A single run of pipe from beginning to end can be shown in one view and is more easily visualized than corresponding orthographic views. Many companies draw piping isometrics to scale, which makes layout and visualization more accurate. Some firms (especially mechanical contractors) do *not* draw piping isometrics to scale. If an isometric is not drawn to scale, it eliminates the need for large sheets of paper and is most often drawn on "B" size media. Dimensions are given on the isometric view and straight lengths of pipe are drawn proportionally to indicate long and short pieces. Note this in the isometric shown in Figure 20–58.

It is easier to check the drawing for interferences and clearances if the isometric is drawn to scale. A bill of materials is often included on the piping isometric, which enables the pipe fitter, checker, and purchasing agent to cross check the drawing with a list of the materials required.

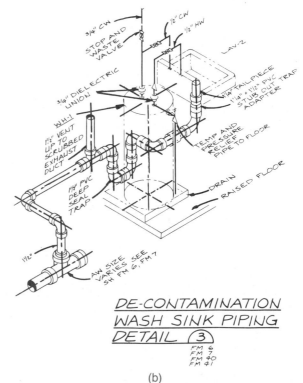

DE-CONTAMINATION
WASH SINK PIPING
DETAIL ③

(b)

Figure 20–58 **(a)** Piping isometric with nonscale proportional dimensioning; **(b)** nonscale double-line piping isometric. *(a) Courtesy Harder Mechanical Contractors, Inc. (b) CH₂M Hill—I.D.C.*

Some companies may illustrate a system or portions of a system in a pictorial manner for purposes of orientation, training, assembly, and installation. It is not often practical for large and complex systems to be shown in isometric form if manual drafting is used. But the development of more powerful CADD systems is enabling companies to create complete pictorially modeled systems.

Piping systems are constructed by pipefitters who must initially weld or thread pipe fittings and pipe together to create pipe *spools.* (See Figure 20–59.) The pipefitters work from assembly instructions called *spool drawings* or spool sheets. The pipe assemblies (spools) are then transported to the job site and installed. A pile of pipe spools awaiting installation is shown in Figure 20–60.

Pipe spools are drawn in isometric or orthographic style on "B" size paper and may or may not be drawn to scale. (See Figure 20–12.) Spool sheets are assembly drawings that contain complete dimensions and a bill of materials (B.O.M.) that indicates the exact size and specifications for each fitting. Extra pipe is usually added to pipe lengths to compensate for errors. If the

drawing is orthogonal it only shows the views required to present all of the lengths of straight pipe and fittings. Valves are never shown on spool drawings because they are installed at the job site when the spools are erected.

CADD Applications

3-D DIGITIZING

Three-dimensional digitizers allow you to pick points on an object in order to input 3-D model data into CADD packages such as AutoCAD and CADKEY. This form of data input is called *reverse engineering,* because the real part is used to construct a model. (See Figure 20–61.)

The 3-D digitizer has an articulated arm that can be moved and rotated in order to pick any point on an existing object or model. The data is transmitted as 3-D points to the CADD software through a cable attached to a serial port on the computer. This process is useful for collecting CADD data of an existing object in order to create a computer model. The computer model can then be revised in order to construct a new part.

This CADD input is also useful for digitizing an existing scale model of an object, building, or industrial site. The digitized data from the model can then be used to create construction drawings and 3-D computer models and renderings.

Figure 20–59 Pipefitter assembling a piping spool. *Courtesy Alaskan Copper Works.*

Figure 20–60 Assembled pipe spools awaiting installation. *Courtesy Alaskan Copper Works.*

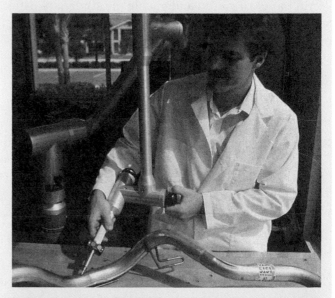

Figure 20–61 The dimensions of an existing pipe can be input to the CADD software using a 3-D digitizer. *Courtesy FARO Technologies, Inc.*

Models

The construction of the model (see Figure 20–62) may take months, even years, making it impossible to ship to the job site during most of the construction phases. Photos of the model are used at the job site for checking purposes. Drawings of the system must still be generated for use in construction.

One of the greatest advantages is the ability to see readily clearances and interferences. (See Figure 20–63.) These can be quickly transferred to the drawings, thus eliminating many errors that might occur if only drawings were used. Some companies may use a three-dimensional digitizing system to locate pipe, equipment, and structural steel on the model. This information is then fed into the CADD system and used to generate drawings.

Equipment on the model is often shown in the form of block shapes, details seldom being required. Equipment is created from standard, inexpensive materials such as rigid foam, plastic, and wood. Piping and structural steel components are available from model supply vendors.

Upon completion, the model is shipped to the job site where it is used in the final stages of the construction project. It is also used to check the design drawings of the system. Operators, maintenance technicians, and engineers can be trained on the workings of the system using the model. Revisions, alterations, and design changes can be accomplished efficiently using the model to check for clearances, and the addition of new equipment and piping locations.

Piping model kits are available for use in piping and blueprint reading classes. (See Figure 20–64.) These kits come with all the parts necessary to construct a small piping and equipment arrangement. These models aid in visualization and enable the student to better construct piping drawings.

Layout Techniques

When constructing scaled piping drawings, use the following layout procedures:

Step 1. Locate all building outlines, concrete foundations, structural steel columns, and equipment. Draw the centerlines of the pipe that connects the equipment. (See Figure 20–65a.)

Step 2. Insert fittings and valves in the pipe centerlines and draw pipe, as shown in Figure 20–65b.

Step 3. Place dimensions and elevations on the drawing, remembering to keep dimensions close to where they apply. Locate pipe specification symbols and text in the pipe runs. Add all notes and textual information required. (See Figure 20–65c.)

Figure 20–62 Complex piping models can require months to build. *Courtesy Engineering Model Associates, Inc.*

Figure 20–63 Designers can readily check for interferences on a piping model. *Courtesy Engineering Model Associates, Inc.*

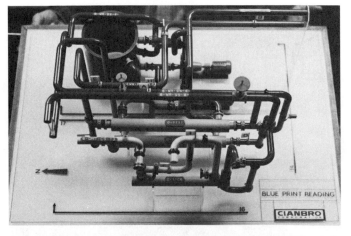

Figure 20–64 Piping model kits are good visualization tools for the classroom. *Courtesy Engineering Model Associates, Inc.*

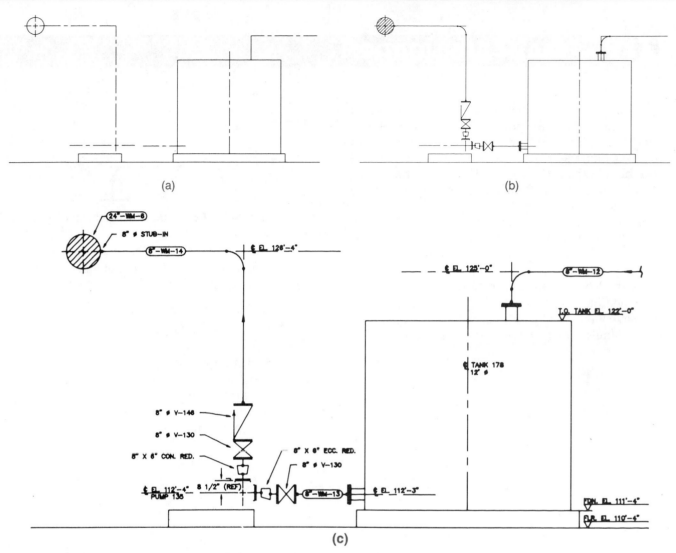

Figure 20–65   **(a)** Equipment outlines and pipe centerlines are located; **(b)** pipe, fittings, and valves are added; **(c)** dimensions, elevations, and text are added to complete the drawing.

PROFESSIONAL PERSPECTIVE

The field of process piping design and drafting can be an excellent opportunity for a person to gain entry into engineering. Most consulting engineering companies will pay for the education of employees who wish to upgrade their skills or work toward an engineering degree. This is often a good method of becoming a designer, technician, or engineer, because you can get valuable job training while working toward a degree. This job training is not possible if you are going to school full time.

A good piping drafter, designer, or engineer is one who is aware of the actual job site requirements and problems. These "field" situations are often considerably different from the layout that is designed in the office. Therefore, if you are planning to work in the industrial piping profession, do your best to work on projects in which you can gain field experience. Field work is especially important when adding new equipment and pipe in an existing facility. There are many stories of inexperienced piping designers who have created a design in the office, and then gone to the field to find a new 4" pipe routed directly through an existing 12" pipe, exactly according to the plan!

MATH APPLICATION

Industrial pipe design often requires the use of fittings such as 45° elbows to route pipe past obstructions. Exact dimensions of all lengths of pipe and fittings must be calculated before the pipe can be drawn and assembled. When 45° elbows are used, the piping designer must use the Pythagorean Theorem to solve for one side of the triangle.

The designer has laid out a run of pipe shown in Figure 20–66. In order to find the true length, or *travel* of the angled run of pipe, a 45° triangle is applied to the pipe. The length of one side of the triangle is determined by finding the difference between the two elevations given, 2'–9". The two adjacent sides each have a length of 2'–9". Solve for the angled side of the triangle, or *hypotenuse*. (See Figure 20–67.)

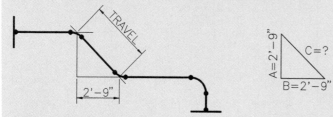

Figure 20–66 Find the true length, or travel, of the angled run of pipe.

The Pythagorean Theorem is stated as follows: The sum of the squares of the two sides is equal to the square of the hypotenuse; or:

$$a^2 + b^2 = c^2$$

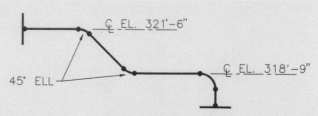

Figure 20–67 Solve for the travel (angled) side of the triangle.

Therefore, the square root of the hypotenuse equals the travel of the angled run of pipe.

This problem is solved as follows:

$$a^2 + b^2 = c^2$$

$$(2'-9")^2 + (2'-9")^2 = c^2$$

$$(33)^2 + (33)^2 = c^2$$

$$1089 + 1089 = c^2$$

$$2172 = c^2$$

$$\sqrt{2172} = c$$

$$c = 46.605$$

$$c = 3.9'(3'-10.8")$$

$$c = 3'-10\ 13/16"$$

CHAPTER

20 INDUSTRIAL PROCESS PIPING TEST

DIRECTIONS

Answer the following questions with short, complete statements.

1. What is the difference between plumbing and piping?
2. In what two forms is piping drawn?
3. What is one of the major uses of cast iron pipe?
4. What type of steel pipe withstands high temperatures?

5. What is steam tracing?
6. What type of pipe is one of the most corrosion proof?
7. What type of pipe is best used for hydraulic lines in automobiles and aircraft?
8. Define *NPS*.
9. What is the meaning of "pipe schedule number"?
10. What designations do the ASTM and ASME use to classify pipe strength?
11. How does butt welding differ from socket welding?

12. What provides the seal in a flanged connection?
13. What type of pipe is commonly used with a bell and spigot connection?
14. What type of pipe is solvent welding used on?
15. Describe the functions of the following pipe fittings: 180° elbow, reducing tee, eccentric reducer, lateral, elbolet, union, and bushing.
16. When is a weld-neck flange used?
17. What is placed between two orifice flanges, and what is its function?
18. Name two types of on/off valves.
19. What is one of the major drawbacks of the globe valve?
20. What type of regulating valve is operated with a one-quarter turn?
21. What type of check valve is used with a globe valve?
22. What type of valve is used to relieve pressure in pipes carrying liquids?
23. Why are detailed representations of equipment not necessary on piping plans?
24. When would the double-line method of pipe drawing be used?
25. How are pipe length dimensions handled in piping sections?
26. What is meant by "fitting-to-fitting"?
27. When would a piping detail drawing be needed?
28. Why are piping isometric drawings often not drawn to scale?
29. What is the relationship between piping isometrics and piping spool drawings?
30. What are the benefits of piping models?
31. What is in store for the future of CADD piping?

MULTIPLE CHOICE

Select the best choice a, b, c, or d that answers or completes each question or statement.

1. Which of the following is a nonscale piping drawing?
   a. plan
   b. elevation
   c. flow diagram
   d. general arrangement

2. Which drawing illustrates a run of pipe in 3-D form?
   a. isolinear
   b. isometric
   c. orthogonal
   d. spool

3. The specific standards that apply to cast pipe are:
   a. AWWA
   b. ANSI
   c. ASTM
   d. ASME

4. Which type of pipe is best used for hydraulic lines in automobiles?
   a. copper
   b. steel tubing
   c. brass
   d. welded steel

5. Underground sewage, water, and gas lines are common applications for _____ connections.
   a. flanged
   b. welded
   c. soldered
   d. bell and spigot

6. A _____ fitting produces a 45° branch to the run pipe.
   a. tee
   b. true Y
   c. lateral
   d. cross

7. What fitting is used to maintain a constant elevation on one side of the pipe?
   a. concentric reducer
   b. cross
   c. lateral
   d. eccentric reducer

8. The _____ flange is used most often on straight pipe.
   a. slip-on
   b. blind
   c. welding neck
   d. stub-end

9. What type of valve is used exclusively for on/off service?
   a. check
   b. globe
   c. butterfly
   d. gate

10. The _____ valve prevents backflow through a pipe.
    a. ball
    b. gate
    c. check
    d. globe

11. The most common and most complex of piping drawings is the _____.
    a. isometric
    b. plan and section
    c. flow diagram
    d. spool

12. Vertical dimensions are given as _____ on a piping section.
    a. local notes
    b. callouts
    c. elevations
    d. reference dimensions

13. The first written information to be placed on a piping plan is the _____.
    a. equipment labels
    b. local notes
    c. pipe specifications
    d. dimensions

14. Piping isometrics are most often drawn on _____-size paper.
    a. C
    b. D
    c. A
    d. B

15. Pipefitters work from _____ drawings to construct the piping assembly.
    a. flow diagram
    b. isometric
    c. spool
    d. plan and section

CHAPTER

20 INDUSTRIAL PROCESS PIPING PROBLEMS

DIRECTIONS

1. Please read problems carefully before you begin working. Your instructor will assign one or more of the following problems. Complete each problem on an appropriately-sized drawing sheet or use the size indicated in the specific instructions.

2. Refer to appendices I and J for fitting and valve dimensions. Consult with your instructor to determine if specific vendor catalogs should be used to obtain dimensions.

3. Construct your drawings using a CADD system, if indicated in course guidelines.

Problem 20–1 Draw the following fittings in double-line representation. Use 8" diameter NPS and draw at a scale of 1/2" = 1'–0".

- 90° elbow
- 45° elbow
- Straight tee
- Concentric reducer
- Eccentric reducer

Draw a front view and each of the four orthographic views (top, bottom, left, and right sides).

Problem 20–2 Draw the fittings listed in Problem 20–1 in single-line representation. Use the same NPS and the same scale.

Problem 20–3 Draw each of the five pipe assemblies shown below (a-e) in double-line form. The view given is the front view. Draw each of the other four orthographic views (top, bottom, left, and right sides). Use the following information when doing this problem:

- Draw at a scale of 3/8" = 1'–0".
- Use pipe diameters indicated.
- Draw the fittings to scale but straight lengths of pipe can be drawn proportional to the sketch.
- Pipe that appears to be going away from the viewer should be drawn "fitting-to-fitting" if indicated with "F-F" on the sketch. This means that there is no straight pipe in the run that is going away from the viewer. Runs that do contain pipe are not shown with "F-F."

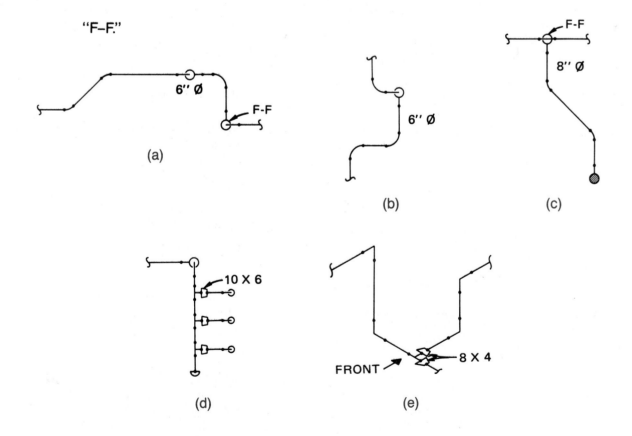

(a)   (b)   (c)   (d)   (e)

Problem 20–4 Draw the assemblies shown in Problem 20–3 in single-line form. Follow the instructions given in Problem 20–3.

Problem 20–5 Redraw the Ground Floor Partial Plan shown in Figure 20–47a on C-size media. Use the following information:

- Draw at a scale of 1/2" = 1'–0".
- Pipe more than 3" in diameter should be drawn in double-line form.
- Use line balloons to indicate pipe specifications.
- Lettering should be 1/8" high.

Use Figure 20–47b–e as references for this drawing.

Problem 20–6 Redraw section E-E (Figure 20–47e) in double-line form at a scale of 1/2" = 1'–0". Use standard dimensions for fittings and valves given in the appendices.

Problem 20–7 Draw a piping detail of the suction piping of pump P-405 in section B-B, Figure 20–47b.

Problem 20–8 Redraw section B-B (Figure 20–47b) in single-line form at a scale of 1/2" = 1'–0".

Problem 20–9 Draw the sections shown in Figure 20–47 single-line form. Use a scale of 3/8" = 1'–0" on C-size media.

Problem 20–10 Redraw section E-E, Figure 20–47e, at a scale of 1/2" = 1'–0". Pipe that is greater than 3" in diameter draw as a single line. Pipe less than 3" diameter, draw as double line.

Problem 20–11 Draw isometric views of the five pipe assemblies shown in Problem 20–3. Draw in single-line form.

Problem 20–12 Draw a piping isometric of lines 1"–HA–101 in section E-E (Figure 20–47e). Use B-size media.

Problem 20–13 Draw a plan and elevation of 8"–SH–009 (Figure 20–58a) in single-line form.

Problem 20–14 Draw a plan and elevation of 8"–SH–009 (Figure 20–58a) in double-line form.

Problem 20–15 Draw the spools required for the isometric 4"–SH–004 shown below. Draw one per sheet of B-size media. Include a bill of materials for each spool. The field weld connection point between the two largest spools is indicated by an "X" just above El. 36'–6".
Courtesy Harder Mechanical Contractors, Inc.

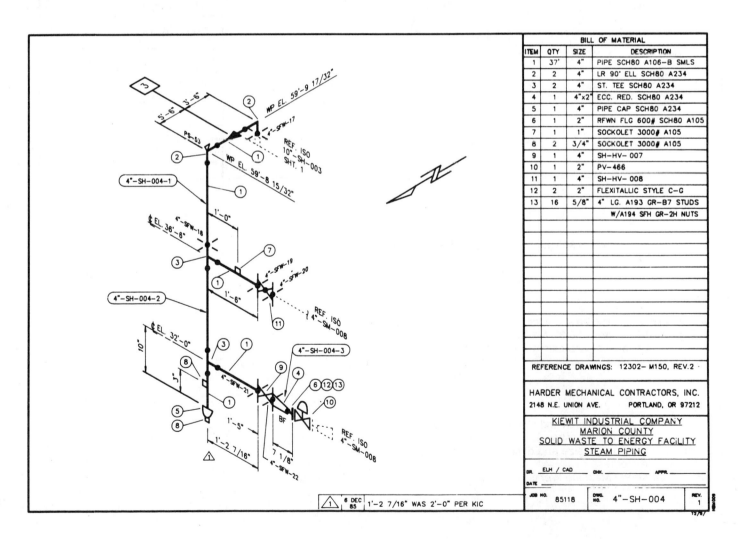

\-		BILL OF MATERIAL	
ITEM	QTY	SIZE	DESCRIPTION
1	37'	4"	PIPE SCH80 A106–B SMLS
2	2	4"	LR 90' ELL SCH80 A234
3	2	4"	ST. TEE SCH80 A234
4	1	4"x2"	ECC. RED. SCH80 A234
5	1	4"	PIPE CAP SCH80 A234
6	1	2"	RFWN FLG 600# SCH80 A105
7	1	1"	SOCKOLET 3000# A105
8	2	3/4"	SOCKOLET 3000# A105
9	1	4"	SH–HV– 007
10	1	2"	PV–466
11	1	4"	SH–HV– 008
12	2	2"	FLEXITALLIC STYLE C–G
13	16	5/8"	4" LG. A193 GR–B7 STUDS
			W/A194 SFH GR–2H NUTS

REFERENCE DRAWINGS: 12302– M150, REV.2

HARDER MECHANICAL CONTRACTORS, INC.
2148 N.E. UNION AVE.   PORTLAND, OR 97212

KIEWIT INDUSTRIAL COMPANY
MARION COUNTY
SOLID WASTE TO ENERGY FACILITY
STEAM PIPING

DR. ELM / CAD   CHK. _____   APPR. _____
DATE _____

| JOB NO. 85118 | DWG. NO. 4"–SH–004 | REV. 1 |

△ 6 DEC 85 1'–2 7/16" WAS 2'–0" PER KIC

Problem 20–16 Draw a plan and elevation of 14"–SL–011 in both single- and double-line forms. Place both drawings on C-size media. Choose a scale that allows both drawings to fit uncrowded on the sheet.
Courtesy Harder Mechanical Contractors, Inc.

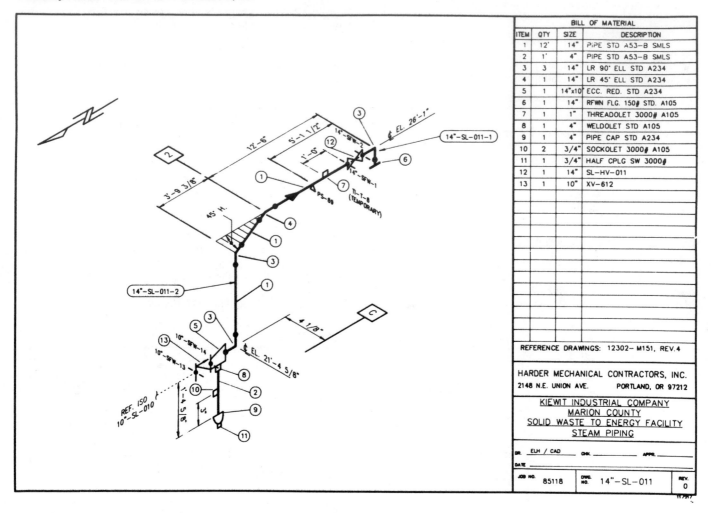

BILL OF MATERIAL			
ITEM	QTY	SIZE	DESCRIPTION
1	12'	14"	PIPE STD A53–B SMLS
2	1'	4"	PIPE STD A53–B SMLS
3	3	14"	LR 90° ELL STD A234
4	1	14"	LR 45° ELL STD A234
5	1	14"x10"	ECC. RED. STD A234
6	1	14"	RFWN FLG. 150# STD. A105
7	1	1"	THREADOLET 3000# A105
8	1	4"	WELDOLET STD A105
9	1	4"	PIPE CAP STD A234
10	2	3/4"	SOCKOLET 3000# A105
11	1	3/4"	HALF CPLG SW 3000#
12	1	14"	SL–HV–011
13	1	10"	XV–612

REFERENCE DRAWINGS: 12302– M151, REV.4

HARDER MECHANICAL CONTRACTORS, INC.
2148 N.E. UNION AVE.   PORTLAND, OR 97212

KIEWIT INDUSTRIAL COMPANY
MARION COUNTY
SOLID WASTE TO ENERGY FACILITY
STEAM PIPING

DR. ELH / CAD   CHK.   APPR.
DATE

JOB NO. 85118 | DWG. NO. 14"–SL–011 | REV. 0

Problem 20–17 Draw the spools required for line 6"–SH–002A (Figure 20–11) on B-size media.

Problem 20–18 Draw the spool for pipe 3"–1C–102 in section E–E (Figure 20–47e) from the 3" V–243 valve at elevation 56'–6" to the 3" V–142 valve at approximately the 64' elevation mark. Draw *no* valves in the spool. Include a bill of materials.

Problem 20–19 Draw a plan and elevation view of the double-line piping isometric drawing shown. Use threaded fittings. Have the instructor assign pipe size and dimensions to the problem. Draw on B-size media.
Courtesy Armstrong Machine Works.

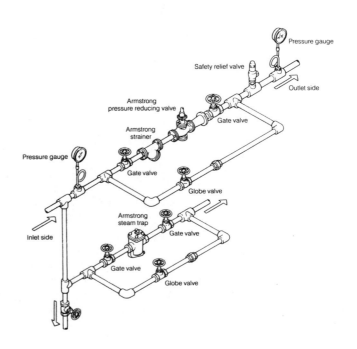

Problem 20–20 Draw the piping assembly shown. The assembly is shown in isometric format and the north arrow is pointing to the lower right of the drawing. Draw the assembly in isometric form, and orient the drawing with the north arrow pointing to the upper right of the drawing. Completely dimension the drawing as shown.

- Use a B-size sheet of vellum or drafting film.
- Draw using a scale of 3/8" = 1'–0" for valves and fittings and no scale for pipe lengths.
- Include the list of materials on your drawing.

Courtesy Willamette Industries, Inc.

Problem 20–21 Draw a plan view of the piping assembly shown in Problem 20–20. Use break lines on the long piece of pipe, so it fits on the drawing sheet. Draw two elevation views of the piping assembly and place them on the same sheet as the plan view. See your instructor for the specific views to draw.

- Use a C-size sheet of vellum or drafting film.
- Draw using a scale of 3/8" = 1'–0".
- Show the single-line pipe and fittings as thick lines.

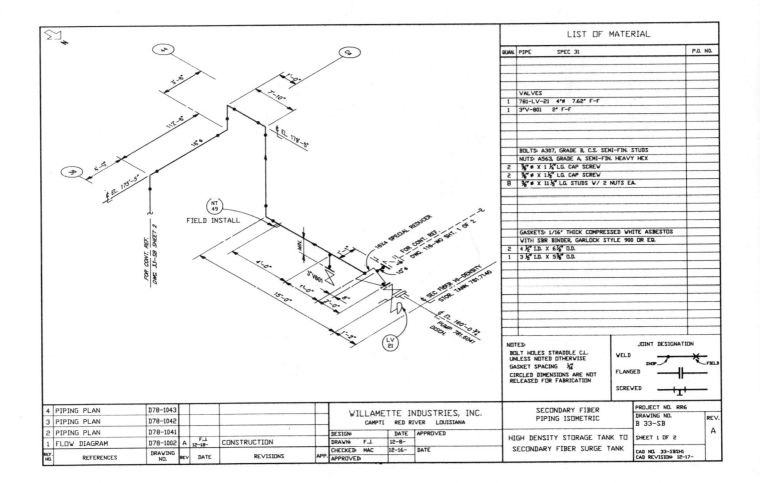

Problem 20–22 Draw the piping plan shown.

- Use D-size vellum.
- Draw using a scale of 1/4" = 1'–0".

- Dimension and label as shown. Place dimensions and labels differently if you can find better locations.

Courtesy Norwest Engineering.

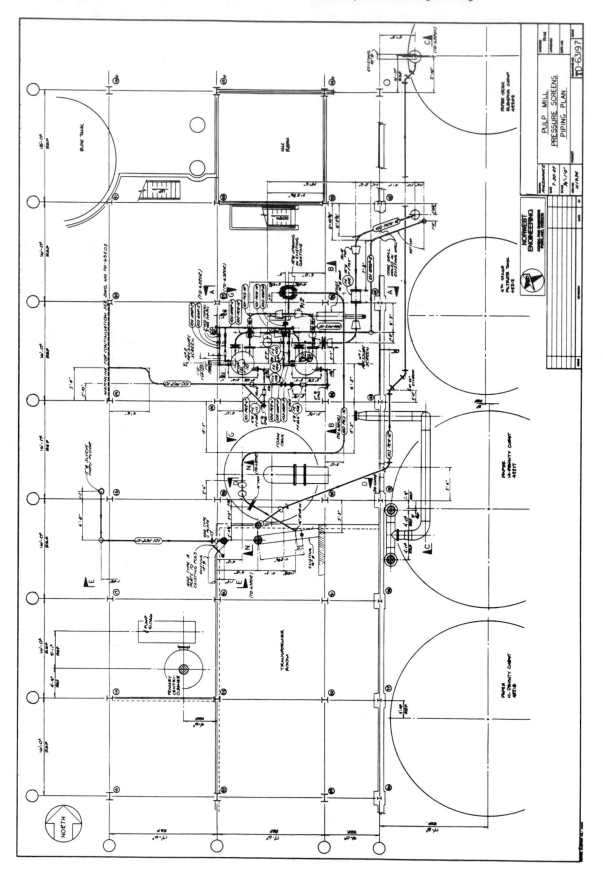

Problem 20–23   Draw the piping plan shown.

- Use D-size vellum.
- Draw using a scale of 1/4" = 1'–0".

- Dimension and label as shown. Place dimensions and labels differently if you can find better locations.

Courtesy Norwest Engineering.

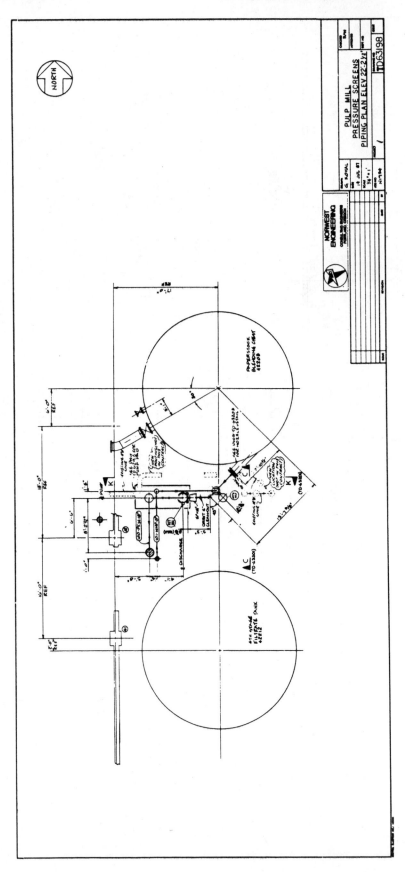

Problem 20–24 Draw the views shown.

- Use D-size vellum.
- Draw using a scale of 1/4" = 1'–0".

- Dimension and label as shown. Place dimensions and labels differently if you can find better locations.

Courtesy Norwest Engineering.

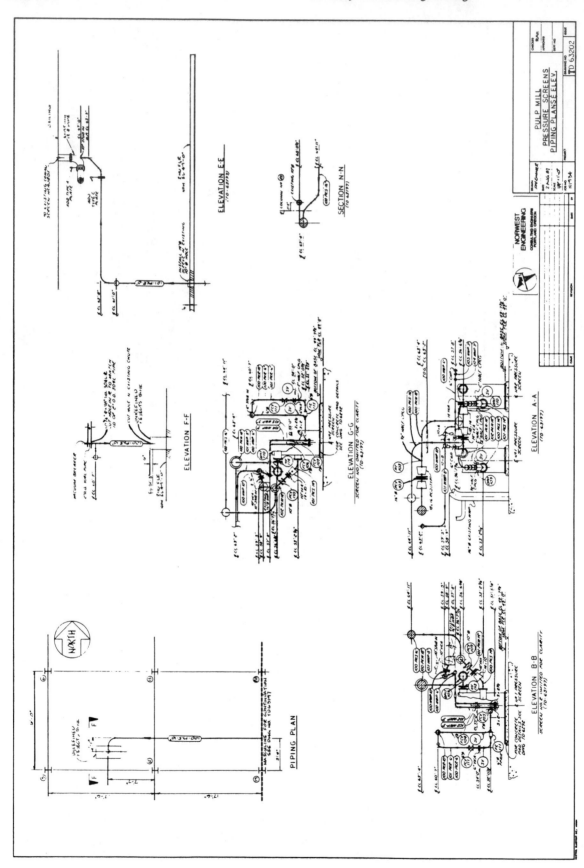

MATH PROBLEMS

Refer to Appendix I for fitting, flange, and valve dimensions to solve the following problems.

1. The drawing shown below is an assembly of pipe, fittings, flanges, and valves. Only one dimension is given for the short, straight length of pipe. This dimension is the length of pipe between two welds. You are to provide all of the missing dimensions. Gaskets are inserted between flanges and valves. All gaskets are 1/16" thick. Use Appendix I to find dimensions for valves and fittings.

    A = _____

    B = _____

    C = _____

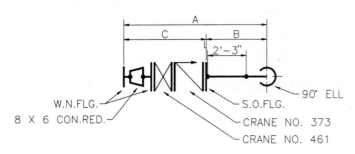

2. You are given a run of pipe shown below. Using the dimensions shown, provide the missing dimensions. Write your answers in the spaces provided.

    A = _____

    B = _____

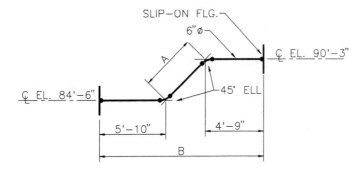

3. Calculate the travel of the angled pipe shown in the figure below. Write your answer in the space provided.

    A = _____

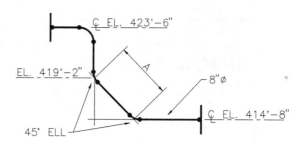

Structural Drafting with Basic Architectural Applications

THE ENGINEERING DESIGN PROCESS

In many situations the drafter prepares formal drawings from engineering calculations and sketches. Registered structural engineers, architects, or designers with experience and training may prepare structural calculations and designs. In many situations the entry-level engineering drafter may not fully understand the calculations, but can interpret the results of the calculations because the engineer or designer normally highlights the solution to be placed on the drawing by clearly underlining or placing a box around, as shown in Figure 21–1. When engineering calculations result in a design sketch, the drafter's job is to convert the sketch to a formal drawing, as shown in Figure 21–2b.

The quality and completeness of the engineering sketch often depends on the experience of the drafter. If the drafter is right out of school, the engineer may have to take a little more time providing detailed information until the drafter gains experience. Once the drafter has had some experience, the engineering sketches may be less complete. For example, if the engineering sketch in Figure 21–2a had omitted the note: 3/16" PLY SHIM EA SIDE GLUE AND NAIL W/4-ROWS 8-10d COMMON, an experienced drafter would have realized that shims are required when two beams of different thicknesses are joined. In some situations, an experienced drafter can work directly from the engineering calculations without sketches. In all situations it is very important for the drafter to

Figure 21–1 Engineering calculations typically contain a problem to be solved, mathematical solutions, and specifications to be placed on the drawing. The drawing note or information is placed in a box or otherwise highlighted.

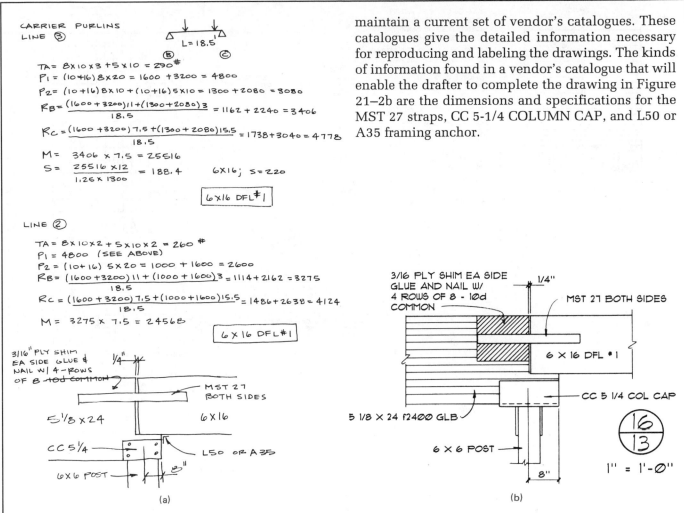

Figure 21–2 (a) A sample of engineer's calculations to determine the loading criteria at a beam connection over a column. (b) The drafter would draw the detail to convey the information of the engineer's calculations and sketch, using proper drafting techniques.

A structural engineer works with architects and building designers to engineer the structural components of a building. Structural engineering is generally associated with commercial steel and concrete buildings, and in some situations, it is also used for the structural design of residential buildings. The structural engineer works with civil engineers to design bridges and other structures related to road and highway construction. There is a wide variety of projects where structural drafting may be involved. Structural drafting techniques are generally the same as mechanical drafting, although a combination of mechanical and architectural methods are used.

Structural engineering drawings and detail drawings show in a condensed form the final results of designing. Drawings, general notes, schedules, and specifications serve as instructions to the contractor. The drawings must be complete and have sufficient detail so no misinterpretation can be made. Structural drawings are usually independent of architectural drawings and other

maintain a current set of vendor's catalogues. These catalogues give the detailed information necessary for reproducing and labeling the drawings. The kinds of information found in a vendor's catalogue that will enable the drafter to complete the drawing in Figure 21–2b are the dimensions and specifications for the MST 27 straps, CC 5-1/4 COLUMN CAP, and L50 or A35 framing anchor.

drawings, such as plumbing or HVAC, or, when necessary, structural drawings are clearly cross-referenced to architectural drawings.

▼ LINE WORK

The lines used in structural drafting are generally the same as those used in mechanical drafting. There are a few exceptions. For example, the object line may be drawn thicker than normal when a shape or feature requires extra emphasis. Object lines may also be thinner than the standard when used on a small scale drawing. (See Figure 21–3.) Dimension lines may be drawn with a space provided for the numeral as in mechanical drafting, or the numeral may be placed above the dimension line as in architectural drafting. The dimension line may be capped on the end with arrowheads, slashes, or dots. (See Figure 21–4.) Cutting-plane lines and symbols are provided one of several ways, as shown in Figure 21–5.

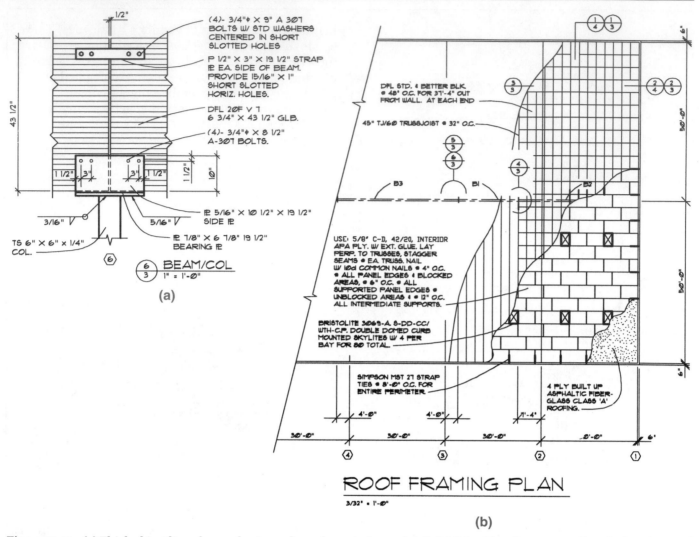

(a)

(b)

Figure 21–3 **(a)** Thick object lines for emphasis used on a beam/column detail; **(b)** thin object lines on small scale drawing used on a panelized roof framing system.

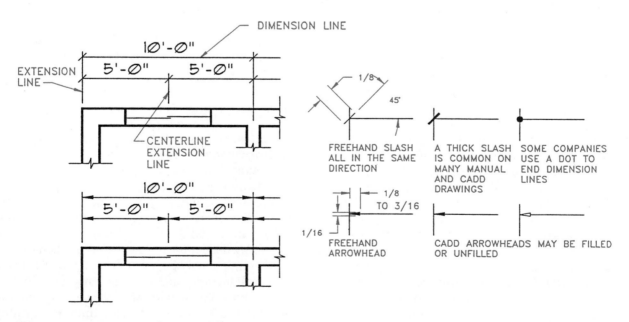

Figure 21–4 Dimension line examples.

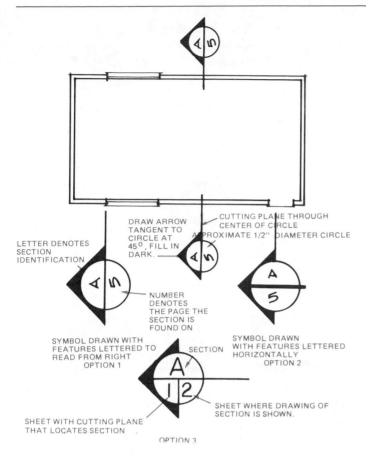

Figure 21–5 Cutting plane line symbols.

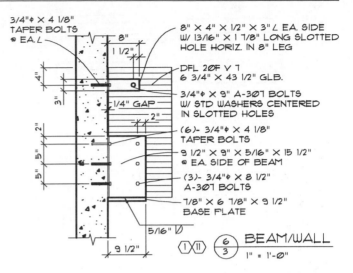

Figure 21–6 Common lettering used in structural drafting used on a beam construction detail.

▼ LETTERING

The quality of lettering in structural drafting is equally as important as it is in other drafting fields. There is a great deal of lettering on structural drawings. A well-developed, legible style of lettering makes the job easier and adds professional quality. Lettering is similar to the Gothic style used in mechanical drafting; however, the structural drafter has more freedom of style, but usually less freedom than is allowed in architectural drafting. The best structural lettering is simple, easy to read, and quick to produce. Structural drafters often prefer slanted letters. Lettering height on drawings is typically 1/8" to 5/32" (3–4 mm) for all lettering except titles, which are 3/16" to 1/4" (5–6 mm) in height. If drawings are to be microfilmed, 5/32" lettering is used. Figure 21–6 shows lettering used in structural drafting. Other lettering rules are presented in Chapter 3.

▼ STRUCTURAL DRAFTING RELATED TO CONSTRUCTION SYSTEMS

Different types of construction methods relate directly to the materials to be used, the area of the country where the construction will take place, the type of structure to be built, and even the office practices of the architect or engineer. The structural drafter should have a knowledge of construction materials and techniques. This chapter provides an introduction to construction techniques and materials. Additional resources should be used for reference as each construction method discussed has volumes of both general and vendor information available. Another valuable way to learn about construction is to visit job sites to talk to builders and see first hand how things are done.

▼ CONCRETE CONSTRUCTION

Concrete is a mixture of Portland cement, sand, gravel, and water. This mixture is poured into *forms* that are built of wood or other materials to contain the mix in the desired shape until it is hard. Concrete is a fundamental material used for building *foundations.* Concrete is also used in commercial applications for wall and floor systems. Residential buildings use concrete foundations with or without steel reinforcing, while commercial buildings usually use steel-reinforced concrete, depending on the structural requirements.

Concrete can either be poured in place at the job site, formed at the job site and lifted into place, or formed off-site and delivered ready to be erected into place. Concrete alone has excellent compression qualities. Steel added to concrete provides tension properties to the material. When concrete is poured around steel bars placed in the forms this is known as *reinforced concrete.* The steel bars are referred to as *rebar.* Steel is the best choice for reinforcing concrete, as its coefficient of thermal expansion is almost the same as cured concrete. The resulting structure has concrete to resist the compressive stress, and steel to resist the tensile stress of the structure.

Steel reinforcing is available in a number of shapes and configurations. The most common steel reinforcing is *plane* and *deformed,* in round or square bars.

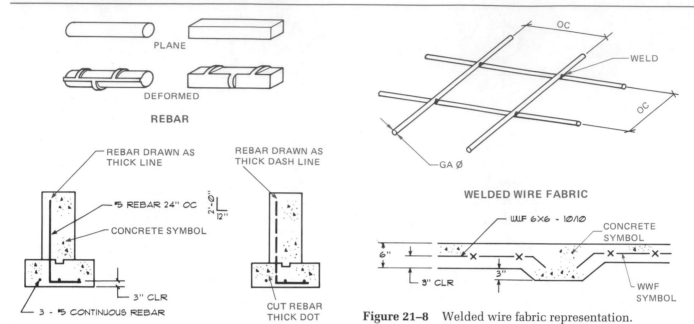

REBAR DRAWING SYMBOLS

Figure 21–7 Representation of reinforcing bars, *rebar*.

Figure 21–8 Welded wire fabric representation.

Deformed steel bars have surface projections that increase the adhesion between the concrete and steel. (See Figure 21–7.) Steel reinforcing bars are sized by number, starting at no. 2, which is 1/8" in diameter and increasing in size at approximately 1/8" intervals. No. 4 rebar is 1/2" in diameter. Another kind of steel reinforcing is wire fabric, which is available as welded wire or expanded metal. Welded wire fabric is commonly used to reinforce concrete slabs. Welded wire fabric is sized by the gage and spacing of the wires. For example, the callout WWF 6 × 6 10/10 means welded wire fabric (WWF), 6 in. on center (6 × 6), #10 gage wires (10/10). (See Figure 21–8.)

Poured-in-place Concrete

Commercial and residential applications for concrete poured in place are similar except that the size of the *casting* (resulting concrete structure) and the amount of reinforcing is generally more extensive in commercial construction. In addition to the foundation and on-grade (ground) floor systems, concrete is often used for walls, columns, and floors above ground. Concrete structures with lateral soil pressure tend to bend the wall inward, thus placing the soil side of the wall in compression and the interior side of the wall in tension. Steel reinforcing is used to reduce this *tensile stress*, as shown in Figure 21–9. Steel reinforced walls and columns are constructed by setting steel reinforcing in place and

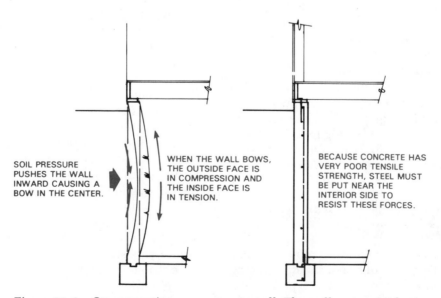

Figure 21–9 Stresses acting on a concrete wall. The wall serves as a beam spanning between each floor, and the soil is the supported load.

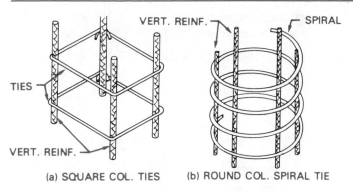

Figure 21–10 Examples of column reinforcing.

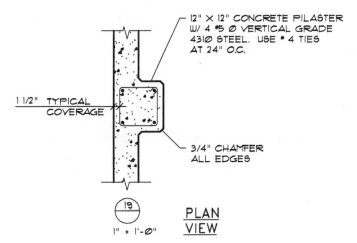

Figure 21–11 Typical reinforcing for a rectangular concrete column.

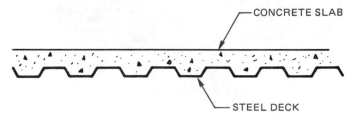

Figure 21–12 Representation of a steel deck system.

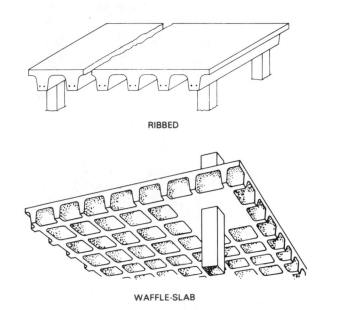

Figure 21–13 Two common concrete floor systems.

then surrounding it by wooden forms to contain the concrete. Once the concrete has been poured and allowed to set (harden), the forms are removed. As a drafter you will be required to draw details showing the sizes of the feature to be constructed and steel placement within the structure. This typically consists of drawing the vertical steel and the horizontal ties. Ties are wrapped around vertical steel in a column or placed horizontally in a wall or slab to help keep the structure from separating when placed under a load. Figure 21–10 shows two examples of column reinforcing. The drawings required to detail the construction of a rectangular concrete column are shown in Figure 21–11.

Concrete is also used on commercial projects to build above-ground floor systems. The floor slab may either be supported by a steel deck or self supported. The steel deck system is typically used on structures constructed with a steel frame. (See Figure 21–12.) Two of the most common poured-in-place concrete floor systems are the *ribbed* and *waffle* floor methods, as shown in Figure 21–13. The ribbed system is used in many office buildings. The ribs serve as floor joists to support the slab, but are actually part of the slab. Spacing of the ribs varies depending on the *span* and the amount and size of reinforcing. The waffle system is used to provide added support for the floor slab and is typically used in the floor systems of parking garages.

Precast Concrete

Precast concrete construction consists of forming the concrete component off-site at a fabrication plant and transporting it to the construction site. Figure 21–14 shows a precast beam being lifted into place. Drawings

Figure 21–14 Precast concrete beams and panels are often formed offsite, delivered to the job site, and then set into place with a crane.

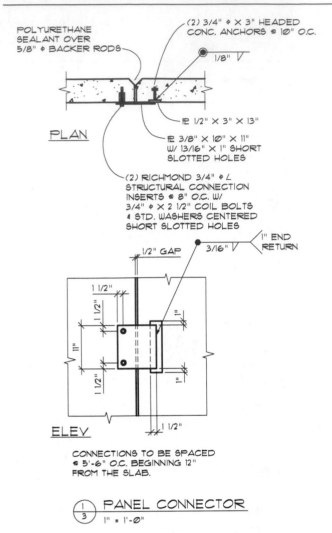

Figure 21–15 Typical panel connection detail.

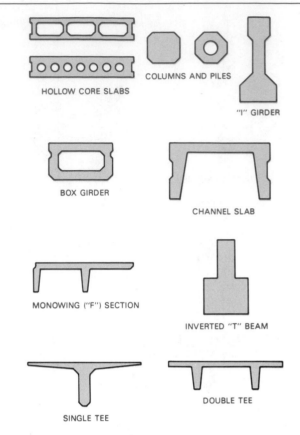

Figure 21–16 Common precast concrete shapes.

for precast components must show how precast members are to be constructed and methods of transporting and lifting the member into place. Precast members often have an exposed metal flange so the member can be connected to other parts of the structure. Common details used for wall connections are shown in Figure 21–15.

Many concrete structures are precast and *prestressed*. Concrete is prestressed by placing steel cables, wires, or bars held in tension between the concrete forms while the concrete is poured around them. Once the concrete has hardened and the forms are removed, the cables act like big springs. As the cables attempt to regain their original shape, compression pressure is created within the concrete. The compression in the concrete helps prevent cracking and deflection. Prestressed concrete members are generally reduced in size in comparison with the same design features of a standard precast concrete member. Prestressed concrete components are commonly used for the structural beams of buildings and bridges. Common prestressed concrete shapes are shown in Figure 21–16.

Tilt-up Concrete

Tilt-up construction is a method using formed wall panels that are lifted or "tilted" into place. Panels may either be formed and poured at or off the job site. Forms for a wall are constructed in a horizontal position and the required steel placed in the form. Concrete is then poured around the steel and allowed to harden. The panel is lifted into place once it has reached its desired hardening and design strength. When using this type of construction, the drafter usually draws a plan view to specify the panel locations, as shown in Figure 21–17. The location and size of steel placement is also important, as shown in Figure 21–18.

Standard Structural Callouts for Concrete Reinforcing

When specifying anchor bolts (AB) on a drawing, give the quantity, diameter, type, length, spacing on center (OC), and projection of the thread out of the concrete. For example:

12-3/4" $\varnothing$ × 12" STD AB 24" OC W/3" PROJ.

When specifying rebar on a drawing, give the quantity (if required), bar size, length (if required), spacing in inches on center, horizontal or vertical, and bend information (if required). Deformed steel rebar is assumed unless otherwise specified. For example:

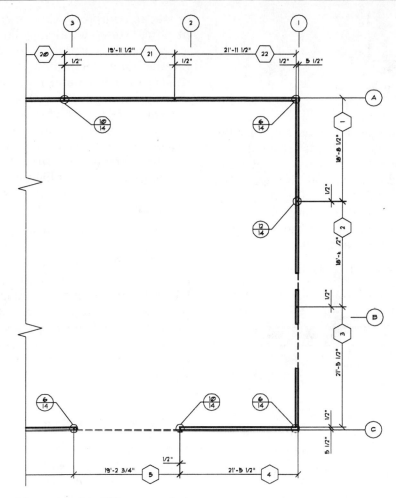

Figure 21–17 Tilt-up panel plan.

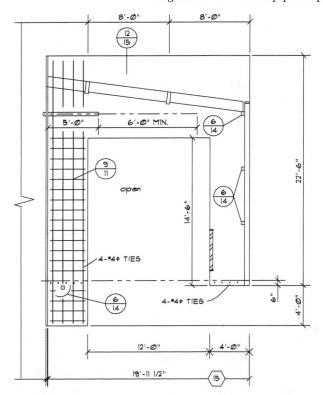

Figure 21–18 Steel locations specified in precast panels.

#4 REBAR AT 12" OC
25 #8 at 24" OC HORIZ AND 16" OC VERT
#6 AT 24" OC EACH WAY (or EW)
#5 AT 16" OC × 12" OC

Dimensions of bends are to be provided in feet and inches without the (') and (") marks given, for example:

#5 AT 16" OC VERT W/90° × 12 BEND

or the bend diagram may be drawn, as shown in Figure 21–19.

When specifying welded wire on a drawing, give the designation WWF, the wire spacing in inches OC, and the wire size. For example:

WWF 6 × 6-6/6

Reinforcing steel rebar lap splices may be shown on the drawing by giving the length of the lap and a lap location dimension. For example:

#5 REBAR 16" OC W/24" MIN SPLICE AT FOOTING

Welded wire fabric lap splices need not be shown as a specific note on the drawing; however, a general note should clarify the amount of allowable splice overlap, in inches, at the cross wires.

Figure 21–19 *Rebar* bend diagram.

Clear distances should be given from the surface of the concrete to the rebar. This dimension is assumed to be to the edge of the rebar or clarified by the abbreviation CLR, as in 3" CLR. If this dimension is designated to the centerline of the rebar, then OC must be specified.

When structural members are embedded in concrete, the size of holes for rebar to pass through should be specified with the rebar callout. For example:

#5 AT 16" OC W/13/16 ∅ HOLES

Recommended hole sizes for rebar passing through steel or timber are as follows:

Bar #	Hole ∅	Bar #	Hole ∅
4	11/16	8	1 1/4
5	13/16	9	1 3/8
6	1	10	1 9/16
7	1 1/8		

When rebar must be driven through timber, a tighter hole tolerance is recommended. For example:

#8 REBAR × 6'–0" at 12" OC W/1-1/8" ∅ HOLES AT TIMBER

This application for given rebar is as follows:

Bar #	Hole ∅	Bar #	Hole ∅
4	9/16	8	1 1/8
5	11/16	9	1 1/4
6	7/8	10	1 7/16
7	1		

Slab thickness, concrete wall, or concrete beam and column cross-sectional dimensions should be given in inches or feet and inches. A complete slab-on-grade callout may read as follows:

4" CONC SLAB ON FIRM UNDISTURBED SOIL
OR 4" SAND FILL

or

6" THICK 3000 PSI CONC SLAB WITH WWF 6 × 6—
W2.9 × W2.9 3" CLR AT 4" MIN 3/4" MINUS FILL
COMPACTED TO 95% OF SOIL DENSITY

Concrete footing thickness should be given in inches or feet and inches. The footing width may be given followed by the thickness all in one note.

Concrete Structural Engineering Drawings

Structural engineering drawings show general information that is required for sales, marketing, engineering, or erection purposes. Concrete structures are drawn in plan (top) view, elevation (side) view, or sectional views at a scale that is dependent on the size of the structure, the amount of detail to be shown, or the size of paper used. Dimensions and notes are used to completely describe the construction characteristics. Concrete material symbols are used where appropriate. Rebar is shown as a very thick line in longitudinal view or as a round dot where the bar appears cut. Drafters in architectural or structural engineering firms typically draw structural drawings as shown in Figure 21–20.

Shop Drawings

Shop drawings are used to break each individual component of a structural engineering drawing down into fabrication parts. Shop drawings are also referred to as fabrication drawings, and are generally drawn by the drafter in the fabrication company. Many large companies do both structural and shop drawings. Depending on the complexity of the structure, the structural and shop drawings may be combined on the same drawing. An example of a shop drawing is shown in Figure 21–21.

▼ CONCRETE BLOCK CONSTRUCTION

Concrete block construction is often used for residential foundations, but is also used in some above-ground construction. In commercial applications, concrete blocks are used to form the wall systems for many types of buildings. Concrete blocks provide a durable construction material and are relatively inexpensive to install and maintain. Blocks are commonly manufactured in nominal size *modules* of 8 × 8 × 16 in., 4 × 8 × 16 in., or 6 × 8 × 16 in. The actual size of the block is smaller than the nominal size so that *mortar* (grout) joints can be included in the final size. Although the building designer determines the size of the structure, it is important that the drafter be aware of the modular principles of concrete block construction. Wall lengths, opening locations, and wall and opening heights must be based on the modular size of the block being used. Failure to maintain the modular layout can result in a tremendous increase in labor costs to cut and lay the blocks.

The drafter's responsibility when working with concrete block structures is also to detail steel reinforcing patterns. Concrete blocks are often reinforced with a wire mesh at every other course of blocks. Where the risk of seismic activity must be considered, concrete block structures are often required to have reinforcing steel placed within the wall to help tie the blocks together. The steel is placed in a block that has a channel or cell running through it. This cell is then filled with grout or concrete to form a *bond beam* within the wall. The bond beam solidifies and ties the block structure together. A typical bond beam concrete block structure is detailed in

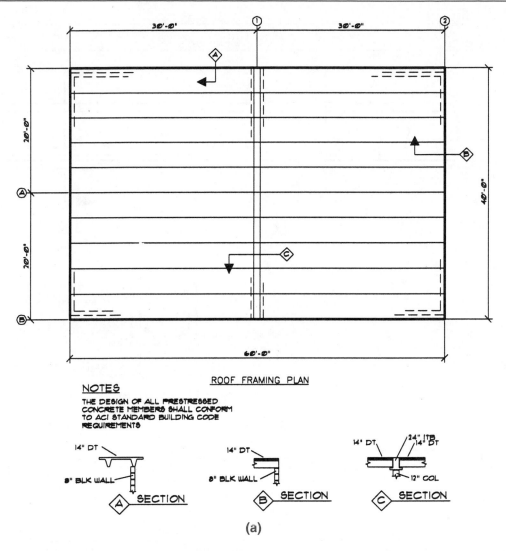

ROOF FRAMING PLAN

NOTES

THE DESIGN OF ALL PRESTRESSED
CONCRETE MEMBERS SHALL CONFORM
TO ACI STANDARD BUILDING CODE
REQUIREMENTS

(a)

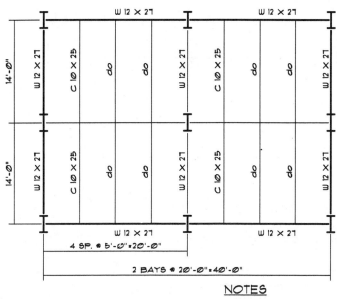

NOTES

ALL BEAMS FLUSH TOP
TOP OF STEEL ELEVATION

(b)

Figure 21–20 **(a)** Precast concrete engineering drawing; **(b)** structural steel engineering drawing.

Figure 21–22. Figure 21–23 (see page 681) shows typical concrete block sizes and reinforcing methods. Openings for windows and doors require steel reinforcing for both poured-in-place and concrete block construction. (See Figure 21–24, page 681.)

When the concrete blocks are required to support a load from a beam, a *pilaster* is often placed in the wall to help transfer the beam loads down the wall to the footing. Pilasters are also used to provide vertical support to the wall when the wall is required to span long distances. Examples of pilasters are shown in Figure 21–25 (see page 682). A structural detail is shown in Figure 21–26 (see page 682).

▼ WOOD CONSTRUCTION

Wood frame construction is typical in residential construction, and is also used in some commercial construction applications, especially for multi-family dwellings and office buildings. Other commercial uses include partition framing, upper-level floor framing, and roof framing. Residential and commercial wood

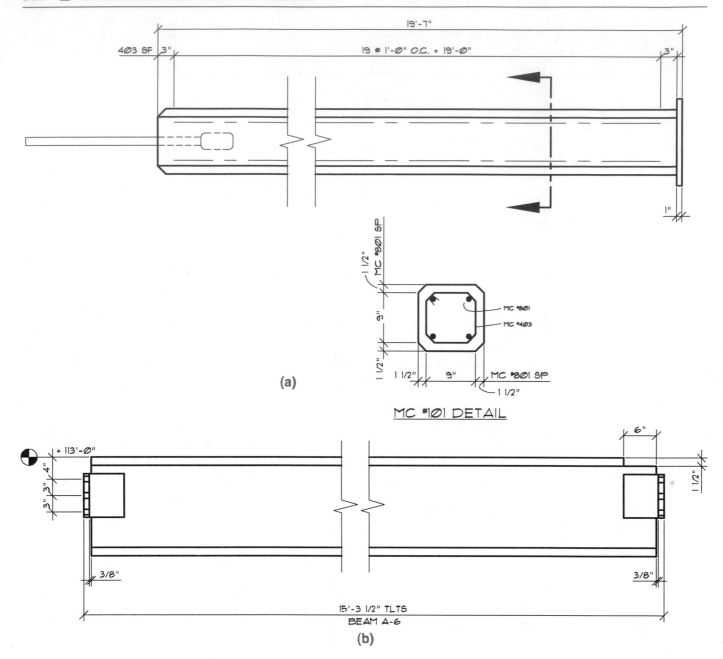

MC #101 DETAIL

(a)

BEAM A-6

(b)

Figure 21–21 **(a)** Precast concrete fabrication detail; **(b)** structural steel fabrication drawing used to show how individual components are to be made.

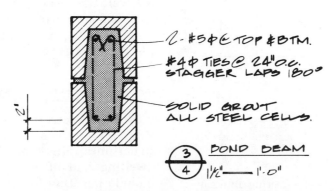

Figure 21–22 Concrete block bond beam reinforcing detail.

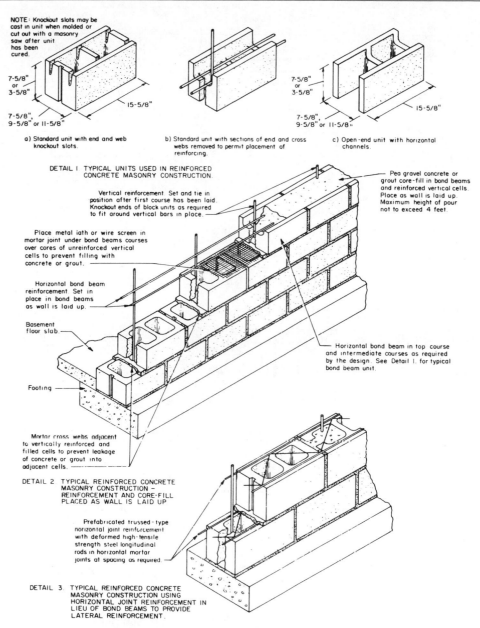

Figure 21–23 Suggested construction details for reinforced concrete masonry foundation walls. *Courtesy National Concrete Masonry Association.*

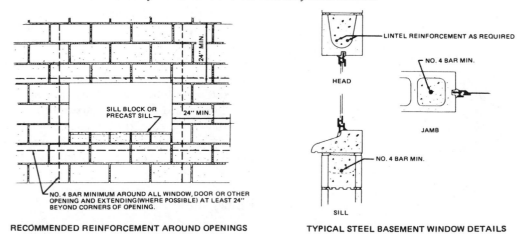

RECOMMENDED REINFORCEMENT AROUND OPENINGS

TYPICAL STEEL BASEMENT WINDOW DETAILS

Figure 21–24 Typical reinforcing and window detail at opening in reinforced concrete masonry. *Courtesy National Concrete Masonry Association.*

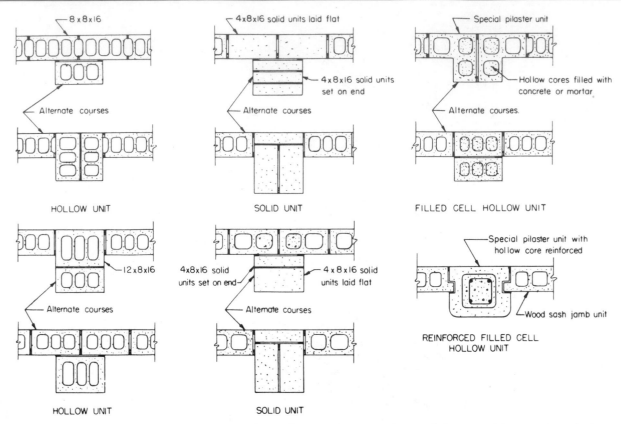

Figure 21–25 Typical concrete masonry pilaster designs. *Courtesy National Concrete Masonry Association.*

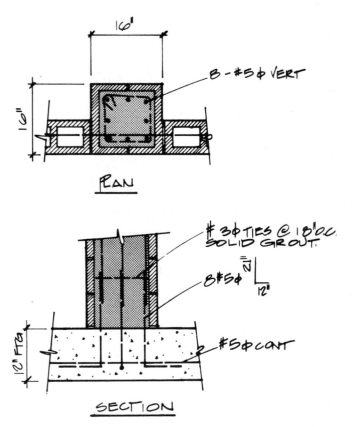

Figure 21–26 Reinforced pilaster detail.

wall construction is essentially the same, with the main difference often being in the type of covering used. In commercial applications, wood walls may require special finishes to meet fire protection requirements, as shown in Figure 21–27.

Joists, trusses, and panelized systems are the most commonly used wood roof framing systems. These systems allow for the members to be placed 24 or 32 in. O.C. Figure 21–3b (see page 672) shows a truss roof system drawing. Panelized roof systems use beams placed 20 to 30 ft apart with smaller beams called *purlins* placed between the main beams on 8 ft centers. Joists that are 2 or 3 in. wide are then placed between the purlins at 24 in. O.C. The roof is then covered with plywood sheathing. Figure 21–28 shows an example of how a panelized roof system is constructed. A roof framing plan for a panelized roof system is shown in Figures 21–3b and 21–29.

When wood frame construction joins concrete or concrete block construction, a rigid connection between the two systems is required. Several methods of connecting wood to concrete block are shown in Figure 21–30 (see page 684).

Heavy Timber Construction

Large wood members are sometimes used for the structural frame of a building. This method of construction is used for appearance, structural purpose, and

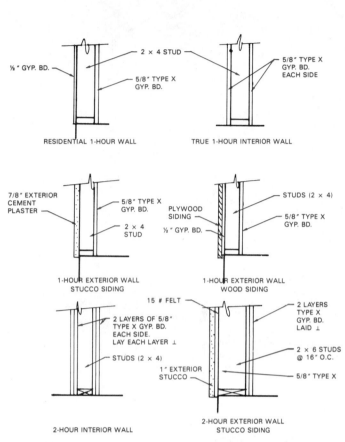

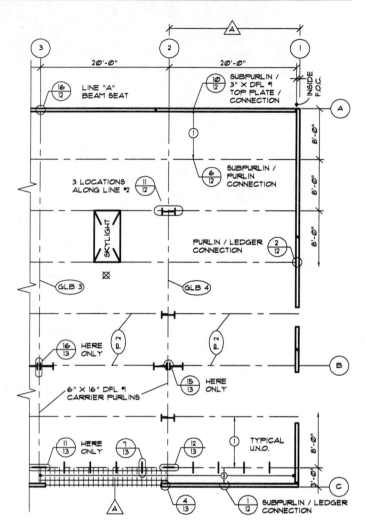

Figure 21–27 Separation walls often require special treatment to achieve the needed fire rating for certain types of construction.

Figure 21–29 A panelized roof and heavy timber framing plan. *Courtesy Structureform Masters, Inc.*

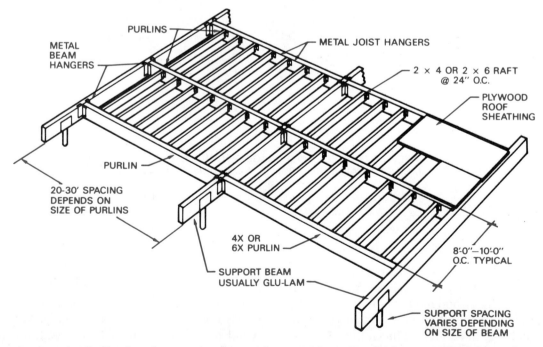

Figure 21–28 A panelized roof system is often used to provide roofing for large area with limited supports.

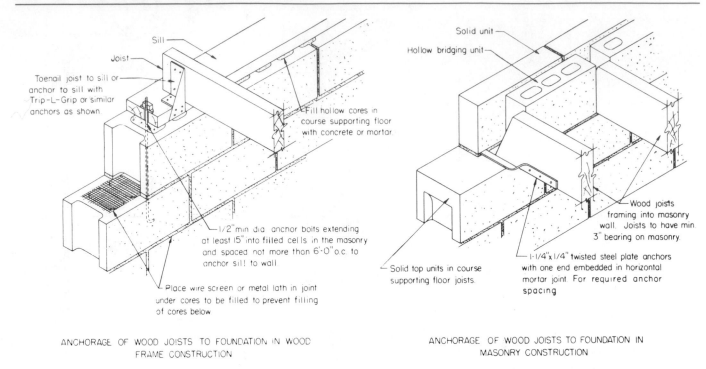

SUGGESTED METHODS OF ANCHORING WOOD JOISTS
BEARING ON CONCRETE MASONRY FOUNDATION WALLS.

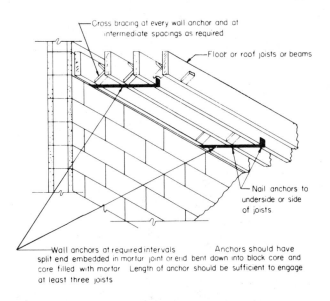

TYPICAL DETAIL FOR ANCHORAGE OF CONCRETE
MASONRY WALLS TO PARALLEL WOOD JOISTS OR BEAMS.

Figure 21–30 Metal angles and straps are typically used to ensure a rigid connection between the wall and floor or roof system. *Courtesy National Concrete Masonry Association.*

when material is available. Heavy timbers have structural advantages for short spans and excellent fire retardant qualities. In a fire, heavy wood members will char on the exposed surfaces while maintaining structural integrity long after a steel beam of equal size has failed. A heavy timber roof framing plan is shown in Figure 21–29.

Laminated Beam Construction

It is difficult and often impossible to produce large-size wood timbers in long lengths. When long-span wood beams are required, the solution is *laminated* (lam) beams, as shown in Figure 21–31. Notice the

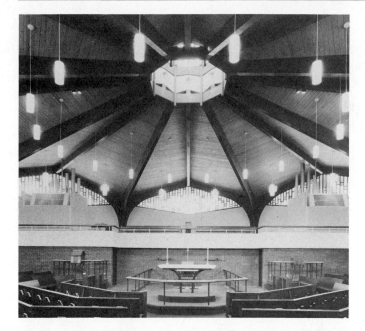

Figure 21–31   Timbers are often used in commercial construction because of their beauty and structural qualities. The structural system of St. Philip's Episcopal Church is formed by glue-laminated Southern pine arches and beams under a sweeping canopy of Southern pine roof decking. *Courtesy Southern Forest Products Association.*

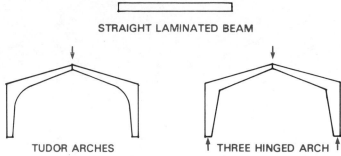

Figure 21–32   Common laminated beam shapes.

items labeled GLB 1, GLB 2, GLB 3, and GLB 4 in Figure 21–29. These are *glu-lam beams* (GLB). The intermediate members labeled P1 and P2 are purlins, as previously discussed. Laminated beams are manufactured from smaller, equally-sized members glued together to form a larger beam. Laminated beams are used to support heavy loads or accommodate long spans and are also used when the wood appearance is important. The common types of laminated beams are the *single span straight, Tudor arch,* and *three-hinged arch* beams. (See Figure 21–32.)

The single span beam is commonly referred to as glu-lam and noted on a drawing with the abbreviation GLB for glu-lam beam. A glu-lam beam can be used to

replace a much larger wood timber due to increased structural qualities. Figure 21–33 compares the strength of common wood members with laminated beams. In some applications, laminated beams have a *camber* or curve built into the beam. The camber is designed into the beam to help resist the loads to be carried.

The Tudor and three-hinged arch members are a post-and-beam system combined into one member. These beams are specified on plans in a method similar to other beams. The drafter's major responsibility when working with either heavy timber construction or laminated beams is in the drawing of connection details. Beam to beam, beam to column, and column to support are among the most common details drawn by drafters. Common types of manufactured connectors used to connect timbers are shown in Figure 21–34. The drafter is required to draw the fabrication details for a connector when the size of the beam does not match existing connectors, as shown in Figure 21–3a (see page 672).

Wood members can be attached to concrete in several ways. Two of the most common methods are by the use of a pocket or seat, as shown in Figure 21–35, or a metal connector.

Standard Wood Structural Callouts

When specifying the callouts for structural wood sizes on a drawing, give the *nominal* (rough, before

SPECIES AND COMMERCIAL GRADE	EXTREME FIBER BENDING F_b	HORIZONTAL SHEAR F_v	COMPRESSION PERPENDICULAR TO GRAIN $F_c \perp$	MODULUS OF ELASTICITY E
DFL #1 22FV4	1350	85	385	1,600,600
DF/DF	2200	165	385	1,700,000
Hem-Fir #1 22F E2	1050	70	245	1,300,000
HF/HF	2200	155	245	1,400,000
SPF 22F-E-1	900	65	265	1,300,000
SP/SP	2200	200	385	1,400,000

Figure 21–33   Comparative values of common framing lumber with laminated beams of equal material. Values based on the *Uniform Building Code.*

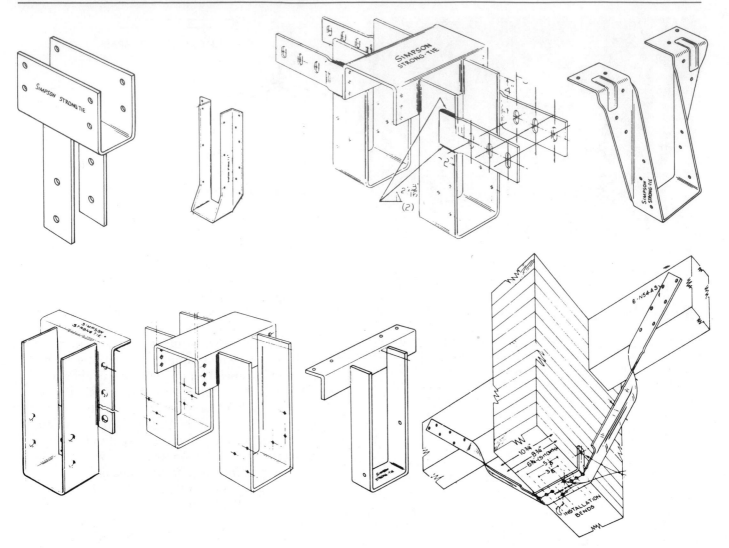

Figure 21–34 Common manufactured metal beam connectors. *Courtesy Simpson Strong-Tie Company, Inc.*

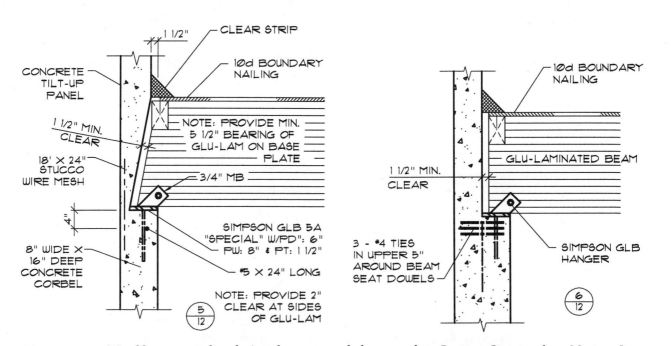

Figure 21–35 Wood beams are often designed to rest on a ledge or pocket. *Courtesy Structureform Masters, Inc.*

planning) cross-sectional dimensions for sawn lumber and timbers. For example:

$$4 \times 12, 6 \times 14$$

The *net* (actual) cross-sectional dimension is given for glu-lams. The net dimensions of a glu-lam are followed by the abbreviation GLB and any other specification. For example:

6-1/8 × 14 GLB f2400
(f2400 = actual units of stress in fiber bending)

If the net cross-sectional dimensions for lumber are required, they may be given in parentheses after the nominal dimensions. If lengths (LG) are required for wood members, they should be given in feet and inches. For example:

6-3/4 × 18 GLB × 24'–0" LG

When wood is used for floor or ceiling joists, rafters or trusses, or stud walls, the member size should be followed by the OC spacing in inches up to 24" OC and feet and inches for over 24" OC. For example, abbreviations are shown in parentheses:

2 × 8 CEILING JOISTS (CJ) 16" OC,
4 × 14 @ 4'–0" OC
2 × 6 STUDS 24" OC,
2 × 12 FLOOR JOISTS (FJ) 12" OC

All lumber and timber on a drawing should be dimensioned to the centerlines unless dimensioning to the face of the member is otherwise required by the engineer or architect. The exception to this rule is dimensioning to the top of a beam or other structural timber.

When specifying plywood on a drawing, give the thickness, group, face veneer grades, identification index (if required), and glue type for each different panel used. Also include nailing, blocking, and edge spacing requirements if specific applications are required. Dimension to the face of the panel when surface location dimensions are required. The following are plywood callout examples:

1. 1/2" CDX 32/16 PLYWOOD SHEATHING.
2. 1/2" CDX SHTG W/10d NAILS 3" OC @ SEAMS AND 8" OC FIELD. (NOTE: *SEAMS* = edges or splices; *FIELD* = within the sheet at supports.)
3. 1/2" GROUP 1, CC, 24/0 EXTERIOR APA PLYWOOD W/8d NAILS 1/4" OC @ EDGES AND @ 6" OC FIELD.
4. 5/8" GROUP 2, UNDERLAYMENT CCPT EXTERIOR APA PLYWOOD W/8d RINGSHANK NAILS @ 3" OC EDGES AND 6" OC FIELD, BLOCK ALL PANEL EDGES PERPENDICULAR TO SUPPORTS. (NOTE: *CC* = CC grade outside veneer, *PT* = plugged and taped patches.)

When lumber decking is used, the lumber size and specification is followed by nailing information, if required. For example:

3 × 8 T&G RANDOM CONTROLLED DECKING W/20d TOENAIL EACH SUPPORT AND 2-30d RINGSHANK FACE NAIL EACH SUPPORT.

Random controlled means various lengths, usually 4'–0" modules placed so there are not two adjacent splices in the same support. T&G = tongue and groove.

▼ STEEL CONSTRUCTION

Steel construction can be divided into three categories: steel studs, prefabricated steel structures, and steel-framed structures.

Steel Studs

Prefabricated steel studs are used in many types of commercial structures. Steel studs offer lightweight, noncombustible, corrosion-resistant framing for interior partitions and load-bearing exterior walls up to four stories high. Steel members are available for use as studs or joists. Members are designed for rapid assembly and are predrilled for electrical and plumbing conduits. The standard 24" spacing reduces the number of studs required by about one-third when compared with wood studs spaced 16" OC. Steel stud widths range from 3-5/8" to 10", but can be manufactured in any width. The material used to make studs ranges from 12- to 20-gage steel depending on the design loads to be supported. Steel studs are mounted in a channel track at the top and bottom of the wall or partition. This channel serves to tie the studs together at the ends. Horizontal bridging is often placed through the predrilled holes in the studs and then welded to the studs to serve as fire blocking within walls and as mid supports. The components of steel stud framing are shown in Figure 21–36.

Prefabricated Steel Structures

Prefabricated or metal buildings, as they are often called, have become a common type of construction for commercial and agricultural structures in many parts of the country. Drafters who are involved in the preparation of drawings for premanufactured structures may work in a structural engineering office or for a building manufacturer. Standardized premanufactured steel buildings are sold as modular units with given spans, wall heights, and lengths in 12' or 20' increments. Most manufacturers provide a wide variety of design options for custom applications that may be required by the client. One advantage of these structures is faster erection time as compared with other construction methods.

The Structural System. The structural system is made up of the frame that supports the walls and roof. There are several different types of structural systems commonly used, as shown in Figure 21–37. The wall system

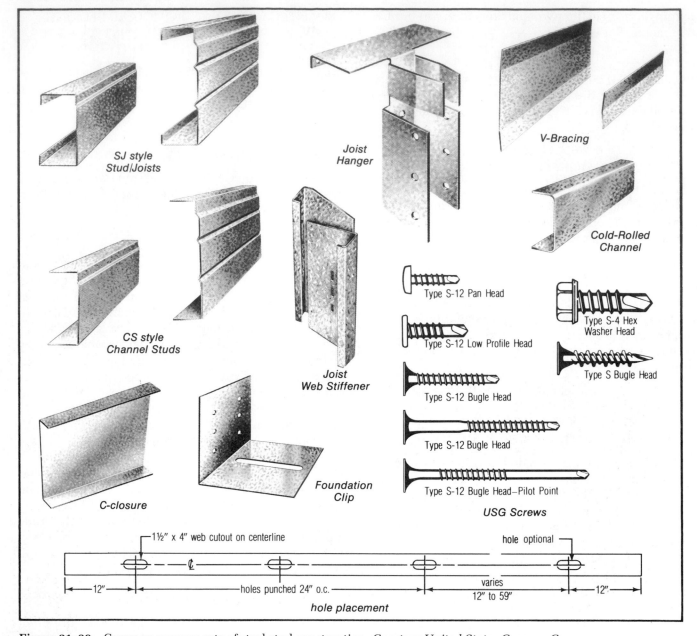

Figure 21–36 Common components of steel stud construction. *Courtesy United States Gypsum Company.*

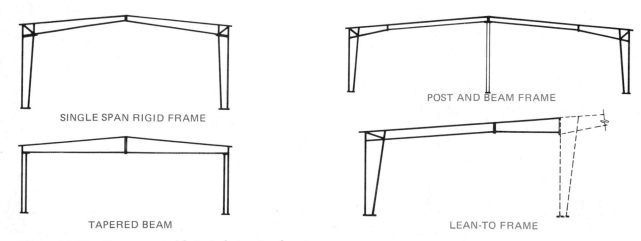

Figure 21–37 Common prefabricated structural systems.

is horizontal *girts* attached to the vertical structure and metal wall sheets attached to the girts. The roof system is horizontal purlins attached to the structure and metal sheets attached to the purlins. (See Figure 21–38.) Steel wall and roof sheets are available from many vendors in a variety of patterns and may be purchased plain, galvanized, or prepainted. A sample pattern design is shown in Figure 21–39.

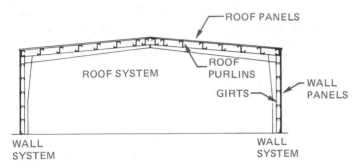

Figure 21–38   Components of the prefabricated structural system.

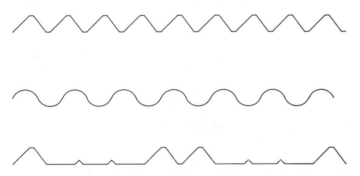

Figure 21–39   Typical cross sections of sheet metal siding and roofing material; many pattern shapes and finish colors are available.

Steel-framed Structures

Steel-framed buildings require structural engineering and shop drawings similar to those used for concrete structures. As a drafter in an engineering or architectural firm, you will most likely be drafting engineering drawings similar to the one shown in Figure 21–40. A drafter working for the steel fabricator prepares shop drawings which are detail drawings of structural steel components, as shown in Figure 21–21.

AISC   Drafters in this field must become familiar with the *Manual of Steel Construction* published by The American Institute of Steel Construction, Inc. (AISC). In addition to the local and national building codes, this manual is one of your prime references, as it helps you determine dimensions and properties of common steel shapes. Another manual that provides information on dimensions for detailing and properties for design work related to steel structural materials is *Structural Steel Shapes,* published by U.S. Steel Corporation.

Common Structural Steel Materials

Structural steels are commonly identified as plates, bars, or shape configurations. *Plates* are flat pieces of steel of various thickness used at the intersection of different members and for the fabrication of custom connectors. Figure 21–42 shows an example of a steel connector that uses top, side, and bottom plates. Plates are typically specified on a drawing by giving the thickness, width, and length in that order. The symbol ℄ is often used to specify plate material. For example:

℄ 1/4 × 6 × 10 (with or without inch marks)

Bars are the smallest of structural steel products and are manufactured in round, square, rectangular, flat, or

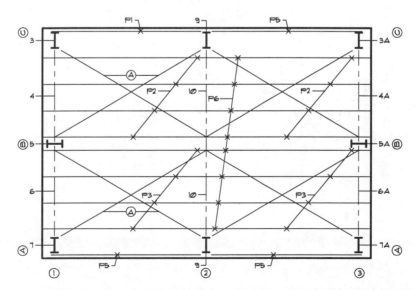

Figure 21–40   Structural steel engineering drawing. *Courtesy Pacific Building Systems.*

CADD Applications

STRUCTURAL DRAFTING

The procedure for preparing structural drawings works especially well with CADD. When a floor plan is drawn, this drawing is used as a base layer upon which each of the other plan views are drawn. For example, the floor plan may be drawn with black lines and labeled as the FLOOR PLAN layer, then the layer is changed and named FOUNDATION PLAN and the color is changed to red. All of the drawing information on the red layer is for the foundation plan. This process of changing layer names and colors continues until all of the drawings are complete. One advantage is that each layer may be turned on or off at the drafter's command. It is the same as tracing a drawing in manual drafting. Time is saved, and the drawings are all extremely accurate. When the drawings are to be plotted, each drawing may be converted to an individual file, or the layer to be plotted is turned on while the others are off. Any combination of plots may be reproduced, with select layers turned on.

There is now available structural CADD software that quickly and accurately draws structural steel shapes to exact specifications in plan, section, or elevation views. These structural steel packages also provide for beam details, integrating cutouts, framing angles, hole groups, welded or bolted connections, and complete dimensioning. An advantage is the capability of such programs to perform tedious standards specification sizing and calculations automatically. You do not have to spend time dimensioning hole patterns, because the program draws bolt or rivet hole patterns to your specifications, automatically displaying on center dimensions, hole diameters, and flange thicknesses. When framing angles are used, the package automatically draws, sizes, positions, dimensions, and notes the angle specifications along with bolt holes and dimensions. These CADD structural packages are often available with a variety of template menu overlays and symbol libraries, as shown in Figure 21–41.

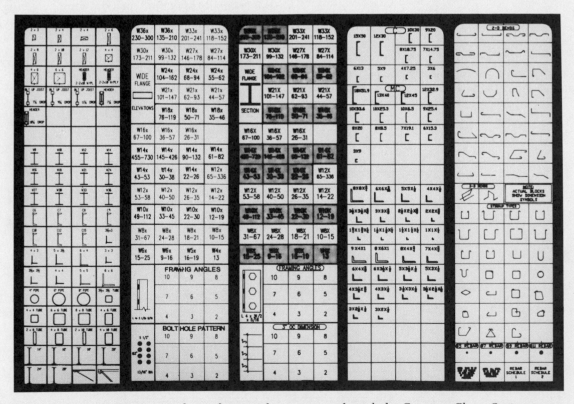

Figure 21–41 CADD structural template overlay menus and symbols. *Courtesy Chase Systems.*

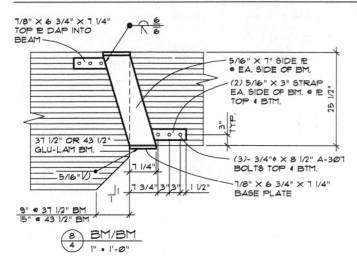

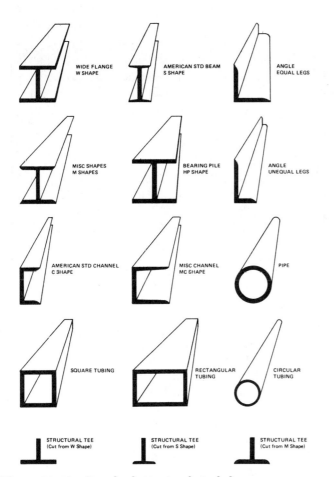

Figure 21–42   Steel plates used to fabricate a beam connector.

Figure 21–43   Standard structural steel shapes.

hexagonal cross sections. Bars are often used as supports or braces for other steel parts or connectors.

Structural steel is also available in several different manufactured shapes, as shown in Figure 21–43. When specifying a steel shape on a drawing, the shape identification letter is followed by the flange width, the "by" sign (×), and the weight in number of pounds per linear foot. For example:

WIDE FLANGE SHAPES

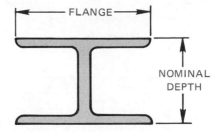

The designation for a wide flange looks like this:

W 12 X 22

Where: **W** is the shape.
       12 is the nominal depth.
       22 is the number of pounds per lineal foot.

CHANNELS

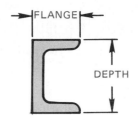

The designation for a channel shape looks like this:

C 6 X 10.5

Where: **C** is the shape
       6 is the depth
       10.5 is the number of pounds per lineal foot.

Figure 21–44   Dimensional elements of the wide flange and channel shapes.

W 12 × 22 or C 6 × 10.5

as shown in Figure 21–44. In the AISC *Manual of Steel Construction,* specific information regarding dimensions for detailing and dimensioning are clearly provided along with typical connection details. The representative page for the W 12 × 22 wide flange and the C 6 × 10.5 channel from Figure 21–44 are shown in Figure 21–45. The W, S, and M shapes all have an I-shaped cross section and are often referred to as "I" beams. The three differ in the width of their flanges. In addition to varied flange widths, the S shape flanges vary in depth.

Angles are structural steel components that have an L shape. The legs of the angle may be either equal or unequal in length but are usually equal in thickness. Channels have a squared C cross-sectional area and are designated with the letter C. Structural tees are produced from W, S, and M steel shapes. Common designations include WT, ST, and MT.

Structural tubing is manufactured in square, rectangular, and round cross-sectional configurations. These

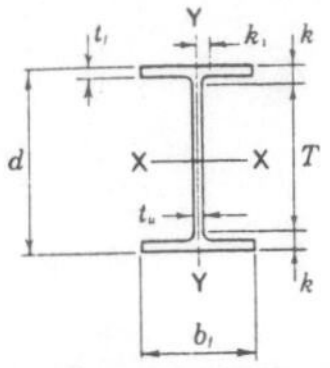

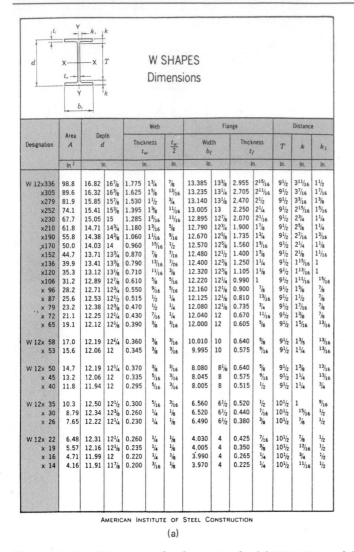

W SHAPES
Dimensions

Designation	Area A	Depth d		Web Thickness t_w		Web $\frac{t_w}{2}$	Flange Width b_f		Flange Thickness t_f		Distance T	Distance k	Distance k_1
	In.²	In.		In.		In.	In.		In.		In.	In.	In.
W 12x336	98.8	16.82	16 7/8	1.775	1 3/4	7/8	13.385	13 3/8	2.955	2 15/16	9 1/2	3 11/16	1 1/2
x305	89.6	16.32	16 3/8	1.625	1 5/8	13/16	13.235	13 1/4	2.705	2 11/16	9 1/2	3 7/16	1 7/16
x279	81.9	15.85	15 7/8	1.530	1 1/2	3/4	13.140	13 1/8	2.470	2 1/2	9 1/2	3 3/16	1 3/8
x252	74.1	15.41	15 3/8	1.395	1 3/8	11/16	13.005	13	2.250	2 1/4	9 1/2	2 15/16	1 5/16
x230	67.7	15.05	15	1.285	1 5/16	11/16	12.895	12 7/8	2.070	2 1/16	9 1/2	2 3/4	1 1/4
x210	61.8	14.71	14 3/4	1.180	1 3/16	5/8	12.790	12 3/4	1.900	1 7/8	9 1/2	2 5/8	1 1/4
x190	55.8	14.38	14 3/8	1.060	1 1/16	9/16	12.670	12 5/8	1.735	1 3/4	9 1/2	2 7/16	1 3/16
x170	50.0	14.03	14	0.960	15/16	1/2	12.570	12 5/8	1.560	1 9/16	9 1/2	2 1/4	1 1/8
x152	44.7	13.71	13 3/4	0.870	7/8	7/16	12.480	12 1/2	1.400	1 3/8	9 1/2	2 1/8	1 1/16
x136	39.9	13.41	13 3/8	0.790	13/16	7/16	12.400	12 3/8	1.250	1 1/4	9 1/2	1 15/16	1
x120	35.3	13.12	13 1/8	0.710	11/16	3/8	12.320	12 3/8	1.105	1 1/8	9 1/2	1 13/16	1
x106	31.2	12.89	12 7/8	0.610	5/8	5/16	12.220	12 1/4	0.990	1	9 1/2	1 11/16	15/16
x 96	28.2	12.71	12 3/4	0.550	9/16	5/16	12.160	12 1/8	0.900	7/8	9 1/2	1 5/8	7/8
x 87	25.6	12.53	12 1/2	0.515	1/2	1/4	12.125	12 1/8	0.810	13/16	9 1/2	1 1/2	7/8
x 79	23.2	12.38	12 3/8	0.470	1/2	1/4	12.080	12 1/8	0.735	3/4	9 1/2	1 7/16	7/8
x 72	21.1	12.25	12 1/4	0.430	7/16	1/4	12.040	12	0.670	11/16	9 1/2	1 3/8	7/8
x 65	19.1	12.12	12 1/8	0.390	3/8	3/16	12.000	12	0.605	5/8	9 1/2	1 5/16	13/16
W 12x 58	17.0	12.19	12 1/4	0.360	3/8	3/16	10.010	10	0.640	5/8	9 1/2	1 3/8	13/16
x 53	15.6	12.06	12	0.345	3/8	3/16	9.995	10	0.575	9/16	9 1/2	1 1/4	13/16
W 12x 50	14.7	12.19	12 1/4	0.370	3/8	3/16	8.080	8 1/8	0.640	5/8	9 1/2	1 3/8	13/16
x 45	13.2	12.06	12	0.335	5/16	3/16	8.045	8	0.575	9/16	9 1/2	1 1/4	13/16
x 40	11.8	11.94	12	0.295	5/16	3/16	8.005	8	0.515	1/2	9 1/2	1 1/4	3/4
W 12x 35	10.3	12.50	12 1/2	0.300	5/16	3/16	6.560	6 1/2	0.520	1/2	10 1/2	1	9/16
x 30	8.79	12.34	12 3/8	0.260	1/4	1/8	6.520	6 1/2	0.440	7/16	10 1/2	15/16	1/2
x 26	7.65	12.22	12 1/4	0.230	1/4	1/8	6.490	6 1/2	0.380	3/8	10 1/2	7/8	1/2
W 12x 22	6.48	12.31	12 1/4	0.260	1/4	1/8	4.030	4	0.425	7/16	10 1/2	7/8	1/2
x 19	5.57	12.16	12 1/8	0.235	1/4	1/8	4.005	4	0.350	3/8	10 1/2	13/16	1/2
x 16	4.71	11.99	12	0.220	1/4	1/8	3.990	4	0.265	1/4	10 1/2	3/4	1/2
x 14	4.16	11.91	11 7/8	0.200	3/16	1/8	3.970	4	0.225	1/4	10 1/2	11/16	1/2

AMERICAN INSTITUTE OF STEEL CONSTRUCTION

(a)

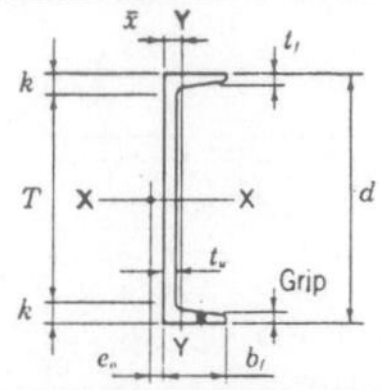

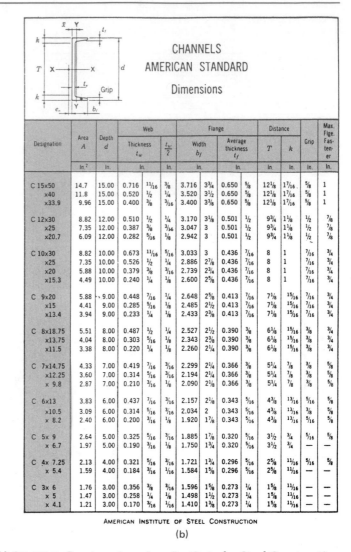

CHANNELS
AMERICAN STANDARD
Dimensions

Designation	Area A	Depth d	Web Thickness t_w		Web $\frac{t_w}{2}$	Flange Width b_f		Flange Average thickness t_f		Distance T	Distance k	Grip	Max. Flge. Fastener
	In.²	In.	In.		In.	In.		In.		In.	In.	In.	In.
C 15x50	14.7	15.00	0.716	11/16	3/8	3.716	3 3/4	0.650	5/8	12 1/8	1 7/16	5/8	1
x40	11.8	15.00	0.520	1/2	1/4	3.520	3 1/2	0.650	5/8	12 1/8	1 7/16	5/8	1
x33.9	9.96	15.00	0.400	3/8	3/16	3.400	3 3/8	0.650	5/8	12 1/8	1 7/16	5/8	1
C 12x30	8.82	12.00	0.510	1/2	1/4	3.170	3 1/8	0.501	1/2	9 3/4	1 1/8	1/2	7/8
x25	7.35	12.00	0.387	3/8	3/16	3.047	3	0.501	1/2	9 3/4	1 1/8	1/2	7/8
x20.7	6.09	12.00	0.282	5/16	1/8	2.942	3	0.501	1/2	9 3/4	1 1/8	1/2	7/8
C 10x30	8.82	10.00	0.673	11/16	5/16	3.033	3	0.436	7/16	8	1	7/16	3/4
x25	7.35	10.00	0.526	1/2	1/4	2.886	2 7/8	0.436	7/16	8	1	7/16	3/4
x20	5.88	10.00	0.379	3/8	3/16	2.739	2 3/4	0.436	7/16	8	1	7/16	3/4
x15.3	4.49	10.00	0.240	1/4	1/8	2.600	2 5/8	0.436	7/16	8	1	7/16	3/4
C 9x20	5.88	9.00	0.448	7/16	1/4	2.648	2 5/8	0.413	7/16	7 1/8	15/16	7/16	3/4
x15	4.41	9.00	0.285	5/16	1/8	2.485	2 1/2	0.413	7/16	7 1/8	15/16	7/16	3/4
x13.4	3.94	9.00	0.233	1/4	1/8	2.433	2 3/8	0.413	7/16	7 1/8	15/16	7/16	3/4
C 8x18.75	5.51	8.00	0.487	1/2	1/4	2.527	2 1/2	0.390	3/8	6 1/8	15/16	3/8	3/4
x13.75	4.04	8.00	0.303	5/16	1/8	2.343	2 3/8	0.390	3/8	6 1/8	15/16	3/8	3/4
x11.5	3.38	8.00	0.220	1/4	1/8	2.260	2 1/4	0.390	3/8	6 1/8	15/16	3/8	3/4
C 7x14.75	4.33	7.00	0.419	7/16	3/16	2.299	2 1/4	0.366	3/8	5 1/4	7/8	3/8	5/8
x12.25	3.60	7.00	0.314	5/16	3/16	2.194	2 1/4	0.366	3/8	5 1/4	7/8	3/8	5/8
x 9.8	2.87	7.00	0.210	3/16	1/8	2.090	2 1/8	0.366	3/8	5 1/4	7/8	3/8	5/8
C 6x13	3.83	6.00	0.437	7/16	3/16	2.157	2 1/8	0.343	5/16	4 3/8	13/16	5/16	5/8
x10.5	3.09	6.00	0.314	5/16	3/16	2.034	2	0.343	5/16	4 3/8	13/16	3/8	5/8
x 8.2	2.40	6.00	0.200	3/16	1/8	1.920	1 7/8	0.343	5/16	4 3/8	13/16	5/16	5/8
C 5x 9	2.64	5.00	0.325	5/16	3/16	1.885	1 7/8	0.320	5/16	3 1/2	3/4	5/16	5/8
x 6.7	1.97	5.00	0.190	3/16	1/8	1.750	1 3/4	0.320	5/16	3 1/2	3/4	—	—
C 4x 7.25	2.13	4.00	0.321	5/16	3/16	1.721	1 3/4	0.296	5/16	2 5/8	11/16	5/16	5/8
x 5.4	1.59	4.00	0.184	3/16	1/16	1.584	1 5/8	0.296	5/16	2 5/8	11/16	—	—
C 3x 6	1.76	3.00	0.356	3/8	3/16	1.596	1 5/8	0.273	1/4	1 5/8	11/16	—	—
x 5	1.47	3.00	0.258	1/4	1/8	1.498	1 1/2	0.273	1/4	1 5/8	11/16	—	—
x 4.1	1.21	3.00	0.170	3/16	1/16	1.410	1 3/8	0.273	1/4	1 5/8	11/16	—	—

AMERICAN INSTITUTE OF STEEL CONSTRUCTION

(b)

Figure 21–45 Dimensional information for **(a)** W12×22, and **(b)** C6×10.5. *Courtesy American Institute for Steel Construction, Manual of Steel Construction.* Additional samples of structural steel dimensions may be found in Appendix H.

members are used as columns to support loads from other members. Tubes are specified by the size of the outer wall, followed by the thickness of the wall.

A variety of templates are available for structural drafting to assist in drawing steel shapes. Many CADD programs are available also to increase structural drafting productivity.

Structural Steel Callouts

Many structural steel materials are specified by shape designation, flange width, and weight in pounds per linear foot. For example:

W 24 × 120

The structural steel shapes that fall into this category are as follows:

W — wide flange shapes
S — American standard beams
M — miscellaneous beam and column shapes
C — American standard channels
MC — miscellaneous channel shapes
WT — structural tees cut from W shapes
ST — structural tees cut from S shapes
MT — structural tees cut from M shapes
T — structural tees
Z — zee shapes
HP — steel H piling

Structural materials that are specified by shape designation, type, diameter or outside dimension, and wall thickness are in the following table. (NOTE: If length dimensions are required they are given at the end of the callout in feet and inches except for plates, which should only be given in inches.)

Designation	Shape	Example	Meaning
P	plate	P 1/2 × 6 × 8	THK × WIDTH × LENGTH
BAR	square bar	BAR 1-1/4□	1-1/4" WIDTH
	round bar	BAR 1-1/4∅	1-1/4"DIAMETER
	flat bar	BAR 2 × 3/8	WIDTH × THICKNESS
PIPE	pipe	PIPE 3 ∅ STD	OD (OUTSIDE DIA) SPEC
TS	structural tubing		
	square	TS 4 × 4 × .250	OUTWIDTH × OUTWIDTH WALL THICKNESS
x	rectangular	TS 6 × 3 × .375	SAME
	round	TS 4 OD × .188	OD × WALL THICKNESS
L	angle		
	unequal leg	L3 × 2 × 7/16	LEG × LEG × THICKNESS
	equal leg	L3 × 3 × 1/2	LEG × LEG × THICKNESS

When plate material is to be bent, the minimum bend radius should be given with the plate callout, and the length of the bend legs are dimensioned on the drawing. For example:

3/8 × 10 W/MIN BEND R 5/8"

Location dimensions for structural steel components should be as shown in Figure 21–46.

▼ COMMON CONNECTION METHODS

Bolting

ASTM Bolts are used for many connections in lumber and steel construction. A bolt specification includes the diameter, length, and strength of the bolt. Washers or plates are specified so the bolt head will not pull through the hole made for the bolt. Bolt strength is classified in accordance with the American Society for Testing Materials (ASTM) specifications. Refer to Chapter 11, Fasteners and Springs, for more information.

Standard Structural Bolt Callouts

When specifying bolts on a drawing, give the quantity, diameter, bolt type if special, length in inches, and ASTM specification. If special washer and nut requirements are present, this should also be specified in the bolt callout. Hexagon head bolts and hexagon nuts are assumed unless either is specified differently in the callout. Examples of bolt callouts are:

> 2-3/4"∅ BOLTS ASTM A503
> 4-1/2"∅ × 10" BOLTS
> 2-5/8"∅ × 6" CARRIAGE BOLTS
> 6-3/4"∅ BOLTS W/MALLEABLE IRON WASHERS
>   AND HEAVY HEX NUTS
> 4-1/2"∅ GALVANIZED BOLTS

Give the hole diameter when holes for standard bolts must be specified on the drawing. In general, holes

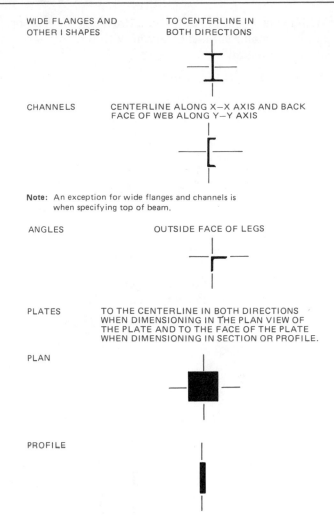

Figure 21–46 Location dimensions for structural components.

should be 1/16" larger in diameter than the specified bolt for standard steel-to-steel, wood-to-wood, or wood-to-steel construction, and 1/8" larger in diameter than the specified bolt for standard steel-to-concrete or wood-to-concrete applications unless otherwise specified by the engineer. For example:

2-3/4"∅ BOLTS FIELD DRILL 13/16"∅ HOLES.

The following are recommended standard hole diameters in inches for given bolt sizes:

Bolt ∅	Standard Hole ∅	Concrete Hole ∅	Oversize Hole ∅
1/2	9/16	5/8	11/16
5/8	11/16	3/4	13/16
3/4	13/16	7/8	15/16
7/8	15/16	1	1 1/16
1	1 1/16	1 1/8	1 1/4
1 1/8	1 3/16	1 1/4	1 7/16
1 1/4	1 5/16	1 3/8	1 9/16
1 3/8	1 7/16	1 1/2	1 11/16
1 1/2	1 9/16	1 5/8	1 13/16

When specifying lag bolts on a drawing, the lead or tap hole diameter should be given with the bolt specification. For example:

2-3/4"∅ × 8 LAG BOLTS W/7/16"∅ LEAD HOLES.

The following are tap hole diameters in inches for lag bolts used in Douglas fir, larch, or Southern pine:

Bolt ∅	Lead Hole ∅
3/8	1/4
1/2	5/16
5/8	7/16
3/4	1/2
7/8	5/8
1	3/4

Bolts on a drawing are located to their centerlines. When counterbores are required, provide the specification to the location of the counterbore. For example:

∅2-3/4" BOLT ∅3-1/4" CBORE × 7/8" DEEP

ASME Y14.5M—1994 symbols for counterbore and depth may also be used (see Chapter 10).

Nails

Nails are used for the fabrication of wood-to-wood members are are sized by the term *penny* and denoted by the letter *d*. Penny is weight classification for nails—the number of pounds per 1000 nails. So, one thousand 16d nails weight 16 pounds. Nails are also sized by diameter when over 60d. Standard nail sizes (2d to 20d) and nail types are shown in Figure 21–47. When specifying nails, the penny weight should be given plus the quantity and spacing if required. The specification for special nails should be given when required. Nailing callout examples include:

4-16d NAILS
5-20d GALV NAILS EA SIDE
10d NAILS 4" OC AT SEAMS
  AND 12" OC IN FIELD

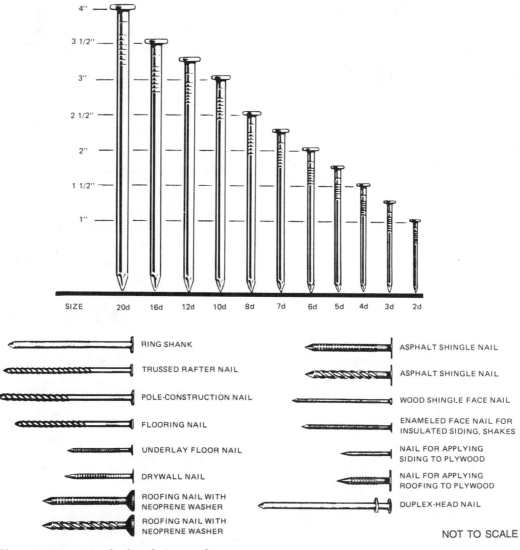

Figure 21–47 Standard nail sizes and types.

30d NAILS ALTERNATELY STAGGERED 12" OC
TOP AND BOTTOM BOTH SIDES
8d RINGSHANK NAILS 6" OC

If pilot holes are required for nailing, the diameter of the pilot hole should be given after the nail callout. For example:

4-20d NAILS W/5/32"∅ HOLES

Verify pilot hole diamaters with manufacturers' recommendations for types of woods and applications.

Welding

Welds are classified according to the type of joint on which they are used. The four common welds used in construction are the fillet, back, plug or slot, and groove welds. A welding symbol is used to designate the type and dimensional specifications of the weld. Review Chapter 12 for an in-depth coverage of welding symbols before doing the drawings in this chapter.

Fabrication Methods

There are an unlimited number of fabrication methods. As a drafter you must become familiar with the fabrications used. Many techniques are typical while others require special design. The drafter either draws from engineering sketches or refers to previously drawn examples. Once a certain amount of experience has been gained, the drafter is able to establish drawings from either written or verbal instructions, or from a particular given situation.

AISC The AISC *Manual of Steel Construction* provides a number of common connection details. Several fabrication details given without specifications or dimensions are shown in Appendix H.

▼ COMPONENTS IN A SET OF STRUCTURAL DRAWINGS

A complete set of structural drawings may have a number of sheets with specific elements of the building provided in a general format or with specific construction elements shown in cross sections or clear details. While not all complete sets of structural drawings contain the same number of sheets or type of information, some of the representative drawings include: a floor plan; a foundation plan and details; a concrete slab plan and details; a roof framing plan and details; a roof drainage plan; building section(s); exterior elevations; a panel plan, elevations, and wall details.

Floor Plan

The *floor plan* is generally designed and drawn by the architect. This floor plan drawing is then used as the layout for the other associated drawing from consulting engineers—for example, the mechanical (HVAC), plumbing, electrical, and structural engineers. In some situations and depending on the structure, the structural engineering firm draws the floor plan as part of the set of drawings. (See Figure 21–48.)

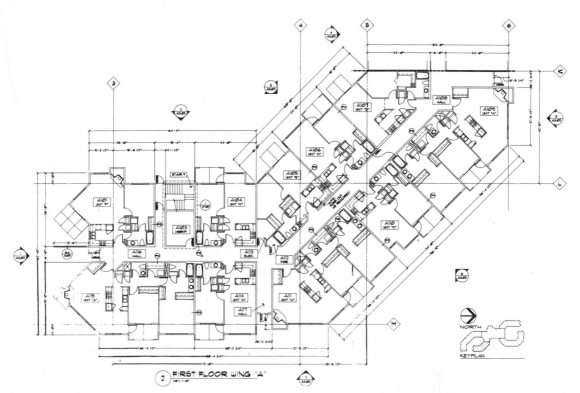

Figure 21–48 CADD floor plan for a set of construction drawings. *Courtesy Soderstrom Architects, PC.*

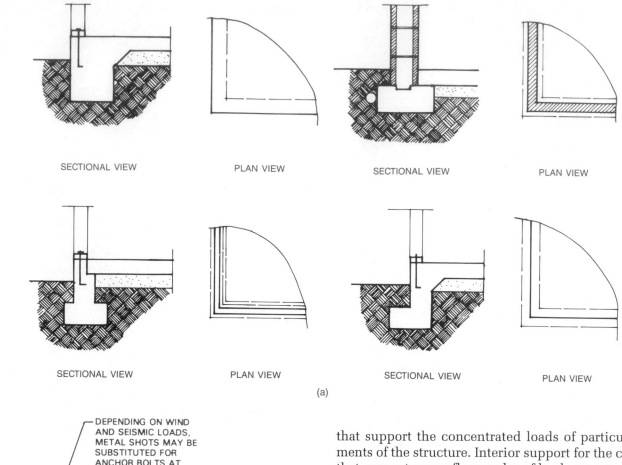

(a)

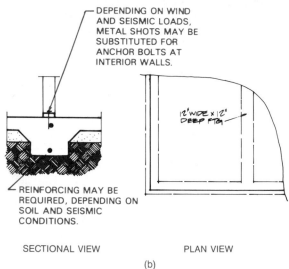

(b)

Figure 21–49 **(a)** Common foundation and slab intersections; **(b)** footing and floor intersections at interior load-bearing wall.

Foundation Plan and Details

The purpose of the *foundation plan* is to show the supporting system for the walls, floor, and roof. The foundation may consist of continuous perimeter footings and walls which support the exterior and main bearing walls of the building. The footings in these locations are rectangular in shape and are continuous concrete supports centered at exterior and interior bearing walls. (See Figure 21–49.) There are also pedestal footings

that support the concentrated loads of particular elements of the structure. Interior support for the columns that support upper floor and roof loads are provided by pedestal footings, as shown in Figure 21–50. Anchor bolts and other metal connectors are shown and located on the foundation plan, as shown in Figure 21–51. A typical foundation plan is shown in Figure 21–52.

The foundation details are drawn to provide information on the concrete foundation at the perimeter walls, and the foundation and pedestals at the center support columns. Possible details include retaining walls, typical exterior foundation and rebar schedule, typical interior bearing wall, and typical pedestal and rebar schedule. Detail drawings are keyed to the plan view using detail markers. Detail markers are usually drawn as a circle of about 3/8" to 1/2" in diameter on the plan view and a coordinating circle of about 3/4" in diameter under the associated detail. Each detail marker is divided in half. The top half contains the detail number and the bottom identifies the page number on which the detail is drawn. Notice the detail markers associated with the example drawings in this chapter. The rebar schedules are charts placed adjacent to the detail that key information about the rebar used to the drawing, as shown in Figure 21–50.

Concrete Slab Plan and Details

The *concrete slab plan* is drawn to outline the concrete that will become the floor(s). Items often found in slab plans include floor slabs, slab reinforcing, expansion

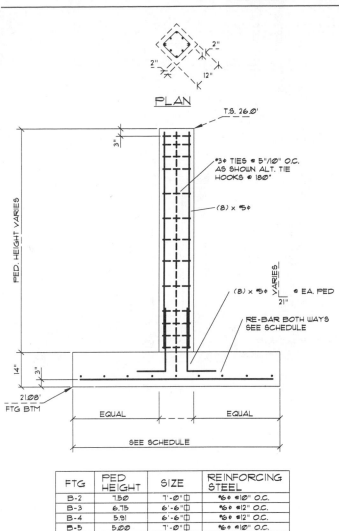

Figure 21–50   Pedestal footing detail.

FTG	PED HEIGHT	SIZE	REINFORCING STEEL
B-2	7.50	7'-0" □	#6 ⌀ @10" O.C.
B-3	6.75	6'-6" □	#6 ⌀ @12" O.C.
B-4	5.91	6'-6" □	#6 ⌀ @12" O.C.
B-5	5.00	7'-0" □	#6 ⌀ @10" O.C.

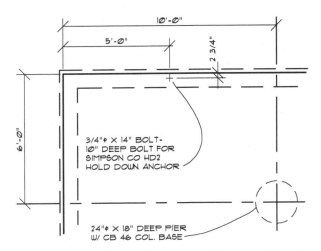

Figure 21–51   Foundation plan showing location of metal connectors.

joints, pedestal footings, metal connectors, anchor bolts, and any foundation cuts for doors or other openings. The openings and other items are located and labeled on the plan, as shown in Figure 21–53. In situations where

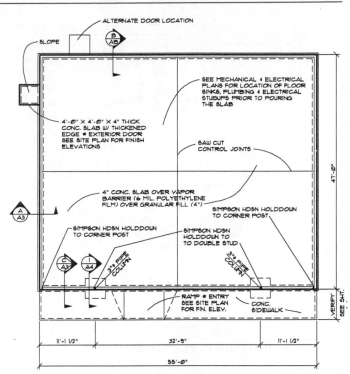

Figure 21–52   A concrete slab foundation is commonly used for commercial structures. *Courtesy The Southland Corporation.*

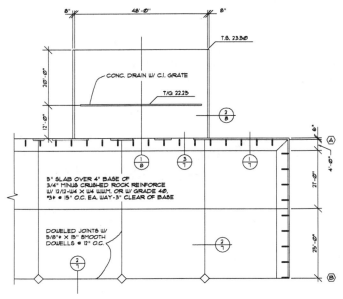

Figure 21–53   A slab on grade plan shows the size and location of all concrete pours plus reinforcing specifications. *Courtesy Structureform Masters, Inc.*

tilt-up construction is used, the opening locations are dimensioned on the elevations.

The concrete slab details are drawn to provide information of the intersections of the concrete slabs. These details include interior and perimeter slab joints as in Figure 21–54, and other slab details. The drawings include such information as slab thicknesses, reinforcing sizes and locations, and slab elevations (heights), as shown in Figure 21–55.

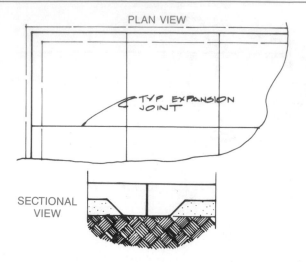

Figure 21–54 Control joints are often placed in large slabs to resist cracking and allow the construction crews manageable areas to pour.

Roof Framing Plan and Details

The purpose of the *roof framing plan* is to show the major structural components in plan view which occur at the roof level. A roof framing plan using truss framing is shown in Figure 21–3b (see page 672). A panelized roof framing plan is shown in Figures 21–3b and 21–29 (see page 683).

Roof framing details are required to show the construction methods used at various member intersections in the building. Details may include the following intersections: wall to beam, beam to column, beam splices and connections, cantilever locations, and roof drains. (See Figure 21–56.)

Roof Drainage Plan

The *roof drainage plan* may be part of a set of structural drawings for some buildings although it may be considered part of the plumbing or piping drawings depending on the particular company's use and interpretation. The purpose of this drawing is to show the elevations of the roof and provide for adequate water drainage generally associated with low slope and flat roofs. (See Figure 21–57.) Some of the terminology associated with roof drainage plans includes: *roof drain* (RD)—a screened opening to allow for drainage; *overflow drain* (OD)—a back-up in case the roof drains fail; *down spout* (DS)—usually a vertical pipe used to transport water from the roof; scupper or *gutter*—a water collector usually on the outside of a wall at the roof level to funnel water from the roof drains to the down spouts.

Building Sections

The *building section* is used to show the relationship between the plans and details previously drawn. This drawing is considered a general arrangement or construction reference, as it is often drawn at a small scale.

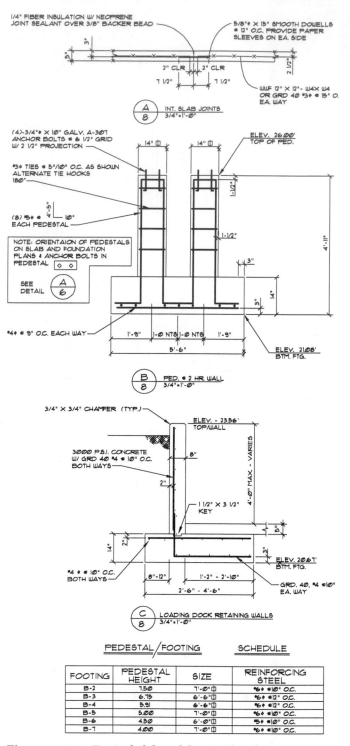

Figure 21–55 Typical slab and footing details.

While some detailed information is provided with regard to building elevations (heights) and general dimensions, the overall section is not intended to provide explanation of building materials. The details more clearly serve this purpose, as they are drawn at a larger scale. Some less complex buildings, however, may show a great deal of construction information on the overall section and use fewer details. (See Figure 21–58, page 700.)

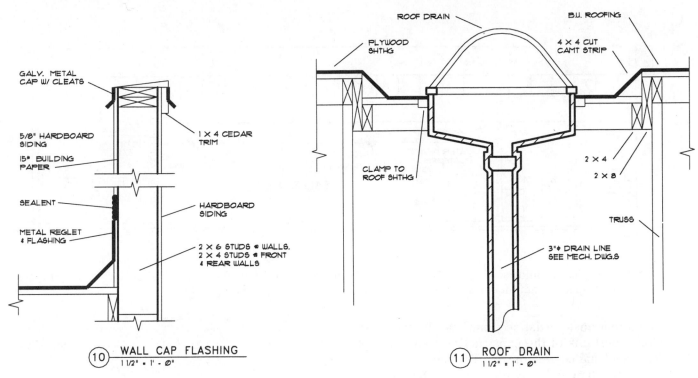

GALV. METAL
CAP W/ CLEATS

5/8" HARDBOARD
SIDING

15# BUILDING
PAPER

SEALENT

METAL REGLET
& FLASHING

1 × 4 CEDAR
TRIM

HARDBOARD
SIDING

2 × 6 STUDS @ WALLS.
2 × 4 STUDS @ FRONT
& REAR WALLS

(10) **WALL CAP FLASHING**
        1 1/2" = 1' - 0"

ROOF DRAIN

PLYWOOD
SHTHG

CLAMP TO
ROOF SHTHG

B.U. ROOFING

4 × 4 CUT
CAMT STRIP

2 × 4

2 × 8

TRUSS

3"∅ DRAIN LINE
SEE MECH. DWG.S

(11) **ROOF DRAIN**
        1 1/2" = 1' - 0"

Figure 21–56   Roof construction details. *Courtesy The Southland Corporation.*

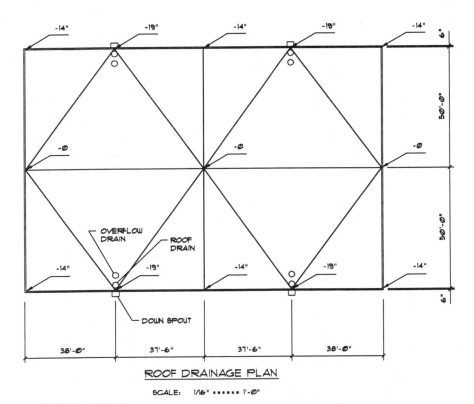

ROOF DRAINAGE PLAN

SCALE: 1/16" •••••• 1'-0"

GENERAL NOTES:

1. ROOF AND OVERFLOW DRAINS SHALL BE LARGE GENERAL PURPOSE TYPE W/
   NON-FERROUS DOMES AND 4"∅ OUTLETS
2. OVERFLOW DRAINS SHALL BE SET W/ INLET 2" ABOVE ROOF DRAIN INLET AND
   SHALL BE CONNECTED TO DRAINS LINES INDEPENDENT FROM ROOF DRAINS.
3. USE A 4"H × 7W SCUPPER W/5" NOMINAL (3 3/4 × 5) RECTANGULAR CORRUGATED
   DOWNSPOUT.
   PROVIDE A 6" × 9" CONDUCTOR HEAD @ TOP OF DOWNSPOUT.

Figure 21–57   Roof drainage plan.

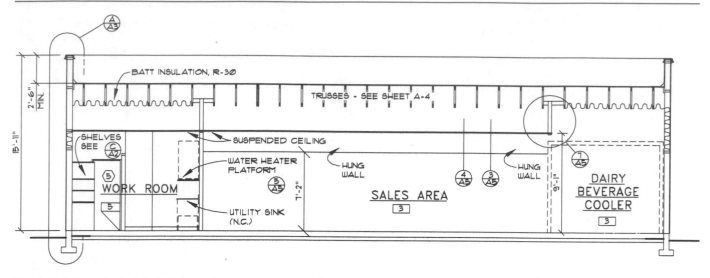

Figure 21–58 Typical building sections for commercial construction are often drawn at a small scale to show major types of construction with specific information shown in details. *Courtesy The Southland Corporation.*

In some situations, partial sections may be useful in describing portions of the construction that may not be effectively handled with the building section and may be larger areas than normally identified with a detailed section, as shown in Figure 21–59.

Exterior Elevations

The *exterior elevations* are drawings that show the external appearance of the building. An elevation is drawn at each side of the building to show the relationship of the building to the final grade, location of openings, wall heights, roof slopes, exterior building materials, and other exterior features. A front elevation is generally the main entry view and is drawn at a 1/4"= 1'–0" scale, depending on the size of the structure. Other elevations may be drawn at a smaller scale. The elevation's scale depends upon the size of the building, the amount of detail shown, and the sheet size used. Many companies prefer to drawn all elevations at the same scale showing an equal amount of detail in all views. An elevation may be omitted if it is the same as another. When this happens, the elevation may be labeled as RIGHT AND LEFT ELEVATION. Elevations are also often labeled by compass orientation, such as SOUTH ELEVATION. The elevations may be drawn showing a great deal of detail, as shown in Figure 21–60. Often, elevations for commercial buildings may be drawn at a small scale representing very little detail, as shown in Figure 21–61 (see page 702). In many situations the drafter may be required to draw interior elevations or details such as the interior finish elevation detail for Farrell's, as shown in Figure 21–62 (see page 702).

Panel Plan, Elevations, and Wall Details

The *panel plan* is used in tilt-up construction to show the location of the panels, as shown in Figure 21–17 (see page 677).

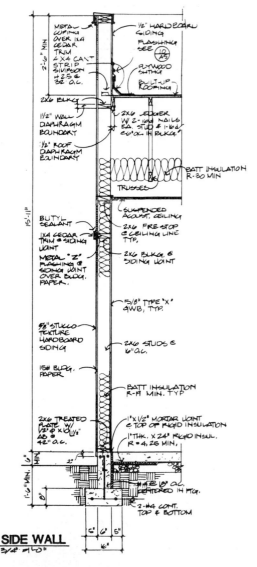

Figure 21–59 Partial sections are used to clarify construction information through various portions of the structure. *Courtesy The Southland Corporation.*

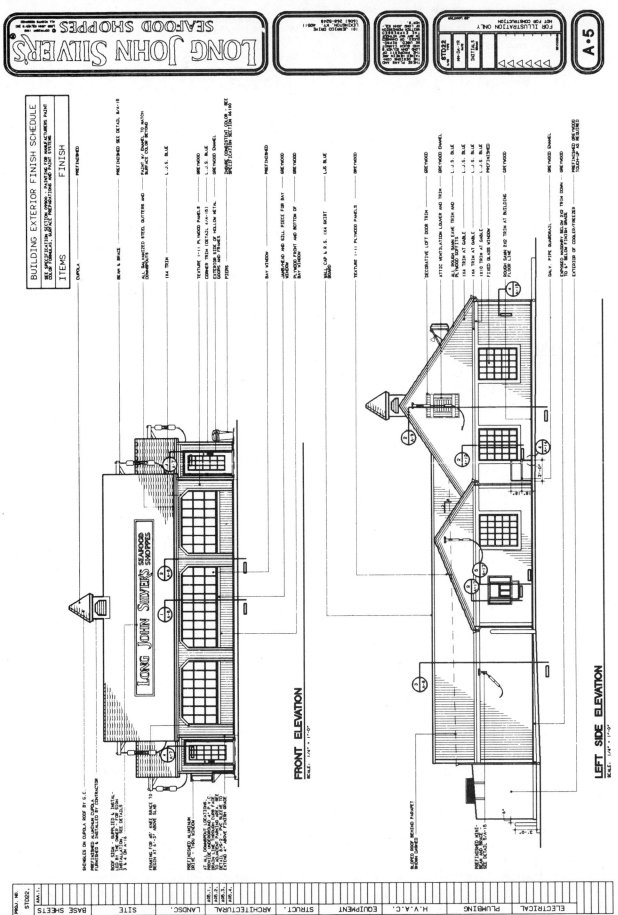

Figure 21–60 CADD elevations of Long John Silver's Seafood Shoppes. *Courtesy Jerrico, Inc.*

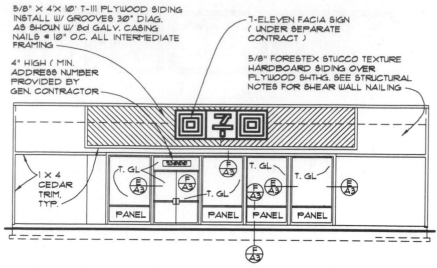

5/8" X 4'X 10' T-III PLYWOOD SIDING
INSTALL W/ GROOVES 30° DIAG.
AS SHOWN W/ 8d GALV. CASING
NAILS @ 10" O.C. ALL INTERMEDIATE
FRAMING

4" HIGH (MIN.
ADDRESS NUMBER
PROVIDED BY
GEN. CONTRACTOR

7-ELEVEN FACIA SIGN
(UNDER SEPARATE
CONTRACT)

5/8" FORESTEX STUCCO TEXTURE
HARDBOARD SIDING OVER
PLYWOOD SHTHG. SEE STRUCTURAL
NOTES FOR SHEAR WALL NAILING

1 X 4
CEDAR
TRIM,
TYP.

Figure 21–61 Elevations for commercial structures, such as this 7-11 store, often require little detail and may be drawn at a small scale. *Courtesy The Southland Corporation.*

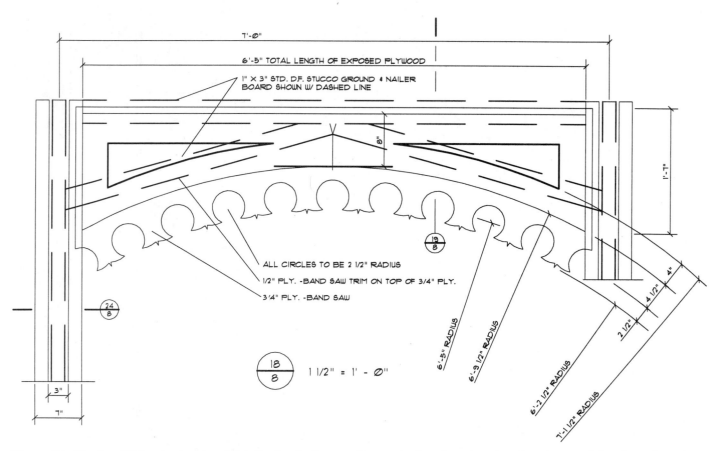

7'-0"

6'-5" TOTAL LENGTH OF EXPOSED PLYWOOD

1" X 3" STD. D.F. STUCCO GROUND & NAILER
BOARD SHOWN W/ DASHED LINE

8"

1'-7"

ALL CIRCLES TO BE 2 1/2" RADIUS

1/2" PLY. -BAND SAW TRIM ON TOP OF 3/4" PLY.

3/4" PLY. -BAND SAW

6'-5" RADIUS

6'-9 1/2" RADIUS

6'-2 1/2" RADIUS

7'-1 1/2" RADIUS

2 1/2"

4 1/2"

4"

3"

7"

1 1/2" = 1' - 0"

Figure 21–62 In addition to structural details, the drafter may be required to draw interior elevations and details such as this finish detail for Farrell's Restaurant. *Courtesy Structureform Masters, Inc.*

Panel elevations are used when the exterior elevations do not clearly show information about items located on or in the walls. Similarly to exterior elevations the panel elevations will show door locations and reinforcing within walls and around openings. Dimensions associated with panel elevations provide both horizontal and vertical dimensions for openings and other features. (See Figure 21–18, page 677.)

Wall details are used to show the connection points of the concrete panels used in tilt-up construction, and connection details at the walls for other types of structures. (See Figure 21–63.)

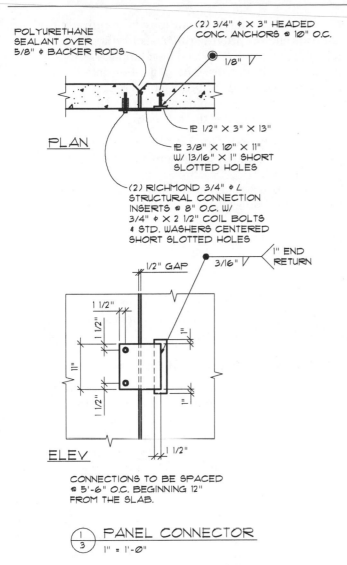

POLYURETHANE
SEALANT OVER
5/8" ⌀ BACKER RODS

(2) 3/4" ⌀ X 3" HEADED
CONC. ANCHORS @ 10" O.C.

1/8"

PLAN

⅊ 1/2" X 3" X 13"

⅊ 3/8" X 10" X 11"
W/ 13/16" X 1" SHORT
SLOTTED HOLES

(2) RICHMOND 3/4" ⌀ L
STRUCTURAL CONNECTION
INSERTS @ 8" O.C. W/
3/4" ⌀ X 2 1/2" COIL BOLTS
⅊ STD. WASHERS CENTERED
SHORT SLOTTED HOLES

1/2" GAP   3/16"   1" END
                         RETURN

1 1/2"

1 1/2"

1"

1 1/2"

ELEV   1 1/2"

CONNECTIONS TO BE SPACED
@ 5'-6" O.C. BEGINNING 12"
FROM THE SLAB.

1/3 PANEL CONNECTOR
     1" = 1'-0"

Figure 21–63 A typical wall connection detail.

▼ PLOT PLAN REQUIREMENTS

A *plot plan,* also known as a lot plan, is a map of a piece of land that may be used for any number of purposes. Plot plans may show a proposed construction site for a specific property. Plots may show topography with contour lines or the numerical value of land elevations may be given at certain locations. Plot plans are also used to show how a construction site will be excavated and are then known as a grading plan. While plot plans can be drawn to serve any number of required functions, they all have some similar characteristics which include showing the following:

○ A legal description of the property based on a survey.
○ Property line bearings and directions.
○ North direction.
○ Roads and easements.
○ Utilities.
○ Elevations.
○ Map scale.

Topography

Topography is a physical description of land surface showing its variation in elevation, known as relief, and locating other features. Surface relief can be shown with graphic symbols that use shading methods to accentuate the character of land or the differences in elevations can be shown with contour lines. Plot plans that require surface relief identification generally use *contour lines.* These lines connect points of equal elevation and help show the general lay of the land.

A good way to visualize the meaning of contour lines is to look at a lake or ocean shore line. When the water is high during the winter or at high tide, a high-water line establishes a contour at that level. As the water recedes during the summer or at low tide, a new lower level line is obtained. This new lines represents another contour. The high-water lines goes all around the lake at one level and the low-water lines goes all around the lake at another level. These two lines represent contours, or lines of equal elevation. The vertical distance between contour lines is known as contour interval. When the contour lines are far apart, the contour interval shows relatively flat or gently sloping land. When the contour lines are close together, then the contour interval shows a land that is much steeper. Drawn contour lines are broken periodically and the numerical value of the contour elevation above sea level is inserted. Figure 21–64 shows sample contour lines. Figure 21–65 shows a graphic example of land relief in pictorial form and contour lines of the same area.

Plot plans do not always require contour lines showing topography. Verify the requirements with the local building codes. In most instances the only contour-related information required is property corner elevations, street elevation at a driveway, and the elevation of the finished floor levels of the structure. Additionally, slope may be defined and labeled with an arrow.

Elements of a Plot Plan

Plot plan requirements vary from one local jurisdiction to the next although there are elements of plot

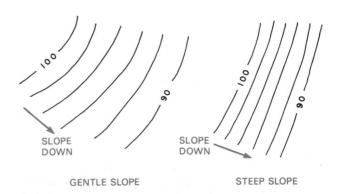

Figure 21–64 Contour lines showing both gentle and steep slopes.

Figure 21–65 Land relief pictorial and contour lines. *Courtesy U.S. Department of Interior, Geological Survey.*

plans that are similar around the country. Guidelines for plot plans can be obtained from the local building official or building permit department. Some agencies, for example, require that the plot plan be drawn on specific size paper such as 8-1/2" × 14". Typical plot plan items include the following:

- Plot plan scale.
- Legal description of the property.
- Property line bearings and dimensions.
- North direction.
- Existing and proposed roads.
- Driveways, patios, walks, and parking areas.
- Existing and proposed structures.
- Public or private water supply.
- Public or private sewage disposal.
- Location of utilities.
- Rain and footing drains, and storm sewers or drainage.
- Topography including contour lines or land elevations at lot corners, street centerline, driveways, and floor elevations.
- Setbacks, front, rear, and sides.
- Specific items on adjacent properties may be required.
- Existing and proposed trees may be required.

Site Analysis Plan

Local requirements for subdivisions should be confirmed as there are a variety of procedures depending upon local guidelines and zoning rules. In areas where zoning and building permit applications require a design

review, a site analysis plan may be required. The site analysis should provide the basis for the proper design relationship of the proposed development to the site and to adjacent properties. The degree of detail of the site analysis is generally appropriate to the scale of the proposed project. A site analysis plan, as shown in Figure 21–66, often includes the following:

- A vicinity map showing the location of the property in relationship to adjacent properties, roads, and utilities.
- Site features such as existing structures and plants on the property and adjacent property.
- The scale.
- North direction.
- Property boundaries.
- Slope shown by contour lines, cross sections, or both.
- Plan legend.
- Traffic patterns.
- Solar site information if solar application is intended.
- Pedestrian patterns.

Planned Unit Development

A creative and flexible approach to land development is a planned unit development. Planned unit developments may include such uses as residential areas, recreational areas, open spaces, schools, libraries, churches, or convenient shopping facilities. Developers involved in these projects must pay particular attention to the impact on local existing developments. Generally the *plats* (site plans) for these developments must include all of the same information shown on a subdivision plat, plus:

1. A detailed vicinity map, as shown in Figure 21–67.
2. Land use summary.
3. Symbol legend.
4. Special spaces such as recreational and open spaces, or other unique characteristics.

Figure 21–68 (see page 706) shows a typical planned unit development plan. These plans, as in any proposed plat plan, may require changes before the final drawings are approved for development.

There are several specific applications for a plot or plat plan. The applied purpose of each will be different, although the characteristics of each type of plot plan may be similar. Local districts have guidelines for the type of plot plan required. Be sure to evaluate local guidelines before preparing a plot plan for a specific purpose. Prepare the plot plan in strict accordance with the requirements in order to receive acceptance.

A variety of plot plan-related templates are available that can help make the preparation of these plans a little easier.

Plot Plans

The plot or site plan for a commercial project resembles the plot plan for a residential project except for its

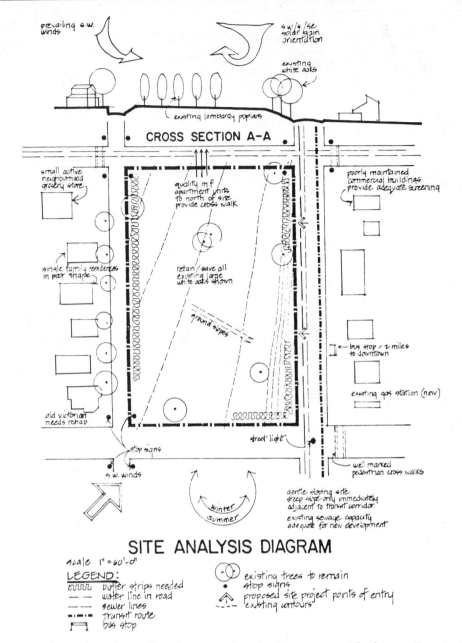

SITE ANALYSIS DIAGRAM

scale 1" = 60'-0"

LEGEND:

~~	buffer strips needed
— · —	water line in road
— — —	sewer lines
— · · — ·	transit route
🚏	bus stop

⊙⊘	existing trees to remain
•	stop signs
⟰	proposed site project points of entry
- - - -	existing contours

Figure 21–66 Site analysis plan. *Courtesy Planning Department, Clackamas County, OR.*

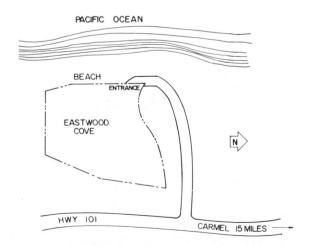

Figure 21–67 Vicinity map.

size. One major difference is the need to plan and draw parking spaces, driveways, curbs, and walkways. In commercial projects, the plot plan is often drawn as part of the preliminary design study. Parking spaces are determined by the square footage of the building and the type of usage the building will receive. Parking requirements often dictate the layout of the floor plan. Figure 21–69 shows an example of a plot plan for an office building.

Grading Plan

Once basic information is placed on the plot plan to describe the structure, the plan is usually given to a civil engineer or licensed surveyor so a grading plan can be prepared. A drafter working for a civil engineer then

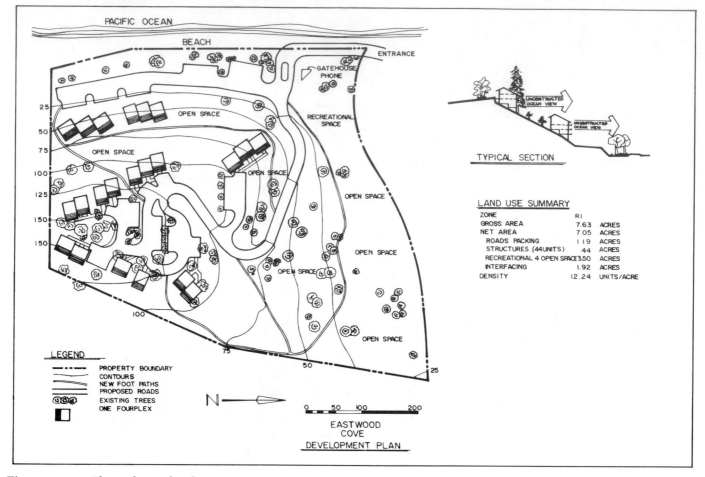

Figure 21–68 Planned unit development plan.

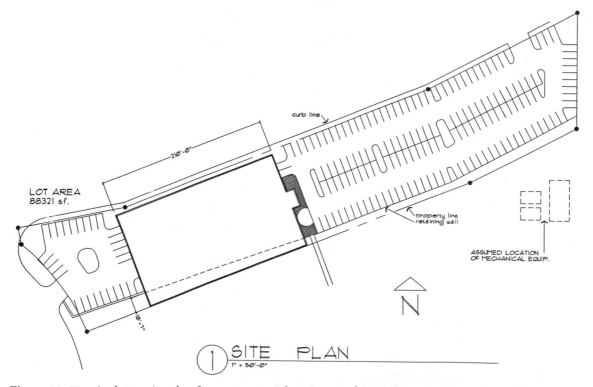

Figure 21–69 A plot or site plan for a commercial project. Parking information is a major addition to a commercial site plan. *Courtesy Soderstrom Architects, PC.*

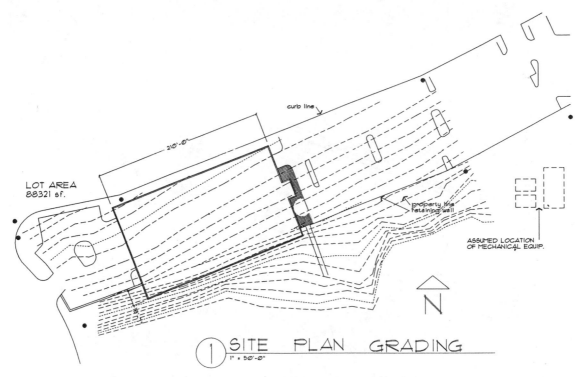

Figure 21–70 Grading information is often placed on an overlay of the plot plan to show cut and fill requirements of a job site. *Courtesy Soderstrom Architects, PC.*

translates field notes into a grading plan as seen in Figure 21–70. Although titled a grading plan, the plan typically contains much more than just grading elevations. Drainage information is often placed on this plan as it relates to the structure, as well as walks and paving areas.

Landscape Plan

The size of the project determines how the landscape plan is drawn. On a small project, a drafter may be given the job of drawing the landscape plan. When this is the case, the drafter must determine suitable plants for the area and the job site, and then specify the plants by their proper Latin name on the plan. On typical commercial projects, the plot plan is given to a landscape architect who specifies the size, type, and location of plantings. A method of maintaining the plantings is also typically shown on the plan as seen in Figure 21–71.

Laying Out the Plan Views

The structural plan views are normally the drawings that begin the complete set. These include the types of drawings already discussed, such as the floor plan, foundation plan, and roof framing plan. After the plan views have been drawn, the sections and details may be properly correlated.

The Floor Plan, Step 1. Begin by laying out the floor plan at a scale that will provide the clearest representation on the sheet size to be used. Normally a scale range of 1/8" = 1'–0" to 1/4" = 1'–0" is used. Draw the outline of the building, leaving adequate space for dimensions and notes, as shown in Figure 21–72.

The Floor Plan, Step 2. Draw all internal walls, partitions, posts, and fixtures. Draw all dimensions. Structural drafting normally uses aligned dimensioning with the dimension numeral placed above the dimension line, and slashes placed where the dimension and extension lines meet, as shown in Figure 21–73. Finally, notes, symbols, and titles are added to complete the drawing.

The Foundation Plan, Step 1. After the floor plan has been drawn, turn off or freeze the layers that are not used to lay out the foundation such as partitions and fixtures. Leave the perimeter wall, bearing wall, and post layers turned on. The floor plan makes an excellent guide for the other plans because it is the base for all other construction.

The Foundation Plan, Step 2. Begin by drawing the foundation walls and footings. Next, add all dimensions, notes, and symbols. Finally, letter the title and scale, as shown in Figure 21–74 (see page 709).

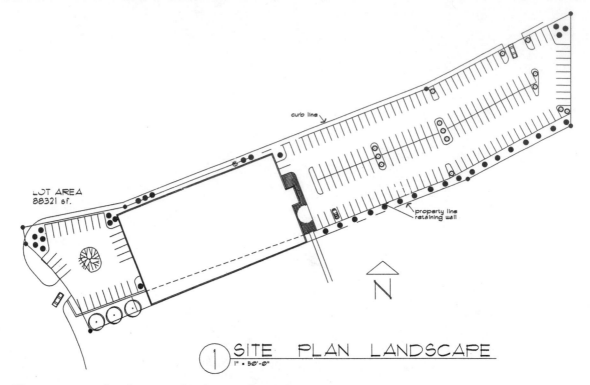

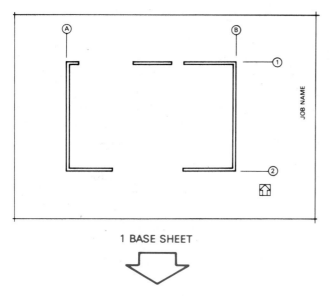

Figure 21–71 The planting or landscape plan is drawn as an overlay to the plot plan to show how the site will be planted and maintained. *Courtesy Soderstrom Architects, PC.*

1 BASE SHEET

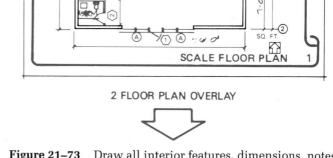

2 FLOOR PLAN OVERLAY

Figure 21–72 Outline the floor plan. *Courtesy Structureform Masters, Inc.*

Figure 21–73 Draw all interior features, dimensions, notes, and titles to complete the floor plan. *Courtesy Structureform Masters, Inc.*

The Roof Framing Plan. The layout of the roof framing plan works in much the same way as the foundation plan. Again, use the floor plan as a basis for the roof plan drawing. This gives you all of the layout features to quickly begin the new drawing. First, place all beams and purlins. Secondly, completely dimension the drawing and add all necessary notes and symbols. Finally, add all schedules and titles, as shown in Figure 21–75.

The Details, Step 1. The detail drawings may be placed with the plan views or on separate sheets. In either case, the details and sections must be clearly correlated to the plans. The scale of the detail drawings depend on the size and complexity of the structure to be drawn. Select a scale that clearly shows all construction details without oversizing the drawing, thus wasting space and drawing time. Common scales for detail

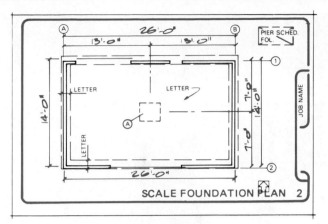

3 FOUNDATION PLAN OVERLAY

Figure 21–74   Lay out the foundation with the floor plan as a guide. *Courtesy Structureform Masters, Inc.*

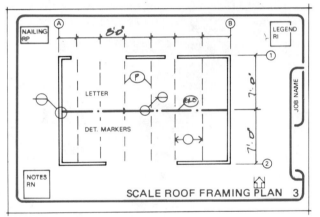

4 ROOF FRAMING PLAN OVERLAY

Figure 21–75   Lay out the roof framing plan with the floor plan as a guide. *Courtesy Structureform Masters, Inc.*

drawing range from 1/2" = 1'–0" to 3" = 1'–0". All the details may be the same scale, or the scale may change depending on the complexity of the drawing. In many cases it is necessary to arrange many details on one sheet or in an area on a sheet. Do this by laying out the details, beginning at the top left corner of the working area. Proceed with additional details placed in rows from left to right. The engineering sketch in Figure 21–2a (see page 671) is used as a guide for the following layout. Start the detail drawing by blocking out the major components of the structure. As you can see in the engineer's sketch, the beam on the left is 24" high and the

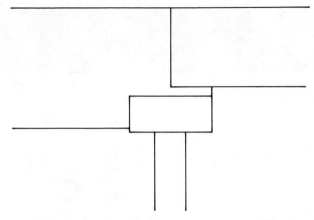

Figure 21–76   Step 1—Block out components of the detail.

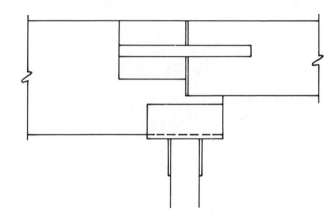

Figure 21–77   Step 2—Draw all construction members.

beam on the right is 16" high. Thus the maximum beam height is 24". If you add another 12" for the post and 12 more inches for the notes, at a scale of 1" = 1'–0" you need a total drawing height of 48", or an actual height of 4" at the selected scale. The width of the detail is up to you. All you need is enough space to show all of the construction members. So, a width of about 4" should work. (See Figure 21–76.)

The Details, Step 2.   Proceed by drawing in all of the construction members to scale. Refer to a vendor's catalog (Figure 21–34) to determine the actual sizes of the prefabricated connectors. Lay out the detail, as shown in Figure 21–77.

The Details, Step 3.   Place all dimensions and notes on the drawing, followed by the title and scale, or the detail identification symbol and scale, as shown in Figure 21–2b (see page 671).

CADD Applications

Structural engineering software programs are available that integrate structural analysis, design, and drafting. A 3-D model is made during the design and drafting process as shown in Figure 21–78. The model may be updated at any time from new design information. The model may be used to automatically create framing plans and elevations as shown in Figure 21–79. The software package also creates construction details for a variety of materials. The construction details may be drawn at any scale and stored for later use. Parametric design of welding symbols and specifications allows the designer to change a variable and automatically update the entire drawing. Detail models as shown in Figure 21–80 may be converted to traditional 2-D views complete with dimensions and specifications.

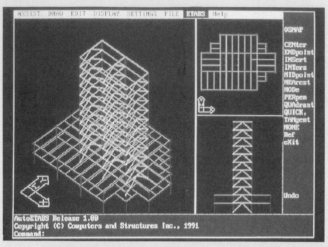

Figure 21–79 Creating framing plans and elevations with the 3-D model. *Courtesy Computers and Structures, Inc.*

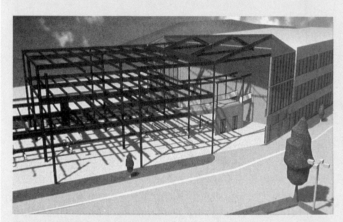

Figure 21–78 A 3-D model is made during the CADD design and drafting process. *Photo courtesy of Softdesk.*

Figure 21–80 3-D CADD model of a structural detail. *Photo courtesy of Softdesk.*

PROFESSIONAL PERSPECTIVE

The drafting procedure for the preparation of structural drawings is often complicated by the amount of information to be contained in them. Structural drawings are normally done on large sheets with as much information as possible placed on each sheet. The first task is to determine the sheet size. This may already be decided as a standard within the company: for example, 22 × 34. The factors that should be considered when selecting sheet size or when drawing on predetermined sheet sizes are:

○ The size of the plan.

○ The number of dimensions and notes needed.

○ The scale used.

○ The number of details to be placed on the sheet.

○ The amount of free space required for possible future revisions.

All of these things must be considered, because you do not want to end up with some sheets that are overcrowded and others that have little information. Use CADD layers such as FL1, FL2, FDN, ROOF, ELEV, and NOTES.

MATH APPLICATION

DESIGNING AN OBSERVATION PLATFORM

Problem: Suppose a building code states that the rise between stair threads cannot exceed 8" for a 12" thread. Find the minimum distance for dimension L and a corresponding angle A in Figure 21–81 of an observation platform.

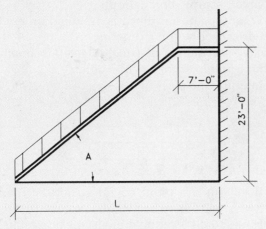

Figure 21–81   Observation platform.

Solution: Geometrical figures are *similar* when they have the same shape. They are *congruent* when they have the same shape *and* size. When two triangles are similar, proportional equations can be written. In this design problem, we have similar triangles: the large staircase and the smaller single tread. (See Figure 21–82.)

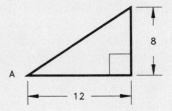

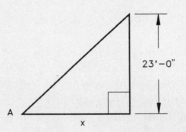

Figure 21–82   Two similar triangles.

Set up a proportion

$$\frac{23}{x} = \frac{8}{12}$$

which gives

$$x = 34.5'$$

Then dimension L is an additional 7'.

$$L = 34.5 + 7 = 41.5' \text{ or } \mathbf{41'–6''}$$

Also, being right triangles, angle A can be found by the application of one trig function:

$$A = \text{Inv tan } \frac{8}{12} = 33.7°$$

CHAPTER 21   STRUCTURAL DRAFTING TEST

DIRECTIONS

Answer the questions with short complete statements or drawings as needed.

1. Describe situations in structural drafting when object lines may be drawn thicker than standard width and when they may be drawn thinner than standard width.

2. Complete the statement: The best structural lettering is _____.

3. Identify the primary components of concrete.

4. Identify the three basic methods of using concrete for building construction.

5. What is the diameter of a #5 steel rebar?

6. What do the parts of the callout 6 × 6 – #10 WWF mean?

    a. 6 × 6 _____

    b. #10 _____

    c. WWF _____

7. Describe the purpose of placing reinforcing steel in poured-in-place concrete.

8. Identify three poured-in-place concrete floor systems.
9. Define *precast concrete.*
10. Define *precast-prestressed concrete.*
11. Define *tilt-up construction.*
12. Give a sample anchor bolt callout.
13. Give a sample rebar callout.
14. How should concrete rebar clear distances be dimensioned?
15. Give a sample slab-on-grade callout.
16. Define *structural engineering drawings.*
17. Define *shop drawings.*
18. What is another name for shop drawings?
19. Describe the modular principles of concrete block construction.
20. How is a bond beam constructed?
21. Define *pilaster.*
22. Describe a panelized roof system.
23. Give a sample timber callout.
24. List the three common types of laminated beams.
25. Give a sample glu-lam callout.
26. Give a sample callout that may be used to label lumber floor joists.
27. How should all lumber and timber be dimensioned on a drawing?
28. Put the following plywood information together into a proper callout:
    a. Exterior APA plywood
    b. 3/4" thick
    c. Group 1
    d. 48/24
    e. A grade veneer one side and C grade other side
    f. Nail w/ 12d 4" OC edges and 12" OC field
29. List at least three advantages of using steel studs.
30. Name at least three types of structural systems commonly used for prefabricated steel structure systems.
31. Define *girts* and *purlins.*
32. Define *plates* as associated with steel materials and give the symbol for plate steel.
33. Describe bars as associated with steel materials.

34. What do the components of the steel shape callout W24 × 55 mean?
    a. W _____
    b. 24 _____
    c. 55 _____
35. Identify the callout symbol for the three "I"-shaped structural materials.
36. Give a sample callout for a C shape.
37. Give a sample callout for flat bar.
38. Give a sample plate callout.
39. Give a sample rectangular tubing callout.
40. Give a sample pipe callout.
41. Give a sample equal leg angle callout.
42. Give a sample bolt callout.
43. What type of bolt head is assumed unless otherwise specified on a drawing?
44. Give the "standard" (steel-to-steel) recommended hole diameter for a 5/8"∅ bolt.
45. Interpret the parts of this note: #5 @ 12" OC EACH WAY.
    a. #5 _____
    b. @ _____
    c. 12" OC _____
    d. EACH WAY _____
46. If lengths are required for structural members, how should they be given?
47. How should location dimensions be established to the following members?
    a. Wide flange and other "I" shapes _____
    b. Channels _____
    c. Angles _____
    d. Plates _____
48. What size hole is recommended for #8 rebar to be driven through timber?

CHAPTER
21 STRUCTURAL DRAFTING PROBLEMS

DIRECTIONS

1. Please read all related instructions before you begin working.
2. Use manual or computer-aided drafting as required by your course guidelines.
3. Use the engineering sketches to prepare a complete set of drawings for the given problems.
4. Do all drawings on 22" × 34" or 24" × 36" sheets unless otherwise specified by your instructor.
5. Use proper sectioning and detailing techniques as correlated to the engineering sketches. Some recommended scales are provided. You may increase or decrease the scale depending on the available space and the complexity of the section or detail.

Sections and details should be drawn at a minimum scale of 3/8" = 1'–0". Judgment should be used if a section or detail requires a lot of information; a larger scale ranging from 3/4" to 1-1/2" = 1'–0" should then be used.

6. It is recommended that you evaluate the entire set of sketches for each problem before beginning to draw. Information needed to draw one problem may be found on the sketch or data for another. The problems are real drafting situations, so you will interpret rough engineering sketches and prepare formal drawings using the best drafting techniques and standards that you have learned.

7. It is suggested that drawing sheets be organized with as much information as possible without crowding or reducing clarity. If additional drawing area is required, it would be better to add another sheet than to overcrowd the drawing.

8. Some problems in this chapter may contain errors, missing information, or slight inaccuracies. This is intentional and is meant to encourage you to apply appropriate problem solving methods, engineering, and drafting standards in order to solve the problems. This is meant to force you to think about each part and how parts fit together in the structure. As in "real-world" projects, the engineering problem should be considered as a basis for your preliminary layouts. Always question inaccuracies in project designs and consult with the proper standards and other sources. In some cases, an error might be the source of engineering changes provided by your instructor; however, this is determined by your specific course objectives. Other situations may require that corrections be made during the development of the original design drawings. This is not intended as a source of frustration, but is considered to be part of the engineering drafter's daily responsibility in project development.

Problems 21–1 through 21–17 are the drawings for a 2400-square-foot storage building.

Problem 21–1 Storage building floor plan

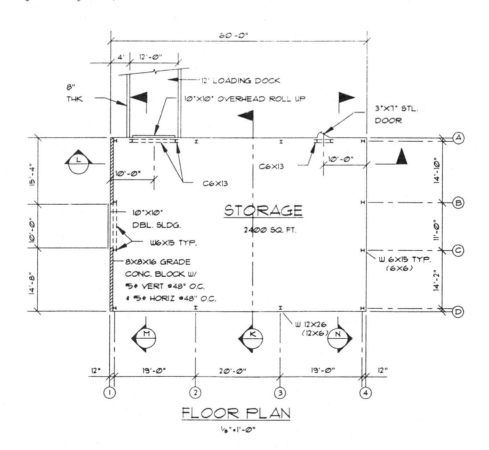

FLOOR PLAN

Problem 21–2   Foundation plan

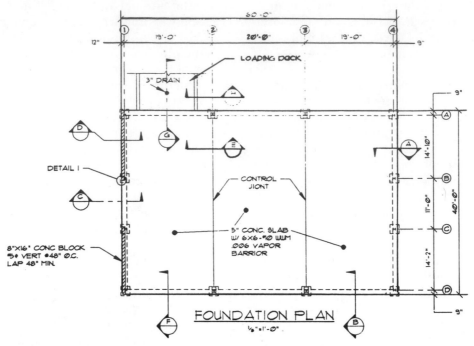

FOUNDATION PLAN
⅛"=1'-0"

GENERAL NOTES:
1. ALL CONC. TO BE 2500 PSI @28 DAYS MIN.
   COMP. STRENGTH.
2. ASSUME SOIL BEAR PRESSURE IS 2000 PSF
3. LAP ALL STEEL 40 DIA. MIN.

Problem 21–3   Section A, Footing section

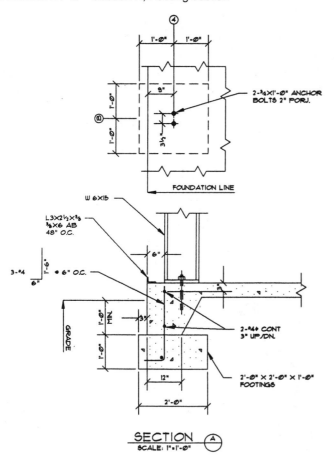

SECTION Ⓐ
SCALE: 1"=1'-0"

Problem 21–4   Section B, Footing section

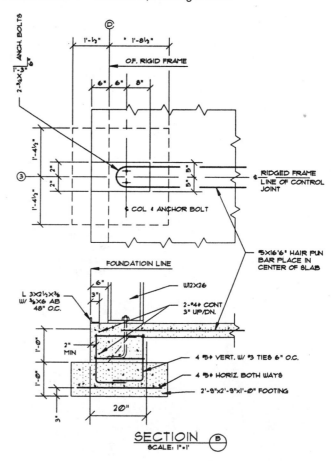

SECTIOIN Ⓑ
SCALE: 1"=1'

Problem 21–5   Section C, Footing and concrete block section; Section D, Footing section; Detail 1, Bond beam detail

Problem 21–6   Section E, Slab footing section; Section F, Slab joint section

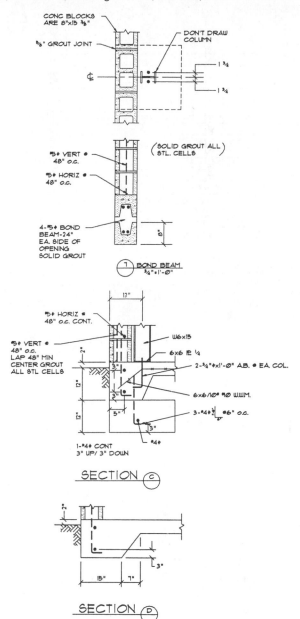

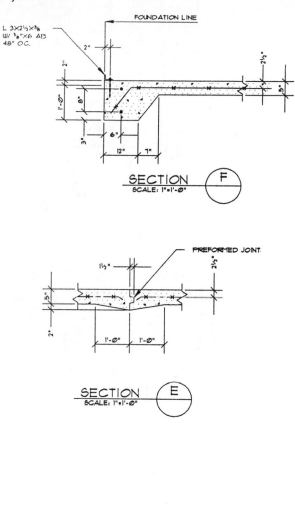

Problem 21–7   Section G, Loading dock ramp section

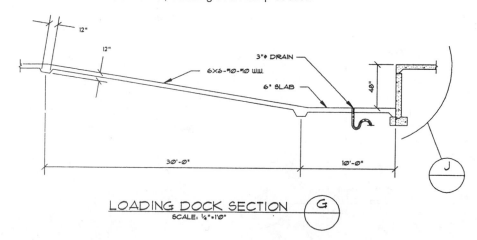

Problem 21–8 Sections H and J, Loading dock wall sections   **Problem 21–9** Roof framing plan

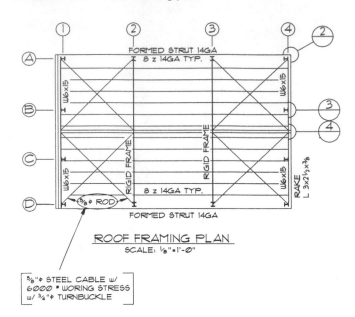

SECTION (H)

SECTION (J)

ROOF FRAMING PLAN
SCALE: ⅛"=1'-0"

⅝"⌀ STEEL CABLE w/
6000# WORING STRESS
w/ ¾"⌀ TURNBUCKLE

Problem 21–10 Section K, Typical building section with detail

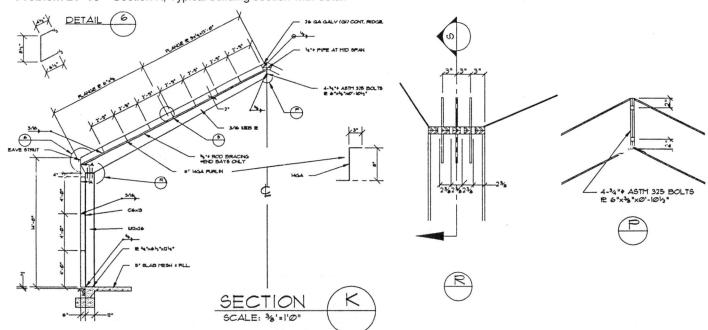

DETAIL (6)

SECTION (K)
SCALE: ⅜'=1'0"

Problem 21–11   Flange detail and connection section

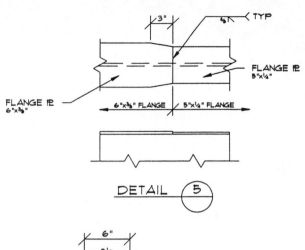

Problem 21–13   Fire wall section

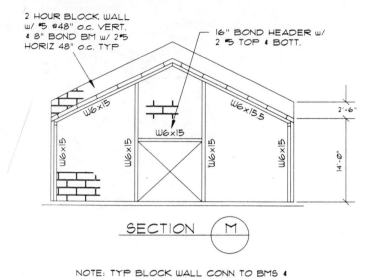

NOTE: TYP BLOCK WALL CONN TO BMS & COLS USE ¾"Φ ANCHOR BOLTS @ 4'-0" o.c.

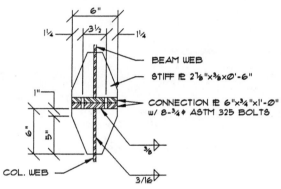

Problem 21–14   Section side wall framing

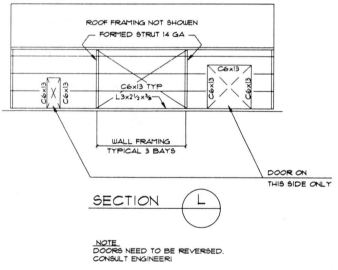

NOTE
DOORS NEED TO BE REVERSED. CONSULT ENGINEER!

Problem 21–12   End wall section

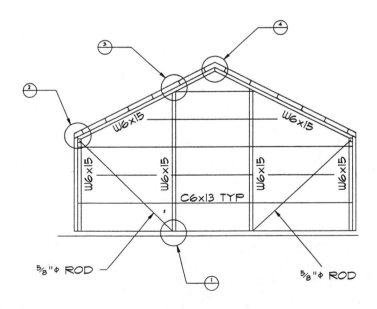

Problem 21–15   Column base and cap connection details

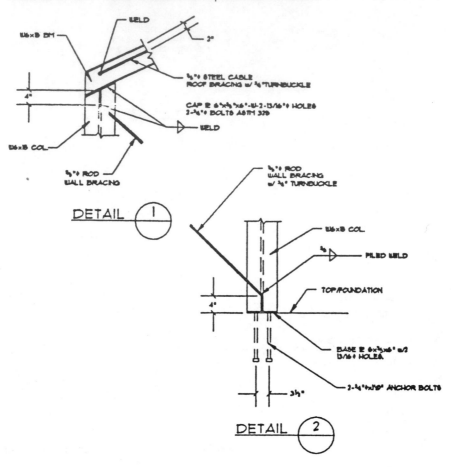

Problem 21–16   Column cap and ridge connection details

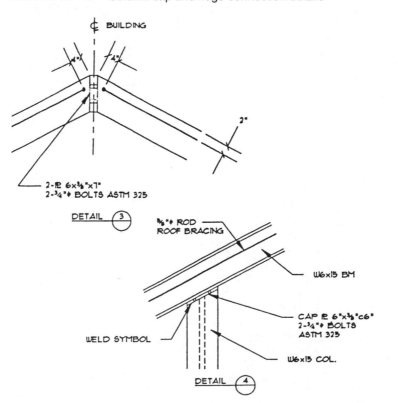

Problem 21–17   Elevations

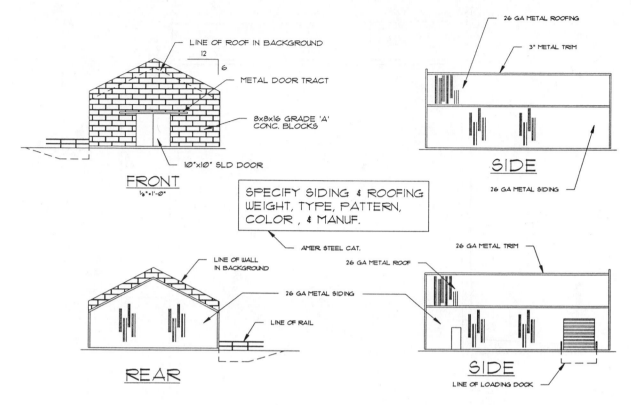

SPECIFY SIDING & ROOFING
WEIGHT, TYPE, PATTERN,
COLOR , & MANUF.

Problems 21–18 and 21–19 are for a warehouse facility.

Problem 21–18   Floor plan and schedules, 3/32" = 1'–0"

NOTE: Alternate problems may be developed using plumbing,

electrical, and HVAC overlays after studying Chapters 20, 23, and 24. Other drawings may be designed to complete a set of plans based on the content of this chapter if appropriate with course guidelines.

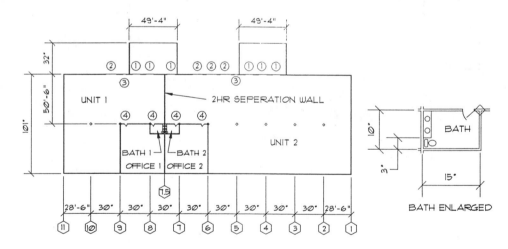

GAR DR OPENERS
    OSCO LDM COMMERCIAL OR EQUIV.
½HP CHAIN DRIVE w/ INSTANT REVERSE
MOTOR, QUICK DISCONNECT ARM, REVERSING
DOOR EDGE, PULL CORD AND 3 BUTTON
COMMAND STATION.

DOOR SCHEDULE

KEY	SIZE	DESCRIPTION
1	8"x8"	KINNEAY COMMERCIAL STEEL INSULATE OVER-HEAD, WEATHER STRIP ALL AROUND, STD 14" R TRACK
2	11"x12"	SAME
3	3"x7"	MESKER FRAME DOUBLE LAMINATED WIRE MESH LITE, ONE HR FIRE RATED, SCHLAGE A53PD HDWARE
4	3"x7"	THERMA-TRU 105H, SCHLAGE A5PD HDWARE

Problem 21–19   Foundation plan, 3/32" = 1'–0"

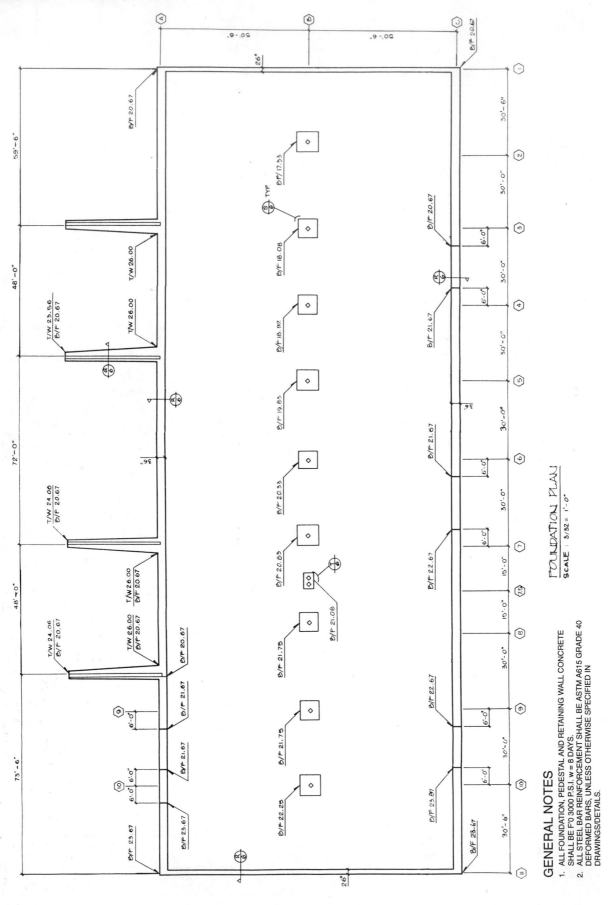

GENERAL NOTES

1. ALL FOUNDATION, PEDESTAL AND RETAINING WALL CONCRETE
   SHALL BE F'0 3000 P.S.I. w = 8 DAYS.
2. ALL STEEL BAR REINFORCEMENT SHALL BE ASTM A615 GRADE 40
   DEFORMED BARS, UNLESS OTHERWISE SPECIFIED IN
   DRAWINGS/DETAILS.

FOUNDATION PLAN
SCALE : 3/32 = 1'-0"

Problem 21–20 Site analysis plan

Use a scale of 1" = 60' to take measurements from the engineering layout shown below. Convert the measurements to a drawing on C-size vellum using a scale of 1" = 20'.

Courtesy Planning Department, Clackamas County, Oregon.

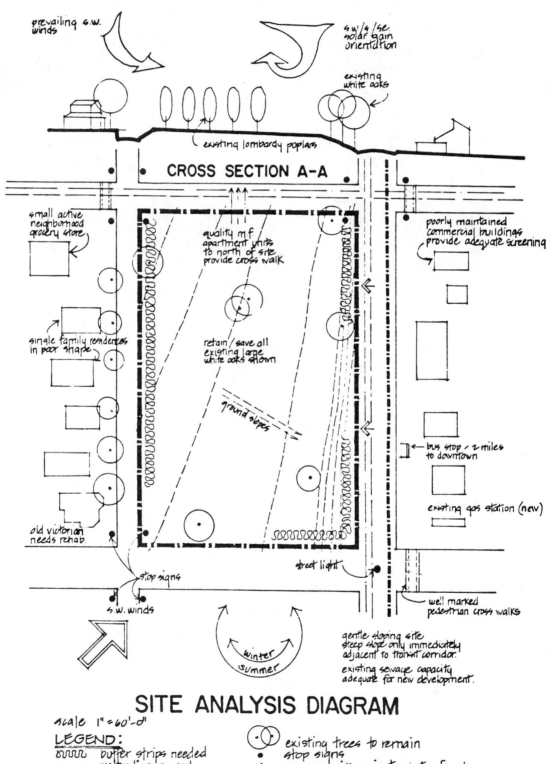

prevailing s.w. winds

s.w/s/se solar gain orientation

existing white oaks

existing lombardy poplars

CROSS SECTION A-A

small active neighborhood grocery store

quality m.f. apartment units to north of site provide cross walk

poorly maintained commercial buildings provide adequate screening

single family residences in poor shape

retain/save all existing large white oaks shown

ground slopes

bus stop · 2 miles to downtown

existing gas station (new)

old victorian needs rehab.

stop signs

street light

s.w. winds

well marked pedestrian cross walks

winter summer

gentle sloping site steep slope only immediately adjacent to transit corridor.

existing sewage capacity adequate for new development.

SITE ANALYSIS DIAGRAM

scale 1"=60'-0"

LEGEND:
ᴠᴠᴠᴠ buffer strips needed
— — — water line in road
– – – sewer lines
–··–··– transit route
⊓ bus stop

⊙ existing trees to remain
• stop signs
⌂ proposed site project points of entry
– – – existing contours

Problem 21–21   Planned unit development

Use the graphic scale shown in the layout below to take mea-surements. Convert to a scale of 1" = 50' on appropriately-sized vellum.

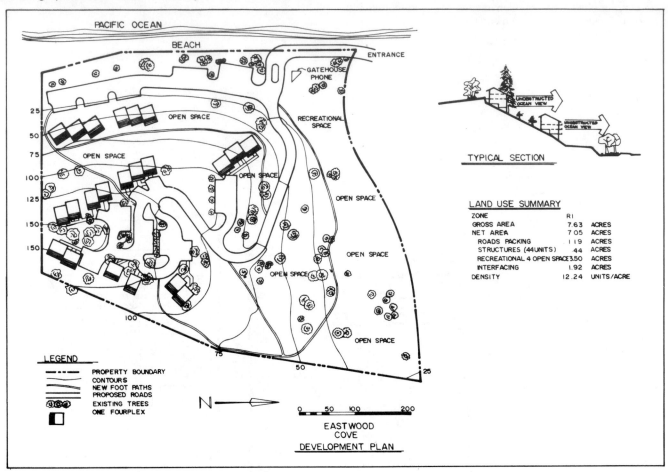

Problem 21–22   Paving site plan

Use the graphic scale shown in the layout below to take mea-surements. Convert to a scale of 1" = 60' on appropriately-sized vellum.

Courtesy Soderstrom Architects, PC.

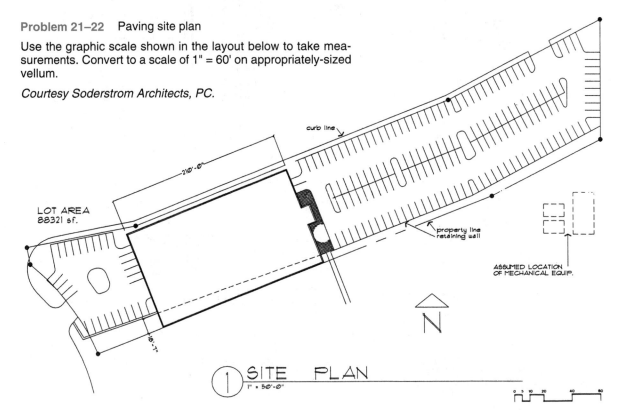

Problem 21–23   Grading plan

Follow the same instructions as in Problem 21–22.

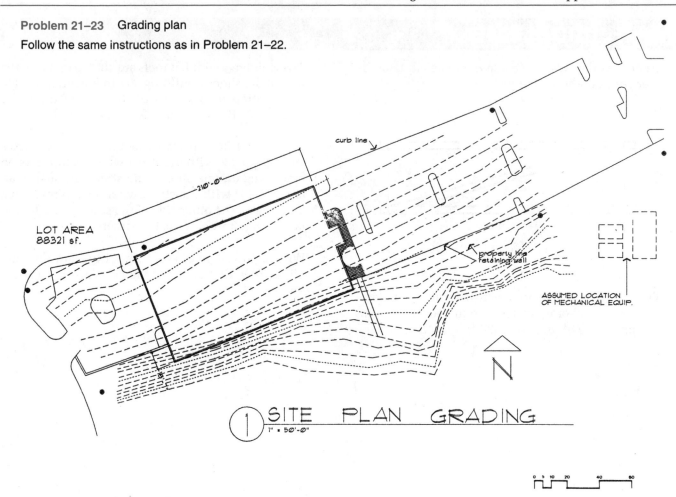

curb line

210'-0"

LOT AREA
88321 sf.

property line
retaining wall

ASSUMED LOCATION
OF MECHANICAL EQUIP.

N

① SITE   PLAN   GRADING
1" = 50'-0"

Problem 21–24   Decorative interior elevations

Use a scale of 3/32" = 1'–0" to measure the layout shown below.
Convert to a 1/4" = 1'–0" scale on appropriately sized vellum.

Courtesy Structureform Masters, Inc.

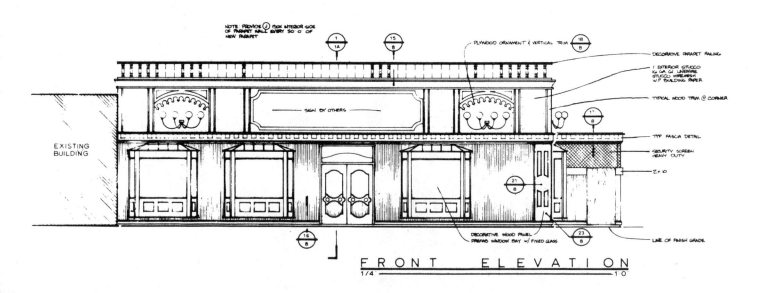

NOTE: PROVIDE Ⓙ BOX INTERIOR SIDE
OF PARAPET WALL EVERY 50'-0' OF
NEW PARAPET

PLYWOOD ORNAMENT & VERTICAL TRIM

DECORATIVE PARAPET RAILING

1 EXTERIOR STUCCO
16 GA GI LINENIRE
STUCCO WIRENESH
W.P. BUILDING PAPER

TYPICAL WOOD TRIM @ CORNER

TYP FASCIA DETAIL

SECURITY SCREEN
HEAVY DUTY

2 x 10

SIGN BY OTHERS

EXISTING
BUILDING

DECORATIVE WOOD PANEL
PREFAB WINDOW BAY W/ FIXED GLASS

LINE OF FINISH GRADE

FRONT   ELEVATION
1/4   1'-0

MATH PROBLEMS

1. Determine the values of angle A and length L for the pier support shown below.

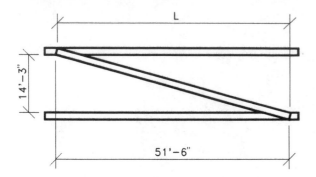

2. Two straight roads intersect to form an angle of 80°. Find the shortest distance from one road to a gas station on the other road 1500 ft from the junction.

3. Two buildings with flat roofs are 160' apart. From the roof of the shorter building, 50' in height, the angle of elevation to the edge of the roof of the taller building is 17°. How high is the taller building?

4. A ladder, with its pivoting foot in the street, makes an angle of 38° with the street when its top rests on a building on one side of the street and makes an angle of 45° with the street when its top rests on a building on the other side of the street. If the ladder is 40' long, how wide is the street?

5. What is the angle of elevation of the sun if a telephone pole that is 60.0' long casts a shadow 72.3' long?

CHAPTER 22

Civil Drafting

LEARNING OBJECTIVES

After completing this chapter, you will:

○ Calculate unknown values of bearings, slopes, and curve dimensions.
○ Draw transit lines for roadway layouts.
○ Reduce survey information from field notes.
○ Prepare site plans and profiles for cut and fill work.
○ Draw plans and profiles from survey information.

THE ENGINEERING DESIGN PROCESS

A new roadway, Valerian Lane, is going to be built from Erika Street to Quincey Avenue. (See Figure 22–1.) The information in the table in Figure 22–2 is given to you from the designer and survey department.

How would you go about executing this project? First, finish the required mathematical calculations for the chart, then lay out Valerian Lane. To complete the unknown portions of the chart and draw the new roadway, you must understand the concepts of survey and civil drafting.

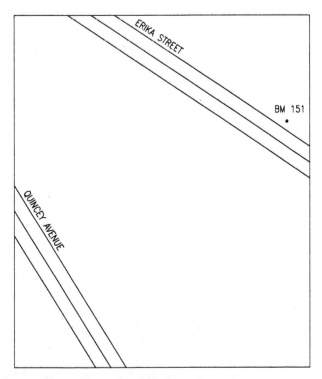

Figure 22–1 Example of roadway layout.

STATION	TRANSIT LINE FOR VALERIAN LANE FROM ERIKA STREET (BEARING N56°16'W) TO QUINCEY AVENUE (BEARING N31°24'W)
0 + 00 Begin Project	245.78'N67°37'W from BM 151 to center line of Erika Street. Then 132.00'S33°44'W to P.C.
STA 1 + 32	P.C. to the left Curve Data R = 200' Δ = 47°23' L = ? D = ?
STA ?	P.R.C. Curve Data R = 170' Δ = 72°15' L = ? D = ?
STA ?	P.T.
STA ? End Project	S58°36'W from STA ? to center line of Quincey Avenue

Figure 22–2 Survey information.

▼ INTRODUCTION TO SURVEY: DIRECTION

Think of the earth as having imaginary lines that partition it into sections for the purpose of exactly pinpointing the area to be described. These lines divide the earth horizontally and vertically into lines of *latitude* (sometimes referred to as parallels) and lines of *longitude* (sometimes referred to as meridians). These two sets of grid lines are called the *graticule* of the earth. Parallels and meridians are shown in Figure 22–3.

Lines of latitude do not intersect. They are approximately the same distance apart (about 69 statute miles). The longest line of latitude is around the earth at the equator and the shortest lines are around the North and South Poles. These parallels are numbered from 0 degrees at the equator to 90 degrees at each Pole as shown in Figure 22–4. When describing a particular point on the earth, a numbered latitude coordinate is given with *N* or *S* to describe north or south of the equator.

Lines of longitude are all about the same length and meet at both Poles. They divide the earth into 360 degrees. Each line of longitude is numbered as shown in Figure 22–5. The zero line of longitude is the *prime meridian,* and the other lines, including those between the ones shown here, are *local meridians.* The most generally accepted prime meridian runs through Greenwich, England, although some countries in a spirit of

CITY	LATITUDE	LONGITUDE
Calcutta, West Bengal, India	22.30 N	88.20 E
Greenwich, England	51.29 N	00
Los Angeles, California, USA	34.00 N	118.15 W
Madrid, Spain	40.25 N	3.43 W
Paris, France	48.52 N	2.20 E
Rio de Janeiro, Brazil	22.53 S	43.17 W
Sydney, Queensland, Australia	16.40 S	139.45 E
Tokyo, Japan	35.40 N	139.45 E
Vancouver, British Columbia, Canada	49.13 N	123.06 W

Figure 22–6   Cities and their coordinates.

nationalism have defined the prime meridian (0 degrees longitude) as going through their major city. Note that while the length of the lines of longitude remains constant, the area between them decreases the closer they get to the Poles.

Each latitude degree and longitude degree is divided even further into sixty equal sections called minutes, and each minute is divided into sixty equal sections called seconds. The symbol for minute is ('), and the symbol for second is ("). Therefore, a latitude of 29 degrees, 15 minutes, 47 seconds to the North of the equator would be shown as 29°15'47"N. A longitude of 124 degrees, 12 minutes, 53 seconds to the east of the prime meridian (at Greenwich, England) would be shown as 124°12'53"E. Figure 22–6 shows some cities and their latitude and longitude coordinates.

Direction

The direction around the earth from a given point in relation to the latitude and longitude coordinates describes a *great circle*. There are an infinite number of great circles. The great circle in Figure 22–7 has a starting point in northern Africa. From this one point, which has coordinates of its own, describe in which direction the great circle goes. To do this, use the true azimuth or bearing.

In order to understand both of these processes, imagine the earth as having an X axis through the equator, and a Y axis through the Poles. Draw smaller portions of the earth on this square grid or plane coordinate system as shown in Figure 22–8.

True Azimuth

The measurement that indicates the direction of a line (in this case, the great circle) is the horizontal angle to the right of the line from north. In other words, the *azimuth* is the angle (going clockwise) between the meridian at the starting point and the great circle. This means that the direction of that particular line is relative to the meridian at a particular point. These horizontal angles are described as the azimuth readings and can mea-

Lines of Latitude (parallels)   Lines of Longitude (meridians)

Figure 22–3   Latitude and longitude.

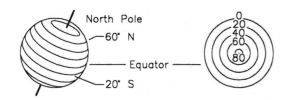

Figure 22–4   Latitude coordinates.

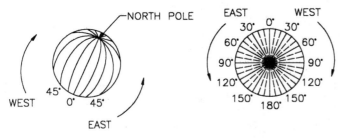

Figure 22–5   Longitude coordinates.

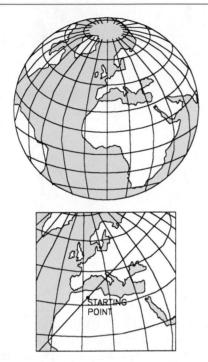

Figure 22–7 Great circle.

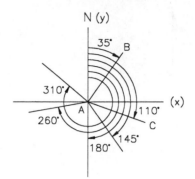

Figure 22–9 Azimuth angles.

sure from 0° to 360°. In Figure 22–9, the azimuth of line AB is 35° and the azimuth of line AC is 110°. All azimuth readings are from a meridian, yet depending on which meridian system is used, a different type azimuth is described:

True azimuth is measured from a true meridian.

Magnetic azimuth is measured from a magnetic meridian.

Grid azimuth is measured from a central meridian on a grid system.

Assumed azimuth is measured from an arbitrary meridian.

Bearings

The *bearing* of a line is basically the same as the azimuth, except it only has values up to 90°, while the azimuth goes through all four quadrants of the grid up to 360°. A bearing is also described by the north or south meridian and by the east or west parallel. For example, a line in the northwest quadrant (using the X and Y coordinates) that has a horizontal angle of 45° counterclockwise from the north meridian would be described as having an azimuth reading of 315°, while its bearing would be N45°W (360° – 45° = 315°). A line 75° to the east from the south meridian has an azimuth of 150° and a bearing of S75°E. Note that a bearing of S75°E has the same angle as a bearing of N75°W. (See Figure 22–10.)

Magnetic Azimuth

The bearings explained so far have been gauged from a true azimuth, or from true north or south. The magnetic north and south are not at the actual North and South Poles where the meridians meet, and where the earth spins on its axis. Figure 22–11 shows the position of the

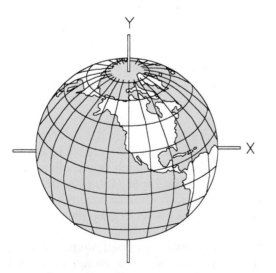

Figure 22–8 Coordinates.

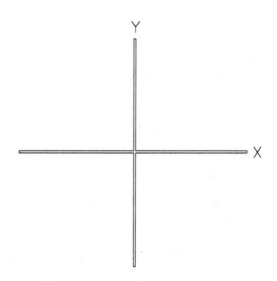

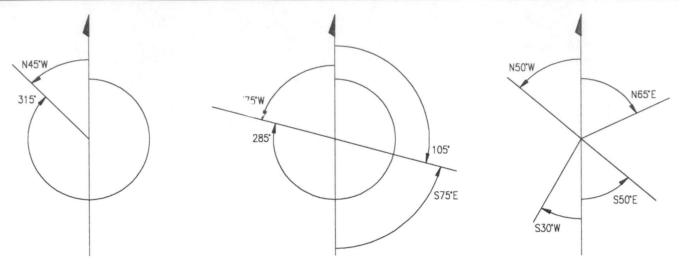

Figure 22–10   Bearings.

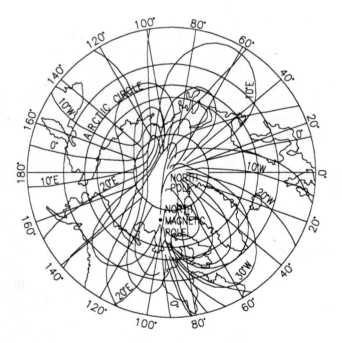

Figure 22–11   Azimuthal equidistant projection.

true poles and the magnetic poles. The needle of the compass points to these magnetic poles. The degree of difference on the compass between magnetic azimuth and true azimuth is called *magnetic declination.* To adjust for this difference, isogonic lines have been established that show how many degrees to the east or west the magnetic pole is from the actual pole. The isogonic lines must be reestablished periodically because the position of the magnetic pole is constantly changing. Therefore, a map showing an arrow for a magnetic pole will also give the year in which the magnetic declination was calculated.

An example of an *isogonic chart* showing isogonic lines or magnetic declination for the United States is shown in Figure 22–12. This chart, which is based by the

U.S. Geological Survey of 1983, shows Chicago, Illinois, as being on the line of zero declination, also known as the *agonic line.* This means that the magnetic North Pole is directly in line with the actual North Pole and the compass reading is the actual North Pole reading, or true azimuth. Los Angeles, California, is approximately 14.3°E of zero declination. This means that when the compass reads 0° it is pointing toward magnetic north, which is 14.3° east of true north. Therefore, the north arrow on any map around Los Angeles would appear as in Figure 22–13.

Grid Azimuth

The grid azimuth is measured from the central meridian in a rectangular grid system. This will be discussed later. It is common in maps to find three north arrows to define true azimuth, magnetic azimuth, and grid azimuth, as shown in Figure 22–14.

▼ SURVEYING

Civil engineering drafting is involved in portraying a portion of the earth. Now that we know about positioning and direction, we can proceed in our understanding of survey.

Traverses

A *traverse* is a series of lines that each have a known length and are connected by known angles. Each traverse line is a *course* and each point where courses intersect is a *traverse station* or *station point.* When traversing, one begins with a *point of beginning* (POB) and proceeds to utilize one of several traverse types.

Open Traverse.   An open traverse does not close upon itself, meaning that the courses do not return to the POB. It is used for route surveys when mapping linear

Figure 22–12 Isogonic chart.

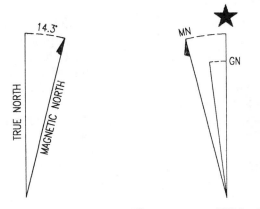

Figure 22–13 Approximate mean declination, 1983.

Figure 22–14 UTM grid and 1983 magnetic north declination at center of sheet.

features such as highways or power lines. An example is shown in Figure 22–15.

Closed Traverse. In a closed traverse a) the courses return to the POB, or b) the courses close on a point different from the POB, but a point that has a known position, called a *control point*. Figure 22–16 shows both a *loop traverse* and a *connecting traverse*. The connecting traverse has known positions at both the POB and at the end of the survey. They are similar in that they each can be checked for accuracy. These traverses are the ones used exclusively for land and construction surveys.

Direct-Angle Traverse. A *direct-angle* traverse is used primarily with a loop traverse. The interior angles are

Figure 22–15 Open traverse.

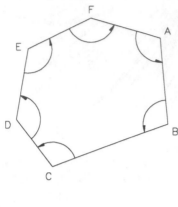
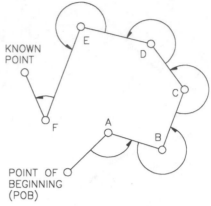

Figure 22–16   Closed traverses.

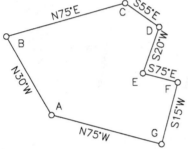

Figure 22–17   Compass bearing traverse.

measured as the traverse proceeds either clockwise or counterclockwise.

Compass-Bearing Traverse.   In a compass-bearing traverse, the bearings of the courses are read directly from the compass. (See Figure 22–17.) When using a compass, one must be aware of the magnetic declination at that particular point.

Deflection-Angle Traverse.   In a deflection-angle traverse, each course veers to the right or left from each station measured. Figure 22–18 shows that an *R* or *L* (right or left) must always be shown with the angle. This

method is used most often when doing route surveys for features such as highways and railroads.

Azimuth Traverse.   An azimuth-traverse uses the azimuth angle to show the direction of the next course. The azimuth reference line is usually the north-south line and can either be the magnetic azimuth or the true azimuth. Remember that the azimuth is always measured clockwise from north. (See Figure 22–19.)

▼ PLOTTING TRAVERSES

When plotting traverses on vellum or Mylar®, a few simple steps are required. First, use a nonreproducible blue pencil until you have the courses plotted. At that point you can draw them permanently in pencil or ink. Begin by drawing a light north-south line on the drawing surface. Mark the POB wherever you want on the vertical line, then draw another light line horizontally through the POB. When plotting a compass-bearing traverse, the next step is to place the center of a protractor on the POB. If the bearing of the course from POB is, for example, N25°E, then make a mark on the drawing surface at the point shown in Figure 22–20. Next, remove the protractor and draw a line through the POB to the N25°E mark. This is the bearing line, which is one characteristic of the course. The other necessary component of the

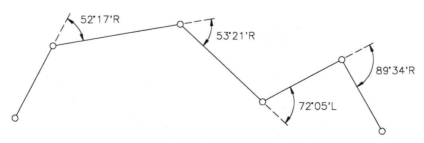

Figure 22–18   Deflection angle traverse.

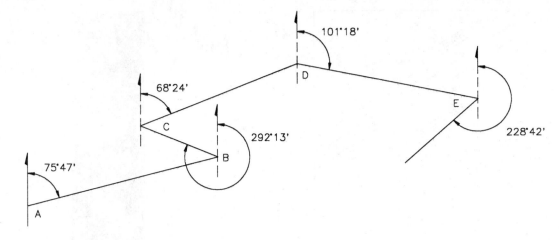

Figure 22–19   Azimuth traverse.

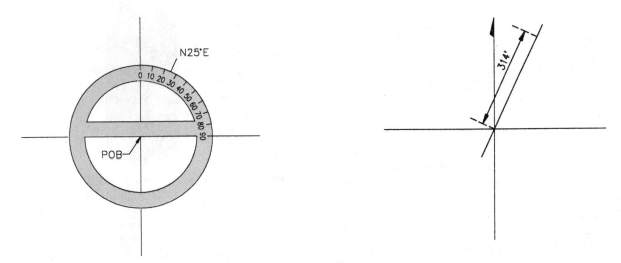

Figure 22–20   Using a protractor.

Figure 22–21   Length.

course is the distance. If that distance is perhaps 314 feet, then measure from the POB 314 feet along the bearing line using a scale that has been previously determined (for example: 1"=100'). Make a mark at that point, as shown in Figure 22–21. The next step is to draw another north-south line through that point and plot the next compass-bearing traverse.

There is a major point to keep in mind when plotting traverses. It was mentioned earlier that a bearing of N25°E and a bearing of S25°W have the identical angle. However, when actually plotting the distance on the course, you must plot in the direction that is given. Otherwise, the 314 feet in Figure 22–21 would be measured to the southwest from the POB instead of to the northeast.

When plotting traverses with a CADD system, the same concepts are kept in mind. Then, depending on the particular system you are using, the layout of the traverse will follow certain steps. For example, a line can begin at the POB then be given information relating to the

length and bearing (azimuth or otherwise). The system will then draw the line automatically.

The direct-angle traverse is drawn by plotting the measured interior angles and by using the lengths given for each course.

A deflection-angle traverse is drawn similarly to the direct-angle traverse except that the length of each course is measured and plotted. Then the deflection angle to the right or the left is measured from that point.

The azimuth traverse is plotted similarly to the compass-bearing traverse except that the complete counterclockwise angle is measured from north. Measure each course from information given, make another north-south line at that point, then plot the next course. (See Figure 22–19.)

In the example at the beginning of the chapter, the POB of the project is a certain length and distance from BM 151. BM refers to *bench mark*; it will be discussed later. Now you know how to find the point of beginning and the first distance of Valerian Lane. (See Figure 22–22.)

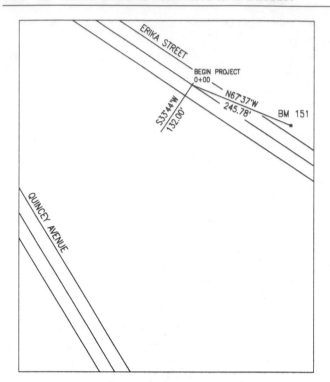

Figure 22–22   Finding the POB.

▼ DISTANCE AND ELEVATION

So far, you have learned how to establish the direction of a line by using the different azimuths and the bearing of the line. Now you need to know how to calculate the distance and the elevation in order to complete your basic knowledge of how to make accurate maps.

Gunter, the mathematician who invented terms such as *cosine* and *cotangent,* also invented a standard of land measurement called *Gunter's chain.* It is 66 feet long, or 1/80th of a mile. It was used extensively in the days before using steel tape became common. Each chain is divided into 100 links, each of which is 7.92 feet long. Because of the widespread and long-term use of Gunter's chain, terminology such as *breaking chain, rear chainman,* and *head chainman,* all of which will be described shortly, is still used.

Distance in modern-day usage is measured several ways. The most common is the *stadia* technique. With this method, a rod, called a Philadelphia rod, is used along with a level, which is an instrument with crosshairs. The Philadelphia rod is 7 feet long, can be extended 12 or 13 feet long (or up to 45 feet long in some models), and is graduated to hundredths of a foot. Figure 22–23 shows a Philadelphia rod. The level is placed on a tripod a certain distance away from the rod. When one looks through the level at the rod, the top stadia crosshair and the bottom stadia crosshair encompass an amount that is then multiplied by 100 to equal the distance. (See Figure 22–24.) In this example, the crosshairs encompass the distance between 10 and 11 feet. When multiplied by 100, the distance to the rod is 100 feet.

Figure 22–23   Philadelphia rod. *Courtesy Chicago Steel Tape.*

This type of surveying technique uses mechanical and optical principles and uses MOD instruments. Another technique is the EDM system, or *Electronic Distance Measuring,* which uses electronic principles. The EDM system is unique in that it uses a reflected beam to gauge distances. (See Figure 22–25.) Examples of theodolites and transits, which are instruments of the EDM system, are shown in Figure 22–26. Theodolites and transits are used instead of compasses because they are more accurate. The magnetic declination is taken into account by an adjustment that can be made on the instrument itself.

The Philadelphia rod and level or transit can also be used to find the elevation of an unknown point. The method is as follows:

1. A known elevation is located, and the rod is placed at that point.
2. The transit (or level) is placed between the rod and the unknown elevation and then leveled.
3. A reading is taken back to the rod. The rod location is called the *backsight.* When this height is added to the known elevation, calculate the height of the instrument or HI.
4. The rod is then placed on the unknown elevation, called the *foresight,* and a reading is taken forward

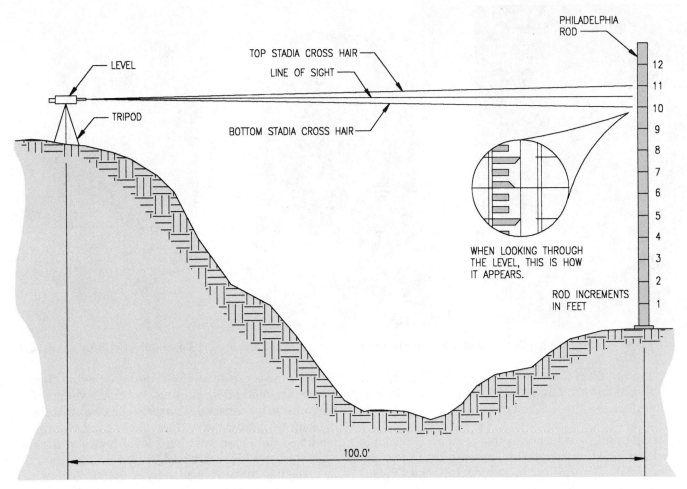

Figure 22–24 Stadia technique.

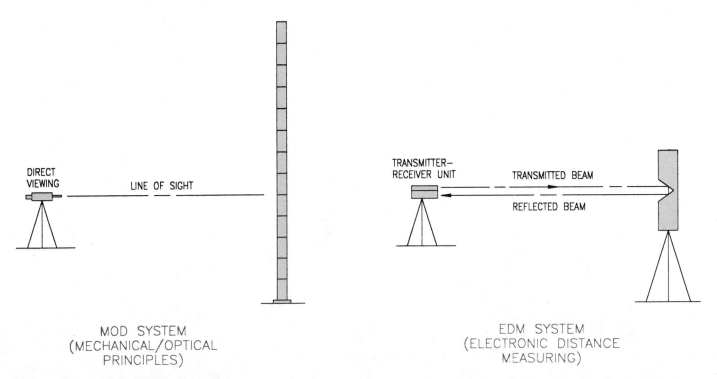

Figure 22–25 MOD and EDM systems.

Figure 22–26 **(a)** Transit (Optical Theodolite). **(b)** Electronic Theodolite. **(c)** Total Station Theodolite. *Courtesy Leica, Inc.*

to the rod from the transit. What is read on the rod is then subtracted from the HI (height of the instrument) to find the unknown elevation.

To find unknown elevations that are a great distance away from the known elevation, several turning points are required. *Turning points* are nothing more than each location that the rod is placed. (See Figure 22–27.)

When distances were first measured using Gunter's chains, the method for finding elevations was similar to what was just described. The difference was that instead of using an instrument, such as a transit, between the rod readings, the chain was held level between the rods. Then the elevations were read, and the chain would then be brought to the next level to be read again. This is what was known as breaking chain, which was referred to ear-

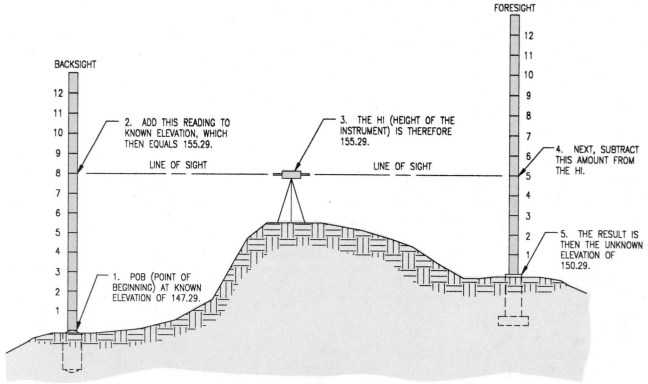

Figure 22–27 Elevation reading.

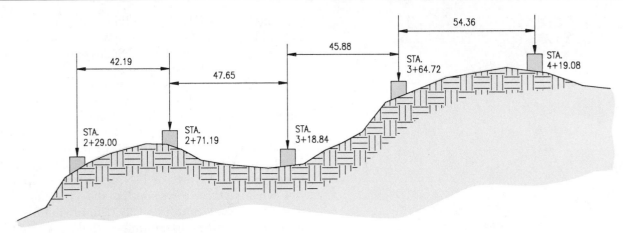

Figure 22–28 Leveling.

lier. The person holding the chain at the backsight was the rear chainman, and the person holding the chain at the foresight was the head chainman. This process was, and still is, called leveling. (See Figure 22–28.)

A known elevation can be read from a bench mark or monument, which is more or less a permanent, fixed object on which the elevation is marked. Often a bench mark is a brass disk about 3 inches in diameter that may have "U.S. GEOLOGICAL SURVEY" engraved on it, along with the elevation above sea level. Another way to describe turning points is temporary bench marks or *stations*, often abbreviated as STA.

When the surveyor begins a survey, the stations, or turning points, are shown in different ways. The distances are sectioned into 100-foot intervals, with the 100s separated from the 10s by a plus (+) sign. For example, the POB is shown as 0 + 00. If the next station is at 57 feet, it is shown as 0 + 57. If the following station is at 244 feet, it is shown as 2 + 44. The beginning of a survey (station 0 + 00) is always positioned according to a known bench mark so that exact locations and elevations can be calculated. A station can also be shown as a turning point. An example would be TP1, meaning turning point #1. BM 35, meaning bench mark #35, may be one of the stations. In Figure 22–29, the surveyor has given these field notes to the draftsperson. The draftsperson then calculates the rest of the information, as shown in Figure 22–30. BM 35 can be double-checked to see if the calculation actually adds up to the posted elevation.

STATION	B.S. (+)	H.I.	F.S. (−)	ELEVATION
BM 34	7.25		11.92	124.13
0 + 75	5.87		10.29	
1 + 50	5.57		12.62	
TP1	7.04		3.44	
2 + 25	11.70		6.54	
BM35 116.75				

Figure 22–29 Leveling field notes.

STATION	B.S. (+)	H.I.	F.S. (−)	ELEVATION
BM 34	7.25	131.38	11.92	124.13
0 + 75	5.87	125.33	10.29	119.46
1 + 50	5.57	120.61	12.62	115.04
TP1	7.04	115.03	3.44	107.99
2 + 25	11.70	123.29	6.54	111.59
BM 35 116.75				116.75

Figure 22–30 Leveling field notes completed by draftsperson.

▼ CIVIL ENGINEERING DRAFTING

Civil engineering drafting in general can be used in a wide variety of applications: highway and traffic design, water and sewage design, park planning, subdivision planning, single-family residence planning, and design, to name a few. When describing parcels or plots of land, there are two primary types of legal descriptions used. One is the *metes and bounds system*, and the other is the *rectangular system*, which is based on the public land system.

Metes and Bounds System

When the United States was first formed, the method used for describing the boundaries of properties was by measuring the length (metes) and bearings (bounds) of the property. The metes and bounds system is the primary form of legal description for nineteen states, including the original thirteen states. These states are not part of the public land system, which is measured by the rectangular system of surveys. Metes and bounds are occasionally used in public land system states as shown in the following example. (See Figure 22–31.)

Beginning at the Northwest corner of that certain tract of land conveyed to James H. Wilcox *et ux* by Deed recorded in Film Volume 95, Page 1488, Deed Records of Yamhill County, Oregon, said point is in the center of County Road No. 87, thence along the

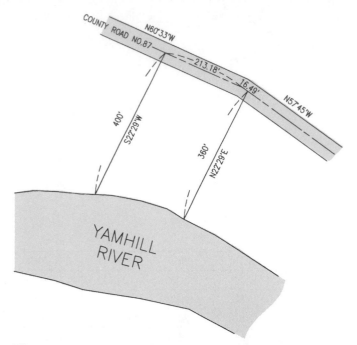

Figure 22–31   Metes and bound tract of land.

center of said County Road No. 87, N57°45'W 16.49 feet; thence N69°33'W 213.18 feet; thence leaving said County Road No. 87, S22°29'W 400.00 feet more or less to a point on the North bank of the Yamhill River; thence Easterly along said North bank of the Yamhill River to the Southwest corner of the aforesaid James H. Wilcox *et ux* tract of land; thence along the West line thereof N22°29'E 360.00 feet more or less to the point of beginning.

Rectangular System

In the late 1700s when the remaining land to the Pacific Ocean was opened up to homesteading, a more complete and universal method of determining land ownership was needed. The government, through the General Land Office (now known as the Bureau of Land Management), measured large area in these public lands and provided a rectangular system of surveys from which ownership could be determined. The public-land states are Alabama, Alaska, Arizona, Arkansas, California, Colorado, Florida, Idaho, Illinois, Indiana, Iowa, Kansas, Louisiana, Michigan, Minnesota, Mississippi, Missouri, Montana, Nebraska, Nevada, New Mexico, North Dakota, Ohio, Oklahoma, Oregon, South Dakota, Utah, Washington, Wisconsin, Wyoming, in addition to 75 million acres of Texas bought by the federal government. Each large area of public land was a single great survey and had a basic reference point where one particular meridian of longitude and one particular parallel of latitude crossed. The meridians in each of these public-land surveys were named separately with names such as the first principal meridian (which is the western boundary of the state of Ohio), Michigan meridan, Choctaw merid-

ian, and the Black Hills meridian. The parallels were all named base line. There are thirty-one sets of great surveys in the contiguous United States and three in Alaska.

With the base line and principal meridian as reference points, each survey is arranged into rows of 6-mile-square blocks (each having an area of 36 square miles) called *townships*. From the base line, each row of townships can be located by referring it to the number and direction from the base line. For example, a township just to the north of the base line would be identified as Township No. 1 North or T1N, and a township in a row of townships three rows south of the base line would be Township No. 3 South or T3S. The townships are also arranged according to columns called *ranges,* and each column of townships can be located from the principal meridian by referring to how far east or west it is from the meridian. For example, a township in a column of townships four columns to the east of the principal meridian would be labelled Range No. 4 East, or R4E. The marked township in Figure 22–32 is T2N, R3W.

Each of the township's thirty-six squares is called a *section*, with the sections numbered as shown in Figure 22–33. Each section is approximately one square mile

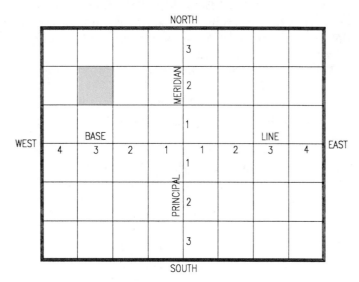

Figure 22–32   Base line and meridian.

6	5	4	3	2	1
7	8	9	10	11	12
18	17	16	15	14	13
19	20	21	22	23	24
30	29	28	27	26	25
31	32	33	34	35	36

Figure 22–33   Township divided into sections.

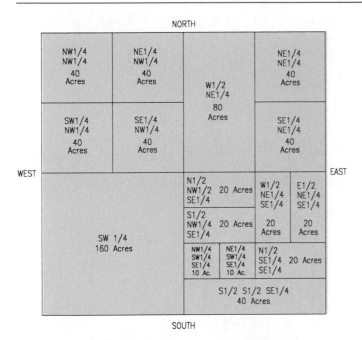

Figure 22-34 Section divided into parcels.

and may be divided into quarter-sections, which are sometimes referred to as quarters. An example would be the Southwest Quarter of Section 19, which can also be written SW 1/4 of Sec. 19. The quarter-sections can be divided further into quarter-quarter-sections, as seen in Figure 22-34. Note that the description of the particular piece of land begins with the smallest portion and proceeds through the largest portion.

▼ BEGINNING A CIVIL ENGINEERING DRAFTING PROJECT

An *engineering map* is a large-scale map that shows the information necessary for planning an engineering project such as a highway layout or utility layout. The information can come from an engineering survey, an aerial survey, or other reliable source, and will include locations, dimensions, and elevations of buildings, roads, water, and utility lines. This type of map can also include contour lines in addition to showing the boundaries of plats. A *plat* is a tract of land (such as a subdivision) showing building lots. Figure 22-35 shows an example of this. Notice also the section, township, and range notations. WM is an abbreviation for Willamette meridian, which is the name of the meridian in this particular great land survey.

Contour Lines

One of the most important symbols in civil engineering drafting is the contour line. *Contours* are simply lines that join points of similar elevation. When looking at a map with contours, one is able to understand the topography of the land. Figure 22-36 (see page 739)

shows the relationship between actual elevations and how the contour lines would appear on the map. Each horizontal plane represents an additional 10 feet in elevation from the horizontal plane below it. The contour interval in this case is 10 (10 feet).

Figure 22-37 (see page 739) shows the contour lines around Mt. St. Helens in Washington State. As you can see, the contour lines are spaced more closely together on steep slopes than they are on gentle slopes. They also point upstream when depicting creeks.

Some of the contours are drawn with a heavier line than others. These are *index contours* and can occur every fourth or fifth contour, depending on the map. Some of the index contours have their elevations labeled on them. *Intermediate contours* are the lighter lines shown between the index contours, and they are not usually labeled with elevations. In the Mt. St. Helens map, find the index contour at 8000 feet of elevation. The next downhill index contour that is labeled is at 7600 feet. This means the index contour between these two is at 7800 feet, and each intermediate contour (there are four between each index contour) shows a difference of 40 feet. The contour interval is, therefore, 40 feet, and this interval will be stated on the map.

Highway Layout

For a *highway layout*, the draftsperson first draws a *plan view* and then a profile of the proposed highway from the information given in the survey notes. The plan view shows existing features such as roads, buildings, and trees in addition to the centerline (transit line) of the proposed highway.

Plan Layout. In order to plot the transit line of the highway, use curve data information like that shown for the sewer layout in Figure 22-35.

Point of curve (PC) is the station at which the curve begins.

Radius (R) means the radius of that particular curve. The center point of the curve can be found by measuring the radius distance perpendicularly from the transit line at the PC. For example, to find the center of a curve with a radius of 265 feet in which the transit lines are known, draw two lines parallel to the transit lines on the inside of the curve. Their intersection will then be the center of the curve. (See Figure 22-38, page 739.)

Point of tangency (PT) denotes the end of the curve.

Delta angle (Δ) is that inclusive angle measurement of the curve from the PC to the PT.

Curve length (L) is the actual length of the curve from the PC to the PT.

Degree of curve (D) is the angle within the delta angle that has a chord of 100 feet along the curve. A chord connects two points on the curve with a straight line.

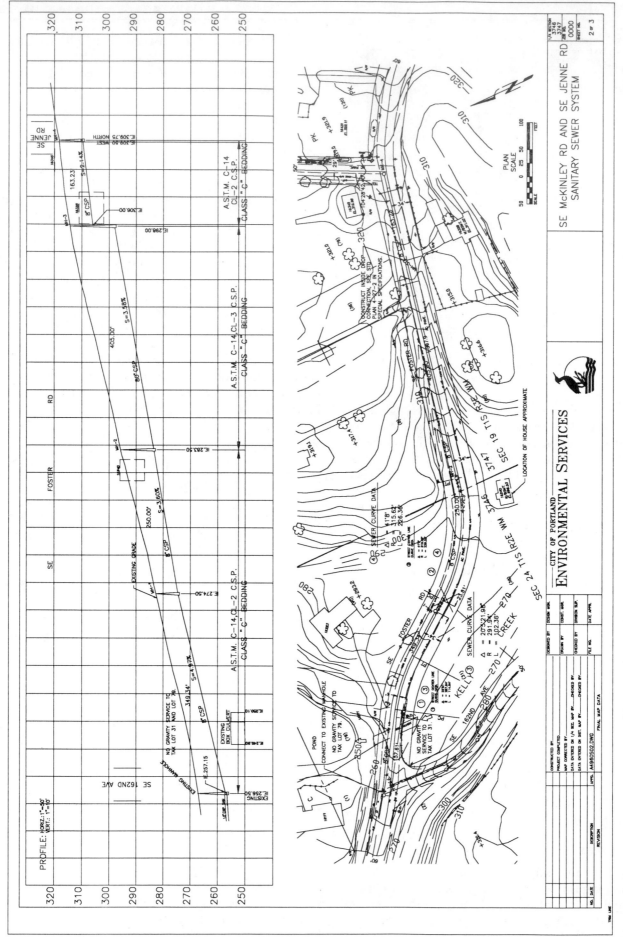

Figure 22-35 Engineering map. *Printed by permission of City of Portland.*

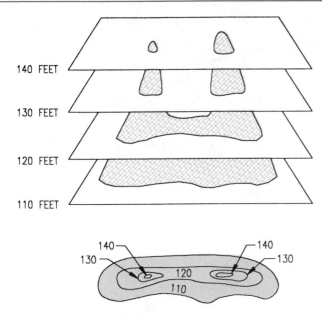

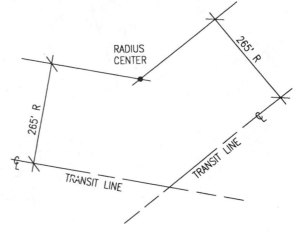

Figure 22–38 Locating radius center of highway curve.

Figure 22–36 Contours.

Point of reverse curve (PRC) is the point where two curves meet and curve in different directions.

The simplified layout in Figure 22–39 shows the relationship of the transit lines with the curve data.

Station points are given at all significant points along the transit line such as the PC, PRC, PT, and other features such as utility crossings and railroad crossings.

Fill in some of the information in the example at the beginning of the chapter. The mathematical equation for the length of the curve is

$$L = (\pi R\Delta) \div (180)$$

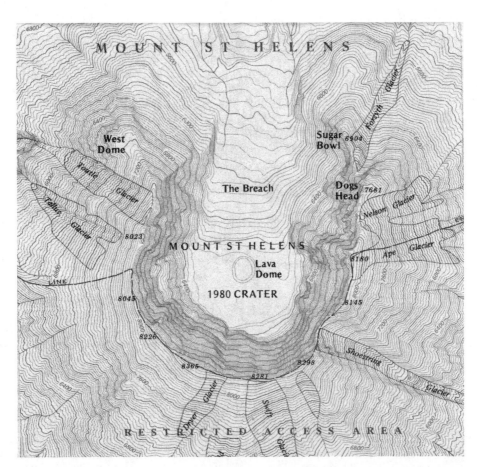

Figure 22–37 Contour intervals of Mount St. Helens. *Printed by permission of U.S. Geological Survey.*

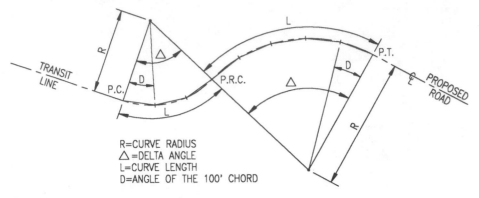

Figure 22–39   A simplified highway layout.

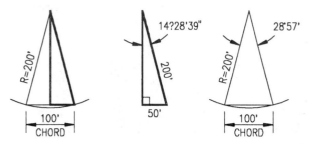

Figure 22–40   Finding delta for Figure 22–1.

Therefore, for the first curve in Figure 22–1, the length would be:

$$[(3.1416)(200)(47°23')] \div 180, \text{ or } 165.40'$$

One way to calculate the degree of curve is to imagine that the 100-foot chord is divided in half. Then, imagine that a triangle is made from the radius length (200 feet), the 50-foot length, and the default length to the center of the curve. (See Figure 22–40.) The sine of the angle at the center of the curve is the opposite side (50 feet) divided by the hypotenuse (200 feet). This number is 0.25. The inverse of the sine will give you the angle, which is 14.477° or 14°28'39". Twice this amount, which would be the amount for the full 100-foot chord, is rounded to 28°57', or the degree of curve.

Since you know the length of the first curve in Figure 22–1 is 165.40', you know that the PRC stationing is 1 + 65.40 added to 1 + 32.00, which equals 2 + 97.40. Other information for Figure 22–1 can now be calculated and drawn. (See Figure 22–41.)

A draftsperson could also receive survey notes that establish a *point of intersection (PI)* of a curve instead of a PC or PT. A point of intersection is where two transit lines would converge if a curve was not taken into account. Figure 22–38 shows an example of this. After the curve center is found, lines are drawn from the center perpendicularly to the transit lines. Where the perpendicular lines intersect the transit lines is the PC and the PT of the curve, and the angle that these two lines include is the delta angle (Δ).

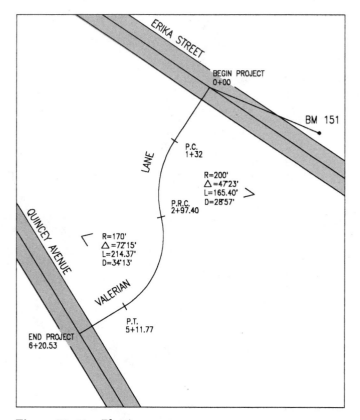

Figure 22–41   Plotting curves.

After plotting the transit line of the proposed highway (using either method), the right-of-ways can then be plotted by measuring from the transit line. In the example in Figure 22–1, plot the right-of-way 20 feet on each side of the transit line. Add station marks at every 1 + 00 station, a north arrow, and a graphic scale. (See Figure 22–42.)

Profile Layout.   A person's profile is what the person's face or body looks like from the side. A *profile* for a map is what that map looks like in elevation if a vertical slice is cut through it. Profiles are derived from the contour information of the map and information from the survey.

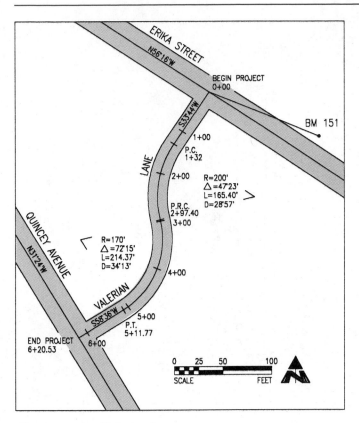

Figure 22-42   Valerian Lane.

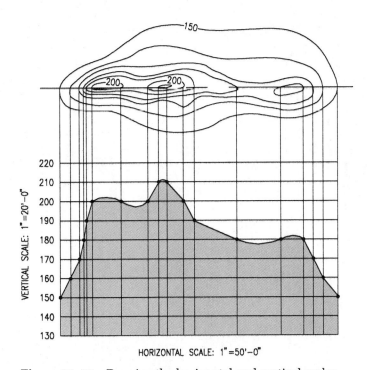

Figure 22-43   Drawing the horizontal and vertical scales.

Figure 22–43 shows a portion of a map with contours and a horizontal line through it. Wherever the horizontal line (or cutting line) crosses a contour line, a perpendicular construction line is drawn. (In this case, it would be a vertical line.) A mark is made at the point where the

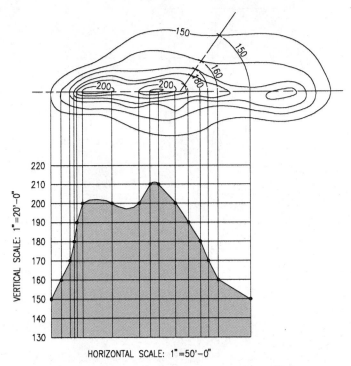

Figure 22-44   Drawing the horizontal and vertical scales.

construction line intersects with the corresponding vertical scale line. The marks are then connected. The resulting profile shown at the base of the figure has the same horizontal scale as the map, with an exaggerated vertical scale. This is common in civil drafting. It helps to visualize the slope of the land, especially in areas that are relatively flat. Horizontal scales are most often 1"=50' or 1"=100' and vertical scales are most often 1"=5' or 1"=10'. Figure 22–44 shows the profile that would be drawn if the cutting line was drawn through the plan at an angle. Remember that the construction lines need to be perpendicular to the cutting line.

Plan and profile drawings are also used when drawing proposed utility services. First, the plan of the particular utility line is plotted and drawn showing right-of-way lines, other services, and manholes, with their corresponding elevations and station points. Then a profile of the service is drawn onto the appropriate vertical scale, usually above the plan on the drawing surface. When drawing a sewer line, for example, if the elevation at the top of MH (manhole) #1 is 76.4 feet at STA 2 + 32, then the manhole would be plotted on the profile at 76.4 vertical feet. If possible, it would be plotted directly above the 2 + 32 station point. In cases in which the utility line makes turns and angles, this is not possible. An example of a plan and profile for a storm sewer layout is shown in Figure 22–45.

Invert elevation (IE) is the lowest elevation of the inside of a sewer pipe at a particular station. The IE of a pipe is shown on the profile; it is information necessary to determine the "grade slope" of the pipe. The *grade*

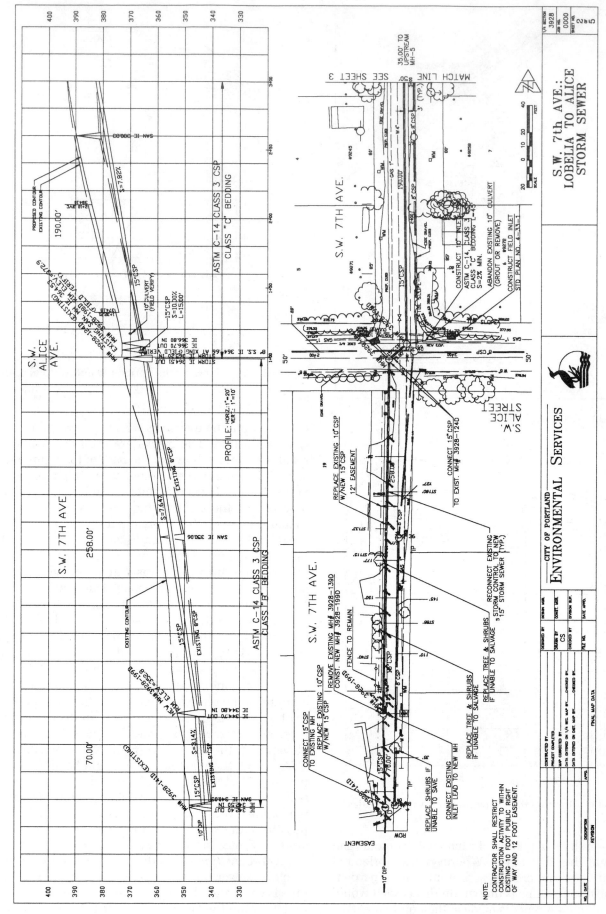

Figure 22–45 Storm sewer plan project sheet. *Printed by permission of City of Portland.*

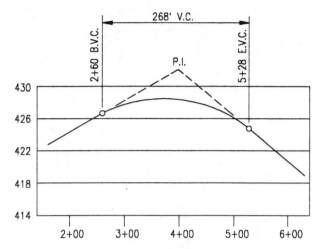

Figure 22–46   Layout of vertical curve.

slope is the percentage number given to show the amount of slope of the pipe (or road) and is shown on the profile. For example, the grade slope shown in Figure 22–45 is 7.64%. Grade slope is the amount of elevation gain (or loss) divided by the distance. So, the elevation difference is 19.71 feet (364.51 – 344.80), and the distance is 258.00 feet. The slope is 0.0764 or 7.64%.

Vertical Curves.   A *vertical curve* is one that is shown on the profile when the road travels over a hill or down a valley then up again. To draft a vertical curve, the draftsperson is first given the elevation points of the road from engineer or survey people, and these points are plotted on the profile grid. Here are a few abbreviations used when plotting vertical curves. (See Figure 22–46.)

BVC (begin vertical curve) is used at the point where the curve begins.

EVC (end vertical curve) is used at the point where the curve ends and the road has an even slope again.

VC (vertical curve) is the horizontal length of the curve from the point where it begins to the point where it ends.

PI (point of intersection) is the point where the two road grades intersect.

Cut and Fill

When designing a road, highway, or building site, portions of earth must sometimes be removed (cut) from hillsides that are too steep and added (filled) to valleys and low spots. The amount of cut and fill needed can be shown on the profile, the plan, or sometimes on both. Examine profile cut and fill in relation to planning a roadway, then in relation to a building site.

Roadway Cut and Fill.   The first step in drafting the cut and fill of a roadway is to plot the roadway on the original contour plan. Next, the necessary *cut and fill ratio* of the road is placed off the edge of the drawing surface

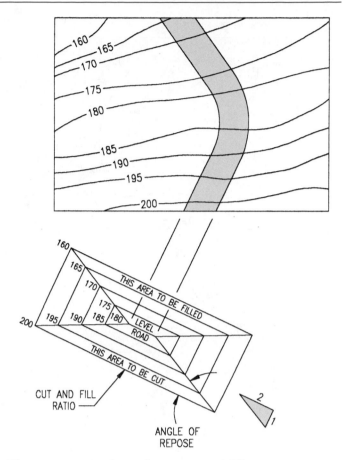

Figure 22–47   Highway layout cut and fill.

in line with the road. This ratio is given to the draftsperson by the designer or engineer and is based on information such as soil conditions. The cut and fill ratio is the relationship of run to rise (*run* being the horizontal length, *rise* being the vertical height), and is called the *angle of repose*. The angle of repose is the design slope of the area on either side of the roadway. In Figure 22–47 the angle of repose is 2:1, and the cut and fill ratio is set up so that the designed roadway will have an elevation of 180 feet. The cut will occur as the elevation increases, and the fill will occur as the elevation decreases.

Next, the desired cut and fill lines from the ratio are projected onto the drawing of the road and parallel to the road. Marks are made at the intersection of the new contour intervals and the original contours. The lines connecting the marks portray the boundaries of the proposed cut and fill around the road. (See Figure 22–48.)

Building Site Cut and Fill.   The first step when planning cut and fill changes for a building site is to plot the original contours of the site and determine where the new contours will be. Deciding where the new contours will be involves knowing if the builder wants the site to be level or wants a slope. The building site in Figure 22–49 needs to be level. In the plan view, the portion of original contours that will change are shown with

THIS AREA TO BE FILLED

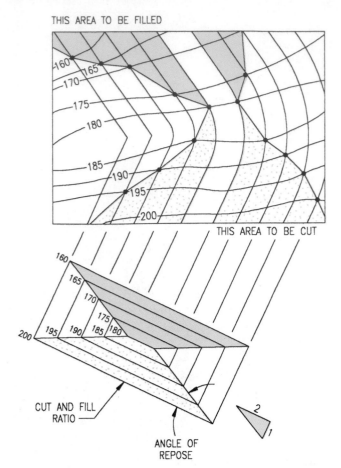

THIS AREA TO BE CUT

CUT AND FILL RATIO

ANGLE OF REPOSE

Figure 22–48   Highway layout cut and fill.

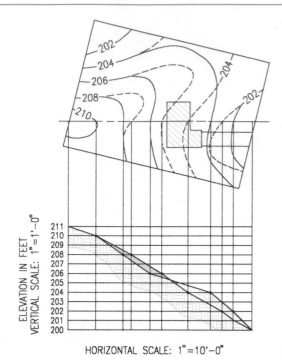

HORIZONTAL SCALE: 1"=10'-0"

Figure 22–49   Building site cut and fill.

lating the resulting areas of cut and fill in the profiles, one can determine the total volume of earth that needs to be moved.

dashed lines. The original contours that will remain unchanged as well as the new contours are shown with thin lines. A cutting line is drawn through the building site. Construction lines, perpendicular from the cutting line, are drawn from both the original contours and the new contours and plotted onto a vertical grid in which the vertical scale (1"=1'-0") is exaggerated from the horizontal scale (1"=10'). The areas that need to be cut or filled are those areas in which there is a discrepancy between the old and the new contours. If the new contour line shows higher in elevation on the profile than the old contour line, that area needs to be filled. If the new contour line is lower in elevation than the old contour line, that area needs to be cut.

The cutting line through the building site shows one section of earth that needs to be moved. By drafting several sections, perhaps 5 or 10 feet apart, and calcu-

PROFESSIONAL PERSPECTIVE

This chapter has touched on the basic portions of civil engineering drafting, with an overview of surveying fundamentals. The civil drafter needs to keep in mind the overall scope of the project when putting it together. The various phases of completing a project require the drafter to work with the different disciplines. It takes a great deal of well-organized thought and planning to synthesize the material and produce a high-quality graphic document. In your work, keep the broad scope of the project in mind.

 **MATH APPLICATION**

When finding the degree of curve (D) for a roadway or utility line, the use of the Law of Cosines is helpful. To use this law, you must know two sides and the included angle, or know all three sides. In finding the degree of curve, we know the three sides: two sides are simply the radius of the curve, and the third side is the 100-foot chord. The Law of Cosines follows:

$$a^2 = b^2 + c^2 - 2bc(\text{cosine } A)$$

in which A = the angle to be determined

a = the side opposite the A angle (the 100-foot chord)

b and c = the radius of the curve

The formula can then be rewritten for our purposes.

$$a^2 = 2b^2 - 2b^2(\cos A) \quad or \quad \cos A = (2b^2 - a^2) \div (2b^2)$$

For example, if the radius of a curve is 145.62 feet, then the formula would appear as follows:

$$\cos A = [2(145.62)^2 - 100^2] \div [2(145.62)^2]$$
$$\cos A = 32,410.3688 \div 42,410.3688$$
$$\cos A = 0.764208$$

Use the 2nd function on your calculator to obtain the inverse cosine, and the angle is 40°9'48".

The use of the simple formula shown in Figure 22–40 will get the same answer. Simply double the angle after you find the sine of the angle using the 50-foot side.

CHAPTER
22 CIVIL DRAFTING TEST

DIRECTIONS

Answer the questions with short complete statements or drawings as needed.

1. Describe the difference between a bearing and a true azimuth. Give specific examples of each.
2. What is the azimuth of a line that has a bearing of S25°W?
3. What is the azimuth of a line that has a bearing of N30°W?
4. What is the bearing of a line that has an azimuth of 80°?
5. What is the bearing of a line that has an azimuth of 237°?
6. Describe isogonic lines. What importance do they have in surveying?
7. If you were to draw an example of a north arrow on a map showing the area around Caracas, South America, what would be the approximate difference in degrees and direction for the magnetic north arrow? (See Figure 22–12.)
8. What is the difference between a closed traverse and a deflection-angle traverse?
9. What is a "chain"?
10. In what two ways can a Philadelphia rod be used?
11. Define *contour lines.*
12. Describe the difference between an EDM system and an MOD system.

13. Given a known elevation at a known distance from a transit, describe how you would find an unknown elevation using foresight and backsight. Give an example with your own arbitrary figures.
14. What is a bench mark?
15. What is meant by the symbol 13 + 79.06?
16. Is Virginia one of the public-land states?
17. Describe the difference between the rectangular system and the metes and bounds system.
18. Is there only one meridian used for land surveys in the United States? Explain.
19. What is the difference between TIN and TIS?
20. When looking at a topographical map (one which shows topography), how would you ascertain the difference between an index contour and an intermediate contour?
21. What information is used in curve data?
22. What is a PI?
23. In a profile layout, why is the vertical scale exaggerated from the horizontal scale?
24. What is the grade slope of a water main in which the IE at MH#1 is 34.88, the IE at MN#2 is 45.52, and the distance between the manholes is 1038.76 feet?
25. Describe the angle of repose. Give an example of how it is used.
26. Describe how new contours are shown on a building site. Also describe how new contours affect the profile of a section of a building site.

27. Describe a great circle.
28. What is the similarity between a connecting traverse and a loop traverse?
29. Describe the difference between a contour map and a profile map.
30. Define the IE. Describe how it is used.
31. What are the four common abbreviations used when plotting a vertical curve? Describe them.

32. Define the prime meridian.
33. How large is a township?
34. How large is a section?
35. What is the number of the section in the lower, left-hand corner of a township?

CHAPTER

22 CIVIL DRAFTING PROBLEMS

DIRECTIONS

Use Mylar®, vellum, or a CADD system for all of the problems that call for a drawing to be created. Include a north arrow and note the scale where applicable.

Problem 22–1 Draw an open traverse at a scale of 1"=20' using the following information:

From POB, traverse 34.23' at S61°42'E, then 15.69' at S44°7'E, then 40.89' at N78°48'E, then 39.33' at N56°14'E.
Be sure to label each course.

Problem 22–2 Give the azimuth angles as well as the bearings of the lines shown. Fill in the values in the following table.

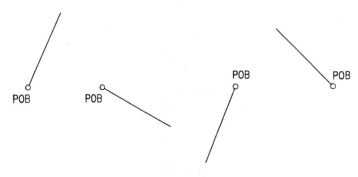

POB   POB   POB   POB

LINE	AZIMUTHAL ANGLE	BEARING
AB		
CD		
EF		
GH		

Problem 22–3 Using either Mylar®, vellum, or a CADD system, draw a traverse at a scale of 1"=20' using the following lengths and bearings. What kind of traverse is this?

21 feet at N40°7'E, 77 feet at S84°9'E, 19 feet at S24°29'E, 48 feet at S39°22'W, 31 feet at S76°9'W, 65 feet at N34°57'W.

○
POB

Problem 22–4 Using either Mylar®, vellum, or a CADD system, transfer the information shown on the following figure. Then draw a traverse at a scale of 1"=20' using the following lengths and azimuth angles. What kind of traverse is this?

23 feet at an angle of 131°40', 11 feet at an angle of 231°8', 36 feet at an angle of 113°24', 49 feet at an angle of 64°39', 27 feet at an angle of 303°50', 36 feet at an angle of 65°33'.

○
POB

○
KNOWN
POINT

Problem 22–5 Using either Mylar®, vellum, or a CADD system, transfer the information shown on the following figure. Then draw a deflection angle traverse at a scale of 1"=40' using the following lengths and bearings. Be sure to use proper notation for the deflections. What type of traverse is this?

48' at S63°46'E, 33' at S37°32'E, 71' at N86°10'E, 23' at S81°4'E, 73' at S73°9'E.

○
POB

○
BENCH
MARK

Problem 22–6 Complete the leveling notes that were gathered by the survey crew. Create a leveling notes table and fill in the HI and elevations for each station.

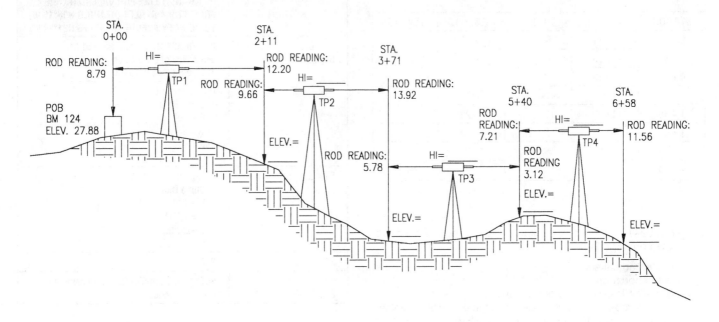

Problem 22–7 Draw the grid as shown on a separate piece of Mylar®, vellum, or on a CADD system. Mark the following townships as shown:

A Township: T2S, R2W
B Township: T3N, R4E
C Township: T1N, R2E
D Township: T2S, R2E
E Township: T1N, R4W

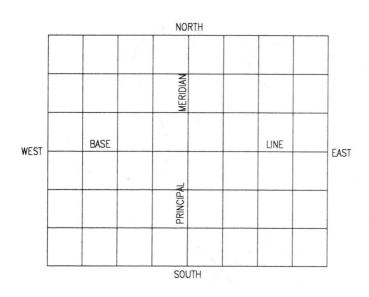

Problem 22–8 Draw the section grid as shown on a separate piece of Mylar®, vellum, or on a CADD system. Mark the following parcels of land as shown:

A Parcel: SW 1/4, SE 1/4
B Parcel: W 1/2, NW 1/4
C Parcel: W 1/2, NE 1/4, SW 1/4
D Parcel: NE 1/4, NE 1/4, NW 1/4
E Parcel: E 1/2, SE 1/4, SE 1/4

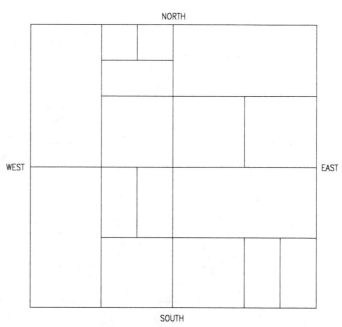

Problem 22–9 Complete the leveling notes that were gathered by the survey crew. Then plot the resulting profile using 1"=10' for the vertical scale and 1"=40' for the horizontal scale.

STATION	B.S. (+)	H.I.	F.S. (−)	ELEVATION
BM 12	17.31		7.46	34.12
0 + 30	6.83		4.87	
0 + 55	9.66		7.72	
1 + 05	14.30		10.13	
1 + 45	12.20		13.29	
1 + 85	7.12		15.82	
2 + 15	10.04		12.73	
2 + 40				BM 13

Problem 22–10 Answer the following questions using the Figure shown on page 749.

What is the vertical scale?
What is the horizontal scale?
What is the degree of curve of the sewer?
Where is the property located in relation to the public land survey?
If the IE at MH-4 is 309.75, and the IE at MH-5 is 313.50, then what is the grade slope?
What is the contour interval?
Printed by permission of City of Portland.

Problem 22–11 Complete the unknown values that were given at the beginning of the chapter.

STATION	TRANSIT LINE FOR VALERIAN LANE FROM ERIKA STREET (BEARING N56°16'W) TO QUINCEY AVENUE (BEARING N31°24'W)
0 + 00 Begin Project	245.78'N67°37'W from BM 151 to center line of Erika Street. Then 132.00'S33°44'W to P.C.
STA 1 + 32	P.C. to the left Curve Data R = 200' Δ = 47°23' L = _____ D = _____
STA _____	P.R.C. Curve Data R = 170' Δ = 72°15' L = _____ D = _____
STA _____	P.T.
STA _____ End Project	S58°36'W from STA _____ to center line of Quincey Avenue

Problem 22–12 Draw a profile from this contour, drawing along line A-A. Use a vertical scale of 1"=10'.

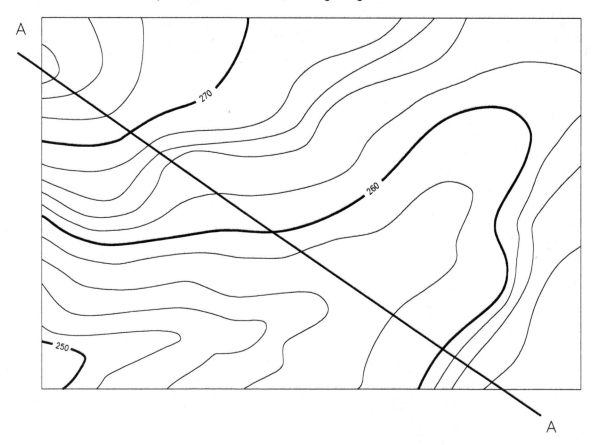

Problem 22–10 (Continued)

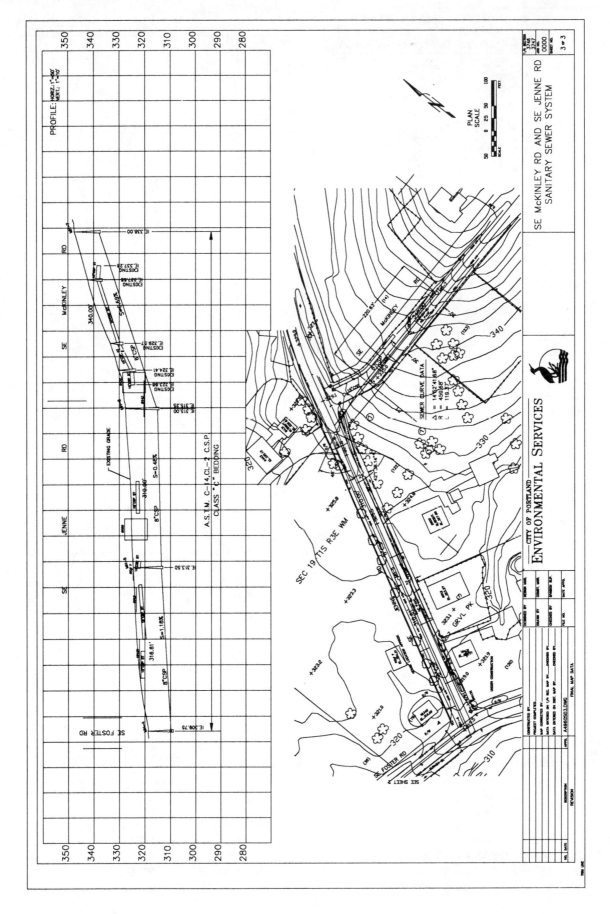

Problem 22–13 Draw a roadway, Jean Road, from the following table. Be sure to include scale (1"=100') and north arrow.

STATION	TRANSIT LINE FOR JEAN ROAD FROM RORY BLVD. (BEARING N53°14'E) TO MURPHY COURT (BEARING N8°15'W)
0 + 00 Begin Project	93.24'S27°42'W from BM 43 to centerline of Rory Blvd. Then 147.93'S80°15'E to P.C.
STA ____	P.C. to the right Curve Data R = 115' Δ = 47.00° L = ____ D = ____
STA ____	P.T., then 50.00'S33°15'E
STA ____	P.C. to the left Curve Data R = 102' Δ = 65.00° L = ____ D = ____
STA ____	P.T., then 149.56'N81°45'E to centerline of Murphy Court
STA ____ End Project	Length and bearing from BM 44: ____

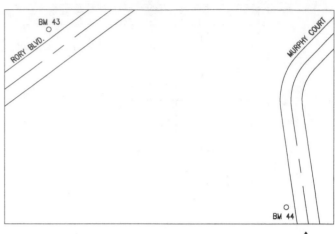

SCALE: 1"=100'

Problem 22–14 A proposed highway is shown. Using an angle of repose of 1:1-1/2 and a contour interval of 10', and assuming the proposed highway will be level with an elevation of 120 feet, draw the proposed cut and fill areas. Use different symbols for the cut and fill.

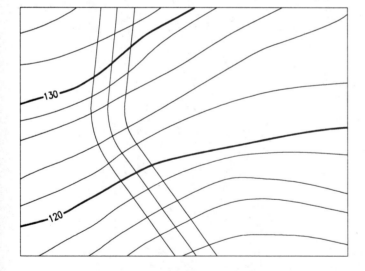

Problem 22–15 The building site needs some contour changes to accommodate the new house being built. Draw a profile line along line A-A, showing the cut and fill needed. Draw the profile at 1"=40' horizontal and 1"10' vertical.

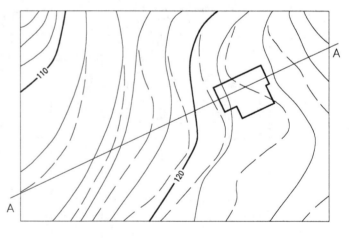

MATH PROBLEMS

1. Find the grade slopes between the water main manholes.

   MH-1 IE = 124.63 STA. 0 + 67.02
   MH-2 IE = 137.44 STA. 2 + 78.14
   MH-3 IE = 145.67 STA. 4 + 50.07
   MH-4 IE = 159.12 STA. 6 + 82.58
   MH-5 IE = 174.88 STA. 9 + 43.32

2. Using the Law of Cosines, find the degree of curve (D) to the nearest second of the following curves.

   R = 1,224.90', Δ = 15°34'
   R = 335.00', Δ = 27°56'
   R = 75.45', Δ = 24°26'
   R = 602.25', Δ = 35°00'

CHAPTER 23

Heating, Ventilating, and Air Conditioning (HVAC)/ Sheet Metal Drafting

LEARNING OBJECTIVES

After completing this chapter, you will:

○ Discuss the purpose and function of HVAC systems.
○ Prepare complete HVAC drawings, including plans, schedules, and details.
○ Draw sheet metal pattern developments and intersections.
○ Calculate and apply bend allowances to sheet metal components.
○ Draw and completely dimension precision sheet metal fabrication drawings.
○ Use an engineering problem as an example for HVAC and sheet metal drawing solutions.

THE ENGINEERING DESIGN PROCESS

Your company produces a wide range of sheet metal products for various customers, ranging from HVAC duct work to precision housing structures for electronic equipment. Your current drawing project is a chassis for an electronic testing device.

With HVAC drawings, it is not usually necessary to provide bend allowances. In this project, however, the final development requires tighter tolerances for all of the components to mount properly within the chassis. Calculating bend allowances enables you to determine the required size of the flat pattern.

While there are a number of methods for determining bend allowance, your company standards reference the tables listed in the *Machinery's Handbook*. By checking the table for the type of metal you are using plus the bend radius and thickness of material you are able to determine the appropriate bend allowances.

Applying the bend allowance to each bend in the flat pattern ensures that when the material is formed into its final shape the desired final dimensions can be achieved. Figure 23–1 shows an example flat pattern layout.

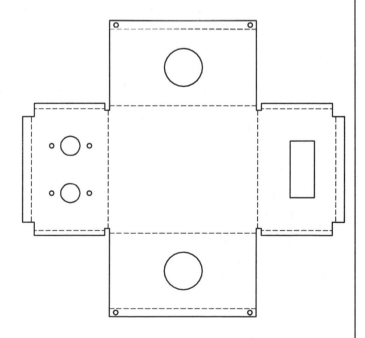

Figure 23–1   Sample flat pattern layout for the engineering problem.

Most residential and commercial structures have *heating, ventilating, and air conditioning (HVAC)* systems. These systems are also commonly known as *mechanical systems*. The purpose of most heating and air conditioning systems is to help maintain a normal comfort zone for the occupants living or working in the structure. In other applications, such as meat lockers, the product may be the reason for environmental controls. The function of ventilating systems is to provide air movement or exchange within the structure. Ventilation is required when excess heat, fumes, moisture, odor, or pollutants must be removed and fresh air replaced. Most residential structures do not use HVAC plans unless the system is complex, or required by the lending or code enforcement agency. The architect of a commercial or public building often consults with an HVAC engineer for the design, drafting, and installation supervision of the HVAC system. The consulting engineer prepares all of the designs and specifications for the project and the drafter either works directly from written specifications or from engineering sketches. In many cases the engineer prepares rough sketches directly on a copy of the floor plans. The HVAC drafter then creates a new drawing using proper lines, symbols, dimensions, and notes. Some companies use an HVAC overlay where the base sheet is the floor plan and another sheet, known as the overlay, is used to draw the HVAC plan.

▼ SHEET METAL DRAFTING

HVAC systems are made up of mechanical equipment such as the furnace, air conditioner or ventilator, and duct work. The *duct* is generally sheet metal pipe designed as the passageway for conveying air from the HVAC equipment to the source. Types of duct shapes include cylindrical, oval, and rectangular. Connectors such as elbows and transition shapes that allow conversion from round to square or rectangular are also used. Any duct shape is possible depending on the application. In most residential applications standard premanufactured duct work is used. In commercial structures the duct work may be premanufactured or custom built in a sheet metal shop. Flat patterns are made when custom sheet metal shapes are required. These patterns are made by a sheet metal layout person or a sheet metal drafter. The flat pattern drawings, known as pattern developments, are transferred to thick paper. The paper pattern is then transferred to flat sheet metal to be formed into the desired duct shape. CADD systems are being used to develop the flat pattern and transfer the pattern to the metal.

Other industries also make sheet metal patterns. Whenever flat metal is bent into shape, a sheet metal pattern is required. Examples are auto body parts, storage facilities, or electronics chassis components.

▼ HVAC SYSTEMS
Central Forced-air Systems

Central forced-air systems, among the most common systems for climate heating and air conditioning, circulate the air from the living spaces through or around heating or cooling devices. A thermostat starts the cycle as a fan forces the air into ducts. These ducts connect to openings called diffusers, or air supply registers, which put warm air (WA) or cold air (CA) in the room. The air enters the room and either heats or cools as needed. Air then flows from the room through another opening called a return-air register (RA) and into the return duct. The return duct directs the air from the room over a heating or cooling device, depending on which is needed. If cool air is required, the return air is passed over the surface of a cooling coil. If warm air is required, the return air is passed over either the surface of a combustion chamber (the part of a furnace where fuel is burned) or a heating coil. The conditioned air is picked up again by the fan and the cycle is repeated. Figure 23–2 shows the air cycle in a forced-air system.

Refrigeration

Refrigeration, the most common type of cooling system, is based on the principle that as a liquid changes to vapor it absorbs large amounts of heat. The boiling point of a liquid can be altered by changing the pressure applied to the liquid so that when a gas is changed to a liquid, it will give up the heat. The basic parts of a refrigeration system are the cooling coil (evaporator), compressor (air pump), condenser (where vaporized refrigerant is liquified), and expansion valve. Common refrigerants boil at low temperatures. Figure 23–3 shows a pictorial diagram of the refrigeration cycle.

Hot Water System

In a hot water system water is heated as it circulates around the combustion chamber of a fuel-fired boiler. The water is then circulated through pipes to radiators or convectors in the rooms. In a one-pipe system one pipe leaves the boiler and runs through the building and back to the boiler as shown in Figure 23–4. In a two-pipe system one pipe supplies heated water to all the outlets, while the other is a return pipe, which carries the water back to the boiler for reheating as shown in Figure 23–5. Hot water systems use a pump, called a circulator, to move the water through the system. The water is kept at a temperature of 150°–180° F in the boiler. When heat is needed, a thermostat starts the circulator pump.

Zoned Control System

A *zoned* system allows for one or more heaters and one thermostat per room. No ductwork is required and

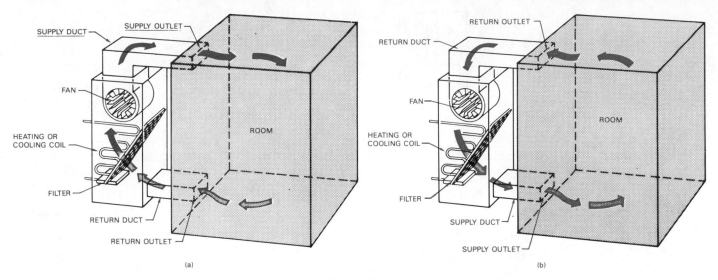

Figure 23–2   **(a)** Down draft forced air system air cycle. **(b)** Up draft forced air system air cycle.

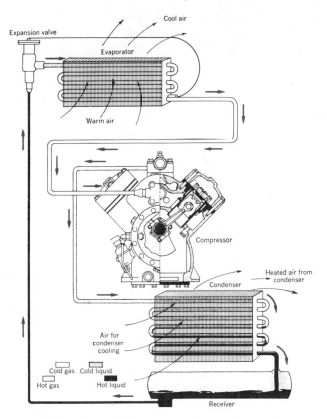

Figure 23–3   Pictorial diagram of refrigeration cycle.

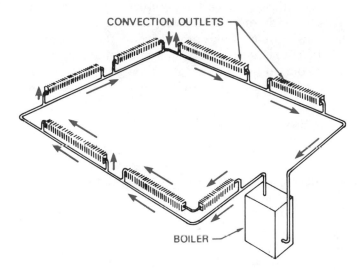

Figure 23–4   One-pipe hot water system.

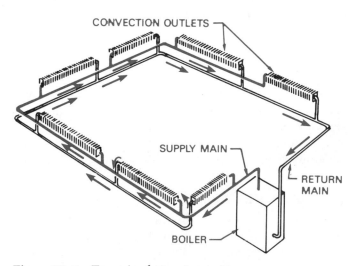

Figure 23–5   Two-pipe hot water system.

only the heaters in occupied rooms need to be turned on. One of the major differences between a zoned and central system is flexibility. A zoned system allows the occupant to determine how many areas are heated and how much energy is used.

Radiant Heat

Radiant heating and cooling systems function on the basis of providing a comfortable environment by means

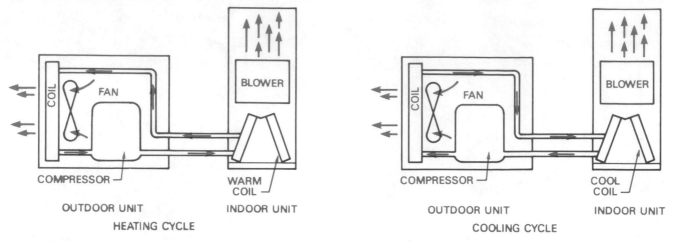

Figure 23–6 Heat pump heating and cooling cycle. *Courtesy Lennox Industries, Inc.*

of controlling surface temperatures and minimizing excessive air movement within the space. Surface-mounted radiant panels provide comfort at a lower thermostat temperature than other systems. Radiant systems vary from oil or gas hot water piping in the ceiling or floor, to electric coils, wiring, or elements at the ceiling.

Heat Pump System

The heat pump is a forced-air central heating and cooling system that operates using a compressor and a circulating liquid gas refrigerant. Heat is extracted from the outside air and pumped inside the structure. The heat pump supplies up to three times as much heat per year for the same amount of electrical consumption. A standard electrical forced-air heating system works best when outside air is above 20°F. In the summer the cycle is reversed and the unit operates as an air conditioner. In this mode the heat is extracted from the inside air and pumped outside. On the cooling cycle the heat pump also acts as a dehumidifier. Figure 23–6 shows how a heat pump works.

▼ VENTILATION

There are a number of reasons why ventilation of a structure or area is necessary. Residential applications include bath, kitchen, and laundry exhaust fans. In commercial applications ventilation may be necessary to exhaust fumes, pollutants, or moisture.

Sources of Pollutants

There are a number of sources of pollutants that make it necessary to plan ventilation systems. Moisture in the form of relative humidity can cause structural damage and health problems such as respiratory troubles

and microbial growth. Each individual can produce up to one gallon of water vapor per day.

Indoor combustion from such items as gas-fired or wood-burning appliances, or fireplaces can generate a variety of pollutants, including carbon monoxide and nitrogen oxides.

Humans and pets can transmit diseases through the air by exhaling a variety of bacterial and viral contaminants.

Tobacco smoke may contribute chemical compounds to the air environment. The pollution can affect smokers and nonsmokers.

Formaldehyde in glues used in construction materials such as plywood, particle board, or even carpet and furniture causes pollution, as do the components of certain insulations. Formaldehyde has been considered a factor in certain diseases, eye irritation, and respiratory problems.

Radon is a naturally occurring radioactive gas that breaks down into compounds that may cause cancer when large quantities are inhaled over a long period of time. Radon may be more apparent in a structure that contains a great deal of concrete, or is located in certain geographic areas of the country. Radon can be scientifically monitored at a nominal cost. Barriers can be built that help reduce the concern of radon contamination.

Household products such as aerosols and crafts materials such as glues and paints can contribute a number of toxic pollutants.

Air-to-air Heat Exchangers

The air-to-air heat exchanger is a heat-recovery ventilation device tht pulls stale, polluted, warm air from the living or working space through a duct system and transfers the heat in that air to the fresh cold air being pulled into the structure. Heat exchangers do not produce heat; they only move heat from one air stream to the other. The heat transfers to the fresh air stream in the

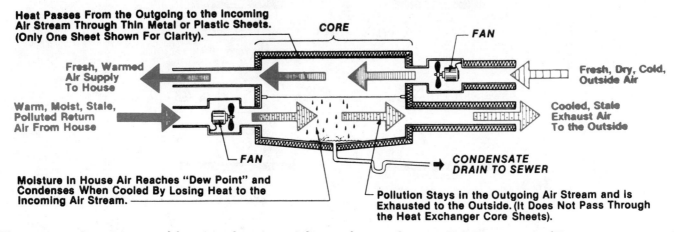

Figure 23–7 Components and function of an air-to-air heat exchanger. *Courtesy U.S. Department of Energy.*

core of the heat exchanger. The core is usually designed to avoid mixing of the two air streams to insure that indoor pollutants are removed. Moisture in the stale air condenses in the core and is drained from the unit. Figure 23–7 shows the basic components and function of an air-to-air heat exchanger. Knowledgeable HVAC engineers, designers, and contractors are able to implement air-to-air heat exchanger technology by the selection of the proper size of a ducted system.

▼ THERMOSTAT

The thermostat is an automatic mechanism for controlling the amount of heating or cooling given by a central or zoned heating or cooling system. The thermostat symbol is shown in Figure 23–8.

The location of the thermostat is an important consideration to the proper functioning of the system. For zoned units there may be thermostats placed in each room or a central panel placed in a convenient location. For central systems there may be one or more thermostats, depending on the layout of the system or the number of units required to service the structure. For example, an office complex may have a split system where the structure is divided into two or more zones. Each individual segment of the system has its own thermostat.

Several factors contribute to the effective placement of the thermostat for a central system. A good common location is near the center of the structure and close to a return air duct. The air entering the return air duct is usually temperate, thus causing little variation in temperature. This is a location where an average temperature

reading can be achieved. There should be no drafts. Avoid locations where sun light or a heat register could cause an unreliable reading. Avoid placement near an exterior opening or on an outside wall. Avoid placement near stairs or a similar traffic area where significant bouncing or shaking could cause the mechanism to alter the actual reading.

▼ HVAC SYMBOLS

There are over a hundred HVAC symbols that may be used in residential and commercial heating plans. Only a few of the possible symbols are typically used in residential HVAC drawings. The symbols are divided into two types: standard heating, ventilating, and air-conditioning symbols and air-conditioning symbols. Figure 23–9 shows some common HVAC symbols. Sheet metal conduit (duct) and air-conditioning templates are available to help speed manual drafting and CADD menus and symbols libraries may be used to customize drafting practices.

ANSI Standard symbols adopted by ANSI are available in the document, "Graphical Symbols for Heating, Ventilating, and Air Conditioning; ANSI Y32.2.4." Standard symbols are often modified to fit individual needs, with a legend placed on the drawing for interpretation.

▼ HVAC DRAWINGS

Drawings for the HVAC system show the size and location of all equipment, duct work, and components with accurate symbols, specifications, notes, and schedules that form the basis of contract requirements for construction. Specifications are documents that accompany the drawings and contain all pertinent written information related to the HVAC system.

Figure 23–8 Thermostat floor plan symbol.

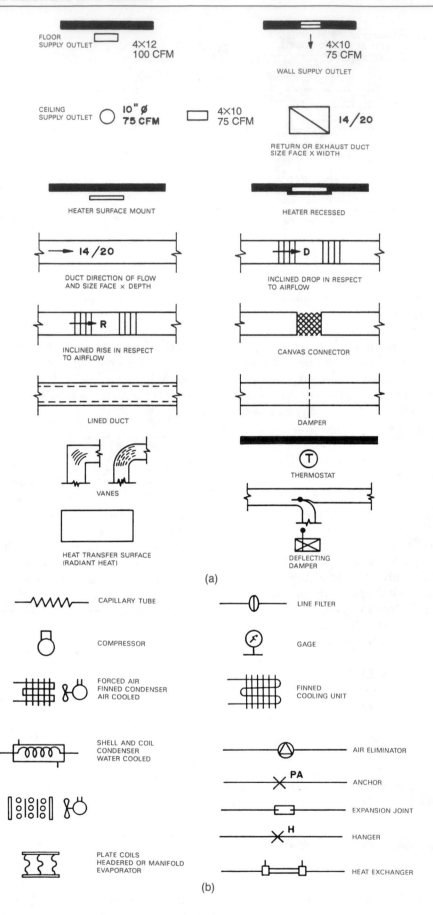

Figure 23–9   Common HVAC symbols.

CADD Applications

HVAC PICTORIALS

CADD applications often make pictorial representations much easier to implement, especially when the HVAC program allows direct conversion from the plan view to the pictorial. The CADD pictorials, known as graphic models, may be used to view the HVAC system from any angle or orientation. Some CADD programs automatically analyze the layout for obstacles where an error in design may result in a duct that does not have a clear path. One of the biggest advantages of the CADD system is that when changes are made in the HVAC plan, these changes are simultaneously corrected on all drawings, schedules, and lists of materials. Figure 23–10 shows a CADD-generated perspective of HVAC duct routing.

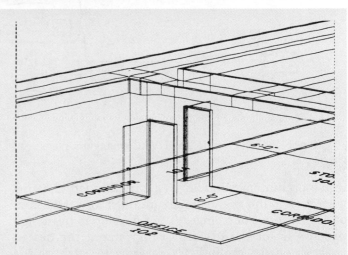

Figure 23–10   CADD-generated perspective of HVAC duct routing. *Courtesy Computervision Corporation.*

Drawings may be prepared by the architect, architectural drafter, or the heating contractor when a complete HVAC layout is necessary. Figure 23–11 shows a heating plan for a residential structure.

For commercial structures the HVAC plan may be prepared by an HVAC engineer as a consultant for the architect. The consulting engineer is responsible for the HVAC design and installation. The engineer determines the placement of all equipment and the location of all duct runs and components. He or she also determines all of the specifications for unit and duct size based on calculations of structure volume, exterior surface areas and construction materials, rate of air flow, and pressure. The engineer may prepare single-line sketches or submit data and calculations to a design drafter who prepares design sketches or final drawings. Drafters without design experience work from engineering or design sketches to prepare formal drawings. A single-line engineer's sketch is shown in Figure 23–12. The next step in the HVAC design is for the drafter to convert the rough sketch

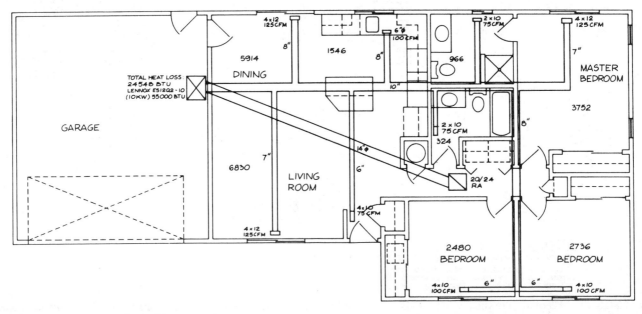

Figure 23–11   Residential heating plan.

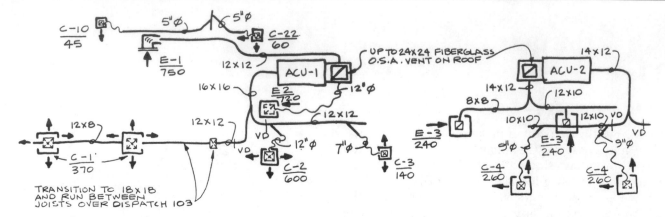

Figure 23–12 Single line HVAC engineer's sketch.

into a preliminary drawing. This preliminary drawing goes back to the engineer and architect for verification and corrections or changes. The final step in the design process is for the drafter to implement the design changes on the preliminary drawing to establish the final HVAC drawing. The final HVAC drawing is shown in Figure 23–13.

Convert an engineering sketch to a formal drawing using manual or CADD in this manner:

1. Draw duct runs using thick (0.7 or 0.9 mm) line widths.
2. Label duct sizes within the duct when appropriate, or use a note with a leader to the duct in other situations.
3. Duct sizes may be noted as 22 × 12 or 22/12, where

the first number, 22, is the duct width and the second numeral, 12, indicates the duct depth.

4. Place notes on the drawing to avoid crowding. Aligned techniques may be used where horizontal notes read from the bottom of the sheet and vertical notes read from the right side of the sheet. Make notes clear and concise.
5. Refer to schedules to get specific drawing information that may not otherwise be available on the sketch.
6. Label equipment to clearly stand out from other information on the drawing either blocked out or bold.

Several examples of duct system elements are shown comparing the engineering sketch and formal drawing in Figure 23–14.

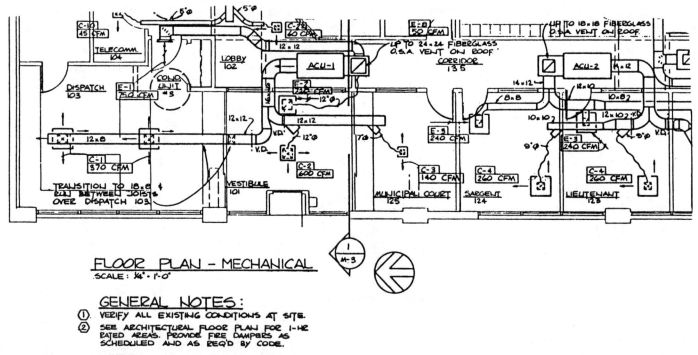

Figure 23–13 HVAC plan from the engineer's sketch shown in Figure 23–12. *Courtesy W. Alan Gold Consulting Mechanical Engineer and Robert Evenson Associates AIA Architects.*

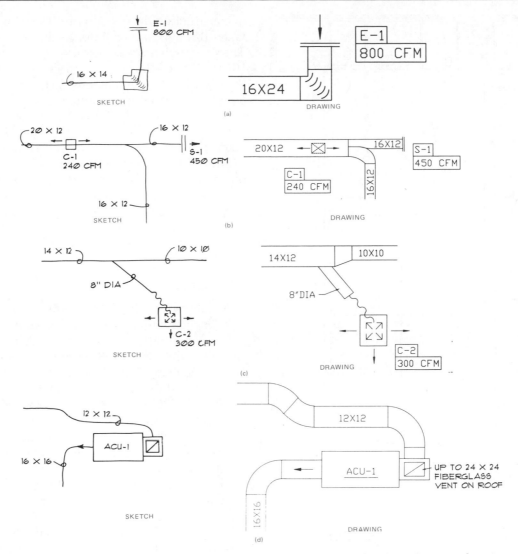

Figure 23–14 Examples showing engineering sketches converted to formal HVAC drawings.

Single- and Double-line HVAC Plans

HVAC plans are drawn over the outline of the floor plan or as an overlay. The floor plan layout is drawn first using thin lines as a base sheet for the HVAC layout and other overlays. The HVAC plan is then drawn using thick lines and notes for contrast with the floor plan. The HVAC plan shows the placement of equipment and duct work. The size (in inches) and shape (with symbols, ∅ = round, □ = square or rectangular) of duct work and system component labeling is placed on the drawing or keyed to schedules. Drawings may be either single line or double line, depending on the needs of the client or how much detail must be shown. Single-line drawings are easier and faster to draw. In many situations they are adequate to provide the equipment placement and duct routing as shown in Figure 23–15. Double-line drawings take up more space and are more time consuming to draw than single line, but they are often necessary when complex systems require more detail as shown in Figure 23–13.

Detail Drawings

Detail drawings are used to clarify specific features of the HVAC plan. Single- and double-line drawings are intended to establish the general arrangement of the system; they do not always provide enough information to fabricate specific components. When further clarification of features is required, detail drawings are made. A detail drawing is an enlarged view(s) of equipment, equipment installations, duct components, or any feature that is not defined on the plan. Detail drawings may be scaled or unscaled and provide adequate views and dimensions for sheet metal shops to prepare fabrication patterns as shown in Figure 23–16.

Section Drawings

Sections or sectional views are used to show and describe the interior portions of an object or structure that would otherwise be difficult to visualize. Section drawings may be used to provide a clear representation

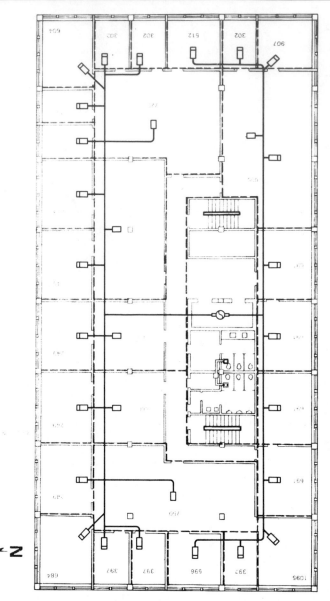

Figure 23–15 Single line ducted HVAC system showing a layout of the proposed trunk and runout ductwork. *Courtesy The Trane Company, La Crosse, WI.*

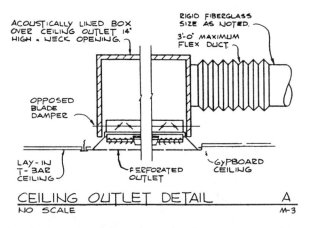

Figure 23–16 Detail drawings. *Courtesy W. Alan Gold Consulting Mechanical Engineer and Robert Evenson Associates AIA Architects.*

of construction details or a profile of the HVAC plan as taken through one or more locations in the building. There are two basic types of section drawings used in HVAC. One method is used to show the construction of the HVAC system in relationship to the structure. In this case the building is sectioned and the duct system is shown unsectioned. This drawing provides a profile of the HVAC system. There may be one or more sections taken through the structure, depending on the complexity of the project. The building structure may be drawn using thin lines as shown in Figure 23–17. Figure 23–17 is a section through the HVAC plan shown in Figure 23–13. The other sectioning method is used to show detail of equipment, or to show how parts of an assembly fit together. (See Figure 23–18.)

Schedules

Numbered symbols that are used on the HVAC plan to key specific items to charts are known as schedules. These schedules are used to describe items such as ceiling outlets, supply and exhaust grills, hardware, and equipment. Schedules are charts of materials or products that include size, description, quantity used, capacity, location, vendor's specification, and any other information needed to construct or finish the system. Schedules aid the drawing by keeping it clear of unnecessary notes. They are generally placed in any convenient area of the drawing field or on a separate sheet. Items on the plan may be keyed to schedules by using a letter and number combination such as C-1 for CEILING OUTLET NO. 1, E-1 for EXHAUST GRILL NO. 1, or ACU-1 for EQUIPMENT UNIT NO. 1. The exhaust grill schedule keyed to the HVAC plan in Figure 23–13 may be set up as a chart as shown in Figure 23–19.

Pictorial Drawings

Pictorial drawings may be isometric or oblique as shown in Figure 23–20 (see page 762). Isometric, oblique, and perspective techniques are discussed in Chapter 17. Pictorial drawings are usually not drawn to scale. They are used in HVAC for a number of applications, such as assisting in visualization of the duct system, and when the plan and sectional views are not adequate to show difficult duct routing. Manually drawn pictorials are time consuming and generally not used unless necessary.

▼ SHEET METAL DRAFTING

Sheet metal drafting is used in any industry where flat material is used for fabrication into desired shapes. One of the most common applications is the HVAC industry, although sheet metal shapes are common in the automotive, electronics, and other related industries. Sheet metal drafting is also called pattern development.

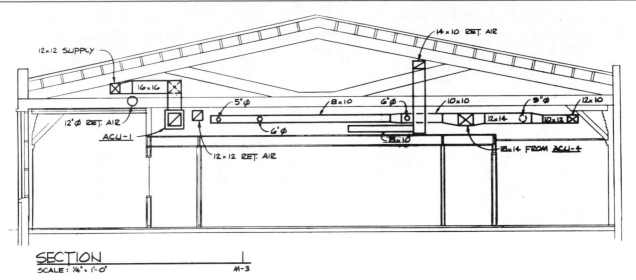

SECTION
SCALE: ¼" = 1'-0" M-3

Figure 23–17 Section drawing. *Courtesy W. Alan Gold Consulting Mechanical Engineer and Robert Evenson Associates AIA Architects.*

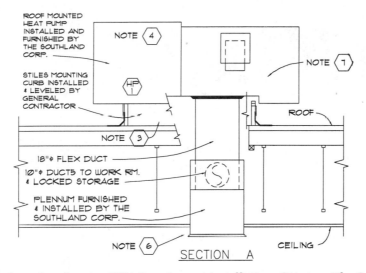

SECTION A

Figure 23–18 Detailed section showing HVAC equipment installation. *Courtesy The Southland Corporation.*

EXHAUST GRILL SCHEDULE

SYMBOL	SIZE	CFM	LOCATION	FIRE DPR.	KEY OP. OPP BLD	KEY OP. EXTR	NO. DPR.	TYPE	REMARKS
				DAMPER TYPE					
E–1	24x12	750	HIGH WALL	✕	✕			4	
E–2	18x18	720	CEILING	✕	✕			2	
E–3	10x10	240	CEILING		✕			1	24x24 PANEL
E–4	10x10	350	CEILING		✕			1	
E–5	12x12	280	CEILING		✕			1	
E–6	10x10	500	CEILING	✕				1	
E–7	6x6	350						2	
E–8	12x6	50			✕			3	
E–9	12x6	200			✕			3	
E–10	12x8	150			✕			3	
E–11	10x10	290			✕			3	
E–12	9x4	160						1	24x24 PANEL
E–13	9x4	75	HIGH WALL	✕	✕			4	

TYPE 1: KRUGER 1190 SERIES STEEL PERFORATED FRAME 23 FOR LAY-IN TILE

TYPE 2: KRUGER 1190 SERIES STEEL PERFORATED FRAME 22 FOR SURFACE MOUNT

TYPE 3: KRUGER EGC-5:1/2"X1/2"X1/2" ALUMINUM GRID.

TYPE 4: KRUGER S80H:35° HORIZ. BLADES 3/4"O.C.

Figure 23–19 A typical HVAC schedule.

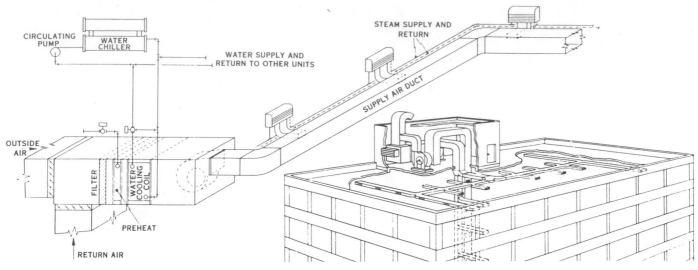

Figure 23–20 Pictorial drawings. *Courtesy The Trane Company, La Crosse, WI.*

CADD Applications

HVAC INDUSTRY

HVAC CADD software is available that allows you to place duct fittings and then automatically size ducts in accordance with common mechanical equipment suppliers' specifications. The floor plan is commonly used as a reference layer and the HVAC plan is a separate layer. The CADD drafter follows these simple steps:

Step 1. The drawing begins with the preliminary layout drawn as the duct centerlines. (See Figure 23–21.)

Step 2. Select supply and return registers from the template menu symbols library and add the symbols to the end of the centerlines where appropriate. (See Figure 23–22.)

Step 3. The program then automatically identifies and records the lengths of individual duct runs, and tags each run. Fittings are located and identified by the type of intersection. (See Figure 23–23.) While all of this draw-

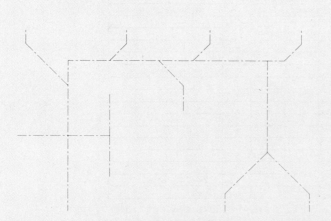

Figure 23–21 The drawing begins with the preliminary layout drawn as duct centerlines. *Courtesy Chase Systems.*

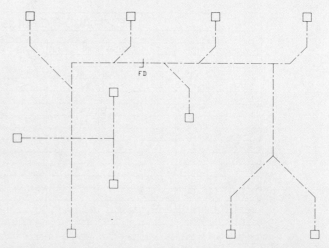

Figure 23–22 Select supply and return registers from a tablet menu. *Courtesy Chase Systems.*

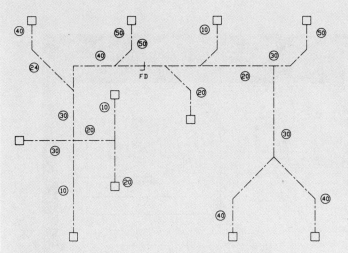

Figure 23-23   Fittings are located and identified by the type of intersection. *Courtesy Chase Systems.*

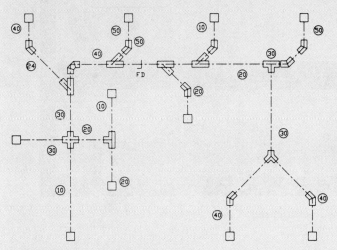

Figure 23-24   HVAC symbols are drawn as accurate double-line symbols exactly to ANSI Y32.2.4 standards. *Courtesy Chase Systems.*

ing information is added to the layout, the computer automatically gathers design information into a file for duct sizing based on a specific mechanical manufacturer's specifications that you select.

Step 4.   After the fitting location and sizes are determined, the program transforms each fitting into accurate double-line symbols exactly to the ANSI Y32.2.4 standard. (See Figure 23-24.)

Step 5.   When the fittings are in place, the program calculates and draws the connecting ducts, adding couplings automatically at the maximum duct lengths. If a transition is needed in a duct run, the program recommends the location and all you have to do is pick a transition fitting from the menu library. See Figure 23-25 for the complete HVAC layout.

An added advantage to using HVAC CADD software is that the program automatically records information, while you draw, to generate a complete bill of materials. The systems that offer you the greatest flexibility and productivity are designed as a parametric package. This type of program allows you to set the design parameters that you want and then the computer automatically draws and details according to these settings. As you draw, information such as the type of fitting, CFM, and gauge, is placed with each fitting. A complete HVAC plan and related schedules are shown in Figure 23-26.

CADD HVAC packages provide a variety of tablet menu overlays that assist in the rapid selection of symbols.

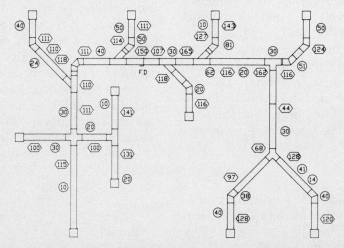

Figure 23-25   The finished HVAC plan. *Courtesy Chase Systems.*   **(Continued)**

CADD Applications (Continued)

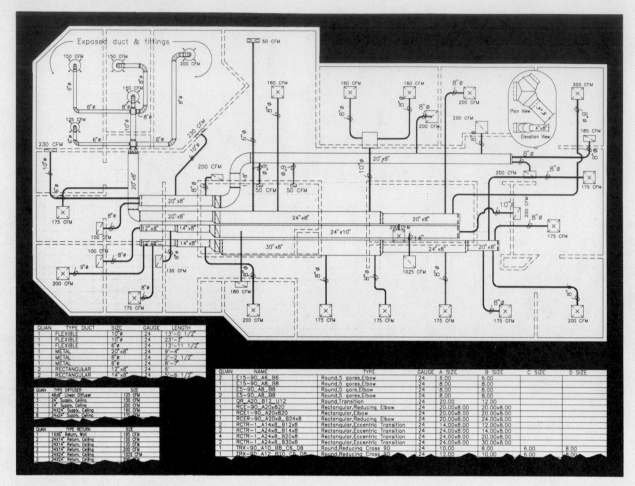

Figure 23–26 The complete HVAC plan and related schedules. *Courtesy Chase Systems.*

▼ PATTERN DEVELOPMENT

The principle of pattern development is based on laying out geometric shapes in true size and shape flat patterns. The fundamental concepts involved in making patterns for basic geometric shapes may be used in the development of any pattern. In most situations a front and top or bottom view should be drawn to help establish true length lines and true size shapes. *The key to pattern development is any line or element used in the development must be in true length.* Use construction lines for all preliminary layout work so errors may be easily erased.

Stretch-out Line

A *stretch-out line* is typically the beginning line upon which measurements are made and the pattern development is established. Carefully observe how the stretch-out lines are established for each of the following developments as this is the first process in making a layout. Specific developments that do not begin with a stretch-out line will be identified. The instruction provided for the following pattern development is broken down into the most basic procedures. Individual shortcuts may be taken after adequate experience has been gained. It is extremely important to maintain a high degree of accuracy. Scale drawings and transfer dimensions very carefully. Another important consideration is to accurately label the elements of the views with numbers and/or letters and then transfer these labels to the pattern. This procedure may seem unnecessary on simple developments, but on complex patterns it is absolutely necessary. It is recommended that beginners label all developments as suggested in the given procedures.

Seams

When sheet metal parts are bent and formed into the desired shape, a seam results where the ends of the pattern come together. The fastening method depends on the kind and thickness of material, on the fabrication

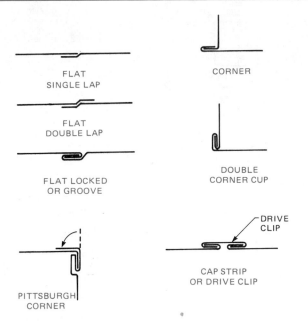

Figure 23–27 Common seams.

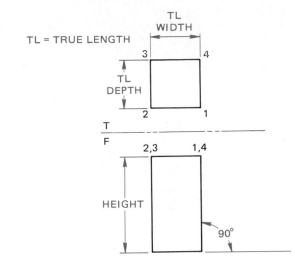

Figure 23–29 Step 1—right prism development.

Figure 23–28 Common hems.

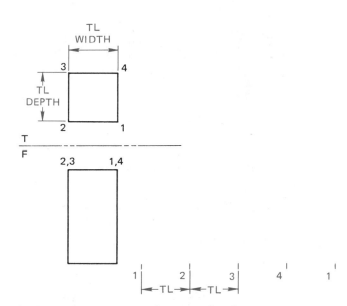

Figure 23–30 Step 2—right prism development.

processes available, and on the end use of the part. Sheet metal components that must hold gases or liquid, or are pressurized may require soldering or welding. Other applications may use mechanical seams, which hold the part together by pressure lapped metal, metal clips, or pop rivets. Some of the most common seams used in the sheet metal fabrication industry are shown in Figure 23–27. Extra material may be required on the pattern to allow for seaming. For the purpose of discussion, in the following procedures and problems, either a single- or double-lap seam is used. For a single-lap seam add the given amount to one side of the pattern. For double-lap seams add the given amount to both sides of the pattern. The corner of a seam may be cut off at an angle (usually 45°) if it interferes with adjacent parts during fastening.

Hemmed edges are necessary when an exposed edge of a pattern must be strengthened. When hems are used, extra material must be added to the pattern on the side of the hem. Figure 23–28 shows some common hems.

Rectangular (Right) or Square Prism

Commonly referred to as a box (open ended in this example), the rectangular or right prism may be developed as follows:

Step 1. Draw the front and top view, label the corners, and establish the stretch-out line off the

base of the front view and perpendicular to the height line. (See Figure 23–29.)

Step 2. Beginning at 1 in the top view use dividers to measure the true length distance from 1 to 2. Transfer this dimension to the stretch-out line, starting at any point near the front view. Continue this process by transferring the distance from 2 to 3, 3 to 4, and 4 to 1 to the stretch-out line. (See Figure 23–30.) You must end at the point you began; point 1 in this example.

Step 3. From each of the points established in Step 2, draw vertical construction lines to meet a horizontal line drawn from the true length (TL) height in the front view. (See Figure 23–31.)

Step 4. Darken in object lines. Notice that thick lines are used where the pattern sides make a bend. Add any required seam material. (See Figure

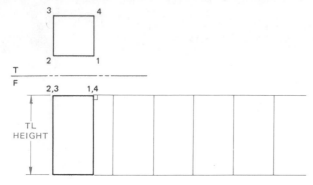

Figure 23–31 Step 3—right prism development.

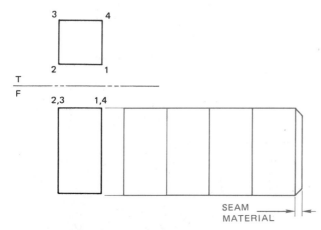

Figure 23–32 Step 4—right prism development.

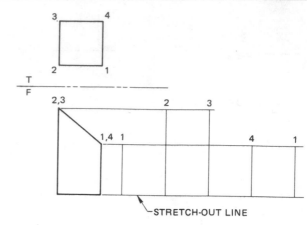

Figure 23–33 Step 1—truncated prism development.

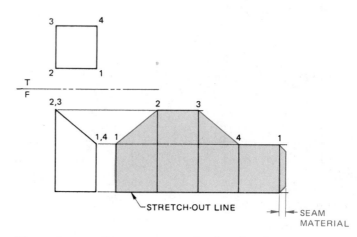

Figure 23–34 Step 2—truncated prism development.

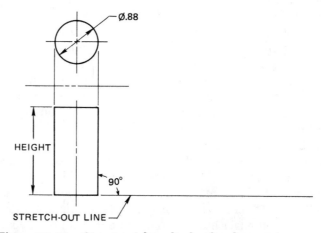

Figure 23–35 Step 1—right cylinder development.

23–32.) The pattern is now ready to be cut out and formed into the given shape or transferred to sheet metal for fabrication.

Truncated Prism

A sheet metal part such as a prism, pyramid, or cone is considered to be *truncated* if a portion is cut off, generally at an angle.

Step 1. Proceed as described in Steps 1 through 3 for a right prism. Begin with the shortest element as the seam, when possible. The shortest seam is stronger, easier to fabricate, and requires less materials. (See Figure 23–33.)

Step 2. Darken all object lines by connecting the ends of the true height elements to form the outline of the object. Darken in all bend lines and add seam material as shown in Figure 23–34.

Regular (Right) Cylinder

One of the most common shapes is the cylinder. The following procedure may be used for the development of any right cylindrical shape:

Step 1. Draw the top and front views. Establish the stretch-out line perpendicular to the side of the front view. The stretch-out line may be placed anywhere next to the front view, but it must be perpendicular to the side. (See Figure 23–35.)

Step 2. Establish the length of the stretch-out line equal to the circumference of the circle with the formula C = πD. The diameter is .88 in., so C = 3.14(π) × .88 = 2.76 in. Now, from the ends of the stretch-out line, draw perpendicular lines that meet a line projected from the

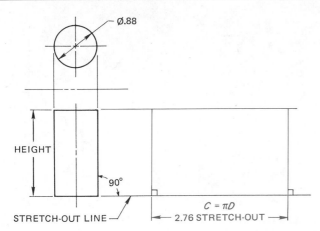

Figure 23–36   Step 2—right cylinder development.

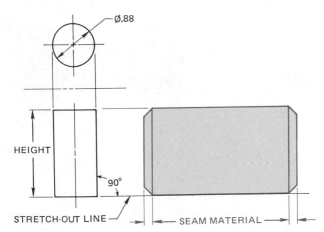

Figure 23–37   Step 3—right cylinder development.

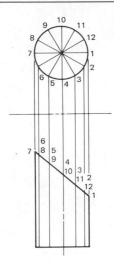

Figure 23–38   Step 1—truncated cylinder development.

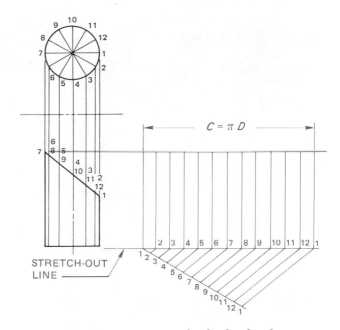

Figure 23–39   Step 2—truncated cylinder development.

true height in the front view as shown in Figure 23–36.

Step 3.   Darken the outline of the pattern and add seam material as shown in Figure 23–37, which shows a double-lap seam.

Truncated Cylinder

Truncated cylinders have many HVAC applications with development procedures similar to the regular cylinder.

Step 1.   Draw the top and front views. Divide the top view into twelve equal parts. More divisions establish better accuracy; less give less accuracy. Twelve divisions have been selected because of ease and effectiveness. Work carefully with a sharp pencil to maintain the best accuracy possible. Number each element in the top view and extend the numbering system into the front view. (See Figure 23–38.)

Step 2.   Establish the stretch-out line perpendicular to the side of the front view and the length with the formula $C = \pi D$ as previously discussed. Divide the stretch-out into twelve equal parts (refer to dividing a line into equal parts, Chapter 4). Number each part from one through twelve, ending at one (the seam). From each part on the stretch-out line, draw a perpendicular construction line equal to the total height of the cylinder. (See Figure 23–39.)

Step 3.   Project the true height of each line segment in the front view to the correspondingly numbered line in the pattern layout. Where these lines intersect, a pattern of points is established. Use an irregular curve to connect the points forming the curved outline of the truncated cylinder pattern. Darken the outline of the pattern and add seams. The division lines

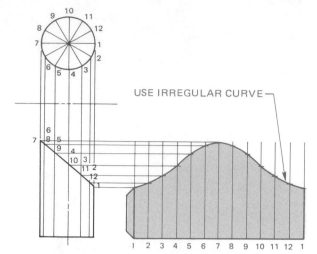

Figure 23–40   Step 3—truncated cylinder development.

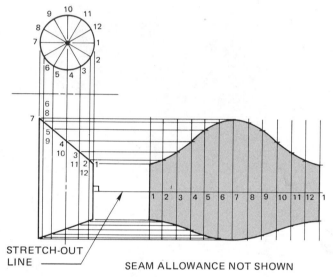

Figure 23–41 Development of cylinder truncated on both ends.

may be left as thin lines to represent a curved object. Construction lines should not have to be erased if properly drawn. (See Figure 23–40.)

When the cylinder is truncated on both ends, the process is the same, with the stretch-out line established perpendicular to the side at any desired location. (See Figure 23–41.)

Cylindrical Elbow

Cylindrical elbows are used to make turns or corners in duct work. The turns may be any number of degrees. Standard cylindrical elbows are made up of any given number of truncated cylinders. Each piece of the elbow may be developed as demonstrated in the previous discussion. Figure 23–42 shows a standard 90° three-piece and four-piece elbow. Each piece of the three-piece elbow is developed as truncated cylinders, with the

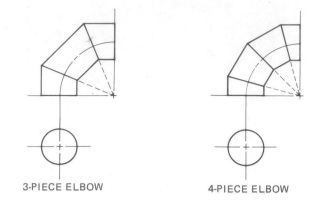

Figure 23–42 Standard 90° elbows.

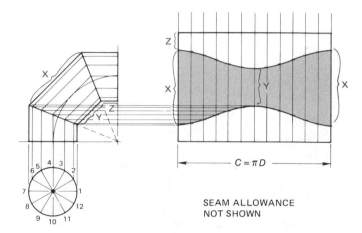

Figure 23–43 Development of a three-piece cylindrical elbow.

patterns developed separately or together by alternating starting elements as shown in Figure 23–43.

Cone

The size of the cone is established by the base diameter and the height and may be developed as follows:

Step 1.   Given the height and base diameter, draw the front and top (or bottom) views. Divide the top (bottom) view into twelve equal parts. Number each part and project these to the cone base in the front view. At the base in the front view, extend each point to the vertex of the cone. (See Figure 23–44.)

Step 2.   The stretch-out line for a right circular cone development is a true circular arc. The radius of the arc is taken from the true length element measured from zero to one in the front view. (See Figure 23–44.) At any convenient location where you have adequate space, draw an arc with the true length radius. (See Figure 23–45.)

Step 3.   Go to the top view and use dividers to establish the increment from point 1 to point 2. Rotate the dividers around the circle, adjusting the

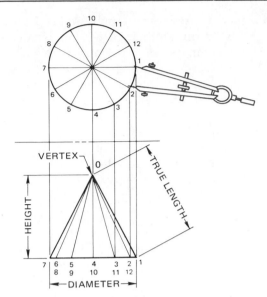

Figure 23–44 Step 1—cone development.

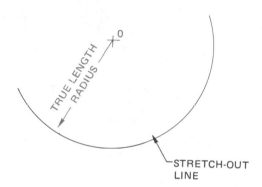

Figure 23–45 Step 2—cone development.

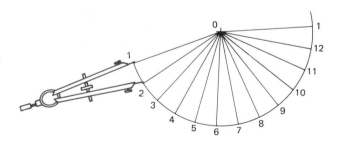

Figure 23–46 Step 3—cone development.

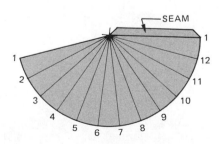

Figure 23–47 Step 4—cone development.

dividers until you have accurately established the increment at least half way around the circle. (See Figure 23–44.) Beginning at any point on the stretch-out line, use the established increment to locate twelve equal spaces. Remember if you begin at "1," you must end at "1." Now, connect each point along the stretch-out line to the vertex "0" with construction lines. (See Figure 23–46.) When using CADD, this procedure may be done quickly and accurately. The AutoCAD MEASURE command may be used to establish the circumference on the stretch-out line followed by the DIVIDE command to find the twelve equal spaces.

Step 4. Darken the outline of the cone pattern and add seam material as necessary. (See Figure 23–47.)

Truncated Cone

The procedure for developing a truncated cone is the same as the regular cone previously described. Always begin by developing a pattern for a full regular cone. After Step 3, the true length of each line segment from the vertex to the truncated line in the front view must be established. Project each point to the true length line. Then measure the true length of each element from the vertex along the true length line as shown in Figure 23–48. Transfer the true length of each individual line to the corresponding line in the development.

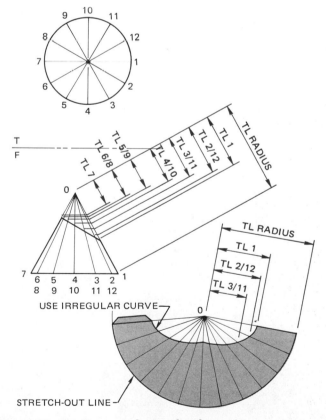

Figure 23–48 Truncated cone development.

(See Figure 23–48.) Finally, darken the outline using an irregular curve, where necessary, and add seam material as shown in Figure 23–48.

Offset Cone

The offset cone has the vertex offset from the center of the base. The procedure is slightly different from the development of a regular cone as the true length of each line element must be found. Refer to Figure 23–49 as you read the following discussion:

Step 1. Draw the top and front views. Divide the top view base circle into twelve equal parts. Project each point to the vertex and establish the same elements in the front view.

Step 2. The true length of elements 0–1 and 0–7 are true length in the front view. Establish the true length diagram for the other elements using the revolution method as shown in color in Figure 23–49.

Step 3. There will be no stretch-out line in this development. Begin the development by laying out the true length of element 0–7 in a convenient location. Take a compass and lay out the true length of 0–6 and 0–8 on each side of 0–7. Set another compass with the radius equal to the distance X in the top view. This measure-

ment is used several times, so keep your compass at this setting. Using the distance X, set the compass point at 7 and draw an arc that intersects the arcs drawn for 0–6 and 0–8. This procedure locates the exact position of elements 0–6 and 0–8. Follow this same method for each of the elements, 0–5 and 0–9, 0–4 and 0–10, 0–3 and 0–11, 0–2 and 0–12, and 0–1 at each end. You now have a series of points along the base that may be connected with an irregular curve.

Step 4. The true lengths of elements 0–1 and 0–7 are in the front view. Use the true length diagram to establish the true lengths of the other elements, 0–2 through 0–12. Transfer these true lengths to the corresponding elements in the development and connect the points with an irregular curve. Darken the outline and add seam material, if required. (See Figure 23–49.)

Pyramid

Pyramids are developed in a manner similar to a cone as described in the following steps:

Step 1. Draw the top and front views and establish the true length of the edge of a side or lateral edge. Set the compass with a radius equal to the true length and draw an arc in a convenient location. This is the stretch-out line. (See Figure 23–50.)

Step 2. Use dividers or a compass to lay out the true lengths from points 1–2, 2–3, 3–4, and 4–1 in the top view and transfer these distances consecutively to the stretch-out line. Connect points 1, 2, 3, 4, and 1 in the development and also connect these points to the vertex "0." Darken all lines and add seam allowances as shown in Figure 23–51.

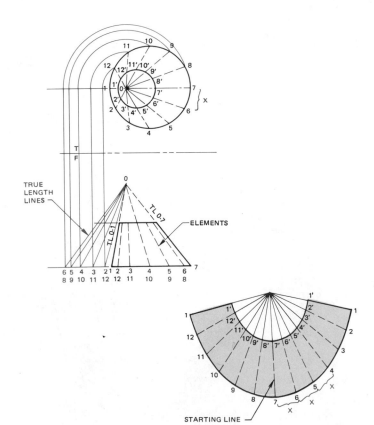

Figure 23–49 Offset cone development.

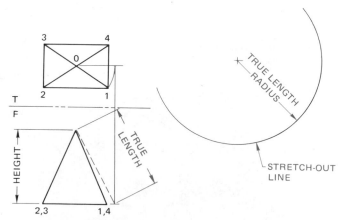

Figure 23–50 Right pyramid development.

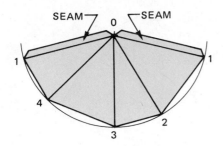

Figure 23–51 Pyramid development.

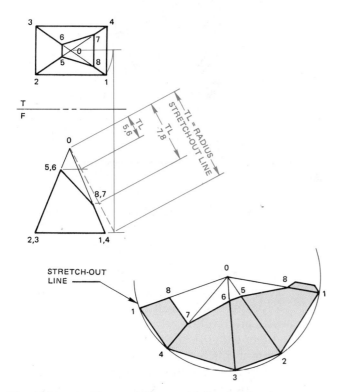

Figure 23–52 Truncated pyramid development.

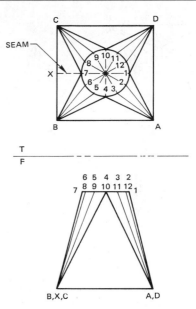

Figure 23–53 Step 1—right transition piece view setup.

Truncated Pyramid

The procedure for the development of a truncated pyramid is the same as for a regular pyramid except that the true length of each element from the vertex to the truncated line must be established and transferred to the pattern as shown in Figure 23–52.

Transition Piece

Also known as a *square to round,* this *transition piece* is a duct component that provides a change in shape from square or rectangular to round. Transition pieces may be designed to fit any given situation, but the pattern development technique is always as follows: (Note: *Accuracy and the use of a number system is very important. Do all layout work using construction lines or a CADD construction layer.*)

Step 1. Draw the top and front views. Divide the circle in the top view into twelve equal parts. Con-

nect the points in each quarter of the circle to the adjacent corner of the square. Project the same corresponding system to the front view. Number and letter each point in each view as shown in Figure 23–53. The transition piece shown in Figure 23–53 is made up of a series of triangles; for example, 1,A,D; 1,2,A; and 2,3,A. The development progresses by attaching the true size and shape of each triangle together in order. This technique is known as *triangulation.* There is no stretch-out line.

Step 2. Each element of the development must be in true length. Establish the true length diagram using revolution as shown in Figure 23–53. Only one set of true length elements is required because the given problem is a right transition piece (symmetrical about both axes).

Step 3. Begin the pattern development on a sheet in an area where there is a lot of space. Start with the true size and shape of triangle 1,A,D. The true lengths of 1–A and 1–D are the same and may be found in the true length diagram. The true length of A–D is in the top view. (See Figure 23–54.) A review of Chapter 4, Geometric Construction, and Chapters 7 and 8, Descriptive Geometry I and II, may be helpful here.

Step 4. Work both ways from triangle 1,A,D in Figure 23–55 to develop the adjacent triangles. From points A and D draw arcs equal to A–2 and D–12, respectively. Set a compass for the distance from 1–2. Keep this compass setting, as it is used several times. With this setting, scribe two arcs from point 1 that intersects the previously drawn arcs. These intersections are points 2 and 12. Connect points 2–A and 12–D.

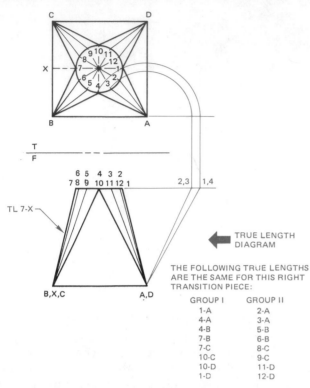

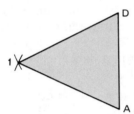

THE FOLLOWING TRUE LENGTHS ARE THE SAME FOR THIS RIGHT TRANSITION PIECE:

GROUP I	GROUP II
1-A	2-A
4-A	3-A
4-B	5-B
7-B	6-B
7-C	8-C
10-C	9-C
10-D	11-D
1-D	12-D

Figure 23–54 Step 2—transition piece development; true length diagram.

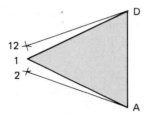

Figure 23–55 Step 3—transition piece development; triangulation, true size and shape of starting triangle.

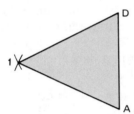

Figure 23–56 Step 4—transition piece development; continue the pattern development both ways from the starting triangle.

Do not connect points 1,2 and 12 yet. (See Figure 23–56.) Do all work with construction lines until complete.

Step 5.   Continue the same procedure, working both ways around the transition piece until the entire development is complete. Every triangle must be included and the pattern will end on both sides with the element 7–X, which is

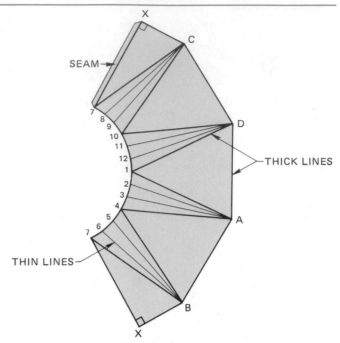

Figure 23–57 Step 5—complete the transition piece development.

true length in the front view. Remember that every line transferred from the views to the pattern must be true length. It is recommended that the numbering system be used to avoid errors. When all points have been established, the outline may be darkened. A series of points form the inside curve. Connect these points with an irregular curve. If any point is out of alignment with the others, you have made an error at this location. Notice in Figure 23–57 that thick lines in the pattern are either outlines or bend lines, while thin lines form a smooth curve or contour. Cut out your pattern and see for yourself how it fits together.

The right transition piece used in the process previously described is the easiest transition to develop because it is symmetrical. If the round were offset as shown in Figure 23–58, then true length diagrams would

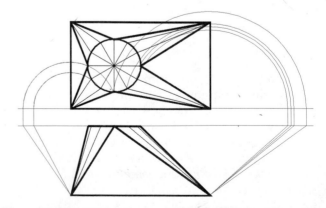

Figure 23–58 Setting up a symetrical offset transition piece; two true length diagrams are required.

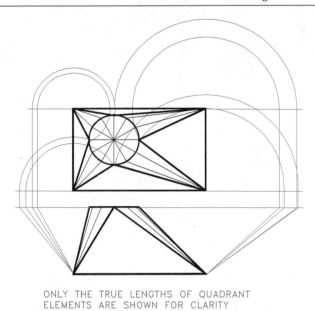

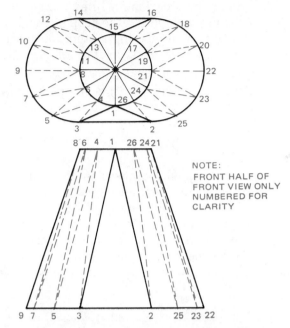

ONLY THE TRUE LENGTHS OF QUADRANT
ELEMENTS ARE SHOWN FOR CLARITY

Figure 23–59 Setting up an offset transition piece; four true length diagrams are required.

Figure 23–61 Setting up the triangulation on a curve-to-curve transition.

be required for both halves. When the round is offset as in Figure 23–59, true length diagrams are necessary for the elements on each quarter of the transition. In these more complex transitions a numbering system and accurate work are very important. Colored lines may be used to help keep true length diagrams clearly separate.

Curve-to-curve Triangulation

Triangulation is a technique that may be used for the development of transition pieces or any other pattern where a series of triangles are used to form the desired shape. Figure 23–60 shows a part that makes a transition from one curved shape to another. In this situation the adjacent curves must be broken down into a series of tri-

angles before the pattern can be developed. The triangles help define one shape in relationship to the other. (See Figure 23–61.) After the views have been drawn, the following procedure is similar to the development of a transition piece:

Step 1. The part is symmetrical, so one set of true length lines must be established. (See Figure 23–62.) Be careful when laying out and num-

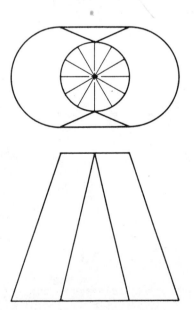

Figure 23–60 Curve-to-curve transition.

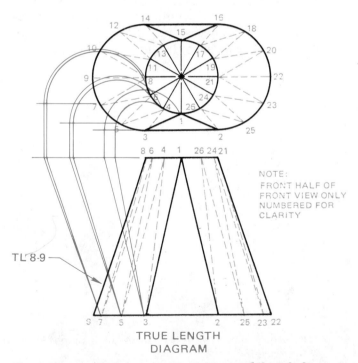

Figure 23–62 Curve-to-curve transition, true length diagram.

bering the true length lines. Some drafters even use colored lines to keep elements coordinated. A numbering system and accuracy are critical for this type of problem.

Step 2. After the true length elements have been established, the development technique is the same as the transition piece. (See Figure 23–63.)

▼ INTERSECTIONS

When one geometric shape meets another, the line of intersection between the shapes must be determined before the pattern of each piece can be developed. The key to determining the intersection between parts is setting up views with numbered true length elements that correlate between views. The points of intersection of line elements at planes may then be projected between views.

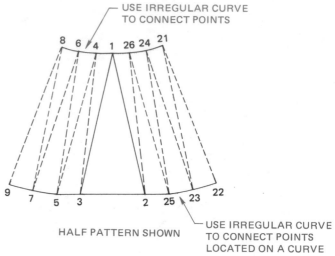

Figure 23–63 Curve-to-curve transition development.

CADD Applications

SHEET METAL PATTERN DESIGN

CADD software is available that automatically produces nearly any sheet metal flat pattern layout you need. Figure 23–64 shows a variety of available shapes and patterns. Full-size plots can also be generated on the computer screen or the information can be transferred to computer numerical control burners, lasers, punch presses, or arc machines so the pattern can be cut or punched, as necessary. The programs can automatically calculate sheet layout for the best material conservation possible. Parametric design of sheet metal fabrication packages allows you to enter specific information about the design; for example, the variables H, Y, X, OX, and D shown on the transition piece in Figure 23–65. After you input the variable information, the program automatically draws the flat pattern on the screen and provides a print file of all the development information for your reference as shown in Figure 23–66.

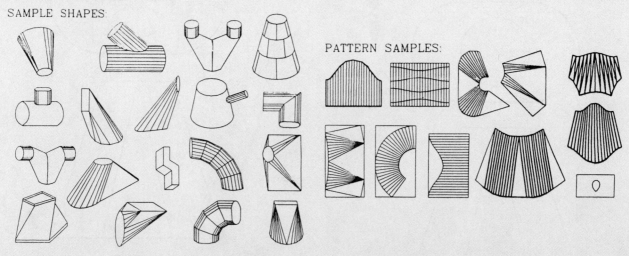

Figure 23–64 A variety of sheet metal shapes and patterns may be drawn on a CADD system. *Courtesy C.A.M. Systems, Inc.*

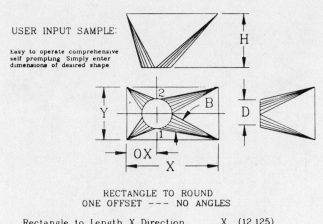

USER INPUT SAMPLE:

Easy to operate comprehensive self prompting Simply enter dimensions of desired shape.

RECTANGLE TO ROUND
ONE OFFSET --- NO ANGLES

Rectangle to Length X Direction	X	(12.125)
Rectangle to Length Y Direction	Y	(10.500)
Height	D	(5.375)
Number of Breaks per Corner	H	(12.562)
Select X or Y Offset	B	(5.0)
Offset in X Direction	OX	(12.125)

(Offset may be in X or Y direction from
lower left corner of the rectangle)

Figure 23–65 Parametric CADD packages are available to design duct fittings by giving variable information. *Courtesy C.A.M. Systems, Inc.*

SAMPLE SCREEN DISPLAY

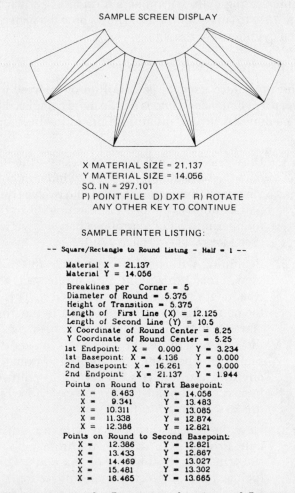

X MATERIAL SIZE = 21.137
Y MATERIAL SIZE = 14.056
SQ. IN = 297.101
P) POINT FILE   D) DXF   R) ROTATE
ANY OTHER KEY TO CONTINUE

SAMPLE PRINTER LISTING:

-- Square/Rectangle to Round Listing - Half - 1 --

Material X = 21.137
Material Y = 14.056

Breaklines per Corner = 5
Diameter of Round = 5.375
Height of Transition = 5.375
Length of First Line (X) = 12.125
Length of Second Line (Y) = 10.5
X Coordinate of Round Center = 8.25
Y Coordinate of Round Center = 5.25
1st Endpoint: X =   0.000   Y = 3.234
1st Basepoint: X =   4.136   Y = 0.000
2nd Basepoint: X = 16.261   Y = 0.000
2nd Endpoint: X = 21.137   Y = 1.944
Points on Round to First Basepoint:
    X =   8.483   Y = 14.058
    X =   9.341   Y = 13.483
    X =   10.311   Y = 13.085
    X =   11.338   Y = 12.874
    X =   12.386   Y = 12.821
Points on Round to Second Basepoint:
    X =   12.386   Y = 12.821
    X =   13.433   Y = 12.867
    X =   14.469   Y = 13.027
    X =   15.481   Y = 13.302
    X =   16.465   Y = 13.665

Figure 23–66 The flat pattern drawing and flat pattern development information provided by the CADD system. *Courtesy C.A.M. Systems, Inc.*

Intersecting Prisms

The line of intersection between intersecting prisms may be found by projecting the true lengths of individual line elements between views. The corresponding points of intersection where lines intersect planes establish a series of points that are connected to form the lines of intersection. Several examples are shown in Figure 23–67. After the lines of intersection are determined, the pattern development may be made of the resulting shape.

Intersecting Cylinders

The line of intersection between intersecting cylinders is determined in a manner that is similar to that for

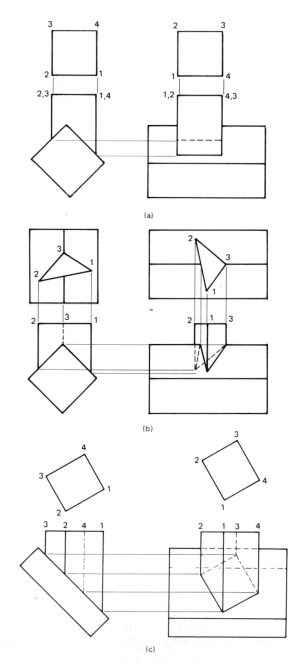

Figure 23–67 Intersecting prisms.

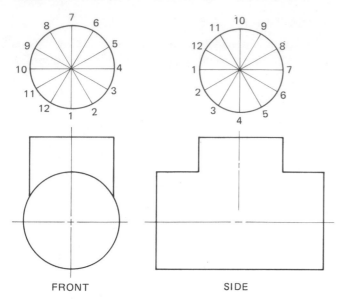

Figure 23–68 Setting up intersecting cylinders.

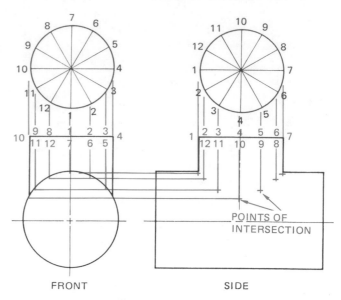

Figure 23–69 Establishing points of intersection for intersecting cylinders.

intersecting prisms. A series of line elements are established on the intersecting cylinder. Points of intersection of these line elements are then plotted. When the points are connected, the line of intersection is formed. The following procedure provides a typical example:

Step 1.   Establish two adjacent views and show the circular view of the intersecting cylinder in both views. Divide the circular views of the intersecting cylinder into twelve equal parts in each view and number the points correspondingly. Be sure that your numbering system correlates between views as shown in Figure 23–68.

Step 2.   Project point 1 in the front view down until it meets the intersecting circle. Then project to the side view until it meets the same corresponding point 1 projected down in the side view. This is the point of intersection for line element 1. Follow the same procedure for each of the other eleven points. When complete, you have a series of points of intersection as shown in Figure 23–69.

Step 3.   Connect the points of intersection in the side view to establish the line of intersection as shown in Figure 23–70. Now pattern developments of each cylinder can be made as discussed earlier in this chapter and as shown in Figure 23–70.

No matter how the intersecting cylinders are arranged, the technique is the same. The circular views are always drawn looking into the intersecting cylinders even when the cylinders intersect at an angle as in Figure 23–71 (see

page 778). When the cylinders are offset, the back half of the intersecting cylinder appears hidden as shown in Figure 23–71. This example also shows how the patterns for the cylinders are drawn.

Cylinder Intersecting Cone

The procedure used to get the line of intersection between a cylinder and cone is a little more complex, but may be done given any situation if the following steps are used:

Step 1.   Draw the front and top views. Draw the circular view of the intersecting cylinder in both views and divide these circles into twelve equal parts. Number the parts of each circle so the numbering system correlates between views. (See Figure 23–72, page 779.)

Step 2.   Beginning with point number 1 in the front view, project this point until it intersects the true length element of the cone, O–X. This is the point of intersection of point 1 in the front view. From this point of intersection, project to the top view until the projection meets the same corresponding point projected from the circle in the top view. This is the point of intersection of point 1 in the top view. (See Figure 23–73, page 779.)

Step 3.   Now the process becomes a little more difficult. Project point 2 from the front view to the true length element of the cone, O–X (points 2 and 12 are on the same projection line if the object is symmetrical). From this point, project point 2 into the top view until inter-

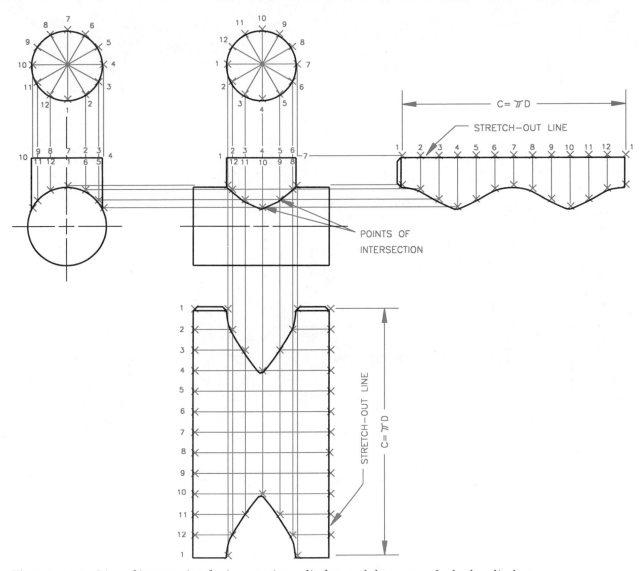

Figure 23–70   Line of intersection for intersecting cylinders and the pattern for both cylinders.

secting O–X. Set a compass with a radius from "O" to this point of intersection and draw an arc. Project points 2 and 12 in the top view until they intersect this arc. These are the points of intersection of points 2 and 12 in the top view. Now, project points 2 and 12 from the top view until they intersect the projection line for 2 and 12 in the front view. This is the point of intersection for 2 and 12 in the front view. (See Figure 23–73.)

Step 4.   Continue this process until each point has been located in each view. (See Figure 23–74, page 779.) Darken the line of intersection by connecting the points of intersection with an irregular curve. Be sure hidden lines are properly represented. (See Figure 23–75, page 779.)

Step 5.   The pattern development for the cone is started by using the true length side of the cone to draw the stretch-out line arc. (This is the edge view.) Look at Figure 23–75 as you follow these steps: Lay out the circumference of the cone base along the stretch-out line, as you learned earlier in this chapter. Then, establish element O–X in the middle of the pattern; this line coincides with O–X in the front and top views. In the front view, project a radius from the vertex (O) and begin where the projection lines from the circle next intersect the true length line (O–X); the points for elements 1 and 7 are where the arcs from the front view cross line O–X in the pattern. Repeat this for each intersection, and extend these radii past the O–X line in the pattern.

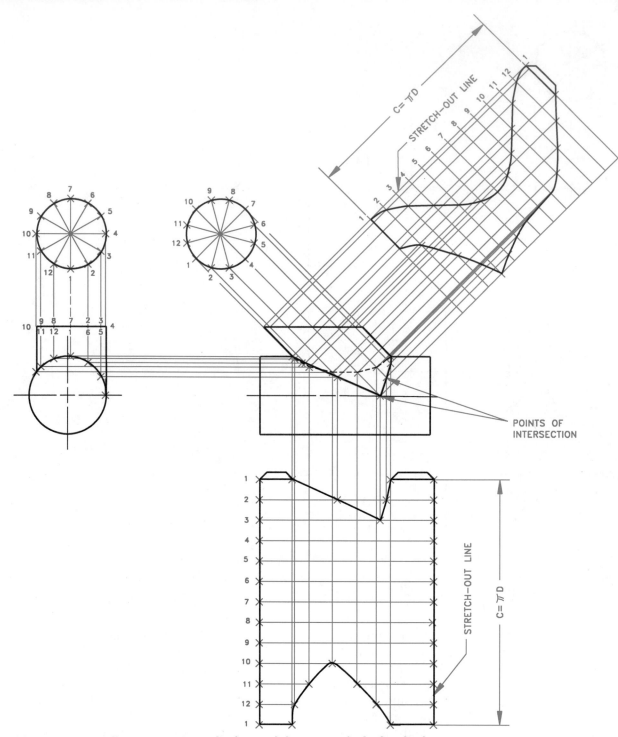

Figure 23–71 Offset intersecting cylinders and the pattern for both cylinders.

In the top view, take measurements from the O–X line at the point where the arc passing through points 2 and 12 intersect. Measure from this intersection to point 2. The distance from O–X to point 2 should be the same as from O–X to point 12 because this object is symmetrical. Take this measurement and transfer it to the

pattern by measuring from the intersection of the 2, 12 radius on the O–X line, and place a point on each side of the O–X line on this radius. Continue this process for each pair of points (3,11; 4,10; 5,9; and 6,8) until each point has been transferred into the pattern. Connect the points in the pattern to establish

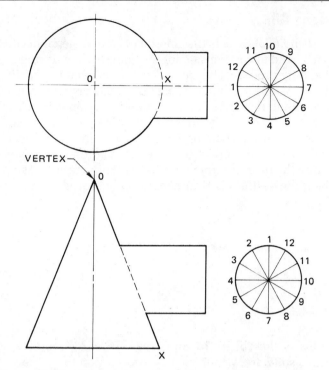

Figure 23–72   Set up intersection between a cone and cylinder.

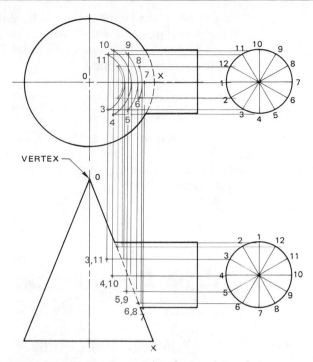

Figure 23–74   Determine points of intersection between cone and cylinder.

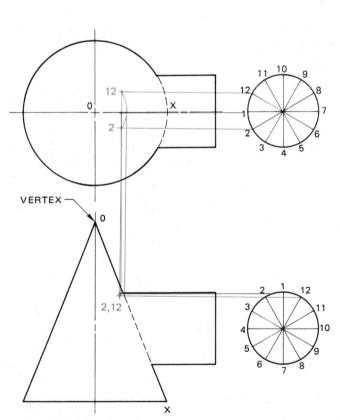

Figure 23–73   Determine points of intersection between cone and cylinder.

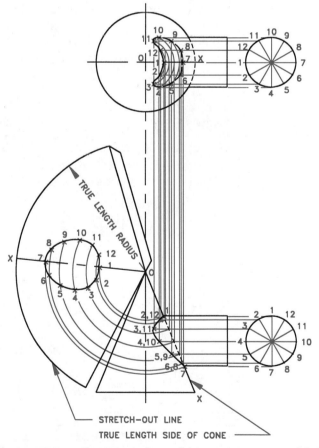

Figure 23–75   Connect points of intersection to establish line of intersection between cone and cylinder and the pattern development for the cone.

the cut-out where the cylinder intersects. Make a pattern development for the cylinder using the same procedure that you have used for other cylinder developments.

▼ MATERIAL BENDING AND CHASSIS LAYOUT

In precision sheet metal applications such as fabrication for electronics chassis components or sheet metal appliance body parts, the condition of material when bent must be taken into consideration.

Bend Allowance

Bend allowance is the amount of extra material needed for a bend to compensate for the compression during the bending process. This consideration is often not critical to HVAC sheet metal bending due to the larger tolerances applied to these components. Bend allowance is important when close tolerances must be held or when thick material must be bent or formed into desired shapes. The purpose of a bend allowance calculation is to determine the overall dimension of the flat pattern, so when bent the desired final dimension is achieved. There are

CADD Applications

PRECISION SHEET METAL FABRICATION

CADD applications are available for developing flat-pattern layouts and converting existing drawings into flat layouts. These packages can automatically calculate bend allowances and notch locations. All you need to enter in the computer for bend allowance calculations is material thickness and type, and bend radius in degrees. The added advantage of some sys-

tems is the ability to interface with CAM applications software, which allows information that is generated at the CADD workstation to be sent to the fabrication shop to run a computer numerical control metal break or punch press. A drawing generated by a CADD precision sheet metal program is shown in Figure 23–76.

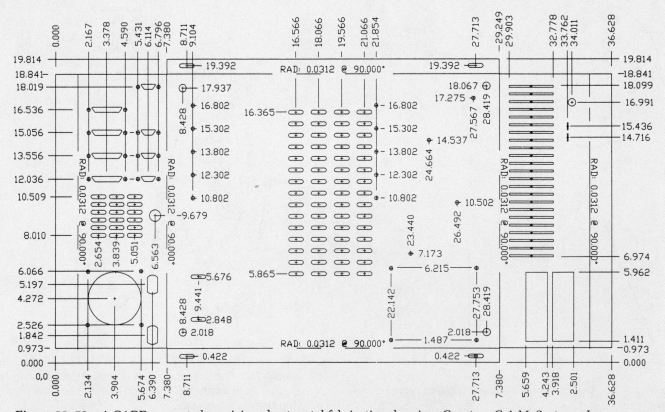

Figure 23–76 A CADD-generated precision sheet metal fabrication drawing. *Courtesy C.A.M. Systems, Inc.*

a number of slightly different methods used to calculate bend allowance. There are different formulas in the *Machinery's Handbook, ASME Handbook,* and in most text books. Many companies use formulas derived from a proven method or by individual experimentation. The amount of bend allowance depends on the material thickness, type of material, bending process, degree of bend, and bend radius. The grain of material, surface condition, and amount of lubrication used also influence bend allowance. The only exact way to establish bend allowance for a specific application is to experiment with the equipment and material to be used. Some companies have made these tests and have developed charts that give the bend allowance for the type of material, material thickness, and bend radius.

When material bends, there is compression on the inside of the bend and stretching on the outside. Somewhere between, there is a neutral zone where neither stretching nor compression occurs; this is called the *neutral axis.* The neutral axis is approximately four-tenths of the thickness from inside of the bend, but this depends on the material and other factors. Information related to calculating the bend allowance and length of the flat pattern is shown in Figure 23–77. The following formula is used to calculate the length of the flat pattern (straight stock before bending):

Length of Flat Pattern = X + Y + Z
X = Horizontal Dimension to Bend = B – R – C
Y = Vertical Dimension to Bend = A – R – C

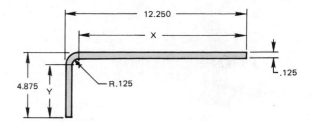

Figure 23–78 Sample bend allowance problem.

Z = Length of Neutral Axis
C = Material Thickness
R = Bend Radius
Z = 2(R + .4C) × π ÷ 4(90° bend)

Given the sheet metal bend shown in Figure 23–78, determine the length of the flat pattern.

X = B – R – C = 12.250 – .125 – .125 = 12.000
Y = A – R – C = 4.875 – .125 – .125 = 4.625
Z = 2(R + .4C) × π ÷ 4 = 2(.125 + .4 × .125)
        × 3.14 ÷ 4 = .275
Length of Flat Pattern = X + Y + Z = 12.000
                        + 4.625 + .275 = 16.900

Bend Relief

A corner with two edges bent in the same direction has internal stresses that may cause a crack at the corner. When necessary, some material at the corner may be cut away to help relieve this stress. This cutout at the corner is called *bend relief.* (See Figure 23–79.)

Chassis and Precision Sheet Metal Layout

Electronic assemblies often require shields, frames, panels, and chassis to be fabricated. Layout drawings are required for the fabrication of these items. Preparing chassis layouts and other precision sheet metal drawings requires close tolerances and bend allowance calculations. The method of dimensioning these parts includes standard unidirectional, arrowless, or tabular dimensioning systems. (Refer to Chapter 10, Dimensioning.) A precision sheet metal pattern may be drawn as a flat pattern, as a finished product, or with the finished part shown and the flat pattern drawn using phantom lines.

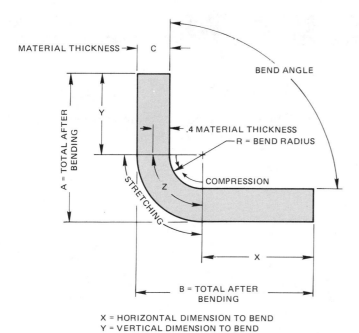

X = HORIZONTAL DIMENSION TO BEND
Y = VERTICAL DIMENSION TO BEND
Z = LENGTH OF NEUTRAL AXIS

Figure 23–77 Bend allowance variables.

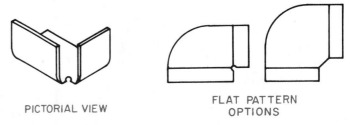

PICTORIAL VIEW

FLAT PATTERN OPTIONS

Figure 23–79 Bend relief.

PROFESSIONAL PERSPECTIVE

This chapter has covered the engineering drafting that takes place in three different, but closely allied, engineering fields. The first is HVAC drafting, the second is sheet metal fabrication, and the third is precision sheet metal fabrication. The first two, HVAC and sheet metal fabrication, are the most closely allied. If you are a drafter for a mechanical (HVAC) engineer, you are likely to draw HVAC plans, details, and schedules. The drawings from your office are sent to a sheet metal fabricator so the duct components can be built and delivered to the construction site. Your colleague in the fabrication industry takes the drawing from the mechanical engineer's office and converts the duct shapes to flat-pattern layouts for fabrication. The third type of drafting deals with sheet metal shapes and parts. Although these components are not used in the HVAC industry, they are used in the precision sheet metal fabrication industry. The dif-

ference is that this industry normally deals with closer tolerances. The types of items made in the precision sheet metal industry are electronics chassis and automotive (cars, trucks, and tractors) sheet metal components. In many cases, HVAC fabrication drawings do not even have dimensions; the pattern itself is used to fabricate the duct shape. On the other hand, the dimensioning and tolerancing for precision sheet metal parts is critical, and arrowless, datum, and tabular dimensioning is often used with close tolerances to achieve an accurate layout.

If you enter the HVAC industry, you should become very familiar with HVAC vendors' catalogs and manuals, and duct design applications engineering information available from mechanical suppliers such as Trane, Apec, and Elite. If you enter the precision sheet metal fabrication industry, you need to have a good understanding of tolerancing, material bending, and precision sheet metal fabrication methods.

MATH APPLICATION

SHEET METAL BEND ANGLES

It is required to find bend angle A for a sheet of metal having the cross section shown in Figure 23–80. The solution to this problem requires working with *two* right triangles because thickness of the material must be taken into account. Figure 23–81 is a drawing of the solution. OML stands for outside mold line. The steps taken follow.

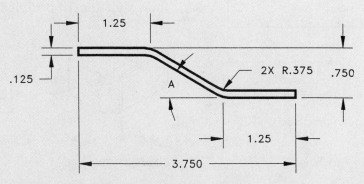

Figure 23–80 Sheet metal with two identical bends.

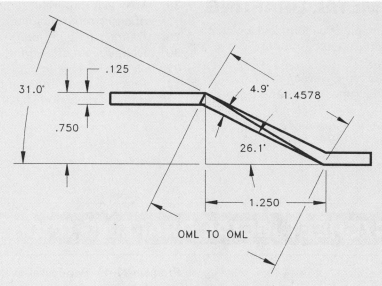

Figure 23–81 Triangles drawn to solve the problem.

1. Find the angle of the large triangle with Inv $\tan\left(\dfrac{.75}{1.25}\right) = 31.0°$.

2. Find the hypotenuse of both triangles with $\sqrt{1.25^2 + .75^2} = 1.4578$.

3. Find the angle of the slender, shaded triangle with Inv $\sin\left(\dfrac{1.25}{1.4578}\right) = 4.9°$

4. Subtract the two angles to find the bend angle.

$$A = 31.0 - 4.9 = \mathbf{26.1°}$$

If the true distance along the bend (OML to OML) is known, then the problem is much simpler and requires working with only one right triangle.

CHAPTER
23 HVAC/SHEET METAL DRAFTING TEST

DIRECTIONS
Answer the questions with short complete statements or drawings as needed.

PART 1. HVAC DRAFTING

1. HVAC systems are also commonly known as _____ systems.
2. What is the general purpose of an HVAC system?
3. Under what general conditions is ventilation required?
4. Define *duct*.
5. Name two types of forced-air systems.
6. What is the basic principle of refrigeration?
7. Name two types of hot water systems.
8. How does a zoned control system differ from a central forced-air system?

9. Describe the basic function of a heat pump system.
10. Identify six sources of pollutants in a structure.
11. Describe the basic function of an air-to-air heat exchanger.
12. Define *thermostat*.
13. List three factors that contribute to the proper placement of a thermostat.
14. Name two types of HVAC plans (based on line technique).
15. What are detail drawings used for in HVAC drafting?
16. What are section drawings used for in HVAC drafting?
17. Describe two types of sectioning practices used in HVAC drafting.
18. When are pictorial drawings used in HVAC drafting?
19. Describe HVAC schedules and their importance.
20. How are schedules keyed to the HVAC plan?

PART 2. SHEET METAL DRAFTING

21. Describe the basic principle of pattern development.
22. What is the key to pattern development?
23. Define *stretch-out line*.
24. Name and sketch four types of seams.
25. Name and sketch three types of hemmed edges.
26. Define *truncated*.
27. A transition piece is also known as a _____.

28. Why is it important to have a working knowledge of descriptive geometry for pattern development?
29. Define *triangulation*.
30. Define *bend allowance*.
31. Define *bend relief*.
32. Describe three different methods that may be used to draw precision sheet metal patterns.

CHAPTER

23 HVAC/SHEET METAL DRAFTING PROBLEMS

DIRECTIONS

1. Please read all related instructions before you begin working. Specific information will be provided for each problem.
2. Use manual or computer-aided drafting as required by your course guidelines.

Problem 23–1   Residential HVAC plan

Given: Residential heating engineering sketch of main floor plan and basement. Do the following on appropriately sized vellum (2 - B size sheets or 1 - C size is recommended):

1. Make a formal double-line HVAC floor plan layout at a 1/4" = 1'-0" scale.
2. Approximate the location of undimensioned items such as windows.
3. Use thin lines for the floor plan and use thick lines for the heating equipment and duct runs.

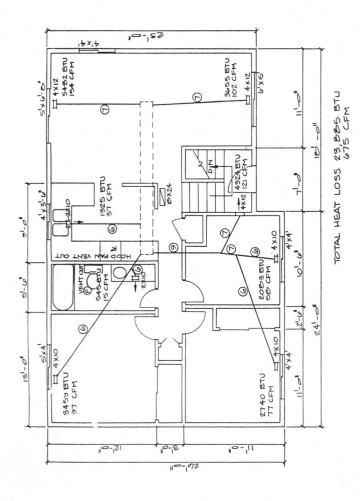

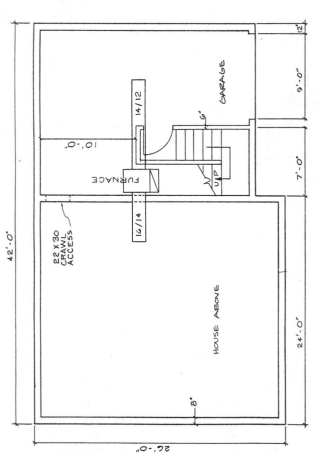

BASEMENT FURNANCE HEATING PLAN

Problem 23–2   Residential air-to-air heat exchanger plan

Given: Residential air-to-air heat exchanger ducting engineering sketch of basement floor plan. Do the following on appropriately sized vellum (one B or C size sheet is recommended):

1. Make a formal single-line air-to-air heat exchanger floor plan layout at a 1/4" = 1'–0" scale.

2. Approximate the location of undimensioned items such as doors.

3. Use thin lines for the floor plan and use thick lines for the air-to-air heat exchanger equipment and duct runs.

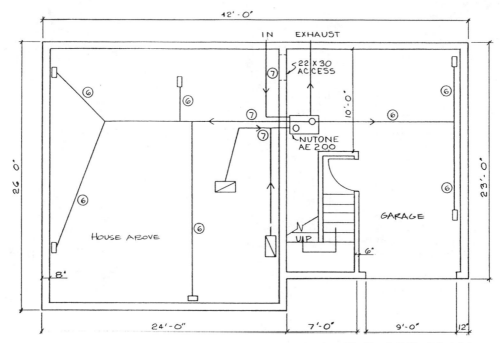

BASEMENT FLOOR PLAN
AIR-TO-AIR HEAT EXCHANGER PLAN

Problem 23–3   Commercial HVAC plan

Given the following:

1. HVAC floor plan engineering layout at approximately 1/16" = 1'–0" The engineer's layout is rough, so round off dimensions to the nearest convenient units. For example, if the dimension you scale reads 24 ft. 3 in. round off to 24 ft. 0 in. The floor plan will not require dimensioning; therefore, the representation is more important then the specific dimensions.
2. Related schedules.
3. Engineer's sketch for exhaust hood.

SCHEDULES				
CEILING OUTLET SCHEDULE				
Symbol	Size	CFM	Damper Type	Panel Size
C-10	9 × 9	230	Key Operated	12 × 12
C-11	8 × 8	185	Key Operated	12 × 12
C-12	6 × 6	40	Key Operated	12 × 12
C-13	6 × 6	45	Key Operated	12 × 12
C-14	6 × 18	300	Fire Damper	24 × 24

SUPPLY GRILL SCHEDULE				
Symbol	Size	CFM	Location	Damper Type
S-1	20 × 8	450	High Wall	Key Operation
S-2	12 × 12	450	High Wall	External Operation

EXHAUST GRILL SCHEDULE				
Symbol	Size	CFM	Location	Damper Type
E-5	18 × 24	1000	Low Wall	No Damper

ROOF EXHAUST FAN SCHEDULE			
Symbol	Area Served	CFM	Fan Specifications
REF-1	Solvent Tank	900	1/4 HP, 12 in. Non-spark Wheel, 1050 Max. Outlet Velocity

Do the following on appropriately sized vellum (D size is recommended; all required items will fit on one sheet with careful planning):

1. Make a formal double-line HVAC floor plan layout at a 1/4" = 1'–0" scale. (Note: You measured the given engineer's sketch at 1/16" = 1'–0".) Now convert the established dimensions to a formal drawing at 1/4" = 1'–0". Approximate the location of the HVAC duct runs and equipment in proportion to the presentation on the sketch.
2. Prepare correlated schedules in the space available. Set up the schedules in a manner *similar* to the examples in Figure 23–19 for layout.
3. Make a detail drawing of the exhaust hood either scaled or unscaled. Make the detail large enough to clearly show the features. Refer to Figure 23–16 for an example of a detail drawing.

(Continued)

Problem 23–3 (Continued)

Note: Do not include notes and dimensions for wall thickness, door sizes, and tangent. Top view of exhaust hood detail is drawn as a transition piece, similar to Figure 23–53.

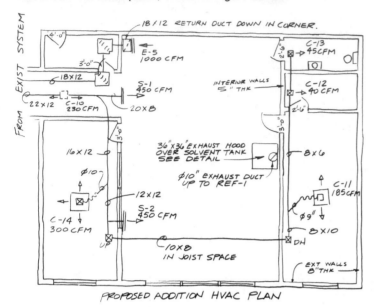

PROPOSED ADDITION HVAC PLAN

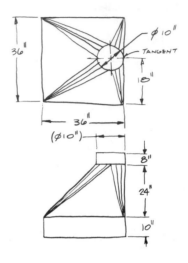

EXHAUST HOOD DETAIL

Problem 23–4 Exhaust duct system

Given: The engineer's sketch and specifications for an exhaust duct system. The sketch displays the top, front, and partial left side views of an exhaust duct system that could be found in any commercial solid-fuel exhaust. The exhaust pickup is rectangular in shape and the discharge throat is cylindrical. The directional path of the system is often obstructed and closely confined for reasons of design and operation of the system.

Do the following on appropriately sized paper:

1. Make a pattern development for each of the five exhaust duct components: (There will be five individual pattern development drawings.)
   A. Truncated cylinder
   B. Truncated cone
   C. Three-piece elbow
   D. Square-to-round transition piece
   E. Rectangular transitional elbow.
2. Use full scale unless otherwise specified by your instructor. A beam compass is necessary for manual drafting or use CADD.
3. Use a 3/8-in. single-lap seam on individual parts and between adjacent parts.
4. Show all layout and construction. Do not erase your construction lines.

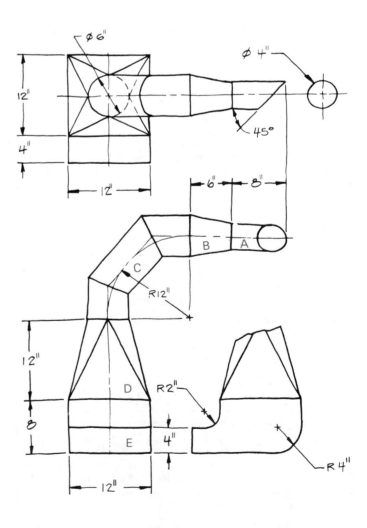

Problem 23–5 Welding booth hood

Given: The engineer's sketch of a fabrication shop's welding booth hood.

Do the following:

1. Make a pattern development drawing of the pyramid-shaped hood and the shroud base at a scale of 1 1/2" = 1'–0" or full scale for CADD.
2. Provide the cutout for the window to be added later. Be careful to find the true location, and true size and shape of the cutout in the pattern.
3. No seam material allowance is required as the seams will be welded.
4. Show all layout and construction.

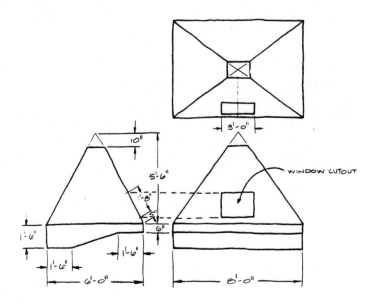

Problem 23–7 Chemistry laboratory hood

Given: The engineer's rough sketch of the chemistry laboratory hood.

Do the following:

1. Make a pattern development drawing of the chemistry laboratory hood, including the top and bottom collars.
2. Use a 3" = 1'–0" scale, or full scale for CADD.
3. No seams required.
4. Show all layout and construction.

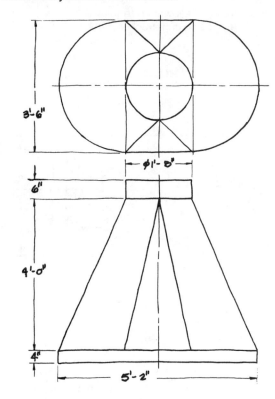

Problem 23–6 Exhaust hood

Given: The exhaust hood detail from Problem 23–3.

Do the following:

1. Make a pattern development drawing on appropriately sized layout for the transition piece, the top collar, and the base collar.
2. Use a half scale or full scale in CADD.
3. Provide a 1-in. single-lap seam for each part and between adjacent parts.
4. Show all layout and construction.

Problem 23–8 Cylindrical duct intersection

Given: The engineer's computer sketch of intersecting cylindrical ducts.

Do the following:

1. Use a 1" = 1'–0" scale, or full scale for CADD.
2. Find the intersection between the cylindrical ducts.
3. Make a pattern development for each cylinder.
4. Use a 1-in. (scale) double-lap seam.
5. Show all layout and construction.

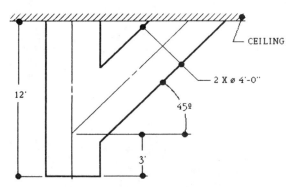

Problem 23–9   Grain hopper

Given the following:

1. Use a 3/4" = 1'–0" scale, or full scale for CADD.
2. The engineer's sketch of the grain hopper.

Do the following:

1. Determine the line of intersection between the cylinder and the cone in both views.
2. Make the resulting pattern development of the cone and side intersecting cylinder.
3. No seam material allowance required.
4. Show all layout and construction.

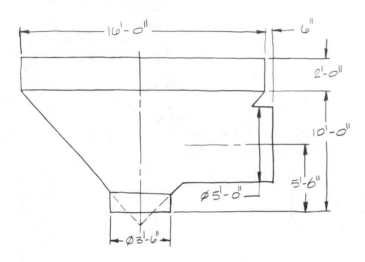

Problem 23–10   Intersection

Given the following drawing, use your scale to double the size shown and draw the object. Determine the line of intersection between the parts.

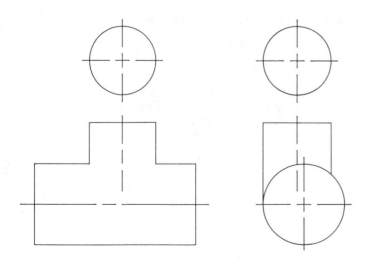

Problem 23–11   Intersection

Given the following drawing, use your scale to double the size shown and draw the object. Determine the line of intersection between the parts.

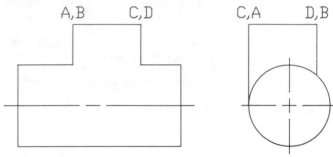

Problem 23–12   Intersection

Given the following drawing, use your scale to double the size shown and draw the object. Complete the left side view and determine the line of intersection between the parts in both views.

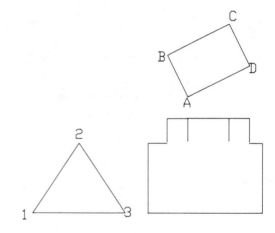

Problem 23–13   Drawing displayed in form view

Draw the following object as shown:

Part Name: Mount Leg Tank. *Courtesy TEMCO.*
Material: .25 A36

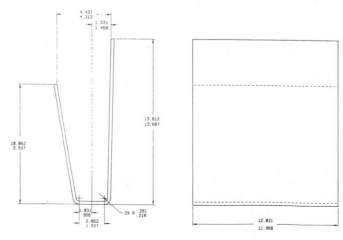

Problem 23–14   Drawing displayed in form view and flat pattern

Draw the following object as shown:

Part Name: Mounting Bracket. *Courtesy TEMCO.*
Material: .25 THK 5086-H32

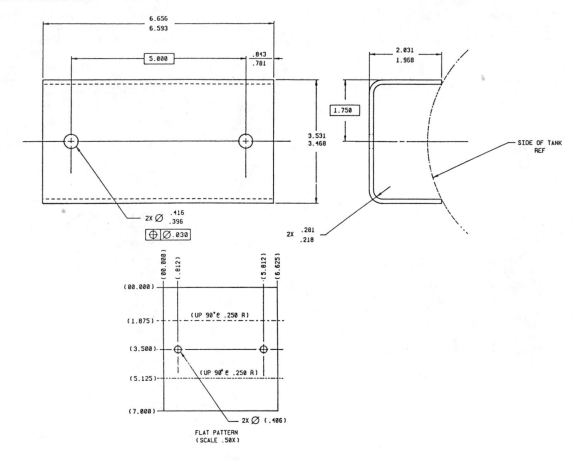

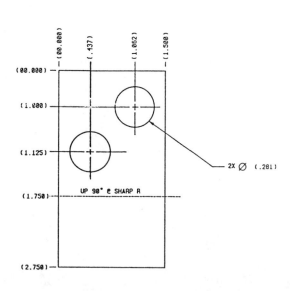

Problem 23–15   Drawing displayed in form view and flat pattern

Draw the following object as shown:

Part Name: Bracket. *Courtesy TEMCO.*
Material: 14 GA GALV CRS

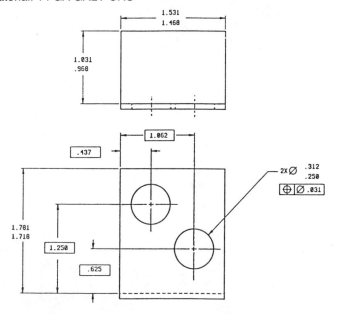

Problem 23–16 Chassis layout

Given: the engineer's rough sketch of a computer component chassis.

Do the following:

1. Make a flat-pattern drawing of the given chassis on properly sized vellum. Full scale is recommended.
2. Use arrowless tabular dimensioning from the given datums.

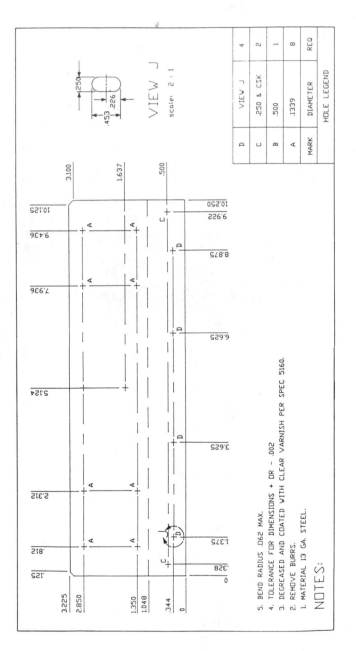

Problem 23–17 Chassis layout

Given: the engineer's rough sketch of a computer component chassis.

Do the following:

1. Make a flat-pattern drawing of the given chassis on properly sized vellum. Full scale is recommended. Establish the flat-pattern layout dimension by bend allowance calculations. Show all math formulas and calculations on another paper.
2. Use arrowless tabular dimensioning from the given datums.

DIMENSIONING TABLE			
Hole Symbol	Hole Diameter	Depth	Quantity
A	.125	Thru	9
B	.250	Thru	2
C	.469	.030	2
D	See View D	Thru	1

3. **Standard dimensioning is needed for the total length of flat pattern and a dimension from one datum to the bend line in the flat pattern.**

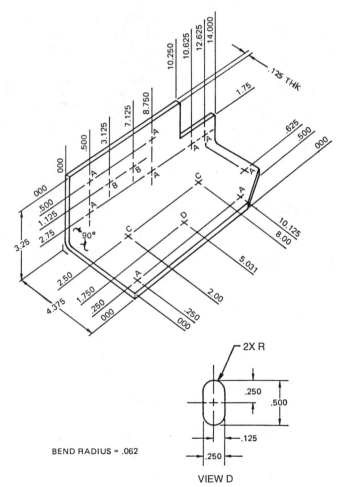

BEND RADIUS = .062

VIEW D

Problem 23–18 Display of part in form view with flat pattern shown as phantom line

Draw the following object as shown:

Part Name: Plate-formed.
Material: HC-112 6 mm THK.
Problem based on original art courtesy Hyster Company.

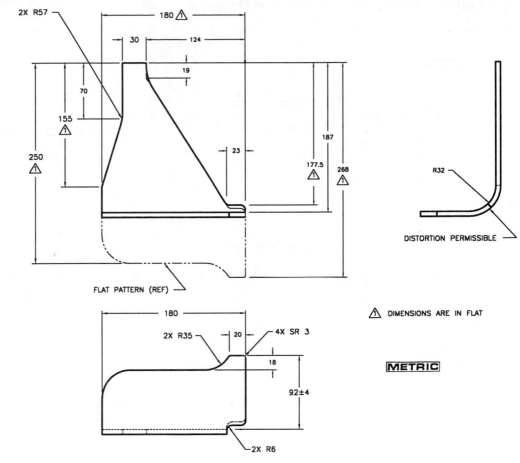

FLAT PATTERN (REF)

DISTORTION PERMISSIBLE

⚠ DIMENSIONS ARE IN FLAT

METRIC

Problem 23–19 Display of part in form view and in flat pattern

Given the following engineer's layout, draw the flat pattern, and the formed view. Use geometric dimensioning and tolerancing as shown.

Part Name: Mounting Bracket.
Material: 11 GA A569
Courtesy TEMCO.

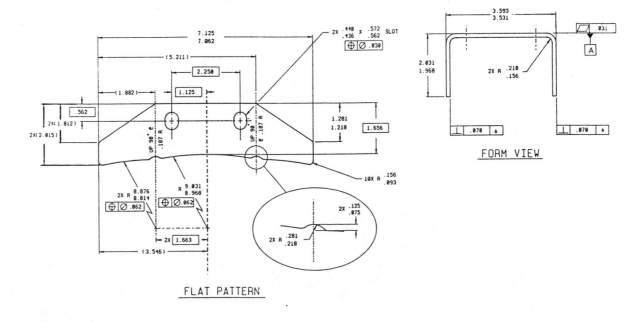

FLAT PATTERN

FORM VIEW

MATH PROBLEMS

Find the bend angle for each of the following figures:

1.

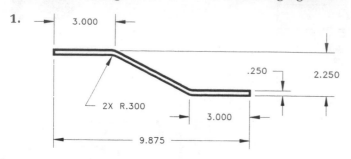

2.

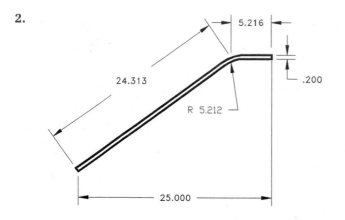

3.

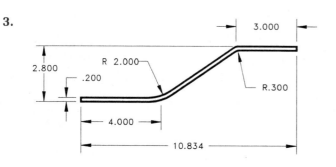

4.

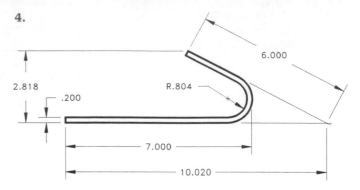

5.

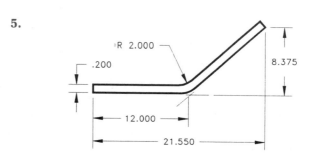

6.

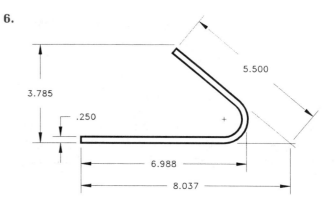

CHAPTER 24

Electrical and Electronic Schematic Drafting

THE ENGINEERING DESIGN PROCESS

In both electronic and electrical drafting, much of your work involves the use of symbols to show the components of your schematic or diagram. Having the proper tools available can greatly simplify your work in this area. If you are using manual drafting techniques, drawing templates are very useful for quick construction of standard symbols. CADD drafting offers additional advantages, however, because once a symbol has been drawn it can be used over and over again very simply. Additionally, CADD symbols can be created that automatically prompt for and place the required notational data in your drawing to properly define the symbol.

In this example, you have been provided a rough sketch of a circuit diagram. (See Figure 24–1.) By utilizing your CADD symbol library and adding the required notation, you are able to create the finished drawing (Figure 24–2) with a minimum of time and effort.

In the fully automated electronics industry, design begins with an engineer's sketch, which goes to an electronics technician who inputs the design into a computer. The computer system automatically evaluates the design and produces a prototype fabrication. The engineer then evaluates the system and makes modifications. After modifications are made and the system is completely tested and approved, the computer layout is sent to the drafter for engineering documentation. The drafter uses the CADD system to check design and drafting standards, evaluate engineering information for documentation accuracy and correct component values. The drafter adds device tables that relate exactly how each component is attached to the board. The entire process is computerized. The drawing then progresses through the system

(Continued)

ENGINEERING DESIGN PROCESS *(Continued)*

to the automatic generation of the circuit boards. The drafter works closely with the circuit board computer program to insure the accuracy of the product. This system may vary from one industry to the next, but as you can see, the electronics drafter is an important link between the engineering department and the final product.

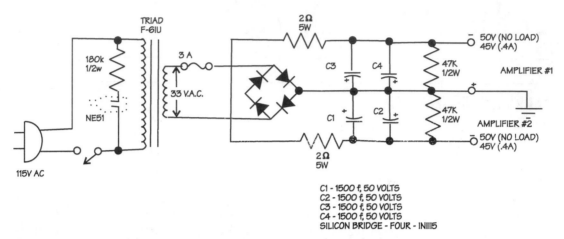

Figure 24–1   Engineer's rough computer sketch of a circuit diagram.

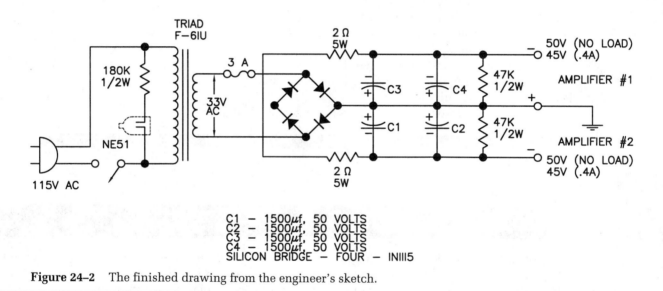

Figure 24–2   The finished drawing from the engineer's sketch.

Electrical drafting deals with concepts and symbols that relate to high-voltage applications from the production of electricity in power plants through distribution to industry and homes. While there is often a fine line between electrical and electronic drafting, *electronic drafting* is more oriented toward the design of electronic circuitry for radios, computers, and other low-voltage equipment.

ANSI   The key to effective communication on electrical drawings is the use of standardized symbols so that anyone who uses the diagram makes the same interpretation. To ensure proper standardization, these engineering drawings and related documents should be prepared in accordance with the American National Standards Institute publication ANSI Y32.2.

▼ FUNDAMENTALS OF ELECTRICAL DIAGRAMS

The purpose of electrical diagrams is to communicate information about the electrical system or *circuit* in a simple, easy-to-understand format of lines and symbols. *Electrical circuits* provide the path for electrical flow from the source of electricity through system components and connections and back to the source. Electrical diagrams are generally not drawn to scale. The

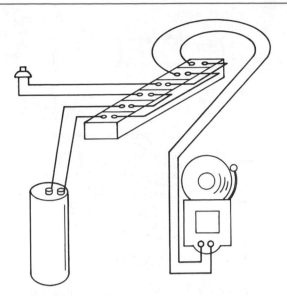

Figure 24–3 Pictorial drawing of a simple doorbell circuit.

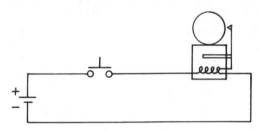

Figure 24–4 Schematic diagram of doorbell circuit shown in Figure 24–3.

responsibility of the drafter is to organize the information in a logical, orderly manner without crowding or large variations in spacing layout.

Pictorial Diagram

Pictorial diagrams represent the electrical circuit as a three-dimensional drawing. This type of diagram provides the most realistic and easy-to-understand representation and may commonly be used in sales brochures, catalogs, service manuals, or assembly drawings. Figure 24–3 shows the pictorial drawing of a simple doorbell circuit.

Schematic Diagram

Schematic diagrams are drawn as a series of lines and symbols that represent the electrical current path and the components of the circuit. Figure 24–4 shows a schematic diagram of the doorbell circuit.

One-line Diagram

The *one-line diagram* is a simplified version of the schematic diagram. Simplified symbols exhibit a minimum of detail of the component and generally no con-

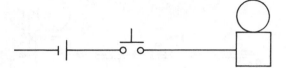

Figure 24–5 One-line diagram of the doorbell circuit shown in Figure 24–3.

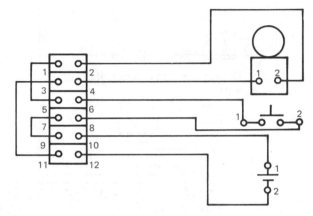

Figure 24–6 Wiring diagram of the doorbell circuit shown in Figure 24–3.

nections at individual *terminals* as shown in Figure 24–5.

Wiring Diagram

The *wiring diagram* is a type of schematic that shows all of the interconnections of the system components. These diagrams are often referred to as point-to-point interconnecting wiring diagrams. The wiring diagram is much more detailed than a standard schematic diagram because it shows the layout of individual wire runs. Figure 24–6 shows a wiring diagram of the doorbell circuit.

Schematic Wiring Diagram

A *schematic wiring diagram* combines the simplicity of a schematic diagram and the completeness of a wiring diagram. The complete circuit is drawn as a series of lines and symbols that represent the electrical current path and the components of the circuit, plus the connection terminals are shown in their proper locations along the circuit. (See Figure 24–7.)

Highway Wiring Diagram

Also known as *highway diagrams*, these drawings are used for fabrication, quality control, and troubleshooting of the wiring of electrical circuits and systems. A highway wiring diagram is a simplified or condensed representation of a point-to-point interconnecting wiring diagram. Highway wiring diagrams may be used when it becomes difficult to draw individual connect lines

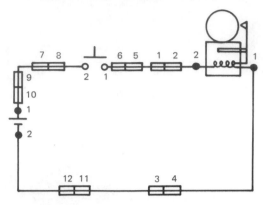

Figure 24–7 Schematic wiring diagram of the doorbell circuit shown in Figure 24–3.

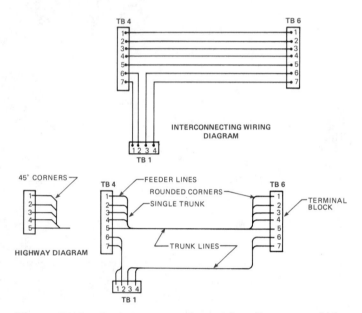

Figure 24–8 An interconnecting wiring diagram and the same electrical system converted to a highway diagram.

because of diagram complexity or when it is not necessary to show all of the wires between terminal blocks.

The wiring lines are merged at convenient locations into main *trunk lines,* called *highways,* that run horizontally or vertically between component symbols. The lines that run from the component to the trunk lines are called *feed lines.* The feed lines are identified by code letter, number, or a combination letter/number at the point where each line leaves the component. By reducing the number of lines drawn, the wiring diagram becomes easier to draw and interpret. Figure 24–8 shows an interconnecting wiring diagram and the same electricity system converted to a highway diagram.

A complete highway diagram has an identification system that guides the reader through the system, with a code at one terminal to the same corresponding terminal on another component. An identification system recommended by the American Standards Association (ASA) is a code made up of the wire destination, terminal number at the destination, wire size, and wire-cov-

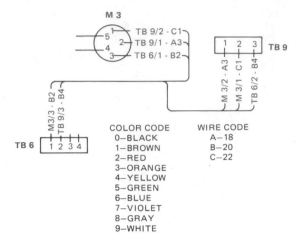

Figure 24–9 Highway wiring diagram.

ering color. Look at Figure 24–9 as you interpret the following code:

Component—M3	Component—TB6
TB6/1-B2	M3/3-B2
TB6 = Destination	M3 = Destination
1 = Terminal at Destination	3 = Terminal at Destination
B = Wire Size	B = Wire Size
2 = Color of Wire Covering	2 = Color of Wire Covering

Wireless Diagram

Wireless diagrams are similar to highway diagrams except that interconnecting lines are omitted as shown in Figure 24–10.

Cable Diagram and Assemblies

Cable diagrams are associated with *multiconductor* systems. A multiconductor is a *cable* or group of insu-

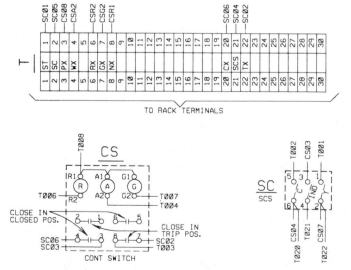

Figure 24–10 CADD-generated wireless diagram. *Courtesy Bonneville Power Administration.*

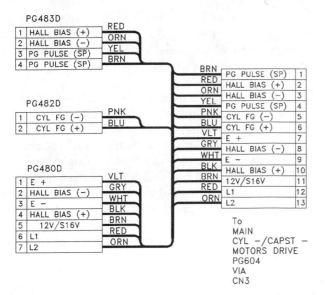

Figure 24–11 Cable diagram. *Reproduced by permission of RCA Consumer Electronics.*

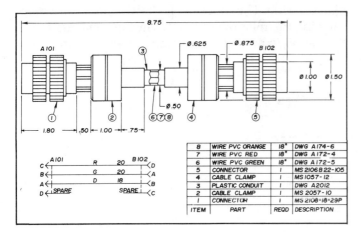

Figure 24–12 Cable assembly.

lated wires that is put together in one sealed assembly. For example, the trunk line shown in the highway diagram in Figure 24–8 or 24–9 could be a multiconductor. Cable systems are made up of insulated conductor wires, protective outer jacket, or some other means of holding the wires together, and connectors at one end or both ends. Cables are used to connect components, equipment assemblies, or systems together. Cable diagrams usually provide circuit destination, conductor size, number of leads, conductor type, and power rating. A cable diagram is shown in Figure 24–11.

Cable assemblies are drawn to scale and include a parts list that is coordinated with the drawing by identification balloons as shown in Figure 24–12.

▼ GENERATION, TRANSMISSION, AND DISTRIBUTION OF ELECTRICITY

Electricity is generated around the world in hydroelectric, coal burning, and nuclear power plants. The elec-

tricity created at the generating station is increased in voltage at a step-up transformer and sent through high-voltage lines to a switching station where the electricity is retransmitted to various locations. Before the electricity can be used, it goes to a substation where a step-down transformer converts the high voltage to a lower voltage for heavy industry or transmission through lines to a distribution substation for further voltage reduction and distribution to light commercial and residential users.

▼ THE ELECTRIC POWER SUBSTATION DESIGN DRAWINGS

A *substation* is the part of the electrical transmission system where electricity is switched or transformed from a very high voltage to a conveniently usable form for distribution to homes or businesses. Substation design drawings are an important part of any power supply system.

One-line Diagrams

The *one-line diagram* is a simple way for electrical engineers and drafters to communicate the design of an electrical power substation as shown in Figure 24–13.

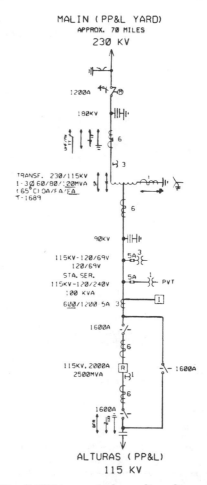

Figure 24–13 CADD-generated one-line diagram. *Courtesy Bonneville Power Administration.*

Elementary Diagrams

Elementary diagrams provide the detail necessary for engineering analysis and operation of the substation equipment by operators or maintenance people.

Elementary diagram, DC (Direct Current) Circuits, show the direct current circuits that operate the relaying and controls for the substation equipment. Elementary diagram, AC (Alternating Current) Circuits, depict the circuits that provide information for both the protective relays and the instruments and meters used by people who work in the substation.

Schematic Diagrams

The *schematic diagram* is used to show the relationship of the components and equipment located in the substation. The schematic diagram is more detailed than the one-line diagram.

Electric Power Systems Schematic Symbols

ANSI Most standards for the preparation of electrical diagrams use symbols in accordance with the ANSI Y32.2 document. Any special variation to symbols should be shown in a legend or described in a general note. Commonly used electrical equipment symbols are shown in Figure 24–14.

Electrical relays are magnetic switching devices.

Nondirectional relays are relays that operate when current flows in either direction. *Directional relays* operate only when current flows in one direction.

Differential relays provide a switching connection between a circuit with two different voltage values. *Pilot relay* systems are controlled by communication devices and are designated by the type of communication circuit or the function of the relay system.

Supervisory control relays are used to check, monitor, or control other relays.

Plot Plan

The entire layout of a substation is shown by a plot-plan drawing. The plot plan is drawn similar to a map, showing the relationship between the elements of the substation in correct orientation to compass direction.

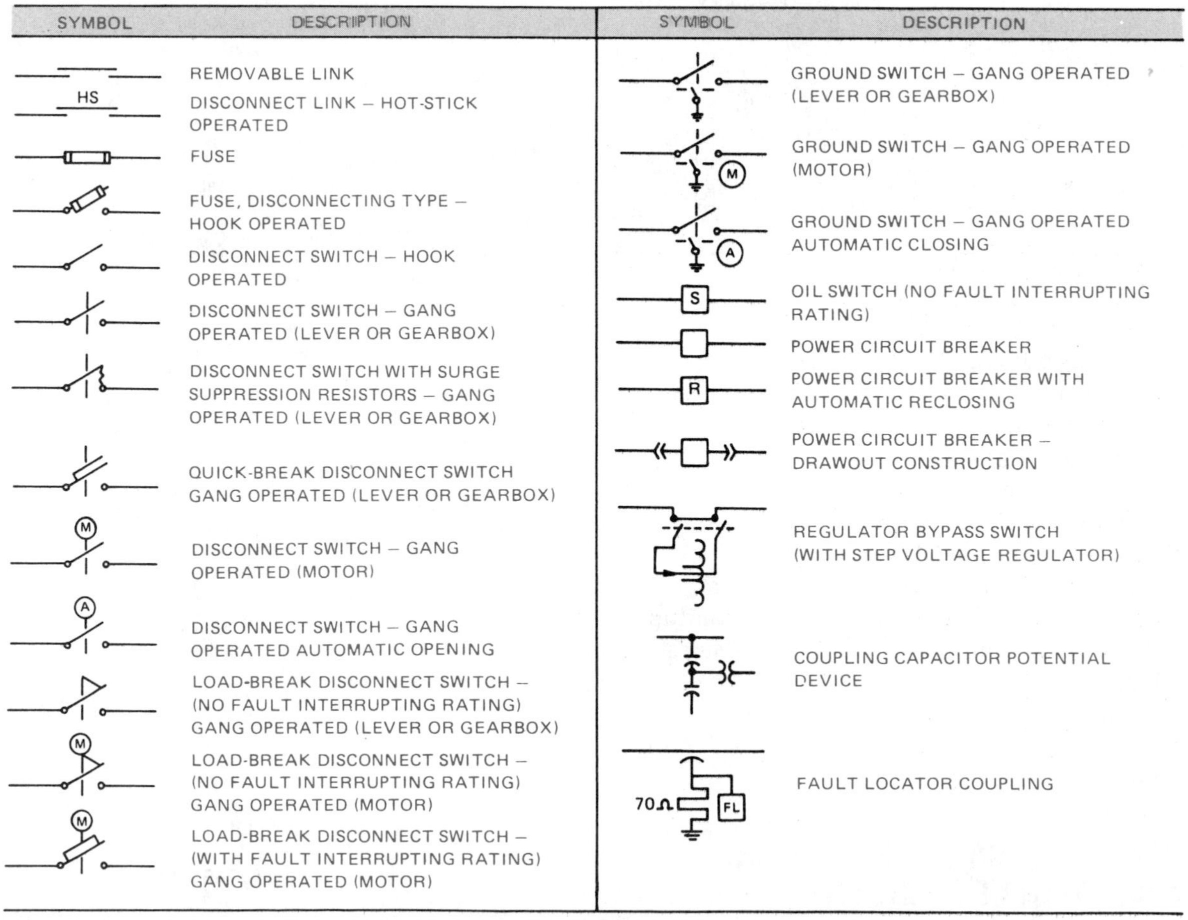

SCHEMATIC SYMBOLS

Figure 24–14   Electric power systems schematic symbols.

NONDIRECTIONAL SYMBOLS

RELAY SYMBOL	RELAY DESCRIPTION
	OVER, NONDIRECTIONAL
	UNDER, NONDIRECTIONAL
	RECEIVING, UNDER, DIRECTIONAL
	OVER, DIRECTIONAL, INITIATING (ARROWS INDICATE TRIPPING DIRECTION)
	RELAY WITH TRIP ATTACHMENT
	GROUND
T	TEMPERATURE
x	DIFFERENTIAL
z	DISTANCE
c	CURRENT (ASSUMED IF LETTER C OMITTED)
V	VOLTAGE
W	POWER
F	FREQUENCY
	BALANCE
	COMMUNICATION CHANNEL (BASIC)
PW	PILOT WIRE
CC	POWER LINE CARRIER CURRENT
MW	MICROWAVE
VHF	RADIO
UHF	RADIO

FUNCTION SYMBOLS

RELAY SYMBOL	RELAY DESCRIPTION
	PHASE OVERCURRENT
	GROUND OVERCURRENT
I	PHASE OVERCURRENT (INSTANTANEOUS)
I	GROUND OVERCURRENT (INSTANTANEOUS)
FD	PHASE OVERCURRENT – FAULT DETECTOR
BFR	PHASE OVERCURRENT – BREAKER FAILURE
MFD	PHASE OVERCURRENT – MULTIPHASE FAULT DETECTOR
DLF	PHASE OVERCURRENT – DEAD-LINE FAULT
V	OVERVOLTAGE
V	UNDERVOLTAGE
VC	VOLTAGE CONTROL (AUTOMATIC)
HLC	HOT-LINE CHECK (VOLTAGE)
F	OVERFREQUENCY
F	UNDERFREQUENCY
S	SYNCHRONISM-CHECK (FREQUENCY)

DIRECTIONAL SYMBOLS

RELAY SYMBOL	RELAY DESCRIPTION
	PHASE OVERCURRENT
	GROUND OVERCURRENT
	PHASE OVERCURRENT (INSTANTANEOUS)
	GROUND OVERCURRENT (INSTANTANEOUS)
c_0	ZERO-SEQUENCE OR POLARIZING OVERCURRENT
c_2	NEGATIVE-SEQUENCE OVERCURRENT
W	POWER
z_1	DISTANCE PHASE – ZONE 1 ONLY
z_{12}	DISTANCE PHASE – ZONES 1 AND 2
z_{123}	DISTANCE PHASE – ZONES 1, 2, AND 3
z_2	DISTANCE PHASE – ZONE 2 ONLY
z_{23}	DISTANCE PHASE – ZONES 2 AND 3
z_3	DISTANCE PHASE – ZONE 3 ONLY
z_{12}	DISTANCE GROUND – (ZONES INDICATED AS ABOVE)
OSB-z	OUT-OF-STEP BLOCKING
OST-z	OUT-OF-STEP TRIPPING

APPLICATIONS – DIFFERENTIAL RELAYS

RELAY SYMBOL	RELAY DESCRIPTION
	CURRENT DIFFERENTIAL (BASIC)
	VOLTAGE DIFFERENTIAL

APPLICATIONS – MISCELLANEOUS RELAYS

RELAY SYMBOL	RELAY DESCRIPTION
SP	SUDDEN PRESSURE
GP	HIGH GAS PRESSURE
GP	LOW GAS PRESSURE
T	HIGH TEMPERATURE
T	LOW TEMPERATURE
R2	AUTOMATIC RECLOSING (R IN POWER CIRCUIT BREAKER SYMBOL. NUMBERS INDICATE 1-, 2-, 3-, OR 4-SHOT.)
	CONNECTION
	NONCONNECTION

DIFFERENTIAL AND MISCELLANEOUS SYMBOLS

RELAY SYMBOL	RELAY DESCRIPTION
PW	PILOT WIRE RELAYS
PW-TT	PILOT WIRE TRANSFER TRIP (INITIATING)
PW-TT	PILOT WIRE TRANSFER TRIP (RECEIVING)
CC-DC	CARRIER CURRENT DIRECTIONAL COMPARISON
CC-PC	CARRIER CURRENT PHASE COMPARISON
CC-TT	CARRIER CURRENT TRANSFER TRIP (INITIATING)
CC-TT	CARRIER CURRENT TRANSFER TRIP (RECEIVING)
CC-TT	CARRIER CURRENT TRANSFER TRIP (INITIATING AND RECEIVING)
MW-PC	MICROWAVE PHASE COMPARISON
MW-TT	MICROWAVE TRANSFER TRIP (INITIATING)
MW-TT	MICROWAVE TRANSFER TRIP (RECEIVING)
MW-TT	MICROWAVE TRANSFER TRIP (INITIATING AND RECEIVING)
VHF-TT	VHF RADIO TRANSFER TRIP (INITIATING)
VHF-TT	VHF RADIO TRANSFER TRIP (RECEIVING)
VHF-TT	VHF RADIO TRANSFER TRIP (INITIATING AND RECEIVING)
UHF-TT	UHF RADIO TRANSFER TRIP (INITIATING)
UHF-TT	UHF RADIO TRANSFER TRIP (RECEIVING)
UHF-TT	UHF RADIO TRANSFER TRIP (INITIATING AND RECEIVING)
MW-GD	MICROWAVE GENERATOR DROPPING
MW-LD	MICROWAVE LOAD DROPPING

PILOT RELAYING SCHEMES

RELAY SYMBOL	RELAY DESCRIPTION
SC	INITIATING – MASTER UNIT (CONTROL CENTER)
SC	RECEIVING – REMOTE UNIT (LOCAL SUBSTATION)
	NOTE: SUPERVISORY CONTROL OF EQUIPMENT IS SHOWN WITH A LETTER DENOTING THE MASTER UNIT LOCATION, AND AN EXPLANATORY NOTE IS MADE ON THE DRAWING. AN EXAMPLE FOLLOWS.
SC-D	RECEIVING – FROM DITTMER CENTER (D)
	THE LAST LETTER OR ABBREVIATION DENOTES THE CENTER LOCATION.

SUPERVISORY CONTROL SYMBOLS

Figure 24–14 (Continued)

SYMBOL	DESCRIPTION
⊗— TM	INITIATING – REMOTE UNIT (LOCAL SUBSTATION)
⊗— TM	RECEIVING – MASTER UNIT (CONTROL CENTER)
⊗— TM-AT	ANALOG TYPE (INITIATING)
⊗— TM-DT	DIGITAL TYPE (INITIATING)
⊗— TM-PPT	PULSE-PRINT TYPE (INITIATING)

TELEMETERING SYMBOLS

Figure 24–14   (Continued)

The scale used for the plot plan generally ranges from 1" = 20' to 1" = 50'. Typical plot-plan symbols are shown in Figure 24–15. The items that are commonly found on a plot plan include:

1. Property boundaries and developed-yard boundary with chain link fence.
2. Primary bearing lines. *Bearing* is direction in relation to the Northwest or Northeast and Southwest or Southeast quadrants of a compass. For example, N 50°30'15"W is a line that is located at an angle of 50°30'15" toward West from North.
3. Service and access roads.
4. Buildings.

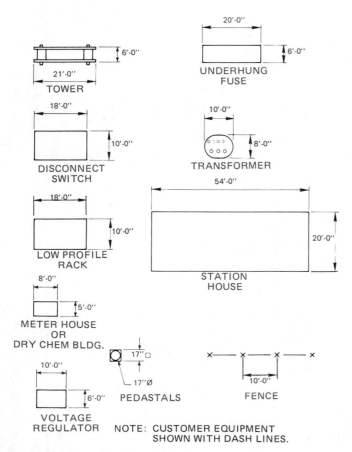

Figure 24–15   Plot-plan symbols.

5. System components such as switches, fuses, transformers, and racks.
6. Complete dimensioning.

A complete plot-plan discussion is found in Chapter 22.

Bus Layout

A *bus* is an aluminum or copper plate or tubing that carries the electrical current. The bus layout drawing is used by construction crews for the construction of the current-carrying portion of the substation. Bus layout drawings show the system in plan (top) view as shown in Figure 24–16, and elevation (side) view as shown in Figure 24–17.

The components are identified with numbered balloons, which key each item to a parts list. Some common bus layout symbols are shown in Figure 24–18.

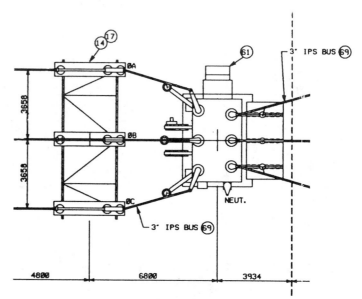

Figure 24–16   CADD-generated bus layout plan. *Courtesy Bonneville Power Administration.*

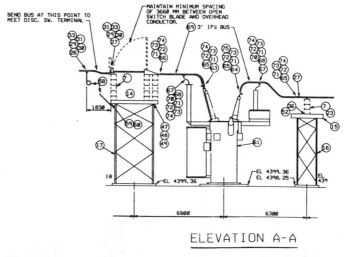

ELEVATION A-A

Figure 24–17   CADD-generated bus elevations. *Courtesy Bonneville Power Administration.*

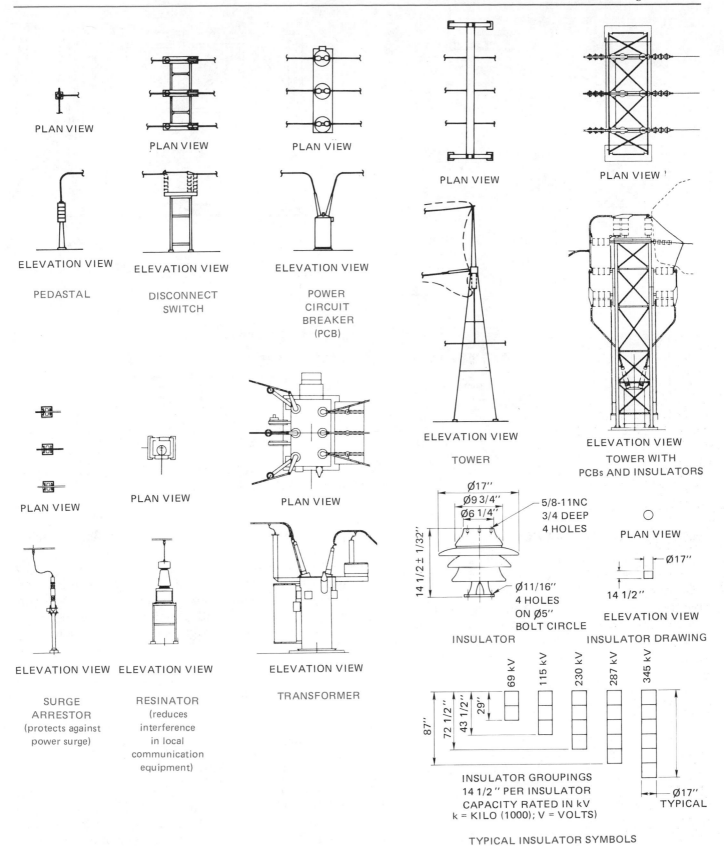

Figure 24–18 Common bus layout symbols.

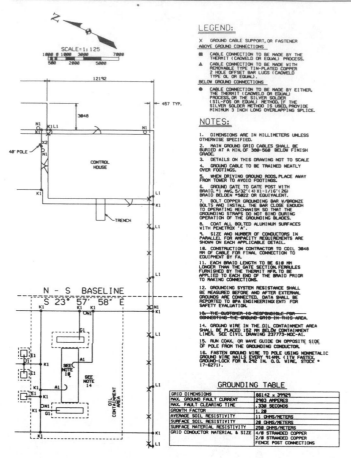

LEGEND:

X   GROUND CABLE SUPPORT, OR FASTENER
ABOVE GROUND CONNECTIONS

■   CABLE CONNECTION TO BE MADE BY THE
     THERMIT (CADWELD OR EQUAL) PROCESS.

▲   CABLE CONNECTION TO BE MADE WITH
     REMOVABLE TYPE TIN-PLATED COPPER
     2 HOLE OFFSET BAR LUGS (CADWELD
     TYPE QL OR EQUAL).

BELOW GROUND CONNECTIONS

●   CABLE CONNECTION TO BE MADE BY EITHER
     THE THERMIT (CADWELD OR EQUAL)
     PROCESS, OR THE SILVER SOLDER
     (SIL-FOS OR EQUAL) METHOD. IF THE
     SILVER SOLDER METHOD IS USED, PROVIDE
     MINIMUM 3 INCH LONG OVERLAPPING SPLICE.

NOTES:

1. DIMENSIONS ARE IN MILLIMETERS UNLESS
    OTHERWISE SPECIFIED.
2. MAIN GROUND GRID CABLES SHALL BE
    BURIED AT A MIN OF 300-500 BELOW FINISH
    GRADE.
3. DETAILS ON THIS DRAWING NOT TO SCALE
4. GROUND CABLE TO BE TRAINED NEATLY
    OVER FOOTINGS.
5. WHEN DRIVING GROUND RODS, PLACE AWAY
    FROM TOWER TO AVOID FOOTINGS.
6. GROUND GATE TO GATE POST WITH
    BRAID #1 AWG. 5/32"(4) X1-1/16"(26)
    BRAID BELDEN #5022 OR EQUIVALENT.
7. BOLT COPPER GROUNDING BAR W/BRONZE
    BOLTS AND INSTALL THE BAR CLOSE ENOUGH
    TO OPERATING MECHANISM SO THAT THE
    GROUNDING STRAPS DO NOT BIND DURING
    OPERATION OF THE GROUNDING BLADES.
8. COAT ALL BOLTED ALUMINUM SURFACES
    WITH PENETROX "A".
9. SIZE AND NUMBER OF CONDUCTORS IN
    PARALLEL FOR AMPACITY REQUIREMENTS ARE
    SHOWN ON EACH APPLICABLE DETAIL.
10. CONSTRUCTION CONTRACTOR TO COIL 3048
    MM OF CABLE FOR FINAL CONNECTION TO
    EQUIPMENT BY FA.
11. EACH BRAID LENGTH TO BE 610 MM
    LONGER THAN THE GATE SECTION FERRULES
    FURNISHED BY THE THERMIT MFR. TO BE
    APPLIED TO EACH END OF THE BRAID PRIOR
    TO MAKING CONNECTIONS.
12. GROUNDING SYSTEM RESISTANCE SHALL
    BE MEASURED BEFORE AND AFTER EXTERNAL
    GROUNDS ARE CONNECTED. DATA SHALL BE
    REPORTED TO BPA ENGINEERING(ENT) FOR
    SAFETY EVALUATION.
13. ~~THE CUSTOMER IS RESPONSIBLE FOR
    CONNECTING THE GROUND GRID IN THIS AREA.~~
14. GROUND WIRE IN THE OIL CONTAINMENT AREA
    SHALL BE PLACED 152 MM BELOW CONTAINMENT
    LINER. SEE CIVIL DRAWING 237773-NDC-A1.
15. RUN COAX, OR WAVE GUIDE ON OPPOSITE SIDE
    OF POLE FROM THE GROUNDING CONDUCTOR.
16. FASTEN GROUND WIRE TO POLE USING NONMETAL
    GROUND WIRE NAILS EVERY 914MM (ITW FASTEX
    GROUND-LOCK FOR 8.292 IN. O.D. WIRE, STOCK #
    17-6271).

GROUNDING TABLE

GRID DIMENSIONS	66142 x 39529
MAX. GROUND FAULT CURRENT	2903 AMPERES
MAX. FAULT CLEARING TIME	.330 SECONDS
GROWTH FACTOR	1.20
AVERAGE SOIL RESISTIVITY	11 OHMS/METERS
SURFACE SOIL RESISTIVITY	20 OHMS/METERS
SURFACE MATERIAL RESISTIVITY	250 OHMS/METERS
GRID CONDUCTOR MATERIAL & SIZE	4/0 STRANDED COPPER 2/0 STRANDED COPPER FENCE POST CONNECTIONS

Figure 24–19 CADD-generated grounding layout. *Courtesy Bonneville Power Administration.*

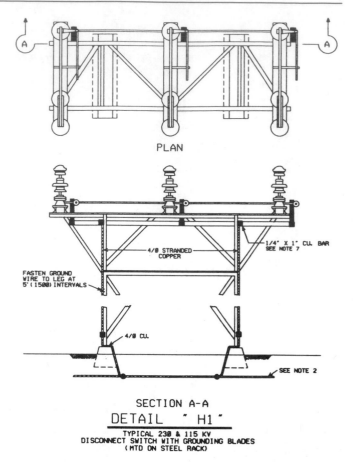

PLAN

SECTION A-A

DETAIL " H1 "

TYPICAL 230 & 115 KV
DISCONNECT SWITCH WITH GROUNDING BLADES
(MTD ON STEEL RACK)

Figure 24–20 CADD-generated grounding details. *Courtesy Bonneville Power Administration.*

Grounding Layout and Details

There is a tremendous hazard in substations due to the possibility of high voltage occurring on pieces of equipment. The voltage at different pieces of equipment varies with the location in the yard and the fault amount available. The amount of potential electrical shock depends on the voltages available during a fault condition. A *fault condition* is a short circuit, which is a zero resistance path for current flow. The voltage difference between the metal parts of the equipment and the ground surface on which a person stands must be mtaintained at a very low level. This is determined by the amount and location of the *ground grid,* which is the grounding system. Grounding layouts are often drawn at a scale of 1" = 10'–0" as shown in Figure 24–19.

Grounding details provide an enlarged view of how a structure or piece of equipment is grounded. Details are keyed to the grounding layout with letters; for example, "DETAIL H1," as shown in Figure 24–20.

Conduit Installation Layout and Details

Each component in a substation has a specific function and contains a different type of electrical equipment. The control facility is a building that interconnects each piece of equipment using multiple conductor cables. (See Figure 24–21.)

Conduit detail drawings coordinate with the layout by providing construction details and the locations of various fittings, junction boxes, and brackets. Details correlate to the layout with callouts that give the detail identification and the page where the detail is located; for example, DETAIL N/2. These details are either drawn at a scale of 1/2" = 1'–0" or may be drawn without scale. (See Figure 24–22.) Various components in the drawing are keyed by balloons to a parts list.

▼ RESIDENTIAL AND COMMERCIAL ELECTRICAL PLANS

The design of the electrical layout is an important part of the total livability of a home or function and safety of a commercial or industrial facility. Careful analysis should be given to the design placement of equipment and furniture, and the planned use of each room. Local and national electrical codes provide specifications for installations and layout. Layout planning should play an important role in conjunction with code guidelines. An evaluation of need and code requirements should be

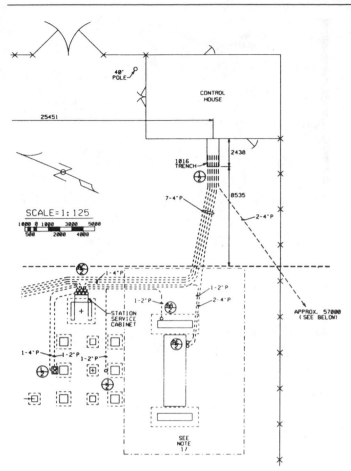

Figure 24–21 CADD-generated conduit installation layout drawing. *Courtesy Bonneville Power Administration.*

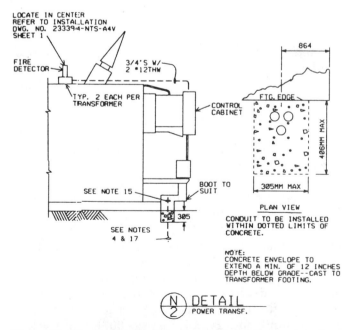

Figure 24–22 CADD-generated conduit detail. *Courtesy Bonneville Power Administration.*

closely compared so the electrical layout is not over or under designed.

Architectural Electrical Symbols

Symbols and lines are used to show the electrical layout in a structure. In residential applications the electrical layout is often part of the floor plan. Commercial electrical plans are commonly an overlay of the base floor-plan sheet. Electrical symbols are generally 1/8 in. in height, where applicable. The electrical plan should be drawn in a clear, concise manner so the layout remains uncluttered. All lettering for switches and other notes should be 1/8 in. high, although a 5/32-in. height is used by some companies. Common electrical symbols are shown in Figure 24–23.

Switch symbols are drawn perpendicular to the wall and are placed to read from the right side or bottom of the sheet. The switch relay dashed line should intersect the symbol at right angles to the wall or the relay may begin next to the symbol. Do not mix methods. Verify the preferred procedure with your instructor or employer. (See Figure 24–24.) Figure 24–25 (see page 805) shows several typical electrical installations with switches to light outlets.

When special characteristics are required, such as a specific size fixture, a location requirement, or any other specification, a local note may be applied next to the electrical symbol to briefly describe the situation. (See Figure 24–26, page 805.) Where a specification affects electrical installations on the entire layout, general notes may be used. Some common errors related to the placement and practice of electrical floor-plan layout are shown in Figure 24–27 (see page 805).

Residential Electrical Plan Examples

Residential electrical plans are generally less complex than commercial plans. Figure 24–28 (see page 805) shows some maximum spacing recommendations for installation of wall outlets in a residential structure. A typical residential electrical layout is shown in Figure 24–29 (see page 806). A typical kitchen layout is shown in Figure 24–30 (see page 806).

Commercial Electrical Plan Examples

Commercial electrical plans often follow much more detailed installation guidelines than residential applications. Electrical plan overlays are commonly used so the only information provided is that of electrical installations.

The electrical circuit switch legs for commercial applications are generally drawn as solid lines rather than dashed, as in residential electrical plans. The electrical circuit lines that continue from an installation to the service distribution panel are terminated next to the fixture and capped with an arrowhead, meaning that the circuit continues to the distribution panel. When multiple arrow-

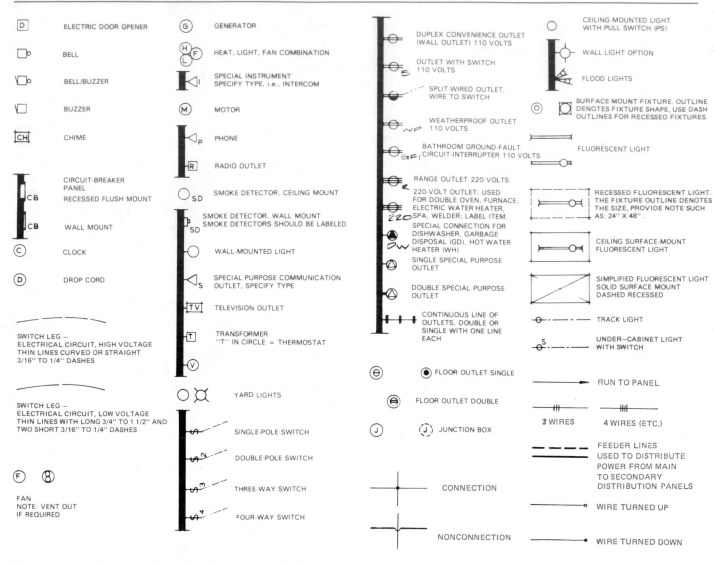

Figure 24–23   Common floor-plan electrical symbols.

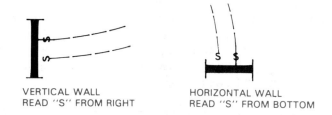

VERTICAL WALL
READ "S" FROM RIGHT

HORIZONTAL WALL
READ "S" FROM BOTTOM

Figure 24–24   Switch symbol placement.

heads are shown, this indicates the number of circuits in the electrical run. (See Figure 24–31, page 806.) For many installations where a number of circuit wires are used, the number of wires is indicated by slash marks placed in the circuit run. The number of slash makes equals the number of wires. (See Figure 24–32, page 806.)

There may be more than one commercial electrical overlay; for example, floor-plan lighting, electrical-plan power supplies, or reflected ceiling plan. The floor-plan lighting layout provides the location and identification of lighting fixtures and circuits. It is usually coordinated with a lighting fixture schedule as shown in Figure 24–33 (see page 806). In some applications a power supply plan is used to show all electrical outlets, junction boxes, and related circuits. (See Figure 24–34, page 806.) The reflected ceiling plan is used to show the layout for the suspended ceiling system as shown in Figure 24–35 (see page 807). Electrical plans for equipment installations may also be needed to supplement the power supply and lighting plans. Figure 24–36 (see page 807) shows the plan for roof installation of equipment. The drafter is also often required to draw schematic diagrams for specific electrical installations as shown in Figure 24–37 (see page 807).

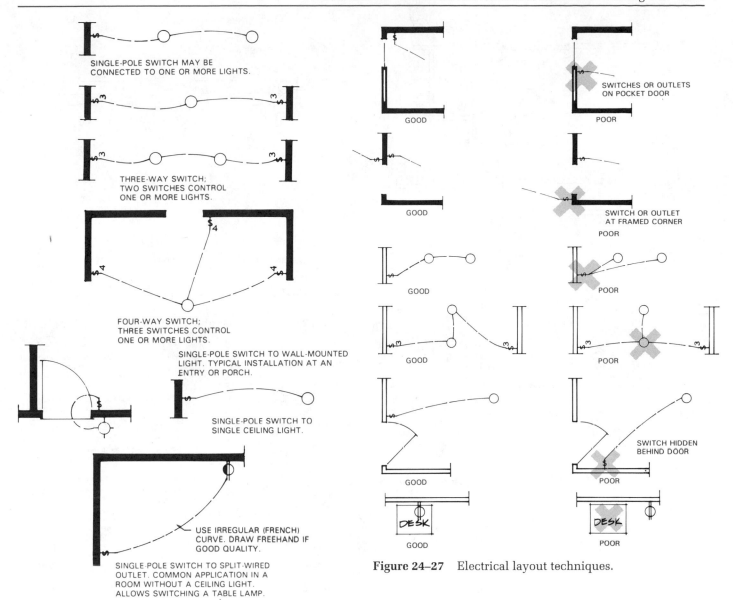

SINGLE-POLE SWITCH MAY BE
CONNECTED TO ONE OR MORE LIGHTS.

THREE-WAY SWITCH;
TWO SWITCHES CONTROL
ONE OR MORE LIGHTS.

FOUR-WAY SWITCH;
THREE SWITCHES CONTROL
ONE OR MORE LIGHTS.

SINGLE-POLE SWITCH TO WALL-MOUNTED
LIGHT. TYPICAL INSTALLATION AT AN
ENTRY OR PORCH.

SINGLE-POLE SWITCH TO
SINGLE CEILING LIGHT.

USE IRREGULAR (FRENCH)
CURVE. DRAW FREEHAND IF
GOOD QUALITY.

SINGLE-POLE SWITCH TO SPLIT-WIRED
OUTLET. COMMON APPLICATION IN A
ROOM WITHOUT A CEILING LIGHT.
ALLOWS SWITCHING A TABLE LAMP.

Figure 24–25 Typical electrical installations.

GOOD

GOOD

GOOD

GOOD

GOOD

DESK
GOOD

SWITCHES OR OUTLETS
ON POCKET DOOR
POOR

SWITCH OR OUTLET
AT FRAMED CORNER
POOR

POOR

POOR

SWITCH HIDDEN
BEHIND DOOR
POOR

DESK
POOR

Figure 24–27 Electrical layout techniques.

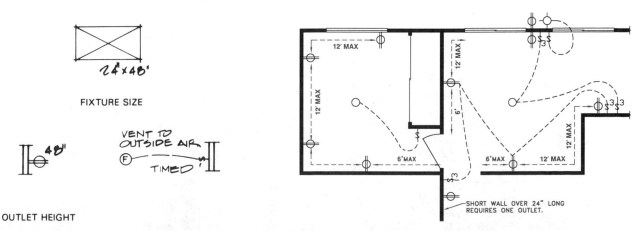

24″×48″

FIXTURE SIZE

48″

VENT TO
OUTSIDE AIR
F
TIMED

OUTLET HEIGHT

Figure 24–26 Special notes for electrical fixtures.

12′ MAX

12′ MAX

12′ MAX

12′ MAX

12′ MAX

6′ MAX

6′ MAX

6′ MAX   12′ MAX

SHORT WALL OVER 24″ LONG
REQUIRES ONE OUTLET.

Figure 24–28 Maximum spacing requirements.

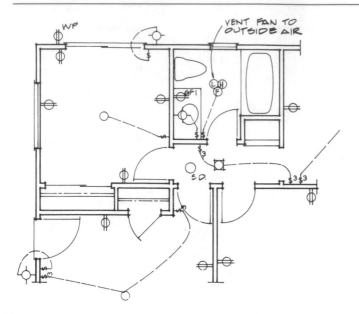

Figure 24–29 Typical electrical layouts.

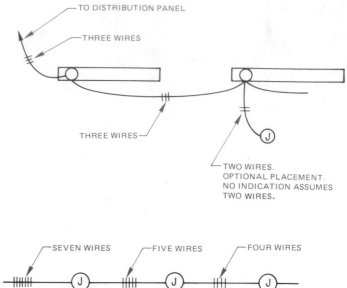

Figure 24–32 Number of wires designated in circuit runs.

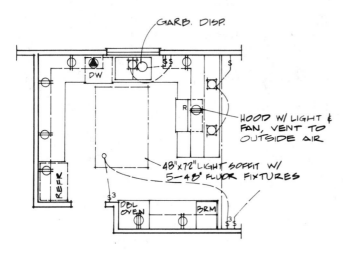

Figure 24–30 Kitchen electrical layout.

LIGHTING FIXTURE SCHEDULE

F1 – Surface mounted 8' open strip fluorescent. Lamps: (1) F96T12/LW/WM (75 watt). Manufacturer: Lithonia PUN 196 – 120V

F2 – Surface mounted 8' open strip fluorescent with damp location label and low temperature ballast. Lamps: (1) F96T12/LW/WM (75 watt). Manufacturer: Lithonia PUN 196 – DL – 120V

F3 – Surface mounted 4' open strip fluorescent. Lamps: (2) F48T12/LW/WM (30 watt). Manufacturer: Lithonia PUN 248 – 120V

F4 – Surface ceiling mounted vapor-tight incandescent with cast guard. Lamp: (1) 100W A19 Manufacturer: Steneo QVCXL – 11GC

F5 – Surface mounted incandescent with prismatic lexan cylinder and dump location label. Lamp: (1) 100W A19 Manufacturer: Marco QB5NP – SA

F6 – Surface wall mounted sodium vapor security flood light. Lexan lens and weather tight. Lamp: 70w Manufacturer: Crousshinds Sc – 711 – 70W HPS

F7 – Recessed ceiling mounted incandescent fan/light combination. Lamp: 10w Manufacturer: Broan Q678

F8 – 16' pole mounted sodium vapor flood area luminaire. Type III distribution flat lens. Bronze finish. Pole to be 16' straight square steel. Coated with paint to match fixture. See detail. Lamp: LU150 – 55 Manufacturer: ELSCO ZCHL – 150 – MPS – 16 – DP – 120 – B2A Alternate: Nu-Art QULT – III – MPS – 150

Figure 24–33 Lighting needs are shown using an overlay of the floor plan and a lighting fixture schedule. *Courtesy The Southland Corporation.*

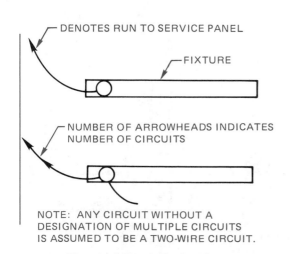

Figure 24–31 Electrical circuit designations.

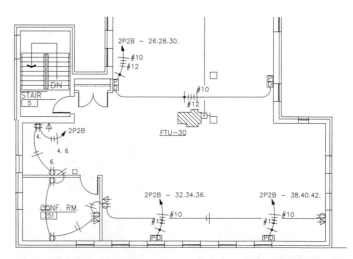

Figure 24–34 Power supply plan. *Courtesy System Design Consultants.*

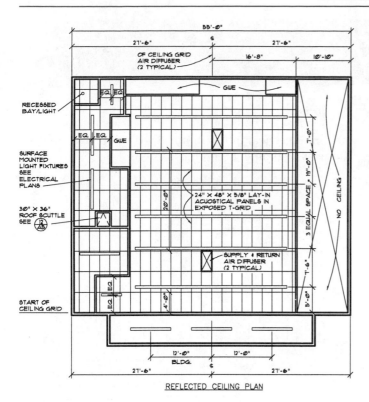

Figure 24–35 A reflected ceiling plan to show the suspended ceiling layout. *Courtesy The Southland Corporation.*

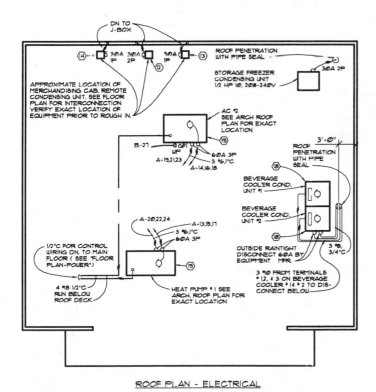

Figure 24–36 To supplement the power supply and lighting plans, a plan showing the electrical needs of equipment on the roof is also drawn. *Courtesy The Southland Corporation.*

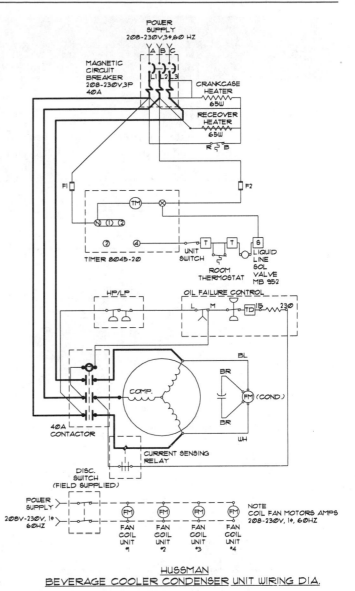

Figure 24–37 Schematic wiring diagrams for specific applications. *Courtesy The Southland Corporation.*

▼ ELECTRONIC SCHEMATIC DRAFTING

Electronics is the control of electrons for use in devices that are dependent on low voltage, amperage, and signal paths. The electronic drafter's responsibility is to convert engineering sketches or instructions to formal drawings, or to revise existing drawings. This task dictates that drawings be prepared in a neat, organized manner using proper symbols. Entry-level drafters can adequately prepare electrical or electronic drawings without knowledge of electricity and the function of electronic components. However, it is advisable for the electronic drafter to become familiar with the fundamentals of electronic technology, and as experience is gained in the industry, in-depth knowledge of product function is important for advancement into design and engineering.

CADD Applications

ELECTRICAL DRAFTING

The use of CADD makes architectural and electrical drafting quick and easy, especially with the variety of custom electrical tablet menu overlays and symbols libraries available. (See Figure 24–38.)

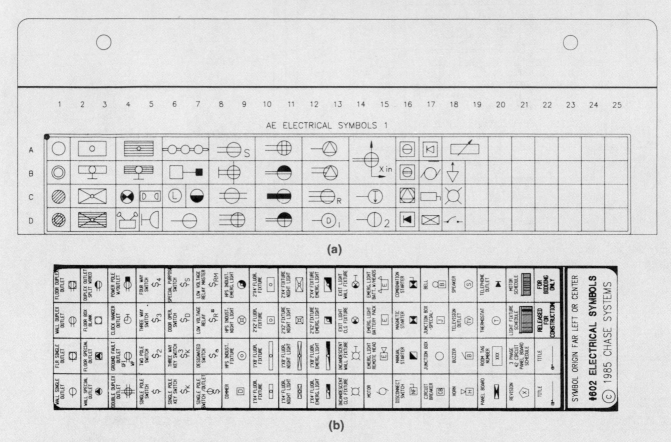

(a)

(b)

Figure 24–38 CADD template overlay symbol libraries for electrical applications. **(a)** *Courtesy Drafting Technology Services, Inc. Bartlesville, OK.* **(b)** *Courtesy Chase Systems.*

▼ ELECTRONIC DIAGRAMS

Many of the same types of diagrams used in electrical drafting are used in electronic drafting. There is only a slight difference between the appearance of electrical and electronic drawings. This text separates the two areas due to the specialized difference that exists between electrical power transmission systems and intricate electronic systems.

Block Diagrams

A *block diagram* outlines the path of a signal through a series of "steps" or "operations." The purpose of these drawings is to provide a quick interpretation of the relationship between the different components in electronic equipment. The steps or operations are shown in "blocks," which summarize a schematic drawing by omitting the details. A square or rectangular block is the usual basic symbol used in block diagrams. The size of the block is determined by the largest title going into the block, such as CONVERTER or AMPLIFIER. The block diagram should start in the upper left corner of the drawing and the components read from left to right, as shown in Figure 24–39. Graphic symbols are often used to demonstrate input and output devices, such as speakers, microphones, or antennas, as shown in Figure 24–40.

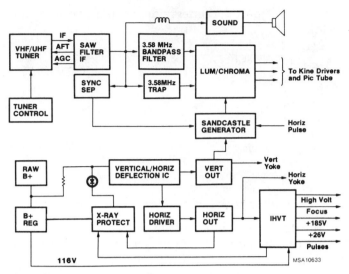

Figure 24–39   Block diagram. *Reproduced by permission of RCA Consumer Electronics.*

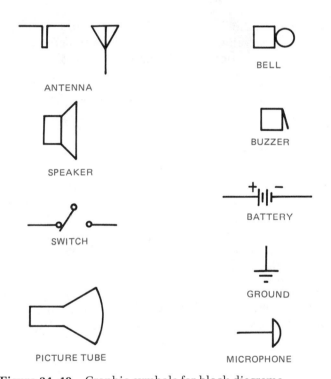

Figure 24–40   Graphic symbols for block diagrams.

Entry-level electronic drafters may be given a schematic diagram where the engineer has clearly defined the components to be placed in the block diagram. Design drafters in this field are able to determine the block diagram elements directly from a given schematic diagram. Figure 24–41 shows a block diagram prepared from a schematic diagram. In some situations, the drafter may prepare a block diagram from a pictorial diagram, as shown in Figure 24–42. Separate or optional components may be represented with a dashed line.

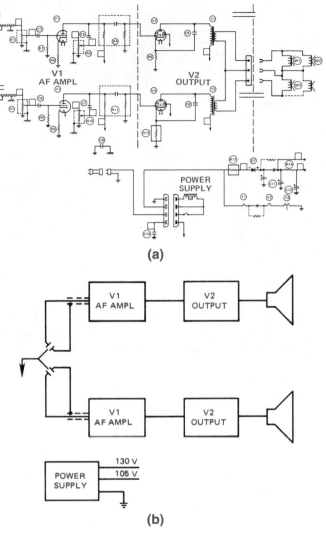

(a)

(b)

Figure 24–41   Block diagram from a given schematic diagram.

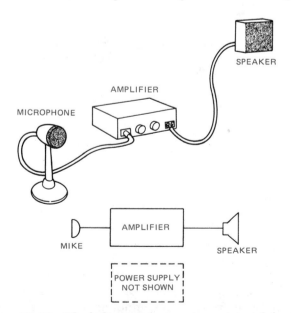

Figure 24–42   Block diagram from a given pictorial drawing and optional components shown with dashed lines.

PROFESSIONAL PERSPECTIVE

The responsibilities of the electrical drafter are varied depending on the electrical needs of the client or employer. This chapter has touched on the basics of electrical drafting as related to electrical power systems, and residential and commercial electrical applications. Engineering drafting related to these fields requires practical experience associated with the specific engineering needs of the company.

In addition to a thorough understanding of company requirements, you must know electrical terminology and know how the components go together. It is important for an engineering drafter in this business to visualize the systems and understand electrical clearance requirements. Schooling in descriptive geometry, communication skills, and problem-solving techniques, along with electrical theory and practices is essential.

When you go to work in this field, you will find that it is extremely important for you to check your work completely, making sure that electrical connections made are necessary. A *schematic* method of checking your own work is recommended. This method involves running a print of your drawing and then marking the print with a colored pencil to compare it to the original engineer's sketch or job assignment. You will find that drawings develop from one to the next. Beginning with the single-line diagram, each stage of project development becomes more specific. The checking process continues as an evolutionary process. With each stage, checking and accuracy become more critical.

Schematic Diagrams

Preparation of a schematic diagram is typically the first stage of product development. *Schematic diagrams* provide the basic circuit connection information for electronic products. Components of the electronic system are often drawn in stages that represent the function of the device. Individual components are labeled with "reference designations," values, and suppliers identification. Reference designations tie the component directly with the schematic drawing. A reference des-

ignation is a letter (or letters) identifying a component, as follows:

COMPONENT	REFERENCE DESIGNATION
Capacitor	C
Inductor	L
Resistor	R
Diode	D
Transistor	Q
Transformer	T
Switch	S
Electron tube	V
Semiconductor	CR

The drafter should know the product well enough to prepare schematics that are uncluttered and easy to read, follow standards, fit the size and shape of the product, and are technically correct. The following discussions provides enough electronic information to help the beginning drafter prepare a schematic.

Basic Electronic Symbols. The symbol and brief definition of the following components is provided only for general information. A full understanding of function would require technical training.

Capacitor—C. A simple *capacitor* consists of two metal plates with wire connectors separated by an insulator. A capacitor in an electronic circuit opposes a change in voltge and stores electronic energy. (See Figure 24–43.)

Coil or Inductor—L. A *coil* or *inductor* is a conductor wound on a form or in a spiral that has inductive properties. *Inductance* is the property in an electronic circuit that opposes a change in current flow or where energy may be stored in a magnetic field, as in a transformer. (See Figure 24–44.)

Resistor—R. *Resistors* are components that resist the flow of electricity. They are used in applications where the circuit requires protection by reducing the flow of

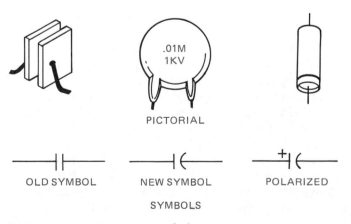

Figure 24–43 Capacitor symbols.

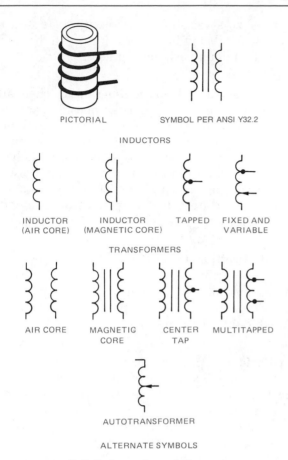

Figure 24–44 Coil (inductor) symbols.

Figure 24–45 Resistor symbol.

current. Resistors are available in fixed or variable values. Variable resistors allow the user to adjust the resistance. An example of a variable resistor is the volume control of a radio. The symbol is shown in Figure 24–45.

Semiconductor—CR. A simplified definition of a *semiconductor* is a device that provides a degree of resistance in an electronic circuit. There are various kinds of semiconductors used in electronics. Specific types of semiconductors are diodes, transistors, and others that are discussed later. Under certain conditions, this group of devices allows the current to pass through them freely, and under other conditions, they block the flow of current. The simplest of these devices is the *diode*. (See Figure 24–46.) The symbols picture the basic characteristics of the device. The arrow points in the direction in which conventional current flows. The arrow and bar are present in some form in almost all semiconductor symbols. The bar end of the diode is referred to as the *cathode* or negative side. The direction the symbol arrow points is important to the drafter and may be the only difference in the symbol for two different com-

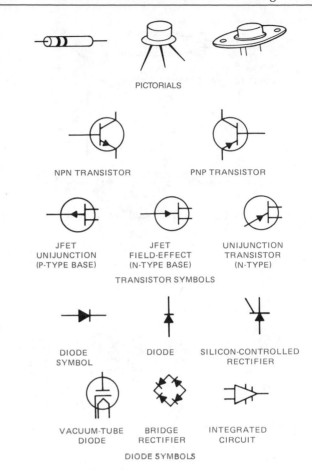

Figure 24–46 Semiconductor symbols.

ponents. Figure 24–46 is not a complete illustration of semiconductor devices, but it does show how slight changes are used to indicate different components.

Symbol Variations. An arrow through or to a symbol changes the component from a *fixed-value* device to a *variable-value* device. (See Figure 24–47.) A screwdriver symbol added to a schematic tells the technician that the component has a variable value by service adjustment, as shown in Figure 24–48.

Another group of symbol variations exists in coils, as shown in Figure 24–49.

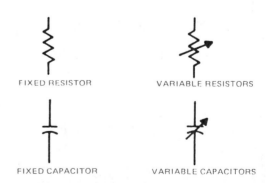

Figure 24–47 Fixed and variable component symbols.

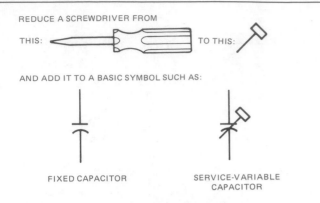

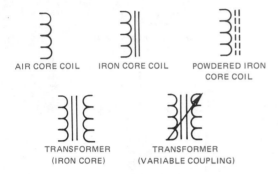

Figure 24–48 Fixed and service-variable component symbols.

Figure 24–49 Coil symbol variations.

As an electronic drafter you may be working from rough engineering sketches.

ANSI You need to be very familiar with standard symbols because engineers usually have operational problems on their minds and not the quality of drawings and sketches. Keep a copy of ANSI Y32.2 or the company standards handy for reference. When CADD is used a symbol library or a template designed to ANSI standards is an excellent drafting aid.

Here are some sample potential communication problems:

1. Know current standards so if the symbol ⊣ ⊢ is on a sketch, you know it is an old symbol of an electrical application for a capacitor.
2. Question symbols with odd details, such as the one in Figure 24–50. Does the line across the bottom of the resistor mean something special or was the engineer's pencil resting on the paper? In this case it could mean either, but be careful, as it could

Figure 24–50 Confusing symbol sketch.

symbolize a variable resistor with a stop on the adjustment that will not allow it to go all the way to zero. If in doubt, ask the engineer.

Do's and Don'ts for Schematic Diagrams. There are a variety of electronic drafting techniques that are more acceptable than others. While there are some methods that the beginning drafter should learn, there may be some situations when common applications may change due to the given situation or a difference between national and company standards. The examples shown in Figure 24–51 are some samples of recommended do's and don'ts.

Electronic Drafting Templates. If you are not using CADD for electronic drafting, you should be using the many available templates. Templates may be purchased that provide virtually all of the possible electronic symbols used in schematic drafting.

Active Devices. An *active device* is an electronic component that contains voltage or current sources, such as a transistor and integrated circuit. Transistors are semiconductor devices in that they are conductors of electricity with resistance to electron flow, are are used to transfer or amplify an electronic signal. The key to the schematic layout is the location of the active devices. When schematics are designed, the active devices are often established first. Follow the engineering sketch as a guide for overall size and layout balance, and begin the schematic diagram by positioning the active devices.

Bias Circuitry. *Bias* is the voltage applied to a circuit element to control the mode of operation. Bias may be fixed, forward, or reversed. The basic circuitry that makes any electronic device function is known as the DC biasing. The configuration of this basic circuit is resistors that are connected to an active device. Resistors are components that maintain resistance to the flow of an electric current. The resistors should be placed as close as possible to the active device without crowding or otherwise destroying the balance of the drawing. Resistors are normally placed vertically in the schematic, as shown in Figure 24–52. Transistors may be drawn one of two basic ways, depending on the bias direction by changing the arrow direction, as shown in Figure 24–53. After the active devices have been established on the drawing sheet, the next step is to add the resistors. There are a few common bias arrangements for transistor circuits, as shown in Figure 24–54 (see page 814).

Coupling Circuitry. After the bias circuitry has been established, the next step is generally the layout of the *coupling* circuitry. This is the point where each schematic becomes unique. The drafter's goal is to provide

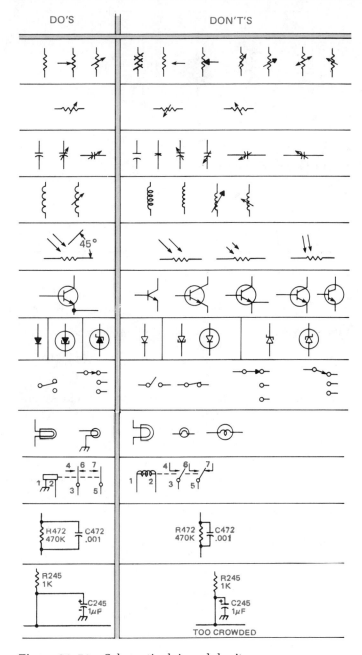

Figure 24–51   Schematic do's and don'ts.

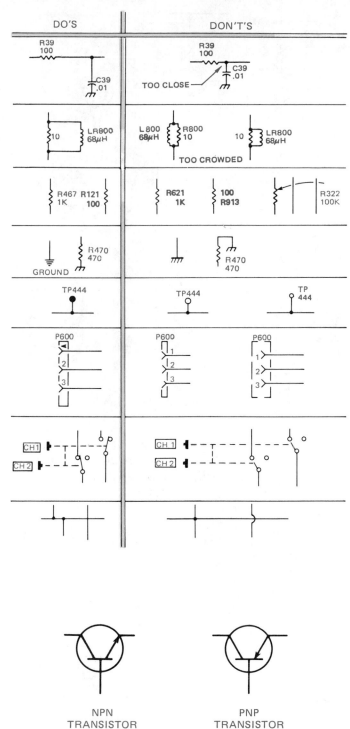

Figure 24–53   Changing the bias direction of a transistor.

adequate space for all of the information while keeping the schematic compact and readable.

Making a Schematic Layout Sketch.   Now that some fundamental guidelines have been provided for the schematic diagram layout, it is time to make a layout sketch. Making this sketch on graph paper helps in estimating the space required for component labels and values. The drafter often works from an engineer's design

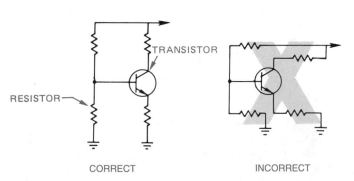

Figure 24–52   Resistors placed in relation to the active device.

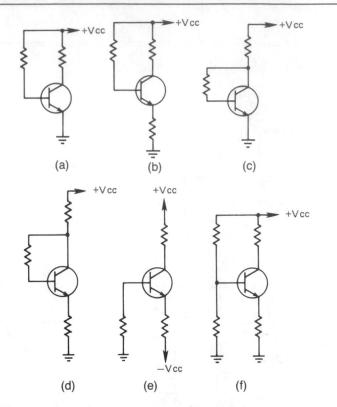

(a)　　　　(b)　　　　(c)

(d)　　　　(e)　　　　(f)

Figure 24–54　Common transistor bias circuit arrangements.

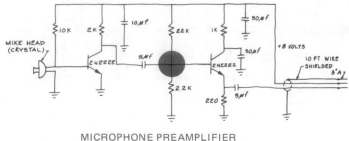

MICROPHONE PREAMPLIFIER

Figure 24–55　Engineer's rough sketch.

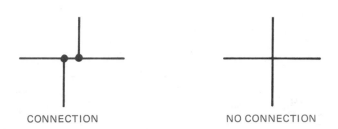

CONNECTION　　　　　　NO CONNECTION

Figure 24–56　Standard connection and nonconnection details.

sketch. This sketch may be rough, but gives the layout characteristics, as shown in Figure 24–55. The engineering sketch may be difficult to read and interpret. For example, notice in Figure 24–55 the screened area where the lines cross. These crossing lines represent a question that you should answer. Is there a connection at this point or do the lines cross without making a connection? In this case, the resistors are part of the universal bias pattern as

previously discussed, so you can conclude that a connection is intended. The standard representations for connections and nonconnections are shown in Figure 24–56. If there is any doubt about how to interpret the engineer's sketch; ask the engineer!

Now, make a layout sketch on graph paper, as shown in Figure 24–57. The blocked out areas next to the symbols represent intended component labeling.

There is more than one correct way to lay out a schematic. The same schematic with an optional layout is shown in Figure 24–58. The basic difference between

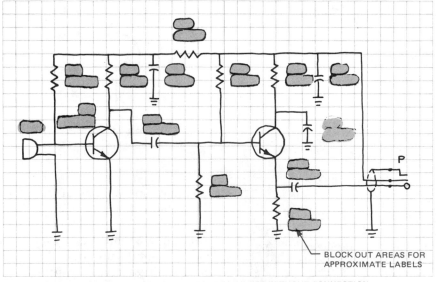

BLOCK OUT AREAS FOR
APPROXIMATE LABELS

NOTE: SOMETIMES THE SKETCH COMES FROM THE ENGINEER WITHOUT CONNECTION DOTS AS SHOWN HERE.

Figure 24–57　Schematic layout sketch.

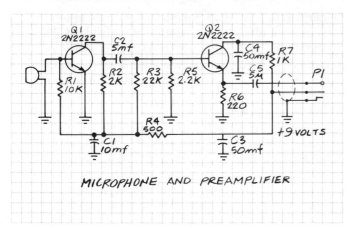

Figure 24–58 Optional layout sketch.

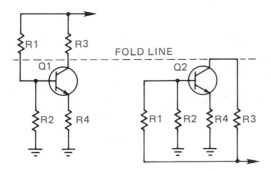

Figure 24–59 Folding down the schematic.

this layout and the previous one is that the components attached to the top side of the active devices have been "folded down," as shown in Figure 24–59. The result of this folded-down form is a long, narrow schematic. This layout can be used to an advantage where the height of the schematic is fixed but the length can vary, as in fold-out pages in instruction manuals. This form gives the advantage of providing a clear space above the active device for labeling or circuit notations.

Labeling and Circuit Notations.   There are two parts to the schematic layout—the symbols and the labeling. Without labeling, the symbols mean nothing more than the component they represent. The designating system has developed around a letter indicating the type of component followed by a number that gives the sequence of the part in the circuit. For example, the first resistor in the upper left of the schematic is designated R1. The next resistor to the right is R2, and so on across the schematic. The same holds true for capacitors, for example, C1, C2, and C3. Some of the other common component reference designators are: Transistor = Q, Inductor = L, Diode = D, Jack = J, and Plug = P. On simple circuits, the number sequence flows from left to right, as shown in Figure 24–60. In more complex schematics there can be two or more levels with each level flowing from left to right and top to bottom, as shown in Figure 24–61. Another method, used in units

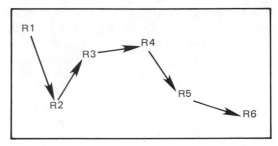

Figure 24–60   Left-to-right component numbering sequence.

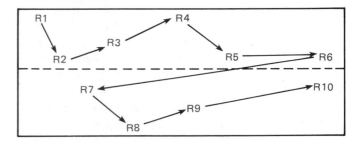

Figure 24–61   Component numbering sequence applied in layers.

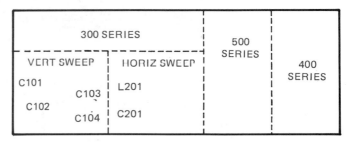

Figure 24–62   Numbering components in groups.

that contain subassemblies (individual groups of components that make up a complete system) or modular construction (component groups made up of equal components), is to assign the components a set of three or four digit numbers. For example, Figure 24–62 shows the vertical sweep module containing numbers from 100 to 199, the horizontal sweep module containing numbers from 200 to 299, the 300 series numbers range from 300 to 399, the 400 series numbers range from 400 to 499, and the 500 series numbers range from 500 to 599. The sequence of number series in schematics may be scrambled, established on the basis of physical layout, or arranged from left to right and top to bottom.

Units Used for Parts Values.   Electronic component values run from extremely small values (0.000000000005) to extremely large values (2,500,000,000,000). In either case, numbers in this form take up too much space on the schematic. The numbers 1500 or 0.0003 would probably be acceptable. A zero should precede the decimal point for values less than one. It has become common practice to move the decimal point in groups of three and then modify the numerical name to indicate the number of places the decimal point has been moved. For example, if you have a 2,200-ohm resistor, you can move the decimal point three places to the left, drop the zeros, and change the numerical name to *kilohms,* which means 1000 ohms. The result is the designation 2.2 kilohms. The word kilohms may be abbreviated "k," thus reducing the notation to 2.2k for a savings of more drafting space and time. The word ohms may be omitted because all resistors are rated in ohms. The following terms are used to rate or size common components:

❍ Resistors = ohms = the unit of measurement of resistance. Symbol = Ω.

❍ Capacitor = farad = the unit of measure of capacitance. *Capacitance* is the property of an electric circuit to oppose a change in voltage. Symbol = *f.*

❍ Inductor or coil = henry = the unit of measurement of inductance. Inductance is the property of a component in an electric circuit that opposes a change in current flow. Symbol = H.

The following chart shows the values achieved by moving decimal points in given numbers:

NUMBER	MOVE DECIMAL	ADD PREFIX	ABBREVIATION
1,000,000,000,000	Left 12 places	1 tera	1 T
1,000,000,000	Left 9 places	1 giga	1 G
1,000,000	Left 6 places	1 mega	1 M
1,000	Left 3 places	1 kilo	1 k
1	No change	1	1
0.001	Right 3 places	1 milli	1 m
0.000001	Right 6 places	1 micro	1 μ
0.000000001	Right 9 places	1 nano	1 n
0.000000000001	Right 12 places	1 pico	1 p

There are some gray areas in this system. For example, there are no firm rules on which conversion for 500,000 is best—0.5 M or 500 k. The number 1200 may be left as is or reduced to 1.2 k.

Military Identification Systems. Schematic diagrams that are drawn to meet military specifications provide a more elaborate component identification system. For example, a conventional part identification may be R304. In the military system, this part is labeled something like 14A2A4R304. This means that component R304 is located in subassembly 4 of subassembly 2A of assembly 14A. Military specifications are available covering every element used in the design, construction, and packaging of electronic equipment.

MIL-STD Some of the Military Standard Specifications (MIL-STD) include: MIL-STD-454, Standard General Requirements for Electronic Equipment, and MIL-STD-681, Identification Coding for Hook-up Wire.

Placement of Identification Numbers and Values. The placement of identification numbers is standard. The reference designator and value numbers are typically placed above horizontal components. (See Figure 24–63.) Other arrangements could cause confusion when components are not adequately spaced as shown in Figure 24–64.

For vertically drawn components the reference designator and the value are commonly placed on the right side of the symbol. (See Figure 24–65.)

The preferred technique in component labeling is placement of part identification at the top of all horizontal symbols and to the right of all vertical symbols, as shown in Figure 24–66.

A final drawing of the schematic used in the sketching layout examples is shown in Figure 24–67.

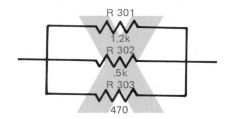

Figure 24–63 Horizontal component labeling.

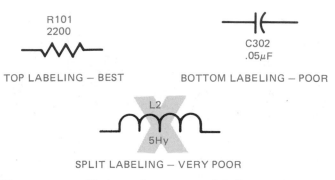

Figure 24–64 An overcrowded schematic using split labeling. Avoid this practice.

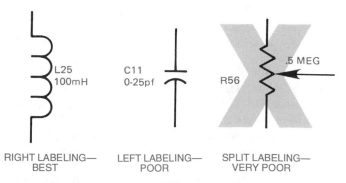

Figure 24–65 Vertical component labeling.

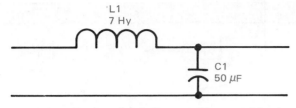

Figure 24–66   Preferred labeling at the top of horizontal components and to the right of vertical components.

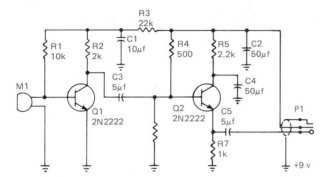

Figure 24–67   Final drawing of schematic used in layout sketch. (Figures 24–57 through 24–59)

Operational Amplifiers and Integrated Circuit Schematics

An *amplifier* (AMP) is a device that allows an input signal to control power and is capable of an output signal greater than the input signal. An *operational amplifier* (OP AMP) is a high gain amplifier created from an integrated circuit. An *integrated circuit* (IC) is an electronic circuit that has been fabricated with extremely small components as an inseparable assembly known as a *chip.* You can become involved with integrated circuit schematics by drawing the schematic of the internal circuitry involved, or drawing the schematic for a system in which integrated circuits or operational amplifiers are used.

Internal Integrated Circuit Schematics.   The internal circuitry of an integrated circuit is shown in Figure 24–68. The only exception is that in the IC diagram, the transistor symbols are not circled. The diagram is also sometimes enclosed by dashed lines in the form of a box. This box represents the package within which the IC is contained. Points are shown and labeled with pin numbers, which are external connectors for attachment to other circuitry. This usually means that all of the components shown in the schematic are made up of one piece of semiconductor material. Thus the definition of integrated circuit—all of the components of the circuit have been put together on one small piece of material. The schematic for an IC can be very simple or so complex that it contains several thousand components.

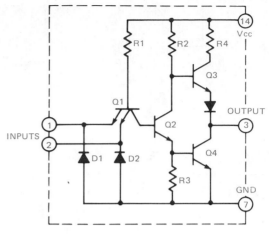

Figure 24–68   The internal components of an integrated circuit.

IC Symbols and Logic Circuits.   In logic circuits, which are computer-oriented circuits, the schematics become a cross between a flow diagram and a schematic diagram. The internal components of an IC are self-contained for a specific function and may not be altered. Therefore, symbols are used that identify the IC. The shape of the symbol tells the technician what the device can be expected to do. The numbers next to the input and output lines or pins tells the reader how to connect the IC to make it work. The basic format of IC symbols is shown in Figure 24–69. The IC symbol that represents the internal IC shown in Figure 24–68 is shown in Figure 24–70. This symbol is called a NAND gate. The *gate*

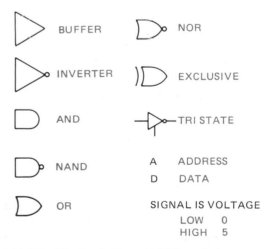

Figure 24–69   The basic format of IC symbols.

Figure 24–70   The IC symbol of the internal IC schematic shown in Figure 24–68.

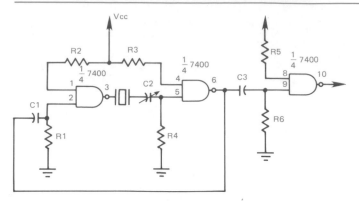

Figure 24–71   A schematic diagram with an IC included.

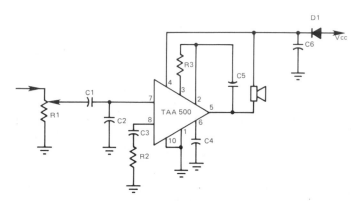

Figure 24–72   Operational amplifier IC.

is the part of the system that makes the electronic circuit operate and permits an output only when a predetermined set of input conditions are met. The code NAND is the predetermined function of the gates. There are also AND, OR, and NOR functions. Simply put, the functions relate to a condition or pattern of conditions whereby an electrical pulse can be received at one pin when the input at another pin or pins is triggered at specific times.

The precise operation of each logic function is required knowledge for electronic technicians and design drafters, but premature for drafting fundamentals and beyond the scope of this study. Your concern at this time is to prepare a well-balanced, accurate drawing from given engineering sketches. Notice how the IC schematic diagram shown in Figure 24–71 compares to previous schematics. The difference is in the IC symbols. While the integrated circuits may be complex in nature, the drafting task becomes easier when ICs are used. For example, the IC labeled TAA-500 in Figure 24–72 is an operational amplifier that contains 11 transistors, 5 diodes, 13 resistors, and 1 capacitor.

Logic diagrams are a type of schematic that are used to show the logical sequence of events in an electrical or electronic system. The basic electronic logic symbols and their functions are shown in Figure 24–73. Part of an electronic logic diagram is shown in Figure 24–74.

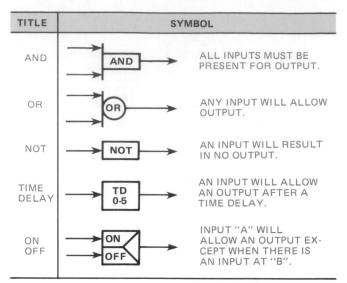

Figure 24–73   Basic electronic logic symbols.

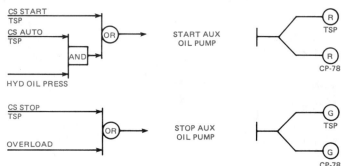

Figure 24–74   Part of an electrical logic diagram.

Large Scale Integration.   The next generation beyond the specific symbols (such as the gates) came about because of the rapid advancement of the electronic industry's ability to put more and more circuits on a single, small IC chip. Many calculators contain only one IC with hundreds of gates. The trend in drawing schematics for this *large scale integration* (LSI) is simply to show the chip as a box. This type of schematic becomes like a block diagram with the name or function identified within the rectangle for each LSI. The LSI pin terminals are shown connected to the rest of the schematic. An LSI schematic diagram is shown in Figure 24–75. Keep in mind that each LSI block in the schematic may contain hundreds or thousands of components.

▼ PRINTED CIRCUIT TECHNOLOGY

Schematic diagrams are designed and drawn to show the location of electronic components using symbols and lines to represent circuit paths or connections. In the production of the actual electronic product, the symbols

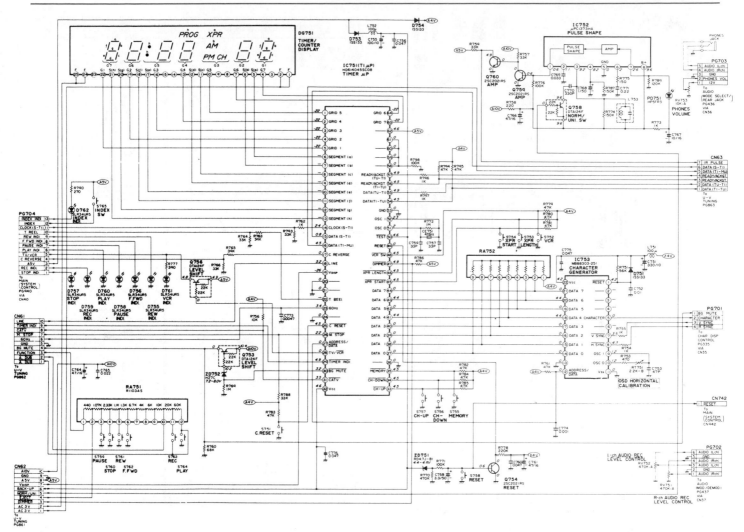

Figure 24–75 A large scale integration (LSI) schematic. *Reproduced by permission of RCA Consumer Electronics.*

shown on the schematic become electronic devices and the lines become wires that connect these devices. At one time, wires were soldered to component terminals and used as the circuit path between components. Today, wires are used as the connection cables that provide the source of current between pieces of electronic equipment. As electronic equipment designs have become increasingly smaller, the internal connection between electronic devices must also take up less space, be easier to install, and be extremely accurate. Printed circuit technology is the answer to these needs.

Schematic and Printed Circuit Board Accuracy

Accuracy and close attention to detail in the preparation of the schematic and printed circuit board are absolutely essential. Everything depends on the accuracy of these two items. The schematic and the printed circuit board cannot have a single mistake, because any mistake affects the master artwork, the drilling drawing, and the assembly and parts lists.

Printed Circuit Design and Layout

Printed circuits (PC) form the interconnection between electronic devices. The base materials for circuit boards are special paper, plastic, glass, or Teflon. The quality of the *printed circuit board* (PCB) begins with the quality of the base material. Depending on the complexity of the electronics, printed circuits are prepared on one side or both sides of the board. In many applications, there are multiple boards or layers for one piece of electronic equipment.

The printed circuits consist of pads and conductor lines that are made of thin conductive material such as copper on a base sheet or board. The *pads* or lands are the circuit termination locations where the electronic devices are attached. The pad normally has a location where a hole is to be drilled in the circuit board for mounting the device. Pads may be placed individually or in patterns

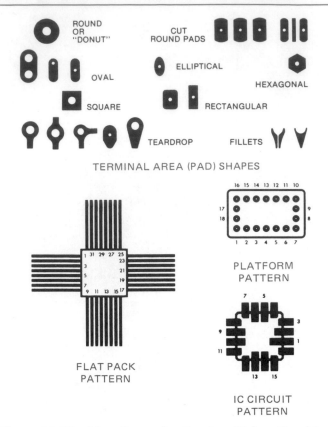

Figure 24–76 Mounting pads. *Courtesy Bishop Graphics, Inc.*

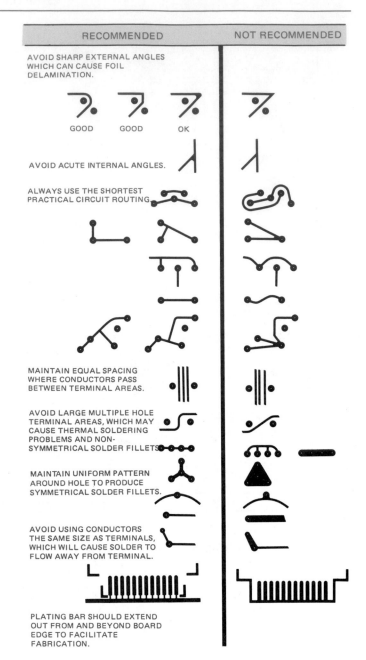

Figure 24–77 Artwork pattern configuration, conductor traces, and taping techniques. *Courtesy Bishop Graphics, Inc.*

depending on the connection characteristics of the device. (See Figure 24–76.) *Conductor traces* (lines) connect the pads to complete the circuit design. Design characteristics of conductors are shown in Figure 24–77. The size depends on the amount of current, the temperature, and the type of board specified. Conductors are .010 in. (0.254 mm) wide for many applications, but may be designed for applications where a width of .031 in. (0.787 mm) or more for large current-carrying requirements. Factors that influence conductor spacing include product requirements and current specifications. For general uses, conductor spacing should be .031 in.

Printed Circuit Board Artwork

The printed circuit artwork is an accurate, scaled, undimensioned drawing used to produce the master pattern from which the actual board is manufactured. The artwork should be prepared from a design layout at an enlarged scale on polyester film using computer-aided design drafting. In addition to all electrical circuitry, the board should contain board edge identification marks; registration, datum, and indexing marks; photo reduction dimensions; scale designation; an identification number; and a revision number. Letters to be etched on the board should be .062 in. (1.57 mm) in height (minimum) and .015 in. (0.38 mm) in thickness (minimum).

Master Pattern Artwork

The *master pattern* is a one-to-one scale circuit pattern that is used to produce a printed circuit board. The master pattern artwork is normally prepared at an enlarged scale so that when it is reduced to a 1:1 ratio, its quality is enhanced.

Grid System. The use of a grid system is essential in laying out and preparing the master pattern artwork. The grid system aids in the placement of pads and conductor traces. A *grid* is a network of equally-spaced parallel lines running vertical and horizontal on a glass or poly-

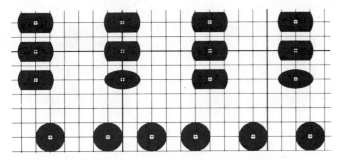

Figure 24–78 Grid system. *Courtesy Bishop Graphics, Inc.*

ester film sheet, as shown in Figure 24–78. Standard spacing is .100 in. (2.54 mm), .050 in. (1.27 mm), or .025 in. (0.635 mm).

MIL-STD MIL-STD 275D specifies any multiple of .005 in. (0.127 mm).

IEC The International Electrotechnical Commission (IEC) specifies 0.5-mm and 0.1-mm increments.

Printed Circuit Scale. The printed circuit board layout should be prepared at an enlarged scale of 2:1 or 4:1 so that possible drafting errors or slight flaws are minimized when the layout is reduced to actual size. The accuracy of the PC layout is also critical. Sometimes it is necessary to prepare the PC layout at a 100:1 scale to insure clarity and accuracy.

Board Size and Number of Layers. Printed circuit boards may be designed in three basic configurations: single layer, multilayer, and multilayer sandwich. *Single-layer* boards contain all printed wiring on one side with the components on the opposite side. *Multilayer* boards have printed circuits on both sides with most of the components on one side and the circuitry on the other. *Multilayer sandwich* boards consist of many thin boards laminated together, with the components on one or both sides of the external layers. As boards increase in a complexity, they also cost more. However, the cost of a single multilayer board may be less than several single-layer boards for the same system.

The circuit board should be just large enough to contain the components and interconnections and remain economical to manufacture. One way to estimate the board size is to randomly place scaled component cutouts, allowing enough space for interconnections. Allow extra space for IC interconnections.

Solder Masks. *Wave solder* has become a common method of attaching components to printed circuit boards. A wave of molten solder is passed over the noncomponent side of the board, making all solder connections. A polymer coating, called a *solder mask* or *solder resist mask,* is applied to the board, covering all conductors except pads, connector lands, and test points. Solder masks are used to prevent the bridging of solder between pads or conductor traces, to cut down on the amount of solder used, and to reduce the weight of the board.

Conductor Width and Spacing. Careful consideration must be given to conductor trace width and spacing during printed circuit design. Width and spacing that is too small can cause service problems in the circuitry. Width and spacing that is too great wastes space and increases costs. Verify minimum width and spacing requirements with product and company specifications. Standard charts and specifications are available; but in general, conductor widths of .050 in. (1.27 mm) or .062 in. (1.57 mm), and minimum spacing between conductors of from .031 in. (0.79 mm) to .050 in. (1.27 mm) are recommended for low-voltage applications.

Component Terminal Holes. A printed circuit board should have a separate mounting hole for each component lead or terminal. Unsupported holes contain no conductive material. To determine the diameter of unsupported holes use the formula:

Minimum hole diameter = Maximum lead diameter + Minimum drill tolerance

Plated-through holes have conductive material plated on the inside wall to form a conductive connection between layers of the circuit board, as shown in Figure 24–79. Plated-through holes are drilled prior to plating. A general note "ALL HOLES TO BE PLATED THROUGH" should be specified. Usually plating thickness is specified as a minimum with an accepted tolerance of − 0 + 100%. For plated holes, the minimum diameter should be greater than the minimum lead size plus .028 in. (0.71 mm). A second drilling process is required if the predrilled holes do not take into account the plating thickness. Drilling twice adds cost to the board. The number of different hole sizes should be kept to a minimum to help save manufacturing costs.

Terminal Pads. There should be a separate terminal pad for each component lead or wire attachment. Terminal pads vary with designer preference and compo-

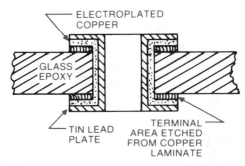

Figure 24–79 Cross section of plated-through hole. *Courtesy Bishop Graphics, Inc.*

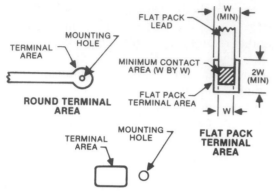

Figure 24–80 Terminal areas. *Courtesy Bishop Graphics, Inc.*

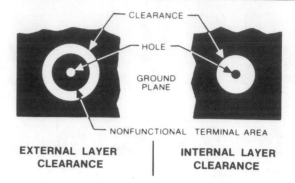

Figure 24–82 Layer clearance at ground planes. *Courtesy Bishop Graphics, Inc.*

nent characteristics. (See Figure 24–76 and 24–80.) The *minimum required annular ring* is the smallest part of the circular strip of conductive material surrounding a mounting hole that meets design requirements.

MIL-STD MIL-STD 275D specifies .015 in. (0.38 mm) minimum for unsupported holes, .005 in. (0.13 mm) minimum for plated-through holes on external layers, and .002 in. (0.05 mm) for internal layers of multilayer sandwich boards.

Ground Planes. A *ground plane* is a continuous conductive area used as a common reference point for circuit returns, signal potentials, shielding, or heat sinks. Ground plane patterns should be broken up so the conductive area is equal to about half of the nonconductive area. (See Figure 24–81.) Clearance should be provided between terminal pads and the ground plane, as shown in Figure 24–82.

Printed Circuit Board Layout

Before starting a printed circuit board layout, you should have a schematic or logic diagram, a parts list, design specifications, component sizes, lead-and-trace pattern and spacing, hole and terminal sizes, grid system, scale, board size and number of layers, and know the manufacturing processes to be used.

Marking Drawings

The circuit board artwork may have a separate sheet called a *marking drawing* or *silk screen* artwork containing component outlines and orientation symbols. This pattern is printed on the component side of the board after etching and plating. The reference designations should be placed so they are visible after component assembly. The component outlines should be located in the same position that the actual components occupy after assembly. The marking drawing should include registration marks that align with the pattern. (See Figure 24–83.)

Drilling Drawings

Drilling drawings are prepared after the master layout is complete. The *drilling* drawing is used to provide size and location dimensions for component and chassis mounting holes and the final dimensions for trimming the board. Drilling drawings are often set up using

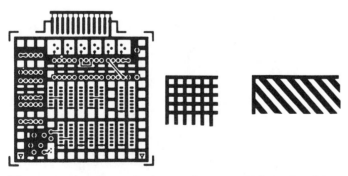

Figure 24–81 Ground planes. *Courtesy Bishop Graphics, Inc.*

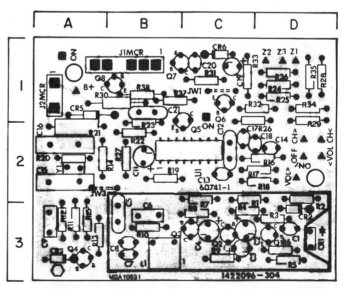

Figure 24–83 Marking drawing. *Reproduced by permission of RCA Consumer Electronics.*

datum arrowless tabular dimensioning in accordance with computer numerical control manufacturing methods. (See Figure 24–84.) Refer to the previous discussion on hole sizes and to Chapter 10 for a more in-depth discussion of dimensioning practices.

Assembly Drawings

The printed circuit board *assembly* drawing is a complete engineering drawing, including components, assembly and fastening or soldering specifications, and a parts list or bill of materials. The parts list should include the: part number, item identification keyed to the assembly, quantity of each part, electronics designations for components, and Federal Code Identification, if required by contract. (See Figure 24–85.)

Printed Circuit Preparation

There are several methods used to prepare the final printed circuit board from a printed circuit layout; however, the two fundamental methods are the additive and the etching processes.

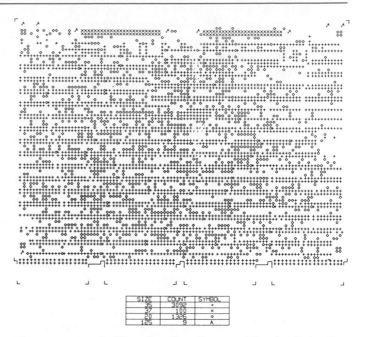

SIZE	COUNT	SYMBOL
35	3092	+
37	100	×
20	1326	○
125	9	A

Figure 24–84   Drilling drawing. *Courtesy Floating Point Systems, Inc.*

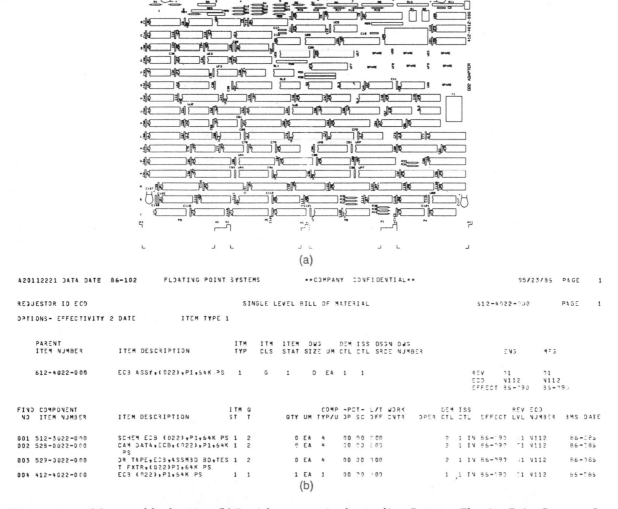

(a)

(b)

Figure 24–85   **(a)** Assembly drawing. **(b)** Partial computerized parts list. *Courtesy Floating Point Systems, Inc.*

Additive Process. The *additive process* takes place on a board that is covered with a chemically etched material that will accept copper. The PCB layout design image is then transferred to the board using a silk-screen or photo printing process. Copper conductor is then deposited on the image using electroless. *Electroless* is the depositing of a metal on another material through the action of an electric current. A solder resist mask is then applied to the copper deposit. This process prevents solder from bridging between closely spaced conductors when components are added to the printed circuit board. The board is then coated with a lacquer to protect the surface from corrosion.

Etched Process. A copper-covered base is used in the *etched process.* (Other conductive material may also be used.) The printed circuit layout is normally prepared at 2:1 or 4:1 scale and then photographically reduced to a photo negative at full scale. The copper-covered base is cleaned and treated with a light-sensitive emulsion. The base is then exposed to light through the negative. During this process, the emulsion hardens and an acid resist is formed at the printed circuit areas. The board is then placed in a chemical solution that etches away the copper coating in areas that are not protected by the acid resist. The board is then rinsed to stop the etching process and a protective coating is added. The resulting printed circuit is ready for components to be added.

Surface-mount Technology (SMT)

Surface-mounted devices (SMD) have emerged from the increasing need to get more electronic components into smaller places. Surface-mounted devices commonly take up to less than one-third the space of feed-through printed circuits. Conventional printed circuit board connectors require plated-through holes for each pin or connection point. Surface-mounted devices save the area, and are especially useful when designing very complex boards. The only drilled holes that are required are for chassis mounting or feed-through between layers.

In surface-mount technology, the traditional component lead-through is replaced with a solder paste to hold or *glue* the components in place on the surface of the printed circuit board. A solder paste template is made that allows for the placement of the paste at required areas. Then the components are automatically or manually placed into the solder paste, although most companies use robotic fabrication. A baking process allows the solder to liquify and then, upon solidification, the components remain soldered to the printed circuit. Some solderless connections have also been designed for surface-mounted devices.

Surface-mount technology is responsible for new types of electrical components and materials used in electronic circuits; for example, leadless plastic and ceramic IC chip housings. There are some problems, however, with surface-mounted devices; they are so dense that solder connections often become difficult, and testing electrical circuits becomes a problem because testing equipment may cause electrical current to flow in tight places that are otherwise defective. In most cases, surface-mounted boards are replaced, rather than serviced, when maintenance is required.

▼ PICTORIAL DRAWINGS

Three-dimensional drawings are frequently made to show pictorial representations of products for display in vendors' catalogues or instruction manuals. Pictorial assembly drawings are often used to show the physical arrangement of components so components are properly placed by factory workers during assembly. These drawings are also helpful for maintenance as they provide a realistic representation of the product.

One of the most effective uses of photodrafting is found in the electronics field. This is where a photo is taken of a complete assembly or product and drafting is used to label individual components. A common method of pictorial representation is the exploded technical illustration often used for parts identification and location, as shown in Figure 24–86.

Another type of pictorial diagram known as a semi-pictorial wiring diagram, uses two-dimensional images of components in a diagram form for use in electrical or electronic applications. The idea is to show components as features that are recognizable by laypersons. This type of pictorial schematic is commonly used in automobile operations manuals, as shown in Figure 24–87.

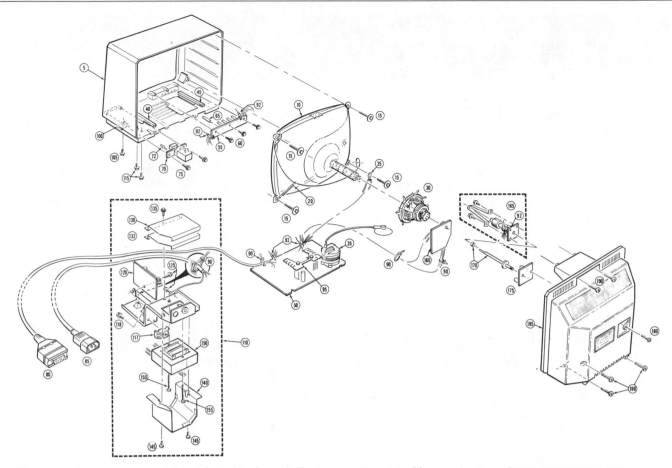

Figure 24–86 Parts pictorial, exploded technical illustration. *Reprinted by permission of HEATH COMPANY.*

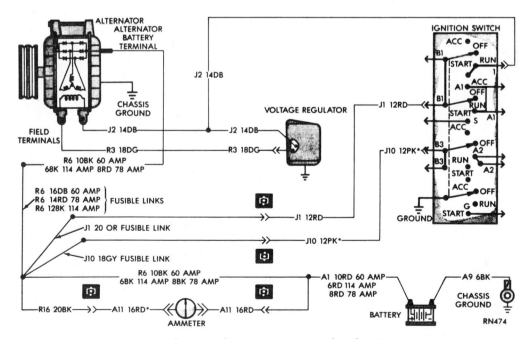

Figure 24–87 Semipictorial wiring diagram. *Courtesy Chrysler Corporation.*

CADD Applications

COMPUTER-AIDED ENGINEERING (CAE)

With electronic symbols established in a CADD template or symbols library, the design and layout of block diagrams, schematic diagrams, and other electronics drawings are greatly enhanced. The implementation of computers in this industry has advanced into what is known as *computer-aided engineering* (CAE), where the CADD function is taken further into the complete design, engineering, and functional analysis of the product. Beginning with the development of a computer-aided schematic design, the CAE program can perform the following functions:

1. It will simulate the circuit operation.

2. It will test the system for possible problems.

3. It can identify and analyze design complexity to help reduce manufacturing costs.

4. Thermal characteristics of the circuitry are evaluated to identify possible overheating situations where heat sinks may be needed or mechanical cooling is required.

5. Arced or 45° corners, rather than sharp corners, are automatically placed in the conductor traces during the routing process.

6. Artwork masters are generated at 1:1 scale for all circuit levels, eliminating costly and error-prone photoreduction, as shown in Figure 24–88.

7. The solder resist mask is generated from the same database as the artwork master. This improves the manufacturability of the board and reduces external changes. (See Figure 24–89.)

8. The component layout that is created as part of the symbols can be extracted to create a marking drawing silk-screen master, as shown in Figure 24–90.

9. The drill template can be generated from the same database as the artwork master and is used to determine plating requirements; generate a drill chart with sizes, quantities, and X-Y coordinate location dimensions; and create a drill pattern or provide direct communication to a computer numerical control (CNC) drilling machine to achieve repeatable board-to-board registration.

10. A description and reference designator may be placed on each symbol and may be used to simultaneously complete assembly drawings, parts lists, and other documentation.

11. A large-format digitizing table may be used to enable the designer to go from rough layout or existing works to tight, on-screen design.

12. Laser-directed photoplotters provide PCB artwork with accuracy tolerances up to 0.4 mil (0.000010 mm) along with image quality and speed that exceeds the highest industry standards.

13. Printed circuit conductor routing may be performed automatically with completion success on most boards of 97–100%. In complex board design, the incomplete part of the trace routing, if any, may be quicked engineered in the manual mode.

14. The designer or engineer may stop the routing process at any time to make component or connection changes, and restart the automatic routing process at the new location.

15. Component locations are designed along with trace routing.

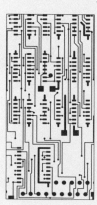

Figure 24–88 Full scale CAE artwork eliminating the need to photoreduce. *Courtesy The Gerber Scientific Instrument Company.*

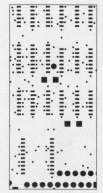

Figure 24–89 The solder resist mask is generated from the same database as the artwork master shown in Figure 24–86. *Courtesy The Gerber Scientific Instrument Company.*

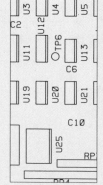

Figure 24–90 Marking drawing created from symbols library used to design the artwork master. *Courtesy The Gerber Scientific Instrument Company.*

PROFESSIONAL PERSPECTIVE

As an entry-level drafter with training in electronics schematic drafting, you are able to make engineering drawing changes and prepare drawings from engineering sketches without specific knowledge of how electronic components and the systems function. However, in order to advance in skill level to designer or engineering technician, you need a thorough understanding of company requirements, and the ability to communicate electrical and electronic terminology, and the knowledge of how the components go together. It is important for an engineering drafter in this business to visualize the systems and understand electrical clearance requirements. A lot of this can be learned on the job because the engineers are responsible for the design, but as you become experienced, you can assist in identifying engineering problems and trouble-shooting systems. Schooling in communication skills and problem-solving techniques, along with electrical and electronic theory and practices, is essential. Helpful college courses in addition to your electronic drafting could include:

- ○   Basic AC/DC electronics.
- ○   Solid state devices.
- ○   Digital logic.
- ○   Basic microprocessors.
- ○   Electrical physics.

When you go to work in this field you will find that it is extremely important for you to check your work completely, making sure that electrical connections made are necessary. Nearly all drafting is done on a CADD system, so it is important for you to fully understand the computer operating systems. When you are bringing electronic symbols from a menu library, such as the sample shown in Figure 24–91, be sure that you leave adequate space to display the symbol, labeling, and circuit notations. Be sure that you set up the schematic in a logical layout sequence so the components, test points, and other items are identified from left to right and top to bottom. You need to know and understand the numbering systems relating to the schematic and the PCB design. This is all part of the knowledge you will gain with experience in the industry. An engineering drafter at Intel warns, "Keep an open mind on new CADD systems and better ways to do things. The hardest part of the job is to constantly learn new systems, because the industry changes so fast. To sum it up, be flexible."

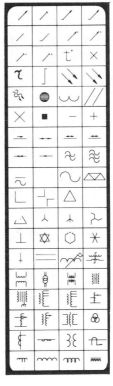

Figure 24–91   Electronic CADD symbols library. *Courtesy Chase Systems.*

 MATH APPLICATION

WORKING WITH POWERS OF TEN

Powers of ten (scientific notation) is used to express the very large and small numbers found in electronics compactly, making multiplying or dividing them much easier. Here are some of the powers of ten.

$$10^4 = 10,000$$
$$10^3 = 1,000$$
$$10^2 = 100$$
$$10^1 = 10$$
$$10^0 = 1$$
$$10^{-1} = .1$$
$$10^{-2} = .01$$
$$10^{-3} = .001$$

A number may be changed to a power of ten form by first writing the digits as a number between 1 and 10, then multiplying it by the appropriate power of ten. Here are some examples to illustrate this:

$$3,400 = 3.4 \times 10^3$$
$$53,000,000 = 5.3 \times 10^7$$
$$.000051 = 5.1 \times 10^{-5}$$
$$.02 = 2 \times 10^{-2}$$

To Multiply: Multiply the numbers out front and add the exponents.

To Divide: Divide the numbers out front and subtract the exponent in the bottom from the exponent on top.

Here are examples to illustrate:

$$(2 \times 10^3)(7 \times 10^8) = 14 \times 10^{11} = 1.4 \times 10^{12}$$
$$(3 \times 10^{-6})(2 \times 10^5) = 6 \times 10^{-6+5} = 6 \times 10^{-1}$$
$$(5 \times 10^{-20})(7 \times 10^{-3}) = 35 \times 10^{-23} = 3.5 \times 10^{-22}$$
$$\frac{6 \times 10^7}{3 \times 10^3} = 2 \times 10^{7-(3)} = 2 \times 10^4$$
$$\frac{1.4 \times 10^{15}}{2 \times 10^{-5}} = .7 \times 10^{15-(-5)} = .7 \times 10^{15+5} = .7 \times 10^{20} = 7 \times 10^{19}$$

These results can be obtained on the calculator using the EE, EXP, or EEX buttons, depending on the make of your calculator. For the last example, many calculators would use this sequence of buttons to obtain the final answer. 1.4 EE 15 ÷ 2 EE 5 +/− =. The calculator may display the answer without the "× 10" like this: 7¹⁹. However, the answer should never be written down and communicated like this because it would be confused with 7 raised to the 19th power, which is an entirely different number! Always include the "× 10" when writing down a power of ten number.

CHAPTER

24 ELECTRICAL AND ELECTRONIC SCHEMATIC DRAFTING TEST

PART 1

DIRECTIONS

Answer the questions with short complete statements or drawings as needed.

QUESTIONS

1. Explain the basic difference between electrical and electronic drafting.
2. Define *electrical circuit*.
3. Describe a pictorial diagram.
4. Draw a simple schematic diagram.
5. Draw a simple one-line diagram.
6. Draw a simple wiring diagram.
7. Draw a simple schematic wiring diagram.
8. Define *terminals*.
9. Briefly describe the transmission and distribution of electricity.
10. Describe the function of a substation.
11. Explain the purpose of a one-line diagram in an electrical power system design.
12. What is the document that provides most electrical symbol standards?

13. Describe the purpose of a substation plot plan.
14. What are the scales most commonly used to draw plot plans?
15. Define *bearing*.
16. Explain the purpose of a bus layout drawing.
17. What is the purpose of a ground layout?
18. What do grounding details show and at what scale are they drawn?
19. Define *conduit*.
20. Describe the purpose of conduit installation details.
21. What is the standard lettering height used on electrical floor plans?
22. How should switch symbols be drawn in relation to the wall upon which they are located?
23. Draw the following electrical symbols:

    Duplex convenience outlet
    Range outlet
    Recessed circuit breaker panel
    Phone
    Light
    Wall-mounted light
    Single-pole switch
    Simplified fluorescent light fixture
    Fan

24. What method is used to describe special characteristics in an electrical plan layout?

PART 2

1. Name the American National Standards Institute document that governs the use of electrical and electronic drafting symbols.
2. Describe the purpose of a block diagram.
3. What is the function of a capacitor?
4. Define *inductance*.
5. Describe the function of a resistor.
6. Define *semiconductor*.
7. Name two components that are considered semiconductors.
8. Draw and label a fixed and variable resistor.
9. Draw and label a circuit connection and a crossover.
10. Define *active devices*.
11. Define *bias*.
12. How should resistors be placed in relationship to the active devices?
13. What is the advantage of a folded-down schematic?
14. Describe the common sequence standard for component identification.

15. Give the standard letter identification for the following components: resistor, capacitor, transistor, tube, diode, inductor, plug, and jack.
16. Reduce the value 4,000 ohms to its simplest form.
17. Reduce the value 0.000000000001 farads to its simplest form.
18. Reduce the value 1,000,000 henrys to its simplest form.
19. Draw an example of a horizontal capacitor (C1, 5µ*f*) and a vertical resistor (R1, 1,000 ohms) connected, and use the most popular labeling technique to show the identification and value of each.
20. Define *amplifier*.
21. Define *integrated circuit*.
22. What does the dashed line box around an internal IC schematic diagram denote?
23. Define *gate*.
24. Why is it possible that an IC schematic may be much easier to lay out and draw than a conventional schematic?
25. What does the abbreviation *LSI* denote?
26. How are electronic pictorials used in industry?
27. What is the main advantage of surface-mounted devices?
28. Printed circuits are used for what purpose?
29. Describe pads in PCB layout.
30. Why are holes drilled at the pad location?
31. Identify three factors that influence the size of conductor traces.
32. Make a general statement describing printed circuit artwork.
33. What scale is the master pattern?
34. What is the function of a grid system in PCB design?
35. Why is manual PCB layout usually prepared at an enlarged scale?
36. Describe the three basic PCB configurations: single layer, multilayer, and multilayer sandwich.
37. Describe a solder mask.
38. Describe unsupported and plated-through holes.
39. What general note should be provided when holes are to be plated prior to drilling?
40. Define *ground plane*.
41. What does the term *dolls* refer to in PCB layout?
42. Describe a marking drawing.
43. Describe a drilling drawing.
44. Identify the components that normally make up a complete assembly drawing.
45. Describe computer-aided engineering as related to electronics engineering.

CHAPTER

24 ELECTRICAL AND ELECTRONIC SCHEMATIC DRAFTING PROBLEMS

PART 1

DIRECTIONS

1. Please read all instructions before you begin working, unless otherwise specified by your instructor.
2. Use manual or computer-aided drafting as required by your course guidelines.
3. Use the attached engineering layouts to prepare each drawing. Refer to chapter coverage for symbol and component dimensions and also refer to previous drawings for information as you progress to each drawing.
4. Prepare well balanced, easy-to-read drawings. Most drawings have no scale.
5. Many of the following problems contain symbols that are duplicated multiple times. These electrical problems demonstrate the power of CADD when you create one symbol and have the ability to use it several times on one or more drawings.
6. Recommended media:

   ○ Polyester film or vellum.
   ○ Ink or CADD ink plot, if available.

7. Additional information is provided for some problems.

Problem 24–1   Switch panel

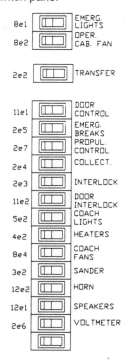

Problem 24–2   Schematic

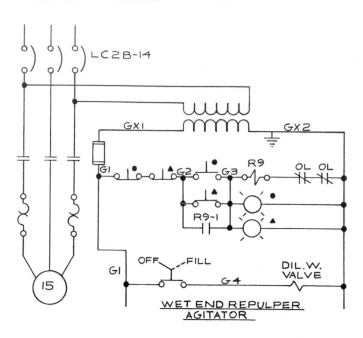

Problem 24–3   Schematic

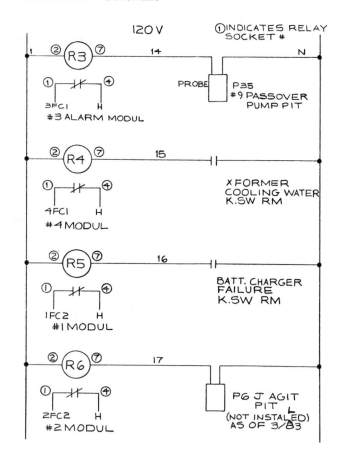

Problem 24–4 Schematic

Problem 24–5 Schematic

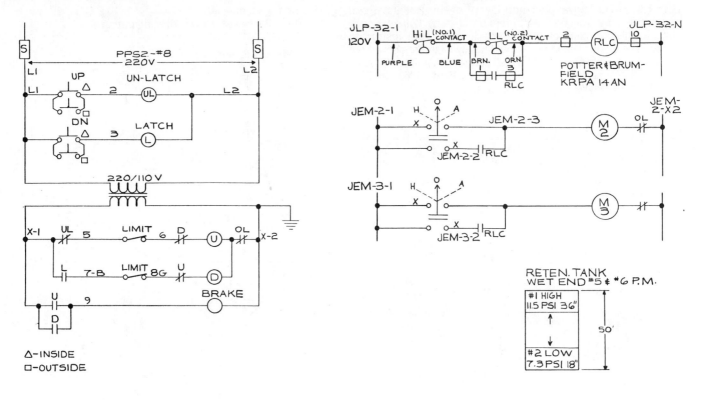

Δ–INSIDE
□–OUTSIDE

Problem 24–6 Schematic

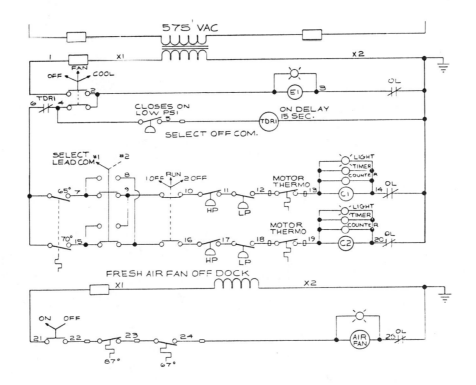

Problem 24–7 One-line diagram

Given the following engineering layout of the Potlatch-Pearl Substation, draw the one-line diagram on 11 in. by 17 in. polyester film or vellum unless otherwise specified by your instructor. There is no scale. The layout should neatly fill the sheet without crowding symbols or notes. Make all connection points ⌀3/32 in. Provide the general note: INTERPRET PER ANSI Y32.2. *Courtesy Bonneville Power Administration.*

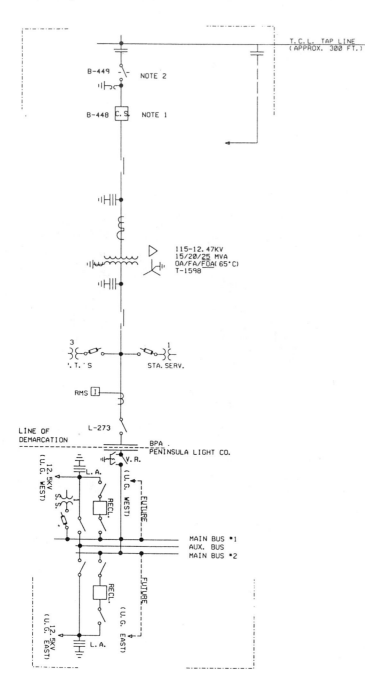

Problem 24–8 Elementary diagram

Metering System.
Courtesy Bonneville Power Administration.

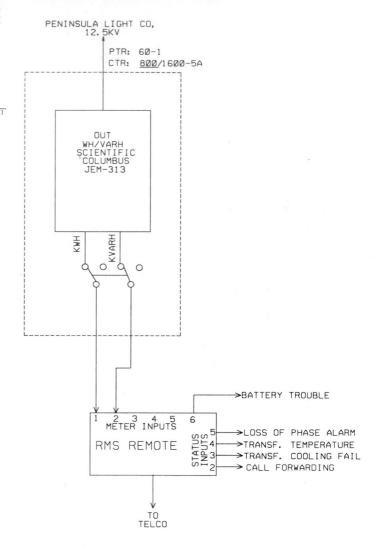

Problem 24–9 Highway diagram

Draw all component outlines and feeder lines 0.5 mm wide and trunk lines 0.9 mm wide.

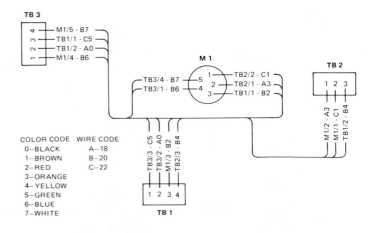

Problem 24–10   Wireless diagram

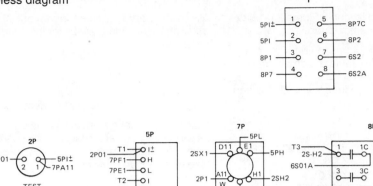

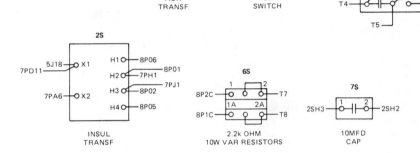

Problem 24–11   Cable diagram

Display Panel-CVS-3 COND. Make all panel outlines and feeder lines 0.5 mm wide and all trunk lines 0.9 mm wide. Do all lettering 1/8 in. high, with component labels 3/16 in. high and titles 1/4 in. high.

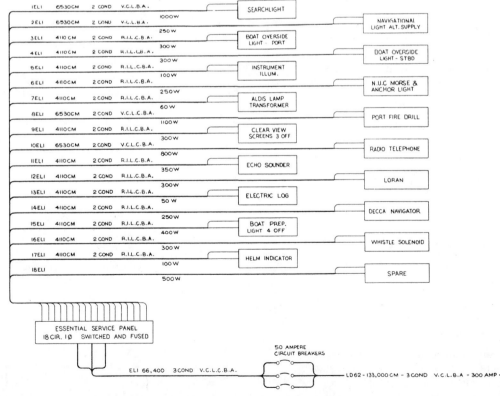

CABLE DIAGRAM FOR INSTRUMENT DISPLAY PANEL - CVS - 3 COND.

Problem 24–12   Cable assembly

Given the cable assembly engineering layout in Figure 24–12, do the following (approximate dimensions not given):

1. Draw the cable assembly, wiring diagram, and list of materials.
2. Use ANSI standard line widths for cable assembly and 0.5 mm lines for wiring diagram.
3. Use 1/8-in. lettering with 1/4-in. high letters in balloons and titles.
4. Make balloons ∅1/2 in.
5. Letter the following general notes:

1. INTERPRET DIMENSIONING AND TOLERANCING PER ANSI Y14.5M—1994.
2. INTERPRET DRAWING PER ANSI Y32.2.
3. DIMENSIONS ARE IN INCHES WITH TOLERANCES: FRACTIONS = ±1/64, .XX = ±.010, .XXX = ±.005.
4. 32 MICRO IN FINISH ON METAL PARTS.
5. MATERIAL BRONZE.
6. USE SPEC 6712 FOR WIRE END PREPARATION.

Problem 24–13   Bus layout plan view

Given the engineering layout shown in Figure 24–16 for Narrows Substation, draw the bus layout using a 3/16" = 1'–0" scale. The following line widths are recommended:

0.35 mm	Extension, dimension, balloon lines (make balloons ∅9/32 in.), all connections
0.50 mm	Component and bus lines, and lettering
0.60 mm	Titles

Draw north arrow S45°W. Bus lines are to be drawn from given dimensions and filled in solid except for connections. Dimensions not given are to be estimated.

Problem 24–14   Bus elevations

Given the partial engineering layout shown in Figure 24–17 for Narrows Substation, draw the bus elevation. Follow all other instructions given in Problem 24–13.

Problem 24–15   Grounding layout

Given the partial engineering layout shown in Figure 24–19 of the Narrows Substation, draw the grounding layout at a scale of 1" = 10'–0" Include notes and grounding table.

Problem 24–16   Ground details

Given the partial engineering layout shown in Figure 24–20 of the Narrows Substation, draw the grounding detail.

Problem 24–17   Conduit installation layout

Given the partial engineering layout shown in Figure 24–21 of the Narrows Substation, draw the conduit installation layout.

Problem 24–18   Conduit installation details

Given the partial engineering layout of the conduit installation detail shown in Figure 24–22 for the Narrows Substation, draw the given views.

Problem 24–19   Residential electrical

Given the typical electrical layouts shown in Figures 24–29 and 24–30, draw each at a scale of 1/4" = 1'–0". Place one next to the other. Estimate dimensions making your drawing about twice the size of the given drawing.

Problem 24–20   Power panel detail

Narrows Substation.
Courtesy Bonneville Power Administration.

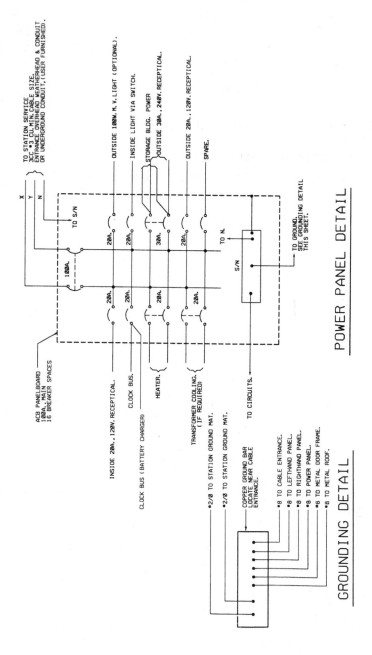

Problem 24–21   Electrical floor plan

Given the partial engineering upper floor power layout shown in Figure 24–34, draw the plan using a 1/4" = 1'–0" scale. Estimate dimensions, making your drawing about three times the size of the given drawing.

Problem 24–22   Reflected ceiling plan

Given the engineering layout shown in Figure 24–35, draw the reflected ceiling plan at a 1/4" = 1'–0" scale. Estimate unknown dimensions.

Problem 24–23   Roof plan—electrical

Given the partial engineering layout shown in Figure 24–36, draw the roof plan–electrical at a 1/4" = 1'–0" scale. Estimate unknown dimensions.

Problem 24–24   Commercial schematic wiring diagram

Given the engineering layout shown in Figure 24–37, draw the beverage cooler condenser unit wiring diagram.

PART 2

DIRECTIONS

1. Follow previous instructions.
2. Use the selected engineering layouts and sketches to prepare each drawing. Keep in mind that engi-

neering sketches may contain slight errors in format and symbol accuracy. Verify proper representation before drawing each symbol.

BLOCK AND SCHEMATIC DIAGRAMS

Problem 24–25   Block diagram

Given the schematic diagram sketch that has been divided into stages, prepare a block diagram on 11 × 17-size vellum or polyester film unless otherwise speclfled by your instructor. Make all lines 0.5 mm wide. Do all lettering 1/8" high with titles 1/4" high. Sketch does not show connection dots.

Problem 24–26   Audio playback block diagram

Given the block diagram engineering layout, draw the block diagram using 11 × 17-size vellum or polyester film unless otherwise specified by your instructor. Do all lottering 1/8" high with titles 1/4" high. Use 0.5 mm thick lines unless otherwise specified on the engineering sketch.
Courtesy RCA Consumer Electronics.

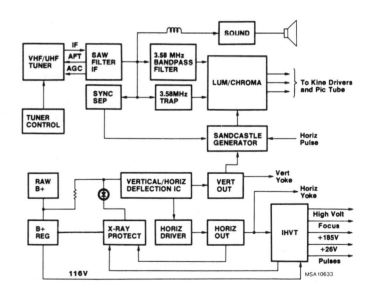

Problem 24–27 Cylinder speed block diagram

Given the block diagram engineering layout, draw the block diagram using 17 × 22-size vellum or polyester film unless otherwise specified by your instructor. Do all lettering 1/8" high with titles 1/4" high. Use 0.5-mm thick lines unless otherwise specified on the engineering sketch.

Courtesy RCA Consumer Electronics.

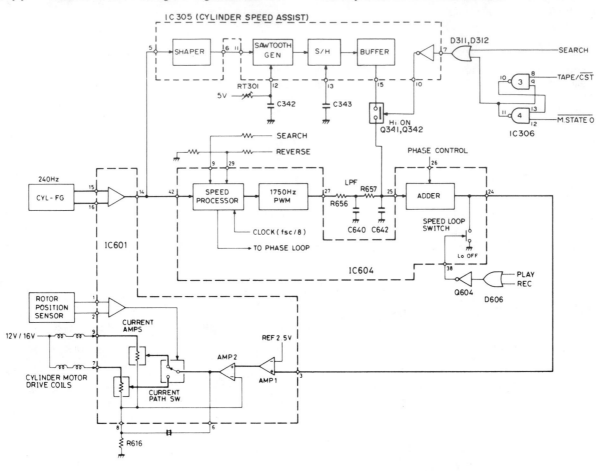

Problem 24–28 Schematic diagram

Given the schematic engineering sketch, make a schematic diagram on 11 × 17-size vellum. Blue grid vellum will be helpful, if available. There is no scale and the drawing must be balanced, uncluttered, and easy to read. Use the following instructions unless otherwise specified by your instructor:

1. Draw the entire schematic with 0.5-mm wide lines.

2. Do all lettering 1/8" high with titles 1/4" high.
3. Follow the left-to-right and top-to-bottom labeling system using coding as described in this chapter: R1, R2, R3. . . , C1, C2. . . .
4. Label horizontal components above and vertical components to the right.

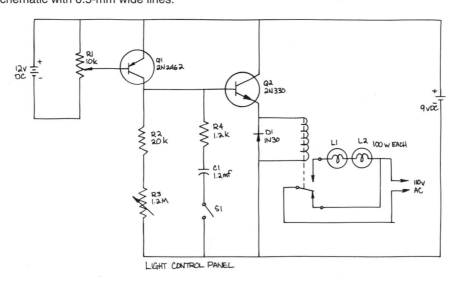

Problem 24–29   Tuner schematic diagram

Given the schematic engineering sketch, make a schematic diagram on 11 × 17-size vellum. Blue grid vellum will be helpful, if available. There is no scale, and the drawing must be balanced, uncluttered, and easy to read. Use the following instructions unless otherwise specified by your instructor.

1. Draw the entire schematic with 0.5-mm wide lines, and dashed stage lines 0.7 mm wide.
2. Do all lettering 1/8" high with titles 1/4" high.

3. Reference designators are not shown; use the left-to-right and top-to-bottom labeling system, using coding as described in this chapter: R1, R2, R3. . . , C1, C2. . . .
4. Label horizontal components above and vertical components to the right.
5. Place connection dots as necessary.

Courtesy RCA Consumer Electronics.

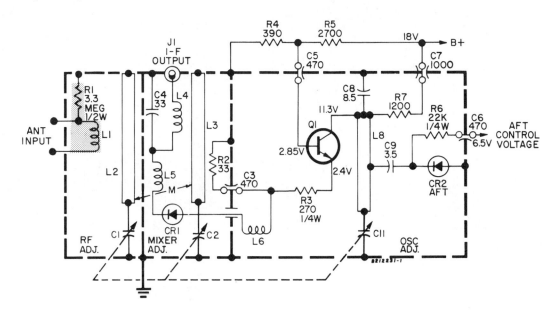

Problem 24–30   Television receiver tuner schematic diagram

Given the schematic engineering sketch, make a schematic diagram on 17 × 22-size vellum. Blue grid vellum will be helpful, if available. There is no scale, and the drawing must be balanced, uncluttered, and easy to read. Use the following instructions unless otherwise specified by your instructor:

1. Draw the entire schematic with 0.5-mm wide lines, and dashed stage lines 0.7 mm wide.

2. Do all lettering 1/8" high with titles 1/4" high.
3. Reference designators are not shown; use the left-to-right and top-to-bottom labeling system, using coding as described in this chapter: R1, R2, R3. . . , C1, C2. . . .
4. Label horizontal components above and vertical components to the right.
5. Place connection dots as necessary.

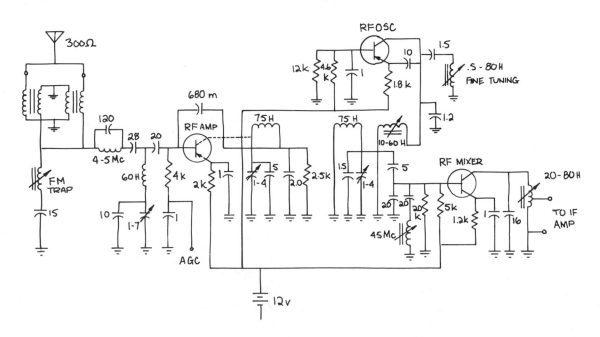

Problem 24–31   Logic diagram

Given the IC schematic engineering layout for the logic diagram, make a schematic drawing on 17 × 22-size vellum. Use the following instructions unless otherwise specified by your instructor:

1.   Do all line work and lettering using 0.5-mm width.

2.   Do all lettering 1/8" high with titles 1/4" high.
3.   All connection points will be ∅3/32" filled in.
4.   Provide the general note: INTERPRET PER ANSI Y 32.2.

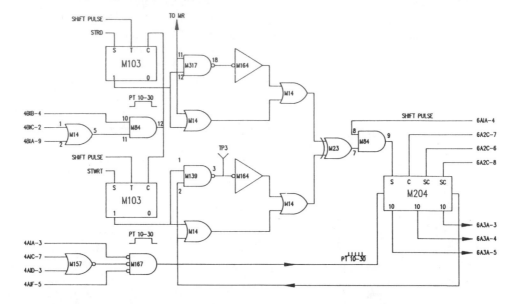

Problem 24–32   Schematic diagram

Given the schematic engineering sketch, make a schematic diagram on 17 × 22-size vellum. Blue grid vellum will be helpful, if available. There is no scale and the drawing must be balanced, uncluttered, and easy to read. Use the following instructions unless otherwise specified by your instructor:

1.   Draw the entire schematic with 0.5-mm wide lines, and dashed stage lines 0.7 mm wide.
2.   Do all lettering 1/8" high with titles 1/4" high.
3.   Reference designators are not shown; use the left-to-right and top-to-bottom labeling system, using coding as described in this chapter: R1, R2, R3. . ., C1, C2. . . .

4.   Label horizontal components above and vertical components to the right.
5.   Place connection dots as necessary.

Notes: UNLESS OTHERWISE SPECIFIED.

1.   Reference designators are for reference only and may not appear on part.
2.   Resistor values are in OHMs.
3.   For PWB/component assembly, see dwg. 100/01.
4.   For parts, see SPL 100/02.
5.   For printed wiring board, see dwg. 100/03.

Courtesy RCA Consumer Electronics.

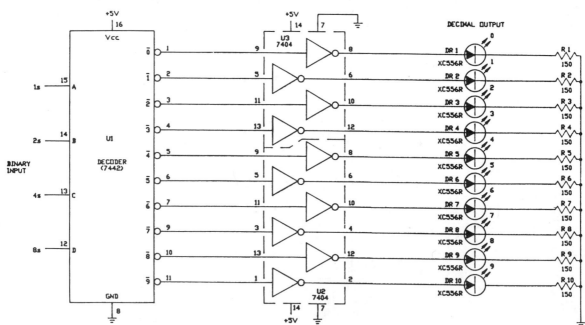

Problem 24–33   Printed circuit board layout

Given the "component side" engineer's sketch of a printed circuit board, make a printed circuit board layout on the "back side" at 2:1 scale. Use the following instructions unless otherwise specified by your instructor:

1. Use Bishop Graphics, Inc. supplies or equivalent or CADD software.
2. Use polyester film base sheet.
3. Use teardrop pads with fillet radius, .200" OD and .032" ID.
4. Use .080 trace line width.

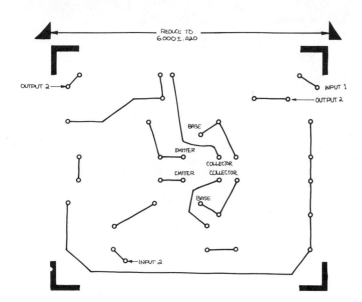

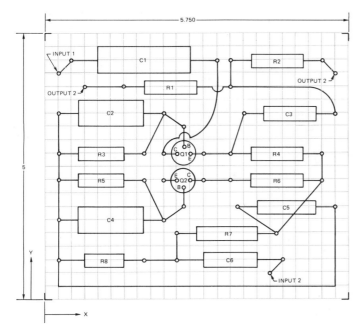

Problem 24–34   Marking drawing

Given the PCB layout sketch from Problem 24–33, make a marking drawing at 1:1 scale on polyester film. Use 5/32" high lettering with ink and mechanical lettering device or rub-on letters, unless otherwise specified by your instructor.

Problem 24–35   Drilling drawing

Given the PCB layout from Problem 24–33 and the attached dimension table, prepare a drilling drawing using tabular dimensioning. Refer to Chapter 10, Dimensioning. The holes may be shown using centerline "cross" at the center locations without drawing the actual circles. The 0,0 location is at the lower left corner of the PCB. Number the holes from top to bottom and left to right as coordinated with the given table. NOTE: Dimensions are given from component side.

HOLE NO.	X	Y	DIAMETER
1	.500	4.500	.032
2	3.250	4.500	.032
3	3.500	4.500	.032
4	5.250	4.500	.032
5	5.500	4.250	.032
6	.750	4.000	.032
7	1.500	4.000	.032
8	3.500	4.000	.032
9	.250	3.500	.032
10	2.250	3.500	.032
11	3.750	3.500	.032
12	5.500	3.500	.032
13	2.625	3.250	.032
14	.250	2.750	.032
15	1.875	2.750	.032
16	2.250	2.750	.032
17	3.000	2.750	.032
18	3.500	2.750	.032
19	5.250	2.750	.032
20	.250	2.250	.032
21	1.875	2.250	.032
22	2.250	2.250	.032
23	3.000	2.250	.032
24	3.500	2.250	.032
25	5.250	2.250	.032
26	2.625	1.750	.032
27	3.625	1.750	.032
28	5.500	1.750	.032
29	.250	1.500	.032
30	2.250	1.500	.032
31	2.500	1.250	.032
32	4.375	1.250	.032
33	.250	.750	.032
34	1.875	.750	.032
35	2.500	.750	.032
36	4.500	.750	.032
37	4.250	.500	.032

PICTORIAL DRAWING

Given the engineer's layout, make a pictorial drawing using any one of the specific shading techniques discussed in Chapter 17, Pictorial Drawings. Scale the engineer's layout directly to determine dimensions. Increase the size of your final drawing. Place all labels on the final drawing in ink using mechanical lettering devices, or prepare a CADD ink plot if available. Use polyester film or vellum.

Problem 24–36   Exploded technical illustration

Given the exploded technical illustration shown in Figure 24–89, select and draw the pictorial of any one subassembly unless otherwise specified by your instructor. Make your drawing larger but proportional to the given drawing.

Problem 24–37 Pictorial schematic

Given the pictorial schematic engineering layout shown in Figure 24–90, make a formal drawing.

MATH PROBLEMS

Convert these numbers to power of ten form:

1. 243
2. 14,000
3. .004

Convert these numbers to regular form:

4. 6.7×10^3
5. 9×10^6
6. 6.2×10^{-3}

Perform these calculations.

7. $(4 \times 10^6)(3 \times 10^{13})$
8. $\dfrac{25 \times 10^{17}}{5 \times 10^{11}}$
9. $(7 \times 10^{-6})(5 \times 10^{-2})$
10. $\dfrac{2.8 \times 10^{-6}}{8.1 \times 10^5}$

CHAPTER 25

Engineering Charts and Graphs

LEARNING OBJECTIVES

After completing this chapter, you will:

○ Design and draw the following types of charts and graphs: rectilinear, surface, column, pie, polar, nomographs, trilinear, flow, distribution, and pictorial.
○ Use a computer-aided design and drafting system to design and draw a chart from engineering specifications and sketches.

THE ENGINEERING DESIGN PROCESS

Given the following statistical data, design a rectilinear chart that will represent the information.

Projected population of the United States from 1985 to 2165. The source is World Bank data, projected assuming constant fertility and migration. The year is followed by the population in millions: 1985—238, 1995—255, 2005—269, 2015—282, 2025—290, 2035—291, 2045—287, 2055—283, 2065—279, 2075—273, 2085—269, 2095—264, 2105—259, 2115—255, 2125—251, 2135—247, 2145—243, 2155—240, 2165—239.

Step 1. Establish the range of units. In this case the lowest unit in years is 1985 years and the highest is 2165 years. The lowest unit in population is 238 million and the highest is 291 million.

Step 2. Determine the vertical scale to accommodate the range. For example, 230 to 300 million as shown in Figure 25–1.

Step 3. Determine the horizontal scale. In this case there is a set of data every ten years. The horizontal scale will be divided into units representing ten-year modules. It is best to make the horizontal scale approximately equal to the vertical units without crowding. In this case there are more horizontal than vertical units, so the ten-year divisions will be closer than preferred; however, the thirty-year divisions will be the main units as shown in Figure 25–2.

Step 4. Plot the points and draw the curve as shown in Figure 25–3.

Step 5. Complete the chart by labeling the title and any required captions. (See Figure 25–4.)

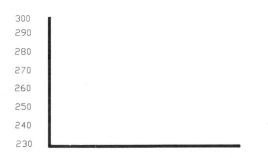

Figure 25–1 Step 2—determine the range for the vertical scale units.

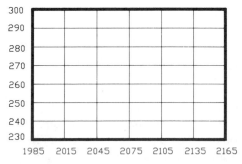

Figure 25–2 Step 3—determine the horizontal scale and establish the grid. ***(Continued)***

ENGINEERING DESIGN PROCESS *(Continued)*

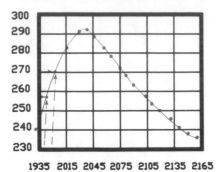

Figure 25–3 Step 4—plot points from given data and draw the curve. Note points are exaggerated for emphasis. Should be plotted very lightly.

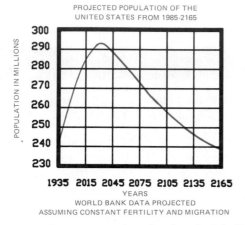

Figure 25–4 Step 5—complete the chart by labeling the title and required captions.

Charts are graphic representations of any measurable data. Charts, graphs, and diagrams are all synonymous with the graphic representation of numerical data. Charts may be more specifically defined as illustrations that give information that would otherwise be arranged in a table (tabular form). Graphs and diagrams are a series of points, a line, a curve, or an area representing the comparison of variables. The advantage of charts and graphs is that technical data may be shown in a manner that quickly and graphically communicates the information.

Professionals in any field can use charts and graphs to graphically explain or demonstrate the results of statistics. *Statistics* is the science of the collection, arrangement, and interpretation of quantifiable information. Statistics has two areas or branches: *descriptive statistics* and *inferential statistics*. Descriptive statistics is the summarizing and organization of data. Inferential statistics includes making inferences (drawing conclusions) about a whole population based on information obtained from a sample. Many engineers, quality-control people, and managers are beginning to understand that to really interpret data, they need the sta-

tistics and a picture of the data in the form of a chart or graph. Many people have difficulty understanding and interpreting statistical data unless the quantities are shown on a chart or graph.

The accurate preparation of charts and graphs requires a degree of knowledge of the topic related to the data and an understanding of where, how, and why the chart is used. There are two fundamental applications for charts: analysis and presentation. *Analytical charts* are used to analyze, examine, and explore data, or the information is used to calculate specific required values. *Presentation charts* are generally more artistic in appearance and are used to demonstrate or present information. Presentation charts are used in advertising and may be pictorial for easy communication of data for lay people. While there may be two classifications of charts, there is a crossover. For example, analytical charts may be used for presentation and presentation charts often contain analytical data.

ASA According to the American Standards Association document ASA Y15.2, the following questions should be answered before a chart design is started:

1. What is the general purpose of the chart?
2. What kind of data are to be presented?
3. What features of the data is the chart to identify?
4. For what audience is the chart intended?
5. What method will be used to show the chart to the audience?

The selection of the specific type of chart is the designer's choice. Not all chart design questions are applied to each chart. Some charts may be designed to focus on one or more of the related design questions. One of the most important factors in chart design is simplicity. Keep the chart as simple as possible by avoiding the presentation of too much information or detail.

Charts may be designed for any purpose to accurately communicate numerical data from a simple line format to a pictorial presentation. The selection of the chart type depends on the audience, the intent of the presentation, and the type of data. The basic types of charts include rectilinear (line) charts, surface charts, column (bar) charts, pie charts, polar charts, nomographs, trilinear charts, flowcharts, distribution charts, and pictorial charts.

▼ RECTILINEAR CHARTS

Rectilinear charts, also known as line charts, are the most common type of chart used for the presentation of analytical data. The rectilinear chart is commonly set up on a horizontal and vertical grid where the horizontal axis or scale represents amounts of time or other significant independent values. The vertical axis identifies the dependent quantities or values related to the horizontal values. The horizontal axis is called the X axis or *abscissa* and the vertical axis is called the Y axis or *ordinate.* The line formed by connecting the data elements is called the *curve.* (See Figure 25–5.) Rectilinear charts are referred to as a slope curve when the data represents points in time that are not backed by mathematical computations. (See Figure 25–6a.) Step curves are established by data that changes abruptly. (See Figure 25–6b.) In some applications a smooth curve is used when the data plotted forms a consistent pattern, or when the average of the calculated points is represented. A smooth curve generally acknowledges an empirical relationship between

the data and the curve. Empirical refers to information that is based on observation and measurement rather than theory. (See Figure 25–6c.) A type of rectangular chart where greatly varying data does not fit into a defined curve is called a rectangular coordinate distribution chart or scatter chart. The term *scatter* is used to describe these charts because the plotted data is scattered around on the chart. The purpose of these charts is to observe the distribution of data and to identify any areas of concentration. (See Figure 25–6d.)

Scales

The selection of horizontal and vertical scales is one of the most important aspects of chart design. There should be a good balance between the scales. If one scale is exaggerated out of proportion, then the resulting curve may be misleading. Scale selection greatly affects chart design. For example, the two charts shown in Figure 25–7 contain the same data; however, the chart in Figure 25–7a has a smaller vertical scale than the chart in Figure 25–7b. Notice the extreme difference in curve representation between the two charts.

When selecting the proper scale, the first consideration is the range of data. The scale range extends from the smallest value to the largest value. Charts often, but not always, begin at zero and extend beyond the largest Y value and are at least as long as the largest X value. The scale should include zero when a comparison of two or more curve magnitudes is made. The scales may not necessarily show zero when the intent of the chart is to show the relationship between one group of units and another, or when the general shape of one curve is compared to another.

The selection of a beginning and ending scale value also depends on the range of values to be displayed. For example, if the display values range from 500 to 1500, then a zero beginning may be too far removed from the first value of 500. When zero does not begin the scale, the first numerical value should be labeled with bold letters or otherwise clearly identified so the reader does not assume a zero beginning. The scale divisions are usually

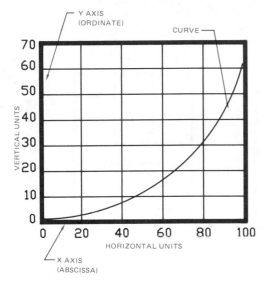

Figure 25–5 Elements of a rectilinear chart.

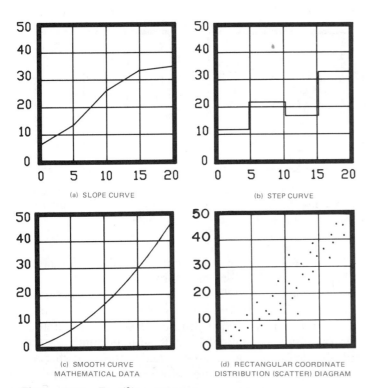

Figure 25–6 Rectilinear curves.

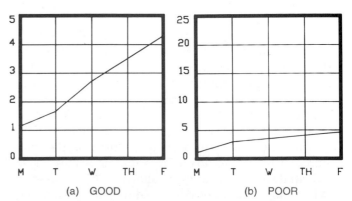

Figure 25–7 Scale selection may greatly affect chart design.

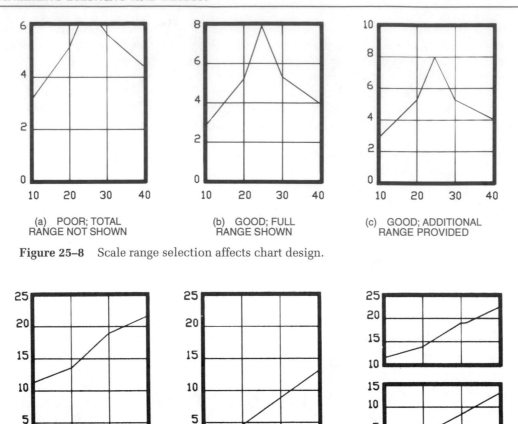

(a) POOR; TOTAL
RANGE NOT SHOWN

(b) GOOD; FULL
RANGE SHOWN

(c) GOOD; ADDITIONAL
RANGE PROVIDED

Figure 25–8 Scale range selection affects chart design.

(a)

(b)

Figure 25–9 **(a)** Charts placed side-by-side for comparison; same scale range. **(b)** Comparative charts stacked to save space.

established as convenient units such as 0,2,4. . . ; or 0,5,10. . . . The scale should also be selected so the curve is well distributed throughout the chart. This requires that the scale range be extended beyond the value limits. (See Figure 25–8.) As previously shown in Figure 25–7b, too much range is also not desirable because it provides an excessive amount of blank area in the chart. In some situations a certain amount of blank space may be necessary to accommodate notes, details, or other information. When this is appropriate, add enough scale units to provide this space.

When one or more charts are compared, the designer should set the scales of all charts the same by starting with the chart that has the greatest range of values. The only potential problem is that some of the curves may be flat; however, this may be necessary to provide a realistic comparison. Figure 25–9 shows two charts placed for comparison. Another method to show the comparison of two or more curves is to place each curve on the same chart. When the curves do not cross, the curve lines may be drawn all the same thickness as shown in Figure 25–10.

When the curves to be compared intersect each other, a technique should be used to differentiate each curve.

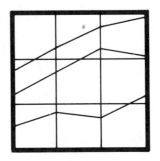

Figure 25–10 Noncrossing curves may be drawn the same thickness for comparison.

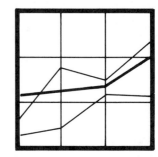

Figure 25–11 Curve comparison using different line thicknesses.

Figure 25–11 shows curve comparison on the same chart by using different line thicknesses. Another common method used to distinguish curves in a comparative chart is to show the curves using different line representations; for example, solid line, dash line, and dotted line as shown in Figure 25–12. It is not normally a good idea to design a chart with unusual patterns or shapes. These types of designs often clutter the chart or may be difficult to differentiate without careful analysis. (See

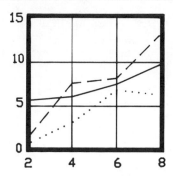

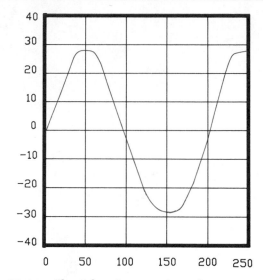

Figure 25–12   A comparative chart using different line representations.

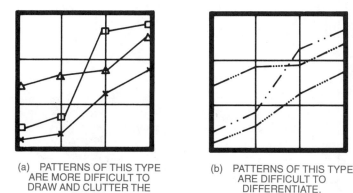

(a) PATTERNS OF THIS TYPE
ARE MORE DIFFICULT TO
DRAW AND CLUTTER THE
CHART.

(b) PATTERNS OF THIS TYPE
ARE DIFFICULT TO
DIFFERENTIATE.

Figure 25–13   Charts designed with unusual patterns or line designations should be avoided.

Figure 25–14   Chart showing negative values.

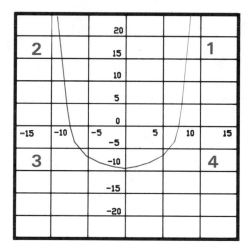

Figure 25–15   Chart set up in quadrants on an X–Y axis.

Figure 25–13.) Another technique for charts with the same horizontal and vertical scales is to prepare the charts separately on clear polyester film and overlay them for comparison.

The previous discussion and examples have shown charts where numerical values along the abscissa and ordinate have been positive, beginning with zero on the X and Y axis. Many charts that are designed from mathematical data do not follow this layout pattern. When numerical values exist as negative quantities, the chart must reflect the quantities below zero as shown in Figure 25–14. Some applications exist where plotted values are negative X and Y, or positive X and Y, or positive X and negative Y, or negative X and positive Y values. When this occurs, the chart will be set up in quadrants as shown in Figure 25–15. Quadrant 1 shows positive X and Y values. Quadrant 2 provides negative X and positive Y values. Quadrant 3 displays negative X and Y values. Quadrant 4 shows positive X and negative Y values.

Ratio Scales

Ratio scales are special scales that are referred to as logarithmic and semilogarithmic scales. The scales that were discussed previously are arithmetic scales where equal distances represent equal amounts. Arithmetic scales are used to display the absolute dimension of val-

ues. Ratio scales represent a different function where the rate of change of the variables is more important than the absolute values or amount of change. Logarithmic charts use a logarithmic scale for both the X and Y axis. Semilogarithmic charts use a logarithmic scale for the Y axis and an arithmetic scale for the X axis. The advantage of ratio scales exists when it is desirable to display changes at one level either relatively larger or smaller than changes at another level, or for showing a pattern of relative changes. Logarithmic scales are established by making the divisions proportional to the logarithms of the arithmetic scales. A comparison of the arithmetic and logarithmic scales is shown in Figure 25–16.

There are a variety of logarithmic scales available. The scales most commonly used for chart design are one-cycle, two-cycle, and three-cycle. (See Figure 25–17.)

Logarithmic charts may be prepared directly on preprinted logarithmic paper or designed to fit any required area. Preprinted log scales may be used to divide the

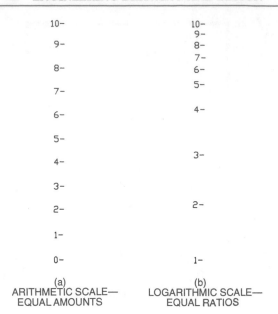

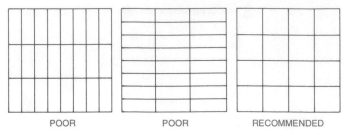

Figure 25–18 Grid spacing.

Figure 25–16 A comparison of the arithmetic and logarithmic scales.

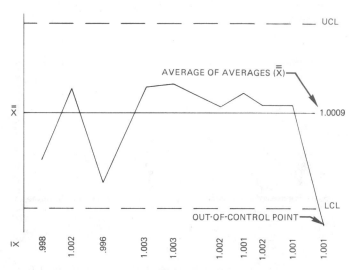

Figure 25–19 A chart showing feature dimensions at inspection intervals.

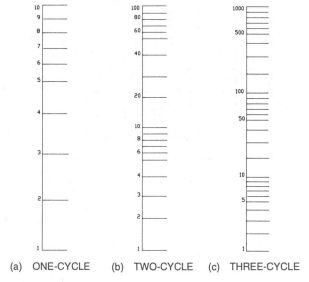

Figure 25–17 Logarithmic scales.

1. A grid with wide horizontal and short vertical increments is used for data that covers a long period of time.

2. A grid with narrow horizontal and high vertical increments is used for data that represents short periods of time or rapid value changes.

3. Most applications work out well if the horizontal and vertical grids are approximately equal. (See Figure 25–18.)

Grid lines are generally thin. The thickness of these lines may increase as the distance between them increases. Close grid lines should usually be drawn very thin.

Certain horizontal or vertical grid lines may be distinguished from other lines by either drawing them thicker or as a dashed line. For example, manufacturing quality control often uses computerized monitoring of dimensional inspections in Statistical Process Control (SPC). When this is done, a chart that shows feature dimensions obtained at inspection intervals is developed. The chart shows the expected limits of sample averages as two dashed, parallel horizontal lines as shown in Figure 25–19.

It is important not to confuse control limits with tolerances—they are not related to each other. The control

given area into proportional divisions by the same technique used when dividing a space into equal parts (Chapter 4), or using scales set up in the CADD program.

Grids

The grid is the horizontal and vertical lines that form the chart layout. The grid lines are determined by the scales and are used to aid in both drawing and reading the chart. Charts and graphs may be prepared on pre-printed grid paper or designed to fit specific needs. The actual grid format depends on the purpose of the chart. In most applications the following guidelines should be considered:

limits come from the manufacturing process as it is operating. For example, if you set out to make a part 1.000 ± .005 in., and you start the run and periodically take five samples and plot the averages (x) of the samples, the sample averages will vary less than the individual parts. The control limits represent the expected variation of the sample averages if the process is stable. If the process shifts or a problem occurs, the control limits signal that change.

Notice in Figure 25–19 that the x values represent the average of each five samples; $\bar{x}$ is the average of averages over a period of sample taking. The upper control limit (UCL) and the lower control limit (LCL) represent the expected variation of the sample averages. A sample average may be "out of control" yet remain within tolerance. During this part monitoring process the out-of-control point represents an individual situation that may not be a problem; however, if samples continue to be measured out of the control limits, the situation must be analyzed and the problem rectified. When this process is used in manufacturing, part dimensions remain within tolerance limits and parts will not be scrapped. Statistical Process Control is discussed further in Chapters 9, 10, and 14.

Line Chart Labeling

There are standard guidelines for the placement of labels and captions. In general, however, labels should be large enough for easy reading and placed to avoid cluttering the chart. The most important feature of the chart should be emphasized by bold display or special effects such as color. For example, the title or a predominant curve may be emphasized. Avoid overemphasizing a particular feature to the extent that other items are lost.

Chart Titles. Chart titles are generally aligned with the left edge of the paper or centered at the top of the chart. Title lettering is higher and bolder than other labels and captions.

Labeling Scales. The scales are usually identified with numerical values next to the corresponding grid lines. Abscissa scale numerals are placed below the horizontal scale and ordinate scales are labeled to the left or right of the chart. Very long or high charts may have scales labeled at both grid limits. The horizontal and vertical units are identified next to the numerical values or below the X values and directly above the column of Y values. Figure 25–20 shows several examples.

Scale identification should be kept as simple as possible. For example, long scale numerals should be shortened using the easiest format for the reader to understand and proper abbreviations should be used when appropriate. (See Figure 25–21.)

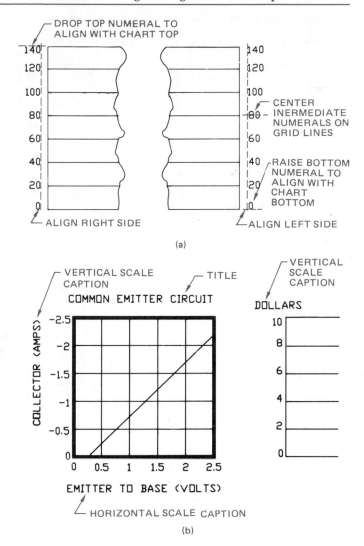

Figure 25–20 Chart labeling.

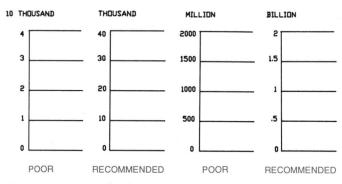

Figure 25–21 Scale identification.

Curve Captions. Curve captions are necessary when two or more curves are displayed on the same chart. Single curves are labeled only if identification is not otherwise apparent by other captions and titles. The lettering for curve labels is generally smaller than titles and larger than general notes. Place curve labels on the face of the chart in areas that clearly avoid crowding or confusion. (See Figure 25–22.)

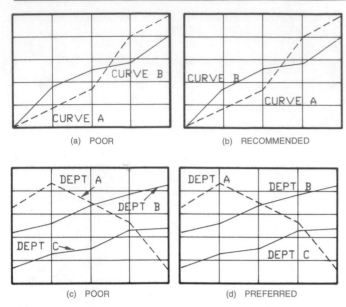

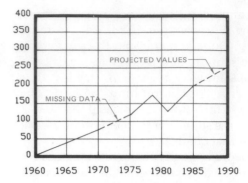

Figure 25–24   Missing and projected data.

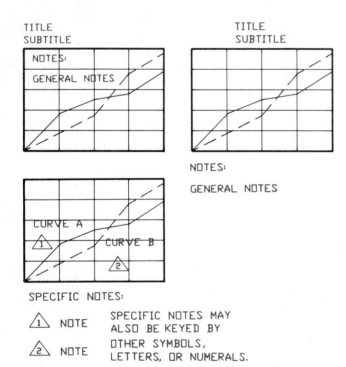

Figure 25–22   Curve captions.

Figure 25–23   Subtitles and notes.

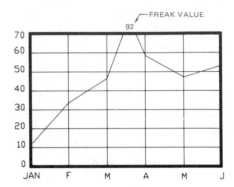

Figure 25–25   Technique for freak values. Avoid this method whenever possible.

Subtitles and Notes.   Subtitles and general notes are subordinate captions that relate to the entire chart. These items may be placed below the title, in the field of the chart, or below the chart, depending on the design influence of the information and the space requirements. Specific notes relate to individual items on the chart and may be placed in the field of the chart or keyed to the chart by identification symbols and placed in the area of general notes. (See Figure 25–23.)

Missing or Projected Data.   When some data is omitted due to missing or unavailable information, the estimated curve representing the missing data should be shown as demonstrated in Figure 25–24.

There are some applications when the function of the chart shows current data and provides an estimate of projected values. This may be done by continuing the curve into a future interval as shown in Figure 25–24.

In some unique situations there may be one or more values that fall well out of the normal distribution of values. When these *freak* values occur, they may completely alter the appearance of the chart unless intentionally left beyond the grid boundaries. When this is done, the numerical designation of the freak value is labeled as shown in Figure 25–25. Freak values should be carefully evaluated for possible error because in actual practice this approach is uncommon.

▼ SURFACE CHARTS

Surface charts, also known as area charts, are designed to show values that are represented by the extent of a shaded area. The only difference between a surface chart and a line chart is that the area between the curve and the X axis or the area between curves is shaded for emphasis. The advantage of surface charts is that they

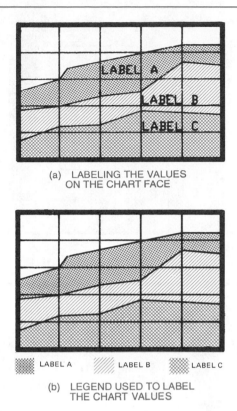

(a)   LABELING THE VALUES
ON THE CHART FACE

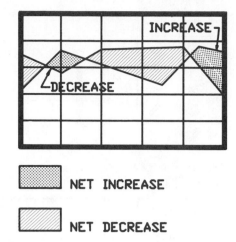

NET INCREASE

NET DECREASE

Figure 25–28   Emphasizing the net difference between two curves.

LABEL A   LABEL B   LABEL C

(b)   LEGEND USED TO LABEL
THE CHART VALUES

Figure 25–26   Surface chart design and layout techniques.

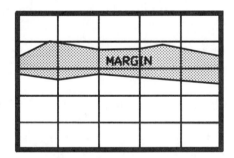

Figure 25–27   Emphasizing the difference between two curves.

more clearly define the difference between curves or the extent of a single curve. Surface charts should be avoided when accurate readings are required, or when greatly irregular layers exist. Surface charts are designed much like line charts, except that transfer films or computer graphics hatching techniques are used to create shading. Some of the common design layout and labeling techniques used with surface charts are shown in Figure 25–26.

One of the best design considerations for surface charts exists when the intent of the chart is to show a margin between two curves. (See Figure 25–27.) Another design characteristic is the implementation of a surface chart that shows the net increase or the net decrease between two sets of values. (See Figure 25–28.)

▼ COLUMN CHARTS

Column charts are commonly referred to as bar charts. A *bar chart* is used to represent numerical values by the height or length of columns. The data presented usually represents total periods of time or total percentages as opposed to various periods of time shown in line charts. Bar charts provide less detail than line charts, but provide a more dramatic and easy-to-understand display for nontechnical readers.

The columns of a bar chart may be placed either horizontally or vertically; the vertical application is most common. The design of the bar chart vertical scale is similar to the technique used for line charts. For vertical column charts the horizontal grid is determined by the range of data and vertical grids are replaced by the columns. The design of the columns should assist and not distract from reading the chart. Chart columns should not be too bold or too thin as shown in Figure 25–29. The column chart shown in Figure 25–29c is used a lot in quality control and is referred to as a *histogram*. The vertical scale is the frequency of occurrence and the horizontal scale is the value measured.

Subdivided column charts are used to show the values to individual components within total quantities of column data. The component values in each column are shaded to stand out as different quantities. The column subdivisions may be labeled on the face of the chart within the designated areas or with leaders pointing to the area, or a legend may be used that differentiates the components. Caution should be exercised when using these charts as they may be deceiving, and hard to read and interpret. (See Figure 25–30.)

Grouped column charts are used to compare two quantities at different intervals. Two columns are attached to show the comparison at the given periods as shown in Figure 25–31.

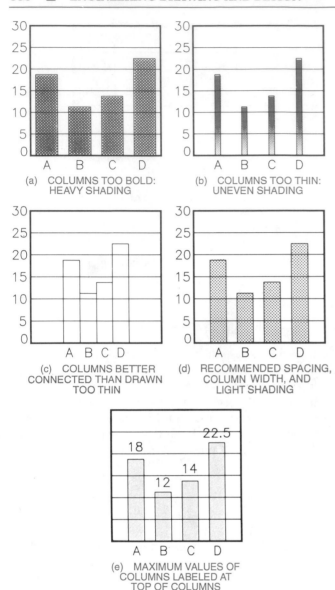

Figure 25-29 Column chart techniques.

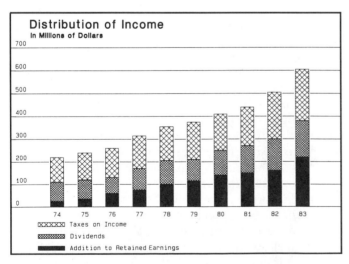

Figure 25-30 Subdivided column chart. *Courtesy Computer Support Corporation.*

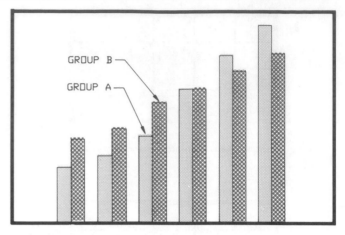

Figure 25-31 Grouped column chart.

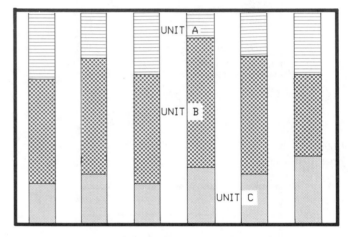

Figure 25-32 One hundred percent column chart.

One hundred percent column charts are used to show the relationship between the distribution of components where the totals are always equal values, such as 100 percent of any given unit. (See Figure 25-32.) These charts are effective in some applications, but may be hard to read. It is usually better to separate the data.

Over-under column charts are used to emphasize the difference between two variables or the difference between a variable and a given standard. The over-under column chart has columns that originate at a common value or standard line as shown in Figure 25-33.

The previously discussed bar charts have been index charts where each of the columns originate at a common base. Another type of bar chart is the range bar, where individual columns represent segments of the total applicable range. *Range column* charts are used to emphasize the difference between values where one value is always higher or larger than the other. Many applications of range bar charts also show the average value for each column. (See Figure 25-34.) These charts show the central tendency (average) of data and the variability (range) of the data. Range bar charts are also used with median values, rather than average values, marked, and with control limits the same as the chart in Figure 25-19.

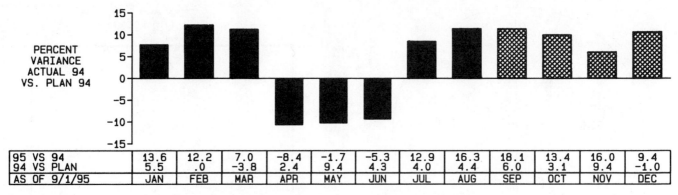

95 VS 94 94 VS PLAN	13.6 5.5	12.2 .0	7.0 -3.8	-8.4 2.4	-1.7 9.4	-5.3 4.3	12.9 4.0	16.3 4.4	18.1 6.0	13.4 3.1	16.0 9.4	9.4 -1.0
AS OF 9/1/95	JAN	FEB	MAR	APR	MAY	JUN	JUL	AUG	SEP	OCT	NOV	DEC

Figure 25–33 Over-under column chart. *Courtesy Computer Support Corporation.*

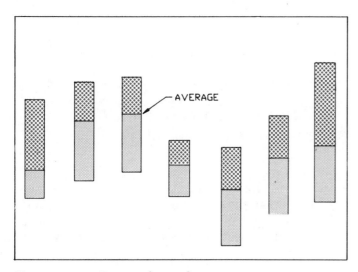

Figure 25–34 Range column chart.

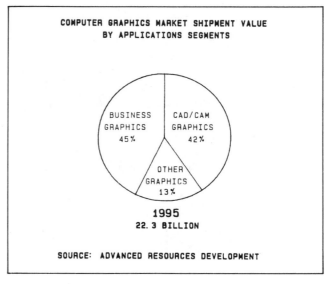

Figure 25–35 Comparative pie charts. *Courtesy MICRO-GRAFX.*

▼ PIE CHARTS

Pie charts are used for presentation purposes and are one of the most popular methods of making graphic representations of data for people to understand. It is easy to understand how the portions of a pie represent quantity. Even without reference to the values of the pieces, readers can quickly determine the relationship between large and small portions. Pie charts are also referred to as 100 percent circle charts because the entire pie or circle represents 100 percent of the total and each piece of pie relates to a percentage of the total. For example, a pie chart may represent 100 percent of 1995 computer graphics market shipment, where the percentage of each segment's shipment is shown as a piece of the pie. (See Figure 25–35.)

Pie Chart Design

Given the following information, design a pie chart that will graphically display the data:

1995 agricultural sales for Washington County:
Nursery crops = 44%
Vegetable crops = 26%

Fruits and berries = 11%
Grain and hay = 8%
Greenhouse crops = 6%
Miscellaneous crops = 5%

Step 1. Draw a circle with a diameter of approximately 4 inches. The diameter of the circle depends on the sheet size and the information to be included. Individual judgment must be used.

Step 2. Convert each category into degrees. There are 360° in the circle representing 100% of the values.
Use the formula 3.6 × % = degrees.
Nursery crops = 3.6 × 44 = 158.4°.
Vegetable crops = 3.6 × 26 = 93.6°.
Fruits and berries = 3.6 × 11 = 39.6°.
Grain and hay = 3.6 × 8 = 28.8°.
Greenhouse crops = 3.6 × 6 = 21.6°.
Miscellaneous crops = 3.6 × 5 = 18°.

Step 3. Using the degrees calculated in Step 2, lay out each piece of the pie on the scale as shown

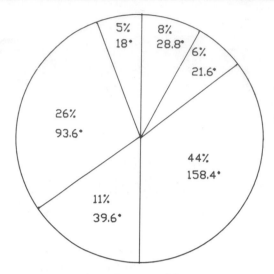

Figure 25–36 Pie chart design and layout.

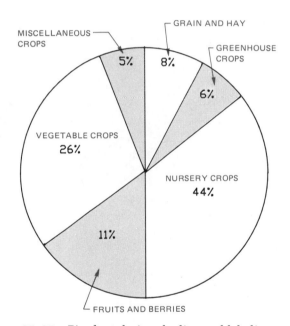

Figure 25–37 Pie chart design shading and labeling.

in Figure 25–36. Notice that at least one of the small pie parts is separated from the group of small parts for ease in labeling later.

Step 4. Add the title, notes, captions, and do any shading to suit the design. (See Figure 25–37.)

▼ POLAR CHARTS

Polar charts are designed by establishing polar coordinate scales. Polar coordinates are points determined by an angle and distance from a center or pole. There are two scales on the polar chart. The first scale is related to the degrees or radians of a circle and these radiate from the pole. The second scale represents distances from the center. Each distance is represented by a concentric

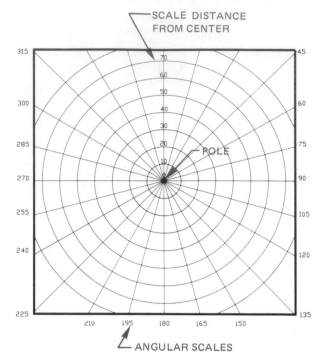

Figure 25–38 Polar coordinate scales.

circle. (See Figure 25–38.) Polar coordinate charts are used to display the effects of data that radiate from a source. The applications may include the range of intensity of a light source or the effective range of communication signals.

▼ NOMOGRAPHS

Nomographs are graphic representations of the relationship between two or more variables of a mathematical equation. Nomographs are used as a quick graphic reference where unknown values of a given equation may be determined by aligning given numerals on a series of scales. There are two general types of nomographs. The most elementary is a *concurrency chart*, also known as a rectangular *Cartesian coordinate* chart. The other type of nomograph is called an alignment chart. Nomographs are generally designed only when the chart will get a great deal of use, otherwise the mathematical solution of unknown values would be the easiest and fastest approach. Nomographs, in general, are difficult and time consuming to design. A nomograph is shown in Figure 25–39.

Concurrency Charts

The *concurrency chart* uses the X and Y axis, as previously described in the discussion on rectilinear charts, to graphically solve values for given mathematical equations. For example, given the equation a + b = c, a series of curves are established on a rectilinear chart where each line represents one of the equation values. (See Figure

Nomogram
Horsepower to Torque Conversion Table

Nomogram Horsepower to Torque Conversion Table

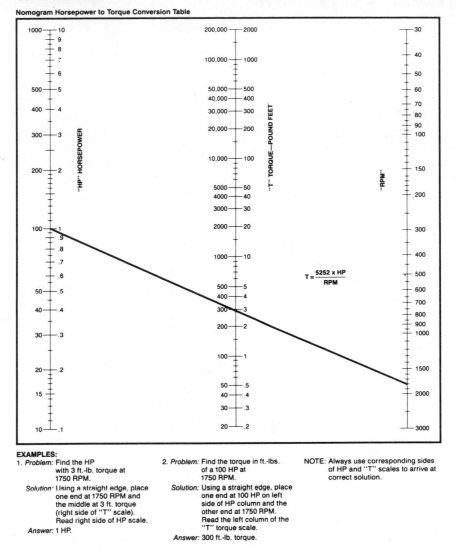

EXAMPLES:

1. *Problem:* Find the HP with 3 ft.-lb. torque at 1750 RPM.

   Solution: Using a straight edge, place one end at 1750 RPM and the middle at 3 ft. torque (right side of "T" scale). Read right side of HP scale.

   Answer: 1 HP.

2. *Problem:* Find the torque in ft.-lbs. of a 100 HP at 1750 RPM.

   Solution: Using a straight edge, place one end at 100 HP on left side of HP column and the other end at 1750 RPM. Read the left column of the "T" torque scale.

   Answer: 300 ft.-lb. torque.

NOTE: Always use corresponding sides of HP and "T" scales to arrive at correct solution.

Figure 25–39   Nomograph. *Courtesy Lovejoy, Inc., Downers Grove, Illinois.*

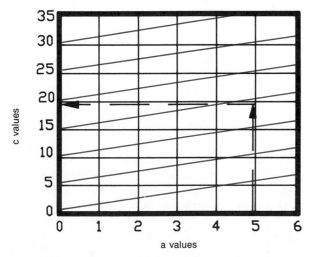

Figure 25–40   Concurrency chart where a + b = c.

25–40.) The chart is used to solve for an unknown value by selecting any two known values. Use Figure 25–40 to solve the formula a + b = c when a = 5 and b = 15. Find 5 on the X axis. Then project directly up to the curve where b = 15. Project to the left and locate the c value at the Y axis. The answer should be 20. This is an extremely simplified version of the concurrency chart; however, the technique is the same for more complex versions.

Alignment Charts

Alignment charts are designed to graphically solve mathematical equation values using three or more scaled lines. The known values are aligned on a combination of scales and the solution is found when the line is extended or aligned with the additional scale. The alignment chart scales are typically vertical lines; however,

CADD Applications

GRAPHICS

The examples displayed in this chapter have been designed and plotted or printed by computer. CADD software is available that transforms numerical data into presentation-quality graphics using any desired format; for example, line, bar, pie, or pictorial charts. Charts may be designed with any grid, scales, line type, shading technique, or text size and style. Programs are available that allow the designer to use a variety of predetermined symbols for two- or three-dimensional applications. Using a CADD system of this type to design charts is the same as having a portfolio of artwork that can be automatically used in any desired configuration for unlimited creativity. With CADD, scales may be automatically reduced, enlarged, or otherwise changed to accommodate any format. (See Figure 25–41.)

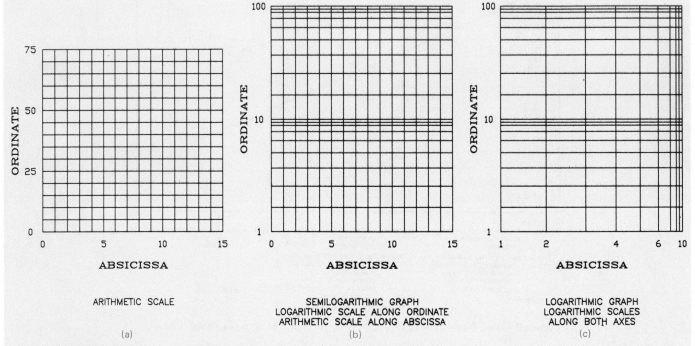

CADD GRAPH PROGRAMS ARE AVAILABLE THAT ALLOW YOU TO AUTOMATICALLY CHANGE THE SCALE OF EITHER AXIS, ENLARGE, REDUCE, OR DESIGN ANY TYPE OF GRAPH OR CHART WITH TEXT, SYMBOLS, AND PICTORIALS.

Figure 25–41 Computer-generated charts automatically change the scales by changing the ordinate and abscissa requirements: **(a)** arithmetic scale; **(b)** semilogarithmic scale; **(c)** logarithmic scale.

other combinations of scale arrangements are used. To demonstrate the alignment chart principle, the equation a + b = c is used again as shown in Figure 25–42. Notice that if a line is drawn between the 5 on the "a" scale and the 10 on the "b" scale, the unknown value of the "c" scale is 15. Therefore, 5 + 10 = 15.

Scales may have equal graduations. These graduations are called *uniform scales*. When the scales resulting from the solution of the equation variables are not uniform, they are called functional scales.

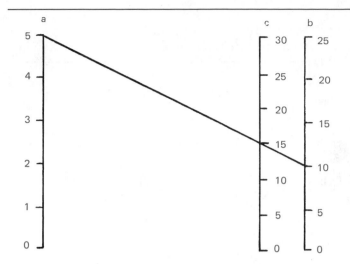

Figure 25–42   Alignment chart where a + b = c.

▼ TRILINEAR CHARTS

Trilinear means consisting of three lines. A *trilinear chart* is designed in the shape of an equilateral triangle. These charts are used to show the interrelationship between three variables on a two-dimensional diagram. The amounts of the three variables, usually presented as percentages, can be represented as shown in Figure 25–43. The corners of the triangle labeled A, B, and C correspond to 100 percent values of A, B, and C. The side of the triangle opposite the corner labeled A represents the absence of A values. Therefore, the horizontal lines across the triangle show increasing percentages of A from 0 percent at the base to 100 percent at the A vertex. In a similar interpretation the percentages of B and C, respectively, are given by the distances opposite the two sides of the vertices at either B or C. From the three scales of the triangle the value corresponding to any point can be read and the total of the three values at any given point is always 100 percent. This is possible because of the

geometric result that the sum of the three perpendicular distances from any point to the three sides of the triangle is equal to the height of the triangle.

▼ FLOWCHARTS

Flowcharts are used to show organizational structure, steps in a series of operations, the flow of something through a system, or the progression of materials through manufacturing processes. Organizational charts show the relationship of the different levels of an organization as shown in Figure 25–44.

Production control or *process* charts are used for planning and coordinating a series of activities such as a marketing planning cycle as shown in Figure 25–45.

▼ DISTRIBUTION CHARTS

Distribution charts are used to display data based on geographical region. The data may be defined by outlining

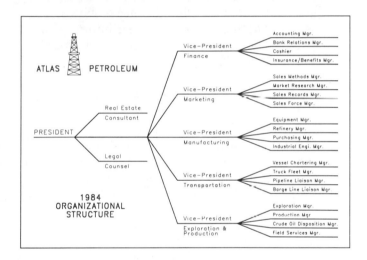

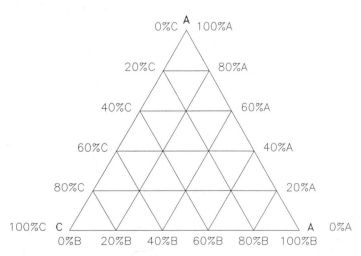

Figure 25–43   Trilinear chart scales.

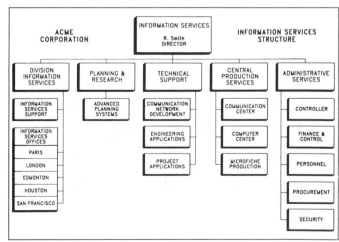

Figure 25–44   Organizational charts. *Courtesy Computer Support Corporation.*

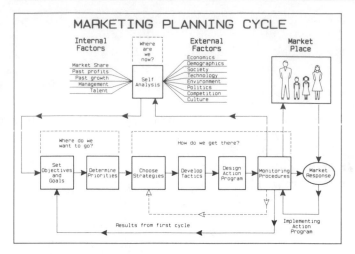

Figure 25–45 Production control or process chart. *Courtesy Computer Support Corporation.*

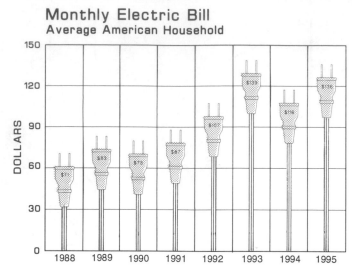

Figure 25–47 Pictorial bar chart. *Courtesy Computer Support Corporation.*

geographic areas. These charts are often in the form of maps to provide locational information. For example, the average frost depth in the United States is shown in Figure 25–46a. Another use of the distribution chart is placing dots or other symbols to define specific locations as shown in Figure 25–46b.

▼ PICTORIAL CHARTS

Any chart type may be designed in a pictorial manner. *Pictorial charts* are used for advertising promotions where the pictorial representation is more important than the data presented. Pictorial charts are used to enhance the meaning of the data using two- or three-

dimensional graphics, photography, or color. Any pictorial representation may be used to help the reader visualize the intent of the chart. Pictorial bar charts may be drawn with the bars shown three dimensionally or with pictorial representations of the product as shown in Figure 25–47. Pie charts may be easily shown pictorially by drawing the pie isometrically as shown in Figure 25–48. The designer may also prepare creative pictorial line charts as shown in Figure 25–49. The chart design, format, and layout may serve any function, using a variety of symbols to pictorially depict recognizable features as shown in Figure 25–50.

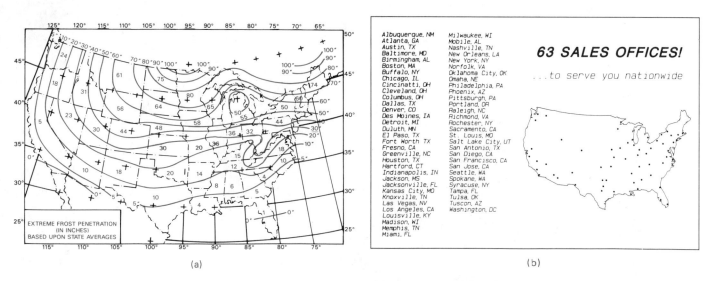

Figure 25–46 Distribution charts. **(a)** Average frost depth in the United States. **(b)** National sales offices. *Part b courtesy Computer Support Corporation.*

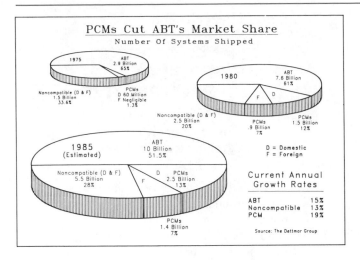

Figure 25–48 Pictorial pie chart. *Courtesy Computer Support Corporation.*

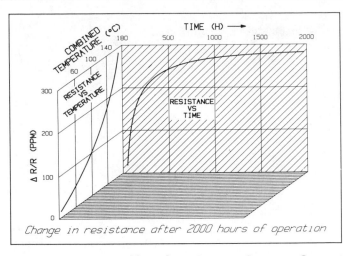

Figure 25–49 Pictorial line chart. *Courtesy Computer Support Corporation.*

Coal Chains and Markets

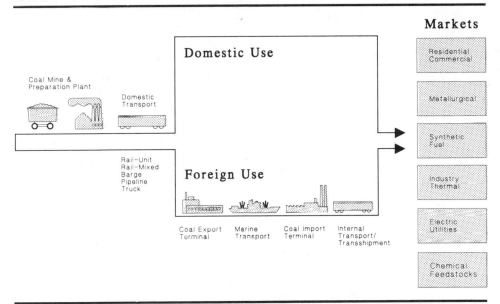

Figure 25–50 Pictorial flowcharts using symbols. *Courtesy Computer Support Corporation.*

CADD Applications

CADD PROGRAMS FOR CHART DESIGN

There are a variety of chart and graph software packages that work within your regular CADD program. For example, Figure 25–51 shows a bar chart application being used inside AutoCAD. Creating schedules for presentations or proposals is easy. All you have to do is add activities and their corresponding dates for automatic display in the drawing. Additionally, the information is stored in a database for editing as needed. Other advantages include adding a title block and inserting an existing drawing to enhance the presentation of your chart. Figure 25–52 shows a bar chart created in this manner. ***(Continued)***

CADD Applications (Continued)

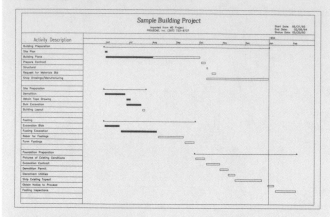

Figure 25–51 A bar chart application being used inside AutoCAD. *Courtesy PROJECAD, Inc.*

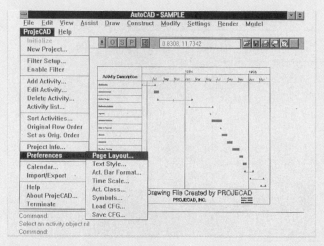

Figure 25–52 A bar chart created using CADD software. *Courtesy PROJECAD, Inc.*

PROFESSIONAL PERSPECTIVE

Following are some professional tips that can make your charts and graphs accurate for the intended audience.

○ Make your graphs and charts clean and simple, avoiding unnecessary detail and information.
○ Do not misrepresent the facts.
○ Keep all sets of data proportional.
○ Avoid distorting the data.
○ Study the background of the data to insure you understand the intended representation.
○ Know the audience and know what they want to have the graph or chart represent. If you display one set of information and they want something else, then your work is useless.
○ Be sure the numbers you are representing are accurate and appropriate.

You should consider taking some classes in statistics, or have someone who understands statistics available to evaluate your designs to insure the data is represented accurately. If you are unprepared or if the data is misrepresented, someone will question the quality of the presentation.

MATH APPLICATION

EQUATION OF A STRAIGHT LINE

CADD computers and all other computers are designed to follow the established math rules when drawing a graph. For a straight line, the equation used is **y = mx + b**, where m is a number for the slope and b is the point where the line crosses the y-axis (ordinate).

Problem: Find the slope of the rectilinear graph shown in Figure 25–53 and write its mathematical formula.

Solution: Mathematical slope is the same as engineering slope; both are rise divided by run. Lines that slant up and to the right have positive slope, those that slant down and to the right have negative slope, and a horizontal line has zero for a slope. From the scaling, the line rises 2 units for every 10 units to the right; so its slope must be 2 ÷ 10, or **.2**. Also, the y-intercept is 2. Then for this graph, m = .2, b = 2; so the equation is **y = .2x + 2.**

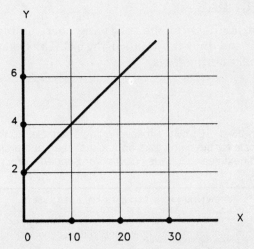

Figure 25–53 Rectilinear graph requiring an equation.

CHAPTER
25 ENGINEERING CHARTS AND GRAPHS TEST

DIRECTIONS

Answer the questions with short complete statements or drawings as needed.

QUESTIONS

1. Define *statistics*.
2. Name and define the two fundamental applications for charts.
3. Identify at least four questions that should be answered before a chart design is started.
4. What is another name for the X axis?
5. What is another name for the Y axis?
6. How does the range of data affect chart scales?
7. Logarithmic and semilogarithmic scales are also referred to as _____ scales.
8. How are grid lines layed out for most applications?
9. What is the general guideline used for chart labeling?
10. When should curve captions be used?
11. Rectilinear charts are also referred to as _____ charts.
12. Surface charts are also known as _____ charts.
13. What is the best design consideration for surface chart use?
14. Column charts are commonly referred to as _____ charts.
15. What are subdivided column charts used to show?
16. What are column charts that show the relationship between the distribution of components where the totals are always equal values called?
17. Define *range bar chart*.
18. What are pie charts used for?
19. How many degrees of a pie chart would a value equal to 36 percent occupy? Show the formula and calculations.
20. What is the purpose of a polar chart?
21. Define *nomograph*.
22. What are the two types of nomographs?
23. Identify at least three applications for flowcharts.
24. For what general purpose are pictorial charts used?

CHAPTER

25 ENGINEERING CHARTS AND GRAPHS PROBLEMS

DIRECTIONS

Given the data for the following chart problems, design and draw the charts using a CADD system or manual drafting tools.

Problem 25–1 Draw the following line chart. Set up the horizontal scale for the nine years ending last year, and project this year and next year. *Courtesy Computer Support Corporation.*

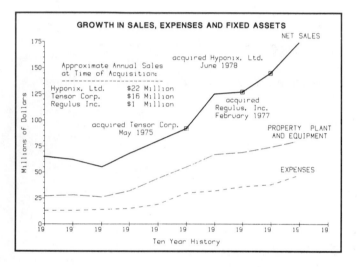

Problem 25–2 Redesign similar to the format shown in Figure 25–30, and draw the following column chart. *Courtesy Computer Support Corporation.*

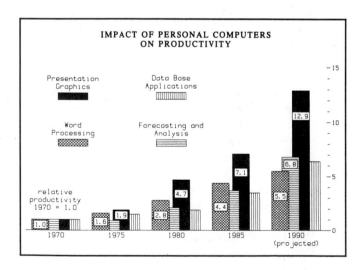

Problem 25–3 Design a rectilinear chart that represents the increase in recommended current amperage with the increase in round solid copper wire gauge size.

AGW GAUGE SIZE	RECOMMENDED MAXIMUM CURRENT IN AMPERES
28	.4
27	
26	.7
25	
24	1.1
23	
22	1.8
21	
20	3
19	
18	4.5
17	
16	7
15	
14	12
13	
12	18
11	
10	30

Problem 25–4 Design a comparative rectilinear chart showing the five-speed transmission diagram for the Porsche 911 Carrera (establish straight line curve average for each year).

ENGINE SPEED (RPM)	SPEED (MPH)				
	1ST GEAR	2ND	3RD	4TH	5TH
1000	10	17	25	30	40
2000	20	34	48	62	78
3000	38	50	74	92	116
4000	38	67	97	122	155
5000	47	84	121	152	192
6000	56	100	145	182	230
7000	66	105	168	212	268

Tire size: 185/70-VR15-215/60 VR15

Note: This diagram shows guiding values, based on a medium effective rolling radius. Slight deviations due to tire tolerance, variations in the rolling radius, wear and skidding on the wheels have not been taken into account.

Problem 25–5 Design a chart similar to Figure 25–19 showing feature dimensions of a part obtained at inspection intervals based on the following information:

1. Average of samples taken at given intervals (x): 2.6248, 2.6252, 2.6250, 2.6250, 2.6244, 2.6248, 2.6251.
2. Upper Control Limit (UCL) = 2.6253.
3. Lower Control Limit (LCL) = 2.6243.
4. Determine the average of averages ($\bar{\bar{x}}$).

Problem 25–6 Design a surface semilogarithmic chart from the following information: Ten years of community college graduates with technical Associate of Science (AS) degrees vs. transfer Associate of Arts (AA) degrees.

YEAR	AS	AA	YEAR	AS	AA
1984-85	266	37	1990-91	232	24
1985-86	214	24	1991-92	233	14
1986-87	225	26	1982-93	242	18
1987-88	225	34	1993-94	263	30
1988-89	269	32	1994-95	270	36
1989-90	265	33			

Problem 25–7 Design a column chart based on the following information: Where top companies reside; more than half of the USA's Fortune 500 companies are headquartered in six states. *Information courtesy USA Today.*

STATE	NUMBER OF FORTUNE 500 HEADQUARTERS
NY	75
IL	49
CA	38
OH	37
CT	35
PA	34

Problem 25–8 Design a column chart that shows net sales with values on the left vertical scale and a step curve to show income per share with values on the right vertical scale. The horizontal scale shows the range of years from 1986 to 1995 for Dial Industries, ten-year summary of operations and financial review.

YEAR	1986	1987	1988	1989	1990
Net Sales in Millions	1400	1500	1600	1825	2150
Income/Share $	1.39	1.56	1.85	1.87	2.06

YEAR	1991	1992	1993	1994	1995
Net Sales in Millions	2275	2500	2900	3225	3475
Income/Share $	2.00	2.36	2.61	1.55	2.62

Problem 25–9 Design an expense/revenue over-under column chart for Precision Manufacturing, Ltd. by country ($ in millions).

	EXPENSE	REVENUE		EXPENSE	REVENUE
United States	343.2	651.7	France	239.4	302.7
England	264.8	466.9	Taiwan	201.6	224.3
Canada	301.6	485.2	Hong Kong	214.7	102.5
Mexico	241.8	413.3	Brazil	225.0	88.2
Australia	213.7	339.9			

Problem 25–10 Design a range bar chart that represents values for surface roughness in microinches and micrometers produced by common processing methods.

PROCESS	RANGE		AVERAGE	
	MICROINCH	MICROMETER	MICROINCH	MICROMETER
Flame cutting	2000-250	50-6.3	1000-500	25-12.5
Snagging	2000-125	50-3.2	1000-250	25-6.3
Sawing	2000-32	50-0.80	1000-63	25-1.6
Planing, shaping	1000-16	25-0.40	500-63	12.5-1.6
Drilling, chemical milling, elect. discharge machining	500-32	12.5-0.80	250-63	6.3-1.6
Milling	1000-8	25-0.20	250-32	6.3-0.80
Broaching, reaming	250-16	6.3-0.40	125-32	3.2-0.80
Electron beam, laser	250-8	6.3-0.40	250-32	6.3-0.80
Electrochemical	500-2	12.5-0.05	125-8	3.2-0.20
Boring, turning	1000-1	25-0.025	250-16	6.3-0.40
Barrel finishing	125-2	3.2-0.05	32-8	0.80-0.20
Electrolytic grinding	32-4	0.80-0.10	20-8	0.60-0.20
Roller burnishing	32-4	0.80-0.10	16-8	0.40-0.20
Grinding	250-1	3.2-0.025	63-4	1.6-0.10
Honing	63-1	1.6-0.025	32-4	0.80-0.10
Electropolish	63-0.5	1.6-0.012	32-4	0.80-0.10
Polishing	32-0.5	0.80-0.012	16-4	0.40-0.10
Lapping	32-0.05	0.80-0.012	16-2	0.40-0.05
Superfinish	32-0.05	0.80-0.012	8-1	0.20-0.025

Note: The ranges indicated are typical of the process; higher or lower values may be obtained under special conditions.

Problem 25–11 Design a pie chart that represents the agricultural sales for Washington County during 1995.

Nursery crops = 44%
Vegetable crops = 26%
Small fruit and berries = 11%
Other crops = 8%
Miscellaneous animals = 6%
Greenhouse crops = 5%

Problem 25–12 Design a pie chart that represents Product Revenue by category for 1995. *Data courtesy Computer Support Corporation.*

Automated process equipment = 28%
Marine products = 20%
Energy services = 17%
Sports products = 13%
Electronic controls = 9%
Specialty products = 6%
Wheel goods = 4%

Problem 25–13 Design a nomograph that may be used to determine grades for the Engineering Drafting I class drafting projects and term tests based on the following assumptions. Total points for drafting projects equal 500 and total points for tests equal 300. Grades are based on the following percentages: A = 100–91%; B = 90–81%; C = 80–71%; D = 70–61%; F = below 60%.

Problem 25–14 Design a flowchart for the chain of command of your school or company.

Problem 25–15   Design a pictorial chart that provides the following information: 25 years in space from 1960 to 1985. More than 200 people—mostly from the USA—have flown in space since Soviet cosmonaut Yuri Gagarin became the first to do so. Tally of space travelers: (Tip: Rockets leaving the earth might be considered.)

    USA citizens = 124
    Soviet citizens = 62
    Foreigners on USA flights = 8
    Foreigners on Soviet flights = 11

Source: Congressional Research Service.

Problem 25–16   Design and draw an alignment chart for the formula a + b = c.

Problem 25–17   Design and draw a distribution chart showing the cities in the United States with populations over 500,000. Research is required.

Problem 25–18   Review your family electric bills over the past year and design and draw a pictorial chart similar to the one shown in Figure 25–47.

Problem 25–19   Research the monthly average rainfall in your state over the past year and design and draw a pictorial chart.

MATH PROBLEMS

1. What are the slope and y-intercept of the line with this equation: $y = 2x - 7$?

2. Which of the following pairs of equations will graph as parallel lines? (Hint: Parallel lines have the same slope.)

    a. $y = 2x - 7$ and $y = 2x + 6$
    b. $y = 5x + 2$ and $y = 3x + 2$
    c. $y = -2x - 3$ and $y = -2x + 4$
    d. $y = 17x + 5$ and $y = 17x - 5$

3. Which of the pairs of equations from Problem 2 cross the y-axis at the same point?

4. Find the slope and equation of the graph shown.

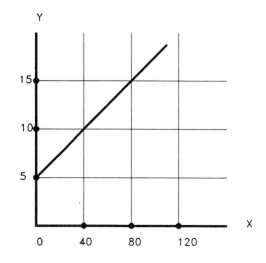

5. Find the equation of the following graph.

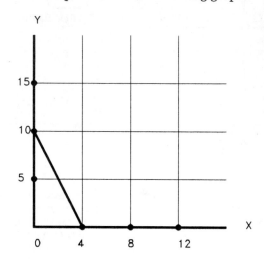

6. Which of the following equations would graph as a horizontal line?

    a. $y = 6x$
    b. $y = 7$
    c. $y = 2x + 5$
    d. $y = -4x - 3$

7. Which of the equations from Problem 6 pass through the origin (0,0)?

▼ APPENDIX A Abbreviations

The following are standard abbreviations commonly used on working drawings:

Accessory	ACCESS.
Accumulate	ACCUM
Adaption	ADAPT.
Addendum	ADD.
Addition	ADD.
Alteration	ALT
Alternate	ALT
Alternating Current	AC
Altitude	ALT
Aluminum	AL
American Iron and Steel Institute	AISI
American Society for Testing Materials	ASTM
American Society of Mechanical Engineers	ASME
American Wire Gage	AWG
Ampere	AMP
Approved	APPD
Approximate	APPROX
Assembly	ASM
Attach	ATT
Authorize	AUTH
Auxiliary	AUX
Average	AVG
Balance	BAL
Battery	BAT.
Bearing	BRG
Bill of material	B/M
Bracket	BRKT
Brass	BRS
British thermal unit	BTU
Bronze	BRZ
Brown and Sharpe	B&S
Bushing	BUSH.
Cadmium	CAD.
Capacity	CAP.
Carbon steel	CS
Cast iron	CI
Cast steel	CS
Casting	CST
Center	CTR
Center of gravity	CG
Centerline	(℄) or CL
Centigrade	C
Centimeter	CM
Chamfer	CHAM
Change	CHG
Check	CHK
Chief	CH
Chromium	CHR
Circular pitch	CP
Circumference	CIRC
Clockwise	CW
Cold drawn steel	CDS
Cold rolled steel	CRS
Company	CO
Composition	COMP
Compression, Compressor	COMP

Concentric	CONC
Condition	COND
Conductor	COND
Conduit	CDT
Connector	CONN
Continue	CONT
Control	CONT
Copper	CU
Corporation	CORP
Correspond	CORRES
Corrosion resistant steel	CRES
Counterbalance	CBAL
Counterbore	CBORE
Counterclockwise	CCW
Counterdrill	CDRILL
Countersink	CSK
Counterweight	CTWT
Cubic	CU
Cubic centimeter	CC
Cubic feet per minute	CFM
Cubic foot	CU FT
Cubic inch	CU IN.
Cycle	CY
Dedendum	DED
Degree	(°) or DEG
Department	DEPT
Detail	DET
Develop	DEV
Deviation	DEV
Diagonal	DIAG
Diameter	DIA, ∅
Diametral pitch	DP
Dimension	DIM.
Direct current	DC
Distance	DIST
Division	DIV
Drawing	DWG
Each	EA
Eccentric	ECC
Effective	EFF
Electric	ELEC
Elevation	ELEV
Engineer	ENGR
Engineering	ENGRG
Equipment	EQUIP.
Equivalent	EQUIV
Estimate	EST
Etcetera	ETC
Extension	EXT
External	EXT
Extrusion	EXTR
Fahrenheit	F
Feet per minute	FPM
Figure	FIG.
Fillister head	FIL HD

Finish	FIN.
Fitting	FTG
Flat	FL
Flat head	FHD
Flexible	FLEX.
Foot	(') or FT
Foot pounds	FT LB
Forging	FORG
Forward	FWD
Front	FRT
Gage (Gauge)	GA
Gallon	GAL
Galvanize	GALV
Gasket	GSKT
Generator	GEN
Grain	GRN
Gravity	G
Grind	GRD
Harden	HDN
Hardware	HDW
Head	HD
Heat treat	HT TR
Height	HGT
Hexagon	HEX
Horizontal	HORIZ
Hot rolled steel	HRS
Hour	HR
Hydraulic	HYD
Illustration	ILLUS
Inch	(") or IN.
Inch ounce	IN. OZ
Inch pounds	IN. LB
Inclusive	INCL
Information	INFO
Inside diameter	ID
Inside radius	IR
Inspection	INSP
Installation	INSTL
Instrument	INST
Insulation	INSL
Interchangeable	INTCHG
Intermediate	INTER.
Internal	INT
Joint	JT
Kilometer	KM
Knock out	KO
Laboratory	LAB
Left hand	LH
Length	LG
Lock washer	LWASH
Longitude, Longitudinal	LONG.
Lower	LWR
Lubricate	LUB
Machine	MACH
Magnesium	MG

Maintenance	MAINT
Malleable	MAL
Manufacture	MFR
Material	MATL
Maximum	MAX
Maximum material condition	MMC
Mechanical	MECH
Medium	MED
Memorandum	MEMO
Mercury	HG
Mile	MI
Miles per hour	MPH
Millimeter	MM
Minimum	MIN
Minute	(') or MIN
Miscellaneous	MISC
Modify	MOD
Molding	MLDG
Mounting	MTG
National	NATL
National Electrical Mfg. Association	NEMA
National Machine Tool Builders Association	NMTBA
Negative	(–) or NEG
Nickel	NI
No drawing	ND
Nominal	NOM
Nonstandard	NONSTD
Number	NO.
Obsolete	OBS
Of true position	OTP
On Center	OC
Operate	OPER
Opposite	OPP
Optional	OPT
Original	ORIG
Ounce	OZ
Outside diameter	OD
Outside radius	OR.
Oval head	OV HD
Overall	OA
Oxygen	OXY
Package, Packing	PKG
Page	P
Part	PT
Parting line	PL
Patent	PAT.
Pattern	PATT
Perpendicular	PERP
Pint	PT
Pitch	P
Pitch circle	PC
Pitch diameter	PD
Plan view	PV
Point on line	POL
Position	POSN
Positive	(+) or POS
Pound	LB
Pounds per square inch	PSI
Preliminary	PRELIM

Pressure	PRESS.
Process	PROC
Product, Production	PROD.
Quality	QUAL
Quantity	QTY
Quart	QT
Quarter	QTR
Rear view	RV
Rectangular	RECT
Reduction	RED.
Reference	REF
Regardless of feature size	RFS
Regular	REG
Reinforce	REINF
Remove	REM
Require	REQ
Required	REQD
Resistor	RES
Reverse	REV
Revision	REV
Revolution	REV
Revolutions per minute	RPM
Right hand	RH
Root diameter	RD
Round	RD
Round head	RD HD
Rubber	RUB.
Screw	SCR
Screw threads	
American National coarse	NC
American National fine	NF
American National extra fine	NEF
American National 8 pitch	8N
American National 12 pitch	12N
American National 16 pitch	16N
American Standard taper pipe	NPT
American Standard straight pipe coupling	NPSC
American Standard taper	NPTF
Unified screw thread coarse	UNC
Unified screw thread fine	UNF
Unified screw thread extra fine	UNEF
Unified screw thread 8 thread	8UN
Unified screw thread 12 thread	12UN
Unified screw thread 16 thread	16UN
Unified screw thread special	UNS
Second	(") or SEC
Section	SECT.
Serial, Series	SER
Sheet	SH
Side view	SV
Sketch	SK
Society of Automotive Engineers	SAE
Special	SPL
Specification	SPEC
Specific gravity	SP GR

Spherical	SPHER
Spot face	SF
Spring	SPR
Square	SQ
Standard	STD
Steel	STL
Support	SUPT
Switch	SW
Symbol	SYM
Symmetrical	SYM
Synthetic	SYN
Tangent	TAN.
Technical	TECH
Teeth	T
Temperature	TEMP
Tensile strength	TS
Terminal	TERM.
Theoretical	THEO
Thickness	THK
Thread	THD
Through	THRU
Tolerance	TOL
Tracer	TCR
Trademark	TM
Transformer	TRANS
Transmission	TRANS
Transverse	TRANSV
True length	TL
True position	TP
True view	TV
Turnbuckle	TRNBKL
Typical	TYP
Ultimate	ULT
United States	US
United States of America Standards Institute	USASI
United States gage	USG
Universal	UNIV
Unless otherwise specified	UOS
Upper	UPR
Vacuum	VAC
Velocity	V
Versus	VS
Vertical	VERT
Volt	V
Volume	VOL
Water line	WL
Watt	W
Weight	WT
Wood	WD
Wood screw	WD SCR
Wrought iron	WI
Yard	YD
Year	YR

▼ APPENDIX B   Tables

TABLE 1   INCHES TO MILLIMETERS

in.	mm	in.	mm	in.	mm	in.	mm
1	25.4	26	660.4	51	1295.4	76	1930.4
2	50.8	27	685.8	52	1320.8	77	1955.8
3	76.2	28	711.2	53	1346.2	78	1981.2
4	101.6	29	736.6	54	1371.6	79	2006.6
5	127.0	30	762.0	55	1397.0	80	2032.0
6	152.4	31	787.4	56	1422.4	81	2057.4
7	177.8	32	812.8	57	1447.8	82	2082.8
8	203.2	33	838.2	58	1473.2	83	2108.2
9	228.6	34	863.6	59	1498.6	84	2133.6
10	254.0	35	889.0	60	1524.0	85	2159.0
11	279.4	36	914.4	61	1549.4	86	2184.4
12	304.8	37	939.8	62	1574.8	87	2209.8
13	330.2	38	965.2	63	1600.2	88	2235.2
14	355.6	39	990.6	64	1625.6	89	2260.6
15	381.0	40	1016.0	65	1651.0	90	2286.0
16	406.4	41	1041.4	66	1676.4	91	2311.4
17	431.8	42	1066.8	67	1701.8	92	2336.8
18	457.2	43	1092.2	68	1727.2	93	2362.2
19	482.6	44	1117.6	69	1752.6	94	2387.6
20	508.0	45	1143.0	70	1778.0	95	2413.0
21	533.4	46	1168.4	71	1803.4	96	2438.4
22	558.8	47	1193.8	72	1828.8	97	2463.8
23	584.2	48	1219.2	73	1854.2	98	2489.2
24	609.6	49	1244.6	74	1879.6	99	2514.6
25	635.0	50	1270.0	75	1905.0	100	2540.0

The above table is exact on the basis: 1 in. = 25.4 mm

TABLE 2   MILLIMETERS TO INCHES

mm	in.	mm	in.	mm	in.	mm	in.
1	0.039370	26	1.023622	51	2.007874	76	2.992126
2	0.078740	27	1.062992	52	2.047244	77	3.031496
3	0.118110	28	1.102362	53	2.086614	78	3.070866
4	0.157480	29	1.141732	54	2.125984	79	3.110236
5	0.196850	30	1.181102	55	2.165354	80	3.149606
6	0.236220	31	1.220472	56	2.204724	81	3.188976
7	0.275591	32	1.259843	57	2.244094	82	3.228346
8	0.314961	33	1.299213	58	2.283465	83	3.267717
9	0.354331	34	1.338583	59	2.322835	84	3.307087
10	0.393701	35	1.377953	60	2.362205	85	3.346457
11	0.433071	36	1.417323	61	2.401575	86	3.385827
12	0.472441	37	1.456693	62	2.440945	87	3.425197
13	0.511811	38	1.496063	63	2.480315	88	3.464567
14	0.551181	39	1.535433	64	2.519685	89	3.503937
15	0.590551	40	1.574803	65	2.559055	90	3.543307
16	0.629921	41	1.614173	66	2.598425	91	3.582677
17	0.669291	42	1.653543	67	2.637795	92	3.622047
18	0.708661	43	1.692913	68	2.677165	93	3.661417
19	0.748031	44	1.732283	69	2.716535	94	3.700787
20	0.787402	45	1.771654	70	2.755906	95	3.740157
21	0.826772	46	1.811024	71	2.795276	96	3.779528
22	0.866142	47	1.850394	72	2.834646	97	3.818898
23	0.905512	48	1.889764	73	2.874016	98	3.858268
24	0.944882	49	1.929134	74	2.913386	99	3.897638
25	0.984252	50	1.968504	75	2.952756	100	3.937008

The above table is approximate on the basis: 1 in. = 25.4 mm, 1/25.4 = 0.039370078740+

TABLE 3   INCH/METRIC—EQUIVALENTS

Fraction	Decimal Equivalent Customary (in.)	Metric (mm)	Fraction	Decimal Equivalent Customary (in.)	Metric (mm)
1/64 — .015625		0.3969	33/64 — .515625		13.0969
1/32 — .03125		0.7938	17/32 — .53125		13.4938
3/64 — .046875		1.1906	35/64 — .546875		13.8906
1/16 — .0625		1.5875	9/16 — .5625		14.2875
5/64 — .078125		1.9844	37/64 — .578125		14.6844
3/32 — .09375		2.3813	19/32 — .59375		15.0813
7/64 — .109375		2.7781	39/64 — .609375		15.4781
1/8 — .1250		3.1750	5/8 — .6250		15.8750
9/64 — .140625		3.5719	41/64 — .640625		16.2719
5/32 — .15625		3.9688	21/32 — .65625		16.6688
11/64 — .171875		4.3656	43/64 — .671875		17.0656
3/16 — .1875		4.7625	11/16 — .6875		17.4625
13/64 — .203125		5.1594	45/64 — .703125		17.8594
7/32 — .21875		5.5563	23/32 — .71875		18.2563
15/64 — .234375		5.9531	47/64 — .734375		18.6531
1/4 — .250		6.3500	3/4 — .750		19.0500
17/64 — .265625		6.7469	49/64 — .765625		19.4469
9/32 — .28125		7.1438	25/32 — .78125		19.8438
19/64 — .296875		7.5406	51/64 — .796875		20.2406
5/16 — .3125		7.9375	13/16 — .8125		20.6375
21/64 — .328125		8.3384	53/64 — .828125		21.0344
11/32 — .34375		8.7313	27/32 — .84375		21.4313
23/64 — .359375		9.1281	55/64 — .859375		21.8281
3/8 — .3750		9.5250	7/8 — .8750		22.2250
25/64 — .390625		9.9219	57/64 — .890625		22.6219
13/32 — .40625		10.3188	29/32 — .90625		23.0188
27/64 — .421875		10.7156	59/64 — .921875		23.4156
7/16 — .4375		11.1125	15/16 — .9375		23.8125
29/64 — .453125		11.5094	61/64 — .953125		24.2094
15/32 — .46875		11.9063	31/32 — .96875		24.6063
31/64 — .484375		12.3031	63/64 — .984375		25.0031
1/2 — .500		12.7000	1 — 1.000		25.4000

TABLE 4   INCH/METRIC—CONVERSION

Measures of Length
1 millimeter (mm) = 0.03937 inch
1 centimeter (cm) = 0.39370 inch
1 meter (m) = 39.37008 inches
= 3.2808 feet
= 1.0936 yards
1 kilometer (km) = 0.6214 mile
1 inch = 25.4 millimeters (mm)
= 2.54 centimeters (cm)
1 foot = 304.8 millimeters (mm)
= 0.3048 meter (m)
1 yard = 0.9144 meter (m)
1 mile = 1.609 kilometers (km)

Measures of Area
1 square millimeter = 0.00155 square inch
1 square centimeter = 0.155 square inch
1 square meter = 10.764 square feet
= 1.196 square yards
1 square kilometer = 0.3861 square mile
1 square inch = 645.2 square millimeters
= 6.452 square centimeters
1 square foot = 929 square centimeters
= 0.0929 square meter
1 square yard = 0.836 square meter
1 square mile = 2.5899 square kilometers

Measures of Capacity (Dry)
1 cubic centimeter (cm^3) = 0.061 cubic inch
1 liter = 0.0353 cubic foot
= 61.023 cubic inches
1 cubic meter (m^3) = 35.315 cubic feet
= 1.308 cubic yards
1 cubic inch = 16.38706 cubic centimeters (cm^3)
1 cubic foot = 0.02832 cubic meter (m^3)
= 28.317 liters
1 cubic yard = 0.7646 cubic meter (m^3)

Measures of Capacity (Liquid)
1 liter = 1.0567 U.S. quarts
= 0.2642 U.S. gallon
= 0.2200 Imperial gallon
1 cubic meter (m^3) = 264.2 U.S. gallons
= 219.969 Imperial gallons
1 U.S. quart = 0.946 liter
1 Imperial quart = 1.136 liters
1 U.S. gallon = 3.785 liters
1 Imperial gallon = 4.546 liters

Measures of Weight
1 gram (g) = 15.432 grains
= 0.03215 ounce troy
= 0.03527 ounce avoirdupois
1 kilogram (kg) = 35.274 ounces avoirdupois
= 2.2046 pounds
1000 kilograms (kg) = 1 metric ton (t)
= 1.1023 tons of 2000 pounds
= 0.9842 ton of 2240 pounds
1 ounce avoirdupois = 28.35 grams (g)
1 ounce troy = 31.103 grams (g)
1 pound = 453.6 grams
= 0.4536 kilogram (kg)
1 ton of 2240 pounds = 1016 kilograms (kg)
= 1.016 metric tons
1 grain = 0.0648 gram (g)
1 metric ton = 0.9842 ton of 2240 pounds
= 2204.6 pounds

TABLE 5   RULES RELATIVE TO THE CIRCLE

To Find Circumference—			
Multiply diameter by	3.1416................Or divide diameter by	0.3183
To Find Diameter—			
Multiply circumference by	0.3183................Or divide circumference by	3.1416
To Find Radius—			
Multiply circumference by	0.15915................Or divide circumference by	6.28318
To Find Side of an Inscribed Square—			
Multiply diameter by	0.7071		
Or multiply circumference by 0.2251................Or divide circumference by	4.4428
To Find Side of an Equal Square—			
Multiply diameter by	0.8862................Or divide diameter by	1.1284
Or multiply circumference by 0.2821................Or divide circumference by	3.545

Square—
A side multiplied by   1.4142   equals diameter of its circumscribing circle.
A side multiplied by   4.443   equals circumference of its circumscribing circle.
A side multiplied by   1.128   equals diameter of an equal circle.
A side multiplied by   3.547   equals circumference of an equal circle.

To Find the Area of a Circle—
Multiply circumference by one-quarter of the diameter.
Or multiply the square of diameter by   0.7854
Or multiply the square of circumference by   .07958
Or multiply the square of $1/2$ diameter by   3.1416

To Find the Surface of a Sphere or Globe—
Multiply the diameter by the circumference.
Or multiply the square of diameter by   3.1416
Or multiply four times the square of radius by 3.1416

TABLE 6   STANDARD LINE TYPES

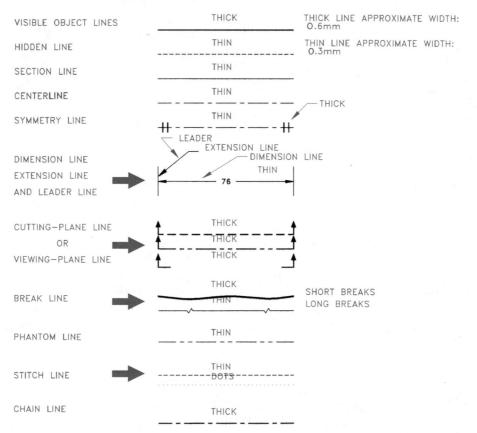

VISIBLE OBJECT LINES — THICK — THICK LINE APPROXIMATE WIDTH: 0.6mm

HIDDEN LINE — THIN — THIN LINE APPROXIMATE WIDTH: 0.3mm

SECTION LINE — THIN

CENTERLINE — THIN — THICK

SYMMETRY LINE — THIN

DIMENSION LINE
EXTENSION LINE
AND LEADER LINE — LEADER — EXTENSION LINE — DIMENSION LINE — THIN — 76

CUTTING–PLANE LINE
OR
VIEWING–PLANE LINE — THICK

BREAK LINE — THICK — THIN — SHORT BREAKS — LONG BREAKS

PHANTOM LINE — THIN

STITCH LINE — THIN — DOTS

CHAIN LINE — THICK

TABLE 7   ASTM AND SAE GRADE MARKINGS FOR STEEL BOLTS AND SCREWS

Grade Marking	Specification	Material
NO MARK	SAE—Grade 1	Low or Medium Carbon Steel
	ASTM—A307	Low Carbon Steel
	SAE—Grade 2	Low or Medium Carbon Steel
	SAE—Grade 5	Medium Carbon Steel, Quenched and Tempered
	ASTM—A 449	
	SAE—Grade 5.2	Low Carbon Martensite Steel, Quenched and Tempered
A 325	ASTM—A 325 Type 1	Medium Carbon Steel, Quenched and Tempered Radial dashes optional
A 325	ASTM—A 325 Type 2	Low Carbon Martensite Steel, Quenched and Tempered
A 325	ASTM—A 325 Type 3	Atmospheric Corrosion (Weathering) Steel, Quenched and Tempered
BC	ASTM—A 354 Grade BC	Alloy Steel, Quenched and Tempered
	SAE—Grade 7	Medium Carbon Alloy Steel, Quenched and Tempered, Roll Threaded After Heat Treatment
	SAE—Grade 8	Medium Carbon Alloy Steel, Quenched and Tempered
	ASTM—A 354 Grade BD	Alloy Steel, Quenched and Tempered
	SAE—Grade 8.2	Low Carbon Martensite Steel, Quenched and Tempered
A 490	ASTM—A 490 Type 1	Alloy Steel, Quenched and Tempered
A 490	ASTM—A 490 Type 3	Atmospheric Corrosion (Weathering) Steel, Quenched and Tempered

(Reprinted from The American Society of Mechanical Engineers—ANSI B18.2.1–1981 (R1992).)

TABLE 8 UNIFIED STANDARD SCREW THREAD SERIES

Primary	Secondary	Basic Major Diameter	Coarse UNC	Fine UNF	Extra fine UNEF	4UN	6UN	8UN	12UN	16UN	20UN	28UN	32UN	Sizes
0		0.0600	—	80	—	—	—	—	—	—	—	—	—	0
	1	0.0730	64	72	—	—	—	—	—	—	—	—	—	1
2		0.0860	56	64	—	—	—	—	—	—	—	—	—	2
	3	0.0990	48	56	—	—	—	—	—	—	—	—	—	3
4		0.1120	40	48	—	—	—	—	—	—	—	—	—	4
5		0.1250	40	44	—	—	—	—	—	—	—	—	—	5
6		0.1380	32	40	—	—	—	—	—	—	—	—	UNC	6
8		0.1640	32	36	—	—	—	—	—	—	—	—	UNC	8
10		0.1900	24	32	—	—	—	—	—	—	—	—	UNF	10
	12	0.2160	24	28	32	—	—	—	—	—	—	UNF	UNEF	12
¼		0.2500	20	28	32	—	—	—	—	—	UNC	UNF	UNEF	¼
⁵⁄₁₆		0.3125	18	24	32	—	—	—	—	—	20	28	UNEF	⁵⁄₁₆
⅜		0.3750	16	24	32	—	—	—	—	UNC	20	28	UNEF	⅜
⁷⁄₁₆		0.4375	14	20	28	—	—	—	—	16	UNF	UNEF	32	⁷⁄₁₆
½		0.5000	13	20	28	—	—	—	—	16	UNF	UNEF	32	½
⁹⁄₁₆		0.5625	12	18	24	—	—	—	UNC	16	20	28	32	⁹⁄₁₆
⅝		0.6250	11	18	24	—	—	—	12	16	20	28	32	⅝
	¹¹⁄₁₆	0.6875	—	—	24	—	—	—	12	16	20	28	32	¹¹⁄₁₆
¾		0.7500	10	16	20	—	—	—	12	UNF	UNEF	28	32	¾
	¹³⁄₁₆	0.8125	—	—	20	—	—	—	12	16	UNEF	28	32	¹³⁄₁₆
⅞		0.8750	9	14	20	—	—	—	12	16	UNEF	28	32	⅞
	¹⁵⁄₁₆	0.9375	—	—	20	—	—	—	12	16	UNEF	28	32	¹⁵⁄₁₆
1		1.0000	8	12	20	—	—	UNC	UNF	16	UNEF	28	32	1
	1¹⁄₁₆	1.0625	—	—	18	—	—	8	12	16	20	28	—	1¹⁄₁₆
1⅛		1.1250	7	12	18	—	—	8	UNF	16	20	28	—	1⅛
	1³⁄₁₆	1.1875	—	—	18	—	—	8	12	16	20	28	—	1³⁄₁₆
1¼		1.2500	7	12	18	—	—	8	UNF	16	20	28	—	1¼
	1⁵⁄₁₆	1.3125	—	—	18	—	—	8	12	16	20	28	—	1⁵⁄₁₆
1⅜		1.3750	6	12	18	—	UNC	8	UNF	16	20	28	—	1⅜
	1⁷⁄₁₆	1.4375	—	—	18	—	6	8	12	16	20	28	—	1⁷⁄₁₆
1½		1.5000	6	12	18	—	UNC	8	UNF	16	20	28	—	1½
	1⁹⁄₁₆	1.5625	—	—	18	—	6	8	12	16	20	—	—	1⁹⁄₁₆
1⅝		1.6250	—	—	18	—	6	8	12	16	20	—	—	1⅝
	1¹¹⁄₁₆	1.6875	—	—	18	—	6	8	12	16	20	—	—	1¹¹⁄₁₆
1¾		1.7500	5	—	—	—	6	8	12	16	20	—	—	1¾
	1¹³⁄₁₆	1.8125	—	—	—	—	6	8	12	16	20	—	—	1¹³⁄₁₆
1⅞		1.8750	—	—	—	—	6	8	12	16	20	—	—	1⅞
	1¹⁵⁄₁₆	1.9375	—	—	—	—	6	8	12	16	20	—	—	1¹⁵⁄₁₆
2		2.0000	4½	—	—	—	6	8	12	16	20	—	—	2
	2⅛	2.1250	—	—	—	—	6	8	12	16	20	—	—	2⅛
2¼		2.2500	4½	—	—	—	6	8	12	16	20	—	—	2¼
	2⅜	2.3750	—	—	—	—	6	8	12	16	20	—	—	2⅜
2½		2.5000	4	—	—	UNC	6	8	12	16	20	—	—	2½
	2⅝	2.6250	—	—	—	4	6	8	12	16	20	—	—	2⅝
2¾		2.7500	4	—	—	UNC	6	8	12	16	20	—	—	2¾
	2⅞	2.8750	—	—	—	4	6	8	12	16	20	—	—	2⅞
3		3.0000	4	—	—	UNC	6	8	12	16	20	—	—	3
	3⅛	3.1250	—	—	—	4	6	8	12	16	—	—	—	3⅛
3¼		3.2500	4	—	—	UNC	6	8	12	16	—	—	—	3¼
	3⅜	3.3750	—	—	—	4	6	8	12	16	—	—	—	3⅜
3½		3.5000	4	—	—	UNC	6	8	12	16	—	—	—	3½
	3⅝	3.6250	—	—	—	4	6	8	12	16	—	—	—	3⅝
3¾		3.7500	4	—	—	UNC	6	8	12	16	—	—	—	3¾
	3⅞	3.8750	—	—	—	4	6	8	12	16	—	—	—	3⅞
4		4.0000	4	—	—	UNC	6	8	12	16	—	—	—	4
	4⅛	4.1250	—	—	—	4	6	8	12	16	—	—	—	4⅛
4¼		4.2500	—	—	—	4	6	8	12	16	—	—	—	4¼
	4⅜	4.3750	—	—	—	4	6	8	12	16	—	—	—	4⅜
4½		4.5000	—	—	—	4	6	8	12	16	—	—	—	4½
	4⅝	4.6250	—	—	—	4	6	8	12	16	—	—	—	4⅝
4¾		4.7500	—	—	—	4	6	8	12	16	—	—	—	4¾
	4⅞	4.8750	—	—	—	4	6	8	12	16	—	—	—	4⅞
5		5.0000	—	—	—	4	6	8	12	16	—	—	—	5
	5⅛	5.1250	—	—	—	4	6	8	12	16	—	—	—	5⅛
5¼		5.2500	—	—	—	4	6	8	12	16	—	—	—	5¼
	5⅜	5.3750	—	—	—	4	6	8	12	16	—	—	—	5⅜
5½		5.5000	—	—	—	4	6	8	12	16	—	—	—	5½
	5⅝	5.6250	—	—	—	4	6	8	12	16	—	—	—	5⅝
5¾		5.7500	—	—	—	4	6	8	12	16	—	—	—	5¾
	5⅞	5.8750	—	—	—	4	6	8	12	16	—	—	—	5⅞
6		6.0000	—	—	—	4	6	8	12	16	—	—	—	6

TABLE 9   DIMENSIONS OF HEX CAP SCREWS (FINISHED HEX BOLTS)

Nominal Size or Basic Product Dia (18)	E Body Dia (8) Max	E Min	F Width Across Flats Basic	F Max	F Min	G Width Across Corners (4) Max	G Min	H Height Basic	H Max	H Min	J Wrenching Height (4) Min	L_T Thread Length 6 in. and Shorter Basic	L_T Over 6 in. Basic	Y Transition Thread Length (10) Max	Runout of Bearing Surface FIM (5) Max
1/4 0.2500	0.2500	0.2450	7/16	0.438	0.428	0.505	0.488	5/32	0.163	0.150	0.106	0.750	1.000	0.250	0.010
5/16 0.3125	0.3125	0.3065	1/2	0.500	0.489	0.577	0.557	13/64	0.211	0.195	0.140	0.875	1.125	0.278	0.011
3/8 0.3750	0.3750	0.3690	9/16	0.562	0.551	0.650	0.628	15/64	0.243	0.226	0.160	1.000	1.250	0.312	0.012
7/16 0.4375	0.4375	0.4305	5/8	0.625	0.612	0.722	0.698	9/32	0.291	0.272	0.195	1.125	1.375	0.357	0.013
1/2 0.5000	0.5000	0.4930	3/4	0.750	0.736	0.866	0.840	5/16	0.323	0.302	0.215	1.250	1.500	0.385	0.014
9/16 0.5625	0.5625	0.5545	13/16	0.812	0.798	0.938	0.910	23/64	0.371	0.348	0.250	1.375	1.625	0.417	0.015
5/8 0.6250	0.6250	0.6170	15/16	0.938	0.922	1.083	1.051	25/64	0.403	0.378	0.269	1.500	1.750	0.455	0.017
3/4 0.7500	0.7500	0.7410	1 1/8	1.125	1.100	1.299	1.254	15/32	0.483	0.455	0.324	1.750	2.000	0.500	0.020
7/8 0.8750	0.8750	0.8660	1 5/16	1.312	1.285	1.516	1.465	35/64	0.563	0.531	0.378	2.000	2.250	0.556	0.023
1 1.0000	1.0000	0.9900	1 1/2	1.500	1.469	1.732	1.675	39/64	0.627	0.591	0.416	2.250	2.500	0.625	0.026
1 1/8 1.1250	1.1250	1.1140	1 11/16	1.688	1.631	1.949	1.859	11/16	0.718	0.658	0.461	2.500	2.750	0.714	0.029
1 1/4 1.2500	1.2500	1.2390	1 7/8	1.875	1.812	2.165	2.066	25/32	0.813	0.749	0.530	2.750	3.000	0.714	0.033
1 3/8 1.3750	1.3750	1.3630	2 1/16	2.062	1.994	2.382	2.273	27/32	0.878	0.810	0.569	3.000	3.250	0.833	0.036
1 1/2 1.5000	1.5000	1.4880	2 1/4	2.230	2.175	2.598	2.480	15/16	0.974	0.902	0.640	3.250	3.500	0.833	0.039
1 3/4 1.7500	1.7500	1.7380	2 5/8	2.625	2.538	3.031	2.893	1 3/32	1.134	1.054	0.748	3.750	4.000	1.000	0.046
2 2.0000	2.0000	1.9880	3	3.000	2.900	3.464	3.306	1 7/32	1.263	1.175	0.825	4.250	4.500	1.111	0.052
2 1/4 2.2500	2.2500	2.2380	3 3/8	3.375	3.262	3.897	3.719	1 3/8	1.423	1.327	0.933	4.750	5.000	1.111	0.059
2 1/2 2.5000	2.5000	2.4880	3 3/4	3.750	3.625	4.330	4.133	1 17/32	1.583	1.479	1.042	5.250	5.500	1.250	0.065
2 3/4 2.7500	2.7500	2.7380	4 1/8	4.125	3.988	4.763	4.546	1 11/16	1.744	1.632	1.151	5.750	6.000	1.250	0.072
3 3.0000	3.0000	2.9880	4 1/2	4.500	4.350	5.196	4.959	1 7/8	1.935	1.815	1.290	6.250	6.500	1.250	0.079

$30° \begin{array}{c} +0 \\ -15 \end{array}$

(Reprinted from The American Society of Mechanical Engineers—ANSI B18.2.1–1981 (R1992).)

TABLE 10 DIMENSIONS OF HEXAGON AND SPLINE SOCKET HEAD CAP SCREWS (1960 SERIES)

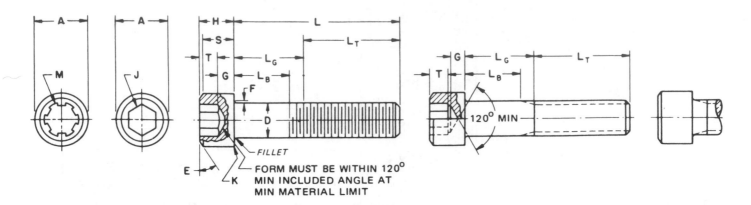

Nominal Size or Basic Screw Diameter		D Body Diameter		A Head Diameter		H Head Height		S Head Side Height	M Spline Socket Size	J Hexagon Socket Size		T Key Engagement	G Wall Thickness	K Chamfer or Radius
		Max	Min	Max	Min	Max	Min	Min	Nom	Nom		Min	Min	Max
0	0.0600	0.0600	0.0568	0.096	0.091	0.060	0.057	0.054	0.060	0.050		0.025	0.020	0.003
1	0.0730	0.0730	0.0695	0.118	0.112	0.073	0.070	0.066	0.072	1/16	0.062	0.031	0.025	0.003
2	0.0860	0.0860	0.0822	0.140	0.134	0.086	0.083	0.077	0.096	5/64	0.078	0.038	0.029	0.003
3	0.0990	0.0990	0.0949	0.161	0.154	0.099	0.095	0.089	0.096	5/64	0.078	0.044	0.034	0.003
4	0.1120	0.1120	0.1075	0.183	0.176	0.112	0.108	0.101	0.111	3/32	0.094	0.051	0.038	0.005
5	0.1250	0.1250	0.1202	0.205	0.198	0.125	0.121	0.112	0.111	3/32	0.094	0.057	0.043	0.005
6	0.1380	0.1380	0.1329	0.226	0.218	0.138	0.134	0.124	0.133	7/64	0.109	0.064	0.047	0.005
8	0.1640	0.1640	0.1585	0.270	0.262	0.164	0.159	0.148	0.168	9/64	0.141	0.077	0.056	0.005
10	0.1900	0.1900	0.1840	0.312	0.303	0.190	0.185	0.171	0.183	5/32	0.156	0.090	0.065	0.005
1/4	0.2500	0.2500	0.2435	0.375	0.365	0.250	0.244	0.225	0.216	3/16	0.188	0.120	0.095	0.008
5/16	0.3125	0.3125	0.3053	0.469	0.457	0.312	0.306	0.281	0.291	1/4	0.250	0.151	0.119	0.008
3/8	0.3750	0.3750	0.3678	0.562	0.550	0.375	0.368	0.337	0.372	5/16	0.312	0.182	0.143	0.008
7/16	0.4375	0.4375	0.4294	0.656	0.642	0.438	0.430	0.394	0.454	3/8	0.375	0.213	0.166	0.010
1/2	0.5000	0.5000	0.4919	0.750	0.735	0.500	0.492	0.450	0.454	3/8	0.375	0.245	0.190	0.010
5/8	0.6250	0.6250	0.6163	0.938	0.921	0.625	0.616	0.562	0.595	1/2	0.500	0.307	0.238	0.010
3/4	0.7500	0.7500	0.7406	1.125	1.107	0.750	0.740	0.675	0.620	5/8	0.625	0.370	0.285	0.010
7/8	0.8750	0.8750	0.8647	1.312	1.293	0.875	0.864	0.787	0.698	3/4	0.750	0.432	0.333	0.015
1	1.0000	1.0000	0.9886	1.500	1.479	1.000	0.988	0.900	0.790	3/4	0.750	0.495	0.380	0.015
1 1/8	1.1250	1.1250	1.1086	1.688	1.665	1.125	1.111	1.012	....	7/8	0.875	0.557	0.428	0.015
1 1/4	1.2500	1.2500	1.2336	1.875	1.852	1.250	1.236	1.125	....	7/8	0.875	0.620	0.475	0.015
1 3/8	1.3750	1.3750	1.3568	2.062	2.038	1.375	1.360	1.237	....	1	1.000	0.682	0.523	0.015
1 1/2	1.5000	1.5000	1.4818	2.250	2.224	1.500	1.485	1.350	....	1	1.000	0.745	0.570	0.015
1 3/4	1.7500	1.7500	1.7295	2.625	2.597	1.750	1.734	1.575	....	1 1/4	1.250	0.870	0.665	0.015
2	2.0000	2.0000	1.9780	3.000	2.970	2.000	1.983	1.800	....	1 1/2	1.500	0.995	0.760	0.015
2 1/4	2.2500	2.2500	2.2280	3.375	3.344	2.250	2.232	2.025	....	1 3/4	1.750	1.120	0.855	0.031
2 1/2	2.5000	2.5000	2.4762	3.750	3.717	2.500	2.481	2.250	....	1 3/4	1.750	1.245	0.950	0.031
2 3/4	2.7500	2.7500	2.7262	4.125	4.090	2.750	2.730	2.475	....	2	2.000	1.370	1.045	0.031
3	3.0000	3.0000	2.9762	4.500	4.464	3.000	2.979	2.700	....	2 1/4	2.250	1.495	1.140	0.031
3 1/4	3.2500	3.2500	3.2262	4.875	4.837	3.250	3.228	2.925	....	2 1/4	2.250	1.620	1.235	0.031
3 1/2	3.5000	3.5000	3.4762	5.250	5.211	3.500	3.478	3.150	....	2 3/4	2.750	1.745	1.330	0.031
3 3/4	3.7500	3.7500	3.7262	5.625	5.584	3.750	3.727	3.375	....	2 3/4	2.750	1.870	1.425	0.031
4	4.0000	4.0000	3.9762	6.000	5.958	4.000	3.976	3.600	....	3	3.000	1.995	1.520	0.031

(Reprinted from The American Society of Mechanical Engineers—ANSI/ASME B18.3–1986 (R1993).)

TABLE 11   DIMENSIONS OF HEXAGON AND SPLINE SOCKET FLAT COUNTERSUNK HEAD CAP SCREWS

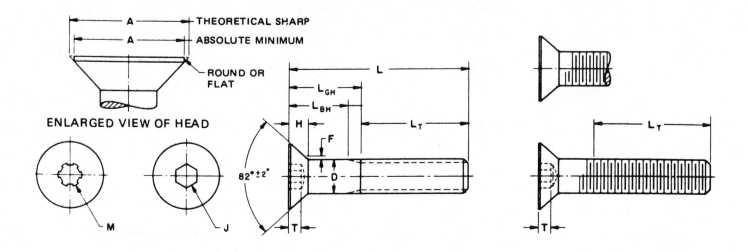

Nominal Size or Basic Screw Diameter		D Body Dia		A Head Diameter		H Head Height		M Spline Socket Size	J Hexagon Socket Size		T Key Engagement	F Fillet Extension Above D Max
		Max	Min	Theoretical Sharp Max	Abs. Min	Reference	Flushness Tolerance		Nom		Min	Max
0	0.0600	0.0600	0.0568	0.138	0.117	0.044	0.006	0.048	0.035		0.025	0.006
1	0.0730	0.0730	0.0695	0.168	0.143	0.054	0.007	0.060	0.050		0.031	0.008
2	0.0860	0.0860	0.0822	0.197	0.168	0.064	0.008	0.060	0.050		0.038	0.010
3	0.0990	0.0990	0.0949	0.226	0.193	0.073	0.010	0.072	1/16	0.062	0.044	0.010
4	0.1120	0.1120	0.1075	0.255	0.218	0.083	0.011	0.072	1/16	0.062	0.055	0.012
5	0.1250	0.1250	0.1202	0.281	0.240	0.090	0.012	0.096	5/64	0.078	0.061	0.014
6	0.1380	0.1380	0.1329	0.307	0.263	0.097	0.013	0.096	5/64	0.078	0.066	0.015
8	0.1640	0.1640	0.1585	0.359	0.311	0.112	0.014	0.111	3/32	0.094	0.076	0.015
10	0.1900	0.1900	0.1840	0.411	0.359	0.127	0.015	0.145	1/8	0.125	0.087	0.015
1/4	0.2500	0.2500	0.2435	0.531	0.480	0.161	0.016	0.183	5/32	0.156	0.111	0.015
5/16	0.3125	0.3125	0.3053	0.656	0.600	0.198	0.017	0.216	3/16	0.188	0.135	0.015
3/8	0.3750	0.3750	0.3678	0.781	0.720	0.234	0.018	0.251	7/32	0.219	0.159	0.015
7/16	0.4375	0.4375	0.4294	0.844	0.781	0.234	0.018	0.291	1/4	0.250	0.159	0.015
1/2	0.5000	0.5000	0.4919	0.938	0.872	0.251	0.018	0.372	5/16	0.312	0.172	0.015
5/8	0.6250	0.6250	0.6163	1.188	1.112	0.324	0.022	0.454	3/8	0.375	0.220	0.015
3/4	0.7500	0.7500	0.7406	1.438	1.355	0.396	0.024	0.454	1/2	0.500	0.220	0.015
7/8	0.8750	0.8750	0.8647	1.688	1.604	0.468	0.025	. . .	9/16	0.562	0.248	0.015
1	1.0000	1.0000	0.9886	1.938	1.841	0.540	0.028	. . .	5/8	0.625	0.297	0.015
1 1/8	1.1250	1.1250	1.1086	2.188	2.079	0.611	0.031	. . .	3/4	0.750	0.325	0.031
1 1/4	1.2500	1.2500	1.2336	2.438	2.316	0.683	0.035	. . .	7/8	0.875	0.358	0.031
1 3/8	1.3750	1.3750	1.3568	2.688	2.553	0.755	0.038	. . .	7/8	0.875	0.402	0.031
1 1/2	1.5000	1.5000	1.4818	2.938	2.791	0.827	0.042	. . .	1	1.000	0.435	0.031

(Reprinted from The American Society of Mechanical Engineers—ANSI/ASME B18.3–1986 (R1993).)

TABLE 12 DIMENSIONS OF HEXAGON AND SPLINE SOCKET SET SCREWS

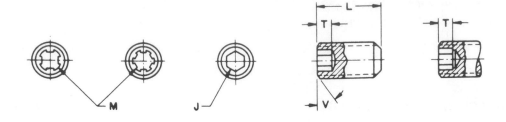

Nominal Size or Basic Screw Diameter		J Hexagon Socket Size		M Spline Socket Size	T Min Key Engagement to Develop Functional Capability of Key		C Cup and Flat Point Diameters		R Oval Point Radius	Y Cone Point Angle 90° ±2° For These Nominal Lengths or Longer; 118° ±2° For Shorter Nominal Lengths
					Hex Socket T_H Min	Spline Socket T_S Min				
		Nom		Nom			Max	Min	Basic	
0	0.0600	0.028		0.033	0.050	0.026	0.033	0.027	0.045	5/64
1	0.0730	0.035		0.033	0.060	0.035	0.040	0.033	0.055	3/32
2	0.0860	0.035		0.048	0.060	0.040	0.047	0.039	0.064	7/64
3	0.0990	0.050		0.048	0.070	0.040	0.054	0.045	0.074	1/8
4	0.1120	0.050		0.060	0.070	0.045	0.061	0.051	0.084	5/32
5	0.1250	1/16	0.062	0.072	0.080	0.055	0.067	0.057	0.094	3/16
6	0.1380	1/16	0.062	0.072	0.080	0.055	0.074	0.064	0.104	3/16
8	0.1640	5/64	0.078	0.096	0.090	0.080	0.087	0.076	0.123	1/4
10	0.1900	3/32	0.094	0.111	0.100	0.080	0.102	0.088	0.142	1/4
1/4	0.2500	1/8	0.125	0.145	0.125	0.125	0.132	0.118	0.188	5/16
5/16	0.3125	5/32	0.156	0.183	0.156	0.156	0.172	0.156	0.234	3/8
3/8	0.3750	3/16	0.188	0.216	0.188	0.188	0.212	0.194	0.281	7/16
7/16	0.4375	7/32	0.219	0.251	0.219	0.219	0.252	0.232	0.328	1/2
1/2	0.5000	1/4	0.250	0.291	0.250	0.250	0.291	0.270	0.375.	9/16
5/8	0.6250	5/16	0.312	0.372	0.312	0.312	0.371	0.347	0.469	3/4
3/4	0.7500	3/8	0.375	0.454	0.375	0.375	0.450	0.425	0.562	7/8
7/8	0.8750	1/2	0.500	0.595	0.500	0.500	0.530	0.502	0.656	1
1	1.0000	9/16	0.562	. . .	0.562	. . .	0.609	0.579	0.750	1 1/8
1 1/8	1.1250	9/16	0.562	. . .	0.562	. . .	0.689	0.655	0.844	1 1/4
1 1/4	1.2500	5/8	0.625	. . .	0.625	. . .	0.767	0.733	0.938	1 1/2
1 3/8	1.3750	5/8	0.625	. . .	0.625	. . .	0.848	0.808	1.031	1 5/8
1 1/2	1.5000	3/4	0.750	. . .	0.750	. . .	0.926	0.886	1.125	1 3/4
1 3/4	1.7500	1	1.000	. . .	1.000	. . .	1.086	1.039	1.312	2
2	2.0000	1	1.000	. . .	1.000	. . .	1.244	1.193	1.500	2 1/4

TABLE 12 (Continued)

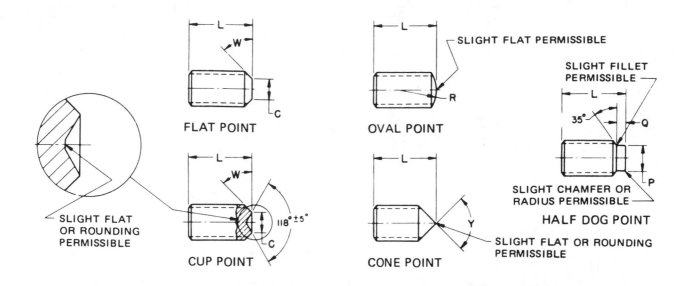

FLAT POINT

OVAL POINT

SLIGHT FLAT OR ROUNDING PERMISSIBLE

CUP POINT

CONE POINT

HALF DOG POINT

Nominal Size or Basic Screw Diameter		P		Q		B			B₁		
		Half Dog Point				Shortest Optimum Nominal Length To Which Column T_H Applies			Shortest Optimum Nominal Length To Which Column T_S Applies		
		Diameter		Length		Cup and Flat Points	90° Cone and Oval Points	Half Dog Point	Cup and Flat Points	90° Cone and Oval Points	Half Dog Point
		Max	Min	Max	Min						
0	0.0600	0.040	0.037	0.017	0.013	7/64	1/8	7/64	1/16	1/8	7/64
1	0.0730	0.049	0.045	0.021	0.017	1/8	9/64	1/8	3/32	9/64	1/8
2	0.0860	0.057	0.053	0.024	0.020	1/8	9/64	9/64	3/32	9/64	9/64
3	0.0990	0.066	0.062	0.027	0.023	9/64	5/32	5/32	3/32	5/32	5/32
4	0.1120	0.075	0.070	0.030	0.026	9/64	11/64	5/32	3/32	11/64	5/32
5	0.1250	0.083	0.078	0.033	0.027	3/16	3/16	11/64	1/8	3/16	11/64
6	0.1380	0.092	0.087	0.038	0.032	11/64	13/64	3/16	1/8	13/64	3/16
8	0.1640	0.109	0.103	0.043	0.037	3/16	7/32	13/64	3/16	7/32	13/64
10	0.1900	0.127	0.120	0.049	0.041	3/16	1/4	15/64	3/16	1/4	15/64
1/4	0.2500	0.156	0.149	0.067	0.059	1/4	5/16	19/64	1/4	5/16	19/64
5/16	0.3125	0.203	0.195	0.082	0.074	5/16	25/64	23/64	5/16	25/64	23/64
3/8	0.3750	0.250	0.241	0.099	0.089	3/8	7/16	7/16	3/8	7/16	7/16
7/16	0.4375	0.297	0.287	0.114	0.104	7/16	35/64	31/64	7/16	35/64	31/64
1/2	0.5000	0.344	0.334	0.130	0.120	1/2	39/64	35/64	1/2	39/64	35/64
5/8	0.6250	0.469	0.456	0.164	0.148	5/8	49/64	43/64	5/8	49/64	43/64
3/4	0.7500	0.562	0.549	0.196	0.180	3/4	29/32	51/64	3/4	29/32	51/64
7/8	0.8750	0.656	0.642	0.227	0.211	7/8	1 1/8	63/64	7/8	1 1/8	63/64
1	1.0000	0.750	0.734	0.260	0.240	1	1 17/64	1 1/8	...	...	...
1 1/8	1.1250	0.844	0.826	0.291	0.271	1 1/8	1 25/64	1 3/16	...	...	...
1 1/4	1.2500	0.938	0.920	0.323	0.303	1 1/4	1 1/2	1 5/16	...	...	...
1 3/8	1.3750	1.031	1.011	0.354	0.334	1 3/8	1 21/32	1 7/16	...	...	...
1 1/2	1.5000	1.125	1.105	0.385	0.365	1 1/2	1 51/64	1 9/16	...	...	...
1 3/4	1.7500	1.312	1.289	0.448	0.428	1 3/4	2 7/32	1 61/64	...	...	...
2	2.0000	1.500	1.474	0.510	0.490	2	2 25/64	2 5/64	...	...	...

(Reprinted from The American Society of Mechanical Engineers—ANSI/ASME B18.3–1986 (R1993).)

TABLE 13   DIMENSIONS OF SLOTTED FLAT COUNTERSUNK HEAD CAP SCREWS

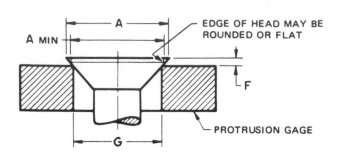

CAP SCREWS

FLAT

Type of Head

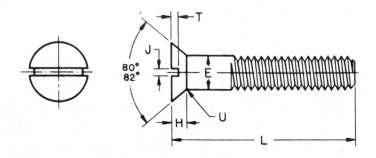

Nominal Size[1] or Basic Screw Diameter		E Body Diameter		A Head Diameter		H[2] Head Height	J Slot Width		T Slot Depth		U Fillet Radius	F[3] Protrusion Above Gaging Diameter		G[3] Gaging Diameter
		Max	Min	Max, Edge Sharp	Min, Edge Rounded or Flat	Ref	Max	Min	Max	Min	Max	Max	Min	
1/4	0.2500	0.2500	0.2450	0.500	0.452	0.140	0.075	0.064	0.068	0.045	0.100	0.046	0.030	0.424
5/16	0.3125	0.3125	0.3070	0.625	0.567	0.177	0.084	0.072	0.086	0.057	0.125	0.053	0.035	0.538
3/8	0.3750	0.3750	0.3690	0.750	0.682	0.210	0.094	0.081	0.103	0.068	0.150	0.060	0.040	0.651
7/16	0.4375	0.4375	0.4310	0.812	0.736	0.210	0.094	0.081	0.103	0.068	0.175	0.065	0.044	0.703
1/2	0.5000	0.5000	0.4930	0.875	0.791	0.210	0.106	0.091	0.103	0.068	0.200	0.071	0.049	0.756
9/16	0.5625	0.5625	0.5550	1.000	0.906	0.244	0.118	0.102	0.120	0.080	0.225	0.078	0.054	0.869
5/8	0.6250	0.6250	0.6170	1.125	1.020	0.281	0.133	0.116	0.137	0.091	0.250	0.085	0.058	0.982
3/4	0.7500	0.7500	0.7420	1.375	1.251	0.352	0.149	0.131	0.171	0.115	0.300	0.099	0.068	1.208
7/8	0.8750	0.8750	0.8660	1.625	1.480	0.423	0.167	0.147	0.206	0.138	0.350	0.113	0.077	1.435
1	1.0000	1.0000	0.9900	1.875	1.711	0.494	0.188	0.166	0.240	0.162	0.400	0.127	0.087	1.661
1 1/8	1.1250	1.1250	1.1140	2.062	1.880	0.529	0.196	0.178	0.257	0.173	0.450	0.141	0.096	1.826
1 1/4	1.2500	1.2500	1.2390	2.312	2.110	0.600	0.211	0.193	0.291	0.197	0.500	0.155	0.105	2.052
1 3/8	1.3750	1.3750	1.3630	2.562	2.340	0.665	0.226	0.208	0.326	0.220	0.550	0.169	0.115	2.279
1 1/2	1.5000	1.5000	1.4880	2.812	2.570	0.742	0.258	0.240	0.360	0.244	0.600	0.183	0.124	2.505

[1] Where specifying nominal size in decimals, zeros preceding decimal and in the fourth decimal place shall be omitted.

[2] Tabulated values determined from formula for maximum H, Appendix III.

[3] No tolerance for gaging diameter is given. If the gaging diameter of the gage used differs from tabulated value, the protrusion will be affected accordingly and the proper protrusion values must be recalculated using the formulas shown in Appendix II.

FOOTNOTES REFER TO ANSI B18.6.2–1972 (R1993).

(Reprinted from The American Society of Mechanical Engineers—ANSI B18.6.2–1972 (R1993).)

TABLE 14 DIMENSIONS OF SLOTTED ROUND HEAD CAP SCREWS

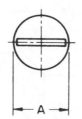

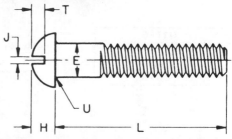

Type of Head

Nominal Size[1] or Basic Screw Diameter		E Body Diameter		A Head Diameter		H Head Height		J Slot Width		T Slot Depth		U Fillet Radius	
		Max	Min	Max	Min	Max	Min	Max	Min	Max	Min	Max	Min
1/4	0.2500	0.2500	0.2450	0.437	0.418	0.191	0.175	0.075	0.064	0.117	0.097	0.031	0.016
5/16	0.3125	0.3125	0.3070	0.562	0.540	0.245	0.226	0.084	0.072	0.151	0.126	0.031	0.016
3/8	0.3750	0.3750	0.3690	0.625	0.603	0.273	0.252	0.094	0.081	0.168	0.138	0.031	0.016
7/16	0.4375	0.4375	0.4310	0.750	0.725	0.328	0.302	0.094	0.081	0.202	0.167	0.047	0.016
1/2	0.5000	0.5000	0.4930	0.812	0.786	0.354	0.327	0.106	0.091	0.218	0.178	0.047	0.016
9/16	0.5625	0.5625	0.5550	0.937	0.909	0.409	0.378	0.118	0.102	0.252	0.207	0.047	0.016
5/8	0.6250	0.6250	0.6170	1.000	0.970	0.437	0.405	0.133	0.116	0.270	0.220	0.062	0.031
3/4	0.7500	0.7500	0.7420	1.250	1.215	0.546	0.507	0.149	0.131	0.338	0.278	0.062	0.031

[1] Where specifying nominal size in decimals, zeros preceding decimal and in the fourth decimal place shall be omitted.

(Reprinted from The American Society of Mechanical Engineers—ANSI B18.6.2–1972 (R1993).)

TABLE 15 DIMENSIONS OF SLOTTED FILISTER HEAD CAP SCREWS

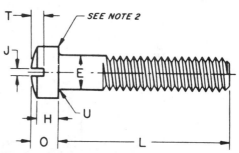

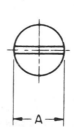

Type of Head

Nominal Size[1] or Basic Screw Diameter		E Body Diameter		A Head Diameter		H Head Side Height		O Total Head Height		J Slot Width		T Slot Depth		U Fillet Radius	
		Max	Min	Max	Min	Max	Min	Max	Min	Max	Min	Max	Min	Max	Min
1/4	0.2500	0.2500	0.2450	0.375	0.363	0.172	0.157	0.216	0.194	0.075	0.064	0.097	0.077	0.031	0.016
5/16	0.3125	0.3125	0.3070	0.437	0.424	0.203	0.186	0.253	0.230	0.084	0.072	0.115	0.090	0.031	0.016
3/8	0.3750	0.3750	0.3690	0.562	0.547	0.250	0.229	0.314	0.284	0.094	0.081	0.142	0.112	0.031	0.016
7/16	0.4375	0.4375	0.4310	0.625	0.608	0.297	0.274	0.368	0.336	0.094	0.081	0.168	0.133	0.047	0.016
1/2	0.5000	0.5000	0.4930	0.750	0.731	0.328	0.301	0.413	0.376	0.106	0.091	0.193	0.153	0.047	0.016
9/16	0.5625	0.5625	0.5550	0.812	0.792	0.375	0.346	0.467	0.427	0.118	0.102	0.213	0.168	0.047	0.016
5/8	0.6250	0.6250	0.6170	0.875	0.853	0.422	0.391	0.521	0.478	0.133	0.116	0.239	0.189	0.062	0.031
3/4	0.7500	0.7500	0.7420	1.000	0.976	0.500	0.466	0.612	0.566	0.149	0.131	0.283	0.223	0.062	0.031
7/8	0.8750	0.8750	0.8660	1.125	1.098	0.594	0.556	0.720	0.668	0.167	0.147	0.334	0.264	0.062	0.031
1	1.0000	1.0000	0.9900	1.312	1.282	0.656	0.612	0.803	0.743	0.188	0.166	0.371	0.291	0.062	0.031

[1] Where specifying nominal size in decimals, zeros preceding decimal and in the fourth decimal place shall be omitted.
[2] A slight rounding of the edges at periphery of head shall be permissible provided the diameter of the bearing circle is equal to no less than 90 per cent of the specified minimum head diameter.

(Reprinted from The American Society of Mechanical Engineers—ANSI B18.6.2–1972 (R1993).)

TABLE 16   DIMENSIONS OF SLOTTED FLAT COUNTERSUNK HEAD MACHINE SCREWS

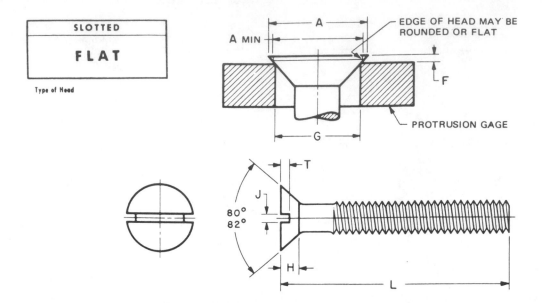

Nominal Size[1] or Basic Screw Diameter		L[2] These Lengths or Shorter are Undercut	A Head Diameter		H[3] Head Height	J Slot Width		T Slot Depth		F[4] Protrusion Above Gaging Diameter		G[4] Gaging Diameter
			Max, Edge Sharp	Min, Edge Rounded or Flat	Ref	Max	Min	Max	Min	Max	Min	
0000	0.0210	—	0.043	0.037	0.011	0.008	0.004	0.007	0.003	*	*	*
000	0.0340	—	0.064	0.058	0.016	0.011	0.007	0.009	0.005	*	*	*
00	0.0470	—	0.093	0.085	0.028	0.017	0.010	0.014	0.009	*	*	*
0	0.0600	1/8	0.119	0.099	0.035	0.023	0.016	0.015	0.010	0.026	0.016	0.078
1	0.0730	1/8	0.146	0.123	0.043	0.026	0.019	0.019	0.012	0.028	0.016	0.101
2	0.0860	1/8	0.172	0.147	0.051	0.031	0.023	0.023	0.015	0.029	0.017	0.124
3	0.0990	1/8	0.199	0.171	0.059	0.035	0.027	0.027	0.017	0.031	0.018	0.148
4	0.1120	3/16	0.225	0.195	0.067	0.039	0.031	0.030	0.020	0.032	0.019	0.172
5	0.1250	3/16	0.252	0.220	0.075	0.043	0.035	0.034	0.022	0.034	0.020	0.196
6	0.1380	3/16	0.279	0.244	0.083	0.048	0.039	0.038	0.024	0.036	0.021	0.220
8	0.1640	1/4	0.332	0.292	0.100	0.054	0.045	0.045	0.029	0.039	0.023	0.267
10	0.1900	5/16	0.385	0.340	0.116	0.060	0.050	0.053	0.034	0.042	0.025	0.313
12	0.2160	3/8	0.438	0.389	0.132	0.067	0.056	0.060	0.039	0.045	0.027	0.362
1/4	0.2500	7/16	0.507	0.452	0.153	0.075	0.064	0.070	0.046	0.050	0.029	0.424
5/16	0.3125	1/2	0.635	0.568	0.191	0.084	0.072	0.088	0.058	0.057	0.034	0.539
3/8	0.3750	9/16	0.762	0.685	0.230	0.094	0.081	0.106	0.070	0.065	0.039	0.653
7/16	0.4375	5/8	0.812	0.723	0.223	0.094	0.081	0.103	0.066	0.073	0.044	0.690
1/2	0.5000	3/4	0.875	0.775	0.223	0.106	0.091	0.103	0.065	0.081	0.049	0.739
9/16	0.5625	—	1.000	0.889	0.260	0.118	0.102	0.120	0.077	0.089	0.053	0.851
5/8	0.6250	—	1.125	1.002	0.298	0.133	0.116	0.137	0.088	0.097	0.058	0.962
3/4	0.7500	—	1.375	1.230	0.372	0.149	0.131	0.171	0.111	0.112	0.067	1.186

[1] Where specifying nominal size in decimals, zeros preceding decimal and in the fourth decimal place shall be omitted.
[2] Screws of these lengths and shorter shall have undercut heads as shown in Table 5.
[3] Tabulated values determined from formula for maximum H, Appendix V.
[4] No tolerance for gaging diameter is given. If the gaging diameter of the gage used differs from tabulated value, the protrusion will be affected accordingly and the proper protrusion values must be recalculated using the formulas shown in Appendix I.
*Not practical to gage.

For additional requirements refer to General Data on Pages 3, 4 and 5.

FOOTNOTES REFER TO ANSI B18.6.3–1972 (R1991).

(Reprinted from The American Society of Mechanical Engineers—ANSI B18.6.3–1972 (R1991).)

TABLE 17 DIMENSIONS OF HEX NUTS AND HEX JAM NUTS

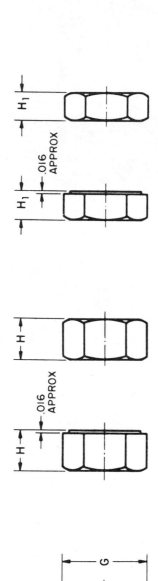

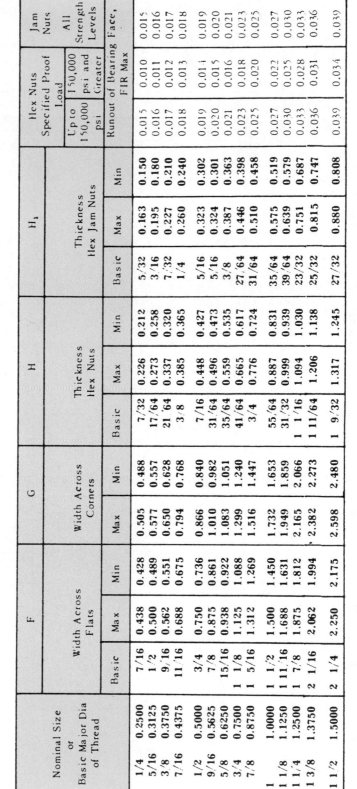

Nominal Size or Basic Major Dia of Thread		F Width Across Flats			G Width Across Corners		H Thickness Hex Nuts			H₁ Thickness Hex Jam Nuts			Runout of Bearing Face, FIR Max		
		Basic	Max	Min	Max	Min	Basic	Max	Min	Basic	Max	Min	Hex Nuts Specified Proof Load Up to 150,000 psi	Hex Nuts 150,000 psi and Greater	Jam Nuts All Strength Levels
1/4	0.2500	7/16	0.438	0.428	0.505	0.488	7/32	0.226	0.212	5/32	0.163	0.150	0.015	0.010	0.015
5/16	0.3125	1/2	0.500	0.489	0.577	0.557	17/64	0.273	0.258	3/16	0.195	0.180	0.016	0.011	0.016
3/8	0.3750	9/16	0.562	0.551	0.650	0.628	21/64	0.337	0.320	7/32	0.227	0.210	0.017	0.012	0.017
7/16	0.4375	11/16	0.688	0.675	0.794	0.768	3/8	0.385	0.365	1/4	0.260	0.240	0.018	0.013	0.018
1/2	0.5000	3/4	0.750	0.736	0.866	0.840	7/16	0.448	0.427	5/16	0.323	0.302	0.019	0.014	0.019
9/16	0.5625	7/8	0.875	0.861	1.010	0.982	31/64	0.496	0.473	5/16	0.324	0.301	0.020	0.015	0.020
5/8	0.6250	15/16	0.938	0.922	1.083	1.051	35/64	0.559	0.535	3/8	0.387	0.363	0.021	0.016	0.021
3/4	0.7500	1 1/8	1.125	1.088	1.299	1.240	41/64	0.665	0.617	27/64	0.446	0.398	0.023	0.018	0.023
7/8	0.8750	1 5/16	1.312	1.269	1.516	1.447	3/4	0.776	0.724	31/64	0.510	0.458	0.025	0.020	0.025
1	1.0000	1 1/2	1.500	1.450	1.732	1.653	55/64	0.887	0.831	35/64	0.575	0.519	0.027	0.022	0.027
1 1/8	1.1250	1 11/16	1.688	1.631	1.949	1.859	31/32	0.999	0.939	39/64	0.639	0.579	0.030	0.025	0.030
1 1/4	1.2500	1 7/8	1.875	1.812	2.165	2.066	1 1/16	1.094	1.030	23/32	0.751	0.687	0.033	0.028	0.033
1 3/8	1.3750	2 1/16	2.062	1.994	2.382	2.273	1 11/64	1.206	1.138	25/32	0.815	0.747	0.036	0.031	0.036
1 1/2	1.5000	2 1/4	2.250	2.175	2.598	2.480	1 9/32	1.317	1.245	27/32	0.880	0.808	0.039	0.034	0.039

(Reprinted from The American Society of Mechanical Engineers—ANSI B18.2.2–1987 (R1993).)

TABLE 18   WOODRUFF KEY DIMENSIONS

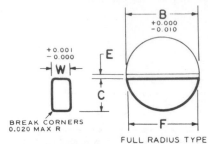

FULL RADIUS TYPE

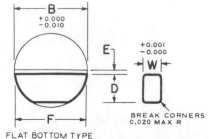

FLAT BOTTOM TYPE

Key No.	Nominal Key Size W × B	Actual Length F +0.000-0.010	Height of Key				Distance Below Center E
			C		D		
			Max	Min	Max	Min	
202	1/16 × 1/4	0.248	0.109	0.104	0.109	0.104	1/64
202.5	1/16 × 5/16	0.311	0.140	0.135	0.140	0.135	1/64
302.5	3/32 × 5/16	0.311	0.140	0.135	0.140	0.135	1/64
203	1/16 × 3/8	0.374	0.172	0.167	0.172	0.167	1/64
303	3/32 × 3/8	0.374	0.172	0.167	0.172	0.167	1/64
403	1/8 × 3/8	0.374	0.172	0.167	0.172	0.167	1/64
204	1/16 × 1/2	0.491	0.203	0.198	0.194	0.188	3/64
304	3/32 × 1/2	0.491	0.203	0.198	0.194	0.188	3/64
404	1/8 × 1/2	0.491	0.203	0.198	0.194	0.188	3/64
305	3/32 × 5/8	0.612	0.250	0.245	0.240	0.234	1/16
405	1/8 × 5/8	0.612	0.250	0.245	0.240	0.234	1/16
505	5/32 × 5/8	0.612	0.250	0.245	0.240	0.234	1/16
605	3/16 × 5/8	0.612	0.250	0.245	0.240	0.234	1/16
406	1/8 × 3/4	0.740	0.313	0.308	0.303	0.297	1/16
506	5/32 × 3/4	0.740	0.313	0.308	0.303	0.297	1/16
606	3/16 × 3/4	0.740	0.313	0.308	0.303	0.297	1/16
806	1/4 × 3/4	0.740	0.313	0.308	0.303	0.297	1/16
507	5/32 × 7/8	0.866	0.375	0.370	0.365	0.359	1/16
607	3/16 × 7/8	0.866	0.375	0.370	0.365	0.359	1/16
707	7/32 × 7/8	0.866	0.375	0.370	0.365	0.359	1/16
807	1/4 × 7/8	0.866	0.375	0.370	0.365	0.359	1/16
608	3/16 × 1	0.992	0.438	0.433	0.428	0.422	1/16
708	7/32 × 1	0.992	0.438	0.433	0.428	0.422	1/16
808	1/4 × 1	0.992	0.438	0.433	0.428	0.422	1/16
1008	5/16 × 1	0.992	0.438	0.433	0.428	0.422	1/16
1208	3/8 × 1	0.992	0.438	0.433	0.428	0.422	1/16
609	3/16 × 1 1/8	1.114	0.484	0.479	0.475	0.469	5/64
709	7/32 × 1 1/8	1.114	0.484	0.479	0.475	0.469	5/64
809	1/4 × 1 1/8	1.114	0.484	0.479	0.475	0.469	5/64
1009	5/16 × 1 1/8	1.114	0.484	0.479	0.475	0.469	5/64
610	3/16 × 1 1/4	1.240	0.547	0.542	0.537	0.531	5/64
710	7/32 × 1 1/4	1.240	0.547	0.542	0.537	0.531	5/64
810	1/4 × 1 1/4	1.240	0.547	0.542	0.537	0.531	5/64
1010	5/16 × 1 1/4	1.240	0.547	0.542	0.537	0.531	5/64
1210	3/8 × 1 1/4	1.240	0.547	0.542	0.537	0.531	5/64
811	1/4 × 1 3/8	1.362	0.594	0.589	0.584	0.578	3/32
1011	5/16 × 1 3/8	1.362	0.594	0.589	0.584	0.578	3/32
1211	3/8 × 1 3/8	1.362	0.594	0.589	0.584	0.578	3/32
812	1/4 × 1 1/2	1.484	0.641	0.636	0.631	0.625	7/64
1012	5/16 × 1 1/2	1.484	0.641	0.636	0.631	0.625	7/64
1212	3/8 × 1 1/2	1.484	0.641	0.636	0.631	0.625	7/64

All dimensions given are in inches.

The key numbers indicate nominal key dimensions. The last two digits give the nominal diameter B in eighths of an inch and the digits preceding the last two give the nominal width W in thirty-seconds of an inch.

Example: No. 204 indicates a key 2/32 × 4/8 or 1/16 × 1/2.
No. 808 indicates a key 8/32 × 8/8 or 1/4 × 1.
No. 1212 indicates a key 12/32 × 12/8 or 3/8 × 1 1/2.

TABLE 18   (Continued)

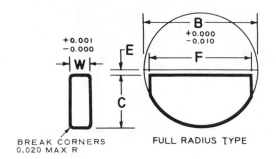

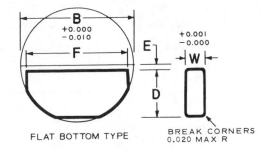

Key No.	Nominal Key Size W × B	Actual Length F +0.000-0.010	Height of Key				Distance Below Center E
			C		D		
			Max	Min	Max	Min	
617-1	3/16 × 2 1/8	1.380	0.406	0.401	0.396	0.390	21/32
817-1	1/4 × 2 1/8	1.380	0.406	0.401	0.396	0.390	21/32
1017-1	5/16 × 2 1/8	1.380	0.406	0.401	0.396	0.390	21/32
1217-1	3/8 × 2 1/8	1.380	0.406	0.401	0.396	0.390	21/32
617	3/16 × 2 1/8	1.723	0.531	0.526	0.521	0.515	17/32
817	1/4 × 2 1/8	1.723	0.531	0.526	0.521	0.515	17/32
1017	5/16 × 2 1/8	1.723	0.531	0.526	0.521	0.515	17/32
1217	3/8 × 2 1/8	1.723	0.531	0.526	0.521	0.515	17/32
822-1	1/4 × 2 3/4	2.000	0.594	0.589	0.584	0.578	25/32
1022-1	5/16 × 2 3/4	2.000	0.594	0.589	0.584	0.578	25/32
1222-1	3/8 × 2 3/4	2.000	0.594	0.589	0.584	0.578	25/32
1422-1	7/16 × 2 3/4	2.000	0.594	0.589	0.584	0.578	25/32
1622-1	1/2 × 2 3/4	2.000	0.594	0.589	0.584	0.578	25/32
822	1/4 × 2 3/4	2.317	0.750	0.745	0.740	0.734	5/8
1022	5/16 × 2 3/4	2.317	0.750	0.745	0.740	0.734	5/8
1222	3/8 × 2 3/4	2.317	0.750	0.745	0.740	0.734	5/8
1422	7/16 × 2 3/4	2.317	0.750	0.745	0.740	0.734	5/8
1622	1/2 × 2 3/4	2.317	0.750	0.745	0.740	0.734	5/8
1228	3/8 × 3 1/2	2.880	0.938	0.933	0.928	0.922	13/16
1428	7/16 × 3 1/2	2.880	0.938	0.933	0.928	0.922	13/16
1628	1/2 × 3 1/2	2.880	0.938	0.933	0.928	0.922	13/16
1828	9/16 × 3 1/2	2.880	0.938	0.933	0.928	0.922	13/16
2028	5/8 × 3 1/2	2.880	0.938	0.933	0.928	0.922	13/16
2228	11/16 × 3 1/2	2.880	0.938	0.933	0.928	0.922	13/16
2428	3/4 × 3 1/2	2.880	0.938	0.933	0.928	0.922	13/16

All dimensions given are in inches.

The key numbers indicate nominal key dimensions. The last two digits give the nominal diameter B in eighths of an inch and the digits preceding the last two give the nominal width W in thirty-seconds of an inch.

Example:

No. 617 indicates a key $6/32 \times 17/8$ or $3/16 \times 2 1/8$
No. 822 indicates a key $8/32 \times 22/8$ or $1/4 \times 2 3/4$
No. 1228 indicates a key $12/32 \times 28/8$ or $3/8 \times 3 1/2$

The key numbers with the -1 designation, while representing the nominal key size have a shorter length F and due to a greater distance below center E are less in height than the keys of the same number without the -1 designation.

(Reprinted from The American Society of Mechanical Engineers—ANSI B17.2–1967 (R1990).)

TABLE 19   WOODRUFF KEYSEAT DIMENSIONS

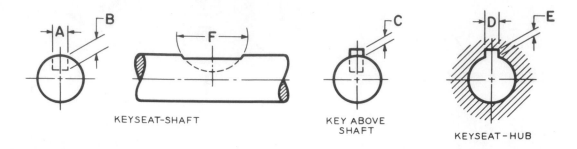

KEYSEAT-SHAFT   KEY ABOVE SHAFT   KEYSEAT-HUB

Key Number	Nominal Size Key	Keyseat — Shaft					Key Above Shaft	Keyseat — Hub	
		Width A*		Depth B	Diameter F		Height C	Width D	Depth E
		Min	Max	+0.005 -0.000	Min	Max	+0.005 -0.005	+0.002 -0.000	+0.005 -0.000
202	1/16 × 1/4	0.0615	0.0630	0.0728	0.250	0.268	0.0312	0.0635	0.0372
202.5	1/16 × 5/16	0.0615	0.0630	0.1038	0.312	0.330	0.0312	0.0635	0.0372
302.5	3/32 × 5/16	0.0928	0.0943	0.0882	0.312	0.330	0.0469	0.0948	0.0529
203	1/16 × 3/8	0.0615	0.0630	0.1358	0.375	0.393	0.0312	0.0635	0.0372
303	3/32 × 3/8	0.0928	0.0943	0.1202	0.375	0.393	0.0469	0.0948	0.0529
403	1/8 × 3/8	0.1240	0.1255	0.1045	0.375	0.393	0.0625	0.1260	0.0685
204	1/16 × 1/2	0.0615	0.0630	0.1668	0.500	0.518	0.0312	0.0635	0.0372
304	3/32 × 1/2	0.0928	0.0943	0.1511	0.500	0.518	0.0469	0.0948	0.0529
404	1/8 × 1/2	0.1240	0.1255	0.1355	0.500	0.518	0.0625	0.1260	0.0685
305	3/32 × 5/8	0.0928	0.0943	0.1981	0.625	0.643	0.0469	0.0948	0.0529
405	1/8 × 5/8	0.1240	0.1255	0.1825	0.625	0.643	0.0625	0.1260	0.0685
505	5/32 × 5/8	0.1553	0.1568	0.1669	0.625	0.643	0.0781	0.1573	0.0841
605	3/16 × 5/8	0.1863	0.1880	0.1513	0.625	0.643	0.0937	0.1885	0.0997
406	1/8 × 3/4	0.1240	0.1255	0.2455	0.750	0.768	0.0625	0.1260	0.0685
506	5/32 × 3/4	0.1553	0.1568	0.2299	0.750	0.768	0.0781	0.1573	0.0841
606	3/16 × 3/4	0.1863	0.1880	0.2143	0.750	0.768	0.0937	0.1885	0.0997
806	1/4 × 3/4	0.2487	0.2505	0.1830	0.750	0.768	0.1250	0.2510	0.1310
507	5/32 × 7/8	0.1553	0.1568	0.2919	0.875	0.895	0.0781	0.1573	0.0841
607	3/16 × 7/8	0.1863	0.1880	0.2763	0.875	0.895	0.0937	0.1885	0.0997
707	7/32 × 7/8	0.2175	0.2193	0.2607	0.875	0.895	0.1093	0.2198	0.1153
807	1/4 × 7/8	0.2487	0.2505	0.2450	0.875	0.895	0.1250	0.2510	0.1310
608	3/16 × 1	0.1863	0.1880	0.3393	1.000	1.020	0.0937	0.1885	0.0997
708	7/32 × 1	0.2175	0.2193	0.3237	1.000	1.020	0.1093	0.2198	0.1153
808	1/4 × 1	0.2487	0.2505	0.3080	1.000	1.020	0.1250	0.2510	0.1310
1008	5/16 × 1	0.3111	0.3130	0.2768	1.000	1.020	0.1562	0.3135	0.1622
1208	3/8 × 1	0.3735	0.3755	0.2455	1.000	1.020	0.1875	0.3760	0.1935
609	3/16 × 1 1/8	0.1863	0.1880	0.3853	1.125	1.145	0.0937	0.1885	0.0997
709	7/32 × 1 1/8	0.2175	0.2193	0.3697	1.125	1.145	0.1093	0.2198	0.1153
809	1/4 × 1 1/8	0.2487	0.2505	0.3540	1.125	1.145	0.1250	0.2510	0.1310
1009	5/16 × 1 1/8	0.3111	0.3130	0.3228	1.125	1.145	0.1562	0.3135	0.1622

TABLE 19   (Continued)

Key Number	Nominal Size Key	Keyseat — Shaft					Key Above Shaft	Keyseat — Hub	
		Width A*		Depth B	Diameter F		Height C	Width D	Depth E
		Min	Max	+0.005 -0.000	Min	Max	+0.005 -0.005	+0.002 -0.000	+0.005 -0.000
610	³⁄₁₆ × 1¼	0.1863	0.1880	0.4483	1.250	1.273	0.0937	0.1885	0.0997
710	⁷⁄₃₂ × 1¼	0.2175	0.2193	0.4327	1.250	1.273	0.1093	0.2198	0.1153
810	¼ × 1¼	0.2487	0.2505	0.4170	1.250	1.273	0.1250	0.2510	0.1310
1010	⁵⁄₁₆ × 1¼	0.3111	0.3130	0.3858	1.250	1.273	0.1562	0.3135	0.1622
1210	³⁄₈ × 1¼	0.3735	0.3755	0.3545	1.250	1.273	0.1875	0.3760	0.1935
811	¼ × 1³⁄₈	0.2487	0.2505	0.4640	1.375	1.398	0.1250	0.2510	0.1310
1011	⁵⁄₁₆ × 1³⁄₈	0.3111	0.3130	0.4328	1.375	1.398	0.1562	0.3135	0.1622
1211	³⁄₈ × 1³⁄₈	0.3735	0.3755	0.4015	1.375	1.398	0.1875	0.3760	0.1935
812	¼ × 1½	0.2487	0.2505	0.5110	1.500	1.523	0.1250	0.2510	0.1310
1012	⁵⁄₁₆ × 1½	0.3111	0.3130	0.4798	1.500	1.523	0.1562	0.3135	0.1622
1212	³⁄₈ × 1½	0.3735	0.3755	0.4485	1.500	1.523	0.1875	0.3760	0.1935
617-1	³⁄₁₆ × 2⅛	0.1863	0.1880	0.3073	2.125	2.160	0.0937	0.1885	0.0997
817-1	¼ × 2⅛	0.2487	0.2505	0.2760	2.125	2.160	0.1250	0.2510	0.1310
1017-1	⁵⁄₁₆ × 2⅛	0.3111	0.3130	0.2448	2.125	2.160	0.1562	0.3135	0.1622
1217-1	³⁄₈ × 2⅛	0.3735	0.3755	0.2135	2.125	2.160	0.1875	0.3760	0.1935
617	³⁄₁₆ × 2⅛	0.1863	0.1880	0.4323	2.125	2.160	0.0937	0.1885	0.0997
817	¼ × 2⅛	0.2487	0.2505	0.4010	2.125	2.160	0.1250	0.2510	0.1310
1017	⁵⁄₁₆ × 2⅛	0.3111	0.3130	0.3698	2.125	2.160	0.1562	0.3135	0.1622
1217	³⁄₈ × 2⅛	0.3735	0.3755	0.3385	2.125	2.160	0.1875	0.3760	0.1935
822-1	¼ × 2¾	0.2487	0.2505	0.4640	2.750	2.785	0.1250	0.2510	0.1310
1022-1	⁵⁄₁₆ × 2¾	0.3111	0.3130	0.4328	2.750	2.785	0.1562	0.3135	0.1622
1222-1	³⁄₈ × 2¾	0.3735	0.3755	0.4015	2.750	2.785	0.1875	0.3760	0.1935
1422-1	⁷⁄₁₆ × 2¾	0.4360	0.4380	0.3703	2.750	2.785	0.2187	0.4385	0.2247
1622-1	½ × 2¾	0.4985	0.5005	0.3390	2.750	2.785	0.2500	0.5010	0.2560
822	¼ × 2¾	0.2487	0.2505	0.6200	2.750	2.785	0.1250	0.2510	0.1310
1022	⁵⁄₁₆ × 2¾	0.3111	0.3130	0.5888	2.750	2.785	0.1562	0.3135	0.1622
1222	³⁄₈ × 2¾	0.3735	0.3755	0.5575	2.750	2.785	0.1875	0.3760	0.1935
1422	⁷⁄₁₆ × 2¾	0.4360	0.4380	0.5263	2.750	2.785	0.2187	0.4385	0.2247
1622	½ × 2¾	0.4985	0.5005	0.4950	2.750	2.785	0.2500	0.5010	0.2560
1228	³⁄₈ × 3½	0.3735	0.3755	0.7455	3.500	3.535	0.1875	0.3760	0.1935
1428	⁷⁄₁₆ × 3½	0.4360	0.4380	0.7143	3.500	3.535	0.2187	0.4385	0.2247
1628	½ × 3½	0.4985	0.5005	0.6830	3.500	3.535	0.2500	0.5010	0.2560
1828	⁹⁄₁₆ × 3½	0.5610	0.5630	0.6518	3.500	3.535	0.2812	0.5635	0.2872
2028	⅝ × 3½	0.6235	0.6255	0.6205	3.500	3.535	0.3125	0.6260	0.3185
2228	¹¹⁄₁₆ × 3½	0.6860	0.6880	0.5893	3.500	3.535	0.3437	0.6885	0.3497
2428	¾ × 3½	0.7485	0.7505	0.5580	3.500	3.535	0.3750	0.7510	0.3810

Width A values were set with the maximum keyseat (shaft) width as that figure which will receive a key with the greatest amount of looseness consistent with assuring the key's sticking in the keyseat (shaft). Minimum keyseat width is that figure permitting the largest shaft distortion acceptable when assembling maximum key in minimum keyseat.

Dimensions A, B, C, D are taken at side intersection.

(Reprinted from The American Society of Mechanical Engineers—ANSI B17.2–1967 (R1990).)

TABLE 20   KEY SIZE VERSUS SHAFT DIAMETER

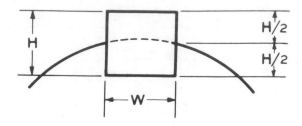

NOMINAL SHAFT DIAMETER		NOMINAL KEY SIZE			NOMINAL KEYSEAT DEPTH	
Over	To (Incl)	Width, W	Height, H		H/2	
			Square	Rectangular	Square	Rectangular
5/16	7/16	3/32	3/32		3/64	
7/16	9/16	1/8	1/8	3/32	1/16	3/64
9/16	7/8	3/16	3/16	1/8	3/32	1/16
7/8	1-1/4	1/4	1/4	3/16	1/8	3/32
1-1/4	1-3/8	5/16	5/16	1/4	5/32	1/8
1-3/8	1-3/4	3/8	3/8	1/4	3/16	1/8
1-3/4	2-1/4	1/2	1/2	3/8	1/4	3/16
2-1/4	2-3/4	5/8	5/8	7/16	5/16	7/32
2-3/4	3-1/4	3/4	3/4	1/2	3/8	1/4
3-1/4	3-3/4	7/8	7/8	5/8	7/16	5/16
3-3/4	4-1/2	1	1	3/4	1/2	3/8
4-1/2	5-1/2	1-1/4	1-1/4	7/8	5/8	7/16
5-1/2	6-1/2	1-1/2	1-1/2	1	3/4	1/2
6-1/2	7-1/2	1-3/4	1-3/4	1-1/2*	7/8	3/4
7-1/2	9	2	2	1-1/2	1	3/4
9	11	2-1/2	2-1/2	1-3/4	1-1/4	7/8
11	13	3	3	2	1-1/2	1
13	15	3-1/2	3-1/2	2-1/2	1-3/4	1-1/4
15	18	4		3		1-1/2
18	22	5		3-1/2		1-3/4
22	26	6		4		2
26	30	7		5		2-1/2

*Some key standards show 1-1/4 in. Preferred size is 1-1/2 in.
All dimensions given in inches.

Shaded areas:

For a stepped shaft, the size of a key is determined by the diameter of the shaft at the point of location of the key, regardless of the number of different diameters on the shaft.

Square keys are preferred through 6½-inch diameter shafts and rectangular keys for larger shafts. Sizes and dimensions in unshaded area are preferred.

If special considerations dictate the use of a keyseat in the hub shallower than the preferred nominal depth shown in Table 1, it is recommended that the tabulated preferred nominal standard keyseat be used in the shaft in all cases.

(Reprinted from The American Society of Mechanical Engineers—ANSI B17.1–1967 (R1989).)

TABLE 21   KEY DIMENSIONS AND TOLERANCES

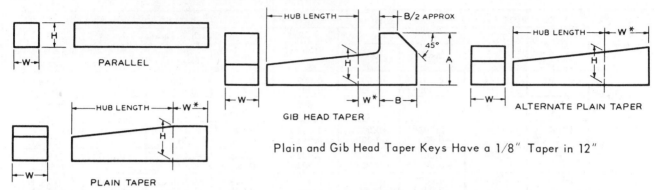

GIB HEAD TAPER

PLAIN TAPER

Plain and Gib Head Taper Keys Have a 1/8″ Taper in 12″

ALTERNATE PLAIN TAPER

KEY			NOMINAL KEY SIZE		TOLERANCE	
			Width, *W*		Width, *W*	Height, *H*
			Over	To (Incl)		
Parallel	Square	Bar Stock	— 3/4 1-1/2 2-1/2	3/4 1-1/2 2-1/2 3-1/2	+0.000 −0.002 +0.000 −0.003 +0.000 −0.004 +0.000 −0.006	+0.000 −0.002 +0.000 −0.003 +0.000 −0.004 +0.000 −0.006
		Keystock	— 1-1/4 3	1-1/4 3 3-1/2	+0.001 −0.000 +0.002 −0.000 +0.003 −0.000	+0.001 −0.000 +0.002 −0.000 +0.003 −0.000
	Rectangular	Bar Stock	— 3/4 1-1/2 3 4 6	3/4 1-1/2 3 4 6 7	+0.000 −0.003 +0.000 −0.004 +0.000 −0.005 +0.000 −0.006 +0.000 −0.008 +0.000 −0.013	+0.000 −0.003 +0.000 −0.004 +0.000 −0.005 +0.000 −0.006 +0.000 −0.008 +0.000 −0.013
		Keystock	— 1-1/4 3	1-1/4 3 7	+0.001 −0.000 +0.002 −0.000 +0.003 −0.000	+0.005 −0.005 +0.005 −0.005 +0.005 −0.005
Taper	Plain or Gib Head Square or Rectangular		— 1-1/4 3	1-1/4 3 7	+0.001 −0.000 +0.002 −0.000 +0.003 −0.000	+0.005 −0.000 +0.005 −0.000 +0.005 −0.000

*For locating position of dimension *H*. Tolerance does not apply. All dimensions given in inches.

(Reprinted from The American Society of Mechanical Engineers—ANSI B17.1–1967 (R1989).)

TABLE 22   GIB HEAD NOMINAL DIMENSIONS

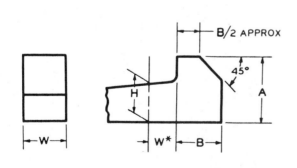

B/2 APPROX

Nominal Key Size Width, *W*	SQUARE			RECTANGULAR		
	H	*A*	*B*	*H*	*A*	*B*
1/8 3/16 1/4	1/8 3/16 1/4	1/4 5/16 7/16	1/4 5/16 3/8	3/32 1/8 3/16	3/16 1/4 5/16	1/8 1/4 5/16
5/16 3/8 1/2	5/16 3/8 1/2	1/2 5/8 7/8	7/16 1/2 5/8	1/4 1/4 3/8	7/16 7/16 5/8	3/8 3/8 1/2
5/8 3/4 7/8	5/8 3/4 7/8	1 1-1/4 1-3/8	3/4 7/8 1	7/16 1/2 5/8	3/4 7/8 1	9/16 5/8 3/4
1 1-1/4 1-1/2	1 1-1/4 1-1/2	1-5/8 2 2-3/8	1-1/8 1-7/16 1-3/4	3/4 7/8 1	1-1/8 1-3/8 1-5/8	7/8 1 1-1/8
1-3/4 2 2-1/2	1-3/4 2 2-1/2	2-3/4 3-1/2 4	2 2-1/4 3	1-1/2 1-1/2 1-3/4	2-3/8 2-3/8 2-3/4	1-3/4 1-3/4 2
3 3-1/2	3 3-1/2	5 6	3-1/2 4	2 2-1/2	3-1/2 4	2-1/4 3

*For locating position of dimension *H*.

For larger sizes the following relationships are suggested as guides
for establishing *A* and *B*.

$$A = 1.8\,H \qquad\qquad B = 1.2\,H$$

All dimensions given in inches.

(Reprinted from The American Society of Mechanical Engineers—ANSI B17.1–1967 (R1989).)

TABLE 23 CLASS 2 FIT FOR PARALLEL AND TAPER KEYS

Type of Key	KEY WIDTH		SIDE FIT			TOP AND BOTTOM FIT			
	Over	To (Incl)	Width Tolerance		Fit Range*	Depth Tolerance			Fit Range*
			Key	Keyset		Key	Shaft Keyseat	Hub Keyseat	
Parallel Square	–	1-1/4	+0.001 −0.000	+0.002 −0.000	0.002 CL 0.001 INT	+0.001 −0.000	+0.000 −0.015	+0.010 −0.000	0.030 CL 0.004 CL
	1-1/4	3	+0.002 −0.000	+0.002 −0.000	0.002 CL 0.002 INT	+0.002 −0.000	+0.000 −0.015	+0.010 −0.000	0.030 CL 0.003 CL
	3	3-1/2	+0.003 −0.000	+0.002 −0.000	0.002 CL 0.003 INT	+0.003 −0.000	+0.000 −0.015	+0.010 −0.000	0.030 CL 0.002 CL
Parallel Rectangular	–	1-1/4	+0.001 −0.000	+0.002 −0.000	0.002 CL 0.001 INT	+0.005 −0.005	+0.000 −0.015	+0.010 −0.000	0.035 CL 0.000 CL
	1-1/4	3	+0.002 −0.000	+0.002 −0.000	0.002 CL 0.002 INT	+0.005 −0.005	+0.000 −0.015	+0.010 −0.000	0.035 CL 0.000 CL
	3	7	+0.003 −0.000	+0.002 −0.000	0.002 CL 0.003 INT	+0.005 −0.005	+0.000 −0.015	+0.010 −0.000	0.035 CL 0.000 CL
Taper	–	1-1/4	+0.001 −0.000	+0.002 −0.000	0.002 CL 0.001 INT	+0.005 −0.000	+0.000 −0.015	+0.010 −0.000	0.005 CL 0.025 INT
	1-1/4	3	+0.002 −0.000	+0.002 −0.000	0.002 CL 0.002 INT	+0.005 −0.000	+0.000 −0.015	+0.010 −0.000	0.005 CL 0.025 INT
	3	Δ	+0.003 −0.000	+0.002 −0.000	0.002 CL 0.003 INT	+0.005 −0.000	+0.000 −0.015	+0.010 −0.000	0.005 CL 0.025 INT

*Limits of variation. CL = Clearance; INT = Interference

Δ To (Incl) 3-1/2 Square and 7 Rectangular key widths.

All dimensions given in inches.

(Reprinted from The American Society of Mechanical Engineers—ANSI B17.1–1967 (R1989).)

TABLE 24 DECIMAL EQUIVALENTS AND TAP DRILL SIZES

FRACTION OR DRILL SIZE	DECIMAL EQUIVALENT	TAP SIZE
NUMBER SIZE DRILLS 80	.0135	
79	.0145	
1/64	.0156	
78	.0160	
77	.0180	
76	.0200	
75	.0210	
74	.0225	
73	.0240	
72	.0250	
71	.0260	
70	.0280	
69	.0292	
68	.0310	
1/32	.0312	
67	.0320	
66	.0330	
65	.0350	
64	.0360	
63	.0370	
62	.0380	
61	.0390	
60	.0400	
59	.0410	
58	.0420	
57	.0430	
56	.0465	
3/64	.0469	0-80
55	.0520	
54	.0550	1-56
53	.0595	1-64, 72
1/16	.0625	
52	.0635	
51	.0670	
50	.0700	2-56, 64
49	.0730	
48	.0760	
5/64	.0781	
47	.0785	
46	.0810	
45	.0820	3-48
44	.0860	3-56, 4-32
43	.0890	4-36
42	.0935	4-40
3/32	.0938	4-48
41	.0960	
40	.0980	

FRACTION OR DRILL SIZE	DECIMAL EQUIVALENT	TAP SIZE
39	.0995	
38	.1015	5-40
37	.1040	5-44
36	.1065	6-32
7/64	.1094	
35	.1100	
34	.1110	
33	.1130	6-36
32	.1160	6-40
31	.1200	
1/8	.1250	
30	.1285	
29	.1360	8-32, 36
28	.1405	8-40
9/64	.1406	
27	.1440	
26	.1470	
25	.1495	10-24
24	.1520	
23	.1540	
5/32	.1562	
22	.1570	10-30
21	.1590	10-32
20	.1610	
19	.1660	
18	.1695	
11/64	.1719	
17	.1730	
16	.1770	12-24
15	.1800	
14	.1820	12-28
13	.1850	12-32
3/16	.1875	
12	.1890	
11	.1910	
10	.1935	
9	.1960	
8	.1990	
7	.2010	1/4-20
13/64	.2031	
6	.2040	
5	.2055	
4	.2090	
3	.2130	1/4-28
7/32	.2188	
2	.2210	
1	.2280	
LETTER SIZE DRILLS A	.2340	

FRACTION OR DRILL SIZE	DECIMAL EQUIVALENT	TAP SIZE
15/64	.2344	
LETTER SIZE DRILLS B	.2380	
C	.2420	
D	.2460	
1/4	.2500	
E	.2570	5/16-18
F	.2610	
17/64	.2656	
G	.2660	
H	.2720	
I	.2770	5/16-24
J	.2810	
9/32	.2812	
K	.2900	
L	.2950	
19/64	.2969	
M	.3020	
5/16	.3125	3/8-16
N	.3160	
O	.3230	
21/64	.3281	
P	.3320	3/8-24
Q	.3390	
11/32	.3438	
R	.3480	
S	.3580	
23/64	.3594	
T	.3680	7/16-14
3/8	.3750	
U	.3770	
V	.3860	
25/64	.3906	7/16-20
W	.3970	
X	.4040	
13/32	.4062	
Y	.4130	
27/64 7/16	.4219	1/2-13
29/64 15/64	.4375	1/2-20
31/32	.4531	
	.4688	
1/2	.4844	9/16-12
	.5000	
33/64 17/32	.5156	9/16-18
35/64	.5312	5/8-11
	.5469	
37/64 9/16	.5625	
	.5781	5/8-18

FRACTION OR DRILL SIZE	DECIMAL EQUIVALENT	TAP SIZE
19/32	.5938	11/16-11
39/64	.6094	
5/8	.6250	11/16-16
41/64	.6406	
21/32	.6562	3/4-10
43/64	.6719	
11/16	.6875	3/4-16
45/64	.7031	
23/32	.7188	
47/64	.7344	
3/4	.7500	
49/64	.7656	7/8-9
25/32	.7812	
51/64	.7969	
13/16	.8125	7/8-14
53/64	.8281	
27/32	.8438	
55/64	.8594	
7/8	.8750	1-8
57/64	.8906	
29/32	.9062	
59/64	.9219	
15/16	.9375	1-12, 14
61/64	.9531	
31/32	.9688	
63/64	.9844	
1	1.0000	1 1/8-7
1 3/64	1.0469	1 1/8-12
1 1/64	1.1094	1 1/4-7
1 1/8	1.1250	1 1/4-12
1 11/64	1.1719	1 1/4-12
1 7/32	1.2188	1 3/8-6
1 1/4	1.2500	1 3/8-12
1 19/64	1.2969	1 1/2-6
1 11/32	1.3438	1 1/2-12
1 3/8	1.3750	
1 27/64	1.4219	1 1/2-12
1 1/2	1.5000	

PIPE THREAD SIZES

THREAD	DRILL	THREAD	DRILL
1/8-27	R	1 1/2-11 1/2	1 47/64
1/4-18	7/16	2-11 1/2	2 7/32
3/8-18	37/64	2 1/2-8	2 5/8
1/2-14	23/32	3-8	3 1/4
3/4-14	59/64	3 1/2-8	3 3/4
1-11 1/2	1 5/32	4-8	4 1/4
1 1/4-11 1/2	1 1/2		

Courtesy The L. S. Starrett Company, Athol, Massachusetts.

TABLE 25   GENERAL APPLICATIONS OF SAE STEELS

Application	SAE No.	Application	SAE No.
Adapters	1145	Chain pins, transmission	4320
Agricultural steel	1070	" " "	4815
" "	1080	" " "	4820
Aircraft forgings	4140	Chains, transmission	3135
Axles, front or rear	1040	" "	3140
" "	4140	Clutch disks	1060
Axle shafts	1045	"	1070
" "	2340	"	1085
" "	2345	Clutch springs	1060
" "	3135	Coil springs	4063
" "	3140	Cold-headed bolts	4042
" "	3141	Cold-heading steel	30905
" "	4063	Cold-heading wire or rod	rimmed*
" "	4340	" " " "	1035
Ball-bearing races	52100	Cold-rolled steel	1070
Balls for ball bearings	52100	Connecting-rods	1040
Body stock for cars	rimmed*	"	3141
Bolts, anchor	1040	Connecting-rod bolts	3130
Bolts and screws	1035	Corrosion resisting	51710
Bolts, cold-headed	4042	"	30805
Bolts, connecting-rod	3130	Covers, transmission	rimmed*
Bolts, heat-treated	2330	Crankshafts	1045
Bolts, heavy-duty	4815	"	1145
"	4820	"	3135
Bolts, steering-arm	3130	"	3140
Brake levers	1030	"	3141
"	1040	Crankshafts, Diesel engine	4340
Bumper bars	1085	Cushion springs	1060
Cams, free-wheeling	4615	Cutlery, stainless	51335
" "	4620	Cylinder studs	3130
Camshafts	1020	Deep-drawing steel	rimmed*
"	1040	" "	30905
Carburized parts	1020	Differential gears	4023
" "	1022	Disks, clutch	1070
" "	1024	"	1060
" "	1320	Ductile steel	30905
" "	2317	Fan blades	1020
" "	2515	Fatigue resisting	4340
" "	3310	" "	4640
" "	3115	Fender stock for cars	rimmed*
" "	3120	Forgings, aircraft	4140
" "	4023	Forgings, carbon steel	1040
" "	4032	"	1045
" "	1117	Forgings, heat-treated	3240
" "	1118	" " "	5140

Application	SAE No.	Application	SAE No.
Forgings, heat-treated	6150	Key stock	1030
Forgings, high-duty	6150	"	2330
Forgings, small or medium	1035	"	3130
Forgings, large	1036	Leaf springs	1085
Free-cutting carbon steel	1111	"	9260
" "	1113	Levers, brake	1030
Free-cutting chro.-ni. steel	30615	"	1040
Free-cutting mang. steel	1132	Levers, gear shift	1030
" "	1137	Levers, heat-treated	2330
Gears, carburized	1320	Lock-washers	1060
" "	2317	Mower knives	1085
" "	3115	Mower sections	1070
" "	3120	Music wire	1085
" "	3310	Nuts	3130
" "	4119	Nuts, heat-treated	2330
" "	4125	Oil-pans, automobile	rimmed*
" "	4320	Pinions, carburized	3115
" "	4615	" "	3120
" "	4620	" "	4320
" "	4815	Piston-pins	3115
" "	4820	"	3120
Gears, heat-treated	2345	Plow beams	1070
Gears, car and truck	4027	Plow disks	1080
"	4032	Plow shares	1080
Gears, cyanide-hardening	5140	Propeller shafts	2340
Gears, differential	4023	" "	2345
Gears, high duty	4640	" "	4140
"	6150	Races, ball-bearing	52100
Gears, oil-hardening	3145	Ring gears	3115
" "	3150	"	3120
" "	4340	"	4119
" "	5150	Rings, snap	1060
Gears, ring	1045	Rivets	rimmed*
" "	3115	Rod and wire	killed*
" "	3120	Rod, cold-heading	1035
" "	4119	Roller bearings	4815
Gears, transmission	3115	Rollers for bearings	52100
" "	3120	Screws and bolts	1035
" "	4119	Screw stock, Bessemer	1111
Gears, truck and bus	3310	" " "	1112
"	4320	" " "	1113
Gear shift levers	1030	Screw stock, open hearth	1115
Harrow disks	1080	Screws, heat-treated	2330
Hay-rake teeth	1095	Seat springs	1095
		Shafts, axle	1045

Application	SAE No.	Application	SAE No.
Shafts, cyanide-hardening	5140	Steel, cold-heading	30905
Shafts, heavy-duty	4340	Steel, free-cutting carbon	1111
" " "	6150	" " "	1113
" " "	4615	Steel, free-cutting chro.-nl.	30615
" " "	4620	Steel, free-cutting mang.	1132
Shafts, oil hardening	5150	"	0000
Shafts, propeller	2340	Steel, minimum distortion	4615
"	2345	" " "	4620
"	4140	" " "	4640
Shafts, transmission	4140	Steel, soft ductile	30905
Sheets and strips	rimmed*	Steering arms	4012
Snap rings	1060	Steering-arm bolts	3130
Spline shafts	1045	Steering knuckles	3141
"	1320	Steering-knuckle pins	4815
" "	2340	" " "	4820
" "	2345	Studs	1040
" "	3115	"	1111
" "	3120	Studs, cold-headed	4042
" "	3135	Studs, cylinder	3130
" "	3140	Studs, heat-treated	2330
" "	4023	Studs, heavy-duty	4815
Spring clips	1060	" "	4820
Springs, coil	1095	Tacks	rimmed*
"	4063	Thrust washers	1060
" "	6150	Thrust washers, oil-harden.	5150
Springs, clutch	1060	Transmission shafts	4140
Springs, cushion	1060	Tubing	1040
Springs, leaf	1085	Tubing, front axle	4140
"	1095	Tubing, seamless	1030
" "	4063	Tubing, welded	1020
" "	4068	Universal joints	1145
" "	9260	Valve springs	1060
" "	6150	Washers, lock	1060
Springs, hard-drawn coiled	1066	Welded structures	30705
Springs, oil-hardening	5150	Wire and rod	killed*
Springs, oil-tempered wire	1066	Wire, cold-heading	rimmed*
Springs, seat	1095	" "	1035
Springs, valve	1060	Wire, hard-drawn spring	1045
Spring wire	1045	" " "	1055
Spring wire, hard-drawn	1055	Wire, music	1085
Spring wire, oil-tempered	1055	Wire, oil-tempered spring	1055
Stainless irons	51210	Wrist-pins, automobile	1020
" "	51710	Yokes	1145
Steel, cold-rolled	1070		

* The "rimmed" and "killed" steels listed are in the SAE 1008, 1010 and 1015 group.   See general description of these steels.

Reprinted by permission from Oberg, Jones, and Horton, Machinery's Handbook, 24th Edition. (New York: Industrial Press, Inc., 1992), table 6, pp. 382–84.

TABLE 26 SURFACE ROUGHNESS PRODUCED BY COMMON PRODUCTION METHODS

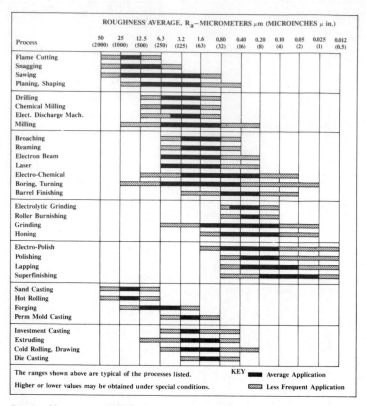

Reprinted by permission, from Oberg, Jones, and Horton, Machinery's Handbook, 24th ed. (New York: Industrial Press, Inc., 1992), figure 5, p. 672.

TABLE 27 ALLOWANCES AND TOLERANCES

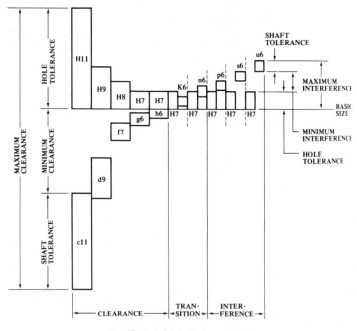

Preferred Hole Basis Fits

Reprinted by permission, from Oberg, Jones, and Horton, Machinery's Handbook, 24th ed. (New York: Industrial Press, Inc., 1992), figure 2, p. 622.

TABLE 27 (Continued)

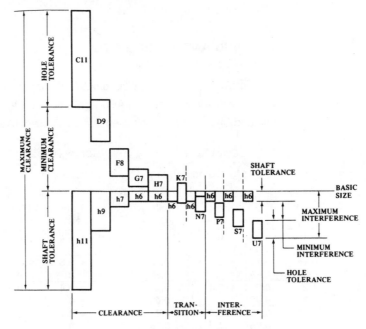

Preferred Shaft Basis Fits

Reprinted by permission, from Oberg, Jones, and Horton, Machinery's Handbook, *24th ed. (New York: Industrial Press, Inc., 1992), figure 3, p. 623.*

ISO SYMBOL		DESCRIPTION
Hole Basis	**Shaft Basis**	
H11/c11	C11/h11	*Loose running* fit for wide commercial tolerances or allowances on external members.
H9/d9	D9/h9	*Free running* fit not for use where accuracy is essential, but good for large temperature variations, high running speeds, or heavy journal pressures.
H8/f7	F8/h7	*Close running* fit for running on accurate machines and for accurate location at moderate speeds and journal pressures.
H7/g6	G7/h6	*Sliding* fit not intended to run freely, but to move and turn freely and locate accurately.
H7/h6	H7/h6	*Locational clearance* fit provides snug fit for locating stationary parts; but can be freely assembled and disassembled.
H7/k6	K7/h6	*Locational transition* fit for accurate location, a compromise between clearance and interference.
H7/n6	N7/h6	*Locational transition* fit for more accurate location where greater interference is permissible.
H7/p6[1]	P7/h6	*Locational interference* fit for parts requiring rigidity and alignment with prime accuracy of location but without special bore pressure requirements.
H7/s6	S7/h6	*Medium drive* fit for ordinary steel parts or shrink fits on light sections, the tightest fit usable with cast iron.
H7/u6	U7/h6	*Force* fit suitable for parts which can be highly stressed or for shrink fits where the heavy pressing forces required are impractical.

Clearance Fits / Transition Fits / Interference Fits

More Clearance / More Interference

[1] Transition fit for basic sizes in range from 0 through 3 mm.

Description of Preferred Fits

Reprinted by permission, from Oberg, Jones, and Horton, Machinery's Handbook, *24th ed. (New York: Industrial Press, Inc., 1992), figure 4, p. 624.*

TABLE 28   AMERICAN NATIONAL STANDARD FITS

RUNNING AND SLIDING FITS

Limits are in thousandths of an inch.

Limits for hole and shaft are applied algebraically to the basic size to obtain the limits of size for the parts.

Data in **boldface** are in accordance with ABC agreements.

Symbols H5, g5, etc., are Hole and Shaft designations used in ABC System (Appendix I).

Nominal Size Range Inches (Over)	(To)	Class RC 1 Limits of Clearance	Hole H5	Shaft g4	Class RC 2 Limits of Clearance	Hole H6	Shaft g5	Class RC 3 Limits of Clearance	Hole H7	Shaft f6	Class RC 4 Limits of Clearance	Hole H8	Shaft f7
0	0.12	0.1 / 0.45	+0.2 / 0	−0.1 / −0.25	0.1 / 0.55	+0.25 / 0	−0.1 / −0.3	0.3 / 0.95	+0.4 / 0	−0.3 / −0.55	0.3 / 1.3	+0.6 / 0	−0.3 / −0.7
0.12	0.24	0.15 / 0.5	+0.2 / 0	−0.15 / −0.3	0.15 / 0.65	+0.3 / 0	−0.15 / −0.35	0.4 / 1.12	+0.5 / 0	−0.4 / −0.7	0.4 / 1.6	+0.7 / 0	−0.4 / −0.9
0.24	0.40	0.2 / 0.6	0.25 / 0	−0.2 / −0.35	0.2 / 0.85	+0.4 / 0	−0.2 / −0.45	0.5 / 1.5	+0.6 / 0	−0.5 / −0.9	0.5 / 2.0	+0.9 / 0	−0.5 / −1.1
0.40	0.71	0.25 / 0.75	+0.3 / 0	−0.25 / −0.45	0.25 / 0.95	+0.4 / 0	−0.25 / −0.55	0.6 / 1.7	+0.7 / 0	−0.6 / −1.0	0.6 / 2.3	+1.0 / 0	−0.6 / −1.3
0.71	1.19	0.3 / 0.95	+0.4 / 0	−0.3 / −0.55	0.3 / 1.2	+0.5 / 0	−0.3 / −0.7	0.8 / 2.1	+0.8 / 0	−0.8 / −1.3	0.8 / 2.8	+1.2 / 0	−0.8 / −1.6
1.19	1.97	0.4 / 1.1	+0.4 / 0	−0.4 / −0.7	0.4 / 1.4	+0.6 / 0	−0.4 / −0.8	1.0 / 2.6	+1.0 / 0	−1.0 / −1.6	1.0 / 3.6	+1.6 / 0	−1.0 / −2.0
1.97	3.15	0.4 / 1.2	+0.5 / 0	−0.4 / −0.7	0.4 / 1.6	+0.7 / 0	−0.4 / −0.9	1.2 / 3.1	+1.2 / 0	−1.2 / −1.9	1.2 / 4.2	+1.8 / 0	−1.2 / −2.4
3.15	4.73	0.5 / 1.5	+0.6 / 0	−0.5 / −0.9	0.5 / 2.0	+0.9 / 0	−0.5 / −1.1	1.4 / 3.7	+1.4 / 0	−1.4 / −2.3	1.4 / 5.0	+2.2 / 0	−1.4 / −2.8
4.73	7.09	0.6 / 1.8	+0.7 / 0	−0.6 / −1.1	0.6 / 2.3	+1.0 / 0	−0.6 / −1.3	1.6 / 4.2	+1.6 / 0	−1.6 / −2.6	1.6 / 5.7	+2.5 / 0	−1.6 / −3.2
7.09	9.85	0.6 / 2.0	+0.8 / 0	−0.6 / −1.2	0.6 / 2.6	+1.2 / 0	−0.6 / −1.4	2.0 / 5.0	+1.8 / 0	−2.0 / −3.2	2.0 / 6.6	+2.8 / 0	−2.0 / −3.8
9.85	12.41	0.8 / 2.3	+0.9 / 0	−0.8 / −1.4	0.8 / 2.9	+1.2 / 0	−0.8 / −1.7	2.5 / 5.7	+2.0 / 0	−2.5 / −3.7	2.5 / 7.5	+3.0 / 0	−2.5 / −4.5
12.41	15.75	1.0 / 2.7	+1.0 / 0	−1.0 / −1.7	1.0 / 3.4	+1.4 / 0	−1.0 / −2.0	3.0 / 6.6	+ / 0	−3.0 / −4.4	3.0 / 8.7	+3.5 / 0	−3.0 / −5.2
15.75	19.69	1.2 / 3.0	+1.0 / 0	−1.2 / −2.0	1.2 / 3.8	+1.6 / 0	−1.2 / −2.2	4.0 / 8.1	+1.6 / 0	−4.0 / −5.6	4.0 / 10.5	+4.0 / 0	−4.0 / −6.5
19.69	30.09	1.6 / 3.7	+1.2 / 0	−1.6 / −2.5	1.6 / 4.8	+2.0 / 0	−1.6 / −2.8	5.0 / 10.0	+3.0 / 0	−5.0 / −7.0	5.0 / 13.0	+5.0 / 0	−5.0 / −8.0
30.09	41.49	2.0 / 4.6	+1.6 / 0	−2.0 / −3.0	2.0 / 6.1	+2.5 / 0	−2.0 / −3.6	6.0 / 12.5	+4.0 / 0	−6.0 / −8.5	6.0 / 16.0	+6.0 / 0	−6.0 / −10.0
41.49	56.19	2.5 / 5.7	+2.0 / 0	−2.5 / −3.7	2.5 / 7.5	+3.0 / 0	−2.5 / −4.5	8.0 / 16.0	+5.0 / 0	−8.0 / −11.0	8.0 / 21.0	+8.0 / 0	−8.0 / −13.0
56.19	76.39	3.0 / 7.1	+2.5 / 0	−3.0 / −4.6	3.0 / 9.5	+4.0 / 0	−3.0 / −5.5	10.0 / 20.0	+6.0 / 0	−10.0 / −14.0	10.0 / 26.0	+10.0 / 0	−10.0 / −16.0
76.39	100.9	4.0 / 9.0	+3.0 / 0	−4.0 / −6.0	4.0 / 12.0	+5.0 / 0	−4.0 / −7.0	12.0 / 25.0	+8.0 / 0	−12.0 / −17.0	12.0 / 32.0	+12.0 / 0	−12.0 / −20.0
100.9	131.9	5.0 / 11.5	+4.0 / 0	−5.0 / −7.5	5.0 / 15.0	+6.0 / 0	−5.0 / −9.0	16.0 / 32.0	+10.0 / 0	−16.0 / −22.0	16.0 / 36.0	+16.0 / 0	−16.0 / −26.0
131.9	171.9	6.0 / 14.0	+5.0 / 0	−6.0 / −9.0	6.0 / 19.0	+8.0 / 0	−6.0 / −11.0	18.0 / 38.0	+8.0 / 0	−18.0 / −26.0	18.0 / 50.0	+20.0 / 0	−18.0 / −30.0
171.9	200	8.0 / 18.0	+6.0 / 0	−8.0 / −12.0	8.0 / 22.0	+10.0 / 0	−8.0 / −12.0	22.0 / 48.0	+16.0 / 0	−22.0 / −32.0	22.0 / 63.0	+25.0 / 0	−22.0 / −38.0

TABLE 28   (Continued)

RUNNING AND SLIDING FITS *(continued)*

Limits are in thousandths of an inch.

Limits for hole and shaft are applied algebraically to the basic size to obtain the limits of size for the parts.

Data in **boldface** are in accordance with ABC agreements.

Symbols H8, e7, etc., are Hole and Shaft designations used in ABC System (Appendix I).

Class RC 5 LoC	Hole H8	Shaft c7	Class RC 6 LoC	Hole H9	Shaft e8	Class RC 7 LoC	Hole H9	Shaft d8	Class RC 8 LoC	Hole H10	Shaft c9	Class RC 9 LoC	Hole H11	Shaft	Nominal Over	To
0.6 / 1.6	+0.6 / −0	−0.6 / −1.0	0.6 / 2.2	+1.0 / 0	−0.6 / −1.2	1.0 / 2.6	+1.0 / 0	−1.0 / −1.6	2.5 / 5.1	+1.6 / 0	−2.5 / −3.5	4.0 / 8.1	+2.5 / 0	−4.0 / −5.6	0	0.12
0.8 / 2.0	+0.7 / −0	−0.8 / −1.3	0.8 / 2.7	+1.2 / 0	−0.8 / −1.5	1.2 / 3.1	+1.2 / 0	−1.2 / −1.9	2.8 / 5.8	+1.8 / 0	−2.8 / −4.0	4.5 / 9.0	+3.0 / 0	−4.5 / −6.0	0.12	0.24
1.0 / 2.5	+0.9 / −0	−1.0 / −1.6	1.0 / 3.3	+1.4 / 0	−1.0 / −1.9	1.6 / 3.9	+1.4 / 0	−1.6 / −2.5	3.0 / 6.6	+2.2 / 0	−3.0 / −4.4	5.0 / 10.7	+3.5 / 0	−5.0 / −7.2	0.24	0.40
1.2 / 2.9	+1.0 / −0	−1.2 / −1.9	1.2 / 3.8	+1.6 / 0	−1.2 / −2.2	2.0 / 4.6	+1.6 / 0	−2.0 / −3.0	3.5 / 7.9	+2.8 / 0	−3.5 / −5.1	6.0 / 12.8	+4.0 / 0	−6.0 / −8.8	0.40	0.71
1.6 / 3.6	+1.2 / −0	−1.6 / −2.4	1.6 / 4.8	+2.0 / 0	−1.6 / −2.8	2.5 / 5.7	+2.0 / 0	−2.5 / −3.7	4.5 / 10.0	+3.5 / 0	−4.5 / −6.5	7.0 / 15.5	+5.0 / 0	−7.0 / −10.5	0.71	1.19
2.0 / 4.6	+1.6 / −0	−2.0 / −3.0	2.0 / 6.1	+2.5 / 0	−2.0 / −3.6	3.0 / 7.1	+2.5 / 0	−3.0 / −4.6	5.0 / 11.5	+4.0 / 0	−5.0 / −7.5	8.0 / 18.0	+6.0 / 0	−8.0 / −12.0	1.19	1.97
2.5 / 5.5	+1.8 / −0	−2.5 / −3.7	2.5 / 7.3	+3.0 / 0	−2.5 / −4.3	4.0 / 8.8	+3.0 / 0	−4.0 / −5.8	6.0 / 13.5	+4.5 / 0	−6.0 / −9.0	9.0 / 20.5	+7.0 / 0	−9.0 / −13.5	1.97	3.15
3.0 / 6.6	+2.2 / −0	−3.0 / −4.4	3.0 / 8.7	+3.5 / 0	−3.0 / −5.2	5.0 / 10.7	+3.5 / 0	−5.0 / −7.2	7.0 / 15.5	+5.0 / 0	−7.0 / −10.5	10.0 / 24.0	+9.0 / 0	−10.0 / −15.0	3.15	4.73
3.5 / 7.6	+2.5 / −0	−3.5 / −5.1	3.5 / 10.0	+4.0 / 0	−3.5 / −6.0	6.0 / 12.5	+4.0 / 0	−6.0 / −8.5	8.0 / 18.0	+6.0 / 0	−8.0 / −12.0	12.0 / 28.0	+10.0 / 0	−12.0 / −18.0	4.73	7.09
4.0 / 8.6	+2.8 / −0	−4.0 / −5.8	4.0 / 11.3	+4.5 / 0	−4.0 / −6.8	7.0 / 14.3	+4.5 / 0	−7.0 / −9.8	10.0 / 21.5	+7.0 / 0	−10.0 / −14.5	15.0 / 34.0	+12.0 / 0	−15.0 / −22.0	7.09	9.85
5.0 / 10.0	+3.0 / 0	−5.0 / −7.0	5.0 / 13.0	+5.0 / 0	−5.0 / −8.0	8.0 / 16.0	+5.0 / 0	−8.0 / −11.0	12.0 / 25.0	+8.0 / 0	−12.0 / −17.0	18.0 / 38.0	+12.0 / 0	−18.0 / −26.0	9.85	12.41
6.0 / 11.7	+3.5 / 0	−6.0 / −8.2	6.0 / 15.5	+6.0 / 0	−6.0 / −9.5	10.0 / 19.5	+6.0 / 0	−10.0 / −13.5	**14.0 / 29.0**	**+9.0 / 0**	**−14.0 / −20.0**	22.0 / 45.0	+14.0 / 0	−22.0 / −31.0	12.41	15.75
8.0 / 14.5	+4.0 / 0	−8.0 / −10.5	8.0 / 18.0	+6.0 / 0	−8.0 / −12.0	12.0 / 22.0	+6.0 / 0	−12.0 / −16.0	**16.0 / 32.0**	**+10.0 / 0**	**−16.0 / −22.0**	25.0 / 51.0	+16.0 / 0	−25.0 / −35.0	15.75	19.69
10.0 / 18.0	+5.0 / 0	−10.0 / −13.0	10.0 / 23.0	+8.0 / 0	−10.0 / −15.0	16.0 / 29.0	+8.0 / 0	−16.0 / −21.0	20.0 / 40.0	+12.0 / 0	−20.0 / −28.0	30.0 / 62.0	+20.0 / 0	−30.0 / −42.0	19.69	30.09
12.0 / 22.0	+6.0 / 0	−12.0 / −16.0	12.0 / 28.0	+10.0 / 0	−12.0 / −18.0	20.0 / 36.0	+10.0 / 0	−20.0 / −26.0	25.0 / 51.0	+16.0 / 0	−25.0 / −35.0	40.0 / 81.0	+25.0 / 0	−40.0 / −56.0	30.09	41.49
16.0 / 29.0	+8.0 / 0	−16.0 / −21.0	16.0 / 36.0	+12.0 / 0	−16.0 / −24.0	25.0 / 45.0	+12.0 / 0	−25.0 / −33.0	30.0 / 62.0	+20.0 / 0	−30.0 / −42.0	50.0 / 100	+30.0 / 0	−50.0 / −70.0	41.49	56.19
20.0 / 36.0	+10.0 / 0	−20.0 / −26.0	20.0 / 46.0	+16.0 / 0	−20.0 / −30.0	30.0 / 56.0	+16.0 / 0	−30.0 / −40.0	40.0 / 81.0	+25.0 / 0	−40.0 / −56.0	60.0 / 125	+40.0 / 0	−60.0 / −85.0	56.19	76.39
25.0 / 45.0	+12.0 / 0	−25.0 / −33.0	25.0 / 57.0	+20.0 / 0	−25.0 / −37.0	40.0 / 72.0	+20.0 / 0	−40.0 / −52.0	50.0 / 100	+30.0 / 0	−50.0 / −70.0	80.0 / 160	+50.0 / 0	−80.0 / −110	76.39	100.9
30.0 / 56.0	+16.0 / 0	−30.0 / −40.0	30.0 / 71.0	+25.0 / 0	−30.0 / −46.0	50.0 / 91.0	+25.0 / 0	−50.0 / −66.0	60.0 / 125	+40.0 / 0	−60.0 / −85.0	100 / 200	+60.0 / 0	−100 / −140	100.9	131.9
35.0 / 67.0	+20.0 / 0	−35.0 / −47.0	35.0 / 85.0	+30.0 / 0	−35.0 / −55.0	60.0 / 110.0	+30.0 / 0	−60.0 / −80.0	80.0 / 160	+50.0 / 0	−80.0 / −110	130 / 260	+80.0 / 0	−130 / −180	131.9	171.9
45.0 / 86.0	+25.0 / 0	−45.0 / −61.0	45.0 / 110.0	+40.0 / 0	−45.0 / −70.0	80.0 / 145.0	+40.0 / 0	−80.0 / −105.0	100 / 200	+60.0 / 0	−100 / −140	150 / 310	+100 / 0	−150 / −210	171.9	200

(Reprinted from The American Society of Mechanical Engineers—ANSI B4.1–1967 (R1987)

TABLE 28 (Continued)

AMERICAN NATIONAL STANDARD PREFERRED METRIC LIMITS AND FITS   ANSI B4.2–1978

Dimensions in mm.

TABLE 2 PREFERRED HOLE BASIS CLEARANCE FITS

BASIC SIZE		LOOSE RUNNING			FREE RUNNING			CLOSE RUNNING			SLIDING			LOCATIONAL CLEARANCE		
		Hole H11	Shaft c11	Fit	Hole H9	Shaft d9	Fit	Hole H8	Shaft f7	Fit	Hole H7	Shaft g6	Fit	Hole H7	Shaft h6	Fit
1	MAX	1.060	0.940	0.180	1.025	0.980	0.070	1.014	0.994	0.030	1.010	0.998	0.018	1.010	1.000	0.016
	MIN	1.000	0.880	0.060	1.000	0.955	0.020	1.000	0.984	0.006	1.000	0.992	0.002	1.000	0.994	0.000
1.2	MAX	1.260	1.140	0.180	1.225	1.180	0.070	1.214	1.194	0.030	1.210	1.198	0.018	1.210	1.200	0.016
	MIN	1.200	1.080	0.060	1.200	1.155	0.020	1.200	1.184	0.006	1.200	1.192	0.002	1.200	1.194	0.000
1.6	MAX	1.660	1.540	0.180	1.625	1.580	0.070	1.614	1.594	0.030	1.610	1.598	0.018	1.610	1.600	0.016
	MIN	1.600	1.480	0.060	1.600	1.555	0.020	1.600	1.584	0.006	1.600	1.592	0.002	1.600	1.594	0.000
2	MAX	2.060	1.940	0.180	2.025	1.980	0.070	2.014	1.994	0.030	2.010	1.998	0.018	2.010	2.000	0.016
	MIN	2.000	1.880	0.060	2.000	1.955	0.020	2.000	1.984	0.006	2.000	1.992	0.002	2.000	1.994	0.000
2.5	MAX	2.560	2.440	0.180	2.525	2.480	0.070	2.514	2.494	0.030	2.510	2.498	0.018	2.510	2.500	0.016
	MIN	2.500	2.380	0.060	2.500	2.455	0.020	2.500	2.484	0.006	2.500	2.492	0.002	2.500	2.494	0.000
3	MAX	3.060	2.940	0.180	3.025	2.980	0.070	3.014	2.994	0.030	3.010	2.998	0.018	3.010	3.000	0.016
	MIN	3.000	2.880	0.060	3.000	2.955	0.020	3.000	2.984	0.006	3.000	2.992	0.002	3.000	2.994	0.000
4	MAX	4.075	3.930	0.220	4.030	3.970	0.090	4.018	3.990	0.040	4.012	3.996	0.024	4.012	4.000	0.020
	MIN	4.000	3.855	0.070	4.000	3.940	0.030	4.000	3.978	0.010	4.000	3.988	0.004	4.000	3.992	0.000
5	MAX	5.075	4.930	0.220	5.030	4.970	0.090	5.018	4.990	0.040	5.012	4.996	0.024	5.012	5.000	0.020
	MIN	5.000	4.855	0.070	5.000	4.940	0.030	5.000	4.978	0.010	5.000	4.988	0.004	5.000	4.992	0.000
6	MAX	6.075	5.930	0.220	6.030	5.970	0.090	6.018	5.990	0.040	6.012	5.996	0.024	6.012	6.000	0.020
	MIN	6.000	5.855	0.070	6.000	5.940	0.030	6.000	5.978	0.010	6.000	5.988	0.004	6.000	5.992	0.000
8	MAX	8.090	7.920	0.260	8.036	7.960	0.112	8.022	7.987	0.050	8.015	7.995	0.029	8.015	8.000	0.024
	MIN	8.000	7.830	0.080	8.000	7.924	0.040	8.000	7.972	0.013	8.000	7.986	0.005	8.000	7.991	0.000
10	MAX	10.090	9.920	0.260	10.036	9.960	0.112	10.022	9.987	0.050	10.015	9.995	0.029	10.015	10.000	0.024
	MIN	10.000	9.830	0.080	10.000	9.924	0.040	10.000	9.972	0.013	10.000	9.986	0.005	10.000	9.991	0.000
12	MAX	12.110	11.905	0.315	12.043	11.950	0.136	12.027	11.984	0.061	12.018	11.994	0.035	12.018	12.000	0.029
	MIN	12.000	11.795	0.095	12.000	11.907	0.050	12.000	11.966	0.016	12.000	11.983	0.006	12.000	11.989	0.000
16	MAX	16.110	15.905	0.315	16.043	15.950	0.136	16.027	15.984	0.061	16.018	15.994	0.035	16.018	16.000	0.029
	MIN	16.000	15.795	0.095	16.000	15.907	0.050	16.000	15.966	0.016	16.000	15.983	0.006	16.000	15.989	0.000
20	MAX	20.130	19.890	0.370	20.052	19.935	0.169	20.033	19.980	0.074	20.021	19.993	0.041	20.021	20.000	0.034
	MIN	20.000	19.760	0.110	20.000	19.883	0.065	20.000	19.959	0.020	20.000	19.980	0.007	20.000	19.987	0.000
25	MAX	25.130	24.890	0.370	25.052	24.935	0.169	25.033	24.980	0.074	25.021	24.993	0.041	25.021	25.000	0.034
	MIN	25.000	24.760	0.110	25.000	24.883	0.065	25.000	24.959	0.020	25.000	24.980	0.007	25.000	24.987	0.000
30	MAX	30.130	29.890	0.370	30.052	29.935	0.169	30.033	29.980	0.074	30.021	29.993	0.041	30.021	30.000	0.034
	MIN	30.000	29.760	0.110	30.000	29.883	0.065	30.000	29.959	0.020	30.000	29.980	0.007	30.000	29.987	0.000

**AMERICAN NATIONAL STANDARD
PREFERRED METRIC LIMITS AND FITS**

ANSI B4.2–1978

Dimensions in mm.

TABLE 28 (Continued)

TABLE 2 PREFERRED HOLE BASIS CLEARANCE FITS (Continued)

BASIC SIZE		LOOSE RUNNING			FREE RUNNING			CLOSE RUNNING			SLIDING			LOCATIONAL CLEARANCE		
		Hole H11	Shaft c11	Fit	Hole H9	Shaft d9	Fit	Hole H8	Shaft f7	Fit	Hole H7	Shaft g6	Fit	Hole H7	Shaft h6	Fit
40	MAX	40.160	39.880	0.440	40.062	39.920	0.204	40.039	39.975	0.089	40.025	39.991	0.050	40.025	40.000	0.041
	MIN	40.000	39.720	0.120	40.000	39.858	0.080	40.000	39.950	0.025	40.000	39.975	0.009	40.000	39.984	0.000
50	MAX	50.160	49.870	0.450	50.062	49.920	0.204	50.039	49.975	0.089	50.025	49.991	0.050	50.025	50.000	0.041
	MIN	50.000	49.710	0.130	50.000	49.858	0.080	50.000	49.950	0.025	50.000	49.975	0.009	50.000	49.984	0.000
60	MAX	60.190	59.860	0.520	60.074	59.900	0.248	60.046	59.970	0.106	60.030	59.990	0.059	60.030	60.000	0.049
	MIN	60.000	59.670	0.140	60.000	59.826	0.100	60.000	59.940	0.030	60.000	59.971	0.010	60.000	59.981	0.000
80	MAX	80.190	79.850	0.530	80.074	79.900	0.248	80.046	79.970	0.106	80.030	79.990	0.059	80.030	80.000	0.049
	MIN	80.000	79.660	0.150	80.000	79.826	0.100	80.000	79.940	0.030	80.000	79.971	0.010	80.000	79.981	0.000
100	MAX	100.220	99.830	0.610	100.087	99.880	0.294	100.054	99.964	0.125	100.035	99.988	0.069	100.035	100.000	0.057
	MIN	100.000	99.610	0.170	100.000	99.793	0.120	100.000	99.929	0.036	100.000	99.966	0.012	100.000	99.978	0.000
120	MAX	120.220	119.820	0.620	120.087	119.880	0.294	120.054	119.964	0.125	120.035	119.988	0.069	120.035	120.000	0.057
	MIN	120.000	119.600	0.180	120.000	119.793	0.120	120.000	119.929	0.036	120.000	119.966	0.012	120.000	119.978	0.000
160	MAX	160.250	159.790	0.710	160.100	159.855	0.345	160.063	159.957	0.146	160.040	159.986	0.079	160.040	160.000	0.065
	MIN	160.000	159.540	0.210	160.000	159.755	0.145	160.000	159.917	0.043	160.000	159.961	0.014	160.000	159.975	0.000
200	MAX	200.290	199.760	0.820	200.115	199.830	0.400	200.072	199.950	0.168	200.046	199.985	0.090	200.046	200.000	0.075
	MIN	200.000	199.470	0.240	200.000	199.715	0.170	200.000	199.904	0.050	200.000	199.956	0.015	200.000	199.971	0.000
250	MAX	250.290	249.720	0.860	250.115	249.830	0.400	250.072	249.950	0.168	250.046	249.985	0.090	250.046	250.000	0.075
	MIN	250.000	249.430	0.280	250.000	249.715	0.170	250.000	249.904	0.050	250.000	249.956	0.015	250.000	249.971	0.000
300	MAX	300.320	299.670	0.970	300.130	299.810	0.450	300.081	299.944	0.189	300.052	299.983	0.101	300.052	300.000	0.084
	MIN	300.000	299.350	0.330	300.000	299.680	0.190	300.000	299.892	0.056	300.000	299.951	0.017	300.000	299.968	0.000
400	MAX	400.360	399.600	1.120	400.140	399.790	0.490	400.089	399.938	0.208	400.057	399.982	0.111	400.057	400.000	0.093
	MIN	400.000	399.240	0.400	400.000	399.650	0.210	400.000	399.881	0.062	400.000	399.946	0.018	400.000	399.964	0.000
500	MAX	500.400	499.520	1.280	500.155	499.770	0.540	500.097	499.932	0.228	500.063	499.980	0.123	500.063	500.000	0.103
	MIN	500.000	499.120	0.480	500.000	499.615	0.230	500.000	499.869	0.068	500.000	499.940	0.020	500.000	499.960	0.000

(Reprinted from The American Society of Mechanical Engineers—ANSI B4.2–1978 (R1994)

**AMERICAN NATIONAL STANDARD
PREFERRED METRIC LIMITS AND FITS**

ANSI B4.2–1978

Dimensions in mm.

TABLE 28 (Continued)

TABLE 4 PREFERRED SHAFT BASIS CLEARANCE FITS

BASIC SIZE		LOOSE RUNNING Hole C11	Shaft h11	Fit	FREE RUNNING Hole D9	Shaft h9	Fit	CLOSE RUNNING Hole F8	Shaft h7	Fit	SLIDING Hole G7	Shaft h6	Fit	LOCATIONAL CLEARANCE Hole H7	Shaft h6	Fit
1	MAX	1·120	1·000	0·180	1·045	1·000	0·070	1·020	1·000	0·030	1·012	1·000	0·018	1·010	1·000	0·016
	MIN	1·060	0·940	0·060	1·020	0·975	0·020	1·006	0·990	0·006	1·002	0·994	0·002	1·000	0·994	0·000
1·2	MAX	1·320	1·200	0·180	1·245	1·200	0·070	1·220	1·200	0·030	1·212	1·200	0·018	1·210	1·200	0·016
	MIN	1·260	1·140	0·060	1·220	1·175	0·020	1·206	1·190	0·006	1·202	1·194	0·002	1·200	1·194	0·000
1·6	MAX	1·720	1·600	0·180	1·645	1·600	0·070	1·620	1·600	0·030	1·612	1·600	0·018	1·610	1·600	0·016
	MIN	1·660	1·540	0·060	1·620	1·575	0·020	1·606	1·590	0·006	1·602	1·594	0·002	1·600	1·594	0·000
2	MAX	2·120	2·000	0·180	2·045	2·000	0·070	2·020	2·000	0·030	2·012	2·000	0·018	2·010	2·000	0·016
	MIN	2·060	1·940	0·060	2·020	1·975	0·020	2·006	1·990	0·006	2·002	1·994	0·002	2·000	1·994	0·000
2·5	MAX	2·620	2·500	0·180	2·545	2·500	0·070	2·520	2·500	0·030	2·512	2·500	0·018	2·510	2·500	0·016
	MIN	2·560	2·440	0·060	2·520	2·475	0·020	2·506	2·490	0·006	2·502	2·494	0·002	2·500	2·494	0·000
3	MAX	3·120	3·000	0·180	3·045	3·000	0·070	3·020	3·000	0·030	3·012	3·000	0·018	3·010	3·000	0·016
	MIN	3·060	2·940	0·060	3·020	2·975	0·020	3·006	2·990	0·006	3·002	2·994	0·002	3·000	2·994	0·000
4	MAX	4·145	4·000	0·220	4·060	4·000	0·090	4·028	4·000	0·040	4·016	4·000	0·024	4·012	4·000	0·020
	MIN	4·070	3·925	0·070	4·030	3·970	0·030	4·010	3·988	0·010	4·004	3·992	0·004	4·000	3·992	0·000
5	MAX	5·145	5·000	0·220	5·060	5·000	0·090	5·028	5·000	0·040	5·016	5·000	0·024	5·012	5·000	0·020
	MIN	5·070	4·925	0·070	5·030	4·970	0·030	5·010	4·988	0·010	5·004	4·992	0·004	5·000	4·992	0·000
6	MAX	6·145	6·000	0·220	6·060	6·000	0·090	6·028	6·000	0·040	6·016	6·000	0·024	6·012	6·000	0·020
	MIN	6·070	5·925	0·070	6·030	5·970	0·030	6·010	5·988	0·010	6·004	5·992	0·004	6·000	5·992	0·000
8	MAX	8·170	8·000	0·260	8·076	8·000	0·112	8·035	8·000	0·050	8·020	8·000	0·029	8·015	8·000	0·024
	MIN	8·080	7·910	0·080	8·040	7·964	0·040	8·013	7·985	0·013	8·005	7·991	0·005	8·000	7·991	0·000
10	MAX	10·170	10·000	0·260	10·076	10·000	0·112	10·035	10·000	0·050	10·020	10·000	0·029	10·015	10·000	0·024
	MIN	10·080	9·910	0·080	10·040	9·964	0·040	10·013	9·985	0·013	10·005	9·991	0·005	10·000	9·991	0·000
12	MAX	12·205	12·000	0·315	12·093	12·000	0·136	12·043	12·000	0·061	12·024	12·000	0·035	12·018	12·000	0·029
	MIN	12·095	11·890	0·095	12·050	11·957	0·050	12·016	11·982	0·016	12·006	11·989	0·006	12·000	11·989	0·000
16	MAX	16·205	16·000	0·315	16·093	16·000	0·136	16·043	16·000	0·061	16·024	16·000	0·035	16·018	16·000	0·029
	MIN	16·095	15·890	0·095	16·050	15·957	0·050	16·016	15·982	0·016	16·006	15·989	0·006	16·000	15·989	0·000
20	MAX	20·240	20·000	0·370	20·117	20·000	0·169	20·053	20·000	0·074	20·028	20·000	0·041	20·021	20·000	0·034
	MIN	20·110	19·870	0·110	20·065	19·948	0·065	20·020	19·979	0·020	20·007	19·987	0·007	20·000	19·987	0·000
25	MAX	25·240	25·000	0·370	25·117	25·000	0·169	25·053	25·000	0·074	25·028	25·000	0·041	25·021	25·000	0·034
	MIN	25·110	24·870	0·110	25·065	24·948	0·065	25·020	24·979	0·020	25·007	24·987	0·007	25·000	24·987	0·000
30	MAX	30·240	30·000	0·370	30·117	30·000	0·169	30·053	30·000	0·074	30·028	30·000	0·041	30·021	30·000	0·034
	MIN	30·110	29·870	0·110	30·065	29·948	0·065	30·020	29·979	0·020	30·007	29·987	0·007	30·000	29·987	0·000

**AMERICAN NATIONAL STANDARD
PREFERRED METRIC LIMITS AND FITS**

ANSI B4.2–1978

TABLE 28   (Continued)

TABLE 4   PREFERRED SHAFT BASIS CLEARANCE FITS (Continued)

Dimensions in mm.

BASIC SIZE		LOOSE RUNNING			FREE RUNNING			CLOSE RUNNING			SLIDING			LOCATIONAL CLEARANCE		
		Hole C11	Shaft h11	Fit	Hole D9	Shaft h9	Fit	Hole F8	Shaft h7	Fit	Hole G7	Shaft h6	Fit	Hole H7	Shaft h6	Fit
40	MAX	40.280	40.000	0.440	40.142	40.000	0.204	40.064	40.000	0.089	40.034	40.000	0.050	40.025	40.000	0.041
	MIN	40.120	39.840	0.120	40.080	39.938	0.080	40.025	39.975	0.025	40.009	39.984	0.009	40.000	39.984	0.000
50	MAX	50.290	50.000	0.450	50.142	50.000	0.204	50.064	50.000	0.089	50.034	50.000	0.050	50.025	50.000	0.041
	MIN	50.130	49.840	0.130	50.080	49.938	0.080	50.025	49.975	0.025	50.009	49.984	0.009	50.000	49.984	0.000
60	MAX	60.330	60.000	0.520	60.174	60.000	0.248	60.076	60.000	0.106	60.040	60.000	0.059	60.030	60.000	0.049
	MIN	60.140	59.810	0.140	60.100	59.926	0.100	60.030	59.970	0.030	60.010	59.981	0.010	60.000	59.981	0.000
80	MAX	80.340	80.000	0.530	80.174	80.000	0.248	80.076	80.000	0.106	80.040	80.000	0.059	80.030	80.000	0.049
	MIN	80.150	79.810	0.150	80.100	79.926	0.100	80.030	79.970	0.030	80.010	79.981	0.010	80.000	75.981	0.000
100	MAX	100.390	100.000	0.610	100.207	100.000	0.294	100.090	100.000	0.125	100.047	100.000	0.069	100.035	100.000	0.057
	MIN	100.170	99.780	0.170	100.120	99.913	0.120	100.036	99.965	0.036	100.012	99.978	0.012	100.000	99.978	0.000
120	MAX	120.400	120.000	0.620	120.207	120.000	0.294	120.090	120.000	0.125	120.047	120.000	0.069	120.035	120.000	0.057
	MIN	120.180	119.780	0.180	120.120	119.913	0.120	120.036	119.965	0.036	120.012	119.978	0.012	120.000	119.978	0.000
160	MAX	160.460	160.000	0.710	160.245	160.000	0.345	160.106	160.000	0.146	160.054	160.000	0.079	160.040	160.000	0.065
	MIN	160.210	159.750	0.210	160.145	159.900	0.145	160.043	159.960	0.043	160.014	159.975	0.014	160.000	159.975	0.000
200	MAX	200.530	200.000	0.820	200.285	200.000	0.400	200.122	200.000	0.168	200.061	200.000	0.090	200.046	200.000	0.075
	MIN	200.240	199.710	0.240	200.170	199.885	0.170	200.050	199.954	0.050	200.015	199.971	0.015	200.000	199.971	0.000
250	MAX	250.570	250.000	0.860	250.285	250.000	0.400	250.122	250.000	0.168	250.061	250.000	0.090	250.046	250.000	0.075
	MIN	250.280	249.710	0.280	250.170	249.885	0.170	250.050	249.954	0.050	250.015	249.971	0.015	250.000	249.971	0.000
300	MAX	300.650	300.000	0.970	300.320	300.000	0.450	300.137	300.000	0.189	300.069	300.000	0.101	300.052	300.000	0.084
	MIN	300.330	299.680	0.330	300.190	299.870	0.190	300.056	299.948	0.056	300.017	299.968	0.017	300.000	299.968	0.000
400	MAX	400.760	400.000	1.120	400.350	400.000	0.490	400.151	400.000	0.208	400.075	400.000	0.111	400.057	400.000	0.093
	MIN	400.400	399.640	0.400	400.210	399.860	0.210	400.062	399.943	0.062	400.018	399.964	0.018	400.000	399.964	0.000
500	MAX	500.880	500.000	1.280	500.385	500.000	0.540	500.165	500.000	0.228	500.083	500.000	0.123	500.063	500.000	0.103
	MIN	500.480	499.600	0.480	500.230	499.845	0.230	500.068	499.937	0.068	500.020	499.960	0.020	500.000	499.960	0.000

(Reprinted from The American Society of Mechanical Engineers—ANSI B4.2–1978 (R1994))

TABLE 28 (Continued)

Table A1 Tolerance Zones for Internal (Hole) Dimensions (A14 through A9 and B14 through B9)   Dimensions in mm.

BASIC SIZE		A14	A13	A12	A11	A10	A9	B14	B13	B12	B11	B10	B9
OVER	0	+0•520	+0•410	+0•370	+0•330	+0•310	+0•295	+0•390	+0•280	+0•240	+0•200	+0•180	+0•165
TO	3	+0•270	+0•270	+0•270	+0•270	+0•270	+0•270	+0•140	+0•140	+0•140	+0•140	+0•140	+0•140
OVER	3	+0•570	+0•450	+0•390	+0•345	+0•318	+0•300	+0•440	+0•320	+0•260	+0•215	+0•188	+0•170
TO	6	+0•270	+0•270	+0•270	+0•270	+0•270	+0•270	+0•140	+0•140	+0•140	+0•140	+0•140	+0•140
OVER	6	+0•640	+0•500	+0•430	+0•370	+0•338	+0•316	+0•510	+0•370	+0•300	+0•240	+0•208	+0•186
TO	10	+0•280	+0•280	+0•280	+0•280	+0•280	+0•280	+0•150	+0•150	+0•150	+0•150	+0•150	+0•150
OVER	10	+0•720	+0•560	+0•470	+0•400	+0•360	+0•333	+0•580	+0•420	+0•330	+0•260	+0•220	+0•193
TO	14	+0•290	+0•290	+0•290	+0•290	+0•290	+0•290	+0•150	+0•150	+0•150	+0•150	+0•150	+0•150
OVER	14	+0•720	+0•560	+0•470	+0•400	+0•360	+0•333	+0•580	+0•420	+0•330	+0•260	+0•220	+0•193
TO	18	+0•290	+0•290	+0•290	+0•290	+0•290	+0•290	+0•150	+0•150	+0•150	+0•150	+0•150	+0•150
OVER	18	+0•820	+0•630	+0•510	+0•430	+0•384	+0•352	+0•680	+0•490	+0•370	+0•290	+0•244	+0•212
TO	24	+0•300	+0•300	+0•300	+0•300	+0•300	+0•300	+0•160	+0•160	+0•160	+0•160	+0•160	+0•160
OVER	24	+0•820	+0•630	+0•510	+0•430	+0•384	+0•352	+0•680	+0•490	+0•370	+0•290	+0•244	+0•212
TO	30	+0•300	+0•300	+0•300	+0•300	+0•300	+0•300	+0•160	+0•160	+0•160	+0•160	+0•160	+0•160
OVER	30	+0•930	+0•700	+0•560	+0•470	+0•410	+0•372	+0•790	+0•560	+0•420	+0•330	+0•270	+0•232
TO	40	+0•310	+0•310	+0•310	+0•310	+0•310	+0•310	+0•170	+0•170	+0•170	+0•170	+0•170	+0•170
OVER	40	+0•940	+0•710	+0•570	+0•480	+0•420	+0•382	+0•800	+0•570	+0•430	+0•340	+0•280	+0•242
TO	50	+0•320	+0•320	+0•320	+0•320	+0•320	+0•320	+0•180	+0•180	+0•180	+0•180	+0•180	+0•180
OVER	50	+1•080	+0•800	+0•640	+0•530	+0•460	+0•414	+0•930	+0•650	+0•490	+0•380	+0•310	+0•264
TO	65	+0•340	+0•340	+0•340	+0•340	+0•340	+0•340	+0•190	+0•190	+0•190	+0•190	+0•190	+0•190
OVER	65	+1•100	+0•820	+0•660	+0•550	+0•480	+0•434	+0•940	+0•660	+0•500	+0•390	+0•320	+0•274
TO	80	+0•360	+0•360	+0•360	+0•360	+0•360	+0•360	+0•200	+0•200	+0•200	+0•200	+0•200	+0•200
OVER	80	+1•250	+0•920	+0•730	+0•600	+0•520	+0•467	+1•090	+0•760	+0•570	+0•440	+0•360	+0•307
TO	100	+0•380	+0•380	+0•380	+0•380	+0•380	+0•380	+0•220	+0•220	+0•220	+0•220	+0•220	+0•220
OVER	100	+1•280	+0•950	+0•760	+0•630	+0•550	+0•497	+1•110	+0•780	+0•590	+0•460	+0•380	+0•327
TO	120	+0•410	+0•410	+0•410	+0•410	+0•410	+0•410	+0•240	+0•240	+0•240	+0•240	+0•240	+0•240
OVER	120	+1•460	+1•090	+0•860	+0•710	+0•620	+0•560	+1•260	+0•890	+0•660	+0•510	+0•420	+0•360
TO	140	+0•460	+0•460	+0•460	+0•460	+0•460	+0•460	+0•260	+0•260	+0•260	+0•260	+0•260	+0•260
OVER	140	+1•520	+1•150	+0•920	+0•770	+0•680	+0•620	+1•280	+0•910	+0•680	+0•530	+0•440	+0•380
TO	160	+0•520	+0•520	+0•520	+0•520	+0•520	+0•520	+0•280	+0•280	+0•280	+0•280	+0•280	+0•280
OVER	160	+1•580	+1•210	+0•980	+0•830	+0•740	+0•680	+1•310	+0•940	+0•710	+0•560	+0•470	+0•410
TO	180	+0•580	+0•580	+0•580	+0•580	+0•580	+0•580	+0•310	+0•310	+0•310	+0•310	+0•310	+0•310
OVER	180	+1•810	+1•380	+1•120	+0•950	+0•845	+0•775	+1•490	+1•060	+0•800	+0•630	+0•525	+0•455
TO	200	+0•660	+0•660	+0•660	+0•660	+0•660	+0•660	+0•340	+0•340	+0•340	+0•340	+0•340	+0•340
OVER	200	+1•890	+1•460	+1•200	+1•030	+0•925	+0•855	+1•530	+1•100	+0•840	+0•670	+0•565	+0•495
TO	225	+0•740	+0•740	+0•740	+0•740	+0•740	+0•740	+0•380	+0•380	+0•380	+0•380	+0•380	+0•380
OVER	225	+1•970	+1•540	+1•280	+1•110	+1•005	+0•935	+1•570	+1•140	+0•880	+0•710	+0•605	+0•535
TO	250	+0•820	+0•820	+0•820	+0•820	+0•820	+0•820	+0•420	+0•420	+0•420	+0•420	+0•420	+0•420
OVER	250	+2•220	+1•730	+1•440	+1•240	+1•130	+1•050	+1•780	+1•290	+1•000	+0•800	+0•690	+0•610
TO	280	+0•920	+0•920	+0•920	+0•920	+0•920	+0•920	+0•480	+0•480	+0•480	+0•480	+0•480	+0•480
OVER	280	+2•350	+1•860	+1•570	+1•370	+1•260	+1•180	+1•840	+1•350	+1•060	+0•860	+0•750	+0•670
TO	315	+1•050	+1•050	+1•050	+1•050	+1•050	+1•050	+0•540	+0•540	+0•540	+0•540	+0•540	+0•540
OVER	315	+2•600	+2•090	+1•770	+1•560	+1•430	+1•340	+2•000	+1•490	+1•170	+0•960	+0•830	+0•740
TO	355	+1•200	+1•200	+1•200	+1•200	+1•200	+1•200	+0•600	+0•600	+0•600	+0•600	+0•600	+0•600
OVER	355	+2•750	+2•240	+1•920	+1•710	+1•580	+1•490	+2•080	+1•570	+1•250	+1•040	+0•910	+0•820
TO	400	+1•350	+1•350	+1•350	+1•350	+1•350	+1•350	+0•680	+0•680	+0•680	+0•680	+0•680	+0•680
OVER	400	+3•050	+2•470	+2•130	+1•900	+1•750	+1•655	+2•310	+1•730	+1•390	+1•160	+1•010	+0•915
TO	450	+1•500	+1•500	+1•500	+1•500	+1•500	+1•500	+0•760	+0•760	+0•760	+0•760	+0•760	+0•760
OVER	450	+3•200	+2•620	+2•280	+2•050	+1•900	+1•805	+2•390	+1•810	+1•470	+1•240	+1•090	+0•995
TO	500	+1•650	+1•650	+1•650	+1•650	+1•650	+1•650	+0•840	+0•840	+0•840	+0•840	+0•840	+0•840

(Reprinted from The American Society of Mechanical Engineers—ANSI B4.2–1978 (R1994)

TABLE 28   (Continued)

Table A13   Tolerance Zones for External (Shaft) Dimensions (a14 through a9 and b14 through b9)   Dimensions in mm.

BASIC SIZE		a14	a13	a12	a11	a10	a9	b14	b13	b12	b11	b10	b9
OVER TO	0 3	−0.270 −0.520	−0.270 −0.410	−0.270 −0.370	−0.270 −0.330	−0.270 −0.310	−0.270 −0.295	−0.140 −0.390	−0.140 −0.280	−0.140 −0.240	−0.140 −0.200	−0.140 −0.180	−0.140 −0.165
OVER TO	3 6	−0.270 −0.570	−0.270 −0.450	−0.270 −0.390	−0.270 −0.345	−0.270 −0.318	−0.270 −0.300	−0.140 −0.440	−0.140 −0.320	−0.140 −0.260	−0.140 −0.215	−0.140 −0.188	−0.140 −0.170
OVER TO	6 10	−0.280 −0.640	−0.280 −0.500	−0.280 −0.430	−0.280 −0.370	−0.280 −0.338	−0.280 −0.316	−0.150 −0.510	−0.150 −0.370	−0.150 −0.300	−0.150 −0.240	−0.150 −0.208	−0.150 −0.186
OVER TO	10 14	−0.290 −0.720	−0.290 −0.560	−0.290 −0.470	−0.290 −0.400	−0.290 −0.360	−0.290 −0.333	−0.150 −0.580	−0.150 −0.420	−0.150 −0.330	−0.150 −0.260	−0.150 −0.220	−0.150 −0.193
OVER TO	14 18	−0.290 −0.720	−0.290 −0.560	−0.290 −0.470	−0.290 −0.400	−0.290 −0.360	−0.290 −0.333	−0.150 −0.580	−0.150 −0.420	−0.150 −0.330	−0.150 −0.260	−0.150 −0.220	−0.150 −0.193
OVER TO	18 24	−0.300 −0.820	−0.300 −0.630	−0.300 −0.510	−0.300 −0.430	−0.300 −0.384	−0.300 −0.352	−0.160 −0.680	−0.160 −0.490	−0.160 −0.370	−0.160 −0.290	−0.160 −0.244	−0.160 −0.212
OVER TO	24 30	−0.300 −0.820	−0.300 −0.630	−0.300 −0.510	−0.300 −0.430	−0.300 −0.384	−0.300 −0.352	−0.160 −0.680	−0.160 −0.490	−0.160 −0.370	−0.160 −0.290	−0.160 −0.244	−0.160 −0.212
OVER TO	30 40	−0.310 −0.930	−0.310 −0.700	−0.310 −0.560	−0.310 −0.470	−0.310 −0.410	−0.310 −0.372	−0.170 −0.790	−0.170 −0.560	−0.170 −0.420	−0.170 −0.330	−0.170 −0.270	−0.170 −0.232
OVER TO	40 50	−0.320 −0.940	−0.320 −0.710	−0.320 −0.570	−0.320 −0.480	−0.320 −0.420	−0.320 −0.382	−0.180 −0.800	−0.180 −0.570	−0.180 −0.430	−0.180 −0.340	−0.180 −0.280	−0.180 −0.242
OVER TO	50 65	−0.340 −1.080	−0.340 −0.800	−0.340 −0.640	−0.340 −0.530	−0.340 −0.460	−0.340 −0.414	−0.190 −0.930	−0.190 −0.650	−0.190 −0.490	−0.190 −0.380	−0.190 −0.310	−0.190 −0.264
OVER TO	65 80	−0.360 −1.100	−0.360 −0.820	−0.360 −0.660	−0.360 −0.550	−0.360 −0.480	−0.360 −0.434	−0.200 −0.940	−0.200 −0.660	−0.200 −0.500	−0.200 −0.390	−0.200 −0.320	−0.200 −0.274
OVER TO	80 100	−0.380 −1.250	−0.380 −0.920	−0.380 −0.730	−0.380 −0.600	−0.380 −0.520	−0.380 −0.467	−0.220 −1.090	−0.220 −0.760	−0.220 −0.570	−0.220 −0.440	−0.220 −0.360	−0.220 −0.307
OVER TO	100 120	−0.410 −1.280	−0.410 −0.950	−0.410 −0.760	−0.410 −0.630	−0.410 −0.550	−0.410 −0.497	−0.240 −1.110	−0.240 −0.780	−0.240 −0.590	−0.240 −0.460	−0.240 −0.380	−0.240 −0.327
OVER TO	120 140	−0.460 −1.460	−0.460 −1.090	−0.460 −0.860	−0.460 −0.710	−0.460 −0.620	−0.460 −0.560	−0.260 −1.260	−0.260 −0.890	−0.260 −0.660	−0.260 −0.510	−0.260 −0.420	−0.260 −0.360
OVER TO	140 160	−0.520 −1.520	−0.520 −1.150	−0.520 −0.920	−0.520 −0.770	−0.520 −0.680	−0.520 −0.620	−0.280 −1.280	−0.280 −0.910	−0.280 −0.680	−0.280 −0.530	−0.280 −0.440	−0.280 −0.380
OVER TO	160 180	−0.580 −1.580	−0.580 −1.210	−0.580 −0.980	−0.580 −0.830	−0.580 −0.740	−0.580 −0.680	−0.310 −1.310	−0.310 −0.940	−0.310 −0.710	−0.310 −0.560	−0.310 −0.470	−0.310 −0.410
OVER TO	180 200	−0.660 −1.810	−0.660 −1.380	−0.660 −1.120	−0.660 −0.950	−0.660 −0.845	−0.660 −0.775	−0.340 −1.490	−0.340 −1.060	−0.340 −0.800	−0.340 −0.630	−0.340 −0.525	−0.340 −0.455
OVER TO	200 225	−0.740 −1.890	−0.740 −1.460	−0.740 −1.200	−0.740 −1.030	−0.740 −0.925	−0.740 −0.855	−0.380 −1.530	−0.380 −1.100	−0.380 −0.840	−0.380 −0.670	−0.380 −0.565	−0.380 −0.495
OVER TO	225 250	−0.820 −1.970	−0.820 −1.540	−0.820 −1.280	−0.820 −1.110	−0.820 −1.005	−0.820 −0.935	−0.420 −1.570	−0.420 −1.140	−0.420 −0.880	−0.420 −0.710	−0.420 −0.605	−0.420 −0.535
OVER TO	250 280	−0.920 −2.220	−0.920 −1.730	−0.920 −1.440	−0.920 −1.240	−0.920 −1.130	−0.920 −1.050	−0.480 −1.780	−0.480 −1.290	−0.480 −1.000	−0.480 −0.800	−0.480 −0.690	−0.480 −0.610
OVER TO	280 315	−1.050 −2.350	−1.050 −1.860	−1.050 −1.570	−1.050 −1.370	−1.050 −1.260	−1.050 −1.180	−0.540 −1.840	−0.540 −1.350	−0.540 −1.060	−0.540 −0.860	−0.540 −0.750	−0.540 −0.670
OVER TO	315 355	−1.200 −2.600	−1.200 −2.090	−1.200 −1.770	−1.200 −1.560	−1.200 −1.430	−1.200 −1.340	−0.600 −2.000	−0.600 −1.490	−0.600 −1.170	−0.600 −0.960	−0.600 −0.830	−0.600 −0.740
OVER TO	355 400	−1.350 −2.750	−1.350 −2.240	−1.350 −1.920	−1.350 −1.710	−1.350 −1.580	−1.350 −1.490	−0.680 −2.080	−0.680 −1.570	−0.680 −1.250	−0.680 −1.040	−0.680 −0.910	−0.680 −0.820
OVER TO	400 450	−1.500 −3.050	−1.500 −2.470	−1.500 −2.130	−1.500 −1.900	−1.500 −1.750	−1.500 −1.655	−0.760 −2.310	−0.760 −1.730	−0.760 −1.390	−0.760 −1.160	−0.760 −1.010	−0.760 −0.915
OVER TO	450 500	−1.650 −3.200	−1.650 −2.620	−1.650 −2.280	−1.650 −2.050	−1.650 −1.900	−1.650 −1.805	−0.840 +2.390	−0.840 −1.810	−0.840 −1.470	−0.840 −1.240	−0.840 −1.090	−0.840 −0.995

(Reprinted from The American Society of Mechanical Engineers—ANSI B4.2–1978 (R1994)

▼ APPENDIX C American National Standards of Interest to Designers, Architects, and Drafters

TITLE OF STANDARD

Abbreviations for Use on Drawings and Text. . . . . ANSI/ASME Y1.1—1989
American National Standard Drafting Practice:
   Metric Drawing Sheet Size and Format . . . . ANSI/ASME Y14.1M—1995
   Decimal Inch Drawing Sheet Size and Format . . . . ANSI/ASME Y14.1—1995
   Line Conventions and Lettering . . . . ANSI/ASME Y14.2M—1992
   Multi and Sectional View Drawings. . . . . ANSI/ASME Y14.3M—1994
   Pictorial Drawing . . . . ANSI/ASME Y14.4—1989 (R1994)
   Revision of Engineering Drawings and Associated Documents. . . . . ANSI/ASME Y14.35M—1992
   Dimensioning and Tolerancing . . . . ANSI/ASME Y14.5M—1994
   Screw Thread Representation, Engineering Drawing and Related Documentation Practice . . . . ANSI Y14.6—1978 (R1993)
   Engineering Drawing and Related Documentation Practices—Screw Thread Representation
     (Metric Supplement) . . . . ANSI Y14.6aM—1981 (R1993)
   Gears and Splines
     Gear Drawing Standards—Part 1, for Spur, Helical, Double Helical, and Rack . . . . ANSI/ASME Y14.7.1—1971 (R1993)
     Gear and Spline Drawing Standards—Part 2, Bevel and Hypoid Gears. . . . . ANSI Y14.7.2—1978 (R1994)
   Castings and Forgings . . . . ANSI/ASME Y14.8M—1989
   Engineering Drawing and Related Documentation Practices—Mechanical Spring Representation . . . . ANSI/ASME Y14.13M—1981 (R1992)
   Electrical and Electronics Diagrams (includes supplements ANSI Y14.15a—1971 and ANSI Y14.15b—1973) . . . . ANSI Y14.15—1966 (R1988)
   Fluid Power Systems and Products—Moving Parts Fluid Controls—Method of Diagramming. . . . . ANSI/(NFPA) T3.28.9R1—1989
   Engineering Drawings and Related Documentation Practices—Optical Parts . . . . ANSI/ASME Y14.18M—1986 (R1993)
   Types and Applications of Engineering Drawings . . . . ANSI/ASME Y14.24M—1989
   Digital Representation for Communication of Product Definition Data . . . . ANSI/US PRO/IPO—100—1993
   Ground Vehicle Drawing Practices . . . . In Preparation
   Chassis Frames—Passenger Car and Light Truck . . . . ANSI/ASME Y14.32.1M—1994
   Parts Lists, Data Lists, and Index Lists. . . . . ANSI/ASME Y14.34M—1989
   Surface Texture Symbols . . . . ANSI Y14.36—1978 (R1993)
Graphic Symbols:
   Electrical Wiring and Layout Diagrams Used in Architecture and Building Construction, Graphic Symbols for . . . . . . ANSI Y32.9—1972 (R1989)
   Plumbing Fixtures for Diagrams Used in Architecture and Building Construction, Graphic Symbols for. . . . . ANSI Y32.4—1977 (R1987)
   Railroad Maps and Profiles, Graphic Symbols for . . . . ANSI Y32.7—1972 (R1987)
   Fluid Power Diagrams, Graphic Symbols for . . . . ANSI Y32.10—1967 (R1987)
   Process Flow Diagrams in the Petroleum and Chemical Industries, Graphic Symbols for . . . . ANSI Y32.11—1961 (R1993)
   Mechanical and Acoustical Elements as Used in Schematic Diagrams, Symbols for . . . . ANSI Y32.18—1972 (R1993)
   Graphical Symbols for Pipe Fittings, Valves, and Piping . . . . ANSI/ASME Y32.2.3—1949 (R1988)
   Heating, Ventilating, and Air Conditioning, Graphic Symbols for. . . . . ANSI Y32.2.4—1949 (R1993)
   Heat-Power Apparatus, Graphic Symbols for. . . . . ANSI Y32.2.6M—1950 (R1993)
   Symbols for Welding, Brazing, and Nondestructive Examination . . . . ANSI/AWS A2.4—1993
Letter Symbols For:
   Glossary of Terms Concerning Letter Symbols . . . . ANSI Y10.1—1972 (R1988)
   Quantities Used in Electrical Science and Electrical Engineering, Letter Symbols for. . . . . ANSI/IEEE 280—1985 (R1992)
   Letter Symbols—Acoustics, Letter Symbols and Abbreviations for Quantities Used in . . . . ANSI/ASME Y10.11—1984
   Chemical Engineering, Letter Symbols for . . . . ANSI Y10.12—1955 (R1988)
   Illuminating Engineering, Letter Symbols for . . . . ANSI Y10.18—1967 (R1977)
   Mathematical Signs and Symbols for Use in Physical Sciences and Technology . . . . ANSI/IEEE 260.3—1993

▼ APPENDIX D   Unified Screw Thread Variations

THE FOLLOWING list is a ready reference of available standard series and selected combinations of Unified screw threads. Each thread is given as part of a proper thread note including major diameter, threads per inch, and series identification.

0–80 UNF	7/16–27 UNS	7/8–28 UN	1-3/8–18 UNEF	1-15/16–6 UN	2-7/8–12 UN
1–64 UNC	7/16–28 UNEF	7/8–32 UN	1-3/8–20 UN	1-15/16–8 UN	2-7/8–16 UN
1–72 UNF	7/16–32 UN	15/16–12 UN	1-3/8–28 UN	1-15/16–12 UN	2-7/8–20 UN
2–56 UNC	1/2–12 UNS	15/16–16 UN	1-7/16–6 UN	1-15/16–16 UN	3–4 UNC
2–64 UNF	1/2–13 UNC	15/16–20 UNEF	1-7/16–8 UN	1-15/16–20 UN	3–6 UN
3–48 UNC	1/2–14 UNS	15/16–28 UN	1-7/16–12 UN	2–4 1/2 UNC	3–8 UN
3–56 UNF	1/2–16 UN	15/16–32 UN	1-7/16–16 UN	2–6 UN	3–10 UNS
4–40 UNC	1/2–18 UNS	1–8 UNC	1-7/16–18 UNEF	2–8 UN	3–12 UN
4–48 UNF	1/2–20 UNF	1–10 UNS	1-7/16–20 UN	2–10 UNS	3–14 UNS
5–40 UNC	1/2–24 UNS	1–12 UNF	1-7/16–28 UN	2–12 UN	3–16 UN
5–44 UNF	1/2–27 UNS	1–14 UNS	1-1/2–6 UNC	2–14 UNS	3–18 UNS
6–32 UNC	1/2–28 UNEF	1–16 UN	1-1/2–8 UN	2–16 UN	3–20 UN
6–40 UNF	1/2–32 UN	1–18 UNS	1-1/2–10 UNS	2–18 UNS	3-1/8–6 UN
8–32 UNC	9/16–12 UNC	1–20 UNEF	1-1/2–12 UNF	2–20 UN	3-1/8–8 UN
8–36 UNF	9/16–14 UNS	1–24 UNS	1-1/2–14 UNS	2-1/16–16 UNS	3-1/8–12 UN
10–24 UNC	9/16–16 UN	1–27 UNS	1-1/2–16 UN	2-1/8–6 UN	3-1/8–16 UN
10–28 UNS	9/16–18 UNF	1–28 UN	1-1/2–18 UNEF	2-1/8–8 UN	3-1/4–4 UNC
10–32 UNF	9/16–20 UN	1–32 UN	1-1/2–20 UN	2-1/8–12 UN	3-1/4–6 UN
10–36 UNS	9/16–24 UNEF	1-1/16–8 UN	1-1/2–24 UNS	2-1/8–16 UN	3-1/4–8 UN
10–40 UNS	9/16–27 UNS	1-1/16–12 UN	1-1/2–28 UN	2-1/8–20 UN	3-1/4–10 UNS
10–48 UNS	9/16–28 UN	1-1/16–16 UN	1-9/16–6 UN	2-3/16–16 UNS	3-1/4–12 UN
10–56 UNS	9/16–32 UN	1-1/16–18 UNEF	1-9/16–8 UN	2-1/4–4-1/2 UNC	3-1/4–14 UNS
12–24 UNC	5/8–11 UNC	1-1/16–20 UN	1-9/16–12 UN	2-1/4–6 UN	3-1/4–16 UN
12–28 UNF	5/8–12 UN	1-1/16–28 UN	1-9/16–16 UN	2-1/4–8 UN	3-1/4–18 UNS
12–32 UNEF	5/8–14 UNS	1-1/8–7 UNC	1-9/16–18 UNEF	2-1/4–10 UNS	3-3/8–6 UN
12–36 UNS	5/8–16 UN	1-1/8–8 UN	1-9/16–20 UN	2-1/4–12 UN	3-3/8–8 UN
12–40 UNS	5/8–18 UN	1-1/8–10 UN	1-5/8–6 UN	2-1/4–14 UN	3-3/8–12 UN
12–48 UNS	5/8–20 UN	1-1/8–12 UNF	1-5/8–8 UN	2-1/4–16 UN	3-3/8–16 UN
12–56 UNS	5/8–24 UNEF	1-1/8–14 UNS	1-5/8–10 UNS	2-1/4–18 UNS	3-1/2–4 UNC
1/4–20 UNC	5/8–27 UNS	1-1/8–16 UN	1-5/8–12 UN	2-1/4–20 UN	3-1/2–6 UN
1/4–24 UNS	5/8–28 UN	1-1/8–18 UNEF	1-5/8–14 UNS	2-5/16–16 UNS	3-1/2–8 UN
1/4–27 UNS	5/8–32 UN	1-1/8–20 UN	1-5/8–16 UN	2-3/8–8 UN	3-1/2–10 UNS
1/4–28 UNF	11/16–12 UN	1-1/8–24 UNS	1-5/8–18 UNEF	2-3/8–12 UN	3-1/2–12 UN
1/4–32 UNEF	11/16–16 UN	1-1/8–28 UN	1-5/8–20 UN	2-3/8–16 UN	3-1/2–14 UNS
1/4–36 UNS	11/16–20 UN	1-3/16–8 UN	1-5/8–24 UNS	2-3/8–20 UN	3-1/2–16 UN
1/4–40 UNS	11/16–24 UNEF	1-3/16–12 UN	1-11/16–6 UN	2-7/16–16 UNS	3-1/2–18 UNS
1/4–48 UNS	11/16–28 UN	1-3/16–16 UN	1-11/16–8 UN	2-1/2–4 UNC	3-5/8–6 UN
1/4–56 UNS	11/16–32 UN	1-3/16–18 UNEF	1-11/16–12 UN	2-1/2–6 UN	3-5/8–8 UN
5/16–18 UNC	3/4–10 UNC	1-3/16–20 UN	1-11/16–16 UN	2-1/2–8 UN	3-5/8–12 UN
5/16–20 UN	3/4–12 UN	1-3/16–28 UN	1-11/16–18 UNEF	2-1/2–10 UNS	3-5/8–16 UN
5/16–24 UNF	3/4–14 UNS	1-1/4–7 UNC	1-11/16–20 UN	2-1/2–12 UN	3-3/4–4 UNC
5/16–27 UNS	3/4–16 UNF	1-1/4–8 UN	1-3/4–5 UNC	2-1/2–14 UNS	3-3/4–6 UN
5/16–28 UN	3/4–18 UNS	1-1/4–10 UNS	1-3/4–6 UN	2-1/2–16 UN	3-3/4–8 UN
5/16–32 UNEF	3/4–20 UNEF	1-1/4–12 UNF	1-3/4–8 UN	1-1/2–18 UNS	3-3/4–10 UN
5/16–36 UNS	3/4–24 UNS	1-1/4–14 UNS	1-3/4–10 UNS	2-1/2–20 UN	3-3/4–12 UN
5/16–40 UNS	3/4–27 UNS	1-1/4–16 UN	1-3/4–12 UN	2-5/8–6 UN	3-3/4–14 UNS
5/16–48 UNS	3/4–28 UN	1-1/4–18 UNEF	1-3/4–14 UNS	2-5/8–8 UN	3-3/4–16 UN
3/8–16 UNC	3/4–32 UN	1-1/4–20 UN	1-3/4–16 UN	2-5/8–12 UN	3-3/4–18 UNS
3/8–18 UNS	13/16–12 UN	1-1/4–24 UNS	1-3/4–18 UNS	2-5/8–16 UN	3-7/8–6 UN
3/8–20 UN	13/16–16 UN	1-1/4–28 UN	1-3/4–20 UN	2-5/8–20 UN	3-7/8–8 UN
3/8–24 UNF	13/16–20 UNEF	1-5/16–8 UN	1-13/16–6 UN	2-3/4–4 UNC	3-7/8–12 UN
3/8–27 UNS	13/16–28 UN	1-5/16–12 UN	1-13/16–8 UN	2-3/4–6 UN	3-7/8–16 UN
3/8–28 UN	13/16–32 UN	1-5/16–16 UN	1-13/16–12 UN	2-3/4–8 UN	4–4 UNC
3/8–32 UNEF	7/8–9 UNC	1-5/16–18 UNEF	1-13/16–20 UN	2-3/4–10 UNS	4–6 UN
3/8–36 UNS	7/8–10 UNS	1-5/16–20 UN	1-7/8–6 UN	2-3/4–12 UN	4–8 UN
3/8–40 UNS	7/8–12 UN	1-5/16–28 UN	1-7/8–8 UN	2-3/4–14 UNS	4–10 UNS
.390–27 UNS	7/8–14 UNF	1-3/8–6 UNC	1-7/8–10 UNS	2-3/4–16 UN	4–12 UN
7/16–14 UNC	7/8–16 UN	1-3/8–8 UN	1-7/8–12 UN	2-3/4–18 UNS	4–14 UNS
7/16–16 UN	7/8–18 UNS	1-3/8–10 UNS	1-7/8–14 UNS	2-3/4–20 UN	4–16 UN
7/16–18 UNS	7/8–20 UNEF	1-3/8–12 UNF	1-7/8–16 UN	2-7/8–6 UN	
7/16–20 UNF	7/8–24 UNS	1-3/8–14 UNS	1-7/8–18 UNS	2-7/8–8 UN	
7/16–24 UNS	7/8–27 UNS	1-3/8–16 UN	1-7/8–20 UN		

▼ APPENDIX E   Metric Screw Thread Variations

THE FOLLOWING list is a ready reference of available standard coarse pitch series ISO metric screw threads. Each thread is given as part of a proper thread note, including metric symbol, major diameter, and thread pitch.

M1 × 0.25	M2.2 × 0.45	M6 × 1	M14 × 2	M30 × 3.5	M52 × 5
M1.1 × 0.25	M2.5 × 0.45	M7 × 1	M16 × 2	M33 × 3.5	M56 × 5.5
M1.2 × 0.25	M3 × 0.5	M8 × 1.25	M18 × 2.5	M36 × 4	M60 × 5.5
M1.4 × 0.3	M3.5 × 0.6	M9 × 1.25	M20 × 2.5	M39 × 4	M64 × 6
M1.6 × 0.35	M4 × 0.7	M10 × 1.5	M22 × 2.5	M42 × 4.5	M68 × 6
M1.8 × 0.35	M4.5 × 0.75	M11 × 1.5	M24 × 3	M45 × 4.5	
M2 × 0.4	M5 × 0.8	M12 × 1.75	M27 × 3	M48 × 5	

▼ APPENDIX F   Dimensioning and Tolerancing Symbols

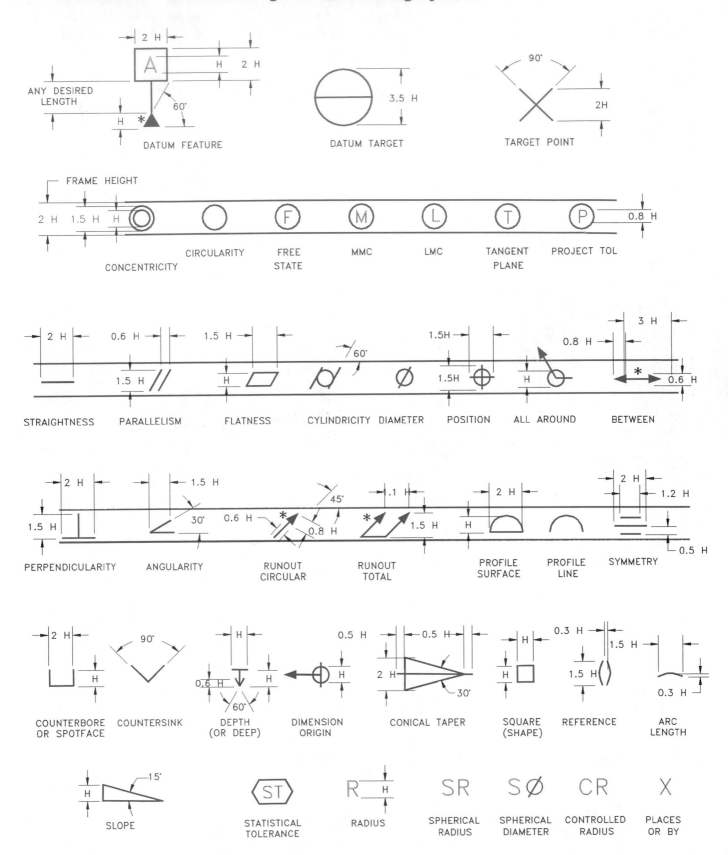

H = LETTER HEIGHT   *MAY BE FILLED OR UNFILLED

▼ APPENDIX G   Designation of Welding and Allied Processes by Letters

Welding and allied processes	Letter designation	Welding and allied processes	Letter designation
adhesive bonding	ABD	resistance spot welding	RSW
arc welding	AW	upset welding	UW
atomic hydrogen welding	AHW	high frequency	UW-HF
bare metal arc welding	BMAW	induction	UW-I
carbon arc welding	CAW	soldering	S
gas	CAW-G	dip soldering	DS
shielded	CAW-S	furnace soldering	FS
twin	CAW-T	induction soldering	IS
electrogas	EGW	infrared soldering	IRS
flux cored arc welding	FCAW	iron soldering	INS
gas metal arc welding	GMAW	resistance soldering	RS
pulsed arc	GMAW-P	torch soldering	TS
short circuiting arc	GMAW-S	wave soldering	WS
gas tungsten arc welding	GTAW	solid-state welding	SSW
pulsed arc	GTAW-P	coextrusion welding	CEW
plasma arc welding	PAW	cold welding	CW
shielded metal arc welding	SMAW	diffusion welding	DFW
stud arc welding	SW	explosion welding	EXW
submerged arc welding	SAW	forge welding	FOW
series	SAW-S	friction welding	FRW
brazing	B	hot pressure welding	HPW
arc brazing	AB	roll welding	ROW
block brazing	BB	ultrasonic welding	USW
carbon arc brazing	CAB	thermal cutting	TC
diffusion brazing	DFB	arc cutting	AC
dip brazing	DB	air carbon arc cutting	AAC
flow frazing	FLB	carbon arc cutting	CAC
furnace brazing	FB	gas metal arc cutting	GMAC
induction brazing	IB	gas tungsten arc cutting	GTAC
infrared brazing	IRB	metal arc cutting	MAC
resistance brazing	RB	plasma arc cutting	PAC
torch brazing	TB	shielded metal arc cutting	SMAC
other welding processes		electron beam cutting	EBC
electron beam welding	EBW	laser beam cutting	LBC
high vacuum	EBW-HV	air	LCB-A
medium vacuum	EBW-MV	evaporative	LBC-EV
nonvacuum	EBW-NV	inert gas	LBC-IG
electroslag welding	ESW	oxygen	LBC-O
flow welding	FLOW	oxygen cutting	OC
induction welding	IW	chemical flux cutting	FOC
laser beam welding	LBW	metal powder cutting	POC
thermit welding	TW	oxyfuel gas cutting	OFC
oxyfuel gas welding	OFW	oxyacetylene cutting	OFC-A
air acetylene welding	AAW	oxyhydrogen cutting	OFC-H
oxyacetylene welding	OAW	oxynatural gas cutting	OFC-N
oxyhydrogen welding	OHW	oxypropane cutting	OFC-P
pressure gas welding	PGW	oxygen arc cutting	AOC
resistance welding	RW	oxygen lance cutting	LOC
flash welding	FW	thermal spraying	THSP
percussion welding	PEW	arc spraying	ASP
projection welding	PW	flame spraying	FLSP
resistance seam welding	RSEW	plasma spraying	PSP
high frequency	RSEW-HF		
induction	RSEW-I		

▼ APPENDIX H Structural Metal Shape Designations

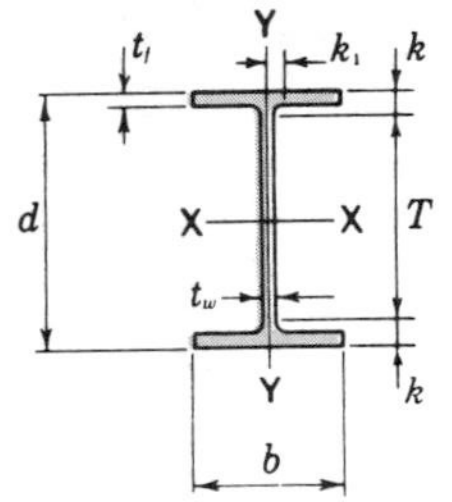

W SHAPES Dimensions

Designation	Area A (In.²)	Depth d (In.)	Depth d (In.)	Web Thickness tw (In.)	tw (In.)	tw/2 (In.)	Flange Width bf (In.)	bf (In.)	Flange Thickness tf (In.)	tf (In.)	Distance T (In.)	Distance k (In.)	Distance k1 (In.)
W 10×112	32.9	11.36	11³/₈	0.755	³/₄	³/₈	10.415	10³/₈	1.250	1¼	7⅝	1⅞	15/16
×100	29.4	11.10	11⅛	0.680	11/16	³/₈	10.340	10³/₈	1.120	1⅛	7⅝	1¾	7/8
× 88	25.9	10.84	10⅞	0.605	⅝	5/16	10.265	10¼	0.990	1	7⅝	1⅝	13/16
× 77	22.6	10.60	10⅝	0.530	½	¼	10.190	10¼	0.870	⅞	7⅝	1½	13/16
× 68	20.0	10.40	10³/₈	0.470	½	¼	10.130	10⅛	0.770	¾	7⅝	1⅜	¾
× 60	17.6	10.22	10¼	0.420	7/16	¼	10.080	10⅛	0.680	11/16	7⅝	1⅜	¾
× 54	15.8	10.09	10⅛	0.370	³/₈	3/16	10.030	10	0.615	⅝	7⅝	1¼	11/16
× 49	14.4	9.98	10	0.340	5/16	3/16	10.000	10	0.560	9/16	7⅝	1³/16	11/16
W 10× 45	13.3	10.10	10⅛	0.350	³/₈	3/16	8.020	8	0.620	⅝	7⅝	1¼	11/16
× 39	11.5	9.92	9⅞	0.315	5/16	3/16	7.985	8	0.530	½	7⅝	1⅛	11/16
× 33	9.71	9.73	9¾	0.290	5/16	3/16	7.960	8	0.435	7/16	7⅝	1¹/16	11/16
W 10× 30	8.84	10.47	10½	0.300	5/16	3/16	5.810	5¾	0.510	½	8⅝	15/16	½
× 26	7.61	10.33	10³/₈	0.260	¼	⅛	5.770	5¾	0.440	7/16	8⅝	⅞	½
× 22	6.49	10.17	10⅛	0.240	¼	⅛	5.750	5¾	0.360	⅜	8⅝	¾	½
W 10× 19	5.62	10.24	10¼	0.250	¼	⅛	4.020	4	0.395	⅜	8⅝	13/16	½
× 17	4.99	10.11	10⅛	0.240	¼	⅛	4.010	4	0.330	5/16	8⅝	¾	½
× 15	4.41	9.99	10	0.230	¼	⅛	4.000	4	0.270	¼	8⅝	11/16	7/16
× 12	3.54	9.87	9⅞	0.190	3/16	⅛	3.960	4	0.210	3/16	8⅝	⅝	7/16

AMERICAN INSTITUTE OF STEEL CONSTRUCTION

W SHAPES Dimensions

Designation	Area A (In.²)	Depth d (In.)	Depth d (In.)	Web Thickness tw (In.)	tw (In.)	tw/2 (In.)	Flange Width bf (In.)	bf (In.)	Flange Thickness tf (In.)	tf (In.)	Distance T (In.)	Distance k (In.)	Distance k1 (In.)
W 8×67	19.7	9.00	9	0.570	9/16	5/16	8.280	8¼	0.935	15/16	6⅛	1 7/16	11/16
×58	17.1	8.75	8¾	0.510	½	¼	8.220	8¼	0.810	13/16	6⅛	1 5/16	11/16
×48	14.1	8.50	8½	0.400	³/₈	3/16	8.110	8⅛	0.685	11/16	6⅛	1 3/16	⅝
×40	11.7	8.25	8¼	0.360	⅜	3/16	8.070	8⅛	0.560	9/16	6⅛	1 1/16	⅝
×35	10.3	8.12	8⅛	0.310	5/16	3/16	8.020	8	0.495	½	6⅛	1	9/16
×31	9.13	8.00	8	0.285	5/16	3/16	7.995	8	0.435	7/16	6⅛	15/16	9/16
W 8×28	8.25	8.06	8	0.285	5/16	3/16	6.535	6½	0.465	7/16	6⅛	15/16	9/16
×24	7.08	7.93	7⅞	0.245	¼	⅛	6.495	6½	0.400	⅜	6⅛	⅞	9/16
W 8×21	6.16	8.28	8¼	0.250	¼	⅛	5.270	5¼	0.400	⅜	6⅝	13/16	½
×18	5.26	8.14	8⅛	0.230	¼	⅛	5.250	5¼	0.330	5/16	6⅝	¾	7/16
W 8×15	4.44	8.11	8⅛	0.245	¼	⅛	4.015	4	0.315	5/16	6⅝	¾	½
×13	3.84	7.99	8	0.230	¼	⅛	4.000	4	0.255	¼	6⅝	11/16	7/16
×10	2.96	7.89	7⅞	0.170	3/16	⅛	3.940	4	0.205	3/16	6⅝	⅝	7/16
W 6×25	7.34	6.38	6⅜	0.320	5/16	3/16	6.080	6⅛	0.455	7/16	4¾	13/16	7/16
×20	5.87	6.20	6¼	0.260	¼	⅛	6.020	6	0.365	⅜	4¾	¾	7/16
×15	4.43	5.99	6	0.230	¼	⅛	5.990	6	0.260	¼	4¾	⅝	⅜
W 6×16	4.74	6.28	6¼	0.260	¼	⅛	4.030	4	0.405	⅜	4¾	¾	7/16
×12	3.55	6.03	6	0.230	¼	⅛	4.000	4	0.280	¼	4¾	⅝	⅜
× 9	2.68	5.90	5⅞	0.170	3/16	⅛	3.940	4	0.215	3/16	4¾	9/16	⅜
W 5×19	5.54	5.15	5⅛	0.270	¼	⅛	5.030	5	0.430	7/16	3½	13/16	7/16
×16	4.68	5.01	5	0.240	¼	⅛	5.000	5	0.360	⅜	3½	¾	7/16
W 4×13	3.83	4.16	4⅛	0.280	¼	⅛	4.060	4	0.345	⅜	2¾	11/16	7/16

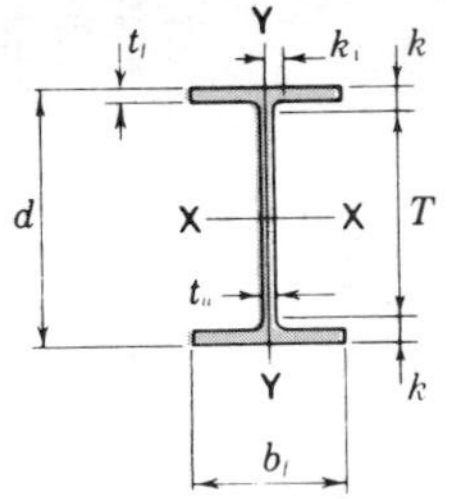

AMERICAN INSTITUTE OF STEEL CONSTRUCTION

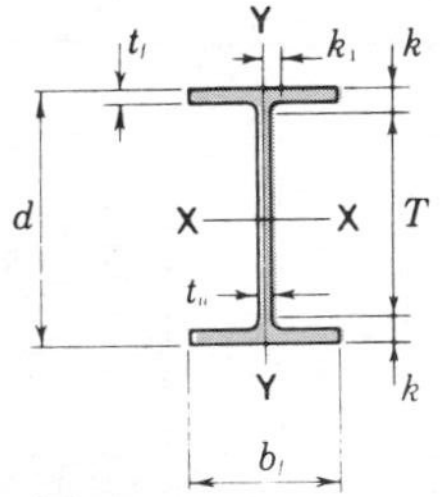

W SHAPES
Dimensions

Designation	Area A (In.²)	Depth d (In.)	Depth d (frac)	Web t_w (In.)	Web t_w (frac)	t_w/2 (In.)	Flange b_f (In.)	Flange b_f (frac)	Flange t_f (In.)	Flange t_f (frac)	T (In.)	k (In.)	k_1 (In.)
W 14×132	38.8	14.66	14⅝	0.645	⅝	5/16	14.725	14¾	1.030	1	11¼	1 11/16	15/16
×120	35.3	14.48	14½	0.590	9/16	5/16	14.670	14⅝	0.940	15/16	11¼	1⅝	15/16
×109	32.0	14.32	14⅜	0.525	½	¼	14.605	14⅝	0.860	⅞	11¼	1 9/16	⅞
×99	29.1	14.16	14⅛	0.485	½	¼	14.565	14⅝	0.780	¾	11¼	1½	⅞
×90	26.5	14.02	14	0.440	7/16	¼	14.520	14½	0.710	11/16	11¼	1⅜	⅞
W 14×82	24.1	14.31	14¼	0.510	½	¼	10.130	10⅛	0.855	⅞	11	1⅝	15/16
×74	21.8	14.17	14⅛	0.450	7/16	¼	10.070	10⅛	0.785	13/16	11	1 9/16	15/16
×68	20.0	14.04	14	0.415	7/16	¼	10.035	10	0.720	¾	11	1½	15/16
×61	17.9	13.89	13⅞	0.375	⅜	3/16	9.995	10	0.645	⅝	11	1 7/16	15/16
W 14×53	15.6	13.92	13⅞	0.370	⅜	3/16	8.060	8	0.660	11/16	11	1 7/16	11/16
×48	14.1	13.79	13¾	0.340	5/16	3/16	8.030	8	0.595	⅝	11	1⅜	11/16
×43	12.6	13.66	13⅝	0.305	5/16	3/16	7.995	8	0.530	½	11	1 5/16	11/16
W 14×38	11.2	14.10	14⅛	0.310	5/16	3/16	6.770	6¾	0.515	½	12	1 1/16	⅝
×34	10.0	13.98	14	0.285	5/16	3/16	6.745	6¾	0.455	7/16	12	1	⅝
×30	8.85	13.84	13⅞	0.270	¼	⅛	6.730	6¾	0.385	⅜	12	15/16	⅝
W 14×26	7.69	13.91	13⅞	0.255	¼	⅛	5.025	5	0.420	7/16	12	15/16	9/16
×22	6.49	13.74	13¾	0.230	¼	⅛	5.000	5	0.335	5/16	12	⅞	9/16

AMERICAN INSTITUTE OF STEEL CONSTRUCTION

W SHAPES
Dimensions

Designation	Area A (In.²)	Depth d (In.)	Depth d (frac)	Web t_w (In.)	Web t_w (frac)	t_w/2 (In.)	Flange b_f (In.)	Flange b_f (frac)	Flange t_f (In.)	Flange t_f (frac)	T (In.)	k (In.)	k_1 (In.)
W 12×336	98.8	16.82	16⅞	1.775	1¾	⅞	13.385	13⅜	2.955	2 15/16	9½	3 11/16	1½
×305	89.6	16.32	16⅜	1.625	1⅝	13/16	13.235	13¼	2.705	2 11/16	9½	3 7/16	1 7/16
×279	81.9	15.85	15⅞	1.530	1½	¾	13.140	13⅛	2.470	2½	9½	3 3/16	1⅜
×252	74.1	15.41	15⅜	1.395	1⅜	11/16	13.005	13	2.250	2¼	9½	2 15/16	1 5/16
×230	67.7	15.05	15	1.285	1 5/16	11/16	12.895	12⅞	2.070	2 1/16	9½	2¾	1¼
×210	61.8	14.71	14¾	1.180	1 3/16	⅝	12.790	12¾	1.900	1⅞	9½	2⅝	1¼
×190	55.8	14.38	14⅜	1.060	1 1/16	9/16	12.670	12⅝	1.735	1¾	9½	2 7/16	1 3/16
×170	50.0	14.03	14	0.960	15/16	½	12.570	12⅝	1.560	1 9/16	9½	2¼	1⅛
×152	44.7	13.71	13¾	0.870	⅞	7/16	12.480	12½	1.400	1⅜	9½	2⅛	1 1/16
×136	39.9	13.41	13⅜	0.790	13/16	7/16	12.400	12⅜	1.250	1¼	9½	1 15/16	1
×120	35.3	13.12	13⅛	0.710	11/16	⅜	12.320	12⅜	1.105	1⅛	9½	1 13/16	1
×106	31.2	12.89	12⅞	0.610	⅝	5/16	12.220	12¼	0.990	1	9½	1 11/16	15/16
×96	28.2	12.71	12¾	0.550	9/16	5/16	12.160	12⅛	0.900	⅞	9½	1⅝	⅞
×87	25.6	12.53	12½	0.515	½	¼	12.125	12⅛	0.810	13/16	9½	1½	⅞
×79	23.2	12.38	12⅜	0.470	½	¼	12.080	12⅛	0.735	¾	9½	1 7/16	⅞
×72	21.1	12.25	12¼	0.430	7/16	¼	12.040	12	0.670	11/16	9½	1⅜	⅞
×65	19.1	12.12	12⅛	0.390	⅜	3/16	12.000	12	0.605	⅝	9½	1 5/16	13/16
W 12×58	17.0	12.19	12¼	0.360	⅜	3/16	10.010	10	0.640	⅝	9½	1⅜	13/16
×53	15.6	12.06	12	0.345	⅜	3/16	9.995	10	0.575	9/16	9½	1¼	13/16
W 12×50	14.7	12.19	12¼	0.370	⅜	3/16	8.080	8⅛	0.640	⅝	9½	1⅜	13/16
×45	13.2	12.06	12	0.335	5/16	3/16	8.045	8	0.575	9/16	9½	1¼	13/16
×40	11.8	11.94	12	0.295	5/16	3/16	8.005	8	0.515	½	9½	1¼	¾
W 12×35	10.3	12.50	12½	0.300	5/16	3/16	6.560	6½	0.520	½	10½	1	9/16
×30	8.79	12.34	12⅜	0.260	¼	⅛	6.520	6½	0.440	7/16	10½	15/16	½
×26	7.65	12.22	12¼	0.230	¼	⅛	6.490	6½	0.380	⅜	10½	⅞	½
W 12×22	6.48	12.31	12¼	0.260	¼	⅛	4.030	4	0.425	7/16	10½	⅞	½
×19	5.57	12.16	12⅛	0.235	¼	⅛	4.005	4	0.350	⅜	10½	13/16	½
×16	4.71	11.99	12	0.220	¼	⅛	3.990	4	0.265	¼	10½	¾	½
×14	4.16	11.91	11⅞	0.200	3/16	⅛	3.970	4	0.225	¼	10½	11/16	½

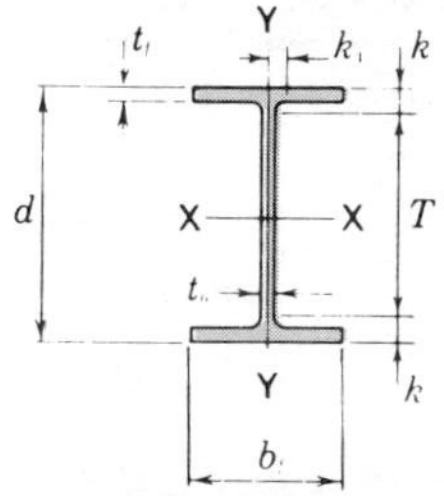

AMERICAN INSTITUTE OF STEEL CONSTRUCTION

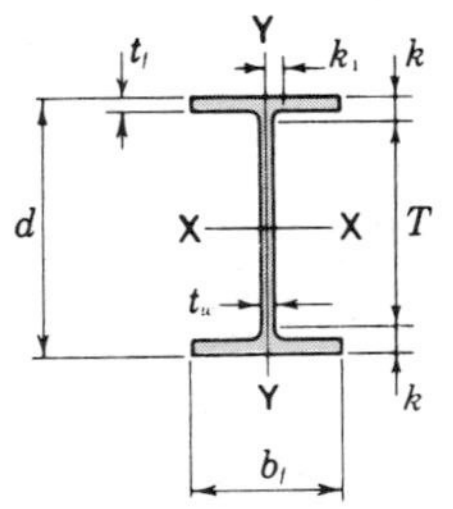

W SHAPES
Dimensions

Designation	Area A	Depth d		Web Thickness t_w		Web $\frac{t_w}{2}$	Flange Width b_f		Flange Thickness t_f		Distance T	Distance k	Distance k_1
	In.^{2}	In.		In.		In.	In.		In.		In.	In.	In.
W 14x730	215.0	22.42	22 3/8	3.070	3 1/16	1 9/16	17.890	17 7/8	4.910	4 15/16	11 1/4	5 9/16	2 3/16
x665	196.0	21.64	21 5/8	2.830	2 13/16	1 7/16	17.650	17 5/8	4.520	4 1/2	11 1/4	5 3/16	2 1/16
x605	178.0	20.92	20 7/8	2.595	2 5/8	1 5/16	17.415	17 3/8	4.160	4 3/16	11 1/4	4 13/16	1 15/16
x550	162.0	20.24	20 1/4	2.380	2 3/8	1 3/16	17.200	17 1/4	3.820	3 13/16	11 1/4	4 1/2	1 13/16
x500	147.0	19.60	19 5/8	2.190	2 3/16	1 1/8	17.010	17	3.500	3 1/2	11 1/4	4 3/16	1 3/4
x455	134.0	19.02	19	2.015	2	1	16.835	16 7/8	3.210	3 3/16	11 1/4	3 7/8	1 5/8
W 14x426	125.0	18.67	18 5/8	1.875	1 7/8	15/16	16.695	16 3/4	3.035	3 1/16	11 1/4	3 11/16	1 9/16
x398	117.0	18.29	18 1/4	1.770	1 3/4	7/8	16.590	16 5/8	2.845	2 7/8	11 1/4	3 1/2	1 1/2
x370	109.0	17.92	17 7/8	1.655	1 5/8	13/16	16.475	16 1/2	2.660	2 11/16	11 1/4	3 5/16	1 7/16
x342	101.0	17.54	17 1/2	1.540	1 9/16	13/16	16.360	16 3/8	2.470	2 1/2	11 1/4	3 1/8	1 3/8
x311	91.4	17.12	17 1/8	1.410	1 7/16	3/4	16.230	16 1/4	2.260	2 1/4	11 1/4	2 15/16	1 5/16
x283	83.3	16.74	16 3/4	1.290	1 5/16	11/16	16.110	16 1/8	2.070	2 1/16	11 1/4	2 3/4	1 1/4
x257	75.6	16.38	16 3/8	1.175	1 3/16	5/8	15.995	16	1.890	1 7/8	11 1/4	2 9/16	1 3/16
x233	68.5	16.04	16	1.070	1 1/16	9/16	15.890	15 7/8	1.720	1 3/4	11 1/4	2 3/8	1 3/16
x211	62.0	15.72	15 3/4	0.980	1	1/2	15.800	15 3/4	1.560	1 9/16	11 1/4	2 1/4	1 1/8
x193	56.8	15.48	15 1/2	0.890	7/8	7/16	15.710	15 3/4	1.440	1 7/16	11 1/4	2 1/8	1 1/16
x176	51.8	15.22	15 1/4	0.830	13/16	7/16	15.650	15 5/8	1.310	1 5/16	11 1/4	2	1 1/16
x159	46.7	14.98	15	0.745	3/4	3/8	15.565	15 5/8	1.190	1 3/16	11 1/4	1 7/8	1
x145	42.7	14.78	14 3/4	0.680	11/16	3/8	15.500	15 1/2	1.090	1 1/16	11 1/4	1 3/4	1

AMERICAN INSTITUTE OF STEEL CONSTRUCTION

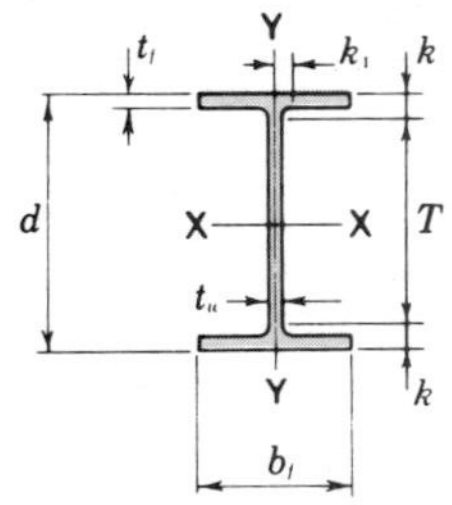

W SHAPES
Dimensions

Designation	Area A	Depth d		Web Thickness t_w		Web $\frac{t_w}{2}$	Flange Width b_f		Flange Thickness t_f		Distance T	Distance k	Distance k_1
	In.^{2}	In.		In.		In.	In.		In.		In.	In.	In.
W 18x119	35.1	18.97	19	0.655	5/8	5/16	11.265	11 1/4	1.060	1 1/16	15 1/2	1 3/4	15/16
x106	31.1	18.73	18 3/4	0.590	9/16	5/16	11.200	11 1/4	0.940	15/16	15 1/2	1 5/8	15/16
x 97	28.5	18.59	18 5/8	0.535	9/16	5/16	11.145	11 1/8	0.870	7/8	15 1/2	1 9/16	7/8
x 86	25.3	18.39	18 3/8	0.480	1/2	1/4	11.090	11 1/8	0.770	3/4	15 1/2	1 7/16	7/8
x 76	22.3	18.21	18 1/4	0.425	7/16	1/4	11.035	11	0.680	11/16	15 1/2	1 3/8	13/16
W 18x 71	20.8	18.47	18 1/2	0.495	1/2	1/4	7.635	7 5/8	0.810	13/16	15 1/2	1 1/2	7/8
x 65	19.1	18.35	18 3/8	0.450	7/16	1/4	7.590	7 5/8	0.750	3/4	15 1/2	1 7/16	7/8
x 60	17.6	18.24	18 1/4	0.415	7/16	1/4	7.555	7 1/2	0.695	11/16	15 1/2	1 3/8	13/16
x 55	16.2	18.11	18 1/8	0.390	3/8	3/16	7.530	7 1/2	0.630	5/8	15 1/2	1 5/16	13/16
x 50	14.7	17.99	18	0.355	3/8	3/16	7.495	7 1/2	0.570	9/16	15 1/2	1 1/4	13/16
W 18x 46	13.5	18.06	18	0.360	3/8	3/16	6.060	6	0.605	5/8	15 1/2	1 1/4	13/16
x 40	11.8	17.90	17 7/8	0.315	5/16	3/16	6.015	6	0.525	1/2	15 1/2	1 3/16	13/16
x 35	10.3	17.70	17 3/4	0.300	5/16	3/16	6.000	6	0.425	7/16	15 1/2	1 1/8	3/4
W 16x100	29.4	16.97	17	0.585	9/16	5/16	10.425	10 3/8	0.985	1	13 5/8	1 11/16	15/16
x 89	26.2	16.75	16 3/4	0.525	1/2	1/4	10.365	10 3/8	0.875	7/8	13 5/8	1 9/16	7/8
x 77	22.6	16.52	16 1/2	0.455	7/16	1/4	10.295	10 1/4	0.760	3/4	13 5/8	1 7/16	7/8
x 67	19.7	16.33	16 3/8	0.395	3/8	3/16	10.235	10 1/4	0.665	11/16	13 5/8	1 3/8	13/16
W 16x 57	16.8	16.43	16 3/8	0.430	7/16	1/4	7.120	7 1/8	0.715	11/16	13 5/8	1 3/8	7/8
x 50	14.7	16.26	16 1/4	0.380	3/8	3/16	7.070	7 1/8	0.630	5/8	13 5/8	1 5/16	13/16
x 45	13.3	16.13	16 1/8	0.345	3/8	3/16	7.035	7	0.565	9/16	13 5/8	1 1/4	13/16
x 40	11.8	16.01	16	0.305	5/16	3/16	6.995	7	0.505	1/2	13 5/8	1 3/16	13/16
x 36	10.6	15.86	15 7/8	0.295	5/16	3/16	6.985	7	0.430	7/16	13 5/8	1 1/8	3/4
W 16x 31	9.12	15.88	15 7/8	0.275	1/4	1/8	5.525	5 1/2	0.440	7/16	13 5/8	1 1/8	3/4
x 26	7.68	15.69	15 3/4	0.250	1/4	1/8	5.500	5 1/2	0.345	3/8	13 5/8	1 1/16	3/4

AMERICAN INSTITUTE OF STEEL CONSTRUCTION

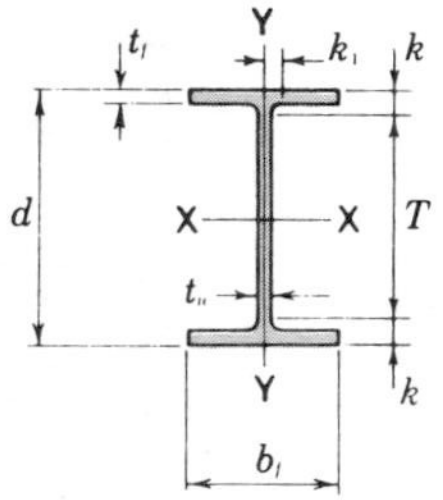

W SHAPES
Dimensions

Designation	Area A	Depth d		Web Thickness t_w		Web $\frac{t_w}{2}$	Flange Width b_f		Flange Thickness t_f		Distance T	Distance k	Distance k_1
	In.^{2}	In.		In.		In.	In.		In.		In.	In.	In.
W 27x178	52.3	27.81	27 3/4	0.725	3/4	3/8	14.085	14 1/8	1.190	1 3/16	24	1 7/8	1 1/16
x161	47.4	27.59	27 5/8	0.660	11/16	3/8	14.020	14	1.080	1 1/16	24	1 13/16	1
x146	42.9	27.38	27 3/8	0.605	5/8	5/16	13.965	14	0.975	1	24	1 11/16	1
W 27x114	33.5	27.29	27 1/4	0.570	9/16	5/16	10.070	10 1/8	0.930	15/16	24	1 5/8	15/16
x102	30.0	27.09	27 1/8	0.515	1/2	1/4	10.015	10	0.830	13/16	24	1 9/16	15/16
x 94	27.7	26.92	26 7/8	0.490	1/2	1/4	9.990	10	0.745	3/4	24	1 7/16	15/16
x 84	24.8	26.71	26 3/4	0.460	7/16	1/4	9.960	10	0.640	5/8	24	1 3/8	15/16
W 24x162	47.7	25.00	25	0.705	11/16	3/8	12.955	13	1.220	1 1/4	21	2	1 1/16
x146	43.0	24.74	24 3/4	0.650	5/8	5/16	12.900	12 7/8	1.090	1 1/16	21	1 7/8	1 1/16
x131	38.5	24.48	24 1/2	0.605	5/8	5/16	12.855	12 7/8	0.960	15/16	21	1 3/4	1 1/16
x117	34.4	24.26	24 1/4	0.550	9/16	5/16	12.800	12 3/4	0.850	7/8	21	1 5/8	1
x104	30.6	24.06	24	0.500	1/2	1/4	12.750	12 3/4	0.750	3/4	21	1 1/2	1
W 24x 94	27.7	24.31	24 1/4	0.515	1/2	1/4	9.065	9 1/8	0.875	7/8	21	1 5/8	1
x 84	24.7	24.10	24 1/8	0.470	1/2	1/4	9.020	9	0.770	3/4	21	1 9/16	15/16
x 76	22.4	23.92	23 7/8	0.440	7/16	1/4	8.990	9	0.680	11/16	21	1 7/16	15/16
x 68	20.1	23.73	23 3/4	0.415	7/16	1/4	8.965	9	0.585	9/16	21	1 3/8	15/16
W 24x 62	18.2	23.74	23 3/4	0.430	7/16	1/4	7.040	7	0.590	9/16	21	1 3/8	15/16
x 55	16.2	23.57	23 5/8	0.395	3/8	3/16	7.005	7	0.505	1/2	21	1 5/15	15/16
W 21x147	43.2	22.06	22	0.720	3/4	3/8	12.510	12 1/2	1.150	1 1/8	18 1/4	1 7/8	1 1/16
x132	38.8	21.83	21 7/8	0.650	5/8	5/16	12.440	12 1/2	1.035	1 1/16	18 1/4	1 13/16	1
x122	35.9	21.68	21 5/8	0.600	5/8	5/16	12.390	12 3/8	0.960	15/16	18 1/4	1 11/16	1
x111	32.7	21.51	21 1/2	0.550	9/16	5/16	12.340	12 3/8	0.875	7/8	18 1/4	1 5/8	15/16
x101	29.8	21.36	21 3/8	0.500	1/2	1/4	12.290	12 1/4	0.800	13/16	18 1/4	1 9/16	15/16
W 21x 93	27.3	21.62	21 5/8	0.580	9/16	5/16	8.420	8 3/8	0.930	15/16	18 1/4	1 11/16	1
x 83	24.3	21.43	21 3/8	0.515	1/2	1/4	8.355	8 3/8	0.835	13/16	18 1/4	1 9/16	15/16
x 73	21.5	21.24	21 1/4	0.455	7/16	1/4	8.295	8 1/4	0.740	3/4	18 1/4	1 1/2	15/16
x 68	20.0	21.13	21 1/8	0.430	7/16	1/4	8.270	8 1/4	0.685	11/16	18 1/4	1 7/16	7/8
x 62	18.3	20.99	21	0.400	3/8	3/16	8.240	8 1/4	0.615	5/8	18 1/4	1 3/8	7/8
W 21x 57	16.7	21.06	21	0.405	3/8	3/16	6.555	6 1/2	0.650	5/8	18 1/4	1 3/8	7/8
x 50	14.7	20.83	20 7/8	0.380	3/8	3/16	6.530	6 1/2	0.535	9/16	18 1/4	1 5/16	7/8
x 44	13.0	20.66	20 5/8	0.350	3/8	3/16	6.500	6 1/2	0.450	7/16	18 1/4	1 3/16	7/8

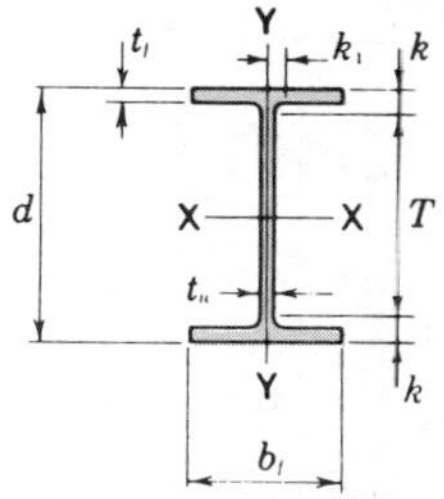

W SHAPES
Dimensions

Designation	Area A	Depth d		Web Thickness t_w		Web $\frac{t_w}{2}$	Flange Width b_f		Flange Thickness t_f		Distance T	Distance k	Distance k_1
	In.^{2}	In.		In.		In.	In.		In.		In.	In.	In.
W 36x300	88.3	36.74	36 3/4	0.945	15/16	1/2	16.655	16 5/8	1.680	1 11/16	31 1/8	2 13/16	1 1/2
x280	82.4	36.52	36 1/2	0.885	7/8	7/16	16.595	16 5/8	1.570	1 9/16	31 1/8	2 11/16	1 1/2
x260	76.5	36.26	36 1/4	0.840	13/16	7/16	16.550	16 1/2	1.440	1 7/16	31 1/8	2 9/16	1 1/2
x245	72.1	36.08	36 1/8	0.800	13/16	7/16	16.510	16 1/2	1.350	1 3/8	31 1/8	2 1/2	1 7/16
x230	67.6	35.90	35 7/8	0.760	3/4	3/8	16.470	16 1/2	1.260	1 1/4	31 1/8	2 3/8	1 7/16
W 36x210	61.8	36.69	36 3/4	0.830	13/16	7/16	12.180	12 1/8	1.360	1 3/8	32 1/8	2 5/16	1 1/4
x194	57.0	36.49	36 1/2	0.765	3/4	3/8	12.115	12 1/8	1.260	1 1/4	32 1/8	2 3/16	1 3/16
x182	53.6	36.33	36 3/8	0.725	3/4	3/8	12.075	12 1/8	1.180	1 3/16	32 1/8	2 1/8	1 3/16
x170	50.0	36.17	36 1/8	0.680	11/16	3/8	12.030	12	1.100	1 1/8	32 1/8	2	1 3/16
x160	47.0	36.01	36	0.650	5/8	5/16	12.000	12	1.020	1	32 1/8	1 15/16	1 1/8
x150	44.2	35.85	35 7/8	0.625	5/8	5/16	11.975	12	0.940	15/16	32 1/8	1 7/8	1 1/8
x135	39.7	35.55	35 1/2	0.600	5/8	5/16	11.950	12	0.790	13/16	32 1/8	1 11/16	1 1/8
W 33x241	70.9	34.18	34 1/8	0.830	13/16	7/16	15.860	15 7/8	1.400	1 3/8	29 3/4	2 3/16	1 3/16
x221	65.0	33.93	33 7/8	0.775	3/4	3/8	15.805	15 3/4	1.275	1 1/4	29 3/4	2 1/16	1 3/16
x201	59.1	33.68	33 5/8	0.715	11/16	3/8	15.745	15 3/4	1.150	1 1/8	29 3/4	1 15/16	1 1/8
W 33x152	44.7	33.49	33 1/2	0.635	5/8	5/16	11.565	11 5/8	1.055	1 1/16	29 3/4	1 7/8	1 1/8
x141	41.6	33.30	33 1/4	0.605	5/8	5/16	11.535	11 1/2	0.960	15/16	29 3/4	1 3/4	1 1/16
x130	38.3	33.09	33 1/8	0.580	9/16	5/16	11.510	11 1/2	0.855	7/8	29 3/4	1 11/16	1 1/16
x118	34.7	32.86	32 7/8	0.550	9/16	5/16	11.480	11 1/2	0.740	3/4	29 3/4	1 9/16	1 1/16
W 30x211	62.0	30.94	31	0.775	3/4	3/8	15.105	15 1/8	1.315	1 5/16	26 3/4	2 1/8	1 1/8
x191	56.1	30.68	30 5/8	0.710	11/16	3/8	15.040	15	1.185	1 3/16	26 3/4	1 15/16	1 1/16
x173	50.8	30.44	30 1/2	0.655	5/8	5/16	14.985	15	1.065	1 1/16	26 3/4	1 7/8	1 1/16
W 30x132	38.9	30.31	30 1/4	0.615	5/8	5/16	10.545	10 1/2	1.000	1	26 3/4	1 3/4	1 1/16
x124	36.5	30.17	30 1/8	0.585	9/16	5/16	10.515	10 1/2	0.930	15/16	26 3/4	1 11/16	1
x116	34.2	30.01	30	0.565	9/16	5/16	10.495	10 1/2	0.850	7/8	26 3/4	1 5/8	1
x108	31.7	29.83	29 7/8	0.545	9/16	5/16	10.475	10 1/2	0.760	3/4	26 3/4	1 9/16	1
x 99	29.1	29.65	29 5/8	0.520	1/2	1/4	10.450	10 1/2	0.670	11/16	26 3/4	1 7/16	1

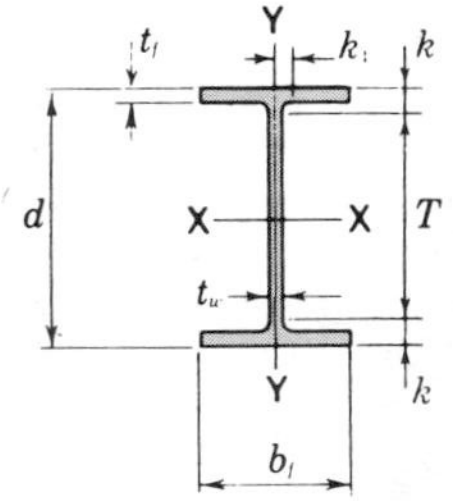

M SHAPES
Dimensions

Designation	Area A	Depth d		Web Thickness t_w		Web $\frac{t_w}{2}$	Flange Width b_f		Flange Thickness t_f		Distance T	Distance k	Grip	Max. Flge. Fas-ten-er
	In.²	In.		In.		In.	In.		In.		In.	In.	In.	In.
M 14x18	5.10	14.00	14	0.215	3/16	1/8	4.000	4	0.270	1/4	12 3/4	5/8	1/4	3/4
M 12x11.8	3.47	12.00	12	0.177	3/16	1/8	3.065	3 1/8	0.225	1/4	10 7/8	9/16	1/4	—
M 10x9	2.65	10.00	10	0.157	3/16	1/8	2.690	2 3/4	0.206	3/16	8 7/8	9/16	3/16	—
M 8x6.5	1.92	8.00	8	0.135	1/8	1/16	2.281	2 1/4	0.189	3/16	7	1/2	3/16	—
M 6x20	5.89	6.00	6	0.250	1/4	1/8	5.938	6	0.379	3/8	4 1/4	7/8	3/8	7/8
M 6x4.4	1.29	6.00	6	0.114	1/8	1/16	1.844	1 7/8	0.171	3/16	5 1/8	7/16	3/16	—
M 5x18.9	5.55	5.00	5	0.316	5/16	3/16	5.003	5	0.416	7/16	3 1/4	7/8	7/16	7/8
M 4x13	3.81	4.00	4	0.254	1/4	1/8	3.940	4	0.371	3/8	2 3/8	13/16	3/8	3/4

AMERICAN INSTITUTE OF STEEL CONSTRUCTION

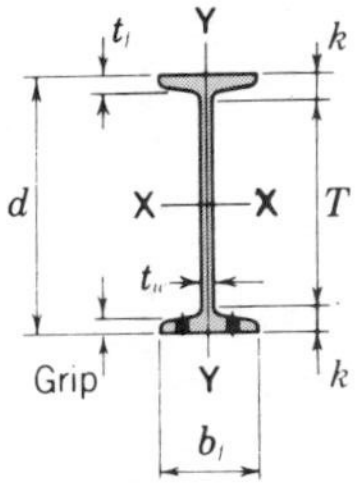

S SHAPES
Dimensions

Designation	Area A	Depth d		Web Thickness t_w		Web $\frac{t_w}{2}$	Flange Width b_f		Flange Thickness t_f		Distance T	Distance k	Grip	Max. Flge. Fas-ten-er
	In.²	In.		In.		In.	In.		In.		In.	In.	In.	In.
S 24x121	35.6	24.50	24 1/2	0.800	13/16	7/16	8.050	8	1.090	1 1/16	20 1/2	2	1 1/8	1
x106	31.2	24.50	24 1/2	0.620	5/8	5/16	7.870	7 7/8	1.090	1 1/16	20 1/2	2	1 1/8	1
S 24x100	29.3	24.00	24	0.745	3/4	3/8	7.245	7 1/4	0.870	7/8	20 1/2	1 3/4	7/8	1
x90	26.5	24.00	24	0.625	5/8	5/16	7.125	7 1/8	0.870	7/8	20 1/2	1 3/4	7/8	1
x80	23.5	24.00	24	0.500	1/2	1/4	7.000	7	0.870	7/8	20 1/2	1 3/4	7/8	1
S 20x96	28.2	20.30	20 1/4	0.800	13/16	7/16	7.200	7 1/4	0.920	15/16	16 3/4	1 3/4	15/16	1
x86	25.3	20.30	20 1/4	0.660	11/16	3/8	7.060	7	0.920	15/16	16 3/4	1 3/4	15/16	1
S 20x75	22.0	20.00	20	0.635	5/8	5/16	6.385	6 3/8	0.795	13/16	16 3/4	1 5/8	13/16	7/8
x66	19.4	20.00	20	0.505	1/2	1/4	6.255	6 1/4	0.795	13/16	16 3/4	1 5/8	13/16	7/8
S 18x70	20.6	18.00	18	0.711	11/16	3/8	6.251	6 1/4	0.691	11/16	15	1 1/2	11/16	7/8
x54.7	16.1	18.00	18	0.461	7/16	1/4	6.001	6	0.691	11/16	15	1 1/2	11/16	7/8
S 15x50	14.7	15.00	15	0.550	9/16	5/16	5.640	5 5/8	0.622	5/8	12 1/4	1 3/8	9/16	3/4
x42.9	12.6	15.00	15	0.411	7/16	1/4	5.501	5 1/2	0.622	5/8	12 1/4	1 3/8	9/16	3/4
S 12x50	14.7	12.00	12	0.687	11/16	3/8	5.477	5 1/2	0.659	11/16	9 1/8	1 7/16	11/16	3/4
x40.8	12.0	12.00	12	0.462	7/16	1/4	5.252	5 1/4	0.659	11/16	9 1/8	1 7/16	5/8	3/4
S 12x35	10.3	12.00	12	0.428	7/16	1/4	5.078	5 1/8	0.544	9/16	9 5/8	1 3/16	1/2	3/4
x31.8	9.35	12.00	12	0.350	3/8	3/16	5.000	5	0.544	9/16	9 5/8	1 3/16	1/2	3/4
S 10x35	10.3	10.00	10	0.594	5/8	5/16	4.944	5	0.491	1/2	7 3/4	1 1/8	1/2	3/4
x25.4	7.46	10.00	10	0.311	5/16	3/16	4.661	4 5/8	0.491	1/2	7 3/4	1 1/8	1/2	3/4
S 8x23	6.77	8.00	8	0.441	7/16	1/4	4.171	4 1/8	0.426	7/16	6	1	7/16	3/4
x18.4	5.41	8.00	8	0.271	1/4	1/8	4.001	4	0.426	7/16	6	1	7/16	3/4
S 7x20	5.88	7.00	7	0.450	7/16	1/4	3.860	3 7/8	0.392	3/8	5 1/8	15/16	3/8	5/8
x15.3	4.50	7.00	7	0.252	1/4	1/8	3.662	3 5/8	0.392	3/8	5 1/8	15/16	3/8	5/8
S 6x17.25	5.07	6.00	6	0.465	7/16	1/4	3.565	3 5/8	0.359	3/8	4 1/4	7/8	3/8	5/8
x12.5	3.67	6.00	6	0.232	1/4	1/8	3.332	3 3/8	0.359	3/8	4 1/4	7/8	3/8	—
S 5x14.75	4.34	5.00	5	0.494	1/2	1/4	3.284	3 1/4	0.326	5/16	3 3/8	13/16	5/16	—
x10	2.94	5.00	5	0.214	3/16	1/8	3.004	3	0.326	5/16	3 3/8	13/16	5/16	—
S 4x9.5	2.79	4.00	4	0.326	5/16	3/16	2.796	2 3/4	0.293	5/16	2 1/2	3/4	5/16	—
x7.7	2.26	4.00	4	0.193	3/16	1/8	2.663	2 5/8	0.293	5/16	2 1/2	3/4	5/16	—
S 3x7.5	2.21	3.00	3	0.349	3/8	3/16	2.509	2 1/2	0.260	1/4	1 5/8	11/16	1/4	—
x5.7	1.67	3.00	3	0.170	3/16	1/8	2.330	2 3/8	0.260	1/4	1 5/8	11/16	1/4	—

AMERICAN INSTITUTE OF STEEL CONSTRUCTION

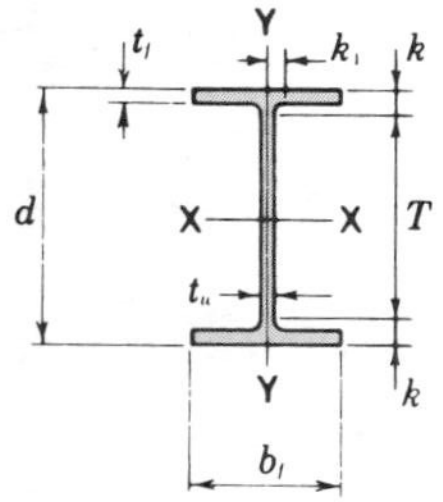

HP SHAPES
Dimensions

Designation	Area A	Depth d		Web: Thickness t_w		Web: $\frac{t_w}{2}$	Flange: Width b_f		Flange: Thickness t_f		Distance: T	Distance: k	Distance: k_1
	In.^{2}	In.		In.		In.	In.		In.		In.	In.	In.
HP 14x117	34.4	14.21	14 1/4	0.805	13/16	7/16	14.885	14 7/8	0.805	13/16	11 1/4	1 1/2	1 1/16
x102	30.0	14.01	14	0.705	11/16	3/8	14.785	14 3/4	0.705	11/16	11 1/4	1 3/8	1
x 89	26.1	13.83	13 7/8	0.615	5/8	5/16	14.695	14 3/4	0.615	5/8	11 1/4	1 5/16	15/16
x 73	21.4	13.61	13 5/8	0.505	1/2	1/4	14.585	14 5/8	0.505	1/2	11 1/4	1 3/16	7/8
HP 13x100	29.4	13.15	13 1/8	0.765	3/4	3/8	13.205	13 1/4	0.765	3/4	10 1/4	1 7/16	1
x 87	25.5	12.95	13	0.665	11/16	3/8	13.105	13 1/8	0.665	11/16	10 1/4	1 3/8	15/16
x 73	21.6	12.75	12 3/4	0.565	9/16	5/16	13.005	13	0.565	9/16	10 1/4	1 1/4	15/16
x 60	17.5	12.54	12 1/2	0.460	7/16	1/4	12.900	12 7/8	0.460	7/16	10 1/4	1 1/8	7/8
HP 12x 84	24.6	12.28	12 1/4	0.685	11/16	3/8	12.295	12 1/4	0.685	11/16	9 1/2	1 3/8	1
x 74	21.8	12.13	12 1/8	0.605	5/8	5/16	12.215	12 1/4	0.610	5/8	9 1/2	1 5/16	15/16
x 63	18.4	11.94	12	0.515	1/2	1/4	12.125	12 1/8	0.515	1/2	9 1/2	1 1/4	7/8
x 53	15.5	11.78	11 3/4	0.435	7/16	1/4	12.045	12	0.435	7/16	9 1/2	1 1/8	7/8
HP 10x 57	16.8	9.99	10	0.565	9/16	5/16	10.225	10 1/4	0.565	9/16	7 5/8	1 3/16	13/16
x 42	12.4	9.70	9 3/4	0.415	7/16	1/4	10.075	10 1/8	0.420	7/16	7 5/8	1 1/16	3/4
HP 8x 36	10.6	8.02	8	0.445	7/16	1/4	8.155	8 1/8	0.445	7/16	6 1/8	15/16	5/8

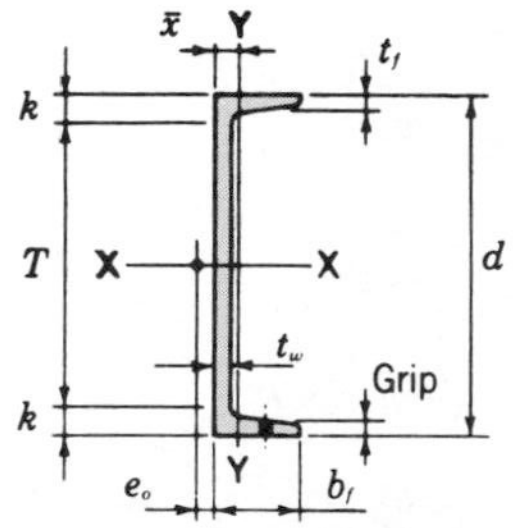

CHANNELS AMERICAN STANDARD
Dimensions

Designation	Area A	Depth d	Web: Thickness t_w		Web: $\frac{t_w}{2}$	Flange: Width b_f		Flange: Average thickness t_f		Distance: T	Distance: k	Grip	Max. Flge. Fas-ten-er
	In.^{2}	In.	In.		In.	In.		In.		In.	In.	In.	In.
C 15x50	14.7	15.00	0.716	11/16	3/8	3.716	3 3/4	0.650	5/8	12 1/8	1 7/16	5/8	1
x40	11.8	15.00	0.520	1/2	1/4	3.520	3 1/2	0.650	5/8	12 1/8	1 7/16	5/8	1
x33.9	9.96	15.00	0.400	3/8	3/16	3.400	3 3/8	0.650	5/8	12 1/8	1 7/16	5/8	1
C 12x30	8.82	12.00	0.510	1/2	1/4	3.170	3 1/8	0.501	1/2	9 3/4	1 1/8	1/2	7/8
x25	7.35	12.00	0.387	3/8	3/16	3.047	3	0.501	1/2	9 3/4	1 1/8	1/2	7/8
x20.7	6.09	12.00	0.282	5/16	1/8	2.942	3	0.501	1/2	9 3/4	1 1/8	1/2	7/8
C 10x30	8.82	10.00	0.673	11/16	5/16	3.033	3	0.436	7/16	8	1	7/16	3/4
x25	7.35	10.00	0.526	1/2	1/4	2.886	2 7/8	0.436	7/16	8	1	7/16	3/4
x20	5.88	10.00	0.379	3/8	3/16	2.739	2 3/4	0.436	7/16	8	1	7/16	3/4
x15.3	4.49	10.00	0.240	1/4	1/8	2.600	2 5/8	0.436	7/16	8	1	7/16	3/4
C 9x20	5.88	9.00	0.448	7/16	1/4	2.648	2 5/8	0.413	7/16	7 1/8	15/16	7/16	3/4
x15	4.41	9.00	0.285	5/16	1/8	2.485	2 1/2	0.413	7/16	7 1/8	15/16	7/16	3/4
x13.4	3.94	9.00	0.233	1/4	1/8	2.433	2 3/8	0.413	7/16	7 1/8	15/16	7/16	3/4
C 8x18.75	5.51	8.00	0.487	1/2	1/4	2.527	2 1/2	0.390	3/8	6 1/8	15/16	3/8	3/4
x13.75	4.04	8.00	0.303	5/16	1/8	2.343	2 3/8	0.390	3/8	6 1/8	15/16	3/8	3/4
x11.5	3.38	8.00	0.220	1/4	1/8	2.260	2 1/4	0.390	3/8	6 1/8	15/16	3/8	3/4
C 7x14.75	4.33	7.00	0.419	7/16	3/16	2.299	2 1/4	0.366	3/8	5 1/4	7/8	3/8	5/8
x12.25	3.60	7.00	0.314	5/16	3/16	2.194	2 1/4	0.366	3/8	5 1/4	7/8	3/8	5/8
x 9.8	2.87	7.00	0.210	3/16	1/8	2.090	2 1/8	0.366	3/8	5 1/4	7/8	3/8	5/8
C 6x13	3.83	6.00	0.437	7/16	3/16	2.157	2 1/8	0.343	5/16	4 3/8	13/16	5/16	5/8
x10.5	3.09	6.00	0.314	5/16	3/16	2.034	2	0.343	5/16	4 3/8	13/16	3/8	5/8
x 8.2	2.40	6.00	0.200	3/16	1/8	1.920	1 7/8	0.343	5/16	4 3/8	13/16	5/16	5/8
C 5x 9	2.64	5.00	0.325	5/16	3/16	1.885	1 7/8	0.320	5/16	3 1/2	3/4	5/16	5/8
x 6.7	1.97	5.00	0.190	3/16	1/8	1.750	1 3/4	0.320	5/16	3 1/2	3/4	—	—
C 4x 7.25	2.13	4.00	0.321	5/16	3/16	1.721	1 3/4	0.296	5/16	2 5/8	11/16	5/16	5/8
x 5.4	1.59	4.00	0.184	3/16	1/16	1.584	1 5/8	0.296	5/16	2 5/8	11/16	—	—
C 3x 6	1.76	3.00	0.356	3/8	3/16	1.596	1 5/8	0.273	1/4	1 5/8	11/16	—	—
x 5	1.47	3.00	0.258	1/4	1/8	1.498	1 1/2	0.273	1/4	1 5/8	11/16	—	—
x 4.1	1.21	3.00	0.170	3/16	1/16	1.410	1 3/8	0.273	1/4	1 5/8	11/16	—	—

AMERICAN INSTITUTE OF STEEL CONSTRUCTION

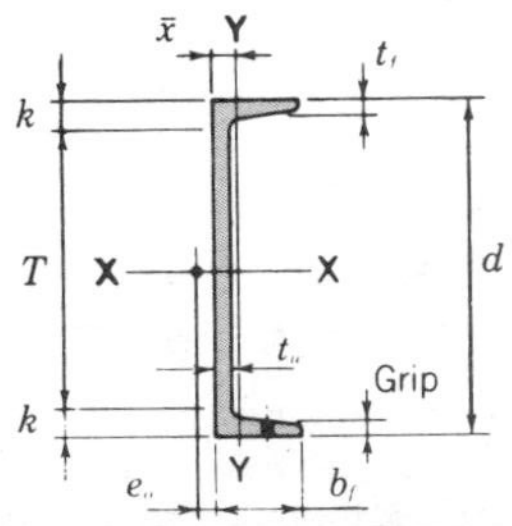

CHANNELS MISCELLANEOUS

Dimensions

Designation	Area A	Depth d	Web: Thickness t_w		Web: $\frac{t_w}{2}$	Flange: Width b_f		Flange: Average thickness t_f		Distance: T	Distance: k	Grip	Max. Flge. Fastener
	In.^{2}	In.	In.		In.	In.		In.		In.	In.	In.	In.
MC 18x58	17.1	18.00	0.700	11/16	3/8	4.200	4 1/4	0.625	5/8	15 1/4	1 3/8	5/8	1
x51.9	15.3	18.00	0.600	5/8	5/16	4.100	4 1/8	0.625	5/8	15 1/4	1 3/8	5/8	1
x45.8	13.5	18.00	0.500	1/2	1/4	4.000	4	0.625	5/8	15 1/4	1 3/8	5/8	1
x42.7	12.6	18.00	0.450	7/16	1/4	3.950	4	0.625	5/8	15 1/4	1 3/8	5/8	1
MC 13x50	14.7	13.00	0.787	13/16	3/8	4.412	4 3/8	0.610	5/8	10 1/4	1 3/8	5/8	1
x40	11.8	13.00	0.560	9/16	1/4	4.185	4 1/8	0.610	5/8	10 1/4	1 3/8	9/16	1
x35	10.3	13.00	0.447	7/16	1/4	4.072	4 1/8	0.610	5/8	10 1/4	1 3/8	9/16	1
x31.8	9.35	13.00	0.375	3/8	3/16	4.000	4	0.610	5/8	10 1/4	1 3/8	9/16	1
MC 12x50	14.7	12.00	0.835	13/16	7/16	4.135	4 1/8	0.700	11/16	9 3/8	1 5/16	11/16	1
x45	13.2	12.00	0.712	11/16	3/8	4.012	4	0.700	11/16	9 3/8	1 5/16	11/16	1
x40	11.8	12.00	0.590	9/16	5/16	3.890	3 7/8	0.700	11/16	9 3/8	1 5/16	11/16	1
x35	10.3	12.00	0.467	7/16	1/4	3.767	3 3/4	0.700	11/16	9 3/8	1 5/16	11/16	1
MC 12x37	10.9	12.00	0.600	5/8	5/16	3.600	3 5/8	0.600	5/8	9 3/8	1 5/16	5/8	7/8
x32.9	9.67	12.00	0.500	1/2	1/4	3.500	3 1/2	0.600	5/8	9 3/8	1 5/16	9/16	7/8
x30.9	9.07	12.00	0.450	7/16	1/4	3.450	3 1/2	0.600	5/8	9 3/8	1 5/16	9/16	7/8
MC 12x10.6	3.10	12.00	0.190	3/16	1/8	1.500	1 1/2	0.309	5/16	10 5/8	11/16	—	—
MC 10x41.1	12.1	10.00	0.796	13/16	3/8	4.321	4 3/8	0.575	9/16	7 1/2	1 1/4	9/16	7/8
x33.6	9.87	10.00	0.575	9/16	5/16	4.100	4 1/8	0.575	9/16	7 1/2	1 1/4	9/16	7/8
x28.5	8.37	10.00	0.425	7/16	3/16	3.950	4	0.575	9/16	7 1/2	1 1/4	9/16	7/8
MC 10x28.3	8.32	10.00	0.477	1/2	1/4	3.502	3 1/2	0.575	9/16	7 1/2	1 1/4	9/16	7/8
x25.3	7.43	10.00	0.425	7/16	3/16	3.550	3 1/2	0.500	1/2	7 3/4	1 1/8	1/2	7/8
x24.9	7.32	10.00	0.377	3/8	3/16	3.402	3 3/8	0.575	9/16	7 1/2	1 1/4	9/16	7/8
x21.9	6.43	10.00	0.325	5/16	3/16	3.450	3 1/2	0.500	1/2	7 3/4	1 1/8	1/2	7/8
MC 10x 8.4	2.46	10.00	0.170	3/16	1/16	1.500	1 1/2	0.280	1/4	8 5/8	11/16	—	—
MC 10x 6.5	1.91	10.00	0.152	1/8	1/16	1.127	1 1/8	0.202	3/16	9 1/8	7/16	—	—

AMERICAN INSTITUTE OF STEEL CONSTRUCTION

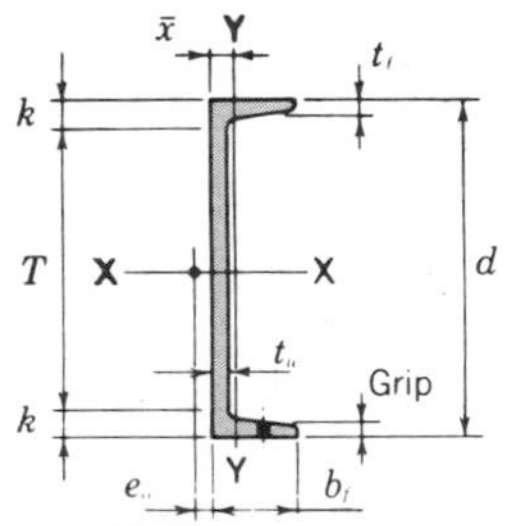

CHANNELS MISCELLANEOUS

Dimensions

Designation	Area A	Depth d	Web: Thickness t_w		Web: $\frac{t_w}{2}$	Flange: Width b_f		Flange: Average thickness t_f		Distance: T	Distance: k	Grip	Max. Flge. Fastener
	In.^{2}	In.	In.		In.	In.		In.		In.	In.	In.	In.
MC 9x25.4	7.47	9.00	0.450	7/16	1/4	3.500	3 1/2	0.550	9/16	6 5/8	1 3/16	9/16	7/8
x23.9	7.02	9.00	0.400	3/8	3/16	3.450	3 1/2	0.550	9/16	6 5/8	1 3/16	9/16	7/8
MC 8x22.8	6.70	8.00	0.427	7/16	3/16	3.502	3 1/2	0.525	1/2	5 5/8	1 3/16	1/2	7/8
x21.4	6.28	8.00	0.375	3/8	3/16	3.450	3 1/2	0.525	1/2	5 5/8	1 3/16	1/2	7/8
MC 8x20	5.88	8.00	0.400	3/8	3/16	3.025	3	0.500	1/2	5 3/4	1 1/8	1/2	7/8
x18.7	5.50	8.00	0.353	3/8	3/16	2.978	3	0.500	1/2	5 3/4	1 1/8	1/2	7/8
MC 8x 8.5	2.50	8.00	0.179	3/16	1/16	1.874	1 7/8	0.311	5/16	6 1/2	3/4	5/16	5/8
MC 7x22.7	6.67	7.00	0.503	1/2	1/4	3.603	3 5/8	0.500	1/2	4 3/4	1 1/8	1/2	7/8
x19.1	5.61	7.00	0.352	3/8	3/16	3.452	3 1/2	0.500	1/2	4 3/4	1 1/8	1/2	7/8
MC 7x17.6	5.17	7.00	0.375	3/8	3/16	3.000	3	0.475	1/2	4 7/8	1 1/16	1/2	3/4
MC 6x18	5.29	6.00	0.379	3/8	3/16	3.504	3 1/2	0.475	1/2	3 7/8	1 1/16	1/2	7/8
x15.3	4.50	6.00	0.340	5/16	3/16	3.500	3 1/2	0.385	3/8	4 1/4	7/8	3/8	7/8
MC 6x16.3	4.79	6.00	0.375	3/8	3/16	3.000	3	0.475	1/2	3 7/8	1 1/16	1/2	3/4
x15.1	4.44	6.00	0.316	5/16	3/16	2.941	3	0.475	1/2	3 7/8	1 1/16	1/2	3/4
MC 6x12	3.53	6.00	0.310	5/16	1/8	2.497	2 1/2	0.375	3/8	4 3/8	13/16	3/8	5/8

AMERICAN INSTITUTE OF STEEL CONSTRUCTION

APPENDIX I Corrosion-Resistant Flanges and Fittings

WELDING FITTINGS — DIMENSIONS

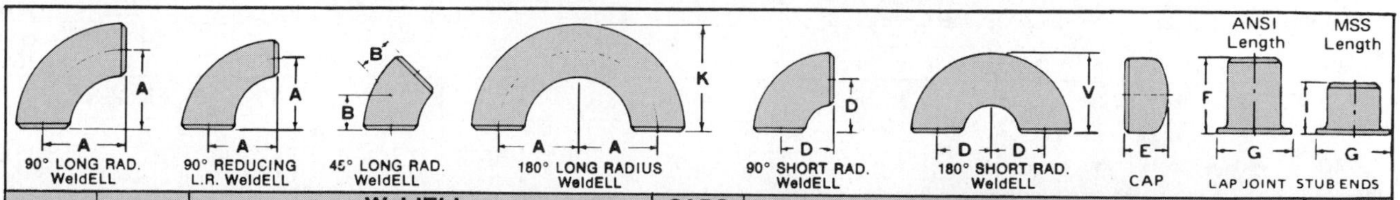

90° LONG RAD. WeldELL
90° REDUCING L.R. WeldELL
45° LONG RAD. WeldELL
180° LONG RADIUS WeldELL
90° SHORT RAD. WeldELL
180° SHORT RAD. WeldELL
CAP
ANSI Length
MSS Length
LAP JOINT STUB ENDS
A
B
D
K
V
E
F
I
G

Nom. Pipe Size	Pipe O.D.	WeldELL A	WeldELL B	WeldELL D	WeldELL K	WeldELL V	CAPS E	STUB ENDS (See note below) G O.D. of Lap	Type A F (Length) ANSI Std.	Type A I (Length) MSS Std.	Type A Corner Radius	Type B, C ▲ I (Length) MSS Std.	Type B, C ▲ Corner Radius	Nom. Pipe Size
1/2	.840	1 1/2	5/8	–	1 7/8	–	1	1 3/8	3	2	1/8	2	1/32	1/2
3/4	1.050	1 1/8		–	1 11/16	–	1	1 11/16	3	2	1/8	2	1/32	3/4
1	1.315	1 1/2	7/8	1	2 3/16	1 5/8	1 1/2	2	4	2	1/8	2	1/32	1
1 1/4	1.660	1 7/8	1	1 1/4	2 3/4	2 1/16	1 1/2	2 1/2	4	2	3/16	2	1/32	1 1/4
1 1/2	1.900	2 1/4	1 1/8	1 1/2	3 1/4	2 7/16	1 1/2	2 7/8	4	2	1/4	2	1/32	1 1/2
2	2.375	3	1 3/8	2	4 3/16	3 3/16	1 1/2	3 5/8	6	2 1/2	5/16	2 1/2	1/32	2
2 1/2	2.875	3 3/4	1 3/4	2 1/2	5 3/16	3 15/16	1 1/2	4 1/8	6	2 1/2	5/16	2 1/2	1/32	2 1/2
3	3.500	4 1/2	2	3	6 1/4	4 3/4	2	5	6	2 1/2	3/8	2 1/2	1/32	3
3 1/2	4.000	5 1/4	2 1/4	3 1/2	7 1/4	5 1/2	2 1/2	5 1/2	6	3	3/8	3	1/32	3 1/2
4	4.500	6	2 1/2	4	8 1/4	6 1/4	2 1/2	6 3/16	6	3	7/16	3	1/32	4
5	5.563	7 1/2	3 1/8	5	10 5/16	7 3/4	3	7 5/16	8	3	7/16	3	1/16	5
6	6.625	9	3 3/4	6	12 5/16	9 5/16	3 1/2	8 1/2	8	3 1/2	1/2	3 1/2	1/16	6
8	8.625	12	5	8	16 5/16	12 5/16	4	10 5/8	8	4	1/2	4	1/16	8
10	10.750	15	6 1/4	10	20 3/8	15 5/8	5	12 3/4	10	5	1/2	5	1/16	10
12	12.750	18	7 1/2	12	24 3/8	18 3/8	6	15	10	6	1/2	6	1/16	12
14	14.000	21	8 3/4	14	28	21	6 1/2	16 1/4	12	–	1/2	–	–	14
16	16.000	24	10	16	32	24	7	18 1/2	12	–	1/2	–	–	16
18	18.000	27	11 1/4	18	36	27	8	21	12	–	1/2	–	–	18
20	20.000	30	12 1/2	20	40	30	9	23	12	–	1/2	–	–	20
24	24.000	36	15	24	48	26	10 1/2	27 1/2	12	–	1/2	–	–	24
30	30.000	45	18 1/2	30	60	45	10 1/2	–	–	–	–	–	–	30

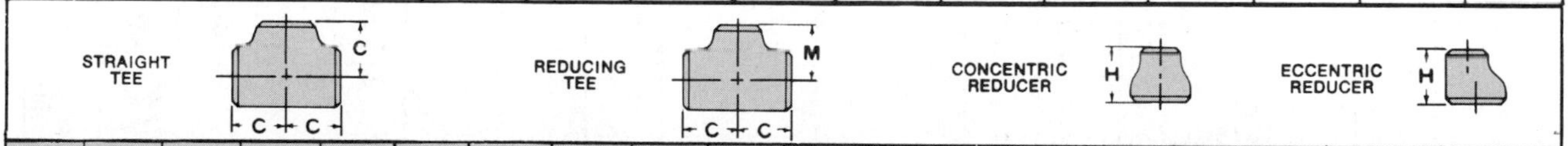

STRAIGHT TEE
C
C C
REDUCING TEE
M
C C
CONCENTRIC REDUCER
H
ECCENTRIC REDUCER
H

Nom. Pipe Size	Outlet	C	M	H
3/4	3/4	1 1/8	. . .	. . .
	1/2	1 1/8	1 1/8	1 1/2
1	1	1 1/2	. . .	. . .
	3/4	1 1/2	1 1/2	2
	1/2	1 1/2	1 1/2	2
1 1/4	1 1/4	1 7/8	. . .	. . .
	1	1 7/8	1 7/8	2
	3/4	1 7/8	1 7/8	2
	1/2	1 7/8	1 7/8	2
1 1/2	1 1/2	2 1/4	. . .	. . .
	1 1/4	2 1/4	2 1/4	2 1/2
	1	2 1/4	2 1/4	2 1/2
	3/4	2 1/4	2 1/4	2 1/2
	1/2	2 1/4	2 1/4	2 1/2
2	2	2 1/2	. . .	. . .
	1 1/2	2 1/2	2 3/8	3
	1 1/4	2 1/2	2 1/4	3
	1	2 1/2	2	3
	3/4	2 1/2	1 3/4	3
2 1/2	2 1/2	3	. . .	. . .
	2	3	2 3/4	3 1/2
	1 1/2	3	2 5/8	3 1/2
	1 1/4	3	2 1/2	3 1/2
	1	3	2 1/4	3 1/2
3	3	3 3/8	. . .	. . .
	2 1/2	3 3/8	3 1/4	3 1/2
	2	3 3/8	3	3 1/2
	1 1/2	3 3/8	2 7/8	3 1/2
	1 1/4	3 3/8	2 3/4	3 1/2

Nom. Pipe Size	Outlet	C	M	H
3 1/2	3 1/2	3 3/4	. . .	. . .
	3	3 3/4	3 5/8	4
	2 1/2	3 3/4	3 1/2	4
	2	3 3/4	3 1/4	4
	1 1/2	3 3/4	3 1/8	4
4	4	4 1/8	. . .	. . .
	3 1/2	4 1/8	4	4
	3	4 1/8	3 7/8	4
	2 1/2	4 1/8	3 3/4	4
	2	4 1/8	3 1/2	4
	1 1/2	4 1/8	3 3/8	4
5	5	4 7/8	. . .	. . .
	4	4 7/8	4 5/8	5
	3 1/2	4 7/8	4 1/2	5
	3	4 7/8	4 3/8	5
	2 1/2	4 7/8	4 1/4	5
	2	4 7/8	4 1/8	5
6	6	5 5/8	. . .	. . .
	5	5 5/8	5 3/8	5 1/2
	4	5 5/8	5 1/8	5 1/2
	3 1/2	5 5/8	5	5 1/2
	3	5 5/8	4 7/8	5 1/2
	2 1/2	5 5/8	4 3/4	5 1/2
8	8	7	. . .	. . .
	6	7	6 5/8	6
	5	7	6 3/8	6
	4	7	6 1/8	6
	3 1/2	7	6	6

Nom. Pipe Size	Outlet	C	M	H
10	10	8 1/2	. . .	. . .
	8	8 1/2	8	7
	6	8 1/2	7 5/8	7
	5	8 1/2	7 1/2	7
	4	8 1/2	7 1/4	7
12	12	10	. . .	. . .
	10	10	9 1/2	8
	8	10	9	8
	6	10	8 5/8	8
	5	10	8 1/2	8
14	14	11	. . .	. . .
	12	11	10 5/8	13
	10	11	10 1/8	13
	8	11	9 3/4	13
	6	11	9 3/8	13
16	16	12	. . .	. . .
	14	12	12	14
	12	12	11 5/8	14
	10	12	11 1/8	14
	8	12	10 3/4	14
	6	12	10 3/8	14
18	18	13 1/2	. . .	. . .
	16	13 1/2	13	15
	14	13 1/2	13	15
	12	13 1/2	12 5/8	15
	10	13 1/2	12 1/8	15
	8	13 1/2	11 3/4	15

Nom. Pipe Size	Outlet	C	M	H
20	20	15	. . .	. . .
	18	15	14 1/2	20
	16	15	14	20
	14	15	14	20
	12	15	13 5/8	20
	10	15	13 1/8	20
	8	15	12 3/4	20
24	24	17	. . .	. . .
	20	17	17	20
	18	17	16 1/2	20
	16	17	16	20
	14	17	16	20
	12	17	15 5/8	20
	10	17	15 1/8	20
30	30	22	. . .	. . .
	24	22	21	24
	20	22	20	24
	18	22	19 1/2	24
	16	22	19	24
	14	22	19	24
36	36	26 1/2	. . .	. . .
	30	26 1/2	25	24
	24	26 1/2	24	24
	20	26 1/2	23	24
	18	26 1/2	22 1/2	24
	16	26 1/2	22	24
42	42	30	. . .	. . .
	36	30	28	24
	30	30	28	24
	24	30	26	24
	20	30	26	24

NOTES:

Fittings having wall thicknesses of 0.065'' are furnished with square ends unless otherwise specified.

STUB ENDS:

Type A for use with Lap Joint Flanges.

Types B and C for use with Slip-On-Flanges.

Types A and B are furnished with square inside corner, gasket surfaces machined and minimum lap thickness equal to nominal wall of barrel.

▲ Dimensions of Type C are same as tabulated for Type B in 5S and 10S thicknesses. Type C is not available in 40S thicknesses. Type C has no fixed corner radius, lap face is not machined. Type C is only available in MSS length.

Type B, usually purchased in the MSS length is also available in the ANSI length in Shedules 10S and 40S.

FORGED FLANGES — DIMENSIONS

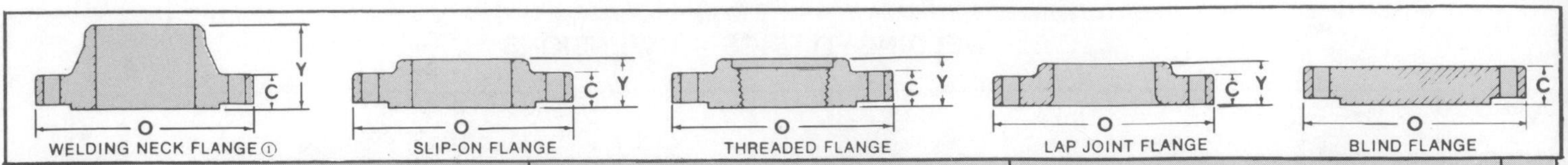
WELDING NECK FLANGE ①
SLIP-ON FLANGE
THREADED FLANGE
LAP JOINT FLANGE
BLIND FLANGE
O
C
Y

150 LB. FLANGES

Nom. Pipe Size	O	C②	Y② Weld Neck	Y② Slip On Thrd.	Y② Lap Joint	Bolt Circle	★ No. and Size of Holes
1/2	3 1/2	7/16	1 7/8	5/8	5/8	2 3/8	4-5/8
3/4	3 7/8	1/2	2 1/16	5/8	5/8	2 3/4	4-5/8
1	4 1/4	9/16	2 3/16	11/16	11/16	3 1/8	4-5/8
1 1/4	4 5/8	5/8	2 1/4	13/16	13/16	3 1/2	4-5/8
1 1/2	5	11/16	2 7/16	7/8	7/8	3 7/8	4-5/8
2	6	3/4	2 1/2	1	1	4 3/4	4-3/4
2 1/2	7	7/8	2 3/4	1 1/8	1 1/8	5 1/2	4-3/4
3	7 1/2	15/16	2 3/4	1 3/16	1 3/16	6	4-3/4
3 1/2	8 1/2	15/16	2 13/16	1 1/4	1 1/4	7	8-3/4
4	9	15/16	3	1 5/16	1 5/16	7 1/2	8-3/4
5	10	15/16	3 1/2	1 7/16	1 7/16	8 1/2	8-7/8
6	11	1	3 1/2	1 9/16	1 9/16	9 1/2	8-7/8
8	13 1/2	1 1/8	4	1 3/4	1 3/4	11 3/4	8-7/8
10	16	1 3/16	4	1 15/16	1 15/16	14 1/4	12-1
12	19	1 1/4	4 1/2	2 3/16	2 3/16	17	12-1
14	21	1 3/8	5	2 1/4	3 1/8	18 3/4	12-1 1/8
16	23 1/2	1 7/16	5	2 1/2	3 7/16	21 1/4	16-1 1/8
18	25	1 9/16	5 1/2	2 11/16	3 13/16	22 3/4	16-1 1/4
20	27 1/2	1 11/16	5 11/16	2 7/8	4 1/16	25	20-1 1/4
24	32	1 7/8	6	3 1/4	4 3/8	29 1/2	20-1 3/8

300 LB. FLANGES

Nom. Pipe Size	O	C②	Y② Weld Neck	Y② Slip On Thrd.	Y② Lap Joint	Bolt Circle	★ No. and Size of Holes
1/2	3 3/4	9/16	2 1/16	7/8	7/8	2 5/8	4-5/8
3/4	4 5/8	5/8	2 1/4	1	1	3 1/4	4-3/4
1	4 7/8	11/16	2 7/16	1 1/16	1 1/16	3 1/2	4-3/4
1 1/4	5 1/4	3/4	2 9/16	1 1/16	1 1/16	3 7/8	4-3/4
1 1/2	6 1/8	13/16	2 11/16	1 3/16	1 3/16	4 1/2	4-7/8
2	6 1/2	7/8	2 3/4	1 5/16	1 5/16	5	8-3/4
2 1/2	7 1/2	1	3	1 1/2	1 1/2	5 7/8	8-7/8
3	8 1/4	1 1/8	3 1/8	1 11/16	1 11/16	6 5/8	8-7/8
3 1/2	9	1 3/16	3 3/16	1 3/4	1 3/4	7 1/4	8-7/8
4	10	1 1/4	3 3/8	1 7/8	1 7/8	7 7/8	8-7/8
5	11	1 3/8	3 7/8	2	2	9 1/4	8-7/8
6	12 1/2	1 7/16	3 7/8	2 1/16	2 1/16	10 5/8	12-7/8
8	15	1 5/8	4 3/8	2 7/16	2 7/16	13	12-1
10	17 1/2	1 7/8	4 5/8	2 5/8	3 3/4	15 1/4	16-1 1/8
12	20 1/2	2	5 1/8	2 7/8	4	17 3/4	16-1 1/4
14	23	2 1/8	5 5/8	3	4 3/8	20 1/4	20-1 1/4
16	25 1/2	2 1/4	5 3/4	3 1/4	4 3/4	22 1/2	20-1 3/8
18	28	2 3/8	6 1/4	3 1/2	5 1/8	24 3/4	24-1 3/8
20	30 1/2	2 1/2	6 3/8	3 3/4	5 1/2	27	24-1 3/8
24	36	2 3/4	6 5/8	4 3/16	6	32	24-1 5/8

400 LB. FLANGES

Nom. Pipe Size	O	C②	Y② Weld Neck	Y② Slip On Thrd.	Y② Lap Joint	Bolt Circle	★ No. and Size of Holes
1/2	3 3/4	9/16	2 1/16	7/8	7/8	2 5/8	4-5/8
3/4	4 5/8	5/8	2 1/4	1	1	3 1/4	4-3/4
1	4 7/8	11/16	2 7/16	1 1/16	1 1/16	3 1/2	4-3/4
1 1/4	5 1/4	13/16	2 5/8	1 1/8	1 1/8	3 7/8	4-3/4
1 1/2	6 1/8	7/8	2 3/4	1 1/4	1 1/4	4 1/2	4-7/8
2	6 1/2	1	2 7/8	1 7/16	1 7/16	5	8-3/4
2 1/2	7 1/2	1 1/8	3 1/8	1 5/8	1 5/8	5 7/8	8-7/8
3	8 1/4	1 1/4	3 1/4	1 13/16	1 13/16	6 5/8	8-7/8
3 1/2	9	1 3/8	3 3/8	1 15/16	1 15/16	7 1/4	8-1
4	10	1 3/8	3 1/2	2	2	7 7/8	8-1
5	11	1 1/2	4	2 1/8	2 1/8	9 1/4	8-1
6	12 1/2	1 5/8	4 1/16	2 1/4	2 1/4	10 5/8	12-1
8	15	1 7/8	4 5/8	2 11/16	2 11/16	13	12-1 1/8
10	17 1/2	2 1/8	4 7/8	2 7/8	4	15 1/4	16-1 1/4
12	20 1/2	2 1/4	5 3/8	3 1/8	4 1/4	17 3/4	16-1 3/8
14	23	2 3/8	5 7/8	3 5/16	4 5/8	20 1/4	20-1 3/8
16	25 1/2	2 1/2	6	3 11/16	5	22 1/2	20-1 1/2
18	28	2 5/8	6 1/2	3 7/8	5 3/8	24 3/4	24-1 1/2
20	30 1/2	2 3/4	6 5/8	4	5 3/4	27	24-1 5/8
24	36	3	6 7/8	4 1/2	6 1/4	32	24-1 7/8

600 LB. FLANGES

Nom. Pipe Size	O	C②	Y② Weld Neck	Y② Slip On Thrd.	Y② Lap Joint	Bolt Circle	★ No. and Size of Holes
1/2	3 3/4	9/16	2 1/16	7/8	7/8	2 5/8	4-5/8
3/4	4 5/8	5/8	2 1/4	1	1	3 1/4	4-3/4
1	4 7/8	11/16	2 7/16	1 1/16	1 1/16	3 1/2	4-3/4
1 1/4	5 1/4	13/16	2 5/8	1 1/8	1 1/8	3 7/8	4-3/4
1 1/2	6 1/8	7/8	2 3/4	1 1/4	1 1/4	4 1/2	4-7/8
2	6 1/2	1	2 7/8	1 7/16	1 7/16	5	8-3/4
2 1/2	7 1/2	1 1/8	3 1/8	1 5/8	1 5/8	5 7/8	8-7/8
3	8 1/4	1 1/4	3 1/4	1 13/16	1 13/16	6 5/8	8-7/8
3 1/2	9	1 3/8	3 3/8	1 15/16	1 15/16	7 1/4	8-1
4	10 3/4	1 1/2	4	2 1/8	2 1/8	8 1/2	8-1
5	13	1 3/4	4 1/2	2 3/8	2 3/8	10 1/2	8-1 1/8
6	14	1 7/8	4 5/8	2 5/8	2 5/8	11 1/2	12-1 1/8
8	16 1/2	2 3/16	5 1/4	3	3	13 3/4	12-1 1/4
10	20	2 1/2	6	3 3/8	4 3/8	17	16-1 3/8
12	22	2 5/8	6 1/8	3 5/8	4 5/8	19 1/4	20-1 3/8
14	23 3/4	2 3/4	6 1/2	3 11/16	5	20 3/4	20-1 1/2
16	27	3	7	4 3/16	5 1/2	23 3/4	20-1 5/8
18	29 1/4	3 1/4	7 1/4	4 5/8	6	25 3/4	20-1 3/4
20	32	3 1/2	7 1/2	5	6 1/2	28 1/2	24-1 3/4
24	37	4	8	5 1/2	7 1/4	33	24-2

900 LB. FLANGES

Nom. Pipe Size	O	C②	Y② Weld Neck	Y② Slip On Thrd.	Y② Lap Joint	Bolt Circle	★ No. and Size of Holes
1/2	4 3/4	7/8	2 3/8	1 1/4	1 1/4	3 1/4	4-7/8
3/4	5 1/8	1	2 3/4	1 3/8	1 3/8	3 1/2	4-7/8
1	5 7/8	1 1/8	2 7/8	1 5/8	1 5/8	4	4-1
1 1/4	6 1/4	1 1/8	2 7/8	1 5/8	1 5/8	4 3/8	4-1
1 1/2	7	1 1/4	3 1/4	1 3/4	1 3/4	4 7/8	4-1 1/8
2	8 1/2	1 1/2	4	2 1/4	2 1/4	6 1/2	8-1
2 1/2	9 5/8	1 5/8	4 1/8	2 1/2	2 1/2	7 1/2	8-1 1/8
3	9 1/2	1 1/2	4	2 1/8	2 1/8	7 1/2	8-1
3 1/2	..	..	..	..	..	..	..
4	11 1/2	1 3/4	4 1/2	2 3/4	2 3/4	9 1/4	8-1 1/4
5	13 3/4	2	5	3 1/8	3 1/8	11	8-1 3/8
6	15	2 3/16	5 1/2	3 3/8	3 3/8	12 1/2	12-1 1/4
8	18 1/2	2 1/2	6 3/8	4	4 1/2	15 1/2	12-1 1/2
10	21 1/2	2 3/4	7 1/4	4 1/4	5	18 1/2	16-1 1/2
12	24	3 1/8	7 7/8	4 5/8	5 5/8	21	20-1 1/2
14	25 1/4	3 3/8	8 3/8	5 1/8	6 1/8	22	20-1 5/8
16	27 3/4	3 1/2	8 1/2	5 1/4	6 1/2	24 1/4	20-1 3/4
18	31	4	9	6	7 1/2	27	20-2
20	33 3/4	4 1/4	9 3/4	6 1/4	8 1/4	29 1/2	20-2 1/8
24	41	5 1/2	11 1/2	8	10 1/2	35 1/2	20-2 5/8

1500 LB. FLANGES

Nom. Pipe Size	O	C②	Y② Weld Neck	Y② Slip On Thrd.	Y② Lap Joint	Bolt Circle	★ No. and Size of Holes
1/2	4 3/4	7/8	2 3/8	1 1/4	1 1/4	3 1/4	4-7/8
3/4	5 1/8	1 7/8	2 3/4	1 3/8	1 3/8	3 1/2	4-7/8
1	5 7/8	1 1/8	2 7/8	1 5/8	1 5/8	4	4-1
1 1/4	6 1/4	1 1/8	2 7/8	1 5/8	1 5/8	4 3/8	4-1
1 1/2	7	1 1/4	3 1/4	1 3/4	1 3/4	4 7/8	4-1 1/8
2	8 1/2	1 1/2	4	2 1/4	2 1/4	6 1/2	8-1
2 1/2	9 5/8	1 5/8	4 1/8	2 1/2	2 1/2	7 1/2	8-1 1/8
3	10 1/2	1 7/8	4 5/8	2 7/8	2 7/8	8	8-1 1/4
3 1/2	..	..	..	..	..	..	..
4	12 1/4	2 1/8	4 7/8	3 9/16	3 9/16	9 1/2	8-1 3/8
5	14 3/4	2 7/8	6 1/8	4 1/8	4 1/8	11 1/2	8-1 5/8
6	15 1/2	3 1/4	6 3/4	4 11/16	4 11/16	12 1/2	12-1 1/2
8	19	3 5/8	8 3/8	5 5/8	5 5/8	15 1/2	12-1 3/4
10	23	4 1/4	10	6 1/4	7	19	12-2
12	26 1/2	4 7/8	11 1/8	7 1/8	8 5/8	22 1/2	16-2 1/8
14	29 1/2	5 1/4	11 3/4	..	9 1/2	25	16-2 3/8
16	32 1/2	5 3/4	12 1/4	..	10 1/4	27 3/4	16-2 5/8
18	36	6 3/8	12 7/8	..	10 7/8	30 1/2	16-2 7/8
20	38 3/4	7	14	..	11 1/2	32 3/4	16-3 1/8
24	46	8	16	..	13	39	16-3 5/8

2500 LB. FLANGES

Nom. Pipe Size	O	C②	Y② Weld Neck	Y② Slip On Thrd.	Y② Lap Joint	Bolt Circle	★ No. and Size of Holes
1/2	5 1/4	1 3/16	2 7/8	1 9/16	1 9/16	3 1/2	4-7/8
3/4	5 1/2	1 1/4	3 1/8	1 11/16	1 11/16	3 3/4	4-7/8
1	6 1/4	1 3/8	3 1/2	1 7/8	1 7/8	4 1/4	4-1
1 1/4	7 1/4	1 1/2	3 3/4	2 1/16	2 1/16	5 1/8	4-1 1/8
1 1/2	8	1 3/4	4 3/8	2 3/8	2 3/8	5 3/4	4-1 1/4
2	9 1/4	2	5	2 3/4	2 3/4	6 3/4	8-1 1/8
2 1/2	10 1/2	2 1/4	5 5/8	3 1/8	3 1/8	7 3/4	8-1 1/4
3	12	2 5/8	6 5/8	3 5/8	3 5/8	9	8-1 3/8
4	14	3	7 1/2	4 1/4	4 1/4	10 3/4	8-1 5/8
5	16 1/2	3 5/8	9	5 1/8	5 1/8	12 3/4	8-1 7/8
6	19	4 1/4	10 3/4	6	6	14 1/2	8-2 1/8
8	21 3/4	5	12 1/2	7	7	17 1/4	12-2 1/8
10	26 1/2	6 1/2	16 1/2	9	9	21 1/4	12-2 5/8
12	30	7 1/4	18 1/4	10	10	24 3/8	12-2 7/8

150 LB. FLANGES ①⑤

Nom. Pipe Size	O	C②	Y② Weld Neck	Y② Slip On Thrd.	Y② Lap Joint	Bolt Circle	★ No. and Size of Holes
1/2	3 1/2	3/8	7/8	9/16	9/16	2 3/8	4-5/8
3/4	3 7/8	3/8	7/8	9/16	9/16	2 3/4	4-5/8
1	4 1/4	3/8	7/8	9/16	9/16	3 1/8	4-5/8
1 1/4	4 5/8	3/8	7/8	5/8	5/8	3 1/2	4-5/8
1 1/2	5	3/8	7/8	5/8	5/8	3 7/8	4-5/8
2	6	7/16	1	3/4	3/4	4 3/4	4-3/4
2 1/2	7	7/16	1	3/4	3/4	5 1/2	4-3/4
3	7 1/2	1/2	1 1/8	7/8	7/8	6	4-3/4
4	9	1/2	1 1/8	7/8	7/8	7 1/2	8-3/4
5	10	9/16	1 1/4	7/8	7/8	8 1/2	8-7/8
6	11	9/16	1 1/4	1 1/4	1 1/4	9 1/2	8-7/8
8	13 1/2	9/16	1 1/4	1 1/4	1 1/4	11 3/4	8-7/8
10	16	11/16	1 3/8	1 1/4	1 1/4	14 1/4	12-1
12	19	11/16	1 3/8	1 1/4	1 1/4	17	12-1

MSS 150 LB. FLANGES ⑤⑥

Nom. Pipe Size	O	C②	Y② Weld Neck	Y② Slip On Thrd.	Y② Lap Joint	Bolt Circle	★ No. and Size of Holes
1/2	3 1/2	5/16	1 7/8	5/8	5/8	2 3/8	4-5/8
3/4	3 7/8	11/32	2 1/16	5/8	5/8	2 3/4	4-5/8
1	4 1/4	3/8	2 3/16	11/16	11/16	3 1/8	4-5/8
1 1/4	4 5/8	13/32	2 1/4	13/16	13/16	3 1/2	4-5/8
1 1/2	5	7/16	2 7/16	7/8	7/8	3 7/8	4-5/8
2	6	1/2	2 1/2	1	1	4 3/4	4-3/4
2 1/2	7	9/16	2 3/4	1 1/8	1 1/8	5 1/2	4-3/4
3	7 1/2	5/8	2 3/4	1 3/16	1 3/16	6	4-3/4
4	9	11/16	3	1 5/16	1 5/16	7 1/2	8-3/4
5	10	3/4	3 1/2	1 7/16	1 7/16	8 1/2	8-7/8
6	11	13/16	3 1/2	1 9/16	1 9/16	9 1/2	8-7/8
8	13 1/2	15/16	4	1 3/4	1 3/4	11 3/4	8-7/8
10	16	1	4	1 15/16	1 15/16	14 1/4	12-1
12	19	1 1/16	4 1/2	2 3/16	2 3/16	17	12-1

NOTES:

1 Always specify bore when ordering.

2 Includes 1/16'' raised face in 150 lb. and 300 lb. Standard. Does not include 1/4'' raised face in 400 lb. and heavier standards.

3 Other types, sizes, and facings on application.

4 For low pressure service up to 150 PSI at 500°F, 225 PSI at 150°F.

5 Drilling and OD match ANSI B16.5 150 lb. steel flange standard, MSS SP-42 150 lb. corrosion resistant value standards and ANSI B16.1 125 lb. cast iron flange standard.

This class flange has flat face.

6 Thicknesses conform to MSS Standards.

★Bolt holes are 1/8'' larger than recommended bolt.

Courtesy Taylor Forge.

▼ APPENDIX J   Threaded Fittings and Threaded Couplings, Reducers, and Caps

Threaded Fittings—Class 2000, 3000 and 6000

DIMENSIONS (Inches)

	NOMINAL PIPE SIZE	1/8	1/4	3/8	1/2	3/4	1	1 1/4	1 1/2	2	2 1/2	3	4
Class 2000	A		7/8	31/32	1 1/8	1 5/16	1 1/2	1 3/4	2	2 3/8	3	3 3/8	4 3/16
	B		29/32	1 1/16	1 5/16	1 9/16	1 27/32	2 1/32	2 1/2	3 1/2	3 11/16	4 5/16	5 3/4
	F		3/4	3/4	1	1 1/8	1 1/4	1 5/16	1 3/8	1 11/16	2 1/16	2 1/2	3 1/8
Class 3000	A	7/8	31/32	1 1/8	1 5/16	1 1/2	1 3/4	2	2 3/8	2 1/2	3 1/4	3 3/4	4 1/2
	B	29/32	1 1/16	1 5/16	1 9/16	1 27/32	2 1/32	2 1/2	3 1/2	3 11/32	4	4 3/4	6
	F	3/4	3/4	1	1 1/8	1 1/4	1 5/16	1 3/8	1 11/16	1 3/4	2 1/16	2 1/2	3 1/8
	H	1 1/4	1 1/4	1 1/2	1 5/8	1 7/8	2 1/4	2 5/8	2 15/16	3 5/16			
	J	7/8	7/8	1	1 1/8	1 3/8	1 3/4	2	2 1/8	2 1/2			
	K		1 7/8	2 1/8	2 9/16	3	3 1/2	3 15/16	4 1/4	5			
	L		2 11/16	3	3 9/16	4 1/8	4 13/16	5 3/8	6 7/16	6 5/8			
Class 6000	A	31/32	1 1/8	1 5/16	1 1/2	1 3/4	2	2 3/8	2 1/2	3 1/4	3 3/4	4 3/16	4 1/2
	B	1 1/16	1 5/16	1 9/16	1 27/32	2 1/32	2 1/2	3 1/32	3 11/32	4	4 3/4	5 3/4	6
	F	3/4	1	1 1/8	1 1/4	1 5/16	1 3/8	1 11/16	1 1/4	2 1/16	2 1/2	3 1/8	3 1/8
	H	1 1/4	1 1/2	1 5/8	1 7/8	2 1/4	2 5/8	2 15/16	3 5/16				
	J		7/8	1	1 1/8	1 3/8	1 3/4	2	2 1/8	2 1/2			
	K			2 9/16	3	3 1/2	3 15/16	4 3/4	5				
	L			3 9/16	4 1/8	4 13/16	5 3/8	6 7/16	6 5/8				

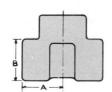
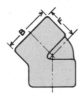

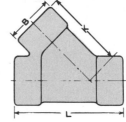
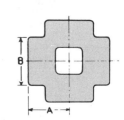

Threaded Couplings, Reducers and Caps—Class 3000 and 6000

DIMENSIONS (Inches)

	NOMINAL PIPE SIZE	1/8	1/4	3/8	1/2	3/4	1	1 1/4	1 1/2	2	2 1/2	3	4
Class 3000	A	1 1/4	1 3/8	1 1/2	1 7/8	2	2 3/8	2 5/8	3 1/8	3 3/8	3 5/8	4 1/4	4 3/4
	B	3/4	3/4	7/8	1 1/8	1 3/8	1 3/4	2 1/4	2 1/2	3	3 5/8	4 1/4	5 1/2
	C	5/8	11/16	3/4	15/16	1	1 3/16	1 5/16	1 9/16	1 11/16	1 13/16	2 1/8	2 3/8
	D	15/16	1	1	1 1/4	1 7/16	1 5/8	1 3/4	1 3/4	1 7/8	2 3/8	2 9/16	2 11/16
Class 6000	A	1 1/4	1 3/8	1 1/2	1 7/8	2	2 3/8	2 5/8	3 1/8	3 3/8	3 5/8	4 1/4	4 3/4
	B	7/8	1	1 1/4	1 1/2	1 3/4	2 1/4	2 1/2	3	3 5/8	4 1/4	5	6 1/4
	C	5/8	11/16	3/4	15/16	1	1 3/16	1 5/16	1 9/16	1 11/16	1 13/16	2 1/8	2 3/8
	D	1	1 1/16	1 1/16	1 5/16	1 1/2	1 11/16	1 13/16	1 7/8	2	2 1/2	2 11/16	2 15/16

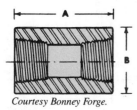
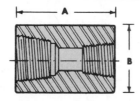

Courtesy Bonney Forge.

▼ APPENDIX K   Valve Specifications

CRANE

No. 438

GATE VALVES CLASS 125
¼″ to 3″

Non-Rising Stem
Screwed Bonnet
Solid Wedge Disc

No. 438, Threaded

RATINGS

Temp. F.	Psi Non-Shock
–20 to 150°	200
200	185
250	170
300	155
350	140
406	125
450	120

Weights and Dimensions

Valve N.P.S.	Weight—Pounds	Dimensions—Inches		
		A	B	C
¼	.6	1.64	3.44	1.75
⅜	.6	1.64	3.44	1.75
½	1.0	1.90	3.75	2.06
¾	1.4	2.14	4.38	2.75
1	2.1	2.47	4.88	2.75
1¼	3.1	3.08	5.63	3.06
1½	4.2	3.11	6.44	3.63
2	6.2	3.39	7.50	4.50
2½	11.3	4.25	9.06	5.00
3	16.0	4.59	9.69	5.00

"B" dimension is with valve open

CRANE

No. 1

No. 2

GLOBE AND ANGLE VALVES CLASS 125
⅛″ to 3″

Screwed Bonnet

Globe
No. 1, Threaded

Angle
No. 2, Threaded

RATINGS

Temp. F.	Psi Non-Shock
–20 to 150°	200
200	185
250	170
300	155
350	140
406	125
450	120

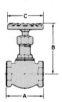

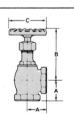

Weights and Dimensions

Valve N.P.S.	Weight—Pounds		Dimensions—Inches			
	No. 1	No. 2	A		B	C
			No. 1	No. 2		
⅛	.3	.3	1.44	.75	2.75	1.75
¼	.4	.4	1.63	.81	3.00	1.75
⅜	.5	.4	1.88	.94	3.25	2.03
½	1.0	.8	2.19	1.13	3.75	2.75
¾	1.4	1.4	2.69	1.38	4.25	2.75
1	2.2	2.0	3.19	1.63	5.00	3.00
1¼	3.3	3.0	3.69	1.81	5.50	3.72
1½	4.7	4.5	4.19	2.06	6.25	4.50
2	7.6	7.6	5.13	2.50	7.50	5.00
2½	11.6	—	6.06	—	7.50	5.00
3	19.5	—	7.06	—	9.25	6.00

"B" dimension is with valve open

CRANE

No. 366E

LIFT CHECK VALVES CLASS 300 ¼″ to 3″

No. 366E Threaded

RATINGS Non-Shock

Temp. F.	Psi ¼″-2″	Psi 2½″-3″
-20-150°	1000	600
200	920	560
250	830	525
300	740	490
350	650	450
400	560	410
450	480	375
500	390	340
550	300	300

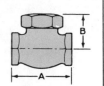

Weights and Dimensions

Valve N.P.S.	Weight—Pounds	Dimensions—Inches	
		A	B
¼	0.4	1.82	1.00
⅜	0.6	2.00	1.12
½	0.9	2.50	1.38
¾	1.5	2.94	1.88
1	2.6	3.50	2.00
1¼	4.2	4.06	2.38
1½	5.4	4.62	2.62
2	10.8	5.75	3.25
2½	15.6	6.88	3.88
3	24.0	8.00	4.50

CRANE

WEDGE GATE VALVES CLASS 125 2″ to 48″

Non-Rising Stem

No. 461

Bronze Trim
No. 460, Threaded
No. 461, Flanged

All Iron
No. 473, Flanged

RATINGS

Temp. F.	Psi, Non-Shock		
	2 to 12″	14 to 24″	30 to 48″
-20 to 150°	200	150	150
200	190	135	115
225	180	130	100
250	175	125	85
275	170	120	65
300	165	110	50
325	155	105	
350	150	100	
375	145	*	
400	140	*	
425	130	*	
450	125	*	

*Use Crane 150-pound steel valves.

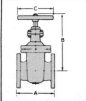

Weights and Dimensions

Valve N.P.S.	Weight—Pounds			Dimensions—Inches			
	460	461	473	Threaded	Flanged		
				A	A	B	C
2	25	30	30	5.38	7.00	11.31	8.0
2½	31	40	40	6.62	7.50	12.40	8.0
3	44	56	56	7.00	8.00	13.25	8.0
4	71	90	90	8.00	9.00	16.31	10.0
5	—	126	—	—	10.00	18.00	10.0
6	—	152	152	—	10.50	20.69	12.0
8	—	260	260	—	11.50	24.12	14.0

Valve N.P.S.	Weight—Pounds 461	Dimensions—Inches		
		A	B	C
10	475	13.00	33.00	20.0
12	680	14.00	36.50	20.0
14	850	15.00	40.50	20.0
16	1280	16.00	47.25	22.0
18	1480	17.00	50.25	24.0
20	1840	18.00	54.75	24.0
24	2860	20.00	65.00	30.0
30	On Request	24.00	76.00	36.0
36	On Request	28.00	85.00	42.0
42	On Request	33.00	106.00	42.0
48	On Request	36.00	111.25	42.0

CRANE

No. 353

GLOBE AND ANGLE VALVES CLASS 125 2″ to 10″

Outside Screw & Yoke Bronze Trim

Globe No. 351, Flanged

Angle No. 353, Flanged

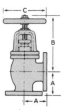

RATINGS

Temp. F.	Psi, Non-Shock	Temp. F.	Psi Non-Shock
–20 to 150°	200	325	155
200	190	350	150
225	180	375	145
250	175	400	140
275	170	425	130
300	165	450	125

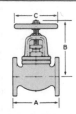

Weights and Dimensions

Valve N.P.S.	Weight—Pounds		Dimensions—Inches				
	351	353	351	353	351	353	C
			A	A	B	B	
2	34	32	8.00	4.00	11.12	11.00	8.00
2½	40	38	8.50	4.25	11.50	11.50	8.00
3	57	54	9.50	4.75	13.25	12.75	9.00
4	95	88	11.50	5.75	15.50	15.00	10.00
5	126	—	13.00	—	17.50	—	10.00
6	176	158	14.00	7.00	19.50	19.50	12.00
8	344	—	19.50	—	25.00	—	16.00
10	570	—	24.50	—	30.50	—	18.00

"B" dimension is with valve open.

CRANE

No. 373

SWING CHECK VALVES CLASS 125 2″ to 24″

Bolted Cap

Bronze Trim No. 372, Threaded No. 373, Flanged

All Iron No. 373½, Flanged

RATINGS

Temp. F.	Psi, Non-Shock	
	Sizes 2″-12″	Sizes 14″-24″
–20 to 150°	200	150
200	190	135
225	180	130
250	175	125
275	170	120
300	165	110
325	155	105
350	150	100
375	145	—
400	140	—
425	130	—
450	125	—

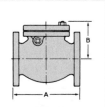

Weights and Dimensions

Valve N.P.S.	Weight—Pounds			Dimensions—Inches		
	372	373	373½	Threaded A	Flanged A	B
2	18	25	25	6.12	8.00	4.50
2½	22	34	34	7.25	8.50	5.38
3	29	44	44	8.00	9.50	5.88
4	54	75	75	9.25	11.50	6.62
5	—	103	—	—	13.00	7.75
6	—	127	127	—	14.00	8.25
8	—	230	230	—	19.50	10.25
10	—	510	510	—	24.50	11.60
12	—	695	695	—	27.50	13.60
14	—	875	—	—	31.00	15.75
16	—	1410	—	—	34.00	17.00
18	—	1540	—	—	38.50	17.40
20	—	1940	—	—	38.50	19.50
24	—	3000	—	—	51.00	20.50

Courtesy Crane Co.

▼ APPENDIX L   Symbols for Pipe Fittings and Valves

SYMBOLS FOR SINGLE LINE PIPE FITTINGS

SYMBOLS FOR BUTT-WELDED SYSTEMS

SYMBOLS FOR VALVES

▼ APPENDIX M   Spur and Helical Gear Data

	SPUR GEAR DATA	SUGGESTED NUMBER OF DECIMAL PLACES
BASIC SPECIFICATIONS	NUMBER OF TEETH	
	DIAMETRAL PITCH	XX.XXXX
	PRESSURE ANGLE	XX°
	STANDARD PITCH DIAMETER	X.XXXX
	TOOTH FORM	
	ADDENDUM	.XXXX
	WHOLE DEPTH	.XXXX
	CALC. CIR. TOOTH THICKNESS ON STD. PITCH CIRCLE	.XXXX MAX. .XXXX MIN.
MANUFACTURING AND INSPECTION	GEAR TESTING RADIUS	X.XXXX MAX. X.XXXX MIN.
	AGMA QUALITY NUMBER	
	MAX. TOTAL COMPOSITE TOLERANCE	.XXXX
	MAX. TOOTH-TO-TOOTH COMPOSITE TOLERANCE	.XXXX
	MASTER GEAR SPECIFICATIONS	
	TESTING PRESSURE (OUNCES)	XX
	DIAMETER OF MEASURING PIN	.XXXX
	MEASUREMENT OVER TWO PINS (FOR SETUP ONLY)	X.XXXX MAX. X.XXXX MIN.
	OUTSIDE DIAMETER	+ .000 X.XXX − .00X
	MAX. ROOT DIAMETER	X.XXX
ENGINEERING REFERENCES	MATING GEAR PART NUMBER	
	NUMBER OF TEETH IN MATING GEAR	
	OPERATING CENTER DISTANCE	X.XXXX MAX. X.XXXX MIN.

	HELICAL GEAR DATA	SUGGESTED NUMBER OF DECIMAL PLACES
BASIC SPECIFICATIONS	NUMBER OF TEETH	
	NORMAL DIAMETRAL PITCH	XX.XXXX
	NORMAL PRESSURE ANGLE	XX°
	HELIX ANGLE	XX.XXXX°
	HAND OF HELIX	L.H. OR R.H.
	STANDARD PITCH DIAMETER	X.XXXX
	TOOTH FORM	
	ADDENDUM	.XXXX
	WHOLE DEPTH	.XXXX
	CALC. NORMAL CIR. TOOTH THICKNESS ON STD. PITCH CIRCLE	.XXXX MAX. .XXXX MIN.
MANUFACTURING AND INSPECTION	GEAR TESTING RADIUS	X.XXXX MAX. X.XXXX MIN.
	AGMA QUALITY NUMBER	
	MAX. TOTAL COMPOSITE TOLERANCE	.XXXX
	MAX. TOOTH-TO-TOOTH COMPOSITE TOLERANCE	.XXXX
	MASTER GEAR SPECIFICATIONS	
	TESTING PRESSURE (OUNCES)	XX
	DIAMETER OF MEASURING PIN	.XXXX
	MEASUREMENT OVER TWO PINS (FOR SETUP ONLY)	X.XXXX MAX. X.XXXX MIN.
	LEAD	
	OUTSIDE DIAMETER	+ .000 X.XXX − .00X
	MAX. ROOT DIAMETER	X.XXXX
ENGINEERING REFERENCES	MATING GEAR PART NUMBER	
	NUMBER OF TEETH IN MATING GEAR	
	OPERATING CENTER DISTANCE	X.XXXX MAX. X.XXXX MIN.

▼ APPENDIX N   Math Instruction

▼ ARITHMETIC

Common Fractions

To add and subtract common fractions, the denominators (bottoms) must be the same.

$$\frac{1}{16} + \frac{3}{16} = \frac{4}{16}$$

Normally, fractions are reduced to lowest form by dividing both the numerator (top) and denominator by the same number. Dividing by 4 reduces the previous fraction.

$$\frac{4}{16} = \frac{1}{4}$$

To multiply fractions, the denominators do not have to be the same. Simply multiply the numerators and denominators separately.

$$\frac{1}{4} \times \frac{2}{3} = \frac{2}{12} = \frac{1}{6}$$

To divide fractions, flip the second fraction upside down (invert the divisor), then multiply.

$$\frac{1}{4} \div \frac{2}{3} = \frac{1}{4} \times \frac{3}{2} = \frac{3}{8}$$

When multiplying and dividing mixed numbers, it is necessary to convert them to fractions first. This is done by multiplying the whole number by the denominator of the fraction and adding that to the top of the fraction for a new numerator.

$$6\frac{3}{8} = \frac{(6 \times 8 + 3)}{8} = \frac{51}{8}$$

For example:

$$2\frac{1}{4} \times 3\frac{2}{3} = \frac{9}{4} \times \frac{11}{3} = \frac{99}{12} = 8\frac{1}{4}$$

The final answer was obtained by dividing the 99 by the 12 with long division.

Decimal Fractions

To convert a common fraction to a decimal fraction, divide the numerator by the denominator. Dividing 3 by 8 shows that

$$\frac{3}{8} = .375$$

To convert a decimal to common fraction form, place the decimal number without the decimal point on top and place its place value, given by this scheme, on the bottom of the fraction.

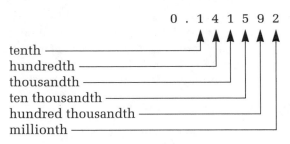

0 . 1 4 1 5 9 2

tenth
hundredth
thousandth
ten thousandth
hundred thousandth
millionth

Examples are:

$$.375 = \frac{375}{1000} = \frac{3}{8}$$

$$.014 = \frac{14}{1000} = \frac{7}{500}$$

$$6.75 = 6\frac{75}{100} = 6\frac{3}{4}$$

With the advent of computers and calculators, it has become important to know how to round decimal answers properly. Look one place to the right of where the decimal is to be rounded. If that digit is a 5 or larger, increase the value of the digit to the left of it by one; if it is a 4 or less, do not increase the left digit. This concept is best grasped by viewing examples.

Rounding to the nearest hundredth:

.275 becomes .28 (due to the 5 one place to the right of the hundredth position)
.273 become .27 (the 3 is "4 or less")
.27499 also becomes .27 (the 9s have no effect on rounding to hundredths)
.279 becomes .28 (the 9 is a "5 or more")

Rounding to the nearest tenth:

.275 becomes .3
.274 also becomes .3 (look only to the 7)
.249 becomes .2 (look only to the 4)
.05 becomes .1
.04 becomes .0
.729 becomes .7

Percentages

Percents are hundredths:

$$5\% = \frac{5}{100} = \frac{1}{20}$$

$$50\% = \frac{50}{100} = \frac{1}{2}$$

Percentages are easiest to calculate in decimal form. To convert a percentage to a decimal, move the decimal point two places to the left. Examples follow.

$$5\% = .05$$
$$50\% = .50$$
$$3.75\% = .0375$$

Success in working percentage problems requires correctly identifying three pieces of the problem called the part, the base, and the rate. The *part* is a portion of any whole amount, the *base* is the whole amount, and the *rate* is the number with the percent sign (%).

To find the part:

1. Convert the rate to decimal form.
2. Multiply the rate by the base.

Example 1: What is 15% of 233?

In this example the rate is 15% and 233 is the whole amount or base. The answer is **34.95** because $15\% \times 233 = .15 \times 233 = 34.95$

Example 2: What is 5.25% of 1000?

The answer is **52.5** because $.0525 \times 1000 = 52.5$

Example 3: What is 125% of 500?

The answer is **625** because $1.25 \times 500 = 625$

To find the base:

1. Convert the rate to decimal form.
2. Divide the part by the rate.

Example 4: 10% of what number is 375?

This problem requires realizing that 375 is a portion of some unknown whole amount. So you know the rate and part and are seeking the base.

The answer is **3750** because $375 \div .10 = 3,750$

Example 5: An old copy machine is known to waste 2% of the copies it makes. How many copies were run if 14 were wasted?

The wasted copies are the part and the entire run is the base, so the answer is $14 \div .02 =$ **700 copies run.**

To find the rate:

1. Divide the part by the base.
2. Convert the answer to percent by moving the decimal point two places to the right.

Example 6: What percent of 500 is 125?

This type of problem requires identifying which number is the base. From the wording, the whole amount is 500 so *that* is the base and 125 is a portion of this whole amount, making it the part. Often, the base is the number immediately following the word *of.* The answer to the problem is **25%** because $125 \div 500 = .25$, which equals 25%.

Example 7: A machine cutting tool has a useful life of 110 hours. If the tool is used 50 hours, what percentage of its useful life is left?

In this problem first find the useful life left in hours: $110 - 50 = 60$. So this problem is the same as asking

what percent of 110 is 60. The solution is **54.5%** because $60 \div 110 = .545$ (after rounding), then $.545 = 54.5\%$.

Example 8: $200 is what percent of $50?

The answer is **400%** because $200 \div 50 = 4 = 400\%$. The moral of this example is that the base is not always the larger number.

Powers and Roots

A *power* is a small, raised number and stands for repeated multiplication.

Example 1: $5^3 = 5 \times 5 \times 5 = 125$. This third power is called a *cube.* The example is read, "5 cubed equals 125."

Example 2: $7^2 = 7 \times 7 = 49$. The second power is the *square,* and this example is read, "7 squared equals 49."

A *root* is the reverse of a power.

Example 3: $\sqrt[3]{125} = 5$. This cube root of 125 equals 5 because 5 cubed is 125.

Example 4: The square root is the most common root. The square root of 49 may be written $\sqrt[2]{49}$, but the index number 2 is usually suppressed for square root. Write $\sqrt{49} = 7$. It equals 7 because 7 squared is 49.

Here are other examples of powers and roots.

$$2^5 = 32 \quad 6^2 = 36 \quad \sqrt[3]{8} = 2 \quad \sqrt{16} = 4$$

All scientific calculators have buttons or sequences of buttons that make finding powers and roots very easy. The booklet that comes with the calculator should be consulted to learn how to do this.

▼ ALGEBRA

Signed Numbers

The *sign* of a number is found immediately to the left of a number. If there is no sign, this means the same as if there is a plus sign. For example, in the expression "4 – 7" the 4 is a positive number and the 7 is a negative number. Multiplication of two numbers is indicated by parentheses around one or both numbers and no sign between the numbers: $(-3)(-6)$ is multiplication but $(-3) + (-6)$ is not multiplication. Here are other examples of multiplication:

$$(2)(-3) = -6$$
$$-7(8) = -56$$
$$(-5)(-6) = 30$$
$$7(4) = 28$$

As these examples show, the rules for signs when multiplying are:

If signs are the same, the answer is positive.

If signs are different, the answer is negative.

Multiplication is also indicated by a raised dot (not to be confused with a decimal point).

$$-4 \cdot -3 = 12$$

The sign rules for division are identical to multiplication. For example, $-9 \div 3 = -3$. Often in algebra, division is indicated as a fraction. The previous division example might be expressed as

$$\frac{-9}{3}$$

The rules for combining (adding and subtracting) signed numbers are:

If signs are the same, add the numbers
and give the answer the common sign.

If signs are different, subtract the two numbers
and give the answer the sign of the bigger.

Example 1: $-1 +3$

This is not a multiplication problem because there are no paraentheses or raised dots; it's a combining problem. The 1 is negative and the 3 is positive, so their signs are different, which calls for subtraction: 1 from 3 leaves 2. The 3 is bigger than the 1 so the answer gets the positive sign of the 3. Thus, $-1 +3 = 2$.

Example 2: $-4 -6$

This is also a combining problem. Both numbers are negative so their signs are the same. Add 4 and 6 and get 10. Give the answer the common negative sign, so $-4 -6 = -10$. It does not matter whether you interpret this problem as "-4 add -6" or "-4 subtract $+6$", but the easiest interpretation is "*combine* -4 and -6". Here are further examples of combining problems:

$$+7 -5 = +2$$
$$+5 -7 = -2$$
$$-3 -1 = -4$$
$$-4 +3 = -1$$

Sometimes parentheses do occur in combining problems involving equations. Such problems require parentheses to first be removed by using the rules of multiplication on the signs around the left parenthesis of each pair. The following example illustrates this.

Example 3: $(-3) + (-6)$

The (-3) has no sign in front so a positive sign can be attached: $+(-3)$. Now the signs around the left parenthesis symbol are $+$ and $-$. These signs are different, and applying the multiplication sign rule for them (different signs answer is negative) resolves them into a single negative $(-)$ sign. Likewise, the signs around the left parenthesis of the 6 resolves them into a single negative $(-)$ sign. So the problem $(-3) + (-6)$ is the same as $-3 -6$. Using the combining rules on this, $-3 -6 = -9$.

Example 4: $-4 - (-5) = -4 +5 = 1$

Example 5: $6 - (7) = +6 - (+7) = +6 -7 = -1$

Sometimes technical publications and applications omit the parentheses and show just the double signs. Solve such combining problems in the same way:

$$-4 - -5 = -4 +5 = 1.$$

Here are further examples of combining problems:

$$-4 - (-4) = -4 +4 = 0$$
$$3 - (-3) = 3 + 3 = 6$$
$$2 + (-7) - (8) = 2 -7 -8 = -5 -8 = -13$$

Evaluating Expressions

When using formulas the first step is to replace (substitute) the letters with numbers. The resulting expression is then solved by applying the rules for signed numbers.

Example 1 (temperature): Find F if C = −10 using the formula

$$F = 1.8 \, C + 32$$

Replacing the letters with numbers gives F = (1.8) (−10) + 32.

In algebra, when a letter is next to a number and there is no + or − sign, multiplication is indicated. This formula calls for C to be multiplied by 1.8: (1.8) (−10) = −18. So we now have F = −18 + 32, which gives F = 14.

Since most formulas involve a mixture of addition, multiplication, and other math operations, it is necessary to follow the worldwide *order of operations* when evaluating expressions.

It is:	1. Work within parentheses first
then do	2. Powers and Roots
then	3. Multiplication and Division
finally	4. Addition and Subtraction

Example 2: $(5 - 1)^2 - 3 = (4)^2 - 3 = 16 - 3 = 13$

Example 3: $17 - 8 \div 4 = 17 - 2 = 15$ (It's tempting to subtract the 8 from the 17 but the division must be done first.)

Solving Simple Equations

There are basically two operations that solve simple equations. One operation involves combining, the other involves multiplying. When a letter (unknown) has a number combined with it, move that number to the other side of the equal sign but also *change its sign*.

Example 1: $x + 5 = 7$

Moving the +5 to the right side of the equation and changing its sign to −5 gives x = 7 − 5. Evaluating x = 2.

Example 2:

$$10 = -4 + y$$
$$10 +4 = +y$$

or just: $\quad 10 +4 = y$

so: $\quad 14 = y$

or: $\quad y = 14$

Example 3:
$$3 + z = 2$$
$$z = 2 - 3$$
$$z = -1$$

Example 4: Solve for x.

$$x + abc = z$$
$$x = z - abc$$

When a letter is being multiplied by a number, *divide both sides of the equation by that number.*

Example 5:
$$5x = 15$$
$$\frac{5x}{5} = \frac{15}{5}$$
$$x = 3$$

Example 6:
$$-2y = 10$$
$$\frac{-2y}{-2} = \frac{10}{-2}$$
$$y = -5$$

Example 7:
$$Ax = B$$
$$\frac{Ax}{A} = \frac{B}{A}$$
$$x = \frac{B}{A}$$

Conversely, if the problem has division, multiply.

Example 8:
$$\frac{x}{2} = 6$$
$$(2)\left(\frac{x}{2}\right) = (6)(2)$$
$$x = 12$$

The solution of larger simple equations involves applying these two basic operations more than once.

Example 9:
$$5x + 7 = 22$$
$$5x = 22 - 7$$
$$5x = 15$$
$$\frac{5x}{5} = \frac{15}{5}$$
$$x = 3$$

Example 10:
$$-2y - 9 = 11$$
$$-2y = 11 + 9$$
$$-2y = 20$$
$$\frac{-2y}{-2} = \frac{20}{-2}$$
$$y = -10$$

To solve an equation with a letter squared, take the square root of both sides of the equation.

Example 11:
$$x^2 = 49$$
$$\sqrt{x^2} = \sqrt{49}$$
$$x = 7 \text{ and } x = -7$$
$$\text{(because } -7 \cdot -7 \text{ also equals 49)}$$

Conversely, to solve an equation with the square root of a letter, square both sides of the equation.

Example 12:
$$\sqrt{y} = 8$$
$$(\sqrt{y})^2 = 8^2$$
$$y = 64$$

Ratio and Proportion

One very common and useful type of equation is called a *proportion*. Each side of a proportion is a fraction or ratio. To solve a proportion by cross-multiplying, multiply the top of one ratio by the bottom of the other and set them equal:

Example 1:
$$\frac{x}{6} = \frac{15}{5}$$
$$(x)(5) = (6)(15)$$
$$5x = 90$$
$$\frac{5x}{5} = \frac{90}{5}$$
$$x = 18$$

An older way to express this example is X:6::15:5. The x and 5 were the *extremes* and the 6 and 15 the *means.*

Example 2: On a drawing, the scale is 1:150. What actual length does a 3.25" line of the drawing represent?
Set up a proportion.

$$\frac{1"}{150"} = \frac{3.25"}{x}$$
$$(1)x = (150)(3.25)$$
$$x = 487.5" \text{ or } 40' -7\tfrac{1}{2}"$$

Example 3: A cylindrical container holds 1000 gallons of oil when filled to a depth of 8 feet. How many gallons are there when the depth is 3½ feet?

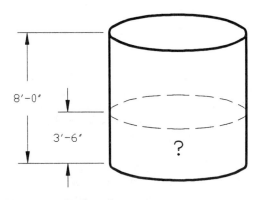

Figure N–1   A cylindrical container.

Set up a proportion.

$$\frac{1000 \text{ gallons}}{8 \text{ feet}} = \frac{x \text{ gallons}}{3.5 \text{ feet}}$$
$$(8)(x) = (1000)(3.5)$$
$$8x = 3500$$
$$x = 437.5 \text{ gallons}$$

▼ GEOMETRY

Two-Dimensional Figures

This section contains formulas for the perimeter (P) and area of common geometric figures the drafter may encounter. *Perimeter* is the straight-line (linear) distance around a figure and *area* is the number of square units that fit within a figure.

Parallelogram

$A + B = 180°$
$P = 2a + 2b$
Area = bh

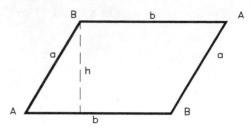

Figure N–5 Parallelogram.

Right Triangle

$A + B = 90°$
$x^2 + y^2 = z^2$
$P = x + y + r$
Area = $\frac{1}{2}xy$

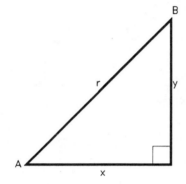

Figure N–2 Right triangle.

General Triangle

$A + B + C = 180°$
$P = a + b + c$
Area = $\frac{1}{2}bh$

Area = $\sqrt{s(s-a)(s-b)(s-c)}$ where s = $\frac{1}{2}(a + b + c)$

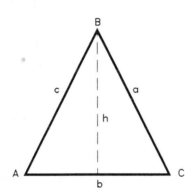

Figure N–3 General triangle.

Rhombus

$P = 4a$

Area = $\frac{1}{2}pq$

Note: The letters and *p* and *q* represent diagonal distances.
$p^2 + q^2 = 4a^2$

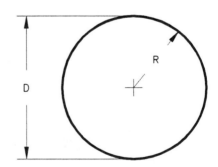

Figure N–6 Rhombus.

Rectangle

$P = 2a + 2b$
Area = ab
$d = \sqrt{a^2 + b^2}$

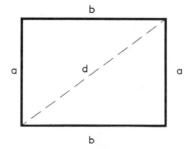

Figure N–4 Rectangle.

Circle

R = radius
D = diameter
$\pi = 3.14159 \ldots$
C = the circle's perimeter, or circumference
$C = \pi D$
Area = πr^2
Area = $\frac{\pi D^2}{4}$

Figure N–7 Circle.

Circle Inscribed within a Right Triangle

$$R = \frac{ab}{a + b + c}$$

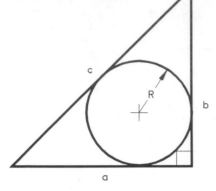

Figure N–8 Circle within a right triangle.

Circle Inscribed within a General Triangle

$$R = \sqrt{\frac{(s-a)(s-b)(s-c)}{s}} \quad \text{where } s = \frac{1}{2}(a + b + c)$$

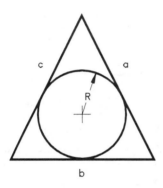

Figure N–9 Circle within a general triangle.

Circle Circumscribed around a Right Triangle

$$R = \frac{1}{2}c$$

Note: The letter *c* is the diameter of the circle as well as the hypotenuse of the right triangle.

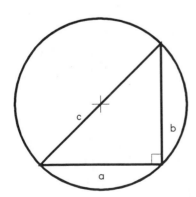

Figure N–10 Circle around a right triangle.

Circle Circumscribed around a General Triangle

$$R = \frac{a}{2 \sin A} = \frac{b}{2 \sin B} = \frac{c}{2 \sin C}$$

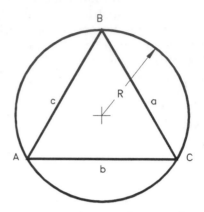

Figure N–11 Circle around a general triangle.

Angle Relationship for the General Triangle Incribed within a Circle

$$B = \frac{1}{2}D$$

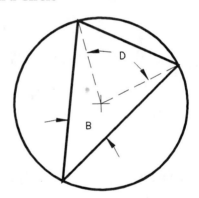

Figure N–12 Angles within an inscribed triangle.

Regular Polygons

n = number of sides
P = nf
$$\theta = \left(\frac{n-2}{n}\right)180°$$
$$f = 2R \sin \frac{180°}{n}$$
$$\text{Area} = \frac{1}{2}nR^2 \sin \frac{360°}{n}$$
$$d = \frac{f}{2 \tan \frac{180°}{n}}$$

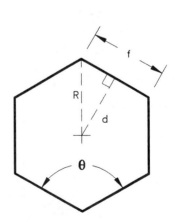

Figure N–13 Regular polygon.

Arc of a Circle

s = length of the arc of the circle (arc length)
Note: θ must be in radians.

$s = R\theta$

$f = 2R \sin \dfrac{\theta}{2}$

$g = R \cos \dfrac{\theta}{2}$

$h = R - g$

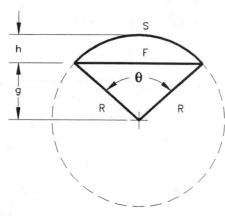

Figure N–14   Arc of a circle.

Ellipse

$P = 2\pi \sqrt{\dfrac{x^2 + y^2}{2}}$

(The perimeter formula is approximate.)
Area = πxy
(The area formula is exact.)

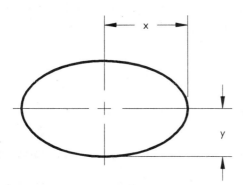

Figure N–15   Ellipse.

Three-Dimensional Figures

This section shows formulas for total surface area (S) in square units and volume (V) in cubic units for common shapes.

Rectangular Solid

$S = 2(wh + lw + lh)$
$V = lwh$

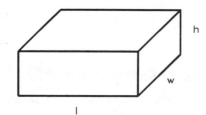

Figure N–16   Rectangular solid.

Cylinder

$S = 2\pi R^2 + 2\pi Rh$
$V = \pi R^2 h$

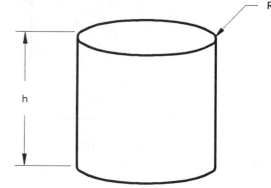

Figure N–17   Cylinder.

Right Circular Cone

$S = \pi R^2 + \pi RL$
$L = \sqrt{R^2 + h^2}$
$V = \dfrac{1}{3} \pi R^2 h$

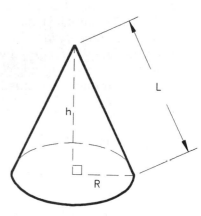

Figure N–18   Right circular cone.

Sphere

$S = 4\pi R^2$
$S = \pi D^2$, where D = 2R
$V = \dfrac{4}{3} \pi R^3$

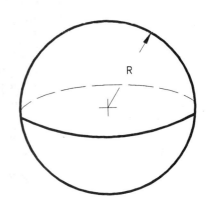

Figure N–19   Sphere.

▼ TRIGONOMETRY

Right Triangle Trigonometry

Trigonometry is based upon the lengths of sides and the angles of right triangles. It is best to treat the *trigonometric function* definitions as simply *working formulas* involving two sides and an angle of the right triangle of

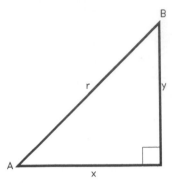

Figure N–20   Right triangle used for trigonometry definitions.

Figure N–20 and to use the formula that involves the information in the problem. Here are the definitions of the trig functions: In the right triangle, the longest side (r) is called the *hypotenuse*. The side (y) is called the *opposite side*, because it is opposite to angle (A). Side (x) is called the *adjacent side.* The first formula says that dividing the length of side y by hypotenuse r gives a number called the *sine of angle A*. (Sine is abbreviated "sin" but still pronounced *sine*.)

Example 1: A very common right triangle is the 3-4-5 right triangle. In Figure N–21 find the sine of angle A.
  Using the sine formula:

$$\text{Sin } A = \frac{y}{r}$$

$$\text{Sin } A = \frac{3}{5} = .6$$

  The usefulness of trigonometry is that it provides a connection between the lengths of sides and measurement of angles. Knowing, for example, that sin A = .6, a calculator can be used to find angle A. For most calculators, entering .6 then pushing INV and SIN (or 2ND and SIN) gives angle A as 36.9°.
  Similarly, *cosine* is the adjacent side divided by the hypotenuse and *tangent* is the opposite side divided by the adjacent side. An additional formula used to solve right triangles involves only the sides. It is the *Pythagorean Theorem*:

$$x^2 + y^2 = r^2$$

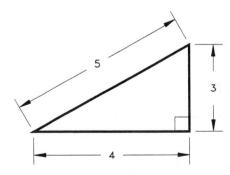

Figure N–21   3-4-5 right triangle.

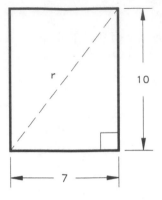

Figure N–22   Rectangular plate.

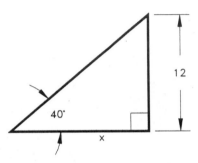

Figure N–23   Right triangle.

Example 2: Find the diagonal distance in Figure N–22 for a rectangular plate.
  Using the formula:

$$x^2 + y^2 = r^2$$
$$7^2 + 10^2 = r^2$$
$$49 + 100 = r^2$$
$$149 = r^2$$
$$r = \sqrt{149} = 12.21"$$

Example 3: Find the length of side x for the right triangle in Figure N–23.
  Because you know angle A and side y, use the tangent formula; only tangent involves the unknown side x and the known angle and side.

$$\tan A = \frac{y}{x}$$
$$\tan 40° = \frac{12"}{x}$$
$$.8391 = \frac{12"}{x}$$
$$.8391x = 12"$$
$$x = \frac{12"}{.8391}$$
$$x = 14.3"$$

Angle Conversion and Arc Length

One degree equals 60 minutes of arc: 1° = 60'
One minute equals 60 seconds of arc: 1' = 60"
Therefore, 1° = 3600"

Example 1: Convert 4°50'35" to decimal degrees (nearest hundredth).

The answer is

$$4 + \frac{50}{60} + \frac{35}{3600} = 4 + .8333 + .0097$$
$$= 4.843 \text{ or } \mathbf{4.84°}$$

Example 2: Convert 10.268° to degrees, minutes, and seconds.

Working with the decimal fraction of the degrees, $.268 \times 60 = 16.08$ for 16 whole minutes. Then, working with the decimal fraction of minutes, $.08 \times 60 = 4.8$ or (rounding off) 5 whole seconds. The complete answer becomes **10°16'5"**.

Here are additional conversion facts:

Degrees and Radians:

$$180° = \pi \text{ radians (where } \pi = 3.14159 \ldots)$$
$$1 \text{ radian} = \text{about } 57.3°$$
$$1° = \text{about } .01745 \text{ radian}$$

Example 3: Convert 30° to radians (nearest thousandth).
The answer is $30 \times .01745 = .5235$ or .524 radian.

Example 4: Convert 2 radians to degrees (nearest tenth).
The answer is $2 \times 57.3 = 114.6°$.

Vectors

A *vector* is a directed line segment, or arrow, with two attributes: *length* (magnitude) and *angle* (direction). Points in the plane may be specified by their x and y coordinates or by magnitude and direction of a vector with its tail at the origin and its head at the point in the plane as shown in Figure N–24.

Vector notation and equations conform to the definitions of the trig functions except that Greek letter θ (theta) is often used for the angle A. Here are the conversion formulas:

Polar to Rectangular (vector form to x-y form):

$$x = r \cos \theta$$
$$y = r \sin \theta$$

The x and y are also called *vector components.*

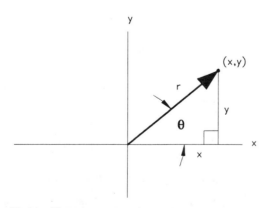

Figure N–24   Vector.

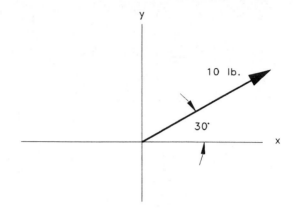

Figure N–25   Vector in polar form.

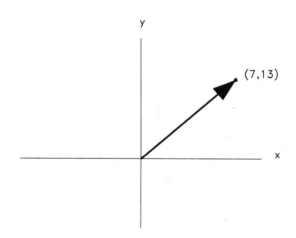

Figure N–26   Vector in rectangular form.

Example 1: Find the components of the vector shown in Figure N–25.

Using the conversion formulas with $r = 10$ and $\theta = 30°$,

$$x = (10)(\cos 30°) = (10)(.8660) = 8.66 \text{ lb, and}$$
$$y = (10)(\sin 30°) = (10)(.5) = 5 \text{ lb}$$

Rectangular to Polar (x-y form to vector form):

$$r = \sqrt{x^2 + y^2}$$
$$\theta = \text{Inv Tan} \frac{y}{x}$$

Example 2: Convert the coordinates (7,13) in Figure N–26 to polar form.

Using the conversion formulas with $x = 7$ and $y = 13$,

$$r = \sqrt{7^2 + 13^2} = \sqrt{49 + 169} = \sqrt{218} = 14.8$$
$$\theta = \text{Inv Tan} \frac{13}{7} = \text{Inv Tan } 1.857 = 61.7°$$

Slope of a Line

The slope-intercept form of the equation of a straight line is: $y = mx + b$. Letter *m* is the slope (rise divided by run of the line) and letter *b* is the y-axis intercept (the point on the vertical axis that the line crosses). For example, Figure N–27 is the graph of the equation $y = 2x + 5$.

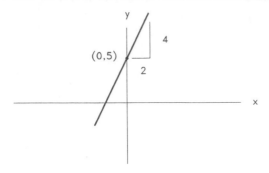

Figure N–27   Graph of y = 2x + 5.

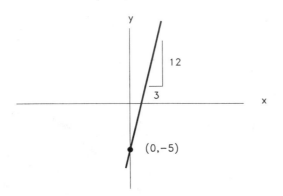

Figure N–28   Graph requiring an equation.

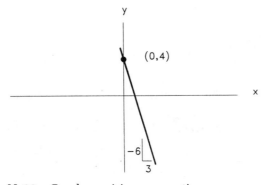

Figure N–29   Graph requiring an equation.

Example 1: What is the equation of the line in Figure N–28?

The slope is 12 ÷ 3 = 4 and the y-intercept is –5, so the equation is y = 4x – 5.

Positive slopes in math slant up and to the right; negative slopes slant down and to the right.

Example 2: What is the equation of the line in Figure N–29?

The slope is –6 ÷ 3 = –2 and the y-intercept is 4, so the equation is y = –2x + 4.

Oblique Triangles

In trig, lowercase letters usually stand for the lengths of sides and capital letters stand for angles, as shown in Figure N–30:

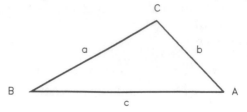

Figure N–30   General oblique triangle.

To solve a triangle means to find all missing angles and sides. The Law of Sines can be used to solve a general triangle when *one* side and *the opposite angle* are known. The Law of Sines is usually expressed as:

$$\frac{a}{Sin\ A} = \frac{b}{Sin\ B} = \frac{c}{Sin\ C}$$

It is really three separate equations; each one is a proportion.

$$\frac{a}{Sin\ A} = \frac{b}{Sin\ B}$$
$$\frac{b}{Sin\ B} = \frac{c}{Sin\ C}$$
$$\frac{c}{Sin\ C} = \frac{a}{Sin\ A}$$

Example 1: Find side b in the triangle show in Figure N–31.

Angle B can be found quickly knowing that the sum of the angles of any triangle is 180°.

$$B = 180 - (80 + 45) = 180 - 125 = 55°$$

Now the Law of Sines can be used:

$$\frac{a}{Sin\ A} = \frac{b}{Sin\ B}$$
$$\frac{14}{Sin\ 45°} = \frac{b}{Sin\ 55°}$$
$$\frac{14}{.7071} = \frac{b}{.8192}$$
$$.7071b = (.8192)(14)$$
$$.7071b = 11.486$$
$$b = 16.2"$$

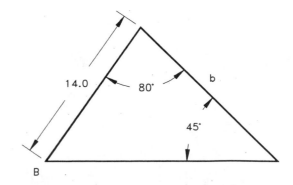

Figure N–31   Oblique triangle with unknown side.

When an oblique triangle has an angle greater than 90°, care must be taken when using the Law of Sines to find that angle. The calculator gives only angles less than 90° when INV SIN is pushed. A theorem from trig class must be used: sin θ = sin (180° − θ), which means the calculator answer must be subtracted from 180° to get the true answer. The next example illustrates this.

Example 2: Find angle A in Figure N–32.

First, convert to inches: 16' − 8" = 200" and 14' − 2" = 170", then using the version of the Law of Sines involving a's and c's:

$$\frac{a}{\text{Sin A}} = \frac{c}{\text{Sin C}}$$
$$\frac{200}{\text{Sin A}} = \frac{170}{\text{Sin } 15°}$$
$$\frac{200}{\text{Sin A}} = \frac{170}{.2588}$$
$$170 \sin A = (200)(.2588)$$
$$170 \sin A = 51.76$$
$$\frac{170 \text{ Sin A}}{170} = \frac{51.76}{170}$$
$$\sin A = .3045$$
$$A = \text{Inv Sin } .3045$$
$$A = 17.7°$$

Angle A is known to be greater than 90°, so A = 180 − 17.7 = **162.3°**. This final subtraction from 180 is taken only when finding an angle greater than 90° with the Law of Sines.

Sometimes a side and opposite angle are not available. In that case, the triangle may be solved with the Law of Cosines. Again, there are three versions of this law.

$$a^2 = b^2 + c^2 - 2bc \cos A$$
$$b^2 = a^2 + c^2 - 2ac \cos B$$
$$c^2 = a^2 + b^2 - 2ab \cos C$$

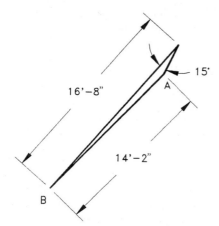

Figure N–32　Oblique triangle with unknown angle.

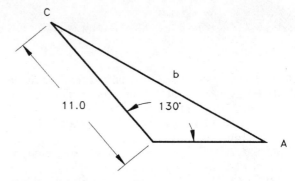

Figure N–33　Oblique triangle with unknown side.

Example 3: Find side b of the triangle in Figure N–33.

Use the version of the Law of Cosines containing angle B.

$$b^2 = 11^2 + 14^2 - 2(11)(14)(\cos 130°)$$
$$b^2 = 121 + 196 - 2(11)(14)(-.6428)$$
$$b^2 = 121 + 196 + 197.98$$
$$b^2 = 514.98$$
$$b = \sqrt{514.98}$$
$$b = 22.7"$$

Example 4: Find angle A of the triangle in Figure N–34.

Use the version of the Law of Cosines containing angle A.

$$60^2 = 100^2 + 87^2 - 2(100)(87) \cos A$$
$$3,600 = 10,000 + 7569 - 17,400 \cos A$$
$$3,600 = 17,569 - 17,400 \cos A$$
$$3,600 - 17,569 = -17,400 \cos A$$
$$-13,969 = -17,400 \cos A$$
$$\frac{-13,969}{-17,400} = \frac{-17,400 \cos A}{-17,400}$$
$$.8028 = \cos A$$
$$A = \text{Inv cos } .8028$$
$$A = 36.6°$$

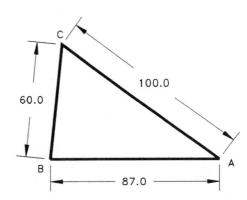

Figure N–34　Oblique triangle with unknown angle.

ACME A thread system used especially for feed mechanisms.

ADDENDUM (SPUR GEAR) The radial distance from the pitch circle to the top of the tooth.

ADDENDUM ANGLE (BEVEL GEAR) The angle subtended by the addendum.

ALIGNED SECTION The cutting plane is staggered to pass through offset features of an object.

ALIGNMENT CHARTS Designed to graphically solve mathematical equation values using three or more scaled lines.

ALLOWANCE The tightest possible fit between two mating parts.

ALLOYS A mixture of two or more metals.

AMPLIFIER (AMP) A device that allows an input signal to control power; capable of having an output signal greater than the input signal.

ANGLE OF REPOSE The run-to-rise ratio of highway cut and fill.

ANNEALING Under certain heating and cooling conditions and techniques, steel may be softened.

APPARENT INTERSECTION This is a condition where lines or planes *look* like they may be intersecting, but in reality they may not be intersecting.

AUXILIARY VIEW A view that is required when a surface is not parallel to one of the principal planes of projection; the auxiliary projection plane is parallel to the inclined surface so that the surface may be viewed in its true size and shape.

AXIS The centerline of a cylindrical feature.

AZIMUTH The clockwise measurement of the angle of a line, measured from the north of its reference meridian.

BACKSIGHT In surveying, the rod reading behind the level toward the point of beginning.

BALL BEARING A friction-reducer where balls roll in two grooved rings.

BASE CIRCLE (CAM) The smallest circle tangent to the CAM follower at the bottom of displacement.

BASE CIRCLE DIAMETER (SPUR GEAR) The diameter of a circle from which the involute tooth is generated.

BASIC DIMENSION A numerical value used to describe the theoretically exact size, profile, orientation, or location of a feature or datum target. It is the basis from which permissible variations are established by tolerances on other dimensions, in notes, or in feature control frames.

BEARING (CIVIL) The measurement of the angle of a line, measured from either the north or the south meridian, whichever is nearer.

BEARING (MECHANICAL) A mechanical device that reduces friction between two surfaces.

BEARING ANGLE The bearing angle of a line is always 90 degrees or less and is identified either from the north or the south.

BEARING OF A LINE The angular relationship of the horizontal projection of the line relative to the compass, expressed in degrees.

BEARING SEAL A rubber, felt, or plastic seal on the outer and inner ring of a bearing. Generally, it is filled with a special lubricant by the manufacturer.

BEARING SHIELD A metal plate on one or both sides of the bearing; serves to retain the lubricant and keep the bearing clean.

BELL AND SPIGOT A pipe connection in which one end of a piece of pipe has a bell-shaped opening and the other end is tapered or notched to fit into the bell.

BELLCRANK A link, pivoted near the center, that oscillates through an angle.

BENCH MARK The name for a known point with a known elevation that is part of the geodetic control system. Another name for bench mark is *monument*.

BEND ALLOWANCE The amount of extra material needed for a bend to compensate for compression during the bending process.

BEND RELIEF Cutting away material at a corner to help relieve stress.

BEVEL The term used to denote the slope of beams, as in structural engineering.

BEVEL GEAR Used to transmit power between intersecting shafts; takes the shape of a frustum of a cone.

BIAS The voltage applied to a circuit element to control the mode of operation.

BILATERAL TOLERANCE A tolerance in which variation is permitted in both directions from the specified dimension.

BIT Binary digit.

BOLT CIRCLE Holes located in a circular pattern.

BORE To enlarge a hole with a single pointed machine tool in a lathe, drill press, or boring mill.

BOSS A cylindrical projection on the surface of a casting for forging.

BOW'S NOTATION A system of notation used to label a vector system. A letter is given to the space on each side of the vector, and each vector is then identified by the two letters on either side of it, read in a clockwise direction.

BROKEN OUT SECTION A portion of a part is broken away to clarify an interior feature; there is no associated cutting plane line.

BUS An aluminum or copper plate or tubing that carries the electrical current.

BUSHING A replaceable lining or sleeve used as a bearing surface.

BUTT WELD A form of pipe manufacture in which the seam of the pipe is a welded flat-faced joint. Also, a form of welding in which two pieces of material are "butted" against each other and welded.

CABINET OBLIQUE DRAWING A form of oblique drawing in which the receding lines are drawn at half scale, and usually at a 45-degree angle from horizontal.

CAD Computer-aided design.

CAD/CAM Computer-aided design/computer-aided manufacturing.

CADD Computer-aided design and drafting.

CAE Computer-aided engineering.

CAM A machine part used to convert constant rotary motion into timed irregular motion.

CAM MOTION The base point from which to begin cam design. There are four basic types of motion: simple harmonic, constant velocity, and cycloidal.

CAPACITOR An electronic component that opposes a change in voltage and storage of electronic energy.

CARBURIZATION A process where carbon is introduced into the metal by heating to a specified temperature range while in contact with a solid, liquid, or gas material consisting of carbon.

CARTESIAN COORDINATE SYSTEM A measurement system based on rectangular grids to measure width, height, and depth (X, Y, and Z).

CASTING An object or part produced by pouring molten metal into a mold.

CAVALIER OBLIQUE DRAWING A form of oblique drawing in which the receding lines are drawn true size, or full scale. Usually drawn at an angle of 45 horizontal degrees.

CENTRAL PROCESSING UNIT (CPU) The processor and main memory chips in a computer. Specifically, the CPU is just the processor, but generally it refers to the computer.

CHAIN DIMENSIONING Also known as point-to-point dimensioning when dimensions are established from one point to the next.

CHAMFER A slight surface angle used to relieve a sharp corner.

CHORDAL ADDENDUM (SPUR GEAR) The height from the top of the tooth to the line of the chordal thickness.

CHORDAL THICKNESS (SPUR GEAR) The straight line thickness of a gear tooth on the pitch circle.

CIM Computer-integrated manufacturing that combines CADD, CAM, and CAE into a controlled system.

CIRCULAR PITCH (SPUR GEAR) The distance from a point on one tooth to the corresponding point on the adjacent tooth, measured on the pitch circle.

CIRCULAR THICKNESS (SPUR GEAR) The length of an arc between the two sides of a gear tooth on the pitch circle.

CLEARANCE (SPUR GEAR) The radial distance between the top of a tooth and the bottom of the mating tooth space.

COIL OR INDUCTOR A conductor wound on a form or in a spiral; contains inductance.

COLD ROLLED STEEL (CRS) The additional cold forming of steel after initial hot rolling; cleans up hot formed steel.

COMMAND A specific instruction issued to the computer by the operator. The computer performs a function or task in response to a command.

COMPRESSIVE Pushing toward the point of currency, as in forces that are compressed.

CONCENTRIC Two or more circles sharing the same center.

CONCURRENT FORCES Forces acting on a common point.

CONE DISTANCE (BEVEL GEAR) The slant height of the pitch cone.

CONSTRUCTION LINES Very lightly drawn, nonreproducing lines used for the layout of a drawing.

CONTOUR INTERVAL The distance in elevation between contour lines.

CONTOUR LINE Denotes a series of connected points at a particular elevation.

COPLANAR FORCES All lie in the same plane.

COUNTERBORE To cylindrically enlarge a hole; generally to allow the head of a screw or bolt to be recessed below the surface of an object.

COUNTERDRILL A machined hole that looks similar to a countersink/counterbore combination.

COUNTERSINK Used to recess the tapered head of a fastener below the surface of an object.

CRANK A link, usually a rod or bar, that makes a complete revolution about a fixed point.

CROWN BACKING (BEVEL GEAR) The distance between the cone apex and the outer tip of the gear teeth.

CRT Cathode Ray Tube.

CURSOR A small rectangle, underline, or set of crosshairs that indicate present location on a video display screen. Also, a hand-held input device used in conjunction with a digitizer.

DATUM A theoretically exact point, axis, or plane derived from the true geometric counterpart of a specified datum feature. The origin from which the location or geometric characteristics of features of a part are established.

DATUM DIMENSIONING A dimensioning system where each dimension originates from a common surface, plane, or axis.

DECLINATION A line that goes downward from its origin; assigned negative values.

DEDENDUM (SPUR GEAR) The radial distance from the pitch circle to the bottom of the tooth.

DEDENDUM ANGLE (BEVEL GEAR) The angle subtended by the dedendum.

DEFAULT An action taken by computer software unless the operator specifies differently.

DETAIL A drawing of an individual part that contains all of the views, dimensions, and specifications necessary to manufacture the part.

DIAMETRAL PITCH A ratio equal to the number of teeth on a gear per inch of pitch diameter.

DIAZO A printing process that produces blue, black, or brown lines on various media (other resultant colors are also produced with certain special products). The print process is a combination of exposing an original in contact with a sensitized material exposed to an ultraviolet light, and then running the exposed material through an ammonia chamber to activate the remaining sensitized image to form the desired print. This is a fast and economical method of making prints commonly used in drafting.

DIGITIZE The act of locating points and selecting commands using an input device (puck or stylus, used with a digitizer tablet).

DIGITIZER An electronically sensitized flat board or tablet that serves as a drawing surface for the input of graphics data. Images can be drawn or traced, and commands and symbols can be selected from a menu attached to the digitizer. *See also* graphics tablet.

DIHEDRAL ANGLE The angle that is formed by two intersecting planes.

DIMETRIC DRAWING A pictorial drawing in which two axes form equal angles with the plane of projection. These can be greater than 90 but less than 180 and cannot have an angle of 120 degrees. The third axis may have an angle less or greater than the two equal axes.

DIP The slope of a stratum.

DISPLACEMENT DIAGRAM A graph; the curve on the diagram is a graph of the path of the cam follower. In the case of a drum cam displacement diagram, the diagram is actually the developed cylindrical surface of the cam.

DOCUMENTATION Instruction manuals, guides, and tutorials provided with any computer hardware/software system.

DOWEL PIN A cylindrical fastener used to retain parts in a fixed position or to keep parts aligned.

DRAFT The taper on the surface of a pattern for castings of the die for forgings, designed to help facilitate removal of the pattern from the mold or the part from the die. Draft is often 7–10 degrees but depends on the material and the process.

DRILLING DRAWING Used to provide size and location dimensions for trimming the printed circuit board.

DRUM CAM A drum, or cylindrical, cam is a cylinder with a groove in its surface. As the cam rotates, the follower moves through the groove, producing a reciprocating motion parallel to the axis of the camshaft.

DRUM PLOTTER A graphics pen plotter that can accommodate continuous feed paper, or in which sheet paper or film is attached to a sheet of flexible material mounted to a drum. The pen moves in one direction and the drum in the other.

DUCT Sheet metal, plastic, or other material pipe designed as the passageway for conveying air from the HVAC equipment to the source.

DUCTILITY The ability to be stretched, drawn, or hammered without breaking.

ECCENTRIC CIRCLE Not having the same center.

ELECTRICAL RELAYS Magnetic switching devices.

ELECTRO-DISCHARGE MACHINING (EDM) A process where material to be machined and an electrode are submerged in a fluid that does not conduct electricity, forming a barrier between the part and the electrode. A high-current, short-duration electrical charge is then used to remove material.

ELECTROLESS The depositing of metal on another material through the action of an electric current.

ELECTRON BEAM (EB) Generated by a heated tungsten filament used to cut or machine very accurate features in a part.

ELEMENT Any line, group of lines, shape, or group of shapes and text that is so defined by the computer operator. *See also* entity.

ELEMENTARY DIAGRAMS Diagrams that provide the detail necessary for engineering analysis and operation or maintenance of substation equipment.

ENGINEERING CHANGE DOCUMENTS Documents used to initiate and implement a change to a production drawing; engineering change request (ECR) and engineering change notice (ECN) are examples.

ENTITY *See* element.

EQUILIBRANT A vector that is equal in magnitude to the resultant and has the opposite direction and sense.

EQUILIBRIUM When a vector system has a resultant of zero, the system is said to be in equilibrium.

EXPLODED ASSEMBLY A pictorial assembly showing all parts removed from each other and aligned along axis lines.

FACE ANGLE (BEVEL ANGLE) The angle between the top of the teeth and the gear axis.

FAULT CONDITION A short circuit that is a zero resistance path for electrical current flow.

FILLET A curve formed at the interior intersection between two or more surfaces.

FIXTURE A device for holding work in a machine tool.

FLANGE A thin rim around a part.

FLAT BED PLOTTER A pen plotter where the drawing surface (bed) is oriented horizontally, and paper or film is attached to the surface by a vacuum or an electrostatic charge. The pen moves in both the x and y directions.

FLOPPY DISK A thin, circular, magnetic storage medium encased in a cover. It comes in 8", 5 1/4", and 3 1/2" sizes.

FLOW CHARTS Used to show organizational structure, steps, or progression in a process or system.

FLOW DIAGRAM A chart-type drawing that illustrates the organization of a system in a symbolic format.

FOLD LINES The reference line of intersection between two reference planes in orthographic projection.

FOLLOWER The cam follower is a reciprocating device whose motion is produced by contact with the cam surface.

FONT A specific type face, such as Helvetica or Gothic.

FORESHORTENED LINE A line that appears shorter than its actual length, because it is at an angle to the line of sight.

FOUR-BAR LINKAGE The most commonly used linkage mechanism. It contains four links; a fixed link called the ground link, a pivoting link called a driver, another pivoted link called a follower, and a link between the driver and follower, called a coupler.

FREE-BODY DIAGRAM A diagram that isolates and studies a part of the system of forces in an entire structure.

FULL SECTION The cutting plane extends completely through the object.

FUNCTION KEYS Extra keys on an alphanumeric keyboard that can be utilized in a computer program to represent different commands and functions. The active commands for function keys may change several times in a program.

GATE The part of an electronic system that makes the electronic circuit operate; permits an output only when a predetermined set of input conditions are met.

GEAR A cylinder or cone with teeth on its contact surface; used to transmit motion and power from one shaft to another.

GEAR RATIO Any two mating gears have a relationship to each other called a gear ratio. This relationship is the same between any of the following: RPMs, number of teeth, and pitch diameters of the gears.

GEAR TRAIN Formed when two or more gears are in contact.

GRADE OF A LINE A way to describe the inclination of a line in relation to the horizontal plane. The percent grade is the vertical rise divided by the horizontal run multiplied by 100.

GRADE SLOPE The percentage given to show amount of slope.

GRAPHICAL KINEMATIC ANALYSIS The process of drawing a particular mechanism in several phases of a full cycle to determine various characteristics of the mechanism.

GRAPHICS TABLET *See* digitizer.

GREAT CIRCLE One of an infinite number of circles from any point on the earth that is described by longitude and latitude.

GUNTER'S CHAIN A 66-foot-long chain that Edmund Gunter invented; it is made up of 100 links. It is used in surveying.

HALF SECTION Used typically for symmetrical objects; the cutting plane line actually cuts through one quarter of the part. The sectional view shows half of the interior and half of the exterior at the same time.

HARDCOPY A paper copy.

HARDWARE The physical computer equipment.

HI Height of instrument. In surveying, the calculation of the level, which is one factor needed to determine elevation.

HIGHWAY DIAGRAM A simplified or condensed representation of a point-to-point, interconnecting wiring diagram for an electrical circuit.

HONE A method of finishing a hole or other surface to a desired close tolerance and fine surface finish using an abrasive.

INCLINATION A line that goes upward from its origin; assigned positive values.

INDUCTANCE The property in an electronic circuit that opposes a change in current flow or where energy may be stored in a magnetic field, as in a transformer.

INTEGRATED CIRCUIT (IC) All of the components in a schematic are made up of one piece of semiconductor material.

INTERSECTING LINES When lines are intersecting, the point of intersection is a point that lies on both lines.

ISOGONIC CHART A chart showing isogonic lines.

ISOGONIC LINES Shows how many degrees to the east or west the magnetic north or south pole is from the true North or South Pole.

ISOMETRIC DRAWING A form of pictorial drawing in which all three drawing axes form equal angles (120 degrees) with the plane of projection.

JIG A device used for guiding a machine tool in the machining of a part or feature.

JOINT The connection point between two links.

JOYSTICK A graphics input device composed of a lever mounted in a small box that allows the user to control the movement of the cursor on the video display screen.

KERF A groove created by the cut of a saw.

KINEMATICS The study of motion without regard to the forces causing the motion.

LAP WELD A form of pipe manufacture in which the seam of the pipe is an angular "lap."

LARGE SCALE INTEGRATION (LSI) More circuits on a single small IC chip.

LASER (LIGHT AMPLIFICATION BY STIMULATED EMISSION OF RADIATION) A device that amplifies focused light waves and concentrates them in a narrow, very intense beam.

LATITUDE The parallels around the earth that do not intersect. The equator is the longest line of latitude.

LAY Describes the basic direction or configuration of the predominant surface pattern in a surface finish.

LAYER An individual aspect of a CADD drawing that makes a complete drawing when combined.

LEAD (WORM THREAD) The distance that the thread advances axially in one revolution of the worm or thread.

LEVELING The process of determining elevation using backsight and foresight.

LEVER A link that moves back and forth through an angle; also known as a rocker.

LIGHT PEN A video display screen input device. It is a light-sensitive stylus connected to the terminal by a wire; enables the user to draw or select menu options directly on the screen.

LINE OF SIGHT An imaginary straight line from the eye of the observer to a point on the object being observed. All lines of sight for a particular view are assumed to be parallel and are perpendicular to the projection plane involved.

LOGIC DIAGRAMS A type of schematic that is used to show the logical sequence in an electronic system.

LONGITUDE The meridians of the earth, which run from the North Pole to the South Pole. The lines of longitude are basically the same length.

MAGNETIC DECLINATION The degree difference between magnetic azimuth and true azimuth.

MALLEABLE The ability to be hammered or pressed into shape without breaking.

MASTER PATTERN A one-to-one scale circuit pattern that is used to produce a printed circuit board.

MAXWELL DIAGRAM A combination vector diagram used to analyze the forces acting in a truss.

MECHANICAL JOINT A pipe connection that is a modification of the "bell and spigot" in which flanges and bolts are used with gaskets, packing rings, or grooved pipe ends providing a seal.

MECHANISM A combination of two or more machine members that work together to perform a specific motion.

METES AND BOUNDS The system of describing portions of land by using lengths and boundaries.

MOUSE A hand-held input device connected to the terminal by a wire. It is moved across a flat surface to control the movement of the cursor on the screen. It rolls on a small ball that sends directional signals to the computer, and may have one or more buttons that serve as function keys.

MULTIVIEW PROJECTION The views of an object as projected upon two or more picture planes in orthographic projection.

NECK A groove around a cylindrical part.

NOMINAL SIZE The designation of the size for a commercial product.

NOMOGRAPHS A graphic representation of the relationship between two or more variables of a mathematical equation.

NONDESTRUCTIVE TESTING (NDS) Tests for potential defects in welds; they do not destroy or damage the weld or the part.

NORMAL PLANE A plane surface that is parallel to any of the primary projection planes.

NORMALIZING A process of heating steel to a specific temperature and then allowing the material to cool slowly by air, bringing the steel to a normal state.

NUMERICAL CONTROL (NC) A system of controlling a machine tool by means of numeric codes that direct the commands for the machine movements; computer numerical control (CNC) is a computer command control of the machine movement.

OBLIQUE DRAWING A form of pictorial drawing in which the plane of projection is parallel to the front surface of the object and the receding angle is normally 45°.

OBLIQUE LINE A straight line that is not parallel to any of the six principal planes.

OBLIQUE PLANE Inclined to all of the principal projection planes.

OFFSET SECTION The cutting plane is offset through staggered interior features of an object to show those features in section as if they were in the same plane.

OPERATIONAL AMPLIFIER (OPAMP) A high-gain amplifier created from an integrated circuit.

OUTSIDE DIAMETER (SPUR GEAR) The overall diameter of the gear; equal to the pitch diameter plus two addendum.

PADS Or lands, are the circuit-termination locations where the electronic devices are attached.

PATTERN DEVELOPMENT Based on laying out geometric forms in true size and shape flat patterns.

PERSPECTIVE DRAWING A form of pictorial drawing in which vanishing points are used to provide the depth and distortion that is seen with the human eye. Perspective drawings can be drawn using one, two, and three vanishing points.

PHILADELPHIA ROD A long pole with numbers and graduated sections that is used in surveying to determine elevation or distance.

PHOTODRAFTING A combination of a photograph(s) with line work and lettering on a drawing.

PICTORIAL DRAWING A form of drawing that shows an object's depth. Three sides of the object can be seen in one view.

PIE CHARTS Used for presentation purposes where portions of a circle represent quantity.

PIERCING POINT A point where a particular line intersects a plane.

PINION GEAR When two gears are mating, the pinion gear is the smaller, usually the driving gear.

PITCH A distance of uniform measure determined at a point on one unit to the same corresponding point on the next unit; used in threads, springs, and other machine parts.

PITCH (WORM) The distance from one tooth to the corresponding point on the next tooth measured parallel to the worm axis; equal to the circular pitch on the worm gear.

PITCH ANGLE (BEVEL GEAR) The angle between an element of a pitch cone and its axis.

PITCH DIAMETER (BEVEL GEAR) The diameter of the base of the pitch cone.

PITCH DIAMETER (SPUR GEAR) The diameter of an imaginary pitch circle on which a gear tooth is designed. Pitch circles of two spur gears are tangent.

PLAIN BEARING Based on a sliding action between the mating parts; also called sleeve or journal bearings.

PLANE A surface that is not curved or warped. It is a surface in which any two points may be connected by a straight line, and the straight line will always lie completely within the surface.

PLAT A tract of land showing building lots.

PLATE CAM A cam in the shape of a plate or disk. The motion of the follower is in a plane perpendicular to the axis of the camshaft.

POB Point of beginning. Any point that has been determined to be the beginning of a survey.

POLAR CHARTS Designed by establishing polar coordinate scales where points are determined by an angle and distance from a center or pole.

POLYESTER DRAFTING FILM A high-quality drafting material with excellent reproduction, durability, and dimensional stability; also known by the trade name Mylar®.

POLYGONS Polygons are enclosed figures such as triangles, squares, rectangles, parallelograms, and hexagons.

PRESSURE ANGLE The direction of pressure between contacting gear teeth. It determines the size of the base circle and the shape of the involute spur gear tooth, commonly 20°.

PRIME CIRCLE (CAM) A circle with a radius equal to the sum of the base circle radius and the roller follower radius.

PRIME MERIDIAN The line of longitude that is given the 0 degree designation, and from which all other lines of longitude are measured.

PRINTED CIRCUITS (PC) Electronic circuits printed on a board that form the interconnection between electronic devices.

PRINTER A device that receives data from the computer and converts it into alphanumeric or graphic printed images.

PROFILE Shows what a section of land (or utility pipe, etc.) looks like in elevation.

PROFILE LINE A profile line is one that is parallel to the profile projection plane; its projection appears in true length in the profile view.

PROJECTION LINE A projection line is a straight line at 90° to the fold line, which connects the projection of a point in a view to the projection of the same point in the adjacent view.

PROJECTION PLANE A projection plane is an imaginary surface on which the view of the object is projected and drawn. This surface is imagined to exist between the object and the observer.

PUBLIC LAND SYSTEM The land that was divided by a rectangular system of surveys, in which the main subdivisions are townships and sections.

QUENCH To cool suddenly by plunging into water, oil, or other liquid.

RACK Basically a straight bar with teeth on it. Theoretically, it is a spur gear with an infinite pitch diameter.

RADIAL MOTION Exists when the path of the motion forms a circle, the diameter of which is perpendicular to the center of the shaft; also known as rotational motion.

RATIO SCALES Special scales that are referred to as logarithmic and semi-logarithmic.

REAM To enlarge a hole slightly with a machine tool called a reamer to produce greater accuracy.

RECTANGULAR SYSTEM OF SURVEYS A system of describing the land that is part of a public land survey. Each one of these public land surveys uses townships, sections, quarter-sections, etc. to describe a particular piece of land.

RECTILINEAR CHARTS Charts that are set up on a horizontal and vertical grid where the vertical axis identifies the quantities or values related to the horizontal values; also known as line charts.

RELIEF A slight groove between perpendicular surfaces to provide clearance between the surfaces for machining.

REMOVED SECTION A sectional view taken from the location of the section cutting plane and placed in any convenient location of the drawing, generally labeled in relation to the cutting plane.

RESISTORS Components that contain resistance to the flow of electric current.

REVOLUTION An alternate method for solving descriptive geometry problems in which the observer remains stationary and the object is rotated to obtain various views.

REVOLVED SECTION A sectional view established by revolving 90 degrees, in place, into a plane perpendicular to the line of sight, generally used to show the cross section of a part or feature that has consistent shape throughout the length.

RIB A thin metal section between parts to reinforce while reducing weight in a part. *See also* web.

RIGHT ANGLE   An angle of 90 degrees.

ROCKER   A link that moves back and forth through an angle; also known as a lever.

ROLLER BEARINGS   A bearing composed of two grooved rings and a set of rollers. The rollers are the friction-reducing element.

ROOT DIAMETER (SPUR GEAR)   The diameter of a circle coinciding with the bottom of the tooth spaces.

ROUND   Two or more exterior surfaces rounded at their intersection.

RPM   Revolutions per minute.

RUNOUTS   Characteristics of intersecting features, determined by locating the line of intersection between the mating parts.

SCHEMATIC DIAGRAMS   Drawn as a series of lines and symbols that represent the electrical current path and the components of the circuit. Provides the basic circuit connection information for electronic products.

SEMICONDUCTORS   Devices that provide a degree of resistance in an electronic circuit; types include diodes and transistors.

SKEW LINES   Lines that are neither parallel nor intersecting.

SLIDER   A link that moves back and forth in a straight line.

SLOPE ANGLE   The angle in degrees that the line makes with the horizontal plane.

SOCKET WELD   A form of pipe connection in which a plain-end pipe is slipped into a larger opening or "socket" of a fitting. One exterior weld is required; thus, no weld material protrudes into the pipe.

SOFTWARE   Computer programs stored on magnetic tape or disk that enable a computer to perform specific functions to accomplish a task.

SOLDER   An alloy of tin and lead.

SOLDER MASK   A polymer coating to prevent the bridging of solder between pads or conductor traces on a printed circuit board.

SOLIDS MODELING   A design and engineering process in which a 3-D model of the actual part is created on the screen as a solid part showing no hidden features.

SPACE DIAGRAM   A drawing of a vector system showing the correct direction and sense but not drawn to scale.

SPLINE   One of a series of keyways cut around a shaft and mating hole; generally used to transfer power from a shaft to a hub while allowing a sliding action between the parts.

SPUR GEAR   The simplest, most common type of gear used for transmitting motion between parallel shafts. Its teeth are straight and parallel to the shaft axis.

STADIA   Technique of measuring distance using a Philadelphia rod and level.

STATION POINT   In surveying, a fixed point from which measurements are made.

STRETCHOUT LINE   Typically, the beginning line upon which measurements are made and the pattern development is established.

SURFACE CHARTS   Designed to show values represented by the extent of a shaded area; also known as area charts.

SURFACE FINISH   Refers to the roughness, waviness, lay, and flaws of a machine surface.

SURFACE MOUNT TECHNOLOGY (SMT)   The traditional component lead through is replaced with a solder paste to hold the electronic components in place on the surface of the printed circuit board and take up to less than one-third of the space of conventional PC boards.

TANGENT   A straight or curved line that intersects a circle or arc at one point only; is always 90° relative to the center.

TAPER   A conical shape on a shaft or hole, or the slope of a plane surface.

TEMPERING   A process of reheating normalized or hardened steel to a specified temperature, followed by cooling at a predetermined rate to achieve certain hardening characteristics.

TENSILE FORCES   Forces that pull away.

TENSILE STRENGTH   Ability to be stretched.

THERMOPLASTIC   Plastic material may be heated and formed by pressure. Upon reheating, the shape can be changed.

THERMOSET   Plastics are formed into permanent shape by heat and pressure and may not be altered after curing.

THERMOSTAT   An automatic mechanism for controlling the amount of heating or cooling given by a central or zoned heating or cooling system.

TILT-UP CONSTRUCTION   Method using formed wall panels that are lifted or tilted into place.

TOLERANCE   The total permissible variation in a size or location dimension.

TRACKBALL   An input device consisting of a smooth ball mounted in a small box. A portion of the ball protrudes above the top of the box and is rotated with the hand to move the cursor on the screen.

TRANSISTORS   Semiconductor devices in that they are conductors of electricity with resistance to electron flow applied and are used to transfer or amplify an electronic signal.

TRANSITION PIECE   A duct component that provides a change from square or rectangular to round; also known as a square to round.

TRANSLATIONAL MOTION   Linear motion.

TRAVERSE   In surveying, a series of lines with directions and lengths that are connected at station points.

TRIANGULATION   A technique used to lay out the true size and shape of a triangle with the true lengths of the sides; used in pattern development on objects such as the transition piece.

TRILINEAR CHART   Designed in the shape of an equilateral triangle; used to show the interrelationship between three variables on a three-dimensional diagram.

TRIMETRIC DRAWING   A type of pictorial drawing in which all three of the principal axes do not make equal angles with the plane of projection.

TRUE AZIMUTH   The azimuth measured from the actual North or South Pole.

TRUE LENGTH OR TRUE SIZE AND SHAPE   When the line of sight is perpendicular to a line, surface, or feature.

TRUE POSITION   The theoretically exact location of a feature established by basic dimensions.

TURNING POINT   In surveying, this is each point at which the Philadelphia rod is placed and measured.

ULTRASONIC MACHINING   A process where a high-frequency mechanical vibration is maintained in a tool designed to a desired shape; also known as impact grinding.

UNDERCUT   A groove cut on the inside of a cylindrical hole.

UNILATERAL TOLERANCE   A tolerance in which variation is permitted in only one direction from the specified dimension.

UPSET   A forging metal used to form a head or enlarged end on a shaft by pressure or hammering between dies.

VALVE   Any mechanism, such as a gate, ball, flapper, or diaphragm, used to regulate the flow of fluids through a pipe.

VECTOR ANALYSIS   A branch of mathematics that includes the manipulation of vectors.

VECTOR DIAGRAM   A drawing of the vector system in which the vectors are drawn with the correct magnitude and sense, and to scale.

VECTOR QUANTITY   A quantity that requires both magnitude and direction for its complete description.

VELLUM   A drafting paper with translucent properties.

VIEWING PLANE LINE   Represents the location of where a view is established.

VISUALIZATION   The process of recreating a three-dimensional image of an object in a person's mind.

WEB   *See* rib.

WHOLE DEPTH (SPUR GEAR)   The full height of the tooth. It is equal to the sum of the addendum and the dedendum.

WIRE FORM   A three-dimensional form in which all edges and features show as lines, thus appearing to be constructed of wire.

WIRELESS DIAGRAM   Similar to highway diagrams except that interconnecting lines are omitted. The interconnection of terminals is provided by coding.

WIRING DIAGRAM   A type of schematic that shows all of the interconnections of the system components, also referred to as a point-to-point interconnecting wiring diagram.

WORKING DEPTH (SPUR GEAR)   The distance that a tooth occupies in the mating space. It is equal to two times the addendum.

WORM GEARS   Used to transmit power between nonintersecting shafts. The worm is like a screw and has teeth similar to the teeth on a rack. The teeth on the worm gear are similar to the spur gear teeth, but they are curved to form the teeth on the worm.

ZERO DECLINATION   Places on the earth where the compass points exactly toward the true North or South Pole.

ZONING   A system of numbers along the top and bottom and letters along the left and right margins of a drawing used for ease of reading and locating items.

INDEX